Surface Diffusion

Metals, Metal Atoms, and Clusters

For the first time, this book unites the theory, experimental techniques, and computational tools used to describe the diffusion of atoms, molecules, and nanoparticles across metal surfaces. Starting with an outline of the formalism that describes diffusion on surfaces, the authors guide the reader through the principles of atomic movement, before moving on to describe diffusion under special circumstances, such as the presence of defects or foreign species. With an initial focus on the behavior of single entities on a surface, later chapters address the movement of clusters of atoms and the interactions between adatoms. While there is a special emphasis on experimental work, attention is paid to the increasingly valuable contributions theoretical work has made in this field. This book has wide interdisciplinary appeal and is ideal for researchers in solid state physics and chemistry, as well as materials science and engineering.

Grażyna Antczak is a Humboldt Fellow in the Solid State Physics Department at Leibniz University, Hannover, Germany. She received her Ph.D. from the Institute of Experimental Physics at the University of Wrocław, Poland, where she is now an adjunct researcher. Dr. Antczak is a Member of the American Physical Society and the American Vacuum Society, and has had 15 publications in scientific journals.

Gert Ehrlich is currently Research Professor in the Department of Materials Science and Engineering at the University of Illinois, Urbana-Champaign. He is internationally recognized as a pioneer in the area of surface diffusion, and he has received numerous scientific honours and awards. Dr. Ehrlich is an active member of various societies, and is a Fellow of the American Physical Society and the New York Academy of Sciences. He has written almost 200 journal articles and has served on several editorial advisory boards.

Surface Diffusion

Metals, Metal Atoms, and Clusters

GRAŻYNA ANTCZAK
Leibniz Universität Hannover, Germany

GERT EHRLICH
University of Illinois, Urbana-Champaign

CAMBRIDGE UNIVERSITY PRESS

CAMBRIDGE UNIVERSITY PRESS
Cambridge, New York, Melbourne, Madrid, Cape Town, Singapore,
São Paulo, Delhi, Dubai, Tokyo

Cambridge University Press
The Edinburgh Building, Cambridge CB2 2RU, UK

Published in the United States of America by Cambridge University Press, New York

www.cambridge.org
Information on this title: www.cambridge.org/9780521899833

© G. Antczak and G. Ehrlich 2010

This publication is in copyright. Subject to statutory exception
and to the provisions of relevant collective licensing agreements,
no reproduction of any part may take place without the written
permission of Cambridge University Press.

First published 2010

Printed in the United Kingdom at the University Press, Cambridge

A catalog record for this publication is available from the British Library

Library of Congress Cataloging in Publication Data
Antczak, Grażyna, 1973–
Surface diffusion : metals, metal atoms, and clusters / Grażyna Antczak, Gert Ehrlich.
  p. cm.
ISBN 978-0-521-89983-3 (hardback)
1. Diffusion. 2. Metals – Surfaces. 3. Surfaces (Physics) I. Ehrlich, Gert. II. Title.
QC176.8.D5A58 2010
530.4′15–dc22

2009050502

ISBN 978-0-521-89983-3 Hardback

Cambridge University Press has no responsibility for the persistence or
accuracy of URLs for external or third-party internet websites referred to
in this publication, and does not guarantee that any content on such
websites is, or will remain, accurate or appropriate.

Contents

	Preface	*page* ix
	List of abbreviations	xi
	List of symbols	xiv
1	**Atomic diffusion on surfaces**	1
	1.1 Diffusivities: an introduction	1
	1.2 Distribution of atomic displacements	7
	1.3 Jump rates	17
2	**Determination of adatom movements**	24
	2.1 Field ion microscopy	24
	2.2 Field electron emission microscopy	29
	2.3 Scanning tunneling microscopy	31
	2.4 FIM measurement of diffusivity	34
	2.5 Displacement distributions	44
	2.6 STM measurements of diffusion	46
	2.7 Other measurement techniques	53
	2.8 Theoretical estimates	56
3	**Atomic events in surface diffusion**	64
	3.1 Adatom binding sites	64
	3.2 Atomic jumps in diffusion	86
	3.3 Long jumps in surface diffusion	116
	3.4 Transient diffusion	146
4	**Diffusion on one-dimensional surfaces**	183
	4.1 Aluminum: Al(110), (311), (331)	184
	4.2 Nickel: Ni(110)	186
	4.3 Nickel: Ni(311) and Ni(331)	193
	4.4 Copper: Cu(110), Cu(311), Cu(331)	194
	4.5 Molybdenum: Mo(211)	200

Contents

4.6	Rhodium: Rh(110), (311), (331)	202
4.7	Palladium: Pd(110), (311), (331)	204
4.8	Silver: Ag(110), (311), (331)	207
4.9	Tungsten: W(211)	211
4.10	Tungsten: W(321)	225
4.11	Iridium: Ir(110), (311), (331)	227
4.12	Platinum: Pt(110), (311), (331)	230
4.13	Gold: Au(110), (311), (331)	235
4.14	Lead: Pb(110)	238
4.15	Tables for 1D Diffusion	238

5 Diffusion on two-dimensional surfaces — 261

5.1	Aluminum: Al(100)	261
5.2	Aluminum: Al(111)	265
5.3	Potassium	269
5.4	Iron: Fe(100), (111), (110)	269
5.5	Nickel: Ni(111)	274
5.6	Nickel: Ni(100)	277
5.7	Copper: Cu(100)	283
5.8	Copper: Cu(111)	295
5.9	Molybdenum: Mo(110), (111), (100)	301
5.10	Ruthenium: Ru(0001)	304
5.11	Rhodium: Rh(111), (100)	304
5.12	Palladium: Pd(100)	309
5.13	Palladium: Pd(111)	312
5.14	Silver: Ag(111)	315
5.15	Silver: Ag(100)	322
5.16	Tantalum: Ta(110)	328
5.17	Tungsten: W(110)	329
5.18	Tungsten: W(100)	341
5.19	Tungsten: W(111)	344
5.20	Rhenium: Re(0001)	347
5.21	Iridium: Ir(111)	348
5.22	Iridium: Ir(100)	353
5.23	Platinum: Pt(111)	355
5.24	Platinum: Pt(100)	361
5.25	Gold: Au(100)	365
5.26	Gold: Au(111)	368
5.27	Lead: Pb(111)	372
5.28	Bismuth: Bi(111)	372
5.29	Tables for 2D diffusion	373
5.30	Predictions and comparisons	402

Contents

6 Diffusion in special environments — 423

- 6.1 Near impurities — 423
- 6.2 To descending step edges — 430
- 6.3 Atom lifetime versus the step-edge barrier — 479
- 6.4 Comparisons — 485
- 6.5 Atom descent over many-layered steps and between facets — 487
- 6.6 To ascending step edges — 498
- 6.7 Diffusion near dislocations — 506

7 Mechanism of cluster diffusion — 517

- 7.1 Via single atom jumps — 520
- 7.2 Concerted displacements — 536
- 7.3 Mechanism of dimer diffusion versus bond length — 546
- 7.4 Kinetic mechanisms of larger clusters — 547
- 7.5 Derivation of the mechanism of large cluster movements — 550

8 Diffusivities of small clusters — 556

- 8.1 Early investigations — 556
- 8.2 Clusters on aluminum surfaces — 561
- 8.3 Clusters on iron surfaces — 567
- 8.4 Clusters on nickel surfaces — 569
- 8.5 Clusters on copper surfaces — 575
- 8.6 Clusters on rhodium surfaces — 593
- 8.7 Clusters on palladium surfaces — 596
- 8.8 Clusters on silver surfaces — 598
- 8.9 Clusters on tantalum surfaces — 601
- 8.10 Clusters on tungsten surfaces — 602
- 8.11 Clusters on rhenium surfaces — 618
- 8.12 Clusters on iridium surfaces — 619
- 8.13 Clusters on platinum surfaces — 634
- 8.14 Clusters on gold surfaces — 647
- 8.15 Comparisons — 651

9 Diffusion of large clusters — 664

- 9.1 Large clusters on fcc(100) surfaces — 664
- 9.2 Large clusters on fcc(111) surfaces — 676
- 9.3 Large clusters on fcc(110) surfaces — 692
- 9.4 Comments and comparisons — 693

10	**Atomic pair interactions**	696
	10.1 Early measurements	698
	10.2 More recent studies	704
	10.3 Summary	730
	Appendix: Preparation of samples for field ion microscopy	735
	Index	743

The color plates are situated between pages 298 and 299.

Preface

Surface diffusion on metals has been a subject of scientific interest for roughly ninety years. During the first forty years of this period it was very hard to do meaningful work because of technical problems – the difficulty of establishing good enough vacuum conditions to maintain a surface clean for measurements. In a few laboratories, mostly industrial, ultrahigh vacuum techniques were already practiced at that time, but this was not the normal course of events. All of this changed after World War II, first with the general adoption of good vacuum practices, and then with the development of more capable techniques for examining kinetic processes that are important on a surface. The first of these techniques was field ion microscopy, invented by Erwin Müller [1,2], the first method to provide a direct view of single atoms on a surface. The next important development was the scanning tunneling microscope, devised by Binnig and Rohrer [3], which established the capability of probing a large scale surface with high resolution. The last major contribution was the progress in theoretical techniques and computer technology, which toward the end of the twentieth century led to the rapid growth of theoretical calculations.

The last forty years have therefore been a time of great progress in our understanding of surface diffusion, especially of metal atoms on metals. These advances have been spread over the scientific literature, and there has been no overview of the entire field, which is what we are trying to provide here. Our primary emphasis will be on experimental work to define the processes participating in surface diffusion. However, theoretical work can now be done so expeditiously that it has provided valuable guidance, and is now being intensively pursued. As such these contributions will also be carefully noted.[1] Surface diffusion has, of course, a long history, dating back to the initiating work of Hamburger [5] in 1918. These early studies have, however, already been reviewed [6], so here we will be concerned with work on surface diffusion under ultra high vacuum (UHV) conditions and on an atomic scale, which began in the 1960s, and has led to the current state of understanding.

The beginnings of modern studies of surface diffusion were greatly influenced by the insights and inspiration of David Turnbull, as well as by the traditions and expertise at General Electric. We have also benefited from the encouragement and suggestions of Ryszard Błaszczyszyn, and were able to draw on the expertise at the Institute of Experimental Physics of the University of Wrocław. Here, at the University of Illinois,

---

[1] For a review of theoretical efforts, see T. Ala-Nissila *et al.* [4].

we have had helpful interactions with Dan Alpert, the man that guided the start of modern ultrahigh vacuum techniques which underlie diffusion studies on surfaces. Above all, GE wants to express his appreciation to his wife for her support and for the time devoted to this effort.

The point of view of this presentation is primarily atomistic, and this was stimulated by the work of J. H. de Boer in his book *The Dynamical Character of Adsorption*, Clarendon Press, Oxford 1953, which had quite an impact on us. It is important to recognize that the term surface diffusion spans topics much broader than what we intend to cover here. Our concern will be concentrated on the behavior of single entities and clusters on a surface. This avoids encountering the interactions between atoms which affect surface diffusion at finite concentrations, and are specific to the particular chemistry of each system. However, with an understanding of surface diffusion gained from experiment and theory, work on interactions between adatoms will be described as well.

Our efforts have greatly benefited from interactions with the various members of the Surface Studies group here over the years, and we express to them our great appreciation. We also want to emphasize again the crucial importance of experimental work, and of the technical support necessary for this. It is therefore a pleasure to give our thanks to the people who primarily provided this support for us: Bob Bales, Jack Gladin, William Lawrence, and Bob MacFarlane. Also important in coming to grips with the subject of surface diffusion was the assistance of Mary Kay Newman, the librarian in the Physics Department, whose help, as well as that of Nicholas Watanabe, has been really appreciated. Finally we want to acknowledge a special debt to Jennifer Lewis, who made it possible for us to continue our work.

References

[1] E. W. Müller, K. Bahadur, Field ionization of gases at a metal surface and the resolution of the field ion microscope, *Phys. Rev.* **102** (1956) 624–631.
[2] E. W. Müller, T. T. Tsong, *Field Ion Microscopy Principles and Applications* (American Elsevier, New York, 1969).
[3] G. Binnig, H. Rohrer, Scanning tunneling microscopy, *Helv. Phys. Acta* **55** (1982) 726–735.
[4] T. Ala-Nissila, R. Ferrando, S. C. Ying, Collective and single particle diffusion on surfaces, *Adv. Phys.* **51** (2002) 949–1078.
[5] L. Hamburger, Ultra-microscopic examinations of very thin metal and salt deposits obtained by evaporation in a high vacuum, *Kolloid Z.* **23** (1918) 177–199.
[6] G. Antczak, G. Ehrlich, The beginnings of surface diffusion studies, *Surf. Sci.* **589** (2005) 52–66.

Abbreviations

A	Type *A* step edge on fcc(111)
AES	Auger electron spectroscopy
AFW	Adams, Foiles, Wolfer
Ass	Assigned
ATVF	Ackland, Tichy, Vitek, Finnis
A-Ex	Adatom catalyzed exchange
B	Type *B* step edge on fcc(111)
CEM	Corrected effective medium method
CEM59	CEM with 59 active atoms
CM	Concerted motion
Coh.	Cohesion approximation
COM	Center of mass
COP	Center of positions
CS	Constrained statics
CY-EAM	EAM of Cai and Ye
CY-EAM1	Extension of CY-EAM
CY-EAM2	Extension of CY-EAM
DFT	Density functional theory
Diam	Diameter
DL	Discommensuration line
D-Ex	Double exchange
EAG	Ercolessi–Adams glue potential
EAM	Embedded atom method
EAM5	Embedded atom method 5
EMT	Effective medium theory
Ener min	Energy minimum
Eq.	Equation
Ex	Exchange
FDB	Foiles, Daw, Baskes
FEM	Field electron emission microscopy
FIM	Field ion microscope or microscopy
Fluct	Fluctuation
F-S	Finnis–Sinclair

List of abbreviations

GGA	Generalized gradient approximation
GP	Glue potential
He-Scat	Helium scattering
^{3}He-SE	^{3}He spin echo
HR	High resolution
HRLEED	High resolution low energy electron diffraction
K	Kelvin
K	Kink
K-K-R	Korringa–Kohn–Rostoker method
LAM	Lonely atom method
LDA	Local density approximation
LDOS	Local density of states
LEED	Low energy electron diffraction
LEEM	Low energy electron microsopy
LEIS	Low energy ion scattering
L-Ex	Long exchange
LF	Leapfrog
LMD	Langevin molecular dynamics
L-J	Lennard Jones
Mag	Magnetic
MAEAM	Modified analytical embedded atom method
MC	Monte Carlo
MD	Molecular dynamics
MD/MC-CEM	Molecular dynamics/Monte Carlo using CEM
MBE	Molecular beam epitaxy
ML	Monolayers
Morse	Morse potential
MS	Molecular statics
MW	Metastable walk
M-Jump	Meta jump
NEB	Nudged elastic band
nn	Nearest neighbor
Nucl	Nucleation theory
OJ	Oh and Johnson
PACS	Perturbed $\gamma - \gamma$ angular correlation studies
PEEM	Photoemission electron microscope
Photo	Photoemission
Pot	Potential
RD	Ring diameter
Rean	Reanalyzed
Refit	Refitted and reanalyzed
Resis	Resistivity
RGL	Rosato, Guillope, Legrand

List of abbreviations

RHEED	Reflection high energy electron diffraction
RS	Rutherford scattering
SC	Sutton–Chen
Scat	Scattering
SEAM	Surface embedded atom method
SEM	Scanning electron microscope
SI	Surface ionization
sim	Simulation
SPA-LEED	Spot profile analysis of low energy electron diffraction
Static	Static barrier
STM	Scanning tunneling microscope or microscopy
T	Temperature
TB	Tight-binding
TDT	Tersoff, Denier van der Gon, and Tromp
TI	Thermodynamic integration
TST	Transition state theory
T-Ex	Triple exchange
Q-Ex	Quadruple exchange
VASP	Vienna *ab initio* simulation package
VC	Voter Chen
VTST	Variational transition state theory
WF	Work function
XPD	X-ray photoelectron diffraction
XPS	X-ray photoelectron spectroscopy
Z	Band occupation
$\parallel$	In-channel
$\perp$	Cross-channel

Symbols

α	Jump rate to nearest-neighbor position at the right for 1D motion, or jump rate to nearest-neighbor position for 2D motion
α_{fh}/α_{hf}	Rate of single jumps from fcc to hcp/hcp to fcc site on fcc(111)
α_M	Morse parameter
α_N/α_L	Exponent describing dependence of diffusivity D on number of atoms N/on radius of island R_r, or island of length L_L
α_{Re}	Rate of short range mechanism of movement for Re-Ir complex
a_ℓ	Lattice spacing
a_S	Atom jump rate along step of type A
A	Island area
A_R	Parameter of repulsive energy
β	Jump rate to next nearest-neighbor position at the right for 1D motion, or jump rate to next nearest-neighbor position for 2D motion
β_{ff}/β_{hh}	Rate of double jumps between fcc/hcp sites on fcc(111)
β_R	Jump rate for rebound jumps
β_{Re}	Long range mechanism of movement for Re-Ir complex
b_S	Atom jump rate along step of type B
χ_c	Energy of condensation on fcc(111) plane
c	Concentration, or rate of dimer jump via horizontal intermediate on bcc(110)
c_0	Concentration at $t = 0$
δ	Jump rate to nearest-neighbor position at the left in 1D movement
δ_F	Fermi-level phase shift
δ_D	Distance between interior and step edge barrier
δ_{xo}	Kronecker delta
δ_x/δ_y	Rate of horizontal/vertical jump on bcc(110)
d_d	Distance
d_{12}	Separation of atom 1 and 2
d_t	Rate of adatom motion on terrace
d_T	Trio perimeter
d_R	Plane diameter
D	Diffusivity
D_o	Prefactor of the diffusivity

List of symbols

D_{0B}	Prefactor of diffusivity over descending step
D_M	Morse parameter
D_{205}	Diffusivity of cluster consisting of 205 atoms
D_γ	Diffusivity calculated with all types of jumps
D^*	Prefactor in diffusivity dependence on cluster size
ε	Jump rate to next-nearest-neighbor position at the left in 1D movement
ε_{LJ}	Energy parameter of L-J potential
$\varepsilon_1/\varepsilon_2/\varepsilon_3$	First/second/third nearest-neighbor pairwise interaction
ε_F	Fermi energy
ε_R	Repulsive pair energy
ε_{AA}	Interaction energy between two similar atoms at nearest-neighbor sites
$\varepsilon_{xx}/\varepsilon_{yy}$	Strain
e	Charge of the electron
E^A/E^B	Activation energy for movement along step A/step B
E_2^{sh}/E_2^{st}	Barrier for dimer shearing / stretching
E_B^i	Band energy
E_D^D/E_D^v	Activation energy for movement obtained from diffusivity/velocity
E_R^i	Repulsive energy between two atoms
E_{ℓ_0}	Energy of two adatoms at nearest-neighbor separation
E_1	Energy of dimer in configuration 1, or binding energy for adatoms at nn separation
E_2/E_3	Binding energy for adatoms in second/third nn separation
E_0	Energy of dimer in configuration 0
$E_{\alpha f}/E_{\alpha h}$	Barrier height for jump out of fcc/hcp site
$E_\alpha/E_\beta/E_{\beta R}$	Activation energy for single/double/rebound jumps
$E_{\delta x}/E_{\delta y}/E_s$	Activation energy for vertical/horizontal/sum of jumps
E_a	Additional step-edge barrier, or activation energy for jump a in dimer movement
E_b	Activation energy for jump b in dimer movement
E_{cb}	Energy of core break up
E_{cc}	Energy of new row nucleation
E_{coh}	Cohesive energy
E_{CJ}	Activation energy for concerted jump
E_e/E_h	Barrier for exchange/hop
E_{eff}	Effective energy barrier
E_e^A/E_e^B	Activation energy for exchange along step A/step B
E_h^A/E_h^B	Activation energy for jump along step A/step B
E_i	Cluster binding energy, or internal energy due to atom i
E_j/E_ℓ	Activation energy for j-/ℓ-type long jump
E_{ij}	Potential energy between atoms i and j
$E_{ij,\ell}$	Energy of two atoms at sites i and j in state ℓ

E_k	Energy of adatom pair in configuration k
E_{kd}/E_{ku}	Energy of down jump/up jump at kink
$E_{b\ell}$	Activation energy for conversion from single to ℓ-type long jump
E_B	Activation energy for overcoming descending step and incorporate
E_{B2}	Activation energy for overcoming descending step at B_2 position and incorporate
E_{He}	Incident energy of helium
E_{LF}	Activation energy for leapfrog event
E_p	Energy of movement along step
E_r	Rebound energy
E_T	Activation energy for diagonal transition around cluster corner
E_{tot}	Total energy
$E(a,b)$	Energy of atom pair at separation $\boldsymbol{R}=(a,b)$
$E(d)$	Pair interaction energy at separation d
$E(s)$	Energy as a function of the displacement s
ΔE	Energy change
ΔE_a	Effective energy for movement of cross-channel dimer from state 0 to state 1
ΔE_b	Effective energy for movement of cross-channel dimer from state 1 to state 0
ΔE_{cs}	Binding energy of core atom relative to adatom at step
ΔE_D	Energy of activation for diffusion
ΔE_ε	Energy width in time of flight spectrum
ΔE_e	Activation energy for cluster movement by atom exchange
ΔE_h	Activation energy for cluster movement by atom hopping
ΔE_{int}	Interaction energy
ΔE_{ks}	Binding energy of kink atom
ΔE_{kt}	Binding energy of kink atom relative to atom on terrace
ΔE_{vib}	Vibrational contribution to energy of activation
$\Delta E(\varepsilon)$	Energy changes during collision
$<E_T>$	Mean kinetic energy
$<\Delta E>_{AT}$	Effective activation energy for atomic motion of dimer
$<\Delta E>_{COM}$	Effective activation energy for center of mass motion of cluster
ϕ	Electron work function
$\phi_{ij}(R_{ij})$	Core–core repulsion between atoms i and j
$\Delta\Phi$	Difference in structural energy between barrier peak and normal position
$f_i(t)$	Auto-correlation function for electron emission fluctuation
$f_j(R_{ij})$	Contribution of electron density of atom i arising from atom j
F	Free energy
F_a	Force
F_e	Electric field
F_f	Rate of atom deposition

List of symbols

F_x	Free energy for atoms x units apart
$F(a,b)$	Free energy of atom pair at separation $\boldsymbol{R} = (a,b)$
$F(\boldsymbol{R})$	Free energy of interaction as a function of the separation $\boldsymbol{R}$
$F(t)$	Fraction of atoms on the surface
$F_i(\rho_i)$	Energy for embedding atom into local density ρ_i
ΔF	Free energy change
ΔF_D	Change in free energy for diffusion
γ	Jump rate to third neighbor position
$\gamma - \gamma$	Angular correlation
γ_s	Formation energy per step atom
Γ	Jump rate
Γ_o	Prefactor for the jump rate
Γ_i	Rate of dissociation of island of size i
Γ_ε	Quasielastic energy width of scattered atoms
g	Geometrical factor
$g(\boldsymbol{R})$	Pair distribution function
$G(t,z)$	Moment generating function of variable z
h	Planck's constant
h_a	Rate of detachment of atom adsorbed at straight edge
h_c	Rate of core breakup
h_e	Rate of straight edge hopping
h_k	Rate of kink escape
h_{ke}/h_{se}	Rate of detachment of atom from kink/from straight edge to terrace
h_r	Rate of conversion of vertical to horizontal dimer, or rate of corner rounding
h_{re}	Rate of detachment of atom from corner to the step edge
H_s^o	Enthalpy of sublimation
$\hbar$	$\dfrac{h}{2\pi}$
i	Critical size of cluster
I	Ionization potential
I_e	Density of the emission current
I_{exp}	Intensity of scattered He atoms
I_{fit}	Best model fit to scattered He atom intensity
I_R	Kinematic RHEED intensity
$\dfrac{I}{I_0}$	Ratio of scattered to incident intensity
$I_n(\tau)$	Modified Bessel function of order n and argument τ
j	Flux across unit length of line
j_B	Atom jump rate over barrier E_B at step edge
j_D	Diffusive flux
j_R	Flux at position R
κ	Ratio of force constants

List of symbols

k	Boltzmann's constant
k_h	Harmonic approximation of escape rate
k_a/k_{ke}	Rate of atom attachment from terrace to straight step/to kink
k_k	Rate of atom attachment from edge to kink
k_{force}	Force constant
k_F	Fermi wave number
λ_{deB}	deBroglie wave length
λ_F	Fermi wave length
λ_x	Jump rate to right, starting from position x
ℓ	Jump length, or quantum state
ℓ_0	Nearest-neighbor spacing
L	Number of sites in one-dimensional plane
L_0	Standard length
L_i	Island separation
L_L	Island length or diameter also side length of square deposit
L_T	Tip to detector distance
μ	Chemical potential
μ_x	Jump rate to left, starting from position x
m	Mass of electron
m_a	Number of deposited atoms, or number of atoms adsorbed per cm^2
m_1, m_2	Number of atoms per unit length
M	Number of atoms adsorbed, or total number of observations
M_S	Number of surface sites
M_T	Magnification of field ion microscope
ν	Attempt frequency of atom
ν_0	Frequency prefactor for diffusion
ν_s	Frequency prefactor for diffusion across descending steps
ν_h	Harmonic approximation attempt frequency for diffusing adatom
$\nu_{0\alpha}/\nu_{0\beta}/\nu_{0\beta R}$	Prefactor for single/double/rebound jumps
$\nu_a/\nu_r/\nu_{ce}/\nu_{all}$	Frequency of single/reinsertion/correlated / all jumps
$\nu_{0\delta x}/\nu_{0\delta y}/\nu_{0s}$	Prefactor for horizontal/vertical/sum of jumps
ν_{0B}	Frequency factor for descending lattice step
ν_a	Relative frequency factor of step edge to terrace diffusion, or frequency for rate a in dimer motion
ν_b	Frequency for rate b in dimer motion
ν_d	Frequency factor
ν_ℓ/ν_j	Frequency factor for ℓ-/j-type jumps
$\nu_{b\ell}$	Frequency factor for conversion from single to longer jump
n	Number of jumps
n_x	Number of islands per site
n_c	Number of charges on the evaporated ion
n_{out}/n_{in}	Number of paths for going out/in over boundary
$<n>$	Number of diffusion events

List of symbols

N	Number of atoms in cluster, size of island, or total number of transitions (jumps)
N_a	Number of atoms simulated
N_{av}/N_{av}^0	Mean island density/initial post-deposition mean island density
N_I/N_{II}	Frequency of occurrence of island in form *I*/form *II*
N_c	Number of atoms in hexagonal form
N_f/N_h	Number of atoms at fcc/hcp sites
N_i/N_t	Number of atoms incorporated/trapped
$N_{\alpha f}/N_{\alpha h}$	Number of hops out from fcc/hcp site to the same kind of site
N_T	Total number of jumps
$N(\bm{R})$	Number of observations of two atoms separated by $\bm{R}$
$N_o(\bm{R})$	Total number of atom pairs at separation $\bm{R}$
$\overline{N}$	Average number of atom jumps
$<\Delta n_1^2>$	Mean-square value of jumps to the right
p	Probability of jump to the right
$p(\bm{R})$	Probablility of finding adatom pair at separation $\bm{R}$
$p_{b\ell}$	Probability of converting from single to long jump
p_{n_1}	Probability of reaching $x = s\ell$ after n_1 jumps
p_x	Probability of atom being at the distance x
$p_{\Delta x}N$	Number of atoms at displacement Δx
P	Probability that material present at $t=0$ will be gone at time τ
P_0/P_1	Probability of being at a site of type 0/type 1
$P_0^{(z)}$	Probability of center of mass being at site of type 0 having started at z
P_{0A}	Probability of finding trimer in configuration *0A*, regardless of position
P_{1D}/P_{2D}	Probability of cluster in 1D/2D configuration
P_b	Probability of atom overcoming step boundary
P_E	Probability of atom occupy edge site
P_{ij}	Probability of finding a pair of atoms at sites i and j
$P(N)$	Term in prefactor for cluster diffusivity accounting for dynamical misfit
$P_f^{(f)}/P_h^{(h)}$	Probablity of atom ending at fcc/hcp site when starting at the same kind of site
θ	Fractional occupation of sites
Θ	Coverage
q	Probability of jump to the left, or distance dependence of hopping integral
q_c	Translational coordinate
q_F	In-surface Fermi wave vector
Q_i	Desorption energy of ion
ρ_i	Electron density of atom i
$\rho(t_f)$	Auto-correlation function
r	Distance between dimer's atoms, or rate of jumps at constant temperature

List of symbols

r_0	Rate of jumps during "zero-time" observations
r_c	Rate of evaporation–condensation mechanism
r_e	Rate of diffusion along cluster perimeter
r_i	Rate of incorpotration to descending step
r_{eq}	Rigid distance between dimer's atoms
r_T	Tip radius
$r1$	Distance of descending step from center
R	Atom deposition rate, or overall rate of jumping
$\boldsymbol{R}$	Adatom–adatom separation vector
$|\boldsymbol{R}|$	Adatom–adatom separation magnitude
R_o	Morse parameter
R_b	Rate of basic jumps, derived from low temperatures
R_c	Cut-off distance for interactions
R_{Fi}	Rate of field ionization
R_ℓ	Rate of long jumps of type ℓ
R_{ij}	Distance between atoms i and j
R_r	Cluster radius or radius of circular deposit
R_s	$\dfrac{E_D}{\Delta H_s^o}$
R_T	Tip to screen distance
R_x	Distance from the center of original distribution
$R1$	Distance of ascending step from center
$<r^2>$	Mean-square displacement in 2D
$<\Delta r^2>$	Fluctuation of displacement in 2D
σ	Interatomic separation at which potential energy vanishes
σ_i	Capture number, relating rate of incorporation to the diffusivity D
σ_x	L-J distance parameter
s	Displacement from initial equilibrium
s_0	Prefactor to $s(T)$
$s(T)$	Ratio of rate of step edge crossing to nearest-neighbor jumps on plane
S_1/S_0	Entropy of dimer in configuration 1/in configuration 0
$\dfrac{s}{s_{tot}}$	Relative distance
S_{av}/S_{av}^0	Mean island size/initial mean island size
ΔS	Change in entropy of system
ΔS_D	Entropy of activation for diffusion
ΔS_{vib}	Vibrational contribution to entropy of activation
τ	Mean lifetime for atom incorporation
τ_0	Prefactor for atom lifetime
τ_c	Lifetime for adatom starting at the center of plane
τ_f	Relaxation time for fluctuation
t_f	Time interval for fluctuation
t	Length of time interval

List of symbols

t_0	Time interval for "zero-time" measurements
t_c	Time interval for diffusion at constant temperature
t_e	Slowly varying functions of $3.79 \times 10^{-4} F_e^{1/2}/\phi$.
T	Temperature
T_d	Temperature for dissociation of cluster
T_D	Temperature for diffusion
T_E	Atom temperature
T_S	Sample temperature
T_m	Melting point
T_R	Temperature for cluster rearrangement
T^*	$\dfrac{kT}{\varepsilon}$
υ	Correction term in field ionization
υ_e	Slowly varying function of $3.79 \times 10^{-4} F_e^{1/2}/\phi$
v_A	Mean velocity in positive direction
ς	Effective hopping integral, or quarto interactions
v	Velocity
V	Voltage
V_0	Effective barrier for non-interacting atoms
$<v_x>$	Average x-component of velocity
ω_0	Angular Debye frequency
ω_d	Angular attempt frequency
$\omega_1/\omega_2/\omega_3$	Frequencies
Ω	Degeneracy
Ω_I/Ω_{II}	Degenerency, number of equal configurations of form *I*/form *II*
W	Free energy change between top and bottom of potential
ξ	Mass of an incident compared with a lattice atom
ξ_1/ξ_2	First/second trio interactions
$<\Delta x^2>$	Fluctuation in displacement x
$<x^2>$	Mean square displacement
$\overline{X}$	Mean diffusion length
$X(N)$	Overall displacement
X	Pair separation measured along channel of W(211) plane
$y(Å)$	Distance perpendicular to step but parallel to surface
$<\Delta y^2>$	Fluctuation in displacement y
$<y^2>$	Mean square displacement
z_A	Partition function of adsorbed material
Z	Canonical partition function
Z_T	Tip sample distance

1 Atomic diffusion on surfaces

One important reason surface diffusion is of interest is that it is so different from diffusion in bulk solids, and is involved in many important processes – among them crystal and film growth as well as evaporation, chemical surface reactions, catalysis, and condensation. However, before becoming intimately involved in the description of surface events, it will be useful to outline the formalism that describes diffusion on surfaces.

1.1 Diffusivities: an introduction

To describe diffusion on a crystal surface it is convenient to adapt the procedures developed for bulk diffusion [1]. The flux J crossing a line of unit length is given by Fick's first law

$$J = -D\, \partial c/\partial x, \tag{1.1}$$

where $\partial c/\partial x$ gives the gradient of the concentration c and D is the diffusivity; the diffusivity establishes the magnitude of the flux in relation to the gradient, and is generally given in units of cm^2/sec. Establishing a known gradient of the concentration on a surface, and measuring the flux J are difficult, and it is therefore useful to transform Eq. (1.1). Consider two parallel lines on a surface, shown in Fig. 1.1, a distance of Δx apart, which is comparable with the jump length ℓ executed in diffusion. The flux into line 1 will be different from that into line 2, as material accumulates in the region between the two lines. If the flux is considered per unit length, then

$$J_1 = J_2 - \partial J/\partial x \Delta x. \tag{1.2}$$

The difference in the flux to the two lines can obviously be written as

$$J_1 - J_2 = -\partial J/\partial x \Delta x, \tag{1.3}$$

that is in terms of the amount of material accumulated, so that

$$J_1 - J_2 = -\partial J/\partial x \Delta x = \partial c/\partial t \Delta x. \tag{1.4}$$

However, from Eq. (1.1) we know that

$$\partial J/\partial x = -\partial (D\, \partial c/\partial x)/\partial x, \tag{1.5}$$

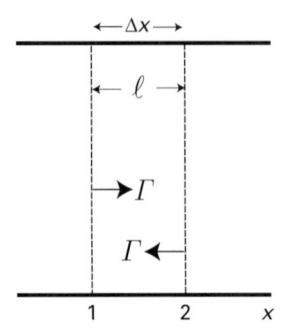

Fig. 1.1 Schematic illustrating atomic jumps at rate Γ per atom in surface diffusion. Jump length = ℓ.

and therefore

$$\partial c/\partial t = \partial(D\, \partial c/\partial x)/\partial x. \tag{1.6}$$

This is Fick's second law, more directly applicable to examining surface diffusion. We will generally consider here diffusion of single atoms over a crystal plane, so that the diffusivity D will not be a function of the concentration, and Eq. (1.6) can therefore be written as

$$\partial c/\partial t = D\, \partial^2 c/\partial x^2. \tag{1.7}$$

Note that we have only considered one-dimensional diffusion, with the flux as well as the gradient along the x-axis, but this will suffice for our problems.

One possible way of looking at diffusion is to deposit a line of m_a atoms on a surface, and to examine how the deposited material spreads out with time. The solution of Eq. (1.7) for this case is

$$c(x,t) = \frac{m_a}{\sqrt{4\pi Dt}} \exp\left(-\frac{x^2}{4Dt}\right), \tag{1.8}$$

where x is the distance normal to the initial deposit, and t indicates the length of the diffusion time interval; the solution is shown in Fig. 1.2a. Equation (1.8) can be readily confirmed by differentiating $c(x,t)$ with respect to both x and t. The boundary conditions here are that at $x=0$, $c \to \infty$ and for $|x| > 0$ as $t \to 0$, $c \to 0$.

If instead of just a line, a part of the entire crystal surface can be covered with adsorbed material, as in Fig. 1.2b; boundary conditions now are that with the border of material at $x=0$, $c=c_o$ for $x > 0$, and $c=0$ for $x < 0$, both with $t=0$. The covered region can be considered as an array of adjacent lines at a separation z. We just integrate the answer in Eq. (1.8) to give

$$c(x,t) = \frac{c_o}{\sqrt{4\pi Dt}} \int_0^\infty \exp\left[-\frac{(x-z)^2}{4Dt}\right] dz. \tag{1.9}$$

If we let $u = \dfrac{x-z}{\sqrt{4Dt}}$, then

1.1 Diffusivities: an introduction

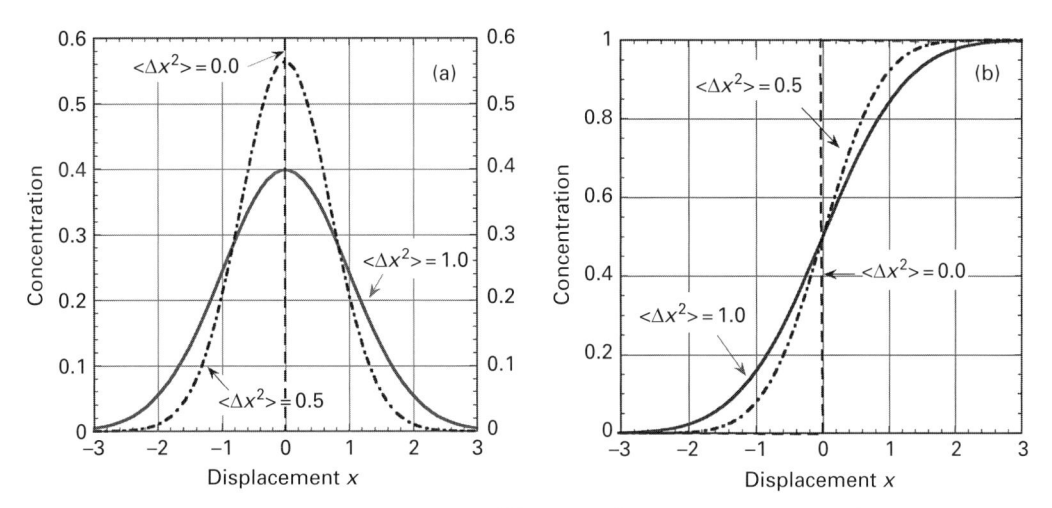

Fig. 1.2 Concentration profiles established in diffusion for different mean-square displacements. (a) Spreading from an initial straight-line deposit. (b) Spreading out of a half-covered surface.

$$c(x,t) = \frac{c_o}{\sqrt{\pi}} \int_{-\infty}^{\frac{x}{\sqrt{4Dt}}} \exp(-u^2) du. \qquad (1.10)$$

Inasmuch as the error function is given by

$$erf(x) = \frac{2}{\sqrt{\pi}} \int_0^x \exp(-u^2) du, \qquad (1.11)$$

and $erf(-\infty) = -1$, the solution can be more simply written as

$$c(x,t) = \frac{c_o}{2}\left[1 + erf\left(\frac{x}{\sqrt{4Dt}}\right)\right]. \qquad (1.12)$$

Spreading will yield a gradually diminishing boundary region, as shown in Fig. 1.2b.

An alternative geometry for observing atomic diffusion is to deposit a circular spot containing m_a atoms and to study spreading from it. The solution to the diffusion equation then is

$$c(R_r, t) = \frac{m_a}{4\pi Dt}\exp\left(-\frac{R_r^2}{4Dt}\right), \qquad (1.13)$$

where R_r gives the distance from the center of the original deposit.

As a final example, consider the spreading of a deposit from the front of a ribbon of width d to the initially clean back. In this finite system we have the initial condition that at $t = 0$, $c = c_o$ for $0 < x < d$, and $c = 0$ for $d < x < w$, where $w > d$. The solution to the diffusion equation under these circumstances, given by Barrer [2], is

$$c(x,t) = c_o\left\{\frac{d}{w} + \frac{2}{\pi}\sum_{n=1}^{\infty}\frac{1}{n}\sin\frac{n\pi d}{w}\cos\frac{n\pi x}{w}\exp\left(-\frac{n^2\pi^2}{w^2}Dt\right)\right\}. \qquad (1.14)$$

By measuring the spreading as a function of time and position, values of $c(x,t)$ can be established experimentally. The diffusivity is then obtained by fitting the appropriate solution to the measured concentration profile. It is clear that here we have covered only the simplest examples useful in studies of surface diffusion. More complicated ones are described by Crank [1]. It should also be noted that under many conditions these approaches are not at all easy to implement.

With a value of the diffusivity in hand, the question arises immediately how to interpret the diffusivity D in terms of the atomic jump processes. We follow here the approach pioneered by Einstein [3]. Consider a surface of unit width, with a uniform concentration gradient in the x direction. Atoms jump in the x direction at the rate Γ per atom, and in the $-x$ direction at the same rate. We now draw two lines perpendicular to the x-axis as was done in Fig. 1.1; the second is separated from the first by a distance equal to the jump length ℓ executed by an atom in diffusion. The rate at which atoms cross line 1 is $m_1\Gamma$, and line 2 is $m_2\Gamma$ where m_1 and m_2 are the number of atoms per unit length. The net rate at which atoms are transferred to the right will be given by

$$m_1\Gamma - m_2\Gamma = \Gamma(m_1 - m_2); \tag{1.15}$$

that is, the flow to the right from line 1 is compensated to some extent by the flow to the left from line 2. The number of atoms m_a can be related to the surface concentration by $m_a = c\ell$, so that the net flux J becomes

$$J = (c_1 - c_2)\ell\Gamma. \tag{1.16}$$

Now $c_1 = c_2 - \ell\partial c/\partial x$, so that

$$J = -\ell^2 \Gamma \partial c/\partial x. \tag{1.17}$$

From Eq. (1.1) it follows that

$$D = \Gamma \ell^2, \tag{1.18}$$

and we see that the diffusivity is just given by the product of the jump rate Γ per atom in one direction times the square of the jump length. For a more realistic view of diffusion, ℓ should of course be taken as the square root of the average of the squares of the individual displacements.

For the jump rate Γ per atom that has entered here we can write the expression available from transition state theory for the rate of overcoming a potential barrier of height W [4],

$$\Gamma = \nu \exp\left(-\frac{W}{kT}\right), \tag{1.19}$$

where ν accounts for the vibrational frequencies of the system, known also as attempt frequency. The diffusivity D can therefore be written as

$$D = \nu \ell^2 \exp\left(-\frac{W}{kT}\right). \tag{1.20}$$

1.1 Diffusivities: an introduction

Here W is really a free energy change ΔF between the top of the potential and the atom in its equilibrium position, but confined to a plane perpendicular to the diffusion path. Since $\Delta F = \Delta E - T\Delta S$, the diffusivity becomes

$$D = v\ell^2 \exp\left(\frac{\Delta S_D}{k}\right) \exp\left(-\frac{\Delta E_D}{kT}\right). \tag{1.21}$$

The usual approximation is that v, ℓ, and ΔS_D are not strongly dependent upon the temperature T, so that a plot of $\ln(D)$ versus $1/T$ will provide us with ΔE_D, the activation energy for diffusion, as the slope and the logarithm of $v\ell^2 \exp\left(\frac{\Delta S_D}{k}\right)$ as the y-intercept of the diffusivity D, with ΔS_D as the change in the entropy in diffusion. It is customary to write

$$v \exp\left(\frac{\Delta S_D}{k}\right) = v_0 \tag{1.22}$$

as we often have some idea about the expected jump length ℓ, and D can be more briefly written as

$$D = D_o \exp\left(-\frac{\Delta E_D}{kT}\right), \text{ with } D_o = v_0 \ell^2. \tag{1.23}$$

where D_o is known as the prefactor for the diffusivity.

A somewhat different formulation has been offered by Kürpick et al. [5], who considered the transfer of an otherwise unconstrained atom from a normal site to the top of the barrier peak, where the degree of freedom in the direction of diffusion is withdrawn from the free energy. She arrived at an expression for the diffusivity as

$$D = D_o \exp\left(-\frac{\Delta \Phi}{kT}\right), \tag{1.24}$$

where $\Delta \Phi$ is the difference in the structural energy of the system between the barrier peak and the normal minimum. The prefactor, itself a function of the temperature T, is given by

$$D_o = \frac{kT}{h} \ell^2 \exp\left(\frac{\Delta S_{vib}}{k}\right) \exp\left(-\frac{\Delta E_{vib}}{kT}\right). \tag{1.25}$$

Here ΔS_{vib} and ΔE_{vib} give the difference between the peak maximum and the minimum in vibrational contributions to the entropy and the internal energy, which have been evaluated by Kürpick [6].

It is useful to establish another connection between diffusivity and jump length, as the standard approaches for evaluating diffusivities are often difficult to carry out. We therefore evaluate the distance covered by a long sequence of N transitions, where $N = 2\Gamma t$. If transition i gives a vector displacement x_i, then the overall displacement $x(N)$ will be

$$x(N) = x_1 + x_2 + x_3 + \cdots + x_N = \sum_{i=1}^{N} x_i. \tag{1.26}$$

On squaring the above we get

$$x^2(N) = \sum_{i=1}^{N} x_i^2 + 2\sum_{i=1}^{N-j}\sum_{j=1}^{N-1} x_i x_{i+j}. \quad (1.27)$$

Averaging $x^2(N)$ causes the second term on the right to disappear: in a random displacement there is no relation between one jump and the next, so positive and negative transitions are equally probable. The mean-square displacement therefore becomes

$$\langle x^2(N) \rangle = \sum_{i=1}^{N} x_i^2 = N\ell^2, \quad (1.28)$$

where the mean-square jump length ℓ^2 is given by

$$\ell^2 = \frac{1}{N}\sum_{i=1}^{N} x_i^2. \quad (1.29)$$

That is, the mean-square displacement is just equal to the total number of jumps times the square of the jump length.

The same result can be obtained in a slightly different way given by Berg [7] for a system of M particles. Consider the x-displacement after n jumps, $x(n)$; this is related to the displacement that has occurred previously by

$$x_i(n) = x_i(n-1) \pm \ell, \quad (1.30)$$

where ℓ is the length of the displacement. For the mean value of $x(n)$ we obtain

$$\langle x(n) \rangle = \frac{1}{M}\sum_{i=1}^{M} [x_i(n-1) \pm \ell] = \frac{1}{M}\sum_{i=1}^{M} x_i(n-1) = \langle x_i(n-1) \rangle. \quad (1.31)$$

The mean location does not change as the number of steps changes, so that a particle starting at $x = 0$ will remain there. The mean value $\langle x(n) \rangle$ is therefore zero. For the square of the displacement we find

$$x_i^2(n) = x_i^2(n-1) \pm 2\ell x_i(n-1) + \ell^2, \quad (1.32)$$

and the mean-square displacement is given by

$$\langle x^2(n) \rangle = \frac{1}{M}\sum_{j=1}^{N} x_j^2(n) = \frac{1}{M}\sum_{j=1}^{N} [x_j^2(n-1) \pm 2\ell x_j(n-1) + \ell^2]. \quad (1.33)$$

However, the second term under the brackets at right disappears, as positive and negative terms balance out, so that

$$\langle x^2(n) \rangle = \langle x^2(n-1) \rangle + \ell^2. \quad (1.34)$$

When $n = 0$, $x(n) = 0$, so that $x^2(1) = \ell^2$, $x^2(2) = 2\ell^2$, and $x^2(n) = n\ell^2$. Therefore,

$$\langle x^2(n) \rangle = n\ell^2, \quad (1.35)$$

as has already been demonstrated. We have previously shown in Eq. (1.18) that $D = \Gamma\ell^2$; but $\langle x(n) \rangle = 0$ and it follows from Eq. (1.28) that $\Gamma\ell^2 = \langle x^2 \rangle / 2t$, so that

$$\langle \Delta x^2 \rangle = \langle x^2 \rangle - \langle x \rangle^2 = 2Dt, \quad (1.36)$$

the Einstein relation. In other words, the diffusivity can be derived directly from measurements of the displacement fluctuation or dispersion <Δx^2>.

We are also going to be interested in the atomic jump processes participating in diffusion. To learn more about these involves just a slight extension – looking in detail at the distribution of atomic displacements. This will be done in various ways in what follows.

1.2 Distribution of atomic displacements

With diffusivities known from measured displacement fluctuations, what can we learn about the atomic jumps contributing to diffusion over a crystal surface? To pursue this question, we will adopt a more detailed view, and explicitly consider diffusion as arising from the movement of atoms or particles; this will involve us in a little more elementary mathematics.

1.2.1 Binomial distributions

Assume that a particle makes a total of N independent, uncorrelated steps, each of length ℓ along an infinite straight line [8]. All jumps take place at the same time interval. We seek the probability p_x that the particle, having started at the origin, will end at position $x = s\ell$, but for the sake of simplicity we will assume the nearest-neighbor jump length to be unity. If the probability of a jump to the right is p and that for a jump to the left is q, then the likelihood of one configuration of n_1 jumps to the right and $N - n_1$ to the left is $p^{n_1} q^{N-n_1}$. Note that $p + q = 1$. How many different independent configurations are there for reaching the endpoint $s\ell$?

The first jump can be assigned in N ways on an empty line, the second in $N-1$, and so on. In total we therefore have $N!$ different choices. However, the selections for the n_1 steps to the right all lead to the same result, and there are $n_1!$ different arrangements for such steps. The same can be said about the $N - n_1$ steps to the left, which can be picked in $(N - n_1)!$ different ways giving the same effect. The total number of different configurations is therefore

$$\frac{N!}{n_1!(N-n_1)!}. \tag{1.37}$$

The probability of reaching the point s after n_1 steps to the right becomes

$$p_{n_1} = \frac{N!}{n_1!(N-n_1)!} p^{n_1} q^{N-n_1}. \tag{1.38}$$

From the binomial theorem we know that

$$(p+q)^N = \sum_{n_1=0}^{N} \frac{N!}{n_1!(N-n_1)!} p^{n_1} q^{N-n_1}. \tag{1.39}$$

Since $p + q = 1$, it is clear that the normalization condition

$$\sum_{n_1=0}^{N} p_{n_1} = 1 \tag{1.40}$$

is satisfied. The average number $\langle n_1 \rangle$ of jumps to the right is given by

$$\langle n_1 \rangle = \sum_{n_1=0}^{N} n_1 p^{n_1} \frac{N!}{n_1!(N-n_1)!} q^{N-n_1}. \quad (1.41)$$

This is easy to evaluate if we remember that

$$n_1 p^{n_1} = p \frac{\partial}{\partial p} p^{n_1}, \quad (1.42)$$

so that

$$\langle n_1 \rangle = \sum_{n_1=0}^{N} p \frac{\partial}{\partial p} p^{n_1} \frac{N!}{n_1!(N-n_1)!} q^{N-n_1}. \quad (1.43)$$

On interchanging the order of summation and differentiation, we have

$$\langle n_1 \rangle = p \frac{\partial}{\partial p} \sum_{n_1=0}^{N} p^{n_1} \frac{N!}{n_1!(N-n_1)!} q^{N-n_1}. \quad (1.44)$$

Taking advantage of the binomial theorem we find that

$$\langle n_1 \rangle = p \frac{\partial}{\partial p} (p+q)^N = Np(p+q)^{N-1} = Np, \quad (1.45)$$

and for the average number $\langle n_2 \rangle$ of jumps to the left

$$\langle n_2 \rangle = N - \langle n_1 \rangle = N(1-p) = Nq. \quad (1.46)$$

For the mean of the second power of the number of jumps to the right, $\langle n_1^2 \rangle$, we proceed in an analogous fashion.

$$\langle n_1^2 \rangle = p \frac{\partial}{\partial p} pN(p+q)^{N-1} = p\left[N(p+q)^{N-1} + pN(n-1)(p+q)^{N-2}\right] \quad (1.47)$$

and

$$\langle n_1^2 \rangle = Np(q+np). \quad (1.48)$$

For the fluctuation of n_1 we therefore find

$$\langle \Delta n_1^2 \rangle = \left\langle (n_1 - \langle n_1 \rangle)^2 \right\rangle = Np(q+np) - N^2 p^2 = Npq. \quad (1.49)$$

We now have many of the interesting quantities for the number of steps n_1 to the right. The value of the position x with ℓ_o unity is given by

$$x = n_1 - n_2 = n_1 - (N - n_1) = 2n_1 - N. \quad (1.50)$$

If the number of steps n_1 to the right is known we also know the position x. The distribution p_x of the position x is therefore the same as for jumps to the right, given by Eq. (1.38). The mean value of x becomes

1.2 Distribution of atomic displacements

$$\langle x \rangle = \langle n_1 \rangle - \langle n_2 \rangle = N(p-q). \tag{1.51}$$

For the displacement we have

$$\Delta x = x - \langle x \rangle = 2n_1 - N - (2\langle n_1 \rangle - N) = 2(n - \langle n_1 \rangle) = 2\Delta n_1 \tag{1.52}$$

so that for the displacement fluctuation

$$\langle \Delta x^2 \rangle = 4Npq. \tag{1.53}$$

The distribution can also be written more clearly in terms of the displacement x by taking advantage of Eq. (1.50). This gives

$$p_x = \frac{N!}{\left(\frac{N+x}{2}\right)!\left(\frac{N-x}{2}\right)!} p^{\left(\frac{N+x}{2}\right)} q^{\left(\frac{N-x}{2}\right)}. \tag{1.54}$$

When jumps to the right occur with the same probability as to the left this reduces to

$$p_x = \frac{N!}{\left(\frac{N+x}{2}\right)!\left(\frac{N-x}{2}\right)!} \left(\frac{1}{2}\right)^N. \tag{1.55}$$

We now have an expression that gives us the number of jumps N in terms of the measured displacements, and from Eq. (1.54) we can also find out the jump rates to the left and right.

1.2.2 Approximation for large values of N

When the total number of jumps N becomes large, evaluation of the probability p_{n_1} given by Eq. (1.38) requires more work, but an approximation can be reached readily. For large values of N, the probability p_{n_1} at the maximum becomes large and n_1 also assumes quite a large value. The condition for the maximum is readily derived by operating on the logarithm of the probability, which is less sensitive to n_1,

$$\ln(p_{n_1}) = \ln(N)! - \ln(n_1)! - \ln(N - n_1)! + n_1 \ln(p) + (N - n_1) \ln(q). \tag{1.56}$$

The necessary condition for the maximum is

$$\frac{d \ln(p_{n_1})}{dn_1} = 0. \tag{1.57}$$

For large values of N, we can resort to Stirlings approximation $\ln(N)! \approx N \ln(N) - N$, so that

$$\frac{d \ln(N)!}{dN} = \ln(N) \tag{1.58}$$

and from Eq. (1.54) we find

$$\frac{d \ln(p_{n_1})}{dn_1} = -\ln(n_1) + \ln(N - n_1) + \ln(p) - \ln(q) = 0. \tag{1.59}$$

It follows that
$$\ln\left(\frac{[(N - <n_1>)p]}{<n_1>q}\right) = 0, \quad (1.60)$$
where $<n_1>$ is the value of n_1 at the maximum. Now
$$(N - \langle n_1 \rangle)p = \langle n_1 \rangle q \quad (1.61)$$
so that
$$Np = \langle n_1 \rangle(p + q) \quad (1.62)$$
and, since $p + q = 1$,
$$\langle n_1 \rangle = Np, \quad (1.63)$$
which agrees with the result already obtained for the binomial distribution in Eq. (1.45).

We can expand the logarithm of the probability p_{n_1} around the maximum as
$$\ln(p_{n_1}) = \ln(p_{\langle n_1 \rangle}) + \frac{d\ln(p)}{dn_1}\Delta n_1 + \frac{1}{2}\frac{d^2\ln(p)}{dn_1^2}\Delta n_1^2 + \cdots. \quad (1.64)$$

The derivatives are all evaluated at the maximum so that the first derivative vanishes, and for the second we get from Eq. (1.59)
$$\frac{d^2\ln(p)}{dn_1^2} = -\frac{1}{n_1} - \frac{1}{N - n_1} = -\frac{N}{n_1(N - n_1)}. \quad (1.65)$$

At the peak, $\langle n_1 \rangle = Np$ and $N - \langle n_1 \rangle = N(1 - p) = Nq$, so that
$$\frac{d^2\ln(p)}{dn_1^2} = -\frac{N}{NpNq} = -\frac{1}{Npq}. \quad (1.66)$$

Note that the second derivative is negative as it must be at a maximum. Inserting the above result into Eq. (1.64) we obtain
$$\ln(p_{n_1}) = \ln(p_{\langle n_1 \rangle}) - \frac{1}{2}\frac{\Delta n_1^2}{Npq} \quad (1.67)$$
and
$$p_{n_1} = B\exp\left(-\frac{\Delta n_1^2}{2Npq}\right), \quad (1.68)$$
where $B = p_{\langle n_1 \rangle}$ is just a constant of proportionality. Now the normalization requires that
$$\int_{-\infty}^{\infty} p_{n_1} dn_1 = 1 \quad (1.69)$$
so we obtain
$$B\int_{-\infty}^{\infty} \exp\left(-\frac{\Delta n_1^2}{2Npq}\right) dn_1 = 1. \quad (1.70)$$

1.2 Distribution of atomic displacements

Substituting $a_k = \dfrac{1}{2Npq}$, we find after integration that

$$B = \frac{1}{\sqrt{2\pi Npq}}. \tag{1.71}$$

Finally we arrive at what is known as the Gaussian distribution

$$p_{n_1} = \frac{1}{\sqrt{2\pi Npq}} \exp\left(-\frac{\Delta n_1^2}{2Npq}\right). \tag{1.72}$$

We already have the average value of n_1. To find the fluctuation $\langle \Delta n_1^2 \rangle$, we note that

$$\langle \Delta n_1^2 \rangle = \langle (n - \langle n_1 \rangle)^2 \rangle = \frac{1}{\sqrt{2\pi Npq}} \int_{-\infty}^{\infty} (n_1 - \langle n \rangle)^2 \exp\left(-\frac{\Delta n_1^2}{2Npq}\right) dn_1. \tag{1.73}$$

Letting $z = \Delta n_1$, we can write

$$\langle z^2 \rangle = \frac{1}{\sqrt{2\pi Npq}} \int_{-\infty}^{\infty} z^2 \exp\left(-\frac{z^2}{2Npq}\right) dz. \tag{1.74}$$

The standard integral gives

$$\int_{-\infty}^{\infty} z^2 \exp\left(-\frac{z^2}{2Npq}\right) dz = \frac{\sqrt{\pi}}{2} (2Npq)^{3/2}, \tag{1.75}$$

so that finally

$$\langle \Delta n_1^2 \rangle = Npq. \tag{1.76}$$

The results here for the mean of n_1 and the fluctuation of n_1 is exactly the same as for the simple binomial with which we started.

We already know from Eq. (1.52) that $\Delta x = 2\Delta n_1$ so that Eq. (1.72) for the distribution can be converted to

$$p_x = \frac{1}{\sqrt{2\pi Npq}} \exp\left(-\frac{\Delta x^2}{8Npq}\right), \tag{1.77}$$

and from Eq. (1.76) we obtain

$$\langle \Delta x^2 \rangle = 4Npq. \tag{1.78}$$

We note here that just as in the binomial distribution, when the number of jumps N is even only even sites are populated, and when N is odd only odd sites are filled, inasmuch as $\Delta x = 2\Delta n_1$.

1.2.3 Motion in continuous time

In the presentation so far we have assumed that a fixed total number of jumps N is made and that these jumps take place regularly in time. Assume now instead that jumps may

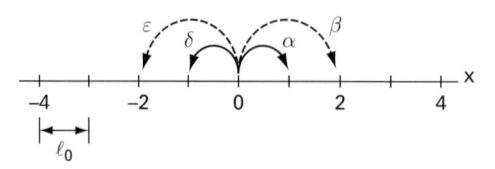

Fig. 1.3 Schematic of atomic processes for one-dimensional random walk. Nearest-neighbor jump length = ℓ_0.

occur at any time, that the rate of jumping is constant, and that jumps occur randomly [9]. In order to anticipate future concerns, we are now also going to expand our horizons to allow not just jumps to nearest-neighbor sites to the right at the rate α and to the left at the rate δ, but also longer jumps spanning two nearest-neighbor distances [10–12]. Double jumps to the right occur at the rate β, and to the left at the rate ε, as indicated in Fig. 1.3. To simplify the presentation, we assume that the nearest-neighbor jump length ℓ_0 is unity.

It is a simple matter to write out the equation for the time rate of change of the distribution function p_x as

$$\frac{dp_x}{dt} = \beta p_{x-2} + \alpha p_{x-1} - (\alpha + \beta + \delta + \varepsilon) p_x + \delta p_{x+1} + \varepsilon p_{x+2}. \tag{1.79}$$

In order to proceed further, it is useful to introduce the moment generating function $G(t, z)$, which is defined by

$$G(t, z) = \sum_{x=-\infty}^{\infty} z^x p_x. \tag{1.80}$$

For the time derivative of this function we find

$$\dot{G}(t, z) = \sum_{x=-\infty}^{\infty} z^x \frac{dp_x}{dt} = \\ \beta \sum_{x=-\infty}^{\infty} z^x p_{x-2} + \alpha \sum_{x=-\infty}^{\infty} z^x p_{x-1} - (\alpha + \beta + \delta + \varepsilon) \sum_{x=-\infty}^{\infty} z^x p_x \\ + \delta \sum_{x=-\infty}^{\infty} z^x p_{x+1} + \varepsilon \sum_{x=-\infty}^{\infty} z^x p_{x+2}. \tag{1.81}$$

This can be written in a more convenient form as

$$\dot{G}(t, z) = \beta z^2 \sum_{x=-\infty}^{\infty} z^{x-2} p_{x-2} + \alpha z \sum_{x=-\infty}^{\infty} z^{x-1} p_{x-1} - (\alpha + \beta + \delta + \varepsilon) \sum_{x=-\infty}^{\infty} z^x p_x \\ + \frac{\delta}{z} \sum_{x=-\infty}^{\infty} z^{x+1} p_{x+1} + \frac{\varepsilon}{z^2} \sum_{x=-\infty}^{\infty} z^{x+2} p_{x+2}. \tag{1.82}$$

A briefer way of presenting this equation is as

$$\dot{G}(t, z) = G(t, z) \left[-(\alpha + \beta + \delta + \varepsilon) + \alpha z + \frac{\delta}{z} + \beta z^2 + \frac{\varepsilon}{z^2} \right]. \tag{1.83}$$

1.2 Distribution of atomic displacements

The generating function is now obtained by integrating over time. Assuming that at time $t = 0$ the atom starts at $x = 0$, so that $G(0, z) = 1$, we find

$$G(t,z) = \exp[-(\alpha+\beta+\delta+\varepsilon)t]\exp\left[\left(\alpha z + \frac{\delta}{z}\right)t\right]\exp\left[\left(\beta z^2 + \frac{\varepsilon}{z^2}\right)t\right]. \quad (1.84)$$

The moments of the x coordinate are immediately accessible through

$$\langle x^n \rangle = \sum_{x=-\infty}^{\infty} x^n p_x = \left(z\frac{\partial}{\partial z}\right)^n G(t,z)_{z=1}. \quad (1.85)$$

The average displacement $\langle x \rangle$ is given by

$$\langle x \rangle = [\alpha + 2\beta - (\delta + 2\varepsilon)]t \quad (1.86)$$

and for the mean-square value we find

$$\langle x^2 \rangle = (\alpha + 4\beta + \delta + 4\varepsilon)t + (\alpha + 2\beta - \delta - 2\varepsilon)^2 t^2. \quad (1.87)$$

Third and fourth moment are given by:

$$\langle x^3 \rangle = \{\alpha + 8\beta - \delta - 8\varepsilon + 3[(\alpha+2\beta)(\alpha+4\beta) - (\delta+2\varepsilon)(\delta+4\varepsilon) - 2(\beta\delta - \alpha\varepsilon)]t \\ + (\alpha + 2\beta - \delta - 2\varepsilon)^3 t^2\}t \quad (1.88)$$

$$\langle x^4 \rangle = \{\alpha + 16\beta + \delta + 16\varepsilon + 12(\beta + \delta + 3\varepsilon)^2 t \\ + 12(\alpha + 2\beta - \delta - 2\varepsilon)(3\beta + \delta + \varepsilon)t + (\alpha + 2\beta - \delta - 2\varepsilon)^2 \\ \times [7 + 6(\alpha + 2\beta - \delta - 2\varepsilon)t + 12(\beta + \delta + 3\varepsilon)t \\ + (\alpha + 2\beta - \delta - 2\varepsilon)^2 t^2]t\}t. \quad (1.89)$$

The fluctuations in x, $\langle \Delta x^2 \rangle = \langle x^2 \rangle - \langle x \rangle^2$ are therefore

$$\langle \Delta x^2 \rangle = (\alpha + 4\beta + \delta + 4\varepsilon)t. \quad (1.90)$$

When jumps to the right take place at the same rate as to the left, as they usually do in diffusion, then

$$\langle x \rangle = 0 \quad (1.91)$$

$$\langle \Delta x^2 \rangle = 2(\alpha + 4\beta)t. \quad (1.92)$$

$$\langle x^3 \rangle = 0 \quad (1.93)$$

$$\langle \Delta x^4 \rangle = 2(\alpha + 16\beta)t + 12(\alpha + 4\beta)^2 t^2. \quad (1.94)$$

Finite values of the odd moments indicate that motion is asymmetric. In Fig. 1.4 is illustrated the average displacement (first moment) together with the third moment of the displacements against the ratio of jump rates to the left compared to the right; here α and β are both equal to 1/sec, δ as well as ε are equal to the ratio of left to right. As the rate of

Atomic diffusion on surfaces

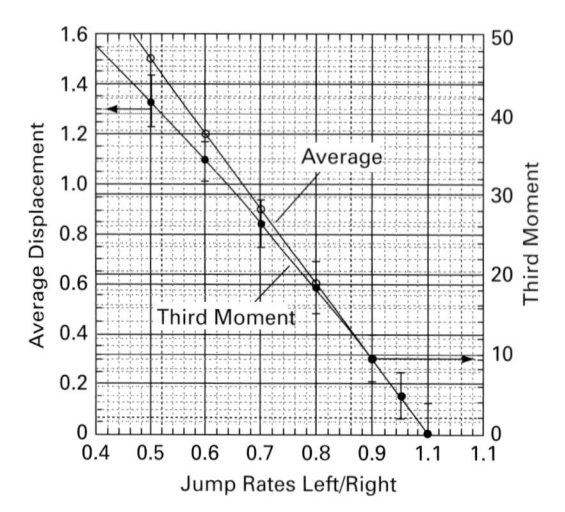

Fig. 1.4 Average as well as third moment of the displacement, as a function of the ratio of jump rates to the left compared with rates to the right. Diffusion time is 1 sec; rates α and β are set at 1/sec, δ as well as ε are equal to the ratio Left/Right. Error bars are estimates for 1000 observations of the third moment.

jumps to the left decreases, the average displacement increases, as the rates to the left are subtracted from rates to the right in Eq. (1.86). The third moment behaves in much the same way, but increases much more rapidly. Estimates of the standard error of the third moment characteristic for 1000 observations are small, so that a finite value for the third moment gives a clear indication of asymmetric movement. In the simplest case, where only single jumps occur, as in the previous sections,

$$\langle \Delta x^2 \rangle = 2\alpha t. \tag{1.95}$$

Deriving the average values of the displacement has been straightforward. Getting the distribution function for the displacements is more involved. As a start in this direction, we rewrite the generating function $G(t,z)$ in Eq. (1.84) as

$$G(t,z) = \exp[-(\alpha+\beta+\delta+\varepsilon)t] \exp\left[t\sqrt{\alpha\delta}\left(z\sqrt{\frac{\alpha}{\delta}} + \frac{1}{z}\sqrt{\frac{\delta}{\alpha}}\right)\right] \times$$

$$\exp\left[t\sqrt{\beta\varepsilon}\left(z^2\sqrt{\frac{\beta}{\varepsilon}} + \frac{1}{z^2}\sqrt{\frac{\varepsilon}{\beta}}\right)\right]. \tag{1.96}$$

At this stage it is useful to recall the Schlömilch [13] relation

$$\exp\left[\frac{\tau}{2}\left(u+\frac{1}{u}\right)\right] = \sum_{n=-\infty}^{\infty} u^n I_n(\tau), \tag{1.97}$$

where $I_n(\tau)$ is the modified Bessel function of order n and argument τ. With this relation we are now able to carry out two transformations:

1.2 Distribution of atomic displacements

$$\exp\left[t\sqrt{\alpha\delta}\left(z\sqrt{\frac{\alpha}{\delta}}+\frac{1}{z}\sqrt{\frac{\delta}{\alpha}}\right)\right] = \sum_{i=-\infty}^{\infty} z^i \left(\frac{\alpha}{\delta}\right)^{\frac{i}{2}} I_i\left(2t\sqrt{\alpha\delta}\right) \quad (1.98)$$

and

$$\exp\left[t\sqrt{\beta\varepsilon}\left(z^2\sqrt{\frac{\beta}{\varepsilon}}+\frac{1}{z^2}\sqrt{\frac{\varepsilon}{\beta}}\right)\right] = \sum_{j=-\infty}^{\infty} z^{2j} \left(\frac{\beta}{\varepsilon}\right)^{\frac{j}{2}} I_j\left(2t\sqrt{\beta\varepsilon}\right). \quad (1.99)$$

The generating function can therefore be written as

$$G(t,z) = \exp[-(\alpha+\beta+\delta+\varepsilon)t]$$
$$\times \sum_{i=-\infty}^{\infty}\sum_{j=-\infty}^{\infty} z^{i+2j} I_i\left(2t\sqrt{\alpha\delta}\right) I_j\left(2t\sqrt{\beta\varepsilon}\right) \left(\frac{\alpha}{\delta}\right)^{\frac{i}{2}} \left(\frac{\beta}{\varepsilon}\right)^{\frac{j}{2}}. \quad (1.100)$$

With the substitution of

$$x = i + 2j \quad (1.101)$$

the generating function transforms to

$$G(t,z) = \exp[-(\alpha+\beta+\delta+\varepsilon)t] \sum_{x=-\infty}^{\infty}\sum_{j=-\infty}^{\infty} z^x I_{x-2j}\left(2t\sqrt{\alpha\delta}\right) I_j\left(2t\sqrt{\beta\varepsilon}\right) \left(\frac{\alpha}{\delta}\right)^{\frac{x-2j}{2}} \left(\frac{\beta}{\varepsilon}\right)^{\frac{j}{2}}. \quad (1.102)$$

From the definition of the moment generating function in Eq. (1.80) we therefore arrive at the probability of finding a displacement x as

$$p_x = \exp[-(\alpha+\beta+\delta+\varepsilon)t] \sum_{j=-\infty}^{\infty} \left(\frac{\alpha}{\delta}\right)^{\frac{x-2j}{2}} \left(\frac{\beta}{\varepsilon}\right)^{\frac{j}{2}} I_j\left(2t\sqrt{\beta\varepsilon}\right) I_{x-2j}\left(2t\sqrt{\alpha\delta}\right). \quad (1.103)$$

When the movement is asymmetric this is clear from the distributions, shown in Fig. 1.5. As the rate of jumps to the left decreases with respect to the jumps to the right, the distribution starts to be asymmetric and gradually is shifted to the right. A diminution of 10 percent in the ratio of jumps to the left with respect to the jumps to the right can be detected. When jumps to the right and left occur at the same rate, so that $\alpha = \delta$ and $\beta = \varepsilon$, the distribution simplifies to

$$p_x = \exp[-2(\alpha+\beta)t] \sum_{j=-\infty}^{\infty} I_j(2\beta t) I_{x-2j}(2\alpha t). \quad (1.104)$$

If only nearest-neighbor transitions are allowed, the modified Bessel function $I_j(2\beta t)$ becomes zero for all orders except zero; in that order it is unity, and the probability can be written as

$$p_x = \exp(-2\alpha t) I_x(2\alpha t). \quad (1.105)$$

1.2.4 Comparisons

The distributions presented here differ significantly. If the mean-square displacement for the binomial or the Gaussian distribution is odd then movement will occur only over odd

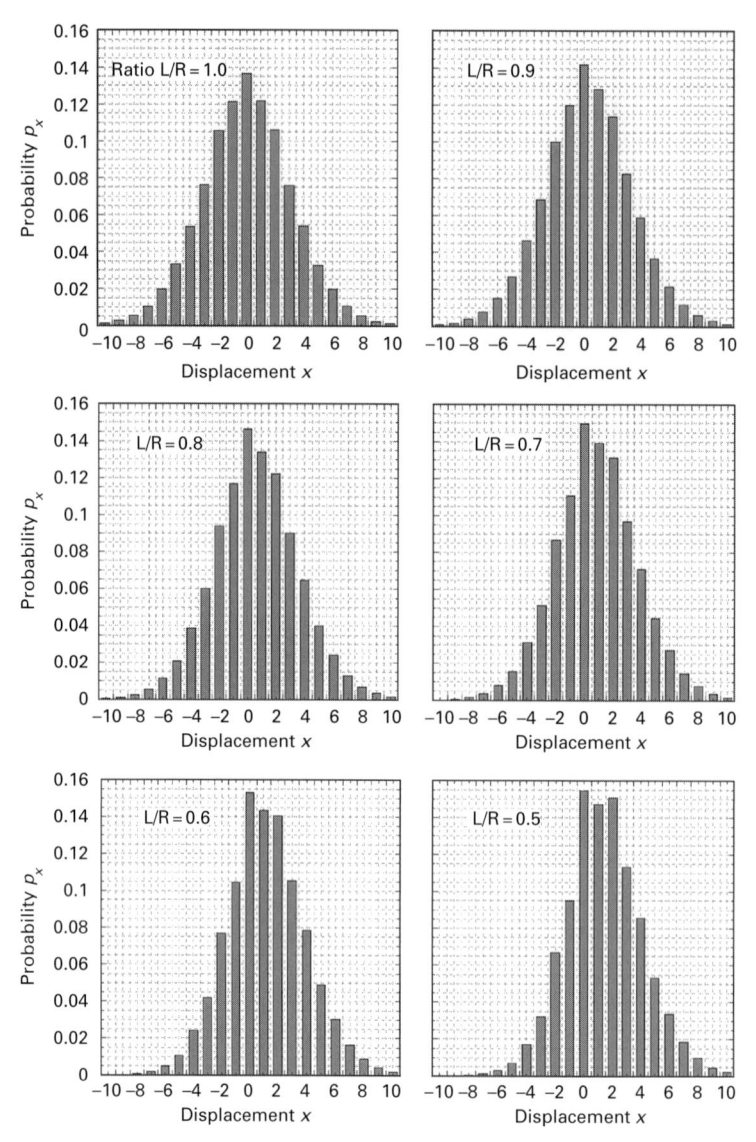

Fig. 1.5 Distribution of displacements for an asymmetric random walk in one dimension. Rates α and β to the right are maintained constant at 1/sec; diffusion interval is always 1 sec.

sites; when the mean-square displacement is even then only the even sites are ever occupied. A random walk in continuous time, however, will populate all kinds of sites. This is most readily seen in plots of the probabilities with single jumps occurring at the same rate right and left, which are shown in Fig. 1.6. It is clear that a measurement of the distribution of probabilities should therefore reveal which expression describes the migration. If only every second site is ever found occupied, then the binomial or the Gaussian distribution is appropriate. However, if the atoms are observed at every site regardless whether they are even or odd then the continuous time description is pertinent.

1.3 Jump rates

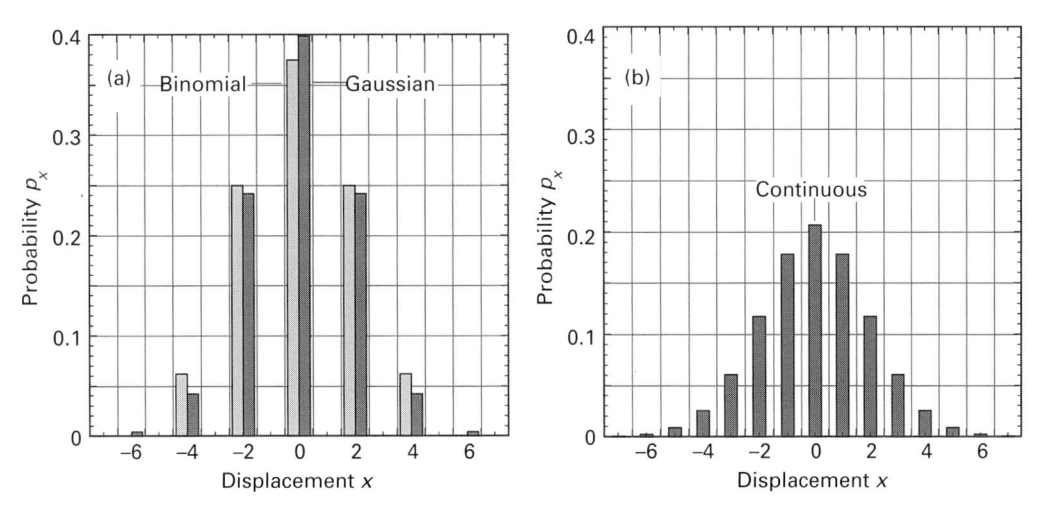

Fig. 1.6 Comparison of binomial and Gaussian distributions in (a) with continuous time distribution in (b), all with the number of jumps equal to 4, and $p = q = 1/2$.

By quantitatively fitting the appropriate distribution to the experimental data, the rates of jump processes participating in the diffusion can be found. The asymmetry of movement, if it occurs, can also be detected from distribution studies. Of course much more data are required to attain a reasonable distribution than just for diffusivities, but this provides the information to determine the rates of atom jumps.

1.3 Jump rates

In previous discussions we have introduced jump rates without any further examination of what they depend upon and how to formulate them. These are matters that we will now examine in more detail, relying on the transition state theory of reaction rates [14,15].

In diffusing over a crystal surface, an atom will experience a potential minimum at the equilibrium sites, which are separated by barriers that tend to localize the atoms. A schematic of such a one-dimensional potential is shown in Fig. 1.7. Of course the lattice is not static; atoms vibrate around their equilibrium positions and the barrier changes during such vibrations. The rate of jumping is given as the product of two terms: the probability that an atom will be at a value of the coordinate q_c at the top of the barrier, divided by its lifetime. The probability of being at the top of the potential barrier we will write as $\varpi(q_c)\delta q_c$, and the lifetime in this activated state is $\delta q_c / v_A$. Here v_A is the mean velocity in the positive direction and q_c is the distance along the path of least potential energy in moving from one site to another, usually referred to as the reaction coordinate.

To describe the system in the simplest terms we assume that atoms at the top of the barrier are in a separate state, the transition state, in equilibrium with the normal atoms at

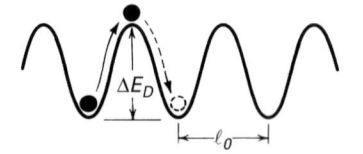

Fig. 1.7 Schematic of potential energy of an adatom undergoing diffusion along a line on a perfect surface.

the equilibrium sites. This of course means that the chemical potentials in the two states are the same,

$$\mu_A = \mu_N, \tag{1.106}$$

where the subscript A designates the active or transition state and N the normal or equilibrium state. For a localized material of non-interacting particles we have

$$\frac{\mu}{kT} = -\ln\frac{z'}{\theta} - \ln(1-\theta), \tag{1.107}$$

where z' is the partition function referred to a standard energy reference. The fractional occupation of surface sites, θ, is given by the ratio of the number of atoms to surface sites, that is by M/M_S. We are interested in the diffusion of individual atoms, so that θ will be vanishingly small and can be neglected compared to unity. The equality of chemical potentials therefore leads to

$$\frac{z'_A}{\theta_A} = \frac{z'_N}{\theta_N}. \tag{1.108}$$

We can write this equation as

$$\frac{\theta_A}{\theta_N} = \frac{z_A}{z_N}\exp\left(-\frac{\Delta E_D}{kT}\right), \tag{1.109}$$

where the partition functions are now evaluated with respect to the bottom of their potentials and ΔE_D indicates the height of the energy barrier; that is the difference in energy between the bottoms of the two potential curves. For one-dimensional motion, the number of sites in the activated state is equal to the number in the equilibrium state, so that

$$\frac{M_A}{M_N} = \frac{z_A}{z_N}\exp\left(-\frac{\Delta E_D}{kT}\right). \tag{1.110}$$

We can also write out the probability of being at the top of the barrier,

$$\varpi(q_c)\delta q_c = \frac{M_A}{M_A + M_N} \sim \frac{M_A}{M_N}, \tag{1.111}$$

where $M_N \gg M_A$.

To calculate the rate of moving from the top of the barrier, we assume that all atoms in a small length δq_c of the coordinate q_c at the saddle are in the activated state. The length δq_c must be so small that the curvature of the potential function is unimportant and motion

1.3 Jump rates

can be considered as a classical translation. Note, however, that δq_c must be large in comparison with the de Broglie wavelength λ_{deB}, where

$$\lambda_{deB} = \frac{h}{mv} \sim \left(\frac{h^2}{2\pi mkT}\right)^{\frac{1}{2}}. \tag{1.112}$$

This condition insures classical behavior for this degree of freedom. We now assume that the partition function for the activated state can be given by the product of the classical partition function for translation, z_{trans}, times a contribution from the remaining degrees of freedom, $z^{\#}$. That is,

$$z_A = z^{\#} z_{trans} = z^{\#} \left(\frac{2\pi mkT}{h^2}\right)^{\frac{1}{2}} \delta q_c. \tag{1.113}$$

From Eq. (1.110) the number of atoms M_A can now be written as

$$M_A = M_N \frac{z^{\#}}{z_N} \left(\frac{2\pi mkT}{h^2}\right)^{\frac{1}{2}} \delta q_c \exp\left(-\frac{\Delta E_D}{kT}\right). \tag{1.114}$$

To find the rate of diffusion, M_A, the number of atoms in the activated state, must be multiplied by the rate at which an atom leaves the activated state to the right. We take this rate v_A as the mean absolute value of the velocity

$$v_A = \frac{1}{2} \langle |v| \rangle \tag{1.115}$$

and

$$\langle |v| \rangle = \frac{\int_{-\infty}^{\infty} |v| \exp\left(-\frac{mv^2}{2kT}\right) dv}{\int_{-\infty}^{\infty} \exp\left(-\frac{mv^2}{2kT}\right) dv}. \tag{1.116}$$

Substituting $a = \frac{m}{2kT}$, we find

$$\int_{-\infty}^{\infty} |v| \exp(-av^2) dv = \frac{1}{a} \tag{1.117}$$

and

$$\int_{-\infty}^{\infty} \exp(-av^2) dv = \sqrt{\frac{\pi}{a}}, \tag{1.118}$$

so that

$$\langle |v| \rangle = \frac{1}{a} \sqrt{\frac{a}{\pi}} = \left(\frac{2kT}{\pi m}\right)^{\frac{1}{2}}. \tag{1.119}$$

Only half of the atoms will be jumping in the right direction, so that the rate at which a single atom moves, the rate constant, is given by

$$\Gamma = \frac{1}{2} M_A \frac{\langle |v| \rangle}{\delta q_c} = \frac{kT}{h} \frac{z^{\#}}{z_N} \exp\left(-\frac{\Delta E_D}{kT}\right). \quad (1.120)$$

To illustrate this result, we evaluate the rate for a simple, if unphysical model: the motion of an atom over a rigid one-dimensional potential. In the partition function for the activated state we no longer consider contributions in the direction of diffusion; only the motion in the normal state remains in z_N, so that

$$\frac{z^{\#}}{z_N} = \frac{1}{z_{vib}} = \frac{1 - \exp\left(-\frac{hv}{kT}\right)}{\exp\left(-\frac{hv}{2kT}\right)} = 2\sinh\left(\frac{hv}{2kT}\right); \quad (1.121)$$

here v is the frequency of vibration of an atom in the normal state. Expanding the hyperbolic sine, we find

$$2\sinh\left(\frac{hv}{2kT}\right) = \left(\frac{hv}{kT}\right)\left[1 + \frac{1}{24}\left(\frac{hv}{kT}\right)^2 + \cdots\right]. \quad (1.122)$$

Provided $\frac{hv}{kT} < 1$, the jump rate Γ in this example becomes

$$\Gamma = \frac{kT}{h} \frac{hv}{kT} \exp\left(-\frac{\Delta E_D}{kT}\right) = v \exp\left(-\frac{\Delta E_D}{kT}\right). \quad (1.123)$$

The rate of jumping is given by the product of v, the frequency of vibration multiplied by a Boltzmann term involving the barrier height ΔE_D. As already mentioned, this is not a realistic model. However, Vineyard [4] has shown that the jump rate can be quite generally written as

$$\Gamma = \frac{\prod_{j=1}^{n} v_j}{\prod_{i=1}^{n-1} v_i} \exp\left(-\frac{\Delta E_D}{kT}\right). \quad (1.124)$$

In the prefactor we have the ratio of the product of n normal frequencies of the system to the product of $n-1$ frequencies when the atom is located at the saddle point.

At this stage it should be emphasized that the dimensionality of the partition functions in Eq. (1.121) is not the same – in the normal state it is higher by one than in the activated state. For a classical system, this is easily changed. We rewrite the partition function for normal atoms as

$$z_N = z_1 z_{vib}. \quad (1.125)$$

1.3 Jump rates

where z_1 is the partition function with the one degree of vibration removed. If the vibrational contribution z_{vib} is assumed to be classical, then it follows that

$$z_{vib} \sim \frac{kT}{h\nu} \qquad (1.126)$$

and

$$\Gamma = \frac{kT}{h}\frac{z^{\#}}{z_1}\frac{h\nu}{kT}\exp\left(-\frac{\Delta E_D}{kT}\right) = \nu\frac{z^{\#}}{z_1}\exp\left(-\frac{\Delta E_D}{kT}\right). \qquad (1.127)$$

We now have an expression for the jump rate Γ that is useful if we can evaluate the two partition functions. It is more convenient for our applications, however, to make a thermodynamic transformation. The standard free energy change ΔF_D in going from the normal state (with one vibration removed) to the activated state is given by

$$\frac{\Delta F_D}{kT} = -\ln\left[\frac{z^{\#}}{z_1}\exp\left(-\frac{\Delta E_D}{kT}\right)\right]. \qquad (1.128)$$

The jump rate Γ for diffusion can therefore be written as

$$\Gamma = \nu\exp\left(-\frac{\Delta F_D}{kT}\right) = \left[\nu\exp\left(\frac{\Delta S_D}{k}\right)\right]\exp\left(-\frac{\Delta E_D}{kT}\right), \qquad (1.129)$$

where ΔS_D gives the entropy and ΔE_D the energy of activation. Here it is customary to set the prefactor Γ_0 in the jump rate to

$$\Gamma_0 = \nu\exp\left(\frac{\Delta S_D}{k}\right) \qquad (1.130)$$

and

$$\Gamma = \Gamma_0\exp\left(-\frac{\Delta E_D}{kT}\right). \qquad (1.131)$$

The displacement fluctuation is therefore

$$\langle \Delta x^2 \rangle = 2\Gamma t \langle \ell^2 \rangle, \qquad (1.132)$$

where $\langle \ell^2 \rangle$ is the mean-square value of the jump lengths. From the Einstein relation, Eq. (1.36), we know that $\langle \Delta x^2 \rangle = 2Dt$. It therefore follows that the diffusivity D is given by

$$D = \Gamma\langle \ell^2 \rangle; \qquad (1.133)$$

that is, the diffusivity D is just equal to the product of the jump rate and the mean-square jump length.

We have so far concentrated on one-dimensional motion. Little new enters when motion is two-dimensional. Consider diffusion by nearest-neighbor jumps on a (100) plane, where $\langle \Delta x^2 \rangle = \langle \Delta y^2 \rangle$. Since

$$\langle \Delta r^2 \rangle = \langle \Delta x^2 \rangle + \langle \Delta y^2 \rangle \qquad (1.134)$$

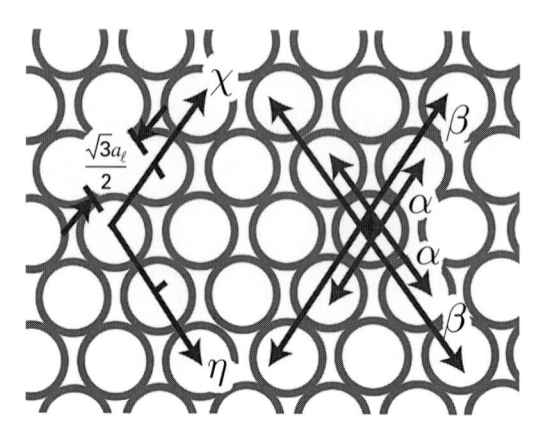

Fig. 1.8 Schematic of atom arrangement on bcc(110) plane, showing jumps to first and second nearest-neighbor sites along <111>.

we now can write

$$\langle \Delta r^2 \rangle = 4Dt. \tag{1.135}$$

On a bcc (110) plane, however, this simple relation does not hold, since only in atomic units is $\langle \Delta x^2 \rangle = \langle \Delta y^2 \rangle$. We therefore again assume that movement occurs by nearest-neighbor jumps but along the close-packed directions <111>, that is along the χ and η coordinates in Fig. 1.8. The relation between coordinates now is

$$x = \chi + \eta \qquad y = \chi - \eta. \tag{1.136}$$

Diffusion along χ and η axes is not correlated, so that

$$\langle \Delta x^2 \rangle = \langle \Delta y^2 \rangle = \langle \Delta \chi^2 \rangle + \langle \Delta \eta^2 \rangle \tag{1.137}$$

and

$$\langle \Delta x^2 \rangle = 4\alpha t, \tag{1.138}$$

with the unit of length being $a_\ell/2$, that is half the lattice constant; α is the rate of jumps along <111>. $\langle \Delta y^2 \rangle$ can be described by the same relation

$$\langle \Delta y^2 \rangle = 4\alpha t, \tag{1.139}$$

but with the unit of length being equal to $a_\ell \sqrt{2}/2$.

On the fcc(111) plane the situation is more complicated, as atoms may sit in either fcc or hcp sites, indicated in Fig. 1.9. Atom jumps, shown in the same figure, occur at the rate α_{fh} from fcc to hcp sites and at the rate α_{hf} for the reverse process. Expressions have been worked out [10] relating the diffusivity to the jump rates,

$$D = \langle \Delta x^2 \rangle / 2t = \frac{\alpha_{fh} \alpha_{hf}}{\alpha_{fh} + \alpha_{hf}} \ell_0^2/2, \tag{1.140}$$

where ℓ_0 is the nearest-neighbor spacing. Evaluation of the diffusivity D in terms of jump processes is therefore straightforward for both one- and two-dimensional processes.

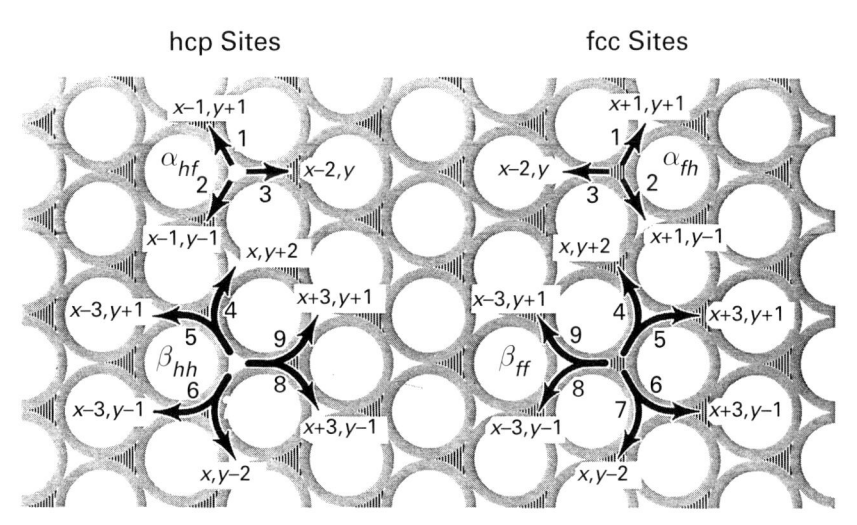

Fig. 1.9 Atom arrangement on fcc(111) plane, showing short as well as long jumps from fcc and hcp sites [10].

References

[1] J. Crank, *The Mathematics of Diffusion*, 2nd edn. (Clarendon Press, Oxford, 1975), Chapter II.
[2] R. M. Barrer, *Diffusion in and through Solids* (Cambridge University Press, London, 1941), p. 14.
[3] A. Einstein, Elementare Theorie der Brownschen Bewegung, *Z. Elektrochem.* **14** (1908) 235–239.
[4] G. H. Vineyard, Frequency factors and isotope effects in solid state rate processes, *J. Phys. Chem. Solids* **3** (1957) 121–127.
[5] U. Kürpick, A. Kara, T. S. Rahman, Role of lattice vibrations in adatom diffusion, *Phys. Rev. Lett.* **78** (1997) 1086–1089.
[6] U. Kürpick, Self-diffusion on (100), (110), and (111) surfaces of Ni and Cu: A detailed study of prefactors and activation energies, *Phys. Rev. B* **64** (2001) 075418 1–7.
[7] H. W. Berg, *Random Walks in Biology* (Princeton University Press, Princeton, 1983), p. 6–11.
[8] S. Chandrasekhar, Stochastic problems in physics and astronomy, *Rev. Mod. Phys.* **15** (1943) 1–89.
[9] G. Ehrlich, Atomic displacements in one- and two-dimensional diffusion, *J. Chem. Phys.* **44** (1966) 1050–1055.
[10] J. D. Wrigley, M. E. Twigg, G. Ehrlich, Lattice walks by long jumps, *J. Chem. Phys.* **93** (1990) 2885–2902.
[11] F. Montalenti, R. Ferrando, Universal law for piecewise dimer diffusion, *Phys. Rev. B* **60** (1999) 11102–11109.
[12] G. Antczak, G. Ehrlich, Asymmetric one-dimensional random walks, *J. Chem. Phys.* **129** (2008) 124702 1–4.
[13] M. Abramowitz, I. A. Stegun, *Handbook of Mathematical Functions* (Dover, New York, 1970), Chapter 9.
[14] S. Glasstone, K. J. Laidler, H. Eyring, *The Theory of Rate Processes* (McGraw-Hill Book Co., New York, 1941).
[15] K. J. Laidler, J. H. Meiser, *Physical Chemistry*, 3rd edn. (Houghton Mifflin, Boston, 1999).

2 Determination of adatom movements

Surface diffusion studies on single adsorbed entities, the focus of our presentation, had to await the development of techniques capable of revealing atoms. This was first accomplished by Müller [1,2] roughly fifty years ago in 1956, with his invention of the field ion microscope (FIM). The natural extension of FIM was the development of the Atom Probe [3] which allowed identification of chemical identities and control of composition for surfaces, but there also were earlier investigative methods, such as field emission microscopy [4], helium scattering [5], contact potential measurements [6] and so on, which provided useful information about surface diffusion. Today there are newer techniques that have been shown to have the capability of revealing atoms. The scanning tunneling microscope (STM), devised by Binnig and Rohrer [7,8] in 1983 is one of them. Less frequently used techniques, such as measurements of work function changes, perturbed $\gamma - \gamma$ angular correlation, or atomic beam scattering will also be mentioned, if only very briefly. Insights into diffusion phenomena on the atomic scale gained with the scanning tunneling microscope are certain to grow in number and importance. Both field ion [9–11] and scanning tunneling microscopy [12–15] have been covered extensively in the literature, and will also be described here in reference to diffusion studies. It should be noted that for examination of diffusion phenomena on clean surfaces by any of these techniques, good vacuum conditions are crucial. The influence of just small amounts of impurities on the movement of atoms, even in the range of 10^{-10} Torr, has recently been recognized [16]. The microscopes must therefore all be built in such a way as to allow bakeout at temperatures of 300 °C or higher, and careful procedures for outgassing must be followed.

2.1 Field ion microscopy

2.1.1 Imaging equipment

A picture of a field ion microscope suitable for examining atom diffusion is shown in Fig. 2.1 [17]. The vital part of the microscope is the tip, welded onto the top of the support loop and positioned opposite to the phosphor screen. An image of the surface is obtained by admitting an imaging gas (helium is the best candidate, but neon can be used as well), at a pressure of $<10^{-3}$ Torr to the microscope, and then raising the positive potential on the tip (or the negative potential on the screen) until the surface field reaches ~ 4.5 V/Å

2.1 Field ion microscopy

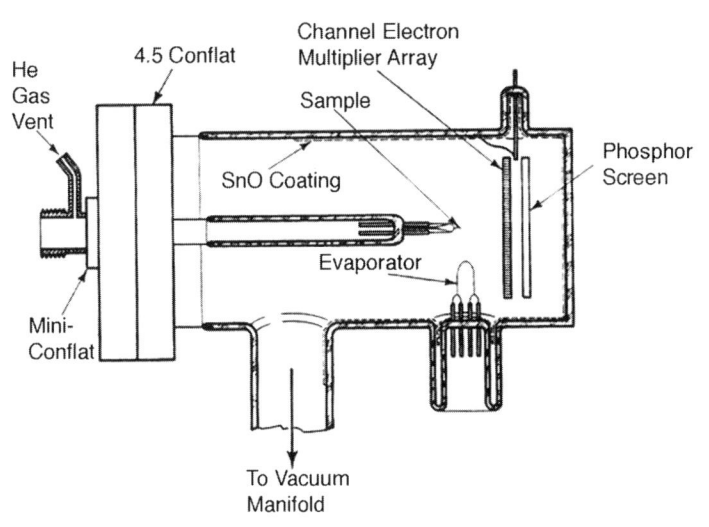

Fig. 2.1 Schematic of a field ion microscope for studies of atom surface diffusion [17].

(for helium gas). Under these conditions, helium atoms are ionized close to the tip, and ions are accelerated by the field to an image intensifier assembly, on the back of which the image is displayed on a fluorescent screen. For imaging, the tip temperature generally is ~20 K, to immobilize atoms adsorbed on the surface; however, cooling with liquid nitrogen (~77 K) is sufficient for some systems. Of course the temperature of the sample has to be higher than the condensation temperature of the image gas. Picture taking is best done with a high-gain TV camera, which captures an image in seconds; the output of the camera is sent to a video recorder or better to a computer for storage. The image is a projection of the curved surface at the end of the tip, consisting of steps, edges, and terraces as well as low energy planes. In most cases the edge atoms of steps and terraces are clearly displayed on the screen. A picture of an Ir(111) surface [18] obtained in the way described is shown in Fig. 2.2a. When large, the low index planes, like fcc(111) and fcc(100) usually remain dark. Only the more protruding atoms are imaged due to differences in the local electric field. The interior of terraces and planes is not resolved directly. However, using an atom as a probe and carefully mapping the sites it occupies gives an insight into this region of the sample. Also, it is not always true that the interior of planes cannot be resolved. When we look at a small plane not close packed, like W(111) [19], all atoms of the plane can be resolved, as in Fig. 2.2b.

The image intensifier assembly in the FIM is a very simple device. A stainless steel ring supports on top a microchannel plate (Photonis USA, Sturbridge, Mass), and on the bottom, at a distance of ~5 mm, a phosphor screen. At a voltage in the range of 1000 V or more on the plate, ions striking the channel plate release electrons which are accelerated down the 25 micron tubes of which the plate is composed and release additional electrons on striking the walls. Thereafter these impinge upon the phosphor to give a much enhanced image of where the ions originated. A single channel plate is adequate for most microscopy; for more demanding enterprises, a double channel plate can be substituted.

26 **Determination of adatom movements**

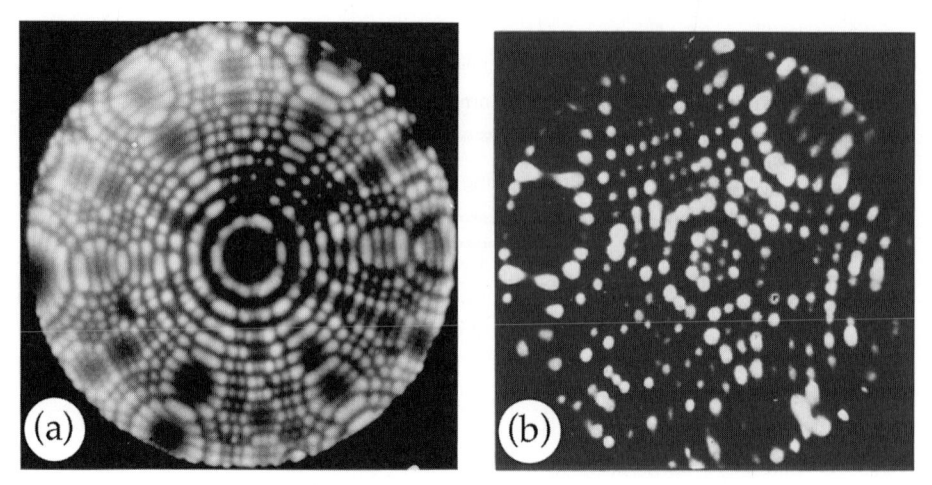

Fig. 2.2 Atom resolution in FIM. FIM image of (a) Ir sample, with an unresolved (111) plane in the center [18], (b) small (111) plane of tungsten in the center is atomically resolved [19].

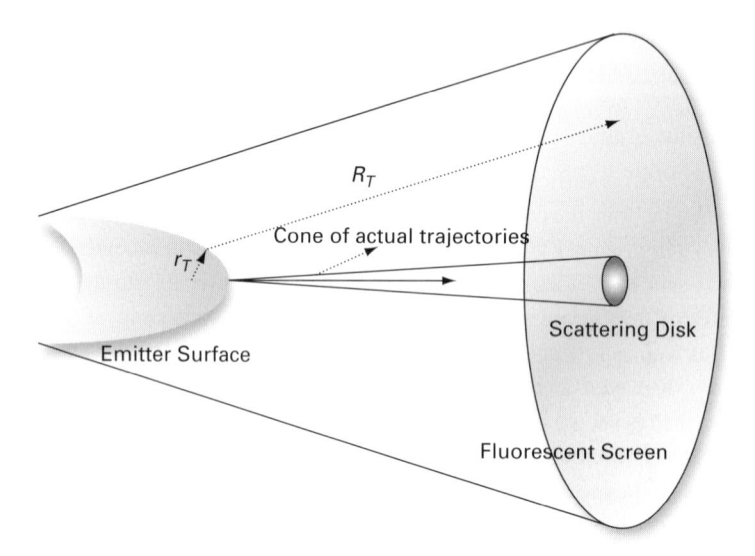

Fig. 2.3 Schematic of a field ion microscope, showing important dimensions.

Cooling for the tip to ~20 K is provided by a mixture of liquid and gaseous helium supplied to the leads on which the sample support is welded. The gas pressure in the microscope is kept as high as possible to enhance the image intensity, but the highest pressure is limited by collisions between the He$^+$ ions from the tip and atoms from the image gas. The mean free path in the gas must be maintained at 10 cm or more to avoid losing information about the image. Typically work is done at pressures in the range 10^{-4} Torr helium. The magnification M_T of the instrument is $M_T \sim R_T/r_T$, the ratio of the distance from tip to screen R_T, compared to the tip radius r_T, shown in Fig. 2.3; the resolution is roughly $\sqrt{kT/eV}$.

2.1 Field ion microscopy

The type of material that is readily imaged has to have a high cohesion to withstand the mechanical force $F_e/8\pi$ exerted by the applied field F_e. Helium gas gives the sharpest images, at a field of ~4.5 V/Å. However, the rate R_{FI} of ionization in a field F_e is roughly proportional to

$$R_{FI} \propto \exp\left[-\frac{4}{3}\sqrt{\frac{2m}{\hbar^2 e^2} \frac{I^{3/2}}{|F_e|}}\right] \quad (2.1)$$

where I is the ionization potential of the gas, m and e are the mass and charge of an electron, and $\hbar = h/2\pi$ where h is Planck's constant. It is clear that the force on the tip is highest for He, with an ionization potential $I = 24.587$ eV. Conditions can be made less strenuous by using neon, with an ionization potential of 21.564 eV, for imaging. Both helium as well as neon can be admitted to the system through heated Vycor tubes, but imaging with neon exerts less of a force on the sample. In this way, images have been obtained for a metal as soft as gold [20,21] as in Fig. 2.4, or palladium [22], and some semiconductors have been investigated as well [23,24]. For diffusion studies, the binding strength of adatoms to the surface also enters, as weakly bound entities are likely to be field desorbed during the imaging process.

Field ion microscopy gives the possibility of cleaning the sample by field evaporation of surface layers, not accessible in many other techniques. The field breaks the bonding between atoms of the first layer and causes their acceleration toward the screen. In this way it is possible to uncover layer after layer of the material, which is basic for the analysis of composition in Atom Probe microscopy, discussed briefly below. To get a well-defined surface, three types of cleaning procedures are applied in field ion

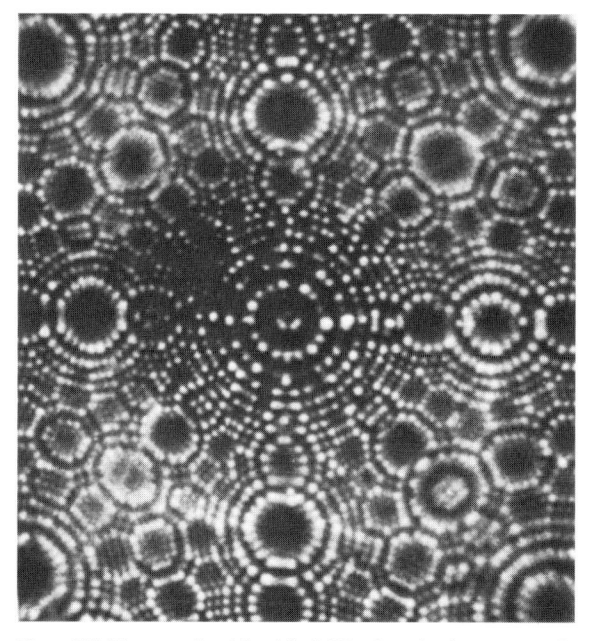

Fig. 2.4 Neon FIM image of gold, with (100) plane in the center (after Averback and Seidman [20]).

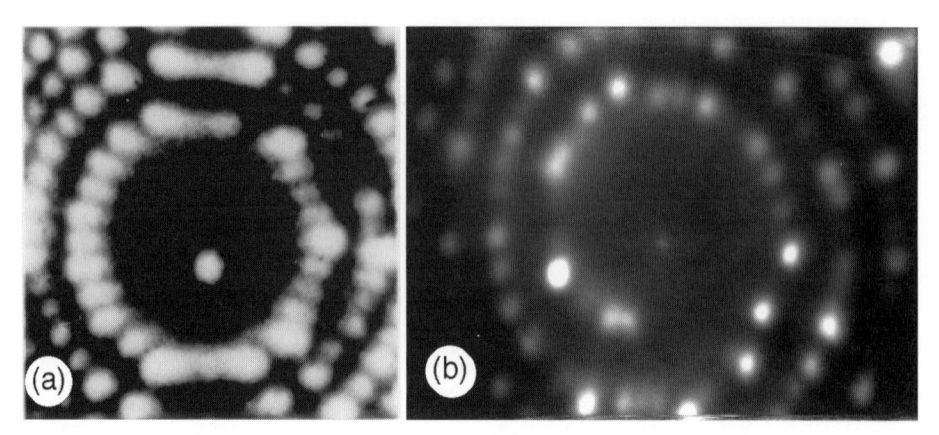

Fig. 2.5 Surface imperfections revealed by FIM. (a) Image of a single Re atom on W(211) plane [58]. (b) Screw dislocation emerging on W(110) plane [26].

microscopy: sputtering by neon [25], heating of the surface, and field evaporation of the outer layer. Sputtering of the tip not only cleans a surface, but can also change the shape and sharpness of the tip. This is accomplished by making the tip negative, causing field emission of electrons from the surface. The electrons on colliding with neon atoms in the gas phase ionize them, and the ions are accelerated to the sample surface, their impact bringing about sputtering. Since the field emission current is direction dependent due to differences in work function from different crystallographic planes, the flux of neon ions striking the tip is also direction dependent and causes changes in the shape of the tip.

FIM allows precise control of the number of adatoms deposited on a surface; these adatoms protrude and their image can be seen directly on the screen, as is clear in Fig 2.5a. The additional adatom deposited at the surface can, in a selective way, be removed from the surface by field evaporation, after migration to the edges of the plane where the field is locally higher. The distance of adatoms from surface imperfections like steps or dislocations [26], shown in Fig. 2.5, can be precisely controlled, making the FIM an ideal tool for examination of single atom events. Unfortunately, use is limited by the strength and conductivity of material as well as the finite size of the planes.

2.1.2 Atom probe microscopy

Field ion microscopy is great in that it delivers information about the location of adatoms on a metallic surface. It does not provide a direct indication of the chemical composition of the atoms imaged, although this can in some instances be inferred indirectly. This limitation was removed by Müller et al. [3], who in 1968 came up with the Atom Probe. In this instrument, a schematic of which is shown in Fig. 2.6 [27], atoms are evaporated from the surface by a field pulse, and the time of flight t to a detector down a flight tube, evacuated like the rest of the system, is measured. From a knowledge of the tip to detector distance L_T and the applied voltage V, the mass-to-charge ratio for the material desorbed is obtained. The velocity v of the evaporated particle is given by

2.2 Field electron emission microscopy

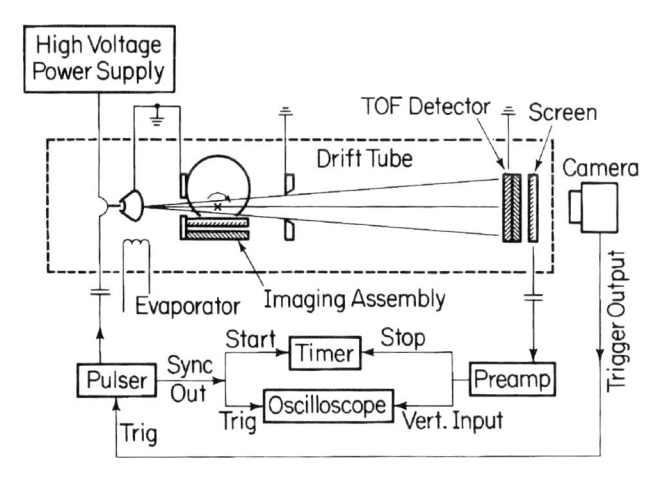

Fig. 2.6 Schematic of Atom Probe [27].

$$v = L_T/t, \quad (2.2)$$

and from the conservation of energy we find that

$$(1/2)mv^2 = n_c eV, \quad (2.3)$$

so that finally

$$\frac{m}{n_c} = 2eV\frac{t^2}{L_T^2}; \quad (2.4)$$

here n_c is the number of charges on the evaporated ion. The Atom Probe therefore allows us to get an indication, not only of the location of an adatom on the surface, but also of its chemical identity. In general, this ability has so far not had important consequences for understanding surface diffusion, except for one experiment: the work of Wrigley [28,29], demonstrating that cross-channel diffusion on Ir(110) actually takes place by the exchange of a deposited adatom with an atom from the substrate.

Descriptions of the instrument and techniques are available in the literature [30–33], and a modern probe for three-dimensional analysis of sample composition has been offered by Imago Scientific Instruments, Madison, Wisconsin as well as by Cameca Science and Metrology Solutions, Gennevilliers, France, so that further details will not be offered here.

2.2 Field electron emission microscopy

Prior to the invention of the field ion microscope, considerable information about surface behavior had been obtained by field emission of electrons, in which electrons instead of ions are emitted from the surface under study when it is subjected to an electric field of ~0.3 V/Å. The density of the emission current I_e (in amps/cm^2) is given by the Fowler–Nordheim relation [34], which we write as

$$I_e = 1.54 \times 10^{-6}(F_e^2/\phi t_e^2)\exp\left[\left(-6.83 \times 10^7 \frac{\phi^{3/2}}{F_e}\right)v_e\right]; \tag{2.5}$$

here ϕ is the work function of the surface in eV, F_e the electric field in V/cm, and t_e as well as v_e are slowly varying functions of $3.79 \times 10^{-4} F_e^{1/2}/\phi$. The addition of atoms to the surface will affect the work function, bringing about a change in the emission current, which is easily detected. Movement of adsorbed material over an emitting surface can therefore be followed, by focusing on the current emitted from a small area as adsorbed material drifts in. The resolution of the technique is limited, however, to ~10 Å or more even on very sharp tips, so that individual atoms cannot easily be visualized [4].

Nevertheless, information about the surface diffusion of gases as well as alkali metals has been obtained by Gomer and coworkers [35] and by Kleint and coworkers [36–38]. There is an alternative available to these kinetic measurements. That is to look at a system in equilibrium, and to examine fluctuations in concentration in a small region on the surface. The amount of material contained in a given area A will change as atoms enter or leave by diffusion, so that measurements of the concentration fluctuations will yield information about the diffusivity. Variations in the number m_a of particles can be represented by the autocorrelation function

$$\rho(t_f) = \frac{\langle \delta m_a(t_f)\delta m_a(0)\rangle}{[\delta m_a(0)]^2}, \tag{2.6}$$

where $\delta m_a(t_f) \equiv m_a(t_f) - m_a(0)$. Smoluchowski [39,40] long ago showed that

$$\langle \delta m_a(t_f)\delta m_a(0)\rangle = \langle m_a\rangle(1 - P). \tag{2.7}$$

Here $1 - P$ is the conditional probability that material present in the small area at time 0 will also be present at time t_f. It is given by

$$1 - P = \frac{1}{4\pi D t_f A}\int\int_{A\ A}\exp\left[-\frac{(R_1 - R_2)^2}{4Dt_f}\right]dR_1 dR_2. \tag{2.8}$$

Integration of the initial position vector R_1 as well as of the final position R_2 is carried out over the area A of the sampled region. For a square with a side of length L_L this gives

$$1 - P = \frac{1}{L_L^2}\left\{L_L erf\left[\frac{L_L}{(4Dt_f)^{1/2}}\right] - \left(\frac{4Dt_f}{\pi}\right)^{1/2}\left[1 - \exp\left(-\frac{L_L^2}{4Dt_f}\right)\right]\right\}^2. \tag{2.9}$$

For a circular area of radius R_r integration yields

$$1 - P = \exp\left(-\frac{R_r^2}{2Dt_f}\right)\left[I_o\left(\frac{R_r^2}{2Dt_f}\right) + I_1\left(\frac{R_r^2}{2Dt_f}\right)\right]. \tag{2.10}$$

Here $I_n(z)$ is the modified Bessel function of order n and argument z.

It is clear that from a measurement of the autocorrelation function it is possible to derive values of the diffusivity D. Adapting emission fluctuations to field emission microscopy

2.3 Scanning tunneling microscopy

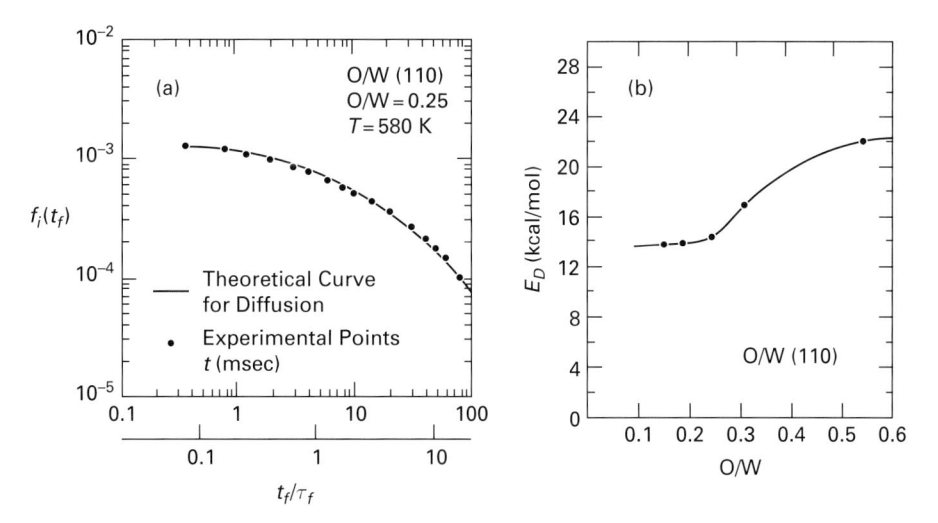

Fig. 2.7 Diffusion of oxygen on W(110), deduced from fluctuations in the field emission current. (a) The fluctuation correlation function in its dependence upon the time. Solid curve gives theoretical fit. τ_f is the relaxation time for fluctuations. (b) Activation energy for oxygen diffusion as a function of surface coverage. Very low coverages are difficult to measure (after Chen and Gomer [44]).

was done by Timm and Van der Ziel [41], by Kleint [42,43] and most successfully by Gomer and coworkers [35], who measured the fluctuation correlation function

$$f_i(t_f) \cong \langle \delta_i(t_f)\delta_i(0)\rangle / i_e^2, \qquad (2.11)$$

where i_e is the average current. An example of this function for the diffusion of oxygen on W(110) is shown in Fig. 2.7a, and a graph of the activation energy is given in Fig. 2.7b [44]. This plot indicates one of the limitations of the technique: measurements at very low concentrations are difficult and have therefore usually not been carried out. Another factor that must be considered, especially in examining diffusion of metal atoms, is that in this powerful technique an electric field to bring about emission of electrons is constantly applied, and this may influence diffusion processes.

2.3 Scanning tunneling microscopy

The basic idea for scanning tunneling microscopy came from Young *et al.* [45], who envisioned a sharp tip, moved parallel to a surface while mounted on piezoelectric supports. By applying a high enough voltage to the surface, electrons would tunnel from the tip, their intensity depending upon the distance of the tip from the surface. Although the equipment was carefully designed and built, its ability to resolve surface features was limited by some lack of rigidity. This was later rectified, and it was possible to bring the tip so close to the surface that electron tunneling occurred [46].

A great advance was made by Binnig and Rohrer [7,47] at IBM Zurich, who built an instrument much like Young's, but considerably more rigid and stable. With this they

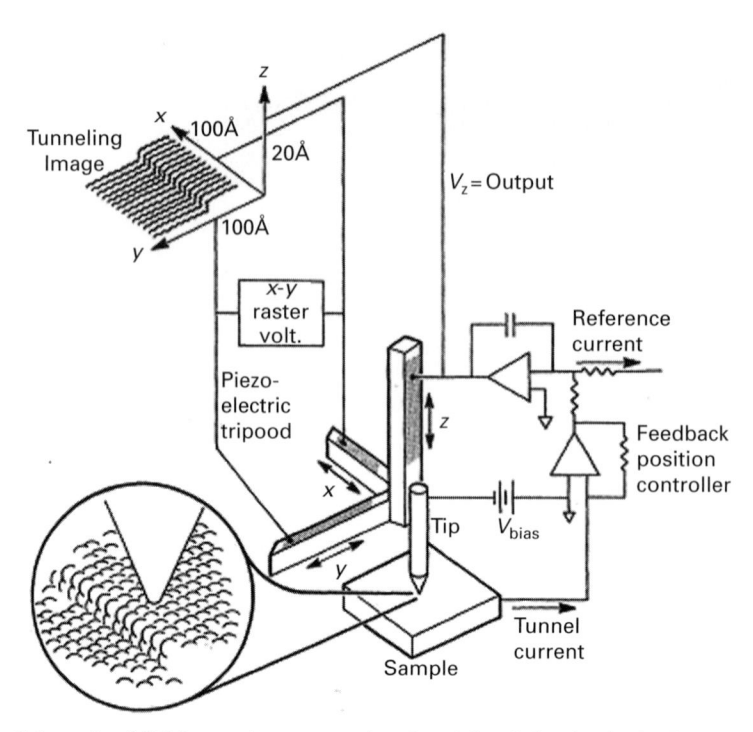

Fig. 2.8 Schematic of STM, scanning a stepped surface (after Golovchenko [48]).

were able to resolve atomic surface details, a feat for which they won the Nobel Prize. A schematic of such a microscope is shown in Fig. 2.8 [48]. The scanning tip is mounted on a tripod of piezoelectrics. The x and y member drive the tip over the surface. The vertical distance to the surface is set by a voltage bias, which induces a tunneling current. This is compared with a reference current, producing a difference voltage which is applied to the z member and adjusts its elevation until the measured current agrees with the set value. The voltage on the z leg is plotted, and provides an indication of the surface structure, in that steps or adatoms will tend to increase the tunneling. The instrument was announced in 1982, caught on rapidly, and is now produced in different versions by several companies, among them Omicron Nanotechnology, Taunusstein, Germany. STM tips are also available commercially, for example from Materials Analytical Services, Raleigh, NC.[1]

Most successful in studies of atomic events has been the fast STM introduced by Aarhus University and produced by Specs Technologies Corporation, Sarasota, Florida; this has already been used for examination of atomic movements on Pt(110)-(1 × 2), shown in Fig. 2.9 [49].

In this figure, the outermost rows are clearly seen as protruding regions, giving greater insight than afforded by the field ion microscope. The STM probes the surface density of states and it is possible to create an image in two regimes, with filled and empty surface

---

[1] For other suppliers, see *Physics Today* buyers guide, August 2009.

2.3 Scanning tunneling microscopy

Fig. 2.9 Image of Pt(110)-(1 × 2) taken with STM after deposition of Pt. Shown are adatoms as well as linear clusters situated in surface channels (after Linderoth *et al.* [49]).

states. For metallic surfaces both images are supposed to overlap and to avoid misunderstandings, comparison of both images is desirable. In studying atomic surface diffusion, it is necessary to pay attention to the distance between the tip and the surface. Too close a tip can cause movement of atoms over the surface or cause changes in potential wells, which will influence the rate of atom jumps. Another thing to consider is charge transfer between tip and surface, which may also influence the movement of atoms and the stability of clusters. In studying single atom diffusion directly by scanning tunneling microscopy, it is necessary to monitor the distance of other atoms to avoid long-range interactions of atoms with each other. For example, Morgenstern and Rieder [50] recently observed a metastable configuration of Cu dimers on the Ag(111) plane caused by the presence of atoms in the vicinity. The relative drift of sample and the tip, which may artificially change diffusion data, must also be carefully followed. The distance from imperfections such as steps or dislocations has to be controlled as well.

STM, like FIM, works only with conducting materials, but is much more successful in looking at soft materials like for example Cu and semiconductors than FIM, due to the lower stress applied to the sample. A major difference between the two techniques is the shape of the sample. In FIM the sample is a small tip with a high curvature, for STM the sample is essentially flat, which makes it possible to work with bigger planes. A number of ways can be used to clean the sample in STM; the combination of heating and sputtering is the most common one. For sputtering, ionization of a gas, for example argon, is accomplished with use of ion guns to bombard the sample surface. Chemical processing is also a possibility for cleaning. One problem of concern is the uniformity of the temperature over the sample; such gradients will induce movement of species.

Neither of the two techniques works with insulator surfaces. Recently, however, atomic resolution was achieved in atomic force microscopy, where the force between tip and surface is monitored instead of the emission current [51]. This technique might be a good candidate for studying atomic surface diffusion on insulator surfaces.

2.4 FIM measurement of diffusivity

2.4.1 Tip preparation

Creating the tip plays a crucial role in both field ion and scanning tunneling microscopy; however, the role of the tip in the two techniques is totally different. In the field ion microscope, the tip is the subject of the investigation, the sample. In the scanning tunneling microscope, the tip is a probe allowing measurements by scanning the flat sample. Crucial for the operation of the field ion microscope is preparing a sharply pointed tip with an outer radius on the order of 100 Å. This is usually accomplished by electro-polishing, and a number of recipes are suggested for different materials in references on electron microscopy. For various materials, etching techniques have been outlined by Godhew [52], for example, and general methods have been given by Müller and Tsong [10] as well as by Miller *et al.* [31]. At Illinois, methods for producing usable FIM tips have been tested by Liu [53], and in the Appendix we reproduce the description of techniques found useful here.

2.4.2 System preparation

It is most straightforward to describe the system and techniques used here at the University of Illinois. At the start, the sample wire, preferably a single crystal specimen of the right orientation and roughly 0.005 in. in diameter, is spot-welded onto the top of the support loop. This is made of tungsten wire, 0.007 in. thick, spot-welded onto the heavy feedthroughs of the sample mount. The support wire has attached to it two fine probe wires, 0.005 in. or less in diameter, and fastened symmetrically on either side of the sample, for determining the temperature. A schematic is given in Fig. 2.10. The loop is

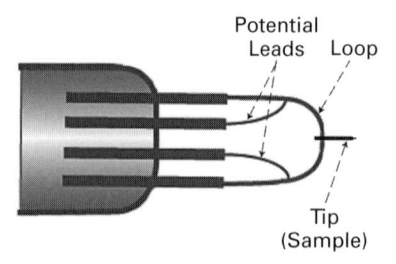

Fig. 2.10 Diagram of crystal support loop for FIM. Outer loop, of 0.007 in tungsten wire, supports and allows heating sample. Probe wires (potential leads) attached to loop measure the temperature in terms of the resistivity.

2.4 FIM measurement of diffusivity

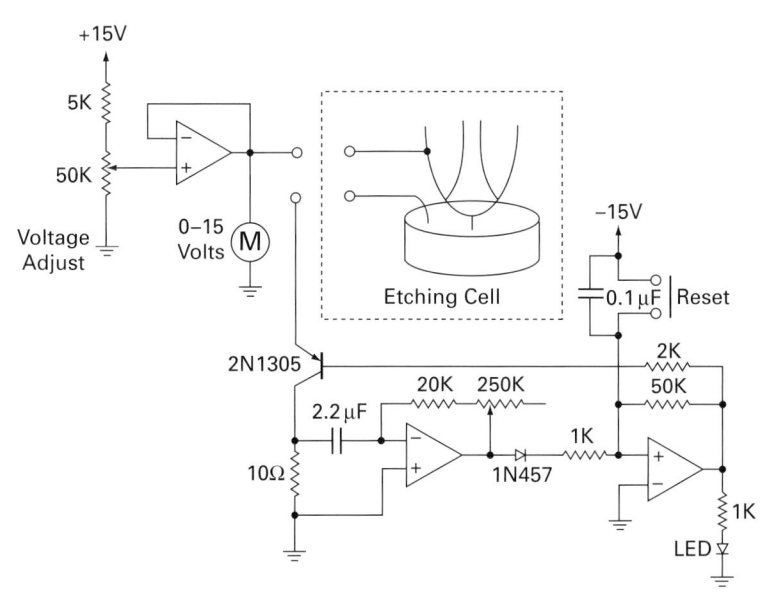

Fig. 2.11 Electronic circuit for etching FIM crystal [17]. Etching current is interrupted automatically to ensure sharp tips.

heated in vacuum by a current of 2.5 amp for 24 hours to allow grain growth and assure constant resistance across the wire, as well as to clean the interior of the loop of gases. The sample wire is then spot-welded to the center of the loop and electropolished with the equipment in Fig. 2.11, which ensures immediate cessation of etching once the bottom part of the wire specimen has dropped off.

After the sample and support loop are washed, the sample support assembly is mounted in the field ion microscope and is pumped out by a three-stage mercury diffusion pump backed by a mechanical pump. A schematic of this system, which is equipped with several getter tubes in which 0.020 in. titanium wire is mounted, is given in Fig. 2.12 [54]. The system as well as the system trap are then baked out overnight at 300 °C. This upper temperature is set so as not to damage the channel plate; without the plate, bakeouts at 415 °C are advantageous. At the end of the bakeout, the trap oven is replaced by a liquid nitrogen dewar and the temperature of the system oven is allowed to fall to 175 °C, at which point electron bombardment outgassing of the ionization gauges and of the shields around the evaporators as well as heating of all the wires is begun for ~5 hours. After cooling the system to room temperature, the bombardment of the channel plate with ~200 eV electrons from one of the tungsten evaporators is started. Eventually, the sample support and sample is heated, and subsequently bombarded by neon ions of ~1 keV energy. These are created by 10^{-6} amp of field emitted electrons colliding with ~2×10^{-4} Torr of neon admitted through a Vycor tube. This sputtering serves to clean as well as sharpen the

36   Determination of adatom movements

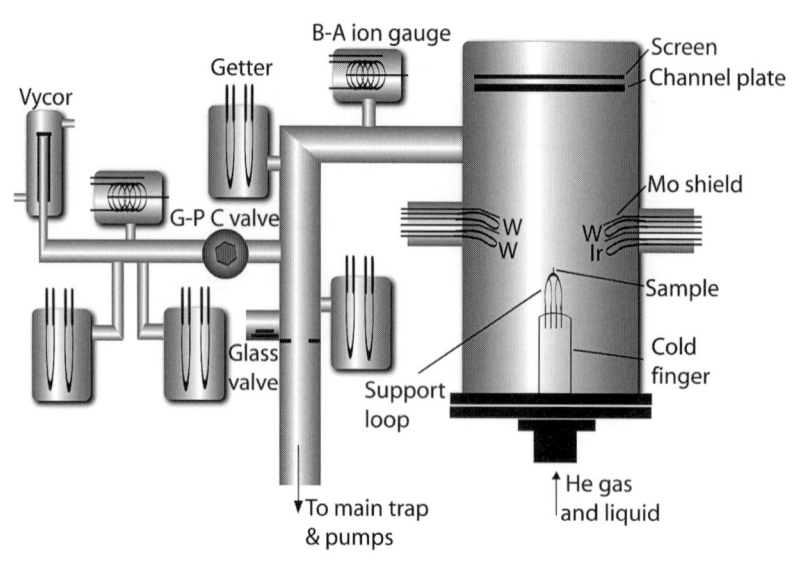

Fig. 2.12   Schematic of field ion microscope and vacuum system for measurements of surface diffusion [54]. Channel plate and Willemite screen serve to enhance image intensity.

tip. The system is usually baked out three times. Thereafter, all filaments are maintained hot (24 hours/7 days), except during measurements, and the channel plate is bombarded by electrons.

When a channel plate is first introduced into a system, really thorough electron bombardment is in order to remove gas trapped in the channels. This may take two to three weeks, interrupted by baking as required, until the pressure is brought to low 10^{-10} Torr. In the final stage of outgassing, the channel plate is bombarded for about five hours by helium ions; gas from a channel plate not well cleaned will impinge on the sample during imaging, so that outgassing is crucial to get reliable information. Evaporation of titanium layers in the getters is very helpful in the final stages of system preparation. Helium or neon gas used for imaging is purified by admission through the Vycor tube in the system.

2.4.3   Atom imaging and analysis

Atoms are deposited on the sample surface from an evaporator wire, previously carefully outgassed, typically 0.006 in. in diameter. For studying self-diffusion of single atoms, some laboratories obtain atoms from the outermost plane, by heating combined with selective field evaporation of atoms at the edges. Single atom diffusion studies in the field ion microscope require only one atom present on the plane to avoid cooperative interactions. In order to study diffusion, the tip is heated to the desired temperature and maintained there by a circuit shown in Fig. 2.13, which senses the voltage across the potential leads on the sample loop and delivers power accordingly. At the end of the diffusion interval the power is turned off and the sample is allowed to cool, usually to ~20 K.

2.4 FIM measurement of diffusivity

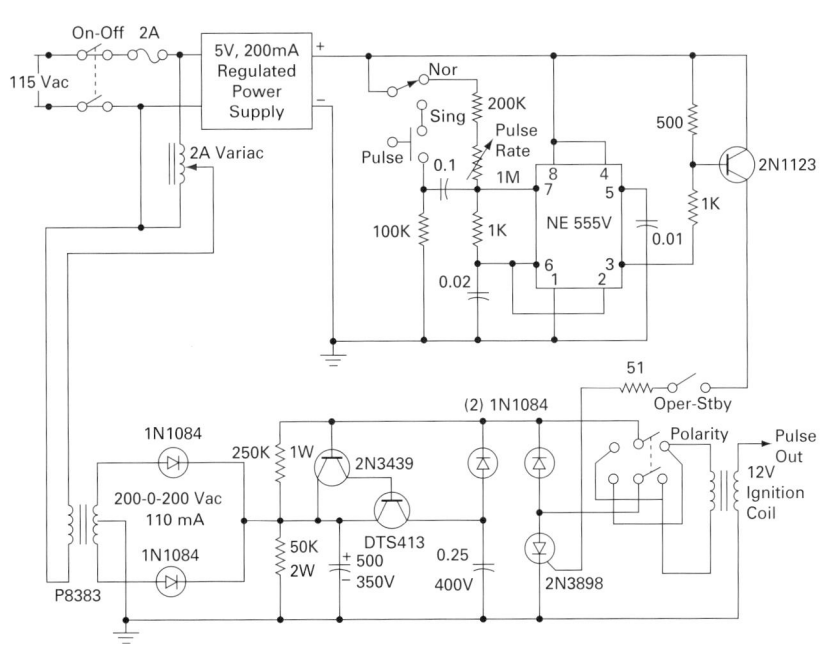

Fig. 2.13 Circuit for control of sample temperature, which measures four-lead resistance [17]. Reference resistor ≈ 30 cm of 16 AWG Manganin wire in parallel with 25 Ω potentiometer.

Heating of the sample has to be done in the absence of an imaging field, since such a field would influence the movement of adatoms. Next to the plane edges the electric field is locally higher than in the middle of the plane, as can be observed by the higher intensity of atom images at positions next to the edge compared with the center, as well as by differences in the voltage necessary to remove an atom from these positions by field evaporation. The differences in the field across a plane would cause directional movement of atoms over the surface if imaging were done at a temperature higher than for movement of adatoms on the plane under study. That is why the sequence of measurement is set up as in Fig. 2.14. First the image of the surface is observed in a high field but a low temperature of ~20 K, where the position of an adatom as well as of the surface atoms is frozen; then the high voltage is turned off and a 15 sec delay is applied before the temperature is raised to the desired value in the absence of any voltage. At the end of the diffusion time interval the temperature is turned off again, and a 15 sec delay is imposed before starting to image the sample in a high field. Imaging in our laboratory involves an RCA TC1430R camera with silicon intensifier target, a reasonably cheap but quite serviceable if old low light-level unit. The image records the state of the surface frozen in after the diffusion interval; that is why speed of recording is not important in these investigations.

As already mentioned, the field ion microscope does not directly yield the positions of surface atoms on low index planes; only the edge atoms protrude and are imaged to reveal the shape of the plane. The first task is therefore to determine the positions of the sites

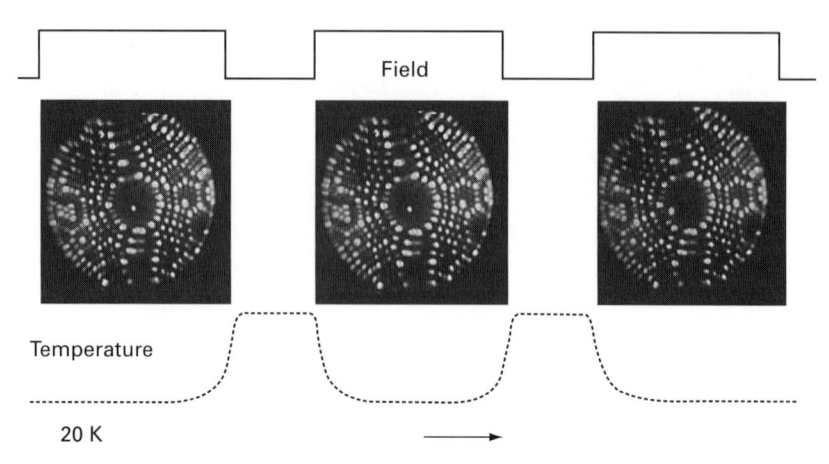

Fig. 2.14 Sequence of FIM images for determining adatom diffusion on W(211) [59]. Image is obtained at a low temperature and a high electric field. Atom motion is induced by elevating temperature, in the absence of applied fields. Atom displacement is determined after shutting off field and lowering temperature.

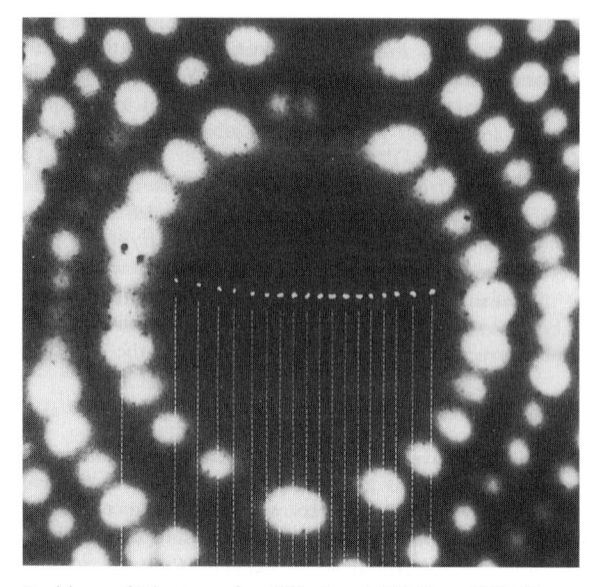

Fig. 2.15 Positions of Rh atom after diffusion at 197 K on W(211), mapped on FIM image of plane [59].

binding atoms on the plane of interest. This is accomplished simply by repeatedly recording the location of an adatom after it has diffused over the surface. Eventually, enough locations are identified to map out all the sites. In Fig. 2.15 is shown a map established for W(211) with a rhodium adatom, which migrates in the individual channels. On a plane on which two-dimensional diffusion occurs, site identification is more tedious [55]. A similar analysis on an Ir(111) surface is shown in Fig. 2.16. With a reasonably small number of observations, sites on the entire surface are mapped and the positions of the surface atoms can be deduced.

2.4 FIM measurement of diffusivity

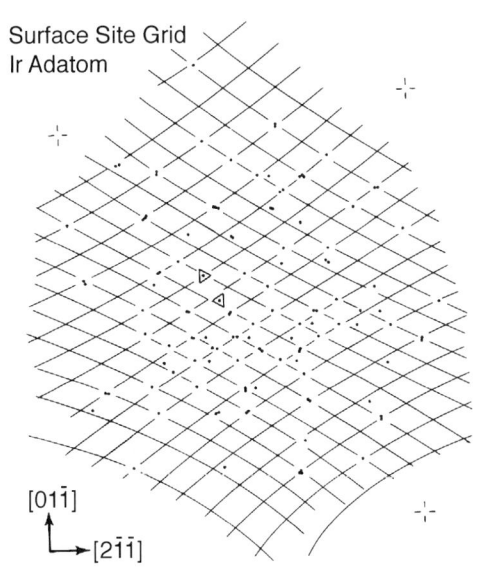

Fig. 2.16 Location of Ir adatom mapped after diffusion on Ir(111) plane at 104 K [55]. Grid is drawn through hcp sites, but occasionally an atom is at an fcc site, toward the left in the surface unit cell.

The lines connecting the mapped points are not always straight; in many cases they are bent in the neighborhood of edges. Similar deviations are observed with both one- and two-dimensional planes. Such deviations of the grid lines are caused by changes in magnification across the plane. We do not expect physical consequences from this fact. The FIM image does not have a linear scale, so distances cannot be obtained directly from the image. This arises from two factors: distortion of the image due to projection onto a flat screen, and changes in the local field across the plane. That is why field ion images do not have scale bars shown as do STM pictures. However, careful mapping of the plane can deliver information about its size in terms of lattice spacings. The uncertainty of such measurements is at worst only ±1 space. Changes in magnification not only alter the orientation of the lines between adsorption sites but can also change distances between the sites [56]. This is seen in Fig. 2.17. Distances between adsorption sites on the image differ depending on the location; they are larger next to cluster edges due to the higher local field there. If two kinds of adsorption sites exist on the surface, as in the case of the Ir(111) plane, then by comparing the structure of the surface with a map of the adsorption places both types can be recognized, as in Fig. 2.16.

Another important matter concerns the temperature of the sample. The temperature of the sample support wire is readily available from the voltage measured across the potential leads and the current through the support loop. This immediately leads to the resistance, which for tungsten, as an example, has been tabulated by Desai et al. [57]. The tip surface is some distance, perhaps as much as 5 mm, away from the support loop, and the potential leads are spotwelded some distance from the top of the loop where the sample is affixed. The sample may therefore have a temperature different from that actually measured. To determine the real surface temperature, the rate of field evaporation is measured with the entire tube equilibrated at room temperature. Thereafter the helium

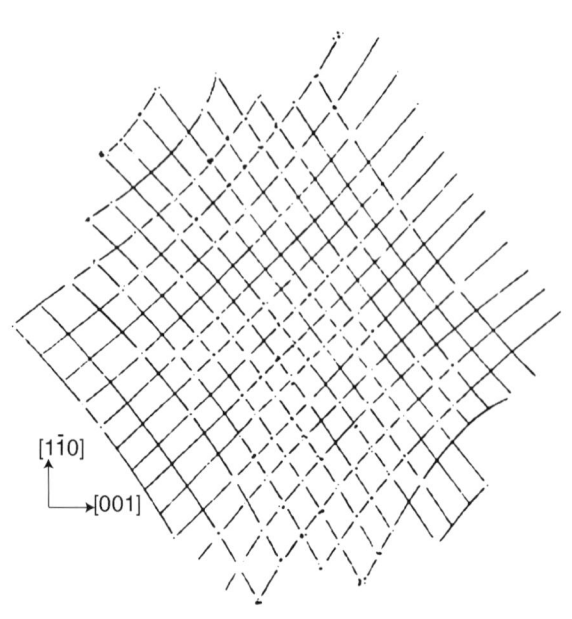

Fig. 2.17 Sites mapped on W(110) plane by diffusing Re atom at 380 K [56]. Grid lines are drawn to connect recorded locations, and clearly reveal change in magnification over the surface.

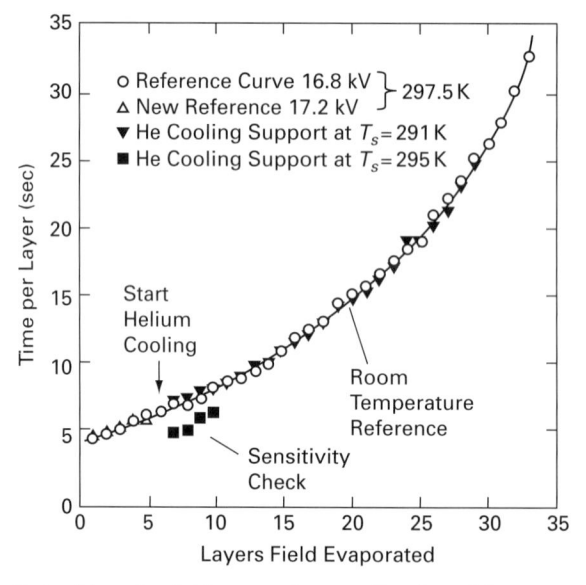

Fig. 2.18 Rate of tungsten field evaporation as a function of the amount evaporated with the system at 297.5 K, and with He cooling to 291K [59].

coolant flow to the leads is started again, and the support is brought to room temperature by adjusting the heating current. The rate of field evaporation, shown in Fig. 2.18, is measured again, and the heating current is changed until the rate matches that found with the system at room temperature. A comparison of the new setting with room temperature gives the magnitude of the correction to be applied to the loop temperature, which

2.4 FIM measurement of diffusivity

typically amounts to ~3 K. This technique, described by Stolt et al. [58] and also by Wang [59], can be applied to any surface that readily field evaporates.

2.4.4 Diffusion measurements

To get data about the diffusivity of atoms on a flat surface we have to design the experiment in a way to get the maximum information, influenced to a minimal extent by other factors, such as the presence of steps or impurities. To minimize the influence of impurities, strict attention is paid to vacuum conditions; however, this is still not enough. The total time of the measurements cannot last more than roughly three hours; thereafter the sample has to be cleaned by field evaporation of at least one layer, and the investigation can then be continued with a new atom deposited on the surface. Flickering of the image and the sudden appearance of new atoms are signs that the vacuum conditions are not satisfactory.

A second factor, which influences the gathering of data, is the finite plane size, and the presence of edges. For different planes the edges can have quite different characteristics, from reflective barriers at the edges, for example on W(110), reflective barriers plus higher binding energy locations close to the edges, as on W(211), to a lack of additional barriers, for example on Ir(100), or the presence of an empty zone next to the edges for Ir(111).[2] Attention has to be paid to make measurements only in the central part of the plane, which has a uniform potential landscape. Finally, keeping all this in mind, the measurements should be designed in such a way to get as many data points as possible in the shortest period of time.

The procedure used in our laboratory divides the area of the plane under study into three parts. The first area is the central part of the plane, where we deposit an atom to start the measurements. The second is the surrounding area, in which we allow the atom to move over the surface, and the third is at the edge; if the atom enters the last region measurements are interrupted, that observation is withdrawn from the data and the atom is allowed to move, without recording, to the central area, sometimes at a raised temperature. Then measurements start again. In this way we obtain the mean-square displacement of an atom on the chosen plane with the minimum of possible influences from the edges. However, even in this procedure there are some effects of the edges on the data; atoms cannot wander infinitely far from their starting position, but they can move in the neighborhood of the edges while not being imaged. We therefore account for the effects of the edges by doing Monte Carlo simulations of the atomic jumps at different rates, on a plane configured just like the actual surface. Jump rates are adjusted until good agreement is obtained with the measured displacement fluctuation. Additionally the influence of edges is controlled by keeping measurements in the area where this influence is low.

Another problem to consider is the non-linear scale of the field ion microscope. It is impossible to come up with distances between the positions of atoms by measuring these on the image, since the magnification across the plane changes with position. What has to be done is to map adsorption sites, which allows us to arrive at distances in terms of surface spacings.

---

[2] More information on this subject is presented in Chapter 3.

The diffusivity D can be derived from measurements of the displacement fluctuation, $<\Delta x^2>$, using the relation (1.36) $<\Delta x^2> = 2Dt$, where t is the time interval for diffusion. The displacement fluctuation can be obtained by measuring each displacement, squaring it, and then averaging, but we emphasize again that this is not the quantity that enters directly in Eq. (1.36). What is required is the displacement fluctuation on an infinite surface. To correct for the fact that the measurements are usually made on a very finite surface, on which the region close to the edges is not included in the displacements, Monte Carlo calculations are done with different jump rates, until a value of the rates is found for which the simulated displacement fluctuation agrees with the experimental value. The simulations are of course done on a plane of the same size and shape as in the experiments. The diffusivity is now accessible from Eq. (1.18) $D = \Gamma \ell^2$, as is the displacement fluctuation on an infinite plane. Roughly 100 observations at each of 7 temperatures are enough to yield activation energies with an uncertainty of ~3%.

There is another matter, however, that must be considered in evaluating diffusion, and that is the sample temperature. Ideally the temperature versus time plot should look like a step function; that is, a constant temperature during the diffusion interval, without transients either at the start or the end. Unfortunately, in the real system it is quite difficult to achieve such a situation. It takes a finite time, the transient time, to reach the set temperature, as indicated in Fig. 2.19. To minimize the transient time we apply a current of 2.5 amp at the beginning of the heating interval and as soon as the sample achieves the set temperature the current is adjusted automatically to the value suitable for this condition. Typically the temperature rises slowly, over a period of ~5 sec at such a setting. We experience a second transient when cooling the sample down from the diffusion temperature to imaging conditions, ~20 K. Once a diffusion interval is over, the heating current is shut down, and the sample cools by conduction, taking more than 10 sec to reach a temperature of ~20 K. At low diffusion temperatures, at which the jump rate is slow, these temperature transients are unimportant – nothing happens during them.

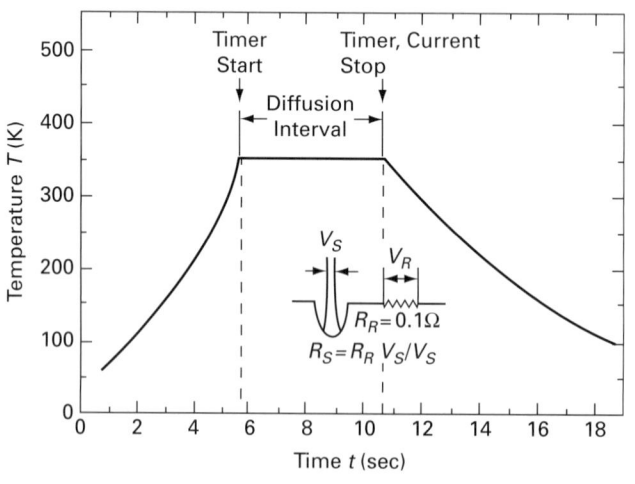

Fig. 2.19   Schematic of transients in the temperature during diffusion observations, obtained by measuring V_S and V_R [59].

2.4 FIM measurement of diffusivity

However, at higher diffusion temperatures atom motion already occurs during the transients, and contributes to the measured displacements. These contributions must be removed, as they have not taken place at the set diffusion temperature. Separate runs are therefore made, in which the heating is cut off as soon as the set diffusion temperature is achieved; the temperature transients are preserved, however, and the displacement fluctuation during these "zero-time" experiments is subtracted from the overall value [60]. The distribution of displacements must also be corrected for effects occurring during transient temperatures, but this is more complicated and is discussed in Section 2.5.

2.4.5 An example: W on W(211)

A system much studied by field ion microscopy is the behavior of tungsten on the (211) plane of tungsten. This plane is made up of close-packed rows of atoms, forming one-dimensional channels along which atoms can move; a diagram of this surface is shown in Fig. 3.1a. Detailed structural studies have been carried out on this plane by argon ion scattering [61], and reveal that there is a small lateral shift of ~0.1 Å of the second-level atoms, in agreement with LEED measurements [62]. An Arrhenius plot of the most recent diffusion results [63] obtained by field ion microscopy is shown in Fig. 2.20. Measurements extended from 260 K to 325 K, a range of 65 degrees. At each temperature above 300 K, 1200 observations were made, supplemented by 1200 "zero-time" determinations; the latter are significant at temperatures above 300 K. From the slope an activation energy of 0.81 ± 0.02 eV was derived, giving a prefactor $D_o = 3.41$ $(\times 2.4^{\pm 1}) \times 10^{-3}$ cm^2/sec [63].

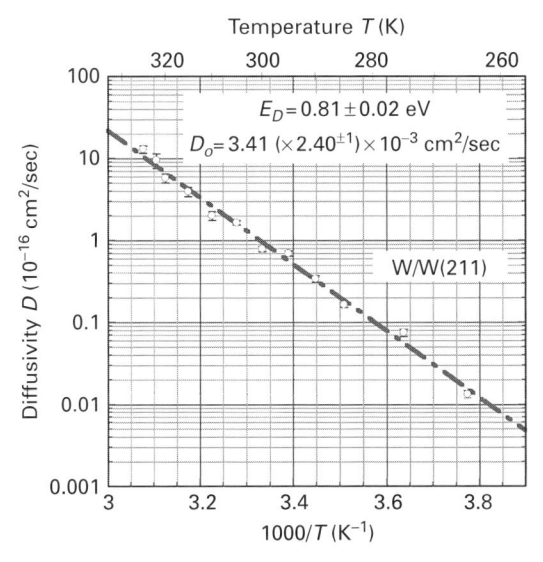

Fig. 2.20 Arrhenius plot for the diffusivity of tungsten atom on W(211) [63].

2.5 Displacement distributions

To determine diffusivities requires only a modest number of observations; roughly 100 at each temperature gives reasonable diffusion characteristics. For obtaining insight into the mechanism of diffusion we need to obtain the distribution of displacements, which involves the same type of measurements, but at least an order of magnitude more data. In the analysis, displacements of different kinds are sorted. For one-dimensional diffusion, the displacements to nearest-neighbor sites α, next-nearest-neighbors β and so on are separated and plotted. For two-dimensional diffusion the different kinds of transitions can also be identified, as in Fig. 2.21 for a bcc(110) surface [64]. After experiments, the actual distribution is simulated by Monte Carlo estimates on a plane of the same size and shape as in the measurements; the different kinds of jumps and their rates are then varied until agreement is reached between simulation and experiment.

To complete the analysis still requires correction for the displacements that occurs during the temperature transients, before and after the steady diffusion temperature is reached. This is again achieved by doing "zero-time" experiments, which we measure for every investigated temperature. We execute this measurement by heating a sample to the desirable temperature and as soon as the sample reaches this temperature, our electronics turns the heating down and the sample is allowed to cool down to 20 K. The "zero-time" distribution consists of jumps that occurred during the transient periods and should be based on roughly the same number of observations as for the main distribution itself. The "zero-time" measurements are analyzed in the same way as regular measurements to determine the jump rates during transients. The rate of jumping r at the set temperature is obtained by subtracting the rate r_o for "zero-time" observations multiplied by the time

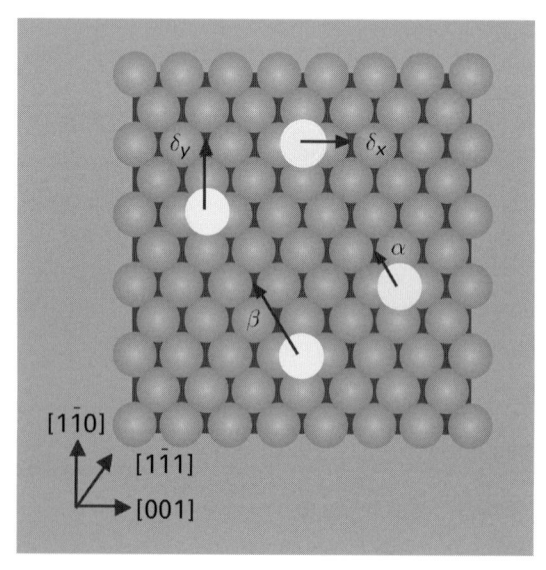

Fig. 2.21 Schematic of possible atom jumps in diffusion over the W(110) surface [64].

2.5 Displacement distributions

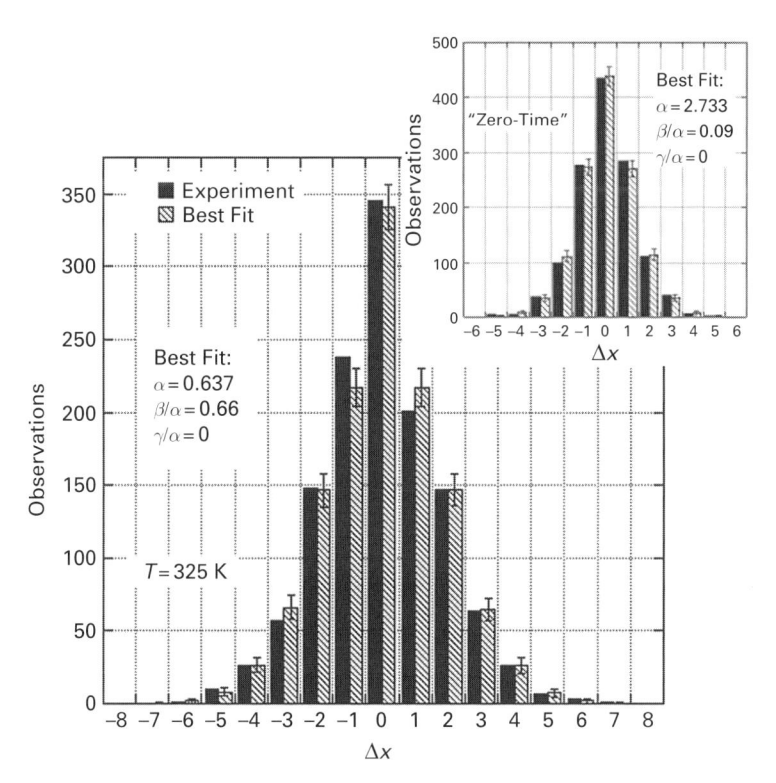

Fig. 2.22 Distribution of displacements for W atom on W(211) plane at 325 K [63]. Inset gives distribution observed during warming and cooling sample.

interval t_o, required for turning the electronics down, from the overall rate R multiplied by the time interval in the overall experiment

$$r = (Rt - r_o t_o)/t_c; \qquad (2.12)$$

here t_c is just the time interval for diffusion at constant temperature.

An example of this type of analysis, which yields jump rates is given in Fig. 2.22. The primary plot shows the displacement distribution determined for tungsten atoms on W(211) at a temperature $T = 325$ K, that is for the one-dimensional distribution along the <111> direction. The second plot, shown in the inset, gives the distribution obtained during "zero-time" experiments at the same diffusion temperature. At elevated temperatures, the "zero-time" measurements have a huge influence on the data and cannot be neglected. The distribution of displacements can of course also be measured in two-dimensional diffusion, as for example on W(110), shown in Fig. 2.23. "Zero-time" measurements are essential. It should be noted that displacement distributions are useful not just in studying adatoms – they can also provide insights into how clusters of several atoms move over the surface.

Here it should be emphasized that what is measured are atomic displacements. The atom starts at a site with which it has equilibrated and ends at another site at which the atom has come to equilibrium; we do not learn anything about any brief intervening

46 Determination of adatom movements

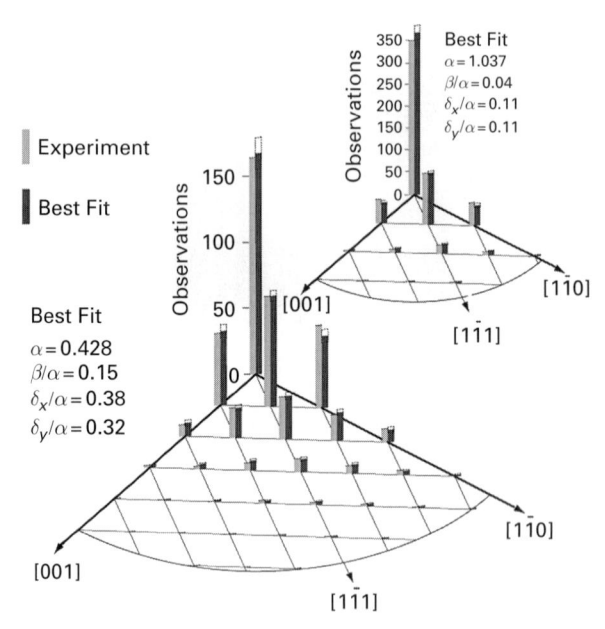

Fig. 2.23 Distribution of displacements for Ir on W(110) in diffusion at 360 K [54]. Inset gives distribution measured in "zero-time" experiments.

touchdowns of the adatom or about the exact path between equilibrium sites, only about the overall displacement. However, by analyzing the temperature dependence of the rates derived from the distribution of displacements we can obtain information about the mechanisms as well as the energetics associated with surface events.

2.6 STM measurements of diffusion

2.6.1 Movement of single atoms investigated by STM

The biggest difference between observing diffusion by FIM compared with STM is that in the latter the tip is scanned at a close distance from the investigated sample, while in FIM the image of the sample is obtained at once. The most direct way of arriving at the diffusivity of single entities by STM is to work with very low coverages, ~0.01 ML, on a surface unimpaired by defects, as in Fig. 2.24b. In this way the movement of an atom can be immediately detected. Observation of different stages in the movement will be limited only by the time of a scan. This problem has been discussed in detail by Morgenstern [65]; here we therefore only give a short description. The problem associated with choosing a scanning speed for the observation of single atoms is illustrated in Fig. 2.24a – it is the interplay between scanning speed and the velocity of the observed adatom. When scanning is slower than the moving adatom the adatom will not be imaged at all. When speed and velocity are comparable then the image of the adatom may be deformed and can be misinterpreted. Ideal for observations is a scanning speed faster than the movement of the adatom. Of course in many cases the movement of adatoms can be

2.6 STM measurements of diffusion

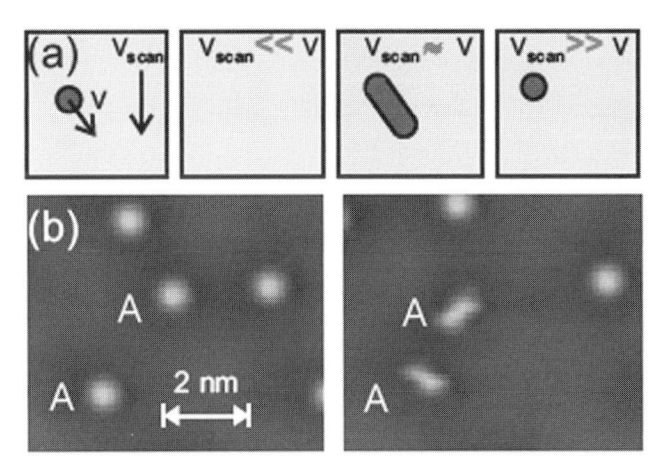

Fig. 2.24 STM scanning speed and visibility. (a) Schematics of atom appearance and scanning speed. When scanning is too slow, nothing is seen. When speed of object and the tip are comparable, a deformed image may arise. Only high speed gives accurate image. (b) Position change of Cu atoms on Ag(111) produces fuzzy image (after Morgenstern [65]).

adjusted by temperature, but in some situations reducing the temperature changes the basic mechanism of movement. Fast scanning is important in imaging two-stage processes, as for example the metastable walk on a Pt(110)-(1 × 2) reconstructed surface. For this metastable walk the two processes are first climbing the {111} walls forming the channels, an energetically demanding step, and second, the very fast movement on the {111} walls. Reducing the temperature in this case may block all movement. The only chance to see the real motion of the atom is to work at a scanning speed faster than the movement of the atom on the {111} walls at a temperature high enough to make the first step proceed, but this has so far not been achieved. Scanning conditions can be adjusted so that a single atom deposited on the surface can be imaged; but the distortion of surface states induced by adsorbed atoms together with elastic and inelastic scattering of electrons by adsorbed adatoms can also be observed [66,67], as is clear in Fig 2.25. This gives us insight into electronic changes at the surface.

The important factor in single atom studies with the STM is the interaction of the adatom and the tip. Since the distance of tip to sample is small, the possibility that the tip will influence a moving atom has to be considered. This can be checked by varying the tip–sample interaction time or by exploring different distances between tip and sample. The possibility of charging the object of observation by injecting electrons from the scanning tip also should be taken into account, especially for semiconducting samples where leakage can be not sufficient [68,69]

Another factor, which has to be taken into account, is thermal drift of the sample; this can influence diffusion as well. That can be a very important factor in the long measurements of slow processes or measurements with big statistics. The atom tracking STM [70] can be very useful in gathering a huge data set. The drawback is lack of control of the distances of atoms and steps or defects.

The first reliable direct information about single atom diffusion by STM was arrived at by Repp et al. [71] on Cu(111); their efforts were followed by Knorr et al. [66]. So far,

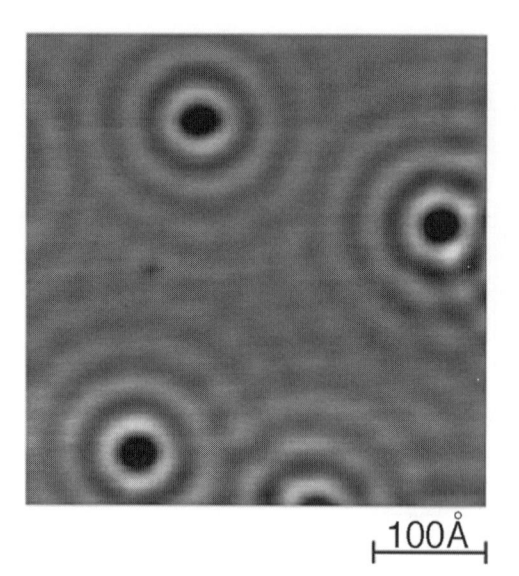

Fig. 2.25 STM image of Al(111) surface with substitutional defects, showing long-range oscillations of the local density of states (after Knorr *et al.* [66]).

direct information about movement of single atoms is rare and has been obtained mostly for fcc(111) surfaces like Cu(111), Ag(111) [72,73]. In addition to all technical challenges, good vacuum conditions are crucial for reliable measurements. The modification of diffusion parameters through the presence of hydrogen in a metal STM chamber should also be taken into account. For this reason most of FIM data were taken in glass microscopes.

2.6.2 Island fluctuation observed by STM

Instead of detecting individual atoms directly, many studies of surface diffusion on metals have relied on predictions from the mean-field theory of nucleation. Assuming that long-range interactions are negligible, the rate at which the concentration n_1 of monomers grows when F_f atoms strike unit area in unit time can be written as [74]

$$dn_1/dt = F_f(1-\Theta) - 2\sigma_1 D_1 n_1^2 - D_1 n_1 \sum_{m=2}^{8} \sigma_m n_m + 2\Gamma_2 n_2 + \sum_{m=3}^{7} \Gamma_m n_m. \qquad (2.13)$$

Here n_i gives the concentration of clusters made up of i atoms, Γ_i is their rate of dissociation, and σ_i is the capture number, which relates the rate of incorporation into a cluster i to the diffusivity of adatoms D_1. Similar relations for higher clusters can also be written out, and for the critical cluster, of size x, we get [75]

$$dn_x/dt = \sigma_4 D_4 n_4^2 + \sigma_5 n_5 \sum_{m=3}^{5} D_m n_m + \sigma_6 n_6 \sum_{m=2}^{6} D_m n_m + \sigma_7 n_7 \sum_{m=1}^{7} D_m n_m. \qquad (2.14)$$

2.6 STM measurements of diffusion

These relations can be integrated numerically to give the density of islands. Venables [76], however, has pointed out a simple alternative: when reevaporation is unimportant, it is possible to arrive at a relation between the concentration of islands n_x and the rate of impingement F_f as well as the diffusivity D. This is

$$n_x \propto \left(\frac{F_f}{D}\right)^{\frac{i}{i+2}} \exp\left[\frac{E_i}{(i+2)kT}\right]. \qquad (2.15)$$

Here i gives the critical size of the cluster, which turns into a stable cluster on adding an atom, and E_i its binding energy. As $E_i = 0$ if $i = 1$, under these circumstances Eq. (2.15) turns into

$$n_x \propto \left(\frac{F_f}{D}\right)^{\frac{1}{3}}. \qquad (2.16)$$

Clearly the most appropriate condition for deriving the diffusivity is to operate at temperatures and impingement rates such that the critical cluster size is one. Detailed conditions for doing such measurements have been given by Brune *et al.* [77]. All that needs to be done is to determine the saturated island density at different temperatures; this yields the prefactor D_o and the activation energy ΔE_D for diffusion.

However, the mean-field theory may not reliably describe island evolution, as long-range interactions are neglected. Failings were already demonstrated some time ago by Rosenfeld *et al.* [78], who found that restricting atom exchange between islands to nearest-neighbor islands gave better agreement with experimental observations for silver islands on Ag(111), as shown in Fig. 2.26.

There is, however, a clear advantage to such island density measurements. In probing a surface with the STM, there is always a possibility of the tip interacting with adatoms and disturbing them unless the distance from tip to surface is high enough. In comparison, the likelihood of disturbing the large clusters is much smaller. Bott *et al.* [79] could not find reasonable energies for Pt self-diffusion on Pt(111) by direct scanning, but observation of islands provided good energetics. However, the interaction of the scanning tip with the surface cannot be excluded completely for larger entities. For example Koch *et al.* [80] found that Ag atoms were extracted from steps on scanning Ag(110).

As already noted, island density measurements have to be done under carefully controlled vacuum conditions. The influence of different kinds of impurities has been investigated and it was noticed, for example, that the presence of hydrogen enhances the diffusivity of Ni atoms and destabilizes islands on Ni(100) by creation of strong Ni-H bonds; the average island separation was also increased [81]. CO changes the number of islands created on Pt(111) due to an increase in the step-edge barrier [16,82]; it also partly blocks the interlayer movement of material, as shown in Fig. 2.27.

The influence of island mobility on island fluctuations was investigated by Kuipers and Palmer [83] who claimed that the density of islands without mobility could be as much as 5 times higher than when there was island mobility present. In the case of island mobility, the size distribution changes shape and should influence analysis of the distribution obtained in STM since information about the growth mechanism then is

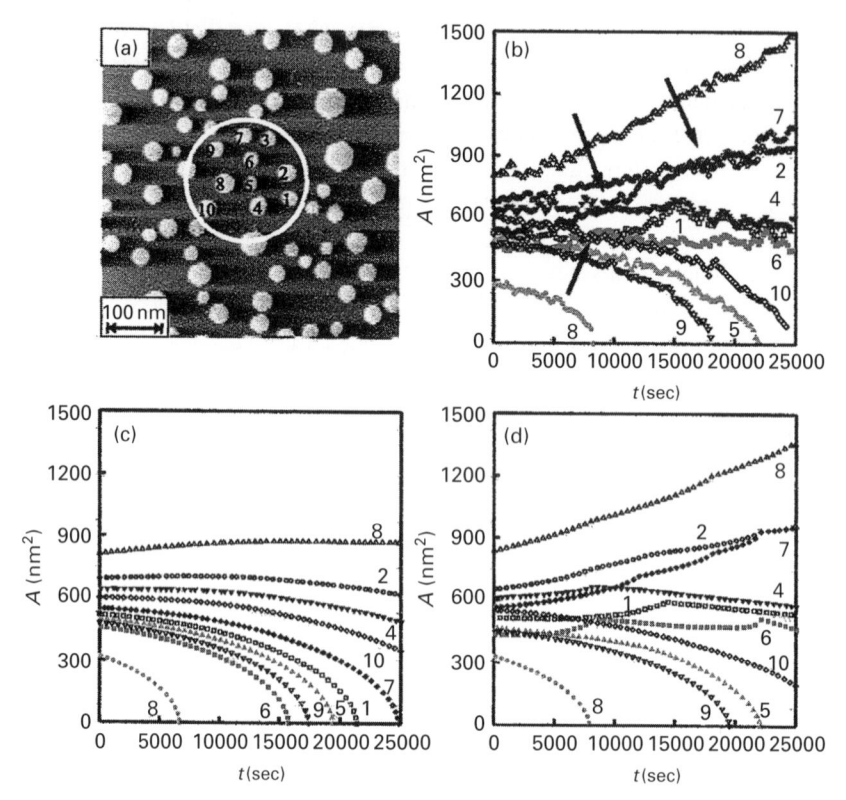

Fig. 2.26 Comparison of mean-field predictions with model allowing atom interchange between nearest-neighbor islands. (a) STM image of Ag islands on Ag(111). (b) Development of island area in circled region with time. (c) Predictions from mean-field theory. (d) Predictions based on exchange between nearest-neighbors (after Rosenfeld et al. [78]).

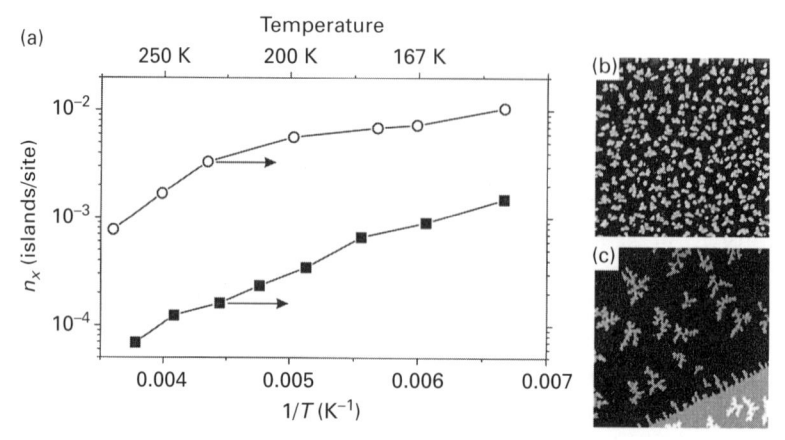

Fig. 2.27 Platinum islands on Pt(111). (a) Island densities. Black squares: STM data [79]. Open circles: similar data, but obtained with high background pressure. (b) STM image after contaminated deposition at 230 K. (c) Image after clean deposition at 225 K (after Michely et al. [82]).

2.6 STM measurements of diffusion

difficult to extract from the ratio D/F_f. The influence of island dissociation and recombination on island fluctuation was tackled by Ratsch et al. [84] in 1996. Based on agreement of their model with experimental data of Stroscio and Pierce [85] they concluded there was a non-negligible influence of the thermal dissociation of doubly coordinated atoms. In 1998 Ratsch and Scheffler [86] listed problems with deriving the diffusivity from the island density using Eq. (2.16). In addition to long-term adatom–adatom and step–adatom interactions, they also pointed out the possibility of long and rebounding jumps at higher temperatures, the possibility of movement of clusters bigger than one, atomic exchange, the non-existence of stable nuclei due to continuous dissociation and recombination of islands, edge movement and funneling.

A severe limitation in deriving the adatom diffusivity from the island density is imposed by interactions between atoms. These extend over long distances, and vitiate the basic assumption of ideal behavior made at the very start of nucleation theory. The implications of interactions on measurements of the island density have been examined by a group in Berlin and another in Göteborg. Fichthorn and Scheffler [87] looked at island formation, and for silver on Ag(111) found that on unstrained Ag(111), interactions are strongly attractive when atoms are located in nearest-neighbor positions. At longer distances, interactions are repulsive, as indicated in Fig. 2.28. In their model, atoms form a ring at the repulsive distance from an adatom, which at low temperatures adatoms cannot penetrate. Only adatoms deposited within the ring can interact with the central atom to form a pair. In the presence of such rings the island density is higher than in the absence of interactions. The net effect of interactions is shown in Fig. 2.29. Ideal nucleation theory gives cluster densities an order of magnitude smaller than in the presence of such repulsive interactions.

This problem has been examined in more detail by Bogicevic et al. [88], as well as Ovesson et al. [89], who made density-functional calculations of interactions between

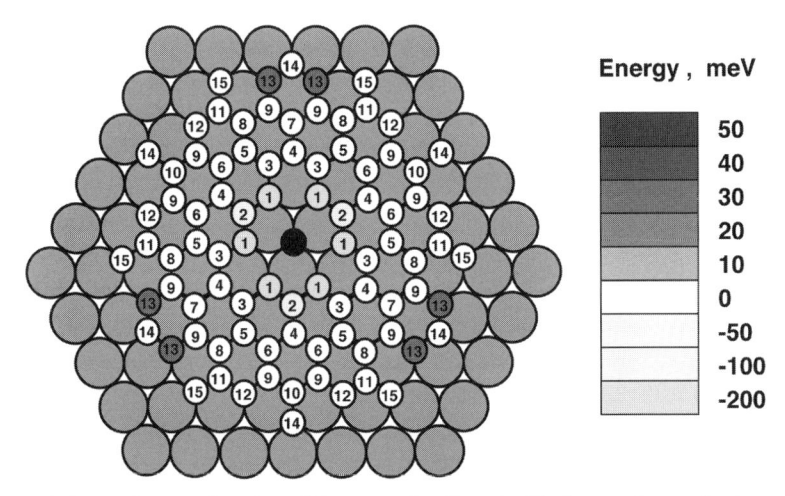

Fig. 2.28 Pair interaction as a function of the separation from the black central atom for Ag on Ag(111) (after Fichthorn and Scheffler [87]). See color plate section for color version of this figure.

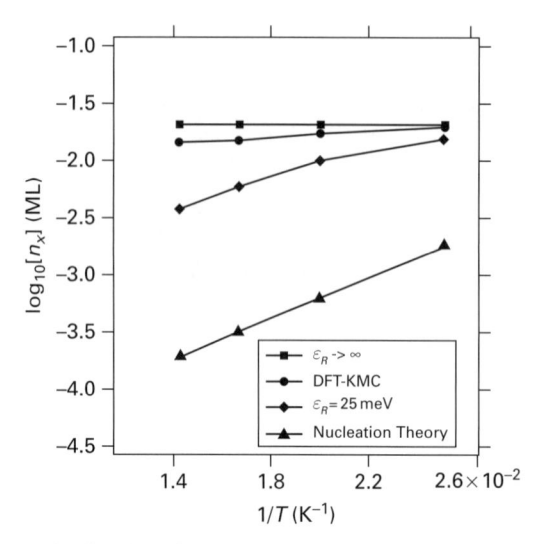

Fig. 2.29 Arrhenius plot of Ag island density on Ag(111) for an impenetrable ring (squares), from density-functional calculations (circles), for repulsive ring with $\varepsilon_R = 25$ meV (diamond) and nucleation theory (triangles) (after Fichthorn and Scheffler [87]).

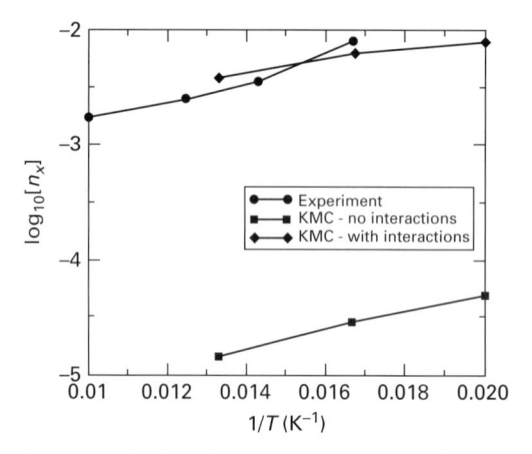

Fig. 2.30 Saturation density of Al islands calculated by DFT on Al(111) and compared with the STM results of Barth et al. [90] (after Ovesson et al. [89]).

copper atoms on Cu(111), as well as for aluminum on Al(111) and on Al(100). With these values, they did Monte Carlo simulations of the growth on these surfaces at low temperatures. Their findings for the island density in Al on Al(111) at different temperatures are given in Fig. 2.30 and calculations with long-range interactions are in good agreement with experiment. In the presence of long-range interactions the island density is increased by two orders of magnitude; for Cu(111) the result is nicely illustrated in Fig. 2.31. It should be noted, however, that the question of the effects of interactions in such experiments is still under discussion and is not yet settled [90–92]. The problem of long-range interactions between atoms in nucleation has recently been addressed by

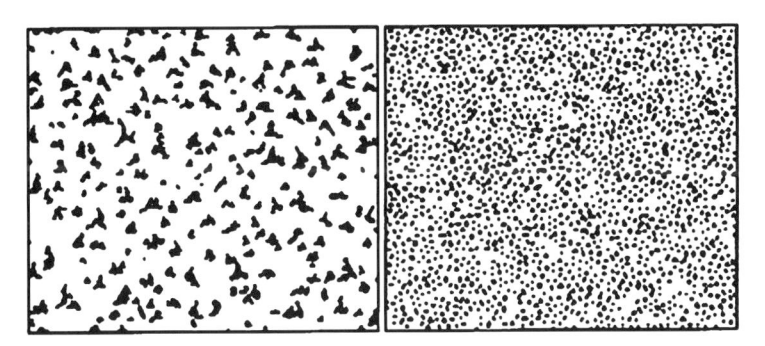

Fig. 2.31 Monte Carlo estimates of the effects of interactions on formation of Cu islands on Cu(111) at 25 K and 0.1 coverage. Left: islands formed without long-range interactions. Right: estimates allowing for interactions show much higher island density (after Bogicevic et al. [88]).

Venables and Brune [93], who have estimated the effects on the capture numbers caused by the presence of a repulsive potential. A straightforward way of deriving diffusivities from the measured island density in the presence of long interactions is not yet available, however.

One thing is certain – for entities diffusing over low barriers, adatom–adatom interactions can severely affect the behavior and can yield much higher island densities than predicted by nucleation theory. Diffusivities derived in this way are therefore somewhat uncertain. Direct observations of atom jumps, on the other hand, are not affected by such problems, and should therefore be preferred. Experimental results about pair interactions observed for a range of systems are reviewed in Chapter 10.

2.7 Other measurement techniques

Almost all surface diffusivities have been measured with the techniques already mentioned – field ion microscopy, field emission microscopy, and scanning tunneling microscopy. However, any technique that can detect adatoms can provide useful information about the movement of adsorbed material over the surface, so there are at least three other methods yielding diffusion data that should be mentioned: work function determinations, perturbed $\gamma - \gamma$ angular correlation, and helium atom scattering. The first of these has been carried out for decades in Naumovets' group [6,94], using Anderson's technique for obtaining concentration profiles, as shown in Fig. 2.32. The electron gun can be moved with a micrometer screw and the electron beam is focused by a magnetic field, giving a resolution as low as 20 micrometers. These measurements have been focused primarily on documenting the concentration dependence of the diffusivity. To attain information about the diffusivity D at a

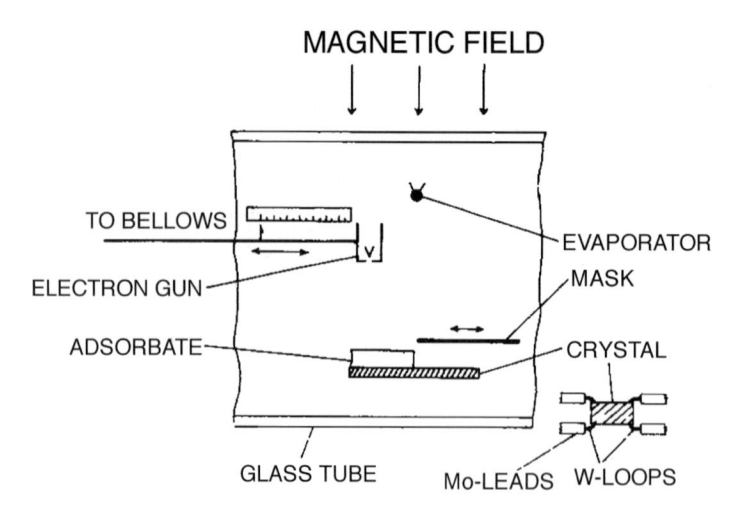

Fig. 2.32 Schematic of experimental arrangement to detect work function changes indicating diffusion over the surface (after Loburets et al. [94]).

concentration c, the entire surface is coated at that concentration. Then a strip at a small additional concentration is deposited, and the spreading of this deposit provides the information to derive the desired diffusivity. The technique of course does not provide information about single atom motion, but work at small coverages is certainly worthwhile.

In perturbed $\gamma - \gamma$ angular correlation studies (PACS) of diffusivities a small quantity, ~10^{-4} ML, usually of radioactive ^{111}In probe atoms is deposited on the surface. Interactions with electric field gradients at the adsorption site split the nuclear level and this leads to emission of three frequencies ϖ_1, ϖ_2, and $\varpi_3 = \varpi_1 + \varpi_2$, which are detected and the delay time between detection of the two γ quanta is recorded. The PACS technique is illustrated in Fig. 2.33. As shown by Klas et al. [95] as well as Krausch et al. [96], the technique eventually makes it possible to identify the nature of the binding sites, and the rate at which they are filled by diffusion over the surface.

The third technique employed to study surface diffusion, helium atom scattering, has been described in detail by Graham [5], and a schematic of the equipment is given in Fig. 2.34. Information about diffusion is obtained by measuring the energy broadening of helium atoms scattered from the adatoms. This scattering is the analogue of neutron scattering used to examine bulk diffusion [97]. One question that still has to be unequivocally answered, however, is if data about the diffusivity of individual metal adatoms can be successfully obtained.

Recently a theory for the determination of atomic jumps by nuclear resonance scattering of synchrotron radiation was proposed by Vogl et al. [98]; however, the technique has not yet been used to draw information about real systems.

2.7 Other measurement techniques

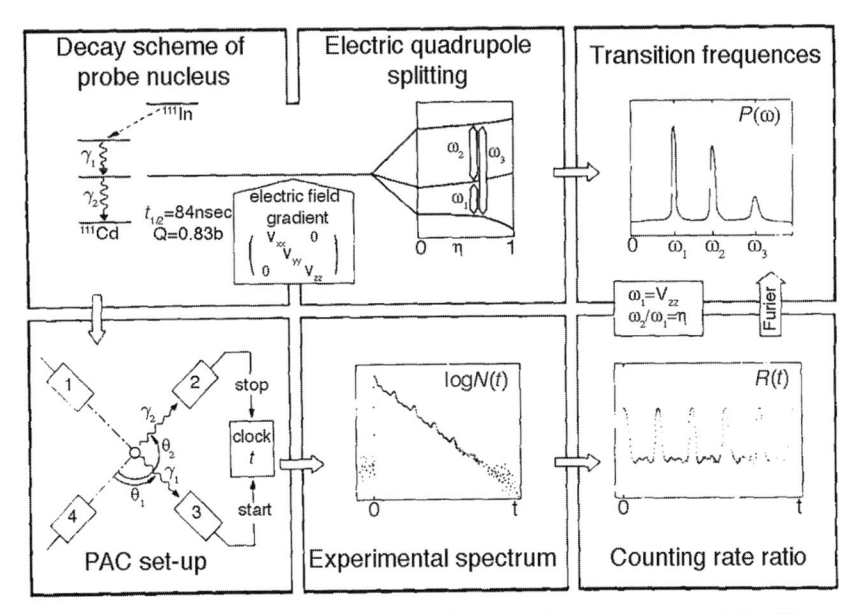

Fig. 2.33 Schematic of settings for perturbed $\gamma - \gamma$ angular correlation measurements (after Klas *et al.* [95]).

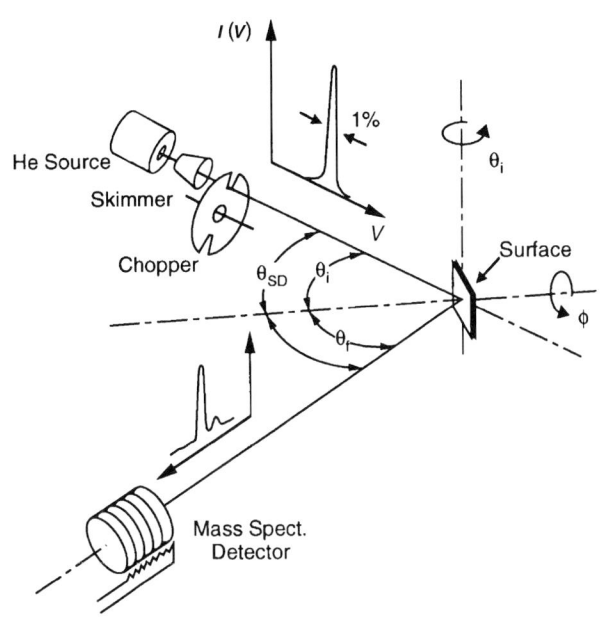

Fig. 2.34 Components of fixed total scattering angle apparatus for helium atom scattering from surfaces. Momentum transfer parallel to the surface is controlled by changing the angle of beam incidence (after Graham [5]).

2.8 Theoretical estimates

The emphasis of our presentation has been on the experimental determination of diffusion processes, and this field has been quite productive. However, in the last 15 or so years, it has become possible to make reasonably quick theoretical estimates of diffusion characteristics, and these now generally outnumber experimental studies. This work has brought with it some important insights, and at least a few words are therefore in order here.

Crucial to the theoretical evaluation of atomic diffusion on surfaces is a knowledge of the interactions between the atoms. A static potential barrier to motion is then obtained from the difference between the energy of the system in the activated state compared to that in a normal site, quantities readily determined from the interactions. For atomic systems, pair potentials have of course been in use for many decades, and the Lennard-Jones (L-J) potential is well known in this field. This relation gives the potential energy between two atoms i and j as

$$E_{ij} = 4\varepsilon_{LJ}\left[\left(\frac{\sigma}{R_{ij}}\right)^{12} - \left(\frac{\sigma}{R_{ij}}\right)^{6}\right], \tag{2.17}$$

where R_{ij} is the distance between the two atoms, ε_{LJ} is the depth of the attractive potential, and σ the separation at which the potential energy goes to zero. Parameters ε_{LJ} and σ can be used to fit the potential to a particular material. A more adaptable form is given by the Morse function, for which

$$E_{ij} = D_M\{\exp\left[-2\alpha_M(R_{ij} - R_o)\right] - 2\exp\left[-\alpha_M(R_{ij} - R_o)\right]\}; \tag{2.18}$$

there are now three parameters D_M, α_M, and R_o which can be obtained by fitting this equation to measured properties such as the lattice constant, the heat of sublimation, elastic constants, compressibilities, etc. Calculations of solid state properties with these relations are quick, but not too meaningful, as they ignore long interactions with other atoms.

To remedy this difficulty, a number of semi-empirical relations have been developed that give rapid calculations but also more trustworthy results. Among the more popular semi-empirical potentials which take account of many-atom effects is that developed by Daw and Baskes [99], the embedded atom method EAM. In density-functional theory, the energy of a crystal can be represented as a functional of the electron density distribution, assumed to be equal to the local density at each site. The total energy E_{tot} can then be written as the sum of all the individual contributions,

$$E_{tot} = \sum_i E_i, \tag{2.19}$$

where E_i is the internal energy attributed to atom i, and is given as the sum of two parts,

$$E_i = \frac{1}{2}\sum_{j \neq i} \phi_{ij}(R_{ij}) + F_i(\rho_i). \tag{2.20}$$

2.8 Theoretical estimates

Here the first term on the right gives the core–core pairwise repulsion, and

$$\rho_i = \sum_{j \neq i} f_j(R_{ij}) \qquad (2.21)$$

is the electron density of atom i arising from the rest of the system; $F_i(\rho_i)$ is the energy for embedding atom i into the local density ρ_i. The two-body potential between atom i and j is written as $f_j(R_{ij})$ and gives the contribution to the electron density of atom i arising from atom j. For calculations, the EAM functions are obtained by fitting to crystal properties, such as the lattice constant, the sublimation energy, elastic constants, etc. Two different sets of functions have been employed in surface calculations. One, designated AFW, was proposed by Adams *et al.* [100]. The other set, the VC potential, was obtained by Voter and Chen [101]. The EAM approximation has the advantage of allowing fast calculations, which are likely to be appropriate for systems in which the d-shell is close to being full.

Other approximations have been developed to give better or faster results, among them the effective-medium method EMT [102,103], the glue model [104], the corrected effective-medium method CEM [105], and a simplification for molecular dynamics and Monte Carlo simulations, MD/MC-CEM [105]. These theories are equivalent, but another type of interaction, the RGL potential, has been developed as a second-moment approximation of the density of states for tight-binding theory [106–108]. The binding energy in this approach is proportional to the square root of the second moment of the density of states, which is represented as a sum of square hopping integrals between adjacent atoms; these depend exponentially on the distance between atoms. From the band energy comes a many-body term which is attractive; this is balanced by introducing a Born–Mayer core repulsion. For an atom i the energy is given by two terms. The band energy E_B^i

$$E_B^i = -\left[\sum_{j, R_{ij} < R_c} \varsigma^2 \exp\left[-2q\left(\frac{R_{ij}}{\ell_o} - 1\right)\right]\right]^{\frac{1}{2}}, \qquad (2.22)$$

where R_{ij} gives the separation between atoms i and j, ℓ_o is the distance between nearest neighbors, R_c is the cut-off distance for the interactions, and q is the distance dependence of the effective hopping integral ς. The repulsive term E_R^i is just

$$E_R^i = \sum_{j, R_{ij} < R_c} A_R \exp\left[-p_R\left(\frac{R_{ij}}{\ell_o} - 1\right)\right]. \qquad (2.23)$$

For the cut-off distance R_c is picked as the distance between second neighbors. There are now four parameters here, A_R, p_R, q, and ς, which are fitted to experimental parameters for the material under study: the lattice parameter, the bulk modulus and elastic constants, as well as the cohesive energy. Another semi-empirical potential, the Sutton–Chen model [109], was obtained in much the same way, with other approximations, and appears in a different format.

Other potentials derived from tight-binding theory by second- or fourth-moment approximations have also been found useful [110–112]. The ultimate to estimating the energetics of diffusion is, of course, an *ab initio* technique [113,114], and this has been refined more recently by Kresse *et al.* [115–117]. The difficulty with the latter is that such estimates require intensive computer work, but they are quite useful in describing diffusion phenomena. Useful simulation packages are now available, like VASP [115–119] or Fireball [120–123].

Most theoretical investigations have explored static energy changes for diffusion; however, it was already known that the dynamics of the lattice might play a role in adatom diffusion as well. So far, there are two contradictory opinions, both based on the same method, EAM. One shows no importance for surface dynamics in the estimation of diffusion prefactors [124] due to compensation effects. The second claims that surface dynamics is important, due to changes in the transition state, and can lead to underestimates as large as a factor of 8 [125,126]. We will leave this problem as an open question; what is worth noting is that a standard prefactor was observed experimentally even if the mechanism of adatom movement changed from single to double jumps or rebound jumps were present [63,127,128]

The extent to which these various approaches are adequate for quantitatively describing diffusion phenomena will appear more clearly in the presentation of calculational efforts in Chapters 3, 4, and 5.

References

[1] E. W. Müller, K. Bahadur, Field ionization of gases at a metal surface and the resolution of the field ion microscope, *Phys. Rev.* **102** (1956) 624–631.

[2] E. W. Müller, Resolution of the atomic structure of a metal surface by the field ion microscope, *J. Appl. Phys.* **27** (1956) 474–476.

[3] E. W. Müller, J. A. Panitz, S. B. McLane, The atom-probe field ion microscope, *Rev. Sci. Instrum.* **39** (1968) 83–86.

[4] R. Gomer, *Field Emission and Field Ionization* (Harvard University Press, Cambridge, 1961).

[5] A. P. Graham, The low energy dynamics of adsorbates on metal surfaces investigated with helium atom scattering, *Surf. Sci. Repts.* **49** (2003) 115–168.

[6] A. G. Naumovets, Y. S. Vedula, Surface diffusion of adsorbates, *Surf. Sci. Rep.* **4** (1985) 365–434.

[7] G. Binnig, H. Rohrer, Scanning tunneling microscopy, *Helv. Phys. Acta* **55** (1982) 726–735.

[8] G. Binnig, H. Rohrer, Scanning tunneling microscopy, *Surf. Sci.* **126** (1983) 236–244.

[9] J. J. Hren, S. Ranganathan, *Field-Ion Microscopy* (Plenum Press, New York, 1968).

[10] E. W. Müller, T. T. Tsong, *Field Ion Microscopy Principles and Applications* (American Elsevier, New York, 1969).

[11] K. M. Bowkett, D. A. Smith, *Field Ion Microscopy* (North-Holland, Amsterdam, 1970).

[12] H.-J. Güntherodt, R. Wiesendanger, Springer Series in Surface Sciences, *Scanning Tunneling Microscopy* (Springer-Verlag, Berlin, 1994).

[13] H.-J. Güntherodt, R. Wiesendanger, Springer Series in Surface Sciences, *Scanning Tunneling Microscopy I* (Springer-Verlag, Berlin, 1992).

[14] R. Wiesendanger, H.-J. Güntherodt, Springer Series in Surface Sciences, *Scanning Tunneling Microscopy II*, Berlin, 1992).

References

[15] R. Wiesendanger, H.-J. Güntherodt, Springer Series in Surface Sciences 29, *Scanning Tunneling Microscopy III* (Springer-Verlag, Berlin, 1993).

[16] M. Kalff, G. Comsa, T. Michely, How sensitive is epitaxial growth to adsorbates? *Phys. Rev. Lett.* **81** (1998) 1255–1258.

[17] D. A. Reed, G. Ehrlich, In-channel clusters: rhenium on W(211), *Surf. Sci.* **151** (1985) 143–165.

[18] S. C. Wang, G. Ehrlich, Self-adsorption sites on a close-packed surface: Ir on Ir(111), *Phys. Rev. Lett.* **62** (1989) 2297–2300.

[19] W. R. Graham, G. Ehrlich, Direct identification of atomic binding sites on a crystal, *Surf. Sci.* **45** (1974) 530–552.

[20] R. S. Averback, D. N. Seidman, Neon gas imaging of gold in the field ion microscope, *Surf. Sci.* **40** (1973) 249–263.

[21] A. van Oostrom, Field ion microscopy, *Acta Electronica* **16** (1973) 59–71.

[22] A. J. Nam, A. Teren, T. A. Lusby, A. J. Melmed, Benign making of sharp tips for STM and FIM: Pt, Ir, Au, Pd, and Rh, *J. Vac. Sci. Technol. B* **13** (1995) 1556–1559.

[23] T. Sakurai, Field ion microscopy of silicon, *Surf. Sci.* **86** (1979) 562–571.

[24] T. Adachi, T. Ariyasu, Field-ion microscopy of the Si(001) surface, *Philos. Mag. A* **66** (1992) 405–414.

[25] J. A. Janssen, J. P. Jones, The sharpening of field emitter tips by ion sputtering, *J. Phys. D.: Appl. Phys.* **4** (1971) 118–124.

[26] G. Antczak, P. Jóźwik, Atom movement on a dislocated surface, *Langmuir* **24** (2008) 9970–9973.

[27] R. S. Chambers, G. Ehrlich, Chemical identification of individual surface atoms in the atom probe, *J. Vac. Sci. Technol.* **13** (1976) 273–276.

[28] J. D. Wrigley, G. Ehrlich, Surface diffusion by an atomic exchange mechanism, *Phys. Rev. Lett.* **44** (1980) 661–663.

[29] J. D. Wrigley, Surface diffusion by an atomic exchange mechanism, Physics Ph.D. Thesis, University of Illinois at Urbana-Champaign, 1982.

[30] M. K. Miller, *Atom Probe Tomography: Analysis at the Atomic Level* (Kluwer Academic, New York, 2000).

[31] M. K. Miller, A. Cerezo, M. G. Hetherington, G. D. W. Smith, *Atom Probe Field Ion Microscopy* (Clarendon Press, Oxford, 1996).

[32] T. T. Tsong, *Atom-probe Field Ion Microscopy: Field Ion Emission, and Surfaces and Interfaces at Atomic Resolution* (Cambridge University Press, Cambridge, 1990).

[33] M. K. Miller, G. D. W. Smith, *Atom Probe Microanalysis: Principles and Applications to Materials Problems* (Materials Research Society, Pittsburgh, PA, 1989).

[34] R. H. Good, E. W. Müller, in: S. Flügge (ed.), Field emission, *Handbuch der Physik*, Vol. XXI–1 (Springer-Verlag, Berlin, 1956), p. 176–231.

[35] R. Gomer, Diffusion of adsorbates on metal surfaces, *Rep. Prog. Phys.* **53** (1990) 917–1002.

[36] J. Beben, C. Kleint, R. Meclewski, Improved adsorbate fluctuation measurements and their explanation by different diffusion mechanisms: I. Arguments in favor of single adatom surface diffusion, *Surf. Sci.* **213** (1989) 438–450.

[37] R. Blaszczyszyn, C. Kleint, Effect of preadsorbed Ni atoms upon the potassium diffusion on the W(112) plane, *Surf. Sci.* **253** (1991) 129–136.

[38] T. Biernat, C. Kleint, R. Meclewski, Surface diffusion of lithium across and along atomic rows on the W(211) plane, *Appl. Surf. Sci.* **67** (1993) 206–210.

[39] M. v. Smoluchowski, Studien über Molekularstatistik von Emulsionen und deren Zusammenhang mit der Brown'schen Bewegung, *Sitzber. Akad. Wiss. Wien Math. Naturw. Kl.* **123** (1914) 2381–2405.

[40] M. v. Smoluchowski, Drei Vorträge über Diffusion, Brownsche Molekularbewegung und Koagulation von Kolloidteilchen, *Physik. Zeitschr.* **17** (1916) 557–571, 585–599.
[41] G. W. Timm, A. Van der Ziel, Noise in field emission diodes, *Physica* **32** (1966) 1333–1344.
[42] C. Kleint, Surface diffusion model of adsorption-induced field emission flicker noise I. Theory, *Surf. Sci.* **25** (1971) 394–410.
[43] C. Kleint, Surface diffusion model of adsorption-induced field emission flicker noise II. Experiments, *Surf. Sci.* **25** (1971) 411–434.
[44] J.-R. Chen, R. Gomer, Mobility of oxygen on the (110) plane of tungsten, *Surf. Sci.* **79** (1979) 413–444.
[45] R. Young, J. Ward, F. Scire, The topographiner: An instrument for measuring surface microtopography, *Rev. Sci. Instrum.* **43** (1972) 999–1011.
[46] E. C. Teague, Room temperature gold-vacuum-gold tunneling experiments, *J. Res. Natl. Bur. Stand.* **91** (1986) 171–233.
[47] G. Binnig, H. Rohrer, C. Gerber, E. Weibel, Surface studies by scanning tunneling microscopy, *Phys. Rev. Lett.* **49** (1982) 57–61.
[48] J. A. Golovchenko, The tunneling microscope: a new look at the atomic world, *Science* **232** (1986) 48–53.
[49] T. R. Linderoth, S. Horch, E. Laegsgaard, I. Stensgaard, F. Besenbacher, Surface diffusion of Pt on Pt(110): Arrhenius behavior of long jumps, *Phys. Rev. Lett.* **78** (1997) 4978–4981.
[50] K. Morgenstern, K.-H. Rieder, Long-range interaction of copper atoms and copper dimers on Ag(111), *New J. Phys.* **7** (2005) 139 1–11.
[51] F. Giessibl, C. F. Quate, Exploring the nanoworld with atomic force microscopy, *Physi. Today* **59** (2006) 44–50.
[52] P. J. Godhew, *Practical Methods in Electron Microscopy* (North-Holland Publishing Company, Amsterdam, 1972).
[53] R. Liu, Chemisorption on perfect surfaces and structural defects, Materials Ph.D. Thesis, University of Illinois at Urbana-Champaign, 1977.
[54] G. Antczak, G. Ehrlich, Long jumps in diffusion of iridium on W(110), *Phys. Rev. B* **71** (2005) 115422 1–9.
[55] S. C. Wang, G. Ehrlich, Imaging and diffusion of individual iridium adatoms on Ir(111), *Surf. Sci.* **224** (1989) L997–1003.
[56] H.-W. Fink, G. Ehrlich, Pair and trio interactions between adatoms: Re on W(110), *J. Chem. Phys.* **81** (1984) 4657–4665.
[57] P. D. Desai, T. K. Chu, H. M. James, C. Y. Ho, Electrical resistivity of selected elements, *J. Phys. Chem. Ref. Data* **13** (1984) 1069–1096.
[58] K. Stolt, W. R. Graham, G. Ehrlich, Surface diffusion of individual atoms and dimers: Re on W(211), *J. Chem. Phys.* **65** (1976) 3206–3222.
[59] S. C. Wang, G. Ehrlich, Adatom diffusion on W(211): Re, W, Mo, Ir and Rh, *Surf. Sci.* **206** (1988) 451–474.
[60] M. F. Lovisa, G. Ehrlich, Quantitative determinations of the temperature dependence of diffusion phenomena in the FIM, *Surf. Sci.* **246** (1991) 43–49.
[61] O. Grizzi, M. Shi, H. Bu, J. W. Rabalais, P. Hochmann, Time of flight scattering and recoiling spectrometry. I. Structure of the W(211) surface, *Phys. Rev. B* **40** (1989) 10127–10146.
[62] H. L. Davis, G.-C. Wang, Registry relaxation of the W(211) surface – a (1 × 1) reconstruction, *Bull. Am. Phys. Soc.* **29** (1984) 221.
[63] G. Antczak, Long jumps in one-dimensional surface self-diffusion: Rebound transitions, *Phys. Rev. B* **73** (2006) 033406 1–4.

References

[64] S.-M. Oh, K. Kyuno, S. J. Koh, G. Ehrlich, Atomic jumps in surface self-diffusion:W on W(110), *Phys. Rev. B* **66** (2002) 233406 1–4.

[65] K. Morgenstern, Fast scanning tunneling microscopy as a tool to understand changes on metal surfaces: from nanostructures to single atoms, *Phys. Stat. Sol.* **242** (2005) 773–796.

[66] N. Knorr, H. Brune, M. Epple, A. Hirstein, M. A. Schneider, K. Kern, Long-range adsorbate interactions mediated by a two-dimensional electron gas, *Phys. Rev. B* **65** (2002) 115420 1–5.

[67] J. Fransson, A. V. Balatsky, Surface imaging of inelastic Friedel oscillations, *Phys. Rev. B* **75** (2007) 195337 1–5.

[68] H. Y. H. Chan, K. Dev, E. G. Seebauer, Vacancy charging on Si(100)-(2 × 1): Consequences for surface diffusion and STM imaging, *Phys. Rev. B* **67** (2003) 035311 1–7.

[69] K. Dev, E. G. Seebauer, Vacancy charging on Si(111)-(7 × 7) investigated by density functional theory, *Surf. Sci.* **538** (2003) L495–499.

[70] B. S. Swartzentruber, Direct measurement of surface diffusion using atom-tracking scanning tunneling microscopy, *Phys. Rev. Lett.* **76** (1996) 459–462.

[71] J. Repp, F. Moresco, G. Meyer, K.-H. Rieder, P. Hyldgard, M. Perrson, Substrate mediated long-range oscillatory interaction between adatoms, *Phys. Rev. Lett.* **85** (2000) 2981–2984.

[72] J. Repp, G. Meyer, K.-H. Rieder, P. Hyldgaard, Site determination and thermally assisted tunneling in homogeneous nucleation, *Phys. Rev. Lett.* **91** (2003) 206102 1–4.

[73] K. Morgenstern, K.-F. Braun, K.-H. Rieder, Direct imaging of Cu dimer formation, motion, and interaction with Cu atoms on Ag(111), *Phys. Rev. Lett.* **93** (2004) 056102 1–4.

[74] H. Brune, Microscopic view of epitaxial metal growth: Nucleation and aggregation, *Surf. Sci. Rep.* **31** (1998) 121–230.

[75] K. Kyuno, G. Ehrlich, Cluster diffusion and dissociation in the kinetics of layer growth: An atomic view, *Phys. Rev. Lett.* **84** (2000) 2658–2661.

[76] J. A. Venables, Nucleation calculations in a pair-binding model, *Phys. Rev. B* **36** (1987) 4153–4162.

[77] H. Brune, G. S. Bales, H. Jacobsen, C. Boragno, K. Kern, Measuring surface diffusion from nucleation island densities, *Phys. Rev. B* **60** (1999) 5991–6006.

[78] G. Rosenfeld, K. Morgenstern, M. Esser, G. Comsa, Dynamics and stability of nanostructures on metal surfaces, *Appl. Phys. A* (1999) 489–496.

[79] M. Bott, M. Hohage, M. Morgenstern, T. Michely, G. Comsa, New approach for determination of diffusion parameters of adatoms, *Phys. Rev. Lett.* **76** (1996) 1304–1307.

[80] R. Koch, J. J. Schulz, K. H. Rieder, Scanning tunneling microscopy artifact and real structure: Steps of Ag(110), *Europhys. Lett.* **48** (1999) 554–560.

[81] K. Haug, Z. Zhang, D. John, C. F. Walters, D. M. Zehner, W. E. Plummer, Effects of hydrogen in Ni(100) submonolayer homoepitaxy, *Phys. Rev. B* **55** (1997) R10233–10236.

[82] T. Michely, J. Krug, in: *Islands, Mounds and Atoms* (Springer-Verlag, Berlin, 2004), Section 2.1.

[83] L. Kuipers, R. E. Palmer, Influence of island mobility on island size distribution in surface growth, *Phys. Rev. B* **53** (1996) R7646–7649.

[84] C. Ratsch, P. Smilauer, A. Zangwill, D. D. Vvedensky, Submonolayer epitaxy without a critical nucleus, *Surf. Sci.* **329** (1995) L599–604.

[85] J. A. Stroscio, D. T. Pierce, Scaling of diffusion-mediated island growth in iron-on-iron homoepitaxy, *Phys. Rev. B* **49** (1994) 8522–8525.

[86] C. Ratsch, M. Scheffler, Density-functional theory calculations of hopping rates of surface diffusion, *Phys. Rev. B* **58** (1998) 13163–13166.

[87] K. Fichthorn, M. Scheffler, Island nucleation in thin-film epitaxy: A first-principles investigation, *Phys. Rev. Lett.* **84** (2000) 5371–5374.

[88] A. Bogicevic, S. Ovesson, P. Hyldgaard, B. I. Lundqvist, H. Brune, D. R. Jennison, Nature, strength, and consequences of indirect adsorbate interactions on metals, *Phys. Rev. Lett.* **85** (2000) 1910–1913.

[89] S. Ovesson, A. Bogicevic, G. Wahnstrom, B. I. Lundqvist, Neglected adsorbate interactions behind diffusion prefactor anomalies on metals, *Phys. Rev. B* **64** (2001) 125423 1–11.

[90] J. V. Barth, H. Brune, B. Fischer, J. Weckesser, K. Kern, Dynamics of surface migration in the weak corrugation regime, *Phys. Rev. Lett.* **84** (2000) 1732–1735.

[91] T. Michely, W. Langenkamp, H. Hansen, C. Busse, Comment on "Dynamics of surface migration in the weak corrugation regime", *Phys. Rev. Lett.* **86** (2001) 2695–2695.

[92] C. Busse, W. Langenkamp, C. Polop, A. Petersen, H. Hansen, U. Linke, P. J. Feibelman, T. Michely, Dimer binding energies on fcc(111) metal surfaces, *Surf. Sci.* **539** (2003) L560–566.

[93] J. A. Venables, H. Brune, Capture numbers in the presence of repulsive adsorbate interactions, *Phys. Rev. B* **66** (2002) 195404 1–16.

[94] A. T. Loburets, A. G. Naumovets, Y. S. Vedula, Surface diffusion of lithium on (011) face of tungsten, *Surf. Sci.* **120** (1982) 347–366.

[95] T. Klas, R. Fink, G. Krausch, R. Platzer, J. Voigt, R. Wesche, G. Schatz, Isolated indium atoms on copper surfaces: a perturbed γ–γ angular correlation study, *Surf. Sci.* **216** (1989) 270–302.

[96] G. Krausch, R. Fink, K. Jacobs, U. Kohl, J. Lohmüller, B. Luckscheiter, R. Platzer, B.-U. Runge, U. Wöhrmann, G. Schatz, Surface and interface studies with perturbed angular correlations, *Hyperfine Interactions* **78** (1993) 261–280.

[97] M. Bée, *Quasielastic Neutron Scattering*, (IOP Publishing, Bristol, 1988).

[98] G. Vogl, M. Sladecek, S. Dattagupta, Probing single jumps of surface atoms, *Phys. Rev. Lett.* **99** (2007) 155902 1–4.

[99] M. S. Daw, M. I. Baskes, Embedded-atom method: Derivation and application to impurities, surfaces, and other defects in metals, *Phys. Rev. B* **29** (1984) 6443–6453.

[100] J. B. Adams, S. M. Foiles, W. G. Wolfer, Self-diffusion and impurity diffusion of fcc metals using the five-frequency model and the Embedded Atom Method, *J. Mater. Res.* **4** (1989) 102–112.

[101] A. F. Voter and S. P. Chen, Accurate interatomic potentials for Ni, Al and Ni$_3$Al, *Mater. Res. Soc. Symp. Proc.* **82** (1987) 175–180.

[102] K. W. Jacobsen, J. K. Nørskov, M. J. Puska, Interatomic interactions in the effective-medium theory, *Phys. Rev. B* **35** (1987) 7423–7442.

[103] K. W. Jacobsen, Bonding in Metallic Systems: An effective-medium approach, *Comments Cond. Mater. Phys.* **14** (1988) 129–161.

[104] F. Ercolessi, M. Parrinello, E. Tosatti, Simulation of gold in the glue model, *Philos. Mag. A* **58** (1988) 213–226.

[105] M. S. Stave, D. E. Sanders, T. J. Raeker, A. E. DePristo, Corrected effective medium method. V. Simplifications for molecular dynamics and Monte Carlo simulations, *J. Chem. Phys.* **93** (1990) 4413–4426.

[106] V. Rosato, M. Guillope, B. Legrand, Thermodynamical and structural properties of f.c.c. transition metals using a simple tight-binding model, *Philos. Mag. A* **59** (1989) 321–336.

[107] M. Guillope, B. Legrand, (110) Surface stability in noble metals, *Surf. Sci.* **215** (1989) 577–595.

References

[108] F. Cleri, V. Rosato, Tight-binding potentials for transition metals and alloys, *Phys. Rev. B* **48** (1993) 22–33.

[109] A. P. Sutton, J. Chen, Long-range Finnis-Sinclair potentials, *Philos. Mag. Lett.* **61** (1990) 139–146.

[110] H. Bulou, O. Lucas, M. Kibaly, C. Goyhenex, Long-time scale molecular dynamics study of Co diffusion on the Au(111) surface, *Comput. Mater. Sci.* **27** (2003) 181–185.

[111] H. Bulou, C. Massobrio, Dynamical behavior of Co adatoms on the herringbone reconstructed surface of Au(111), *Superlattices Microstruct.* **36** (2004) 305–313.

[112] W. Xu, J. B. Adams, Fourth moment approximation to tight binding: Application to bcc metals, *Surf. Sci.* **301** (1994) 371–385.

[113] P. Hohenberg, W. Kohn, Inhomogeneous electron gas, *Phys. Rev.* **136** (1964) B 864–871.

[114] W. Kohn, L. J. Sham, Self-consistent equations including exchange and correlation effects, *Phys. Rev.* **140** (1965) A 1133–1138.

[115] G. Kresse, J. Hafner, *Ab initio* molecular dynamics for liquid metals, *Phys. Rev. B* **47** (1993) 558–561.

[116] G. Kresse, J. Hafner, *Ab initio* molecular dynamics simulation of the liquid-metal-amorphous-semiconductor transition in germanium, *Phys. Rev. B* **49** (1994) 14251–14269.

[117] G. Kresse, J. Furthmüller, Efficient iterative schemes for *ab initio* total-energy calculations using a plane-wave set, *Phys. Rev. B* **54** (1996) 11169–11186.

[118] G. Kresse, J. Furthmüller, Efficiency of ab-initio total energy calculations for metal and semiconductors using a plane-wave basis set, *Comput. Mater. Sci.* **6** (1996) 15–50.

[119] G. Kresse, J. Joubert, From ultrasoft pseudopotentials to the projector augmernted-wave method, *Phys. Rev. B* **59** (1999) 1758–1775.

[120] A. A. Demkov, J. Ortega, O. F. Sankey, M. P. Grumbach, Electronic structure approach for complex silicas, *Phys. Rev. B* **52** (1995) 1618–1630.

[121] O. F. Sankey, D. J. Niklewski, *Ab initio* multicenter tight-binding model for molecular-dynamics simulations and other applications in covalent systems, *Phys. Rev. B* **40** (1989) 3979–3995.

[122] P. Jelinek, H. Wang, J. P. Lewis, O. F. Sankey, J. Ortega, Multicenter approach to the exchange-correlation interactions in *ab initio* tight-binding methods, *Phys. Rev. B* **71** (2005) 235101 1–9.

[123] J. P. Lewis, K. R. Glaesemann, G. A. Voth, J. Fritsch, A. A. Demkov, J. Ortega, O. Sankey, Further developments in the local-orbital density-functional-theory tight-binding method, *Phys. Rev. B* **64** (2001) 195103 1–10.

[124] H. Yildirim, A. Kara, T. S. Rahman, Orgin of quasi-constant pre-exponential factors for adatom diffusion of Cu and Ag surfaces, *Phys. Rev. B* **76** (2007) 165421 1–10.

[125] L. T. Kong, L. J. Lewis, Surface diffusion coefficient: Substrate dynamics matters, *Phys. Rev. B* **77** (2008) 165422 1–5.

[126] L. T. Kong, L. J. Lewis, Transition state theory of the preexponential factors for self-diffusion on Cu, Ag, and Ni surfaces, *Phys. Rev. B* **74** (2006) 073412 1–4.

[127] G. Antczak, G. Ehrlich, Jump processes in surface diffusion, *Surf. Sci. Rep.* **62** (2007) 39–61.

[128] G. Antczak, G. Ehrlich, Long jump rates in surface diffusion: W on W(110), *Phys. Rev. Lett.* **92** (2004) 166105 1–4.

3 Atomic events in surface diffusion

With background information about the kinetic and experimental aspects of surface diffusion in hand we will now turn our attention to the atomistics of diffusion about which much has been learned through modern instrumentation. The usual picture of surface diffusion, which seems to agree at least qualitatively with experiments on diffusivities, is that atoms carry out random jumps between nearest-neighbor sites. Is this picture correct? Has it been tested in reasonable experiments? What can be said about the jumps which move an atom in surface diffusion? What is the nature of the sites at which atoms are bound? These are among the topics that will be considered at length now.

In these matters the geometry of the various surfaces studied plays an important role, and at the very start we therefore show hard-sphere models of planes that will emerge as significant. Presented first are channeled surfaces, bcc (211) and (321) in Fig. 3.1, as well as fcc(110), (311), and (331) in Fig. 3.2. These are followed in Fig. 3.3 by fcc(100) and (111) and bcc(100), (110), and (111) planes in Fig. 3.4. Their structures are sometimes quite similar, but their diffusion behavior may be quite different.

3.1 Adatom binding sites

3.1.1 Location

Where on a surface are metal adatoms bonded? That is an important question for understanding how atoms progress over a surface in diffusion. LEED has been very important in providing information about the geometry of adsorbed layers [1]. Lately STM has also given us some direct insights into the geometry of adsorption layers [2]. We, however, require information about binding of individual adatoms, and for tackling this question the field ion microscope is ideally suited, since it has excellent sensitivity and horizontal resolution, even though its ability to determine vertical positions is limited. STM is quite a suitable technique as well, but one has to be very careful not to influence the magnitude of the potential well or the position of the adatom through the closeness of the scanning tip, as illustrated in Fig. 3.5 [3]. The FIM has no such limitation; however, observations of adsorption take place in high electric fields, above $\sim 10^8$ V/cm, which may have an effect on atom location. In any event, this effect would not be local, as in the STM.

3.1 Adatom binding sites

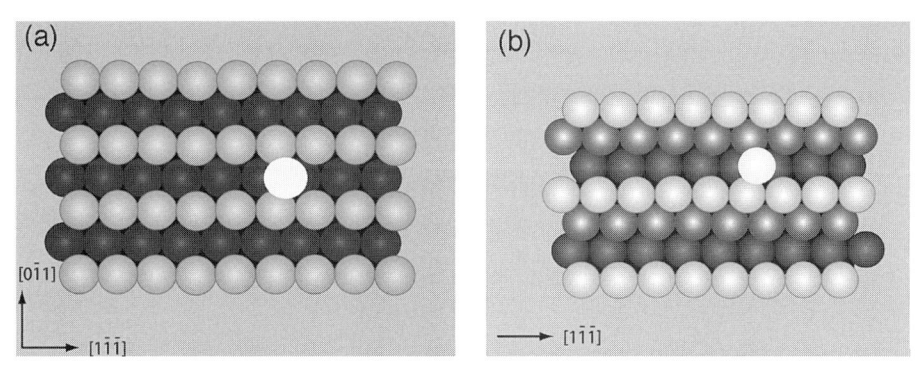

Fig. 3.1 Hard-sphere models of channeled bcc crystal surfaces. (a) bcc(211) and (b) bcc(321).

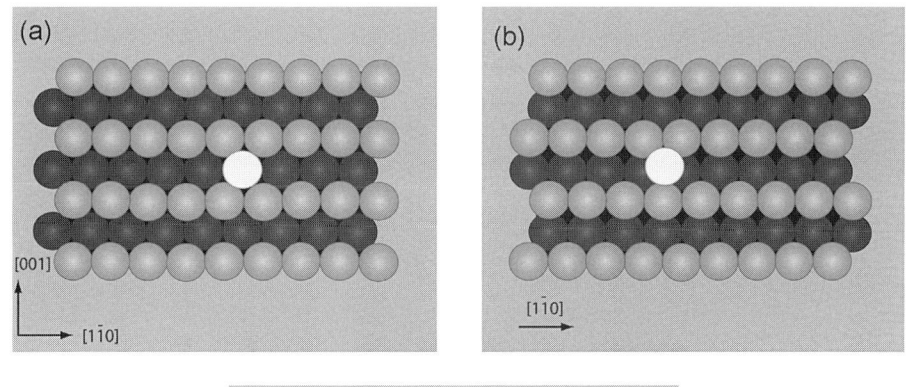

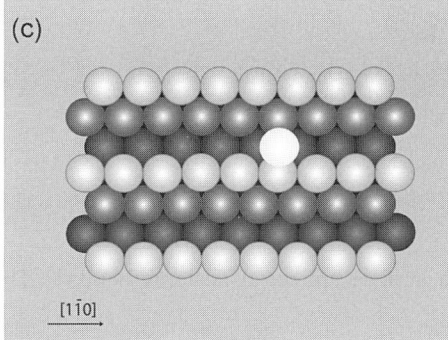

Fig. 3.2 Hard-sphere models of channeled fcc crystal surfaces. (a) fcc(110), (b) fcc(311), and (c) fcc(331).

The first attempt at defining atomic positions was made by Graham and Ehrlich [4] in 1974, who looked at the (111) plane of tungsten. A micrograph with this plane atomically resolved in the center is shown in Fig. 3.6, and a hard-sphere model in Fig. 3.4c. That the structure of the (111) plane conformed to the model in Fig. 3.4 was first established by imaging the surface, then field evaporating a layer, imaging again, until three layers had been removed. The fourth layer should correspond in position to the first, and this was proven in Fig. 3.6b.

66   **Atomic events in surface diffusion**

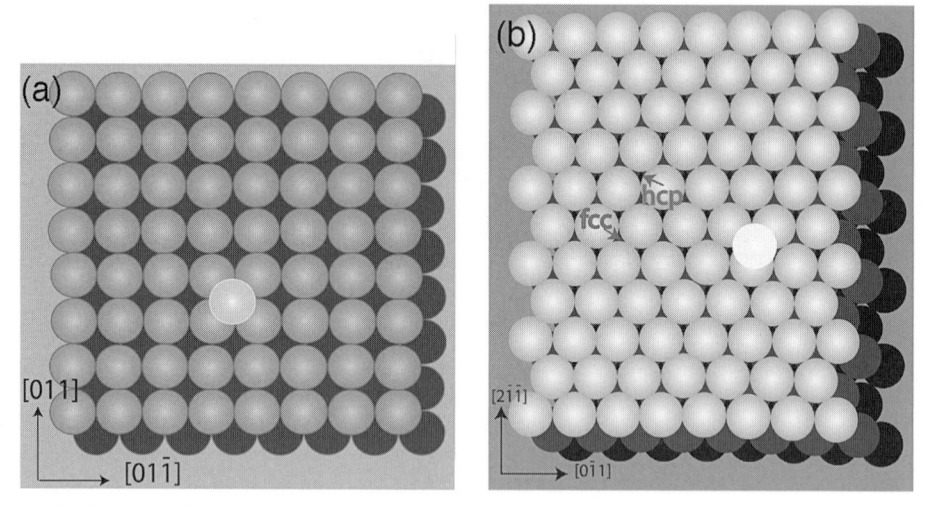

Fig. 3.3   Hard-sphere models of fcc surfaces. (a) fcc(100), (b) fcc(111).

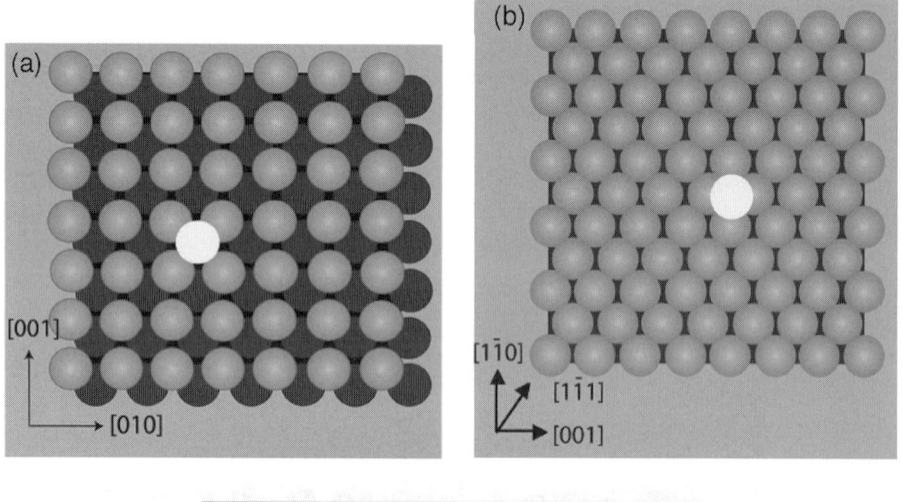

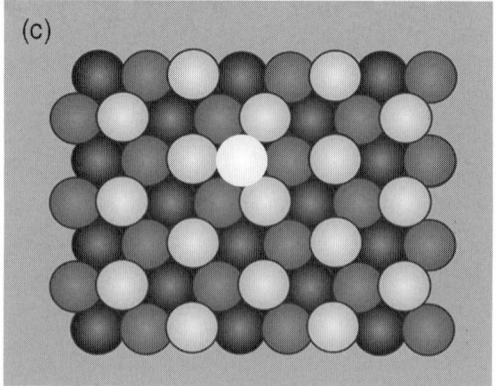

Fig. 3.4   Hard-sphere models of bcc surfaces. (a) bcc(100), (b) bcc(110), (c) bcc(111).

3.1 Adatom binding sites

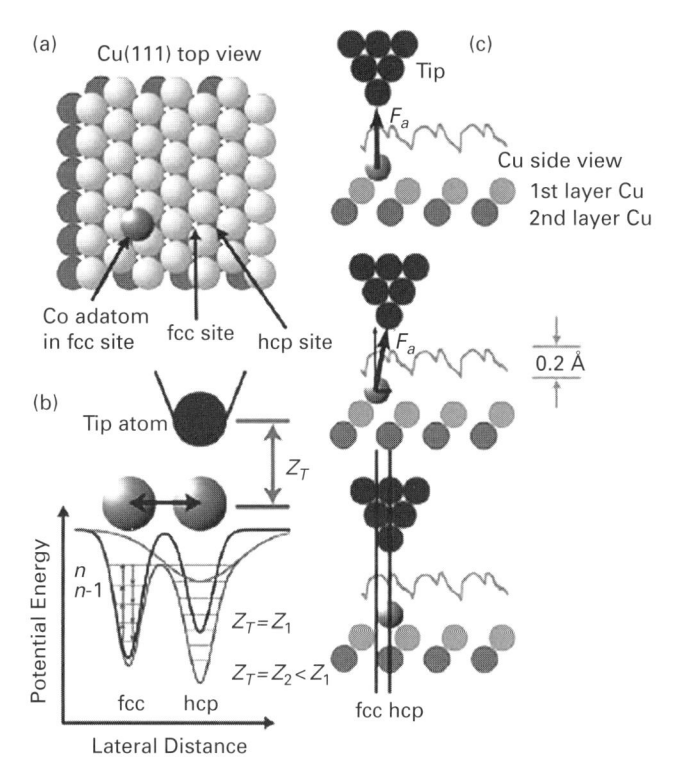

Fig. 3.5 Scanning of adatom location on a Cu(111) surface using STM. (a) View of Cu(111) plane with Co adatom in an fcc site. (b) Effect of probing tip on potential confining adatom. (c) Schematic showing effect of tip on height trace and on atom location (after Stroscio and Celotta [3]).

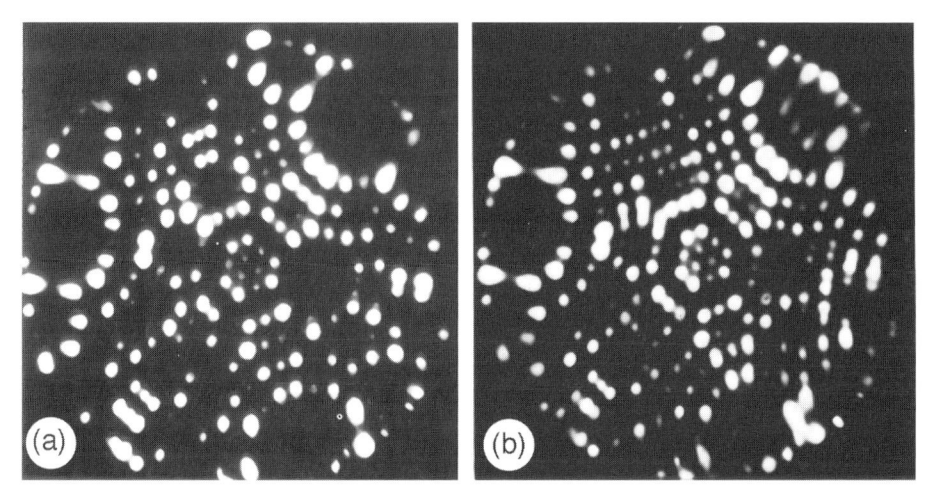

Fig. 3.6 Field ion image of [111] oriented tungsten tip, field evaporated to reveal individual tungsten atoms of (111) plane [4]. (a) Image of first layer. (b) Superposition of first layer of W(111) plane, that is Fig. 3.6a, upon fourth layer, showing agreement of positions.

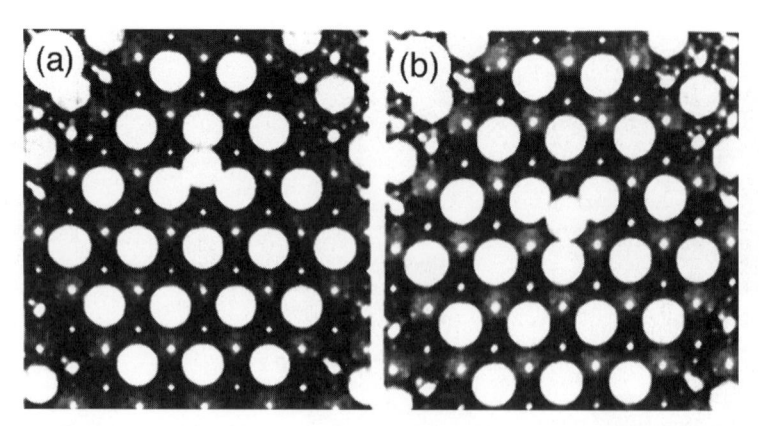

Fig. 3.7 Hard-sphere models of bcc(111) plane, showing (a) lattice site and (b) fault site for binding an adatom.

Where on this surface does a tungsten atom sit? On W(111), there are two obvious possibilities, lattice and fault sites, shown on hard-sphere models in Fig. 3.7. To be able to define the position of an adatom on such a plane, tungsten was evaporated onto a large surface, on which the adatom image is well defined and sharp. However, on such a large surface, the substrate atoms are not revealed, only the atoms at the plane edges are apparent. The atom position is therefore recorded, and the atom is then removed by field evaporation. The (111) plane is cautiously field evaporated until it is so small that all the atoms in it are clearly visible, as in Fig. 3.8. The adatom position is then ascertained by superposing the first image upon the last. In a total of 175 sequences, 165 atoms were found at lattice sites, the rest in fault sites. Warming to 480 K moved atoms from fault to lattice sites. The lattice site was estimated as 0.5 eV energetically more favorable than the fault site.

The next step was taken in 1978 by Flahive and Graham [5]. Because of concerns about the superposition technique, they did direct low-field imaging of nickel atoms on W(111). In roughly 200 experiments, the location of the nickel atom always corresponded to an equilibrium lattice site. Flahive and Graham [6] then turned their attention to the binding sites of tungsten atoms on three different planes – W(111), W(211), and W(321). For W(111), observations were made on small planes, consisting of ~ 24 surface atoms, in which individual lattice atoms were discernable. In a total of 220 experiments in which a tungsten atom was deposited on the (111) plane, 71% showed atoms at lattice sites, as in Fig. 3.9. The remaining 29% were found in various sites at the edges. They did not observe atoms on fault or bridge sites. Flahive and Graham also carefully examined the location of tungsten adatoms on W(211), and concluded "that the sites occupied by a tungsten adatom on the W(211) surface exhibit the same symmetry as the underlying substrate." On the W(321) plane, they found that 70% of the deposited atoms ended up at more strongly binding sites F, while the remainder were in weaker sites I, illustrated in Fig. 3.10. They identified the stronger sites as being positions in the trough of the channel, while the weaker ones were on the flatter portion, on the side of the plane, as in Fig. 3.11. Checking adatom diffusion after deposition, they found that ~ 30% of the

3.1 Adatom binding sites

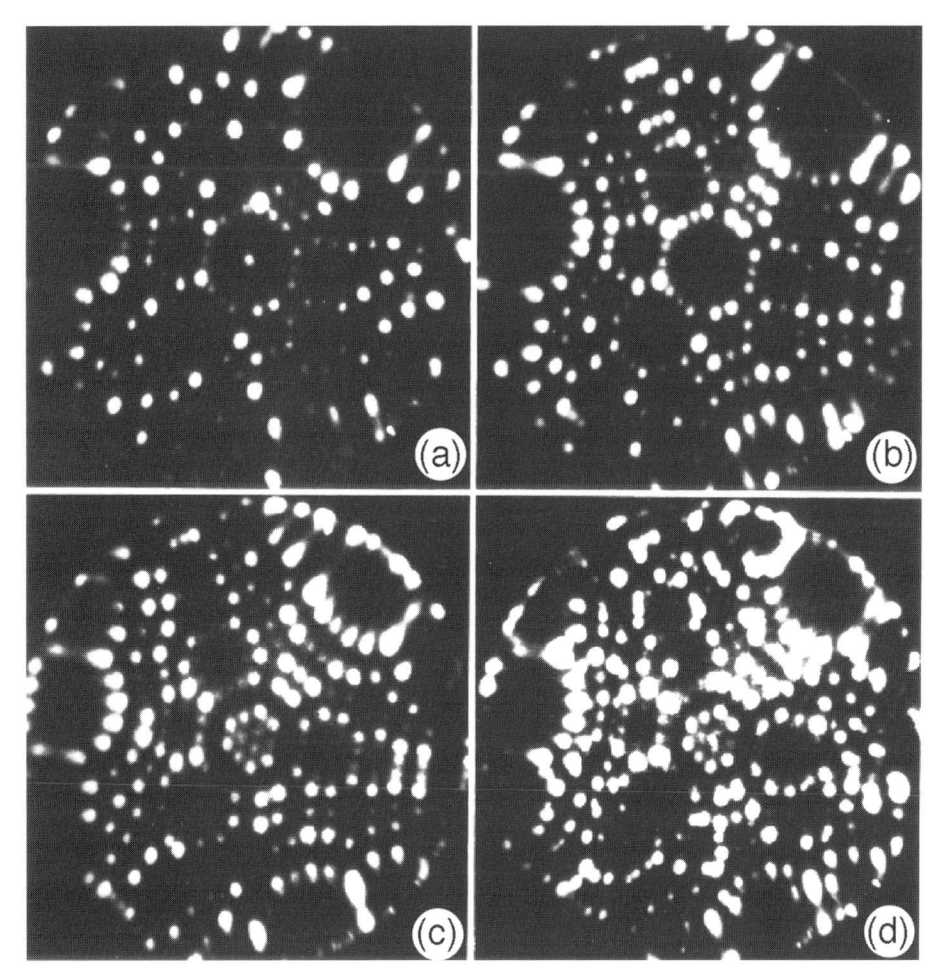

Fig. 3.8 Steps in determining adatom position on W(111) plane [4]. (a) W atom has been deposited on (111) plane. (b) Field evaporation has removed adatom. (c) (111) plane has been reduced in size by field evaporation to make individual atoms visible. (d) Superposition of (a) on (c) gives adatom location with respect to substrate.

adatoms moved transverse to the channel direction. The atom deposited at a shallow site started moving around 225 K and was very likely to descend to a deep site, where it stayed immobilized up to 300 K. Subsequent theoretical predictions by Xu and Adams [7] found the same adsorption sites as experiments on this plane. They estimated the energy difference between the two sites as 0.93 eV.

Identification of binding sites on W(110), the most densely packed plane in the bcc lattice, was done by Fink [8–10]. He allowed a rhenium atom to diffuse over the surface, and recorded the location of the sites at which the atom stopped. A plot revealed an arrangement expected for sites if an atom is held halfway between four lattice atom sites (sometimes named fourfold sites). This is confirmed by a more detailed view of a smaller area of a tungsten (110) surface, shown in Fig 3.12b. Of course the possibility exists that

70 Atomic events in surface diffusion

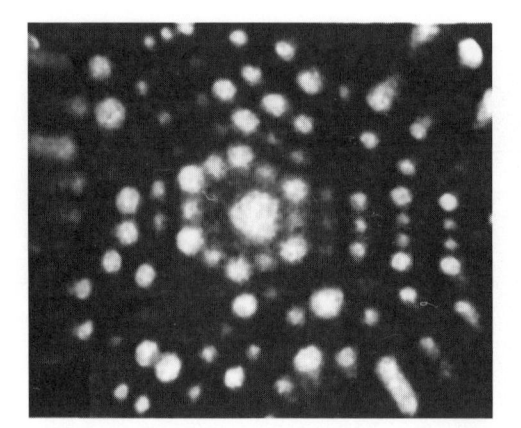

Fig. 3.9 FIM image of tungsten adatom at 12 K at a normal binding site on W(111) plane (after Flahive and Graham [6]).

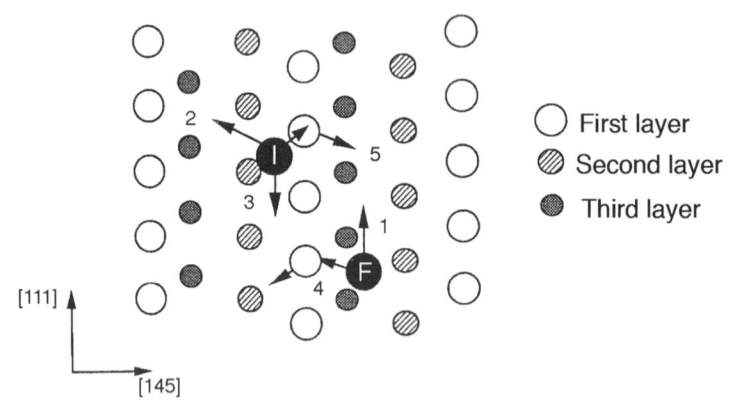

Fig. 3.10 Binding sites and diffusion paths of adatom on W(321) obtained from calculations with modified fourth-moment approximation. (1), (2), and (3) designate transitions inside channel. Cross-channel diffusion is shown in (4) and (5). I stands for shallow and F for deep site (after Xu and Adams [7]).

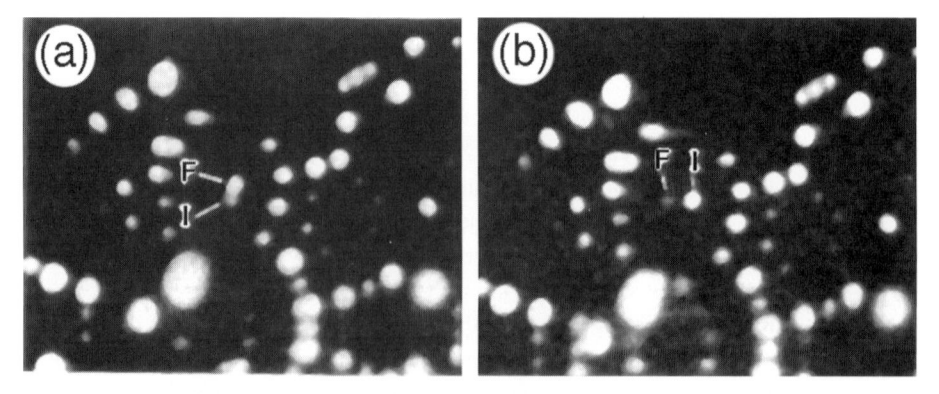

Fig. 3.11 Tungsten adatom diffusion on W(321) plane. (a) Superposition showing transition from an initial shallow site I to a deep site F on heating to 225 K. (b) Superposition demonstrating motion along row of deep sites after heating at 300 K (after Flahive and Graham [5]).

3.1 Adatom binding sites

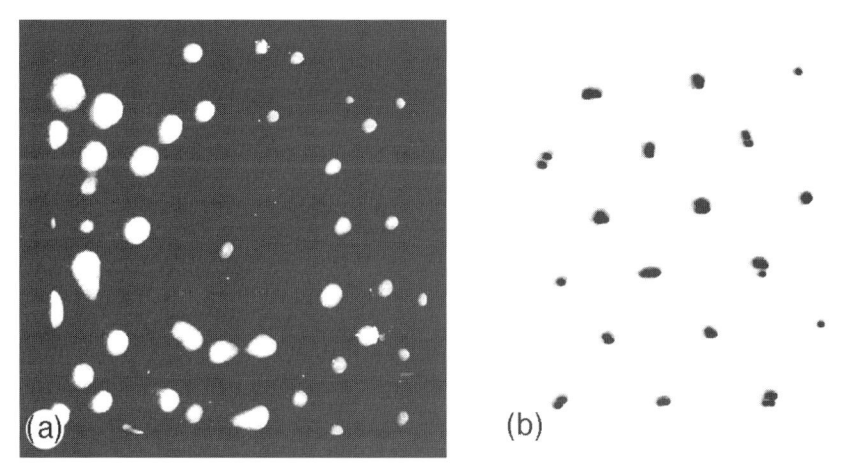

Fig. 3.12 FIM determination of binding sites for Re adatom on W(110) plane [9]. (a) Image of single adatom on W(110). (b) Location of adatom on (110) plane after diffusion over the surface at 368 K.

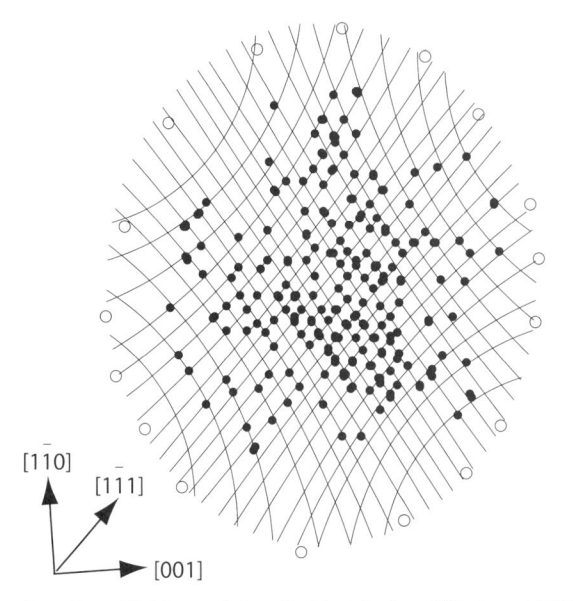

Fig. 3.13 Location of iridium adatom (in black) after diffusing at 350 K on W(110) plane [11]. Open circles show edge atoms.

the location of the adatom is a result of the electric field applied in order to image the surface. However, there are no indications of small variations in the atom locations from one image to the next, as might be expected if they were ordinarily held at some other position on the surface. Later on adsorption sites have been reexamined a number of times and for tungsten as well as for iridium atoms in Fig. 3.13 the sites agreed with the previously identified adsorption places [11]. In the past, two separate adsorption sites were identified on this plane in theoretical work, indicated in Fig. 3.14: a stable one in a

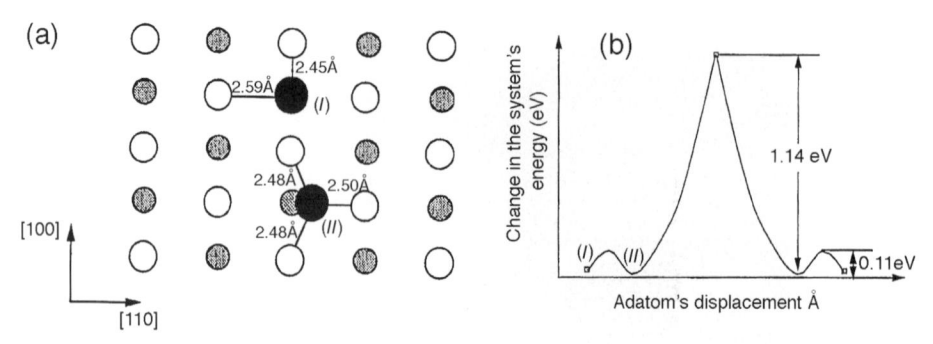

Fig. 3.14 Tungsten adatom on W(110) plane, derived from fourth-moment approximation to tight binding. (a) Four fold binding site indicated by *I*, three fold site by *II*. (b) Energy change with tungsten atom jumping (after Xu and Adams [7]).

three fold hollow *II*, the other metastable in a four fold hollow position *I* [7], with an energy difference around 0.11 eV and a lateral distance of 0.74 Å. This was not observed in experiments, however, nor in calculations of diffusion for iron [12,13] or manganese [14] on the same surface. Recent theoretical investigations by Fijak *et al*. [15] also show the existence of only one adsorption site in the movement of W on W(110). At this point it is worth emphasizing that the map of adsorption sites indicates only those places at which atoms are equilibrated; it does not show places occupied while an atom is in transit. The existence of a very metastable location will, however, change the kinetics of atom movement between stable locations.

The picture is somewhat more complicated for adsorption on the fcc(111) plane. In 1980, Tung and Graham [16] looked at nickel atoms deposited on a Ni(111) surface at ~25 K. They discovered that atoms were field evaporated at two different voltages, ~4.8 kV and ~ 6.5 kV, suggesting the presence of two distinct binding sites. They ascribed the higher voltage peak to atoms desorbed from fcc sites, and the lower one to hcp sites. Although determination of the binding sites was not possible, this work was the first to establish that atom binding could take place at two distinct locations on an fcc(111) plane.

Some time later, iridium was studied on Ir(111) by Wang and Ehrlich [17]. As is clear from the model of the fcc(111) plane in Fig. 3.3b, there are two somewhat similar sites, designated as fcc and hcp, at which an atom may possibly be bound. An adatom at an fcc site sits above an empty space in the second layer of lattice atoms, whereas at an hcp site the adatom is located directly above a second layer atom in the lattice. The field ion microscope image of Ir(111) does not provide any information about the atomic arrangement of this surface; this region is black, as in Fig. 2.2a. An iridium atom placed on the surface, however, is clearly revealed, as is evident in Fig. 3.15 and its locations can be mapped out. A mapping of binding sites, if only fcc sites are occupied, is given in Fig. 3.16a. If, on the contrary, only hcp sites are filled in adsorption, then the site grid is that shown in Fig. 3.16b. The two grids are identical, but if occasionally both types of sites are occupied, then everything changes. If fcc sites are favored in adsorption, but occasionally an hcp site is filled, then this atom will appear at the right side of the unit

3.1 Adatom binding sites

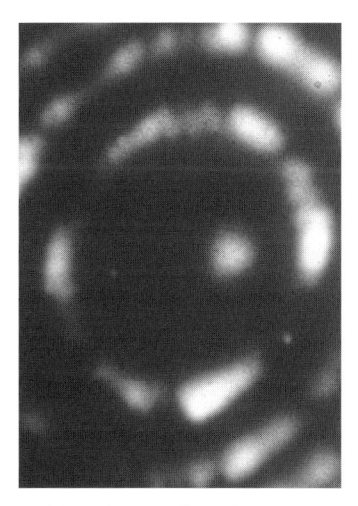

Fig. 3.15 Field ion image of Ir adatom close to the center of Ir(111) plane [18]. Adatom image is triangular, not round.

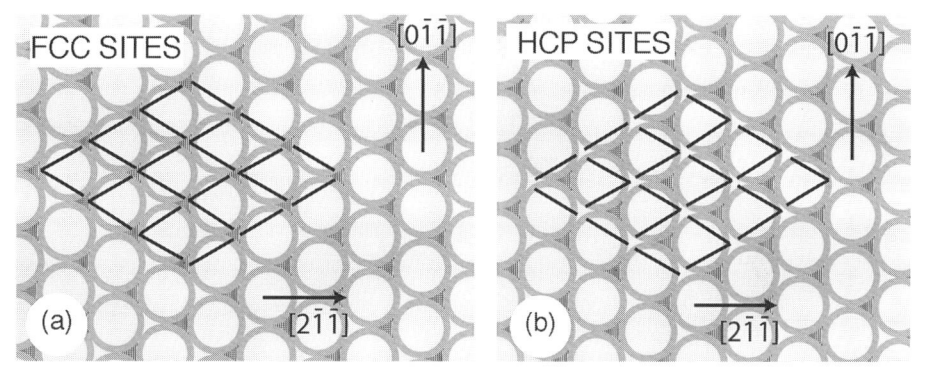

Fig. 3.16 Schematics of surface grids for adatom sites on fcc(111) plane [17]. (a) Adatom in fcc sites. (b) Adatom in hcp sites.

cell, as is clear from Fig. 3.16a. If, on the other hand, atoms prefer occupying an hcp site most of the time, an additional atom position at the left of the unit cell will be visible. A test of the distribution has been made and is shown in Fig. 3.17. For iridium atoms, the occasional atom position is seen on the left side of the unit cell. The conclusion is therefore that iridium atoms prefer to occupy hcp sites.

Somewhat later, Wang and Ehrlich [18] discovered that the field ion image spots from iridium adatoms were not round, but were instead triangular in shape, as had been found for tungsten atoms on W(111) [6] in Fig. 3.9. The orientation of the image triangle depended upon the nature of the site at which the atom was bound. With direction [2$\bar{1}\bar{1}$] as always to the right, image triangles point to the right for atoms at hcp sites and to the left for fcc sites. It is even possible to recognize the type of adsorption site without creating a time-consuming map, even when both types of adsorption places are equally

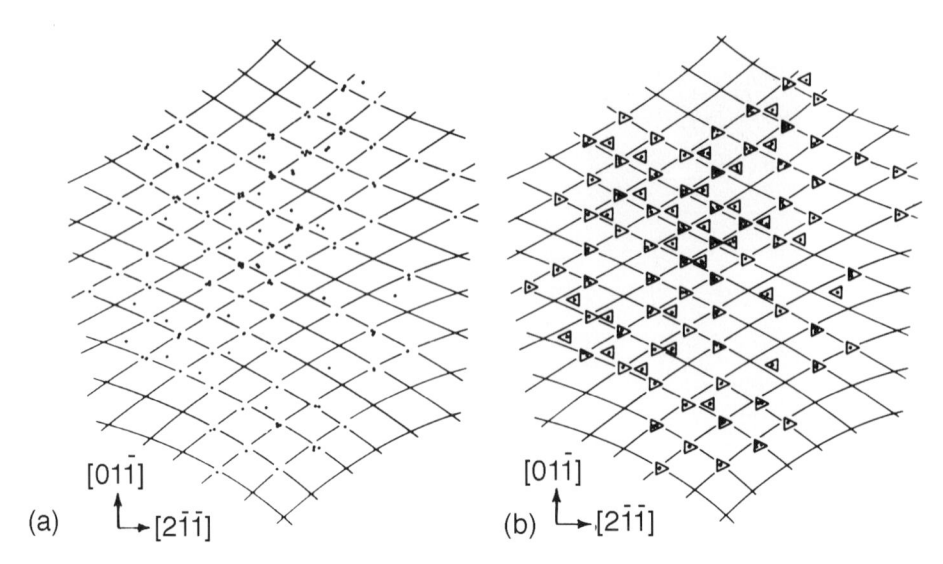

Fig. 3.17 Ir atom locations on Ir(111) [19]. (a) Map after 225 3 sec diffusion intervals at 104 K. (b) Orientation of atom image is indicated for each observation in (a); shown is fcc grid.

occupied. It is also feasible to track the movement of atoms from hcp to fcc sites and back again during diffusion, as in Fig. 3.18.

Other atoms have also been tested on Ir(111) [19]. In these tests the hcp grid was established by first observing the diffusion of an iridium atom. The preferences of another atom can then be readily determined. Tungsten atoms were found to sit in hcp sites, as does rhenium, but palladium sits at fcc sites, as is demonstrated in Fig. 3.19. The images of W, Re and Pd are also triangular, but the orientation depends on the chemical identity of the atom, as shown in Fig. 3.19. For palladium atoms the orientation is the same as for Ir, but for W and Re the orientation is reversed. The physics underlying the orientation of the image triangle is not yet understood, but the image certainly helps in detecting where an atom sits. Of course at the start the surface has to be mapped to establish the orientation associated with each site.

The difference in occupation of the two adsorption sites was roughly estimated at different temperatures and for Ir the hcp site was found to be around 0.043 eV stronger than the fcc site; for Re hcp was around 0.13 eV stronger than fcc; for W the difference was 0.17 eV, but for Pd fcc was stronger than hcp by roughly 0.04 eV [20]. The probability that an atom will be at an fcc site having first struck the surface at an fcc site is [21]

$$P_f^{(f)} = \{N_{ah} + N_{af}\exp[-3(N_{af} + N_{ah})]\}/(N_{af} + N_{ah}). \quad (3.1)$$

Furthermore, the probability of an atom ending at an hcp site when it started on such a site is given by

$$P_h^{(h)} = \{N_{af} + N_{ah}\exp[-3(N_{af} + N_{ah})]\}/(N_{af} + N_{ah}). \quad (3.2)$$

3.1 Adatom binding sites

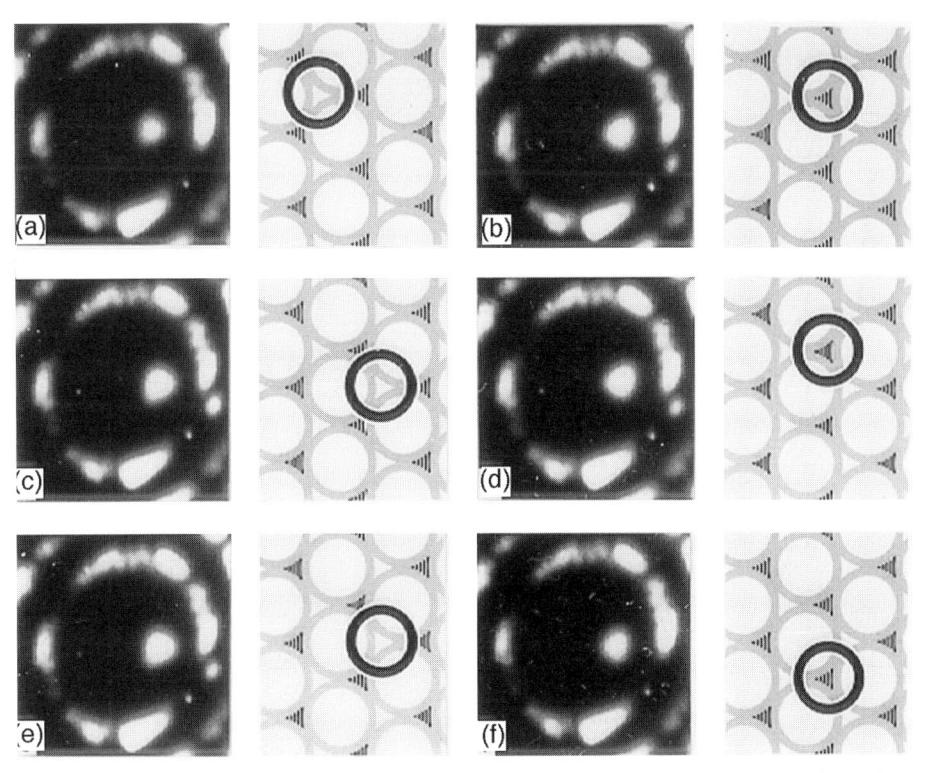

Fig. 3.18 Observations of Ir adatom on Ir(111) diffusing over surface at 93 K [18]. Location of adatom is indicated by schematic to the right of the micrograph.

Here N_{af} is the number of hops out of fcc sites, and N_{ah} the number out from hcp sites. To obtain a random distribution of positions the two probabilities have to be equal. The difference in occupation of the two adsorption sites can also be estimated from

$$\ln(N_f/N_h) = -\Delta F/kT = \Delta S/k - \Delta E/kT. \quad (3.3)$$

The energy difference obtained in this way is presented schematically in Fig. 3.20. A detailed analysis of the temperature dependence yields more reliable values of the energy difference; for Ir this was 0.0216±0.0015 eV, for W 0.188±0.014 eV, and for Re 0.141±0.009 eV. From the plot of N_f/N_h against reciprocal temperature in Fig. 3.21 we find that the entropy of iridium at an fcc site is (0.64±0.17)k higher than at an hcp site [22].

In 1990, Brune et al. [23] observed STM images of an Al(111) surface with carbon apparently on the surface. The carbon is evident in Fig. 3.22 as a bright protrusion surrounded by three bright aluminum lattice atoms. By determining the location of the aluminum atoms in the second layer and extrapolating it to the first they were able to establish that carbon sits in an hcp site on the surface.

Liu et al. [24] in 1991 carried out calculations for diffusion on a variety of surfaces relying on EAM potentials. As part of this effort they also calculated the energy difference between atom binding at hcp and fcc sites on fcc(111) surfaces. Their results

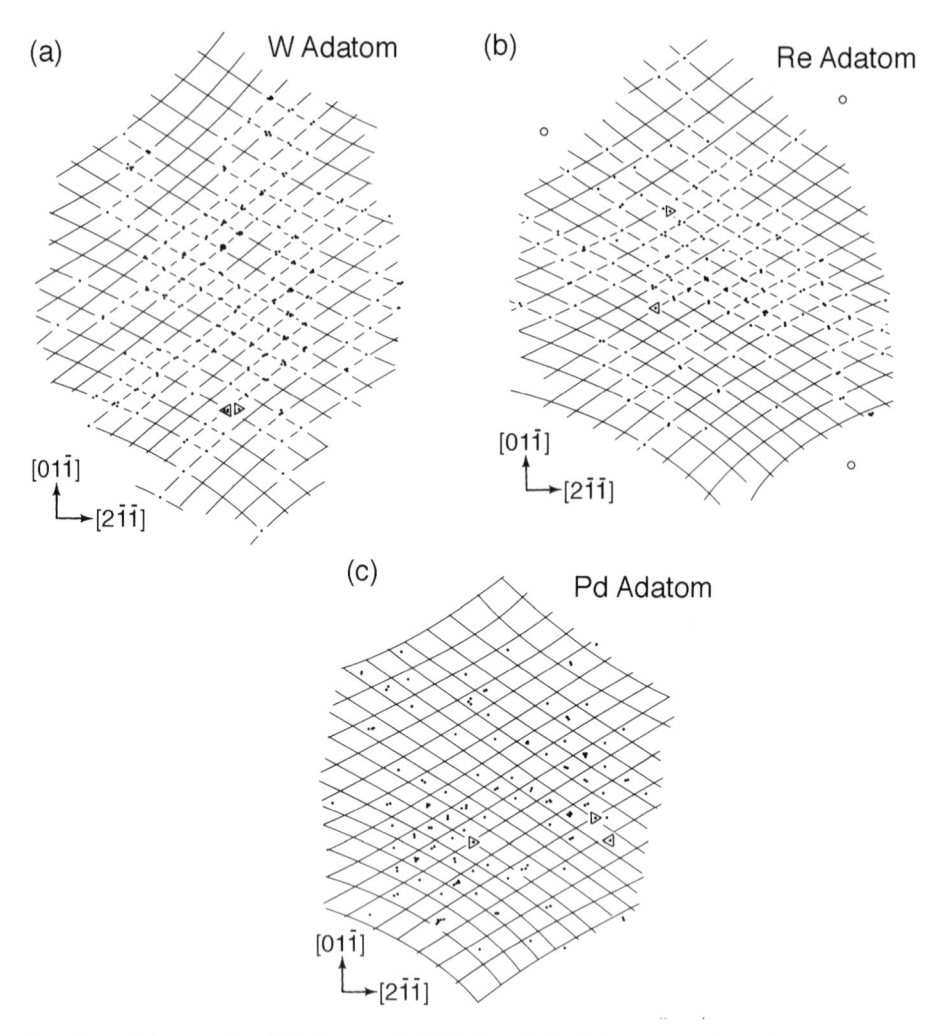

Fig. 3.19 Location of adatoms after diffusion over Ir(111) plane [19]. (a) Tungsten adatom is usually seen at hcp sites. (b) Re adatom is found at hcp sites on the surface. (c) Pd adatom is imaged at fcc sites.

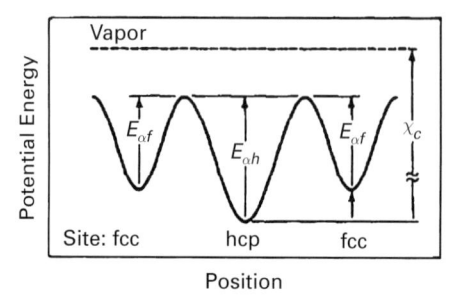

Fig. 3.20 Schematic potential curve for adatom moving over fcc(111) surface [20]. χ_c gives energy of condensation on (111) plane.

3.1 Adatom binding sites

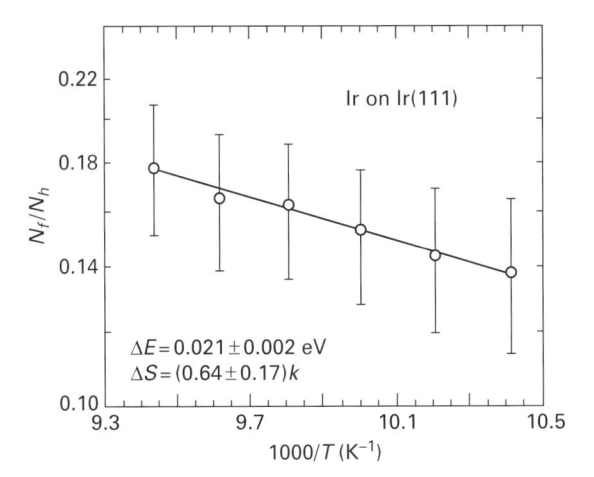

Fig. 3.21 Ratio of Ir adatoms at fcc and hcp sites on Ir(111) against reciprocal temperature, yielding an energy difference of 0.021±0.0015 eV [22].

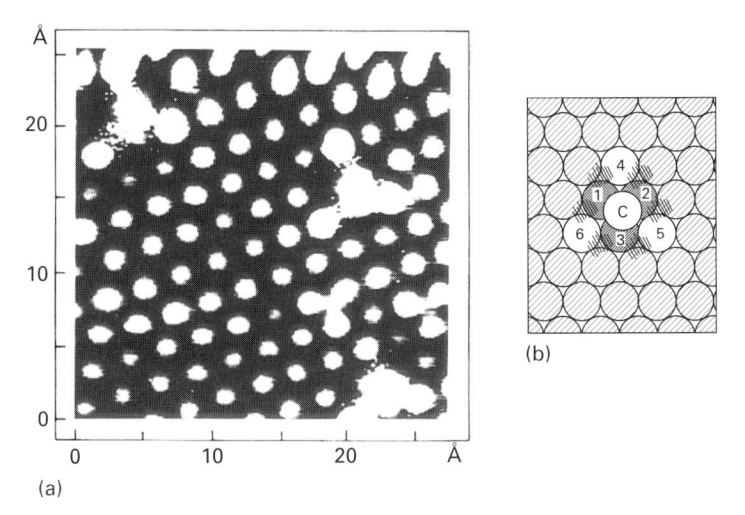

Fig. 3.22 Image of Al(111) surface with carbon atoms on it. (a) STM image with three bright carbon atoms. (b) Schematic of carbon adatom with three Al atoms surrounding it (after Brune *et al.* [23]).

are listed in Table 3.1. For all the surfaces examined using VC parametrization[1], except for platinum, binding was a little more favorable at fcc sites. This finding for platinum is not in agreement with the experiments of Gölzhäuser [25,26].

Feibelman *et al.* [27] looked briefly at platinum atoms adsorbed on Pt(111), and they were able to see triangular image spots from the adatoms, oriented away from [2$\bar{1}\bar{1}$]. They inferred that platinum preferred binding at fcc sites. In 1996 this was confirmed by Gölzhäuser and Ehrlich [25,26], who mapped the binding sites of platinum atoms

---

[1] The authors also claimed that AFW parameters were not suitable for investigation of the stability of binding sites.

Table 3.1 Binding energy at fcc compared to hcp sites on fcc(111) planes, evaluated with EAM interactions [24].

Metal	$(E_{min}^{hcp} - E_{min}^{fcc})^{VC}$
Ni	0.017
Cu	0.008
Al	0.007
Ag	0.006
Au	0.002
Pd	0.005
Pt	−0.0007

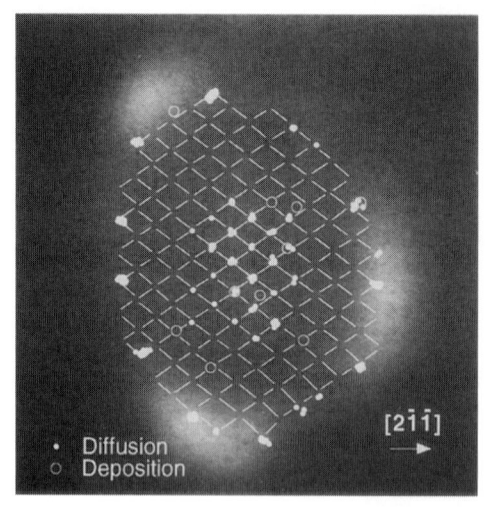

Fig. 3.23   Map of sites for platinum atoms after diffusion on Pt(111) plane with a diameter of 13 spacings [26]. Solid circles indicate observation after diffusion, open circles after atom deposition on surface at ~20 K.

diffusing on Pt(111) and found the map shown in Fig. 3.23. The locations after diffusion are indicated by white dots, and the grid is drawn through them. When plotting the location of atoms deposited on the surface at ~20 K, additional spots are found, indicated by empty circles. These are occasionally inside a unit cell and to the right. This demonstrates that platinum ordinarily sits in fcc sites, with image triangles oriented away from $[2\bar{1}\bar{1}]$ as in Fig. 3.24, except on condensation, when hcp positions may also be occupied.

Density-functional calculations, done by Mortensen *et al.* [28] in 1996 for atoms self-adsorbed on their own fcc(111) surfaces, showed that for Ni, Pd, and Pt the binding energy at fcc sites was 0.05 eV, 0.13 eV, and 0.14 eV higher than for hcp sites. For platinum these results are in agreement with the experiments of Gölzhäuser [25].

The fcc(111) plane seems to have been quite widely examined for adsorption of different species. Fcc and hcp adsorption sites for the movement of Cu on Cu(111) were investigated by Repp *et al.* in 2003 [29] using the scanning tunneling microscope

3.1 Adatom binding sites

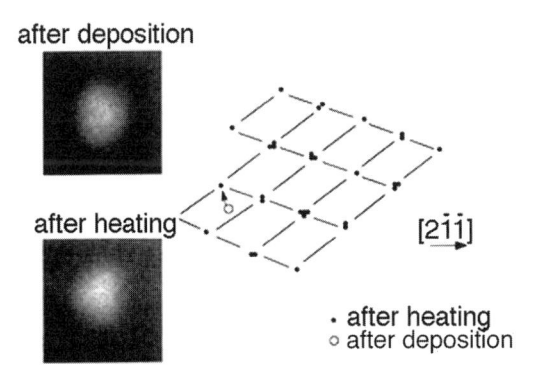

Fig. 3.24 Platinum adatoms on Pt(111) plane [26]. Left: Image of adatom after deposition and after equilibration. Right: Distribution of adatoms over (111) plane. Deposition site does not fit the grid found after diffusion.

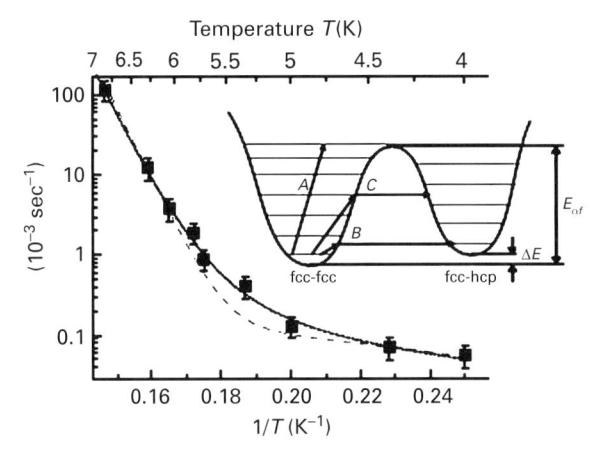

Fig. 3.25 Arrhenius plot for Cu dimer on Cu(111) plane jumping from fcc-fcc configuration to fcc-hcp. Thermal activation is given by process A, thermally assisted tunneling to the ground state by B, and for intermediate energies by C. Dashed line gives rate estimates for these two processes (after Repp *et al.* [29]).

supplemented by *ab initio* calculations using the DACAPO code. From the hcp-to-fcc and fcc-to-hcp hopping rates, they deduced a lower limit of 0.004 eV for the energy difference between the two sites. From the difference in the temperatures for the start of diffusion out of the sites they came up with an upper limit of 0.008 eV, with binding at fcc sites the stronger. Calculations led to a 0.006 eV stronger binding on fcc sites. They also concluded that monomers, dimers, and trimers prefer sitting in fcc sites on the Cu(111) surface. For dimers, they claimed that movement was enabled by thermally associated tunneling, illustrated in Fig. 3.25. For dimers the energy difference between the two sites derived from experiment was only 1.3±0.5 meV, against 21 meV from DFT.

Marinica *et al.* [30] made use of the EAM potential proposed by Mishin *et al.* [31] in working on diffusion over Cu(111). The change in the potential energy during atom migration is shown in Fig. 3.26. Just as Repp *et al.* [29] had observed in their

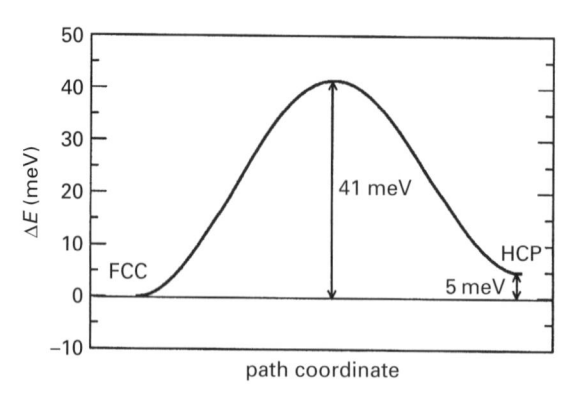

Fig. 3.26 Potential energy plot for a Cu adatom diffusing from an fcc to an hcp site on Cu(111) plane, estimated from EAM potential (after Marinica et al. [30]).

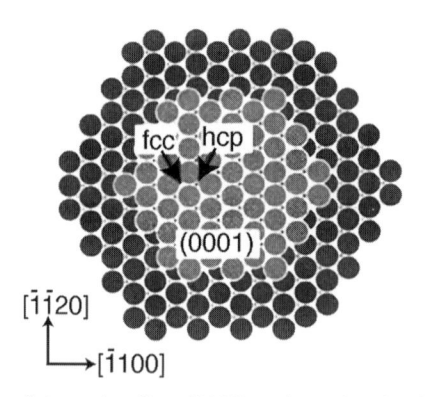

Fig. 3.27 Schematic of hcp (0001) surface, showing location of hcp and fcc sites [33].

experiments, the fcc was found as the more favorable adsorption site. Recently Morgenstern and Rieder [32] looked for movement of Cu on Cu(111) and at 21 K, atoms were observed at hcp sites 207 times, and at fcc sites 7274 times. Based on this, the energy difference between the two sites was estimated as 5.5±1.0 meV.

Fcc and hcp sites also exist on the (0001) plane of an hcp crystal, illustrated in Fig. 3.27. The grids formed by the positions of atoms recorded after diffusion were determined by Goldstein and Ehrlich [33] for both rhenium and tungsten atoms on the Re(0001) plane, as shown in Fig. 3.28. In successive atom layers fcc sites are positioned on top of each other, but hcp sites alternate. Resolution in these experiments was not good enough to discern this, however. All that emerged was that only one type of site was occupied in adsorption.

From this survey it is clear that on fcc(111) planes both fcc as well as hcp sites are occupied, but the extent of adsorption varies with the particular metal. It is also clear that it would be interesting to do more, to establish binding patterns for other atoms on fcc and hcp surfaces. On the bcc(110) plane only one type of location for adsorption has so far been observed: at fourfold sites on this surface. Two types of adsorption sites were found on W(321), with the deep one much more favorable.

3.1 Adatom binding sites

(a) Re adatoms　　**(b) W adatoms**

[1̄1̄20]
[1̄100]

• Adatom binding site
○ Edge of plane

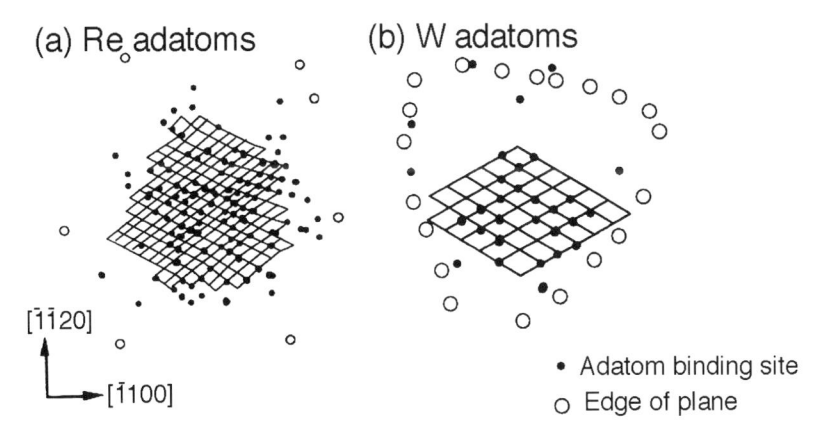

Fig. 3.28 Maps of adatom locations after diffusion on Re(0001) [33]. Only one type of binding site is occupied.

3.1.2 Distribution of binding sites

Although the nature of binding sites has been explored, there is at least one question that still remains – are these sites uniformly strong over an entire plane? This question was first probed by Fink [34] in 1984 for tungsten atoms on the W(211) plane; he did this by determining single adatom positions visited after diffusion in the absence of applied fields at 363 K. The distribution of positions occupied by a single adatom in experiments on a plane with twelve sites is shown in Fig. 3.29. In the central portion of the plane the population of sites by the adatom is uniform but subject to small fluctuations expected for a limited number of observations. Two sites from the plane edges, however, clearly have different properties, and show a much larger population of adatoms. At equilibrium, the probability of being at a site i compared to the probability of being at a reference site o is just

$$\frac{P_i}{P_o} = \exp\left(-\frac{F_i}{kT}\right), \tag{3.4}$$

where F_i is the free energy of the system with the atom at site i compared to site o. The free energy for a tungsten atom observed on W(211) is shown in Fig. 3.30, and for rhenium on a somewhat larger plane of tungsten with 16 adsorption sites in Fig. 3.31. Interactions for both atoms are different at the two types of edges of the W(211) plane compared to the center. The behavior for the two adatoms is also different. For tungsten atoms the free energy goes down as the atom gets close to the edge; for rhenium instead, at both the side toward (100) and toward (111), the free energy decreases but rises again at the last site from the edge. The important conclusion for crystal growth is that the effect of the edges extends only over a short distance, so that the central part of the plane has uniform binding. It is also worth noticing that on W(211), the influence of the edges on atomic movement is not localized at one site next to the edge, but is instead spread over 3 atomic spaces;

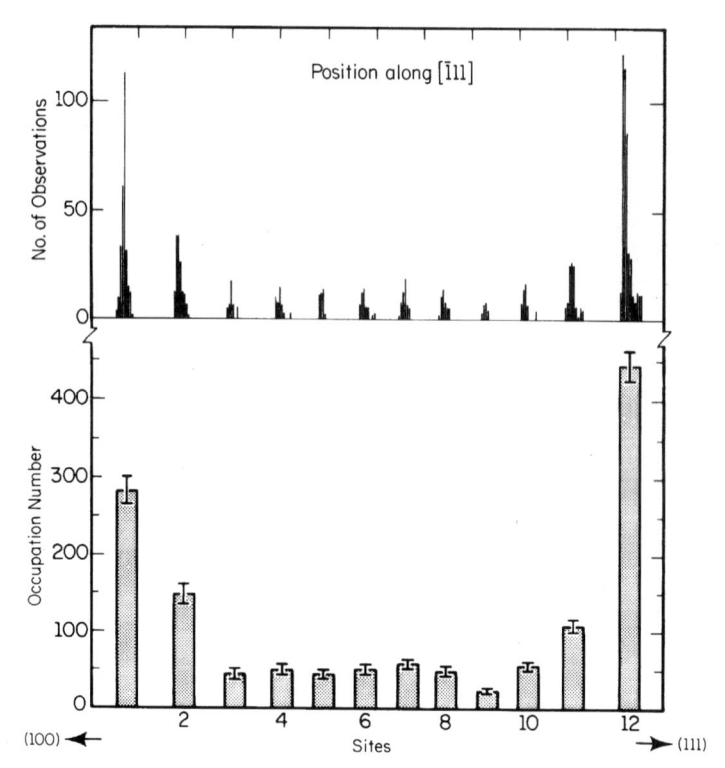

Fig. 3.29 Distribution of a tungsten adatom on W(211) plane after equilibration at 363 K [34]. Top: Frequency of adatom locations, revealing 12 sites for binding. Bottom: Distribution over binding sites in 1340 observations.

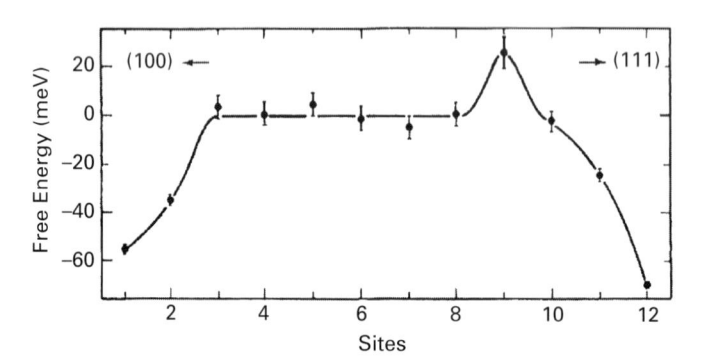

Fig. 3.30 Free energy of adatom binding for a tungsten adatom on W(211) plane of 12 sites [34].

the situation does not change for bigger planes. Similar results were observed for Si on W(211), as shown in Fig. 3.32, and for Ni and Pd [35].

Kellogg [36] looked at the distribution of platinum atoms along a rhodium (331) plane with 18 atomic sites at 191 K in 230 atomic movements. Opposite to what was seen on the W(211) plane, he did not observe any preferred sites on this plane even in the

3.1 Adatom binding sites

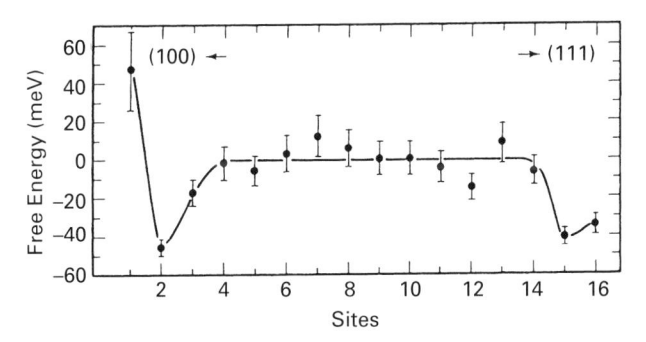

Fig. 3.31 Free energy of Re adatom on W(211) plane of 16 sites [34].

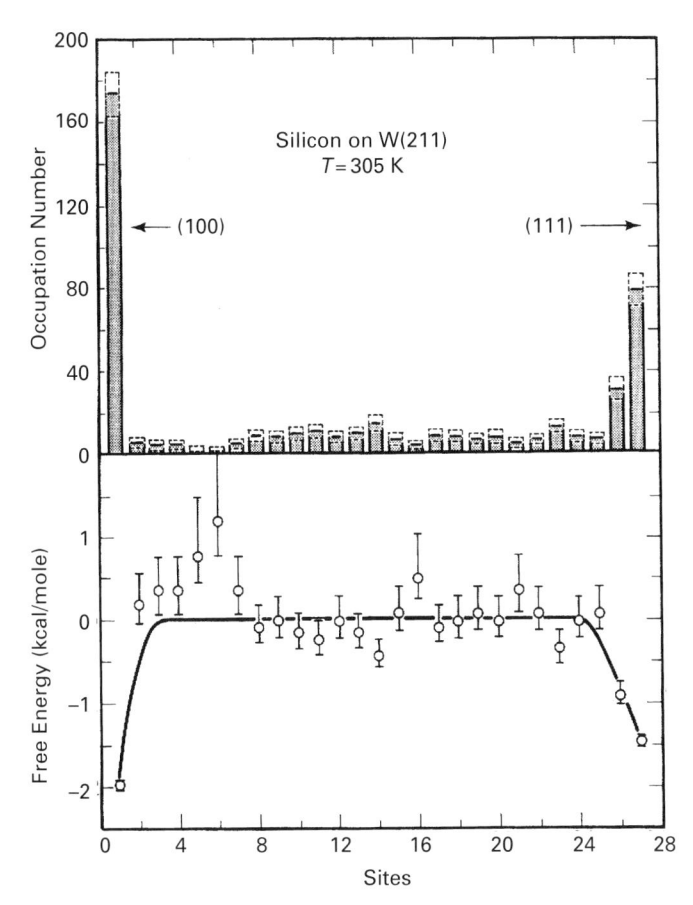

Fig. 3.32 Distribution of Si and free energy on sites on W(211) derived in FIM observations [35]. Sites 0 and 28 are the locations of the step bounding the plane.

neighborhood of edges, as is indicated in Fig. 3.33. A higher population of atoms due to the step-edge barrier was observed for atoms on W(110), but only at one atomic spacing from the edge. Investigations also showed that tungsten, iridium, rhenium, and palladium atoms reflected from the edge, but platinum atoms did not [35].

84   Atomic events in surface diffusion

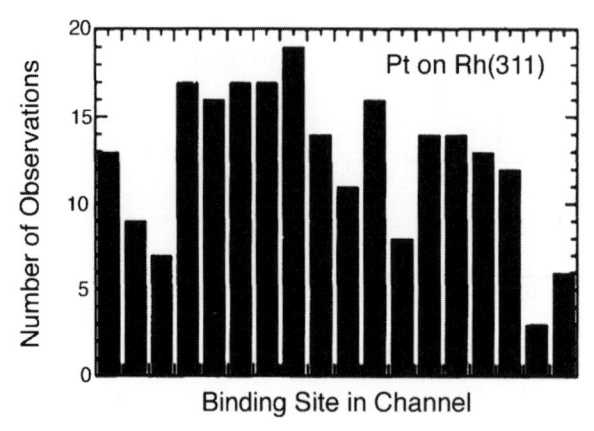

Fig. 3.33   Distribution of Pt adatoms on Rh(311) plane of 18 sites after equilibration at 191 K. Sites at plane edge are not preferentially occupied (after Kellogg [36]).

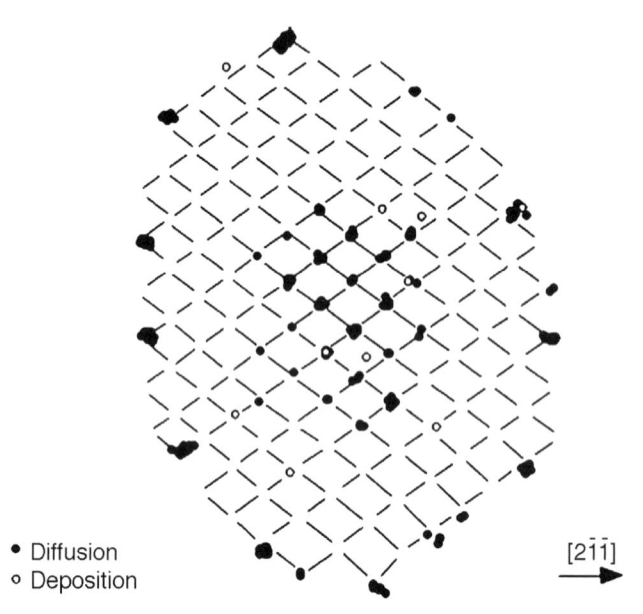

Fig. 3.34   Sites occupied by platinum adatoms on Pt(111) cluster of ~145 atoms [25]. Filled circles indicate location after diffusion, open circles give position after deposition on surface at ~20 K.

Some time later, binding sites for platinum atoms on Pt(111) were explored by Gölzhäuser [25,26]. In Fig. 3.34 are shown his results on a cluster with a diameter of ~ 13 nearest-neighbor spacings, with solid dots representing locations after diffusion, and open circles after deposition at ~ 20 K. Only the center of the plane is occupied after diffusion, leaving empty the zone to the plane edge, ~ 3 spacings wide. This empty zone can be occupied, but only by atoms deposited on the cold surface from the vapor; on warming, these atoms move to the plane edge or the center. The width of the empty zone is not particularly dependent on the cluster size; that is

3.1 Adatom binding sites

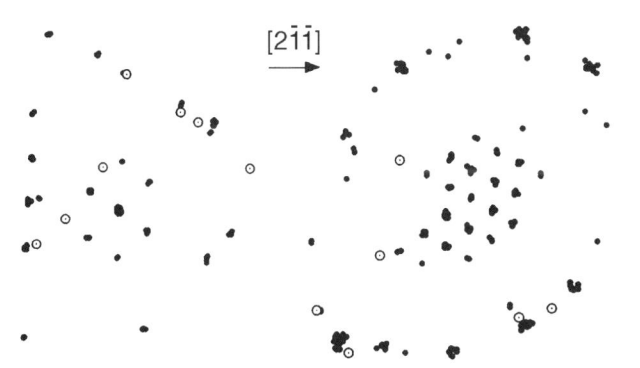

Fig. 3.35 Maps of sites occupied by Pt adatoms on Pt(111) planes of different sizes [25]. Filled circles show position after diffusion, open circles indicate location after deposition. Left image gives map for a plane 8 atomic spaces wide, while the right is 16 spaces wide. Both planes show empty zone of roughly comparable width.

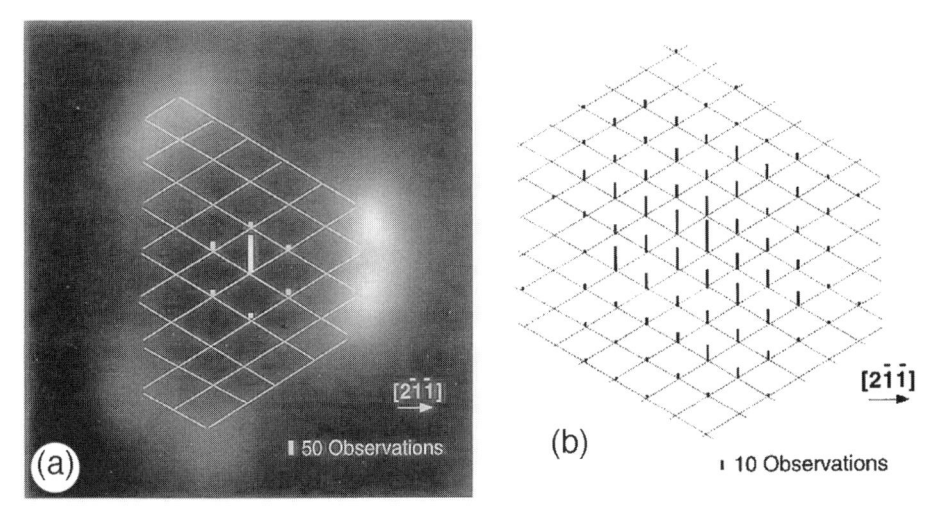

Fig. 3.36 Distribution of Pt adatom over Pt(111) plane [26]. (a) On cluster with diameter of 7 nearest-neighbor spacings. (b) On cluster with diameter of 16 nearest-neighbor spacings, showing gradually diminishing population at boundary between center and empty zone.

apparent from Fig. 3.35, which shows atoms on two clusters, one with a diameter of 8 spacings, the other at the right with 16 spacings. The empty zone has much the same width. When the width of the cluster is diminished to 7 spacings, only the very central site lies outside of the empty zone. However, an atom deposited at this site is not limited to only this site; at $T > 80$ K it can still move, and populates adjacent sites at the boundary with the empty zone, as shown in Fig. 3.36a. The central site is most frequently populated and from the time spent in the center compared to adjacent sites the difference in energy was derived as 0.015 ± 0.002 eV. This suggests that the transition from center to the empty zone is gradual. This gradual transition is not limited to

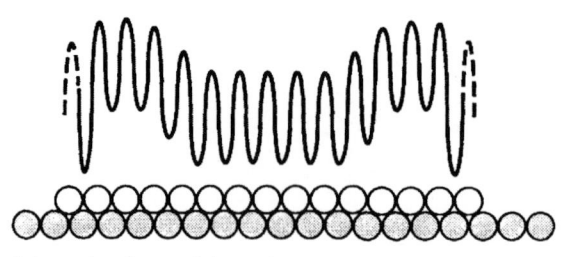

Fig. 3.37 Schematic of potential seen by Pt adatom diffusing radially on Pt(111) cluster [26].

small clusters; a diminishing population is also seen in Fig. 3.36b, for a cluster with a diameter of 16 nearest neighbors.

For platinum atoms on Pt(111), an estimate of the potential energy changes over the surface is given in Fig. 3.37. Although the characteristics of atom motion in the empty zone are not well established, the potential energy in this region is definitely higher. The cause for the empty zone has not been clearly established, but may be due to surface stress, which is high for Pt(111). The cluster atoms may relax their positions, so that the strain at the center may differ from that at the edges.

The phenomenon of empty zones is not limited to platinum; it has also been observed for Ir(111) [37,38]. As appears from Fig. 3.38, the empty zone for iridium atoms on Ir(111) is again ~ 3 spacings wide, and does not vary sensitively with cluster size. For a cluster with a width of 7 spacings shown in Fig. 3.38b, only the central few sites are occupied after diffusion. When the size is reduced to 5 spacings in Fig. 3.38c, the empty zone covers the entire cluster and there are no equilibrium sites of low energy left; all the cluster sites are occupied by atoms after diffusion. Recently, an empty zone was also suggested present on the Cu(111) surface [39]. The empty zone is not a general phenomenon, however. On tungsten, the entire (110) plane up to the edges is filled by atoms diffusing at 340 K, as is clear in the diagram in Fig. 3.39. It must be recognized, of course, that in diffusivity measurements, only observations in the central region of a plane provide information about movement on the flat surface.

This survey clearly shows that the presence of steps can influence the adatom potential landscape in a number of ways, and this influence does not have to be short ranged.

3.2 Atomic jumps in diffusion

At the time of the first observations of individual atom movement in the field ion microscope, in the years 1966 to 1970, diffusivity was seen as a simple phenomenon. Adatoms were believed to jump randomly between nearest-neighbor sites on the surface. This view was consistent with what was emerging from experimental measurements of the temperature dependence of the diffusivity D; these yielded diffusivity prefactors D_o of roughly 10^{-3} cm^2/sec. From Eqs. (1.22) and (1.23) it appears that the prefactor is given by

3.2 Atomic jumps in diffusion

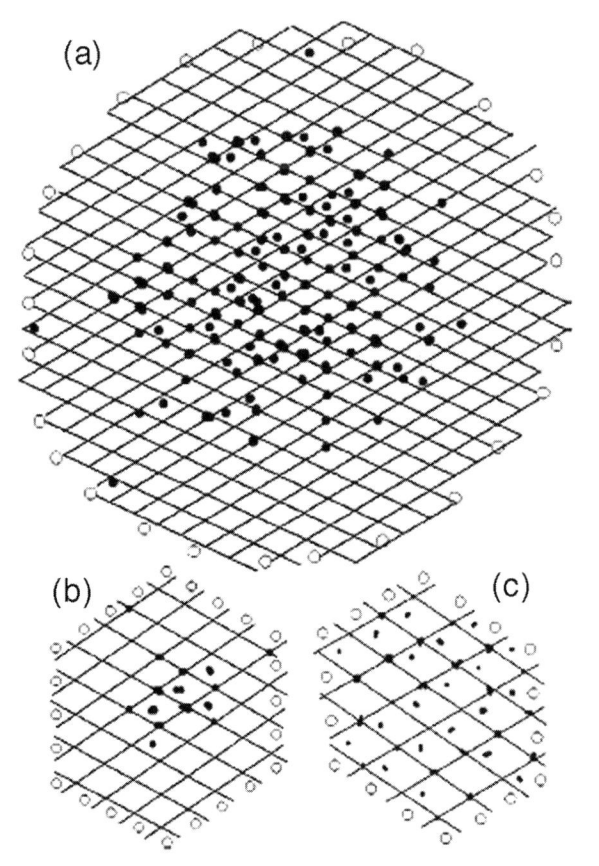

Fig. 3.38 Location of Ir adatoms on Ir(111) clusters of different sizes [38]. Solid circles give position of adatoms, open ones indicate edge atoms. Shown is grid for hcp sites. (a) Distribution on cluster with diameter of 19 spacings after diffusion at 120 K. (b) Ir distributions over Ir$_{61}$ after equilibration at 120 K. (c) Ir adatoms cover entire Ir$_{37}$ cluster, with a diameter of only 5 spacings, after equilibration at 115 K.

$$D_o = v\ell^2 \exp(\Delta S_D/k), \tag{3.5}$$

where ΔS_D is the entropy of activation, v the attempt frequency of the adatom, and ℓ the jump length. If we ignore the entropy of activation and pick a typical value of 10^{12} sec^{-1} for the vibrational frequency, then D_o emerges on the order of 10^{-3} cm^2/sec, as the nearest-neighbor jump length on metals is around 3 Å. These rough estimates suggested that this simple picture was correct in describing the movement of adatoms on the surface.

Detailed information about surface diffusion dimensionality on tungsten, and later on rhodium surfaces, also fit well into this simple picture. On planes such as W(211) and W(321) shown in Fig. 3.1, as well as Rh(110), Rh(311), and Rh(331), in Fig. 3.2, diffusion was one-dimensional and occurred along the channels that characterize these surfaces. This is evident, for example, from a detailed plot of atomic positions seen for Rh during diffusion on W(211), in Fig. 2.15. A look at the hard-sphere surface models

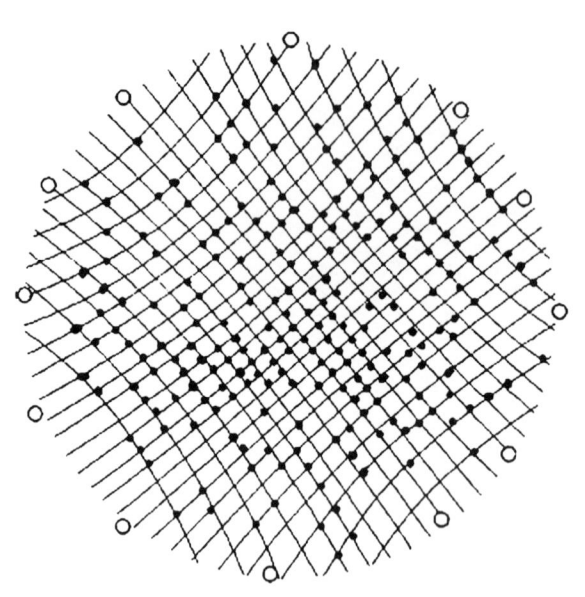

Fig. 3.39 Location of tungsten adatom on W(110) plane after equilibration at 340 K [38]. Black dots indicate tungsten adatom, open circle shows edge atoms of W(110) plane. Entire surface is available for adatoms.

makes it clear the jumps over the sidewalls of the channels would require much more energy than along the direction of the channels, which should be the preferred path. This simple view, of being able to predict the directionality in surface diffusion merely by looking at an atomic model, did not, however, have a long lifetime.

3.2.1 Diffusion mechanisms: atom exchange on fcc(110) planes

The simple picture of diffusion was destroyed by the work of Bassett and Webber [40], who in 1978 examined diffusion of platinum, iridium, and gold atoms on platinum surfaces. The most interesting results were obtained on the (311), (331), and (110) planes of platinum. On the former two, diffusion occurred as expected along the channels. On the (110) plane, however, diffusion was observed not only along the channels but also across them, making the diffusion two-dimensional. For diffusion along the channels, Bassett and Webber found an activation energy of 0.84±0.1 eV, and a prefactor of 8×10^{-3} cm^2/sec, whereas motion across the channels required an activation energy of 0.78±0.1 eV and a prefactor of 1×10^{-3} cm^2/sec. For iridium atoms, comparable results were found, but gold diffused along the channels of the platinum substrate. To explain the unexpected two-dimensional motion, Bassett and Webber considered possibilities such as impurities, but finally opted for two models. Fluctuations could occur in the location of the atoms making up the channel walls, opening up holes through which the adatom could move into an adjacent channel. A more likely alternative was that a fluctuation caused an atom from the channel wall to move aside, as in Fig. 3.40, allowing an adatom to move into the empty site; the atom from the lattice then took its place in an adjacent

3.2 Atomic jumps in diffusion

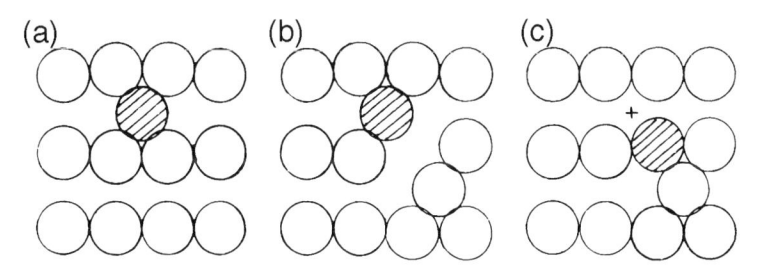

Fig. 3.40 Mechanism for cross-channel diffusion of adatom on Pt(110) proposed by Bassett and Webber. (a) Adatom in channel. (b) Vacancy forms in channel wall. (c) Adatom has moved into vacancy, lattice atom is now in channel site (after Bassett and Webber [40]).

channel and continued the diffusion. Bassett and Webber were unable to go beyond that, and warned that "further experimental and theoretical studies are required to clarify the nature of the inter-channel diffusion."

This followed rapidly. In 1979, Halicioglu [41] as well as Halicioglu and Pound [42] did calculations with Lennard-Jones potentials of the energetics of diffusion on channeled fcc planes. For Pt(110) they found cross-channel motion in which an adatom interacts with an atom from the channel wall to form a dumbbell-shaped pair. This could decompose in different ways, moving a lattice atom into an adjacent channel or else returning the adatom to its original channel.

Shortly thereafter, De Lorenzi *et al.* [43,44] carried out detailed molecular dynamics simulations on fcc planes again bonded by Lennard-Jones potentials. On the (100) surface they found significant non-nearest-neighbor transitions at $T = 0.49 T_m$, where T_m denotes the melting point. However, the most interesting results were for the (110) plane, which showed that atom motion occurred both along and across the channels, as in Fig. 3.41. For cross-channel motion they proposed an exchange of an adatom with a lattice atom, giving the model in Fig. 3.42, in which the transition state at the saddle point is made up of a lattice atom–adatom pair. From their simulations they deduced a lower activation energy for exchange than for in-channel motion.

The work by Bassett and Webber [40] on Pt(110) stimulated much interest, so further simulations of cross-channel diffusion were also being carried out without, however, changing the overall picture very much. Mruzik and Pound [45] did more elaborate molecular dynamics simulations on several planes of a Lennard-Jones fcc crystal. Diffusion on (100) and (311) planes involved random jumping between nearest-neighbor positions. On (110), cross-channel motion again occurred by the exchange mechanism.

Molecular dynamics estimates for gold and iridium adatoms were done by Garofalini and Halicioglu [46] on the Pt(110) surface with Lennard-Jones potentials at low and high temperatures. At low temperatures, both iridium and gold atoms diffused along the <110> channels; at higher temperatures gold was found creating a dumbbell with a channel atom, but instead of exchanging rapidly returned to its original channel in which it continued to move. Overall movement was thus really one-dimensional as seen

Fig. 3.41 Adatom trajectories in diffusion on fcc(110) plane at $0.4T_m$. Transitions across the $[1\bar{1}0]$ channels are apparent. Based on Lennard-Jones potentials (after DeLorenzi et al. [44]).

previously by Bassett and Webber [40] in the FIM. With iridium, exchange occurred, a platinum atom ending up in an adjacent channel and continuing to move.

A direct experimental demonstration of atomic exchange during cross-channel motion on fcc(110) was soon provided by Wrigley and Ehrlich [47] in 1980, who used the atom probe [48] to determine the nature of the atom making a cross-channel jump. They worked on the unreconstructed (110) plane of iridium. As a first step, the atom probe was calibrated by depositing adatoms on the surface maintained at 50 K, and then field desorbing them without heating the surface, while measuring the charge-to-mass ratio. The results for iridium as well as tungsten atoms are shown in Fig. 3.43. The next step was the crucial one. A tungsten atom was deposited on Ir(110), and the surface was heated until cross-channel motion occurred. The atom that appeared in the adjacent channel was then field evaporated and its charge-to-mass ratio determined. As indicated in Fig. 3.43, the atom measured after a cross-channel transition was iridium, a lattice atom. As a further test of the exchange mechanism, the adatom was removed from the surface, and the first lattice plane was field evaporated. After a cross-channel event, a tungsten atom was generally found in the surface layer. For planes on which cross-channel motion had *not* taken place, examination of the surface did *not* reveal a tungsten atom, as shown in Fig. 3.44. In 1982 Wrigley [49] examined the temperature dependence of cross-channel movement in self-diffusion on Ir(110), shown in Fig 3.45. He obtained an activation energy of 0.74 ± 0.09 eV and a prefactor 1.4 $(\times 16^{\pm 1})\times 10^{-6}$cm^2/sec. At that point it was not clear if such a low prefactor was characteristic of the exchange mechanism.

That cross-channel motion involves atom exchange was now established, and at much the same time a new study showing cross-channel movement appeared. Tung and

3.2 Atomic jumps in diffusion

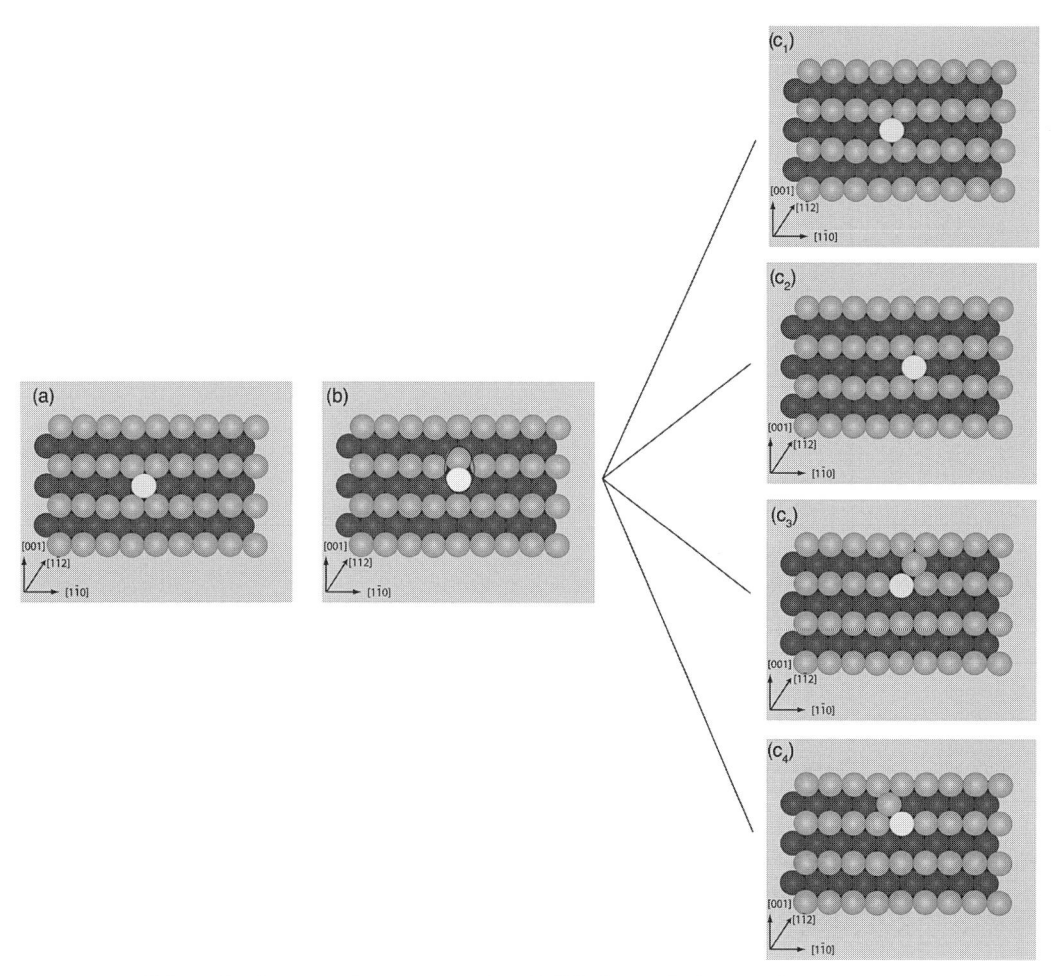

Fig. 3.42 Schematic of atom exchange process in self-diffusion on fcc(110) surface: (a) Atom in equilibrium position. (b) At the saddle point, atom pair sits as a dumbbell (indicated by ellipse) across [1$\bar{1}$0] row of substrate atoms. (c$_1$) – (c$_4$) After diffusion, atoms distributed over allowed sites. See color plate section for color version of this figure.

Graham [16] in 1980 undertook quite a difficult task – they looked at self-diffusion on nickel surfaces. On the Ni(110) plane results depended very much on the pretreatment of the surface, as is clear from Fig. 3.46. Thermal treatment gave quite different diffusivities from what was found on a specimen field evaporated in hydrogen, but after both types of treatment the surfaces showed two-dimensional diffusion. The prefactors D_o were very low on the thermally treated sample, 10^{-7} to 10^{-9} cm^2/sec, with in-channel diffusion over a barrier of 0.23±0.04 eV and over 0.32±0.05 eV for cross-channel motion. With hydrogen exposure, in-channel movement took place over an activation energy of 0.30±0.06 eV and a more reasonable prefactor of 10^1 cm^2/sec. For cross-channel events on a hydrogen fired surface the barrier was only 0.25±0.06 eV and a prefactor of 10^{-1} cm^2/sec. In contrast, diffusion on Ni(311) and Ni(331) always occurred along

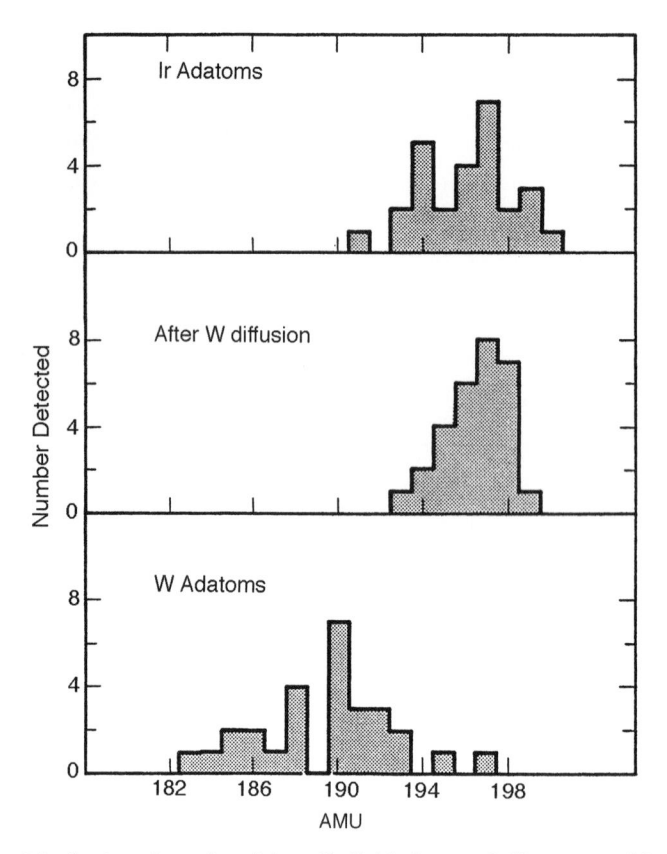

Fig. 3.43 Distribution of atomic weights of individual atoms field evaporated from Ir(110) surface [47]. Top: After deposition of Ir atom. Bottom: After deposition of W atom. Center: After deposition of W atom, followed by heating and observing cross-channel diffusion event; material desorbed is iridium.

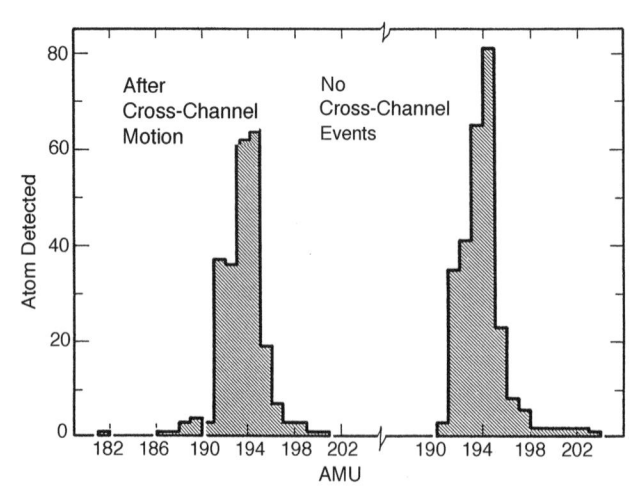

Fig. 3.44 Atomic weight distribution of material field evaporated from first Ir(110) layer [47]. Left: After cross-channel diffusion has occurred, a W atom is detected in the first substrate layer. Right: When no cross-channel motion was detected, only iridium is desorbed.

3.2 Atomic jumps in diffusion

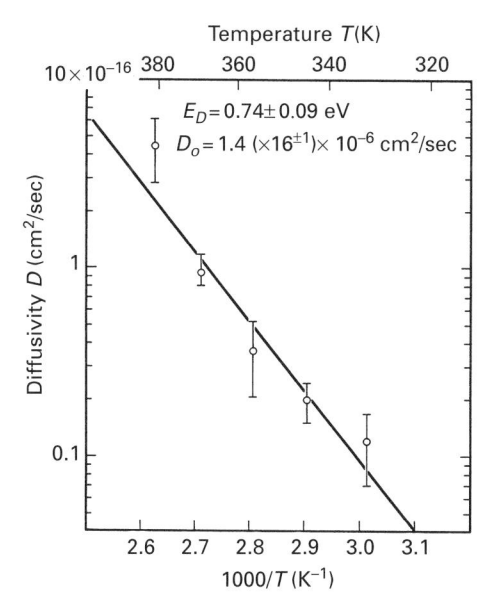

Fig. 3.45 Early Arrhenius plot, based on FIM observations, for the diffusivity of Ir on Ir(110), involving cross-channel jumps (after Wrigley [49]).

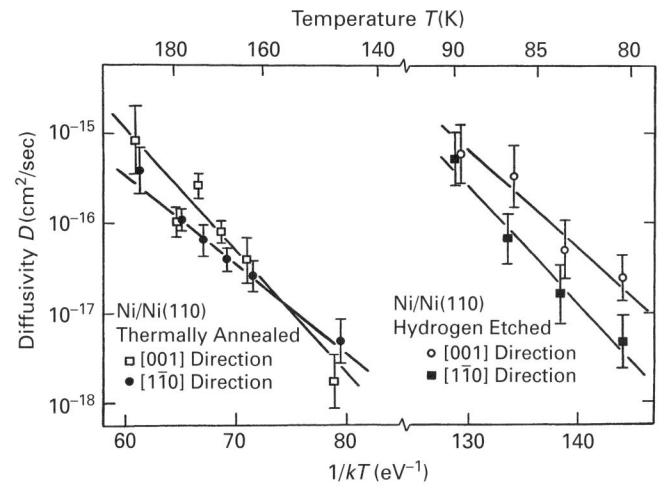

Fig. 3.46 Arrhenius plot for in- and cross-channel diffusion of Ni on Ni(110) plane. Left: Substrate after thermal treatment. Right: Nickel substrate has been hydrogen fired and field evaporated prior to experiments (after Tung and Graham [16]).

the channels. Tung [50] also examined self diffusion on Al(110), and found that at 154 K movement was two-dimensional with an activation energy of ~ 0.43 eV. It should be noted that these results for the (110) plane of Ni and Al are quantitatively uncertain. No new insights into cross-channel diffusion were offered, but they demonstrated quite unequivocally that cross-channel movement occurred on these surfaces.

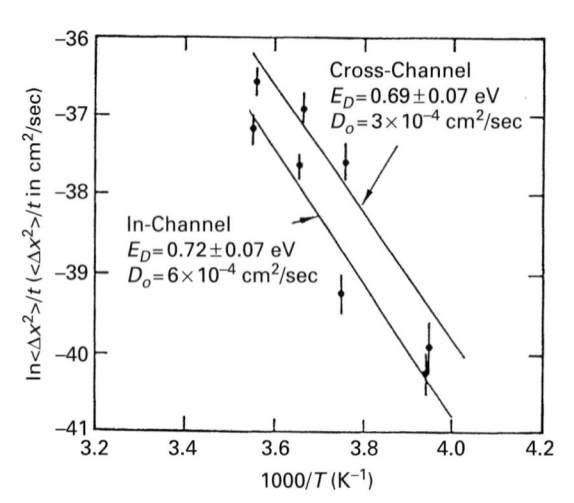

Fig. 3.47 Self-diffusion of Pt on Pt(110) surface observed in FIM. In-channel diffusion: $E_D=0.72\pm0.07$ eV, $D_o=6\times10^{-4}$ cm^2/sec. Cross-channel diffusion: $E_D=0.69\pm0.07$ eV, $D_o=3\times10^{-4}$ cm^2/sec (after Kellogg [51]).

Although it was now established that on fcc(110) planes surface diffusion could be two-dimensional, not much quantitative information was available about diffusion kinetics. Exchange was not observed on every fcc(110) plane; it did not occur, for instance, on Rh(110). In 1986 experiments resumed, however, and Kellogg [51] again looked at self-diffusion on Pt(110) and measured rates of in- and cross-channel motion at different temperatures, as shown in Fig. 3.47. For in-channel diffusion he found a barrier of 0.72±0.07 eV and a prefactor of 6×10^{-4} cm^2/sec; for cross-channel movement the diffusion barrier was 0.69±0.07 eV, with a prefactor of 3×10^{-4} cm^2/sec, both energies lower than first reported by Bassett and Webber [40]. Most important was the fact that the prefactors for exchange were similar to standard values.

Further experimental tests were delayed for some years. Finally, in 1991, Chen and Tsong [52] carried out detailed measurements of self-diffusion on the (110) plane of iridium, and again observed cross-channel motion. In addition to the usual observations, they also quantitatively determined the distribution of displacements, shown in Fig. 3.48. Here x indicates the direction of in-channel motion and y cross-channel motion. From the temperature dependence of the mean-square displacements they arrived at the Arrhenius plot in Fig. 3.49, giving a lower activation energy of 0.71±0.02 eV for cross-channel movement than for movement along the channels; for this the diffusion barrier was 0.80 ±0.04 eV. The prefactors were $6\times10^{-2.0\pm1.8}$ cm^2/sec for cross-channel jumps and $4\times10^{-3.0\pm0.8}$ cm^2/sec for in-channel movement.

What is of particular interest in this study is the distribution of displacements. Apart from the number of atoms at the starting position, with coordinates (0,0), the largest number was found at position (1,1), resulting from a cross-channel transition in the <112> directions with 80% of the total. Atoms very rarely were seen in the (1,0) position, as a result of movements in the <100> directions. From the dumbbell saddle configuration arrived at by all previous theoretical studies, we expect the same number of

3.2 Atomic jumps in diffusion

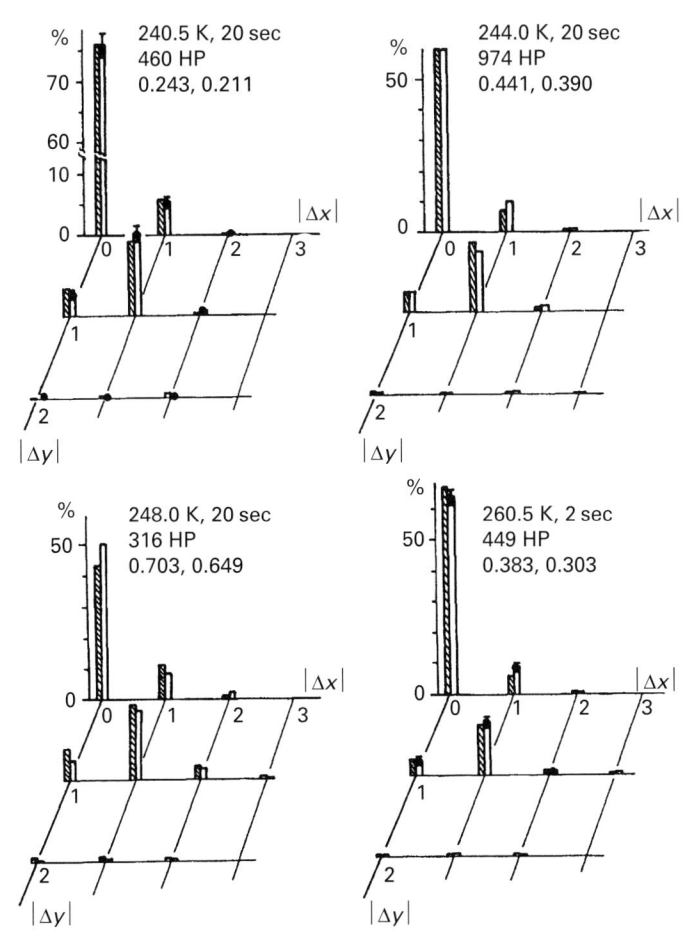

Fig. 3.48 Distribution of displacements in diffusion of Ir on Ir(110) observed in FIM at different temperatures. Transitions in direction of moving atom are favored, contrary to theoretical predictions. HP = heating period; last two numbers are mean-square displacement along x and y directions (after Chen and Tsong [52]).

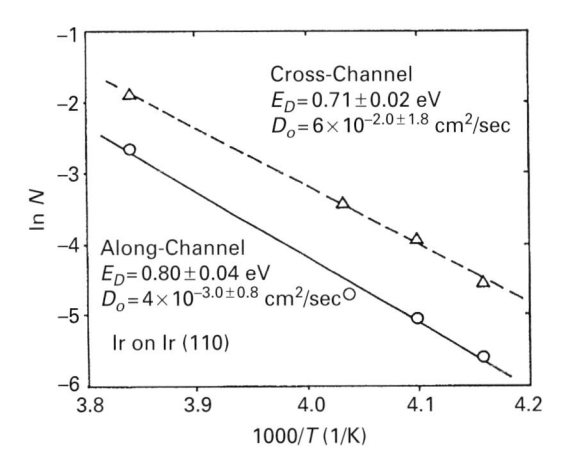

Fig. 3.49 Self-diffusion of Ir on Ir(110) plane examined in FIM. Both in-channel and cross-channel motion is observed (after Chen and Tsong [52]).

transitions at (1,1) and (1,0), as the intermediate has the same probability of decomposing up or down in the neighboring channel. The results of these experiments were not accidental, however. Soon thereafter another examination was done by Kellogg [53] on platinum atoms on the (110) plane of nickel. What is worth noting is that the nickel surface was not cleaned in hydrogen; instead a combination of thermal cleaning, sputtering and evaporation was used. In the field ion microscope, Kellogg observed cross-channel movement of platinum atoms, and was able to recognize the identity of adatoms during atom exchange without resorting to the Atom Probe. The platinum adatom was found to field evaporate at the same voltage as the nickel substrate underneath did; nickel adatoms, on the other hand, were removed at 75–85% of the voltage for evaporating the substrate, making it easy to distinguish between the two species. The size of the image spot also turned out different: large for Pt, small for Ni. The image spot, seen after diffusion in Fig. 3.50b is small, and the adatom was field evaporated at a relatively low voltage, suggesting that the adatom on the surface after heating was nickel, not platinum. Additionally after removing the outermost layer, as in Fig. 3.50c, a platinum atom was left behind, establishing that atom exchange had indeed taken place. Also interesting were the observations of the location of the atom after cross-channel movement. Here the diagram in Fig. 3.2a is useful. Kellogg found that in 10 out of 16 observations the atom moved in the <112> direction. That is, 62.5 % of the transitions were diagonally forward; the atom was never observed to translate in the <001> direction. In the rest of the

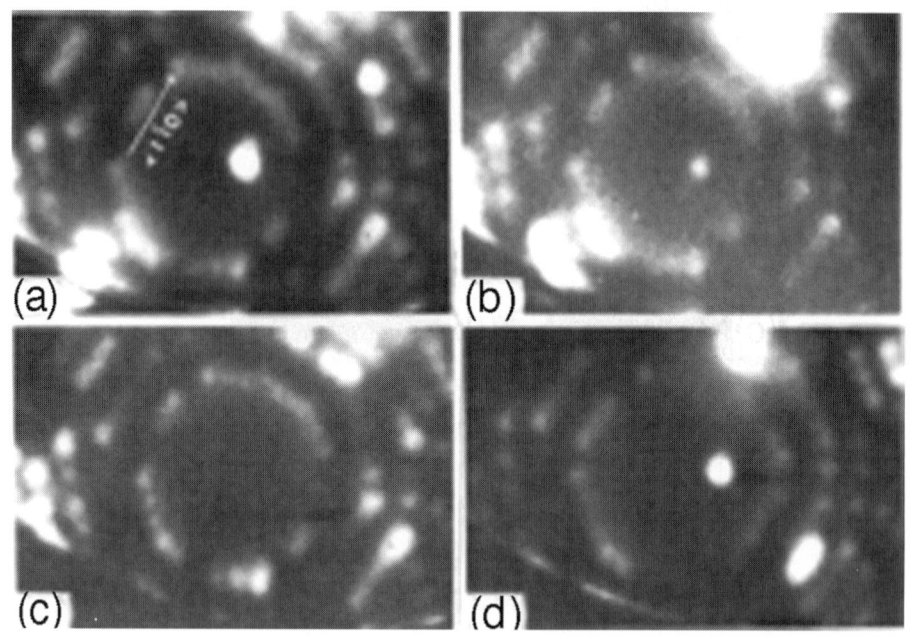

Fig. 3.50 Neon field ion images showing exchange of Pt adatom with Ni from Ni(110) surface. (a) Single Pt adatom on Ni(110) plane. (b) After one minute at 112 K, adatom has changed to Ni, judged from lower desorption field. (c) Adatom has been removed by field evaporation. (d) After removal of one layer of nickel, Pt adatom is again apparent (after Kellogg [53]).

3.2 Atomic jumps in diffusion

observations atoms ended up in the original row. Kellogg associated this finding with the channel wall maintaining its cohesion with the adatom, an idea that was not confirmed, however, in earlier molecular dynamics studies. This result is very similar to the findings of Chen and Tsong [52]. Although these authors give no indication of being aware of previous theoretical studies, their results demonstrate that the picture of a dumbbell-like dimer intermediate, previously the unanimous result of theoretical endeavors, does not describe the real situation. More work is clearly needed to explain the mechanism of cross-channel transitions.

Another year thereafter, Chen et al. [54] looked at Re diffusion on Ir(110), and were able to distinguish Ir from Re adatoms in terms of the shapes of the image spots: the image of Re was circular while Ir was elongated. They evaporated a Re atom onto the cold iridium, as in Fig. 3.51b, and then heated to ~ 256 K, moving the image spot one

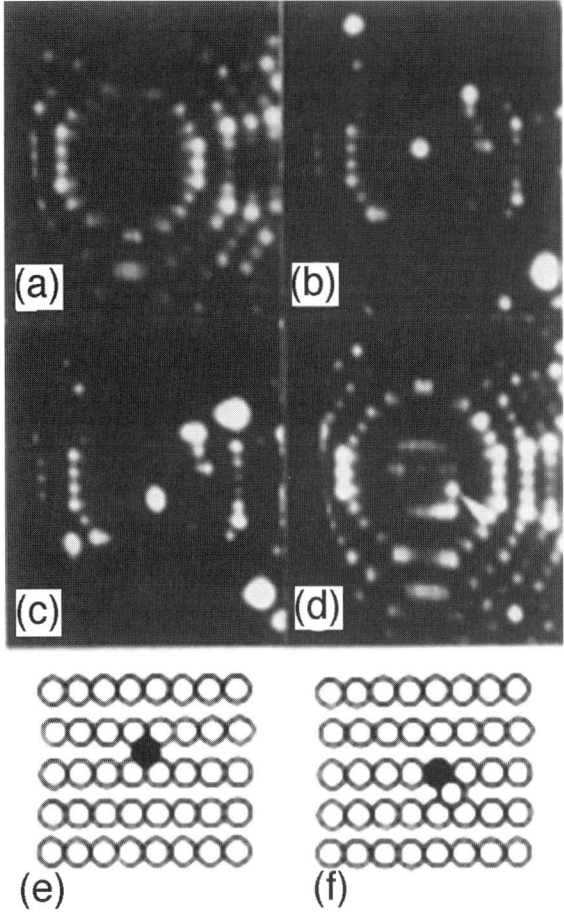

Fig. 3.51 Exchange between Re adatom and Ir(110) substrate illustrated in field ion images. In (b), Re atom has been deposited on the clean Ir(110) surface in (a), giving a round image spot, a schematic of which is in (e). After one minute at 255 K, oblong atom, Ir, is found in (c), also illustrated in (f). Substituted Re atom is revealed on partial field evaporation, in (d) (after Chen et al. [54]).

channel over. The image shape changed, as is clear in Fig. 3.51c. This came about through an atomic replacement of the Ir. After removing the outer layer of the substrate by field evaporation, they found in Fig. 3.51d a Re atom sitting on the surface – exchange had taken place with the iridium substrate; a schematic of the exchange is given in Fig. 3.51e–f.

For proof of cross-channel diffusion it is not necessary to rely only on FIM observations. In 2002, Pedemonte *et al.* [55] did helium atom scattering experiments on Ag(110) at 750 K and higher. They found indications of in- as well as cross-channel movement, but the energetics of diffusion were not probed. Furthermore, beginning in the 1990s, much calculational work was started, usually resorting to semi-empirical potentials, characterizing diffusion on metal surfaces. These efforts are reviewed in Chapters 4 and 5. Here we just note that such work has been reported for (110) planes of aluminum [24,56–58], nickel [24,59–61], copper [24,60–66], palladium [24,60,61,67], silver [24,61,63,65,67,68], iridium [61,69,70], platinum [24,60,61,71,72], gold [24,60,61,65], and lead [73]. In most studies the activation energy for diffusion along the channels proved to be lower than across the channels. This despite the fact that in experiments on Al, Ni, Ir, and Pt the opposite was established.

From all the work done so far it is clear that cross-channel diffusion occurs on some fcc(110) surfaces by exchange of the adatom with an atom from the lattice. On other fcc(110) surfaces, however, such as Rh(110), cross-channel transitions do not occur. Right now it is not clear why one particular system undergoes exchange while another one does not, and exactly how exchange occurs is not yet established. Nevertheless, atom exchange is an established fact. It should be noted that the relative energetics of cross- versus in-channel diffusion vary from one material to the next, but the prefactors appear normal and of comparable magnitude. A new mechanism of surface diffusion beyond simple atom hops has been established.

3.2.2 Diffusion mechanisms: atom exchange on fcc(100) planes

While work on atom exchange in diffusion on fcc(110) surfaces was still being actively pursued, DeLorenzi and Jacuzzi [74] in 1985 were continuing their efforts to find out more about diffusion, but this time on bcc surfaces. This, of course, required a new interaction potential, and for this they picked one developed by Price [75,76] to describe sodium, and examined atom movement on several planes. The one of primary interest here is the (100) plane. In diffusion on this surface an atom is expected to move from a normal site, located at the center of four substrate atoms, to an adjacent equilibrium site, in a direction at right angles to the boundary for the four lattice sites. What was found in their molecular dynamics simulations was quite different, as shown in Fig. 3.52: short, straight arrows indicate normal jumps between nearest-neighbor sites, curved arrows emphasize exchange events in which an adatom replaces a lattice atom. The consequence is a diagonal transition quite different from expectations. In addition to the normal nearest-neighbor transitions, the adatom was found to exchange with an atom from the substrate, which then became an adatom at a site located diagonally from the starting position. The exchange event found here is

3.2 Atomic jumps in diffusion

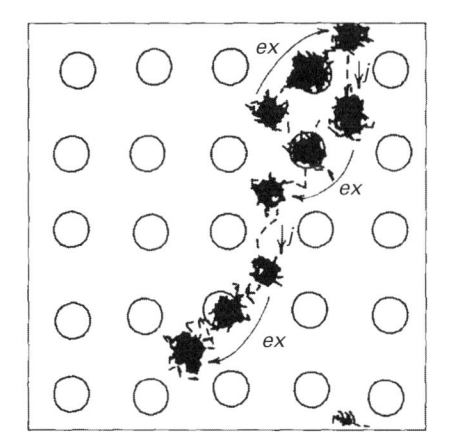

Fig. 3.52 Atom migration on bcc (100) surface, shown in molecular dynamics simulations modeled by Price potential [75,76]. Straight arrows *j* indicate nearest-neighbor jumps, curved arrows *ex* show adatom exchange with substrate (after DeLorenzi and Jacucci [74]).

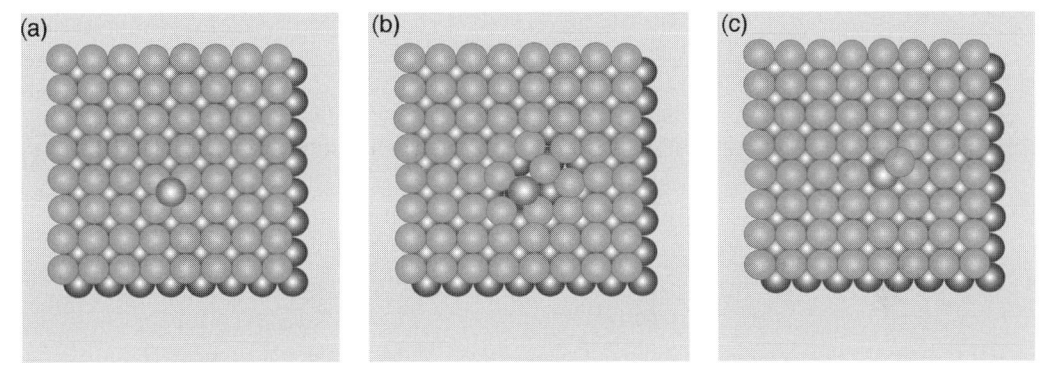

Fig. 3.53 Schematic showing exchange between adatom and substrate atom on (100) plane: (a) Adatom at equilibrium site. (b) Adatom and lattice atom in transition state. (c) Adatom incorporated into lattice; atom from lattice has turned into adatom. See color plate section for color version of this figure.

modeled in Fig. 3.53. The original adatom moves into the outermost layer of the lattice, and the displaced lattice atom takes over on the surface. Quoting from the work of DeLorenzi and Jacuzzi, they found that "In addition to conventional nearest-neighbor jumps between surface sites, the adatom undergoes migration events reminiscent of exchange processes of the intersticialcy in the bulk. In these events, atom A belonging to the surface layer is replaced by atom B originally constituting the adatom. As a result, atom A ends up as an adatom located at a site displaced from the one originally occupied by the point defect." "This is the first observation, in simulations or real experiments, of the occurrence of exchange events between adatom and a substitutional atom as a quantitatively important process contributing to atomic diffusion on isotropic crystal surfaces." The stage was set now to explore whether diffusion by atom exchange would be important in a real two-dimensional system.

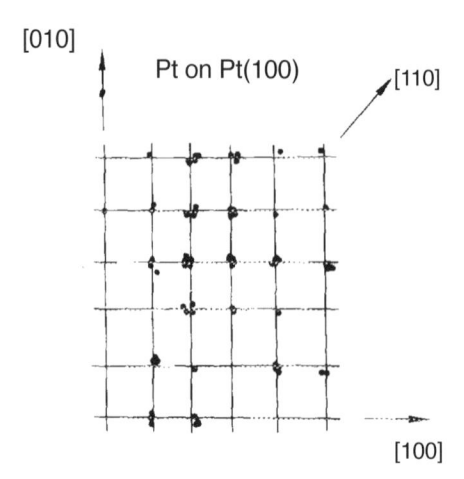

Fig. 3.54 Map of sites on Pt(100) at which Pt adatom was found after diffusion at 175 K. A c(2×2) net is formed, indicative of transitions by atom replacement (after Kellogg et al. [80]).

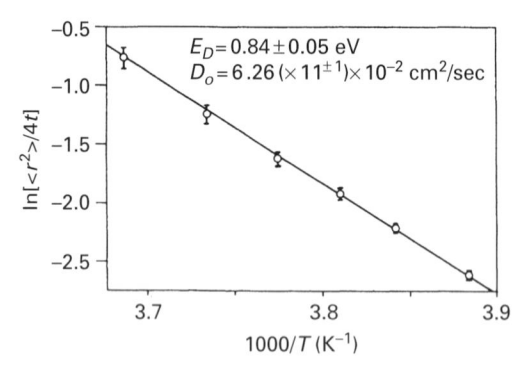

Fig. 3.55 Arrhenius plot for diffusion of Ir adatom on Ir(100) observed in field ion microscope (after Chen and Tsong. [78]).

Nothing was done for several years to explore such exchange experimentally, but finally, five years later, Kellogg and Feibelman [77] examined Pt(100), and Chen and Tsong [78] more extensively studied diffusion on Ir(100). They looked at the displacements executed in diffusion. In normal jumps, atoms leap between nearest-neighbor sites. This is not what was observed, however. Instead, atoms were found to move diagonally, creating a c(2 × 2) net of sites as shown in Fig. 3.54. Kellogg and Feibelman estimated an activation energy of 0.47 eV for diffusion from the mean-square displacement at 175 K, assuming a prefactor of 1×10^{-3} cm^2/sec. On Ir(001), Chen and Tsong actually made measurements at different temperatures to arrive at an activation energy for diffusion of 0.84±0.05 eV and a prefactor of $6.26(\times 11^{\pm 1}) \times 10^{-2}$ cm^2/sec, all from the Arrhenius plot in Fig. 3.55. On both surfaces diffusion occurred as envisioned earlier by DeLorenzi and Jacuzzi [74], through exchange between the adatom and an atom from the substrate. The important end result of these measurements was the first experimental realization of atom exchange on a homogeneous two-dimensional surface.

3.2 Atomic jumps in diffusion

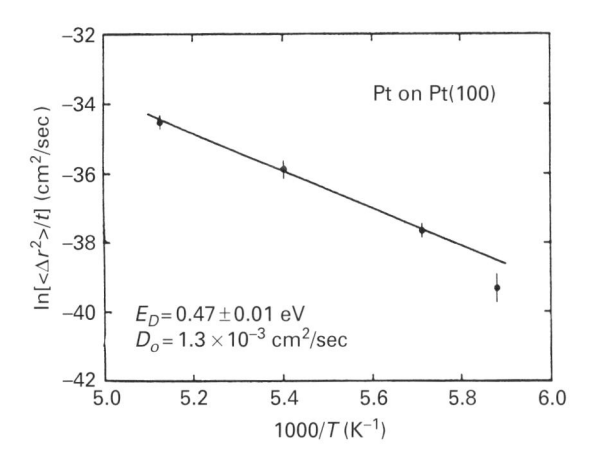

Fig. 3.56 Dependence of Pt atom diffusivity on Pt(100), observed in FIM, upon $1/T$ (after Kellogg [79]).

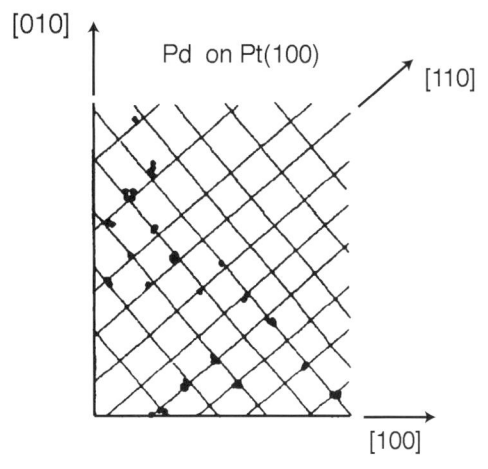

Fig. 3.57 Binding sites for Pd on Pt(100) after diffusion at 265 K, indicating ordinary hopping (after Kellogg et al. [80]).

There now followed a period of high activity studying (100) planes. Kellogg [79] further looked at diffusion on the (100) plane of platinum in 1991. He arrived at an activation energy for diffusion of 0.47±0.01 eV and a prefactor $D_o = 1.3(\times 10^{\pm 1}) \times 10^{-3}$ cm^2/sec from the Arrhenius plot of the rate in Fig. 3.56. The most important result here was to confirm that in self-diffusion on Pt(100), atoms describe a c(2 × 2) pattern of sites, characteristic of motion by atom exchange.

For other atoms studied by Kellogg et al. [80] on Pt(100) the experimental situation was not as good. Detailed measurements at different temperatures were not possible, but for diffusion of palladium an activation energy of 0.70±0.01 eV was deduced assuming a prefactor of 10^{-3} cm^2/sec. For nickel as well as platinum atoms, indications were that diffusion occurred by exchange, but for palladium ordinary jumps were involved, as

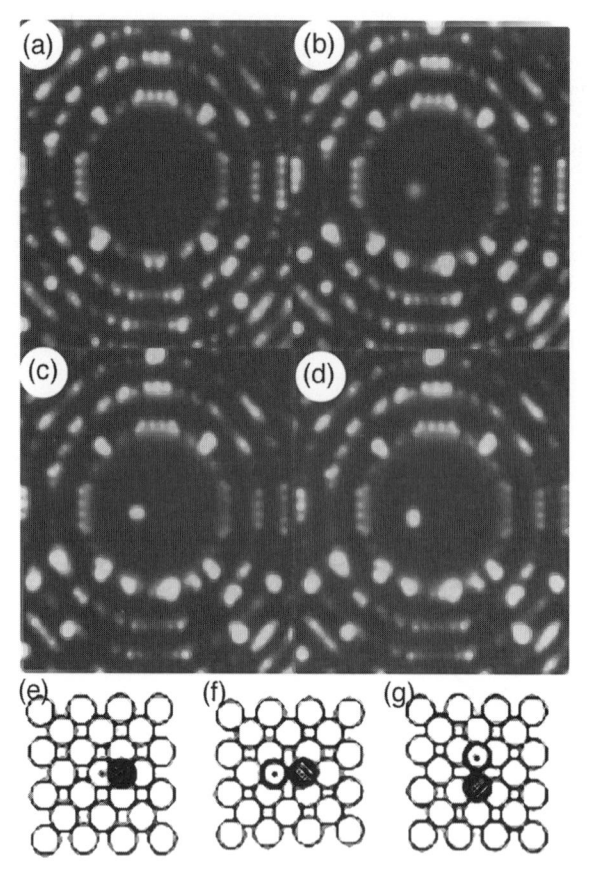

Fig. 3.58 Interactions of Re atom with Ir(100) substrate observed in FIM. (a) Clean Ir(100) surface. (b) Re atom deposited on Ir surface. (c) Re-Ir dimer forms after heating to 230 K. (d) Dimer reorients after heating to 265 K. (e)–(f) Schematics of atom transitions (after Tsong and Chen [81]).

indicated by the (1 × 1) site map in Fig. 3.57. Note the much higher activation energy for diffusion of palladium compared to platinum, which is undoubtedly due to the difference in diffusion mechanisms.

In 1992 Chen et al. [54,81] examined the behavior of rhenium on various iridium surfaces. They found that when a rhenium adatom, deposited on Ir(100), is heated above ~ 220 K as in Fig. 3.58, an iridium atom from the lattice is displaced to the surface and forms a dimer above the vacancy. Created in this way the dimer changes its orientation during annealing at 265 K. Eventually, further heating above 280 K brings about dissociation of the dimer, with the rhenium atom entering the lattice, while the iridium atom diffuses over the surface by atom exchange, a very direct demonstration of what happens in the exchange process.

Kellogg [82] also looked at the behavior of platinum atoms on the Ni(100) surface. He was able to distinguish nickel from platinum atoms by the electric field required for field desorption: a nickel atom deposited on the plane was evaporated from the surface below the best voltage for imaging in the FIM at 77 K, whereas platinum atoms were

3.2 Atomic jumps in diffusion

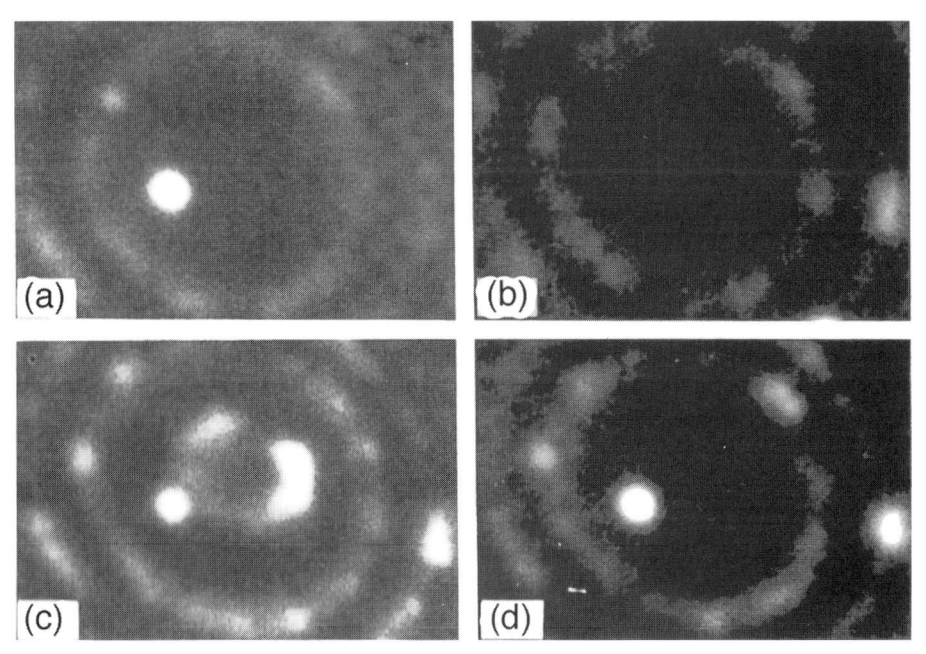

Fig. 3.59 Neon field ion images of Ni(100) undergoing replacement with Pt adatom. (a) Single Pt atom deposited on (100) plane at 77 K. (b) On heating to 250 K, atom disappears. (c) Surface after partial field evaporation. (d) On complete field evaporation of plane, presence of Pt in the layer is detected (after Kellogg [82]).

maintained on the surface, and field desorbed just below evaporation of the substrate. In an experiment, a platinum atom was evaporated onto the surface at 77 K. After heating to 250 K the platinum atom disappeared, but after field evaporating the top nickel layer a platinum atom was again found there, as illustrated in Fig. 3.59, indicating an exchange event.

Friedl *et al.* [83] examined self-diffusion on Ir(100) and again found a c(2 × 2) net for atom displacements. Instead of atom exchange they proposed a new idea – that the surface reconstructed, creating a c(2 × 2) sub-lattice for a diffusing atom, but this explanation did not survive closer examination. The same map of adsorption places should be observed for both Ir and Rh atoms if they move on a c(2 × 2) reconstructed sub-lattice. Fu and Tsong [84] tested this in 1999 and again found a c(2 × 2) net in self-diffusion on Ir(100). Rh atoms, however, gave a (1 × 1) net in diffusion on Ir(100), characteristic of movement by hopping.

It is important to recognize that quite a number of calculations have been carried out for diffusion on (100) planes. For Al(100) more than ten have been reported [24,56,58,85–92], with half giving atom exchange as the low energy process. Out of nineteen estimates for Ni(100) [24,92–104] only eight considered exchange as an option and fewer than half of these suggested an exchange process. Much more has been done on copper [24,62,86,89,92–94,96,98–102,104–125], but only half the studies considered exchange as an option and only three studies gave atom exchange as the low energy

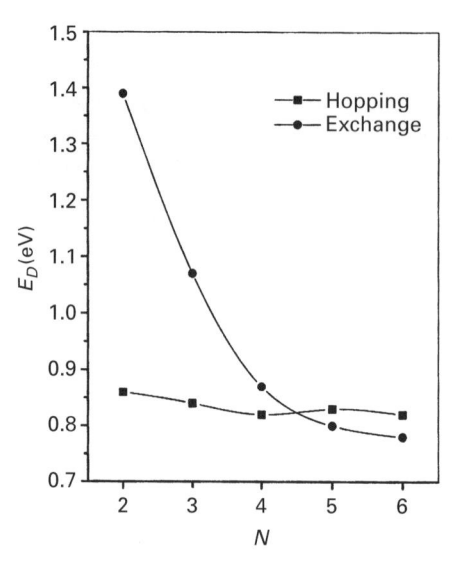

Fig. 3.60 Dependence of activation energy for diffusion of Ni on Ni(100) plane upon the size of the unit cell $(N \times N)$ used in the *ab initio* calculations. Barrier to atom hopping is essentially constant; for adatom exchange it decreases strongly with size of cell (after Chang et al. [97]).

event. For Pd(100) [24,60,92–95,98–100,103,104] half indicated hopping was the likely diffusion process, but increasing the size of the cell used in the calculations started to favor exchange [94,98]. For Ag(100) [24,60,92–96,98–104,106,126–131], one effort considered atom exchange and indicated it as the favored process. The rest arrived at hopping as the leading mechanism. Relatively little calculational work has been done on Ir(100) [70,127], Pt(100) [24,132,133], and Au(100) [24,95,134,135]. However, all these gave atom exchange as the preferred mechanism of diffusion. One interesting fact was noted: in the theoretical studies, exchange depended on the size of the cell used in the simulations, as is apparent from Fig. 3.60.

The theoretical calculations do not provide a clear indication of what is going on at (100) planes, but from the experiments it is evident that atom exchange is an important process in surface diffusion: in self-diffusion on Ir(100), Pt(100), Ni(100) and probably also on Au(100), diffusion occurs through exchange between the adatom and an atom from the substrate. It is still not certain why one atom diffuses by exchange and another by hopping. Feibelman [136] in 1990 first pointed to the covalent bonding in Al as stabilizing the transition state to atom exchange. Diffusion by atom exchange was correlated to the relaxation of atoms around the adatom by Kellogg et al. [80]. Tensile surface stress was believed to play a significant role in diffusion by atom exchange on fcc(100) planes by Yu and Scheffler [126]. However, detailed density-functional calculations for the (100) planes of Rh, Ir, Pd, and Pt by Feibelman and Stumpf [127] found no such clear relation. They felt that Kellogg's proposal of substrate atom relaxation made sense. One thing is clear – predictions from experiments about diffusion by atom exchange are difficult.

3.2.3 Diffusion mechanisms: multiple exchange

The processes considered so far have all occurred through single events; however, this might not always be the case at elevated temperatures. Under such conditions more complicated mechanisms may start to play a non-negligible role. This has so far not been explored in experiments, which are difficult to do, but molecular dynamics simulations are feasible and have revealed interesting phenomena. The first of these was carried out by Black and Tian [105], who did molecular dynamics simulations of copper atoms on Cu(100), with interactions based on the embedded atom method. Their work was done for different diffusion temperatures. At 600 K they found atom exchange processes, but most interesting were simulations at 900 K. "A surface atom entered the substrate, and an atom not adjacent to the entering atom popped out of the substrate. The strain introduced by the entering atom then can be thought of as propagating along a row of substrate copper atoms." Eventually an atom emerged from the substrate crystal at some distance from the original entry point and relieved the strain; a schematic is shown in Fig. 3.61. This appears to have been the first observation of multiple exchange events.

A year later, Cohen [106] did molecular dynamics simulations on fcc(100) surfaces, again using embedded atom potentials. With an aluminum atom on the Al(100) plane at 500 K, she observed that an adatom embedded in the surface of the lattice, and "traveled two binding sites down the fcc(100) row and over one site." The adatom moved on the surface like "a knight in chess." Similar movement was also found for Ag, Au, Cu, Pd, Pt, and Ni, but the barrier for this new mechanism was much higher than for diffusion by the usual atom exchange, as is clear from the Arrhenius plot in Fig. 3.62. Cohen estimated that except for Ni(100), this new exchange process would be observable at temperatures

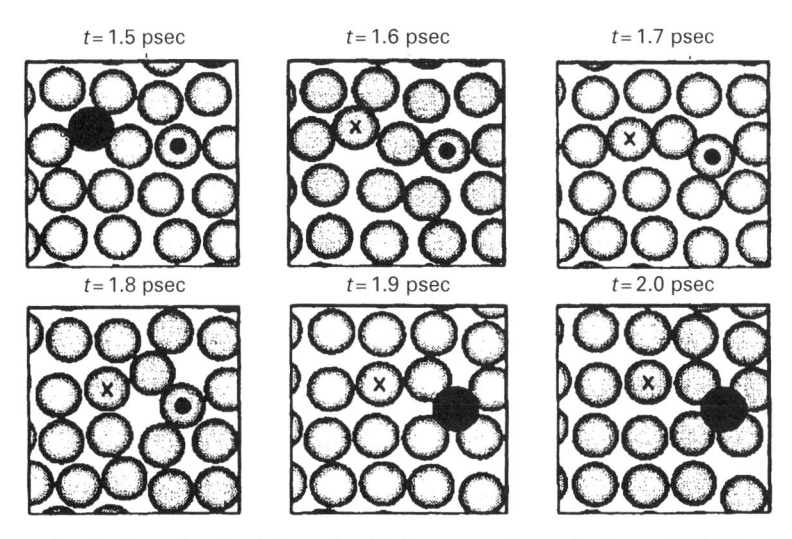

Fig. 3.61 Molecular dynamics simulation of multiple atom exchange for Cu on Cu(100) at 900 K. Black adatom moves into substrate, bringing about eventual emergence of substrate atom at a distance from original point of entry (after Black and Tian [105]).

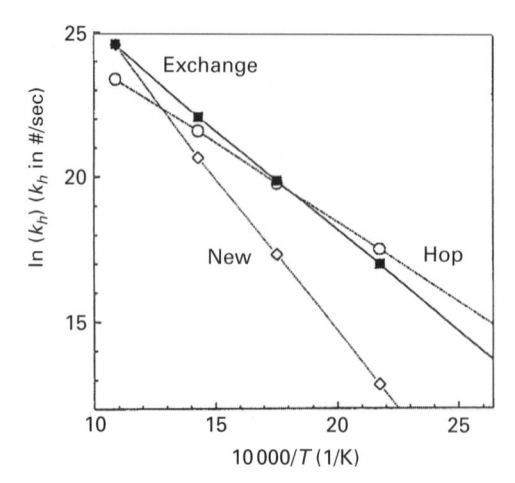

Fig. 3.62 Arrhenius plot for three different mechanisms of diffusion for Ag on Ag(110) derived in molecular dynamics simulations. *New* refers to multiple displacement exchange process, which has a higher activation energy and therefore only contributes at elevated temperatures (after Cohen [106]).

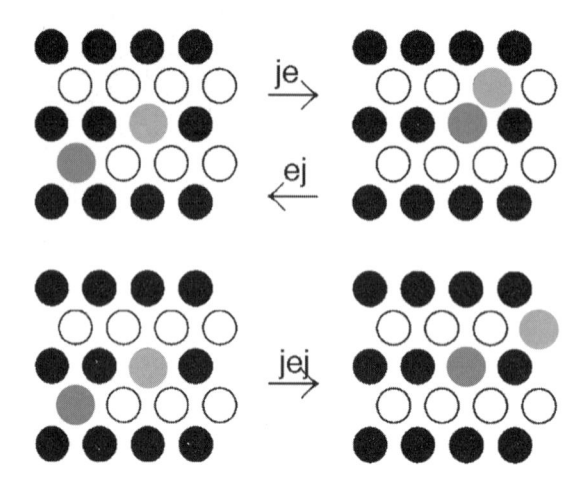

Fig. 3.63 Schematics of movements of Ag adatom on Ag(110) plane obtained in molecular dynamics simulations with RGL potentials at 600 K. Top: jump-exchange and exchange–jump events. Bottom: jump–exchange–jump. Dark grey shows atom in initial position, light grey indicates substrate atom participating in exchange (after Ferrando [137]).

above half the melting point. For nickel, temperatures would have to be even higher, but at high temperatures this multiple process certainly becomes competitive.

A new type of jump process was uncovered by Ferrando [137] in 1996 who looked at molecular dynamics of Ag adatoms diffusing on Ag(110) using RGL potentials. In the vicinity of 600 K, that is at a fairly high temperature, he found the events illustrated in Fig. 3.63: an atom moving in a one-dimensional channel would after a jump carry out an exchange with an atom of the substrate, a jump–exchange event. Also observed in the simulations was the reverse process, exchange followed by a jump, as well as

3.2 Atomic jumps in diffusion

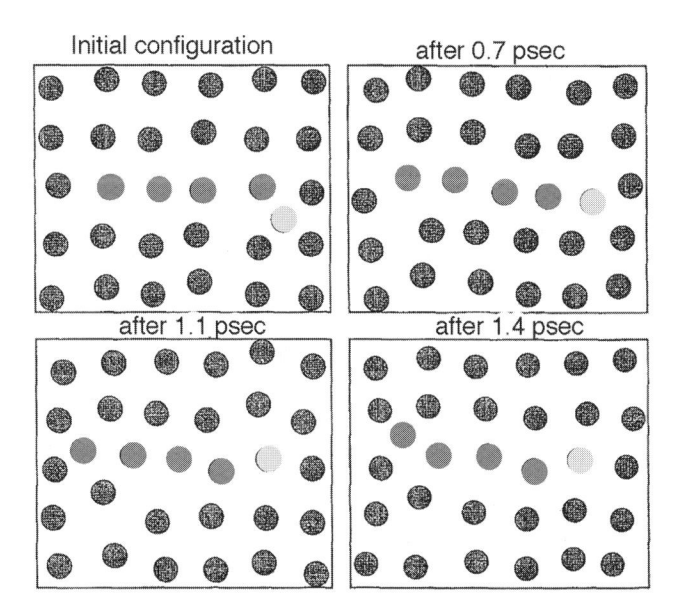

Fig. 3.64 Molecular dynamics simulation with RGL interactions of self-diffusion on Cu(100) at 950 K. Adatom squeezes into the surface layer, and eventually a surface atom pops out at a considerable distance from the first entry (after Evangelakis and Papanicolaou [110]).

jump–exchange–jump events, also illustrated in Fig. 3.63. Correlated movement was also observed on the Au(110) surface [65].

Other high temperature processes were pursued at greater length by Evangelakis and Papanicolaou [110] in 1996. They again did molecular dynamics simulations for a copper atom deposited on the Cu(100) surface, with atoms interacting via RGL type potentials. The simulations were carried out at elevated temperatures, above 700 K, and double-hop transitions were observed. At higher temperatures, they also found exchange processes in which the adatom entered the lattice and a lattice atom emerged nearby. A multiple exchange event is illustrated in Fig. 3.64. The adatom enters the surface layer and pushes aside a row of atoms, the last of which, four spacings away, finally emerges and transforms into an adatom. Evangelakis and Papanicolaou were able to follow these and similar processes at different temperatures, and obtained the Arrhenius plots in Fig. 3.65, which yielded the following diffusion characteristics: single jumps occurred over a barrier of 0.43±0.02 eV with a prefactor of $3.4 \times 10^{-3\pm0.2}$ cm^2/sec, for double jumps the values were 0.71±0.05 eV and $38 \times 10^{-3\pm0.5}$ cm^2/sec. For simple atom exchange the barrier was 0.70±0.04 eV, and the prefactor $42 \times 10^{-3\pm0.3}$ cm^2/sec, for double exchange 0.70±0.06 eV and $45 \times 10^{-3\pm0.6}$ cm^2/sec, for triple exchange 0.82±0.08 eV and $104 \times 10^{-3\pm1.0}$ cm^2/sec, and finally for quadruple exchange 0.75±0.12 eV and $86 \times 10^{-3\pm0.9}$ cm^2/sec. What is especially interesting here is that for multiple processes the activation energies as well as prefactors are all roughly comparable, with both substantially higher than for single atom jumps, so that multiple processes only occur at elevated temperatures.

Calculations for self-diffusion on Pt(100) were carried out by Zhuang and Liu [133], who relied on EAM potentials [138–140]. They made estimates for diffusion by the

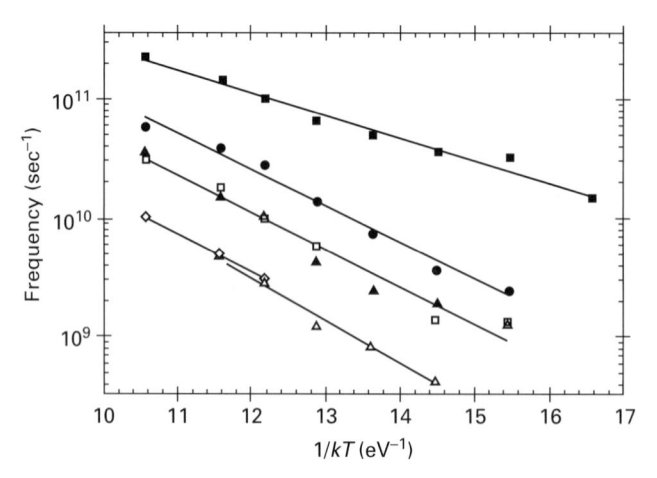

Fig. 3.65 Arrhenius plot for frequency of jumps in self-diffusion on Cu(100). Filled squares – single jumps; open squares – double jumps. Filled circles – simple exchange; filled and open triangles – double and triple exchange. Rhombic squares – quadruple exchange (after Evangelakis and Papanicolaou [110]).

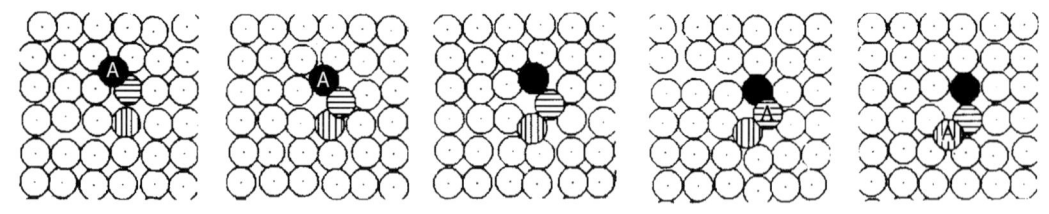

Fig. 3.66 Model of ad-dimer diffusion mechanism on fcc (100) plane. Adatom penetrates into the lattice and dislodges two lattice atoms to form an ad-trimer state. When this decomposes, the third atom is left on the surface as an adatom (after Zhuang and Liu [133]).

standard exchange process, illustrated in Fig. 3.53, in which an adatom penetrates into the lattice and displaces a lattice atom to the surface. Zhuang and Liu also envisioned a more elaborate process, shown in Fig. 3.66. The adatom now enters the lattice, and displaces two adjacent lattice atoms to the surface to form an ad-trimer state. The atom in the center finally returns to the lattice, leaving the third one on the surface as an adatom in this ad-dimer diffusion. The end result is identical to what would transpire in a sequence of two standard exchanges, but the barrier to the new process proved to be somewhat lower, 0.26 eV, than for the conventional exchange, which required 0.28 eV. There are as yet no experiments demonstrating this ad-dimer mechanism, but it is something that should definitely be probed.

Surprising results were obtained by Haftel and Rosen [141] in simulations using SEAM [138] in a molecular dynamics study to observe Au atoms condensing on Ag(111). They used constrained molecular statics, in which they moved the diffusing atom from the initial to the final step in small increments, relaxing all other atoms in every step. Their findings for the diffusion parameters on this close-packed surface are summarized in Table 3.2. Interesting here are the double exchange events occurring on the

3.2 Atomic jumps in diffusion

Table 3.2 Diffusion barriers and prefactors for Au on Ag(111) obtained in SEAM simulations [141].

Diffusion Type		E_D (eV)	$v_D(\times 10^{12} \text{sec}^{-1})$
Hopping	On terrace	0.102	1.19
	Over B step	0.716	
Exchange	On terrace	0.421–0.805	
	Over A step	0.051–0.067	
	Over B step	0.047	
Double Exchange	On terrace	0.212	2.09
Adatom catalyzed double exchange	On terrace	0.047–0.083	2.1

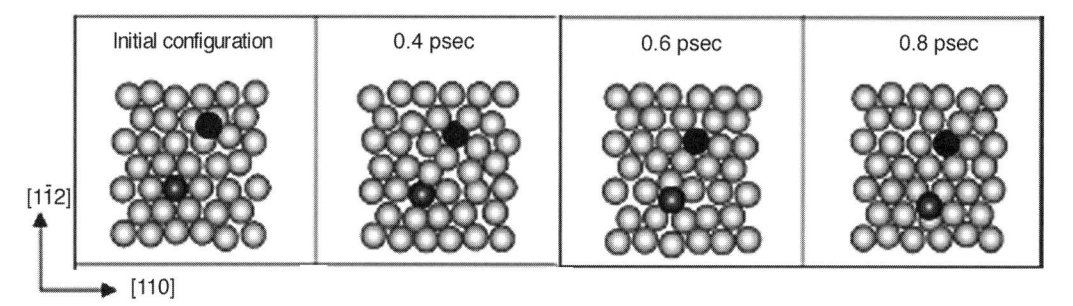

Fig. 3.67 Schematic of exchange on Cu(111) of Cu adatom, in black, with lattice atom, pushing lattice atoms along <110> until one moves to the surface at some distance from original entry of adatom. Events at high temperature and tensile stress found in EAM calculations (after Wang *et al.* [144]).

terraces as well as adatom catalyzed double exchange, which seems to be the lowest energy process. Haftel and Rosen claim that the ballistic energy of the colliding adatom, which they estimate as 0.11 eV, was able to catalyze double exchange, an effect rather unlikely in light of the studies presented in Section 3.4.

The very detailed previous examination of surface processes by Evangelakis and Papanicolaou was rediscovered some years later by Xiao *et al.* [142,143]. Working with embedded atom potentials and the nudged elastic band technique to do simulations on Cu(100) and Pt(100), they observed multiple atom exchange in strained layers and referred to these events as crowdions. An atom emerging from this crowdion can of course contribute to the diffusion process.

Shortly thereafter, Wang *et al.* [144] also published EAM calculations of self-diffusion on Cu(111). At high temperatures, $\geq$ 800 K, and tensile strains 2% or larger, diffusion by atom exchange became more probable even on this closely packed plane, but still with hopping as the leading mechanism. Especially interesting was the observation, shown in Fig. 3.67, of atom exchange, with a lattice atom being pushed out to the surface at some distance from the original exchange event. This complicated mechanism is only observed at high temperatures or under conditions of large strain.

Action-derived molecular dynamics [145] estimates based on RGL interactions have also been made by Kim *et al.* [146] for a multiple exchange process on Cu(100). A Cu adatom penetrates partly into the first lattice layer, eventually causing a lattice atom to

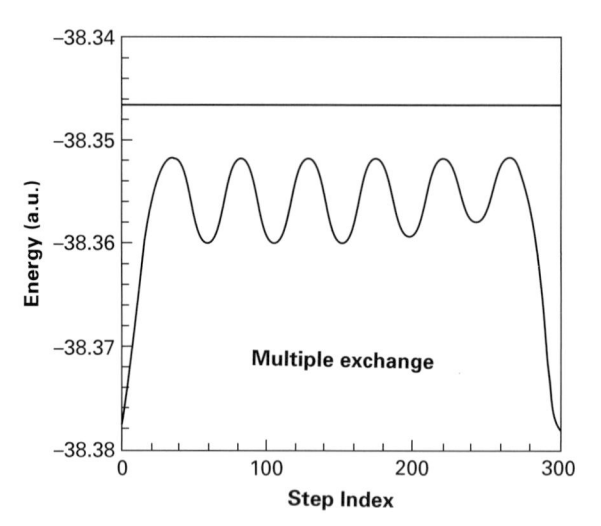

Fig. 3.68 Energy of multiple exchange process for Cu adatom on Cu(100). 1 a.u. = 27.211396 eV (after Kim et al. [146]).

emerge six spacings away from the point of entry. The energetics for this process, in Fig. 3.68, are interesting, showing six peaks of the same height, roughly 0.8 eV and large compared to the barrier for single atom jumps, but comparable to the result of Evangelakis and Papanicolaou [110]. It would be interesting to know the prefactor for this process.[2]

What is clear from these various simulations is that multiple exchange events or crowdions may under the right conditions contribute significantly to diffusion. It must be recognized, however, that so far these are just the product of simulations. What still needs to be done now is to verify all this experimentally, which will require tests at elevated temperatures, or on stressed systems.

3.2.4 Diffusion on strained surfaces

Up to this point, atom jumps have been considered as occurring on ordinary, thermally equilibrated surfaces. What happens, however, when a surface is strained? How does that affect atom diffusion? That is an effect first considered by Brune et al. [147], who did STM experiments on silver diffusion on both unstrained and strained Ag(111) surfaces. A strained surface was obtained by depositing silver on a Pt(111) plane, a layer shown to be under 4.2% compressive strain [148]. To determine the diffusivity they then measured the saturation island density. The Arrhenius plots they found are shown in Fig. 3.69a for Ag(111) as well as Pt(111), and on one monolayer of silver on Pt(111). On Ag(111), the activation energy for diffusion was 0.091 ± 0.010 eV, with a frequency prefactor of $2 \times 10^{11 \pm 0.5}$ sec^{-1}. On the silver monolayer on Pt(111) the diffusion barrier

---

[2] Kim et al. [146] have also commented on long jumps, but are clearly not well informed on the state of such studies.

3.2 Atomic jumps in diffusion

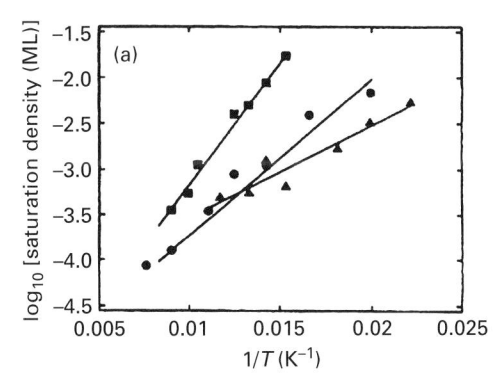

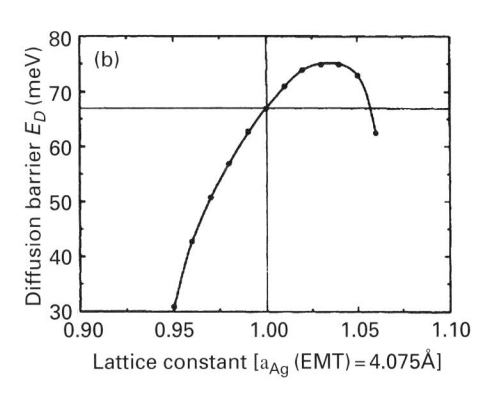

Fig. 3.69 Diffusion studies for Ag atoms on strained and unstrained surfaces. (a) Arrhenius plots for Ag diffusivity, derived from saturation island density. Square dots: Ag on Pt(111). Triangular dots: Ag on silver monolayer on Pt(111). Round dots: Ag on Ag(111). (b) Barrier to silver self-diffusion on Ag(111) as a function of strain, calculated with EMT (after Brune *et al.* [147]).

diminished to 0.060 ± 0.010 eV, with a frequency prefactor $1 \times 10^{9 \pm 0.6}$ sec^{-1}, and for Ag on Pt(111) the activation energy was high, 0.157 ± 0.010 eV, and the prefactor turned out to be $1 \times 10^{13 \pm 0.4}$ sec^{-1}.

The low barrier for diffusion on the silver monolayer could have been caused by electronic interactions with the underlying platinum. To check this, Brune *et al.* carried out EMT calculations for diffusion as a function of the lattice strain, shown in Fig. 3.69b. As the strain changed, the barrier to diffusion changed significantly from a reduction of the barrier for compressive strain to an increase when the strain is tensile. High tensile strains led to a maximum in the barrier followed by a strong decrease. The authors also showed that changes in the diffusion were mainly caused by changes in the binding energy at the equilibrium sites. Apart from the detailed effects, it is clear from these estimates that strain is an important parameter in surface diffusion,[3] with compression smoothing the surface structure.

Two years later, in 1997, Ratsch *et al.* [149] did density-functional estimates aimed specifically at the studies by Brune *et al.* [147]. For Ag diffusion on Pt(111), Ratsch *et al.* found a diffusion barrier of 0.150 eV, for Ag on Ag(111) the barrier was 0.081 eV, and for Ag on a monolayer of silver on Pt(111) the barrier was reduced to 0.063 eV, results in good agreement with the values from experiments. For silver self-diffusion on Ag(111) the effect of surface strain was calculated and is compared in Fig. 3.70 with the EMT calculations of Brune *et al.* [147]; in the former the diffusion barrier is quite linear, unlike the curved results obtained in EMT. What brings about the change in the diffusion barrier is revealed in Fig. 3.70b, which shows the energy of an adatom in the transition state, that is at a bridge site, compared to a normal fcc site. The difference between the two increases as the strain rises, with the binding energy at the fcc site increasing more rapidly. This

---

[3] This was verified by Mortensen *et al.* [28] who used density-functional theory to estimate the effect of strain on the self-diffusion barrier for fcc metals. The derivative of the barrier height with respect to strain, in units of meV/%, turned out as follows: Ni, 2.8; Pd, 3.5; Pt, 3.6; Cu, 3.6; Ag, 3.8; and Au, 2.8.

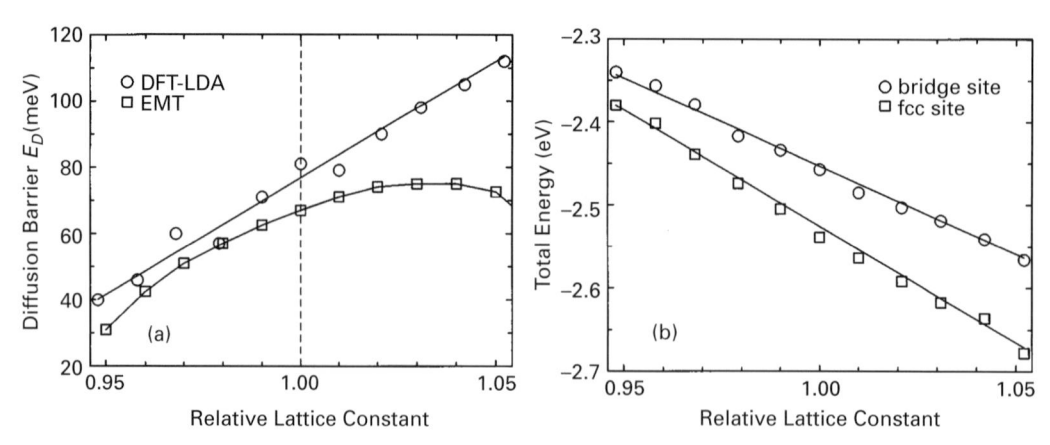

Fig. 3.70 Diffusion of Ag on Ag(111) as a function of strain. (a) Comparison of EMT results from Brune *et al.* [147] with DFT calculations. (b) Energy of Ag atom in transition state (bridge site) and in normal fcc site on Ag(111) plane, as affected by lattice strain (after Ratsch *et al.* [149]).

altered the corrugation of the potential energy surface, making it more pronounced in expansion and flatter during compression. The energy difference between fcc and hcp sites was also estimated and turned out to be 0.04 eV for an Ag atom on a Pt(111) surface but only 0.01 eV for Ag on one monolayer of Ag on Pt(111).

At much the same time as the investigations of Brune *et al.* [147], Meyer *et al.* [150] looked at the growth of nickel films on Ru(0001) with the STM, and found an increase in the mobility of Ni atoms as the number of nickel layers on ruthenium increased. The diffusion barrier for Ni was estimated to diminish by 0.30 eV in going from one nickel monolayer to three Ni layers. Meyer *et al.* pointed out that "the reason for the changing Ni mobility with layer height is not clear." However, based on what is known at this moment, it is likely that strain could also be responsible for these observations.

In 1997, Yu and Scheffler [151] employed density-functional calculations to diffusion on fcc(100) surfaces. They evaluated the diffusion barriers for atom hopping as well as for undergoing exchange as a function of the relative lattice constant, that is the surface lattice constant compared to that of the normal bulk material. As is clear from their work in Fig. 3.71, atom hopping is the favored process for the unstrained Ag(100) surface or under compressive stress; however, exchange becomes the important event under high tensile stress. Yu and Scheffler indicated that "while the significant tensile surface stress of Au(100) pulls the dimer of the exchange transition state 'into' the surface, lowers the energy of the transition state, and enables exchange diffusion, the surface stress at Ag(100) is too weak to have a significant effect." "Smaller lattice constants correspond to a reduced corrugation of the surface potential, thus diffusion energy barriers are reduced. In contrast, for a stretched surface the corrugation increases, and the adsorption energy at the four-fold coordinated site increases." It is interesting to note that the barrier height on (100) is again linear in the relative lattice constant, even out to high strains.

Diffusion on strained surfaces was discussed by Schroeder and Wolf [152], also in the same year, but using Lennard-Jones potentials throughout, which limits the applicability to real systems. The effect of strain on the barrier for diffusion on a simple cubic (100)

3.2 Atomic jumps in diffusion

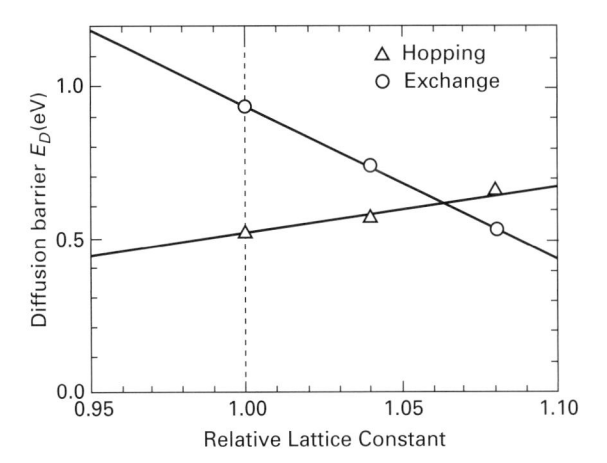

Fig. 3.71 Self-diffusion barrier of Ag on Ag(100), derived in density-functional calculations, in its dependence on strain (after Yu and Scheffler [151]).

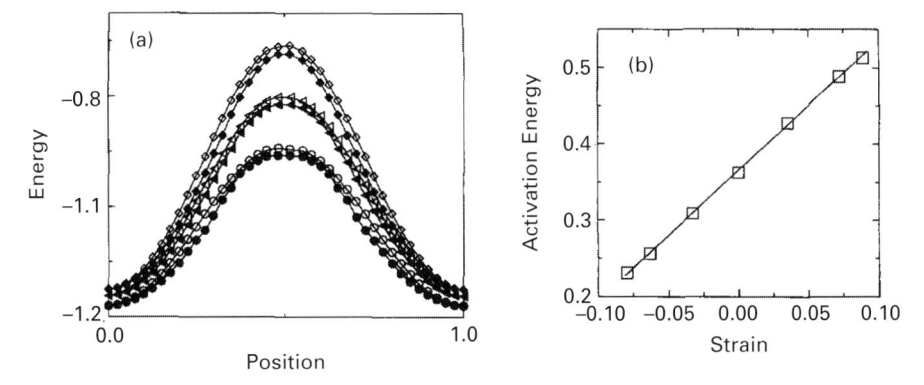

Fig. 3.72 Effect of strain on self-diffusion on a (100) plane. (a) Potential energy change in self-diffusion on a Lennard-Jones (100) plane for different strains. Middle curves: unstrained. Upper curves: tensile strain, $\varepsilon_{xx} = \varepsilon_{yy} = 0.05$. Lower curves: compressive strain, $\varepsilon_{xx} = \varepsilon_{yy} = -0.05$. Filled symbols: global relaxation. Open symbols: frozen lattice. (b) Dependence of self-diffusion barrier on Lennard-Jones (100) plane upon strain (after Schroeder and Wolf [152]).

surface is shown in Fig. 3.72a, with the upper curve calculated for tensile strain $\varepsilon_{xx} = \varepsilon_{yy} = 0.05$, and the bottom one for a compressive strain of 0.05. The central curves give values for the normal system as a function of the x-coordinate of an adatom as it moves from one binding site to an adjacent one. The activation energy as a function of surface strain is plotted in Fig. 3.72b for diffusion on the (100) plane of a simple cubic crystal, increasing as strain becomes more tensile. According to Schroeder and Wolf, changes in the barrier height are mostly due to changes in the saddle point, rather than the changing in the strength of the binding energy. Much the same effects were found for fcc as well as bcc(100) planes. For an fcc(100) crystal, the binding energy weakly depended on the strain, so again changes in the barrier are mostly due to changes in the saddle point. However, the situation differs for a bcc(100) surface, where changes in the strength of binding caused by strain are significant. This is due to the presence, in the second lattice layer, of an atom just below the binding site of the adatom, which contributes to binding energy changes.

More recently, Liu et al. [153] used glue potentials [154,155] to examine the influence of local strain on the surface self-diffusion of Au on the $(23 \times \sqrt{3})$-Au(111) reconstructed surface. A schematic of this surface is given in Fig. 3.73a; B and D indicate fcc- and hcp-like regions, A, C, and E are transition regions. For self-diffusion on the unreconstructed Au(111) this method yielded a value of 0.7 eV, which the authors

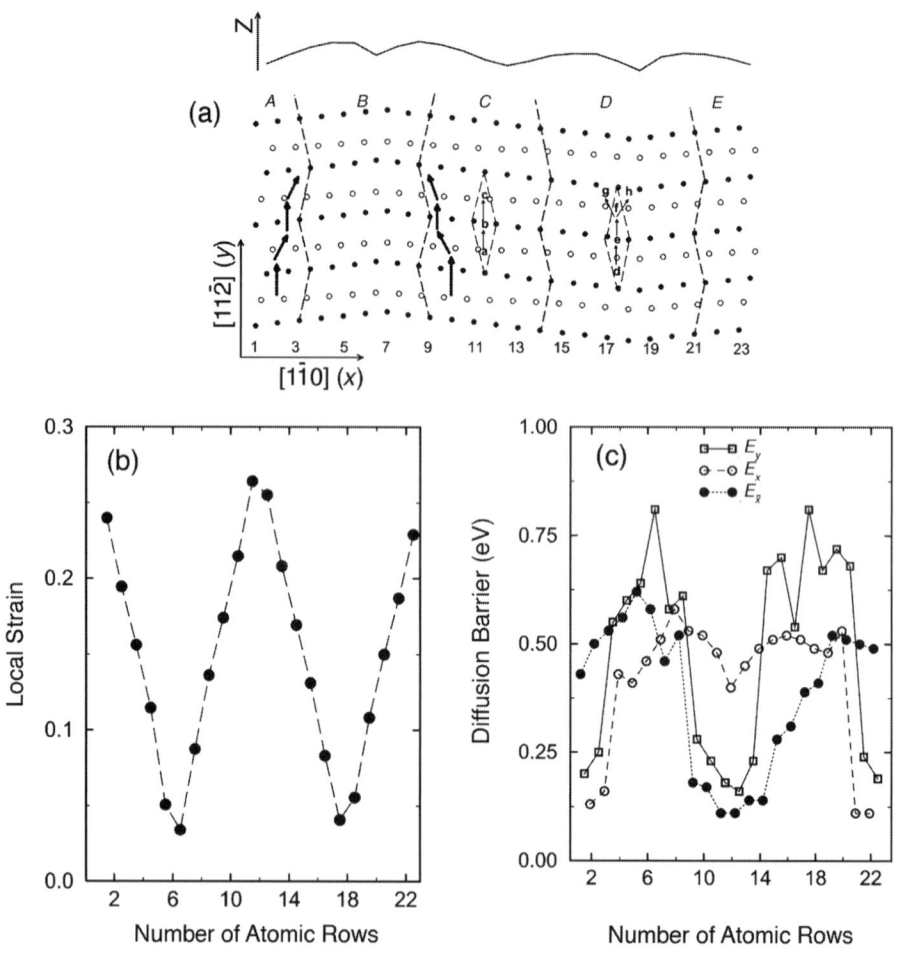

Fig. 3.73 Diffusion on the reconstructed $(23 \times \sqrt{3})$-Au(111) surface. (a) Schematic of $(23 \times \sqrt{3})$-Au(111) surface. B and D denote fcc-and hcp-like regions, A, C, and E indicate transition regions. Preferred directions for diffusion shown by thick arrows. (b) Estimate of strain in different regions of the surface. (c) Self-diffusion barrier at different positions estimated using glue potential (after Liu et al. [153]).

3.2 Atomic jumps in diffusion

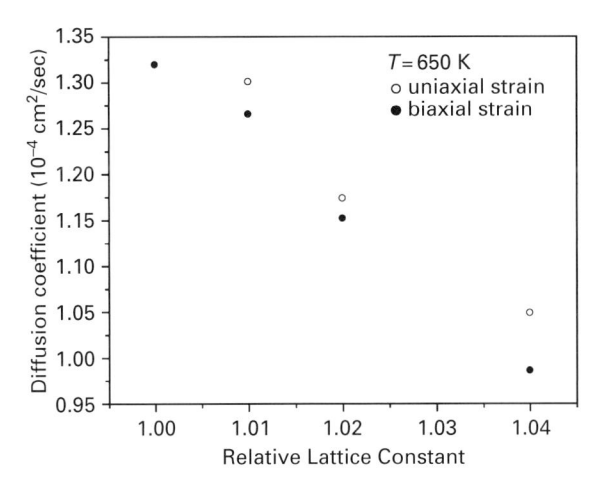

Fig. 3.74 Diffusivity of Cu adatom on Cu(111) plane, calculated with EAM potentials in its dependence on strain (after Wang et al. [144]).

recognized as too high. Only general features can therefore be compared. The binding energy at fcc sites is higher than at hcp sites, and based on this observation the authors concluded that strain does not change the energy sequence. An estimate of the local strain at various locations is plotted in Fig. 3.73b and the calculated diffusion barrier is indicated in Fig. 3.73c. Here $E_{\bar{x}}$ stands for the barrier in the direction f–g shown in Fig. 3.73a, and E_x is for the direction f–h. It is clear that diffusion is fastest in the transition regions A, C, and E, where the strain is high, so these regions create easy paths for adatom movement.

A year later, in 2003, Wang et al. [144] performed molecular dynamics simulations for diffusion on Cu(111) using EAM potentials from Mishin et al. [31]. The strong effect of both uniaxial as well as biaxial strain on the diffusivity is clear in Fig. 3.74, with tensile strain diminishing the diffusivity. High values of the tensile strain also bring about diffusion via atom exchange, absent on a normal Cu(111) surface.

In 2004 Xiao et al. [143] looked at adatom diffusion on a strained Pt(100) surface, investigating the influence of the final size of the computational cell on the energy barrier. As is clear from Fig. 3.75a, they found that the dependence of the energy for exchange on strain was cell-size dependent, and associated this with the displacement of atoms over longer distances during adatom exchange, as in Fig 3.75b. If such displacements are not EAM artifacts, then the size of the cell used can cause systematic overestimation of the energy for atom exchange. The change in hopping energy versus strain for differently sized cells is illustrated in Fig. 3.76a; for small cells the expected trend of changes with increasing strain is reversed compared to bigger cells. A comparison of the energy change for hopping and exchange with strain on a big 10×20 cell is presented in Fig. 3.76b, the barrier rising for atom hops and decreasing for exchange events.

Although the studies of strain effects on surface diffusion have relied on a variety of different techniques, one thing is quite clear: strain exerts a big effect on diffusion kinetics, and is something to take advantage of in guiding diffusion. More experimental

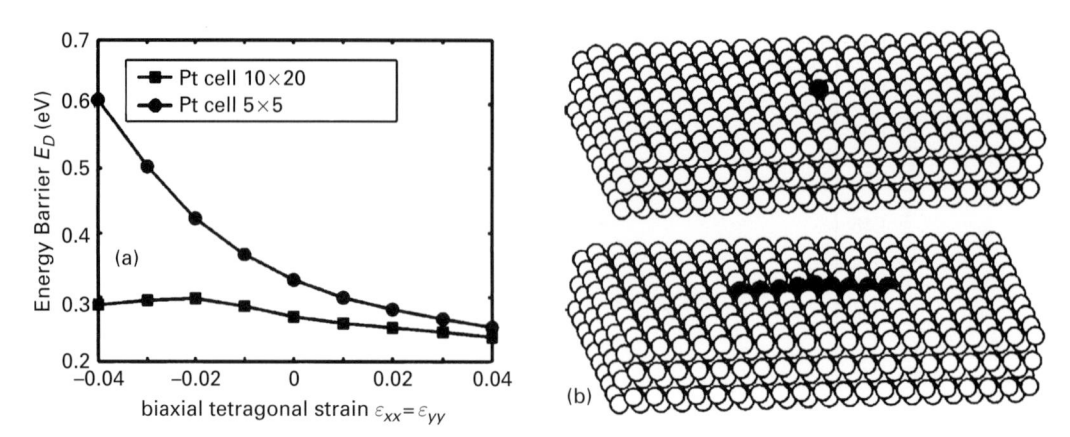

Fig. 3.75 Effect of strain on the self-diffusion barrier on Pt(100) calculated with EAM. (a) Dependence of exchange energy on size of computational cell. (b) 10 × 20 unit cell used in computation, showing initial configuration at the top, and transition state for unstrained surface at the bottom (after Xiao et al. [143]).

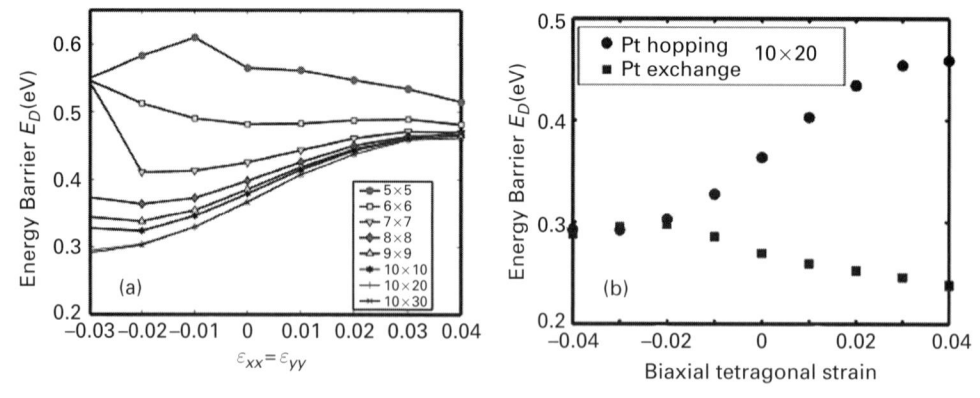

Fig. 3.76 Diffusion of Pt atom on strained Pt(100) surface. (a) Diffusion barrier for atom hopping in its dependence on the computational cell size. (b) Effect of biaxial strain on barrier to hopping and also to exchange (after Xiao et al. [143]).

information will be useful in this area to start deriving information about absolute values of diffusion barriers, not just about trends.

3.3 Long jumps in surface diffusion

3.3.1 Theoretical studies

Even in the 1970s the picture that in diffusion atoms jump randomly between nearest-neighbor sites on the surface still prevailed. However, there gradually appeared theoretical efforts countering this view. In 1979, Tully et al. [156] made ghost particle simulations for diffusion on a (100) plane of a Lennard-Jones crystal at temperatures below the melting point T_m. As shown in Fig. 3.77, they realized that the mean jump length

3.3 Long jumps in surface diffusion

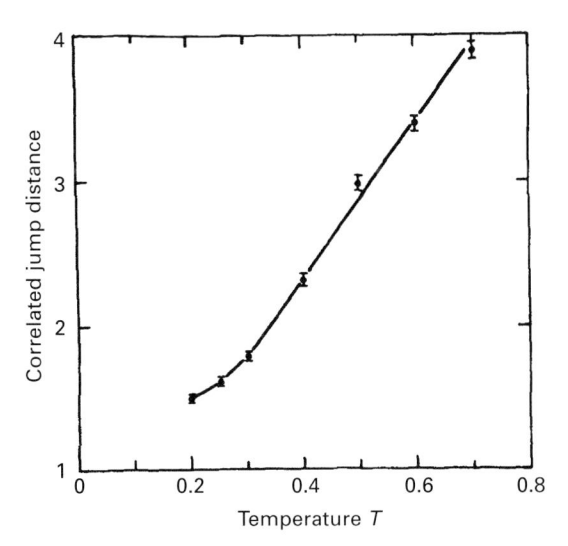

Fig. 3.77 Temperature dependence of the length of correlated jumps in diffusion of adatom on Lennard-Jones fcc(100) surface (after Tully *et al.* [156]).

increased from 1.5 spacings at a low temperature to 3.5 spacings at high values. At much the same time extensive simulations, again with Lennard-Jones potentials, were done by DeLorenzi *et al.* [43,44]. They examined several fcc crystal planes, but most interesting were their results for the (100). At $T \sim 0.5T_m$ long jumps were observed, sometimes between sites three or more spacings apart, as is clear from their trajectories in Fig. 3.78. On the fcc(110) plane they found that diffusion was two-dimensional, occurring both along and across the channels, with exchange between the adatom and an atom from the lattice. Adatom behavior on the close-packed (111) plane at $T \sim 0.44T_m$ proved to be quite different, as seen in Fig. 3.79. Rather than hopping between sites, the adatoms carried on close to free two-dimensional motion.

The same year, Mruzik and Pound [45] also did molecular dynamics estimates of diffusion on an fcc Lennard-Jones crystal. On the (111) plane at $T^* = 0.1$, where $T^* = kT/\varepsilon_{LJ}$, and ε_{LJ} is the energy parameter in the Lennard-Jones potential, diffusion occurred essentially between nearest-neighbor positions, but at higher temperatures, at $T^* = 0.45$, atom motion resembled free flight over the surface, as already shown in Fig. 3.79. The atoms spend little time at the binding sites and long jumps occur. The situation is different on the (311) plane, where atom motion is entirely along the channels on the surface, depicted in Fig. 3.2b. For the (110) plane, diffusion at $T^* = 0.45$ occurred by exchange between the adatom and the lattice. The picture developed by Mruzik and Pound is essentially the same as that of DeLorenzi *et al.* [43,44], but more detailed.

Lennard-Jones interactions were again used by Voter and Doll [157] in 1985 for treating transitions longer than nearest-neighbor spacings in transition-state theory calculations of diffusion on Rh(100). However, longer jumps were found only at quite high temperatures, around 1000 K. In the meantime, DeLorenzi and Jacuzzi [74] had continued their work, and in the same year published their estimates for bcc surfaces,

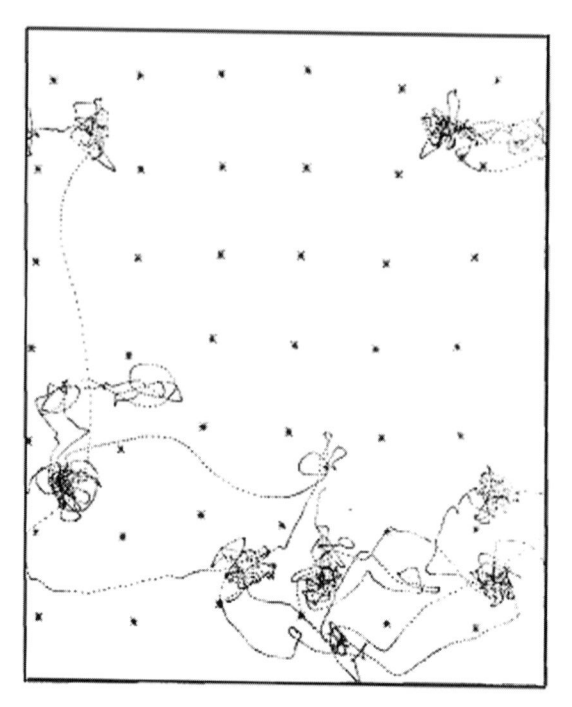

Fig. 3.78 Adatom trajectories on Lennard-Jones (100) surface at $T = 0.49T_m$, revealing non-nearest-neighbor transitions (after DeLorenzi et al. [44]).

using a metallic potential originally intended for sodium [75,76]. All simulations were done at $T \simeq 0.4T_m$. As already pointed out previously, they noted exchange events on the (211) plane, and more importantly, on the (100) face. Most interesting in the present context were the simulations on the (110) surface, the densest on the bcc lattice. There DeLorenzi and Jacuzzi observed single jumps, in Fig. 3.80a, but of greater interest, double jumps in Fig. 3.80b, and even more complicated transitions.

The simulation effort was rounded out a few years later by the work of Sanders and DePristo [158] in 1992, who made molecular dynamics calculations on the fcc(111) surface, approximating atomic interactions by the corrected effective medium theory, but also by Morse and Lennard-Jones potentials. What they found was not surprising. At 800 K, the time an atom spent at a given site, was, as suggested in Fig. 3.81, very small, indicating that the atom did not fully equilibrate there.

Ferrando et al. [159] in 1993 set about to solve the problem of a Brownian particle moving over a one-dimensional periodic potential in an external field. Long jumps are expected to appear when exchange of energy between the adatom and the lattice becomes poor and Ferrando et al. were able to calculate the probability of jumps longer than a nearest-neighbor spacing as a function of the dissipation Δ in traveling over the tops of the barriers, given by

$$\Delta = \frac{4\ell_0 (mE_D)^{1/2}}{\pi kT} \eta. \tag{3.6}$$

3.3 Long jumps in surface diffusion

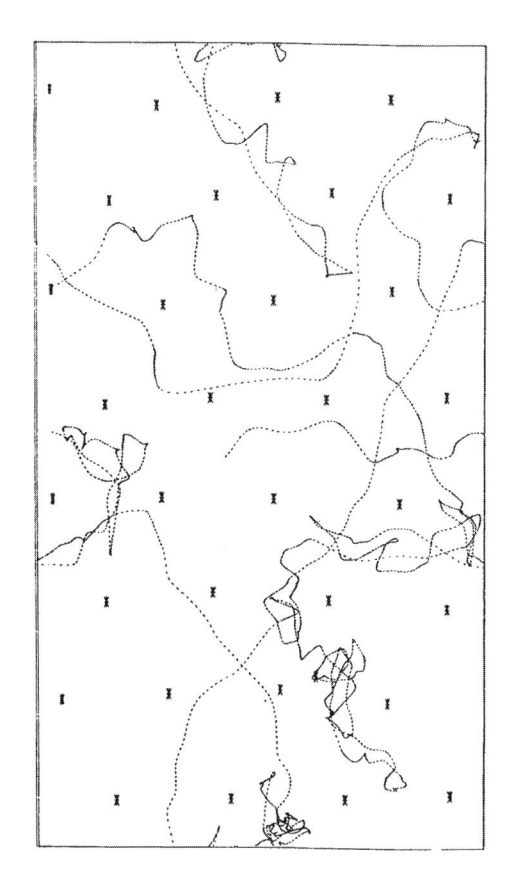

Fig. 3.79 Delocalized adatom trajectories on fcc(111) surface at $0.44 T_m$, obtained by simulations with metallic potential (after DeLorenzi et al. [44]).

Here η is the coefficient of friction between the diffusing atom and the lattice, and the probability of long jump decreases as the dissipation increases. This approach to the diffusion of a particle has been extended by Georgievskii and Pollak [160,161] and by Chen and Ying [162], who arrived at the probability of a jump of many nearest-neighbor spacings in its dependence upon the magnitude of the barrier. What turned out to be of interest is the small effect of temperature upon the length of the jumps.

In 1994, Ellis and Toennies [163] made an interesting discovery. They carried out molecular dynamics simulations for the diffusion of Na on Cu(100), using Morse potentials, and arrived at the Arrhenius plot in Fig. 5.24, giving an activation energy of 0.051 ± 0.006 eV and a frequency factor of $(0.53 \pm 0.2) \times 10^{12}$ sec^{-1}. More importantly, they also discovered that for every nearest-neighbor transition of the Na atom, long jumps occurred more than 0.7 times at 200 K, and just a little less at 300 K. Somewhat later, Evangelakis and Papanicolaou [110] observed double jumps on Cu(100) at temperatures above 750 K, as was already mentioned in Section 3.2.3.

In 1996, Ferrando [137] examined simulations of self diffusion on Ag(110) and observed that 90% of the atom jumps were single transitions at any temperature; the

120   **Atomic events in surface diffusion**

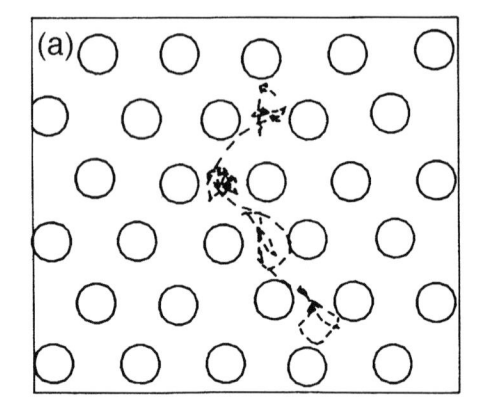

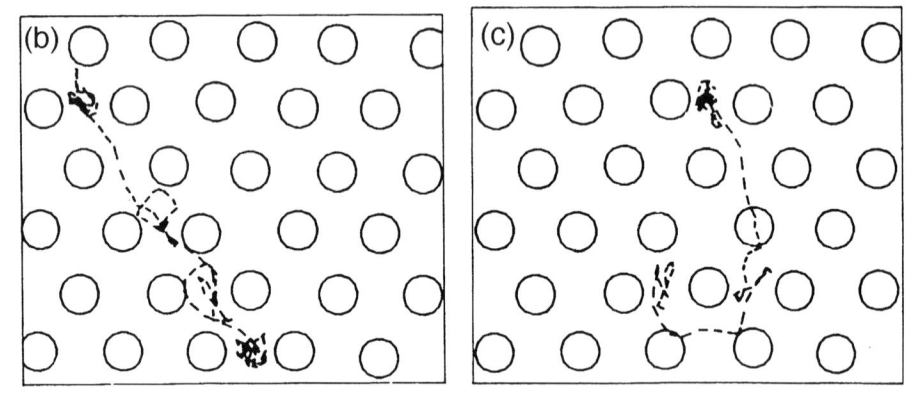

Fig. 3.80 Adatom trajectories on bcc (110) plane modeled with Price potential [75,76] during 200 psec at $\sim 0.4 T_m$. (a) Single jump. (b) Double jump. (c) Complicated trajectories (after DeLorenzi and Jacucci [74]).

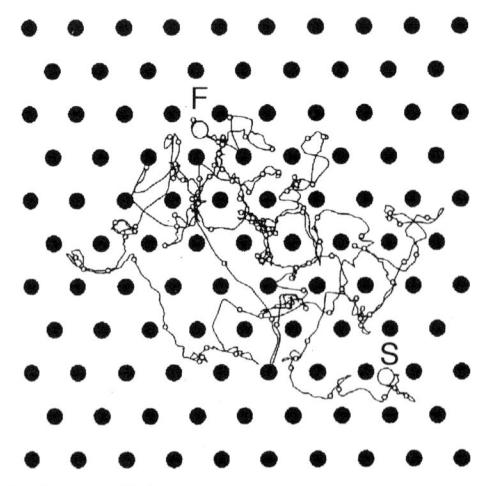

Fig. 3.81 Trajectory of Rh atom on Rh(111) plane at 800 K, obtained in molecular dynamics simulations with MD/MC-CEM (after Sanders and DePristo [158]).

rest were mostly doubles. A few years later, Ferrando extended his study to diffusion on (110) surfaces of Au, Ag, as well as Cu [65]. Despite similarities in the energetics of diffusion on Au and Ag, and a slightly lower barrier for Cu, the behavior of long jumps proved to be quite different on the different surfaces. At 450 K, long jumps were absent on gold, there were 3% long jumps on silver, and 6% on copper. For copper, the fraction of long jumps increased to 15% at 600 K, but never got to this value on the Ag(110) surface. In the meantime Azzouz *et al.* [164] derived a master equation for atoms hopping on a crystal, and found that for atoms weakly interacting with the surface transitions beyond nearest-neighbor sites became more likely, in agreement with the work of Ellis and Toennies [163].

In 2004, Ferron *et al.* [165] carried out molecular dynamics simulations relying on EAM interactions. They saw long jumps as well as rebound (recrossing) jumps in self-diffusion on Cu(111) and claimed that at 500 K, 95% of jumps were correlated; at 100 K this decreased to 50%. They continued their simulations using the DYNAMO code and atomic interactions derived from EAM potentials [166]. A number of correlated jumps, including atomic long events as well as ballistic transitions were identified at temperatures ranging from 7% to 55% of the melting temperature of copper, with the likelihood of correlations increasing with increasing temperature. Also observed was a linear increase of the characteristic jump length with temperature. The average distance covered by adatoms increased faster than expected for a random walk between nearest-neighbor sites.

The influence of long jumps on island morphology on fcc(100) surfaces was studied by Beausoleil *et al.* [167] with Monte Carlo simulations based on bond-counting energetics. They considered only long jumps forward, ignoring rebound jumps, and found that due to long jumps island coalescence is slowed down, adatom density as well as island density increased, but the average size of islands is lowered.

These simulations, although quite different from each other, have all shown that at elevated temperatures long jumps, that is jumps in which the adatom does not equilibrate at a nearest-neighbor site, instead spanning two, three, or longer nearest-neighbor distances, can be expected to take place at higher temperature. What these temperatures are, and what the type of jump is, does not emerge, as the interatomic forces were only roughly approximated. The emphasis therefore turns to experiments: how can long jumps in single atom diffusion be detected?

3.3.2 Distribution of displacements

Long jumps cannot be seen directly in experiments, due to the lack of an experimental technique fast enough to probe jumps in real time and at the same time to have control of a single adatom. One possible way to distinguish experimentally between single and long jumps is measurement of the distribution of displacements. Usually information about diffusivities is obtained from measurements of the displacement fluctuations, $\langle \Delta x^2 \rangle$. These fluctuations are related to the diffusivity D through the Einstein relation (Eq. 1.36), $\langle \Delta x^2 \rangle = 2Dt$. For diffusion via single jumps between nearest-neighbor sites the diffusivity D can be written from Eq. 1.18 as $D = \Gamma_1 \ell_0^2$, so that

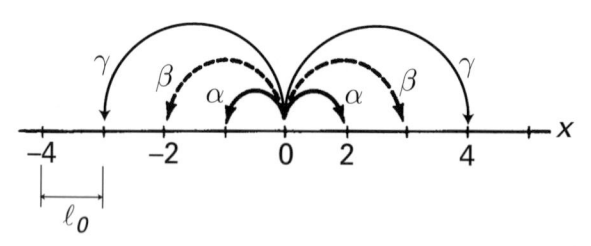

Fig. 3.82 Schematic of single, double, and triple transitions in one-dimensional symmetric atom motion.

$$\langle \Delta x^2 \rangle = 2\Gamma_1 t \ell_0^2, \tag{3.7}$$

where Γ_1 is the rate of nearest-neighbor transitions and ℓ_0 is the nearest-neighbor jump length. Suppose now that diffusion is one-dimensional and involves not only single jumps but also jumps of arbitrary length ℓ_i, as illustrated in Fig. 3.82. Then

$$\langle \Delta x^2 \rangle = 2t \sum_i \Gamma_i \ell_i^2, \tag{3.8}$$

where the summation extends over all types of jumps i and their rates Γ_i. The diffusivity is therefore given by

$$D = \sum_i \Gamma_i \ell_i^2. \tag{3.9}$$

It is evident that long jumps should affect the diffusivity, but this by itself does not allow a determination of what the rate of long jumps amounts to. In terms of transition-state theory, the diffusivity has been written as Eq. (1.21) $D = \nu \ell^2 \exp\left(\frac{\Delta S_D}{k}\right) \exp\left(-\frac{\Delta E_D}{kT}\right) = D_o \exp\left(-\frac{\Delta E_D}{kT}\right)$. From a plot of the natural log of the diffusivity D against $1/kT$ we can arrive at the activation energy, which henceforth will be written as just E_D, and also the prefactor D_o. The value of the prefactor will depend upon the frequency ν, the mean-square jump length ℓ^2, and the entropy of activation ΔS_D. These contributions to the prefactor D_o are not easy to sort out, so that ℓ^2 is not directly accessible in this way.

However, the displacements measured to get information about the diffusivity D can also provide further insights if a sufficient number of measurements are available. It is then possible to ascertain the distribution of displacements, shown in Fig. 3.83 for one-dimensional diffusion with a mean-square displacement of two. Expressions for the distribution have already been derived in Section 1.2.3. If, in addition to nearest-neighbor transitions there are also double jumps, spanning the distance between next-nearest-neighbors, then p_x, the probability of being at a distance x from the origin, is given by (Eq. 1.104) $p_x = \exp[-2(\alpha + \beta)t] \sum_{j=-\infty}^{\infty} I_j(2\beta t) I_{x-2j}(2\alpha t)$, where α is the rate of single jumps, β the rate of double transitions, and $I_i(z)$ is the modified Bessel function of order i and argument z. Suppose that in this migration the mean-square displacement is again

3.3 Long jumps in surface diffusion

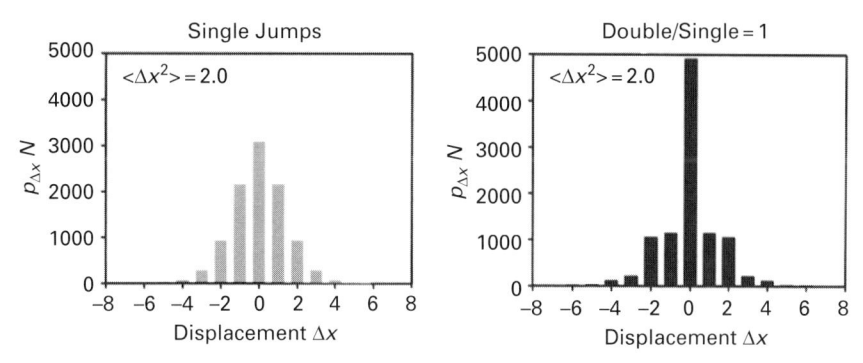

Fig. 3.83 Effect of double jumps on the distribution of atomic displacements in diffusion with a mean-square displacement of two.

two, but double jumps occur at the same rate as singles. This distribution is also shown in Fig. 3.83 at the right, and is obviously quite different from what we have seen for single jumps, at the left. The number of zero displacements is much higher, and the adatom is less frequently seen at longer distances from the origin. By analyzing the distribution of displacements we obviously can gain information about the jump rates in diffusion.

In practice there are of course some problems. Equation (1.104) gives the expression for the distribution of displacements on an infinite line of sites. In the field ion microscope, however, the individual planes are small, usually less than 60 Å in diameter, so that this equation is not really applicable. Furthermore, close to the edges of the plane, we have seen that two or three sites may have properties different from the interior and have to be excluded. To deduce jump rates from the measured distribution we therefore have to resort to Monte Carlo simulations. These are carried out on a plane of the same size, shape, and structure as the plane in the experiments. In the simulations the adatom is allowed to make all predicted kinds of jumps and jump rates are varied until the best agreement is found between simulated and experimental distribution.

It should also be noted that here we have been concerned entirely with one-dimensional diffusion, but this is not a real limitation. In diffusion on a two-dimensional surface, the distribution along x and y can usually be analyzed separately. There are still some corrections to be made, but these have already been discussed in Chapter 2.

3.3.3 Early experiments

The first attempt to experimentally probe the distribution of displacements during diffusion in order to elicit information about jumps to other than nearest-neighbor sites was made in 1989 by Wang *et al.* [168] on a one-dimensional system: W(211). Distributions of displacements were measured in the FIM over a range of temperatures for Re, Mo, Ir, and Rh. In Fig. 3.84 are shown distributions for rhenium at 300 K for two different time intervals and for molybdenum at 262 K in Fig. 3.85. The best fit to these distributions was derived with a ratio of double-to-single jumps equal to zero, that is without any long jumps involved. However, it is clear that the continuous time

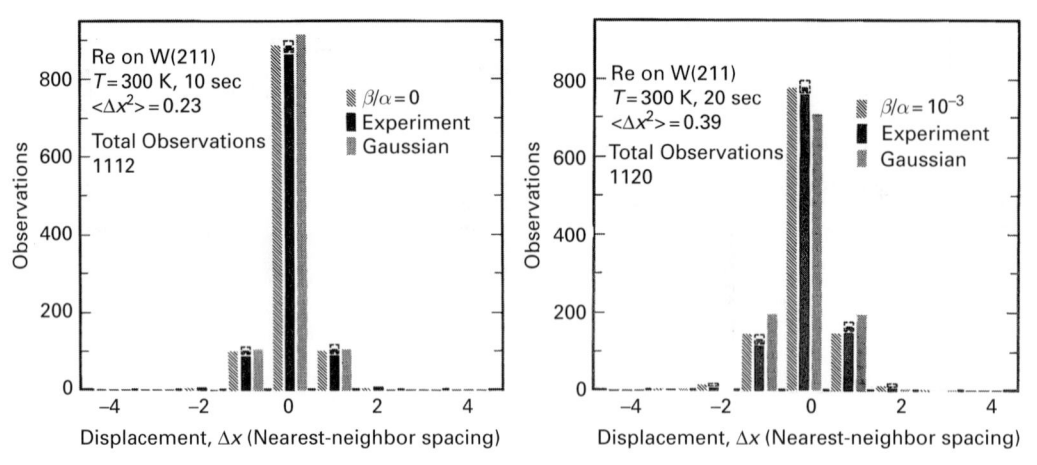

Fig. 3.84 Distribution of Re adatom displacements at 300 K on W(211), obtained from field ion observations [168]. Best fit to experiments obtained with an entirely negligible contribution from long jumps.

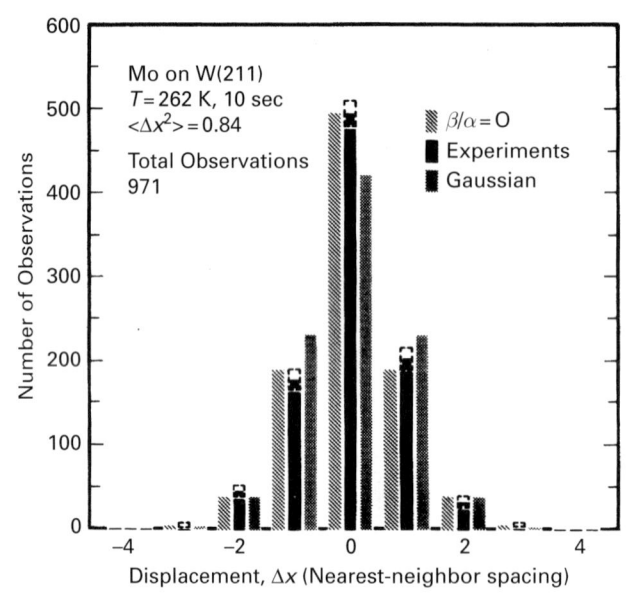

Fig. 3.85 Distribution of Mo adatom displacements on W(211) at 262 K [168]. Only nearest-neighbor transitions are found.

description of diffusion outlined in the previous section correctly describes atomic behavior – atoms migrated between nearest-neighbor sites. Only for iridium and rhodium diffusion, in Fig. 3.86, were small contributions from double jumps found, but these were too small to be really significant. These early studies were important in gaining experience, but also in confirming the simplest picture of diffusion.

The next significant step in the search for long jumps was taken by Senft [169–171]. She decided that long jumps were more likely to occur if energy exchanges between

3.3 Long jumps in surface diffusion

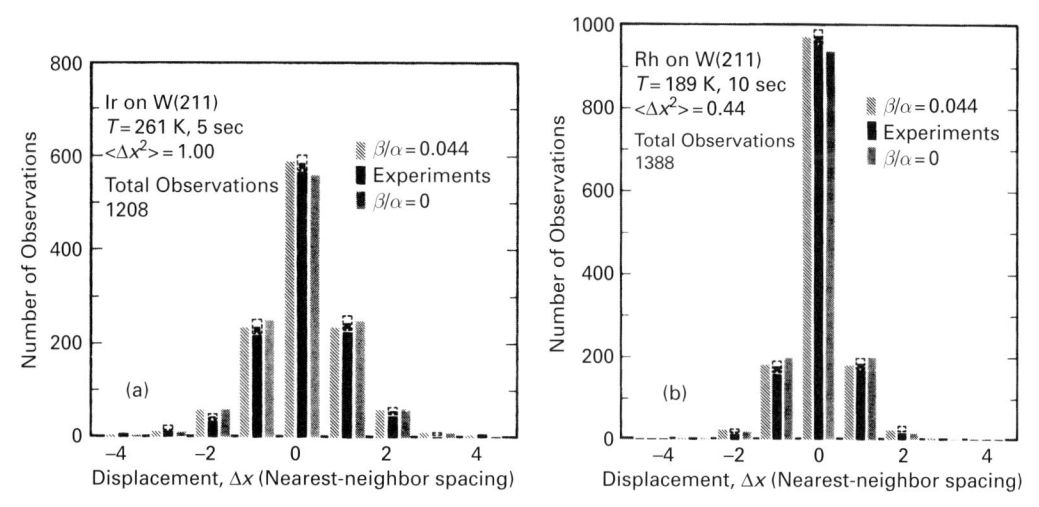

Fig. 3.86 Distribution of displacements for Ir and for Rh adatoms on W(211) plane, suggesting a tiny contribution of double jumps [168].

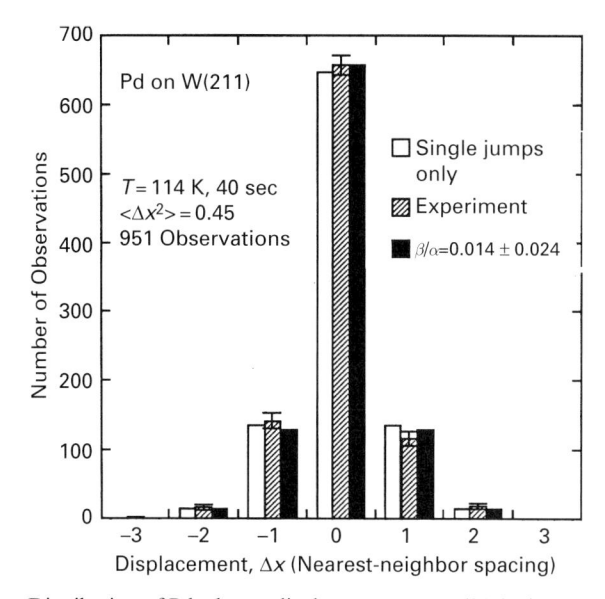

Fig. 3.87 Distribution of Pd adatom displacements on W(211) plane at 114 K, showing essentially single jumps only (after Senft [169]).

adatom and lattice were small, as then the atom might bounce after the first impact and continue farther on. She therefore picked palladium and nickel to study on W(211), as these metals have low barriers to diffusion, 0.314±0.006 eV for palladium and 0.46±0.01 eV for nickel, that should lead to long jumps. At 114 K, the distribution for displacements of palladium is shown in Fig. 3.87. Essentially only single jumps contribute to the distribution at this temperature; the ratio of double-to-single jumps amounted to only ~ 1%, too small to be significant. Measurements at 122 K revealed

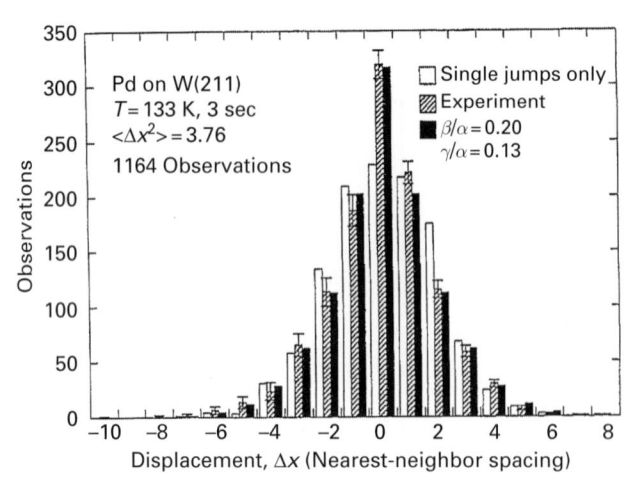

Fig. 3.88 Distribution of Pd atom displacements on W(211) at 133 K. Best fit with double/single jumps equal to 0.20, and triple/single of 0.13 (after Senft [171]).

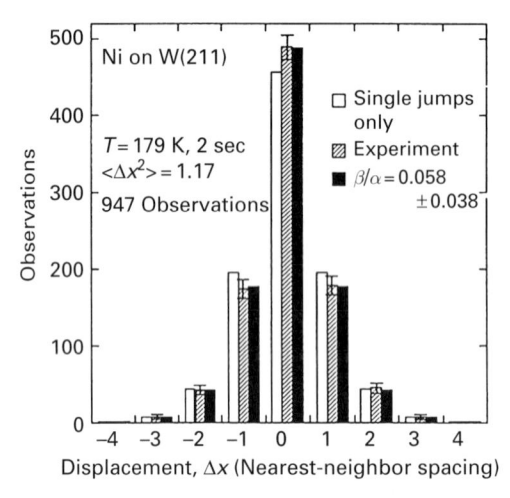

Fig. 3.89 Distribution of Ni adatom displacements on W(211) plane at 179 K. Best fit of observations with double/single jumps equal to 0.058 (after Senft [170]).

essentially the same situation, but everything changed dramatically at a slightly higher temperature of 133 K, as indicated by the distribution in Fig. 3.88. Now the experiments differed significantly from the distribution for only single jumps. The best fit was obtained for a ratio of double-to-single jumps of 0.20±0.06, and even triple jumps contributed, at a ratio of 0.13±0.04 to singles. This marked the first time that long jumps had been found in a real system. And it happened at quite a low temperature, at less than one twenty-fifth the melting point of tungsten. For the diffusion of nickel, Senft also found long jumps, as is clear from the diagram at 179 K in Fig. 3.89. In this system, the ratio of double-to-single jumps turned out to be 0.058±0.038. The behavior of tungsten self-diffusing on W(211), studied by Senft is quite different. As shown in

3.3 Long jumps in surface diffusion

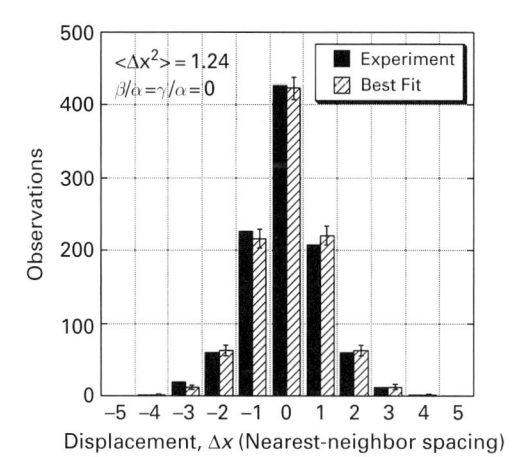

Fig. 3.90 Displacement distribution of W adatom on W(211) at 307 K. Best fit to field ion experiments derived with negligible contribution of β double or longer jumps (after Senft [170]).

Fig. 3.90 at 307 K, these experiments are best represented by a distribution with a quite negligible number of double jumps; only nearest-neighbor transitions mattered. However, Senft did not investigate higher temperatures for a tungsten adatom.

What is especially interesting about the results for palladium is the very strong dependence of the frequency of long jumps on temperature. A drop of only eleven degrees, to 122 K, makes all transitions other than single jumps vanish. Shortly before this work, Ferrando et al. [159] worked out a theory for the random motion of an atom over a periodic potential in a thermal bath subject to frictional forces, and came up with a description for the probability of long jumps. However, in this theory the occurrence of long transitions has only a small dependence upon the temperature, and is therefore not in accord with Senft's findings of a strong sensitivity to temperature, as if long jumps were characterized by a larger activation energy.

With the existence of long jumps established, at least in one-dimensional diffusion, attention now turned to the detailed temperature dependence. Linderoth et al. [172] in 1997 studied this in detail for the diffusion on the reconstructed Pt(110)-(1 × 2) surface, a model of which, showing wide channels, is given in Fig. 3.91. Linear clusters and single atoms can easily be seen on the STM image of this surface in Fig. 2.9. Motion was examined with the STM after deposition of a sub-monolayer quantity of platinum. An example of the distribution of displacements found at 375 K is shown in Fig. 3.92, for which the ratio of double-to-single jumps was reported as 0.095. From measurements at temperatures over an ~ 60 K range they came up with the Arrhenius plot in Fig. 3.93, which gave them an activation energy of 0.81±0.01 eV with a prefactor $v_0 = 10^{10.7 \pm 0.2}$ sec^{-1} for single jumps, and a slightly higher barrier of 0.89±0.06 eV and roughly the same prefactor, $10^{10.9 \pm 0.8}$ sec^{-1}, for what were presumed to be double jumps. This would represent the first estimate of the barrier to long jumps, but the situation changed rapidly.

The next year, Montalenti and Ferrando [173] did molecular dynamics simulations on the (1 × 2) reconstructed Au(110) plane, modeled in terms of the RGL potential

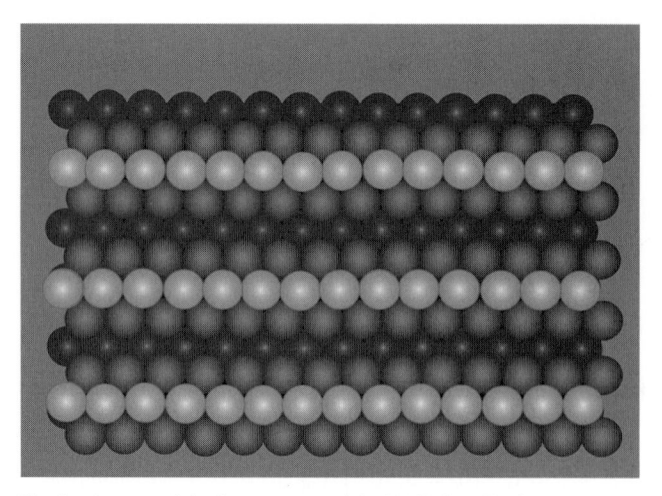

Fig. 3.91 Hard-sphere model of reconstructed fcc(110)-(1 × 2) plane.

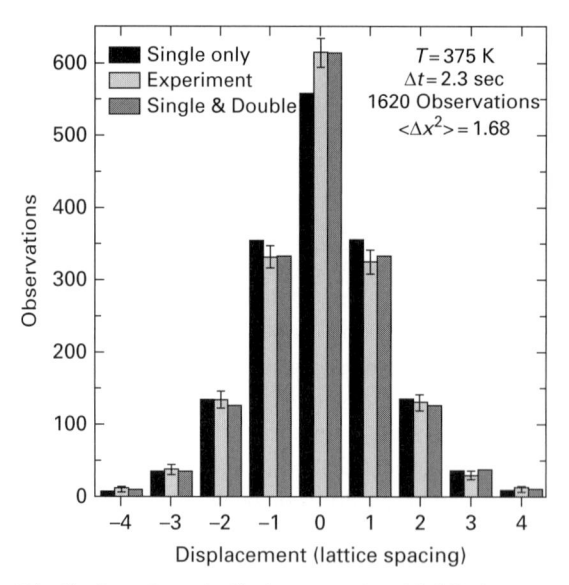

Fig. 3.92 Distribution of atomic displacements in self-diffusion on Pt(110)-(1 × 2) plane. Best fit obtained with ratio of "double" to single jumps of 0.095 (after Linderoth et al. [172]).

[174,175]. What they found is shown in Fig. 3.94. Jumps could take place in a straight line on the bottom of the channels; these could span one, two or three spacings. What they also observed was that atoms could carry out a metastable walk, in which an atom jumps up the (111) channel walls and continues diffusion there. The number of these metastable walks exceeded the number of long jumps at the bottom at all temperatures, from 450 K to 625 K, and ~ 20% of the transitions over two spacings occurred by metastable walks. Although Montalenti and Ferrando made no inferences concerning the results of Linderoth et al. [172], this happened the next year.

3.3 Long jumps in surface diffusion

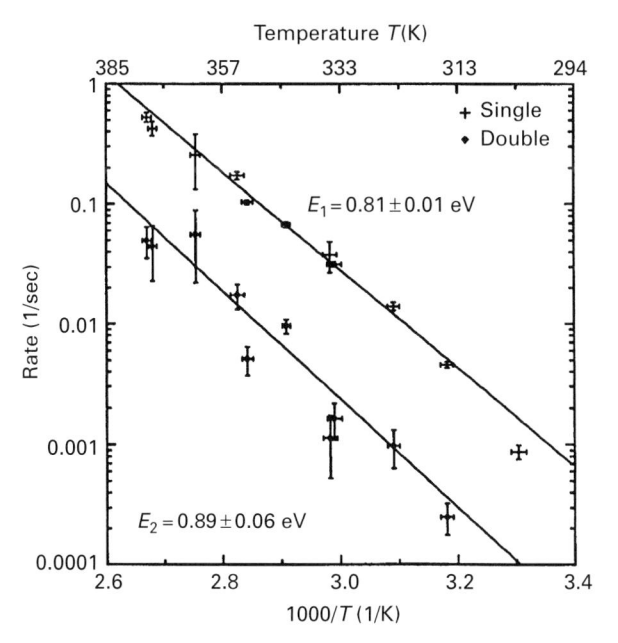

Fig. 3.93 Arrhenius plot for single and double jump rates presumed to occur in self-diffusion on Pt(110)-(1 × 2) (after Linderoth et al. [172]).

Lorensen et al. [176] in 1999 carried out density-functional calculations of the energy barriers in diffusion on the reconstructed Pt(110)-(1 × 2). They again found that not only could atoms jump in a straight line along the bottom of the channels, but, as shown in Fig. 3.95, they could also carry out metastable walks on the (111) facets of the channel sides, as had been reported earlier by Montalenti and Ferrando [173], with energetics indicated in Fig. 3.96. The conclusion of this work was that metastable walks are a likelier explanation than straight-line long jumps along the bottom for the displacement distribution reported by Linderoth et al. [172]. Based on this study, it appears that so far mostly single jumps were observed in self-diffusion on the Pt(110)-(1 × 2) reconstructed surface.

3.3.4 Later studies

At this stage, long jumps had been experimentally found only on channeled surfaces, on W(211). In 2002, however, Oh et al. [177] looked at the diffusion of palladium on W(110), to check if long jumps were limited to channeled planes and to provide a comparison with previous work. What had to be determined first was what sort of jumps might be expected on this surface. Possibilities, already depicted in Fig. 2.21, are single jumps to nearest-neighbor positions along the <111> direction at the rate α, double jumps to next-nearest-neighbor positions in the same direction at the rate β, and horizontal jumps at the rate δ_x to the closest adsorption site along direction <100>, as well as vertical transitions in direction <110> at the rate δ_y.

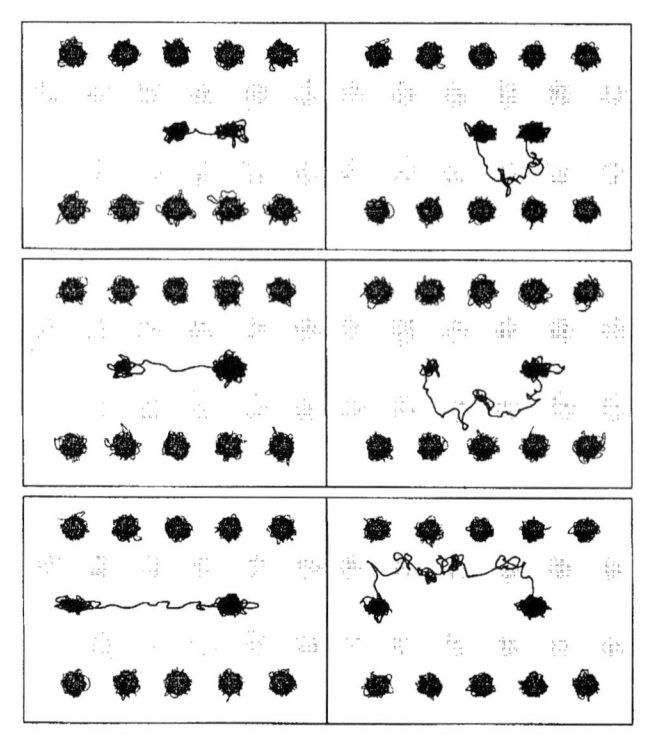

Fig. 3.94 Trajectories of Au adatom on reconstructed Au(110) surface with <110> rows missing; obtained in molecular dynamics simulations using RGL interactions at 450 K. Left column: in-channel jumps single, double, and triple. Right column: metastable transitions single, double, and triple (after Montalenti and Ferrando [173]).

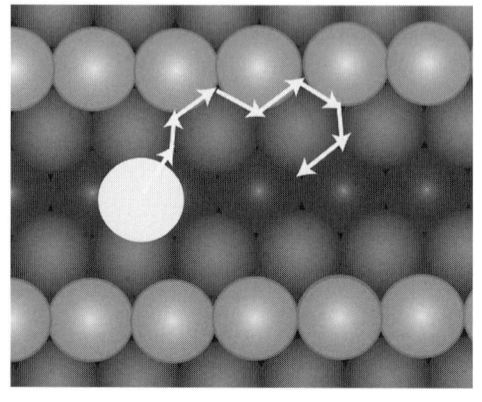

Fig. 3.95 Hard-sphere model of Pt(110)-(1 × 2) surface, showing metastable excursion of adatom on channel walls.

The Arrhenius plot for palladium on W(110) in Fig. 3.97 gives no indication of anything unusual – the experiments are nicely fitted by one straight line. At low temperatures, at $T \leq 185$K, there are only small contributions from long jumps. By 210 K, however, for the distribution shown in Fig. 3.98, before any correction for

3.3 Long jumps in surface diffusion

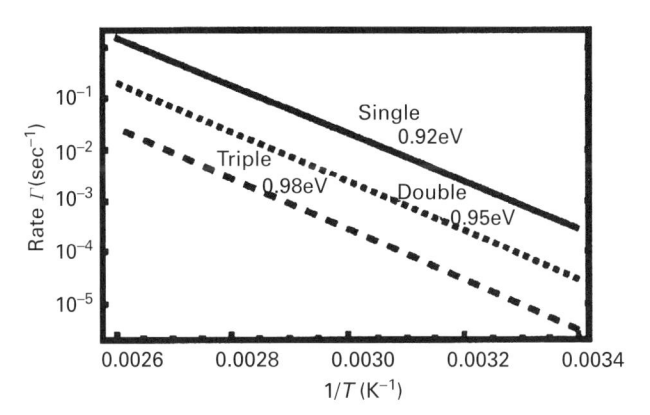

Fig. 3.96 Arrhenius plots for single length, double length, and triple length excursions of Pt adatom on Pt(110)-(1×2), obtained by density-functional calculations (after Lorensen *et al.* [176]).

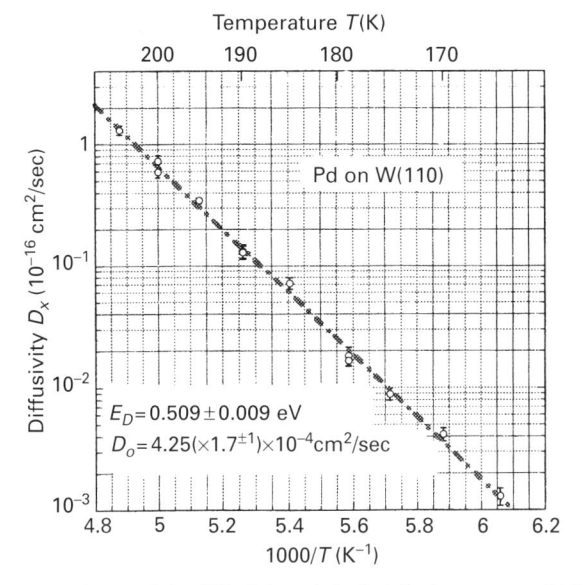

Fig. 3.97 Dependence of the diffusivity of single Pd adatom on W(110) upon the reciprocal temperature [177]. Data corrected for edge effects as well as for migration during transients.

transients in the temperature, the best fit is obtained with significant contributions of double as well as vertical jumps. The final evaluation, after correction for transients, gave $\beta/\alpha = 0.12 \pm 0.06$ and $\delta_y/\alpha = 0.11 \pm 0.10$, where the errors are only rough estimates. These experiments for the first time demonstrated that various long jumps made significant contributions in two-dimensional diffusion even at relatively low temperatures, at less than $0.06 T_m$. What is worth noting is the huge difference between vertical and horizontal transitions for this system, which made it possible to derive a significant energy difference between the two kinds of jumps, as is illustrated in Fig. 3.99, that amounted to ~0.6 eV.

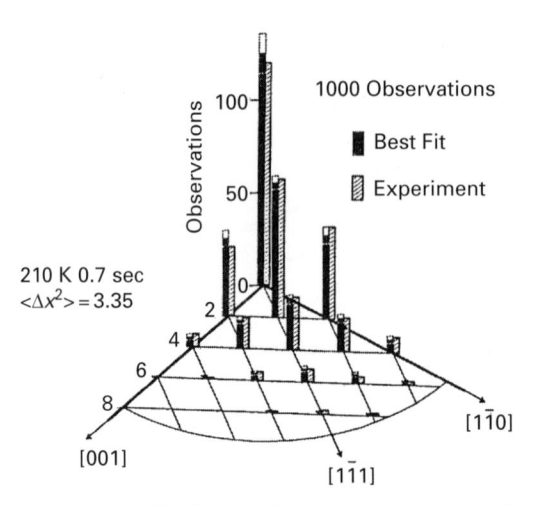

Fig. 3.98 Displacement distribution of Pd adatom on W(110) plane at 210 K [177]. Double/single jump rate β/α is 0.12, δ_y/α is 0.11.

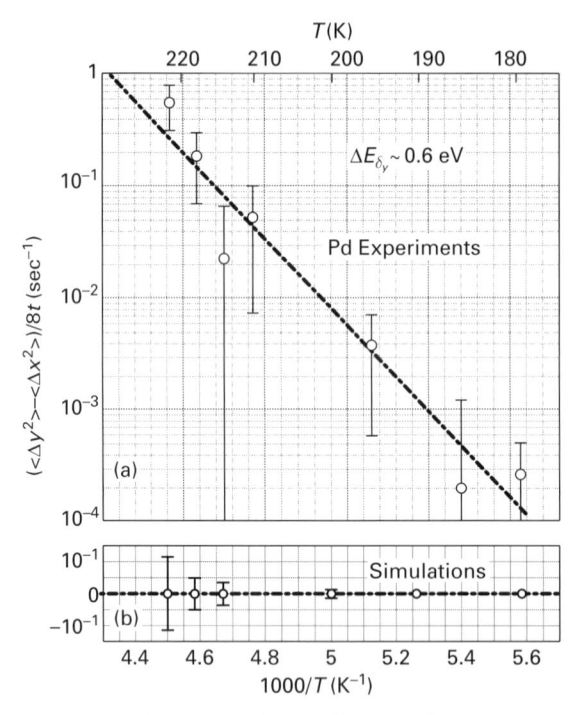

Fig. 3.99 Temperature dependence of the difference in the mean-square displacement of Pd on W(110) along x-and y-axis [177]. In simulations in (b) $\langle \Delta x^2 \rangle = \langle \Delta y^2 \rangle$.

What was particularly surprising is that the vertical transitions are different from the horizontal transitions. How do they occur? Looking at Fig. 3.100, it is clear that atoms could jump directly along <110> and <001> directions, moving over a lattice atom, or else, a more likely alternative is that the atom starts along the energetically easiest

3.3 Long jumps in surface diffusion

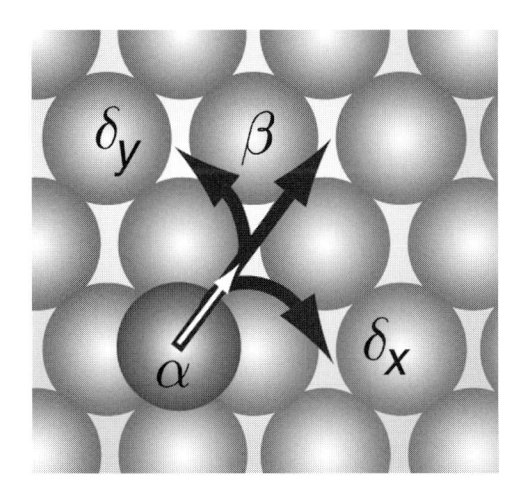

Fig. 3.100 Schematic of jump rates on W(110) plane [179]. Basic jump is α transition along <111>. At elevated temperatures, this can proceed beyond nearest-neighbor end point, giving β, δ_x, or δ_y.

direction, which is <111>, and is then turned aside by interactions with the lattice. A turn in one direction produces a vertical jump. A turn in the other direction, to give a horizontal displacement, involves a larger deviation from the original direction of the jump and is therefore less likely.

The same year Oh *et al.* [178] extended work to the self-diffusion of tungsten on the (110) plane of tungsten, and were able to demonstrate strong contributions from long jumps for this system. The trends were similar to what had previously been found for palladium diffusion. At 350 K, only vertical transitions made a contribution beyond nearest-neighbor single jumps, with a ratio $\delta_y/\alpha \sim 0.1$. At 364 K the situation changed completely, as is apparent in Fig. 3.101. The ratio of rates β/α now amounted to 0.22, $\delta_x/\alpha = 0.36$, and $\delta_y/\alpha = 0.43$. Long jumps constitute roughly half the total number of transitions. It must be stressed here, however, that at elevated temperatures, adatom displacements take place during the rise in temperature before the temperature set for diffusion is reached, and also during the period at the end, after the heating current is shut off. The distribution of displacements occurring during these transients has to be separately determined in "zero-time" experiments, and the final rates are derived taking advantage of Eq. (2.12).

The important question now remained, what are the energetics of long jumps? This question was addressed by Antczak [179] in 2004, who again worked with single tungsten atoms on the W(110) plane. The Arrhenius plot for the diffusion of tungsten is given in Fig. 3.102. Again there is nothing unusual in this plot, with an activation energy of 0.92±0.02 eV along the <001> direction and almost the same barrier, 0.93±0.01 along the vertical <110>. In the Arrhenius plot there is no clear indication of the long jumps which are known to be present. To get a true picture of what is happening, the distributions of displacements at several temperatures, as

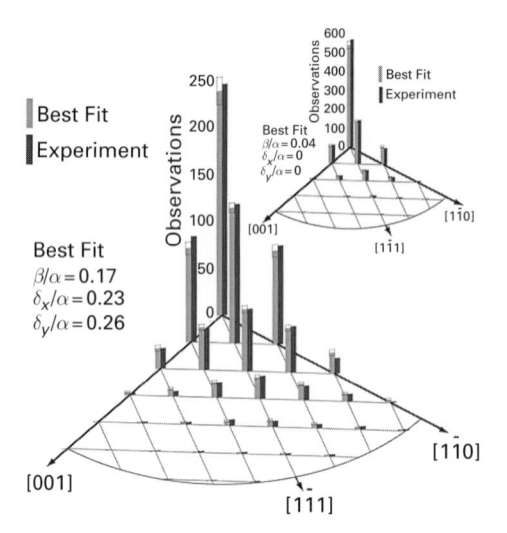

Fig. 3.101 Distribution of tungsten displacements on W(110) at 364 K [179]. Best fit before corrections for "zero-time" effects obtained with significant contributions from long jumps. Inset gives distribution during temperature transients. Rates not corrected for effects from transients.

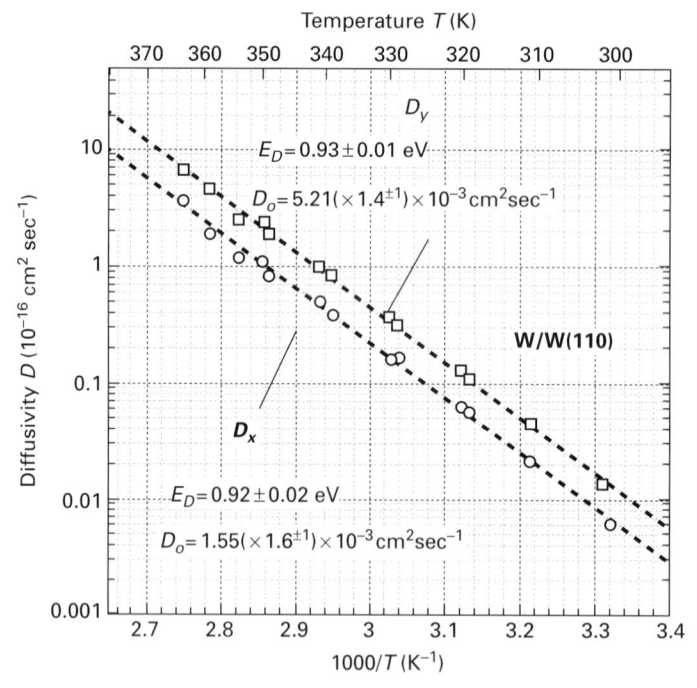

Fig. 3.102 Arrhenius plot for diffusivities of W atom on W(110) along <100> and <110> direction [179]. Best fit is obtained with straight lines.

well as the distributions during temperature transients consisting of at least 1000 observations each were measured. In order to understand what is going on here it is best to look at plots of the rates against $1/T$. In Fig. 3.103 is shown the rate of single α jumps along <111> directions. At low temperatures, the natural log of the single

3.3 Long jumps in surface diffusion

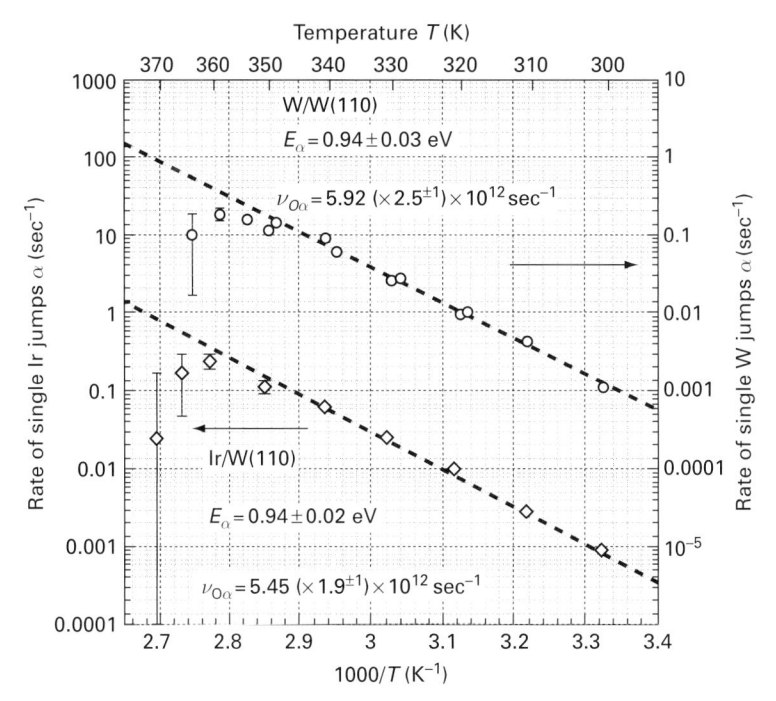

Fig. 3.103 Arrhenius plots for rates of single α jumps of W and Ir adatom on W(110) [11,179]. Single jump rate is fitted at low temperatures. At elevated temperatures, rate drops significantly below this straight line.

jump rate α follows nicely along a straight line. Above 340 K the measurements begin to fall below the Arrhenius curve, and at 365 K the rate α has actually turned down.

This plot shows clearly that the number of single jumps is influenced by the presence of other jumps in the movement over the surface. This is not what happens with the rate of double jumps β, in Fig. 3.104, or the horizontal and vertical rates, in Fig. 3.105 and 3.106 respectively. They all show a linear dependence on reciprocal temperature over the investigated temperature range.

We now have available in Table 3.3 estimates for energies and prefactors for all the tungsten jump rates. What is surprising is that the energy required to make long jumps is at least 30% higher than for singles, and the prefactors obtained for long jumps from the intersection of the Arrhenius plots are 10^3 higher than the standard factors. We can better understand what is going on here by abandoning the notion that the various jumps take place independently. At low temperatures, only nearest-neighbor transitions are possible, but that changes as the temperature is raised and correlations between different types of jumps become important. A possible explanation of the correlation is illustrated in Fig. 3.100. An adatom moving toward a nearest-neighbor site at a higher temperature (having a higher thermal energy) may continue to move, either to a next-nearest-neighbor position along <111> giving a β jump, or the adatom path can deviate to the left to create a δ_y transition, or to the right, to form δ_x. In such a model the total

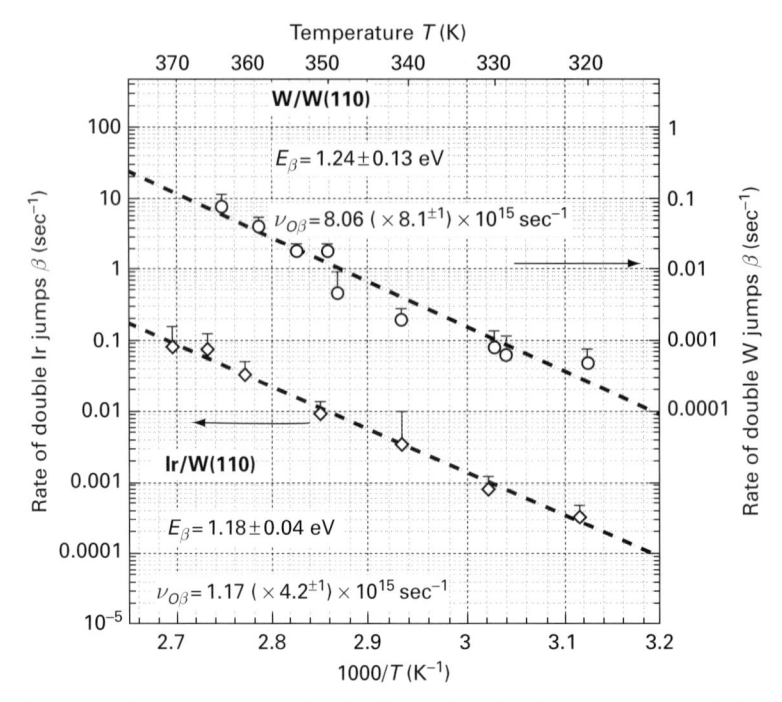

Fig. 3.104 Arrhenius plots for the rate of double β jumps for W and Ir adatoms on W(110) [11,179].

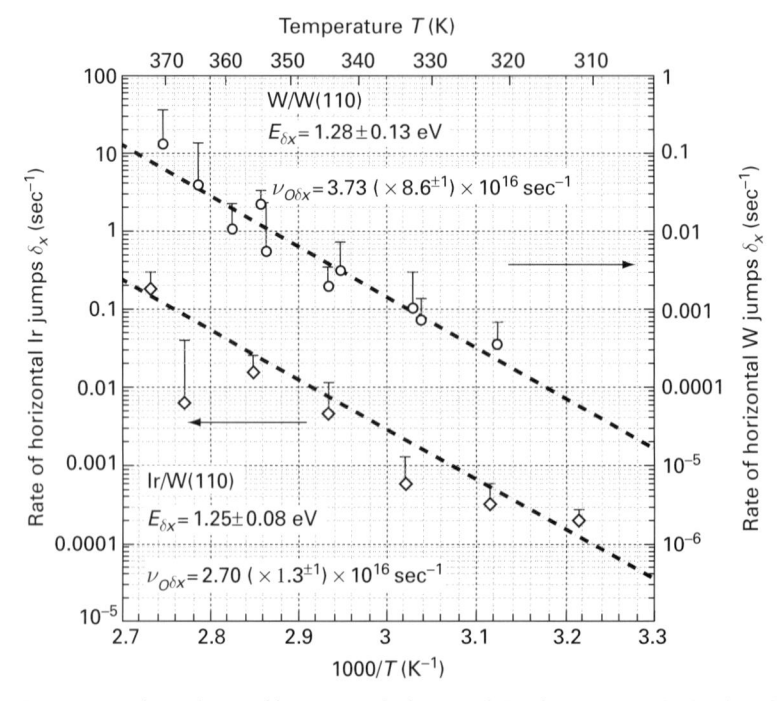

Fig. 3.105 Temperature dependence of jump rates δ_x for Ir and W adatoms on W(110) plane [11,179].

3.3 Long jumps in surface diffusion

Table 3.3 Rate parameters for W adatoms on the W(110) plane [179].

Rate	Activation Energy E_D (eV)	Frequency Prefactor v_0(sec^{-1})
α (low T)	0.94±0.03	5.92($\times 2.5^{\pm 1}$)× 10^{12}
β	1.24±0.13	8.06($\times 8.1^{\pm 1}$)× 10^{15}
δ_x	1.28±0.13	3.78($\times 8.6^{\pm 1}$)× 10^{16}
δ_y	1.37±0.13	1.00($\times 4.4^{\pm 1}$)× 10^{19}

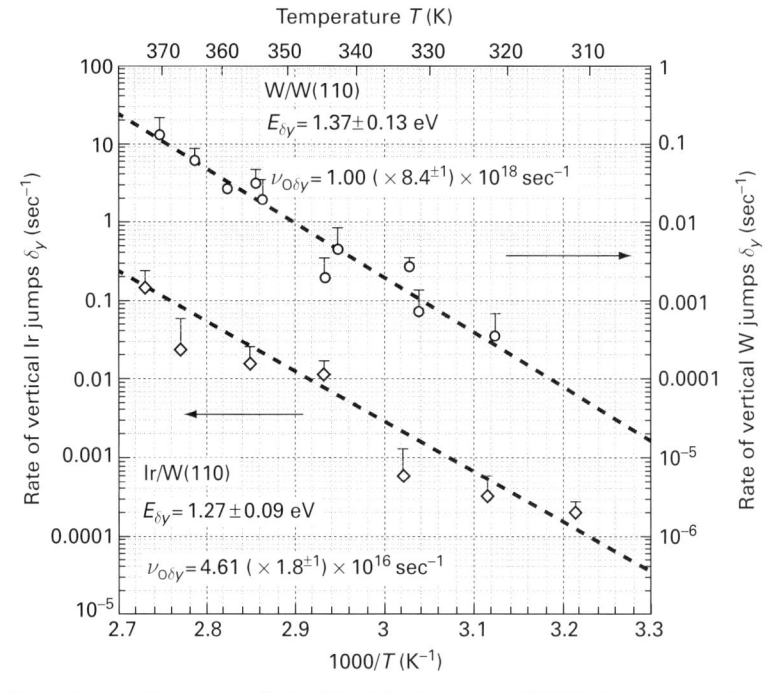

Fig 3.106 Dependence of jump rates δ_y for W and Ir adatoms on W(110) plane upon reciprocal temperature [11,179].

number of jumps the adatom makes at any temperature is given by the number of basic jumps R_b in the <111> direction. The different types of jumps originate from an attempt to make a transition to a nearest neighbor. The rate of long jumps R_ℓ can therefore be written as

$$R_\ell = p_{b\ell} R_b, \tag{3.10}$$

where R_b is the rate of basic jumps, derived by extrapolation of the low temperature rate for single jumps to higher temperatures. The probability $p_{b\ell}$ of converting from a single jump can be represented as

$$p_{b\ell} = v_\ell \exp\left(-\frac{E_\ell}{kT}\right) / \sum_j v_j \exp\left(-\frac{E_j}{kT}\right) = v_{b\ell} \exp\left(-\frac{E_{b\ell}}{kT}\right), \tag{3.11}$$

where $E_{b\ell}$ is the activation energy for this conversion, and j labels the various transitions possible. This makes the high activation energies and frequency factors for long jumps understandable. The activation energy E_ℓ for a given rate R_ℓ is now just the sum of the activation energies for the basic jump plus the transition, and the prefactor is the product of the individual prefactors. Furthermore, it is also clear why the single jump rate α diminishes at higher temperatures. Every time a long jump takes place, it diminishes the number of single jumps, since every long jump is a continuation of a single jump to a next-nearest-neighbor position, as illustrated in Fig. 3.100. Finally, if all the long jumps are related to nearest-neighbor transitions, in the way described above, then an Arrhenius plot of the sum of all the rates should plot as a straight line. As shown in Fig. 3.107, this is what actually happens, and the activation energy as well as prefactor are in reasonable agreement with the values found previously for singles alone at low temperatures as well as from the temperature dependence of the diffusivity. This all shows that at temperatures up to 370 K, the second-range transitions (double, vertical and horizontal jumps) replace single jumps in diffusion. It is likely that such a scenario will occur for longer transitions as well and that longer jumps will be eliminating shorter ones and taking over the leading role in transport at a certain temperature, gradually changing discrete motion into a continuous one along the channels on the surface.

To check if this behavior is unique to self-diffusion, Antczak [11] has also looked at diffusion of iridium on W(110). As shown in Fig. 3.108, the activation energy for iridium

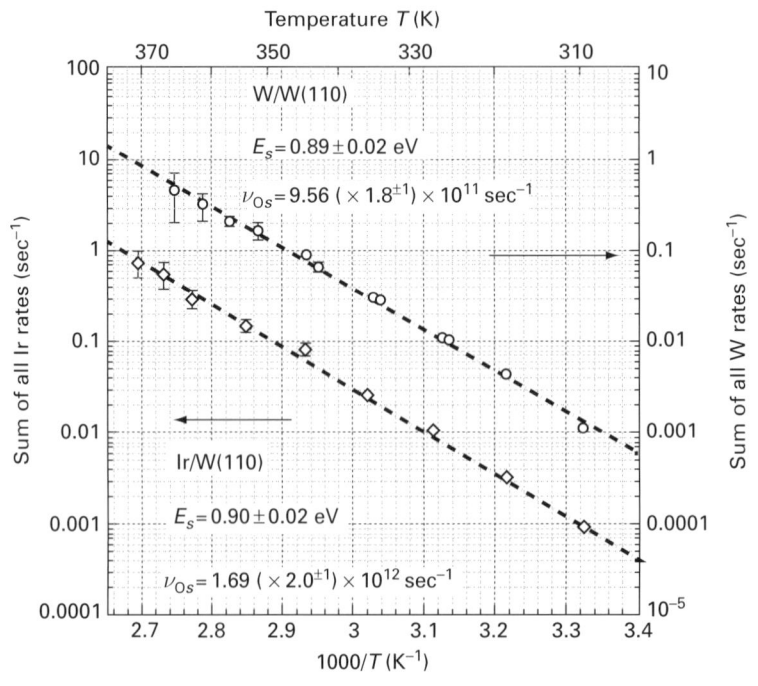

Fig. 3.107 Arrhenius plot of the sum of all jump rates for W and also Ir adatoms on W(110), giving a straight line fit [11,179].

3.3 Long jumps in surface diffusion

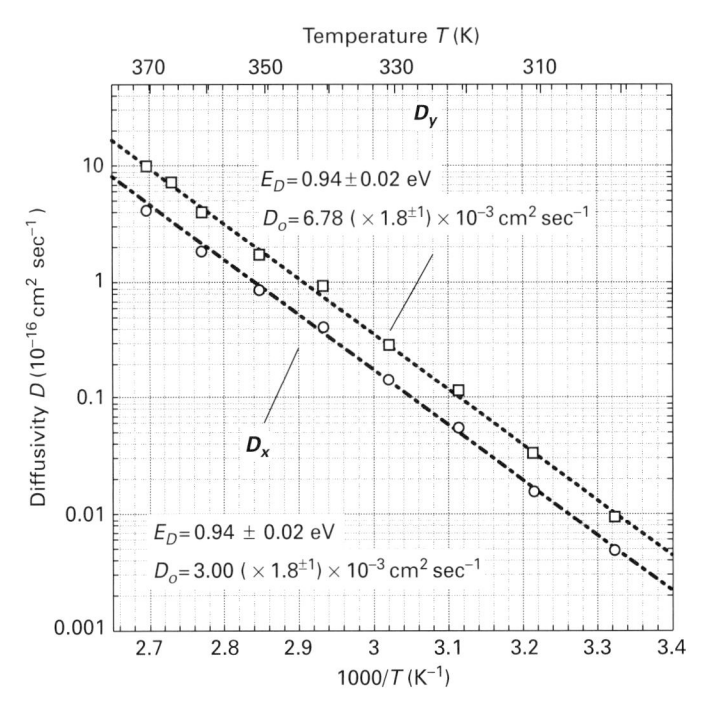

Fig. 3.108 Arrhenius plot for Ir adatom diffusivities on W(110) plane along <100> and <110> [11].

diffusion is very similar to tungsten.[4] The activation energy for motion along <001> and <110> is now the same, 0.94±0.02 eV, and the prefactor along <110> is roughly a factor of two larger than along <001>, as it should be since the inter-atomic spacing in the <110> direction is $\sqrt{2}$ larger. The distribution measured at a low temperature, at 301 K, shown in Fig. 3.109, is essentially due to single jumps between nearest-neighbor positions. As the temperature is raised, the contribution from long jumps increases, as already seen previously. In Fig. 2.23 is shown the displacement distribution at 360 K, together with the distribution obtained for "zero-time" measurements to catch transitions during temperature transients. The contribution from long jumps has now become significant, with a best fit to the experiments obtained for $\beta/\alpha = 0.15, \delta_x/\alpha = 0.38$, and $\delta_y/\alpha = 0.32$. This trend continues as the temperature is raised further. All the jump parameters for Ir are given in Table 3.4.

If the jump rates for iridium are compared with those previously established for tungsten on the same plane they are found to be quite similar, as can be seen in Figs. 3.103–3.107. Activation energies as well as prefactors for long jumps of iridium are again high. The single jump rate α, shown in Fig. 3.103, again decreases as the diffusion temperature reaches higher values, and just as previously, the sum of all the jump rates plots linearly on an Arrhenius plot in Fig. 3.107, with an activation energy very close to that of single jumps obtained at low temperatures and also derived from the

---

[4] The possibility of exchange was ruled out by measurements of the evaporation voltage after diffusion.

Table 3.4 Rate parameters for Ir adatoms on the W(110) plane [11].

Rate	Activation Energy E_D (eV)	Frequency Prefactor $v_0 (\text{sec}^{-1})$
α(low T)	0.94±0.02	5.46($\times 1.9^{\pm 1}$)× 10^{12}
β	1.18±0.04	1.17($\times 4.2^{\pm 1}$)× 10^{15}
δ_x	1.25±0.08	2.70($\times 1.3^{\pm 1}$)× 10^{16}
δ_y	1.27±0.09	4.61($\times 1.8^{\pm 1}$)× 10^{16}

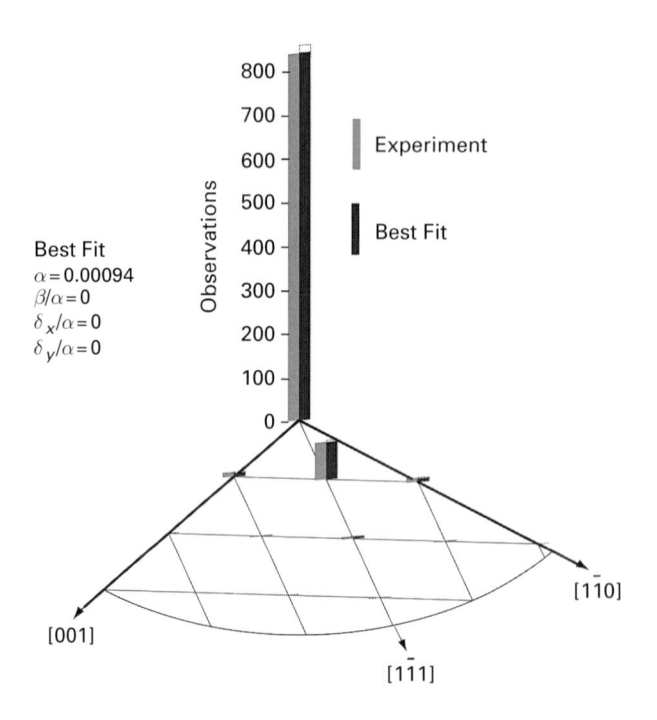

Fig. 3.109 Distribution of Ir atom displacements on W(110) at 301 K, revealing only nearest-neighbor jumps [11].

diffusivity. The one big difference is that the vertical δ_y and the horizontal δ_x jump rates are now essentially equal.

It turns out that the difference in the activation energy required for vertical and horizontal jumps on W(110) depends on the chemical identity of the adatom moving over the surface. The energy difference as a function of the atomic mass of the adatom is plotted in Fig. 3.110 and gives a linear dependence [180]. Lighter atoms like Pd very strongly feel the anisotropy of the W(110) surface, while heavier atoms like Ir are not

3.3 Long jumps in surface diffusion

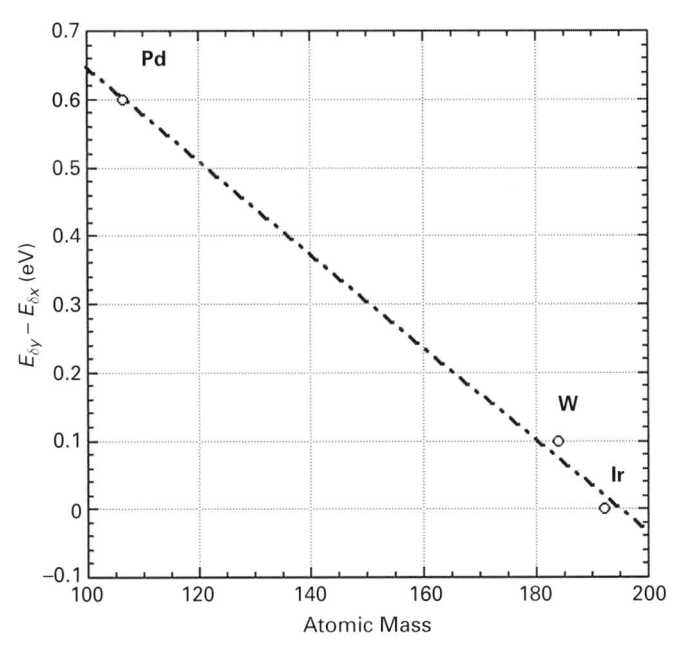

Fig. 3.110 Difference in activation energies for transitions along y- and x-axes of W(110) as a function of adatom mass [180].

much influenced. So far information is available for only three types of adatoms; further investigation is definitely necessary.

On W(110), in addition to single jumps, a second range of transitions including β double, vertical δ_y, and horizontal δ_x was detected and it was demonstrated that this second-range transition started to play a leading role in the movement of atoms over the surface at quite a low temperature, around 1/10 of the melting point. It is just possible that at higher temperatures, longer transitions will in turn start to be present in the movement of adatom.

Both iridium and tungsten atoms show much the same behavior in diffusion on W(110). Can this similarity be transcribed to other surfaces as well? To check this, self-diffusion has also been examined by Antczak [181] on the one-dimensional W(211) plane, on which a tungsten atom diffuses in a single <111> channel. Displacement distributions were measured at increasing temperatures, and were later corrected in "zero-time" experiments. At 295 K only atom jumps between nearest-neighbor sites were detected. However, at 300 K and higher temperatures, jumps spanning two spacings along <111> became more important, as is clear from Fig. 2.22. Up to a temperature of 325 K no triple jumps were detected, but they could be active at higher temperatures. The rates of single jumps, adjusted for "zero-time" results, are shown in Fig. 3.111; what is striking here is that for temperatures above 300 K, the rate falls very rapidly below the values extrapolated from lower temperatures. For the lower temperatures, the activation energy for single jumps is 0.84±0.06 eV, with a frequency prefactor of 2.2 $(\times 11.3^{\pm 1}) \times 10^{13}$ sec^{-1}, close to what was derived from the temperature dependence of

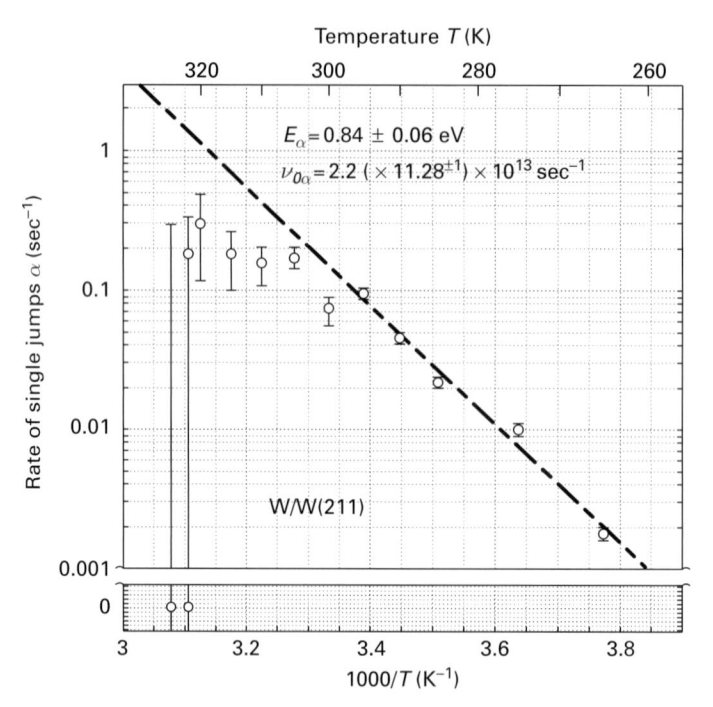

Fig. 3.111 Arrhenius plot for α single jumps of W on W(211) [181]. At $T > 320$ K, rate α has decreased so much it is no longer discernable.

the diffusivity, shown in Fig. 2.20. The rate of double jumps, plotted in Fig. 3.112 for the different temperatures, gives a reasonable straight line, with an activation energy of 1.44 ±0.13 eV and a much larger prefactor, $7.9(\times 127^{\pm 1}) \times 10^{21}$ sec^{-1}, effects similar to what was found on W(110). As was already discovered in the self-diffusion of tungsten atoms on W(110) above 350 K, the single jump rate α descends below that extrapolated from low temperatures, while the other rates β, δ_x, and δ_y do not – they follow normal Arrhenius plots. The explanation for diffusion on W(211) is similar to that for W atoms on the W(110) plane – the transitions are not independent and all arise from one basic jump process. At higher temperatures single jumps are gradually replaced by longer transitions.

The sum of all the rates on W(110) with both tungsten and iridium adatoms yielded good Arrhenius plots, with characteristics similar to what was derived for diffusivities. When the sum of single and double jumps detected on W(211) is plotted against $1/T$, as in Fig. 3.113, the situation is seen to be different: at higher temperatures the sum falls significantly below the line representing the total number of jumps extrapolated from low temperatures. Double jumps evidently do not replace single jumps completely. As indicated in the model in Fig. 3.114, an atom jumping at a nearest-neighbor site can do three things. It can equilibrate at that site, giving a single transition. It may also continue to a site two spacings away, in a double jump, or else it can rebound and return to the origin, to the starting position. Such rebounds have been detected in molecular dynamics

3.3 Long jumps in surface diffusion

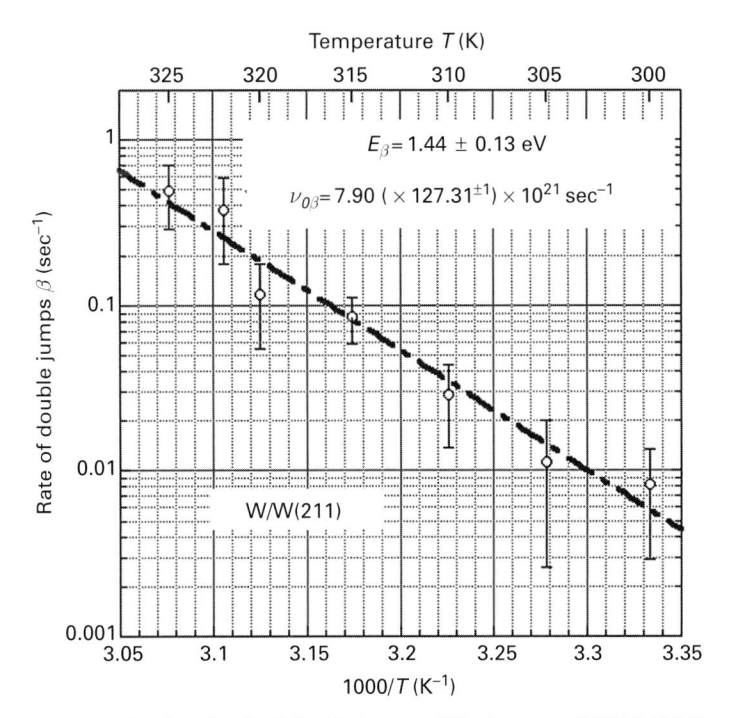

Fig. 3.112 Normal Arrhenius plot for β double jumps of W adatom on W(211) [181].

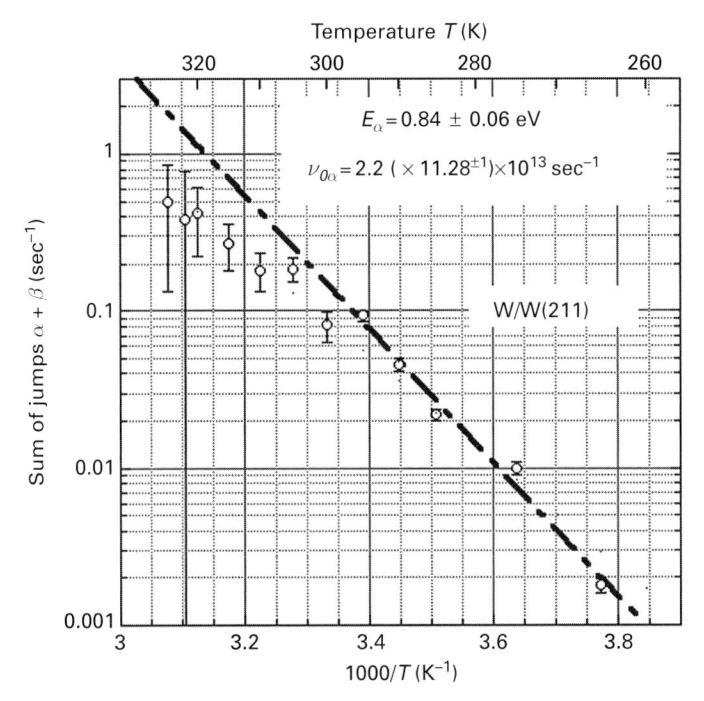

Fig. 3.113 Arrhenius plot for the sum of all jump rates measured for W adatom on W(211); for temperatures ≥ 310 K, rate falls below linear plot [181].

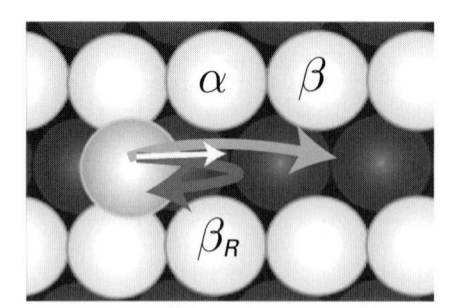

Fig. 3.114 Schematic of different types of jump processes for an adatom on W(211) [181]. β_R denotes rebound transition.

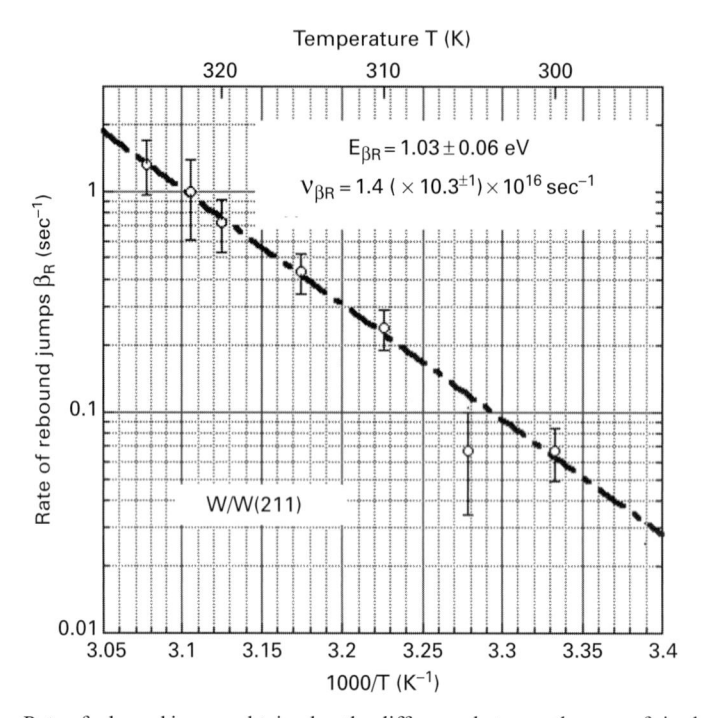

Fig. 3.115 Rate of rebound jumps, obtained as the difference between the sum of single plus double jumps and total jump rate extrapolated from low temperatures [183].

simulations of diffusion on a bcc(211) as well as on fcc(111) planes [158,165,182] and do not result in a displacement but they cost energy. That is why it is important to include them in an atomistic description of diffusion.

These rebound transitions would not be seen directly in the distribution of displacements as their displacements are zero. Nevertheless, Antczak [183] has for the first time ever also been able to determine the characteristics of the rebound transition. They can be found as the difference between the total rate of jumps, obtained by extrapolating the jump rate from low temperatures, and the sum of single plus double transitions. This rate of rebounds is plotted in Fig. 3.115 as a function of the reciprocal temperature; rebounds

3.3 Long jumps in surface diffusion

Table 3.5 Rate parameters for W adatoms on the W(211) plane [183].

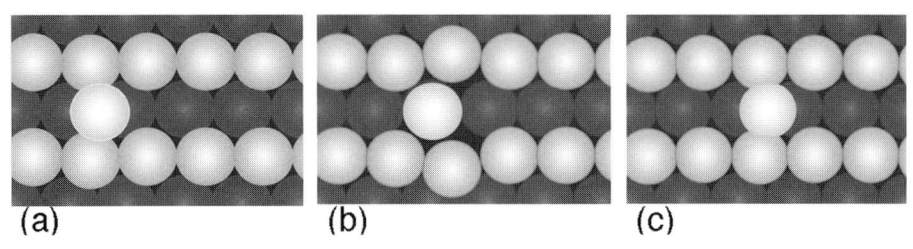

Rate	Activation Energy E_D (eV)	Frequency Prefactor $v_0(\sec^{-1})$
α(low T)	0.84 ± 0.06	2.2(× 11.3$^{\pm 1}$)× 10^{13}
β	1.44 ± 0.13	7.9(× 127$^{\pm 1}$)× 10^{21}
β_R	1.03 ± 0.06	1.40(× 10.3$^{\pm 1}$)× 10^{16}

Fig. 3.116 Model showing adatom movement on W(211) as lattice atoms at the sides move out of the way [183].

occur with an activation energy of 1.03±0.06 eV and a frequency prefactor of 1.40 (× 10.3$^{\pm 1}$)× 10^{16} sec^{-1}. That is, the characteristics for rebounds lie between those of single and double jumps on W(211). The existence of rebound jumps is proof of how important the dynamic of the lattice are: rebounds are caused by the correlation between adatom movement and the vibrations of the lattice. That is also clear from Fig. 3.116, showing how an atom moves down a channel as lattice atoms at the sides open up. The energetics of all jumps in self-diffusion on W(211) up to 325 K are presented in Table 3.5, but the nature of the jumps will probably change as the temperature is increased.

Here we just want to note briefly that long jumps have also been observed for entities larger than adatoms. The first experimental indications of long jumps of bigger clusters with an open structure – Re$_5$ diffusing on the W(110) surface – was obtained by Fink in 1985 [184]. They also were seen for Ir$_{19}$ clusters on Ir(111) surface by Wang *et al.* [185] and later for decacyclene and hexa-tert-butyl decacyclene by Schunack *et al.* [186]. Indications of long jumps were similarly found in molecular dynamics simulations of diffusion along straight steps of Cu(111) by Marinica *et al.* [187], shown in Fig. 3.117, but this has not yet been experimentally investigated.

What is clear from this survey is that long jumps play an important part in surface diffusion, especially at elevated temperatures. Observations of long jumps turn out not to be limited to certain surface structures. The temperature for detecting these transitions is relatively low, so that the existence of long jumps has to be taken into account as one of the significant events on the surface. Nearest-neighbor jumps, which are widely used in modeling surface processes, disappear almost completely at relatively low temperatures. That at least is the case for diffusion on tungsten. What must still be established now is that similar effects occur on other materials, but this is a task for the future.

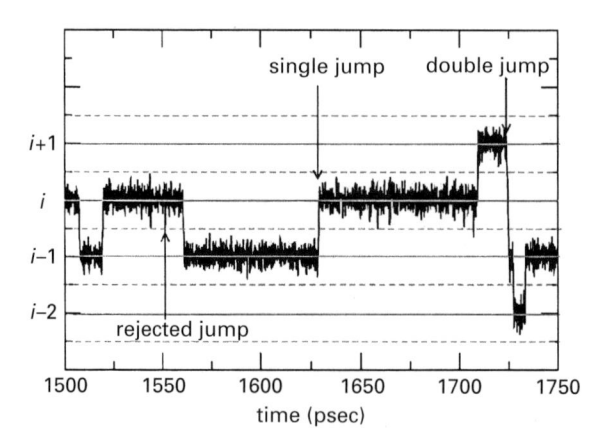

Fig. 3.117 Time evolution of the position of a Cu adatom at a step on Cu(111) plane, showing single, double, and rejected jumps. Results from simulations based on EAM potentials (after Marinica *et al.* [187]).

3.4 Transient diffusion

So far we have concentrated on the atomic processes significant in diffusion in a surface layer in thermal equilibrium. This, however, is not the only way in which diffusion may occur. When atoms strike a surface from the vapor and are trapped there, the energy of condensation still has to be transferred to the lattice. Part of this energy may activate the trapped atom to diffuse over the surface, in what traditionally is referred to as transient diffusion. This is a subject that has long been of interest, as transient diffusion could be important in reactions at low temperatures, where thermal diffusion is stopped.

3.4.1 Transient motion of adatoms

Experiments to detect the diffusion of atoms after striking a cold surface are not easy and require modern surface analytical techniques, but there were rough, early studies made to explore this phenomenon. The first attempts [188–191] to more seriously come to grips with these problems, around 1960, were usually theoretical in nature. Estimates were made for atom collisions with the end of a linear chain of atoms at rest, that is, with a one-dimensional crystal. A discussion of the behavior of atoms after trapping was given by McCarroll [192], who, as shown in Fig. 3.118, determined the time such an atom took to come to equilibrium. It was found that after only three vibrations, 98% of the energy of condensation had been transferred to the one-dimensional lattice. Of course these one-dimensional models provide only a crude pattern for the properties of real systems, but experiments began to get started. In 1963, a tungsten emitter tip in a field ion microscope was shadowed from the side with tungsten atoms, as depicted in Fig. 3.119, and was then imaged [190,191,193]. For extensive transient diffusion, a fairly uniform layer over the surface would be expected. Instead what was found is shown in Fig. 3.120. Some atoms penetrated into the (110) region, but only a few. In view of the poorly defined geometry of the evaporator in relation to the emitter, the only thing clear is that transient diffusion of tungsten on a tungsten surface has at best quite a limited range.

3.4 Transient diffusion

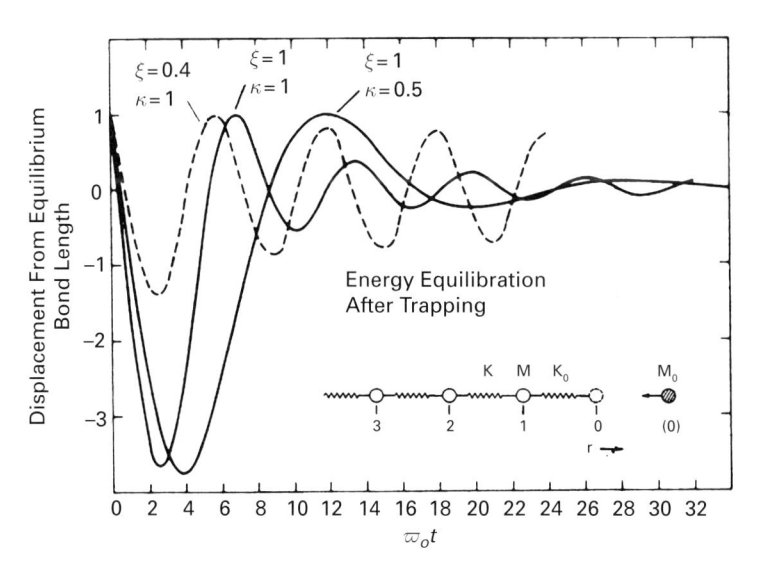

Fig. 3.118 Time for equilibration of energy in collision of different gas atoms with one-dimensional lattice [191,192]. ξ is the ratio of the mass for an incident compared with a lattice atom, κ is ratio of force constants.

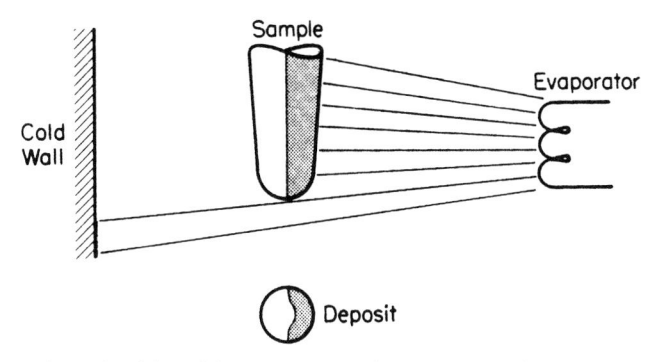

Fig. 3.119 Schematic of deposition arrangement for metal atoms in FIM [193].

Further experiments on tungsten were undertaken by Gurney *et al.* [194] in 1965. They again shadowed with tungsten atoms one side of a tungsten emitter tip, kept at a low temperature, achieving a coverage of one to two layers. Field ion micrographs and field evaporation showed small piles of atoms, as in Fig. 3.121. The nature of the deposits was analyzed in detail by Young and Schubert [195], using Monte Carlo simulations for the deposits shown in Fig. 3.122. Their models assumed two different scenarios: atoms condensing at the first site struck, shown in Fig. 3.122a, and condensation following two random jumps, as in Fig. 3.122b. Statistical estimates for both showed a lowering of the density of atoms in upper layers for transient hopping, contrary to the experimental observations. They concluded, that tungsten atoms condensing on a tungsten surface at a temperature below 77 K are bound in the first potential well found.

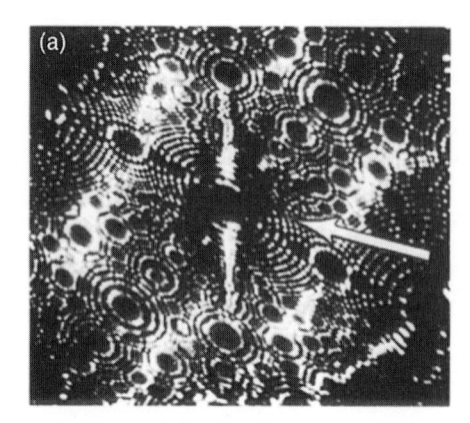

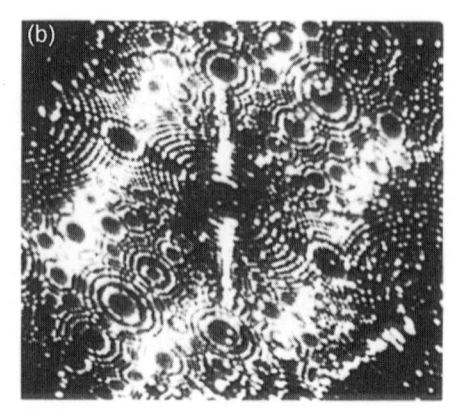

Fig. 3.120 FIM images of sideways deposition of W atoms on W(110)-oriented tip at ~20 K [190]. (a) Clean surface, with direction of atoms shown by arrow. (b) After atoms have been deposited from source at ~3000 K.

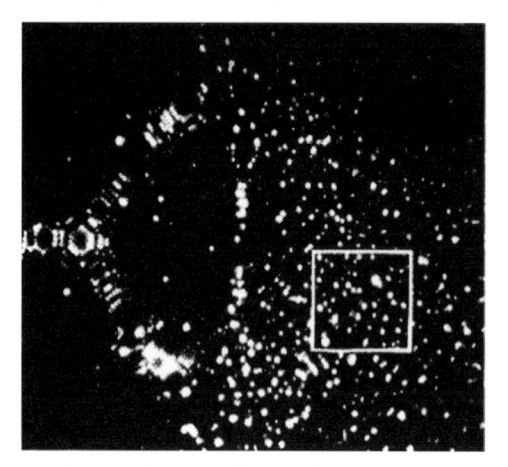

Fig. 3.121 FIM image of tungsten deposit after some field evaporation. White frame indicates area analyzed in Monte Carlo simulations (after Gurney *et al.* [194]).

This finding was bolstered in 1980 when Flahive and Graham [6] deposited tungsten atoms onto a W(111) surface in a field ion microscope. In 220 adsorption observations on a plane of 24 sites the atom was deposited 156 times into a lattice site and 64 times close to the edges. Adatoms were not found at fault sites. Flahive and Graham analyzed their findings with respect to edge atoms and compared them with the distribution obtained for random deposition of adatoms. As it is clear from the results in Table 3.6, a random distribution very well matched experimental observations. Such a random distribution is expected if the colliding atoms settled at the first site encountered.

Some time later, in 1985, Tully [196] turned his attention to the interaction of xenon atoms with a Pt(111) surface. Xenon, of course, interacts only weakly with the surface, primarily through van der Waals forces, and can be expected to behave differently from strongly binding tungsten atoms. Tully did stochastic trajectory calculations which

3.4 Transient diffusion

Table 3.6 Distribution of W atoms deposited on W(111) surface [6].

Adatom surrounding surface atoms which are (111) edge atoms	0	1	2
Observations	48	38	70
Frequency	0.31	0.24	0.45
Random distribution	0.42	0.23	0.35

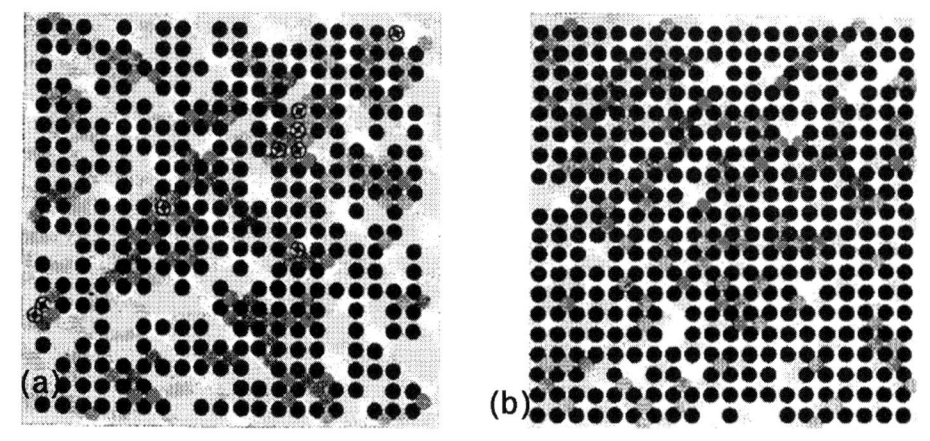

Fig. 3.122 Results of Monte Carlo simulation of tungsten deposition. (a) Condensation on impact. (b) Condensation after two random jumps following impact. Black circles indicate position of first layer obtained in simulations (after Young and Schubert [195]).

established that around 40% of the impinging xenon atoms were trapped briefly on the surface at a temperature of 773 K. The population of trapped atoms decayed exponentially due to thermal desorption, as shown in Fig 3.123a, the z-component of the energy being very quickly quenched. But exchange of energy between atom and lattice is clearly seen in Fig. 3.123b. The trapped atoms retain some of their velocity component parallel to the surface, and drift ~ 200 Å along the surface. In this instance, energy exchange between the incoming atom and the surface was obviously limited and there seemed to be considerable transient diffusion.

In the same year, Schneider et al. [197] carried out molecular dynamics simulations with a Lennard-Jones system to describe epitaxial growth at different temperatures of a (111) substrate. Even at temperatures approaching zero they found the deposited layer to consist of nicely ordered close-packed islands, as in Fig. 3.124. The conclusion drawn from these results was that atom diffusion took place on deposition, even at very low temperatures. However, Schneider et al. [197] had just one movable layer in their simulation, and it later turned out that this was responsible for their findings. Sanders et al. [198] discovered that for such a model the adatom was subject to multiple kicks by the substrate, which does not describe the nature of the deposition correctly.

Attention returned to experiments the next year with the work of Fink [199,200] on the condensation of tungsten atoms. In a field ion microscope, he attempted to deposit a

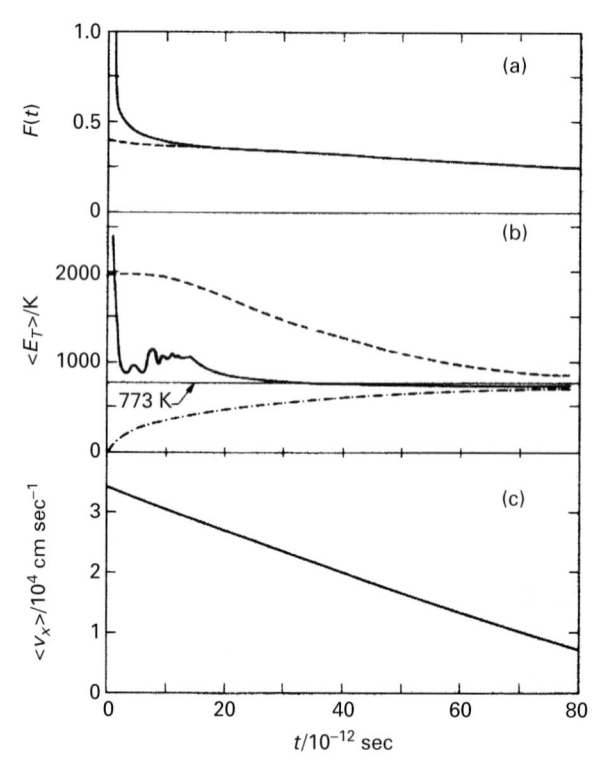

Fig. 3.123 Thermalization of Xe incident at 45° on Pt(111) at 773 K. (a) Fraction of atoms on the surface as a function of time. (b) Average kinetic energy in x-direction (dashed), in y-direction (dot-dashed) and along z, the surface normal, for atoms on the surface. (c) Average x-component of velocity of adsorbed Xe (after Tully [196]).

single atom on a (111) plane of tungsten that had been reduced to just three atoms. It turned out that deposition took place readily, as illustrated in Fig. 3.125. If a tungsten atom trapped on the surface after deposition were to undergo lateral translations, such deposition would be highly unlikely; atoms striking the small (111) plane would move over the edges without any chance of returning. Fink's conclusion therefore was that the condensing tungsten atom settled into the first site struck. It must be noted, however, that a barrier at the step edge could have helped to confine the atom.

In 1989 there appeared an important paper by Egelhoff and Jacob [201] which created a great deal of interest. They looked at reflection high energy electron diffraction (RHEED) oscillations during growth at a low temperature of 77 K for copper and iron on Cu(100), as well as Ag, Cu, Fe, and Mn on Ag(100).[5] If condensation occurs at the first location struck, no layer growth was expected. Egelhoff and Jacob found oscillations, as shown in Fig. 3.126, and concluded that quasi layer-by-layer growth was achieved even at 77 K, where thermally activated diffusion was not possible. "The deposited atoms," they claimed, "must be using their latent heat of condensation

---

[5] Oscillations were found even earlier by Koziol *et al.* [202], who examined the deposition of nickel on the W(110) plane at 100 K.

3.4 Transient diffusion

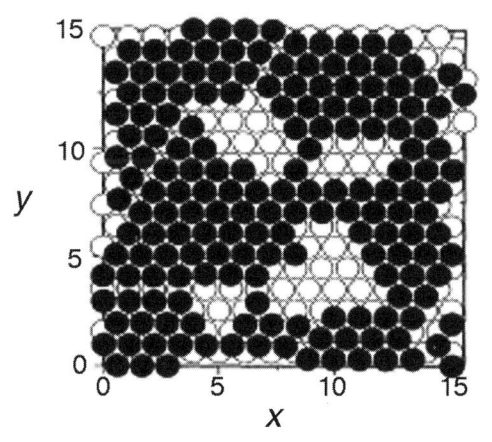

Fig. 3.124 Location of atoms in first layer of Lennard-Jones system after deposition of particles on surface at $T=0$ K of movable substrate (after Schneider *et al.* [197]).

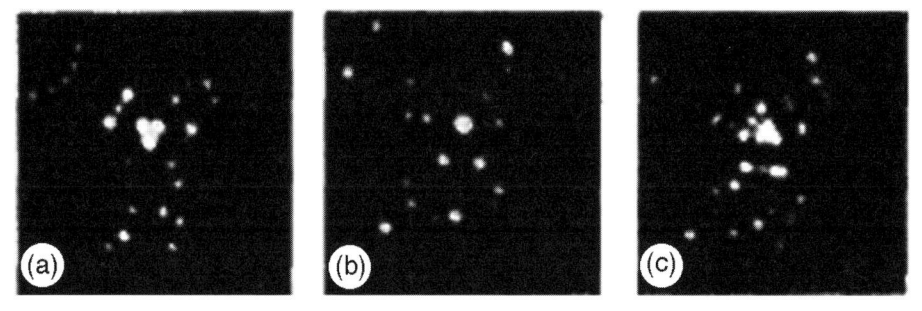

Fig. 3.125 FIM images of W atom deposition onto a W trimer on W(111) plane. (a) Trimer formed by field evaporation. (b) W atom has been deposited on trimer, and now is the only atom seen in center. (c) Superposition of (b) on image after removal of W adatom. Additional deposited atoms are now visible (after Fink [199]).

(≈ 3 eV) in lieu of thermally activated diffusion to overcome the energy barrier to surface diffusion." In light of what had previously been found for the immediate localization of tungsten atoms during condensation, this result was surprising indeed. It must be noted, however, that LEED intensity oscillations were also found by Flynn-Sanders *et al.* [203] for palladium on Pd(100) even at 100 K, as shown in Fig. 3.127.

Evans *et al.* [204] in 1990 offered an ingenious proposal to account for layer-by-layer growth previously observed at low temperatures by Egelhoff and Jacob [201]. Rather than invoking transient diffusion to account for the necessary mobility, their model involved "downward funneling to complete adsorption sites of all impinging atoms at $T=0$ K," as illustrated in Fig. 3.128. "Such downward funneling," they reported, "might correspond to channeling down the (111) faces of micropyramids which developed during film growth." With this model they obtained Bragg peaks oscillating when the number of layers increased, as found in the various experiments mentioned. At this stage the overall conclusion was that transient diffusion played at best a very small role in condensation of metal atoms on a metal surface.

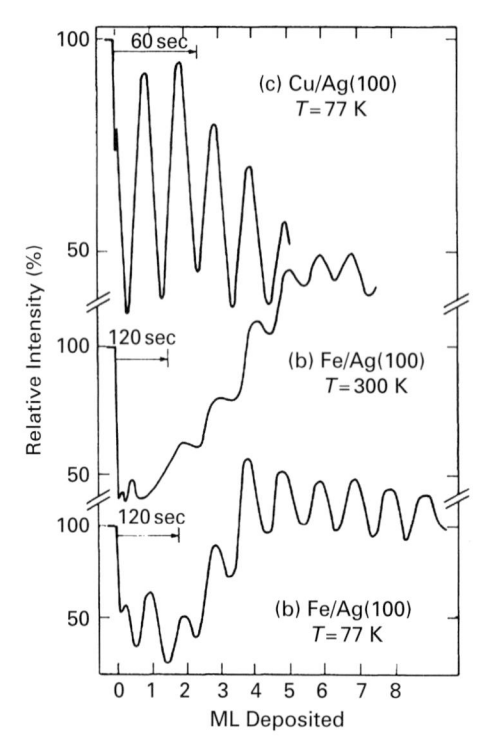

Fig. 3.126 Oscillations in RHEED intensity from layers of various metals deposited on low temperature substrates (after Egelhoff and Jacob [201]).

This impression was strengthened in 1991 by the work of Wang [20], who looked at the condensation of iridium, rhenium, tungsten, and palladium atoms on the Ir(111) plane, using the field ion microscope to identify the location of adsorbed atoms. The lowest surface temperature in these experiments was ~ 20 K, but deposition was also studied for increasingly warmer surfaces. As has already been pointed out, on the fcc(111) plane, two types of binding sites exist, fcc and hcp, at which atoms have slightly different binding energies. For iridium atoms, binding is 0.022 eV stronger at hcp than fcc sites, for rhenium the difference is 0.14 eV, and for tungsten it amounts to 0.19 eV [22]. The distribution of the different atoms over the sites of the (111) plane found in the experiments is shown in Table 3.7. The atom, at a temperature T_E, striking the surface at a temperature T_S, is found at sites ordinarily not significantly populated during thermal motion. It is clear that at a low temperature of ~ 20 K, the distribution over the surface is random and without any preference for the more strongly binding sites, even if such preference is observed during diffusion. The conclusion was "that rapid localization at the point of first impact is fairly common for strongly bound metal atoms, even on as smooth a surface as Ir(111)."

Layer-by-layer growth at low temperature stimulated more refined calculations of the dynamics of atom condensation on metals. Sanders and DePristo [205] simulated condensation on fcc(100) layers at 80 K, relying on corrected effective medium potentials.

3.4 Transient diffusion

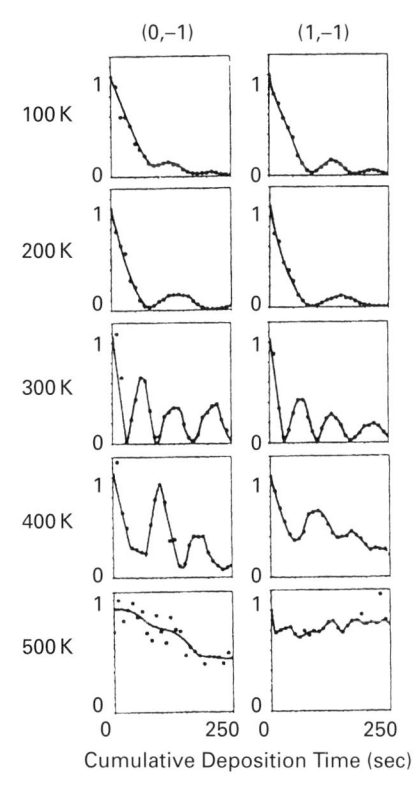

Fig. 3.127 Oscillations of Bragg intensity from Pd layer on Pd(100) as a function of the amount deposited. Shown are results for two diffraction spots under out-of-phase conditions (after Flynn-Sanders et al. [203]).

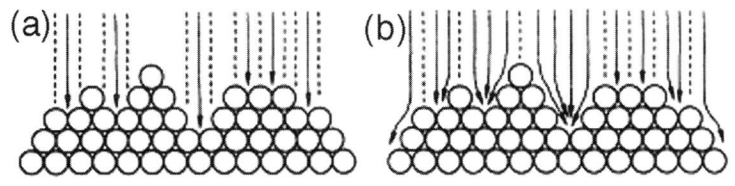

Fig. 3.128 Schematic diagram showing (a) random deposition and (b) downward funneling deposition on a surface (after Evans et al. [204]).

One of the first things they checked was the importance of the number of movable atom layers in the model of the substrate crystal. They discovered that on increasing the number of substrate layers with movable atoms to three, the trapping of incoming atoms happened usually at the first site encountered on their fcc(100) substrate. For nickel on Ni(100), rhodium on Rh(100), palladium on Pd(100), platinum on Pt(100), and gold on Au(100) they reported the absence of transient diffusion, as is suggested by the examples in Fig. 3.129. Only for silver on Ag(100), with quite a small calculated diffusion barrier, were there any signs of atoms beyond the unit cell at which the

Table 3.7 Distribution of deposited metal atoms over sites on Ir(111) [20].

Atom	T_E (K)	T_S (K)	Number deposited	At fcc sites Random	Actual	%
Ir	2500	20	346	173±9	178	51.4
$E_{af} \sim 0.24^a$		80	104	52±5	54	51.9
$E_{ah} \sim 0.28$		100	102	51±5	20	19.6
		104	503	252±11	71	14.1
Re	3000	20	507	254±11	253	49.9
$E_{af} \sim 0.39$		120	319	160±9	158	49.5
$E_{ah} \sim 0.52$		140	158	79±6	26	16.5
		200	106	53±5	0	0
		215	265	133±8	0	0
W	3200	20	248	124±8	120	48.4
$E_{af} \sim 0.30$		100	213	107±7	103	48.4
$E_{ah} \sim 0.48$		200	285	143±8	0	
Pd	1700	20	216	108±7	116	53.7
$E_{af} \sim 0.17$		65	175	88±7	175	100
$E_{ah} \sim 0.13$						

[a] Unit = eV; E_{af} – activation energy for jump out of fcc site; E_{ah} – activation energy for jump out of hcp site

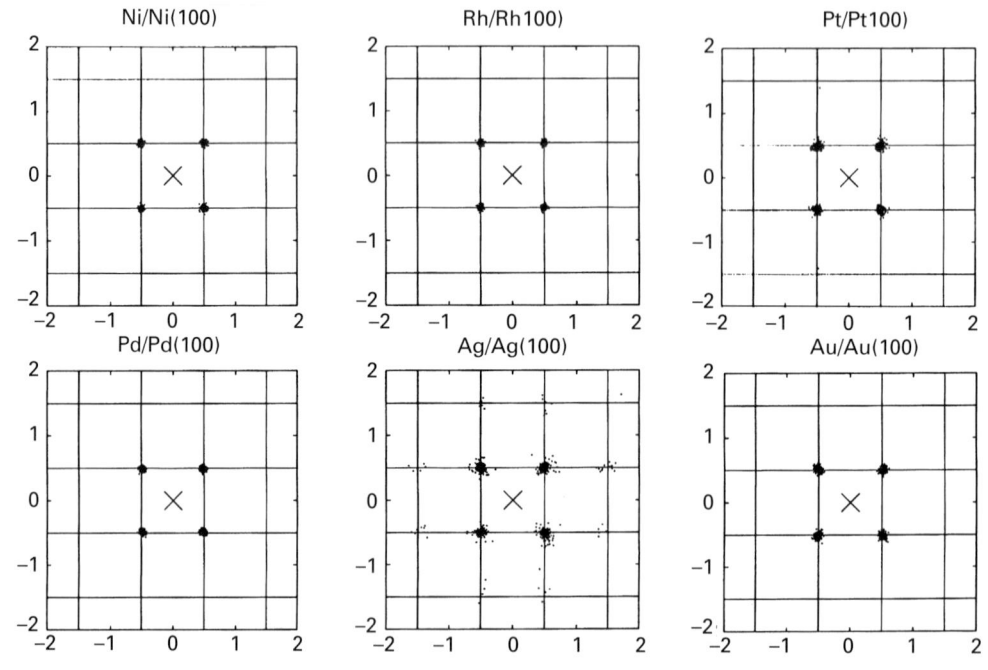

Fig. 3.129 Disposition of atoms condensed on (100) planes of different metals at 80 K. Calculations done with MD/MC-CEM potentials show no transient diffusion (after Sanders and DePristo [205]).

3.4 Transient diffusion

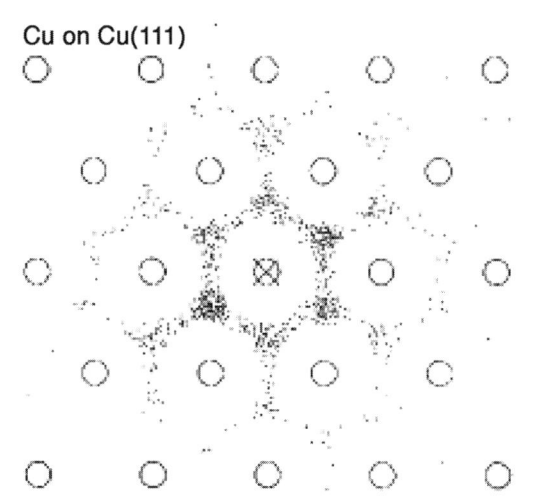

Cu on Cu(111)

Fig. 3.130 Distribution of Cu atoms deposited on Cu(111) plane at 80 K. Only 67% of gas atoms initially concentrated over central cell end in central cell on the surface. Simulations done with MD/MC-CEM potentials. Results suggest some transient motion (after Sanders et al. [198]).

incoming atoms had been aimed. Sanders et al. [198] somewhat later did similar calculations for condensation on the energetically much smoother fcc(111) plane, again at 80 K. They report "no transient mobility for the Cu/Cu(111) system," but their results, obtained with the corrected effective medium theory, revealed that 67% of the condensing atoms settled in the first unit cell struck. Of the rest, 21% were found in the adjacent unit cells, 11% two units away, and the remainder three. Their distribution is shown in Fig. 3.130.

In 1992, Weiss and Eigler [206] reported a very important experiment they had carried out on the adsorption of xenon on Pt(111). Their work was done in a low temperature scanning tunneling microscope with the (111) surface at 4 K and illuminated with a sharply collimated beam of xenon atoms at a temperature between 100 and 300 K. At a low coverage of 0.01 monolayers, the xenon atoms appeared at steps at the edges of the plane, as in Fig. 3.131a. The authors concluded that the atoms move over hundreds of angstrom units over the surface to reach the steps and lose their excess energy. On increasing the Xe coverage to 0.03 ML, Xe islands such as in Fig. 3.131b began to appear. After moving an island away a depression could be observed in the Pt(111) surface associated with the defect on which nucleation started. This finding definitively demonstrated transient diffusion for weakly binding xenon atoms, just as predicted seven years earlier by Tully [196]. It is of interest to note that this long diffusion path is tied to the smoothness of the Pt(111) plane. No such effect was reported for xenon on Ni(110) [207] kept at 4 K, made up of close-packed rows of atoms.

A number of epitaxial growth experiments have also been carried out, and Ernst et al. [208] found oscillations in the scattering intensity with the amount of copper deposited on Cu(100) down to 100 K. Much the same was reported by Bedrossian et al. [209] for silver deposits on Ag(100), where oscillations were again found at 100 K. In both papers transient diffusion as well as downward funneling was mentioned as a possible

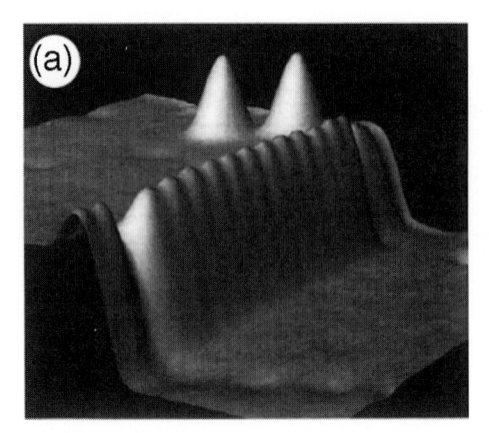

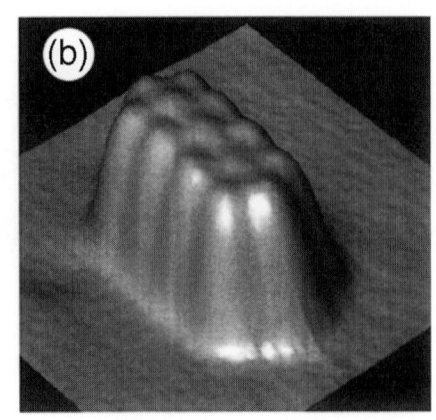

Fig. 3.131 STM image of Pt(111) bombarded by xenon atoms. (a) Step decorated with Xe atoms that had been deposited on the surface. Vertical scale is expanded. (b) Image of Xe island on Pt(111) obtained with STM (after Weiss and Eigler [206]).

explanation for the apparent crystal growth. No final decision was reached about which one is really responsible for these oscillations.

Blandin and Massobrio [210] in 1992 looked at the collision of silver dimers, Ag_2, with the Pt(111) surface in simulations relying on EAM interactions. Their results for the displacement of Ag atoms from the point of impact for dimers colliding with an energy of 0.5 eV and also 5 eV are shown in Fig. 3.132. Even for 5 eV impacts, the distance from the point of impact was at most 5 Å, indicating the absence of transient motion. It is interesting to note that on colliding with the surface the dimer had a considerable likelihood of dissociating. With the surface at 150 K, the fraction of Ag_2 dissociating for 0.5 eV impacts was 0.29.[6]

Condensation on a Lennard-Jones fcc(111) plane, with the lattice atoms initially at rest, was looked at in greater detail by DeLorenzi [211], who found condensed atoms out through the third shell of sites surrounding the aiming point. This result is illustrated in Fig. 3.133. The details of the energy transfer process in condensation were also examined. Atoms that impinged directly on atomic binding sites generally condensed right there, but collisions on top of a surface atom sent the incoming atom far out, as suggested in Fig. 3.134. The process of energy transfer, illustrated in Fig. 3.135, is different in the two cases. After collision with a binding site the energy wave created on impact is quite gentle; the energy of the impinging atom is mostly transferred between three surface atoms. The rebound energy E_r the adatom gets back from the surface is small, not enough to cause further movement. The situation changes in the collision of an adatom with a surface atom. The head-on-collision influences mostly one surface atom and the atom immediately underneath it. The rebound energy E_r the adatom receives in the lattice response to the impact is bigger, big enough to cause the movement of the atom over a few atomic distances, as shown in Fig. 3.133. The conclusion from all these simulation studies was that on the energetically smooth fcc(111) plane the only transient diffusivity

---

[6] At 300 K, dimers dissociate on impact but recombine during further movement on the surface.

3.4 Transient diffusion

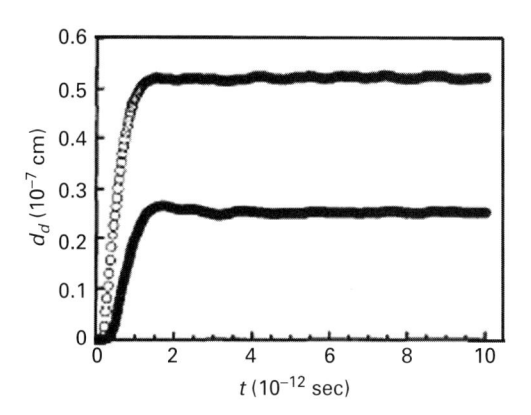

Fig. 3.132 Average distance between point of impact of Ag$_2$ with Pt(111) and resting place on the surface, established in EAM calculations. Impact kinetic energy: open circles 5 eV, full circles 0.5 eV (after Blandin and Massobrio [210]).

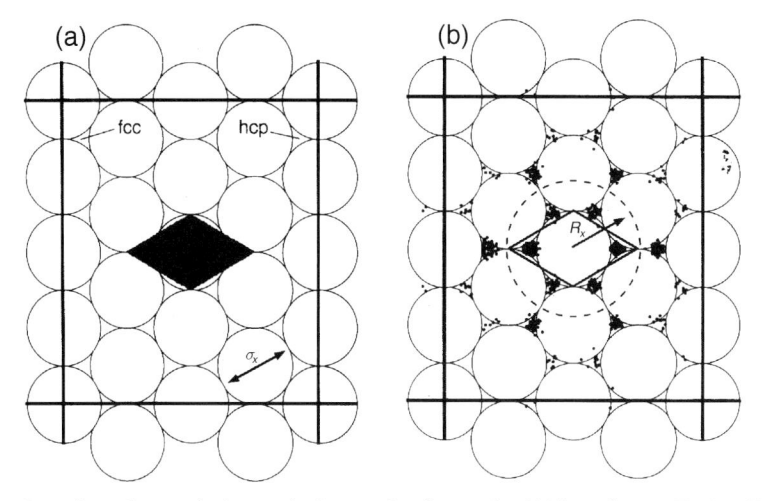

Fig. 3.133 Location of atoms before and after condensing on fcc (111) surface at $T = 0$ K [211]. Results obtained from molecular dynamics simulations with Lennard-Jones potentials.

observed can result from head-on collisions; even there, atoms are confined close to the site of the initial impact, but it is necessary to remember that this investigation was on a Lennard-Jones crystal, not for a specific material.

Experiments on the condensation of iridium atoms on the surroundings of small Ir clusters formed on the Ir(111) plane were carried out by Wang and Ehrlich [212] in 1993. Even when the surface temperature was kept at ∼ 20 K, a zone empty of any deposited atoms was observed, such as that around the Ir$_{59}$ cluster shown in Fig. 3.136. Rather than invoking transient diffusion as an explanation, this empty zone was attributed to changes in the potential binding adatoms close to a cluster

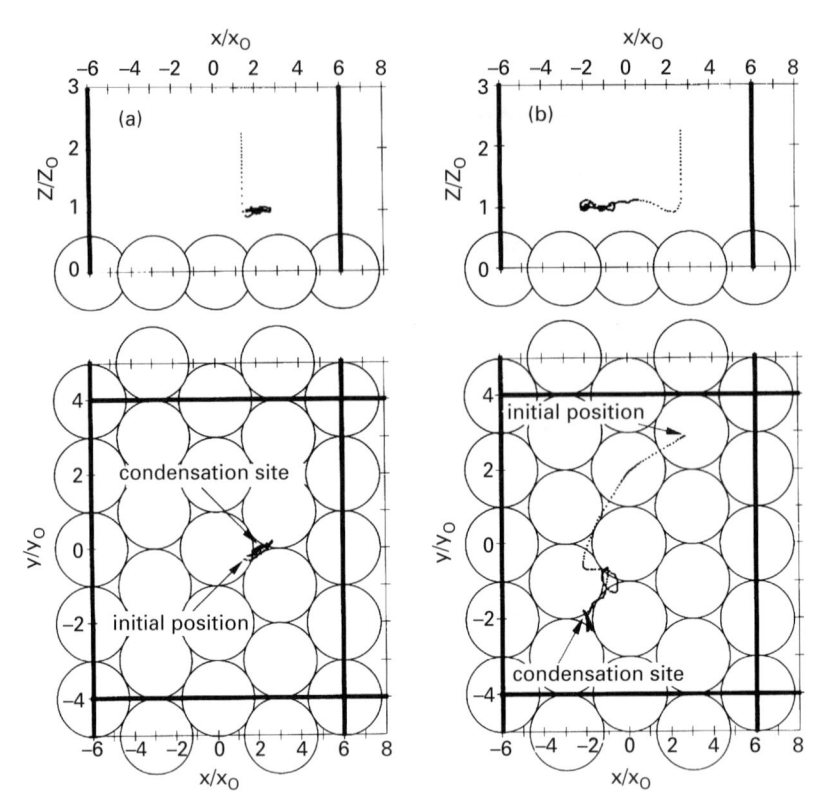

Fig. 3.134 Trajectories for atoms impinging on fcc(111) plane in two situations [211]. (a) Collision at hcp site. (b) Collision on top of surface atom.

edge, as shown in Fig. 3.137. This was described as unphysical by Kellogg [213]. However, such distortions have been reported in calculations by Liu and Adams [214], Stumpf and Scheffler [85], Villarba and Jónsson [71], as well as by Jacobsen et al. [215], eliminating the possibility that transient diffusion occurs for atoms colliding with the cold surface close to the cluster. The problem of the empty zone will be described more extensively in Chapter 6. It should be noted, however, that in STM experiments on nickel films deposited on Au(110)-(1 × 2) at 130 K, no indication of empty zones was found by Hitzke et al. [216]. Recently Smirnov et al. [39] relied on *ab initio* calculations and kinetic MC simulations to suggest that the presence of an empty zone around a cluster is associated with quantum confinement of surface electrons.

Vandoni et al. [217] in 1994 carried out studies of helium atom scattering while a monolayer of 0% to 10% silver was deposited on Pd(100). The surface was maintained at 80 K, a temperature at which Ag adatoms were not mobile. As shown in Fig. 3.138, the experimental results for the measured intensity could not be fitted over the entire range of concentrations deposited on the assumption of complete accommodation of the Ag at the first site struck. Agreement could be obtained assuming six random consecutive jumps prior to permanent adsorption at a site. It should be noted,

3.4 Transient diffusion

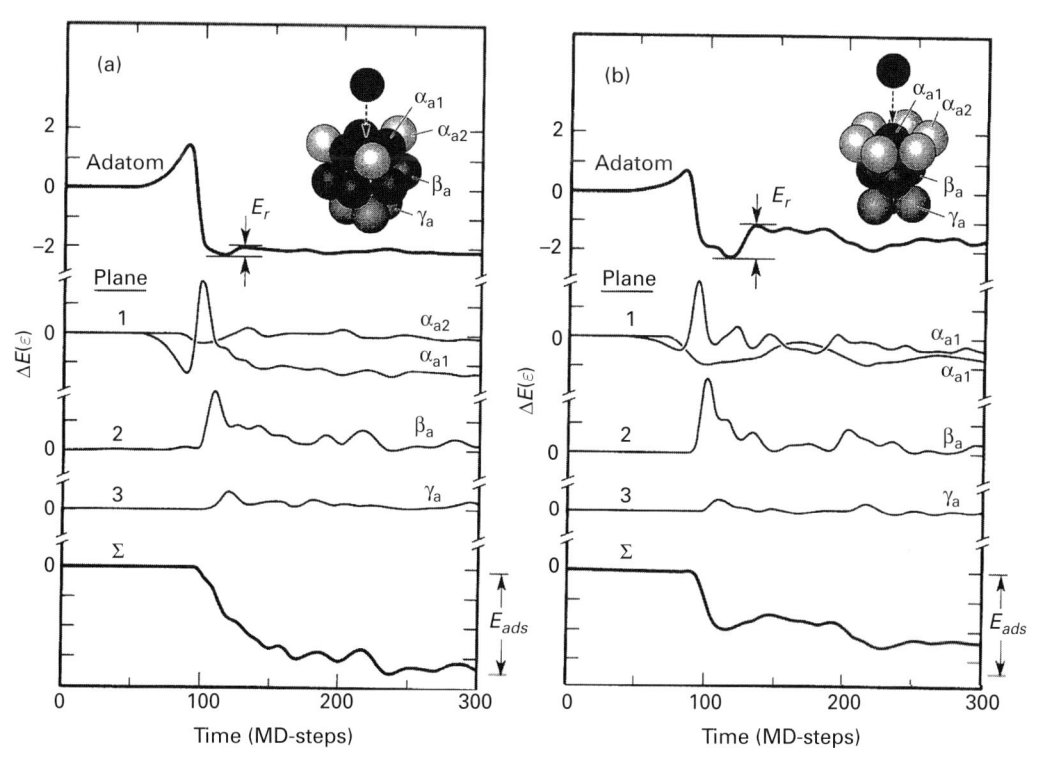

Fig. 3.135 Energy changes of colliding and lattice atoms during collision [211]. (a) Collision at hcp site. (b) Collision with a lattice atom.

however, that up to a coverage of 2% good agreement was obtained for instantaneous condensation. As an alternative model, Vandoni *et al.* assumed that an atom landing close to an already deposited atom or cluster would join up. To obtain good agreement with the measured intensity it had to be assumed, as shown in Fig. 3.139, that an atom landing one site remote would combine with its neighbor despite the low temperature. The authors conclude, however, that "the lowering of the diffusion barrier due to the presence of an adsorbate combined with a transient mobility seems to be the most probable explanation for these experimental findings." A year later the same authors [218] tackled the same problem once more. This time, in addition to MC simulations they used EAM potentials to carry out molecular dynamics simulations. They performed 50 adatom depositions with an energy from 0.16 to 2.98 eV. Some 6% of the adatoms moved towards already adsorbed adatoms, and no transient mobility of atoms was observed. However, these results did not explain their experimental data. They claimed that it was probably due to an inaccurate potential or an overestimation of energy dissipation in EAM. From molecular dynamics simulations they concluded that a 6.1 Å capture zone existed around preadsorbed adatoms and that lowering of the potential barrier combined with transient mobility was the most probable scenario.

160   Atomic events in surface diffusion

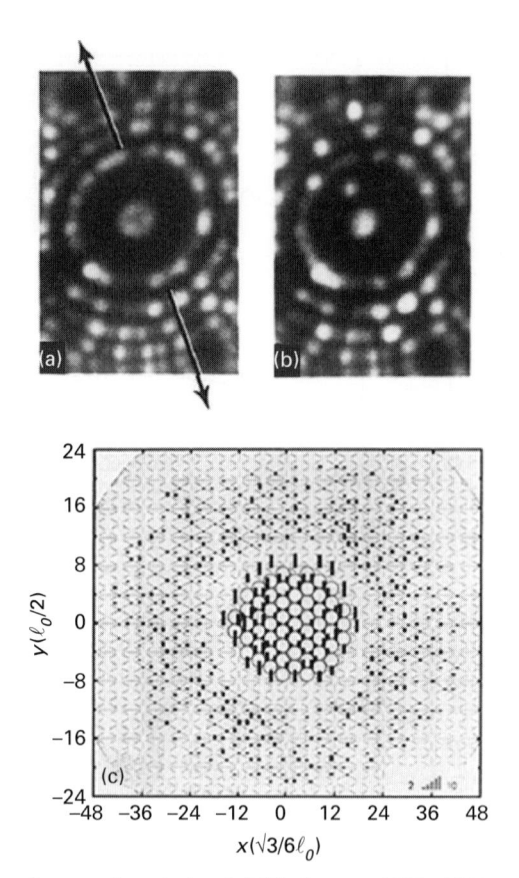

Fig. 3.136   Ir atoms deposited on Ir(111) plane at ~20 K with Ir$_{59}$ cluster at the center [212]. (a) FIM image with island in place on Ir(111) plane. Arrow shows direction of incoming atom. (b) Single Ir atom has been deposited on surface. (c) Distribution of deposited atoms around central island, showing empty ring.

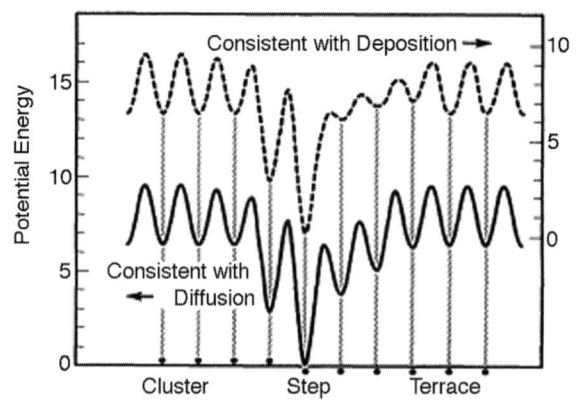

Fig. 3.137   Potential for adatom migrating over Ir(111) plane with a central cluster on it [212]. Bottom curve is adequate for diffusion results, top gives more severe changes required for results in condensation experiments.

3.4 Transient diffusion

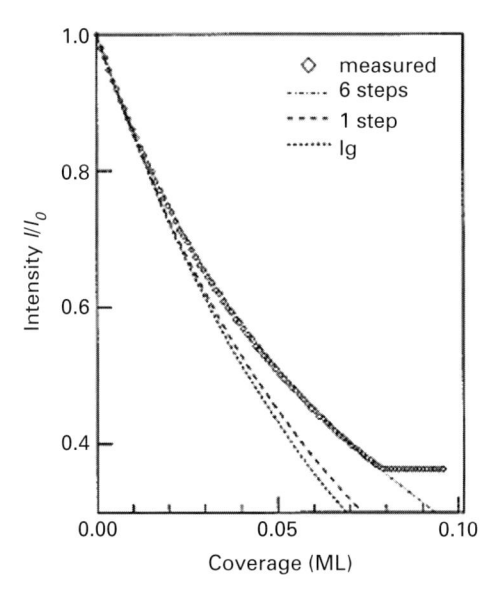

Fig. 3.138 Intensity of reflected helium for Ag deposits on Pd(100). Best fit obtained with six random transient steps (after Vandoni et al. [217]).

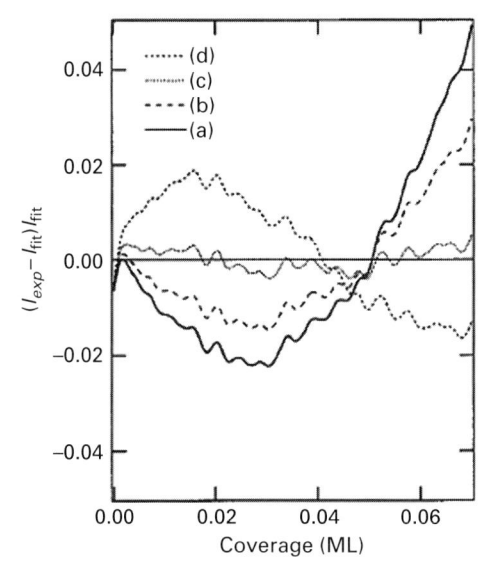

Fig. 3.139 Difference between best model fit and experimental intensity for Ag deposited on Ag(100) at 80 K. (a) Random condensation. (b) Capture zone one nearest neighbor. (c) Zone of one nearest and next-nearest neighbor. (d) Zone of two nearest-neighbor steps (after Félix et al. [218]).

One year after Vandoni et al., Gilmore and Sprague [219] did interesting simulations for deposition of Cu and Ag atoms on Ag(001), relying on embedded atom potentials as well as potentials by Haftel et al. [139]. In their work they found no evidence for transient diffusion of atoms condensing on a bare surface. However, interactions with adatoms and islands brought about atom spreading, conveying the impression of transient motion. A few years earlier Gilmore and Sprague [220] looked at the distribution of silver atoms impinging on a Ag(111) surface with energies between 0.1 and 10 eV. For low energies, ~0.1 eV, they did observe a random distribution of atoms on the surface, indicating no transient mobility. For higher energies, from 1 to 10 eV, ballistic displacements started occurring. These have also been observed by Villarba and Jónsson [221] for Pt on Pt(111). Ballistic exchange from Ir atoms colliding with Rh(100) was later found by Kellogg [213].

The introduction of the scanning tunneling microscope now led to a number of interesting and sometimes confusing studies which dealt with transient diffusion upon dissociation of diatomic molecules on a surface. We will discuss these later. Our main concern here is transient diffusion during adsorption of metal atoms, and we will for the moment continue this emphasis. In 1997 Hitzke et al. [216] used the STM to observe a random distribution of Ni atoms on the Au(110)-(1 × 2) reconstructed surface. They deposited 0.05 ML at 130 K and 0.2 ML at 180K, with immediate quenching to 130 K. Then they studied the lateral and vertical distribution of Ni atoms in the middle of the terrace and counted the number of monomers, dimers, trimers and linear chains as well as the spacing between them. They also did Monte Carlo simulations to compare with the experimental distribution. Based on this comparison they concluded that atoms could on average make no more than 0.5 hops during energy dissipation–energy exchange between the incoming atom and the surface was very efficient. The deposition of iron on Ag(100) was reported by Canepa et al. [222] in 1997, using thermal energy helium atom diffraction. Based on island size in the first layer they concluded that transient diffusion of one hop was consistent with their data. However, their data are a bit puzzling since they also found evidence for intermixing, from which they deduced exchange as the mechanism of movement over the surface.

At the same time, Yue et al. [223,224] did molecular dynamics simulations of the low temperature growth of copper on the Cu(100) plane, relying on hybrid tight-binding-like potentials. They looked for transient diffusion as the amount of copper deposited on the surface was increased. A number of possible scenarios for creating the distributions shown in Fig. 3.140 was considered. These were 1 – direct deposition with lateral movement less than 1 spacing; 2 – direct deposit with lateral movement less than 2; 3 – downward funneling with a distance 0.35 to 0.7 lattice spacings; 4 – downward lattice funneling with a distance of 0.7 to 1 lattice spacing; 5 – downward funneling with a distance from 1 to 1.5 lattice spacings; 6 – impact cascade diffusion when atom takes the place of kicked atom; 7 – impact cascade diffusion when atom does *not* take the place of the kicked atom. As seen in Fig. 3.140, transient diffusion became significant only after a few atoms had been deposited. Their conclusion – "transient motion primarily stems from the impacting atoms interacting with other adatoms already on the surface" – confirmed the results previously obtained by Gilmore and Sprague [219].

At this stage it should be noted that whether or not transient diffusivity is important has little influence on experiments to measure diffusivities, as these are generally done on thermally equilibrated systems. The situation is different, however, for

3.4 Transient diffusion

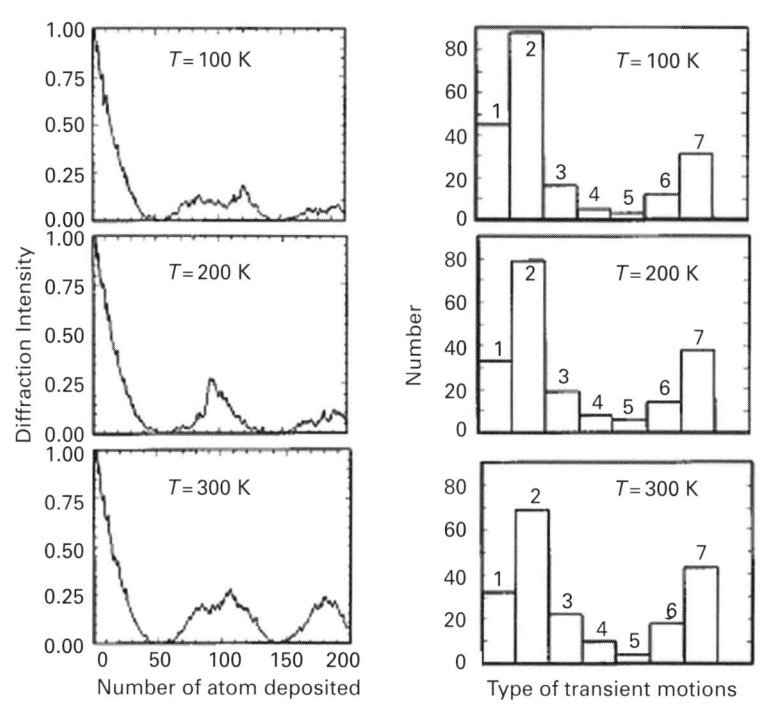

Fig. 3.140 Simulations of deposition of Cu atoms with 0.1 eV energy on Cu(100). Left: Diffraction intensity as a function of the number of deposited atoms. Right: Number of different transient motions occurring during deposition. Transient diffusion occurs only after some atoms have been condensed (after Yue et al. [224]).

experiments in which the saturation island density is measured to learn about the diffusivity; here atoms are constantly colliding with the surface, and atoms excited during the collision process could produce elevated values for the diffusivity. This has been checked by Brune et al. [225] for the deposition of silver on Pt(111) at 35 K, at which temperature ordinary diffusion is effectively stopped. An STM image of the surface in Fig. 3.141a reveals individual spots, most of which correspond to atoms; the mean size of islands formed is plotted in Fig. 3.141b against the coverage. If transient mobility were active over one spacing, the mean island size would be 2.3 atoms at a coverage of 0.1 monolayer; the experiments, however, showed a size of 1.2±0.3 atoms, close to the value expected in the absence of any type of diffusion. Michely and Krug [226] also reported an STM survey of a Pt(111) surface at 23 K on which Pt atoms had been deposited. The number of image spots again corresponded to the number expected for a random distribution without any transient diffusion.

In looking at all the work on transient diffusion of metal atoms it is important to note that interest in this topic was really stimulated by experiments indicating crystal growth at low temperatures. All the studies, including both experimental and theoretical efforts, that have examined the impact of individual metal atoms with a surface are in agreement – transient diffusion is a negligible phenomenon on an empty surface. The situation can be

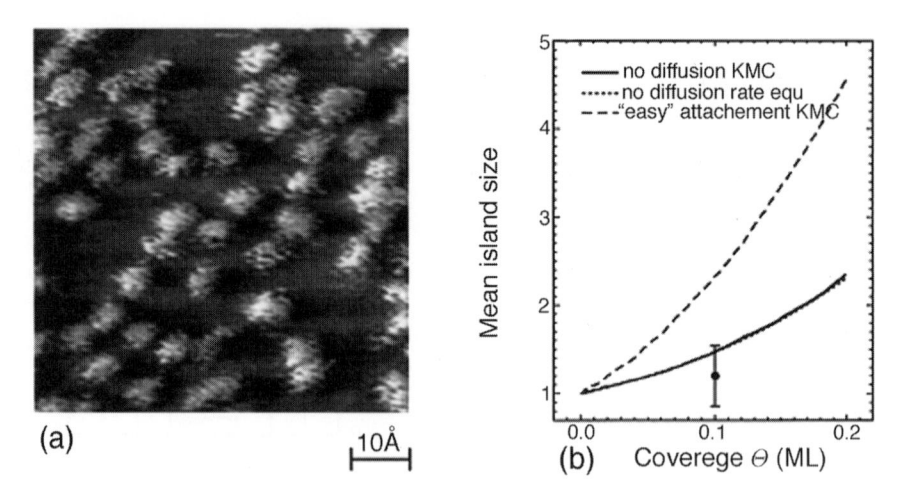

Fig. 3.141 Silver deposition on Pt(111) surface. (a) STM image of silver deposited on Pt(111) surface at 35 K shows predominantly atoms. (b) Condensation without any diffusion compared with transient diffusion over one site, "easy attachment." Results indicate absence of transient motion (after Brune et al. [225]).

more complicated for weakly bound gases like Xe, which are likely to show some transient motion. Simulations have also suggested that some mobility may arise on a surface at a finite coverage in collisions or interactions between incoming and adsorbed atoms or island. The subject is therefore reasonably well understood.

3.4.2 Transient motion of dissociating molecules

For the condensation of metal atoms on metal surfaces the conclusion from all the available experiments is that energy transfer between incoming atoms and the lattice is good so that transient diffusion does not play a significant role. Because of the great interest that has arisen in what happens during dissociation of diatomics on a surface we will, however, discuss this related topic at least briefly. Traditionally, on dissociation of a molecule on a surface, the atoms were supposed to end up in adjacent nearest-neighbor sites. Is this what really happens? It is conceivable that some of the energy released in the dissociation may translationally impel the atoms over the surface to positions far removed from each other. But it is also possible that the atoms start repelling each other after reaching a certain separation. The possibility of transient diffusion was first examined by Brune et al. [227,228], who looked with a scanning tunneling microscope at the distribution of oxygen atoms on Al(111) after adsorption of O_2 at 300 K, a temperature at which oxygen atoms are immobile. After the chemisorption process, rather than atom pairs they found a random distribution of atoms on the surface, shown in Fig. 3.142, with an average separation between adsorbed atoms of 80 Å, suggesting the presence of translationally "hot" oxygen atoms. The histograms of the size distribution in Fig. 3.143 for several exposures of oxygen were also consistent with the random creation of islands from single atoms. This study started a number of investigations.

3.4 Transient diffusion

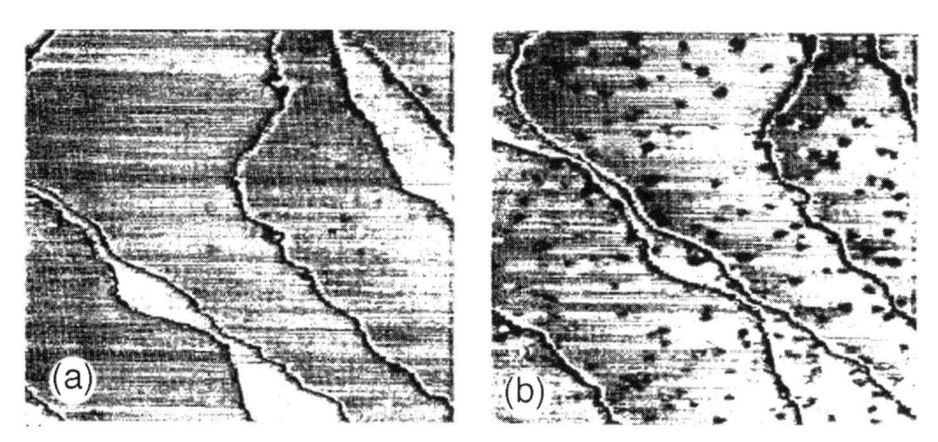

Fig. 3.142 STM images of O_2 adsorption on Al(111) surface. (a) Clean surface. (b) After exposure to six monolayers of O_2, black dots show atoms and small oxygen islands (after Brune et al. [228]).

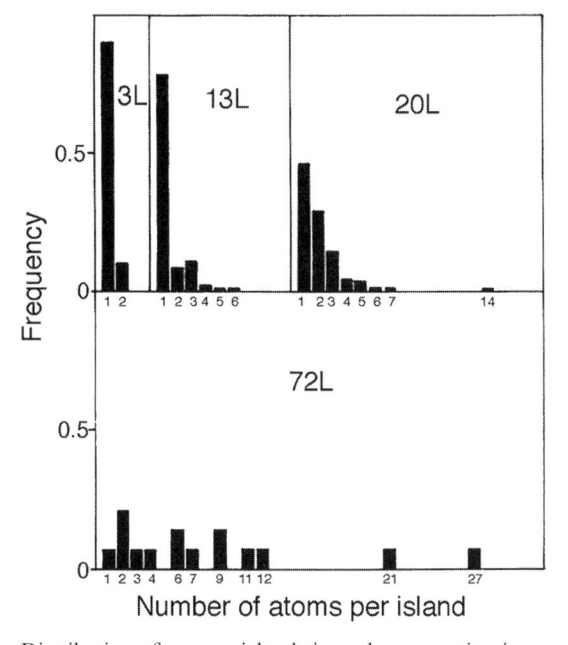

Fig. 3.143 Distribution of oxygen island sizes, demonstrating increased collisions of adatoms with increasing coverage. No preference is indicated for pairs of adjacent adatoms (after Brune et al. [228]).

The findings by Brune et al. [227,228], of very large spacings of O adatoms in dissociation of O_2 on aluminum led Engdahl and Wahnström [229] to do molecular dynamics simulations using an effective medium approximation for interactions. They looked at impacts with two different energies, 3.5 eV and 9.5 eV, with the direction of impact in both cases parallel to the surface. The average displacement between two atoms formed by dissociation was reported as 16.1 Å for an initial energy of 3.5 eV,

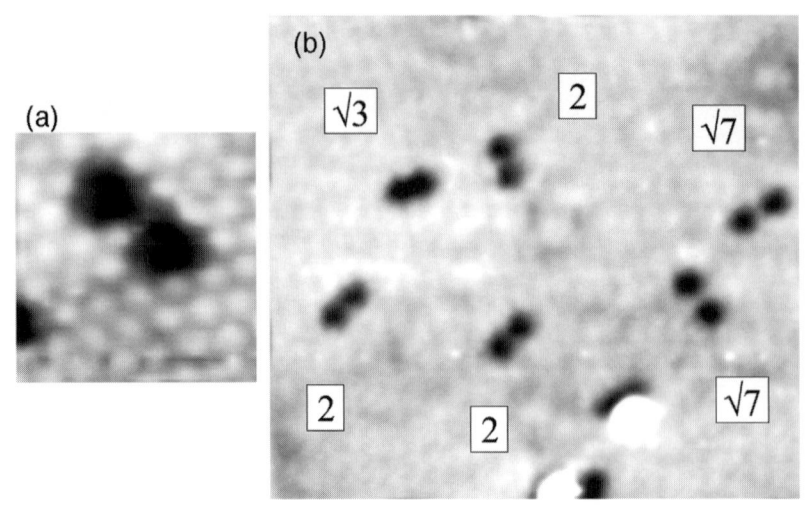

Fig. 3.144 STM images of oxygen adsorption on Pt(111) at ~160 K. (a) Two monolayers of O_2. (b) 1.2 monolayers of O_2. Figures give distance between oxygen atoms (after Wintterlin *et al.* [231]).

considerably smaller than found in the experiments. For an initial energy of 9.5 eV the separation on dissociation was 52 Å, still lower than in the experiment. The separation in a completely random distribution of atoms should be equal to 35.6 Å for a coverage of 0.0014, which is consistent with that of Brune *et al.*, so the value of 80 Å is a bit suspicious. The higher energy used in the simulations to come closer to the experimental findings of Brune *et al.* is likely to be too high compared to the experiments.

In 1995, Jacobsen *et al.* [230] looked briefly at oxygen on Al(111) using density-functional theory calculations. They suggested that a cannon-like trajectory for oxygen atoms, in which an atom spends most of its time far from the surface, could possibly account for long distances between atoms in dissociation. A year later, Wintterlin *et al.* [231] studied the dissociation of O_2 on the Pt(111) plane, using the STM to ascertain the distribution of oxygen atoms over the surface at 163 K; thermal motion of atoms was found around 200 K. An image of the surface with atoms on it is shown in Fig. 3.144. Pairs were found with quite a small inter-atomic separation, with oxygen occupying three fold fcc sites. A chart of the distribution giving a statistical account is shown in Fig. 3.145; the average distance between O atoms is roughly two lattice spacings. They also observed agglomeration of oxygen atoms at only 60 K, indicating mobility. The authors point out "that dissociation of O_2 on Pt(111) actually creates 'hot' atoms … This leads, however, to only a very limited transient motion."

At the same time, Wahnström *et al.* [232] did dynamical simulations for adsorption of oxygen on Al(111) using a model potential fitted to first principle data. They did not find any evidence for the large transient mobility seen by Brune *et al.* [228]. In this investigations atoms were separated by 1 to 3 atomic spaces, results similar to what was found in experiments on the adsorption of oxygen on Pt(111) [231].

3.4 Transient diffusion

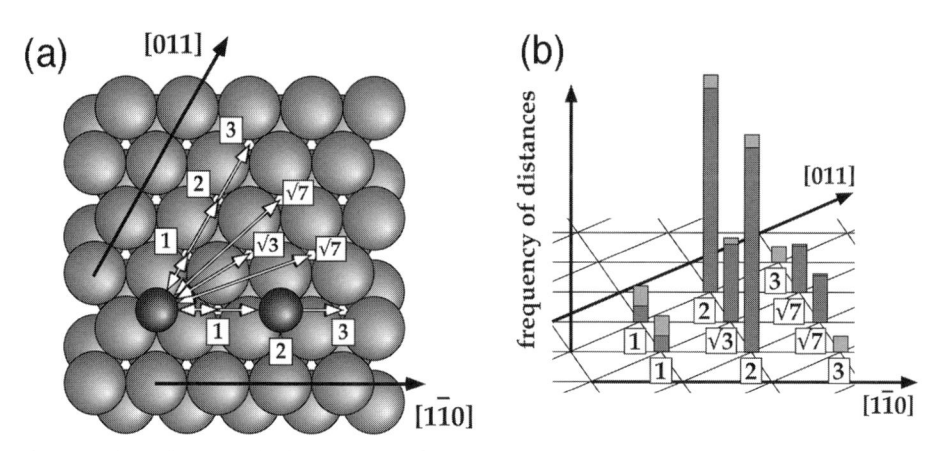

Fig. 3.145 Oxygen adsorption on Pt(111). (a) Model of Pt(111) with oxygen after dissociation. (b) Distribution of inter-pair distances between oxygen atoms (after Wintterlin *et al.* [231]).

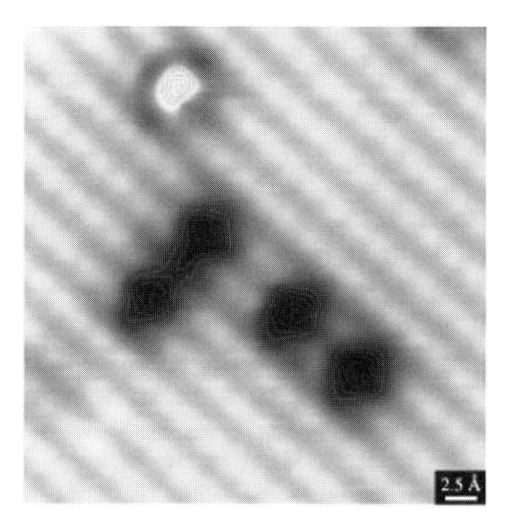

Fig. 3.146 STM image at high resolution of Cu(110), showing both atomic and molecular oxygen (after Briner *et al.* [233]).

The study of oxygen adsorption continued with the work of Briner *et al.* [233], who now examined the adsorption of O_2 on Cu(110) in a low-temperature STM. The oxygen impinged on the sample at an angle of only ~ 5° and with an energy near 50 meV. They found atom pairs at an inter-atomic spacing of 2 nearest neighbors at 4 K, usually aligned along the channels of the plane, as in Fig. 3.146. Precursor molecules were also seen at the surface but oriented across the channels. The distances they observed were roughly what Wintterlin *et al.* found on Pt(111) [231].

At the same time dissociation of O_2 on Pt(111) at temperatures below 100 K was also studied by Stipe *et al.* [234] in several ways using the STM. When dissociation was initiated by photons, electrons, or by heating, the distance between two O adatoms ranged

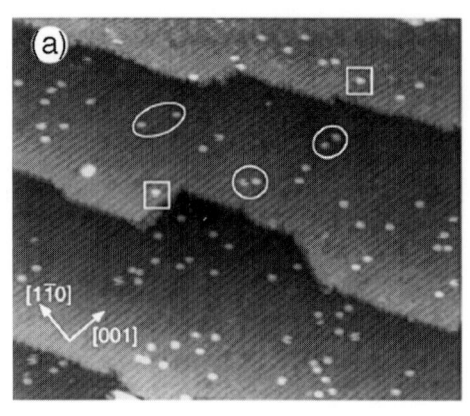

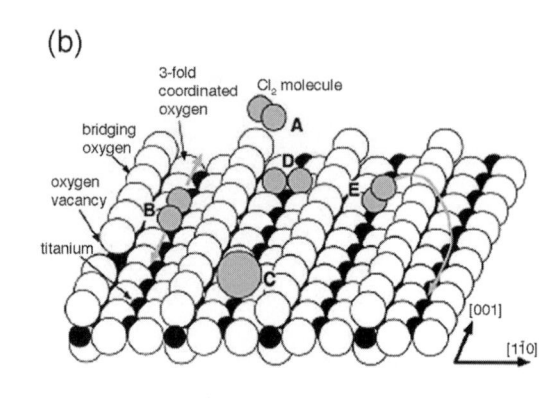

Fig. 3.147 Interaction of Cl_2 with $TiO_2(110)$. (a) STM image of $TiO_2(110)$ surface, exposed to 0.07 Langmuir Cl_2 at 300 K. Chlorine atoms indicated by white squares, Cl-Cl pairs by circle. (b) Model of $TiO_2(110)$ surface exposed to Cl_2. Ti indicated by black spheres. *A* shows gas phase chlorine molecule, *B* and *D* show flat molecules, *C* gives Cl^-. Upright Cl_2 is labeled *E* (after Diebold et al. [235]).

from 1 to 3 lattice spacings, with the latter favored. The results are again in agreement with the previous studies by Wintterlin et al. [231] indicating a lack of transient mobility. In somewhat similar work, Diebold et al. [235] looked at the dissociation of Cl_2 on $TiO_2(110)$. Although the substrate is not a metal, the findings are relevant here. The Cl atoms were observed in the STM and were found separated by an average distance of 26 Å, as indicated in Fig. 3.147a. Diebold et al. found only 23% pairs in the same row, 42% one row apart, and 25% two rows apart; 10% of the chlorine did not have a partner. To account for this, the authors propose that dissociation occurred from an upstanding molecule, in which the upper atom is emitted in a cannon-ball-like path, as suggested in Fig. 3.147b, rather than transient motion connected with the impact of the molecule. The cannon-ball mechanism of dissociation was already proposed by Jacobsen et al. [230] to account for the long distances seen by Brune et al. [227,228] between oxygen atoms on Al(111). On Ni(110), pairs of chlorine atoms were observed by Fishlock et al. [236] at a distance around 3.5 Å, aligned along the [001] direction, but it is not clear if transient motion influenced this distance.

Dissociation of oxygen molecules on Rh(110) was examined with the STM by Hla et al. [237]. Oxygen atoms were found at a separation of 3.3 Å along the [001] direction, slightly smaller than the [001] lattice constant. Hla et al. concluded that atoms were sitting at adjacent asymmetric short bridge sites. Their findings did not support creation of "hot" atoms in molecular dissociation.

The early studies of O_2 dissociation on Al(111) were so startling that work on this system has continued, with rather surprising findings. Schmid et al. [238] in 2001 examined an aluminum (111) surface after O_2 adsorption at different temperatures and found the images in Fig. 3.148a. Oxygen struck the sample at an angle of 60°. Schmid et al. measured the distance of 420 pairs on the surface, and found that atom pairs were reasonably close to each other, separated by no more than $\sqrt{7}$ times the nearest-neighbor

3.4 Transient diffusion

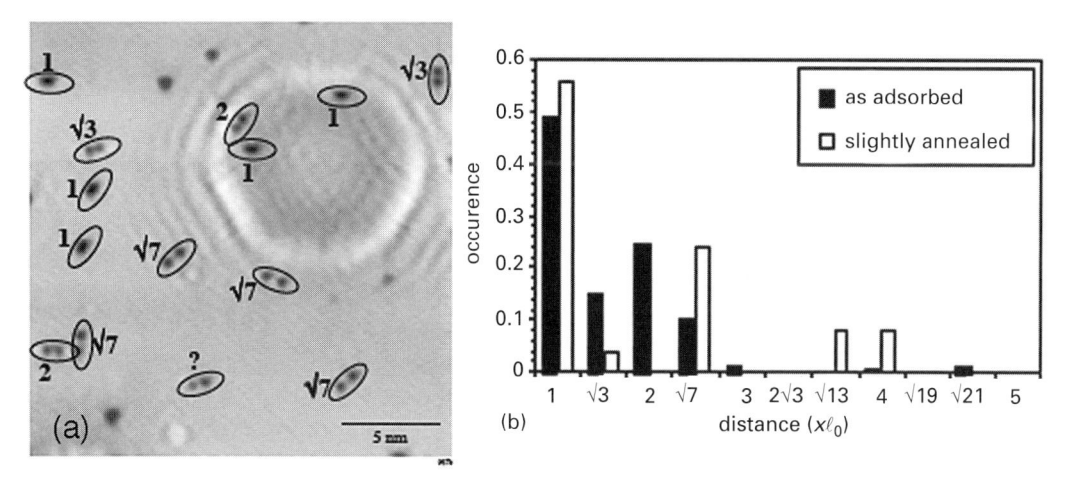

Fig. 3.148 Interaction of O$_2$ with Al(111) surface. (a) Al(111) surface imaged with STM at 80 K, after O$_2$ exposure at 150 K. Oxygen pairs are labeled with interatomic distance (in terms of Al nearest-neighbor spacing ℓ_0). (b) Pair separations of oxygen atoms on Al(111) plane (after Schmid et al. [238]).

spacing of aluminum. A graph of the observed distribution in Fig. 3.148b shows the average inter-atomic distance as 5 Å. After annealing to ~ 250 K the number of nearest-neighbor oxygen pairs increased, but there were also some larger separations observed. The behavior of O$_2$ dissociation on Al(111) now appears very similar to oxygen dissociation on Pt(111) – transient mobility is low. This study did not show a preferential orientation of the pairs, and the question remains, does the direction of the incident beam matter?

The behavior of O$_2$ on the Al(111) plane is not yet clear. Komrowski et al. [239] looked at the surface with a scanning tunneling microscope, but also examined the gas phase with resonant enhanced multi-photon ionization. They dosed oxygen at an angle of 45° with the sample. On the surface they found single O atoms, O atoms at adjacent sites, or small islands of three atoms. The three atom islands were attributed to consecutive adsorption at adjacent surface sites, while the isolated single O adatoms were assumed due to abstractive chemisorption. The oxygen molecule sat on the surface in an end-on geometry, in which the upper atom is lost to the vacuum in dissociation. The important point, however, is that the ejected O atoms were detected in the gas phase. Abstractive chemisorption was proposed earlier for fluorine atom adsorption on Si(100)-(2 × 1) surface [240]. There this mechanism was explained by the presence of dangling bonds.

The overall result of the more recent investigations is that atom pairs produced by dissociation of diatomic O$_2$ are close together, and show little diffusion of atoms excited in the dissociation event. Recently, however, in 2001, Schintke et al. [241] have studied the dissociative adsorption of O$_2$ on Ag(100). At a low concentration of 0.13% of a monolayer, with the surface at 50 K, they obtained the STM image in Fig. 3.149. This shows atom pairs at two kinds of separations: the smaller one has a separation of 20 Å (half of the oxygen pairs), the larger one 40 Å (one-third). The rest is unpaired, which might be a result of

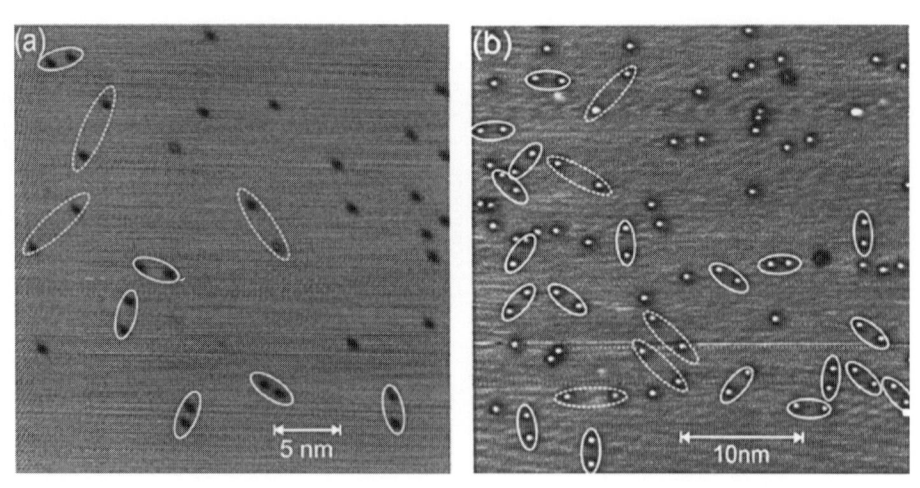

Fig. 3.149 STM images of oxygen adsorbed on Ag(100). (a) 0.13% of a monolayer on the surface. (b) 0.5% of a monolayer adsorbed. Pairs are indicated by white lines (after Schintke et al. [241]).

Fig. 3.150 Location of oxygen atoms produced by thermal dissociation on Pd(111) at 160 K. Inset shows molecules perpendicular and parallel to line from impurity, in white, to the center of the pair (after Rose et al. [242]).

abstractive dissociation with desorption of the second atom from the surface. These spacings are much larger than inter-atomic distances reported recently in dissociation, and the authors suggest two alternative mechanisms that may account for the observations. In the first, if the dissociation energy is released equally, the inter-atomic separation will be smaller than if the energy is distributed unequally for the two adatoms. An alternative mechanism attributes the smaller separation to dissociation from a molecular precursor, the longer separation to dissociation from the gas phase. So far the actual process responsible for the long distances has not been identified, but O_2 on Ag(100) appears as the only example in which atoms from the dissociation are widely separated. This is especially surprising given that Ag(100) is an energetically rough surface.

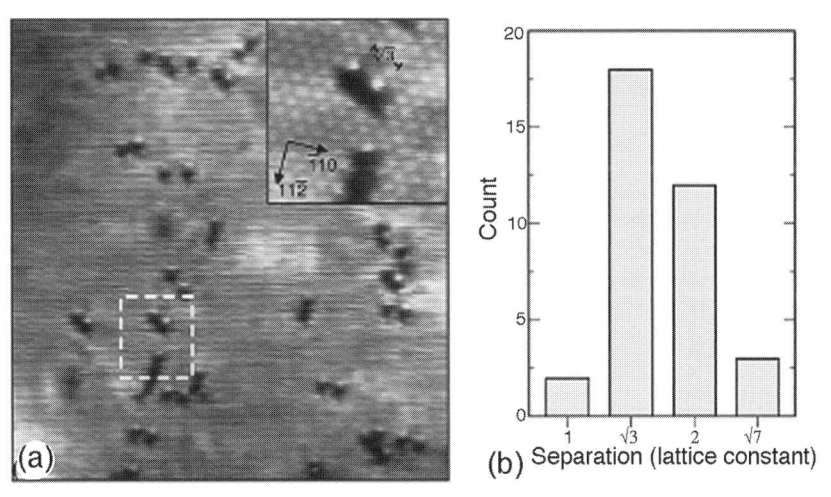

Fig. 3.151 Oxygen atoms produced on Pd(111) by electron bombardment at 0.1 V bias and 10 nA current. (a) Pairs of oxygen atoms. Separation in higher resolution inset is $\sqrt{3}$ spacings along $<11\bar{2}>$. (b) Pair distribution obtained in several experiments (after Rose *et al.* [242]).

In 2004, Rose *et al.* [242] looked at the dissociation of oxygen molecules on a Pd(111) surface under thermal conditions at 160 K. At this temperature dissociation occurs readily, but migration is not active yet. In most cases atoms were separated by distances of $\sqrt{3}$ or 2 times the lattice constant, as shown in Fig. 3.150, but the pair distribution was strongly influenced by the presence of impurities in the Pd crystal. Dissociation caused by injection of electrons from the tip at 100 mV and 10 nA was also examined. Rose *et al.* observed pairs at distances 1, $\sqrt{3}$, 2 or $\sqrt{7}$ times the lattice constant of Pd, as in Fig. 3.151. There was no indication of any significant transient motion in this system.

It is clear that transient diffusion in molecular dissociation is a complicated, not terribly well understood event. It is difficult to separate effects associated only with dissociation from transient motion caused by impact with the surface. Dissociation of molecular oxygen generally leads to only brief diffusion of the dissociating atoms, something on the order of two spacings. For metal atoms, however, the situation is much more straightforward. The general impression is that transient diffusion is insignificant.

References

[1] P. R. Watson, M. A. Van Hove, K. Herman, Atlas of surface structures, based on the NIST structure database (SSB), *J. Phys. Chem. Ref. Data Monograph* **5**, 1994.

[2] K. Morgenstern, Fast scanning tunneling microscopy as a tool to understand changes on metal surfaces: from nanostructures to single atoms, *Phys. Stat. Sol.* **242** (2005) 773–796.

[3] J. A. Stroscio, R. J. Celotta, Controlling the dynamics of a single atom in lateral atom manipulation, *Science* **306** (2004) 242–247.

[4] W. R. Graham, G. Ehrlich, Direct identification of atomic binding sites on a crystal, *Surf. Sci.* **45** (1974) 530–552.

[5] P. G. Flahive, W. R. Graham, Surface site geometry and diffusion characteristics of single Ni atoms on W(111), *Thin Solid Films* **51** (1978) 175–184.

[6] P. G. Flahive, W. R. Graham, The determination of single atom surface site geometry on W(111), W(211) and W(321), *Surf. Sci.* **91** (1980) 463–488.

[7] W. Xu, J. B. Adams, W single adatom diffusion on W surfaces, *Surf. Sci.* **319** (1994) 58–67.

[8] H.-W. Fink, Atomistik der Monolagenbildung: Direkte Beobachtung an Rhenium auf Wolfram, Physics Ph.D. Thesis, Technical University of Munich, 1982.

[9] H.-W. Fink, G. Ehrlich, Pair and trio interactions between adatoms: Re on W(110), *J. Chem. Phys.* **81** (1984) 4657–4665.

[10] H.-W. Fink, in: *Diffusion at Interfaces – Microscopic Concepts*, M. Grunze, H. J. Kreuzer, J. J. Weimer (eds.), Direct observation of atomic motion on surfaces, (Springer-Verlag, Berlin, 1988), p. 75–91.

[11] G. Antczak, G. Ehrlich, Long jumps in diffusion of iridium on W(110), *Phys. Rev. B* **71** (2005) 115422 1–9.

[12] D. Spisák, J. Hafner, Diffusion of Fe atoms on W surfaces and Fe/W films and along surface steps, *Phys. Rev. B* **70** (2004) 195426 1–13.

[13] D. Spisák, J. Hafner, Diffusion mechanisms for iron on tungsten, *Surf. Sci.* **584** (2005) 55–61.

[14] S. Dennler, J. Hafner, First-principles study of ultrathin Mn films on W surfaces. II. Surface diffusion, *Phys. Rev. B* **72** (2005) 214414 1–9.

[15] R. Fijak, L. Jurczyszyn, G. Antczak, Adatom self-diffusion on W(110), in preparation (2009).

[16] R. T. Tung, W. R. Graham, Single atom self-diffusion on nickel surfaces, *Surf. Sci.* **97** (1980) 73–87.

[17] S. C. Wang, G. Ehrlich, Self-adsorption sites on a close-packed surface: Ir on Ir(111), *Phys. Rev. Lett.* **62** (1989) 2297–2300.

[18] S. C. Wang, G. Ehrlich, Imaging and diffusion of individual iridium adatoms on Ir(111), *Surf. Sci.* **224** (1989) L997–1003.

[19] S. C. Wang, G. Ehrlich, Determination of atomic binding sites on the fcc(111) plane, *Surf. Sci.* **246** (1991) 37–42.

[20] S. C. Wang, G. Ehrlich, Atom condensation on an atomically smooth surface: Ir, Re, W, and Pd on Ir(111), *J. Chem. Phys.* **94** (1991) 4071–4074.

[21] G. Ehrlich, An atomic view of crystal growth, *Appl. Phys. A* **55** (1992) 403–410.

[22] S. C. Wang, G. Ehrlich, Atomic behavior at individual binding sites: Ir, Re, and W on Ir(111), *Phys. Rev. Lett.* **68** (1992) 1160–1163.

[23] H. Brune, J. Wintterlin, G. Ertl, R. J. Behm, Direct imaging of adsorption sites and local electronic bond effects on a metal surface: C/Al(111), *Europhys. Lett.* **13** (1990) 123–128.

[24] C. L. Liu, J. M. Cohen, J. B. Adams, A. F. Voter, EAM study of surface self-diffusion of single adatoms of fcc metals Ni, Cu, Al, Ag, Au, Pd, and Pt, *Surf. Sci.* **253** (1991) 334–344.

[25] A. Gölzhäuser, G. Ehrlich, Atom movement and binding on surface clusters: Pt on Pt(111) clusters, *Phys. Rev. Lett.* **77** (1996) 1334–1337.

[26] A. Gölzhäuser, G. Ehrlich, Direct observation of platinum atoms on Pt(111) clusters, *Z. Phys. Chem.* **202** (1997) 59–74.

[27] P. J. Feibelman, J. S. Nelson, G. L. Kellogg, Energetics of Pt adsorption on Pt(111), *Phys. Rev. B* **49** (1994) 10548–10556.

[28] J. J. Mortensen, B. Hammer, O. H. Nielsen, K. W. Jacobsen, J. K. Nørskov, in: *Elementary Processes in Excitations and Reactions on Solid Surfaces*, A. Okiji, H. Kasai, K. Makoshi

(eds.), Density functional theory study of self-diffusion on the (111) surfaces of Ni, Pd, Pt, Cu, Ag and Au, (Springer-Verlag, Berlin, 1996), p. 173–182.

[29] J. Repp, G. Meyer, K.-H. Rieder, P. Hyldgaard, Site determination and thermally assisted tunneling in homogeneous nucleation, *Phys. Rev. Lett.* **91** (2003) 206102 1–4.

[30] M.-C. Marinica, C. Barreteau, M.-C. Desjonqueres, D. Spanjaard, Influence of short-range adatom-adatom interactions on the surface diffusion of Cu on Cu(111), *Phys. Rev. B* **70** (2004) 075415 1–14.

[31] Y. Mishin, M. J. Mehl, D. A. Papaconstantopoulos, A. F. Voter, J. D. Kress, Structural stability and lattice defects in copper: *Ab initio*, tight binding, and embedded-atom calculations, *Phys. Rev. B* **63** (2001) 224106 1–16.

[32] K. Morgenstern, K.-H. Rieder, Long-range interaction of copper adatoms and copper dimers on Ag(111), *New J. Phys.* **7** (2005) 139 1–11.

[33] J. T. Goldstein, G. Ehrlich, Atom and cluster diffusion on Re(0001), *Surf. Sci.* **443** (1999) 105–115.

[34] H.-W. Fink, G. Ehrlich, Lattice steps and adatom binding on W(211), *Surf. Sci.* **143** (1984) 125–144.

[35] G. Ehrlich, Atomic events at lattice steps and clusters: A direct view of crystal growth processes, *Surf. Sci.* **331/333** (1995) 865–877.

[36] G. L. Kellogg, Diffusion of individual platinum atoms on single-crystal surfaces of rhodium, *Phys. Rev. B* **48** (1993) 11305–11312.

[37] T.-Y. Fu, H.-T. Wu, T. T. Tsong, Energetics of surface atomic processes near a lattice step, *Phys. Rev. B* **58** (1998) 2340–2346.

[38] S.-M. Oh, K. Kyuno, S. C. Wang, G. Ehrlich, Step-edge versus interior barriers to atom incorporation at lattice steps, *Phys. Rev. B* **67** (2003) 075413 1–7.

[39] A. S. Smirnov, N. N. Negulyaev, L. Niebergall, W. Hergert, A. M. Saletsky, V. S. Stepanyuk, Effect of quantum confinement of surface electrons on an atomic motion on nanoislands: *Ab initio* calculation and Kinetic Monte Carlo simulations, *Phys. Rev. B* **78** (2008) 041405(R) 1–4.

[40] D. W. Bassett, P. R. Webber, Diffusion of single adatoms of platinum, iridium and gold on platinum surfaces, *Surf. Sci.* **70** (1978) 520–531.

[41] T. Halicioglu, An atomistic calculation of two-dimensional diffusion of a Pt adatom on a Pt(110) surface, *Surf. Sci.* **79** (1979) L346–348.

[42] T. Halicioglu, G. M. Pound, A calculation of the diffusion energies for adatoms on surfaces of FCC metals, *Thin Solid Films* **57** (1979) 241–245.

[43] G. DeLorenzi, G. Jacucci, V. Pontikis, in: *Proc. ICSS-4 and ECOSS-3, Cannes*, D. A. Degras, M. Costa (eds.), Simulation par la dynamique moléculaire de la diffusion des adatomes et adlacunes sur les terrasses (100), (110) et (111) des gaz rares dans les solides, 1980, p. 54.

[44] G. DeLorenzi, G. Jacucci, V. Pontikis, Diffusion of adatoms and vacancies on otherwise perfect surfaces: A molecular dynamics study, *Surf. Sci.* **116** (1982) 391–413.

[45] M. R. Mruzik, G. M. Pound, A molecular dynamics study of surface diffusion, *J. Phys. F* **11** (1981) 1403–1422.

[46] S. H. Garofalini, T. Halicioglu, Mechanism for the self-diffusion of Au and Ir adatoms on Pt(110) surface, *Surf. Sci.* **104** (1981) 199–204.

[47] J. D. Wrigley, G. Ehrlich, Surface diffusion by an atomic exchange mechanism, *Phys. Rev. Lett.* **44** (1980) 661–663.

[48] E. W. Müller, J. A. Panitz, S. B. McLane, The atom-probe field ion microscope, *Rev. Sci. Instrum.* **39** (1968) 83–86.

[49] J. D. Wrigley, Surface diffusion by an atomic exchange mechanism, Physics Ph.D. Thesis, University of Illinois at Urbana-Champaign, 1982.

[50] R. T. Tung, Atomic structure and interactions at single crystal metal surfaces, Physics Ph.D. Thesis, University of Pennsylvania, Philadelphia, 1981.

[51] G. L. Kellogg, Field-ion microscope observations of surface self-diffusion and atomic interactions on Pt, *Microbeam Analysis* **1986** (1986) 399–402.

[52] C. L. Chen, T. T. Tsong, Self-diffusion on the reconstructed and nonreconstructed Ir(110) Surfaces, *Phys. Rev. Lett.* **66** (1991) 1610–1613.

[53] G. L. Kellogg, Direct observations of adatom-surface-atom replacement: Pt on Ni(110), *Phys. Rev. Lett.* **67** (1991) 216–219.

[54] C. L. Chen, T. T. Tsong, L. H. Zhang, Z. W. Yu, Atomic replacement and adatom diffusion: Re on Ir surfaces, *Phys. Rev. B* **46** (1992) 7803–7807.

[55] L. Pedemonte, R. Tatarek, G. Bracco, Surface self-diffusion at intermediate temperature: The Ag(110) case, *Phys. Rev. B* **66** (2002) 045414 1–5.

[56] P. A. Gravil, S. Holloway, Exchange mechanisms for self-diffusion on aluminum surfaces, *Surf. Sci.* **310** (1994) 267–272.

[57] R. Stumpf, M. Scheffler, *Ab initio* calculations of energies and self-diffusion on flat and stepped surfaces of aluminum and their implications on crystal growth, *Phys. Rev. B* **53** (1996) 4958–4973.

[58] Y.-J. Sun, J.-M. Li, Self-diffusion mechanisms of adatom on Al(001), (011) and (111) surfaces, *Chin. Phys. Lett.* **20** (2003) 269–272.

[59] C. L. Liu, J. B. Adams, Diffusion mechanisms on Ni surfaces, *Surf. Sci.* **265** (1992) 262–272.

[60] P. Stoltze, Simulation of surface defects, *J. Phys.: Condens. Matter* **6** (1994) 9495–9517.

[61] U. T. Ndongmouo, F. Hontinfinde, Diffusion and growth on fcc(110) metal surfaces: a computational study, *Surf. Sci.* **571** (2004) 89–101.

[62] L. Hansen, P. Stoltze, K. W. Jacobsen, J. K. Nørskov, Self-diffusion on Copper Surfaces, *Phys. Rev. B* **44** (1991) 6523–6526.

[63] C. Mottet, R. Ferrando, F. Hontinfinde, A. C. Levi, A Monte Carlo simulation of submonolayer homoepitaxial growth on Ag(110) and Cu(110), *Surf. Sci.* **417** (1998) 220–237.

[64] G. A. Evangelakis, D. G. Papageorgiou, G. C. Kallinteris, C. E. Lekka, N. I. Papanicolaou, Self-diffusion processes of copper adatom on Cu(110) surface by molecular dynamics simulations, *Vacuum* **50** (1998) 165–169.

[65] F. Montalenti, R. Ferrando, Jumps and concerted moves in Cu, Ag, and Au(110) adatom self-diffusion, *Phys. Rev. B* **59** (1999) 5881–5891.

[66] S. Durukanoglu, O. S. Trushin, T. S. Rahman, Effect of step-step separation on surface diffusion processes, *Phys. Rev. B* **73** (2006) 125426 1–6.

[67] L. S. Perkins, A. E. DePristo, Self-diffusion of adatoms on fcc(110) surfaces, *Surf. Sci.* **317** (1994) L1152–1156.

[68] F. Hontinfinde, R. Ferrando, A. C. Levi, Diffusion processes relevant to the epitaxial growth of Ag on Ag(110), *Surf. Sci.* **366** (1996) 306–316.

[69] K.-D. Shiang, C. M. Wei, T. T. Tsong, A molecular dynamics study of self-diffusion on metal surfaces, *Surf. Sci.* **301** (1994) 136–150.

[70] C. M. Chang, C. M. Wei, S. P. Chen, Modeling of Ir adatoms on Ir surfaces, *Phys. Rev. B* **54** (1996) 17083–17096.

[71] M. Villarba, H. Jónsson, Diffusion mechanisms relevant to metal crystal growth: Pt/Pt(111), *Surf. Sci.* **317** (1994) 15–36.

References

[72] U. T. Ndongmouo, F. Hontinfinde, R. Ferrando, Numerical study of the stability of (111) and (331) microfacets on Au, Pt, and Ir(110) surfaces, *Phys. Rev. B* **72** (2005) 115412 1–8.

[73] M. Karimi, G. Vidali, I. Dalins, Energetics of the formation and migration of defects in Pb(110), *Phys. Rev. B* **48** (1993) 8986–8992.

[74] G. DeLorenzi, G. Jacucci, The migration of point defects on bcc surfaces using a metallic pair potential, *Surf. Sci.* **164** (1985) 526–542.

[75] D. L. Price, K. S. Singwi, M. P. Tosi, Lattice dynamics of alkali metals in the self-consistent screening theory, *Phys. Rev. B* **2** (1970) 2983–2999.

[76] D. L. Price, Effects of a volume-dependent potential on equilibrium properties of liquid sodium, *Phys. Rev. A* **4** (1971) 358–363.

[77] G. L. Kellogg, P. J. Feibelman, Surface self-diffusion on Pt(001) by an atomic exchange mechanism, *Phys. Rev. Lett.* **64** (1990) 3143–3146.

[78] C. Chen, T. T. Tsong, Displacement distribution and atomic jump direction in diffusion of Ir atoms on the Ir(001) surface, *Phys. Rev. Lett.* **64** (1990) 3147–3150.

[79] G. L. Kellogg, Temperature dependence of surface self-diffusion on Pt(001), *Surf. Sci.* **246** (1991) 31–36.

[80] G. L. Kellogg, A. F. Wright, M. S. Daw, Surface diffusion and adatom-induced substrate relaxations of Pt, Pd, and Ni atoms on Pt(001), *J. Vac. Sci. Technol. A* **9** (1991) 1757–1760.

[81] T. T. Tsong, C. L. Chen, Atomic replacement and vacancy formation and annihilation on iridium surfaces, *Nature* **355** (1992) 328–331.

[82] G. L. Kellogg, Surface diffusion of Pt adatoms on Ni surfaces, *Surf. Sci.* **266** (1992) 18–23.

[83] A. Friedl, O. Schütz, K. Müller, Self-diffusion on iridium (100). A structure investigation by field-ion microscopy, *Surf. Sci.* **266** (1992) 24–29.

[84] T.-Y. Fu, T. T. Tsong, Structure and diffusion of small Ir and Rh clusters on Ir(001) surfaces, *Surf. Sci.* **421** (1999) 157–166.

[85] R. Stumpf, M. Scheffler, Theory of self-diffusion at and growth of Al(111), *Phys. Rev. Lett.* **72** (1994) 254–257.

[86] J.-M. Li, P.-H. Zhang, J.-L. Yang, L. Liu, Theoretical study of adatom self-diffusion on metallic fcc{100} surfaces, *Chin. Phys. Lett.* **14** (1997) 768–771.

[87] O. S. Trushin, P. Salo, M. Alatalo, T. Ala-Nissila, Atomic mechanisms of cluster diffusion on metal fcc(100) surfaces, *Surf. Sci.* **482–485** (2001) 365–369.

[88] N. I. Papanicolaou, V. C. Papathanakos, D. G. Papageorgiou, Self-diffusion on Al(100) and Al(111) surfaces by molecular-dynamics simulation, *Physica B* **296** (2001) 259–263.

[89] T. Fordell, P. Salo, M. Alatalo, Self-diffusion on fcc (100) metal surfaces: Comparison of different approximations, *Phys. Rev. B* **65** (2002) 233408 1–4.

[90] S. Valkealahti, M. Manninen, Diffusion on aluminum-cluster surfaces and the cluster growth, *Phys. Rev. B* **57** (1998) 15533–15540.

[91] S. Ovesson, A. Bogicevic, G. Wahnstrom, B. I. Lundqvist, Neglected adsorbate interactions behind diffusion prefactor anomalies on metals, *Phys. Rev. B* **64** (2001) 125423 1–11.

[92] P. M. Agrawal, B. M. Rice, D. L. Thompson, Predicting trends in rate parameters for self-diffusion on FCC metal surfaces, *Surf. Sci.* **515** (2002) 21–35.

[93] L. S. Perkins, A. E. DePristo, Self-diffusion mechanisms for adatoms on fcc (100) surfaces, *Surf. Sci.* **294** (1993) 67–77.

[94] L. S. Perkins, A. E. DePristo, The influence of lattice distortion on atomic self-diffusion on fcc(001) surfaces: Ni, Cu, Pd, Ag, *Surf. Sci.* **325** (1995) 169–176.

[95] G. Boisvert, L. J. Lewis, A. Yelon, Many-body nature of the Meyer-Neidel compensation law for diffusion, *Phys. Rev. Lett.* **75** (1995) 469–472.

[96] J. Merikoski, I. Vattulainen, J. Heinonen, T. Ala-Nissila, Effect of kinks and concerted diffusion mechanisms on mass transport and growth on stepped metal surfaces, *Surf. Sci.* **387** (1997) 167–182.

[97] C. M. Chang, C. M. Wei, J. Hafner, Self-diffusion of adatoms on Ni(100) surfaces, *J. Phys.: Condens. Matter* **13** (2001) L321–328.

[98] C. M. Chang, C. M. Wei, Self-diffusion of adatoms and dimers on fcc(100) surfaces, *Chin. J. Phys.* **43** (2005) 169–175.

[99] P. G. Flahive, W. R. Graham, Pair potential calculations of single atom self-diffusion activation energies, *Surf. Sci.* **91** (1980) 449–462.

[100] D. E. Sanders, A. E. DePristo, Predicted diffusion rates on fcc (001) metal surfaces for adsorbate/substrate combinations of Ni, Cu, Rh, Pd, Ag, Pt, Au, *Surf. Sci.* **260** (1992) 116–128.

[101] K.-D. Shiang, Molecular dynamics simulation of adatom diffusion on metal surfaces, *J. Chem. Phys.* **99** (1993) 9994–10000.

[102] Z.-P. Shi, Z. Zhang, A. K. Swan, J. F. Wendelken, Dimer shearing as a novel mechanism for cluster diffusion and dissociation on metal (100) surfaces, *Phys. Rev. Lett.* **76** (1996) 4927–4930.

[103] H. Mehl, O. Biham, I. Furman, M. Karimi, Models for adatom diffusion on fcc (001) metal surfaces, *Phys. Rev. B* **60** (1999) 2106–2116.

[104] S. Y. Davydov, Calculation of the activation energy for surface self-diffusion of transition-metal atoms, *Phys. Solid State* **41** (1999) 8–10.

[105] J. E. Black, Z.-J. Tian, Complicated exchange-mediated diffusion mechanisms in and on a Cu(100) substrate at high temperatures, *Phys. Rev. Lett.* **71** (1993) 2445–2448.

[106] J. M. Cohen, Long range adatom diffusion mechanism on fcc (100) EAM modeled materials, *Surf. Sci. Lett.* **306** (1994) L545–549.

[107] C. Lee, G. T. Barkema, M. Breeman, A. Pasquarello, R. Car, Diffusion mechanism of Cu adatoms on a Cu(001) surface, *Surf. Sci.* **306** (1994) L575–578.

[108] M. Karimi, T. Tomkowski, G. Vidali, O. Biham, Diffusion of Cu on Cu surface, *Phys. Rev. B* **52** (1995) 5364–5374.

[109] G. Boisvert, L. J. Lewis, Self-diffusion of adatoms, dimers, and vacancies on Cu(100), *Phys. Rev. B* **56** (1997) 7643–7655.

[110] G. A. Evangelakis, N. I. Papanicolaou, Adatom self-diffusion processes on (001) copper surfaces by molecular dynamics, *Surf. Sci.* **347** (1996) 376–386.

[111] O. S. Trushin, K. Kokko, P. T. Salo, W. Hergert, M. Kotrla, Step roughening effect on adatom diffusion, *Phys. Rev. B* **56** (1997) 12135–12138.

[112] G. Boisvert, N. Mousseau, L. J. Lewis, Surface diffusion coefficients by thermodynamic integration: Cu on Cu(100), *Phys. Rev. B* **58** (1998) 12667–12670.

[113] J. B. Adams, Z. Wang, Y. Li, Modeling Cu thin film growth, *Thin Solid Films* **365** (2000) 201–210.

[114] R. Pentcheva, Ab initio study of microscopic processes in the growth of Co on Cu(001), *Appl. Phys. A* **80** (2005) 971–975.

[115] U. Kürpick, A. Kara, T. S. Rahman, Role of lattice vibrations in adatom diffusion, *Phys. Rev. Lett.* **78** (1997) 1086–1089.

[116] P. Wynblatt, N. A. Gjostein, A calculation of relaxation, migration and formation energies for surface defects in copper, *Surf. Sci.* **12** (1968) 109–127.

[117] C.-L. Liu, Energetics of diffusion processes during nucleation and growth for the Cu/Cu(100) system, *Surf. Sci.* **316** (1994) 294–302.

References

[118] P. V. Kumar, J. S. Raul, S. J. Warakomski, K. A. Fichthorn, Smart Monte Carlo for accurate simulation of rare-event dynamics: Diffusion of adsorbed species on solid surfaces, *J. Chem. Phys.* **105** (1996) 686–695.

[119] Q. Xie, Dynamics of adatom self-diffusion and island morphology evolution at a Cu(100) surface, *Phys. Stat. Sol. B* **207** (1998) 153–170.

[120] O. Biham, I. Furman, M. Karimi, G. Vidali, R. Kennett, H. Zeng, Models for diffusion and island growth in metal monolayers, *Surf. Sci.* **400** (1998) 29–43.

[121] M. O. Jahma, M. Rusanen, A. Karim, I. T. Koponen, T. Ala-Nissila, T. S. Rahman, Diffusion and submonolayer island growth during hyperthermal deposition on Cu(100) and Cu(111), *Surf. Sci.* **598** (2005) 246–252.

[122] H. Yildirim, A. Kara, S. Durukanoglu, T. S. Rahman, Calculated pre-exponential factors and energetics for adatom hopping on terraces and steps of Cu(100) and Cu(110), *Surf. Sci.* **600** (2006) 484–492.

[123] U. Kürpick, Self-diffusion on (100), (110), and (111) surfaces of Ni and Cu: A detailed study of prefactors and activation energies, *Phys. Rev. B* **64** (2001) 075418 1–7.

[124] U. Kürpick, T. S. Rahman, Vibrational free energy contribution to self-diffusion on Ni(100), Cu(100) and Ag(100), *Surf. Sci.* **383** (1997) 137–148.

[125] U. Kürpick, T. S. Rahman, Diffusion processes relevant to homoepitaxial growth on Ag(100), *Phys. Rev. B* **57** (1998) 2482–2492.

[126] B. D. Yu, M. Scheffler, Anisotropy of growth of the close-packed surfaces of silver, *Phys. Rev. Lett.* **77** (1996) 1095–1098.

[127] P. J. Feibelman, R. Stumpf, Adsorption-induced lattice relaxation and diffusion by concerted substitution, *Phys. Rev. B* **59** (1999) 5892–5897.

[128] A. V. Evteev, A. T. Kosilov, S. A. Solyanik, Atomic mechanisms and kinetics of self-diffusion on the Pd(001) surface, *Phys. Solid State* **46** (2004) 1781–1784.

[129] B. D. Yu, M. Scheffler, *Ab initio* study of step formation and self-diffusion on Ag(100), *Phys. Rev. B* **55** (1997) 13916–13924.

[130] R. C. Nelson, T. L. Einstein, S. V. Khare, P. J. Reus, Energetics of steps, kinks, and defects on Ag{100} and Ag{111} using the embedded atom method, and some consequences, *Surf. Sci.* **295** (1993) 462–484.

[131] Z. Chvoj, C. Ghosh, T. S. Rahman, M. C. Tringides, Prefactors for interlayer diffusion on Ag/Ag(111), *J. Phys.: Condens. Matter* **15** (2003) 5223–5230.

[132] R. M. Lynden-Bell, Migration of adatoms on the (100) surface of face-centered-cubic metals, *Surf. Sci.* **259** (1991) 129–138.

[133] J. Zhuang, L. Liu, Adatom self-diffusion on Pt(100) surface by an ad-dimer migrating, *Science in China A* **43** (2000) 1108–1113.

[134] G. Boisvert, L. J. Lewis, Self-diffusion on low-index metallic surfaces: Ag and Au(100) and (111), *Phys. Rev. B* **54** (1996) 2880–2889.

[135] J. E. Müller, H. Ibach, Migration of point defects at charged Cu, Ag, and Au(100) surfaces, *Phys. Rev. B* **74** (2006) 085408 1–10.

[136] P. J. Feibelman, Diffusion Path for an Al Adatom on Al(001), *Phys. Rev. Lett.* **65** (1990) 729–732.

[137] R. Ferrando, Correlated jump-exchange processes in the diffusion of Ag on Ag(110), *Phys. Rev. Lett.* **76** (1996) 4195–4198.

[138] M. I. Haftel, Surface reconstruction of platinum and gold and the embedded-atom method, *Phys. Rev. B* **48** (1993) 2611–2622.

[139] M. I. Haftel, M. Rosen, T. Franklin, M. Hettermann, Molecular dynamics observations of the interdiffusion and Stranski-Krastanov growth in the early film deposition of Au on Ag(100), *Phys. Rev. Lett.* **72** (1994) 1858–1861.

[140] M. I. Haftel, M. Rosen, Molecular-dynamics description of early film deposition of Au on Ag(110), *Phys. Rev. B* **51** (1995) 4426–4434.

[141] M. I. Haftel, M. Rosen, New ballistically and thermally activated exchange processes in the vapor deposition of Au on Ag(111): a molecular dynamics study, *Surf. Sci.* **407** (1998) 16–26.

[142] W. Xiao, P. A. Greaney, D. C. Chrzan, Adatom transport on strained Cu(001): Surface crowdions, *Phys. Rev. Lett.* **90** (2003) 146102 1–4.

[143] W. Xiao, P. A. Greaney, D. C. Chrzan, Pt adatom diffusion on strained Pt(001), *Phys. Rev. B* **70** (2004) 033402 1–4.

[144] Y. X. Wang, Z. Y. Pan, Z. J. Li, Q. Wei, L. K. Zang, Z. X. Zhang, Effect of tensile strain on adatom diffusion on Cu(111) surface, *Surf. Sci.* **545** (2003) 137–142.

[145] D. Passerone, M. Parrinello, Action-derived molecular dynamics in the study of rare events, *Phys. Rev. Lett.* **87** (2001) 108302 1–4.

[146] S. Y. Kim, I.-H. Lee, S. Jun, Transition-pathway models of atomic diffusion on fcc metal surfaces. I. Flat surfaces, *Phys. Rev. B* **76** (2007) 245407 1–15.

[147] H. Brune, K. Bromann, H. Röder, K. Kern, J. Jacobsen, P. Stoltze, K. Jacobsen, J. Nørskov, Effect of strain on surface diffusion and nucleation, *Phys. Rev. B* **52** (1995) R 14380–14383.

[148] H. Brune, H. Röder, C. Boragno, K. Kern, Strain relief at hexagonal-close-packed interfaces, *Phys. Rev. B* **49** (1994) 2997–3000.

[149] C. Ratsch, A. P. Seitsonen, M. Scheffler, Strain dependence of surface diffusion: Ag on Ag(111) and Pt(111), *Phys. Rev. B* **55** (1997) 6750–6753.

[150] J. A. Meyer, J. Vrijmoeth, H. A. van der Vegt, E. Vlieg, R. J. Behm, Importance of the additional step-edge barrier in determining film morphology during epitaxial growth, *Phys. Rev. B* **51** (1995) 14790–14793.

[151] B. D. Yu, M. Scheffler, Physical origin of exchange diffusion on fcc(100) metal surfaces, *Phys. Rev. B* **56** (1997) R15569–15572.

[152] M. Schroeder, D. E. Wolf, Diffusion on strained surfaces, *Surf. Sci.* **375** (1997) 129–140.

[153] Y. B. Liu, D. Y. Sun, X. G. Gong, Local strain induced anisotropic diffusion on $(23 \times \sqrt{3})$-Au(111) surface, *Surf. Sci.* **498** (2002) 337–342.

[154] F. Ercolessi, M. Parinello, E. Tosatti, Au(100) reconstruction in the glue model, *Surf. Sci.* **177** (1986) 314–328.

[155] F. Ercolessi, M. Parrinello, E. Tosatti, Simulation of gold in the glue model, *Phil. Mag. A* **58** (1988) 213–226.

[156] J. C. Tully, G. H. Gilmer, M. Shugard, Molecular dynamics of surface diffusion. I. The motion of adatoms and clusters, *J. Chem. Phys.* **71** (1979) 1630–1642.

[157] A. F. Voter, J. D. Doll, Dynamical corrections to transition state theory for multistate systems: Surface self-diffusion in the rare-event regime, *J. Chem. Phys.* **82** (1985) 80–92.

[158] D. E. Sanders, A. E. DePristo, A non-unique relationship between potential energy surface barrier and dynamical diffusion barrier: fcc(111) metal surface, *Surf. Sci.* **264** (1992) L169–176.

[159] R. Ferrando, R. Spadacini, G. E. Tommei, Kramer's problem in periodic potentials: Jump rate and jump lengths, *Phys. Rev. E* **48** (1993) 2437–2451.

[160] Y. Georgievskii, E. Pollak, Semiclassical theory of activated diffusion, *Phys. Rev. E* **49** (1994) 5098–5102.

References

[161] Y. Georgievskii, E. Pollak, Long hops of an adatom on a surface, *Surf. Sci.* **355** (1996) L366–370.

[162] L. Y. Chen, S. C. Ying, Solution of the Langevin equation for rare event rates using a path integral formalism, *Phys. Rev. B* **60** (1999) 16965–16971.

[163] J. Ellis, J. P. Toennies, A molecular dynamics simulation of the diffusion of sodium on a Cu(001) surface, *Surf. Sci.* **317** (1994) 99–108.

[164] M. Azzouz, H. J. Kreuzer, M. R. A. Shegelski, Long jumps in surface diffusion: A microscopic derivation of the jump frequencies, *Phys. Rev. Lett.* **80** (1998) 1477–1480.

[165] J. Ferrón, L. Gómez, J. J. de Miguel, R. Miranda, Nonstochastic behavior of atomic surface diffusion on Cu(111) down to low temperatures, *Phys. Rev. Lett.* **93** (2004) 166107 1–4.

[166] J. Ferrón, R. Miranda, J. J. de Miguel, Atomic jumps during surface diffusion, *Phys. Rev. B* **79** (2009) 245407 1–9.

[167] A. Beausoleil, P. Desjonquères, A. Rocheford, Effects of long jumps, reversible aggregation, and Meyer-Neldel rule on submonolayer epitaxial growth, *Phys. Rev. E* **78** (2008) 021604 1–18.

[168] S. C. Wang, J. D. Wrigley, G. Ehrlich, Atomic jump lengths in surface diffusion: Re, Mo, Ir, and Rh on W(211), *J. Chem. Phys.* **91** (1989) 5087–5096.

[169] D. C. Senft, G. Ehrlich, Long jumps in surface diffusion: One-dimensional migration of isolated adatoms, *Phys. Rev. Lett.* **74** (1995) 294–297.

[170] D. C. Senft, Long Jumps in Surface Diffusion on Tungsten(211), Materials Science Ph.D. Thesis, University of Illinois at Urbana-Champaign, Urbana, 1994.

[171] D. C. Senft, Atomic jump length in surface diffusion: Experiment and theory, *Appl. Surf. Sci.* **94/95** (1996) 231237.

[172] T. R. Linderoth, S. Horch, E. Laegsgaard, I. Stensgaard, F. Besenbacher, Surface diffusion of Pt on Pt(110): Arrhenius behavior of long jumps, *Phys. Rev. Lett.* **78** (1997) 4978–4981.

[173] F. Montalenti, R. Ferrando, Competing mechanisms in adatom diffusion on a channeled surface: Jumps versus metastable walks, *Phys. Rev. B* **58** (1998) 3617–3620.

[174] V. Rosato, M. Guillope, B. Legrand, Thermodynamical and structural properties of f.c.c. transition metals using a simple tight-binding model, *Philos. Mag. A* **59** (1997) 321–336.

[175] F. Cleri, V. Rosato, Tight-binding potentials for transition metals and alloys, *Phys. Rev. B* **48** (1993) 22–33.

[176] H. T. Lorensen, J. K. Nørskov, K. W. Jacobsen, Mechanism of self-diffusion on Pt(110), *Phys. Rev. B* **60** (1999) R5149–5152.

[177] S.-M. Oh, S. J. Koh, K. Kyuno, G. Ehrlich, Non-nearest-neighbor jumps in 2D diffusion: Pd on W(110), *Phys. Rev. Lett.* **88** (2002) 236102 1–4.

[178] S.-M. Oh, K. Kyuno, S. J. Koh, G. Ehrlich, Atomic jumps in surface self-diffusion:W on W(110), *Phys. Rev. B* **66** (2002) 233406 1–4.

[179] G. Antczak, G. Ehrlich, Long jump rates in surface diffusion: W on W(110), *Phys. Rev. Lett.* **92** (2004) 166105 1–4.

[180] G. Antczak, G. Ehrlich, Jump processes in surface diffusion, *Surf. Sci. Rep.* **62** (2007) 39–61.

[181] G. Antczak, Long jumps in one-dimensional surface self-diffusion: Rebound transitions, *Phys. Rev. B* **73** (2006) 033406 1–4.

[182] G. DeLorenzi, *Dynamics of Atom Jumps on Surfaces*, 1989, www.lanl.gov/orgs/t/tl/surface_diffusion.

[183] G. Antczak, Kinetics of atom rebounding in surface self-diffusion, *Phys. Rev. B* **74** (2006) 153406 1–3.

[184] H.-W. Fink, G. Ehrlich, Rhenium on W(110): Structure and mobility of higher clusters, *Surf. Sci.* **150** (1985) 419–429.

[185] S. C. Wang, U. Kürpick, G. Ehrlich, Surface diffusion of compact and other clusters: Ir$_x$ on Ir(111), *Phys. Rev. Lett.* **81** (1998) 4923–4926.
[186] M. Schunack, T. R. Linderoth, F. Rosei, E. Laegsgaard, I. Stensgaard, F. Besenbacher, Long jumps in the surface diffusion of large molecules, *Phys. Rev. Lett.* **88** (2002) 156102 1–4.
[187] M.-C. Marinica, C. Barreteau, D. Spanjaard, M.-C. Desjonquères, Diffusion rates of Cu adatoms on Cu(111) in the presence of an adisland nucleated at fcc or hcp sites, *Phys. Rev. B* **72** (2005) 115402 1–16.
[188] N. B. Cabrera, The structure of crystal surfaces, *Discussions Faraday Soc.* **28** (1959) 16–22.
[189] R. W. Zwanzig, Collision of a gas atom with a cold surface, *J. Chem. Phys.* **32** (1960) 1173–1177.
[190] G. Ehrlich, in: *Metal Surfaces: Structure, Energetics and Kinetics*, Adsorption and surface structure, (ASM, Metals Park, Ohio, 1963), p. 221–258.
[191] G. Ehrlich, An atomic view of adsorption, *Brit. J. Appl. Phys.* **15** (1964) 349–364.
[192] B. McCarroll, G. Ehrlich, Trapping and Energy Transfer in Atomic Collisions with a Crystal Surface, *J. Chem. Phys.* **38** (1963) 523–532.
[193] G. Ehrlich, in: *Proc. 9th Internat'l Vacuum Cong. and 5th Internat'l Conf. on Solid Surfaces, Invited Speakers' Volume*, J. L. deSegovia (ed.), Layer growth – an atomic picture, (ASEVA, Madrid, 1983), p. 3–16.
[194] T. Gurney, F. Hutchinson, R. D. Young, Condensation of tungsten on tungsten in atomic detail: Observations with the field-ion microscope, *J. Chem. Phys.* **42** (1965) 3939–3942.
[195] R. D. Young, D. C. Schubert, Condensation of tungsten on tungsten in atomic detail: Monte Carlo and statistical calculations vs experiment, *J. Chem. Phys.* **42** (1965) 3943–3950.
[196] J. C. Tully, Stochastic-trajectory simulations of gas–surface interactions: Xe on Pt(111), *Faraday Discuss. Chem. Soc.* **80** (1985) 291–298.
[197] M. Schneider, A. Rahman, I. Schuller, Role of relaxation in epitaxial growth: A molecular-dynamics study, *Phys. Rev. Lett.* **55** (1985) 604–606.
[198] D. E. Sanders, D. M. Halstead, A. E. DePristo, Metal/metal homoepitaxy on fcc(111) and fcc(001) surfaces: Deposition and scattering from small islands, *J. Vac. Sci. Technol. A* **10** (1992) 1986–1992.
[199] H.-W. Fink, Mono-atomic tips for scanning tunneling microscopy, *IBM J. Res. Develop.* **30** (1986) 460–465.
[200] H.-W. Fink, Point source for ions and electrons, *Physica Scripta* **38** (1988) 260–263.
[201] W. F. Egelhoff, I. Jacob, Reflection high-energy electron diffraction (RHEED) oscillations at 77 K, *Phys. Rev. Lett.* **62** (1989) 921–924.
[202] C. Koziol, G. Lilienkamp, E. Bauer, Intensity oscillations in reflection high-energy electron diffraction during molecular beam epitaxy of Ni on W(110), *Appl. Phys. Lett.* **51** (1987) 901–903.
[203] D. K. Flynn-Sanders, J. W. Evans, P. A. Thiel, Homoepitaxial growth on Pd(100), *Surf. Sci.* **289** (1993) 75–84.
[204] J. W. Evans, D. E. Sanders, P. A. Thiel, A. E. DePristo, Low-temperature epitaxial growth of thin metal films, *Phys. Rev. B* **41** (1990) 5410–5413.
[205] D. E. Sanders, A. E. DePristo, Metal/metal homoepitaxy on fcc(001) surfaces: Is there transient mobility of adsorbed atoms? *Surf. Sci.* **254** (1991) 341–353.
[206] P. S. Weiss, D. M. Eigler, Adsorption and accommodation of Xe on Pt{111}, *Phys. Rev. Lett.* **69** (1992) 2240–2243.
[207] D. M. Eigler, E. K. Schweizer, Positioning single atoms with a scanning tunnelling microscope, *Nature* **344** (1990) 524–526.

[208] H. J. Ernst, F. Fabre, J. Lapujoulade, Growth of Cu on Cu(100), *Surf. Sci.* **275** (1992) L682–684.
[209] P. Bedrossian, B. Poelsema, G. Rosenfeld, L. C. Jorritsma, N. N. Lipkin, G. Comsa, Electron density contour smoothing for epitaxial Ag islands on Ag(100), *Surf. Sci.* **334** (1995) 1–9.
[210] P. Blandin, C. Massobrio, Diffusion properties and collisional dynamics of Ag adatoms and dimers on Pt(111), *Surf. Sci.* **279** (1992) L219–224.
[211] G. DeLorenzi, G. Ehrlich, Energy transfer in atom condensation on a crystal, *Surf. Sci. Lett.* **293** (1993) L900–907.
[212] S. C. Wang, G. Ehrlich, Atom condensation at lattice steps and clusters, *Phys. Rev. Lett.* **71** (1993) 4174–4177.
[213] G. L. Kellogg, Experimental observation of ballistic atom exchange on metal surfaces, *Phys. Rev. Lett.* **76** (1996) 98–101.
[214] C.-L. Liu, J. B. Adams, Diffusion behavior of single adatoms near and at steps during growth of metallic thin films on Ni surfaces, *Surf. Sci.* **294** (1993) 197–210.
[215] J. Jacobsen, K. W. Jacobsen, P. Stoltze, J. K. Nørskov, Island shape-induced transition from 2D to 3D growth for Pt/Pt(111), *Phys. Rev. Lett.* **74** (1995) 2295–2298.
[216] A. Hitzke, M. B. Hugenschmidt, R. J. Behm, Low temperature Ni atom adsorption on the Au(110)-(1×2) surface, *Surf. Sci.* **389** (1997) 8–18.
[217] G. Vandoni, C. Félix, R. Monot, J. Buttet, W. Harbich, Neighbor driven mobility of silver adatoms on Pd(100) measured by thermal helium scattering, *Surf. Sci.* **320** (1994) L63–67.
[218] C. Félix, G. Vandoni, W. Harbich, J. Buttet, R. Monot, Surface mobility of Ag on Pd(100) measured by specular helium scattering, *Phys. Rev. B* **54** (1996) 17039–17050.
[219] C. M. Gilmore, J. A. Sprague, A molecular dynamics study of transient processes during deposition on (001) metal surfaces, *J. Vac. Sci. Technol. A* **13** (1995) 1160–1164.
[220] C. M. Gilmore, J. A. Sprague, Molecular-dynamics simulation of the energetic deposition of Ag thin films, *Phys. Rev. B* **44** (1991) 8950–8957.
[221] M. Villarba, H. Jónsson, Atomic exchange processes in sputter deposition of Pt on Pt(111), *Surf. Sci.* **324** (1995) 35–46.
[222] M. Canepa, S. Terreni, P. Cantini, A. Campora, L. Mattera, Initial growth morphology in a heteroepitaxial system at low temperatures: Fe on Ag(100), *Phys. Rev. B* **56** (1997) 4233–4242.
[223] Y. Yue, Y. K. Ho, Z. Y. Pan, Low-temperature epitaxy and transient diffusion mechanisms on Cu(100), *Europhys. Lett.* **40** (1997) 453–457.
[224] Y. Yue, Y. K. Ho, Z. Y. Pan, Molecular-dynamics study of transient-diffusion mechanisms in low-temperature epitaxial growth, *Phys. Rev. B* **57** (1998) 6685–6688.
[225] H. Brune, G. S. Bales, H. Jacobsen, C. Boragno, K. Kern, Measuring surface diffusion from nucleation island densities, *Phys. Rev. B* **60** (1999) 5991–6006.
[226] T. Michely, J. Krug, in: *Islands, Mounds and Atoms*, (Springer-Verlag, Berlin, 2004), Section 2.1.
[227] H. Brune, J. Wintterlin, R. J. Behm, G. Ertl, Surface migration of "hot" adatoms in the course of dissociative chemisorption of oxygen on Al(111), *Phys. Rev. Lett.* **68** (1992) 624–626.
[228] H. Brune, J. Wintterlin, J. Trost, G. Ertl, J. Wiechers, R. J. Behm, Interaction of oxygen with Al(111) studied by scanning tunneling microscopy, *J. Chem. Phys.* **99** (1993) 2128–2148.
[229] C. Engdahl, G. Wahnström, Transient hyperthermal diffusion following dissociative chemisorption: a molecular dynamics study, *Surf. Sci.* **312** (1994) 429–440.
[230] J. Jacobsen, B. Hammer, K. W. Jacobsen, J. K. Nørskov, Electronic structure, total energies, and STM images of clean and oxygen-covered Al(111), *Phys. Rev. B* **52** (1995) 14954–14962.

[231] J. Wintterlin, R. Schuster, G. Ertl, Existence of a "hot" atom mechanism for the dissociation of O_2 on Pt(111), *Phys. Rev. Lett.* **77** (1996) 123–126.

[232] G. Wahnström, A. B. Lee, J. Strömquist, Motion of "hot" oxygen adatoms on corrugated metal surfaces, *J. Chem. Phys.* **105** (1996) 326–336.

[233] B. G. Briner, M. Doering, H.-P. Rust, A. M. Bradshaw, Mobility and trapping of molecules during oxygen adsorption on Cu(110), *Phys. Rev. Lett.* **78** (1997) 1516–1519.

[234] B. C. Stipe, M. A. Rezaei, W. Ho, Atomistic studies of O_2 dissociation on Pt(111) induced by photons, electrons, and by heating, *J. Chem. Phys.* **107** (1997) 6443–6447.

[235] U. Diebold, W. Hebenstreit, G. Leonardelli, M. Schmid, P. Varga, High transient mobility of chlorine on TiO_2(110): Evidence for "cannon-ball" trajectories of hot adsorbates, *Phys. Rev. Lett.* **81** (1998) 405–408.

[236] T. W. Fishlock, J. B. Pethica, F. H. Jones, R. G. Egdell, J. S. Foord, Interaction of chlorine with nickel (110) studied by scanning tunnelling microscopy, *Surf. Sci.* **377–379** (1997) 629–633.

[237] S. W. Hla, P. Lacovig, G. Comelli, A. Baraldi, M. Kiskinova, R. Rosei, Orientational anisotropy in oxygen dissociation on Rh(110), *Phys. Rev. B* **60** (1999) 7800–7803.

[238] M. Schmid, G. Leonardelli, R. Tscheliesznig, A. Biedermann, P. Varga, Oxygen adsorption on Al(111): low transient mobility, *Surf. Sci.* **478** (2001) L355–362.

[239] A. J. Komrowski, J. Z. Sexton, A. C. Kummel, M. Binetti, O. Weisze, E. Hasselbrink, Oxygen abstraction from dioxygen on the Al(111) surface, *Phys. Rev. Lett.* **87** (2001) 246103 1–4.

[240] Y. L. Li, D. P. Pullman, J. J. Yang, A. A. Tsekouras, D. B. Gosalvez, K. B. Laughlin, Z. Zhang, M. T. Schulberg, D. J. Gladstone, M. McGonigal, S. T. Ceyer, Experimental verification of a new mechanism for dissociative chemisorption: Atom abstraction, *Phys. Rev. Lett.* **74** (1995) 2603–2606.

[241] S. Schintke, S. Messerli, K. Morgenstern, J. Nieminen, W.-D. Schneider, Far-ranged transient motion of "hot" oxygen atoms upon dissociation, *J. Chem. Phys.* **114** (2001) 4206–4209.

[242] M. K. Rose, A. Borg, J. C. Dunphy, T. Mitsui, D. F. Ogletree, M. Salmeron, Chemisorption of atomic oxygen on Pd(111) studied by STM, *Surf. Sci.* **561** (2004) 69–78.

4 Diffusion on one-dimensional surfaces

Quite a number of diffusivity measurements have been reported during the last 40 years, not all of them of the same quality. Our aim here will be to outline the data currently existing for diffusivities, which allow us to describe diffusion of single atoms on metal surfaces.

The quantity that needs to be known in order to calculate the flux of atoms diffusing over the surface is the diffusivity D, which establishes the relation between the flux J and the concentration gradient $\partial c/\partial x$, according to Eq. (1.1) $J = -D\partial c/\partial x$. The diffusivity has to be known at different temperatures, and is therefore written as Eq. (1.23), $D = D_o \exp\left(-\dfrac{\Delta E_D}{kT}\right)$, where D_o, the diffusivity prefactor, is usually considered a constant that can be related to the frequency prefactor by Eq. (1.22) and (1.23), $D_o = \nu_0 \ell^2$; here ℓ^2 is the mean-square jump length, and ν_0 is the prefactor for atomic jumps. Our emphasis in this chapter will be largely upon the activation energy for diffusion, which from now on we write more simply as E_D instead of ΔE_D, and on the prefactor D_o.

Many systems have been examined by a variety of techniques, and with different degrees of reliability. What we will therefore do is to outline the facts reported for diffusivities of single atoms in the order in which the substrate appears in the periodic table. We will also comment on the limitations of the techniques used and the advantages of the approach employed. The presentation is divided into two parts: diffusivities on one-dimensional, channeled surfaces are treated first; only then will we examine the data on two-dimensional surfaces. For channeled surfaces we avoid some of the complications that arise on two-dimensional planes, and diffusion is therefore easier to discuss. We will not touch mass transport phenomena of many atoms at all, nor will we be concerned here in detail with the mechanism of diffusion or with atomic jump processes; that has already been discussed at length in the previous chapter. Just the empirical description of the diffusivities, and sometimes a comparison of different results will be attempted. We will not examine all studies; we tend to downplay measurements at but a single temperature, from which the diffusivity is obtained by assuming a value for the prefactor D_o. The aim of diffusion studies is the assignment of values for both the prefactor and the diffusion barrier, because at the moment these are the quantities that are generally difficult to calculate reliably.

In the survey that follows, the emphasis is on experimental results, and measurements of diffusivities by a variety of different techniques are presented. However, we must note a caution here. Most, but not all, scanning tunneling microscopy studies have measured the saturation island density to arrive at the diffusivity. There are several advantages to this approach, but also a big defect: this approach relies on the predictions of nucleation theory, which are based on the assumption that long-range atomic interactions can be ignored. This approximation is known to be incorrect at least for some systems, and may well vitiate results.

In recent years theoretical estimates of diffusivities have become popular, and have been attempted using different approaches. The results from these efforts will be presented in conjunction with experiments, but perhaps not as carefully as the experimental studies. In order to keep the presentation simple, we will, as already pointed out, start our survey by focusing on diffusion on one-dimensional, that is channeled planes.

In examining the work done to characterize atom diffusion on such surfaces, it will be important to keep in mind the geometry of the planes discussed. To make this task easier we have at the very start of the previous chapter shown hard-sphere models of the important channeled planes – the bcc(211) and (321) in Fig. 3.1, as well as fcc(110), (311), and (331) in Fig. 3.2. Their atomic arrangements are similar, but nevertheless their properties are quite different.

4.1   Aluminum: Al(110), (311), (331)

The first measurements of self-diffusion on aluminum were done by Tung [1] in 1981, who looked at (110), (311), and (331) planes in the field ion microscope. He determined the temperatures for the onset of diffusion in only ~20 trials, and by assuming a prefactor for the diffusivity of 10^{-3} cm^2/sec was able to arrive at the following diffusion barriers, all in eV: $(110)_{\parallel} = 0.43$, $(110)_{\perp} = 0.43$, $(311) = 0.48$, $(331) = 0.46$. These three planes are all made up of channels, and for (311) and (331) diffusion took place along the channels. On the (110) plane, however, diffusion was two-dimensional, and occurred at roughly the same rate in the direction of the channels and transverse to them. The cross-channel movement on this plane is probably an indication of atom exchange as the diffusion mechanism for this system. Till today it is the one and only attempt to experimentally investigate single atom surface diffusion for this system. Even if the statistics are not great, the measurements are very valuable.

Liu et al. [2] began calculational efforts by evaluating the self-diffusion barriers on several planes of aluminum, relying on embedded atom, EAM, potentials.[1] On the (110) plane, the activation energy for jumps along the channels was found to be 0.26 eV with a diffusivity prefactor D_o of 1.8×10^{-3} cm^2/sec using AFW parameters [3], while for cross-channel processes the barrier proved to be 0.30 eV with a prefactor of 6.0×10^{-2} cm^2/sec. In the VC approximation [4], cross-channel moves required an activation energy E_D of only 0.15 eV with a prefactor 2.4×10^{-2} cm^2/sec and, from this approach, exchange seems to be the likely mechanism for diffusion on this plane. However, AFW parametrization gave the opposite indication – hopping was more favorable, leaving the mechanism of movement uncertain for the Al(110) plane. On the (311) plane the diffusion barrier for

---

[1] The abbreviations commonly used in theoretical evaluations are briefly described in Section 2.8.

4.1 Aluminum: Al(110), (311), (331)

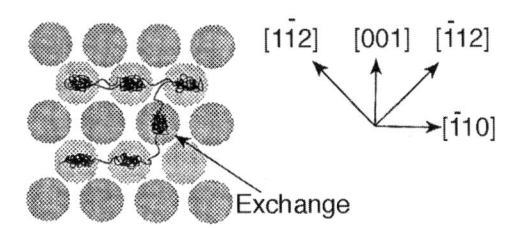

Fig. 4.1 Trajectories of aluminum atom for diffusion on Al(110) surface calculated using effective medium potentials (after Gravil and Holloway [5]).

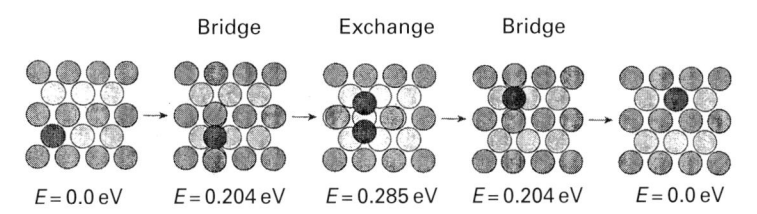

Fig. 4.2 Models for cross-channel motion by Al atom exchange on Al(110). Barrier heights arrived at with EMT potentials (after Gravil and Holloway [5]).

AFW parameters was 0.20 eV with a prefactor 2.0×10^{-3} cm^2/sec; with VC potentials the activation energy was 0.24 eV and the prefactor turned out to be 6.7×10^{-3} cm^2/sec. On (331) the barrier was 0.27 eV, with a prefactor of 2.0×10^{-3} cm^2/sec (AFW potentials), and 0.24 eV and 5.4×10^{-3} cm^2/sec (VC potentials). Only hopping was considered on these planes, however, it is the most likely mechanism and experimental findings seem to confirm that.

Effective medium theory was used by Gravil and Holloway [5] in molecular dynamics simulations, relying on a realistic many-body potential for aluminum diffusion on Al(110). They observed hopping along the channels with an activation energy of 0.204 eV, as well as exchange processes with a higher barrier, 0.285 eV, as indicated in Figs 4.1 and 4.2; the prefactor for the diffusivity was not calculated, but again hopping seems to be slightly more favorable.

Stumpf and Scheffler [6] did density-functional theory estimates for self-diffusion on Al(110). Jumps along the channels occurred over a barrier of 0.33 eV, while cross-channel diffusion over a much higher activation energy of 0.62 eV proceeded by atom exchange. Again the prefactor for diffusivity was not derived. Agrawal *et al.* [7] relied on Lennard-Jones potentials for their work, which gave them a rather too big self-diffusion barrier of 0.67 eV with a prefactor of 3.5×10^{-3} cm^2/sec for in-channel transitions. Cross-channel transitions, by jumps not exchange, occurred over a much higher barrier of 1.19 eV with a prefactor of 6.9×10^{-3} cm^2/sec. A year later, calculations for self-diffusion of aluminum on Al(110) were also done by Sun and Li [8], relying on the Vienna *ab-initio* local-density-functional theory simulation package [9–11]. They considered three mechanisms of motion: in-channel jumps, cross-channel jumps, and atom exchange, with activation energies of 0.38 eV, 0.83 eV, and 0.50 eV respectively. No information was given for the prefactor of the diffusivity. From the theoretical results presented it appears that diffusion should proceed along channels, and atom exchange would only become more significant at elevated temperatures. However, this prediction seems to be contradicted by experiments where cross-channel motion was clearly observed [1].

The situation with aluminum appears to be confused: for the (110) plane, activation energies derived in experiments are unreliable due to limited statistics and do not agree with calculations with the exception of the work by Sun and Li [8], but calculations are not too close together either. What is worth noting, however, is that direct observation of cross-channel motion in the FIM provided proof of exchange on this surface. For (331) and (311) planes experiments are not in accord with the calculated energetics either, but it is clear that movement is one-dimensional on these surfaces, suggesting that diffusion most likely proceeds by in-channel hopping.

4.2   Nickel: Ni(110)

Nickel is a comparatively soft metal, which field evaporates under the conditions for imaging with helium gas and makes FIM investigation difficult. However, because of its simple band structure and its technological significance, Tung and Graham [12] studied self-diffusion of nickel in 1980 on Ni(110) and (311) as well as (331) planes using the FIM. Imaging was done with neon, and after careful preparation to reduce defects it became possible to image adsorbed atoms. Arrhenius plots for diffusion were obtained with over 30 observations at each temperature and are shown in Fig 4.3. The results are unusual, especially on the (110) plane. As was already indicated in Fig. 3.46, rather different behavior was found on thermally treated samples and those heated in hydrogen and then field evaporated. The authors favored the former. On the (110) plane after thermal treatment the activation energy for self-diffusion was 0.23 ± 0.04 eV along the channels, but with a very suspicious prefactor for the diffusivity of 1×10^{-9} cm^2/sec. Cross-channel motion also occurred, with a slightly higher barrier of 0.32 ± 0.05 eV and with a low prefactor $D_o = 10^{-7}$ cm^2/sec.

The low prefactors have raised concerns, and the data for Ni(110) have been reworked by Liu et al. [2], who presumed a prefactor of 10^{-3} cm^2/sec. This yielded activation

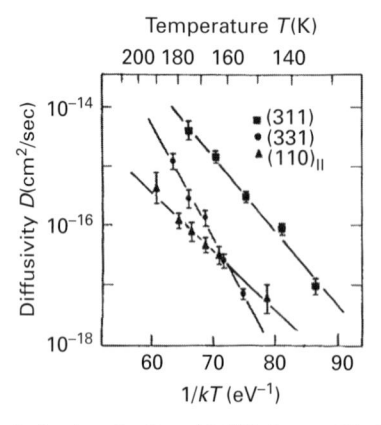

Fig. 4.3   Arrhenius plot for self-diffusion on Ni planes examined by field ion microscopy. Motion shown on (110) plane is along the channels (after Tung and Graham [12]).

4.2 Nickel: Ni(110)

energies of 0.45 eV both along the channels as well as across them, erasing the difference of 0.09 eV Tung and Graham reported between in- and cross-channel movement on the (110) plane. It should be noted, however, that the results on Ni(110) after hydrogen treatment look rather more reasonable, although a bit on the high side. The activation energy for in-channel motion was 0.30 ± 0.06 eV with $D_o = 10^1$ cm^2/sec, whereas for cross-channel diffusion the activation energy was lower, 0.26 ± 0.06 eV, with $D_o = 10^{-1}$ cm^2/sec.

In 1991 and 1992 Kellogg [13,14] reported on the diffusion of platinum adatoms over Ni(110) and (311) planes. The surfaces were treated by a combination of thermal cleaning, sputtering and field evaporation, but not hydrogen. For Ni(110), he obtained an estimate for the activation energy of 0.28 eV from the temperature for the onset of diffusion, assuming $D_o = 1 \times 10^{-3}$ cm^2/sec. From monitoring the voltage required for adatom field evaporation it was obvious that diffusion occurred by platinum atoms exchanging with nickel in the sides of the channels, as was shown in Fig. 3.50.

Scattering of helium atoms from a Ni(110) surface was done by Graham et al. [15,16] in 1997, who measured the broadening of the energy distribution of helium atoms scattered from diffusing nickel. Measurements were done at temperatures from 900 K to 1200 K, and yielded an activation energy of 0.536 ± 0.040 eV. This figure was for atoms jumping along the rows, with 80% single jumps and 20% double jumps. For cross-channel diffusion they deduced a lower barrier of 0.424 ± 0.040 eV, as shown in Fig. 4.4. Since this value is not too far away it suggests coexistence of both mechanisms, in agreement with FIM findings. It must be emphasized that the barriers here are slightly higher than what has been found from previous FIM measurements at lower temperature. This may not be due to the presence of double jumps – it could just be for single jump movement of individual adatoms influenced by interactions. The technique is not capable of controlling atom distances directly.

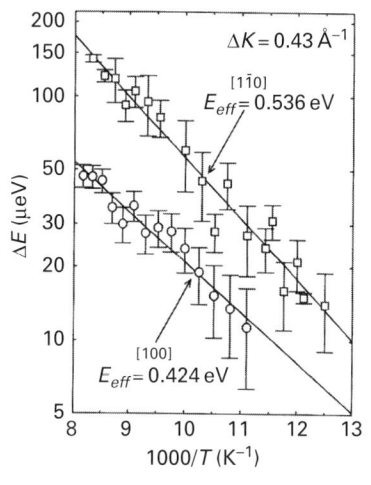

Fig. 4.4 Arrhenius plot for diffusion of Ni on Ni(110) derived from quasi-elastic He atom scattering. Note that cross-channel diffusion along [100] has a lower activation energy than in-channel motion (after Graham et al. [15]).

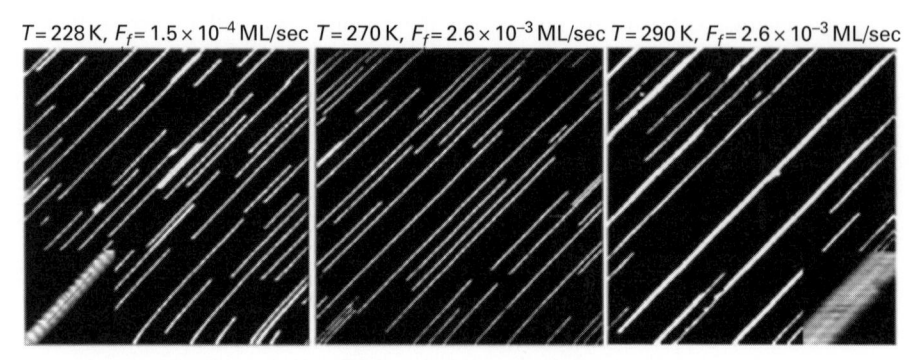

Fig. 4.5 STM images of 0.09 ML nickel deposit on Ni(110), under different deposition conditions, showing narrow deposits along channels. Insets reveal island width, which at low temperature is monatomic, but wider at 290 K (after Memmel *et al.* [17]).

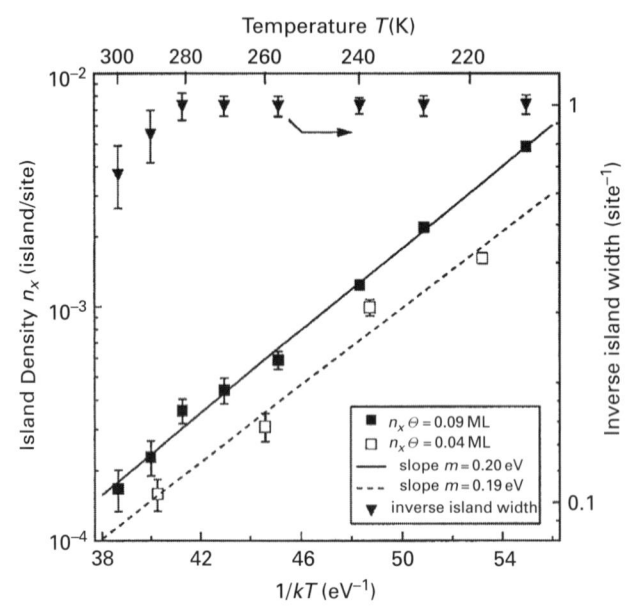

Fig. 4.6 Dependence of nickel island density on Ni(110) as a function of inverse temperature, yielding a diffusion barrier of 0.060 eV (after Memmel *et al.* [17]).

In 2005, scanning tunneling microscopy observations of the growth of nickel films on Ni(110) were made by Memmel *et al.* [17] over a number of temperatures for coverages up to 0.1 ML. Strongly elongated islands, shown in Fig. 4.5, were observed, revealing the anisotropy of the system. They found that despite the channels on the Ni (110) plane, nucleation was quasi-isotropic with a critical cluster size of one. From the temperature dependence of the island density shown in Fig. 4.6, they arrived at a diffusion barrier of 0.60 eV, in reasonable accord with the work of Graham [15] but not with single atom FIM findings, which indicates that barriers might be influenced by long-range interactions.

It is important to note that a slight anisotropy of diffusion on Ni (110) was already observed in 1967 by Maiya and Blakely [18] who found a very small difference between

4.2 Nickel: Ni(110)

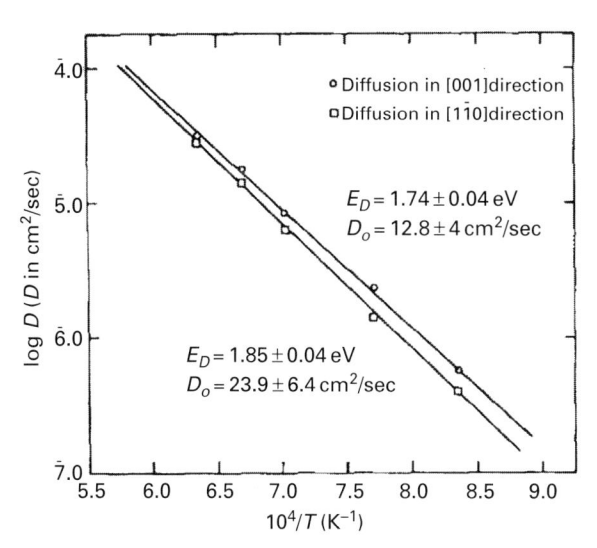

Fig. 4.7 Arrhenius plot for massive self-diffusion on Ni(110) plane, showing both in-channel and cross-channel motion (after Maiya and Blakely [18]).

movement along <110> compared with movement in <100>, shown in Fig. 4.7. This movement is not single atom diffusion, but the almost isotropic values of the activation energy were probably the first indication of atom exchange on this plane. Early diffusion studies were also done by Bonzel and Latta [19], who in 1978 carried out mass transfer experiments measuring the decay of a sinusoidal profile and compared these with theoretical investigations based on pairwise interactions. These also are not single atom studies, but the results, in Fig. 4.8 are interesting. At temperatures below 1000 K there appear to be two modes of diffusion, one along the <110> direction, that is along the channels, over a barrier of 0.76 eV, and a much higher activation energy process of 1.95 eV across them. Even if this value cannot be compared with single atom diffusion, as it is affected by interactions, these are among the early measurements showing anisotropic movement on this channeled surface of nickel.

Theoretical studies on nickel surfaces started in 1980 with the work of Flahive and Graham [20] who investigated Ni as well as other fcc and bcc metals using Morse potentials. Their findings are summarized in Fig. 4.9a and b, and will also be useful for other surfaces. However, it is worth noting that Morse potentials ignore many-body interactions, which is probably the reason for the poor agreement of this data with later studies – values calculated this way are definitely inaccurate.

Theoretical estimates with many-body potentials using EAM interactions for self-diffusion on nickel surfaces were first done by Liu et al. [2,21] in 1991. On Ni(110), jumps along the channels occurred over a barrier of 0.44 eV for AFW potentials, with a prefactor of 2.3×10^{-3} cm^2/sec; with VC potentials the activation energy was lower, 0.39 eV with a prefactor $D_o = 4.0 \times 10^{-3}$ cm^2/sec. For cross-channel motion the barriers were higher, 0.49 eV with a prefactor of 3.7×10^{-3} cm^2/sec (AFW), and 0.42 eV with the prefactor 2.8×10^{-2} cm^2/sec (VC). They also compared their results with investigations by Lennard-Jones and Morse potentials, and found no agreement; many-body effects

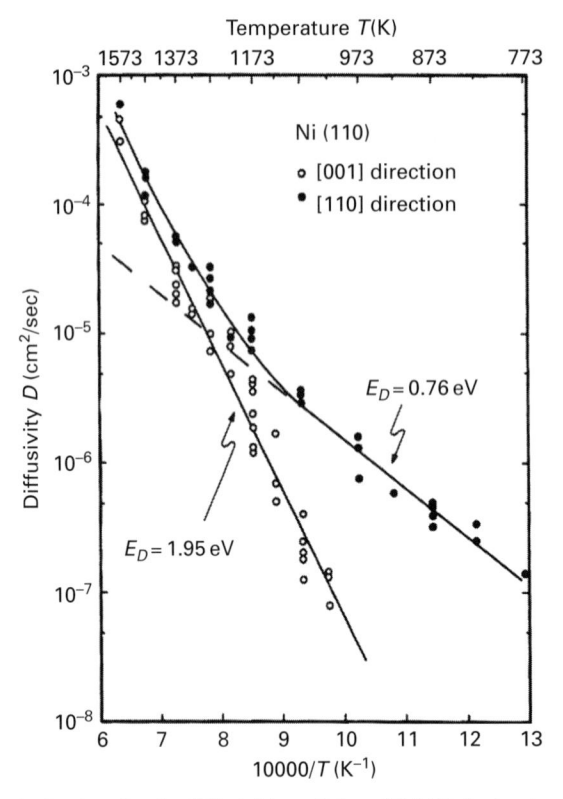

Fig. 4.8 Arrhenius plot for diffusivities of Ni on Ni(110) for in-channel [1$\bar{1}$0] and cross-channel [001] motion, measured in mass-transfer experiments. Strong anisotropy is clear at lower temperatures (after Bonzel and Latta [19]).

clearly play an important role on Ni surfaces. In 1993 Liu and Adams [22] investigated diffusion over steps on nickel using EAM potentials; for movement on the flat Ni(110) surface they once more obtained a value similar to their previous work: 0.41 eV for in-channel and 0.49 eV for cross-channel diffusion.

In 1992 Rice et al. [23] did calculations for self-diffusion on nickel surfaces using EAM with four different interatomic potentials: Foiles, Baskes, and Daw [24] (FBD), Voter and Chen [4] (VC), Oh and Johnson [25] (OJ), and Ackland, Tichy, Vitek and Finnis [26] (ATVF). The potentials differ mostly in the interpretation of the electron density surrounding the atom and the embedding function. The VC approach is similar to FBD but pairwise interactions are given by Morse potentials. Estimates were made over a huge range of temperatures, 50–1600 K. For in-channel motion on Ni(110) they obtained the following values of the activation energy, in eV: 0.50 (FBD), 0.53 (VC), 0.44 (OJ) and 0.64 (ATVF). The spread in values is ~0.2 eV, about 40% of the barrier. The values obtained from VC potentials do not agree with the previous findings of Liu [2]: 0.53 eV compared with 0.39 eV for VC potentials and 0.50 eV versus 0.44 eV with AFW parameters. Prefactors from different potentials are in quite reasonable agreement, all are in the range of standard values: 3.3 (FBD), 3.8 (VC), 3.6 (OJ), 4.5 (ATVF), all multiplied by 10^{-3} cm^2/sec. Rice et al. also investigated cross-channel movement, but only by hopping, not exchange; it

4.2 Nickel: Ni(110)

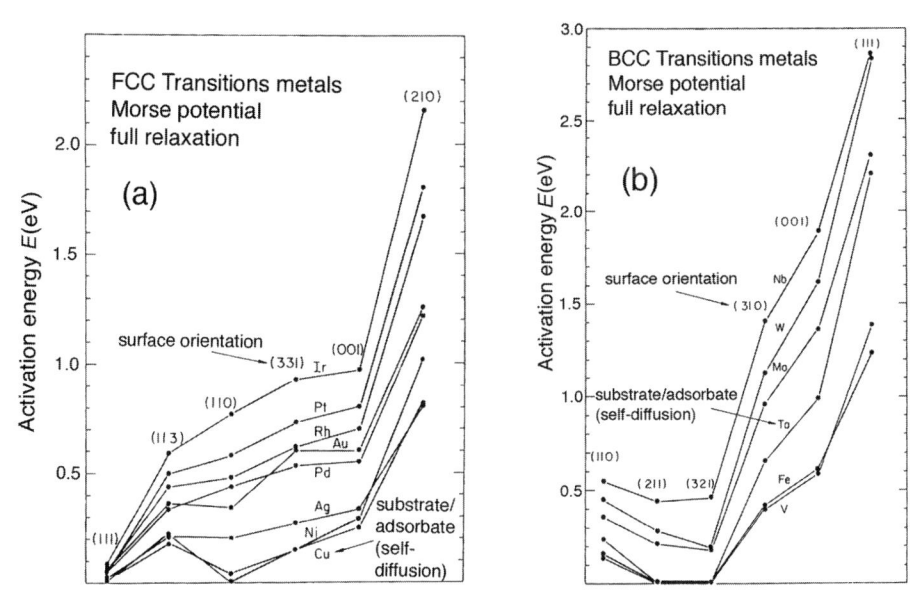

Fig. 4.9 Activation energies in self-diffusion of different single metal atoms, calculated using Morse potentials with complete substrate relaxation. (a) Fcc transition metals. (b) Bcc transition metals (after Flahive and Graham [20]).

required a much higher energy, with the following values, all in eV: 1.30 (FBD), 1.67 (VC), 1.01 (OJ), and 1.53 (ATVF). The diffusivity prefactors for hopping across the channels are slightly higher but in the same range of magnitude as for in-channel motion: 6.5×10^{-3} cm^2/sec (FBD), 7.5×10^{-3} cm^2/sec (VC), 6.5×10^{-3} cm^2/sec (OJ), and 7.7×10^{-3} cm^2/sec (ATVF). The spread of the Arrhenius plots for the different potentials is presented in Fig. 4.10. The authors did not consider exchange in their investigations, despite the clear indication of this mechanism from the experiments of Tung and Graham [12] as well as from the calculations of Liu et al. [2].

In 1994, Stoltze [27] examined self-diffusion on Ni(110) relying on EMT potentials. He estimated a barrier of 0.407 eV for in-channel jumps and 0.564 eV for cross-channel exchange while cross-channel jumps cost 1.157 eV. Again in-channel jumps seem to be the most favorable mechanism based on this study. The same year Perkins and DePristo [28] published calculations using MD/MC-CEM potentials of the activation energy for in-channel hopping as well as for atom exchange. For self-diffusion the activation energy for jumping decreased from 0.273 eV for 13 active atoms to 0.243 eV for 113 active atoms. For atom exchange similar changes were found – the barrier diminished from 0.427 eV to 0.392eV, showing that the size of the supercell used for the theoretical description of the surface influenced the results obtained for the activation energy. From CEM they obtained values of 0.18 eV for the hopping barrier and 0.35 eV for exchange. Six years later, Haug and Jenkins [29] investigated diffusion of Ni on the Ni (110) surface, working with a slab of 146 active atoms and using the EAM5 potential of Wonchoba et al. [30]. For isolated Ni atoms they came out with a barrier of 0.39 eV for hopping and 0.42 eV for exchange. What is more surprising is that the presence of hydrogen lowered the energy to 0.38 eV for both hopping and exchange. Hydrogen

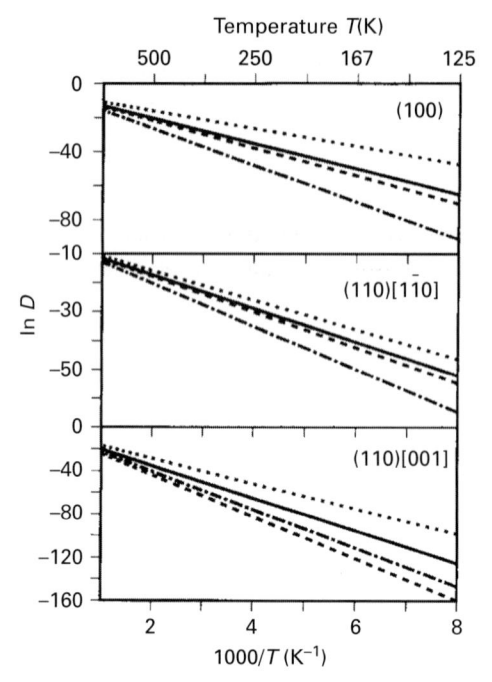

Fig. 4.10 Arrhenius plots for self-diffusion on Ni(100) and Ni(110) calculated with different EAM potentials. Solid line – FBD; long-dashed – VC; short-dashed – OJ; dot-dashed – ATVF (after Rice et al. [23]).

seemed to influence the activation energy for exchange more than for hopping, making both mechanisms equally probable. The authors associated this findings with weak Ni-H bonding, which stabilized the transition state for Ni hopping. The time constants were 1.4×10^{-5} sec for hopping and 5.6×10^{-5} sec for exchange.

More extensive calculations using EAM potentials to understand the physical meaning of prefactors for atom hopping were done by Kürpick [31]. For Ni(110) she found an activation energy of 0.39 eV with a prefactor $(1.4 \pm 0.1) \times 10^{-3}$ cm^2/sec for in-channel jumps, agreeing with prior studies. Her investigation concluded that the usual prefactor for the diffusivity should be observed for single atom movement. Agrawal et al. [7] relied on Lennard-Jones interactions to come up with an activation energy of 0.86 eV and a prefactor of 2.5×10^{-3} cm^2/sec for atom jumps along channels. They estimated a barrier of 1.68 eV for cross-channel jumps with a prefactor of 4.0×10^{-2} cm^2/sec. The barrier given here for in-channel jumps is clearly much too high, but the EAM estimates agree not too badly with the FIM measurements.

Ndongmouo and Hontinfinde [32] did calculations based on RGL potentials and found a barrier of 0.32 eV for in-channel jumping, and a somewhat larger value, 0.38 eV, for cross-channel exchange processes. Cross-channel jumps required overcoming a much higher activation energy of 1.33 eV. Kong and Lewis [33] in 2006 did EAM calculations of the self-diffusion prefactor, summing up all vibrational contributions. For motion along the channels making up the plane they found a temperature independent value of 4.0×10^{-3} cm^2/sec for atom hops, twice higher than one obtained from the local approximation [31], however, in the same magnitude range. Recently, Kim et al. [34] have done molecular dynamics studies of diffusion using RGL interactions. The barrier for in-channel jumps amounted to 0.301 eV, and the same value was found for cross-channel exchange suggesting coexistence of both mechanisms. Cross-channel jumps required a much higher energy of 1.026 eV. For a Cu adatom carrying out in-channel diffusion the barrier was 0.334 eV, for Pd 0.426 eV, for Ag 0.334 eV, for Pt 0.308 eV, and for Au 0.440 eV. The authors did not discuss their findings in detail. For Pt adatoms on Ni (110) Kim did not consider exchange as a mechanism, even though it was shown in experiments [13,14] that exchange not hopping was a leading mechanism.

The calculated results for diffusion on Ni(110) are in reasonable agreement with the somewhat uncertain experimental findings, but clearly, more reliable experiments are desirable. However, the existence of the exchange mechanism for diffusion on Ni(110) seems to be documented. Even though it appears to be slightly more energetically demanding, both processes are clearly present in experiments. There is still a huge lack of data for hetero-diffusion on this plane, the only experiments having been done by Kellogg.

4.3   Nickel: Ni(311) and Ni(331)

Compared with Ni(110), other fcc channeled planes, such as (331) and (311) have gained much less attention. The first investigation of these planes of nickel was done by Tung and Graham in 1978 [35] and in 1980 [12]. Arrhenius plots for diffusion on (311) and (331) planes were shown in Fig 4.3 but are based on only ~ 30 observations per temperature. Self-diffusion on the (311) plane took place over a barrier of 0.30 ± 0.03 eV, again with a low prefactor of 1.9×10^{-6} cm^2/sec. The barrier to diffusion on (331) was higher, 0.45 ± 0.03 eV with a prefactor of 2.3×10^{-3} cm^2/sec. For both these planes the movement of atoms is one-dimensional, an indication that movement is by hopping rather than by exchange with a lattice atom. Doubting the physical meaning of the low prefactor, Liu et al. [2] recalculated the data for these planes presuming a prefactor of 10^{-3} cm^2/sec. On the (311) plane, the recalculated barrier amounted to 0.37 eV, compared to 0.30 eV in the original study; the value for the (331) plane stayed at 0.45 eV after recalculation.

In 1992 Kellogg [14] reported on the diffusion of platinum adatoms over the Ni(311) plane. Diffusion observed at three temperatures took place by in-channel movement, over a barrier of 0.38 eV, estimated by assuming a value for the prefactor. As expected no exchange was observed on this surface.

Theoretical studies on this surfaces started as well in 1980 with the work of Flahive and Graham [20] using Morse potentials. Their findings, summarized in Fig. 4.9a and b, are definitely too low, as their calculations considered only pair interactions.

Many-body theoretical estimates using EAM interactions were done by Liu et al. [2,21]. Diffusion on (311) took place along the channels with an activation energy of 0.34 eV and a prefactor of 1.4×10^{-3} cm^2/sec for AFW potentials, and a barrier of 0.38 eV and a prefactor of 4.4×10^{-3} cm^2/sec using VC potentials. On the (331) plane the barriers were 0.45 eV and 0.46 eV for AFW and VC potentials, and the prefactors amounted to 1.4×10^{-3} and 4.2×10^{-3} cm^2/sec. The comparison with investigations by Lennard-Jones and Morse potentials gave no agreement, indicating that many-body effects clearly play an important role on Ni surfaces.

Merikoski et al. [36] in 1997 carried out EMT investigations for diffusion on the Ni(311) plane. The activation energy for in-channel hopping was 0.351 eV, very close to previous findings. The barrier to going over the edge by atom exchange at a straight step was more favorable than just hopping over the edge: 1.302 eV compared to 1.373 eV.

Both from experiments and theoretical investigations it is clear that movement on these planes proceeds by hopping, and the diffusion parameters seem to be in quite reasonable agreement. However, except for one experiment, there is no information about hetero-diffusion on these planes.

4.4 Copper: Cu(110), Cu(311), Cu(331)

4.4.1 Self-diffusion

The only experiment which derived values for the self-diffusion of copper on Cu(110) at small coverages was done by Robinson et al. [37] in 1996. Copper island growth on Cu(110) was measured in quasi-elastic helium atom scattering for a coverage of 0.5 ML, at which interactions are, however, very likely to influence the data. From the temperature dependence they were able to arrive at an activation energy of 0.84 ± 0.04 eV for diffusion across the channels on this plane, a value much higher than expected. They also were able to derive in the same way a value for in-channel movement, but since this turned out to be higher than for cross-channel motion, they disregarded it, based on number of bonds broken, as unphysical. Today we know that counting bonds is not a reliable method for diffusion estimates, and the lower value for cross-channel movement might be correlated with the existence of the exchange mechanism on this plane. Unfortunately, copper surfaces are not easily imaged in the FIM, but a considerable amount of work has been done on theoretical calculations for self-diffusion on copper. In fact, copper surfaces have received the most intense theoretical attention among metals. It is interesting that the main theoretical efforts have turned in a direction where there is essentially no experimental data available for comparison.

Calculational work seems to have been started by Wynblatt and Gjostein [38], who depended on Morse potentials to represent atomic interactions. For in-channel motion on Cu(110) they calculated a very small barrier height of 0.059 eV. In 1980 Flahive and Graham [20] also did Morse calculations for copper; their data are shown in Fig. 4.9. Their value for the diffusion barrier on Cu(110) is also very small, but slightly higher than the barriers obtained for Cu(311) and Cu(331). Such small values are presumably an indication of the importance of many-body interactions. Liu et al. [2] used EAM interactions. With AFW parameters their barrier for in-channel motion was 0.23 eV, with a prefactor of 8×10^{-4} cm^2/sec; with VC parameters their results were 0.28 eV for the barrier and 4.4×10^{-3} cm^2/sec for the prefactor. In cross-channel motion by atom exchange they came up with a slightly higher barrier, 0.30 eV, with AFW potentials; the prefactor was 3.2×10^{-2} cm^2/sec. With VC interactions they estimated a barrier of 0.31 eV for atom exchange with a prefactor of 2.7×10^{-2} cm^2/sec. Surprisingly this

4.4 Copper: Cu(110), Cu(311), Cu(331)

very first many-body calculation agrees well with one done 16 years later in 2008 [50]. Liu *et al.* also looked at diffusion on the (311) and (331) planes, where movement was one-dimensional. On Cu(311) they found a barrier of 0.26 eV with a prefactor of 1.2×10^{-3} cm^2/sec (AFW) and 0.28 eV and a prefactor of 3.1×10^{-3} cm^2/sec (VC). On Cu(331), they estimated a barrier of 0.28 eV and a prefactor of 1.0×10^{-3} cm^2/sec using AFW potentials; with VC interactions these figures were 0.33 eV and 3.0×10^{-3} cm^2/sec.

Using EMT, Hansen *et al.* [39] offered calculations for motion along the channels on Cu(110), with a barrier of roughly 0.18 eV, and across channels by atom exchange requiring an energy around 0.27 eV. Perkins and DePristo [28] used the CEM method in their work. They investigated the influence of the number of active atoms in the calculations of the energy. For in-channel diffusion on Cu(110) with 113 active atoms they found a barrier of 0.260 eV, and 0.485 eV for cross-channel motion by exchange of atoms using MD/MC-CEM. When the number of active atoms decreased to 13 the energy increased to 0.294 eV for hopping and 0.489 eV for exchange. From CEM they got much lower values of 0.08 eV for hopping and 0.09 eV for exchange in a system with 59 active atoms, which are rather too low. Calculations were also made by Stoltze [27] with effective medium theory. He found a barrier to in-channel diffusion of 0.292 eV and 0.419 eV for atom exchange to give cross-channel motion. Cross-channel atom jumping required overcoming a barrier of 0.826 eV. Karimi *et al.* [40] again relied on EAM interactions with AFW parameters and found a barrier to in-channel motion of 0.24 eV; for exchange with atom transfer to the adjacent channel they obtained 0.30 eV, but what was surprising in this study is that for exchange with atom transfer to the same channel the barrier was much higher, 0.87 eV. That may explain the preferential direction for exchange obtained from the displacement distribution on Ir(110) [41]; however, the physics of this process is not completely understood. Jumping across channels is definitely an unfavorable event with a barrier of ~1.15 eV.

In 1997 Merikoski *et al.* [36] looked at atomic processes next to steps using molecular dynamics based on effective medium theory. For hopping on Cu(311) they found a barrier of 0.232 eV. Hopping across to an adjacent channel required 0.889 eV while exchange needed only 0.822 eV. Once more hopping was favorable on fcc(311) surfaces.

Evangelakis *et al.* [42] examined the movement of Au on Cu(110) in the same year. For comparison they also determined the barriers for self-diffusion: 0.47 eV for hopping and 0.63 eV for exchange. One year later, the same authors derived much smaller values [43], again using RGL interactions, a second-moment approximation of the tight-binding model, to do molecular dynamics simulation of self-diffusion; their results are shown in Fig. 4.11. For temperatures up to 750 K they observed jumps along the channels; from the slope of their Arrhenius plot they derived a barrier of 0.25 eV; cross-channel motion via atom exchange took place over a barrier of 0.30 eV. Hopping is a slightly more favored mechanism in this range of temperatures. Above 850 K the frequency of both mechanisms has saturated and diffusion involves more complicated and concerted movements of hopping and exchange. No information about frequency prefactors was derived.

RGL interactions were utilized in calculations by Mottet *et al.* [44] for self-diffusion on Cu(110). They found a barrier of 0.23 eV for in-channel motion and 0.29 for cross-channel movement by atom exchange. The prefactor for the diffusivity was not

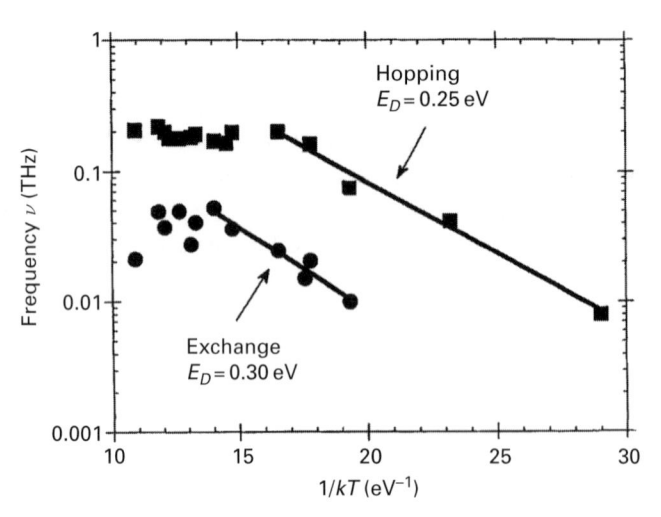

Fig. 4.11 Arrhenius plots for hopping and cross-channel exchange in self-diffusion on Cu(110), based on molecular dynamics simulations using RGL potentials (after Evangelakis *et al.* [43]).

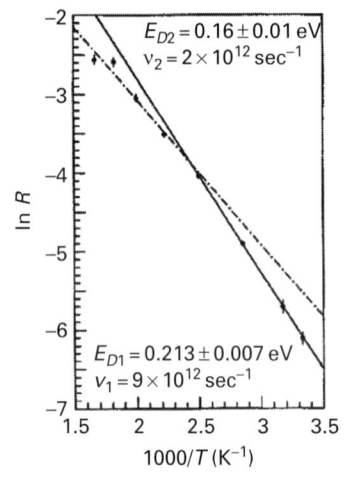

Fig. 4.12 Dependence of the diffusivity of Cu on Cu(110) upon the reciprocal temperature. Results obtained by molecular dynamics simulation with RGL potentials. Low temperature results (solid line) and high temperature (dash-dotted) are plotted separately, with activation energies of 0.213 ± 0.007 eV and 0.16 ± 0.01 eV respectively (after Montalenti and Ferrando [45]).

calculated. This study was continued by Montalenti and Ferrando [45] and the same values were reported. However, a detailed plot of the jump rates for temperatures from 300 K to 600 K revealed that a single straight line Arrhenius plot did not do justice to the data, which divided into two parts, below and above 400 K, as shown in Fig. 4.12. The low temperature results yielded an activation energy of 0.213 ± 0.007 eV, with a frequency prefactor of 9×10^{12} sec^{-1}; in the higher temperature range, the barrier was smaller, 0.16 ± 0.01 eV with a prefactor of 2×10^{12} sec^{-1}. At the highest temperature of 600 K, double jumps amounted to 14% of the total. It is interesting to note that double jumps caused a decrease of the diffusivity not an increase! It is also worth noting the standard frequency prefactor for both temperature ranges.

Prévot et al. [46] also resorted to RGL calculations of the diffusion barriers on Cu(110). For in-channel motion of Cu atoms they found an activation energy of 0.251 eV, whereas Cu-Cu exchange involved a barrier of 0.284 eV. The same year Wang and Adams [47] worked on kinetic Monte Carlo modeling of growth on copper surfaces. For estimates of diffusion barriers they used the EAM developed by Daw and Baskes. On the Cu(110) surface they obtained a value of 0.23 eV for jumps inside the channel and 0.30 eV for perpendicular movement by exchange. Again applying RGL potentials to the problem of self-diffusion by hopping on the (110) plane of copper, Kürpick [31] found an activation energy of 0.25 eV with a prefactor of $(1.1 \pm 0.1) \times 10^{-3}$ cm^2/sec.

Agrawal et al. [7] have made similar estimates, but with Lennard-Jones potentials that give only two-body interactions. For in-channel motion they found an activation energy of 0.76 eV and a prefactor of 7.5×10^{-3} cm^2/sec; for jumps between neighboring channels the barrier was significantly higher, 1.31 eV, with a prefactor of 2.6×10^{-2} cm^2/sec. As usual for studies with only pair interactions, the in-channel value for the diffusion barrier is too high. Ndongmouo and Hontinfinde [32] again used RGL interactions in their simulations and for in-channel motion attained a barrier of 0.23 eV. For cross-channel atom exchange the barrier was 0.27 eV, and for jumps between neighboring channels 1.07 eV. In molecular dynamics simulations also relying on RGL potentials, Kim et al. [34] found a barrier of 0.241 eV for in-channel motion, 0.323 eV for cross-channel exchange, and 1.020 eV for cross-channel jumps.

Calculations with EAM potentials were done by Yildirim et al. [48], who found the barrier to in-channel motion was 0.23 eV; the prefactor at 300 K amounted to 6.29×10^{-4} cm^2/sec. Jumping across a channel required 1.146 eV, with a prefactor of 9.97×10^{-4} cm^2/sec; exchange was not considered in this investigation. The same energetics were reported again by Durukanoglu et al. [49] for in-channel and cross-channel hopping. Atom exchange required an activation energy of 0.34 eV, so that in-channel hopping was the preferred process. The prefactor for hopping did not change much with temperature, becoming 6.39×10^{-4} cm^2/sec at 600 K. This work has been challenged, however, by Kong and Lewis [33,50], who used the same EAM potentials as Yildrim et al. but evaluated the total density of states. For the barrier height along the channels they found a value of 0.23 eV, for cross-channel jumps the value was 1.146 eV, both values are in exact agreement with the results of Yildrim et al. For the prefactor for an atom diffusing at 300 K they calculated a value of 4.0×10^{-3} cm^2/sec by summing over all vibrational frequencies compared to 1.4×10^{-3} cm^2/sec if only local vibrations are taken into account. The same values were measured at 600 K. Kong and Lewis concluded that the local approximation was not always appropriate in calculating the prefactor.

Recently, Stepanyuk et al. [51] investigated growth of cobalt nanostructures on a Cu(110) surface. For Cu in-channel movement they derived a value of 0.26 eV from molecular statics, exchange required only slightly more − 0.3 eV. Exchange of Cu close to an embedded Co atom required 0.1 eV less.

It is regrettable that no reliable experimental work has been done to determine the self-diffusion characteristics of Cu(110). From the work of Robinson et al. the exchange mechanism is likely to be present on the surface. The calculated values for the self-diffusion coefficients are in reasonable shape – the results of the various efforts using reasonable interactions between the atoms are close together. It is clear that many-body interactions are necessary to consider diffusion. Based on the theoretical studies, the exchange mechanism seems to be a bit more energetically demanding for diffusion, but the difference in barriers is not big. A combination of both may be possible but will need direct experimental proof, but the work of Robinson et al. seems to point in this direction.

4.4.2 Hetero-diffusion

Several investigations have been done of silver diffusion on copper surfaces. However, the concentrations of silver in these studies were quite appreciable and it is not clear that information about single atom phenomena was determined and that interactions were insignificant. Among these reports is the work of Roulet [52], Ghaleb and Peraillon [53], and Kürpick et al. [54]. What is worth noting, however, is the suggestion [53] of an exchange movement for Ag on Cu(110), based on observations of the anisotropy of diffusion independent of temperature.

Work has been reported for diffusion of lead on Cu(110) by Prévot et al. [55] using Rutherford backscattering spectrometry. In their studies a square of the surface was covered with ~ 0.1 monolayer of deposited lead, and the spreading of this deposit, at temperatures from 500 K to 800 K was then measured, as shown in Fig. 4.13. The diffusion coefficients along and across the channels at the different temperatures is given in Fig. 4.13b, yielding for motion along the channels the result

$$D = 1.5 \times 10^{-2} \exp(-0.57\,\text{eV}/kT)\text{cm}^2/\text{sec}. \tag{4.1}$$

For cross-channel movement the diffusivity was

$$D = 6.2 \times 10^{-3} \exp(-0.57\,\text{eV}/kT)\,\text{cm}^2/\text{sec}. \tag{4.2}$$

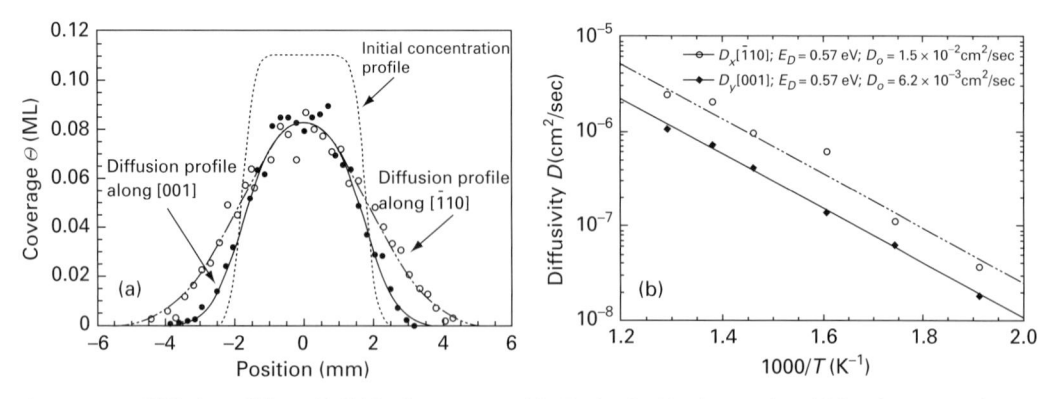

Fig. 4.13 Diffusion of Pb on Cu(110) plane measured by Rutherford backscattering. (a) Lead concentration profiles on Cu(110) plane. In-channel motion is more rapid than cross-channel diffusion. (b) Arrhenius plots for Pb diffusivities on Cu(110). Activation energies are the same in the two directions, but prefactor is larger for in-channel motion (after Prévot et al. [55]).

4.4 Copper: Cu(110), Cu(311), Cu(331)

The activation energies are the same, but the prefactor along the channel is higher than across. To rationalize this behavior, Prévot et al. came up with a model in which at temperatures above 250 K the lead atoms are embedded in the outermost layer; for lead to diffuse, it must first emerge from being embedded and change to an adatom by atom exchange. This exchange occurs with a copper atom, and not with another lead, as the diffusivity is not dependent on the coverage of lead atoms. The lead adatoms then diffuse along the channels until they again embed.

This model was soon tested by Prévot et al. [46], who carried out static as well as molecular dynamics studies using RGL inter-atomic potentials. From static energy calculations they found an activation energy of 0.212 eV for lead diffusion along the channels; for reinsertion of the lead by exchange with a copper lattice atom the barrier was 0.281 eV. The activation energy for emergence of a lead atom by exchange with a copper adatom was obtained as 0.590 eV, quite close to the experimental value of 0.57 eV. For diffusion of copper atoms along the channels the barrier was 0.251 eV, and for exchange with copper 0.284 eV. Molecular dynamics simulations for lead atoms at temperatures between 400 K and 700 K showed that multiple jumps appeared frequently and it was possible to obtain the rate of single jumps

$$\nu_a = 4.9 \times 10^{11} \exp(-0.14 \pm 0.02 \, \text{eV}/kT) \, \text{sec}^{-1}, \qquad (4.3)$$

the rate of correlated jumps

$$\nu_{ce} = 8.9 \times 10^{11} \exp(-0.18 \pm 0.02 \, \text{eV}/kT) \, \text{sec}^{-1}, \qquad (4.4)$$

and finally the rate for returning lead to the outer layer

$$\nu_r = 3.7 \times 10^{12} \exp(-0.27 \pm 0.03 \, \text{eV}/kT) \, \text{sec}^{-1}. \qquad (4.5)$$

These results are shown in the Arrhenius plots in Fig. 4.14a. The value of the barrier for reinsertion from the Arrhenius plot is close to the static one: 0.27 eV compared to

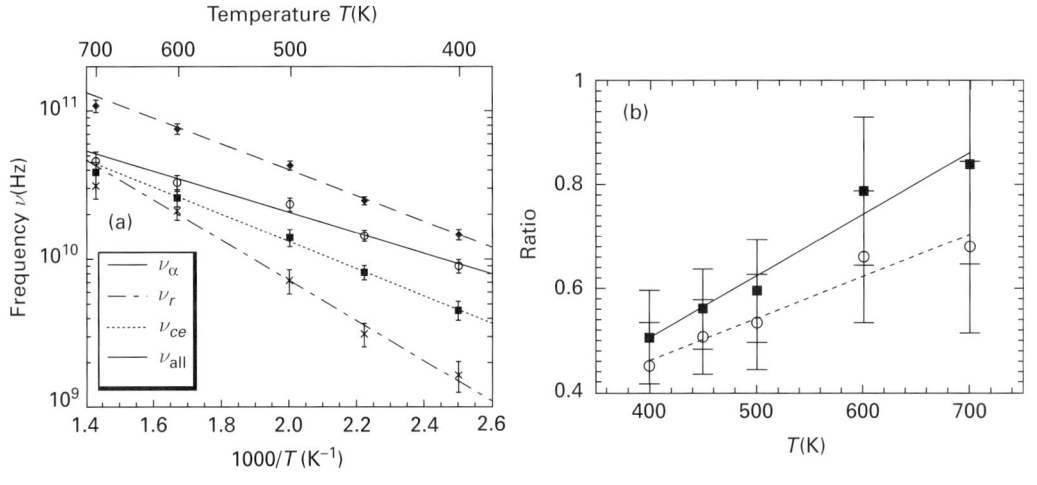

Fig. 4.14 (a) Dependence of jump rates for Pb on Cu(110) upon temperature, obtained with RGL potentials. Shown are single jumps, reinsertions, correlated events, all events. (b) Ratio of multiple (open circles) and of correlated events (full squares) to single jumps for in-channel diffusion (after Prévot et al. [46]).

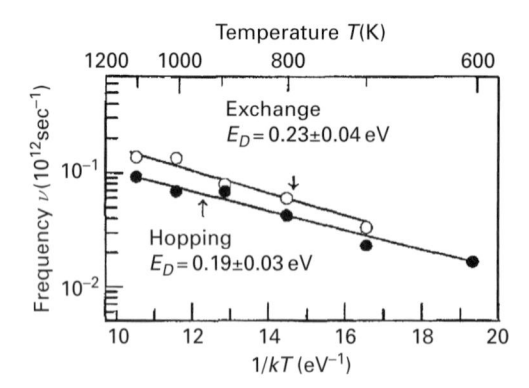

Fig. 4.15 Arrhenius plot for the jump frequency of Au atoms on Cu(110) by hopping along the channels and by atom exchange. Data derived by molecular dynamics simulations with RGL potentials (after Evangelakis et al. [42]).

0.281 eV. But the barrier for single jumps (0.14 eV compared to 0.212 eV for the static value) is quite different. The authors attribute this difference to not taking into account multiple jumps and other events. They also made molecular dynamics estimates and found long transitions in addition to single jumps. As shown in Fig. 4.14b, increasing the temperature increased the number of multiple jumps, but it should be noted that the temperatures were quite high.

Molecular dynamics simulations of the diffusion of gold atoms on Cu(110) were done using RGL interactions by Evangelakis et al. [42]. They observed both atom hopping along the channels and exchange with lattice atoms. The frequency of the diffusion processes is plotted in Fig. 4.15 against $1/kT$, yielding a diffusion barrier of 0.19 ± 0.03 eV for hopping and 0.23 ± 0.04 eV for exchange. It must be noted, however, that in the temperature range from 700 K to ~1100 K, exchange occurs more frequently than hopping. In a similar way, Kim et al. [34] arrived at a barrier height of 0.244 eV for in-channel diffusion of Ni on Cu(110), 0.318 eV for Pd, 0.254 eV for Ag, 0.376 eV for Pt, and 0.333 eV for Au. Exchange was not taken into account. The last value for gold is quite different from that derived by Evangelakis et al. [42], despite the similarity of the methods, but otherwise there are not a sufficient number of measurements available to allow a comparison.

Recently Stepanyuk et al. [51] looked at the movement of cobalt on Cu(110) with molecular statics and with VASP. From molecular statics they derived a value of 0.29 eV for in-channel jumps, while exchange required overcoming a barrier of 0.30 eV; perpendicular jumps were unlikely with an energy higher than 1 eV. From VASP the diffusion barriers were slightly higher, 0.35 eV for in-channel motion and 0.32 for exchange.

4.5 Molybdenum: Mo(211)

Unfortunately there is no direct experimental information available about single atom movement on the molybdenum (211) surface. The only information concerns spreading material over the surface and monitoring the work function changes. A lithium circular

4.5 Molybdenum: Mo(211)

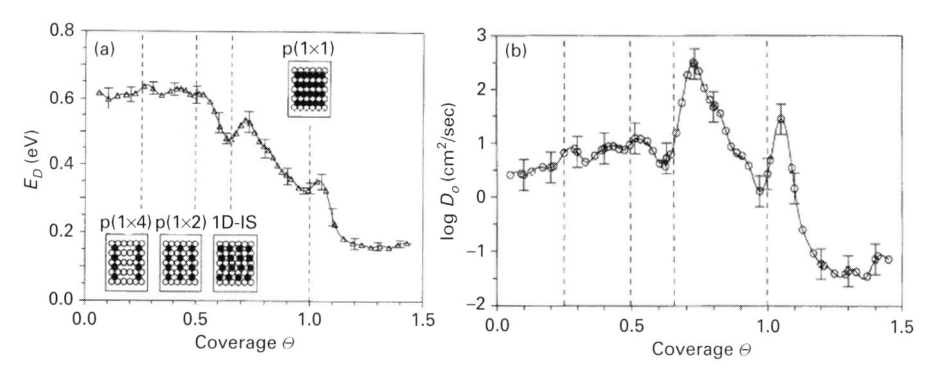

Fig. 4.16 Dependence of the diffusion characteristics for Li in-channel diffusion on Mo(112) upon coverage. (a) Activation energy for diffusion. (b) Prefactor D_o (after Naumovets et al. [56]).

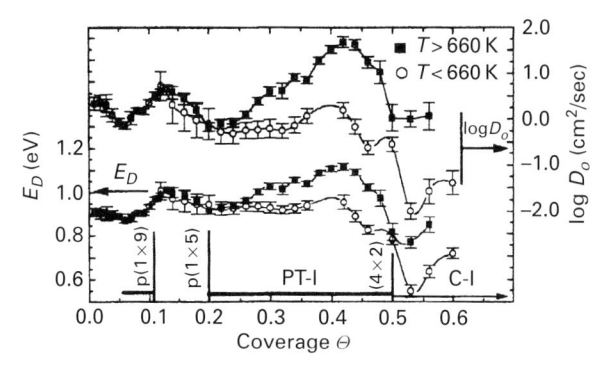

Fig. 4.17 Activation energy and prefactor for diffusion of Sr on Mo(211) channels as a function of coverage (after Loburets et al. [57]).

spot was deposited on the surface by Naumovets et al. [56] and a strong anisotropy of spreading was observed. The results were analyzed using the Boltzmann–Matano method. Coverages as low as 0.02 ML were investigated, which might yield information about single atom movement. For low coverage, the activation energy for motion along channels was around 0.6 eV, and was coverage independent up to 0.5 ML, as shown in Fig. 4.16a. In Fig. 4.16b the dependence of the prefactor for the diffusivity on the coverage is presented; at low concentrations the prefactor D_o amounted to 2.5 cm²/sec.

At the same time, Loburets et al. [57] studied diffusion of Sr on Mo(112) using the same technique. The activation energy and prefactor for the diffusivity are shown as a function of coverage in Fig. 4.17, with the former at 0.9 eV at the lowest concentration. The same method was also used by Loburets et al. [58] to investige diffusion of a submonolayer of Dy on Mo (211). However, for Dy the activation energy increased in going to low coverages, reaching 1.4 eV, as shown in Fig. 4.18. This method does not account for controlling defects on the surface, so it is impossible to get information about the mechanism of surface diffusion.

In 1999, Loburets [59] studied the diffusion of copper on the (211) surface of molybdenum, again measuring changes in the electronic work function of the substrate. Work appears to have been done at reasonably low concentrations, <0.1 ML, so that

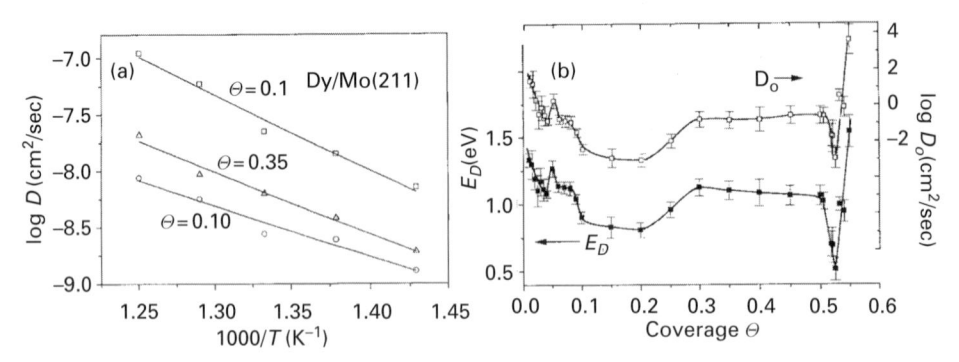

Fig. 4.18 Diffusion of Dy on Mo(112) plane. (a) Arrhenius plots for the diffusivity at different coverages. (b) Diffusion characteristics in their dependence on coverage (after Loburets et al. [58]).

individual adatom behavior may be dominant in the activation energy of 0.45 eV and the prefactor $D_o = 1.2 \times 10^{-3}$ cm^2/sec.

Unfortunately this plane was not investigated in FIM or STM studies, and there are also no theoretical investigations available for a comparison.

4.6 Rhodium: Rh(110), (311), (331)

Measurements of diffusion on rhodium by Ayrault [60] in 1974 came not too long after the initial work on tungsten surfaces, and should therefore be treated carefully. However, measurements were done with a single atom deposited on each plane. Arrhenius plots for these systems are shown in Fig. 4.19. On the (110) plane, diffusion of rhodium occurred over a barrier of 0.60±0.03 eV and a prefactor of 3×10^{-1} cm^2/sec. Diffusion over (311) involved a smaller activation energy, 0.54±0.04 eV and a prefactor $D_o = 2 \times 10^{-3}$ cm^2/sec. On the (331) plane, the barrier to diffusion was 0.64±0.04 eV, higher than for the other two planes, and the prefactor amounted to 1×10^{-2} cm^2/sec. What is interesting is that on all these surfaces, diffusion was always in the direction of the channels, and occurred by simple jumps. Unfortunately no other measurements were done later on for self-diffusion of rhodium, so these data stand alone.

Almost twenty years later, Kellogg [61] also with the FIM examined the movement of a platinum atom on these rhodium planes. Only on Rh(311) was an Arrhenius plot given, Fig. 4.20, with an activation energy of 0.44±0.05 eV and a prefactor $D_o = 1.6 \times 10^{-4\pm 0.05}$ cm^2/sec. Movement on this plane proceeded by simple hopping along the channels. The activation energy, but also the prefactor, are significantly lower than for the self-diffusion of rhodium. On (331) the activation energy, based on four temperatures from 256 K to 278 K, was given as 0.72±0.02 eV, presuming a prefactor of 10^{-3} cm^2/sec. On (110), the activation energy for cross-channel transitions was estimated as ~0.65 eV from the sudden loss of Pt atoms, which shows that the movement proceeded by exchange not by hopping. Cross-channel movement was certain, but the extent of in-channel motion was not clear. This study showed that the exchange mechanism is not forbidden for rhodium surfaces even if self-diffusion proceeds by hopping.

4.6 Rhodium: Rh(110), (311), (331)

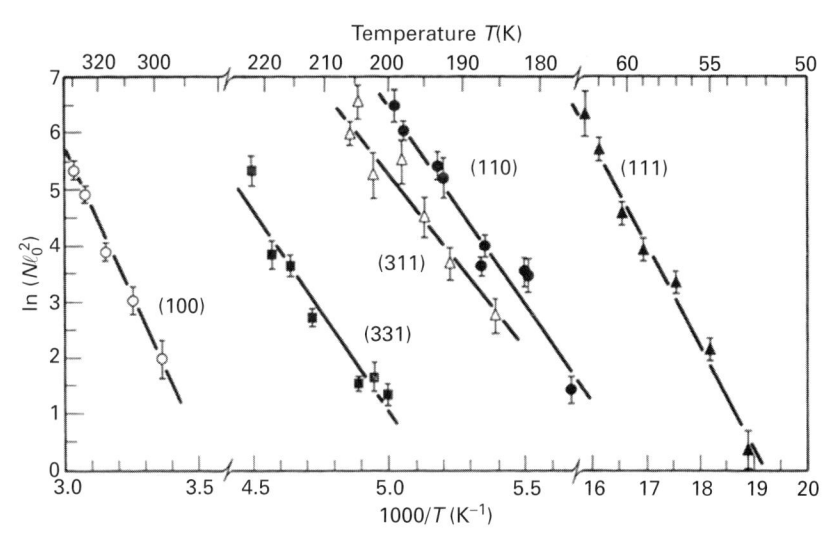

Fig. 4.19 Self-diffusion of rhodium adatoms on Rh planes determined in FIM [60]. On (111) and (100) planes motion is two-dimensional.

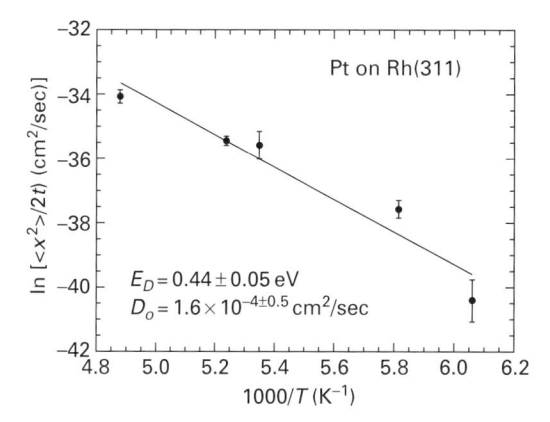

Fig. 4.20 Dependence of Pt atom diffusivity on Rh(311) plane upon the reciprocal temperature. Results from FIM observations of atom displacements (after Kellogg [61]).

In 1980 Flahive and Graham [20] used Morse potentials to calculate diffusion energies shown in Fig. 4.9. Approximation with Morse potentials gives a result a bit lower than the experimental value. Calculations for self-diffusion on Rh(110) were done by Liu et al. [2], who found a barrier of 1.24 eV for hops along the channel with a prefactor of 8×10^{-3} cm^2/sec using Lennard-Jones potentials. With the same potentials they calculated a barrier of 0.64 eV on the (311) with a prefactor of 5.4×10^{-3} cm^2/sec, and on the (331) plane the results were 1.31 eV and 7.4×10^{-3} cm^2/sec. The first and last barriers appear rather high. The barriers calculated with Morse potentials were much smaller and amounted to 0.48 eV, 0.44 eV, and 0.62 eV on the (110), (311), and (331), and are in better accord with experiments. Unfortunately, Liu et al. did not use a many-body EAM potential for rhodium. Agrawal et al. [7], who also utilized Lennard-Jones interactions, came up with an activation energy of 1.20 eV and a prefactor $D_o = 4.4 \times 10^{-3}$ cm^2/sec for self-diffusion along the channels on Rh(110), but this is clearly much too high a value. Their estimates for cross-channel jumps were 2.06 eV for the barrier and 1.0×10^{-2} cm^2/sec for the prefactor. It appears that the Lennard-Jones potential is unsuitable for diffusion on Rh surfaces.

The first and so far only calculations for self-diffusion of rhodium on Rh(110) with RGL many-body potentials was done by Ndongmouo and Hontinfinde [32]. They calculate an activation energy of 0.78 eV for jumps along the channels, rather higher than experiments, and 0.91 for cross-channel motion by atom exchange. Jumps across channel required going over a barrier of 1.80 eV.

Theoretical results for channeled rhodium planes are not in good shape. The experiments, while not really definitive, appear reasonable. The calculations with Morse potentials or from RGL are in reasonable accord, but the other estimates are quite far from the experimental results and more work here is clearly desirable.

4.7 Palladium: Pd(110), (311), (331)

Although experimental self-diffusion studies have not been made on Pd(110), the growth of copper has been examined by Bucher et al. [62], who studied the saturation density of copper islands at a coverage lower than 0.1 ML over a range of temperatures. Their results, in Fig. 4.21, show two different processes. At temperatures below 350 K,

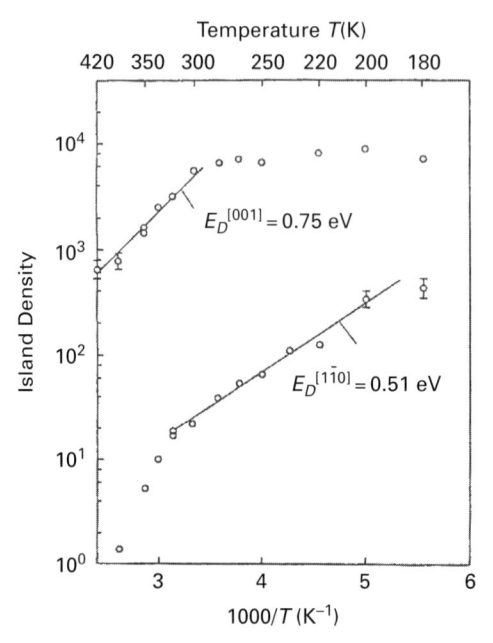

Fig. 4.21 Copper island density on Pd(110) plotted against reciprocal temperature for both transverse density (upper curve) and overall density, giving activation energies for diffusion (after Bucher et al. [62]).

4.7 Palladium: Pd(110), (311), (331)

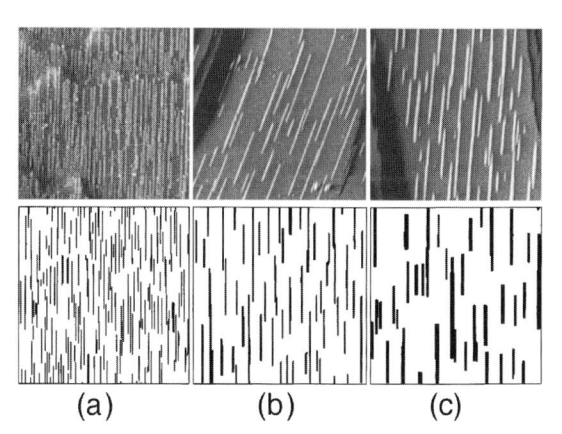

Fig. 4.22 Comparison of STM images of Cu islands on Pd(110) with simulations at (a) 265 K, (b) 300 K, and (c) 320 K. In-channel barrier height was 0.3 eV, cross-channel barrier 0.45 eV (after Li *et al.* [65]).

one-dimensional islands form along the channels, and Bucher *et al.* derived an activation energy for in-channel diffusion of 0.51 ± 0.05 eV. Above 300 K, the island density across the channel direction, shown by the upper curve in Fig. 4.21, yielded an Arrhenius plot with a barrier of 0.75 ± 0.07 eV. However, the values may not represent single atom movement, due to possible long-range interactions at the investigated coverage.

Bucher *et al.* also did calculations with EAM interactions [62–64] to find barriers for diffusion of Cu by hopping along and across the channels of 0.32 eV and 1.25 eV; the latter value was recognized as not realistic. Instead the authors suggested a series of steps including Cu-Pd exchange with an overall barrier height of 0.57 eV. What is interesting is that exchange of a Pd adatom with a Cu atom embedded in a row of palladium atoms costs only 0.42 eV; Pd-Pd exchange required the same energy as Cu-Pd exchange, 0.57 eV. Hopping of a Pd adatom which emerged after exchange cost 0.33 eV. Massobrio and Fernandez [63] examined diffusion of palladium atoms on Pd(110) relying on EAM potentials and calculated an activation energy of 0.28 eV for in-channel hopping, 0.42 eV for cross-channel exchange, and a very high barrier of 1.56 eV for cross-channel jumps.

In 1997 Li *et al.* [65] developed a kinetic model to describe the temperature dependence of island shapes observed in the STM. For islands of Cu on Pd(110) comparisons of experiment and model predictions are shown in Fig. 4.22. From the island densities plotted in Fig. 4.23 they derived activation barriers of 0.3 eV for diffusion along the channel and 0.45 eV for movement across channels. They also established that corner rounding was very important in the investigation of island shape; corner rounding slowed the overall process due to its high activation energy of 0.65 eV.

In 1980 Flahive and Graham [20] used Morse potentials to calculate the activation energy for self-diffusion on palladium. Their results are summarized in Fig. 4.9. Taking into account only pair interactions, they arrived at values a bit higher than later estimates: for Pd(110) around 0.45 eV, Pd(311) ~0.35 eV, and Pd(331) roughly 0.55 eV. Calculations of self-diffusion on Pd(110) were made with EAM many-body potentials by Liu *et al.* [2]. For in-channel motion the activation energy was 0.28 eV and the prefactor 4×10^{-4} cm^2/sec using AFW potentials. With VC potentials they found a diffusion barrier of 0.30 eV and a

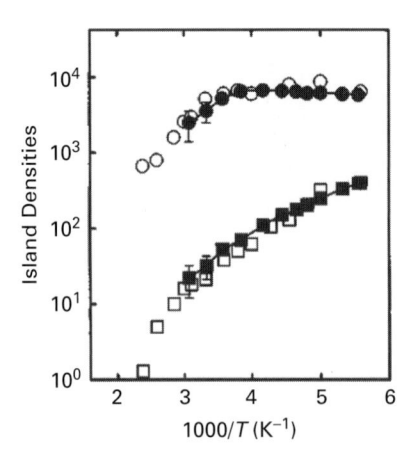

Fig. 4.23 Copper island densities on Pd(110) surface in their dependence upon the reciprocal temperature. Linear cross-channel islands indicated by circles, total island density by squares; experimental results plotted with open circles, solid symbols give simulated findings (after Li *et al.* [65]).

prefactor of 3.5×10^{-3} cm^2/sec. For cross-channel motion by atom exchange, the activation energy was 0.42 eV with a prefactor 3.3×10^{-2} cm^2/sec (AFW), and 0.34 eV with a prefactor of 2.4×10^{-2} cm^2/sec (VC). Liu *et al.* also evaluated diffusivities on the (311) and (331) planes. On the first, the activation energy was 0.37 eV and the prefactor 1.2×10^{-3} cm^2/sec (AFW), and 0.41 eV with a prefactor of 3.1×10^{-3} cm^2/sec (VC). On the (331) plane the barrier was 0.33 eV with 8×10^{-4} cm^2/sec (AFW) and 0.37 eV and a prefactor of 2×10^{-3} cm^2/sec (VC).

Using CEM and MD/MC-CEM methods, Perkins and DePristo [28] calculated the barrier to in-channel as well as to cross-channel movement, by exchange, on Pd(110). For in-channel diffusion the barrier calculated with CEM was 0.30 eV, and 0.33 for atom exchange; using MD/MC-CEM they found a similar activation energy of 0.28 eV for in-channel motion and 0.38 eV for cross-channel movement with 113 atoms active. The barrier was around 0.04 eV higher using only 13 atoms active on the surface: 0.32 eV for hopping and 0.42 eV for exchange. Prefactors for the diffusivity were not investigated.

Stoltze [27], with effective medium theory, arrived at a barrier for self-diffusion along the channels of 0.366 eV, and 0.599 eV for cross-channel exchange; cross-channel jumps required a higher energy of 0.776 eV. Depending on Lennard-Jones potentials, Agrawal *et al.* [7] found a barrier to Pd diffusion along the channels of Pd(110) of 0.72 eV with a prefactor $D_o = 1.8 \times 10^{-3}$ cm^2/sec; for cross-channel jumps the activation energy was 1.35 eV, with a prefactor of 9.2×10^{-3} cm^2/sec. These estimates are of course unrealistic.

Ndongmouo and Hontinfinde [32] used RGL potentials in their calculations for Pd atoms on Pd(110) to arrive at an in-channel barrier of 0.39 eV and 0.54 eV for cross-channel motion by atom exchange. For jumps between adjacent channels the activation energy was higher, 0.99 eV. No studies of prefactors were performed. Similar barriers have been obtained recently by Kim *et al.* [34] using molecular dynamics simulations and RGL interactions. For in-channel self-diffusion they found a barrier of 0.380 eV, 0.551 eV for exchange, and 0.965 eV for cross-channel jumps. In-channel diffusion of

4.8 Silver: Ag(110), (311), (331)

Ni involved a barrier of 0.369 eV, 0.315 eV for Cu, 0.333 eV for Ag, 0.520 eV for Pt, and 0.392 eV for Au. Again no information about prefactors is available in this study, nor about the possibility of the exchange mechanism for hetero-diffusion.

The theoretical estimates available for self-diffusion on Pd(110) are quite close together, but regrettably there is *no* information available from experiments. Values for the movement of Cu on Pd(110) surfaces obtained from MC analysis of STM data by Li *et al.* agree nicely with results from EAM by Bucher *et al.* [62] and RGL by Kim *et. al.* [34].

4.8 Silver: Ag(110), (311), (331)

The mobility of silver atoms on the (110) plane of silver has been examined in a variety of ways. Morgenstern *et al.* [66] looked at the decay of Ag islands on Ag(110) in STM studies, together with KMC simulations relying on molecular dynamics estimates of the diffusion energetics with RGL potentials. From these simulations they found values of 0.279 eV for the barrier to in-channel diffusion and 0.394 for cross-channel movement. Corner rounding was also considered in these investigations.

In 2001, De Giorgi *et al.* [67] did scanning tunneling microscope studies of homo-epitaxial growth on Ag(110). The saturation island density, examined for a deposit of 0.16 monolayers (ML) at temperatures from 140 K to 200 K, is shown in Fig. 4.24. Together with KMC simulations developed earlier by Mottet *et al.* [44], shown in Fig. 4.25, they came up with an in-channel barrier to self-diffusion of 0.30 ± 0.03 eV. A cross-channel barrier was estimated from island rotation at low temperature to be at least 0.05 eV higher than for in-channel diffusion. Frequency prefactors were fixed at a value of 10^{13} sec^{-1}. However, the results may not represent single atom movement due to the high coverage and interactions between atoms.

This work was followed by a study from Pedemonte *et al.* [68] in 2002, who did quasi-elastic scattering experiments using a helium atom beam having a most probable energy of 0.0073 eV. From the width of the quasi-elastic peak shape at surface temperatures from 600 to 750 K, they were able to find the diffusivity at the different temperatures, as shown in Fig. 4.26. A barrier to diffusion of 0.19 ± 0.05 eV was found for movement along the

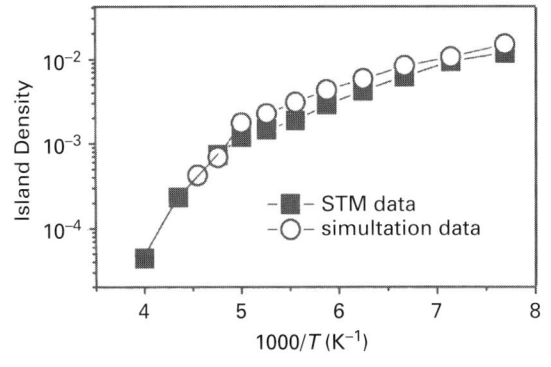

Fig. 4.24 Ag island density on Ag(110) plane as a function of reciprocal temperature. From KMC simulations with RGL potentials, in-channel diffusion barrier was 0.30 ± 0.03 eV (after De Giorgi *et al.* [67]).

208 Diffusion on one-dimensional surfaces

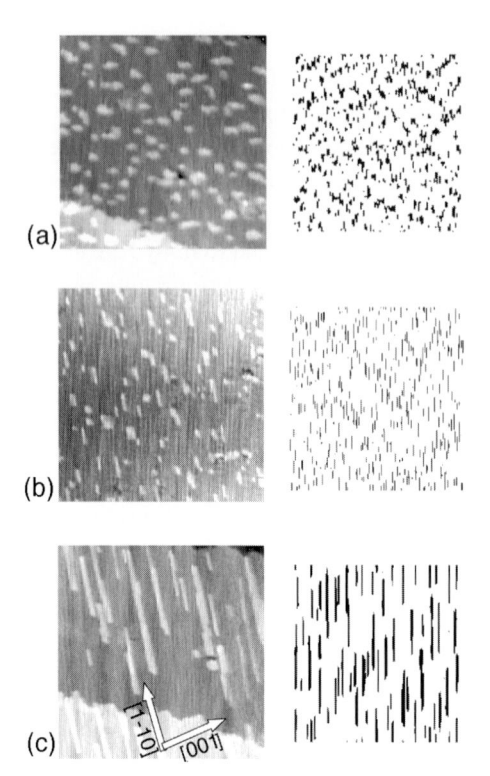

Fig. 4.25 Comparison of STM images of Ag islands on Ag(110) with KMC simulations at (a) 140 K (50 × 50 nm^2), (b) 170 K (100 × 100 nm^2), and (c) 210 K (100 × 100 nm^2), showing good agreement between the two (after De Giorgi et al. [67]).

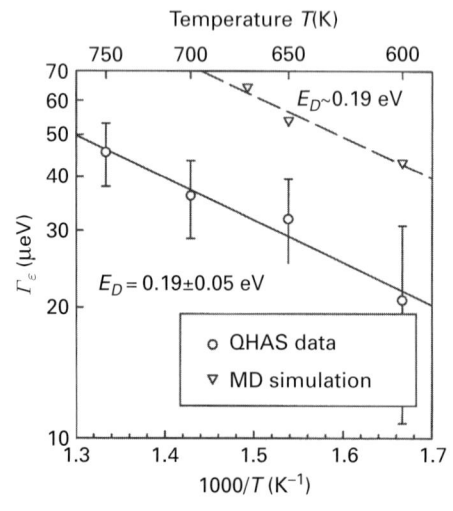

Fig. 4.26 Arrhenius plot of the diffusivity of Ag atom on Ag(110), determined by quasi-elastic He atom scattering. Simulations from Ferrando [76] (after Pedemonte et al. [68]).

4.8 Silver: Ag(110), (311), (331)

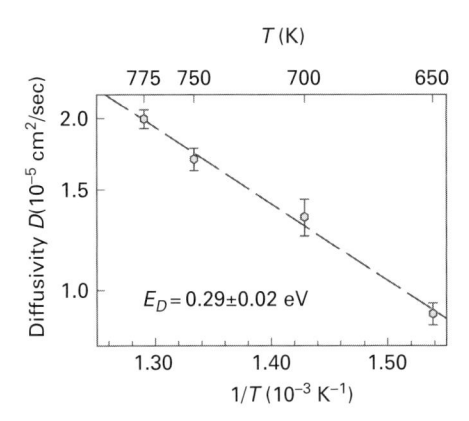

Fig. 4.27 Dependence of Ag diffusivity on Ag(110) plane upon reciprocal temperature, derived from energy resolved scattering of He atoms. Barrier for in-channel motion 0.29 ± 0.02 eV (after Pedemonte et al. [69]).

channels, that is in the <110> direction. Movement of atoms in this range of temperatures was by single jumps only. This technique does not provide the possibility of controlling the number of adatoms on the surface, nor their distance or information about defects, so the work has to be treated carefully.

Further studies were published by Pedemonte et al. [69,70], who continued their helium scattering work. The experiments on Ag(110) were extended to temperatures from 650 K to 775 K, and they arrived at a higher self-diffusion barrier of 0.26 ± 0.03 eV along the channels. Motion across the channels, in the <001> direction, was almost undetectable even at 775 K. Pedemonte et al. still claimed that movement proceeded by single jumps along the channels on the surface, but did not provide an explanation of what gave a higher activation energy at higher temperatures. Additional scattering studies were done by Pedemonte et al. [71], with the upper limit at 800 K, to arrive at the Arrhenius plot of silver diffusion along the channels in Fig. 4.27; the barrier proved to be even higher than in the previous study of the same group, 0.29 ± 0.02 eV. Although diffusion by cross-channel transitions was essentially undetected up to 750 K, such transitions were identified at 800 K and it was concluded that the atom was moving by exchange. The energetics of atom exchange were not determined. One possible reason for these observations could be a change in the coverage of adatoms at the surface with increasing temperature, as the concentration was never specified in this study. If this is the case the activation energy in the lowest temperature range should be closest to single atom movement at the surface, but other factors might play a role as well.

Quite interesting data were reported for gold atoms deposited on the Ag(110) surface. Not only was exchange proposed by Johnson [72], but the burying of gold atoms with hole creation was observed by STM [73,74]. These effects were not observed for silver atoms on this surface.

A significant number of attempts have been made to calculate the diffusion characteristics on silver. The first were by Flahive and Graham [20] using Morse potentials, summarized in Fig. 4.9. Liu et al. [2] used many-body EAM potentials and estimated an activation energy for self-diffusion along the channels on Ag(110) of 0.32 eV with a prefactor of 1.0×10^{-3} cm^2/sec, using AFW parameters. In the VC approximation the barrier proved to be 0.25 eV and the prefactor 2.7×10^{-3} cm^2/sec. For cross-channel motion the barriers were higher, 0.42 eV, and the prefactor 4.0×10^{-2} cm^2/sec (AFW) and 0.31 eV and 2.5×10^{-2} cm^2/sec (VC). Calculations were also made for other planes. On the (311) surface the diffusion barrier was 0.26 eV and the prefactor 1.0×10^{-3} cm^2/sec (AFW parameters), and again 0.26 eV and 3.0×10^{-3} cm^2/sec (VC parameters). For the (331) plane the barriers were derived as 0.34 eV with a prefactor of 1.0×10^{-3} cm^2/sec (AFW) and 0.29 eV and a prefactor of 2.0×10^{-3} cm^2/sec (VC).

Perkins and DePristo [28] also made diffusion estimates on Ag(110). With their MD/MC-CEM method they reported an in-channel diffusion barrier of 0.248 eV, and 0.328 eV for cross-channel exchange. Findings from CEM did not differ much; they obtained 0.26 eV and 0.34 eV, in reasonable agreement with the work above. They also looked at the influence of the number of active atoms taken into account in the simulations; when this number was only 13, the activation energies were higher; for in-channel hopping the barrier was 0.29 eV and for exchange 0.371 eV. In the same year, Stoltze [27] in simulations with EMT potentials found an activation energy of 0.291 eV for jumps along the channels, 0.561 eV for atom exchange, and 0.639 eV for jumps across the channel.

Hontinfinde et al. [75] used RGL interactions to again estimate the self-diffusion barriers on Ag(110). For in-channel single jumps they found an activation energy of 0.28 eV, in very good agreement with the results of Pedemonte et al. [71]. The barrier for cross-channel exchange was 0.38 eV. Cross-channel hops required a much higher energy of 0.82 eV. Ferrando [76] extended this study to investigate diffusion mechanisms. He claimed that at any temperature single transitions represented 90% of the jumps, and that the longer jumps were doubles. Around 600 K more complicated mechanisms started to play a role as well, and a sudden decrease in the number of single jumps was observed at higher temperatures. The same values were again given by Montalenti and Ferrando [45], who also considered the presence of long jumps; at 600 K, 8.4% of the transitions were double jumps, but at 450 K the percentage diminished to 2.8. In 1999 Rusponi et al. [77] looked at vacancy and adatom islands created by ion sputtering of Ag(110) over a range of temperatures to explain the shape evolution of the islands. They used KMC simulations, but for the starting values of the energy did quenched molecular dynamics simulations based on RGL potentials. They used the same diffusion barriers on Ag(110) as previous investigators, 0.28 eV and 0.38 eV for in-channel and cross-channel diffusion.

Calculations of the activation energy of self-diffusion were made by Agrawal et al. [7] using Lennard-Jones potentials. They found a high barrier of 0.59 eV and a prefactor $D_o = 2.6 \times 10^{-3}$ cm^2/sec for in-channel motion, and 1.10 eV with a prefactor of 6.9×10^{-3} cm^2/sec for cross-channel events, quite far away from previous investigations.

Merikoski et al. [36] studied self-diffusion on the Ag(311) surface using EMT potentials. For hopping along the channels, they found a barrier of 0.220 eV; cross-channel exchange required overcoming a barrier of 0.727 eV, whereas cross-channel jumps turned out to have a comparable energy to exchange -0.704 eV.

Nie et al. [78] have done density-functional calculations for the self-diffusion of silver atoms on Ag(110)-(2 × 2). On the bare surface in-channel diffusion of Ag occurred over a barrier of 0.266 eV. For cross-channel jumps the barrier was much higher, amounting to

0.757 eV, but exchange required only ~0.4 eV, so that in-channel motion was clearly preferred. Vacancy formation cost about 0.275 eV. Ag diffusion on an Sb covered Ag(110) surface required 0.275 eV for in-channel movement and 0.673 eV for cross-channel movement; the presence of Sb had only a little effect on diffusion of silver atoms. However it is worth noting that a reconstruction of the clean silver (110) surface was *not* observed in experiments.

In 2006, the same year, Kong and Lewis [33] calculated the kinetic prefactor for diffusion of Ag atoms along the channels of the (110) plane using EAM potentials. The value for atom jumps turned out to be 2.9×10^{-3} cm^2/sec when vibrational contributions were summed over all atoms of the crystal rather than using the local density approximation. For self-diffusion on Ag(110), Kim *et al*. [34] used molecular dynamics simulations and RGL interactions to arrive at a barrier of 0.277 eV for in-channel jumps, 0.388 eV for atom exchange, and 0.818 eV for cross-channel jumps, in good agreement with the work of Hontinfinde *et al*. [75]. In hetero-atom in-channel motion the barrier for diffusion of Ni was 0.268 eV, for Cu 0.249 eV, for Pd 0.337 eV, for Pt 0.492 eV, and 0.360 eV for Au.

For Ag(110), experiments and reasonable calculations appear to be in sensible agreement, but most of the calculations were done using RGL potentials. There is also a lack of data about hetero-diffusion on these planes. No direct single atom investigations have been carried out, and that might have revealed a systematic error due, for example, to interactions. However, one thing seems to be clear from both experiment and theory – self-diffusion of silver proceeds mostly by simple hopping. The exchange mechanism is, however, likely to contribute at higher temperatures.

4.9   Tungsten: W(211)

The (211) plane of tungsten is a surface intensively studied in diffusion experiments, but almost completely neglected by theory. Many different atoms have been examined on this plane, a model of which is given in Fig. 3.1 showing the <111> rows which define the path of atoms moving over the surface. Note that the second layer in this model is not located symmetrically with respect to the first around $[0\bar{1}1]$. The structure of this plane has been examined by low-energy electron diffraction [79], as well as by low-energy ion scattering [80]. These experiments have found that the first-to-second layer spacing is shortened by 9.3% (0.12 Å) and there is a lateral shift of 3.6% along the direction of the channels. However, it was shown recently that such asymmetry of the W(211) surface does not result in asymmetric diffusion [81]. To get an impression of how diffusion occurs, we will examine the available data.

4.9.1   Self-diffusion: FIM studies

Pioneering work in diffusion studies [82] was carried out on W(211) with field ion microscopy in 1966, yielding rather odd results: a prefactor on the order of 10^{-7} cm^2/sec or less, and a low activation energy of 0.56 eV, as in Fig. 4.28. These were the first ever direct measurements of atom movement on a surface. It may be that the presence of more

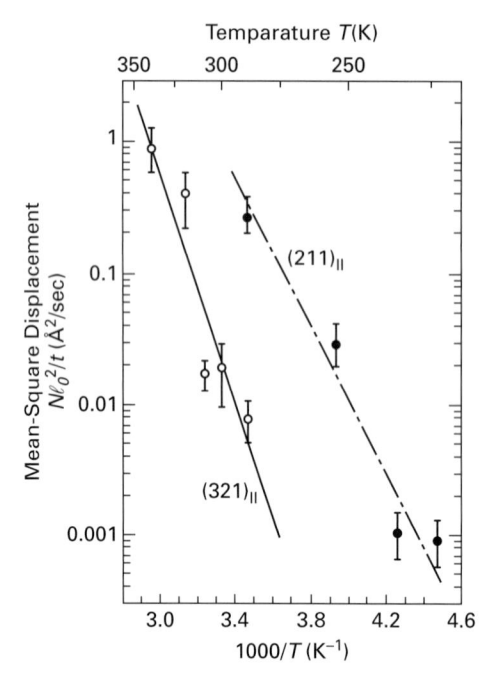

Fig. 4.28 Arrhenius plots for the in-channel diffusivities of W atoms on W(211) and W(321) planes [82]. Deduced was a rather low value of 0.56 eV on W(211), and 0.87 eV for W(321); presence of many atoms on (211) may have caused low value.

than one atom on the plane interfered with the measurements. That may also have been true for the later studies by Bassett and Parsley [83] using the same technique; in 1970 they reported a diffusion barrier of 0.56 eV and another low prefactor of 3.8×10^{-7} cm^2/sec. These were very early determinations, and later studies revealed more consistent behavior.

Graham and Ehrlich [84] in 1975 reexamined the previous values of Ehrlich and Hudda [82] and from FIM observations reported a prefactor $D_o = 3 \times 10^{-4}$ cm^2/sec and an activation energy of 0.76 ± 0.07 eV. In a careful examination of this system by Flahive and Graham with an improved temperature calibration [85], an activation energy of 0.85 ± 0.05 eV and a prefactor of $1.5(\times 10^{\pm 1.5}) \times 10^{-2}$ cm^2/sec was reported and is shown in Fig. 4.29. They also demonstrated that for planes with between 17–11 sites, the activation energy only barely depended on the size of the plane, but decreased between 11–9 sites, as indicated in Fig. 4.30. Here it should be noted that a reanalysis of Flahive and Graham's data yielded the same barrier as before and a prefactor of $19.0(\times 7.3^{\pm 1}) \times 10^{-3}$ cm^2/sec [86], shown in Fig. 4.31.

These results are in reasonable agreement with the work of Wang [86], who examined a number of different adatoms on W(211); for tungsten adatoms he found a barrier of 0.82 ± 0.02 eV and a prefactor $7.7(\times 1.9^{\pm 1}) \times 10^{-3}$ cm^2/sec. A comparison of the reanalyzed Flahive and Wang data is shown in Fig. 4.31. The value of the activation energy has turned out to be reproducible in further investigations with very good statistics.

More recently diffusion on this plane has been studied by Senft looking for long jumps [87,88], whose results are given in Fig. 4.32. These were the first studies corrected for

4.9 Tungsten: W(211)

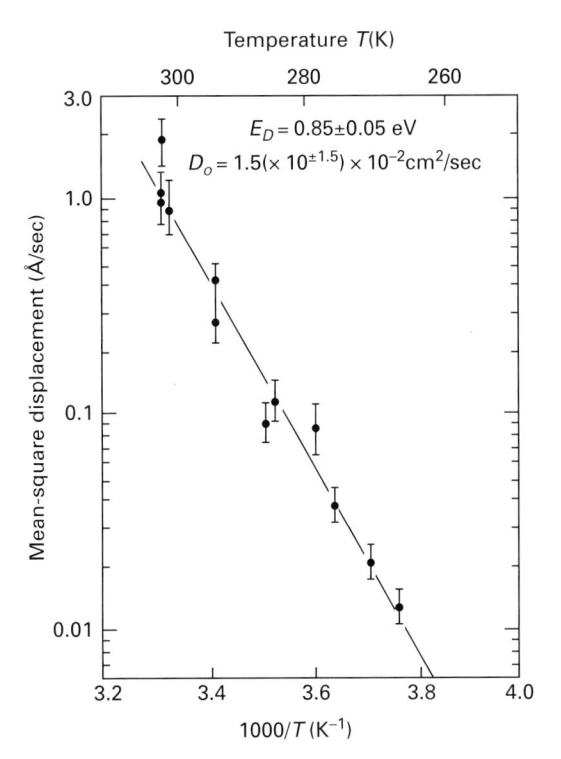

Fig. 4.29 Dependence of in-channel diffusivity for W adatoms on W(211) upon reciprocal temperature, determined by FIM (after Flahive and Graham [85]).

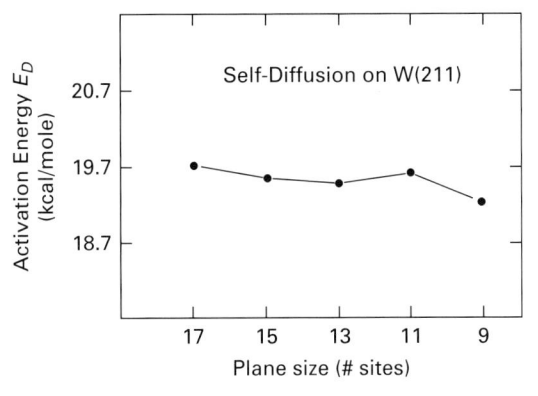

Fig. 4.30 Dependence of the activation energy determined for W(211) upon the number of sites on the plane. Measured energy is essentially constant for planes with 17 to 11 sites (after Flahive and Graham [85]).

diffusion occurring during temperature transients, while the sample is warming to the diffusion temperature and afterwards, during cooling. Her results, an activation energy of 0.81 ± 0.02 eV and a prefactor of $2.5(\times 3.3^{\pm 1}) \times 10^{-3}$ cm^2/sec, fit in with the previous work of Wang and of Flahive and Graham, without problems even if diffusion during temperature transients were not taken into account in the previous studies. It is likely that in the temperature range investigated by Flahive and Graham transient motion was not important.

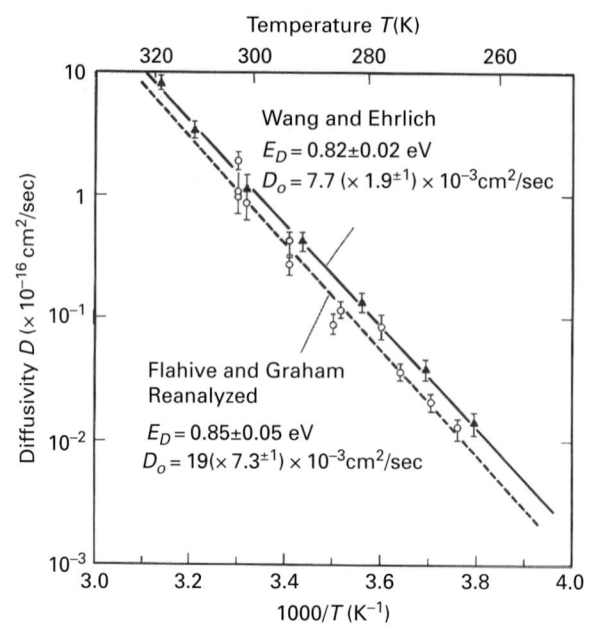

Fig. 4.31 Comparison of different measurements of diffusion on W(211) [86]. Activation energies: 0.82 ± 0.02 eV for Wang, 0.85 ± 0.05 eV for Flahive and Graham [85].

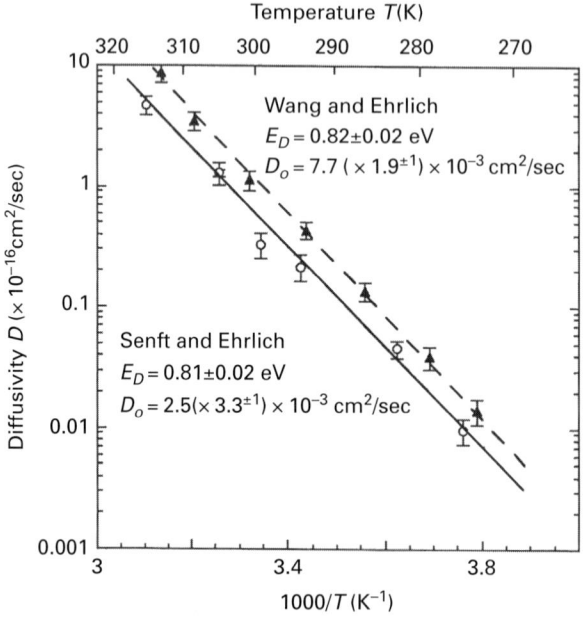

Fig. 4.32 Comparison of W diffusivities on W(211) measured by Senft [88] and by Wang and Ehrlich [86].

4.9 Tungsten: W(211)

With the aim of exploring the mechanism of diffusion at high temperatures, Antczak [89] did more recent diffusion measurements over a 60 K range of temperatures and with huge statistics. An activation energy was found of 0.81 ± 0.02 eV and a prefactor of $3.41(\times 2.40^{\pm 1}) \times 10^{-3}$ cm^2/sec, as shown in Fig. 2.19; this is in nice agreement with previous work. It should be noted that all the more recent studies give a prefactor of $\sim 10^{-3}$ cm^2/sec, which is expected if the effective vibrational frequency is $\sim 10^{12}$ sec^{-1}. Antczak observed the presence of longer as well as rebound transitions at around 300 K, and a gradual withdrawal of single jumps caused by longer transitions. The energetics of all kinds of observed long jumps were derived and data are presented in Chapter 3.3.4. She did not find any changes in diffusivity due to these alterations, and the mechanism of movement in self-diffusion up to 325 K on W(211) was established.

4.9.2 Self-diffusion: FEM studies

At the same time as these FIM measurements, field electron microscopy was carried out by Tringides and Gomer [90]. In 1986 they measured self-diffusion on a (211) plane, relying on fluctuations of the field emission current through a probe hole in the form of a long slit. They found that diffusion, in the temperature range 568–694 K, occurred both along the <111> channels, but also across them. Along the channels, the activation energy amounted to 0.57 eV, with a prefactor $D_o = 7.1 \times 10^{-7}$ cm^2/sec; across the channels, the figures were 0.27 eV and 1.8×10^{-9} cm^2/sec, as shown in Fig. 4.33. Even at the low temperature of 264 K, these figures imply that 10 jumps across the channels occur per second. However, cross-channel transitions have not been seen by field ion microscopy on this plane up to 340 K [89]. In the work of Tringides and Gomer [90], diffusion occurred with the tip biased negatively, and it is possible that the field induced cross-channel motion. Kellogg [91] in 1993 studied the effect of electric fields on the diffusion of platinum atoms on Pt(100), and found that a

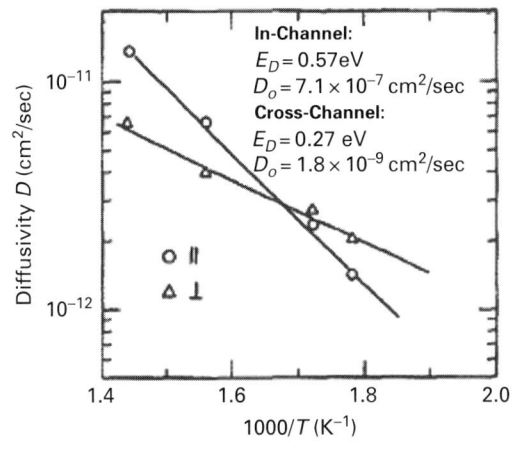

Fig. 4.33 Dependence of in-channel and cross-channel self-diffusivity on W(211) upon reciprocal temperature, determined from fluctuations in field emission current (after Tringides and Gomer [90]).

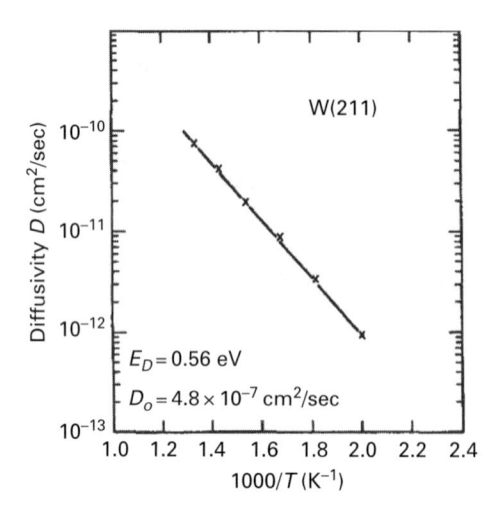

Fig. 4.34 Arrhenius plot for W atoms on W(211), deduced from field emission fluctuations (after Gong and Gomer [92]).

negative field promoted exchange diffusion, and that might be an explanation for these findings.

As part of a larger study on thermal roughening of steps in 1988, Gong and Gomer [92] again looked at diffusion on the (211) plane of tungsten by measuring the fluctuations in the field emission current. They used a circular probe, so it was impossible to distinguish between in- and cross-channel movement. Their results, in Fig. 4.34, indicate a diffusion barrier of 0.56 eV and a prefactor $D_o = 4.8 \times 10^{-7}$ cm^2/sec. The authors suggested that the field applied to obtain field emission affected the activation energy.

Choi and Gomer [93] extended these measurements in 1990; they covered a larger temperature range, from 500 K to more than 800 K, compared to the earlier work that extended from 613 K to 700 K. Their detector slit also was smaller. For diffusion along the channels they found an activation energy of 0.73 ± 0.02 eV and a prefactor $D_o = (3-2) \times 10^{-5}$ cm^2/sec. As shown in Fig 4.35a, there is a sharp drop in diffusion starting at a temperature of 826 K, but at 850 K the diffusivity rises again. In this last range, the activation energy proved to be 0.95 ± 0.02 eV with a prefactor of 5×10^{-5} cm^2/sec. The results for diffusion across channels in the temperature range 526 K to 752 K yielded a barrier of 0.29 eV and a prefactor of 4.2×10^{-9} cm^2/sec, shown in Fig. 4.35b, close to the earlier studies by Tringides and Gomer [90]. At temperatures above 752 K, cross-channel motion occurred over a barrier of 1.04 eV, with a prefactor of 5×10^{-4} cm^2/sec. Choi and Gomer [93] interpreted the motion at lower temperatures as an exchange between an adatom with a lattice atom, and at high temperatures as diffusion over the channels. At the moment it is not really clear what is going on, but it is unlikely that single atom motion is being measured.

More recently, Gong [94] has reported similar measurements, but with a round probe hole, which does not allow distinguishing in-channel and cross-channel diffusion. For the diffusion of tungsten on W(211) he found an activation energy of 0.56 eV and a prefactor $D_o = 4.8 \times 10^{-7}$ cm^2/sec, results and data exactly the same as in previous work [92]. The

4.9 Tungsten: W(211)

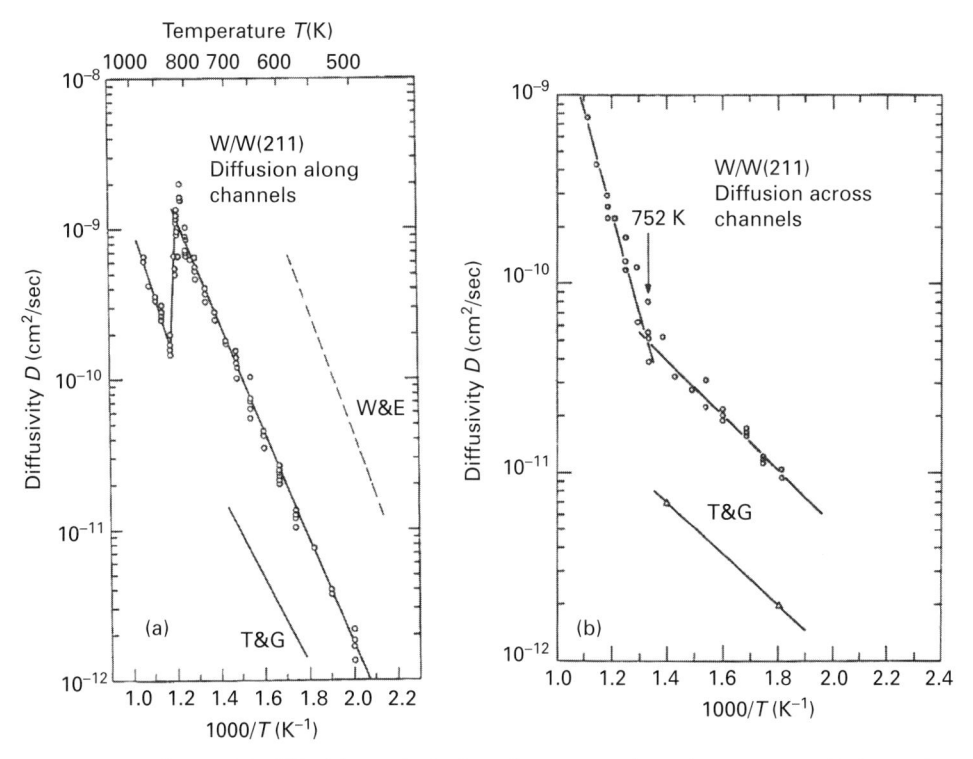

Fig. 4.35 Dependence of W diffusivity on W(211) surface upon reciprocal temperature. (a) Comparison of in-channel results from Choi and Gomer [93] with data from Tringides and Gomer [90] and with Wang and Ehrlich [86]. (b) Comparison of cross-channel measurements with the work of Tringides and Gomer [90] (after Choi and Gomer [93]).

FEM findings for in-channel motion agree with early field ion microscopic observations of diffusion on W(211), where several atoms might have interacted to produce such low values. Since with this technique there is no control in the experiments over the number of atoms on the plane, this is very likely the explanation for the data. Based on the lack of agreement between FIM and FEM data it is reasonable to conclude that FEM is not a technique really suitable for single atom self-diffusion studies. In FIM, movement of a single adatom can be followed directly, and the presence of a second adatom or defect can be detected. The electric field also does not influence the movement, since the state of the surface is frozen during imaging, so only thermal motion is detected.

4.9.3 Self-diffusion: theoretical studies

There has been only little theoretical effort to estimate self-diffusion energetics on tungsten with many-body potentials. However, investigations with pair potentials started quite early. The first estimates of the surface migration barrier were done by Ehrlich and Kirk based on Morse potentials; they ranged from 0.78 eV for a realistic (fluctuating) surface, 1.09 eV for a surface with relaxed atomic positions, and 0.95 eV for a rigid surface [95]. Other early calculations were made using Morse function interactions by Wynblatt and Gjostein [96];

they arrived at a barrier of 0.432 eV for jumps, considerably below the current experimental values. Later studies were done by Flahive and Graham [20]. Using Morse potentials again, they estimated a self-diffusion barrier of 0.28 eV for in-channel motion on a fully relaxed surface, and 0.91 eV for an unrelaxed surface, with the data presented in Fig 4.9. Values on the relaxed surface obtained from Morse potentials were much too low compared with experiments. Lennard-Jones potentials were used by Doll and McDowell [97] in 1982 and from their calculation they arrived at an activation energy of 0.887 ± 0.224 eV. In 1984 Voter and Doll [98] continued studies with Lennard-Jones potentials and found a barrier of 0.952 ± 0.182 eV and a prefactor of 2.8×10^{-3} cm^2/sec using Monte Carlo TST; from simple TST they obtained 0.822 eV and a prefactor of 1.6×10^{-3} cm^2/sec. Values obtained from L-J potentials are not too far from the experimental data.

The only work so far with many-body potentials was done in 1994. Xu and Adams [99] did work on tungsten using the fourth-moment approximation to tight-binding theory. For in-channel diffusion they calculated an activation energy of 0.79 eV with a prefactor of 1.41×10^{-3} cm^2/sec; for cross-channel diffusion by exchange their activation energy turned out quite high, 2.0 eV with a prefactor of 9.4×10^{-4} cm^2/sec. The estimate for in-channel diffusion is quite close to the results from experiments.

4.9.4 Hetero-diffusion: nickel, molybdenum, rhodium

Diffusion of nickel was examined in FIM studies on W(211) by Senft [88], who found an activation energy of 0.46 ± 0.01 eV and a prefactor of $1.9(\times 2.7^{\pm 1}) \times 10^{-3}$ cm^2/sec, as shown in Fig. 4.36. This study was quite careful and statistics were quite reasonable, so it is probably close to the real value for Ni. There are also indications of long jumps in diffusion at 179 K, as discussed in Chapter 3.

The first examination of molybdenum diffusion on W(211) was done by Bassett and Parsley [83] in 1970 with a few atoms on the plane using FIM. They reported an activation

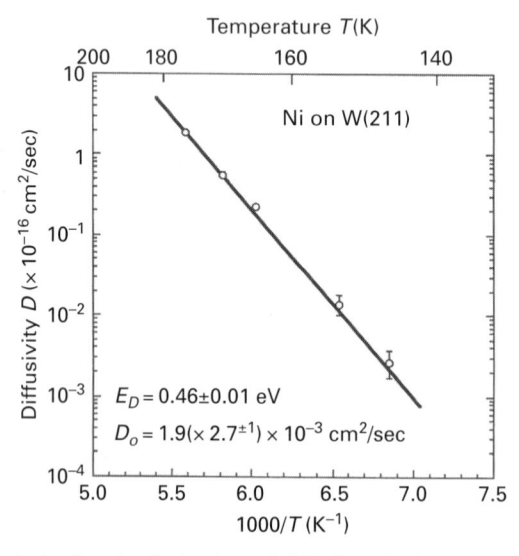

Fig. 4.36   Arrhenius plot for in-channel diffusion of Ni on W(211) (after Senft [88]).

4.9 Tungsten: W(211)

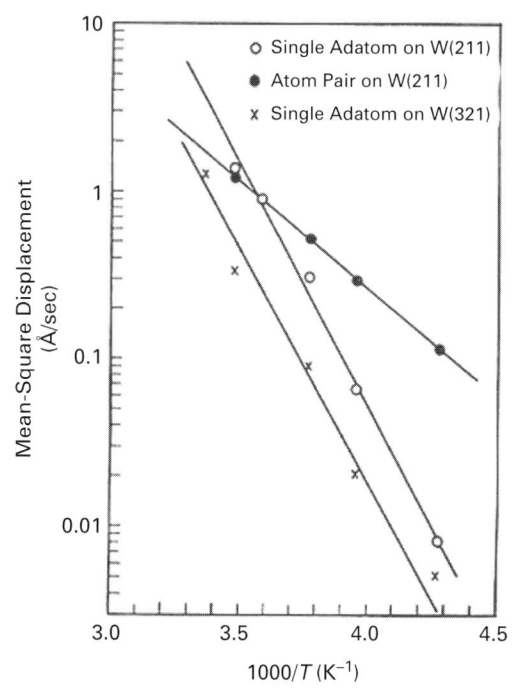

Fig. 4.37 Dependence of diffusivity for Mo on W(211) and W(321) upon reciprocal temperature. For Mo atoms on W(211) a barrier of 0.57 eV was obtained in observations with FIM and 0.55 eV for W(321) (after Sakata and Nakamura [100]).

energy for diffusion of 0.55 eV, with prefactor of 2.4×10^{-6} cm^2/sec. In 1975, Sakata and Nakamura [100] carried out a variety of studies on W(211) also in an FIM, and for the diffusion of Mo atoms they found an activation energy of 0.57 eV, with a prefactor of 9.3×10^{-7} cm^2/sec. These results were based on only ~ 20 measurements at each of five temperatures, and are shown in Fig. 4.37. The agreement with the earlier work is very good, both in the low activation barrier and the prefactor, but this low value is open to considerable doubt. Wang [86] later again used field ion microscopy and studied diffusion at nine temperatures, with one hundred observations at each, and found quite a different result: an activation energy of 0.71 ± 0.01 eV, and a prefactor of $2.0(\times 1.6^{\pm 1}) \times 10^{-3}$ cm^2/sec. The disagreement with earlier work is significant, and the latest measurement appears more reliable, with better statistics and with only a single atom on the plane. The movement of Mo atoms at 262 K proceeded by single jumps only [101] for the investigated temperature range, as is clear from Fig. 3.85.

Diffusion of rhodium on the (211) plane of tungsten was examined by Wang [86] in the FIM. Results obtained at ten temperatures gave an activation energy of 0.536 ± 0.001 eV and a prefactor of $3.3(\times 1.5^{\pm 1}) \times 10^{-3}$ cm^2/sec, derived from the Arrhenius plot in Fig. 4.38. Investigations by Wang et al. [101] at temperatures of 189 K and 192 K gave negligible amounts of double transitions, and higher temperatures were not investigated. No other studies have been reported for this system.

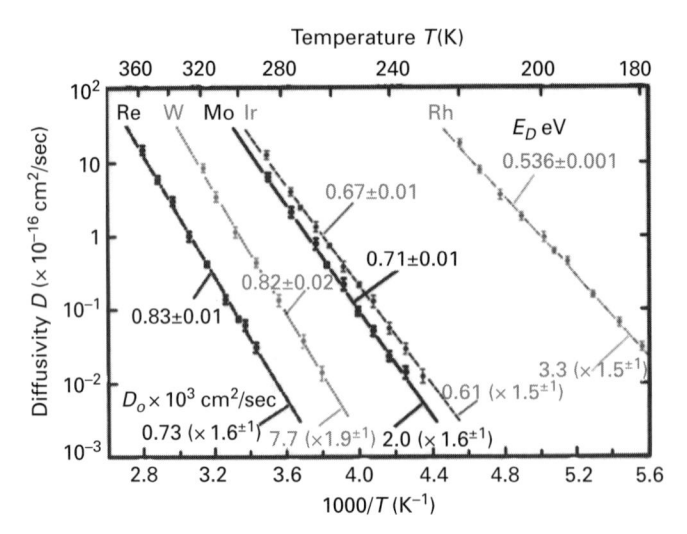

Fig. 4.38 Dependence of adatom in-channel diffusion on W(211) upon reciprocal temperature, obtained from FIM measurements of atom displacements [86]. Uncertainties derived by simulation.

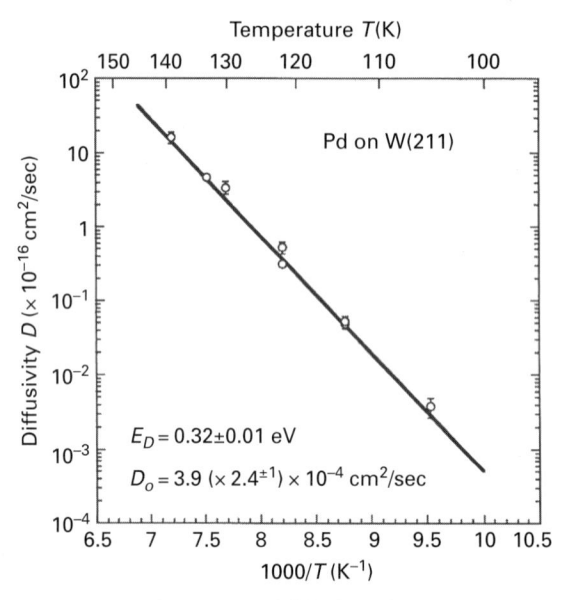

Fig. 4.39 Arrhenius plot for in-channel diffusion of Pd on W(211), obtained from observations in FIM (after Senft [88]).

4.9.5 Hetero-diffusion: palladium, rhenium, iridium

Senft [87,102] utilized the FIM to examine the movement of palladium, as shown in Fig. 4.39, and reported a barrier of 0.32 ± 0.01 eV and a prefactor of $3.9(\times 2.4^{\pm 1}) \times 10^{-4}$ cm^2/sec. She found a significant number of double and triple jumps for this system at a temperature of 133 K, as described in detail in Chapter 3. It should be noted that despite differences in the activation energies for the different

4.9 Tungsten: W(211)

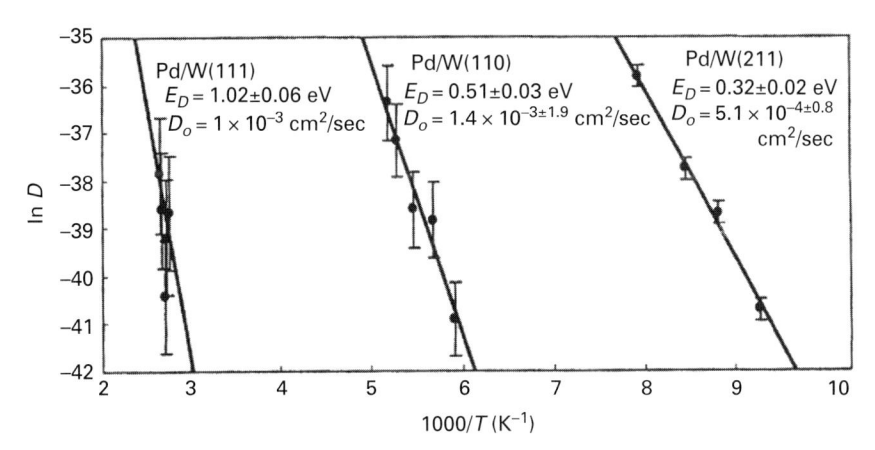

Fig. 4.40 Diffusion of palladium on W(211), W(111), and W(110), shown in Arrhenius plots obtained from FIM observations (after Fu et al. [103]).

types of jumps, the prefactor in the diffusivity is still around 10^{-3} cm^2/sec. In 2002, Fu et al. [103] looked at diffusion of palladium on tungsten surfaces also using FIM, as shown in Fig. 4.40. Movement on W(211) was measured at four temperatures, and an activation energy for diffusion of 0.32 ± 0.02 eV was found, with a prefactor of $5.1 \times 10^{-4.0 \pm 0.8}$ cm^2/sec, in excellent agreement with the results of Senft.

Movement of Re has been the first adatom probed by Bassett and Parsley on W(211) [104] in this multi-atom study in an FIM. They found an activation energy for diffusion of 0.88 eV, and a prefactor of 1.1×10^{-2} cm^2/sec. They also checked the effect of an electric field of 0.9 V/Å on the diffusivity and observed an increase of 0.009 eV in the activation energy and 0.7×10^{-2} cm^2/sec in the prefactor, changes that appeared insignificant. However, observations were done with a few atoms present on the surface. Subsequent work was reported by Stolt et al. [105] in 1976, and later still by Wang [86] in 1988, both with field ion microscopy. With approximately 100 observation per temperature in the range 290 K to 351 K, Stolt et al. found an activation energy of 0.86 ± 0.03 eV and a prefactor of $2.2(\times 2.8^{\pm 1}) \times 10^{-3}$ cm^2/sec from the Arrhenius plot in Fig. 4.41, whereas Wang reported an activation energy of 0.83 ± 0.01 eV and a prefactor of $0.73(\times 1.6^{\pm 1}) \times 10^{-3}$ cm^2/sec. These measurements, done without "zero-time" compensation, are in reasonable agreement. It should be noted that at these diffusion temperatures, atom jumps during temperature transients may have been negligible. Analysis of the distribution of displacements for temperatures 300 and 317 K did not detect the presence of double jumps [101]. The first study, with many atoms present on the plane, seems not to lie too far from later measurements with only one atom, which might indicate weak interactions between the atoms present in different channels.

Bassett and Parsley [83] studied the motion of iridium on W(211) with several atoms on the surface using the FIM technique; they found an activation energy of 0.58 eV and a prefactor of 2.7×10^{-5} cm^2/sec. In a brief comparison study of atom and dimer diffusion, Reed [106] reported an activation energy of 0.53 ± 0.05 eV with a prefactor of 5×10^{-7} cm^2/sec for single adatoms, also measured by FIM, as shown in Fig. 4.42. These values appear too low. A rather higher activation energy of 0.67 ± 0.01 eV and

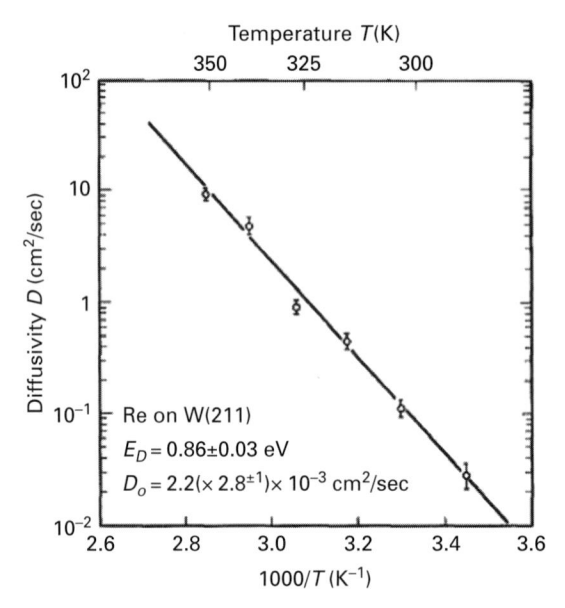

Fig. 4.41 Arrhenius plot for in-channel diffusivity of Re atom on W(211), derived from FIM measurements [105].

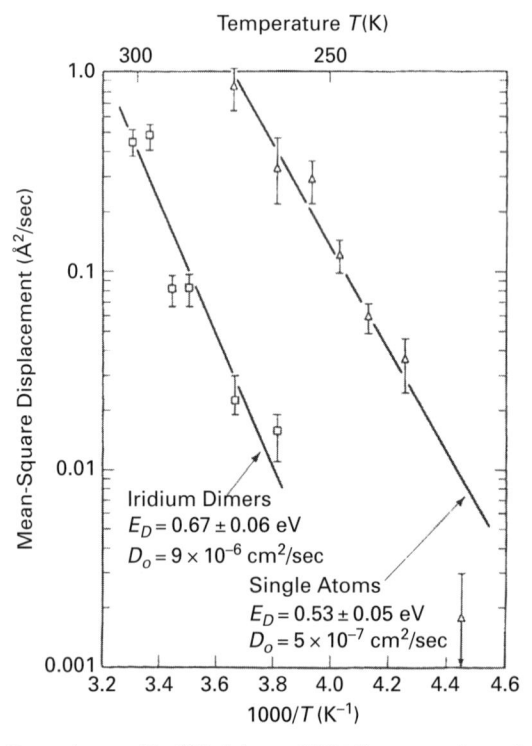

Fig. 4.42 Dependence of Ir diffusivity on W(211) upon reciprocal temperature [106]. Results useful only for comparison of atom and dimer behavior, not for actual kinetics.

4.9 Tungsten: W(211)

prefactor of $0.61(\times 1.5^{\pm 1}) \times 10^{-3}$ cm^2/sec was found by Wang [86] in extensive experiments. These utilized the FIM and were shown previously in Fig. 4.38, which involved roughly 100 observations at each of eleven temperatures. The earlier studies by Reed [106] are useful comparing single atom and dimer motion, but should be ignored in examining details of single atom diffusion. Investigation of the type of jumps Ir atoms make at temperatures of 250 K, 261 K, and 272 K, showed a small contribution of double jumps. The ratio of double to single jumps was 0.004, 0.044 and 0.027, too low to claim they were significant [101], however, still missing are "zero-time" corrections.

4.9.6 Hetero-diffusion: miscellaneous adatoms

There are a number of studies which have concentrated on only one or two different adatoms; these are grouped together here. In 1970, Bassett and Parsley [107] examined diffusion of tantalum on W(211) in the FIM. They reported a diffusion barrier of 0.48 eV and a prefactor of 0.9×10^{-7} cm^2/sec. These values are undoubtedly too low, possibly influenced by interactions with other adatoms, and should be reexamined.

Field emission measurements of gold adsorption on W(211) were made quite early by Jones and Jones [108] in 1976. They measured the rate at which gold swept over this plane, and derived an activation energy of 0.19 eV. They arrived at this value at a high coverage, half of a monolayer, and it is clear that information about single atom diffusion was not generated.

A different type of diffusion study was carried out by Bayat and Wassmuth [109], who examined the movement of potassium on W(211) with a surface ionization microscope. The amount of potassium on the surface was extremely small, less than 10^{-3} of a monolayer, so these measurements should correspond to the limit of zero concentration. Studies were done for quite high temperatures from 840 K to 1400 K, and not only diffusion but also evaporation took place. The latter was examined separately, using the relation for the lifetime τ on the surface of

$$\tau = \tau_0 \exp(Q_i/kT); \qquad (4.6)$$

for Q_i, the desorption energy of potassium ions, a value of 2.52 ± 0.04 eV was found from the data in Fig. 4.43a.

The root mean-square displacements along and across the channels were given as

$$\sqrt{\langle x_\parallel^2 \rangle} = (x_\parallel) \exp[(Q_i - E_\parallel)/2kT],$$
$$\sqrt{\langle x_\perp^2 \rangle} = (x_\perp) \exp[(Q_i - E_\perp)/2kT]. \qquad (4.7)$$

The studies of diffusion are shown in Fig. 4.43b, and gave a barrier to in-channel motion of $E_\parallel = 0.46 \pm 0.08$ eV and to cross-channel motion of $E_\perp = 0.76^{+0.10}_{-0.08}$ eV. The prefactor for both processes proved to be roughly the same, 0.3 cm^2/sec.

Diffusion of lithium on W(211) was examined by Biernat et al. [110] who measured fluctuations in the field electron emission current. They looked at coverages starting from 0.4 ML, and observed a strong coverage dependence of the activation energy as well as the prefactor. For 0.4 ML, they found an activation energy in the <111>

Diffusion on one-dimensional surfaces

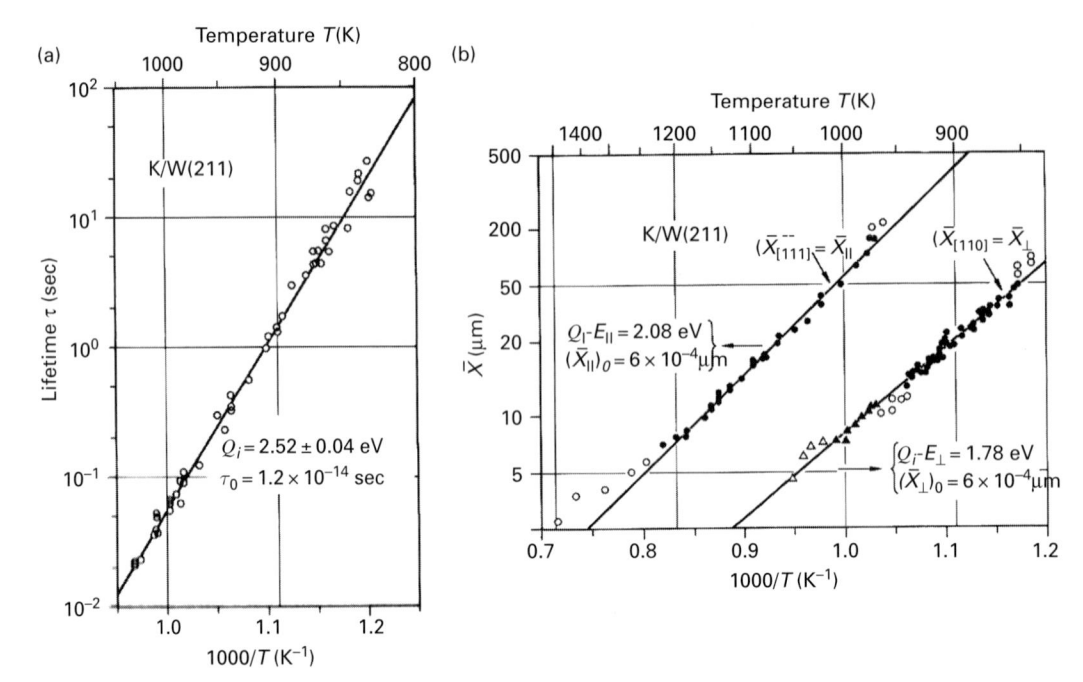

Fig. 4.43 (a) Lifetime of potassium on W(211) as a function of reciprocal temperature, derived from surface ionization microscopy. (b) Arrhenius plot for in- and cross-channel diffusion of K on W(211), giving a barrier of 0.46 ± 0.08 eV for in-channel motion and $0.76^{+0.10}_{-0.08}$ for cross-channel diffusion (after Bayat and Wassmuth [109]).

direction equal to 0.49 eV, with a prefactor of 2.0×10^{-3} cm^2/sec. At low temperature the movement was strictly one-dimensional, at higher temperature two-dimensional movement became apparent. At a coverage of 0.6 ML, they obtained an in-channel barrier of 0.46 eV and a prefactor of 1.2×10^{-3} cm^2/sec, as well as a cross-channel activation energy of 0.54 eV and a prefactor of 5.8×10^{-3} cm^2/sec. However, concentrations were so high that diffusion for single atoms was strongly affected by interactions.

Strontium on W(211) was investigated in Naumovets' group by monitoring changes in work function during spreading over the surface [111]. Coverages down to about 0.05 ML were studied, and the activation energies for diffusion are shown in Fig. 4.44, with a barrier of 0.87 eV in the most dilute state. Unfortunately this technique cannot control defects at the surface, which might influence the diffusion barrier and are likely to be responsible for the high prefactor of ~ 0.4 cm^2/sec.

In 1999, Loburets [59] studied the diffusion of copper on W(211) by measuring work function changes. Regrettably his lowest concentrations of Cu were ~ 1ML, so that no information about individual atom behavior could be derived from his work.

Although it would be desirable to have more data for hetero-atom diffusion, the experiments on single adatoms are in reasonably good shape.

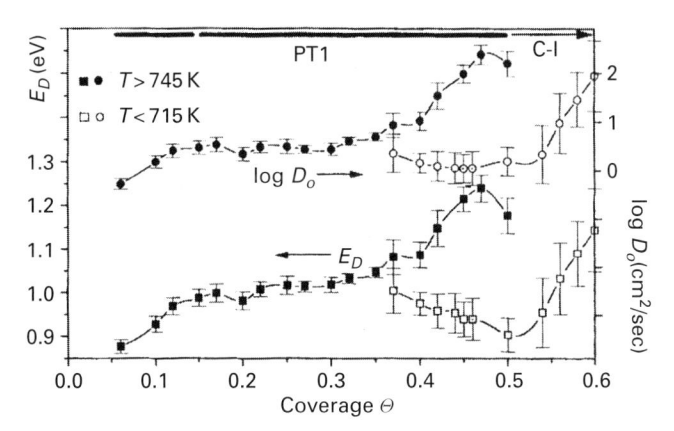

Fig. 4.44 Diffusion characteristics for strontium on W(211) at different surface coverages (after Loburets et al. [111]).

4.10 Tungsten: W(321)

4.10.1 Self-diffusion

Much less has been done to characterize diffusion on the (321) plane of tungsten, a schematic of which was shown in Fig. 3.1. What is significant here is that this plane is again made up of channels; however, they are wider and deeper than on W(211). All studies except for those by Gong and Gomer [92] utilized field ion microscopy. The first measurements were done by Ehrlich and Hudda [82] in 1966. For tungsten adatoms migrating on W(321) they found a barrier of 0.87 eV and a prefactor of 1×10^{-3} cm^2/sec; their data are shown in Fig. 4.28. It should be noted that these measurements were done with a few atoms on the plane. In 1970, Bassett and Parsley [107] examined the mobility of tantalum, tungsten, and rhenium, again with multi-atom deposits. For tungsten they arrived at a barrier of 0.84 eV and a prefactor of 1.2×10^{-3} cm^2/sec.

Graham [84] reported new determinations for self-diffusion, with a barrier of 0.81 ± 0.078 eV and a prefactor of 1×10^{-4} cm^2/sec, but this time with only one atom on the plane giving more reliable results. Some observations of tungsten adatoms were reported by Nishigaki and Nakamura [112], who made a few observations of in-channel self-diffusion at 330 K, and found an activation energy of 0.89 eV. In 1980 Flahive and Graham [85] explored shallow and deep sites on W(321) shown in Fig. 3.11. For movement from shallow to deep sites they found a barrier of 0.65 eV, assuming a standard prefactor; for motion along the deep sites the barrier was ~0.87 eV.

Gong and Gomer [92] in 1988 looked at self-diffusion on the W(321) plane, measuring fluctuations in the field electron emission current; results, including activation energies, are given in Fig. 4.45. The low values for the prefactors are highly suspicious. Fast diffusion was two-dimensional, slow diffusion was one-dimensional. It shows once more that FEM is really not a suitable tool for diffusion measurement of single atoms.

First calculations for diffusion on this plane were done by Ehrlich and Kirk [95] in 1968. Using Morse potentials they found energy barriers of 0.96 eV, 1.09 eV, and 0.91 eV

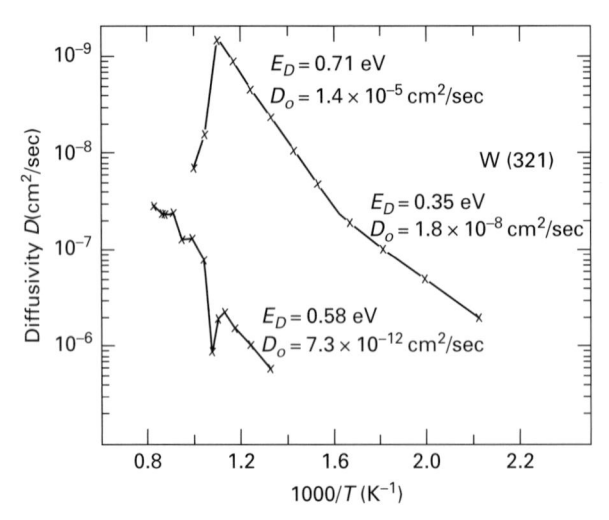

Fig. 4.45 Arrhenius plots for W atoms on W(321) derived from measurements of fluctuations in field emission current. Fast diffusion with barriers of 0.35 eV and 0.706 eV is two-dimensional (after Gong and Gomer [92]).

for rigid, relaxed, and fluctuating surfaces respectively, quite close to the experimental results. Early theoretical estimates for tungsten diffusion on W(321) were made with Morse potentials by Flahive and Graham [20]. For in-channel movement they calculated a barrier of 0.13 eV on a fully relaxed surface, much lower than what was found by experiment; on a surface with no relaxation the barrier increased to 1.07 eV. Xu and Adams [99] much later, in 1994, used interactions derived from tight-binding theory by the fourth-moment method. For in-channel movement by hopping between deep sites they estimated a barrier of 0.73 eV and a prefactor of 1.36×10^{-3} cm^2/sec. Other transitions, between shallow sites and from deep to shallow sites required appreciably more energy. However, atom exchange in moving from shallow to deep sites, shown in Fig. 3.11, involved an activation energy of only 0.23 eV and a prefactor of 3.4×10^{-4} cm^2/sec. The in-channel barrier is not too far from experiments. The exchange from shallow to deep sites predicted here was not observed experimentally by Flahive and Graham [20].

Agreement of the diffusivities in the different measurements of the diffusion of tungsten atoms (except for the field emission studies) is surprisingly good, considering that the early work was done in different laboratories with different equipment. Even values measured with a few atoms on the plane do not stand out, probably due to negligible interactions for adatoms separated by a channel of the plane. Comparison of these findings with results for tungsten on W(211) is interesting, since activation energies and prefactors are quite close. The theoretical estimates with many-body potentials are definitely in order for this surface.

4.10.2 Hetero-diffusion

In 1969 Bassett and Parsley [104] looked at the diffusion of Re atoms on W(321) in the FIM, and found an activation energy of 0.88 eV and a prefactor of 0.5×10^{-3} cm^2/sec.

They also tested the effect of an electric field on the diffusion. With a field of 0.9 V/Å they noted an increase of 0.06 eV in the activation barrier and 1.5×10^{-3} cm²/sec for the prefactor, changes that did not appear significant. A year later, Bassett and Parsley [107] examined the mobility of tantalum, tungsten, and rhenium on this plane, again with multi-atom deposits. For Ta they reported a diffusion barrier of 0.67 eV and a prefactor of 1.9×10^{-5} cm²/sec. For rhenium adatoms their results were a diffusion barrier of 0.88 eV and a prefactor of 4.8×10^{-4} cm²/sec. These prefactors appear somewhat low, possibly influenced by interactions with other adatoms, and should be looked at again.

Sakata and Nakamura [100] examined diffusion of molybdenum adatoms on W(321) also in an FIM, for which they gave an activation energy of 0.55 eV and a prefactor $D_o = 1.2 \times 10^{-7}$ cm²/sec, as shown in the Arrhenius plot in Fig. 4.37. This is orders of magnitude below the usual prefactor values. Only one atom was on the plane in these experiments, but the statistics were rather limited.

From this survey it is clear that there is no reliable information about hetero-atom movement on the W(321) plane.

4.11 Iridium: Ir(110), (311), (331)

On Ir(110), Wrigley [113] in 1980 used the Atom Probe to establish that in a majority of the observations, motion of iridium atoms occurred across the <110> rows rather than just along them. He also looked at the temperature dependence of the cross-channel diffusion of iridium atoms [114], and reported an Arrhenius plot, shown in Fig. 3.45; this yielded an activation energy of 0.74 ± 0.09 eV and $D_o = 1.4(\times 16^{\pm 1}) \times 10^{-6}$ cm²/sec. Measurements were done at only five temperatures, and in view of the low value of the prefactor must be considered with caution. He also looked at tungsten atoms moving over the iridium surface and using the Atom Probe proved that the exchange mechanism took place on this plane; he did not examine the energetics of this movement, however.

Eleven years later, Chen and Tsong [41,115] using field ion microscopy reported detailed self-diffusion measurements on a field evaporated Ir(110) surface, just as above, but also on the (1 × 2) reconstructed plane. Their results are displayed in Fig. 3.49 and 4.46. On the simple Ir(110) plane, the activation energy for self-diffusion by in-channel motion was 0.80 ± 0.04 eV, with a prefactor of $4 \times 10^{-3 \pm 0.8}$ cm²/sec. For cross-channel motion an activation energy of 0.71 ± 0.02 eV was found, as in Fig. 3.49. This energy, derived from measurements at only four temperatures, is in agreement with the previous results of Wrigley. The prefactor of $6 \times 10^{-2 \pm 1.8}$ cm²/sec is much more sensible, however. On the reconstructed (110) plane, with much deeper channels, diffusion occurred only along the <110> direction and over a higher barrier, 0.86 ± 0.03 eV, with a prefactor of $1.2 \times 10^{-3 \pm 1}$ cm²/sec. There is the possibility of a metastable walk on this surface, but the authors did not explore the mechanism of movement.

Chen et al. [116], again using the FIM, did studies of rhenium atoms on Ir(110) and Ir(311). On the latter plane, rhenium atoms diffused along the channels above 250 K and

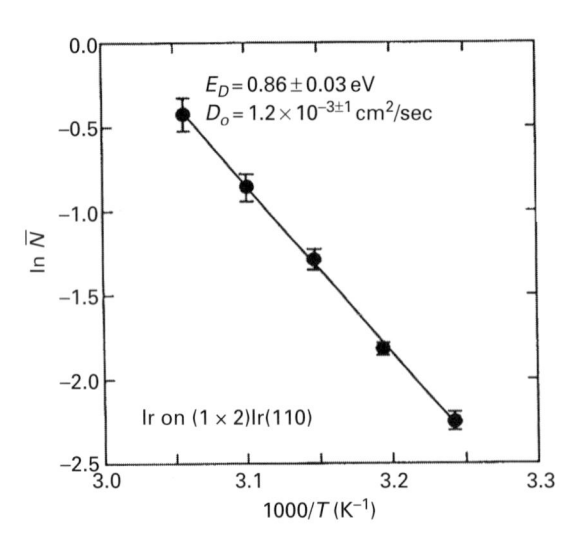

Fig. 4.46 Arrhenius plot for diffusivity of Ir on Ir(110)-(1 × 2) plane, giving higher activation energy than on (1 × 1) plane (after Chen and Tsong [41]).

did not undergo any atom exchanges. On Ir(110), after depositing a rhenium atom and heating to 256 K, there clearly was an exchange with an atom from the channel walls, and an iridium atom appeared in the neighboring channel. Rhenium atoms did not migrate along the channels; the energetics of the processes were not measured, however. In 1996, Fu et al. [117] examined self-diffusion on Ir(311) and Ir(331) planes with the FIM. The activation energy found for the first was 0.72 ± 0.02 eV, with a prefactor of $1 \times 10^{-3 \pm 0.4}$ cm^2/sec. On the (331) plane, the energy barrier, derived from the Arrhenius plot in Fig. 4.47, was 0.91 ± 0.03 eV with a prefactor of $1 \times 10^{-1.9 \pm 0.5}$ cm^2/sec. Movement of the adatom was one-dimensional.

A number of calculations have been made for the self-diffusion of iridium. That started with the calculations of Flahive and Graham [20] using Morse potentials, with the results shown in Fig. 4.9. For Ir(311) they got a value around 0.6 eV, for Ir(110) ~0.75 eV and for Ir(331) ~0.95 eV. Shiang et al. [118] worked with both EAM and RGL interaction potentials. They found an activation energy of 0.92 eV for exchange with EAM potentials and 0.75 eV with RGL, quite a big difference between values from these two methods. For diffusion by atom jumps along the channels, EAM gave an activation energy of 0.70 eV and with RGL 0.83 eV. Jumps from one channel to an adjacent one were found to require 2.58 eV in calculations using the RGL approximation. It turns out that EAM predicts a lower value of the activation energy for jumps compared with exchange, while RGL predicts the opposite sequence. For both methods, however, the differences in activation energies for both mechanisms are not that big, indicating coexistence of both mechanisms, in agreement with experiment.

In 1996, Chang et al. [119] continued with embedded atom calculations for single atom self-diffusion on Ir(110). They found an activation energy of 0.70 eV for hopping along the channels, while exchange in both directions <112> and <001> occurred over the same barrier of 0.81 eV, results not too far from experiments. Similar to the previous

4.11 Iridium: Ir(110), (311), (331)

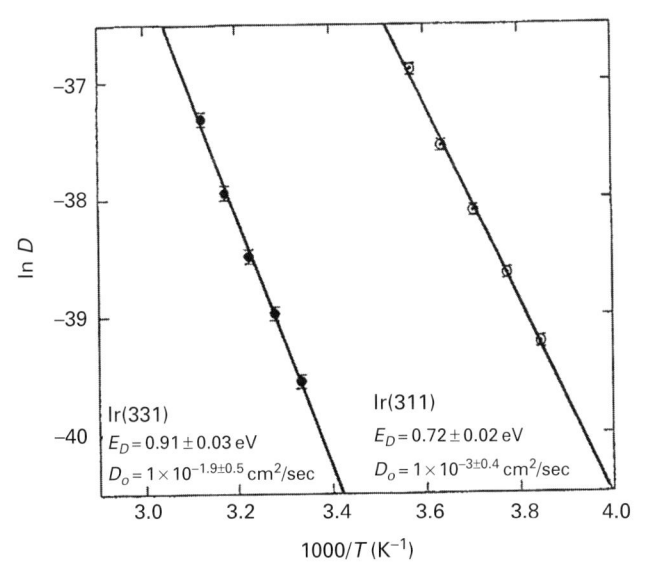

Fig. 4.47 Dependence of diffusivity for Ir atoms on Ir(331) and Ir(311), obtained in observations with FIM (after Fu et al. [117]).

EAM study [118] hopping seemed to be slightly more favorable, opposite to what was observed in experiment, where exchange was a bit more frequent event. Hopping into the adjacent channel required at least 2.5 eV. The lack of a preference for the direction of exchange does not agree with the experimental findings of Chen et al. [41] for the same system.

Agrawal et al. [7] utilized Lennard-Jones potentials to calculate the barrier to jumps of iridium atoms on Ir(110). For cross-channel jumps, they estimated a very high activation energy of 2.47 eV and a prefactor of 1.1×10^{-2} cm^2/sec, but for in-channel transitions the barrier was very high, 1.48 eV, with a prefactor 4.7×10^{-3} cm^2/sec. No investigations of atomic exchange were performed. Calculations with more reliable RGL potentials were done by Ndongmouo and Hontinfinde [32], who came up with a barrier of 1.22 eV for diffusion along the (1×2) reconstructed plane; for the (1×1) surface, the barrier was lower, 1.15 eV, but still high. Cross-channel exchange events were estimated from the surface energy to occur over a barrier of 1.60 eV, and jumps between adjacent channels required an activation barrier of 2.21 eV. All values are much higher than the experimental data as well as previous estimates with many-body potentials.

The experimental results for self-diffusion on Ir(110) seem to be reasonable. However, the calculations, even when done for the same interactions, do not appear reproducible. The work of Shiang [118] with RGL interactions is close to agreement with experiments, the studies by Ndongmouo and Hontinfinde [32], with the same potentials, are far away. Only Agrawal et al. [7] made an effort to provide prefactors for the diffusivity, but all their data stand apart from experimental findings, probably as they included only pair interactions. Theory as well as experiment are in agreement about the coexistence of two mechanisms: hopping and exchange.

4.12   Platinum: Pt(110), (311), (331)

4.12.1   Experimental data

Bassett and Webber [120] were the first to study diffusion on platinum surfaces, studies which gave very interesting and important results. Using field ion microscopy, FIM, on the (110) surface, they found in 1978 that atom movement occurred along the direction of the channels, as expected, with an activation energy of 0.84 ± 0.1 eV and a prefactor of 8×10^{-3} cm^2/sec. The surprise, however, was that diffusion also occurred across the channels, with a lower barrier of 0.78 ± 0.1 eV and a prefactor of 1×10^{-3} cm^2/sec. They also did calculations using Morse potentials and obtained 0.64 eV for the in-channel diffusion barrier and 1.97 for inter-channel jumps. We can see that Morse potentials are not very useful for explaining cross-channel diffusion. However, this study for the first time ever showed the existence of a cross-channel mechanism in diffusion on the surface.

In addition, studies were made of platinum diffusion on the channeled (311) and (331) planes. On these, motion always occurred only along the <110> channels. On the (311), the diffusivity was obtained by assuming a prefactor of $(kT/h)\ell^2$; this gave an activation energy of 0.69 ± 0.2 eV. On the (331) plane the activation energy was higher, 0.84 ± 0.1 eV, with a prefactor of 4×10^{-4} cm^2/sec. From calculations with Morse potentials they obtained 0.50 eV for the (311) barrier and 0.77 eV for diffusion on the (331) plane. The values are of the same order as the experiments for these planes.

Measurements of platinum motion on Pt(110) repeated eight years later by Kellogg [121] for both in-channel as well as cross-channel movement in the FIM gave similar results. For the former he found a barrier of 0.72 ± 0.07 eV and a prefactor of 6×10^{-4}; for cross-channel motion these values were 0.69 ± 0.07 eV and 3×10^{-4} cm^2/sec. As already shown in Fig. 3.47, Arrhenius plots were based on only four temperatures, and within the limits of the experiments in-channel and cross-channel jumps occurred at the same rate, what meant that on this plane both mechanisms coexisted. Kellogg [122] in 1986 also looked at self-diffusion on the platinum (311) plane with the FIM. He reported a diffusion barrier of 0.60 ± 0.03 eV and a prefactor of 1.9×10^{-4} cm^2/sec, as shown in Fig. 4.48, in reasonable agreement with the results of Bassett and Webber [120]. No indication of cross-channel diffusion was found, only in-channel motion occurred.

We do want to mention the work of Preuss et al. [123], who did very interesting studies of self-diffusion on single crystal Pt(110), although they did not measure single atom motion. They looked at the decay of a platinum deposit by analyzing surface profiles occurring by diffusion at temperatures above 1200 K, and as shown in Fig. 4.49, came to the conclusion that diffusion occurred both along the surface channels and across them. The activation energy for movement along the channels proved to be 1.70 eV and across the channels 3.2 eV, much above the results from single atom experiments.

A few results for other adatoms have been reported. Bassett and Webber [120], again using field ion microscopy, estimated a barrier of 0.80 ± 0.15 eV for diffusion of iridium along channels on Pt(110), and the same value, 0.80 ± 0.15 eV for cross-channel motion. These figures were estimated assuming a value for the prefactor. Morse potential

4.12 Platinum: Pt(110), (311), (331)

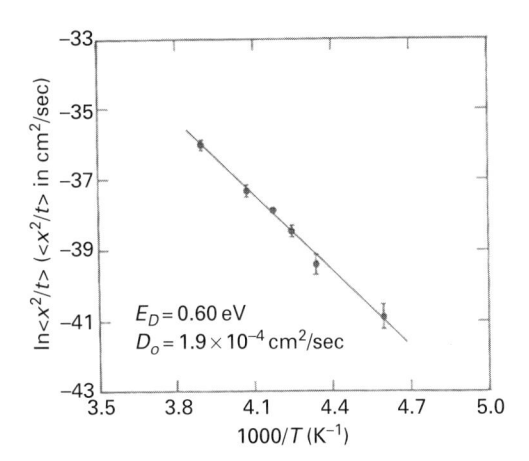

Fig. 4.48 Arrhenius plot for the diffusivity of Pt on Pt(311) plane examined in FIM (after Kellogg [122]).

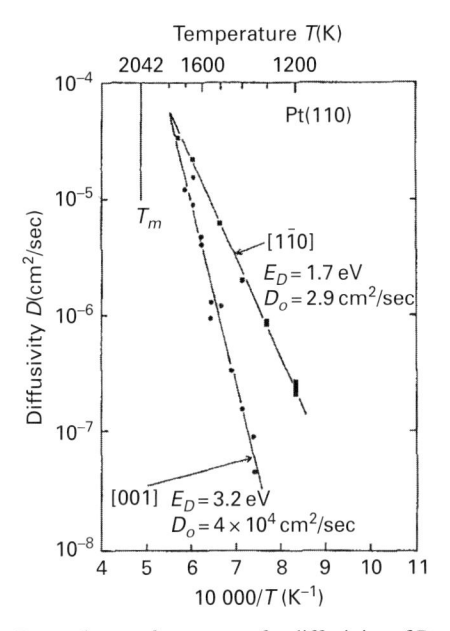

Fig. 4.49 Dependence of mass-transfer diffusivity of Pt on Pt(110) upon reciprocal temperature for in-channel [1$\bar{1}$0] and cross-channel [001] motion (after Preuss *et al.* [123]).

calculations for the same system predicted a barrier of 0.66 eV for jumps along the channels. On the (311) plane the activation energy from experiments was 0.74 ± 0.15 eV, from calculations a value of 0.54 eV was derived. On Pt(331), movement of iridium required overcoming a barrier of 0.81 eV from calculations, and there were no experimental data derived.

For gold on Pt(110), diffusion occurred along the channels over a barrier of 0.63 ± 0.15 eV, again estimated by assuming a prefactor. Calculations with Morse potential gave a barrier of 0.57 eV. What is worth noting is that gold behaves quite differently than Ir or Pt on Pt (110): for gold only one-dimensional movement was observed on this

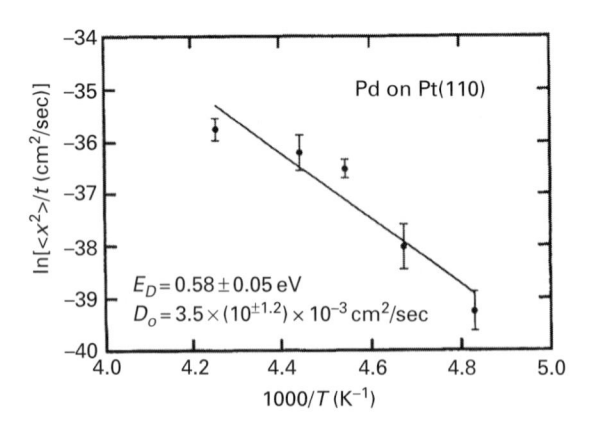

Fig. 4.50 Arrhenius plot for the diffusion of Pd atom on Pt(110) plane, determined by FIM (after Kellogg [124]).

plane. On Pt(311), the barrier measured for gold movement was 0.56 ± 0.1 eV, and a prefactor of 3×10^{-1} cm^2/sec while calculations predicted 0.41 eV. On Pt(331), according to calculations, movement of gold required overcoming a barrier of 0.68 eV.

Field ion microscope studies of palladium atoms diffusing on the Pt(110) plane were made by Kellogg [124]. As shown in Fig. 4.50, in the temperature range 207–235 K he found an activation energy of 0.58 ± 0.05 eV and a prefactor of $3.5 \times (10^{\pm 1.2}) \times 10^{-3}$ cm^2/sec for atoms hopping along the channels. Unfortunately, the statistics were limited, resulting in significant scatter of the data. Palladium is similar to gold in moving in only one dimension, and only by jumps, not by exchange.

Linderoth et al. [125,126] in 1997 examined self-diffusion on the Pt(110)-(1 × 2) plane, shown in Fig. 3.93; it is a Pt (110) surface with every second <110> row of platinum removed. They observed atomic jumps using the STM, with a tunnel junction resistance of 100 MΩ, at which the effect of the probing tip on the adatoms was negligible. Diffusion here occurred entirely along the channels forming this plane, and from measurements of the distribution of displacements, in Fig. 3.92, it was possible to establish that in addition to jumps between nearest-neighbor positions, longer transitions also occurred. From an Arrhenius plot of the jump rates, shown in Fig. 3.93, they were able to arrive at an activation energy of 0.81 ± 0.01 eV with a frequency prefactor of $10^{10.7 \pm 0.2}$ sec^{-1} for single jumps.

The situation with what were called "double" jumps, which had an activation energy of 0.89 ± 0.06 eV and a frequency factor of $10^{10.9 \pm 0.8}$ sec^{-1}, is more complicated, as these turned out not to be just double length transitions along the channel bottoms. A year after this work, Montalenti and Ferrando [127,128] worked on the missing row reconstructed Au(110) plane, doing molecular dynamics simulations with energetics from RGL potentials. They found that the adatom is not confined to the bottom of the trough, but can instead run up the channel sides and diffuse rapidly there until it returns again to the bottom. The implication of these findings are clear – such diffusion should also occur on the reconstructed Pt(110) plane.

The next year, in 1999, Lorensen et al. [129] came out with density-functional calculations for diffusion on this surface, shown in Fig. 4.51. They found that in addition

4.12 Platinum: Pt(110), (311), (331)

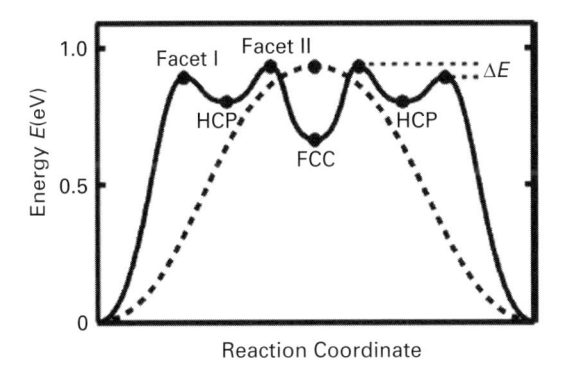

Fig. 4.51 Potential diagram for Pt diffusion on Pt(110) plane calculated with density-functional method in generalized gradient approximation. Dashed curve is for direct jump along channel (after Lorensen et al. [129]).

to atoms following a direct path along the bottoms of the troughs, platinum atoms could indeed jump onto the facets and continue their motion there, as was already shown in Fig. 3.95. An Arrhenius plot in Fig. 3.96 for their jump estimates showed that transitions along the facets could occur with an energy slightly higher than for single transitions along the troughs. Here it should also be noted that reconstruction of the (110) plane increased the activation energy for diffusion by 0.09 eV compared with the ordinary (110) surface studied by Kellogg [121].

4.12.2 Theoretical studies

Observations by Bassett and Webber [120] of cross-channel movement triggered further calculational efforts. The second of these was done by Halicioglu and Pound [130,131] using Lennard-Jones potentials. For diffusion of Pt by exchange on Pt (110) they found a barrier of 1.07 eV. For Ir and Au they obtained 1.01 and 1.02 eV respectively, the former for two-dimensional exchange, the latter for in-channel jumps. They also looked at movement on the Pt(311) plane and obtained barrier heights of 0.656 eV, 0.734 eV, and 0.486 eV for Pt, Ir, and Au atoms respectively. In 1980, Flahive and Graham [20] used Morse potentials to calculate activation energies, shown in Fig. 4.9. For self-diffusion on Pt(110) they got a value around 0.6 eV, for Pt(311) ~0.5 eV, and for Pt(331) ~0.7 eV.

Liu et al. [2] were quite early in making estimates for activation energies of self-diffusion on Pt(110) with EAM potentials. For in-channel motion evaluated with AFW interactions they obtained a low activation energy of 0.25 eV and a prefactor of 4×10^{-4} cm^2/sec; in the VC approximation the results were 0.53 eV and 1.4×10^{-3} cm^2/sec. Cross-channel motion by atom exchange occurred over higher barriers, 0.43 eV and a prefactor of 7×10^{-3} cm^2/sec using AFW potentials, and 0.68 eV with 1.5×10^{-2} cm^2/sec for the prefactor calculated with the VC potentials. On the (311) plane the diffusion characteristics for in-channel motion were a barrier of 0.43 eV with a prefactor 8×10^{-4} cm^2/sec (AFW), and 0.63 eV and 2.8×10^{-3} cm^2/sec (VC). For movement in the channels of the (331) plane, the AFW approximation gave a barrier of 0.28 eV with a prefactor of 6×10^{-4} cm^2/sec, and with VC potentials 0.54 eV and 8.5×10^{-4} cm^2/sec.

Liu et al. [2] also looked at diffusion with Leonard-Jones potentials, which brought much higher values. For in-channel movement on Pt(110) a barrier of 1.26 eV with a prefactor of 6×10^{-3} cm^2/sec and for cross-channel motion 1.34 eV and a prefactor 3.4×10^{-2} cm^2/sec. On the Pt(311) plane the barrier was 0.65 eV with a prefactor of 4×10^{-3} cm^2/sec, on Pt(331) 1.33 eV and 5.5×10^{-3} cm^2/sec. From L-J potentials only the value for Pt(311) was close to results obtained from many-body potentials or experiments, showing the importance of many-body interactions for diffusion studies.

Estimates of platinum diffusion on Pt(110) were later made by Villarba and Jónsson [132], again with EAM potentials. For in-channel motion the barrier was estimated as 0.46 eV, and slightly higher, 0.58 eV, for cross-channel exchange. On the (311) plane the activation energy for diffusion was 0.57 eV, and along the (331) plane 0.53 eV. With Morse potentials they obtained a barrier of 0.53 eV along the channels of the (110) plane and 0.63 eV across them. For diffusion on the (311) plane the Morse potential estimate for the barrier was 0.49 eV and 0.71 eV on the (331) plane. Prefactors were not investigated in this study. Relying on EMT interactions, Stoltze [27] calculated a self-diffusion barrier of 0.420 eV for in-channel jumps, and a much higher value of 0.809 eV for cross-channel diffusion by atom exchange. Cross-channel jumps were a bit more energetic, with a barrier of 0.945 eV, but again without any prefactors.

Lorensen et al. [129] did density-functional estimates of the activation energy for diffusion on the reconstructed Pt(110)-(1 × 2). For diffusion along the bottom of the channel they obtained a barrier of 0.94 eV. When the adatom jumped onto the sidewalls, as illustrated in Fig. 3.95, the barrier was close to the same value, so that both paths were possible, as shown in Fig. 4.51. Feibelman [133] in the year 2000 made estimates of diffusion on Pt(110)-(1 × 2) using the VASP *ab initio* simulation [9–11]. For adatom diffusion via a metastable walk he arrived at a barrier of 0.93 eV, slightly higher than the value of 0.89 eV determined in experiments by Linderoth et al. [125,126].

Calculations with Lennard-Jones potentials for self-diffusion on Pt(110) were carried out by Agrawal et al. [7]. For platinum atom jumps along the channels, they derived a very high activation energy of 1.21 eV and a prefactor of 3.8×10^{-3} cm^2/sec; for jumps across the channels they obtained a barrier of 2.08 eV, with a prefactor of 7.1×10^{-3} cm^2/sec. These finding are not surprising at all, since it was already shown by Liu et al. [2] 11 years earlier that L-J potentials are not suitable for describing diffusion on this surface.

Ndongmouo and Hontinfinde [32] worked with the more reliable RGL interactions and on the reconstructed (110)-(1 × 2) plane found a barrier to self-diffusion of 0.60 eV, and 0.54 eV on the (1 × 1) surface. For cross-channel atom exchange on the (1 × 1) Pt (110) plane the barrier proved to be 0.78 eV, and for jumps to a neighboring channel 1.38 eV, the former in good agreement with the experiments.

Different diffusion barriers were obtained from molecular dynamics simulations using RGL potentials by Kim et al. [34]. For in-channel self-diffusion they found an energy of 0.490 eV was required, for cross-channel exchange the barrier was 0.779, and for cross-channel jumping 1.247 eV. The value for cross-channel motion by exchange seems to be quite high compared with the in-channel barrier. Also determined were the barriers for in-channel hetero-motion on Pt(110) – 0.463 eV for Ni, 0.363 eV for Cu, 0.402 eV for Pd, 0.376 eV for Ag, and 0.386 eV for Au. The value for gold is much too low compared with the experimental findings of Bassett and Webber [120], the value for Pd is lower than the experimental results of Kellogg [124].

Not too much experimental data about self-diffusion on Pt(110) is available, but the two existing reports are not that far apart and it is clear that both mechanisms, hopping and exchange, participate on this surface. The various theoretical efforts are quite different, and even the RGL as well as EMT estimates differ significantly and favor jumping over exchange, which is not seen in the experiments. One thing is clear – many-body potentials must be used in studying platinum surfaces.

4.13 Gold: Au(110), (311), (331)

Hardly any experimental work has been done to measure atomic diffusivities on Au(110). Günther et al. [134], in 1997, carried out STM studies of atom hopping on Au(110). From observations at two temperatures and a coverage of 0.2 ML, they found that on the (1 × 2) reconstructed surface the activation energy for self-diffusion ranged from 0.40 eV to 0.44 eV, with a frequency prefactor of 10^{12}–10^{13} sec^{-1}. On the (1 × 1) surface, prepared from the reconstructed plane after atoms had locally filled in the wide troughs, the minimum barrier height was lowered to 0.38 eV. Hitzke et al. [135] in the same year looked briefly at diffusion of Ni on Au(110)-(1 × 2) using the STM. They examined a surface covered with a 0.22 ML film of Ni at 180 K, and deduced a barrier of 0.52 ± 0.1 eV along <1$\bar{1}$0>, on the assumption of a frequency prefactor of 10^{13} sec^{-1}. However, in both experimental studies movements are likely to be influenced by adatom interactions.

There have, however, been a number of attempts to calculate the diffusion characteristics. In 1980 Flahive and Graham [20] did calculation using Morse potentials with results shown in Fig. 4.9. Early estimates of self-diffusion on Au(110) were made by Roelofs et al. [136] in 1991, who did Monte Carlo simulations with interactions similar to Morse potentials. From an Arrhenius plot of in-channel motion they found an activation energy of only 0.140 eV; for cross-channel movement by atom exchange the barrier was 0.305 eV. Four years later Roelofs et al. [137] continued estimates of diffusion, this time based on EAM interactions. The activation energy for in-channel motion was given as 0.27 eV. For the cross-channel exchange process the barrier was 0.35 eV, with a frequency prefactor of ~0.76 × 10^{12} sec^{-1}.

Liu et al. [2] in 1991 used EAM interactions and transition state theory to reach an in-channel barrier of 0.25 eV and a prefactor of 6 × 10^{-4} cm^2/sec with AFW parameters; for VC potentials the values were 0.34 eV and 1.6 × 10^{-3} cm^2/sec. In cross-channel motion the activation energies were 0.40 eV with a prefactor of 2.5 × 10^{-2} cm^2/sec (AFW) and 0.42 eV and 1.3 × 10^{-2} cm^2/sec (VC). They also did calculations for Au(311) and Au(331) where movement is likely to be one-dimensional. For the (311) plane using AFW potentials they obtained 0.35 eV and a prefactor of 8 × 10^{-4} cm^2/sec, from VC potentials the barrier to diffusion was 0.42 eV with a prefactor of 3.6 × 10^{-3} cm^2/sec. For the (331) plane, the barriers were 0.26 eV and the prefactor 6 × 10^{-4} cm^2/sec (AFW) and 0.34 eV and 9.6 × 10^{-4} cm^2/sec (VC). These are the only determinations for the two planes, making comparison impossible.

With effective medium theory potentials, Stoltze [27] calculated a self-diffusion barrier of 0.268 eV for in-channel jumps on Au(110); for cross-channel exchange the activation energy proved to be higher, 0.554 eV, and the jump across the channel was only slightly more energetic at 0.67 eV.

Montalenti and Ferrando [127] in 1998 made an important discovery in molecular dynamics simulations of self-diffusion on the reconstructed Au(110)-(1 × 2) surface. Representing interactions in the RGL approximation, they found that, as shown in Fig. 3.95, the atom could jump from the bottom of the channel to the (111)-like sidewalls, where diffusion continued until the atom again descended. From Arrhenius plots of the diffusion rate they found a barrier of 0.29 ± 0.01 eV for single jumps, and 0.43 ± 0.03 eV for double jumps, both along the bottom of the troughs. In the metastable excursions on the wall, they discovered a comparable activation energy of 0.37 ± 0.02 eV for single jumps, and 0.43 ± 0.05 eV for double jumps. Jumps along the channel bottoms and along the sides were therefore both likely to contribute to diffusion. They also looked at the static energy and for single jumps they obtained a barrier of 0.31 eV. Climbing a wall required 0.4 eV, but the opposite motion needed only 0.06 eV; movement on the (111) wall required overcoming a barrier of 0.10 eV.

Long jumps in diffusion were explored by Montalenti and Ferrando [45,128] in molecular dynamics simulations using RGL potentials. For in-channel motion of Au on Au(110) they found a barrier of 0.28 eV, and 0.46 eV for cross-channel motion with atom exchange; cross-channel jumps cost much more, with a barrier of 0.7 eV. For diffusion by in-channel jumps on the reconstructed Au(110)-(1 × 2) surface, the barrier was higher, 0.31 eV; cross-channel exchange cost 0.66 eV, cross-channel jumps were comparable to exchange at 0.63 eV. The effective barrier for metastable walks was 0.44 eV. An Arrhenius plot, in Fig. 4.52, showed a nice straight line, and in a detailed examination of jump statistics longer jumps amounted to less than 1% of the total at 450 K.

In 2002 Agrawal et al. [7] used MC variational transition state theory and Lennard-Jones potentials to calculate the different barriers on Au(110). They found a barrier of 0.86 eV for in-channel movement, much higher than the usual estimate, and a prefactor

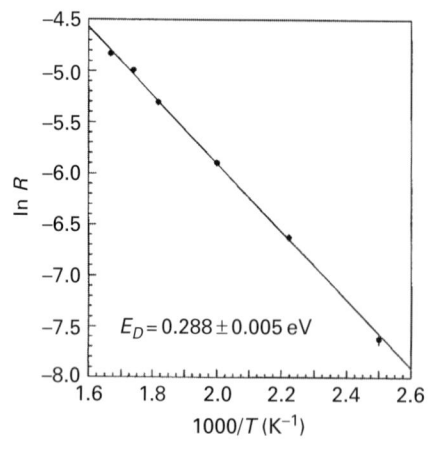

Fig. 4.52 Dependence of Au atom diffusivity on Au(110)-(1 × 2) surface upon reciprocal temperature, derived from molecular dynamics simulations with RGL potentials. Static barrier 0.31 eV, from Arrhenius plot 0.288 ± 0.005 eV (after Montalenti and Ferrando [45]).

4.13 Gold: Au(110), (311), (331)

6.3×10^{-3} cm^2/sec; for cross-channel movement they calculated a barrier of 1.49 eV with a prefactor of 2.8×10^{-2} cm^2/sec.

Ndongmouo and Hontinfinde [32,138] used RGL potentials to calculate a barrier for self-diffusion jumps on Au(110)-(1 × 2) of 0.31 eV. For Au (1 × 1) they found a smaller barrier of height 0.28 eV. For cross-channel diffusion by exchange the barrier was 0.46 eV, and for cross-channel jumps it proved to be higher, 0.69 eV. The same RGL potentials were used by Kim *et al*. [34] in molecular dynamics estimates of diffusion on Au(110). For in-channel motion the barrier was 0.274 eV, for exchange 0.468 eV, and 0.700 eV for cross-channel jumps, close to the previous estimates. In hetero-diffusion of Ni the barrier proved to be 0.313 eV, for Cu 0.243 eV, for Pd 0.281 eV, and Pt 0.393 eV.

Molecular dynamics simulations of nickel adatom diffusion on Au(110)-(1 × 2) were done by Fan and Gong [139] using embedded atom potentials from Johnson [140,141]. Diffusion can of course proceed in different ways, along the troughs of the channels making up this plane, or by moving up to the channel facets and diffusing there. This is demonstrated by the atom trajectories shown in Fig. 4.53. For in-trough motion the activation energy was 0.21 eV; for motion to a position on the sidewalls the barrier was comparable, 0.20 eV. However diffusion in the opposite direction, from the wall to the bottom, required only 0.04 eV. Diffusion on the wall occurred over a small barrier of 0.05 eV between points *C* and *B* as well as *B* and *C* in Fig. 4.54, and 0.08 eV for *C* to *F*. Diffusion on the sidewalls took place with a much smaller activation energy, and can be expected to play a significant role.

Calculated diffusion characteristics for Au(110) are in reasonable agreement with each other, but more experimental information would certainly be desirable, as the available

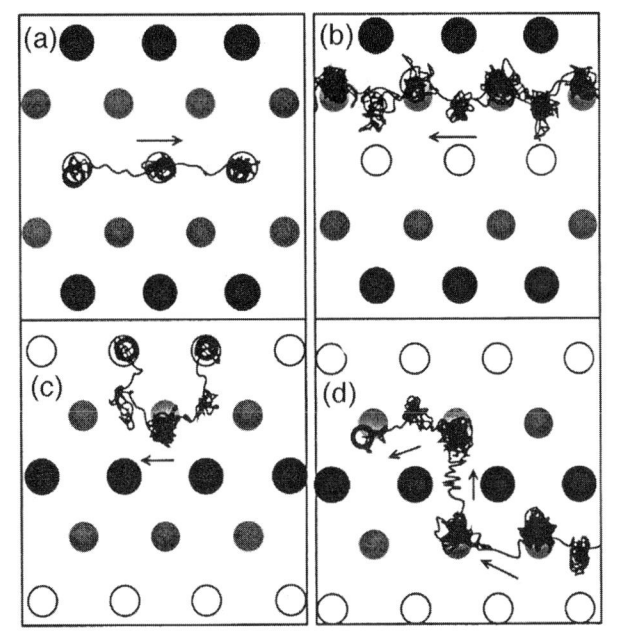

Fig. 4.53 Ni atom trajectories on Au(110)-(1 × 2) surface, obtained by molecular dynamics simulations with Johnson potentials [140,141]. (a) Single in-channel jump. (b) Jumps on channel walls. (c) Diffusion over local sidewall minima. (d) Diffusion from one channel to the next (after Fan and Gong [139]).

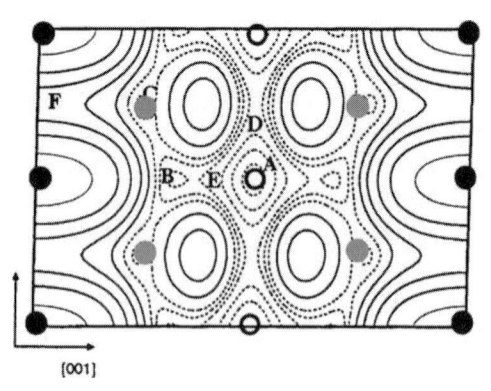

Fig. 4.54 Potential energy for Ni atom on Au(110)-(1 × 2) surface [139]. Filled circles show atoms at top of channel, open circles atoms at channel bottom. Minimum energy locations *A, B, C*; saddle points *E, D* (after Fan and Gong [139]).

experimental data differ from calculated values. Also information about hetero-diffusion is quite scarce.

4.14 Lead: Pb(110)

Quasi-elastic helium scattering studies were carried out by Frenken *et al.* [142] on the Pb (110) surface at close to the melting point. The diffusivity was deduced from the energy distribution of the helium atoms scattered from the surface. The activation energy for diffusion was derived as 0.65 ± 0.2 eV, with a prefactor $D_o = 26$ cm^2/sec, and the diffusivity along and across the surface channels was roughly the same. After additional studies at temperatures from ~300 K to ~560 K, Frenken *et al.* [143] came up with a diffusion barrier of 1.0 ± 0.3 eV and a prefactor of 6.2×10^4 cm^2/sec for cross-channel motion. For motion along the channels they tentatively estimated an activation energy of 1.0 eV and a prefactor of 1.8×10^5 cm^2/sec. These numbers all look to be too large by comparison with other materials, and it does not appear that single atom motion was being measured. The activation energy seems to be not just for atom motion, but may also have a contribution from the energy for forming vacancies.

Calculations of the self-diffusion barrier on Pb(110) have been made by Karimi *et al.* [144] using EAM interactions. For diffusion along the channels they arrived at a barrier of only 0.215 eV; for cross-channel motion by exchange the barrier was higher, 0.454 eV. Adequate experiments to test the validity of these estimates are not available.

It is clear that more experimental as well as calculational work needs to be done to achieve insight into the characteristics of self-diffusion on Pb(110).

4.15 Tables for 1D Diffusion

To provide perspective as well as a ready reference and also to allow quick comparisons, much of the diffusion data is summarized here in tabular form. To ease access, the material is subdivided according to the substrate plane for which results are presented.

Table 4.1 Fcc(110)

System	Mechanism	Experiment E_D (eV)	D_o (cm²/sec)	Theory E_D (eV)	D_o (cm²/sec)	v_0 (sec⁻¹)	Method	Ref.
Al/Al(110)	∥ Jump	0.43	†10⁻³				FIM	[1]
				0.26	1.8×10⁻³		EAM AFW	[2]
				0.204			EMT	[5]
				0.33			DFT	[6]
				0.67	3.5×10⁻³		L-J	[7]
				0.38			VASP	[8]
	⊥Ex	0.43	†10⁻³				FIM	[1]
				0.30	6.0×10⁻²		EAM AFW	[2]
				0.15	2.4×10⁻²		EAM VC	[2]
				0.285			EMT	[5]
				0.62			DFT	[6]
				0.50			VASP	[8]
	⊥ Jump			1.19	6.9×10⁻³		L-J	[7]
				0.83			VASP	[8]
Ni/Ni(110)	∥ Jump	0.23±0.04	1×10⁻⁹				FIM‡	[12]
		0.30±0.06	10¹				FIM‡‡	[12]
				0.45	†10⁻³		Refit	[2]
		0.536±0.040					He Scat	[15,16]
		0.60					STM	[17]
				0.44	2.3×10⁻³		EAM AFW	[2]
				0.39	4.0×10⁻³		EAM VC	[2]
				0.50	3.3×10⁻³		EAM FBD	[23]
				0.53	3.8×10⁻³		EAM VC	[23]
				0.44	3.6×10⁻³		EAM OJ	[23]
				0.64	4.5×10⁻³		EAM ATVF	[23]
				0.41			EAM	[22]
				0.407			EMT	[27]
				0.243			CEM¥	[28]
				0.18			CEM	[28]
				0.39			EAM5	[29]

Table 4.1 (cont.)

System	Mechanism	Experiment E_D (eV)	Experiment D_o (cm²/sec)	Theory E_D (eV)	Theory D_o (cm²/sec)	v_0 (sec⁻¹)	Method	Ref.
				0.39	$(1.4 \pm 0.1) \times 10^{-3}$		EAM	[31]
				0.86	2.5×10^{-3}		L-J	[7]
				0.32			RGL	[32]
				0.301			RGL	[34]
	⊥ Ex	0.32 ± 0.05	10^{-7}				FIM‡	[12]
		0.26 ± 0.06	10^{-1}				FIM‡‡	[12]
				0.45	†10^{-3}		Refit	[2]
		0.424 ± 0.040					He Scat	[15,16]
				0.49	3.7×10^{-3}		EAM AFW	[2]
				0.42	2.8×10^{-2}		EAM VC	[2]
				0.564			EMT	[27]
				0.392			CEM*	[28]
				0.35			CEM	[28]
				0.42			EAM5	[29]
				0.38			RGL	[32]
				0.301			RGL	[34]
	⊥ Jump			1.30	6.5×10^{-3}		EAM FBD	[23]
				1.67	7.5×10^{-3}		EAM VC	[23]
				1.01	6.5×10^{-3}		EAM OJ	[23]
				1.53	7.7×10^{-3}		EAM ATVF	[23]
				1.157			EMT	[27]
				1.68	4.0×10^{-2}		L-J	[7]
				1.33			RGL	[32]
				1.026			RGL	[34]
Cu/Ni(110)	∥ Jump			0.334			RGL	[34]
Pd/Ni(110)	∥ Jump			0.426			RGL	[34]
Ag/Ni(110)	∥ Jump			0.334			RGL	[34]
Pt/Ni(110)	⊥ Ex	0.28	†10^{-3}				FIM	[13]
	∥ Jump			0.308			RGL	[34]
Au/Ni(110)	∥ Jump			0.440			RGL	[34]
Cu/Cu(110)	∥ Jump			0.23	8×10^{-4}		EAM AFW	[2]

	0.28	4.4×10^{-3}	EAM VC	[2]
	0.18		EMT	[39]
	0.26		CEM¥	[28]
	0.08		CEM59	[28]
	0.292		EMT	[27]
	0.24		EAM AFW	[40]
	0.23		RGL	[44]
	0.47		RGL	[42]
	0.25		RGL	[43]
	0.251		RGL	[46]
	0.25	$(1.1 \pm 0.1) \times 10^{-3}$	RGL	[31]
	0.76	7.5×10^{-3}	L-J	[7]
	0.23		RGL	[32]
	0.23	6.29×10^{-4}	EAM	[48]
	0.241		RGL	[34]
	0.23	4.0×10^{-3}	EAM	[50]
	0.26		MS	[51]
∥ Jumpÿ	0.16 ± 0.01	2×10^{12}	RGL	[45]
a	0.213 ± 0.007	9×10^{12}	RGL	[45]
⊥ Move	0.84 ± 0.04		He Scat	[37]
⊥ Ex	0.30	3.2×10^{-2}	EAM AFW	[2]
	0.31	2.7×10^{-2}	EAM VC	[2]
	0.27		EMT	[39]
	0.485		CEM¥	[28]
	0.09		CEM59	[28]
	0.419		EMT	[27]
	0.30		EAM AFW	[40]
	0.29		RGL	[44]
	0.63		RGL	[42]
	0.30		RGL	[43]
	0.284		RGL	[46]
	0.27		RGL	[32]
	0.34		EAM	[49]
	0.323		RGL	[34]
	0.30		MS	[51]

Table 4.1 (cont.)

System	Mechanism	Experiment E_D (eV)	D_o (cm²/sec)	Theory E_D (eV)	D_o (cm²/sec)	v_0 (sec⁻¹)	Method	Ref.
	⊥ Jump			0.826			EMT	[27]
				1.15			EAM	[40]
				1.31	2.6×10^{-2}		L-J	[7]
				1.07			RGL	[32]
				1.146	9.97×10^{-4}		EAM	[48]
				1.020			RGL	[34]
				1.146	7.4×10^{-3}		EAM	[50]
Co/Cu(110)	∥ Jump			0.29			MS	[51]
				0.35			VASP	[51]
	⊥ Ex			0.30			MS	[51]
				0.32			VASP	[51]
				>1.0			MS	[51]
Ni/Cu(110)	⊥ Jump			0.244			VASP	[34]
Pd/Cu(110)	∥ Jump			0.318			RGL	[34]
Ag/Cu(110)	∥ Jump			0.254			RGL	[34]
Pt/Cu(110)	∥ Jump			0.376			RGL	[34]
Au/Cu(110)	∥ Jump			0.19 ± 0.03			RGL	[42]
				0.333			RGL	[34]
Pb/Cu(110)	⊥ Ex			0.23 ± 0.04			RGL	[42]
	∥ Jump			0.212			RGL	[46]
	α			0.14			RGL-Arrh	[46]
	⊥ Ex			0.281			RGL	[46]
Rh/Rh(110)	∥ Jump	0.60 ± 0.03	3×10^{-1}				FIM	[60]
				1.24	8×10^{-3}		L-J	[2]
				0.48			Morse	[2]
				1.20	4.4×10^{-3}		L-J	[7]
				0.78			RGL	[32]
	⊥ Ex			0.91			RGL	[32]
	⊥ Jump			2.06	1.0×10^{-2}		L-J	[7]
				1.80			RGL	[32]
Pt/Rh(110)	⊥ Ex	~0.65					FIM	[61]

Pd/Pd(110)	∥ Jump		0.28		EAM	[63]
			0.45		Morse	[20]
			0.28	4×10^{-4}	EAM AFW	[2]
			0.30	3.5×10^{-3}	EAM VC	[2]
			0.30		CEM	[28]
			0.28		CEM$^\yen$	[28]
			0.366		EMT	[27]
			0.72	1.8×10^{-3}	L-J	[7]
			0.39		RGL	[32]
			0.38		RGL	[34]
	⊥ Ex		0.42		EAM	[63]
			0.42	3.3×10^{-2}	EAM AFW	[2]
			0.34	2.4×10^{-2}	EAM VC	[2]
			0.33		CEM	[28]
			0.38		CEM$^\yen$	[28]
			0.599		EMT	[27]
			0.54		RGL	[32]
			0.551		RGL	[34]
			0.776		EMT	[27]
	⊥ Jump		1.56		EAM	[63]
			1.35	9.2×10^{-3}	L-J	[7]
			0.99		RGL	[32]
			0.965		RGL	[34]
Ni/Pd(110)	∥ Jump		0.369		RGL	[34]
Cu/Pd(110)	∥ Jump	0.51 ± 0.05			STM	[62]
			0.32		EAM	[62]
		0.3			STM	[65]
			0.315		RGL	[34]
	⊥ Move	0.75 ± 0.07			STM	[62]
	⊥ Ex	0.45	0.57		EAM	[62]
					STM	[65]
	⊥ Jump		1.25		EAM	[62]
Ag/Pd(110)	∥ Jump		0.333		RGL	[34]
Pt/Pd(110)	∥ Jump		0.520		RGL	[34]
Au/Pd(110)	∥ Jump		0.392		RGL	[34]
Ag/Ag(110)	∥ Jump	0.279			STM	[66]

Table 4.1 (cont.)

System	Mechanism	Experiment E_D (eV)	Experiment D_o (cm^2/sec)	Theory E_D (eV)	Theory D_o (cm^2/sec)	v_0 (sec^{-1})	Method	Ref.
Ag/Ag(110)-(2 × 2)		0.30 ± 0.03					STM	[67]
				0.32	1.0×10^{-3}		EAM AFW	[2]
				0.25	2.7×10^{-3}		EAM VC	[2]
				0.248			CEM*	[28]
				0.26			CEM	[28]
				0.291			EMT	[27]
				0.28			RGL	[75]
				0.59	2.6×10^{-3}		L-J	[7]
				0.277			RGL	[34]
	∥ Jump[a]	0.19 ± 0.05					He Scat	[68]
	∥ Jump[b]	0.26 ± 0.03					He Scat	[69,70]
	∥ Jump[c]	0.29 ± 0.02					He Scat	[71]
	⊥ Ex	0.394					STM	[66]
				0.42	4.0×10^{-2}		EAM AFW	[2]
				0.31	2.5×10^{-2}		EAM VC	[2]
				0.328			CEM*	[28]
				0.34			CEM	[28]
				0.561			EMT	[27]
				0.38			RGL	[75]
				0.388			RGL	[34]
	⊥ Move	~0.35					STM	[67]
	⊥ Jump			0.639			EMT	[27]
				0.82			RGL	[75]
				1.10	6.9×10^{-3}		L-J	[7]
				0.818			RGL	[34]
Ag/Ag(110)-(2 × 2)	∥ Jump			0.266			DFT	[78]
	⊥ Ex			~0.4			DFT	[78]
	⊥ Jump			0.757			DFT	[78]
Ni/Ag(110)	∥ Jump			0.268			RGL	[34]
Cu/Ag(110)	∥ Jump			0.249			RGL	[34]
Pd/Ag(110)	∥ Jump			0.337			RGL	[34]

System	Direction	Value	Value2	Energy	Method	Ref
Pt/Ag(110)	∥ Jump			0.492	RGL	[34]
Au/Ag(110)	∥ Jump			0.360	RGL	[34]
Ir/Ir(110)	⊥ Ex	0.74 ± 0.09	$1.4(\times 16^{\pm 1}) \times 10^{-6}$		FIM	[114]
		0.71 ± 0.02	$6 \times 10^{-2 \pm 1.8}$		FIM	[41]
				0.92	EAM	[118]
				0.75	RGL	[118]
				0.81	EAM	[119]
	∥ Jump	0.80 ± 0.04	$4.0 \times 10^{-3 \pm 0.8}$	1.60	RGL	[32]
				~0.75	FIM	[41]
				0.70	Morse	[20]
				0.83	EAM	[118]
				0.70	RGL	[118]
				1.48	EAM	[119]
				1.15	L-J	[7]
	⊥ Jump			2.58	RGL	[32]
				2.5	RGL	[118]
				2.47	EAM	[119]
			4.7×10^{-3}	2.21	L-J	[7]
					RGL	[32]
Ir/Ir(110)–(1×2)	∥ Jump	0.86 ± 0.02	$1.2 \times 10^{-3 \pm 1}$		FIM	[41]
				1.22	RGL	[32]
Pt/Pt(110)	∥ Jump	0.84 ± 0.1	8×10^{-3}		FIM	[120]
		0.72 ± 0.07	6×10^{-4}		FIM	[121]
				0.64	Morse	[120]
				0.6	Morse	[20]
			4×10^{-4}	0.25	EAM AFW	[2]
			1.4×10^{-3}	0.53	EAM VC	[2]
			6×10^{-3}	1.26	L-J	[2]
				0.46	EAM	[132]
				0.53	Morse	[132]
				0.42	EMT	[27]
			3.8×10^{-3}	1.21	L-J	[7]
				0.54	RGL	[32]
	⊥ Ex	0.78 ± 0.1	1×10^{-3}	0.49	RGL	[34]
					FIM	[120]

Table 4.1 (cont.)

System	Mechanism	Experiment E_D (eV)	Experiment D_o (cm²/sec)	Theory E_D (eV)	Theory D_o (cm²/sec)	v_0 (sec⁻¹)	Method	Ref.
		0.69 ± 0.07	3 × 10⁻⁴				FIM	[121]
				0.43	7 × 10⁻³		EAM AFW	[2]
				0.68	1.5 × 10⁻²		EAM VC	[2]
				1.34	3.4 × 10⁻²		L-J	[2]
				1.07			L-J	[2]
				0.58			EAM	[130]
				0.63			Morse	[132]
				0.809			Morse	[132]
				0.78			EMT	[27]
	⊥ Jump			0.779			RGL	[32]
				1.97			RGL	[34]
				0.945			Morse	[120]
				2.08	7.1 × 10⁻³		EMT	[27]
				1.38			L-J	[7]
				1.247			RGL	[32]
							RGL	[34]
Pt/Pt(110)–(1×2)	‖ Jump	0.81 ± 0.01				10^(10.7±0.2)	STM*	[125]
				0.94			DFT	[129]
				0.60			RGL	[32]
	MW	0.89 ± 0.06				10^(10.9±0.9)	STM*	[125]
				0.93			VASP	[133]
				~0.94			DFT	[129]
	α Jump			0.92			DFT	[129]
	β Jump			0.95			DFT	[129]
	γ Jump			0.98			DFT	[129]
Ni/Pt(110)	‖ Jump			0.463			RGL	[34]
Cu/Pt(110)	‖ Jump			0.363			RGL	[34]
Pd/Pt(110)	‖ Jump	0.58 ± 0.05	3.5 × 10⁻³±¹·²				FIM	[124]
	‖ Jump			0.402			RGL	[34]
Ag/Pt(110)	‖ Jump			0.376			RGL	[34]
Ir/Pt(110)	‖ Jump	0.80 ± 0.15£	†10⁻⁵				FIM	[120]
				0.66			Morse	[120]

System	Process	Col A	Col B	Col C	Method	Ref.
	⊥ Ex	0.80±0.15£			FIM	[120]
	∥ Jump	0.63±0.15£		†10⁻⁷	L-J	[130]
Au/Pt(110)			1.01		FIM	[120]
	∥ Jump		0.57		Morse	[120]
			1.02		L-J	[130]
			0.386		RGL	[34]
Au/Au(110)	∥ Jump		0.14		Morse	[136]
			0.25	6×10^{-4}	EAM AFW	[2]
			0.34	1.6×10^{-3}	EAM VC	[2]
			0.268		EMT	[27]
		0.38	0.27		EAM	[137,145]
			0.28		STM	[134]
			0.86	6.3×10^{-3}	RGL	[45,128]
	⊥ Ex		0.28		L-J	[7]
			0.274		RGL	[32]
			0.305		RGL	[34]
			0.40	2.5×10^{-2}	Morse	[136]
			0.42	1.3×10^{-2}	EAM AFW	[2]
			0.554		EAM VC	[2]
			0.35		EMT	[27]
			0.46	$\sim 0.76 \times 10^{12}$	EAM	[137,145]
			0.46		RGL	[45,128]
	⊥ Jump		0.468		RGL	[32]
			0.67		RGL	[34]
			0.7		EMT	[27]
			1.49	2.8×10^{-2}	RGL	[45,128]
			0.69		L-J	[7]
			0.70		RGL	[32]
Au/Au(110)–(1×2)	∥ Jump	0.40–0.44	0.31	10^{12}–10^{13}	STM	[134]
	∥ α Jump		0.29±0.01		RGL	[32]
			0.31		RGL^S	[45,128]
	β Jump		0.43±0.03		RGL	[127]
	Meta α		0.37±0.02		RGL	[127]

Table 4.1 (cont.)

System			Experiment		Theory			
	Mechanism	E_D (eV)	D_o (cm²/sec)	E_D (eV)	D_o (cm²/sec)	v_0 (sec⁻¹)	Method	Ref.
	Meta β			0.43±0.05			RGL	[127]
	MW			0.44			RGL	[45,128]
	⊥ Ex			0.66			RGL	[45,128]
	⊥ Jump			0.63			RGL	[45,128]
Ni/Au(110)	∥ Jump			0.313			RGL	[34]
Ni/Au(110)-(1×2)	∥ Jump	0.52±0.1				†10¹³	STM	[135]
				0.21			EAM	[139]
	M-Jump			0.29			EAM	[139]
Cu/Au(110)	∥ Jump			0.243			RGL	[34]
Pd/Au(110)	∥ Jump			0.281			RGL	[34]
Pt/Au(110)	∥ Jump			0.393			RGL	[34]
Pb/Pb(110)		0.65±0.2	26				He Scat	[142]
				0.215			EAM	[144]
	∥ Jump[d]	1.0	1.8×10⁵				He Scat	[143]
	⊥ Ex[d]	1.0±0.3[d]	6.2×10⁴				He Scat	[143]
	⊥ Ex			0.454			EAM	[144]

* direct measurement; ‡ after thermal treatment; ‡‡ after hydrogen treatment; CEM¥ – MD/MC CEM; [a] Temperature range 600–750 K; [b] Temperature range 650–775 K; [c] Temperature range 600–800 K; [d] Temperature range 300–560 K; † assumed value; £ assumed $kT\ell^2/2h$; [s] static estimation; [y] above 400 K.

Table 4.2 Fcc(311)

System	Mechanism	Experiment E_D (eV)	Experiment D_o(cm²/sec)	Theory E_D (eV)	Theory D_o(cm²/sec)	v_0 (sec⁻¹)	Method	Ref.
Al/Al(311)	∥ Jump	0.48	†10⁻³				FIM	[1]
				0.20	2.0 × 10⁻³		EAM AFW	[2]
				0.24	6.7 × 10⁻³		EAM VC	[2]
Ni/Ni(311)	∥ Jump	0.30 ± 0.03	1.9 × 10⁻⁶				FIM	[12]
				0.37	†10⁻³		Refit	[2]
				0.34	1.4 × 10⁻³		EAM AFW	[2]
				0.38	4.4 × 10⁻³		EAM VC	[2]
	⊥ Ex			0.351			EMT	[36]
	⊥ Jump			1.302			EMT	[36]
	∥ Jump			1.373			EMT	[36]
Pt/Ni(311)	∥ Jump	0.38	†10⁻³				FIM	[14]
Cu/Cu(311)	∥ Jump			0.26	1.2 × 10⁻³		EAM AFW	[2]
				0.28	3.1 × 10⁻³		EAM VC	[2]
				0.232			EMT	[36]
	⊥ Ex			0.889			EMT	[36]
	⊥ Jump			0.822			EMT	[36]
Rh/Rh(311)	∥ Jump	0.54 ± 0.04	2 × 10⁻³	0.64	5.4 × 10⁻³		FIM	[60]
				0.44			L-J	[2]
							Morse	[2]
Pt/Rh(311)	∥ Jump	0.44 ± 0.05	(1.6 ± 0.5) × 10⁻⁴				FIM	[61]
Pd/Pd(311)	∥ Jump			~0.35			Morse	[20]
				0.37	1.3 × 10⁻³		EAM AFW	[2]
				0.41	3.1 × 10⁻³		EAM VC	[2]
Ag/Ag(311)	∥ Jump			0.26	3.0 × 10⁻³		EAM VC	[2]

Table 4.2 (cont.)

System	Mechanism	Experiment E_D (eV)	Experiment D_o(cm^2/sec)	Theory E_D (eV)	Theory D_o(cm^2/sec)	v_0 (sec^{-1})	Method	Ref.
	$\perp$ Ex			0.26	1.0×10^{-3}		EAM AFW	[2]
	$\perp$ Jump			0.22			EMT	[36]
	$\perp$ Jump			0.727			EMT	[36]
	∥ Jump			0.704			EMT	[36]
Ir/Ir(311)		0.72 ± 0.02	$1 \times 10^{-3 \pm 0.4}$				FIM	[117]
				~ 0.6			Morse	[20]
Pt/Pt(311)	∥ Jump	$0.69 \pm 0.2^{£}$	$^{†}10^{-6}$				FIM	[120]
				0.50			Morse	[120]
		0.60 ± 0.03	1.9×10^{-4}				FIM	[122]
				~ 0.50			Morse	[20]
				0.656			L-J	[130]
				0.43	8×10^{-4}		EAM AFW	[2]
				0.63	2.8×10^{-3}		EAM VC	[2]
				0.65	4×10^{-3}		L-J	[2]
				0.57			EAM	[132]
				0.49			Morse	[132]
Ir/Pt(311)	∥ Jump	$0.74 \pm 0.15^{£}$		0.54			FIM	[120]
				0.734			Morse	[120]
Au/Pt(311)	∥ Jump	0.56 ± 0.15	3×10^{-1}				Morse	[130]
				0.41			FIM	[120]
				0.486			Morse	[120]
Au/Au(311)	∥ Jump			0.35	8×10^{-4}		Morse	[130]
				0.42	3.6×10^{-3}		EAM AFW	[2]
							EAM VC	[2]

$^{£}$ assumed $kT\ell^2/2h$; † assumed value.

4.15 Tables for 1D Diffusion

Table 4.3 Fcc(331)

System	Mechanism	Experiment E_D (eV)	D_o (cm^2/sec)	Theory E_D (eV)	D_o (cm^2/sec)	v_0 (sec^{-1})	Method	Ref.
Al/Al(331)	∥ Jump	0.46	†10^{-3}				FIM	[1]
				0.27	2.0 × 10^{-3}		EAM AFW	[2]
				0.24	5.4 × 10^{-3}		AFW VC	[2]
Ni/Ni(331)	∥ Jump	0.45 ± 0.03	2.3 × 10^{-3}				FIM	[12]
				0.45	†10^{-3}		Recal.	[2]
				0.45	1.4 × 10^{-3}		EAM AFW	[2]
				0.46	4.2 × 10^{-3}		EAM VC	[2]
Cu/Cu(331)	∥ Jump			0.28	1.0 × 10^{-3}		EAM AFW	[2]
				0.33	3.0 × 10^{-3}		EAM VC	[2]
Rh/Rh(331)	∥ Jump	0.64 ± 0.04	1 × 10^{-2}				FIM	[60]
				1.31	7.4 × 10^{-3}		L-J	[2]
				0.62			Morse	[2]
Pt/Rh(331)	∥ Jump	0.72 ± 0.02	†10^{-3}				FIM	[61]
Pd/Pd(331)	∥ Jump			0.33	8 × 10^{-4}		EAM AFW	[2]
				0.37	2 × 10^{-3}		EAM VC	[2]
				0.55			Morse	[20]
Ag/Ag(331)	∥ Jump			0.34	1.0 × 10^{-3}		EAM AFW	[2]
				0.29	2.0 × 10^{-3}		EAM VC	[2]
Ir/Ir(331)	∥ Jump	0.91 ± 0.03	1 × 10$^{-1.9 \pm 0.5}$				FIM	[117]
				0.95			Morse	[20]
Pt/Pt(331)	∥ Jump	0.84 ± 0.1	4 × 10^{-4}				FIM	[120]
				~0.7			Morse	[20]
				0.77			Morse	[120]
				0.28	6 × 10^{-4}		EAM AFW	[2]
				0.54	8.5 × 10^{-4}		EAM VC	[2]
				1.33	5.5 × 10^{-3}		L-J	[2]
				0.53			EAM	[132]
				0.71			Morse	[132]
Ir/Pt(331)	∥ Jump			0.81			Morse	[120]
Au/Pt(331)	∥ Jump			0.68			Morse	[120]
Au/Au(331)	∥ Jump			0.26	6 × 10^{-4}		EAM AFW	[2]
				0.34	9.6 × 10^{-4}		EAM VC	[2]

† assumed value.

Table 4.4 Bcc (211)

System	Mecha-nism	Experiment E_D (eV)	D_o (cm^2/sec)	Theory E_D (eV)	D_o (cm^2/sec)	v_0 (sec^{-1})	Method	Ref.
Li/Mo(211)	‖ Jump	0.6	2.5				WF	[56]
Cu/Mo(211)	‖ Jump	0.45	1.2×10^{-3}				WF	[59]
Sr/Mo(211)	‖ Jump	0.9	1.8				WF	[57]
Dy/Mo(211)	‖ Jump	1.4	10^2				WF	[58]
W/W(211)	‖ Jump	0.85 ± 0.05	$19.0(\times 7.3^{\pm 1}) \times 10^{-3}$				Rean	[86]
		0.82 ± 0.02	$7.7(\times 1.9^{\pm 1}) \times 10^{-3}$				FIM	[86]
		0.81 ± 0.02	$2.5(\times 3.3^{\pm 1}) \times 10^{-3}$				FIM	[87,88]
		0.81 ± 0.02	$3.41(\times 2.4^{\pm 1}) \times 10^{-3}$				FIM	[89]
		0.56	4.8×10^{-7}				FEM	[92]
		0.73 ± 0.02	$(3-2) \times 10^{-5}$				FEM	[93]
				0.78			Morse	[95]
				0.432			Morse	[96]
				0.28			Morse	[20]
				0.887 ± 0.224			L-J	[97]
				0.952 ± 0.182	2.8×10^{-3}		L-J MC	[98]
				0.822	1.6×10^{-3}		L-J TST	[98]
				0.79	1.41×10^{-3}		4th Moment	[99]
	‖ Jump^e	0.95 ± 0.02	5×10^{-5}				FEM	[93]
	⊥ Move	0.29	4.2×10^{-9}				FEM	[93]
	⊥ Move^e	1.04	5×10^{-4}				FEM	[93]
	⊥ Ex			2.00	9.4×10^{-4}		4th Moment	[99]
	α Jump	0.84 ± 0.06				$2.2(\times 11.28^{\pm 1}) \times 10^{13}$	FIM	[89]
	β Jump	1.44 ± 0.13				$7.90(127.3^{\pm 1}) \times 10^{21}$	FIM	[89]
	β_R Jump	1.03 ± 0.06				$1.4(\times 10.3^{\pm 1}) \times 10^{16}$	FIM	[89]
Li/W(211)	‖ Move	0.49^f	2.0×10^{-3}				FEM	[110]
K/W(211)	‖ Move	0.46 ± 0.08	~0.3				SI	[109]
	⊥ Move	$0.76^{+0.10}_{-0.08}$	~0.3				SI	[109]
Sr/W(211)	‖ Move	0.87	0.4				WF	[111]
Ni/W(211)	‖ Jump	0.46 ± 0.01	$1.9(\times 2.7^{\pm 1}) \times 10^{-3}$				FIM	[88]

Mo/W(211)	∥ Jump	0.57	9.3×10^{-7}	FIM	[100]
		0.71 ± 0.01	$2.0(\times 1.6^{\pm 1}) \times 10^{-3}$	FIM	[86]
Rh/W(211)	∥ Jump	0.536 ± 0.001	$3.3(\times 1.5^{\pm 1}) \times 10^{-3}$	FIM	[86]
Pd/W(211)	∥ Jump	0.32 ± 0.01	$3.9(\times 2.4^{\pm 1}) \times 10^{-4}$	FIM	[87,102]
		0.32 ± 0.02	$5.1 \times 10^{-4.0 \pm 0.8}$	FIM	[103]
Ta/W(211)	∥ Jump	0.48	0.9×10^{-7}	FIM**	[107]
Re/W(211)	∥ Jump	0.86 ± 0.03	$2.2(\times 2.8^{\pm 1}) \times 10^{-3}$	FIM	[105]
		0.83 ± 0.01	$0.73(\times 1.6^{\pm 1}) \times 10^{-3}$	FIM	[86]
Ir/W(211)	∥ Jump	0.53 ± 0.05	5.0×10^{-7}	FIM	[106]
		0.67 ± 0.01	$0.61(\times 1.5^{\pm 1}) \times 10^{-3}$	FIM	[86]
Au/W(211)		$0.19^{\mathfrak{c}}$		FEM	[108]

** multi atom measurement; ^{e} High temperature; ^{f} coverage 0.4 ML; $^{\mathfrak{c}}$ coverage 0.5 ML.

Table 4.5 BCC(321)

System	Mechanism	E_D (eV) Experiment	D_o (cm²/sec)	E_D (eV) Theory	D_o(cm²/sec)	v_0 (sec⁻¹)	Method	Ref.
W/W(321)	∥ Jump	0.81±0.078	1×10⁻⁴				FIM	[84]
		0.89					FIM	[112]
				0.91			Morse	[95]
				0.13			Morse	[20]
				0.73	1.36×10⁻³		4th Moment	[99]
	F->I	0.65	†10⁻³				FIM	[85]
	I->I	~0.87	†10⁻³				FIM	[85]
	2-D	0.35	1.8×10⁻⁸				FEM	[92]
	2-D	0.71	1.4×10⁻⁵				FEM	[92]
	⊥ Ex F->I			0.23	3.4×10⁻⁴		4th Moment	[99]
Re/W(321)	∥ Jump	0.88	0.5×10⁻³				FIM**	[104]
		0.88	4.8×10⁻⁴				FIM**	[107]
Ta/W(321)	∥ Jump	0.67	1.9×10⁻⁵				FIM**	[107]
Mo/W(321)	∥ Jump	0.55	1.2×10⁻⁷				FIM	[100]

** multi atom measurement; † assumed value.

References

[1] R. T. Tung, Atomic structure and interactions at single crystal metal surfaces, Physics Ph.D. Thesis, University of Pennsylvania, Philadelphia, 1981.
[2] C. L. Liu, J. M. Cohen, J. B. Adams, A. F. Voter, EAM study of surface self-diffusion of single adatoms of fcc metals Ni, Cu, Al, Ag, Au, Pd, and Pt, *Surf. Sci.* **253** (1991) 334–344.
[3] J. B. Adams, S. M. Foiles, W. G. Wolfer, Self-diffusion and impurity diffusion of fcc metals using the five-frequency model and the Embedded Atom Method, *J. Mater. Res.* **4** (1989) 102–112.
[4] A. F. Voter, S. P. Chen, Accurate interatomic potentials for Ni, Al and Ni₃Al, *Mater. Res. Soc. Symp. Proc.* **82** (1987) 175–180.
[5] P. A. Gravil, S. Holloway, Exchange mechanisms for self-diffusion on aluminum surfaces, *Surf. Sci.* **310** (1994) 267–272.
[6] R. Stumpf, M. Scheffler, *Ab initio* calculations of energies and self-diffusion on flat and stepped surfaces of aluminum and their implications on crystal growth, *Phys. Rev. B* **53** (1996) 4958–4973.
[7] P. M. Agrawal, B. M. Rice, D. L. Thompson, Predicting trends in rate parameters for self-diffusion on FCC metal surfaces, *Surf. Sci.* **515** (2002) 21–35.
[8] Y.-J. Sun, J.-M. Li, Self-diffusion mechanisms of adatom on Al(001), (011) and (111) surfaces, *Chin. Phys. Lett.* **20** (2003) 269–272.
[9] G. Kresse, J. Hafner, *Ab initio* molecular dynamics for liquid metals, *Phys. Rev. B* **47** (1993) 558–561.
[10] G. Kresse, J. Hafner, *Ab initio* molecular dynamics simulation of the liquid-metal-amorphous-semiconductor transition in germanium, *Phys. Rev. B* **49** (1994) 14251–14269.

[11] G. Kresse, J. Furthmüller, Efficient iterative schemes for *ab initio* total-energy calculations using a plane-wave set, *Phys. Rev. B* **54** (1996) 11169–11186.

[12] R. T. Tung, W. R. Graham, Single atom self-diffusion on nickel surfaces, *Surf. Sci.* **97** (1980) 73–87.

[13] G. L. Kellogg, Direct observations of adatom-surface-atom replacement: Pt on Ni(110), *Phys. Rev. Lett.* **67** (1991) 216–219.

[14] G. L. Kellogg, Surface diffusion of Pt adatoms on Ni surfaces, *Surf. Sci.* **266** (1992) 18–23.

[15] A. P. Graham, W. Silvestri, J. P. Toennies, in: *Surface Diffusion: Atomistic and Collective Processes, NATO ASI Series B: Physics*, M. C. Tringides (ed.), Elementary processes of surface diffusion studied by quasielastic atom scattering (Plenum Press, New York, 1997), 565–580.

[16] A. P. Graham, The low energy dynamics of adsorbates on metal surfaces investigated with helium atom scattering, *Surf. Sci. Rep.* **49** (2003) 115–168.

[17] N. Memmel, E. Laegsgaard, I. Stensgaard, F. Besenbacher, Quasi-isotropic scaling behavior on an anisotropic substrate: Ni/Ni(110), *Phys. Rev. B* **72** (2005) 085411 1–5.

[18] P. S. Maiya, J. M. Blakely, Surface self-diffusion and surface energy of nickel, *J. Appl. Phys.* **38** (1967) 698–704.

[19] H. P. Bonzel, E. E. Latta, Surface self-diffusion on Ni(110): Temperature dependence and directional anisotropy, *Surf. Sci.* **76** (1978) 275–295.

[20] P. G. Flahive, W. R. Graham, Pair potential calculations of single atom self-diffusion activation energies, *Surf. Sci.* **91** (1980) 449–462.

[21] C.-L. Liu, J. B. Adams, Diffusion mechanisms on Ni surfaces, *Surf. Sci.* **265** (1992) 262–272.

[22] C.-L. Liu, J. B. Adams, Diffusion behavior of single adatoms near and at steps during growth of metallic thin films on Ni surfaces, *Surf. Sci.* **294** (1993) 197–210.

[23] B. M. Rice, C. S. Murthy, B. C. Garrett, Effects of surface structure and of embedded-atom pair functionals on adatom diffusion on fcc metallic surfaces, *Surf. Sci.* **276** (1992) 226–240.

[24] S. M. Foiles, M. I. Baskes, M. S. Daw, Embedded-atom-method functions for the fcc metals Cu, Ag, Ni, Pd, Pt, and their alloys, *Phys. Rev. B* **33** (1986) 7983–7991.

[25] D. J. Oh, R. A. Johnson, Simple embedded atom method model for fcc and hcp metals, *J. Mater. Res.* **3** (1988) 471–478.

[26] G. J. Ackland, G. Tichy, V. Vitek, M. W. Finnis, Simple N-body potentials for the noble metals and nickel, *Philos. Mag. A* **56** (1987) 735–756.

[27] P. Stoltze, Simulation of surface defects, *J. Phys.: Condens. Matter* **6** (1994) 9495–9517.

[28] L. S. Perkins, A. E. DePristo, Self-diffusion of adatoms on fcc(110) surfaces, *Surf. Sci.* **317** (1994) L1152–1156.

[29] K. Haug, T. Jenkins, Effects of hydrogen on the three-dimensional epitaxial growth of Ni (100), (110), and (111), *J. Phys. Chem. B* **104** (2000) 10017–10023.

[30] S. E. Wonchoba, W. H. Hu, D. G. Truhlar, Surface diffusion of H on Ni(100): Interpretation of the transition temperature, *Phys. Rev. B* **51** (1995) 9985–10002.

[31] U. Kürpick, Self-diffusion on (100), (110), and (111) surfaces of Ni and Cu: A detailed study of prefactors and activation energies, *Phys. Rev. B* **64** (2001) 075418 1–7.

[32] U. T. Ndongmouo, F. Hontinfinde, Diffusion and growth on fcc(110) metal surfaces: a computational study, *Surf. Sci.* **571** (2004) 89–101.

[33] L. T. Kong, L. J. Lewis, Transition state theory of the preexponential factors for self-diffusion on Cu, Ag, and Ni surfaces, *Phys. Rev. B* **74** (2006) 073412 1–4.

[34] S. Y. Kim, I.-H. Lee, S. Jun, Transition-pathway models of atomic diffusion on fcc metal surfaces. I. Flat surfaces, *Phys. Rev. B* **76** (2007) 245407 1–15.

[35] R. T. Tung, W. R. Graham, Single atom self-diffusion on Ni(331), *J. Chem. Phys.* **68** (1978) 4764–4765.

[36] J. Merikoski, I. Vattulainen, J. Heinonen, T. Ala-Nissila, Effect of kinks and concerted diffusion mechanisms on mass transport and growth on stepped metal surfaces, *Surf. Sci.* **387** (1997) 167–182.

[37] I. K. Robinson, K. L. Whiteaker, D. A. Walko, Cu island growth on Cu(110), *Physica B* **221** (1996) 70–76.

[38] P. Wynblatt, N. A. Gjostein, A calculation of relaxation, migration and formation energies for surface defects in copper, *Surf. Sci.* **12** (1968) 109–127.

[39] L. Hansen, P. Stoltze, K. W. Jacobsen, J. K. Nørskov, Self-diffusion on copper surfaces, *Phys. Rev. B* **44** (1991) 6523–6526.

[40] M. Karimi, T. Tomkowski, G. Vidali, O. Biham, Diffusion of Cu on Cu surface, *Phys. Rev. B* **52** (1995) 5364–5374.

[41] C. L. Chen, T. T. Tsong, Self-diffusion on the reconstructed and nonreconstructed Ir(110) surfaces, *Phys. Rev. Lett.* **66** (1991) 1610–1613.

[42] G. A. Evangelakis, G. C. Kallinteris, N. I. Papanicolaou, Molecular dynamics study of gold adatom diffusion on low-index copper surfaces, *Surf. Sci.* **394** (1997) 185–191.

[43] G. A. Evangelakis, D. G. Papageorgiou, G. C. Kallinteris, C. E. Lekka, N. I. Papanicolaou, Self-diffusion processes of copper adatom on Cu(110) surface by molecular dynamics simulations, *Vacuum* **50** (1998) 165–169.

[44] C. Mottet, R. Ferrando, F. Hontinfinde, A. C. Levi, A Monte Carlo simulation of submonolayer homoepitaxial growth on Ag(110) and Cu(110), *Surf. Sci.* **417** (1998) 220–237.

[45] F. Montalenti, R. Ferrando, Jumps and concerted moves in Cu, Ag, and Au(110) adatom self-diffusion, *Phys. Rev. B* **59** (1999) 5881–5891.

[46] G. Prévot, C. Cohen, D. Schmaus, V. Pontikis, Non-isotropic surface diffusion of lead on Cu(110): a molecular dynamics study, *Surf. Sci.* **459** (2000) 57–68.

[47] Z. Wang, Y. Li, J. B. Adams, Kinetic lattice Monte Carlo simulation of facet growth rate, *Surf. Sci.* **450** (2000) 51–63.

[48] H. Yildirim, A. Kara, S. Durukanoglu, T. S. Rahman, Calculated pre-exponential factors and energetics for adatom hopping on terraces and steps of Cu(100) and Cu(110), *Surf. Sci.* **600** (2006) 484–492.

[49] S. Durukanoglu, O. S. Trushin, T. S. Rahman, Effect of step-step separation on surface diffusion processes, *Phys. Rev. B* **73** (2006) 125426 1–6.

[50] L. T. Kong, L. J. Lewis, Surface diffusion coefficient: Substrate dynamics matters, *Phys. Rev. B* **77** (2008) 165422 1–5.

[51] V. S. Stepanyuk, N. N. Negulyaev, A. M. Saletsky, W. Hergert, Growth of Co nanostructures on Cu(110): Atomistic scale simulations, *Phys. Rev. B* **78** (2008) 113406 1–4.

[52] C. A. Roulet, Diffusion en surface de l'argent sur le plan (001), (111), (110) et des surfaces vicinales du cuivre, *Surf. Sci.* **36** (1973) 295–316.

[53] D. Ghaleb, B. Perraillon, Anisotropy of surface diffusion of silver on (331) and (110) clean copper surface at low temperature, *Surf. Sci.* **162** (1985) 103–108.

[54] U. Kürpick, G. Meister, A. Goldmann, Diffusion of Ag on Cu(110) and Cu(111) studied by spatially resolved UV-photoemission, *Appl. Surf. Sci.* **89** (1995) 383–392.

[55] G. Prévot, C. Cohen, J. Moulin, D. Schmaus, Surface diffusion of Pb on Cu(110) at low coverage: competition between exchange and jump, *Surf. Sci.* **421** (1999) 364–376.

[56] A. G. Naumovets, M. V. Paliy, Y. S. Vedula, A. T. Loburets, N. B. Senenko, Diffusion of Lithium and Strontium on Mo(112), *Prog. Surf. Sci.* **48** (1995) 59–70.

References

[57] A. T. Loburets, N. B. Senenko, A. G. Naumovets, Y. S. Vedula, Surface diffusion of strontium on the molybdenum (112) plane, *Phys. Low-Dim. Struct.* **10/11** (1995) 49–56.

[58] A. T. Loburets, A. G. Naumovets, Y. S. Vedula, Diffusion of dysprosium on the (112) surface of molybdenum, *Surf. Sci.* **399** (1998) 297–304.

[59] A. T. Loburets, Surface diffusion and phase transitions in copper overlayers on the (211) surfaces of molybdenum and tungsten, *Metallofizika I Noveishie Tekhnologii* **21** (1999) 47–51.

[60] G. Ayrault, G. Ehrlich, Surface self-diffusion on an fcc crystal: An atomic view, *J. Chem. Phys.* **60** (1974) 281–294.

[61] G. L. Kellogg, Diffusion of individual platinum atoms on single-crystal surfaces of rhodium, *Phys. Rev. B* **48** (1993) 11305–11312.

[62] J. P. Bucher, E. Hahn, P. Fernandez, C. Massobrio, K. Kern, Transition from one- to two-dimensional growth of Cu on Pd(110) promoted by cross-exchange migration, *Europhys. Lett.* **27** (1994) 473–478.

[63] C. Massobrio, P. Fernandez, Cluster adsorption on metallic surfaces: Structure and diffusion in the Cu/Pd(110) and Pd/Pd(110) systems, *J. Chem. Phys.* **102** (1995) 605–610.

[64] P. Fernandez, C. Massobrio, P. Blandin, J. Buttet, Embedded atom method computations of structural and dynamic properties of Cu and Ag clusters adsorbed on Pd(110) and Pd(100): evolution of the most stable geometries versus cluster size, *Surf. Sci.* **307–309** (1994) 608–613.

[65] Y. Li, M. C. Bartelt, J. W. Evans, N. Waelchli, E. Kampshoff, K. Kern, Transition from one- to two-dimensional island growth on metal (110) surfaces induced by anisotropic corner rounding, *Phys. Rev. B* **56** (1997) 12539–12543.

[66] K. Morgenstern, E. Laegsgaard, I. Stensgaard, F. Besenbacher, Transition from one-dimensional to two-dimensional island decay on an anisotropic surface, *Phys. Rev. Lett.* **83** (1999) 1613–1616.

[67] C. De Giorgi, P. Aihemaiti, F. Buatier de Mongeot, C. Boragno, R. Ferrando, U. Valbusa, Submonolayer homoepitaxial growth on Ag(110), *Surf. Sci.* **487** (2001) 49–54.

[68] L. Pedemonte, R. Tatarek, G. Bracco, Self-diffusion on Ag(110) studied by quasielastic He-atom scattering, *Surf. Sci.* **502–503** (2002) 341–346.

[69] L. Pedemonte, R. Tatarek, M. Vladiskovic, G. Bracco, Anisotropic self-diffusion on Ag(110), *Surf. Sci.* **507–510** (2002) 129–134.

[70] L. Pedemonte, G. Bracco, Surface disordering of Ag(110) studied by a new high resolution scattering apparatus, *Surf. Sci.* **513** (2002) 308–314.

[71] L. Pedemonte, R. Tatarek, G. Bracco, Surface self-diffusion at intermediate temperature: The Ag(110) case, *Phys. Rev. B* **66** (2002) 045414 1–5.

[72] R. A. Johnson, Gold on silver{110}: an embedded-atom-method study, *Modelling Sim. Mater. Sci. Eng.* **2** (1994) 985–994.

[73] S. Rousset, S. Chiang, D. E. Fowler, D. D. Chambliss, Intermixing and three-dimensional islands in the epitaxial growth of Au on Ag(110), *Phys. Rev. Lett.* **69** (1992) 3200–3203.

[74] E. S. Hirschorn, D. S. Lin, E. D. Hansen, Atomic burrowing and hole formation for Au growth on Ag(110), *Surf. Sci.* **323** (1995) L299–304.

[75] F. Hontinfinde, R. Ferrando, A. C. Levi, Diffusion processes relevant to the epitaxial growth of Ag on Ag(110), *Surf. Sci.* **366** (1996) 306–316.

[76] R. Ferrando, Correlated jump-exchange processes in the diffusion of Ag on Ag(110), *Phys. Rev. Lett.* **76** (1996) 4195–4198.

[77] S. Rusponi, C. Boragno, R. Ferrando, F. Hontinfinde, U. Valbusa, Time evolution of adatom and vacancy clusters on Ag(110), *Surf. Sci.* **440** (1999) 451–459.

[78] J. L. Nie, H. Y. Xiao, X. T. Zu, F. Gao, First-principles study of Sb adsorption on Ag(110) (2 × 2), *Chem. Phys.* **326** (2006) 583–588.
[79] H. L. Davis, G.-C. Wang, Registry relaxation of the W(211) surface – a (1 × 1) reconstruction, *Bull. Am. Phys. Soc* **29** (1984) 221.
[80] O. Grizzi, M. Shi, H. Bu, J. W. Rabalais, P. Hochmann, Time of flight scattering and recoiling spectrometry. I. Structure of the W(211) surface, *Phys. Rev. B* **40** (1989) 10127–10146.
[81] G. Antczak, G. Ehrlich, Asymmetric one-dimensional random walks, *J. Chem. Phys.* **129** (2008) 124702 1–4.
[82] G. Ehrlich, F. G. Hudda, Atomic view of surface self-diffusion: Tungsten on tungsten, *J. Chem. Phys.* **44** (1966) 1039–1049.
[83] D. W. Bassett, M. J. Parsley, Field ion microscope studies of transition metal adatom difusion on (110), (211) and (321) tungsten surfaces, *J. Phys. D* **3** (1970) 707–716.
[84] W. R. Graham, G. Ehrlich, Surface self-diffusion of single atoms, *Thin Solid Films* **25** (1975) 85–96.
[85] P. G. Flahive, W. R. Graham, The determination of single atom surface site geometry on W(111), W(211) and W(321), *Surf. Sci.* **91** (1980) 463–488.
[86] S. C. Wang, G. Ehrlich, Adatom diffusion on W(211): Re, W, Mo, Ir and Rh, *Surf. Sci.* **206** (1988) 451–474.
[87] D. C. Senft, G. Ehrlich, Long jumps in surface diffusion: One-dimensional migration of isolated adatoms, *Phys. Rev. Lett.* **74** (1995) 294–297.
[88] D. C. Senft, Long jumps in surface diffusion on tungsten(211), Materials Ph.D. Thesis, University of Illinois at Urbana-Champaign, Urbana, 1994.
[89] G. Antczak, Long jumps in one-dimensional surface self-diffusion: Rebound transitions, *Phys. Rev. B* **73** (2006) 033406 1–4.
[90] M. Tringides, R. Gomer, Diffusion anisotropy of oxygen and of tungsten on the tungsten (211) plane, *J. Chem. Phys.* **84** (1986) 4049–4061.
[91] G. L. Kellogg, Electric field inhibition and promotion of exchange diffusion on Pt(001), *Phys. Rev. Lett.* **70** (1993) 1631–1634.
[92] Y. M. Gong, R. Gomer, Thermal roughening on stepped tungsten surfaces I. The zone (011) – (112), *J. Chem. Phys.* **88** (1988) 1359–1369.
[93] D.-S. Choi, R. Gomer, Diffusion of W on a W(211) plane, *Surf. Sci.* **230** (1990) 277–282.
[94] Y. M. Gong, Surface self-diffusion studies on the W(112) plane by the field emission method, *Surf. Sci.* **266** (1992) 30–34.
[95] G. Ehrlich, C. F. Kirk, Binding and field desorption of individual tungsten atoms, *J. Chem. Phys.* **48** (1968) 1465–1480.
[96] P. Wynblatt, N. A. Gjostein, A calculation of migration energies and binding energies for tungsten adatoms on tungsten surfaces, *Surf. Sci.* **22** (1970) 125–136.
[97] J. D. Doll, H. K. McDowell, Theoretical studies of surface diffusion: Self-diffusion in the bcc (211) System, *Surf. Sci.* **123** (1982) 99–105.
[98] A. F. Voter, J. D. Doll, Transition state theory description of surface self-diffusion: Comparison with classical trajectory results, *J. Chem. Phys.* **80** (1984) 5832–5838.
[99] W. Xu, J. B. Adams, W single adatom diffusion on W surfaces, *Surf. Sci.* **319** (1994) 58–67.
[100] T. Sakata, S. Nakamura, Surface diffusion of molybdenum atoms on tungsten surfaces, *Surf. Sci.* **51** (1975) 313–317.
[101] S. C. Wang, J. D. Wrigley, G. Ehrlich, Atomic jump lengths in surface diffusion: Re, Mo, Ir, and Rh on W(211), *J. Chem. Phys.* **91** (1989) 5087–5096.

References

[102] D. C. Senft, Atomic jump length in surface diffusion: Experiment and theory, *Appl. Surf. Sci.* **94/95** (1996) 231–237.

[103] T.-Y. Fu, L.-C. Cheng, Y.-J. Hwang, T. T. Tseng, Diffusion of Pd adatoms on W surfaces and their interactions with steps, *Surf. Sci.* **507–510** (2002) 103–107.

[104] D. W. Bassett, M. J. Parsley, The effect of an electric field on the surface diffusion of rhenium adsorbed on tungsten, *Brit. J. Appl. Phys. (J. Phys. D)* **2** (1969) 13–16.

[105] K. Stolt, W. R. Graham, G. Ehrlich, Surface diffusion of individual atoms and dimers: Re on W(211), *J. Chem. Phys.* **65** (1976) 3206–3222.

[106] D. A. Reed, G. Ehrlich, Chemical specificity in the surface diffusion of clusters: Ir on W(211), *Philos. Mag.* **32** (1975) 1095–1099.

[107] D. W. Bassett, M. J. Parsley, Field ion microscopic studies of transition metal adatom diffusion on (110), (211) and (321) tungsten surfaces, *J. Phys. D* **3** (1970) 707–716.

[108] J. P. Jones, N. T. Jones, Field emission microscopy of gold on single-crystal planes of tungsten, *Thin Solid Films* **35** (1976) 83–97.

[109] B. Bayat, H.-W. Wassmuth, Directional dependence of the surface diffusion of potassium on tungsten (112), *Surf. Sci.* **133** (1983) 1–8.

[110] T. Biernat, C. Kleint, R. Meclewski, Surface diffusion of lithium across and along atomic rows on the W(211) plane, *Appl. Surf. Sci.* **67** (1993) 206–210.

[111] A. T. Loburets, A. G. Naumovets, N. B. Senenko, Y. S. Vedula, Surface diffusion and phase transitions in strontium overlayers on W(112), *Z. Phys. Chem.* **202** (1997) 75–85.

[112] S. Nishigaki, S. Nakamura, FIM observation of interactions between W atoms on W surfaces, *Jpn. J. Appl. Phys.* **14** (1975) 769–777.

[113] J. D. Wrigley, G. Ehrlich, Surface diffusion by an atomic exchange mechanism, *Phys. Rev. Lett.* **44** (1980) 661–663.

[114] J. D. Wrigley, Surface diffusion by an atomic exchange mechanism, Physics Ph.D. Thesis, University of Illinois at Urbana-Champaign, 1982.

[115] C. L. Chen, T. T. Tsong, Self-diffusion on reconstructed and nonreconstructed Ir surfaces, *J. Vac. Sci. Technol. A* **10** (1992) 2178–2184.

[116] C. L. Chen, T. T. Tsong, L. H. Zhang, Z. W. Yu, Atomic replacement and adatom diffusion: Re on Ir surfaces, *Phys. Rev. B* **46** (1992) 7803–7807.

[117] T.-Y. Fu, Y.-R. Tzeng, T. T. Tsong, Self-diffusion and dynamic behavior of atoms at step edges of iridium surfaces, *Phys. Rev. B* **54** (1996) 5932–5939.

[118] K.-D. Shiang, C. M. Wei, T. T. Tsong, A molecular dynamics study of self-diffusion on metal surfaces, *Surf. Sci.* **301** (1994) 136–150.

[119] C. M. Chang, C. M. Wei, S. P. Chen, Modeling of Ir adatoms on Ir surfaces, *Phys. Rev. B* **54** (1996) 17083–17096.

[120] D. W. Bassett, P. R. Webber, Diffusion of single adatoms of platinum, iridium and gold on platinum surfaces, *Surf. Sci.* **70** (1978) 520–531.

[121] G. L. Kellogg, Field-ion microscope observations of surface self-diffusion and atomic interactions on Pt, *Microbeam Analysis* **1986** (1986) 399–402.

[122] G. L. Kellogg, Surface self-diffusion of Pt on the Pt(311) plane, *J. Physique* **47** (1986) C2 – 331–336.

[123] E. Preuss, N. Freyer, H. P. Bonzel, Surface self-diffusion on Pt(110): Directional dependence and influence of surface-energy anisotropy, *Appl. Phys. A* **41** (1986) 137–143.

[124] G. L. Kellogg, Diffusion of Pd adatoms and stability of Pd overlayers on the (011) surface of Pt, *Phys. Rev. B* **45** (1992) 14354–14357.

[125] T. R. Linderoth, S. Horch, E. Laegsgaard, I. Stensgaard, F. Besenbacher, Surface diffusion of Pt on Pt(110): Arrhenius behavior of long jumps, *Phys. Rev. Lett.* **78** (1997) 4978–4981.

[126] T. R. Linderoth, S. Horch, E. Laegsgaard, I. Stensgaard, F. Besenbacher, Dynamics of Pt adatoms and dimers on Pt(110)-(1 × 2) observed directly by STM, *Surf. Sci.* **402–404** (1998) 308–312.

[127] F. Montalenti, R. Ferrando, Competing mechanisms in adatom diffusion on a channeled surface: Jumps versus metastable walks, *Phys. Rev. B* **58** (1998) 3617–3620.

[128] F. Montalenti, R. Ferrando, An MD study of adatom self-diffusion on Au(110) surfaces, *Surf. Sci.* **433–435** (1999) 445–448.

[129] H. T. Lorensen, J. K. Nørskov, K. W. Jacobsen, Mechanism of self-diffusion on Pt(110), *Phys. Rev. B* **60** (1999) R5149–5152.

[130] T. Halicioglu, G. M. Pound, A calculation of the diffusion energies for adatoms on surfaces of FCC metals, *Thin Solid Films* **57** (1979) 241–245.

[131] T. Halicioglu, An atomistic calculation of two-dimensional diffusion of a Pt adatom on a Pt(110) surface, *Surf. Sci.* **79** (1979) L346–348.

[132] M. Villarba, H. Jónsson, Diffusion mechanisms relevant to metal crystal growth: Pt/Pt(111), *Surf. Sci.* **317** (1994) 15–36.

[133] P. J. Feibelman, Ordering of self-diffusion barrier energies on Pt(110)-(1 × 2), *Phys. Rev. B* **61** (2000) R2452–2455.

[134] S. Günther, A. Hitzke, R. J. Behm, Low adatom mobility on the (1 × 2)-missing-row reconstructed Au(110) surface, *Surf. Rev. Lett.* **4** (1997) 1103.

[135] A. Hitzke, M. B. Hugenschmidt, R. J. Behm, Low temperature Ni atom adsorption on the Au(110)-(1 × 2) surface, *Surf. Sci.* **389** (1997) 8–18.

[136] L. D. Roelofs, J. I. Martin, R. Sheth, Competition between direct and concerted movements in surface diffusion with application to the Au(110) surface, *Surf. Sci.* **250** (1991) 17–26.

[137] L. D. Roelofs, B. J. Greenblatt, N. Boothe, Kinetic prefactors for concerted-mode diffusion: a realistic calculation – Au/Au(110), *Surf. Sci.* **334** (1995) 248–256.

[138] U. T. Ndongmouo, F. Hontinfinde, R. Ferrando, Numerical study of the stability of (111) and (331) microfacets on Au, Pt, and Ir(110) surfaces, *Phys. Rev. B* **72** (2005) 115412 1–8.

[139] W. Fan, X. G. Gong, Simulation of Ni cluster diffusion on Au(110)-(1 × 2) surface, *Appl. Surf. Sci.* **219** (2003) 117–122.

[140] R. A. Johnson, Analytical nearest-neighbor model for fcc metals, *Phys. Rev. B* **37** (1988) 3924–3931.

[141] R. A. Johnson, Alloy models with the embedded atom method, *Phys. Rev. B* **39** (1989) 12554–12559.

[142] J. W. M. Frenken, J. P. Toennies, C. Wöll, Self-diffusion at a melting surface observed by He scattering, *Phys. Rev. Lett.* **60** (1988) 1727–1730.

[143] J. W. M. Frenken, B. J. Hinch, J. P. Toennies, C. Wöll, Anisotropic diffusion at a melting surface studied with He-atom scattering, *Phys. Rev. B* **41** (1990) 938–946.

[144] M. Karimi, G. Vidali, I. Dalins, Energetics of the formation and migration of defects in Pb(110), *Phys. Rev. B* **48** (1993) 8986–8992.

[145] L. D. Roelofs, E. I. Martir, in: *The Structure of Surfaces III*, S. Y. Tong, K. Takayanagi, M. A. Van Hove, X. D. Xie (eds.), Microscopic kinetics of the (1 × 2) missing row reconstruction of the Au(110) surface, (Milwaukee, Wisconsin, 1991), 248–252.

5 Diffusion on two-dimensional surfaces

In this chapter, we continue the task started in the last one: we will list diffusion characteristics determined on a variety of two-dimensional surfaces. For better orientation, ball models of fcc(111) and (100) planes have already been shown in Fig. 3.3, together with the (110), (100), and (111) planes of the bcc lattice in Fig. 3.4. In experiments it has been observed that on fcc(111) planes there are two types of adsorption sites: fcc (sometimes called bulk or stacking) and hcp (also referred to as surface or fault sites); these sites, indicated in Fig. 3.3b, can have quite different energetic properties. Two types of adsorption sites also exist on bcc(111) and hcp(0001) structures. However, on bcc(110) as well as on bcc(100) and fcc(100) planes, only one type of adsorption site has so far been observed, which makes it easier to follow adsorption on these surfaces.

5.1 Aluminum: Al(100)

Experimental work on two-dimensional surfaces of aluminum was long in coming, and was preceded by considerable theoretical work, with which we therefore begin here. The first effort, by Feibelman [1] in 1987, was devoted to the Al(100) surface, and relied on local-density-functional theory (LDA–DFT). The primary aim of the work was to examine the binding energy of atom pairs, but he also estimated a barrier of 0.80 eV for the diffusive hopping of Al adatoms. Two years later, Feibelman [2] investigated surface diffusion which takes place on Al(100) by exchange of an adatom with one from the substrate. His findings are often cited as the first evidence for the occurrence of the exchange mechanism on a two-dimensional surface, but this process had previously been shown to occur in simulations by DeLorenzi and Jacuzzi in 1985 [3] on a bcc(100) plane of sodium. Feibelman arrived at a barrier to atom jumps over the surface of 0.65 eV, compared with 0.20 eV for diffusion by exchange of an aluminum adatom with an atom from the substrate lattice. The value for hopping differed from his previous findings.

Liu et al. [4] in 1991 resorted to the embedded atom method (EAM) in order to evaluate the self-diffusion characteristics. They used two parametrizations: Adams, Foiles, and Wolfer [5] (AFW) and Voter–Chen [6]. The exchange mechanism on Al(100) was investigated but yielded contradictory results. With AFW parameters, ordinary hopping was more favorable, with a diffusion barrier of 0.4 eV and a prefactor $D_o = 2 \times 10^{-3}$ cm^2/sec, compared with a barrier of 0.69 eV for exchange. Using VC potentials the

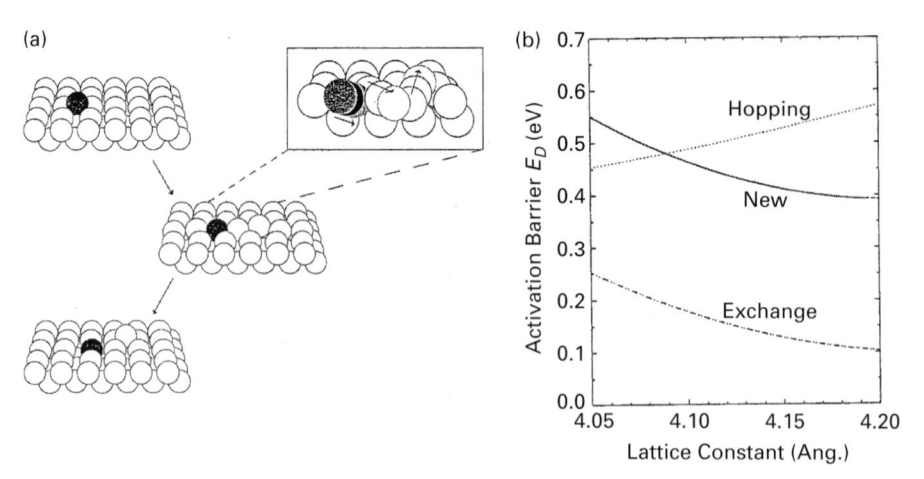

Fig. 5.1 (a) Model of adatom diffusion on fcc(100) plane, in which adatom enters the substrate and eventually displaces a lattice atom some distance away from the point of entry. (b) Activation energy for Al by three different diffusion mechanisms on Al(100) plane, as a function of the lattice constant. Dependence of hopping process is the opposite of the other two (after Cohen [7]).

exchange process was favorable over hopping, with an activation energy of 0.25 eV and a prefactor of $D_o = 4 \times 10^{-2}$ cm^2/sec compared with 0.46 eV for hopping. Calculations were also made with Lennard-Jones interactions and yielded a barrier of 0.60 and a prefactor of 1.0×10^{-2} cm^2/sec for hopping, while for exchange the barrier was much higher, 2.28 eV and a prefactor of 9×10^{-2} cm^2/sec.

Three years later the same system was examined by Cohen [7] using EAM potentials. The activation energy for hopping was higher than for exchange in this system, 0.45–0.58 eV for hopping and 0.1–0.24 eV for exchange, depending on the lattice constant, as in Fig 5.1b. She made interesting observations about the mechanism for diffusion, other than hopping or atom exchange, which also seem to have been seen earlier by Black and Tian [8]. In molecular dynamics simulations at 500 K, an adatom embedded in the lattice, and moved down the binding sites in the surface in an fcc row and over one, as shown in Fig. 5.1. The barrier was much higher (around 0.39–0.55 eV) than for standard atom exchange so the new mechanism of movement starts to be active at temperatures around half the melting point of the substrate. The same year the self-diffusion mechanism of aluminum adatoms was probed in molecular dynamics simulations by Gravil and Holloway [9] using effective medium potentials. They found that atom exchange with a barrier of only 0.271 eV predominated over simple atom hopping over a barrier of 0.692 eV because covalent bonding stabilized the intermediate state. The same year Stumpf and Scheffler [10] used density-functional theory to look at self-diffusion on aluminum surfaces and on the (100) plane they found an activation energy of 0.35 eV for exchange.

Calculations with the discrete variational method in the local-density-functional approximation were made in 1997 by Li et al. [11]. For diffusion by atom exchange along <100> they calculated a barrier of 0.27 eV, while atom hopping in the same direction required much more energy, 0.81 eV. The barrier to bridge hopping along

5.1 Aluminum: Al(100)

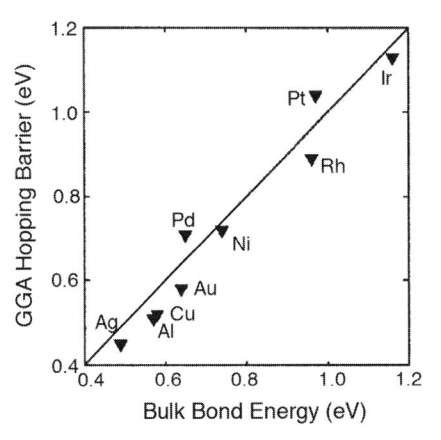

Fig. 5.2 Plot of self-diffusion barriers by atom hopping versus bond energy in bulk (equal to one sixth the cohesive energy) (after Feibelman [13]).

<110> occurred over a barrier of 0.74 eV, making the exchange process the lowest energy event. A year later, Valkealahti and Manninen [12] performed calculations with EMT potentials for aluminum atoms migrating on polyhedral clusters using molecular dynamics. On a (100) surface, the activation energy for hops amounted to 0.30 eV. Exchange was observed only above 500 K, but it must be emphasized that these are values on clusters. Feibelman [13], in 1999, calculated the barrier to hopping of aluminum atoms, this time with the generalized gradient approximation, and found a value of 0.51 eV. He proposed that the barrier for hopping over the surface should be roughly one sixth the cohesive energy of the bulk and, as shown in Fig. 5.2, demonstrated that this worked quite well for self-diffusion barriers estimated in the generalized gradient approximation. A similar approach had previously been used by Wang [14] in 1989, who compared the measured activation energies for diffusion on W(211) to sublimation energies in an attempt to achieve insight into the diffusion process, and also by Kief and Egelhoff [15]. However this rule seems not to work for self-diffusion on Re(0001) compared to self-diffusion on W(110) [16] nor for diffusion by the exchange mechanism.

The Ercolessi–Adams glue (EAG) potential [17] was used by Trushin et al. [18] in calculations of aluminum atom diffusion to obtain an energy barrier of 0.53 eV for atom hopping over the surface and 0.56 eV for atom exchange. The difference in energy was so small that they could not predict which mechanism was the favorable one. They also observed more complicated mechanisms, like three-atom exchange, with an activation energy of 0.85 eV and multi-particle exchange processes with barriers higher than 1 eV. Also in 2001, Ovesson et al. [19] carried out density-functional calculations for self-diffusion by hopping and using transition-state theory they found an energy barrier of 0.57 eV. Papanicolaou et al. [20] made estimates relying on the second-moment approximation to tight-binding theory, that is RGL. The barrier for hopping was very low, 0.20 eV, and the prefactor amounted to 2.4×10^{-3} cm^2/sec. They claimed that, as

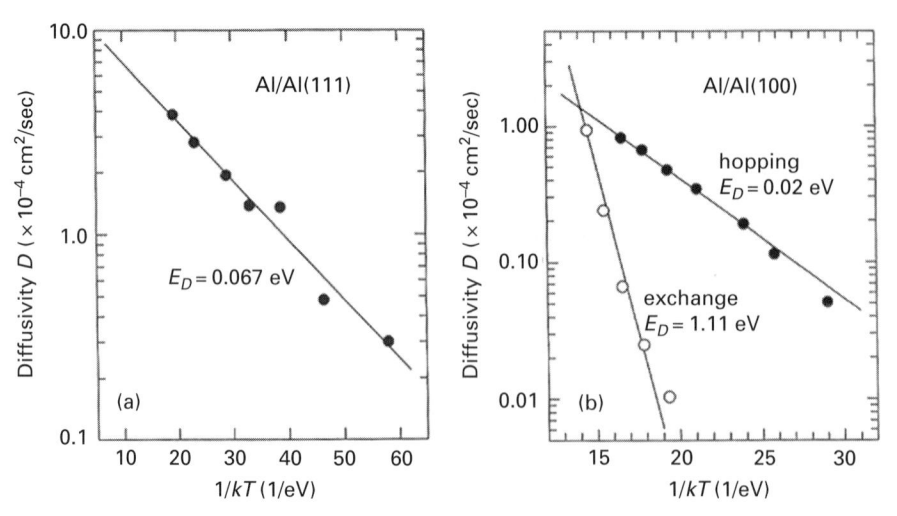

Fig. 5.3 Arrhenius plots for self-diffusion on aluminum surfaces, examined by molecular dynamics simulations with RGL potentials. (a) Diffusion on Al(111) surface. (b) Diffusion both by hopping and atom exchange on Al(100) plane (after Papanicolaou et al. [20]).

illustrated in Fig. 5.3b, hopping was dominant up to 800 K. The barrier to diffusion by atom exchange was by comparison huge, amounting to 1.11 eV with a huge prefactor of $9.4 \times 10^2 \text{cm}^2/\text{sec}$, different from previous studies where exchange was the leading mechanism on this plane.

In 2002, Fordell et al. [21] resorted to *ab initio* calculations with the VASP code [22–24] using the nudged elastic band method implemented for self-diffusion. In the local-density approximation [25], jumping occurred over a barrier of 0.52 eV; with the generalized gradient method (GGA) it was 0.50 eV. The barrier to atom exchange proved to be considerably lower for both approximations: 0.23 eV for LDA and 0.14 for GGA. At roughly the same time, Agrawal et al. [26] carried out detailed calculations for diffusion, without any concern for many-atom effects. They worked with Lennard-Jones potentials to do Monte Carlo simulations for several crystal faces of fcc metals. On the Al(100), the values were 0.60 eV and 6.1×10^{-3} cm^2/sec. It must be noted here that exchange was not considered at all as a mechanism of diffusion. The value reported is out of line.

A year later, density-functional theory estimates were carried out by Sun and Li [27] using the Vienna *ab initio* package with the generalized gradient approximation and a 4×4 surface unit cell for self-diffusion of aluminum. The diffusion occurred by exchange, with an activation energy of 0.17 eV; hopping required overcoming a much larger barrier of 0.46 eV. These are values comparable with previous findings of Fordell et al. obtained with the same method.

Chang and Wei [28] in 2005 again used *ab initio* density-functional theory. For hopping they arrived at a barrier of 0.67 eV; for diffusion by atom exchange the barrier was much lower, 0.13 eV. They also explored the effect of the size of the unit cell used in the calculations on the activation energy, as previously done by Perkins and DePristo [29] and found that for exchange the unit cell size can have a huge effect on the magnitude of the energy barrier. For exchange, a (5 × 5) cell was desirable to ensure relaxation and to get values comparable with experiment.

Diffusion on aluminum surfaces has received considerable theoretical attention, but not much experimental work has been reported examining atom movement. Only theoretical results are available for self-diffusion on Al(100) – there have been no experiments done – and these results range quite widely from 0.2 eV to 0.8 eV for hopping and from 0.14 eV to 0.69 eV for exchange. There is no unanimity about the mechanism of diffusion, whether it occurs by hops or in an exchange process, but exchange over a barrier around 0.25 eV seems to be a likely event. Additionally, there is a possibility that long exchange events will occur on this surface at higher temperatures. No investigations of hetero-diffusion on aluminum, either experimental or theoretical, have been done so far.

5.2 Aluminum: Al(111)

Compared to the Al(100) surface, the Al(111) is much smoother and has two kinds of adsorption sites, fcc and hcp, shown in Fig. 3.3b. These may have different energetic characteristics. Investigations on this plane also started with theoretical studies, and experimental work caught up a bit later.

Theoretical studies of self-diffusion began with the embedded atom method (EAM) work of Liu *et al.* [4] in 1991. Using the parametrization of Adams, Foiles, and Wolfer [5] (AFW) they obtained a barrier of 0.074 eV and $D_o = 9 \times 10^{-4}$ cm^2/sec; using the Voter–Chen [6] approximation (VC), the barrier was lower, 0.054 eV and $D_o = 1.6 \times 10^{-3}$ cm^2/sec. An estimate relying on the Lennard-Jones potentials came out with a diffusion barrier of 0.11 eV. Liu *et al.*'s efforts were followed by Stumpf and Scheffler [10,30] in 1994 with density-functional theory calculations. For normal hopping they reported an activation energy of 0.04 eV for jumps between hcp sites on the surface. They claimed that for Al the hcp site is a stable binding site, with a difference in energy between hcp and fcc site of 0.04 eV. DFT was also used in 1998 by Bogicevic *et al.* [31] to determine the same diffusion barrier of 0.04 eV. The same year, Valkealahti and Manninen [12] did calculations with EMT potentials for aluminum atoms migrating on polyhedral clusters using molecular dynamics. On an Al(111) surface of a polyhedral, the activation energy for hops proved to be 0.02 eV. So far only Liu *et al.* [4] bothered to determine prefactors for the diffusivity; later studies gave only activation energies. The determinations from DFT are exactly the same; EAM seems to give a slightly higher value, but also not far away.

At this stage there began the first efforts at experimental determinations for diffusion on Al(111) by Barth *et al.* [32], using the STM to determine the saturation density of islands. From measurements at eight temperatures, from 60 to 180 K, shown in Fig. 5.4, they derived an activation energy for aluminum self-diffusion of 0.042 ± 0.004 eV, in reasonable agreement with previous theoretical work. Their prefactor, however, had the unusually small value of $D_o = 2 \times 10^{-9 \pm 0.25}$ cm^2/sec. The authors attributed this to a general effect, which lowers the prefactor when diffusion takes place in "the weak

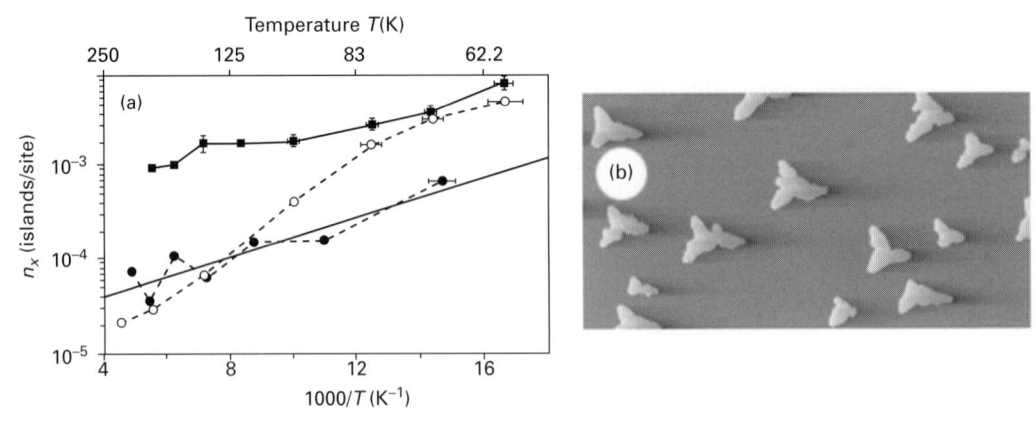

Fig. 5.4 Aluminum islands on Al(111). (a) Plot of aluminum island density on Al(111) against reciprocal temperature. Solid circles: Al evaporated from BN tube, 0.11 ML coverage, deposition rate $7.2 \pm 0.7 \times 10^{-3}$ ML/sec. Open circles: Evaporation from Al_2O_3 tube, 0.10 ML, $1.05 \pm 0.06 \times 10^{-2}$ ML/sec. Full squares: results of Barth et al. [32]. (b) STM image after deposition at 183 K (after Michely et al. [33]).

corrugation regime", over a barrier of less than 0.1 eV. The low value of the prefactor is quite suspicious, however, and leads to concerns about the experiments and about the effect of atomic interactions on the underlying nucleation theory.

The findings of Barth et al. [32] were soon challenged by Michely et al. [33], who repeated island density measurements on Al(111), with the aluminum evaporated both from an Al_2O_3 tube as well as a tube of boron nitride; comparison with results as well as data from Barth et al. [32] is shown in Fig. 5.4. The second arrangement gave a much cleaner deposit, but yielded results quite different from what had been reported by Barth et al. [32]. Although they recognized their experiments still needed further work, Michely et al. reported an activation energy for diffusion on clean Al(111) of 0.06 eV, and a frequency prefactor of $v_0 = 1 \times 10^{11}$ sec^{-1}. They suspected that the low value of the prefactor obtained by Barth et al. was connected with coadsorption of impurities rather than with the weak corrugation of the potential energy of the adatom. Later, in 2003, Busse et al. [34] made an estimate of 0.04 eV for the diffusion barrier, with a prefactor of 5×10^{11} sec^{-1} again based on measurements of the island density in Fig. 5.5, so that at least the activation energies proposed by the two groups are roughly the same.

Chang et al. [35] used density-functional calculations for self-diffusion on Al(111) and found again a barrier of 0.04 eV; with embedded atom potentials the result was similar, 0.03 eV. Ovesson et al. [19] once more carried out density-functional calculations for adatom movement by hopping. Using transition-state theory, they found an energy barrier of 0.042 eV and a frequency prefactor of 4×10^{12} sec^{-1}. They attributed the low prefactor obtained in the experiments by Barth et al. [32] to long-range interactions neglected in nucleation theory rather than to the adsorption of impurities claimed by Michely et al. [33].

A substantial number of diffusivities have been derived from mean-field nucleation theory by plotting the logarithm of the island density at saturation against the reciprocal

5.2 Aluminum: Al(111)

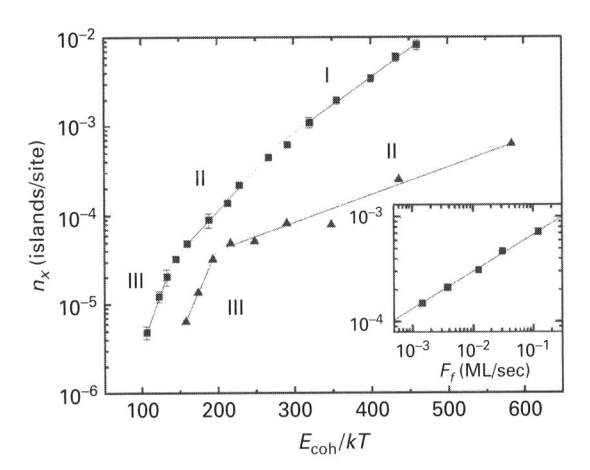

Fig. 5.5 Arrhenius plot for island density n_x on Ir(111) (solid squares) and Al(111) (solid triangles). Inset gives island density versus rate of deposition on Ir(111) (after Busse et al. [34]).

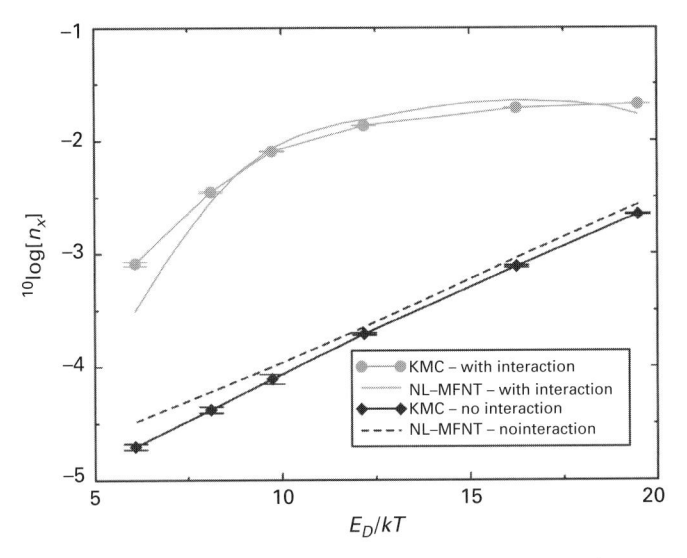

Fig. 5.6 Saturated density of aluminum islands on Al(111) in its dependence upon the reciprocal temperature at 0.05 coverage and an impingement rate of 0.03 ML/sec. Data from simulations by kinetic Monte Carlo methods (after Ovesson [36]).

temperature. However, this approach has come under suspicion, as classical nucleation theory ignores interactions between metal atoms which are known to be strong. Ovesson [36] has therefore worked out a nucleation theory allowing for adatom–adatom interactions. A comparison of the island density plotted against the reciprocal temperature both with and without interactions is shown in Fig. 5.6, where the mean-field approximation is also tested against kinetic Monte Carlo simulations. Markedly larger densities are obtained with interactions (patterned on results for Al on Al(111)) than without. It will be interesting to see if this procedure can be tested for ordering diffusivities in interacting systems.

A slightly different approximation was undertaken by Papanicolaou et al. [20], who made estimates of the diffusion of aluminum atoms on the Al(111) plane relying on the second-moment approximation to tight-binding theory (RGL). From the Arrhenius plot shown in Fig. 5.3a. they arrived at an activation energy of 0.067 eV with a prefactor of $D_o = 1.4 \times 10^{-3}$ cm^2/sec. This value is the result of averaging the two barriers corresponding to fcc and hcp positions. However, from a trajectory analysis, only single hops were detected. Agrawal et al. [26] worked with Lennard-Jones potentials to do Monte Carlo simulations for several crystal faces of fcc metals. For aluminum adatoms diffusing on the (111) plane they found an activation energy of 0.13 eV and a prefactor $D_o = 3.2 \times 10^{-4}$ cm^2/sec. This study shows that pair potentials are not suitable to describe movement on this plane. Density-functional theory estimates using the Vienna *ab initio* package have been carried out again for diffusion of aluminum on Al(111) planes by Sun and Li [27]. Aluminum adatoms jump across the bridge between two substrate atoms with an activation energy of only 0.04 eV. Exchange on this plane involved a large barrier of 0.63 eV, making it unlikely to happen.

Polop et al. [37] as well as Busse et al. [34] reported further work on the saturation island density of Al on Al(111), but came to the conclusion that because of uncertainties in the diffusivity of aluminum dimers, the experiments did not allow an unequivocal decision about the diffusion characteristics of single aluminum atoms. They adopted the value of 0.04 eV for the diffusion barrier, obtained in density-functional calculations, with a frequency prefactor of $5 \times 10^{11 \pm 0.5}$ sec^{-1}. The last work on Al(111) was reported by Polop et al. [38] in 2005. They looked at the island density under carefully controlled clean conditions, but also in the presence of contamination. Island nucleation on the clean surface as well as on a surface with oxygen clusters is compared in Fig. 5.7. For clean Al(111) the activation energy for self-diffusion was found as 0.07 ± 0.01 eV, with a

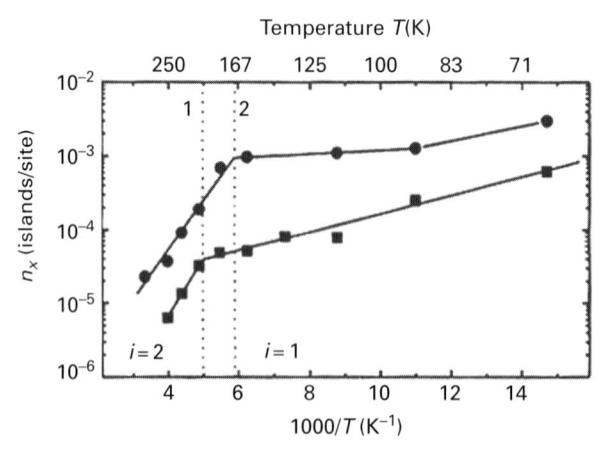

Fig. 5.7 Density of aluminum islands on Al(111) as a function of reciprocal temperature, grown on clean surface (solid squares) and on surface with 0.024 ± 0.002 oxygen clusters. On the clean surface the activation energy for diffusion was 0.07 ± 0.01 eV (after Polop et al. [38]).

frequency prefactor $v_0 = 5 \times 10^{11 \pm 0.5}$ sec^{-1}. On the surface contaminated with oxygen the diffusion barrier decreased to 0.03 eV and the prefactor was 8×10^6 sec^{-1}. Since in previous experiments by Barth et al. [32] the activation energy was 0.042 ± 0.004 eV, with a frequency prefactor of $8 \times 10^{6 \pm 0.25}$ sec^{-1}, Polop et al. took the low values reported for the diffusion barrier and the prefactor as a general indication of contamination.

The activation energies calculated for self-diffusion on Al(111) scatter between 0.02 eV and 0.07 eV, excluding the results of Agrawal et al. [26] which are completely out of range. Experimental studies of the diffusion barrier are in the same range, from 0.042 eV to 0.070 eV, with the latter experimentally preferred. Direct measurement of adatom displacements would also be very desirable. The experimental findings are in good agreement with most of the calculational results. There is no information about hetero-diffusion on this plane.

5.3 Potassium

In 1997 Fuhrmann and Hulpke [39] did helium scattering experiments on a Ni(100) surface covered with 4 ML of potassium. The activation energy for diffusion of single potassium atoms on such a layer was derived as 0.063 ± 0.015 eV, with a prefactor of $D_o = (9.5 \pm 0.9) \times 10^{-4}$ cm^2/sec. At 300 K, desorption contributed to the measurements but single jumps were favored in diffusion on the surface. On lowering the temperature to 285 K, the percentage of long jumps increased; jumps covering two or three spacings were observed, quite a surprising result since the presence of long jumps is usually associated with increased temperatures. Fuhrmann and Hulpke also claimed that their diffusion barrier was an average value from a number of different crystallographic directions, since they detected a rich distribution of domains in the system. It is difficult to judge how their results correspond to single atom movement, as well as the possible influence of strain and defects on movement.

5.4 Iron: Fe(100), (111), (110)

Measurements of the self-diffusion of iron on Fe(100) were made by Stroscio et al. [40,41], using the scanning tunneling microscope and reflection high energy electron diffraction. They measured the temperature dependence of the density of islands formed in the temperature range 20–250 °C. The data were taken at a starting coverage of 0.07 ± 0.016 monolayers and interpreted according to classical nucleation theory; it is shown in Fig. 5.8. It was assumed that only individual adatoms can diffuse in the investigated temperature range and that the capture number for islands was unity; this assumption could affect the diffusivity by a factor less than ten. They arrived at an activation energy for diffusion of 0.45 ± 0.08 eV, with a prefactor of $D_o = 7.2 \times 10^{-4}$ cm^2/sec.

Pfandzelter et al. [42] later studied the submonolayer growth of iron on Fe(100) by examining the scattering of a grazing angle beam of 25 keV He$^+$ ions. From the temperature dependence of the saturation density of islands, shown in Fig. 5.9, they

270   Diffusion on two-dimensional surfaces

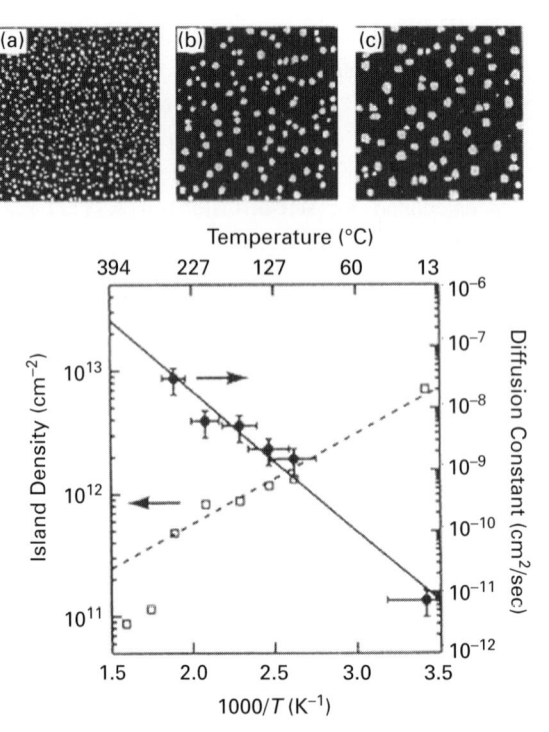

Fig. 5.8 Growth of iron on Fe(100) surface. STM images of single layer islands grown at (a) 20 °C, (b) 108 °C, and (c) 163 °C. (d) Arrhenius plot of the island density and the diffusivity for Fe on Fe(100). Coverage 0.07 ± 0.016 ML, at a flux of $1.4 \pm 0.3 \times 10^{13}$ atoms/cm^2 sec (after Stroscio et al. [40]).

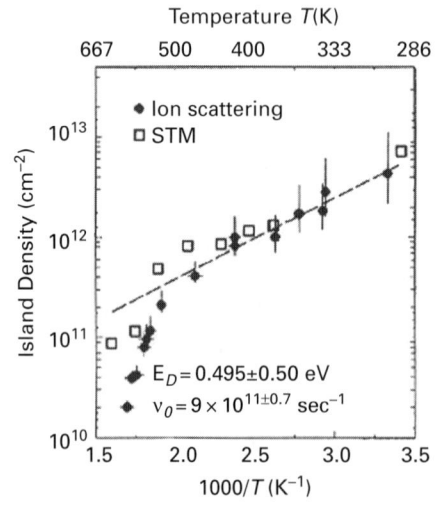

Fig. 5.9 Saturation density of Fe islands on Fe(100) in its dependence upon reciprocal temperature. Open squares: STM results from Stroscio. Solid circles: from ion scattering (after Pfandzelter et al. [42]).

5.4 Iron: Fe(100), (111), (110)

Table 5.1 Temperature variation of the island–island separation L_i determined by RHEED and STM [43].

T (K)	L_i(RHEED) (Å)	L_i(STM) (Å)
340	40	38
375	48	47
430	68	66
490	162	162
600	292	291

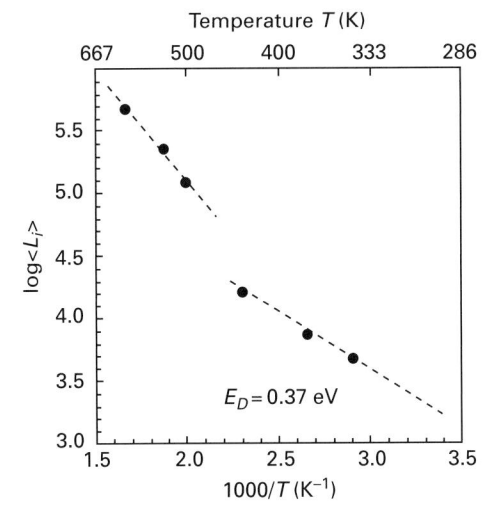

Fig. 5.10 Dependence of the separation of iron islands on Fe(100) upon the reciprocal temperature, yielding an activation energy of 0.37 eV at low temperatures (after Dulot et al. [43]).

derived an activation energy of 0.495 ± 0.050 eV and a frequency prefactor of $9 \times 10^{11 \pm 0.7}$ sec^{-1} for diffusion of iron atoms, in good agreement with the prior work of Stroscio et al. [40].

In 2003, Dulot et al. [43] examined the deposition of iron on Fe(100) using RHEED as well as STM, and were able to determine the island density achieved in growth at substrate temperatures ranging from 340 K to 600 K. They deduced the island–island distance L_i by analyzing the shape of diffraction peaks in RHEED, as well as measuring the distance with scanning tunneling microscopy STM. The data are presented in Table 5.1, and the agreement is excellent. From the temperature dependence of the island separation shown in Fig. 5.10, they arrived at an activation energy for surface diffusion of 0.37 eV at temperatures below 450 K. The agreement with the work of Stroscio et al. [40] is not too bad.

In the year 2000, Köhler et al. [44] performed STM measurements as well as Monte Carlo simulations with Finnis–Sinclair [45] potentials to better understand homoepitaxy on the bcc Fe(110) plane. As part of this effort they also calculated the energy barriers to atom diffusion. For motion by jumps on the rigid surface the barrier in the <111>

direction was 0.38 eV, and along <001>, over a surface atom, it proved to be much higher, 0.88 eV. On fully relaxing the iron substrate, the barrier to diffusion along <111> became much smaller, amounting to only 0.27 eV. Seven years later, Chan et al. [46] looked at the movement of Fe with EAM potentials and MD simulations. Single adatom movement proceeded along <111> directions with an energy of 0.4 eV and a prefactor 1.27×10^{-2} cm²/sec. The potential barrier is symmetric, with a saddle point at 0.3 eV, and has a local minimum 0.02 eV deep in the middle between two stable adsorption places. Note the disagreement between the energy given for the movement of monomer and the height of the barrier, about which the authors did not comment. The data turn out to be comparable with the findings of Köhler et al. [44]. Unfortunately, no experimental estimates of the barrier on Fe(110) appear to be available.

No good theoretical estimates for self-diffusion of iron have been made, but Flahive and Graham [47], using Morse potentials, calculated a barrier of 1.4 eV for self-diffusion on Fe(111), 0.6 eV on Fe(100), and ~0.25 eV for Fe(110), as shown in Fig. 4.9. Kief and Egelhoff [15] much later made approximate determinations by scaling the barrier for other systems with the cohesive energy. For self-diffusion on Fe(111) they obtained a barrier of 0.18 eV by comparing the iron surface with Ir(111) [48], for which the diffusion energy was known. Since the structure of the Ir(111) plane (a very smooth surface) is quite different from that of Fe(111) (a bcc metal and quite rough), such estimates can be misleading. To obtain the activation energy for iron diffusion on Fe(100) as 0.46 eV, scaling from copper self-diffusion on Cu(100) was used. In 1999, Davydov [49] estimated the activation energy for self-diffusion on the Fe(110) surface as 0.65 eV, based on his cohesion approach.

More reliable estimates of diffusion characteristics were done by Spisák and Hafner [50,51] in 2004. They looked at the movement of Fe atoms on 1 ML of Fe covering a W(100) surface with first-principles VASP calculations. The shortest bridge hopping along <011> cost 0.4 eV. For jumps to next-nearest-neighbor sites along the <001> direction the barrier was only 0.5 eV. Exchange was a bit more expensive, requiring 0.7 eV. It should, however, be mentioned that 1ML of iron on a W(100) surface is very likely influenced by strain, which will change the diffusion. The authors did not discuss this factor.

Self-diffusion on the Fe(100) plane has been investigated using molecular dynamics simulations by Chamati et al. [52]. To perform this study, the authors first constructed a new EAM potential, which was fitted to measured quantities as well as first-principles calculations both for bcc α-iron and fcc γ-iron. From analyses of atom trajectories in the temperature range 600–950 K they found that atom movement took place by three mechanisms: diffusion by diagonal exchange, as in the first row of Fig. 5.11, by atoms hopping as in the second row, and finally non-diagonal exchange, shown in the third row of Fig. 5.11. The temperature dependence of all three processes is shown in Fig. 5.12. For diagonal exchange the barrier was 0.66 eV, with a prefactor of 113×10^{-3} cm²/sec; for hopping a barrier of 0.92 eV was found with a prefactor of 206×10^{-3} cm²/sec, and finally for non-diagonal exchange an activation energy of 0.99 eV with a prefactor of 306×10^{-3} cm²/sec was calculated. Using the nudged elastic band (NEB) method, Chamati et al. also calculated static barriers of 0.60 eV, 0.84 eV, and 0.97 eV for diagonal exchange, atom hopping, and non-diagonal exchange. They also observed long and correlated jumps as well as diffusion by concerted exchange, but the frequency was

5.4 Iron: Fe(100), (111), (110)

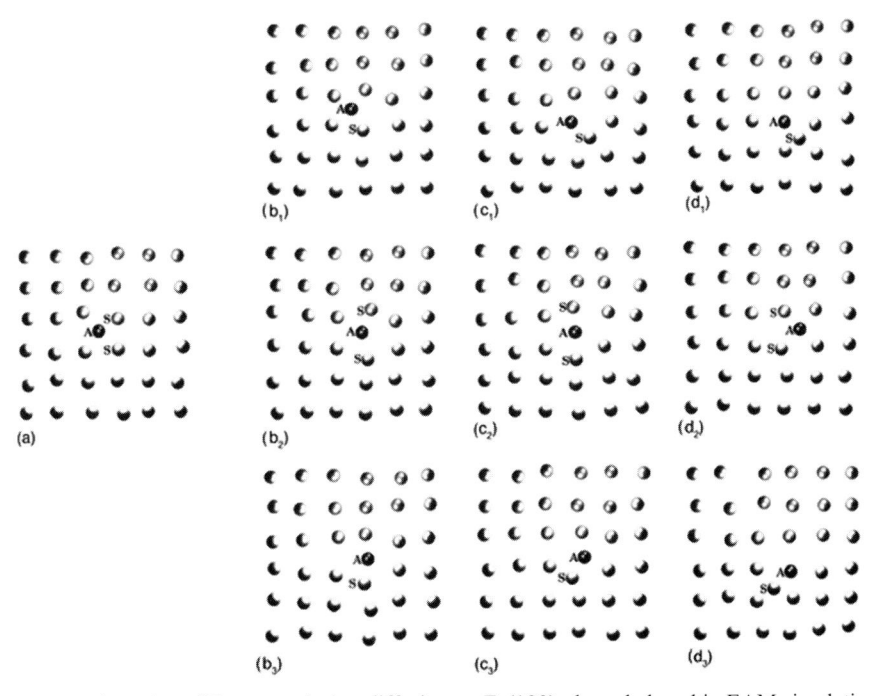

Fig. 5.11 Atom trajectories of Fe atoms during diffusion on Fe(100) plane deduced in EAM simulations. Top row: diagonal atom exchange. Middle row: diffusion by atom hopping. Bottom row: non-diagonal exchange mechanism (after Chamati et al. [52]).

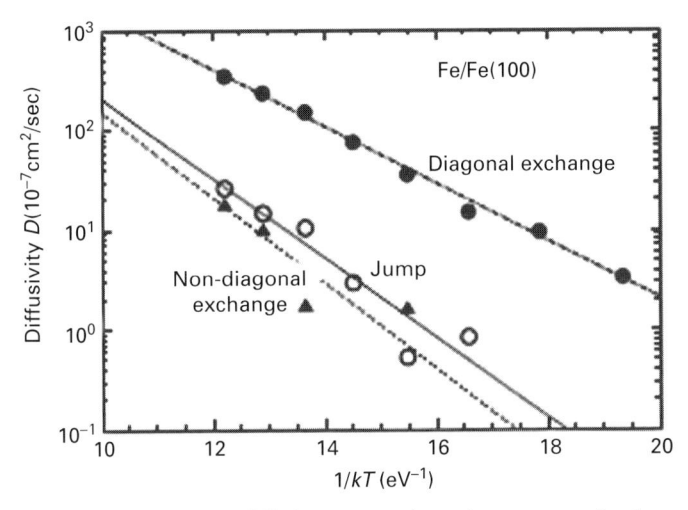

Fig. 5.12 Dependence of Fe atom diffusion upon reciprocal temperature for three mechanisms of diffusion. Diagonal atom exchange has the lowest activation energy of 0.66 eV (after Chamati et al. [52]).

too low to arrive at the characteristics of these processes. It must be noted, however, that all the results are too high compared with what has been obtained in experiments.

Recently, Shvets et al. [53] looked at nanowedge formation of iron on Mo(110). As shown in Fig. 5.13, they calculated a hopping barrier for Fe on Fe(110) at different strains

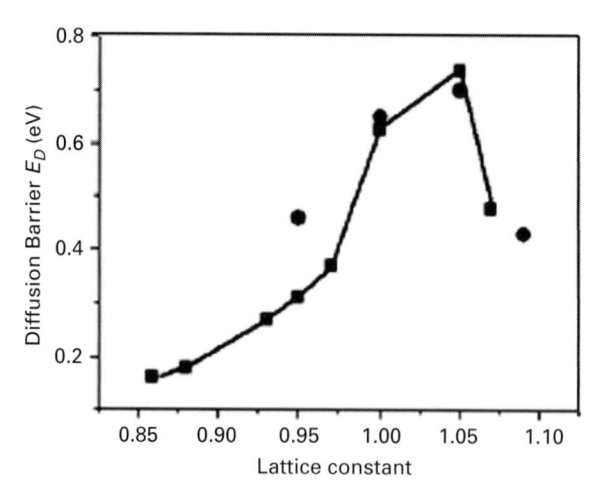

Fig. 5.13 Effect of strain on barrier to diffusion of Fe on Fe(110) (after Shvets et al. [53]).

using density-functional theory in the GGA approximation, relying on the CASTEP numerical code to obtain diffusion barriers. They worked with a four layer thick slab and a 2 × 2 cell – two layers were fixed and two relaxed. The barrier was around 0.65 eV on the surface without strain. Their arrangement did not take into account the presence of the Mo substrate underneath the Fe layer, and the simulation settings were very limited. It is unlikely they obtained reliable values for this system.

Iron (111) and (110) definitely need more work, both experimental and theoretical. For iron (100), experiments seem to be in agreement, but theory still needs effort. There is no data on hetero-diffusion for these substrates.

5.5 Nickel: Ni(111)

5.5.1 Self-diffusion

Tung and Graham [54] were the first to undertake a technically quite difficult project – examination of self-diffusion on nickel surfaces using the field ion microscope, with the surface cooled to ~20 K. On the (111) planes, schematics of which were shown in Fig. 3.3, activation energies for diffusion were derived from the onset temperature, assuming the prefactor to be $(kT/2h)\ell^2$; here k stands for Boltzmann's constant, h for Planck's constant, and ℓ for the jump length. This yielded a somewhat uncertain value of 0.33 eV for the barrier on the (111) surface. They also looked at adsorption places on the Ni(111) surface and definitely recognized two types, with slightly different binding energies; both sites were occupied after thermal treatment.

Twenty years later, Fu and Tsong [55] again looked at Ni(111), also with FIM but with a sample pre-annealed before tip formation. With the sample cooled to 30 K during measurements they derived an activation energy of 0.22 ± 0.02 eV and a prefactor $D_o = 1 \times 10^{-5.2 \pm 0.7}$ cm^2/sec, based on measurements at four temperatures shown in Fig. 5.14. The agreement between the two studies is acceptable, taking into account

5.5 Nickel: Ni(111)

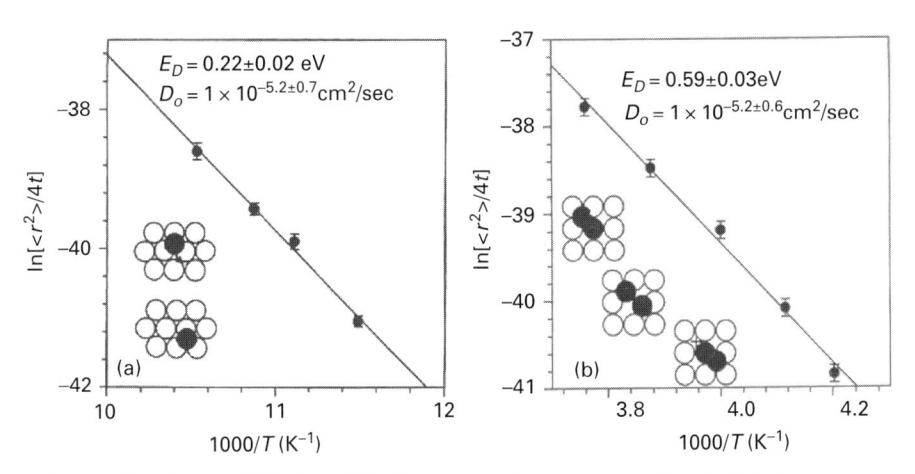

Fig. 5.14 Arrhenius plots for self-diffusion of Ni adatom on nickel, as measured in the field ion microscope (FIM). (a) On Ni(111) plane. (b) On Ni(100) surface. Insets give indication of diffusion mechanism (after Fu et al. [55]).

the preliminary character of the first studies. It should be noted that no attempt to determine the binding sites on Ni(111) was made.

Although there have been only two experimental studies of the diffusion characteristics of nickel atoms on nickel (111), more than ten efforts have been made to calculate these diffusivities. Flahive and Graham [47], using Morse potentials, calculated a barrier for self-diffusion on Ni(111) of ~ 0.01 eV, as in Fig. 4.9, relatively far away from experimental determinations. The first effort with many-atom interactions was done by Liu et al. [4] in 1991, and involved embedded atom calculations for atom hopping, using both AFW [5] and VC [6] parametrizations. They obtained a barrier of only 0.056 eV and a prefactor of 5×10^{-4} cm^2/sec with AFW, and 0.063 eV and 6.2×10^{-4} cm^2/sec with VC parameters. These values are again considerably smaller than what has been found in the experiments. Liu et al. also used a Lennard-Jones potential for estimates. On the (111) plane they found a barrier of 0.15 eV with a prefactor of 9.0×10^{-4} cm^2/sec.

Comparisons of the diffusion barriers on nickel surfaces were obtained by Rice et al. [56] with different EAM potentials. Rates at different temperatures were calculated using transition-state theory with the FBD [57] as well as the VC approximations, but also with the potentials of Oh and Johnson [58] as well as of Ackland et al. [59]. Static estimates were made, giving values of 0.06 eV with FBD parameters, 0.07 eV with VC, 0.05 eV for OJ, and 0.05 eV for ATVF. All these values are much too low compared with experiments.

Shiang [60] performed molecular dynamics estimates using the Sutton–Chen (SC) and also the Morse potential to estimate the self-diffusion characteristics of Ni atoms. Using the SC potential he arrived at an activation energy of 0.086 eV and a prefactor $D_o = 8.74 \times 10^{-4}$ cm^2/sec from an Arrhenius plot of the rate of diffusion at different temperatures, shown in Fig 5.15. From calculations with Morse potentials, which are also shown in Fig. 5.15, the values for the activation energies were 0.016 eV and for the prefactor 5.04×10^{-4} cm^2/sec. It is clear that many-body interactions play an important role, the activation energies being much higher with Sutton–Chen potentials.

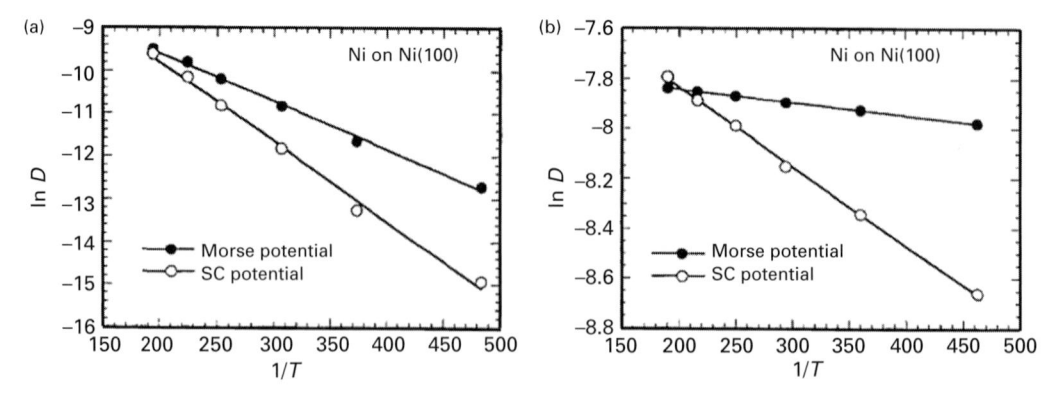

Fig. 5.15 Arrhenius plots for Ni diffusion on nickel surfaces. (a) On Ni(100): $E_D = 0.530$ eV, $D_o = 2.87 \times 10^{-3}$ cm^2/sec from many-body potential, and 0.298 eV from Morse potental. (b) On Ni(111): $E_D = 0.086$ eV, $D_o = 8.47 \times 10^{-4}$ cm^2/sec from many-body potential, and 0.016 eV from Morse potential (after Shiang et al. [60]).

By resorting to effective medium theory, Stoltze [61] arrived at an activation energy of 0.068 eV for self-diffusion. The value is again on the low side compared with experiment. In 1996, Li and DePristo [62] used corrected effective medium theory and molecular dynamics to calculate an activation energy for diffusion of 0.036 eV. Density-functional estimates were done by Mortensen et al. [63] using both the local-density and generalized gradient approximation; in the former they found a barrier of 0.16 eV, in the latter 0.11 eV, both on the static lattice. The uncertainty of the energy prediction was estimated as 0.02 eV. They also looked at the differences between fcc and hcp sites; the former seemed to be 0.05 eV stronger than hcp. Chang et al. [35] relied on embedded atom potentials to arrive at an activation energy of 0.06 eV for diffusion of nickel atoms on Ni(111). In 2000, Haug and Jenkins [64] examined the effect of hydrogen on epitaxial growth. They also looked at self-diffusion using EAM potentials [65,66] and found again a barrier of 0.05 eV for hopping between fcc sites of the plane. The presence of hydrogen lowered the barrier to 0.02 eV. The exchange mechanism required much higher energy of 2.15 eV and hydrogen reduced this barrier to 1.55 eV. It was still too high for this mechanism to participate during growth.

Using the embedded atom approximation for potentials, Kürpick [67] has done transition-state calculations for self-diffusion on nickel and copper surfaces. For nickel she found a diffusion barrier of 0.063 eV, and a prefactor of $(1.8\pm0.1)\times10^{-4}$ cm^2/sec. Agrawal et al. [26] depended on Lennard-Jones potentials in their estimates for the hopping barrier on Ni(111) and obtained a value of 0.14 eV and a prefactor of 1.7×10^{-4} cm^2/sec, in very good agreement with previous findings with the same potential. More recently, Bulou and Massobrio [68] utilized molecular dynamics simulations in the second-moment approximation to tight-binding. For self-diffusion of Ni they arrived at a barrier of 0.049 eV for hopping, and a much higher value, in excess of 2 eV, for atom exchange. The barrier for hopping is again much smaller than the experimental value. Kong and Lewis [69] calculated the prefactor to self-diffusion on Ni(111) at 300 K relying on EAM potentials by summing the vibrational contributions over all atoms, and found a value of 2.99×10^{-4} cm^2/sec starting from an fcc site and 2.89×10^{-4} cm^2/sec starting from an hcp site. These prefactors essentially did not change over the next 300 K. Kim et al. [70], using action-derived molecular dynamics [71] with RGL potentials, have recently estimated a self-diffusion barrier of 0.061 eV for nearest-neighbor jumps, and 1.633 eV for exchange, the former in agreement with previous calculations.

Estimates for self-diffusion on Ni(111) are quite consistent. Exchange has been found to involve overcoming a very high barrier, and is therefore unlikely to occur. The activation energies calculated at different laboratories are usually reasonably close together. Different adsorption sites available on Ni(111) have been recognized. The most significant concern, however, is the disagreement between calculational efforts and experiments, which have yielded a much higher activation energy than the calculations. Additional experimental effort might be in order.

5.5.2 Hetero-diffusion

Only one experimental investigation was done for hetero-diffusion on the Ni(111) plane. Diffusion of platinum atoms was examined on Ni(111) by Kellogg [72] using the field ion microscope. Platinum atoms were mobile on Ni(111) while imaging at 77 K, so little data was obtained, but there were no indications of motion by atom exchange. Kellogg suggested a diffusion barrier of 0.2–0.3 eV.

Kim et al. [70] also made estimates with RGL potentials for hopping of a number of hetero-atom diffusion events on Ni(111). For Cu, the barrier was 0.050 eV, for Pd 0.045 eV, for Ag 0.038 eV, for Pt 0.018 eV, and for Au 0.038 eV.

It must be noted that hetero-diffusion has not been carefully examined in theoretical calculations. Only for Pt on a Ni(111) surface are data available for comparison of experimental and theoretical barriers, and these are an order of magnitude apart: ~0.2 eV from experiment versus 0.018 eV from calculations. However, the value from FIM observed by Kellogg [72] is only an estimate, and might be too high. More thorough studies would certainly be advantageous for this system.

5.6 Nickel: Ni(100)

5.6.1 Self-diffusion

Tung and Graham [54] examined a nickel surface using the field ion microscope, with the surface cooled to ~ 20 K. Imaging of the (100) planes, schematics of which were shown in Fig. 3.3, was quite challenging. Nevertheless, an activation energy for diffusion of 0.63 eV was derived from the onset temperature, assuming the prefactor to be $(kT/2h)\ell^2$.

Two decades later, Fu and Tsong [55], also used the FIM but pre-annealed their wire before electro-etching the tip for longer life and for removing impurities. With the sample cooled to 30 K they derived an activation energy of 0.59 ± 0.03 eV and a prefactor equal to $1 \times 10^{-5.2 \pm 0.6}$ cm^2/sec, shown in Fig. 5.14b. The agreement between the two studies for the (100) plane is surprisingly good. Fu and Tsong also showed that diffusion

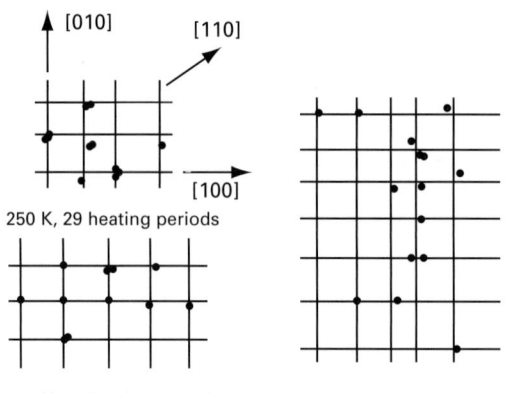

Fig. 5.16 Maps of sites occupied by Ni atoms diffusing on Ni(100) plane, indicating diffusion by atom exchange (after Fu and Tsong [55]).

occurred by exchange between the adatom and a lattice atom; this was deduced from the map of sites visited by adatoms, displayed in Fig. 5.16.

Work function changes on a Ni(100) surface were measured at different temperatures by Schrammen and Hölzl [73] after deposition of nickel atoms to reasonably low coverages (1/80 and 1/40 of a monolayer). Monte Carlo simulations were carried out to fit the measured work function increase with temperature and they found an activation energy for atom jumps between nearest-neighbor sites of 0.60 ± 0.02 eV, assuming a frequency prefactor of 10^{12} sec^{-1}. This result is in excellent agreement with the field ion microscopic observations of diffusion on this plane.

In 1995 Bartelt et al. [74] analyzed island densities on Ni(100), on which half a monolayer of nickel had been deposited at 300 K by Kopatzki et al. [75]; they found an activation energy for diffusion of more than ≈ 0.43 eV and a frequency prefactor of $\approx 10^{12}$ sec^{-1}.

There has been quite a lot of theoretical activity for predicting self-diffusion characteristic on Ni(100), starting with Flahive and Graham [47]. They used Morse potentials and calculated a barrier for self-diffusion on Ni(100) of ~ 0.3 eV, shown in Fig. 4.9, quite far from the experimental determinations. Many-atom interactions were taken into account by Liu et al. [4] in 1991, with embedded atom calculations using both AFW [5] and VC [6] parametrizations. They found for hopping an activation energy of 0.63 eV and a prefactor $D_o = 1.6 \times 10^{-3}$ cm^2/sec with AFW parameters; with VC parameters these values turned out higher, 0.68 eV for the barrier and 5.4×10^{-3} cm^2/sec for the prefactor. The barrier to atom exchange came out to be 0.93 eV using AFW potentials, and 1.15 eV with a prefactor of 4×10^{-2} cm^2/sec for VC interactions, much higher than the activation energy for hopping. These estimates do not reflect on the experimental findings about the mechanism. Liu et al. also used a Lennard-Jones potential for estimates and obtained a barrier of 0.80 eV, with a prefactor of 6.9×10^{-3} cm^2/sec.

Transition-state estimates using the corrected effective medium theory were done for hopping of nickel atoms by Sanders and DePristo [76], who found a barrier of 0.61 eV

5.6 Nickel: Ni(100)

with a prefactor of 3.2×10^{-3} cm^2/sec. From molecular dynamics they arrived at a diffusion energy of 0.63 eV; exchange was not considered. Comparisons of the diffusion barriers on Ni(100) obtained with different EAM potentials were done by Rice *et al.* [56]. They used transition-state theory with the FBD and VC approximations, the Oh and Johnson potential (OJ) [58] as well as the potential of Ackland *et al.* (ATVF) [59]. The results were as follows: activation energy 0.64 eV, D_o 3.5×10^{-3} cm^2/sec; 0.70 eV, 3.7×10^{-3} cm^2/sec; 0.45 eV, 3.5×10^{-3} cm^2/sec; 0.93 eV, 6.4×10^{-3} cm^2/sec, respectively. Values comparable with previous findings were obtained with the first two potentials. A comparison is presented in Fig. 4.10. What emerged from this study is that not all potentials describe nature correctly, and the scatter in the data obtained from different approaches can be quite big.

Shiang [60] utilized molecular dynamics with the Sutton–Chen potential and estimated a barrier of 0.530 eV, with the prefactor equal to 2.87×10^{-3} cm^2/sec. He compared this finding with data from Morse potentials, plotted in Fig. 5.15; the value for the activation energy was 0.298 eV with a prefactor of 1.06×10^{-3} cm^2/sec. As already shown for Ni(111), many-body interactions play an important role for Ni(100) as well, the activation energies being much higher with Sutton–Chen potentials. Shiang did not consider exchange as a possible mechanism of movement.

One year later Perkins and DePristo [77] also made estimates with the effective medium theory to describe inter-atomic energies. They found an activation barrier of 0.68 eV for hopping compared with 0.65 eV for atom exchange, but in estimates using MD/MC-CEM reached a lower value of 0.62 eV for hopping and 1.4 eV for exchange. They reanalyzed this data a year later [29], allowing the number of movable atoms in the lattice to vary. The hopping barrier from effective medium theory remained essentially constant, varying from 0.68 eV to 0.67 eV as the number of movable atoms rose from 13 to 226. For exchange, however, the activation energy changed dramatically from 0.65 eV to 0.47 eV. Similar effects were found in calculations with MD/MC-CEM: the hopping barrier changed from 0.62 eV to 0.60 eV, and for exchange from 1.4 eV to 1.0 V. From this investigation it is not clear which mechanism is favorable, jumping or exchange, since two methods contradict each other.

In molecular dynamics simulations with embedded atom potentials, Cohen [7] found long atom exchange events in self-diffusion on a number of metals, among them Ni(100). The barrier to these long exchanges was 1.42 eV; for ordinary hopping the activation energy was much lower, 0.74 eV, and for the usual atom exchange 0.85 eV. Exchange could occur at elevated temperatures such as 1000 K, as the barriers for hopping and exchange are reasonably close together, suggesting the coexistence of the mechanisms. The same year, Stoltze [61] utilized the effective medium theory to arrive at an activation energy of 0.558 eV for atom hopping; he did not consider exchange on this plane. Boisvert *et al.* [78] did molecular dynamics simulations based on embedded atom potentials for nickel atoms jumping and undergoing exchange; calculations were done on a layer of 72 atoms. For atom hopping they arrived at a barrier of 0.67 eV, with a frequency prefactor of 3.71×10^{12} sec^{-1}. For atom exchange the barrier was much higher, 1.29 eV, with a frequency prefactor of 7.28×10^{12} sec^{-1}. The latter barrier may have been affected by the small size of the substrate surface, but the activation energies are so different that jumping should still be the favored mechanism. Shi et al. [79] also relied on embedded atom potentials to evaluate the activation energies for self-diffusion on three fcc(100) planes, with 12 layer slabs consisting of 128 atoms in each layer and three bottom ones fixed. For single atom movement on Cu(100), they found a barrier of 0.503 eV, on Ag(100) 0.478 eV, and on Ni(100) the barrier was 0.632 eV, in good agreement with similar efforts.

Kürpick and Rahman [80] used embedded atom potentials with a cell of (10 × 10) atoms and 10 layers. They arrived at a barrier of 0.63 eV and a prefactor of 3.4×10^{-4} cm^2/sec with FBD [57] parameters, and 0.68 eV together with $D_o = 1.3 \times 10^{-3}$ cm^2/sec in the VC approximation. Atom exchange during diffusion was not considered. Later Kürpick [67] reported the same value of the diffusion barrier but with a prefactor of $(3.7 \pm 0.1) \times 10^{-3}$ cm^2/sec. Merikoski et al. [81] used EMT potentials to calculate the barriers from the adiabatic energy surface for different diffusion mechanisms. They found an activation energy of 0.631 eV for atom hopping compared to 0.844 eV for atom exchange. In ab initio density-functional estimates, Feibelman [13] obtained an activation energy of 0.72 eV for self-diffusion by hopping on paramagnetic Ni(100).

Embedded atom method potentials were again used by Mehl et al. [82] to calculate the hopping barrier. They worked with 100 atoms in each layer and 20 layers, finding a value of 0.63 eV, in excellent agreement with previous theoretical data. In 1999, Davydov [49] examined self-diffusion on Ni(100), relying on the cohesion based approach of Wills and Harrison [83]. For the activation energy he derived a value of 0.20 eV, which is clearly much too low. Haug et al. [84,85] looked at the influence of hydrogen on homo-epitaxial growth on Ni(100). They worked with 77 dynamically active atoms and estimated the barrier for self-diffusion by hopping, using EAM potentials, as 0.61 eV; exchange was not considered.

Chang et al. [28,86] carried out ab initio density-functional calculations and discovered that atom exchange occurred over a barrier comparable to hopping: 0.78 eV for exchange compared to 0.82 for hopping. As had earlier been done by Perkins and DePristo [29], they also examined the effect of the number of atoms in the unit cell of the surface on the activation energy calculated. The results were plotted in Fig. 3.60, and it is clear that for atom hopping there is essentially no dependence on the number of surface atoms; this is not true for diffusion by atom exchange, where it is necessary to have a large region in which atoms can relax their positions. For Ni(100) the barrier for exchange changed from 1.39 eV to 0.78 eV, making exchange the more favorable mechanism than hopping.

Agrawal et al. [26] depended on Lennard-Jones potentials in their estimates for the hopping barrier and found it to be 0.75 eV, with a prefactor of 2.5×10^{-3} cm^2/sec. No consideration was given to the possibility of an exchange mechanism or many-body effects. In 2006, Kong and Lewis [69] reported the kinetic prefactor for self-diffusion on Ni(100) calculated with EAM potentials by summing the vibrations over all atoms. They found a value of 5.3×10^{-3} cm^2/sec at 300 K, and 5.2×10^{-3} cm^2/sec at 600 K. From molecular dynamics simulations using RGL interactions, Kim et al. [70] derived a barrier of 0.376 eV for atom hopping and a much bigger value of 1.304 eV for exchange. The value for hopping is surprisingly low, and not in agreement with either experiment or previous theoretical estimates.

5.6 Nickel: Ni(100)

It appears that independent of the technique used, calculated estimates for the self-diffusion barrier on Ni(100) are reasonably close together and to experiments, but the estimates are not very satisfactory – they leave uncertain the mechanism of diffusion, which in experiments was established to be atom exchange. Less than half of the theoretical contributions have actually made calculations of the energetics for exchange. These results have a large spread, ranging from a barrier of 0.47 eV to 1.304 eV. The spread may be tied to the effect noted by Perkins and DePristo [29], as well as Chang et al. [86], that a considerable number of adjustable atoms must be included in the surface to accommodate strain. Values for the hopping energy of nickel on Ni(100), excluding work with pair interactions and the cohesive approximation of Davydov [49], are in much better shape; they are usually fairly close together.

5.6.2 Hetero-diffusion

There have been a few attempts to understand diffusion of foreign species on nickel surfaces. However, only three experiments were attempted to look at hetero-diffusion. The first, in 1996, was done by Müller et al. [87] in Kern's laboratory in Lausanne. They relied on nucleation theory to study the deposition of copper on the (100) plane of nickel, and utilized the scanning tunneling microscope as a tool to characterize the details of surface structure. Their results, obtained at a coverage of 0.1 monolayer, are shown in Fig. 5.17. For a critical nucleus size of one, implying stable dimers, they derived a diffusion barrier of 0.351 ± 0.017 eV and an attempt frequency $v_0 = 4 \times 10^{11 \pm 0.3}$ sec^{-1}. In the temperature range studied here, Müller et al. concluded that diffusion did not occur by atom exchange on the surface but rather by hopping.

Diffusion of platinum atoms was examined on nickel surfaces by Kellogg [72] using the field ion microscope. From the disappearance of adatoms from the (100) and the observation of atoms in the surface layer resistant to field evaporation, he concluded that diffusion occurred by atom exchange. Assuming a prefactor of 1×10^{-3} cm^2/sec, Kellogg estimated an activation energy of 0.69 eV.

In 1993 Krausch et al. [88] reported data from perturbed γ–γ angular correlation spectroscopy (PACS) which used ^{111}In as a probe. They deposited probe material in the range 10^{-4} ML on the surface at 77 K and monitored changes in frequency as the probe scanned over the deposit. For diffusion of In on Ni(100) they reported an activation energy of 0.31 ± 0.03 eV. No information about the mechanism was deduced from this study.

There were also a few theoretical investigations of hetero-diffusion on Ni(100). The first was done by Sanders and DePristo in 1992 [76] using MD/MC-CEM. Activation energies and frequency prefactors were: for copper 0.62 eV and 5.2×10^{12} sec^{-1}; for rhodium 0.67 eV and 4.5×10^{12} sec^{-1}; for palladium 0.55 eV and 4.2×10^{12} sec^{-1}; for silver 0.42 eV and 10×10^{12} sec^{-1}; for platinum 0.55 eV and 3.7×10^{12} sec^{-1}; and for gold 0.48 eV and 2.9×10^{12} sec^{-1}. It should be noted that the activation energy found here for Cu diffusion is much larger than the value reported in experiments by Müller et al. [87], and the study considered only hopping as a possible mechanism of movement.

The next attempt to establish the energetics, this time both for hopping and exchange, was made by Perkins and DePristo [89] who also looked at hetero-diffusion with 226

282   Diffusion on two-dimensional surfaces

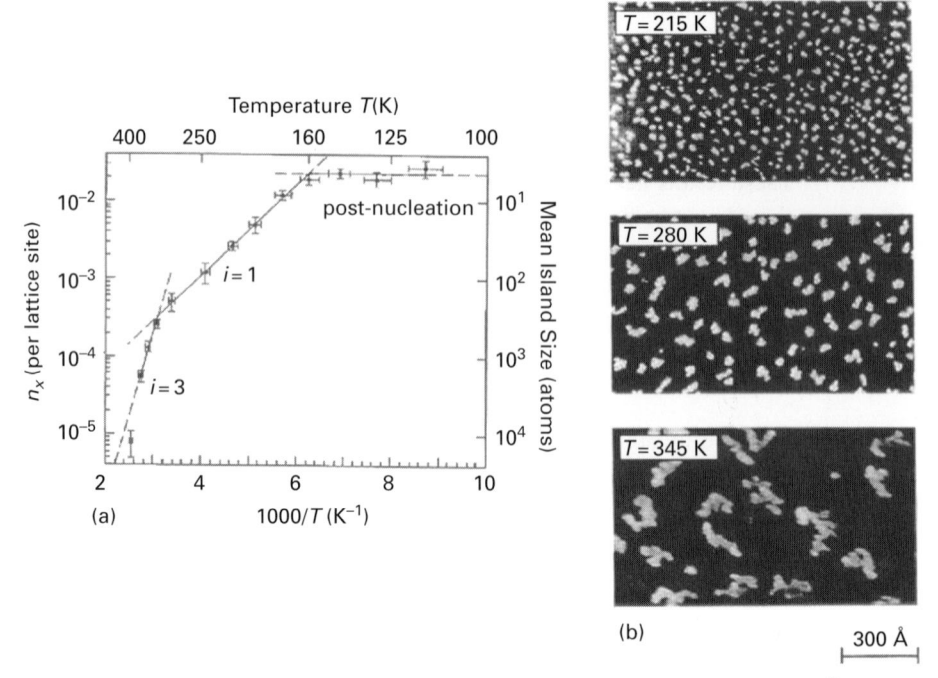

Fig. 5.17 Deposition of Cu on Ni(100) plane at 0.1 ML coverage and a deposition rate of 1.34×10^{-3} ML/sec. (a) Arrhenius plot of saturation island density against reciprocal temperature. (b) STM images of copper growth at different temperatures (after Müller et al. [87]).

movable atoms in the first layer. They obtained the following diffusion barriers from MD/MC-CEM calculations: for hopping of rhodium 0.65 eV, for exchange of a rhodium atom with Ni 0.85 eV; for hopping of palladium atom 0.64 eV, for exchange of palladium 1.10 eV; for hopping of copper 0.45 eV while for exchange of Cu 1.23 eV; for hopping of silver 0.44 eV, for exchange 1.47 eV. The frequency prefactors, expressed as multiples of 10^{12} sec^{-1}, were: for Rh hopping 8.28, for Rh exchange 9.37; for Pd hopping 9.77, for Pd exchange 11.93; for Cu hopping 4.15, for Cu exchange 3.76; for Ag hopping 7.80, for Ag exchange 4.45. The characteristics obtained for diffusion of copper on Ni(100) are in significantly better agreement with the experimental results of Müller et al. [87] than earlier estimates by Sanders and DePristo [76]. Unfortunately there are no experimental values with which to compare the rest of the data. One thing worth noting is that Perkins and DePristo [89] worked with a bigger cell, which should give a better prediction of the barrier for exchange.

Kim et al. [70] estimated the following barriers from molecular dynamics simulations with RGL interactions: Cu 0.407 eV, Pd 0.492 eV, Ag 0.347 eV, Pt 0.706 eV and Au 0.489 eV. These values are always not in good agreement with the work of Perkins and DePristo [89], but are quite close to Kellogg's [72] findings for platinum. Values obtained for copper are close to the experimental results of Müller et al. [87].

It must be noted that hetero-diffusion has not been carefully examined in theoretical calculations. Based on Perkins and DePristo's [89] data, it seems that hopping is a more

5.7 Copper: Cu(100)

5.7.1 Self-diffusion: experimental work

Diffusion measurements, which may be quite far from values for single atom movement on the copper (100) plane, started in 1987 and were done by de Miguel et al. [90]. They carried out thermal energy He atom scattering experiments on copper deposits of at least 0.5 monolayers evaporated onto the (100) plane of copper, and determined the mean terrace size over a range of temperatures. From this they were able to deduce an activation energy for diffusion of 0.40 eV, with a prefactor of $D_o = 1.4 \times 10^{-4}$ cm^2/sec. De Miguel et al. [91] also gave alternative procedures for deriving the diffusion characteristics: from the stationary intensities at different temperatures, where the scattered intensity no longer changes, and also from the temperature at which no oscillations are found in the specular intensity. Their results are plotted in Fig. 5.18 and give an activation energy of 0.48 eV and $D_o = 2.3 \times 10^{-3}$ cm^2/sec, in reasonable agreement with the previous results.

Somewhat later, in 1992, Ernst et al. [92] examined the nucleation of copper on Cu(100) using He-atom beam scattering as a probe. From measurements of the mean separation between islands as a function of temperature at a coverage of 0.5 monolayers, they arrived at a much lower estimate for the diffusion barrier of 0.28 ± 0.06 eV, with a

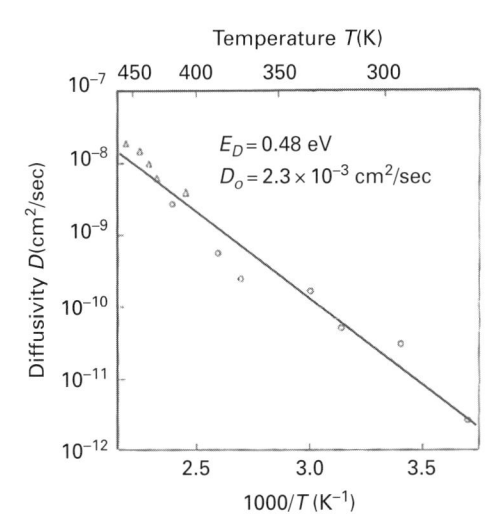

Fig. 5.18 Arrhenius plot for self-diffusion of Cu on Cu(100) derived from thermal energy He atom scattering. Open circles: from stationary intensities. Open triangles: from temperature above which no oscillations are seen (after de Miguel et al. [91]).

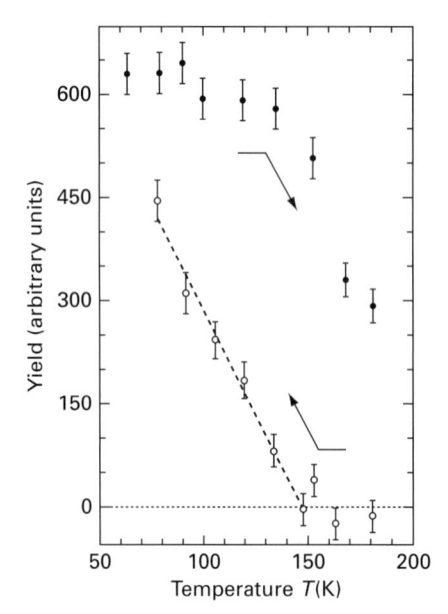

Fig. 5.19 Plots for increasing and decreasing temperature dependence of copper adatom yield under ion beam irradiation of Cu(100), giving 145 K as temperature at which adatoms become mobile (after Breeman and Boerma [94]).

prefactor $\approx 10^{-5}$ cm^2/sec. On reanalyzing this data with a different scaling factor, Brune et al. [93] obtained a value of 0.34 ± 0.07 eV for the activation energy of diffusion.

The same year, Breeman and Boerma [94] produced copper adatoms on Cu(100) terraces, with a small coverage of indium for calibration purposes, by bombarding the crystal with ions (LEIS). The surface mobility was deduced from the number of atoms appearing after irradiation. This of course also created vacancies on the surface, which can influence the mobility, but it was assumed that all vacancies were eliminated at lattice steps. The yield of copper at different temperatures was measured to determine the conditions for the start of surface diffusion, as shown in Fig. 5.19. Assuming a frequency prefactor $v_0 = 1 \times 10^{13}$ sec^{-1}, they arrived at a value of 0.39 ± 0.06 eV for the diffusion barrier, quite close to what de Miguel et al. [90] found.

In 1995, Dürr et al. [95] also looked at the diffusion of copper on Cu(100), using high resolution LEED. From measurements of the island separation, shown in Fig. 5.20 at a coverage of 0.3 monolayers and for temperatures from 180 to 260 K, they arrived at an activation energy of 0.36 ± 0.03 eV relying on mean-field nucleation theory. At higher temperatures detachment of atoms from islands is possible. It must be noted that these results were obtained at a rather high coverage where interactions are likely to play a role, and that the technique used is not sensitive enough to control the state of the surface, so results for single atom movement are somewhat uncertain.

In 1999 Laurens et al. [96] looked at the Cu(17,1,1) plane with PACS techniques; this plane consists mostly of (100) terraces. From the onset temperature for self-diffusion of Cu atoms they derived an activation energy on Cu(100) as 0.36 eV, in agreement with previous results, which are all reasonably close together.

5.7 Copper: Cu(100)

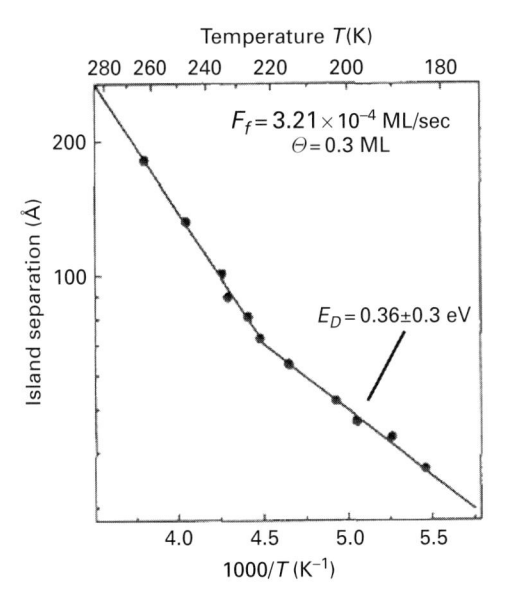

Fig. 5.20 Arrhenius plot of Cu island separation on Cu(100) as a function of reciprocal temperature (after Dürr et al. [95]).

None of the above data are really ideal for drawing conclusions about single atom diffusion, since coverages are much too high and interactions may influence movement on the surface. Unfortunately, the mechanism of motion also did not emerge from these studies.

5.7.2 Self-diffusion: theoretical work

Many theoretical estimates of the self-diffusion parameters for Cu(100) have been made, starting very early, in 1968, with the work of Wynblatt and Gjostein [97]; they did calculations relying on Morse potentials to arrive at a self-diffusion barrier of 0.248 eV. Also using Morse potentials, Flahive and Graham [47] found a self-diffusion barrier of 0.27 eV, as already shown in Fig. 4.9. The first many-body calculations on Cu(100) were carried out by Liu et al. [4], who derived a barrier to hopping of 0.38 eV, evaluated using AFW potentials, with a prefactor of 1.2×10^{-3} cm^2/sec; with VC potentials they arrived at a barrier of 0.53 eV and a prefactor of 5.2×10^{-3} cm^2/sec. The first of these estimates is in good agreement with experiments. For exchange on this plane the barrier was much bigger, 0.72 eV (AFW) and 0.79 eV (VC) with a prefactor $D_o = 2 \times 10^{-2}$ cm^2/sec. Hansen et al. [98] depended upon effective medium theory potentials to obtain a rather low barrier of ~ 0.22 eV. This, it should be noted, was for diffusion by atom exchange; the hopping barrier proved much higher, 0.43 eV. Hansen et al. [99] continued their work in 1993 and evaluated the free energy of activation, both by atom hopping and by exchange. Free energies were evaluated for a range of temperatures, but we confine our attention to ~ 300 K. Here the barrier to atom jumps was 0.398 eV, and only ~ 0.200 eV for diffusion by atom exchange, as shown in Fig. 5.21. The prefactors for the two processes were 4.7×10^{-4} cm^2/sec and 8.5×10^{-4} cm^2/sec.

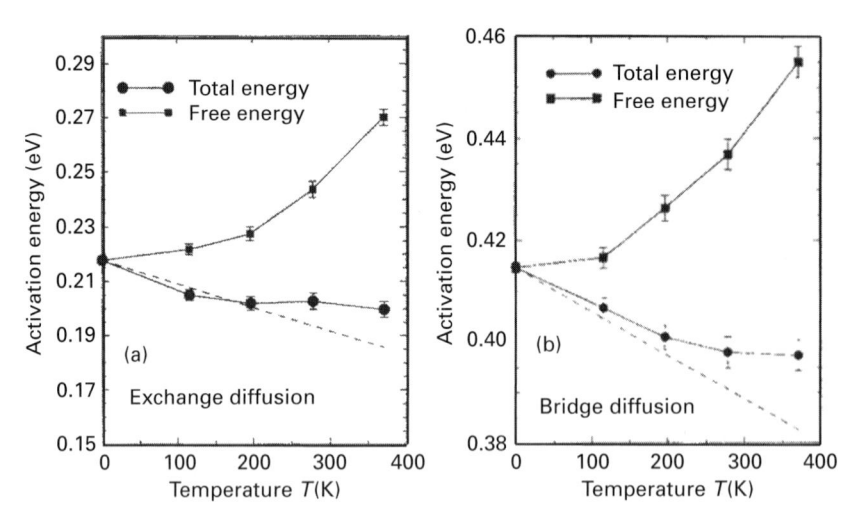

Fig. 5.21 Temperature dependence of the activation energy and free energy of adatom self-diffusion on Cu(100) by (a) atom exchange and (b) atom hopping; calculations by effective medium theory (after Hansen et al. [99]).

Transition-state estimates of the rate of self-diffusion were done a year later by Sanders and DePristo [76] using MD/MC-CEM theory to arrive at interaction energies. They found a barrier of 0.66 eV with a frequency prefactor of 5.5×10^{12} sec^{-1}. In this work only atom hopping was examined; the possibility of atom exchange in diffusion was not considered. Perkins and DePristo [77] continued their previous study using the same method with 13 atoms active at the surface. They found a barrier to hopping of 0.52 eV, with a much higher value of 1.1 eV for exchange. Using EAM potentials in the VC approximation they arrived at much the same barrier, 0.53 eV, and 0.79 eV for atom exchange. In the same paper they also made estimates with CEM potentials and found an activation energy of 0.47 eV for hopping compared to 0.43 eV for atom replacement. From the CEM method they concluded that replacement was the preferred mechanism in self-diffusion on Cu(100) and that the kinetic energy correlation term was crucial in the correct description of the exchange mechanism. However in 1995, they [29] reanalyzed their previous results with an increase in the number of movable atoms in the outer layer from 13 to 226. For hopping the barrier changed only little, from 0.52 eV to 0.51 eV using MD/MC-CEM, and from 0.47 eV to 0.46 eV with CEM. However, in diffusion by atom exchange the activation energy was altered significantly, from 1.1 eV to 0.77 eV with MD/MC-CEM and from 0.43 eV to 0.18 eV using CEM, seemingly making atom exchange the favored mechanism. Their data are still not conclusive since CEM gave a lower value for exchange, while MD/MC-CEM favored hopping.

Molecular dynamics simulations using embedded atom potentials were carried out by Black and Tian [8]. Their work was done for different diffusion temperatures. At 600 K they found atom exchange processes, but most interesting were simulations at 900 K. An adatom entered the substrate, and a lattice atom not adjacent to the entering atom came out of the substrate. The strain introduced by the entering atom was believed to propagate

5.7 Copper: Cu(100)

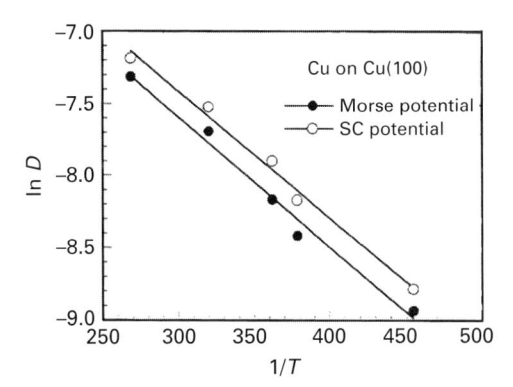

Fig. 5.22 Dependence of Cu adatom diffusivity on Cu(100) plane upon reciprocal temperature. Calculations by Sutton–Chen potentials yielded a diffusion barrier of 0.243 eV and a prefactor of 8.62×10^{-3} cm^2/sec. Using Morse potentials, the values were 0.249 eV and a prefactor of 7.80×10^{-3} cm^2/sec (after Shiang [102]).

along a row of substrate copper atoms. Eventually an atom emerged from the substrate at some distance from the original point of entry. Unfortunately they did not derive the energetics for this process, but this study showed that the exchange mechanism should contribute to movement on this surface, even if most of theoretical studies concluded that exchange was more demanding than hopping. Embedded atom estimates were made by Tian and Rahman [100], who reported a hopping barrier of 0.49 eV; only exchange with edge atoms was investigated, and surprisingly this turned out to be comparable to terrace hopping.

Many-body potentials introduced by Sutton and Chen [101] were used by Shiang [102] to do molecular dynamics simulations shown in Fig. 5.22. He calculated a barrier of 0.243 eV and a prefactor of 8.62×10^{-3} cm^2/sec. Using Morse potentials, without concern for many-body effects, he arrived at almost the same barrier of 0.249 eV and a prefactor of 7.80×10^{-3} cm^2/sec. It is possible that for Cu(100) many-body effects are not so important as for Ni. Stoltze [61] estimated an activation energy of 0.425 eV for jumps of copper atoms on Cu(100) using effective medium potentials. Liu [103] depended on embedded atom interactions in arriving at an energy barrier of 0.45 eV for hopping.

Lee et al. [104] performed *ab initio* calculations to find a barrier of 0.69 eV for hopping of copper adatoms, and a barrier of 0.97 eV for atom exchange. Karimi et al. [82,105, 106] made estimates for the diffusion of copper atoms relying on EAM potentials. For hopping they obtained a barrier of 0.48 eV, compared with 0.80 eV for diffusion by atom exchange. Effective medium theory was employed by Merikoski and Ala-Nissila [107] to arrive at a barrier of 0.399 eV for hopping. Subsequently Merikoski et al. [81] also worked out an activation energy of 0.996 eV for exchange. In 1996, Shi et al. [79] used EAM potentials to establish an activation energy of 0.503 eV for hopping; atom exchange was not considered. Diffusion of copper on Cu(100) was examined with a smart Monte Carlo technique by Kumar et al. [108]. Using MD/MC-CEM potentials, they arrived at an activation energy of 0.52 ± 0.04 eV and a prefactor of $1.6 \pm 0.3 \times 10^{-3}$ cm^2/sec and showed that this value was independent of the time step used.

Using an RGL type potential [109,110], Evangelakis and Papanicolaou [111] resorted to molecular dynamics simulations to find a barrier of 0.43 ± 0.02 eV, with $D_o = 3.4 \times 10^{-3 \pm 0.2}$ cm^2/sec for hopping, compared with 0.70 ± 0.04 eV and $D_o = 42 \times 10^{-3 \pm 0.3}$ cm^2/sec for atom exchange. An additional important development was that they also arrived at the characteristics of long transitions: for double jumps a barrier of 0.71 ± 0.05 eV and a prefactor of $38 \times 10^{-3 \pm 0.5}$ cm^2/sec; double exchange 0.70 ± 0.06 eV with $45 \times 10^{-3 \pm 0.6}$ cm^2/sec; triple exchange 0.82 ± 0.08 eV and $104 \times 10^{-3 \pm 1.0}$ cm^2/sec; as well as quadruple exchange 0.75 ± 0.12 eV and $86 \times 10^{-3 \pm 0.9}$ cm^2/sec. Their study confirmed the earlier work of Black and Tian [8] as well as Cohen [7]. An example of such a long-range transition seen at 950 K is shown in Fig. 3.64. It was possible to gain this detailed information as data were obtained at high temperatures ranging from 700 K to 1100 K.

Boisvert and Lewis [112] worked with the *ab initio* density-functional method using the generalized gradient approximation. For atom jumps they found a barrier of 0.52 ± 0.05 eV, and for exchange 0.96 ± 0.1 eV. However, their surface cell was only 3×3, and they used a block of four layers, which could have affected the energy estimates for atomic exchange. With molecular dynamics using EAM interactions, they found a barrier of 0.49 ± 0.01 eV for hopping with a prefactor of $20(\times e^{\pm 0.2}) \times 10^{12}$ sec^{-1} and 0.70 ± 0.04 eV and a prefactor of $437(\times e^{\pm 0.7}) \times 10^{12}$ sec^{-1} for atom exchange. A year later thermodynamic integration was used by Boisvert et al. [113] to evaluate transition-state barriers. They obtained almost the same value of 0.51 ± 0.02 eV with a frequency prefactor of $18.2^{+4.03}_{-3.29} \times 10^{12}$ sec^{-1} for atom hopping. For atom exchange the barrier was slightly higher at 0.74 ± 0.02 eV, with a frequency prefactor $6.65^{+5.47}_{-3.00} \times 10^{14}$ sec^{-1}.

Relying on EAM potentials and transition-state theory, Kürpick and Rahman [80,114] evaluated the diffusion characteristics of copper on Cu(100) at 600 K using both FBD [57] and VC [6] parameters. For the former the diffusion barrier amounted to 0.51 eV and the prefactor was 3.2×10^{-4} cm^2/sec; with the VC approximation a barrier of 0.53 eV was obtained with a prefactor of 3.3×10^{-4} cm^2/sec. The barriers for the expanded system at 600 K were very similar, 0.53 eV with a prefactor of 3.8×10^{-4} cm^2/sec from FBD parameters and 0.55 eV and 4.7×10^{-4} cm^2/sec from VC. Two years later Kürpick and Rahman [115] resorted to the same methods to arrive at the same values of the activation energies but with different prefactors. With FBD parameters the prefactor was 8.7×10^{-4} cm^2/sec, using VC potentials it turned out to be 9.0×10^{-4} cm^2/sec.

Li et al. [11] relied on the discrete variational method with the local-density-functional approximation to calculate activation energies for different types of transitions. For atom hopping along <110> the barrier proved to be 0.93 eV, for atom exchange it was much higher, 2.81 eV, and for hops over atom tops along <100> it was 1.25 eV. Hopping was clearly the lowest energy process, but the barriers calculated here are much higher than in other studies. Trushin et al. [116,117] used EAM potentials to find an activation energy of 0.49 eV for atom jumps and 0.69 eV for atom exchange.

Xie [118] used the embedded atom method to calculate interactions for molecular dynamics simulations and he arrived at a barrier of 0.56 eV for hopping over an unrelaxed surface, and claimed that relaxation changed the barrier by less than 0.1 eV; no estimates were made for atom exchange. Davydov [49] in 1999 arrived at a barrier of 0.01 eV for atom hopping using his cohesion approximation. This value is entirely out of line compared

5.7 Copper: Cu(100)

to other estimates. Zhuang and Liu [119] used RGL potentials to look at exchange and jump processes. For simple exchange they obtained a barrier of 0.70 ± 0.04 eV, from static calculations the value was 0.85 eV. However, the barrier to simple atom hopping proved to be much lower, 0.44 eV from an Arrhenius plot.

Adams et al. [120,121] used EAM potentials in the AFW approximation to evaluate the activation energy. Their work yielded a barrier of 0.50 eV for atom hopping and 0.75 eV for atom exchange, with the former process clearly preferred. They also showed that bond counting did not properly account for surface diffusion effects. Four years later, Wang et al. [122] did molecular statics calculations using embedded atom potentials and arrived at an activation energy of 0.48 eV for adatom hopping. The exchange mechanism led to a higher value of 0.69 eV. Kürpick [67] used a slightly different approach than previously; with RGL potentials and transition-state calculations she found a barrier of 0.44 eV and a prefactor of $(2.5 \pm 0.1) \times 10^{-3}$ cm^2/sec for hopping over the surface. In the same year, Fordell et al. [21] carried out *ab initio* calculations and in the LDA approximation they arrived at a barrier to atom hopping of 0.67 eV, and for GGA the activation energy was 0.55 eV. Exchange barriers proved to be much higher, 0.94 eV for LDA and 0.81 eV for GGA.

Lennard-Jones potentials were used by Agrawal et al. [26] to calculate an activation energy of 0.59 eV with a prefactor of 2.3×10^{-3} cm^2/sec for copper adatoms hopping, but completely ignored all many-body effects. However Liu et al. [4] already showed in 1991 that many-body effects do not play a crucial role in the case of copper. Pentcheva [123] relied on density-functional theory to look at the diffusion of copper adatoms and she found a barrier of 0.49 eV for jumps of copper, and a much larger barrier of 1.02 eV for diffusion by atom exchange.

Activation energies were evaluated by Chang and Wei [28] using the VASP simulation routines [22–24]. For atom hopping they arrived at a barrier of 0.64 eV, but found a much higher value of 0.94 eV for atom exchange. Effective medium estimates were used by Jahma et al. [124] to look at diffusion and submonolayer growth by kinetic Monte Carlo simulations. For self-diffusion they briefly mentioned a barrier of ~0.4 eV, and assumed hopping as the leading mechanism. Another calculation of atom hopping with embedded atom potentials was made by Yildrim et al. [125] with EAM potentials using the local-density approximation. They found an activation energy of 0.505 eV for jumps, and a prefactor at 300 K of 7.29×10^{-4} cm^2/sec; at 600 K this changed very little to 7.45×10^{-4} cm^2/sec. This local-density approximation has been challenged by Kong and Lewis [126], who used the same EAM potentials but calculated the prefactor by summing over all frequencies. They obtained precisely the same activation energy for diffusion. For the prefactor to the diffusivity, however, they obtained a different value, 5.3×10^{-3} cm^2/sec at 300 K, which changed to 5.8×10^{-3} cm^2/sec at 600 K, and they concluded that the local-density approximation was not in general valid.

The activation energy was calculated by Müller and Ibach [127] using the local-density approximation; they arrived at a value of 0.67 eV for the barrier to hopping, compared to 1.02 eV for atom exchange. For self-diffusion by atom jumps Kim et al. [70] estimated a barrier of 0.477 eV, and for exchange 0.708 eV, both from molecular dynamics simulations relying on RGL potentials. These results are in reasonable accord with much of the prior calculations, but higher than most experimental values.

The characteristics for self-diffusion on Cu(100) are not in a very good state. Four experiments have been done, and an average barrier of 0.37 ±0.03 eV emerges from these. The results are reasonably close together, but it must be noted that all the studies have been based on indirect deductions, without any direct observations of atom motion. Theoretical estimates of atom hopping are in worse shape. There are roughly thirty of these and the more reasonable ones for hopping range from 0.243 eV to 0.69 eV, with an average value of 0.49 ±0.06 eV. In more than a dozen of these evaluations, estimates were made of atom motion by exchange with the lattice; in these, reasonable energies ranged from 0.18 eV to 1.02 eV, with a mean value of 0.69 ±0.30. Early theoretical studies indicated exchange as the primary mechanism of movement, the latest definitely favor hopping. There is not enough experimental information to verify which prediction is correct. At the moment it is therefore not clear which mechanism is primarily responsible for diffusion, and it appears that further work, mostly experimental, would be desirable.

5.7.3 Hetero-diffusion

Some experiments on the diffusion of indium on Cu(100) were done by Breeman and Boerma [128] using low energy neon ion scattering. They looked at the fraction of In atoms deposited on different sites at various deposition temperatures. With 0.013 ±0.002 monolayers of indium on the surface, they were able to determine the temperature at which indium became mobile as being between 77 K and 88 K. Assuming an attempt frequency of 10^{13} sec^{-1}, they arrived at an activation energy of 0.24 ±0.03 eV for atom hopping, and were unable to account for their observations assuming diffusion occurred by atom exchange. In 1993 Krausch et al. [88] reported diffusion data obtained using perturbed γ–γ angular correlation spectroscopy (PACS). This technique used ^{111}In in the range 10^{-4} ML as a probe. Indium was deposited on the surface at 77 K, and atom movement as well as changes in frequency were monitored. The activation energy for movement of In on Cu(100) was reported as 0.28 ±0.05 eV. Diffusion into the bulk required a huge energy of 2.05 ±0.05 eV. Movement of ^{111}In on Cu(17,1,1), quite close to the (100) plane, was also investigated by Rosu et al. [129] in 2001, again using the PACS technique. Coverage of indium was less than 10^{-3} ML and an activation energy of 0.20 ±0.03 eV was found. Diffusion of indium atoms was also studied by Van Siclen [130] using EAM interactions. For diffusion, the energy barrier was estimated as 0.32 eV. Diffusion by atom exchange involved much more energy, requiring 1.01 eV, in agreement with the experimental findings of Breeman and Boerma [128], but their value of the hopping energy was slightly higher.

Cohen et al. [131] looked at the spreading of a two-dimensional submonolayer of Pb deposited on Cu(100) as a function of coverage. Measurements were done by Rutherford backscattering. From the temperature dependence of the diffusivity at low coverages they deduced a barrier of 0.68 ±0.1 eV for isolated jumps of Pb atoms.

Graham et al. [132] also examined the diffusion of sodium atoms on Cu(100) using high resolution scattering of helium atoms. From measurements of the peak width as a function of the temperature they were able to come up with the Arrhenius plot in

5.7 Copper: Cu(100)

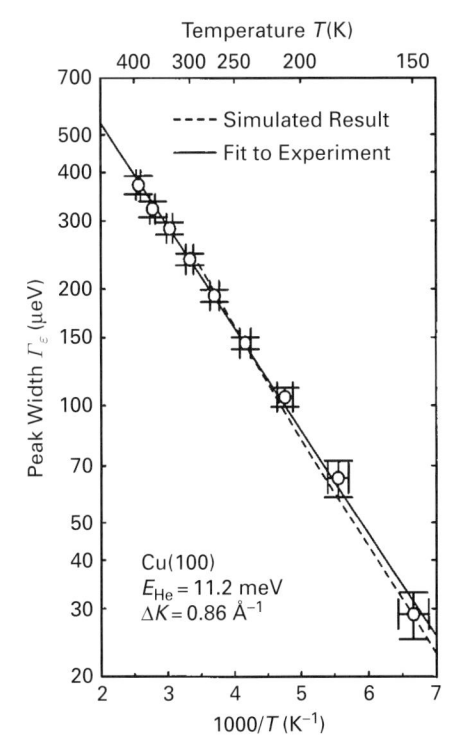

Fig. 5.23 Dependence of quasi-elastic peak width upon reciprocal temperature in He scattering from Na atoms at 0.028 ML coverage on Cu(001). Best fit gives activation energy of 0.0525 ± 0.0009 eV. Predicted slope 0.065 ± 0.006 eV (after Graham et al. [132]).

Fig. 5.23, yielding an activation energy of 0.0525 ± 0.0009 eV. The dashed line indicates the curve predicted by molecular dynamics simulations for temperatures $T \leq 280$ K with a barrier of 0.065 ± 0.006. These results were obtained for a sodium coverage of 0.028, which may yield the behavior of individual adatoms. The data are in excellent agreement with earlier measurements obtained in the same group by Ellis and Toennies [133]. This yielded an activation energy of 0.051 ± 0.006 eV and a frequency prefactor of $(0.53 \pm 0.2) \times 10^{12}$ sec^{-1} at a higher sodium concentration of 0.1 layer, as shown in Fig. 5.24. In 1997 Graham et al. [134] made an attempt to find the diffusivity D in the limit when the wave vector $\Delta K \to 0$. They claimed that the experimental sensitivity in this region was too low for direct extraction of the diffusivity, and instead extrapolated via the calculated quasi-elastic width to the limit $\Delta K \to 0$ and used the zero frequency Laplace transform of the velocity autocorrelation function of Na atoms. By this procedure they found $D_o = (6.3 \pm 0.6) \times 10^{-3}$ cm^2/sec and $E_D = 0.075$ eV. It was claimed that the difference between the earlier value of the barrier and the present one arose because the quasi-elastic peak had contributions from diffusive motion and T-mode vibrations at finite ΔK. The real barrier corresponded to 0.075 eV, not 0.052 eV as suggested previously. They also predicted a value of 0.065 ± 0.006 eV for the effective barrier for diffusion.

In 1999, Cucchetti and Ying [135] carried out molecular dynamics simulations for the movement of Na on the Cu(100) surface using their own model Hamiltonian. Their

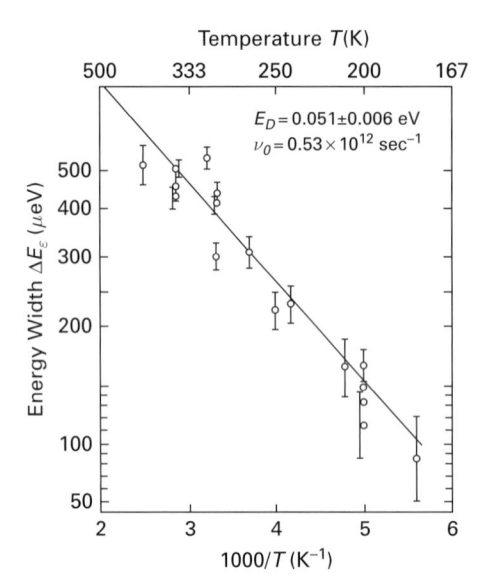

Fig. 5.24 Arrhenius plot of He peak width in scattering from Na atoms on Cu(001) plane, yielding an activation energy of 0.051 eV (after Ellis and Toennies [133]).

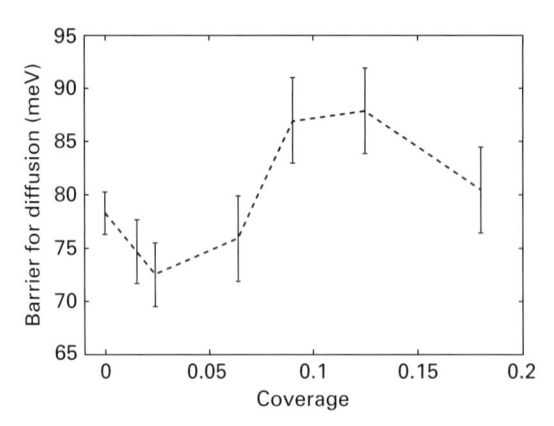

Fig. 5.25 Activation energy for tracer diffusion of Na on Cu(001) surface calculated for different coverages (after Cucchetti et al. [135]).

results for tracer diffusion are shown in Fig. 5.25. The value of the barrier at a very low coverage was around 0.077 eV, very close to the findings of Graham et al. [134]. Recently the system was investigated again with ^{3}He spin echo measurements by Alexandrowicz et al. [136] at lower coverages of 0.02 ML and 0.08 ML; the lower coverage can probably be associated with interaction-free movement. Their data are shown in Fig. 5.26. The activation energy deduced at the lower coverage was 0.058 ± 0.007 eV, rather close to the earlier estimates of Graham et al. [132], and diffusion was by atom hopping. What is more important is that they came out with a value of 0.042 ± 0.005 eV for motion perpendicular to the surface, although there was no previous evidence of atoms getting away from the surface during jumps, and this value was obtained for a higher coverage

5.7 Copper: Cu(100)

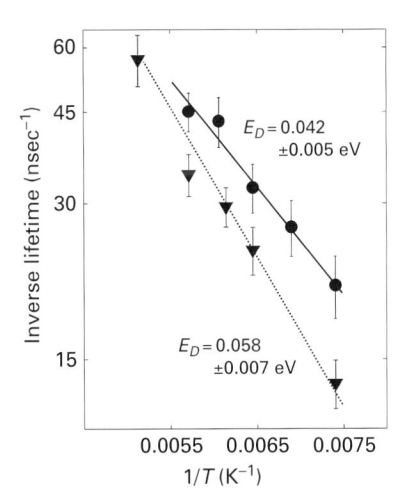

Fig. 5.26 Arrhenius plots for the inverse lifetime of Na on Cu(001), obtained in ^{3}He spin echo experiments. Inverted triangles: 2D motion, coverage 0.02 ML; circles: perpendicular movement, coverage 0.08 ML (after Alexandrowicz et al. [136]).

of 0.08 ML. The same group continued investigations of Cs on Cu(100) at a coverage of 0.014 and 0.044 ML [137]. From their experiments they deduced a value of 0.31 ± 0.02 eV for the diffusion barrier at a coverage of 0.044 ML; from Langevin molecular dynamic simulations they obtained a value of 0.24 ± 0.04 eV, and observed multiple jumps during movement of Cs over the surface.

Sanders and DePristo [76] studied hetero-diffusion in some detail relying on MD/MC-CEM interactions. They first did molecular dynamics simulations and compared the results with estimates using transition-state theory (TST) and found good agreement. With this established, they worked with hetero-atoms on Cu(100), and arrived at the following characteristics from TST: nickel diffusion barrier of 0.64 eV and a frequency prefactor 5.2×10^{12} sec^{-1}; rhodium 0.71 eV and 4.6×10^{12} sec^{-1}; palladium 0.58 eV and 4.6×10^{12} sec^{-1}; silver 0.36 eV and 6.9×10^{12} sec^{-1}; platinum 0.61 eV and 4.1×10^{12} sec^{-1}; gold 0.56 eV and 3.1×10^{12} sec^{-1}. No consideration was given to the possibility that atom exchange might contribute to diffusion. However, the frequency prefactors all seem to lie in the right range, $\sim 10^{12}$ sec^{-1}.

Further work was done by Perkins and DePristo [89], who still continued with MD/MC-CEM and transition-state theory, but now also worked out what happens when atom exchange contributes. Their results for a surface consisting of 226 movable atoms were as follows for the hopping characteristics: rhodium 0.72 eV and 12.43×10^{12} sec^{-1}; nickel 0.64 eV and 9.44×10^{12} sec^{-1}; palladium 0.73 eV and 17.57×10^{12} sec^{-1}; silver 0.48 eV and 9.07×10^{12} sec^{-1}. For diffusion by atom exchange their results were rather different: rhodium 0.42 eV and 10.61×10^{12} sec^{-1}; nickel 0.57 eV and 9.00×10^{12} sec^{-1}; palladium 0.59 eV and 15.58×10^{12} sec^{-1}; silver 0.91 and 7.5×10^{12} sec^{-1}. It appears that for Rh, Ni, and Pd, diffusion by exchange of atoms is really the favored process, but not for Ag. Values for Rh and Ni hopping agreed very well with the previous findings of Sanders and DePristo [76].

Gold atoms on Cu(100) were examined by Evangelakis et al. [138] with RGL potentials; from the Arrhenius plot in Fig. 5.27 they found an activation energy of 0.58 ± 0.02 eV for atom exchange. The barrier estimated from energy minimization was 0.57 eV for exchange, and 0.64 eV for atom hopping.

Pentcheva [123] did density-functional theory calculations in the WIEN97 implementation for diffusion of both copper and also cobalt atoms on Cu(100). For hopping of cobalt the barrier proved to be 0.61 eV, and 1.00 eV for atom exchange, as is illustrated in Fig. 5.28. However, for non-magnetic cobalt adatoms the hopping barrier proved to be much higher, 0.92 eV. Both copper and cobalt appear to diffuse by atoms hopping over the surface. First-principles calculations with RGL potentials for movement of Co on Cu(100) as well as on Co_{36} strained islands were performed by Stepanyuk et al. [139].

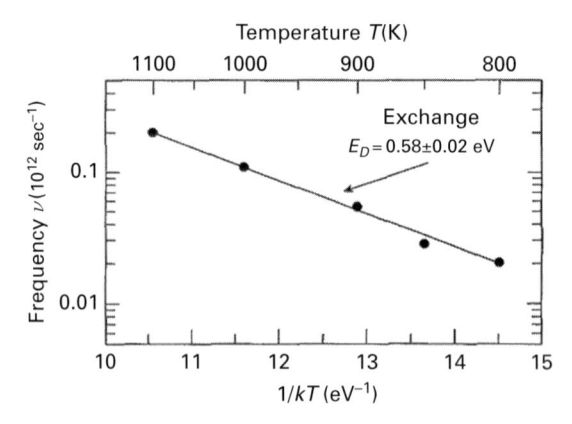

Fig. 5.27 Arrhenius plot for diffusion of Au atom on Cu(100) plane evaluated using RGL potentials (after Evangelakis et al. [138]).

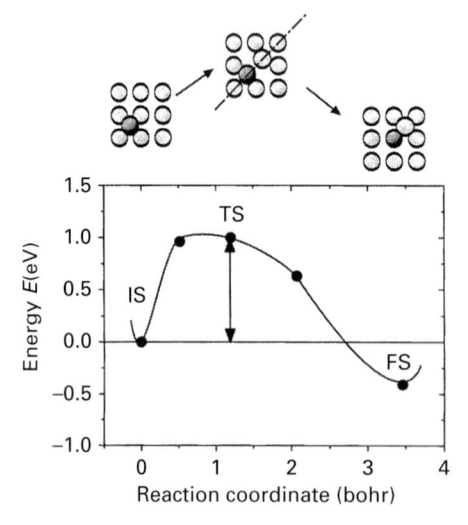

Fig. 5.28 Potential diagram for exchange of Co adatom with lattice atom on Cu(100), calculated by density-functional theory. Adatom hopping is preferred (after Pentcheva [123]).

Movement of Co was preferred by hopping, with an energy of 0.66 eV, while exchange required 0.86 eV. Diffusion on Co$_{36}$ islands required 0.58 eV by hopping and 0.90 eV by exchange. Movement along the islands was, however, a very low energy process with a barrier of 0.20 eV. What is worth noting is that the misfit between Co and Cu is about 2%, so a Co$_{36}$ island is under compressive strain. The amount of such strain decreases with the size of the island, but is present for Co$_{100}$ islands as well. It is interesting that such an island creates a depression in the Cu(100) surface, so not only the Co$_{36}$ island, but also the substrate underneath is under strain. Such depression becomes smaller for an island under 200 atoms.

Kim et al. [70] also made barrier estimates for hetero-atom diffusion using molecular dynamics simulations based on RGL interactions. For Ni on Cu(100) the barrier to atom hopping was 0.439 eV, for Pd 0.566 eV, for Ag 0.393 eV, for Pt 0.874 eV, and for Au 0.554 eV; results rather different from what was obtained in previous attempts.

It appears there is quite a lot of information available about hetero-diffusion on Cu(100), but existing data are difficult to compare and judge.

5.8 Copper: Cu(111)

5.8.1 Self-diffusion

The first experimental work which delivered the characteristics for diffusion on Cu(111) was done by Wulfhekel et al. [140] in 1996. They measured the saturation density of copper islands at a coverage of 0.1 monolayer using thermal energy helium scattering, and assumed that dimers were stable in the investigated range of temperatures. From the data in Fig. 5.29 they arrived at an activation energy for self-diffusion of $0.03^{+0.01}_{-0.005}$ eV, but this may not be the value for single atom movement since the coverage was rather high. In 1994, Henzler et al. [141] analyzed spot profile LEED and at a coverage of 0.5 ML derived an activation energy of 0.064 eV from the island density, assuming a stable nucleus equal to one. For a Cu layer covered with 1 ML of Pb, movement required overcoming a barrier of 0.083 eV. It is unlikely, however, that these data are for the

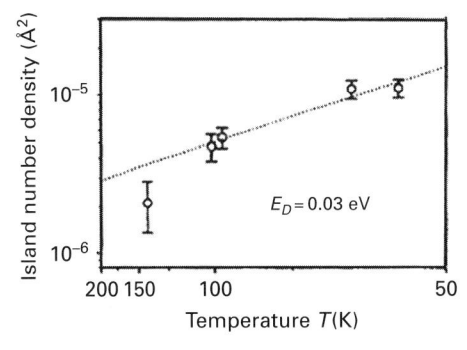

Fig. 5.29 Saturation density of Cu islands on Cu(111) determined by thermal energy He scattering (after Wulfhekel et al. [140]).

movement of single atoms not influenced by interactions, since the coverage was high. In 2000 Schlösser et al. [142] looked at vacancy island movement with a fast STM and modeled their findings with EMT potentials using Stoltze's [61] program, to find for terrace self-diffusion a barrier of 0.057 eV.

Excellent direct studies of atom jumps in self-diffusion on the Cu(111) plane have been carried out by two groups. In 2000, Repp et al. [143] reported STM observations of copper atoms on Cu(111), aimed at examining interactions between the adatoms. As part of that investigation, they also looked at the rate of atom jumps at a coverage of 0.003 ML. From the temperature dependence in the range from 9 to 21 K they arrived at a diffusion barrier of 0.037 ± 0.005 eV and a frequency prefactor of $5 \times 10^{13 \pm 1}$ sec^{-1}. What is noteworthy here, besides the fact that the results were obtained directly from the measured rate of atom jumps and not from nucleation theory, is the small size of the diffusion barrier, which in the past has been connected with an unusually small prefactor [32]. The present results disprove such a connection.

Two years later, Knorr et al. [144] published a more detailed study of the same system. From STM observations of isolated copper adatoms they found the rate of jumping at different temperatures around 13 K, as shown in Fig. 5.30, and arrived at a diffusion barrier of 0.040 ± 0.001 eV and a frequency prefactor $\nu_0 = 1 \times 10^{12 \pm 0.5}$ sec^{-1}. Diffusion was observed between only one type of adsorption site, presumably an fcc site.

In 2003 Repp et al. [145] carried out further studies using the scanning tunneling microscope supplemented by *ab initio* calculations using the DACAPO code. They were able to establish an hcp-to-fcc hopping rate 75 times higher than for fcc-to-hcp, which established a lower limit of 0.004 eV for the energy difference between the two sites. From the difference in the temperatures for the start of diffusion out of the sites they came up with an upper limit of 0.008 eV, with binding at fcc sites the stronger. Calculations led

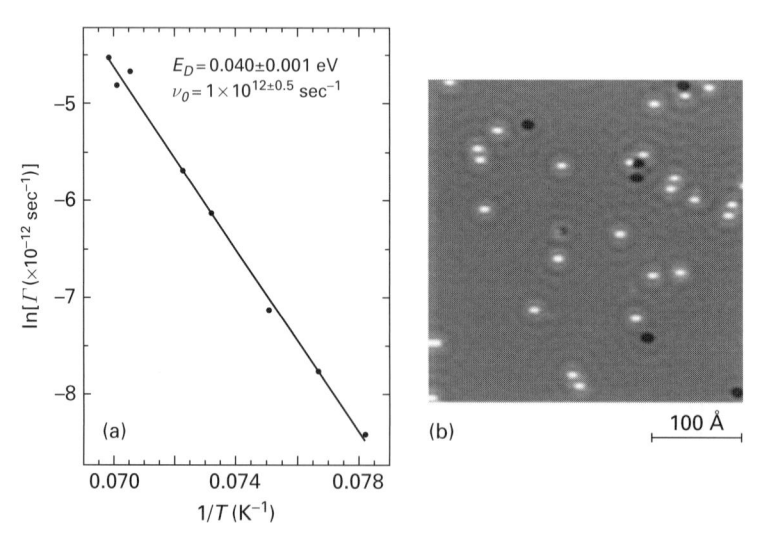

Fig. 5.30 Self-diffusion of Cu on Cu(111) measured with STM. (a) Arrhenius plot yields an energy barrier 0.040 ± 0.001 eV and a frequency prefactor $\nu_0 = 1 \times 10^{12 \pm 0.5}$ sec^{-1}. (b) Excerpt from sequence of STM measurements tracing diffusion at 13.5 K, 0.0014 ML (after Knorr et al. [144]).

5.8 Copper: Cu(111)

to a 0.006 eV stronger binding on fcc sites, as well as a diffusion barrier of 0.050 eV with a prefactor of 1×10^{12} sec^{-1}.

Agreement between the measurements is good, and lots of theoretical estimates have been made, even before any experiments had been done. Wynblatt and Gjostein [97] did energy calculations with Morse potentials and obtained a value between 0.023 eV and 0.032 eV for the barrier to self-diffusion. Also using Morse potentials, Flahive and Graham [47] calculated a barrier close to 0 eV, as in Fig. 4.9. The first estimates with a many-atom potential, EAM, were carried out by Liu et al. [4] in 1991. They found a barrier of 0.026 eV with AFW potentials and 0.044 eV when using VC. In these two cases the prefactors were 3×10^{-4} cm^2/sec and 4.6×10^{-4} cm^2/sec.

Hansen et al. [98] in the same year made estimates with EMT potentials and arrived at a hopping barrier of ∼ 0.12 eV. They also showed that exchange was unfavorable on this plane, requiring around 0.5 eV. The many-body potential of Sutton and Chen [101] was used by Shiang [102] to evaluate the activation energy from an Arrhenius plot, shown in Fig. 5.31. He found a barrier of 0.059 eV with a prefactor of 6.4×10^{-4} cm^2/sec. Surprisingly enough the barrier obtained using Morse potentials, 0.064 eV with a prefactor of 5.24×10^{-4} cm^2/sec, was much the same.

In 1993, Kief and Egelhoff [15] evaluated the barrier for self-diffusion on Cu(111) as 0.14 eV by taking the results for Ir(111) and scaling according to the cohesive energies, but this value is much too high. Using the same sort of potentials as Hansen et al. [98] had used earlier, Stoltze [61] found a lower value, 0.053 eV. Karimi et al. [105] came up with a self-diffusion barrier of 0.028 eV using EAM, exchange cost 1.12 eV. Li and DePristo [62] arrived at a barrier of 0.039 eV with corrected effective medium theory.

Kallinteris et al. [146], in 1996, did molecular dynamics simulations using the semi-empirical RGL potential [110,147]. As shown in Fig. 5.32, they covered a range of temperatures, and found a diffusion barrier of 0.041 ± 0.002 eV as well as a prefactor

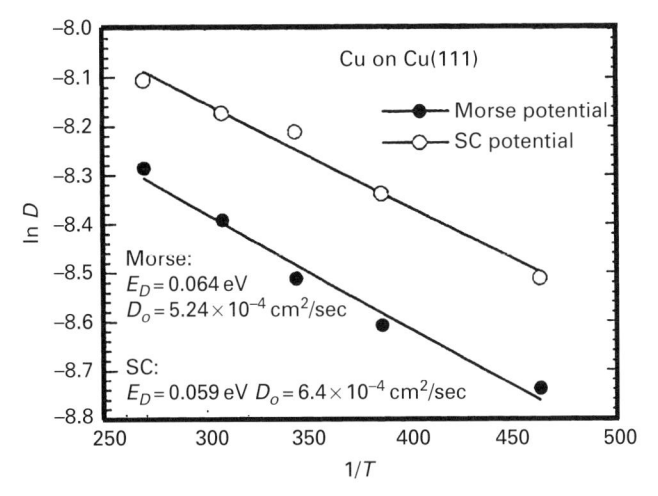

Fig. 5.31 Arrhenius plot for self-diffusion on Cu(111), calculated with Sutton–Chen and Morse potentials to yield a diffusion barrier of 0.059 eV (after Shiang [102]).

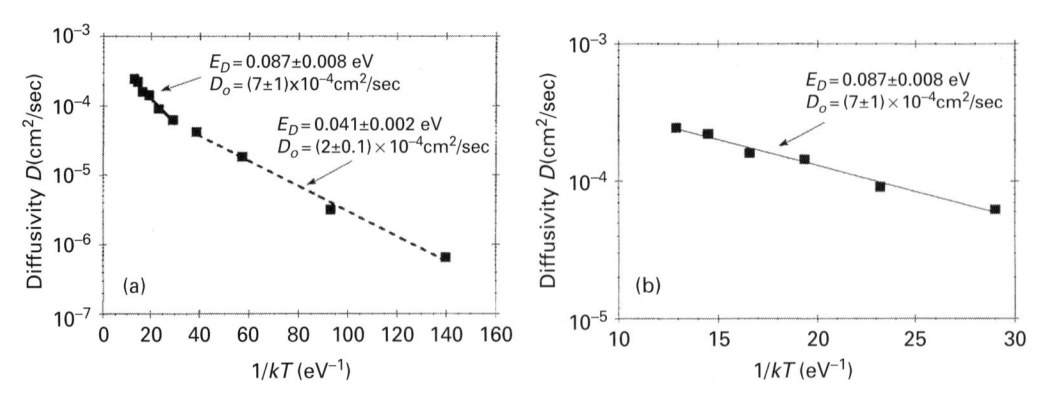

Fig. 5.32 Self-diffusion of Cu adatoms on Cu(111) plane as a function of the reciprocal temperature, derived by molecular dynamics simulations with RGL potentials. (a) Overall temperature range. (b) Only high temperatures (after Kallinteris et al. [146]).

$D_o = (2 \pm 0.1) \times 10^{-4}$ cm^2/sec at lower temperatures from 80 K to 300 K. For direction <110> the barrier was 0.043 ± 0.001 eV, perpendicular to the <110> direction it was almost the same, 0.040 ± 0.001 eV. Agreement with experimental results is excellent. From their Arrhenius plot at higher temperatures, from 400 K to 900 K, they arrived at a barrier of 0.087 ± 0.008 eV and a prefactor $D_o = (7 \pm 1) \times 10^{-4}$ cm^2/sec. In direction <110> the energy barrier was 0.082 ± 0.007 eV, and in the perpendicular direction 0.10 ± 0.01 eV, so no anisotropy for diffusion was observed. At temperatures higher than 1000 K exchange started to be present with an energy of 1.19 eV.

Using EAM interactions, Breeman et al. [148] calculated the activation energies for copper atoms hopping from fcc to hcp sites as 0.13 eV, and 0.11 eV for the reverse jumps. Mortensen et al. [63] estimated the barrier to self-diffusion as 0.16 ± 0.02 eV with the local-density approximation of density-functional theory and 0.12 ± 0.02 eV with the generalized gradient approximation. These estimates are rather too high.

EAM potentials were also employed by Trushin et al. [116] in molecular statics estimates of the activation energy for self-diffusion on copper. On Cu(111), the barrier was 0.029 eV. Estimates of the barrier to self-diffusion were made in 1999 by Davydov [49] using his cohesion approximation, and he found a value of 0.05 eV. Although some of his estimates have come far from most calculations, in this instance agreement is quite reasonable with the experimental value of 0.04 eV [143,144]. Bogicevic et al. [149] carried out density-functional estimates with a cell $12 \times 4 \times 4$ consisting of 192 atoms and obtained a barrier of 0.05 eV; they also pointed out that the barrier changed by a factor of two if the surface was not relaxed. With embedded atom interactions, Chang et al. [35] once more found the barrier to self-diffusion to be 0.05 eV. Relying on RGL potentials, Kürpick [67] estimated a diffusion barrier of 0.042 eV with a prefactor of $1.2 \pm 0.1 \times 10^{-4}$ cm^2/sec, one of not too many theoretical studies which derived the prefactor as well. Larsson [150] studied island diffusion on Cu(111) with Monte Carlo simulations, relying on bond energies derived in the literature. For the activation energy of single copper adatom diffusion he obtained a barrier of 0.09 eV, much higher than the experimental value.

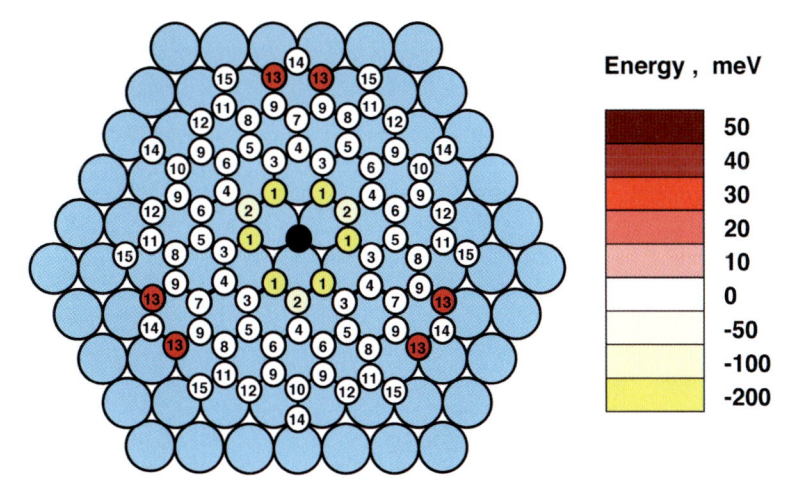

Fig. 2.28 Pair interaction as a function of the separation from the black central atom for Ag on Ag(111) (after Fichthorn and Scheffler [87]).

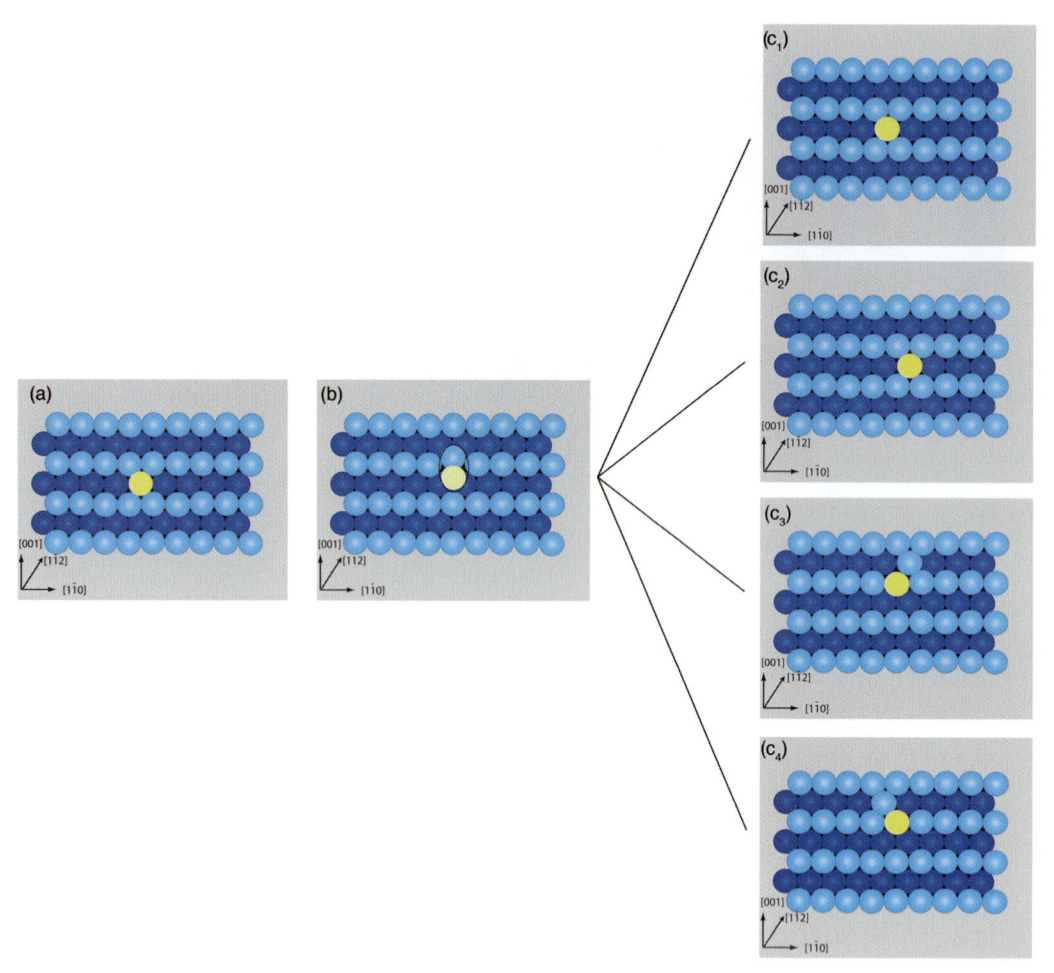

Fig. 3.42 Schematic of atom exchange process in self-diffusion on fcc(110) surface: (a) Atom in equilibrium position. (b) At the saddle point, atom pair sits as a dumbbell (indicated by ellipse) across [1$\bar{1}$0] row of substrate atoms. (c$_1$) – (c$_4$) After diffusion, atoms distributed over allowed sites.

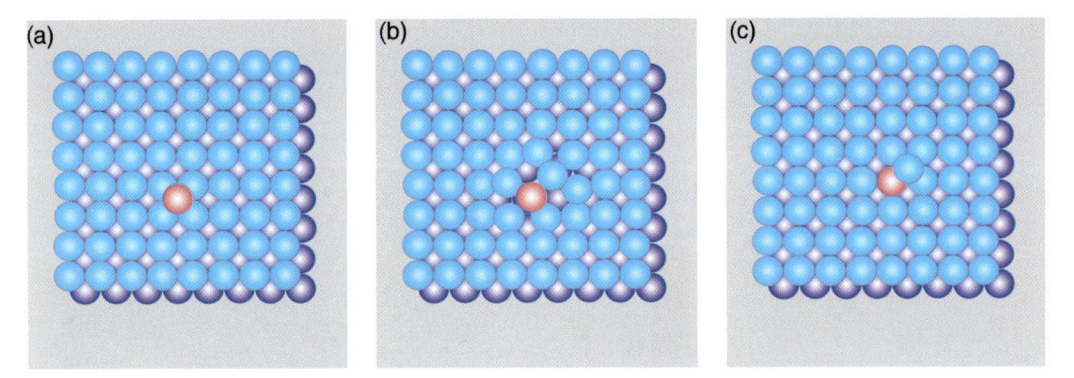

Fig. 3.53 Schematic showing exchange between adatom and substrate atom on (100) plane: (a) Adatom at equilibrium site. (b) Adatom and lattice atom in transition state. (c) Adatom incorporated into lattice; atom from lattice has turned into adatom.

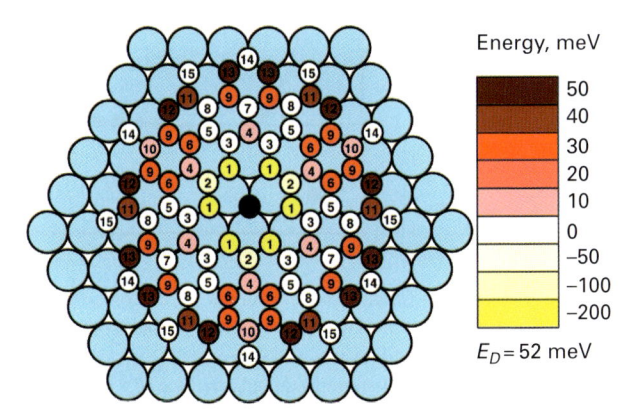

Fig. 10.18 Dependence of pair interaction energy on the separation from the central adatom for Ag on strained Ag(111) surface. Results obtained from density-functional theory calculations (after Fichthorn and Scheffler [39]).

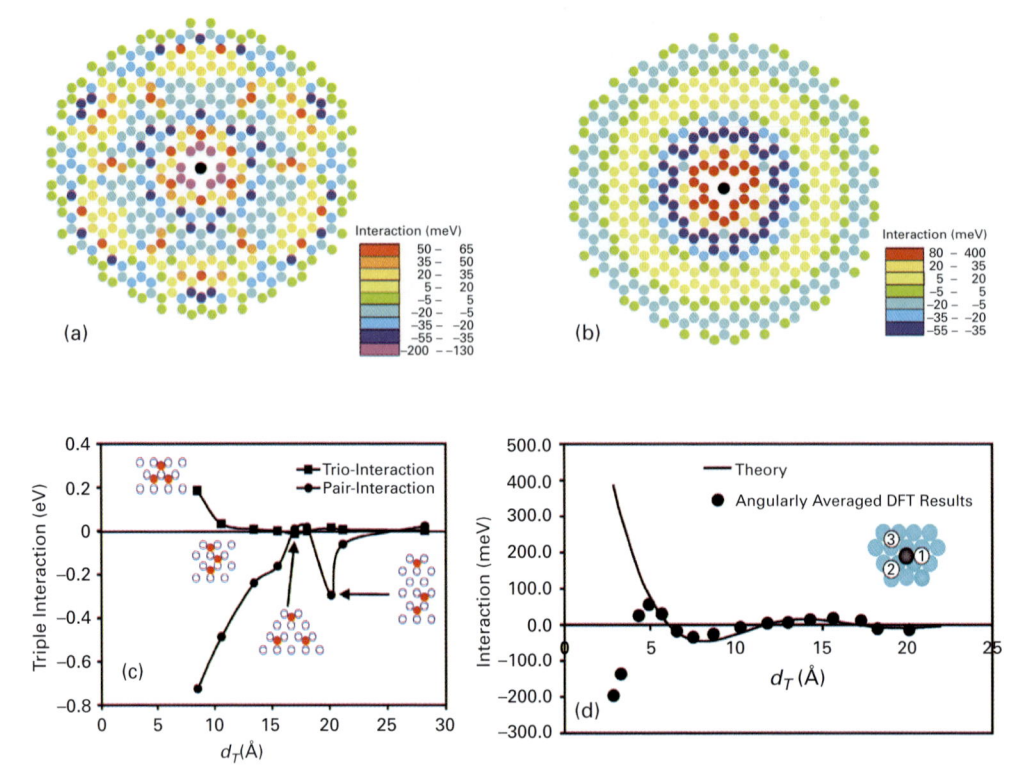

Fig. 10.28 Ag-Ag interactions on Ag(111) plane. (a) Interactions between a Ag adatom and a central Ag on a strained Ag(111) plane as a function of the interatomic separation. Results obtained by density-functional calculations. Lattice constant of silver was 4.16% compressed. (b) Energies from Eq. (10.22) found in four-layer slab in density-functional calculations ($k_F = 0.28$ Å^{-1}, $\varepsilon_F = 0.662 eV$, $\delta_F = 0.45\pi$) for central Ag adatom interacting with a second Ag in different positions. (c) Interactions of three adatoms on Ag(111) in their dependence on trio-perimeter d_T. Total pair-energy indicated by circles. The squares show trio effects. (d) Comparison of interactions from Eq. (10.22) with angularly averaged DFT interactions from (b) (after Luo and Fichthorn [61]).

5.8 Copper: Cu(111)

Agrawal et al. [26] in 2002 used Lennard-Jones potentials to calculate a barrier of 0.11 eV and a prefactor of 0.14×10^{-3} cm^2/sec for hopping. Just as for other systems, this estimate is much too high. In 2003 Wang et al. [151] looked at the influence of strain on the mechanism of movement on Cu(111), using MD simulations based on EAM potentials [152]. They worked at three temperatures, 650 K, 800 K, and 950 K, and demonstrated that the diffusivity decreased as a function of increasing uniaxial and biaxial stress, as shown in Fig. 3.74. On analyzing a trajectory they found both hopping as well as atom exchange – it turned out that stress can change the mechanism of movement.

In the same year, Huang et al. [153] used EAM interactions to establish the potential barrier to movement from fcc to hcp sites as 0.04 eV, in good agreement with the experiments. This work was continued in 2004 [122] and for jumps in the opposite direction, from hcp to fcc, a barrier of 0.03 eV was derived. For direct movement from fcc to fcc sites without passing over an intermediate hcp site, the barrier was huge, 0.37 eV. Based on those findings they concluded that the movement over the surface proceeded by a sequence of fcc-hcp-fcc jumps.

Marinica et al. [154] made use of the EAM potential proposed by Mishin et al. [152] to find an energy barrier of 0.041 eV for copper atoms diffusing from fcc to hcp sites. The change in the potential energy during atom migration was previously shown in Fig. 3.26. Two things are of interest. The diffusion barrier is in good agreement with the experiments of Repp et al. [143] and Knorr et al. [144]. Furthermore, the fcc sites are slightly more favorable, by 0.005 eV, than the hcp sites. Also with EAM potentials, Ferrón et al. [155] did molecular dynamics simulations on a 14 layer slab with 270 atoms in each layer at different temperatures, from 100 K to roughly 800 K. For coordinated jumps, including long-distance transitions and rebounds, they found a good Arrhenius plot, shown in Fig. 5.33, giving a barrier of 0.036 ± 0.002 eV and $D_o = 1.73 \times 10^{-4}$ cm^2/sec. From

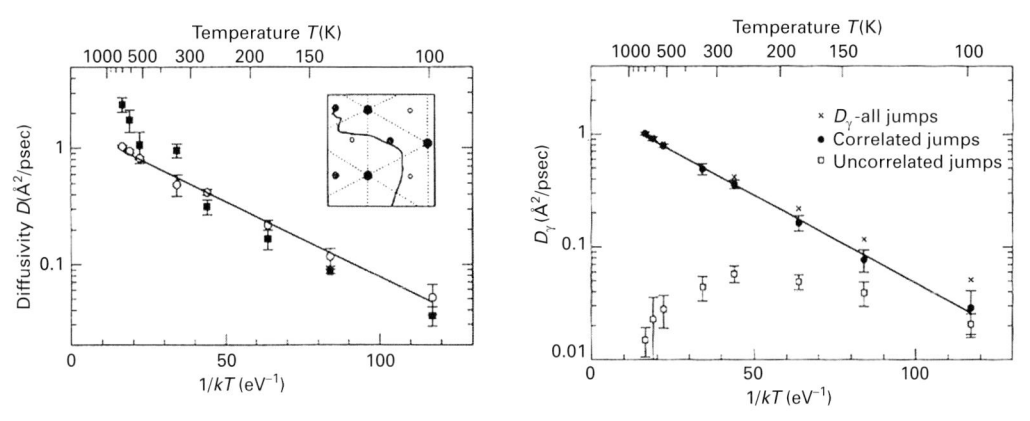

Fig. 5.33 Self-diffusion of Cu atoms on Cu(111) evaluated with EAM potentials. (a) Solid squares: from displacement and time. Open circles: from frequency of hopping. Solid line gives barrier of 0.030 ± 0.001 eV and $D_o = 1.56 \times 10^{-4}$ cm^2/sec. Inset shows surface cells and atom trajectory. (b) Solid circles: correlated jumps. Open squares: uncorrelated jumps. Solid line is Arrhenius plot for the former, yielding a barrier of 0.036 ± 0.002 eV and $D_o = 1.73 \times 10^{-4}$ cm^2/sec (after Ferrón et al. [155]).

observations of the mean-square displacement as a function of the temperature they derived an activation energy of 0.030 ± 0.001 eV and $D_o = 1.56 \times 10^{-4}$ cm^2/sec.

Bulou and Massobrio [68] calculated a self-diffusion barrier of 0.043 eV using a second-moment approximation to tight-binding. They also evaluated a barrier of 1.455 eV for self-diffusion by atom exchange. A kinetic Monte Carlo method and effective medium energetics were used by Jahma et al. [124] to evaluate a self-diffusion barrier of 0.026 eV, which of course is much too small. Durukanoglu et al. [117] used the nudged elastic band method and relied on EAM potentials to evaluate the diffusion barriers of single atoms on a flat Cu(111) surface using the local-density approximation. They calculated much too small a value of 0.01 eV for hopping; for exchange on this plane they found a much higher value of 1.42 eV. This work has been challenged by Kong and Lewis [69,126], who relied on the same EAM potentials, but calculated the diffusion parameters by summing over all atoms of the crystal. At an fcc site of the (111) plane they arrived at a diffusion barrier of 0.030 eV, and a prefactor of 1.84×10^{-4} cm^2/sec; at an hcp site the barrier was 0.027 eV, and the prefactor 2.06×10^{-4} cm^2/sec. They concluded that the local-density approximation was not generally valid for diffusion calculations.

Kim et al. [70] have recently given barriers to atom jumps of 0.043 eV, and 1.352 eV for exchange, derived in molecular dynamics simulations with RGL interactions, in very reasonable agreement with experiments. Yang et al. [156] studied diffusion of clusters on a Cu(111) surface with MD and EAM potentials and for single atom movement derived a value of 0.06 eV and a prefactor of 3.01×10^{-4} cm^2/sec.

Since 1996 the theoretical estimates are in general agreement with the experimental results, and it appears that for self-diffusion on Cu(111) the data are in excellent shape. It also seems that self-diffusion on Cu(111) is quite insensitive to long-range interactions – even calculations with Morse and Lennard-Jones potentials give reasonable results. Atom motion clearly proceeds by hopping; exchange involves huge barriers, and seems unlikely on close-packed surfaces. Only on stressed surfaces is there a chance for exchange. What is not quite clear, however, is how the two kinds of adsorption sites, fcc and hcp, influence the diffusion of adatoms.

5.8.2 Hetero-diffusion

It should also be noted that there has been some work on hetero-diffusion. Van Siclen [130] performed calculations of the activation energy for hopping of indium on Cu(111) using EAM potentials and found a low value of 0.02 eV for the barrier. Exchange was unlikely, requiring 1.26 eV. The difference between binding at fcc and hcp sites was only 0.004 eV. In 1996 Breeman et al. [157] used embedded atom calculations to do Monte Carlo simulations for the same system and arrived at a higher activation energy of 0.08 eV.

RGL potentials were used by Evangelakis et al. [138] to calculate diffusivities of gold atoms on Cu(111), with atoms hopping between fcc and hcp sites. From the Arrhenius plot shown in Fig. 5.34 they arrived at a barrier of 0.08 ± 0.01 eV and a prefactor of 3×10^{-4} cm^2/sec for temperatures below 700 K. They also calculated the barrier for atom hopping by energy minimization and found a slightly smaller value of 0.061 eV, which they ascribed to including all types of jumps in the migration energy.

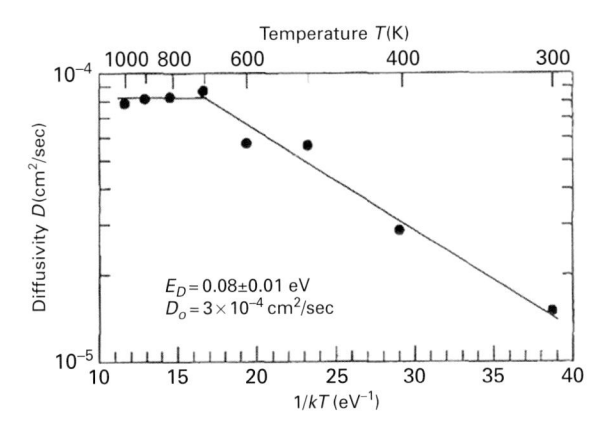

Fig. 5.34 Arrhenius plot for the diffusivity of gold on Cu(111), obtained using RGL potentials. Prefactor is 3×10^{-4} cm^2/sec (after Evangelakis et al. [138]).

In 1997 Padilla-Campos and Toro-Labbé [158] looked at the diffusion of the alkali metals Li, Na, and K on Cu(111) using density-functional methods and the local-density approximation. Their data are highly doubtful, as they had only three layers in their simulation. They were able to distinguish between fcc and hcp sites. However their findings are a bit puzzling. Movement of Li from fcc to hcp involved a barrier of 0.20 eV with a prefactor of 6.49×10^{-5} cm^2/sec, from an hcp to a fcc site 0.054 eV and a prefactor of 4.54×10^{-5} cm^2/sec. For Na, movement between fcc to hcp required overcoming a barrier of 0.30 eV with a prefactor 5.32×10^{-5} cm^2/sec, and for an hcp to an fcc transition the barrier was 0.13 eV and the prefactor 5.21×10^{-5} cm^2/sec. For potassium the energy barrier from fcc to hcp was 0.20 eV and the prefactor 2.02×10^{-5} cm^2/sec; the opposite way cost 0.05 eV, with a prefactor 1.83×10^{-5} cm^2/sec.

Kim et al. [70] relied on molecular dynamics simulations and RGL interactions to estimate the following barriers for atom hops on Cu(111): Ni, 0.045 eV, Pd 0.041 eV, Ag 0.034 eV, Pt 0.009 eV, and Au 0.039 eV. The last barrier is much lower than the value obtained by Evangelakis et al. [138].

The conditions of lead and copper alloying on the Cu(111) surface were studied by Anderson et al. [159] using the STM. To understand their measurement they calculated the diffusion barriers with VASP using the NEB method with the GGA approximation. For movement of lead atoms by hopping they estimated a barrier of ~0.03 eV while for exchange they calculated a much higher value of 1.17 eV. They concluded that intermixing in this system proceeded by vacancy mediated exchange.

Hetero-diffusion on this plane definitely needs much more attention, both in experiments as well as theory, as there are no reports of experimental studies.

5.9 Molybdenum: Mo(110), (111), (100)

Surprisingly there is only little information available, either experimental or theoretical, about self-diffusion on the densely packed (110) surface of molybdenum. This despite

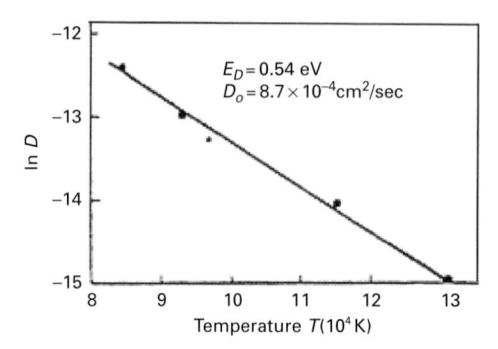

Fig. 5.35 Diffusivity of Cu on Mo film as a function of reciprocal temperature, as measured by secondary ion emission; based on the work of Abramenkov et al. [161].

the fact that molybdenum is a neighbor of tungsten, for which much has been done. However, there are some data that have been obtained for hetero-diffusion. Studies were done by Dhanak and Bassett [160] who examined rhodium diffusion with field ion microscopy and found an activation energy of 0.62 ± 0.06 eV, with a prefactor of 1×10^{-3} cm^2/sec. This is a direct measurement of the movement of one atom deposited on the plane, which makes it quite reliable.

Abramenkov et al. [161] looked at the spreading of copper atoms on a molybdenum ribbon of unspecified surface orientation, and under quite poor vacuum conditions ($\sim 10^{-6}$ Torr). From measurements at four temperatures they obtained an Arrhenius plot, shown in Fig. 5.35, which yielded an activation energy for diffusion of 0.54 eV and a prefactor $D_o = 8.7 \times 10^{-4}$ cm^2/sec, but it is clear that this result cannot be compared with single atom data.

Saadat [162] a few years later did a field ion microscopic study of the motion of gallium, indium, and tin on Mo(110). There are only a few notes about the experimental conditions in this paper, but vacuum conditions seem to have been excellent. The field ion images look very good, but are a bit confusing, as the voltage on the tip is claimed to be very low (3.2 kV), but the photograph is of quite a large surface with an appreciable radius. Measurements of the movement of single atoms deposited at the surface at five temperatures gave an Arrhenius plot, shown in Fig. 5.36, with an activation energy of 0.25 eV for gallium with a prefactor $D_o = 2 \times 10^{-5}$ cm^2/sec, 0.30 eV for indium with $D_o = 3 \times 10^{-5}$ cm^2/sec, and 0.50 eV for diffusion of tin atoms with $D_o = 4 \times 10^{-5}$ cm^2/sec.

Jubert et al. [163] studied the nucleation of iron on Mo(110), using the scanning tunneling microscope to measure the island density formed at different temperatures at a coverage of 0.12 ML. They arrived at a diffusion barrier of 0.1 ± 0.05 eV and a frequency factor of $v_0 = 8 \times 10^{11 \pm 1}$ sec^{-1}. By comparison with films formed in molecular beam epitaxy they concluded that in pulsed laser deposition the diffusivity was four orders of magnitude higher, probably due to the higher energy of the impinging species.

Movement of barium atoms on Mo(110) was investigated by Vedula et al. [164,165] who measured changes in the work function during spreading. For the lowest coverage

5.9 Molybdenum: Mo(110), (111), (100)

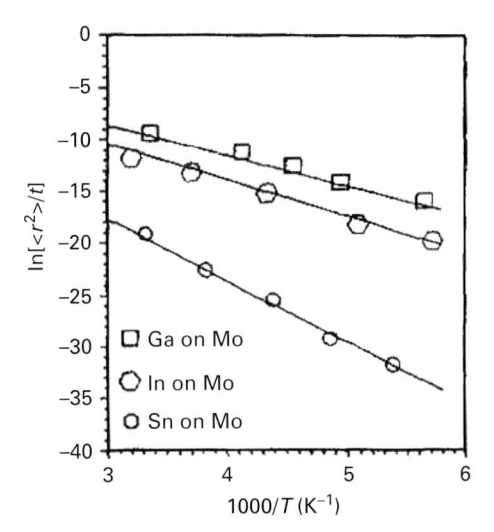

Fig. 5.36 Arrhenius plot for the diffusivity of Ga, In, and Sn on Mo(110) plane, as measured by field ion microscopy. Diffusion barriers for Ga 0.25 eV, In 0.30 eV, and Sn 0.50 eV (after Sadaat [162]).

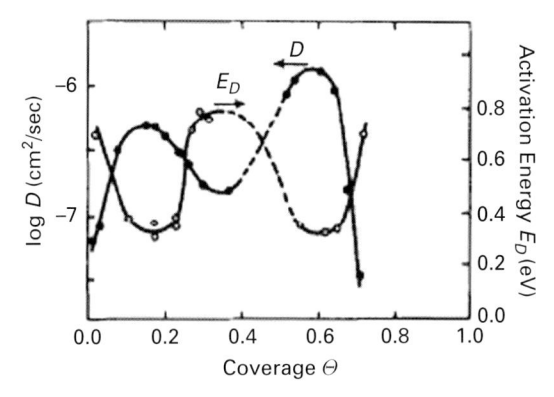

Fig. 5.37 Diffusion characteristics for barium on Mo(110), determined from work function changes. Activation energy and diffusivity are given as a function of the coverage (after Naumovets et al. [165]).

of 10^{13} atoms/cm^2 the value of the activation energy for diffusion amounted to 0.7 eV as indicated in Fig. 5.37 and the prefactor was 20 cm^2/sec. These values probably do not reflect single atom movement. Not much else is known about diffusion on this plane, and more experimental work is clearly called for.

The only two calculations for molybdenum surfaces were done quite early in the game, using Morse potentials without many-body interactions [47]. Flahive and Graham calculated a barrier of 0.4 eV for self-diffusion on Mo(110), a huge value of 1.9 eV for Mo(111) and for Mo(100) 1.4 eV, as shown in Fig. 4.9. Davydov, using the cohesion energy [49], estimated a diffusion barrier of 1.20 eV on Mo(110), a rather high value.

5.10 Ruthenium: Ru(0001)

In 1996, Hamilton [166], in his study of the dislocation mechanism of cluster diffusion, briefly mentioned an EAM value of 0.057 eV for movement of a single silver adatom on the ruthenium (0001) surface. The activation energy was in agreement with the bridge site energy and the prefactor was in the standard range; the mechanism of movement was not investigated.

5.11 Rhodium: Rh(111), (100)

5.11.1 Self-diffusion

For growth on the face-centered cubic (111) plane, atoms must be added to the fcc sites, but also available are very similar hcp binding sites. However, on rhodium, no attempts have been made to determine which of these is filled by adatoms. The (111) surface is very smooth; rapid diffusion can be expected, and that was shown to be the case by the work of Ayrault [167] in Fig. 4.19, who from studies in 1974 by FIM at seven temperatures determined an activation energy of 0.16 ± 0.02 eV and a prefactor of 2×10^{-4} cm^2/sec on (111). The study was done with one adatom deposited on the surface and looked directly at the displacements, making the work quite reliable.

In 1996, Tsui et al. [168] looked at the density of rhodium islands formed on Rh(111) at different temperatures and at a coverage of 0.1 ML. They used the STM and nucleation theory to analyze the data, assuming a critical size equal to one. From these measurements, shown in Fig. 5.38, they arrived at an activation energy of 0.18 ± 0.06 eV for diffusion. This is in quite good agreement with the work of Ayrault [167], and suggests that in this instance at least classical nucleation theory is valid.

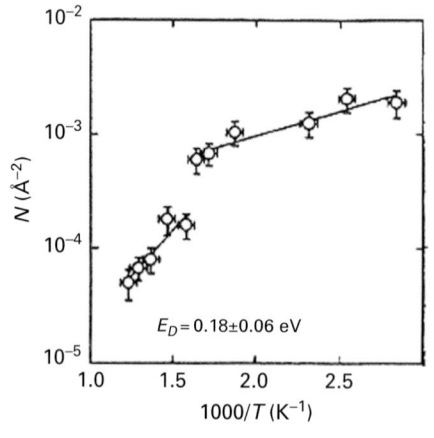

Fig. 5.38 Diffusion of Rh on Rh(111). Saturation density of islands as a function of reciprocal temperature, leading to a barrier of 0.18 ± 0.06 eV (after Tsui et al. [168]).

5.11 Rhodium: Rh(111), (100)

The structure of the (100) surface is quite different from (111), with deep recesses which can hold an adatom. On this surface an activation energy of 0.87 ± 0.07 eV, and a prefactor of 1×10^{-3} cm^2/sec was reported by Ayrault [167] and the Arrhenius plot is shown in Fig. 4.19. These are quite early direct FIM studies, with only one adatom on the surface, but the results still seem to be valid.

Somewhat more work has been done on the Rh(100) plane. Self-diffusion of Rh has been looked at briefly by Kellogg [169], also by FIM. From the mean-square displacement at 300 K, he derived a barrier of 0.83 ± 0.05 eV by assuming a standard value for the prefactor; this is reasonably close to the more detailed measurements by Ayrault [167]. Atoms moved by hopping on this surface; no exchange was observed. In 1994 Kellogg again looked at single atom motion and reported a barrier of 0.84 eV, assuming a standard prefactor. His measurements were done at one temperature, 296 K, and consisted of 70 observations [170]. The agreement in these studies is remarkable.

Theoretical estimates of diffusion on rhodium surfaces began quite early. Using Morse potentials, Flahive and Graham [47] calculated a barrier of 0.05 eV for Rh(111) and 0.7 eV for Rh(100), as shown in Fig. 4.9. In 1983, McDowell and Doll [171,172] used a Lennard-Jones potential to do molecular dynamics simulations for diffusion. On Rh(100) they found a barrier of 0.87 ± 0.16 eV with a prefactor $D_o = 4.06 \times (e^{\pm 0.85}) \times 10^{-3}$ cm^2/sec. For Rh(111) they reported an activation energy of 0.27 ± 0.026 eV and a prefactor of $7.10 \times (e^{\pm 0.25}) \times 10^{-4}$ cm^2/sec. Later, Voter and Doll [173] compared various evaluations of the activation energy for diffusion on Rh(111) and Rh(100) using Lennard-Jones potentials. With Monte Carlo simulation and transition-state theory (MCTST) they found a barrier of 0.224 ± 0.003 eV on Rh(111), with a prefactor of $5.4(\times 1.37^{\pm 1}) \times 10^{-4}$ cm^2/sec for a fit to low temperatures only. When they analyzed a wider range of temperatures they obtained a barrier of 0.20 ± 0.01 eV with a prefactor $2.91(\times 1.16^{\pm 1}) \times 10^{-4}$ cm^2/sec for the (111) plane [174]. In molecular dynamics simulations the results were 0.29 ± 0.03 eV for the diffusion barrier on (111) and $7.1(\times 1.28^{\pm 1}) \times 10^{-4}$ cm^2/sec for the prefactor. On Rh(100), these quantities were 1.03 ± 0.01 eV and $6.1(1.5^{\pm 1}) \times 10^{-3}$ cm^2/sec for low temperatures [171]; for a wider temperature range they got 0.92 ± 0.12 eV for the barrier with a prefactor of $3.7(\times 1.78^{\pm 1}) \times 10^{-3}$ cm^2/sec. A comparison of the data with the experiments of Ayrault [167] for low temperatures is shown in Fig 5.39; estimates over a wider range of data are given in Fig. 5.40. Voter also did a simple TST calculation, but the data are very similar to the MC investigations. In 1986, Voter [175] again used Lennard-Jones potentials to model cluster self-diffusion on Rh(100). Resorting to transition-state theory he came up with an activation energy of 1.094 eV for single Rh atoms.

In 1991 Lynden-Bell [176] looked at self-diffusion on Rh(100) relying on molecular dynamics simulations using Sutton–Chen potentials. For hopping she arrived at an activation energy of 0.388 ± 0.002 eV, far from the experimental values, and a prefactor $5.6 \times 10^{-3}(\times 2^{\pm 1})$ cm^2/sec. Sanders and DePristo [76,177], in one of the first large applications of a many-body potential, used corrected effective medium theory to arrive at a self-diffusion barrier of 0.80 eV with a jump prefactor of 4.4×10^{12} sec^{-1} and $D_o = 3 \times 10^{-3}$ cm^2/sec on Rh(100); on Rh(111) the barrier proved to be 0.035 eV with a prefactor of 6.2×10^{-4} cm^2/sec. Using molecular dynamics they found 0.092 ± 0.033 eV for the barrier and a prefactor $4.6 \pm 2.3 \times 10^{-4}$ cm^2/sec on Rh(111). They also gave estimates

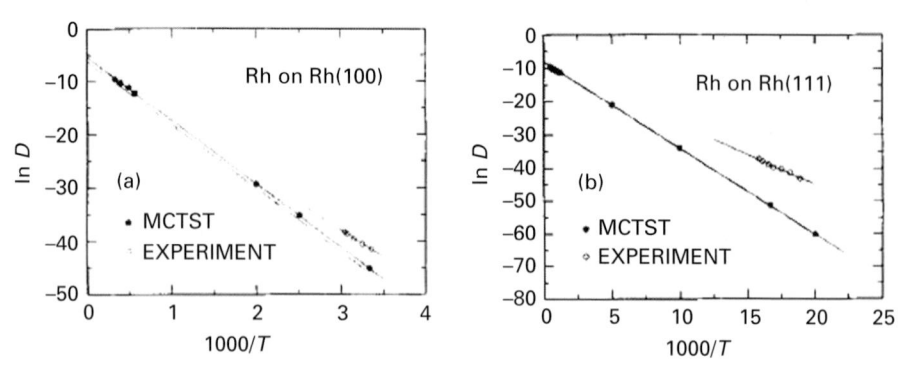

Fig. 5.39 Arrhenius plot for the diffusivity of Rh on Rh surfaces, calculated with Lennard-Jones potentials and compared with the experiments of Ayrault [167] (after Voter and Doll [173]).

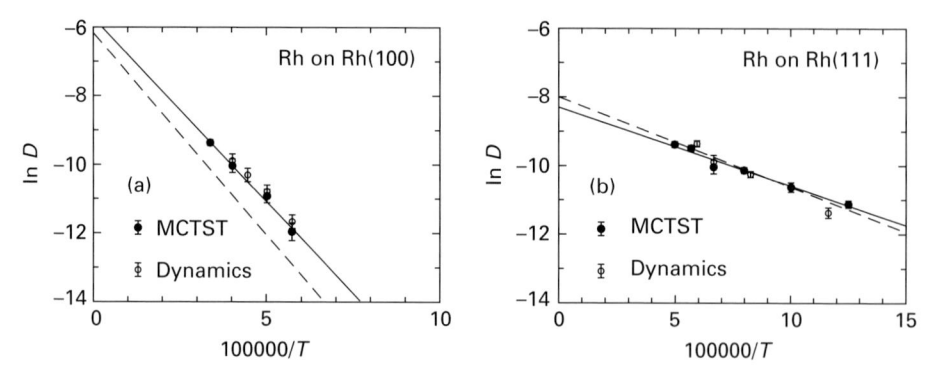

Fig. 5.40 Diffusion of Rh on Rh planes in its dependence upon reciprocal temperature, as obtained from L-J calculations. (a) On Rh(100), diffusion barrier of 1.03 ± 0.01 eV was derived from MCTST and data for dynamics [172]. (b) On Rh(111), a barrier of 0.29 ± 0.03 eV was obtained (after Voter and Doll [174]).

with Morse and Lennard-Jones potentials on this plane by doing molecular dynamics. For the first of the two potentials they gave for the barrier 0.118 ± 0.043 eV and $D_o = (7.1 \pm 4.5) \times 10^{-4}$ cm^2/sec; with Lennard-Jones potentials they found a barrier of 0.234 eV with $D_o = 4.5 \times 10^{-4}$ cm^2/sec, previously also calculated by Mruzik and Pound [178].

Using molecular dynamics simulations with the Sutton–Chen potential, Shiang [60,102] achieved a barrier of 0.832 eV on Rh(100) with a prefactor of 1.36×10^{-3} cm^2/sec. With Morse potentials he found 0.799 eV and 2.13×10^{-3} cm^2/sec. On Rh(111) these quantities proved to be 0.106 eV for the barrier and 6.09×10^{-4} cm^2/sec for the prefactor. Using Morse potentials the barrier obtained was 0.089 eV and the prefactor 6.40×10^{-4} cm^2/sec, as shown in Fig. 5.41. It appears there is no big difference between barriers calculated with many-body and pair potentials.

Kumar et al. [108] used a smart Monte Carlo technique with Lennard-Jones potentials to arrive at a barrier of 0.27 ± 0.01 eV and a prefactor of $1.1 \pm 0.2 \times 10^{-5}$ cm^2/sec for rhodium on Rh(111). The barrier was quite large compared with experimental findings, and no information about movement between fcc and hcp sites was derived. Feibelman

5.11 Rhodium: Rh(111), (100)

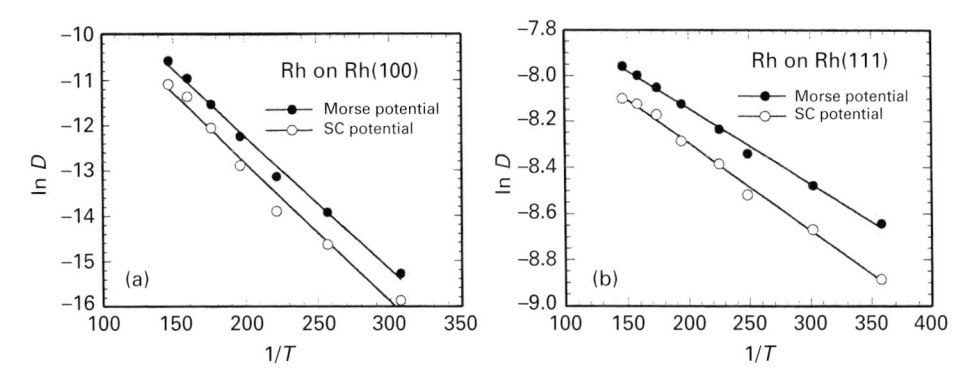

Fig. 5.41 Diffusion of Rh adatoms on Rh planes, with the temperature dependence calculated by the Sutton–Chen as well as with the Morse potential. (a) On Rh(100), barrier is 0.832 eV from SC and 0.799 eV from Morse. (b) On Rh(111) the barrier was lower, 0.106 eV from SC and 0.089 eV from Morse calculations (after Shiang [60,102]).

and Stumpf [179], working with density-functional methods on Rh(100), found a self-diffusion barrier of 0.89 eV for atom hopping and > 1.11 eV for atom exchange with the substrate. This barrier is in excellent agreement with the experiments of Ayrault [167]. For Rh(111), Davydov [49] estimated a barrier of 0.26 eV, and 0.44 eV for the (100) plane, in both cases using his cohesion approximation. It is not unusual for this approach to describe the nature of adatom movements incorrectly. Máca et al. [180,181] calculated diffusion energies in various environments and arrived at an energy barrier of 0.15 eV for self-diffusion on Rh(111) using RGL potentials. Agrawal et al. [26] with Lennard-Jones potentials found a diffusion barrier of 1.02 eV and a prefactor $D_o = 3.0 \times 10^{-3}$ cm^2/sec on Rh(100), and 0.21 eV and $D_o = 1.9 \times 10^{-4}$ cm^2/sec for Rh(111). Chang and Wei [28] in 2005 obtained a barrier of 0.99 eV for atom hopping on Rh(100) using ab initio density-functional methods. For atom exchange the barrier was considerably higher, 1.20 eV.

The theoretical estimates for diffusion on Rh(111) have a significant spread, making a choice difficult, but for the (100) plane they are quite close together and in reasonable accord. All investigations seem to agree that the mechanism of movement on the Rh(100) surface is atom hopping, not exchange. On Rh(111), however, not much attention has been paid to the two kinds of adsorption sites, fcc and hcp. The experimental findings for this material seem to be in quite good shape, even if made relatively early, partly because direct measurements are available.

5.11.2 Hetero-diffusion

Kellogg [182,183] in 1993 briefly looked at platinum on Rh(111) in the field ion microscope, and found Pt to be mobile at 77 K, the temperature at which he did imaging. From this he gave an upper limit of 0.22 eV for the barrier, assuming a prefactor $D_o = 1 \times 10^{-3}$ cm^2/sec. FIM measurements at an imaging temperature close to 20 K will be desirable to get more precise estimates. Kellogg [182] also looked more intensively at the diffusion of platinum atoms on Rh(100). Although based on only four temperatures,

Table 5.2 Diffusion on Rh(100) [89].

	Frequency Prefactor ($\times 10^{12}$ sec^{-1})		Diffusion Barrier (eV)	
Adatom	Bridge Hops	Exchange	Hopping	Exchange
Ni	5.82	6.50	0.75	1.21
Pd	4.07	8.64	0.73	1.46
Cu	2.46	3.62	0.53	1.44
Ag	3.31	4.14	0.51	1.91

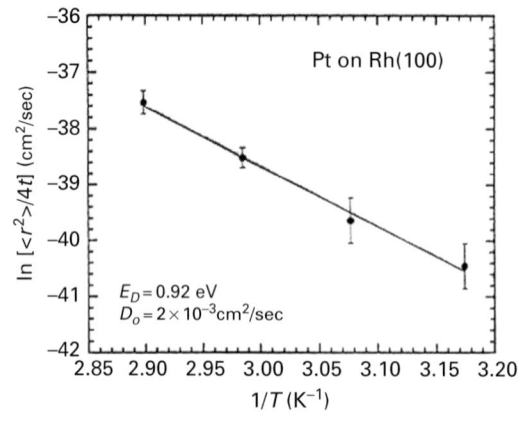

Fig. 5.42 Arrhenius plot for the diffusivity of Pt on Rh(100), obtained by field ion microscopy (after Kellogg [183]).

the activation energy of 0.92 ± 0.13 eV and prefactor of $2.0 \times 10^{-3 \pm 1.9}$ cm^2/sec derived from the Arrhenius plot in Fig. 5.42 appear reasonable. The displacements of platinum during diffusion show a (1 × 1) map, suggesting that diffusion occurred by hopping. This is different from the behavior of iridium atoms, which diffuse on the Rh(100) surface by atom exchange [169].

In 1992, Sanders and DePristo [76] carried out transition-state theory calculations of hetero-diffusion on the Rh(100) plane, examining hopping of Ni, Cu, Pd, Ag, Pt, and Au with potentials from MD/MC-CEM. Activation energies were as follows: nickel 0.76 eV, copper 0.74 eV, palladium 0.66 eV, silver 0.50 eV, platinum 0.66 eV, and gold 0.55 eV. The frequency prefactors were all in the range 10^{12} sec^{-1}. The multiples of this factor were as follows: nickel 5.5; copper 5.4; palladium 4.1; silver 4.9; platinum 3.7; gold 2.7. The value obtained for platinum movement is lower than that experimentally derived by Kellogg [182]. Two years later, Perkins and DePristo [89], relying on the same approach, compared hopping with atom exchange. The results are listed in Table 5.2. These estimates were made with 226 movable atoms in the first three layers, 81 in the first and third, and 64 in the second, but despite that hopping appeared as the favored mechanism for diffusion. Exchange was experimentally found [169] for diffusion of iridium atoms, but Perkins and DePristo unfortunately did not do calculations for this system. As is clear from this survey the hetero-diffusion effort has been much too small.

5.12 Palladium: Pd(100)

5.12.1 Self-diffusion

The first measurements of self-diffusion on Pd(100) were made by Flynn-Sanders et al. [184], who in 1993 did LEED studies of homo-epitaxial growth. From the temperature for the onset of diffusion they estimated a barrier ranging from 0.52 eV to 0.61 eV, assuming a frequency prefactor $v_0 = 10^{13}$ sec^{-1}. A second analysis of the LEED data yielded 0.59 eV to 0.63 eV, assuming a frequency prefactor of 5×10^{12} sec^{-1} [185]. From the ring diameter in the nucleation and growth region together with Monte Carlo simulations a barrier between 0.51 eV and 0.62 eV was deduced, assuming a critical size equal to one. The data were reexamined by Evans and Bartelt [186], who arrived at a barrier of ≈ 0.65 eV assuming $v_0 = 5 \times 10^{12}$ sec^{-1}. The mechanism of movement was not derived in these studies.

Using Morse potentials, Flahive and Graham [47] calculated a barrier of 0.55 eV, as shown in Fig. 4.9. Calculations with many-atom potentials were started by Liu et al. [4], who relied on the embedded atom method and arrived at a barrier for atom hopping of 0.71 eV and a prefactor of 1.2×10^{-3} cm^2/sec in the AFW approximation; with VC potentials they reported a barrier of 0.74 eV and a prefactor $D_o = 6.0 \times 10^{-3}$ cm^2/sec. They also investigated the exchange mechanism on this plane and found slightly lower activation energies, 0.61 eV (AFW) and 0.59 eV with a prefactor of 3×10^{-2} cm^2/sec (VC).

Sanders and DePristo [76] in 1992 relied on MD/MC-CEM and transition-state theory to evaluate a barrier of 0.61 eV with a frequency prefactor of 4.2×10^{12} sec^{-1}, but considered only atom hopping. Using CEM estimates, Perkins and DePristo [77] found a static energy barrier of 0.64 eV, and 0.75 eV by molecular dynamics. The barrier for diffusion by adatom exchange proved considerably higher, 1.03 eV with CEM and 1.38 eV using MD/MC-CEM. It must be noted here, however, that in the first substrate layer only 13 atoms were movable, 9 in the first layer and 4 in the second, and this may have contributed to the high barriers evaluated for atom exchange. In 1995 Perkins and DePristo [29] continued work on the same system with a larger number of movable atoms and found that on increasing the number of movables from 13 (appropriate for the previous work) to 226, the activation energy for the exchange mechanism decreased from 1.38 eV to 0.96 eV with MD/MC-CEM potentials and from 1.03 eV to 0.70 eV using CEM. The hopping mechanism was still slightly favored energetically, but the difference was now much smaller; the size of the used cell is crucial in attaining a correct description of nature, as is apparent in Fig. 5.43.

Stoltze [61] calculated activation energies for hopping in self-diffusion from effective medium theory and derived a barrier of 0.503 eV. Boisvert et al. [78] also did embedded atom calculations and reported an activation energy of 0.51 eV, with a prefactor of 3.41×10^{13} sec^{-1} for atom exchange and 0.7 eV and a prefactor 5.51×10^{13} sec^{-1} for atomic jumps. This study suggests exchange as the leading mechanism. However, the surface layer of the substrate had only 72 atoms in it. Also with embedded atom potentials, Mehl et al. [82] found a barrier of 0.71 eV in self-diffusion by hopping; exchange was not considered. Feibelman and Stumpf [179] did density-functional

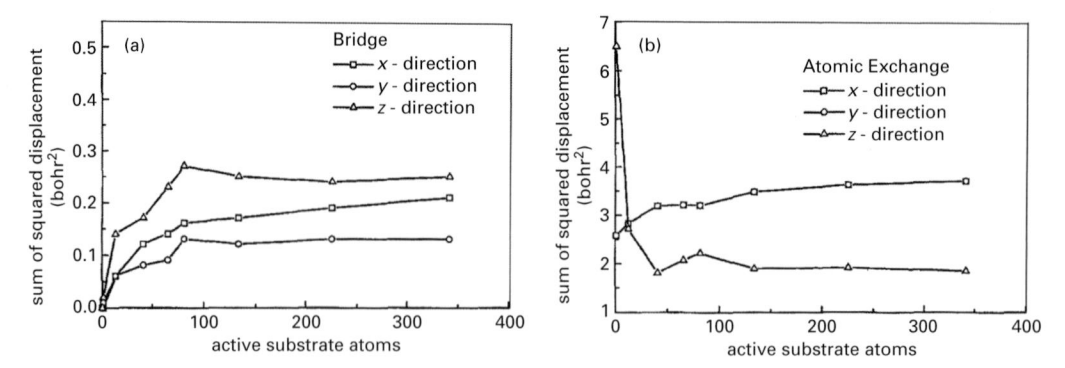

Fig. 5.43 Sum of squared displacements in self-diffusion as a function of the number of atoms active for Pd(100); calculations by corrected effective medium method. (a) For ordinary atom hopping. (b) For diffusion by atom exchange (after Perkins and DePristo [29]).

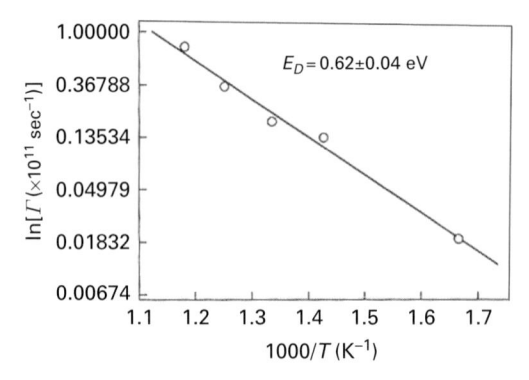

Fig. 5.44 Arrhenius plot for the self-diffusion of Pd on Pd(100), calculated by embedded atom method (after Evteev et al. [187]).

calculations for self-diffusion on Pd(100). They found a barrier of 0.71 eV for atom jumps, and 0.82 eV for diffusion via atom exchange. Using his cohesion approximation, Davydov [49] estimated an activation energy of only 0.15 eV, a value which appears much too low.

Agrawal et al. [26] worked with Lennard-Jones potentials and found a barrier for jumps of 0.60 eV with a prefactor of 1.1×10^{-3} cm^2/sec. Evteev et al. [187] relied on EAM potentials to do molecular dynamics simulations. From the temperature dependence of the number of elementary events they obtained the Arrhenius plot of Fig. 5.44 and derived an activation energy of 0.62 ± 0.04 eV. Exchange was not considered. For hopping, Chang and Wei [28] calculated an energy barrier of 0.87 eV using the VASP [22–24] simulation package. For atom exchange, the barrier proved to be comparable, 0.85 eV, so that it is difficult to figure out which mechanism is favorable, or whether both are present at the surface. Recently, Kim et al. [70] carried out molecular dynamics simulations using RGL potentials to find the barrier to jumps amounted to 0.621 eV, whereas exchange required 0.725 eV.

5.12 Palladium: Pd(100)

The results for diffusion on palladium, both from experiments and from calculations, are not in any sort of agreement. Self-diffusion barriers range from 0.5 eV to 0.75 eV for hopping and from 0.6 eV to 0.9 eV for atom exchange. At the moment it is difficult to say which mechanism is preferred in diffusion or if they coexist at the surface. Better experiments are clearly in order here.

5.12.2 Hetero-diffusion

In 1989, Evans et al. [188] looked at the growth of a Pt film on Pd(100) with LEED. From the temperature for the onset of diffusion, around 150 K, they deduced an activation energy around 0.43 eV, assuming a frequency prefactor of 10^{13} sec^{-1}. There were no attempts made to detect the mechanism of movement. Félix et al. [189] carried out helium scattering experiments on silver deposits on Pd(100). The intensity specularly reflected was measured to determine the particle size distribution, assumed to arise from the random motion of the silver atoms, as a function of time and temperature. From the Arrhenius plot of the jump rate, shown in Fig. 5.45, they derived an activation energy for silver diffusion of 0.37 ± 0.03 eV, with a somewhat small frequency prefactor of 8×10^9 sec^{-1}. They also calculated a static barrier with embedded atom theory and found a much higher value of 0.66 eV. A year earlier, the same authors [190] reported for the same system a barrier of 0.35 ± 0.03 eV and a frequency prefactor 2.7×10^9 sec^{-1}. Massobrio et al. [191] did molecular dynamics simulations using EAM potentials to explain the He scattering results for the Ag/Pd(100) system. For atomic movement they found a hopping barrier in the <110> direction of 0.6 eV and an exchange barrier in the <100> direction of 0.75 eV. Exchange was clearly not the favored mechanism.

Sanders and DePristo [76] in 1992 relied on MD/MC-CEM and transition-state theory to evaluate barriers for hetero-diffusion on Pd(100), with the following results: activation energy for nickel 0.71 eV, frequency prefactor 5.2×10^{12} sec^{-1}; copper 0.72 eV, 5.3×10^{12} sec^{-1}; rhodium 0.75 eV, 4.3×10^{12} sec^{-1}; silver 0.42 eV, 8.0×10^{12} sec^{-1}; platinum

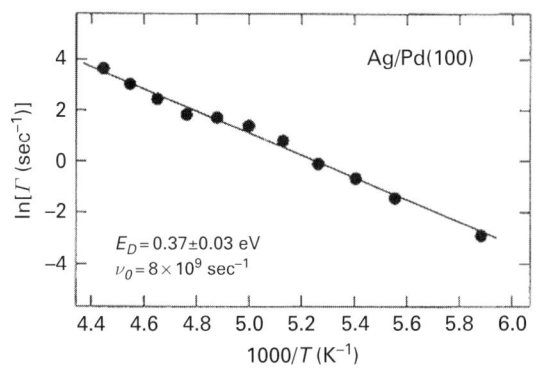

Fig. 5.45 Diffusivity of Ag adatom on Pd(100) in its dependence upon reciprocal temperature, determined by thermal atom scattering (after Félix et al. [189]).

Table 5.3 Diffusion on Pd(100) [89].

Adatom	Frequency Prefactor ($\times 10^{12}$ sec^{-1})		Diffusion Barrier (eV)	
	Bridge Hops	Exchange	Hopping	Exchange
Rh	7.85	13.76	0.75	0.70
Ni	7.01	9.74	0.70	0.78
Cu	2.63	3.98	0.52	0.98
Ag	4.96	6.38	0.50	1.34

0.64 eV, 4.0×10^{12} sec^{-1}; gold 0.54 eV, 2.8×10^{12} sec^{-1}. Perkins and DePristo [89] reconsidered diffusion on Pd(100) and now arrived at the activation energies listed in Table 5.3. In this table the frequency prefactors for the exchange process were roughly twice as high as for hopping, but still in the range 10^{12} sec^{-1}. Only for rhodium adatoms were conditions favorable for diffusion by atom exchange. The barrier for silver on Pd(100) appears to be a bit higher than the experimental findings by Félix et al. [189] but comparable with the results of Massobrio et al. [191]. It must be noted here, however, that in the first substrate layer of Perkins and DePristo [89] 226 atoms were movable.

Estimates for hetero-atom diffusion were made by Kim et al. [70] using molecular dynamics based on RGL potentials. The values were Ni 0.566 eV, Cu 0.558 eV, Ag 0.459 eV, Pt 0.986, and Au 0.590 eV. These barriers are not in very good agreement with the work of Perkins and DePristo [89].

5.13 Palladium: Pd(111)

5.13.1 Self-diffusion

Experimental studies of self-diffusion on Pd(111) were made in 2000. Steltenpohl and Memmel [192,193], used the STM to observe nucleation and growth for 0.08 of a monolayer of palladium deposited at 210 K and at 420 K. Their results for the island density at various temperatures, shown in Fig. 5.46, were interpreted in accord with classical nucleation theory. This led to a diffusion barrier of 0.35 ± 0.04 eV. More surprising is the jump prefactor, $v_0 = 6 \times 10^{16 \pm 0.6}$ sec^{-1}, four orders of magnitude above the normally expected value. The authors relate these results to the electronic structure of Pd(111), but these findings still need to be checked. Unfortunately, this study stands alone and no other direct measurements of adatom movement are available for palladium surfaces.

Theoretical investigations with Morse potentials were carried out by Flahive and Graham [47], who estimated a barrier of 0.05 eV, as shown in Fig. 4.9. Calculations with many-atom potentials started in 1991 by Liu et al. [4], who relied on the embedded atom method with AFW and VC potentials. The results obtained were 0.031 eV for the barrier and 5×10^{-4} cm^2/sec for the prefactor (AFW), and 0.059 eV together with 4.5×10^{-4} cm^2/sec (VC).

5.13 Palladium: Pd(111)

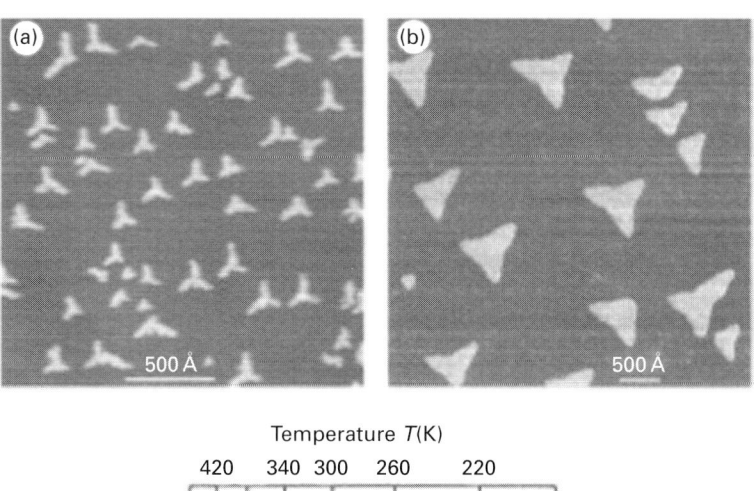

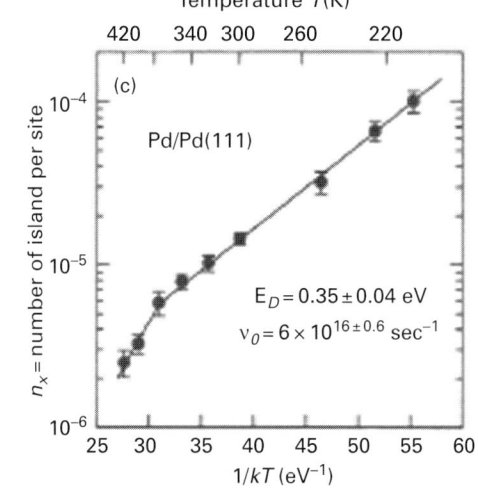

Fig. 5.46 Self-diffusion of Pd atoms on Pd(111) examined by STM. Images of 0.08 ML deposited at (a) 210 K and (b) 420 K. (c) Dependence of diffusivity upon reciprocal temperature (after Steltenpohl and Memmel [192,193]).

Stoltze [61] used effective medium theory to estimate a barrier of 0.104 eV. Mortensen et al. [63] used density-functional theory to arrive at a barrier of 0.21 eV with LDA and 0.17 eV relying on GGA. They also found that binding at an fcc site was 0.13 eV stronger than at an hcp. In 1996, Li and DePristo [62] with corrected effective medium theory found for hopping on this plane an activation energy of 0.034 eV. Davydov [49] estimated an activation energy of 0.10 with his cohesive approach. Chang et al. [35] did calculations using embedded atom interactions and came up with a barrier of 0.04 eV, a value similar to the previous findings.

Lennard-Jones potentials were used by Agrawal et al. [26] to estimate a diffusion energy of 0.14 eV and a prefactor 1.6×10^{-4} cm^2/sec. At much the same time Papanicolaou [194] carried out molecular dynamics simulations at temperatures from 150 K to 900 K for self-diffusion on Pd(111), relying on RGL potentials. The results are given in the Arrhenius plot of Fig. 5.47. Here the data at temperatures less than 500 K are of primary

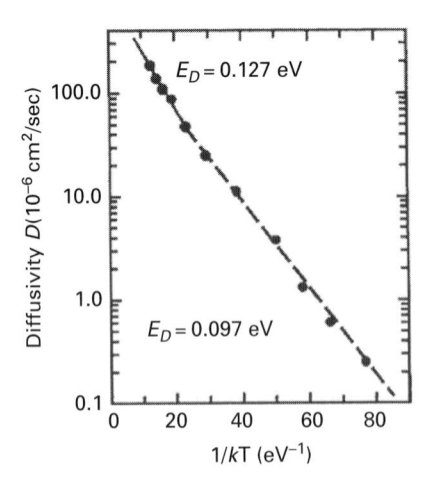

Fig. 5.47 Dependence of diffusivity of Pd on Pd(111) upon reciprocal temperature, calculated using RGL potentials. Data at ≤ 500 K refer to nearest-neighbor jumps (after Papanicolau et al. [194]).

interest, and refer to nearest-neighbor hops. These data lead to a diffusion barrier of 0.097 eV and a prefactor $D_o = 4.4 \times 10^{-4}$ cm^2/sec. This barrier is much lower than the value from experiments. At higher temperatures, 500–900 K, he found a barrier of 0.127 eV and a prefactor 9.1×10^{-4} cm^2/sec, which he attributed to long and correlated jumps. Recently, Kim et al. [70] carried out molecular dynamics simulations using RGL potentials to find that jumps of Pd atoms occurred over a barrier of 0.109 eV, exchange required an energy of 1.315 eV. As in all their other estimates, Kim et al. did not distinguish between fcc and hcp sites on the fcc(111) plane.

The state of the Pd(111) plane is not good. Theoretical estimates for the activation energy lie in the range from 0.03 eV to 0.21 eV, and it is difficult to conclude anything. No information about how the existence of two adsorption sites, fcc and hcp, influence movement is given in the the literature. It is clear that further work is needed to provide a reasonable perspective about diffusion on palladium.

5.13.2 Hetero-diffusion

Hunger and Haas [195] have examined the diffusion of ^{111}In on Pd(111) at different temperatures using perturbed angular correlation spectroscopy. Indium atoms were already mobile at 100 K. Assuming a frequency prefactor of 10^{12} sec^{-1}, they arrived at an activation energy of 0.23 ± 0.03 eV for the diffusion of ^{111}In at a concentration of 10^{-4} monolayers. Due to the very low coverage it is likely that this study detected single atom movement, but unfortunately the technique is not capable of defect control.

Diffusion of potassium on Pd(111) at a coverage of < 0.04 monolayer was studied using photoemission electron microscopy by Ondrejcek et al. [196]. They arrived at a value of < 0.130 eV, assuming a frequency factor of 1.7×10^{14} sec^{-1}. The diffusion of potassium in the limit of zero coverage has been examined on Pd(111) by Snábl et al. [197] in 1996, who used photoelectron emission microscopy to measure the

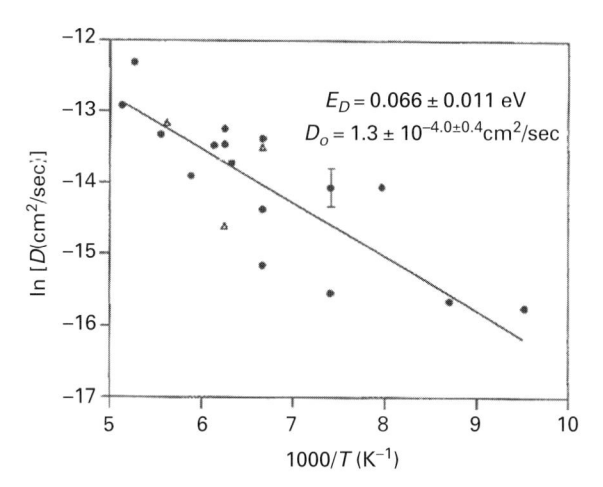

Fig. 5.48 Surface diffusivity of potassium atoms on Pd(111) as a function of reciprocal temperature, at a coverage <0.05 ML. Triangles: from concentration profiles. Closed circles: from contour plots (after Snábl et al. [197]).

potassium concentration. Their results for a coverage lower than 0.05 ML, in Fig. 5.48, lead to an activation energy for diffusion of 0.066 ± 0.011 eV, and a prefactor $D_o = 1.3 \times 10^{-4 \pm 0.4}$ cm^2/sec. The authors ascribe what they consider as a large uncertainty in the diffusion barrier to the presence of defects on the surface not detected by their instrument.

Estimates for jumps in hetero-atom diffusion were made by Kim et al. [70] with molecular dynamics based on RGL potentials. The barrier for Ni was 0.087 eV, for Cu 0.091 eV, for Ag 0.074 eV, for Pt 0.127 eV, and for Au 0.110 eV.

It appears, however, that not enough experimental work has been done to come to any firm conclusions about hetero-diffusion on palladium.

5.14 Silver: Ag(111)

5.14.1 Self-diffusion

Among the first attempts to study self-diffusion on silver was the work of Stark [198], who measured the resistivity changes of thin, (111) oriented films of silver on which silver atoms were deposited. From measurements over a range of temperatures he found an activation energy of 0.58 eV for self-diffusion, a value much higher than what has been reported later and unlikely to represent single atom movement.

Jones et al. [199] studied the deposition of silver on a W(110) surface in 1990. In their experiments, two layers of silver were formed first in the (111) orientation; only then did nucleation and growth of silver islands occur. This was examined with UHV scanning electron microscopy and interpreted by nucleation theory using kinetic rate equations to yield an activation energy for surface diffusion of 0.15 ± 0.10 eV. In calculations with effective medium theory they found a barrier of 0.12 eV. Venables [200] also analyzed measurements by Spiller et al. [201] on nucleation on top of two silver layers deposited on a W(110) surface and derived a barrier of ~0.1 eV. No possible strain factor was taken into account in the above studies.

A few years later, in 1995, Luo et al. [202] carried out studies of silver on the (111) plane of silver, using SPA-LEED to measure the density of islands at different surface temperatures. From the average terrace size for 0.5 monolayers of silver and nucleation theory, with the critical nucleus size equal to one, they deduced a diffusion barrier of roughly 0.18 eV from their data, shown in Fig. 5.49; the coverage may, however, have been too big for single atom diffusion. Two years earlier, Henzler et al. [141] reported SPA-LEED measurements of the temperature dependence of the island density on Ag(111) at quite a high coverage of 0.5 ML, which yielded a barrier of 0.12 eV.

Rosenfeld et al. [203] also examined the epitaxy of silver on Ag(111), using thermal energy He atom scattering from 0.05 monolayers of silver. Their results, in Fig. 5.50,

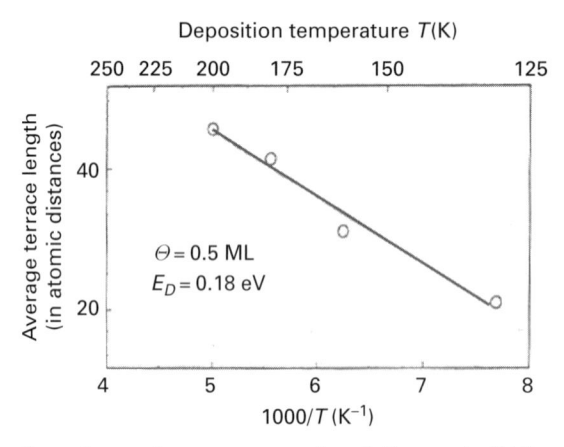

Fig. 5.49 Dependence of average terrace size of silver on Ag(111) upon reciprocal temperature (after Luo et al. [202]).

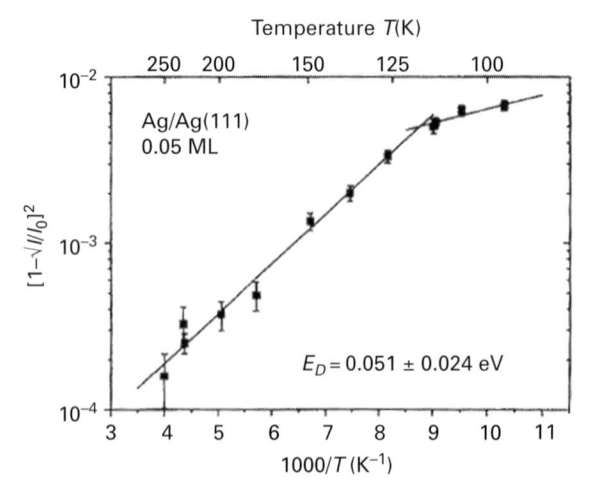

Fig. 5.50 Arrhenius plot of quantity proportional to density of Ag islands obtained from He intensity scattered from Ag(111) (after Rosenfeld et al. [203]).

5.14 Silver: Ag(111)

interpreted according to nucleation theory, led to a diffusion barrier of 0.051 ± 0.024 eV. At roughly the same time, Brune et al. [204] looked at this system with a STM. Relying on the nucleation theory expression for the island density, they arrived at a value of 0.097 ± 0.010 eV for the barrier to diffusion on Ag(111) and a frequency prefactor of $v_0 = 2 \times 10^{11 \pm 0.5}$ sec^{-1}. For silver deposition on one monolayer of silver on Pt(111), Brune et al. [204] made the same sort of measurements, and found a diffusion barrier of 0.060 ± 0.010 eV and a frequency prefactor of $v_0 = 1 \times 10^{9 \pm 0.6}$ sec^{-1} as shown in Fig. 3.69. They attributed the lower value of the activation energy to the effect of strain arising when silver was deposited on Pt(111), but did not comment about the frequency factor, which was at least two orders of magnitude too low. Brune et al. [204], in explaining findings from saturation island density measurements, ran calculations with EMT interactions and arrived at a lower value of 0.067 eV for silver atoms on silver (111) (compared with the 0.097 eV they derived from experiments). For silver on one monolayer of silver on Pt(111) their experiments indicated a barrier of 0.060 ± 0.010 eV, whereas the calculations yielded a value of 0.050 eV.

In 2000, Schlösser et al. [142] examined island diffusion with a variable temperature STM, and to explain the experimental finding employed the ARTwork computer program of Stoltze [61] using EMT potentials. For movement of a single Ag atom on the Ag(111) surface their calculations led to a barrier of 0.068 eV. Cox et al. [205] looked at the shape of silver islands on Ag(111) with the scanning tunneling microscope and modeled the data by Monte Carlo simulations relying on EAM interactions. In their model, the atom sits in an fcc site and moves between fcc sites, with an hcp site intervening. On a silver layer of 0.3 monolayers the activation energy for self-diffusion by hopping was 0.1 eV with a prefactor of 10^{11} sec^{-1}. This value is comparable with the findings of Brune et al. [204].

There is a substantial body of theoretical work dealing with the diffusion of silver atoms on silver surfaces started with the work of Flahive and Graham [47]. Using Morse potentials, they calculated a barrier of ~ 0.01 eV, shown in Fig. 4.9. Rilling et al. [206] in 1990 used EAM potentials to evaluate the energy changes occurring during atom jumps; they found a barrier height of 0.058 eV with a frequency prefactor of 1.4×10^{12} sec^{-1}. Further work with EAM potentials was done by Liu et al. [4]. The barrier for AFW calculations was 0.059 eV and a prefactor of 5×10^{-4} cm^2/sec, whereas using VC values the barrier was smaller, 0.044 eV, with a prefactor 4.1×10^{-4} cm^2/sec.

Sanders and DePristo [177] made estimates in 1992 using the MD/MC-CEM approximation and came up with a barrier of 0.023 eV and a prefactor of 6.2×10^{-4} cm^2/sec in transition-state theory; with molecular dynamics they found the activation energy for diffusion to be 0.055 ± 0.011 eV and a prefactor of $(6.9 \pm 2.2) \times 10^{-4}$ cm^2/sec. Using Morse potentials the results for the barrier were much smaller, 0.005 eV for transition-state theory, with a prefactor of 5.5×10^{-4} cm^2/sec, and 0.039 ± 0.009 eV with a prefactor of $4.7 \pm 1.4 \times 10^{-4}$ cm^2/sec from molecular dynamics.

Nelson et al. [207] carried out calculations using EAM, but with FBD set Universal 3 [57]. The barrier proved to be 0.06 eV. The calculations were done on a static lattice and considered only hopping. Stoltze [61] relied on EMT, gave an activation energy for self-diffusion of 0.064 eV. Full-potential linear-muffin-tin-orbital estimates were made by Boisvert et al. [208] and found an energy barrier of 0.14 ± 0.02 eV, higher than previously reported values. They claimed the difference between fcc and hcp sites was very small and insignificant, justifying why they assumed that both sites were equal. In the same year, Boisvert et al. [78] used EAM and molecular dynamics simulations to validate the Meyer–Neidel rule for jumps on the Ag(111) surface; they reported a much lower barrier of 0.055 eV and a prefactor of 2.29×10^{12} sec^{-1}, as shown in Fig. 5.51. Molecular dynamics simulations were made in 1995 using RGL potentials by Ferrando and Tréglia [209], who arrived at an activation energy of 0.068 eV and a prefactor of 2.6×10^{-4} cm^2/sec from the Arrhenius plot in Fig. 5.52; the difference in the potential energy at fcc and hcp sites was 0.009 eV.

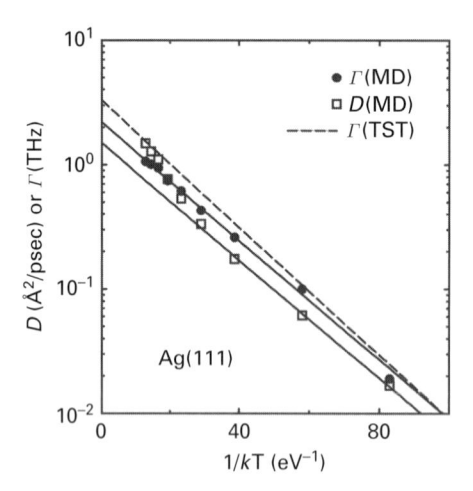

Fig. 5.51 Arrhenius plot for diffusivity of Ag adatom on Ag(111), obtained by molecular dynamics simulations using EAM potentials. Activation energy was 0.055 eV (after Boisvert et al. [359]).

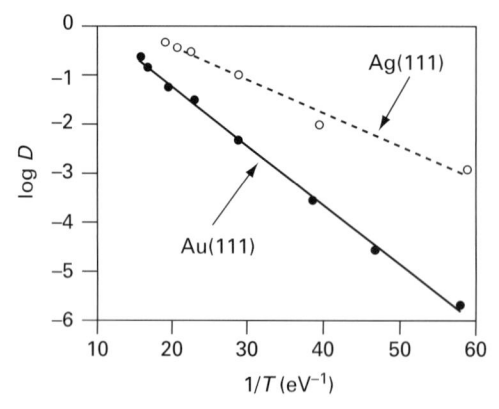

Fig. 5.52 Dependence of the diffusivity of Ag on Ag(111) and Au on Au(111), calculated with RGL interactions, upon reciprocal temperature. Diffusion barrier of 0.12 eV for Au and 0.068 eV for Ag (after Ferrando and Tréglia [209]).

5.14 Silver: Ag(111)

Li and DePristo [62] also made estimates using MD/MC-CEM theory, to obtain a low barrier of 0.02 eV. Calculations with density-functional theory were reported by Mortensen et al. [63] in 1996. They found a barrier of 0.14 eV with the local-density approximation and 0.10 eV using the generalized gradient method, but did not see any difference for fcc and hcp sites. In 1997 Ratsch et al. [210] used density-functional calculations in the local-density approximation to arrive at a barrier of 0.081 eV for jumps. For movement of Ag atoms on one ML of Ag on Pt(111) they found a barrier of 0.063 eV, in good agreement with the calculations of Brune et al. [204], although they worked with a very small (2 × 2) cell. A year later, Ratsch and Scheffler [211], using functional-density theory, reported a slightly higher barrier of 0.082 eV and a frequency prefactor of 0.82×10^{12} sec^{-1}. They also looked at silver moving on one monolayer of silver deposited on Pt(111) and found an activation energy of 0.06 eV and a frequency prefactor of 5.9×10^{12} sec^{-1}.

Papanicolaou et al. [212] used molecular dynamics simulations with a many-body potential from the second-moment approximation to the tight-binding model over a wide range of temperatures. They found that movement proceeds by adatom hopping on the surface, but there are two temperature regimes. Their results are presented in Fig. 5.53. At low temperatures of 200 K to 500 K, the activation energy for moving an adatom was 0.069 ± 0.001 eV; at higher temperatures of 500 K to 900 K the activation energy increased to 0.098 ± 0.006 eV. The static barrier for the lower temperature range was 0.063 eV. The increase in activation energy at higher temperatures was associated with longer and correlated jumps on the surface. However, the recent experimental study of Antczak [213] on W(110) showed an increasing number of long jumps with increasing temperature, but did not reveal an increase in the effective activation energy due to a correlation between the number of single and long jumps.

In 1999, with his cohesion approximation, Davydov [49] estimated a self-diffusion barrier of 0.04 eV. A year later, Chang et al. [35] also worked out the same barrier of 0.04 eV using embedded atom interactions. In 2000 Baletto et al. [214] looked at the movement of silver atoms on a silver cluster, a Wulff polyhedron with (111) and (100) planes; they relied on molecular dynamics simulations using RGL potentials. One

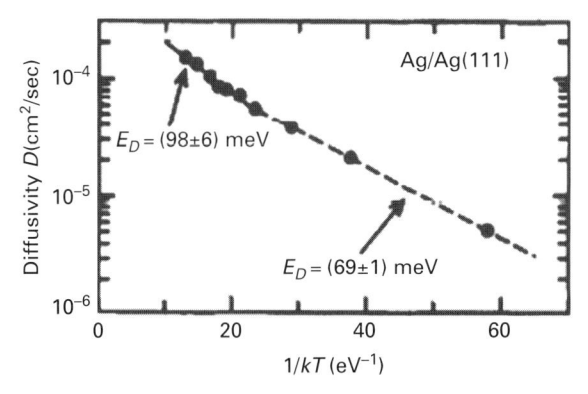

Fig. 5.53 Arrhenius plot for the diffusivity D of Ag atoms on Ag(111), derived by molecular dynamics simulations using RGL potentials (after Papanicolaou et al. [212]).

polyhedron for which diffusion was studied consisted of 201 atoms, with a (100) plane of 9 atoms. The activation energy for movement from fcc to hcp sites on the (111) plane of the Ag polyhedron was found as 0.07 eV.

Depending on Lennard-Jones potentials, Agrawal et al. [26] made estimates for a range of fcc materials. On Ag(111) the barrier was only 0.10 eV with a prefactor 1.4×10^{-4} cm^2/sec. Chvoj et al. [215] relied on EAM interactions to do Monte Carlo simulations to arrive at an activation energy of 0.061 eV for atom hopping. Kong and Lewis [69] in 2006 calculated the diffusivity prefactor D_o relying on EAM potentials and summing vibrations over all atoms in the system. Starting from an fcc site they obtained 2.78×10^{-4} cm^2/sec, from an hcp site the prefactor was 2.80×10^{-4} cm^2/sec, both independent of temperature from 300 to 600 K. Molecular dynamics simulations based on RGL interactions were carried out by Kim et al. [70] to obtain diffusion barriers on Ag(111). Atom jumps on this surface required an energy of 0.064 eV, whereas exchange occurred over a much higher barrier of 1.126 eV, results of the same sort as obtained in previous calculations.

The one thing clear about self-diffusion on the (111) plane of silver is that there is no close agreement between the different experimental results that have been obtained for this system. That is disturbing, since all the results relied on nucleation theory. However, a value around 0.1 eV seems to be the most probable. Unfortunately no direct measurements of hopping rates were done on this plane to clarify the situation. The theoretical results also scatter widely, from 0.14 eV to 0.02 eV, and appear to be lower than what seems to emerge from the experiments, 0.18 eV to 0.05 eV, after excluding the results of Stark. However the range of theoretical and experimental findings is similar.

5.14.2 Hetero-diffusion

The effect of antimony on the growth mode of silver on Ag(111) was examined in detail in 1994 by Vrijmoeth et al. [216], in a continuation of a study begun earlier. From the island density, which increased exponentially with the amount of antimony deposited, they concluded that antimony increases the activation energy of silver diffusion linearly. The increase in the diffusion barrier above that on the clean (111) surface is suggested in Fig. 5.54, but they did not derive values of the activation energy in this paper. However, they continued their study, and in 1998 [217] came out with an equation describing the change in the activation energy for diffusion as a function of antimony coverage Θ

$$E_D = E_{D0} + 1.7\Theta. \tag{5.1}$$

Movement of gold atoms on a Ag(111) surface was investigated by Haftel and Rosen [218] using molecular dynamics with EAM potentials. Hopping of Au on Ag(111) required an energy of 0.102 eV and showed a frequency prefactor of 1.3×10^{12} sec^{-1}, evaluated in the harmonic approximation to the attempt frequency

5.14 Silver: Ag(111)

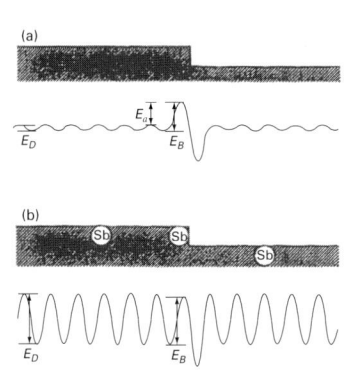

Fig. 5.54 Potentials affecting interlayer surface transport. (a) Step edge barrier E_B limits interlayer transport. (b) Higher diffusion potentials lower step-edge effects, stimulating layer-by-layer growth (after Vrijmoeth et al. [216]).

$$E(s) = 1/2 m\omega_d^2 s^2. \tag{5.2}$$

Here $E(s)$ gives the energy as a function of s, the displacement from the initial equilibrium, m is the mass of the atom diffusing, and ω_d gives the angular attempt frequency. The frequency factor can also be estimated based on the number of diffusion events $<n>$ in the interval t

$$<n>(t) = N_a v_h \int_0^t dt \exp[-E_D/kT(t)], \tag{5.3}$$

where N_a is the number of atoms simulated, v_h is the harmonic aproximation attempt frequency of the diffusing atom. The frequency based on this equation was 1.19×10^{12} sec^{-1} and the barrier amounted to 0.102 eV, the same as in their previous derivation. For exchange the energy barrier was bigger, 0.421–0.805 eV, with a frequency factor of 1.5×10^{12} sec^{-1}, which makes this event unlikely. However, an atom-catalyzed double exchange (already discussed in Chapter 3) on the terrace cost only 0.047–0.083 eV with frequency factors of 2.1×10^{12} sec^{-1} based on Eq. (5.2) and 1.9×10^{10} sec^{-1} based on Eq. (5.3).

The diffusion of copper adatoms on Ag(111) was examined by Morgenstern et al. [219] using scanning tunneling microscopy. Atoms moved primarily over distances comparable with a lattice spacing. An energy difference of 0.0055 ± 0.0010 eV was found between fcc and hcp sites. A plot of the diffusivity against $1/T$ is given in Fig. 5.55, and yields a diffusion barrier of 0.065 ± 0.009 eV and a frequency prefactor of $1 \times 10^{12 \pm 0.5}$ sec^{-1} [220]. This is one of not many direct studies of adatom movement by STM. From molecular dynamics simulations using potentials from effective medium theory, but disregarding long-range interactions, they obtained a value of 0.08 eV.

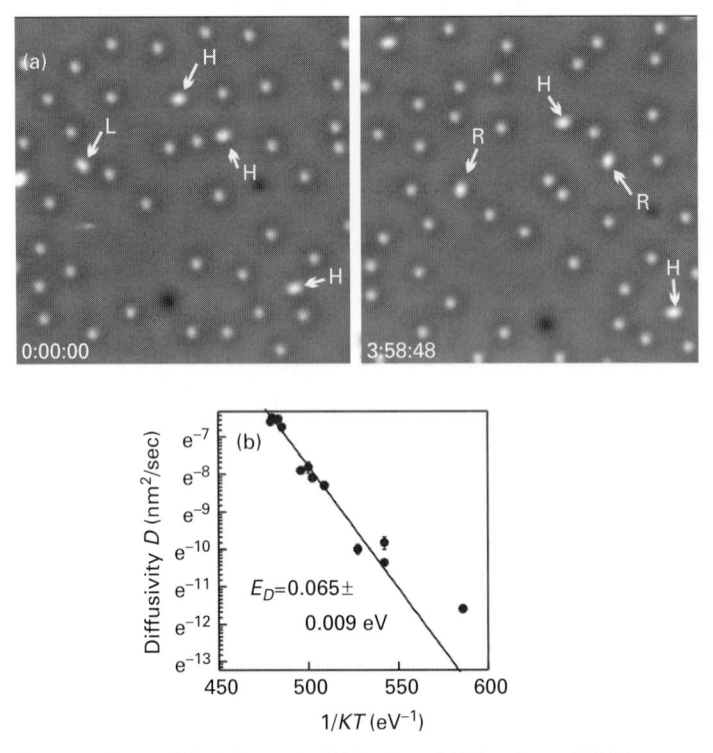

Fig. 5.55 Copper adatom diffusivity on Ag(111) plane. (a) Time lapse STM images at 21.5 K. (b) Arrhenius plot of diffusivity, giving an activation energy of 0.065 ± 0.009 eV (after Morgenstern et al. [219]).

A study of the interactions between cerium adatoms on Ag(111) was done by Silly et al. [221] using the STM. In the course of this work they also measured the time at 4.8 K for a cerium atom to execute a jump to be 0.3 sec; assuming a frequency prefactor of 10^{12} sec^{-1} they arrived at a diffusion barrier of 0.011 eV for cerium on Ag(111).

The molecular dynamics simulations based on RGL interactions of Kim et al. [70] recently yielded the following barriers: Ni 0.071 eV, Cu 0.078 eV, Pd 0.087 eV, Pt 0.040 eV, and Au 0.088 eV. The results for Cu and Au are not far away from previous studies.

There clearly is not enough information available for hetero-diffusion on Ag(111) for any trends to emerge.

5.15 Silver: Ag(100)

5.15.1 Self-diffusion

Investigations of the self-diffusion on Ag(100) started with the study of Venables in 1987 [200] who reanalyzed the data of Hartig et al. [222] for the island density of silver deposited on a Ag layer on top of a Mo(100) surface. Measurements were done at temperatures from 150 °C to 600 °C with scanning electron microscopy in combination

5.15 Silver: Ag(100)

with RHEED and AES. Venables derived a barrier of 0.45 eV for movement of silver on this surface. The effect of the possible influence of strain on atom movement was not disscused.

Some studies of silver diffusing on Ag(100) were done by Langelaar et al. [223] in 1996. They irradiated the surface with 6 keV Ne$^+$ ions, which create Ag adatoms on the (100) surface. As the surface temperature for irradiation increased from 160 K, the yield of silver adatoms was essentially constant. Langelaar et al. determined 160 K to be the temperature at which migration began. Assuming a value of the frequency prefactor of 1×10^{12} sec^{-1}, they arrived at an activation energy of 0.40 ± 0.05 eV. To allow for an uncertainty of a factor of 10 in their assumed frequency prefactor, they increased the error estimate for the barrier by 0.03 eV. The irradiation in this study created vacancies as well as adatoms, and it is not clear how this may have affected adatom motion.

Zhang et al. [224] also examined the homo-epitaxy in 0.1 monolayers of silver on Ag(100), studying nucleation by scanning tunneling microscopy. Comparing experiments with Monte Carlo simulations without atomic interactions, they arrived at an activation energy for surface diffusion of 0.330 ± 0.005 eV, assuming a prefactor $v_0 = 1 \times 10^{12}$ sec^{-1}. Next year another study of this system by Zhang et al. [225] appeared. From an STM examination and a Monte Carlo simulation of the homo-epitaxy using a lattice gas model they arrived at a higher value of the activation energy of 0.38 eV, assuming a frequency prefactor $v_0 = 10^{13}$ sec^{-1}; this was subsequently revised up, to a barrier of ≈ 0.40 eV [226].

Bardotti et al. [227] did high resolution LEED measurements for the submonolayer growth of silver at 0.3 of a monolayer on Ag(100). From the splitting of the diffraction profiles over a range of temperatures from 170 K to 295 K, they were able to arrive at the mean island separation, which gave them an activation energy for silver diffusion ranging from 0.37 ± 0.06 eV to 0.45 ± 0.06 eV. Comparison of these results with previous STM studies led them to a value of 0.40 ± 0.04 eV for the diffusion barrier, assuming a frequency prefactor 3×10^{13} sec^{-1}.

Work with perturbed angular correlation of ^{111}Ag atoms on Ag(100) was carried out by Rosu et al. [228]. From the terrace length distribution, and an estimate of 100 hops to reach a lattice step, they reached a diffusion barrier of 0.35 ± 0.03 eV assuming a frequency prefactor of $10^{12 \pm 1}$ sec^{-1}. For diffusion by atom exchange they estimated an activation energy of 0.45 ± 0.03 eV. In 1999 Laurens et al. [96] looked at the same system with the same technique. One hundredth of a monolayer of ^{111}Ag was deposited on a Ag(100) surface. The adatoms were immobile below 120 K. Steps started to be populated at 130 K, indicating the onset temperature for diffusion and leading to an activation energy of 0.37 ± 0.03 eV. The barrier for exchange was measured as 0.48 ± 0.03 eV, all in agreement with previous findings. The low coverage was probably suitable to learn about single atom motion.

The formation of two-dimensional silver islands in 0.1 monolayers of silver on a Ag(100) surface was followed by Frank et al. [229]. From STM studies of island densities shown in Fig. 5.56 they found a diffusion barrier of 0.40 eV and a frequency prefactor of 5×10^{12} sec^{-1}. These results are close to the work of Langelaar et al. [223], but all except the last study are based on an assumed value for the prefactor, so a better

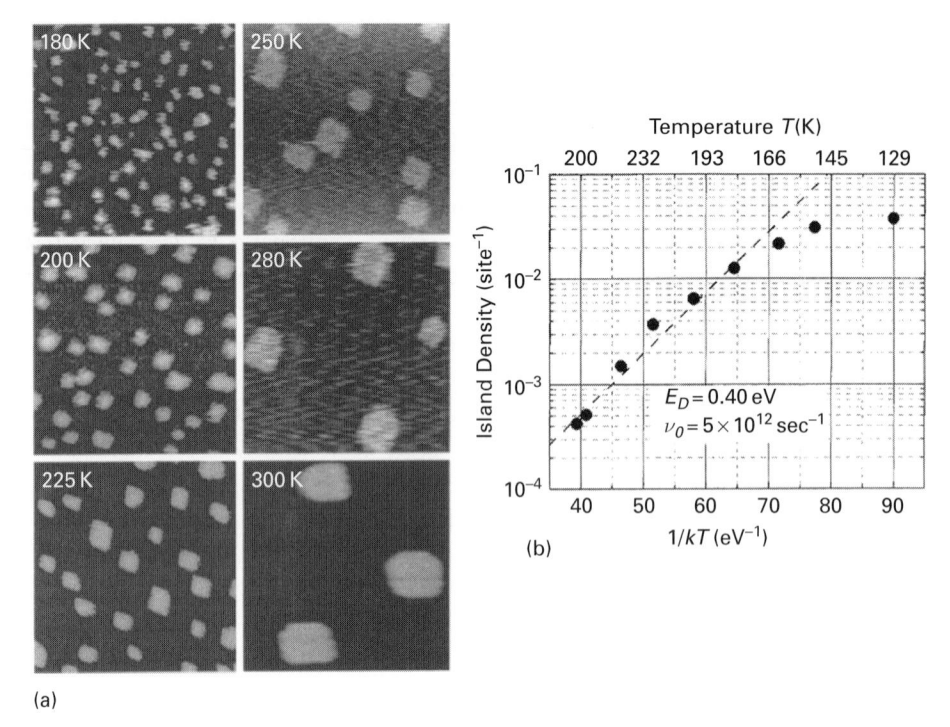

Fig. 5.56 Deposition of Ag on Ag(100). (a) STM images of growth at different temperatures. Coverage 0.1 ML, deposition rate 0.006 ML/sec. (b) Arrhenius plot of island density yields an activation energy for diffusion of 0.40 eV (after Frank et al. [229]).

examination here would not be amiss. However, the experimental barrier heights for self-diffusion on Ag(100) are all between ~0.3 eV and ~0.4 eV, which is a hopeful sign that they reflect adatom movement on this surface.

Theoretical work for the diffusion of silver atoms on silver surfaces started with the work of Flahive and Graham [47], using Morse potentials to give a barrier of 0.35 eV as shown in Fig. 4.9. Voter [230] was among the first to make an estimate of the barrier to diffusion by hopping allowing for many-atom effects. Resorting to EAM potentials between atoms and allowing 43 surface atoms to be movable he arrived at a value of 0.49 eV for the activation barrier. For a vacancy the diffusion barrier was only 0.46 eV. Further work with EAM potentials was done by Liu et al. [4]. They arrived at an activation energy of 0.48 eV with AFW parameters yielding a prefactor of 1.2×10^{-3} cm^2/sec, and the same barrier but with a prefactor of 3.9×10^{-3} cm^2/sec using VC curves. For atom exchange they found higher barriers, 0.75 eV from AFW and 0.60 eV from VC with a prefactor of 2×10^{-2} cm^2/sec.

Sanders and DePristo [76] reported a barrier of 0.24 eV and a frequency prefactor of 1.0×10^{12} sec^{-1} for hopping with the MD/MC-CEM approximation, a value lower than previous findings. However, this changed with the next study in the same group. Perkins and DePristo [77] came up with an activation energy of 0.45 eV for hopping; the exchange mechanism seems to have been unfavorable on this surface with a high barrier

5.15 Silver: Ag(100)

of 1.01 eV. They also did CEM calculations to find a barrier of 0.41 eV for hopping and 0.58 eV for atom exchange. Two years later, Perkins and DePristo [29] investigated the influence the number of movables had on the value of the activation energy for hopping as well as exchange. Their results for silver were similar to what was found for other substrates: the barrier to hopping did not depend sensitively on the number of atoms movable, but for atom exchange it did very strongly. With calculations based on MD/MC-CEM interactions, increasing movables from 13 to 226 lowered the barrier to exchange of silver from 1.01 eV to 0.70 eV, and with CEM interactions from 0.58 eV to 0.41 eV. It is clearly difficult to discern which process is favored.

Nelson et al. [207] made estimates using EAM, but with FBD set Universal3 [57], and arrived at a barrier of 0.479 eV. The calculations, done on a static lattice, considered only hopping. Stoltze [61] relied on EMT and found an activation energy for self-diffusion of 0.365 eV. Cohen [7] carried out molecular dynamics simulations relying on EAM interactions, and identified a "new" diffusion process in which an adatom entered the substrate and a lattice atom subsequently emerged a few spacings away. The temperature dependence for these events, as well as for hopping and the usual exchange, were shown in the Arrhenius plot in Fig. 3.62. Exchange begins to be important above ~ 600 K, but long-range exchange only becomes significant at 1000 K.

Full-potential linear-muffin-tin-orbital estimates were made by Boisvert et al. [208], who found an energy barrier for diffusion on Ag(100) of 0.50 ± 0.03 eV, higher than previously reported values. Continuation of this study with molecular dynamics simulations relying on the EAM [78] gave for hopping values of 0.48 eV and a prefactor of 1.53×10^{13} sec^{-1}. For exchange they arrived at a much higher activation energy of 0.78 eV and a prefactor of 3.92×10^{14} sec^{-1}.

Yu and Scheffler [231,232] used plane-wave pseudo-potential calculations and density-functional methods in a 3×3 cell to arrive at an activation energy of 0.52 eV in the local-density approximation and 0.45 eV using the generalized gradient approach. For diffusion via the exchange mechanism they found a bigger barrier, 0.93 eV from LDA and 0.73 eV using GGA. Shi et al. [79] used the embedded atom method (EAM) to calculate the activation energy to be 0.478 eV. Effective medium theory was used by Merikoski et al. [81] to come up with a hopping barrier of 0.367 eV; for diffusion by atom exchange they found a higher barrier of 0.614 eV. Kürpick et al. [114] calculated a barrier for hopping of 0.48 eV with both EAM potentials, FBD and VC. For the system at 600 K, Kürpick and Rahman [80] reported a slight increase, to 0.49 eV using FBD potentials and 0.50 eV with potentials of type VC; prefactors were 3.4×10^{-4} cm^2/sec (FBD) and 5.5×10^{-4} cm^2/sec (VC). The next year Kürpick and Rahman [233] continued work with EAM potentials to calculate the same activation energy as above for atom hopping on Ag(100) with a prefactor of 8.1×10^{-4} cm^2/sec for FBD potentials and 2.3×10^{-3} cm^2/sec with VC parameters in the harmonic aproximation. The result in the quasi-harmonic approximation at 600 K was as shown a year earlier [80], but with prefactors of 9.3×10^{-4} and 1.5×10^{-3} cm^2/sec for FBD and VC potentials, respectively. No temperature dependence of the prefactor was detected.

In 1998, Papanicolaou et al. [212] used molecular dynamics with a tight-binding RGL model and for bridge hopping they found an energy barrier of 0.43 ± 0.04 eV and

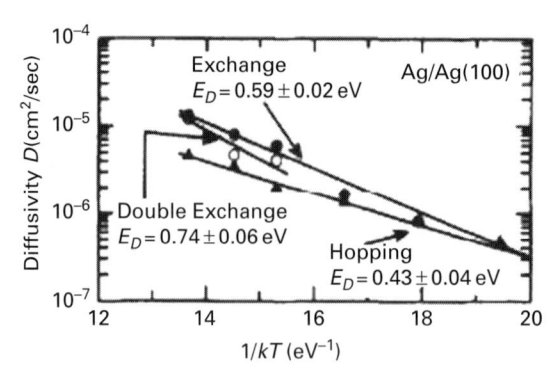

Fig. 5.57 Arrhenius plots for hopping and exchange frequencies of Ag on Ag(100), obtained from molecular dynamics simulations with RGL potentials (after Papanicolaou et al. [212]).

$D_o = (1.4 \pm 0.6) \times 10^{-3}$ cm^2/sec. Exchange cost 0.59 ± 0.02 eV, with a prefactor for the diffusivity of $(40 \pm 10) \times 10^{-3}$ cm^2/sec, one range higher than for hopping; the static barrier for hopping amounted to 0.46 eV. Double exchange required an activation energy of 0.74 ± 0.06 eV, and was characterized by a prefactor of $(300 \pm 200) \times 10^{-3}$ cm^2/sec, ten times higher than for simple exchange and 100 times higher than for hopping. The exchange mechanism started to be important for this surface at a temperature around 600 K. Their data for the transition frequency at different temperatures are illustrated in Fig. 5.57.

Calculations were done by Zhuang and Liu [119] using potentials from SEAM [234]. From rate measurements over a range of temperatures they arrived at an activation energy of 0.26 ± 0.04 eV, rather lower than the static barrier of 0.39 eV they derived for simple exchange. They concluded that diffusion involved other strain-induced exchanges. Mehl et al. [82] relied on EAM potentials to come up with a barrier of 0.48 eV; they considered only hopping in a 3×3 cell. With his cohesion approximation, Davydov [49] estimated a self-diffusion barrier of 0.06 eV on (100), unusually low.

Baletto et al. [214] looked at diffusion on a silver Wulff polyhedron with the nudged elastic band method and RGL. On this polyhedron a (100) plane consisted of nine atoms. On such a plane they found that diffusion proceeded by hopping with a barrier of 0.43 eV, not really different from the barrier on a big, flat plane.

Depending on Lennard-Jones potentials, Agrawal et al. [26] found the barrier for diffusion as 0.52 eV with a prefactor of 2.2×10^{-3} cm^2/sec. Chang and Wei [28] estimated a hopping barrier of 0.60 eV using *ab initio* VASP simulations, and 0.76 eV for diffusion by atom exchange. The results obtained by Müller and Ibach [127] of 0.62 eV for the hopping barrier using a local-density approximation are close to this value. They also made an estimate of 1.00 eV for the barrier to self-diffusion by atom exchange. Kim et al. [70] have recently done molecular dynamics simulations with RGL interactions to find a barrier of 0.467 eV for atom jumps in self-diffusion, and 0.624 eV by exchange.

The scatter in the barriers calculated for self-diffusion on Ag(100) is really not too bad, independent of the technique. The values for hopping range from 0.36 eV to 0.62 eV, excluding the results of Davydov [49], and correspond not too badly with the barriers from experimental studies, 0.3 eV to 0.4 eV. It is clear that many-body potentials are required to study this system. Atoms move by hopping rather than exchange on the Ag(100) plane, but a lot of calculations have not looked into exchange as a possible mechanism of diffusion. However, as shown by Perkins and DePristo [29], the number of substrate atoms considered as movable in simulations has a large effect on the barrier to exchange, and this factor should be more carefully considered. In any event, further experiments, especially direct motion investigations on Ag(100), would be quite desirable.

5.15.2 Hetero-diffusion

Considerable work has been done on hetero-diffusion on the (100) plane of silver. Fink et al. [235,236] looked at diffusion of indium atoms on Ag(100) using perturbed angular correlation spectroscopy with ^{111}In, at a low concentration of 10^{-4} monolayers. For the activation energy they arrived at a value of 0.31 ± 0.03 eV, assuming a frequency prefactor of 10^{12} sec^{-1}.

In 1996, Patthey et al. [237] examined diffusion of palladium atoms over the Ag(100) surface. The coverage was estimated to be 0.1 ML. From measurements of the time rate of change of the photoelectron spectra at 135 K they arrived at a barrier of 0.43 ± 0.05 eV for palladium diffusion on this plane. To reach this value it was assumed that the frequency factor is the same as at higher temperatures, where subsurface diffusion is presumed to occur. Using EAM potentials, Patthey et al. calculated an activation energy of 0.53 eV for exchange of palladium with a silver atom from the lattice, which compares favorably with the measured barrier. Based on this they assumed that movement proceeded by exchange rather than by hopping.

Calculations for hetero-diffusion by hopping on Ag(100) were done by Sanders and DePristo [76]. Their work, based on MD/MC-CEM interactions and transition-state theory, yielded the following results: nickel barrier 0.71 eV, frequency prefactor 5.0×10^{12} sec^{-1}; copper 0.73 eV, 5.6×10^{12} sec^{-1}; rhodium 0.74 eV, 4.5×10^{12} sec^{-1}; platinum 0.59 eV, 3.7×10^{12} sec^{-1}; gold 0.54 eV, 3.2×10^{12} sec^{-1}. For diffusion of palladium the barrier was 0.58 eV, with a prefactor of 5.0×10^{12} sec^{-1}, a value for the activation energy somewhat higher than the 0.43 eV found in experiments by Patthey et al. [237]. These experiments suggested that atom exchange, not considered in these calculations, would be rather favorable. Two years later this was looked at by Perkins and DePristo [89], who did work with 81 movable surface atoms in the first layer. Their results for the diffusion barrier by hopping compared to exchange are given in Table 5.4. Prefactors were in the range 10^{12} sec^{-1}, but the values for atom hopping were at least twice as large as for exchange and are listed in Table 5.4. The barrier for palladium atom exchange agreed quite nicely with the experiments. According to this study in hetero-diffusion, exchange was the favored mechanism on the Ag(100) surface, except for copper atoms; there the barriers for hopping and exchange were too close together to reach a decision.

Canepa et al. [238] looked at the growth of iron on a Ag(100) surface with ion scattering spectroscopy. From observations of the alloying process they concluded that exchange diffusion would explain their findings, but this process was not checked in detail.

Table 5.4 Diffusion on Ag(100) [89].

Adatom	Frequency Prefactor ($\times 10^{12}$ sec^{-1})		Diffusion Barrier (eV)	
	Bridge Hops	Exchange	Hopping	Exchange
Rh	9.69	3.43	0.73	0.26
Ni	18.24	7.29	0.70	0.33
Pd	16.08	4.01	0.76	0.38
Cu	34.02	14.44	0.55	0.52

In 2005 Caffio *et al.* [239] examined the growth of Ni layers on Ag(100) with a number of techniques, LEIS, XPS, STM, and XPD. To understand their findings they employed EAM simulations with a 13 × 13 unit cell and a slab of 5 layers; for hopping of nickel on the Ag(100) surface they found a barrier of 0.61 eV, big compared with the barrier of 0.19 eV for exchange. Their barriers are lower then the findings of Perkins and DePristo [89], but the difference between hopping and exchange is in the same range. The movement of an Ag atom by exchange away from an embedded Ni atom at the surface cost 0.59 eV.

Molecular dynamics simulations by Kim *et al.* [70], based on RGL potentials, yielded the following barriers for hetero-atom hopping on Ag(100): Ni 0.529 eV, Cu 0.569 eV, Pd 0.674 eV, Pt 1.189 eV, and Au 0.662 eV. Unfortunately, no estimates for exchange were given. The especially high value of the activation energy for Pt atoms suggests that a different mechanism might be in order. However, such a high energy barrier was not seen previously by Sanders and DePristo [76].

Not enough information is available about hetero-atom diffusion on Ag(100) to discern any trends.

5.16 Tantalum: Ta(110)

Only brief measurements are available for the diffusion of palladium on the Ta(110) surface. Based on only forty-five observations in an FIM at each of three temperatures, Schwoebel and Kellogg [240] arrived at an estimate of 0.49 ± 0.02 eV for the barrier to hopping, assuming a prefactor of 2×10^{-3} cm^2/sec.

Davydov [49], with his cohesion approximation, came up with a self-diffusion barrier of 1.19 eV on Ta(110); like many of his results, this study is likely to not predict diffusion correctly.

Antczak and Blaszczyszyn [241] looked at the spreading of palladium at two monolayers over a tantalum tip in field emission experiments. They reported an activation energy of 1.45 eV, but this is likely to be the energy of moving over vicinals of the (110) plane rather than over the smooth surface. Furthermore, the layer was really too concentrated to derive single atom information.

Using Morse potentials, Flahive and Graham [47] estimated a barrier of 2.2 eV for self-diffusion on Ta(111), 1.0 eV on Ta(100), and 0.45 eV on Ta(110), as indicated in Fig. 4.9. The last value is in good agreement with experiments.

5.17 Tungsten: W(110)

There is almost a complete lack of information about diffusion on this surface, and more work clearly needs to be done.

5.17 Tungsten: W(110)

5.17.1 Self-diffusion experiments

The first ever measurements of single atom self-diffusion [242] were done in field ion microscopic experiments on the (110) plane of tungsten in 1966. An Arrhenius plot of these measurements is given in Fig. 5.58. Unfortunately the measurements were done with a few atoms on the plane and later on it was demonstrated that the other atoms on the surface could influence the movement observed. Ayrault improved the data analysis [167] and came up with a value of 0.92 ± 0.05 eV for the diffusion barrier, with a prefactor of 2.6×10^{-3} cm^2/sec, which agrees nicely with the value accepted nowadays. A few years later, Bassett and Parsley [243] extended these studies, but still with a few atoms present on the plane, again using field ion microscopy. They determined an activation energy equal to 0.86 eV, with a prefactor $D_o = 2.1 \times 10^{-3}$ cm^2/sec. More introductory experiments, but with one atom on the plane were done by Cowan and Tsong [244], who reported a diffusion barrier of 0.77 ± 0.06 eV and a value of $D_o = (4.5 \pm 0.2) \times 10^{-4}$ cm^2/sec. Both the activation energies and the prefactors are somewhat low in these measurements. Kellogg et al. [245] in 1978 corrected these results for an error in the temperature calibration and arrived at an activation energy for diffusion of 0.90 ± 0.07 eV as well as a prefactor of $6.2(\times 13^{\pm 1}) \times 10^{-3}$ cm^2/sec.

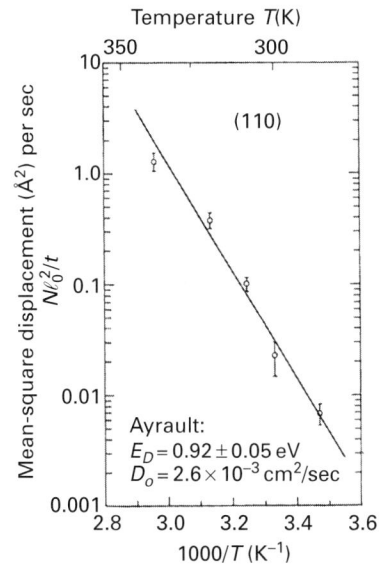

Fig. 5.58 Arrhenius plot for diffusivity of tungsten adatom on W(110) surface [242]. Barrier eventually determined as 0.92 ± 0.05 eV [167].

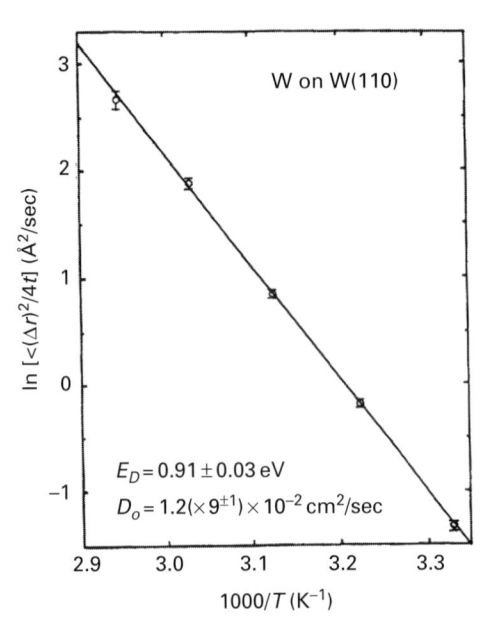

Fig. 5.59 Arrhenius plot for diffusion of tungsten adatom on W(110) plane obtained by laser heating of FIM sample (after Chen et al. [246]).

In 1990, Chen and Tsong [246] used laser heating to look at self-diffusion of tungsten. Their plot for the results is shown in Fig. 5.59 and yielded an activation energy equal to 0.91 ± 0.03 eV and a prefactor of $1.2(\times 9^{\pm 1}) \times 10^{-2}$ cm^2/sec. These parameters are in reasonable accord with currently accepted results.

Recently the diffusion of tungsten on W(110) has been examined by Antczak [247] with huge statistics of 1200 observations at each temperature. From the Arrhenius plot shown in Fig. 3.102 a barrier of 0.92 ± 0.02 eV was found for motion along <001> and 0.93 ± 0.01 eV along <110>, equal to each other within the limit of error. The prefactors were $1.61(\times 1.6^{\pm 1}) \times 10^{-3}$ cm^2/sec and $5.21(1.4^{\pm 1}) \times 10^{-3}$ cm^2/sec, with the y-component larger than along x, because of the larger separation between sites in this direction. They also showed that movement on this plane proceeded along <111> directions. In addition to single jumps, detected at low temperatures, with an activation energy of 0.94 ± 0.03 eV, at temperatures above 350 K three types of longer transitions were detected: vertical, horizontal, and double. Double jumps proceeded with an energy of 1.24 eV, verticals with 1.37 eV, and horizontals over a barrier of 1.28 eV. The details are described in Chapter 3.

Diffusivities of tungsten atoms on the W(110) plane appear to be in very good shape.

5.17.2 Self-diffusion simulations

The first theoretical investigations for self-diffusion on W(110) were done by Ehrlich and Kirk [248] who used Morse potentials to derive a barrier of 0.47 eV on a rigid surface and 0.44 eV on the relaxed surface. The effort was continued by Wynblatt and Gjostein [249] who again used Morse potentials to estimate diffusion barriers and arrived at an

5.17 Tungsten: W(110)

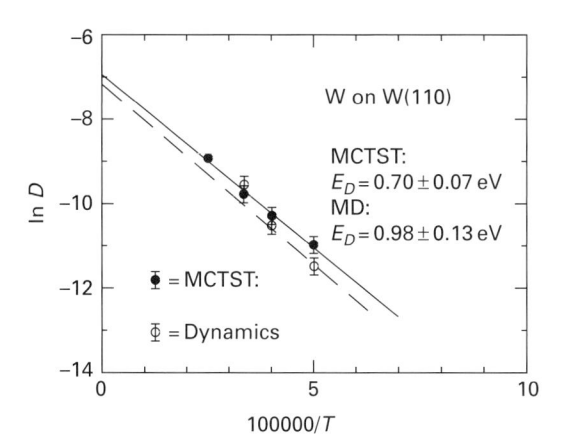

Fig. 5.60 Diffusion calculations for W on W(110) plane with one free layer (after Voter and Doll [174]).

activation energy for self-diffusion of 0.54 eV. Later Flahive and Graham [47] also did simulations again with Morse potentials to describe atomic interactions, they found a self-diffusion barrier of 0.46 eV. Banavar et al. [250] in 1981 used a Fokker–Planck formalism and Morse potentials to come up with a barrier of 0.52 eV and a prefactor of 5.4×10^{-4} cm^2/sec. When they adjusted the Morse parameters to reproduce the measured barrier, a prefactor of 1.3×10^{-3} cm^2/sec was obtained. In 1992 Doll and McDowell [251] used molecular dynamics simulations with L-J potentials to find a barrier of 0.982 eV and a prefactor of $3.6 \times 10^{-3}(\exp^{\pm 0.57})$ cm^2/sec. Two years later, Voter and Doll [174] employed transition-state theory with Lennard-Jones potentials and obtained a value of 0.70 ± 0.07 eV for the barrier and $1.1 \times 10^{-3} \times (1.31^{\pm 1})$ cm^2/sec for the prefactor. They compared this with the molecular dynamics investigation of Doll and McDowell [251], shown in Fig. 5.60. It should be noted, of course, that many-body effects are important and should really be included for this system.

In 1994 Xu and Adams [252] examined diffusion on several planes of tungsten with a modified fourth-moment approximation to tight-binding theory. For self-diffusion on W(110) they gave an activation barrier of 1.14 eV, with a prefactor of 1.11×10^{-3} cm^2/sec; this barrier is only ~24 % larger than the experimental value, and much better than what was found earlier with Morse potentials. The value they obtained for atom exchange appears to be an unusually high 2.94 eV. What is of concern in this study is that they found two types of adsorption sites, never seen in experiments, and one which was seen in experiments they claimed was metastable.

Recently MD and EAM potentials have been used to determine values for cluster movement by Chen et al. [253]. For single atom movement the authors derived a value of 0.89 eV and a prefactor of 6.92×10^{-3} cm^2/sec. Unfortunately, as a comparison Chen et al. used the incorrect value of 0.86 eV from Bassett and Parsley [243] instead of 0.92 eV [247]. Exchange is unlikely with an activation energy of 3.11 eV. The stable adatom adsorption site was fourfold, as observed in experiment.

Using his cohesion method, Davydov [49,254] came up with an activation energy for self-diffusion of 1.22 eV, which is above the value from experiments. With the same

Table 5.5 Activation energies (in eV) for diffusion on W(110), from calculations based on cohesion theory [254]. Z_s gives occupation of band.

Adatom	$Z_s=1.5$	Adatom	$Z_s=1.5$	Adatom	$Z_s=1.5$	$Z_s=2$
Sc	0.87	Y	0.90	Lu	0.94	0.46
Ti	1.03	Zr	1.07	Hf	1.11	0.79
V	1.08	Nb	1.14	Ta	1.23	1.13
Cr	1.01	Mo	1.19	W	1.22	1.27
Mn	0.95	Tc	1.10	Re	1.12	1.18
Fe	0.85	Ru	0.86	Os	1.01	1.05
Co	0.76	Pd	0.71	Ir	0.86	0.85
Ni	0.65	Ag	0.54	Pt	0.72	0.62
Cu	0.53			Au	0.54	0.28

approach [83,255,256] he also looked at the movement of transition metal atoms on the W(110) plane, as shown in Table 5.5. For most of the data there are no experimental comparisons available, but for tungsten the self-diffusion barrier is too high, as are the values for nickel, palladium, tantalum, and rhenium. The barrier for iridium is rather on the low side, but platinum seems to be in good agreement.

The theoretical results so far available for W(110) are somewhat disappointing, but experiments on self-diffusion are in a very good state. It turns out that the only mechanism of movement observed on W(110) is hopping; atoms move over this plane without interchanging with the first surface layer. At low temperatures hopping proceeds between nearest neighbors while increasing temperature causes longer jumps.

5.17.3 Hetero-diffusion: lithium, sodium, potassium

In 1982, Loburets et al. [257] examined the diffusion of Li on W(110) by studying changes in the work function of the surface. A range of lithium coverages was examined, as shown in Fig. 5.61, the lowest being 0.025 ML, but the surface was not probed for defects. From the Arrhenius plot at the lowest Li level they arrived at an activation energy of 0.11 eV and a prefactor $D_o = 1.78 \times 10^{-3}$ cm^2/sec.

Morin [258] studied sodium at coverages of 0.2 to 3×10^{14} atoms/cm^2 on the W(110) plane using field emission fluctuation measurements. Two different temperature regimes were noted, of which only the upper region, ≥ 300 K, will be mentioned. From the temperature dependence of the diffusivity Morin found a diffusion barrier of 0.28 eV and a prefactor of 2.0×10^{-8} cm^2/sec, shown in Fig. 5.62. The small value of the prefactor is highly suspicious, and it is not clear that the sodium concentration was low enough to yield information about single atom behavior.

Diffusion of potassium on tungsten was examined with field emission microscopy by Schmidt and Gomer [259], but this work did not focus on behavior on one specific crystal plane. Subsequent studies were done by Naumovets [260] and also by Meclewski [261],

5.17 Tungsten: W(110)

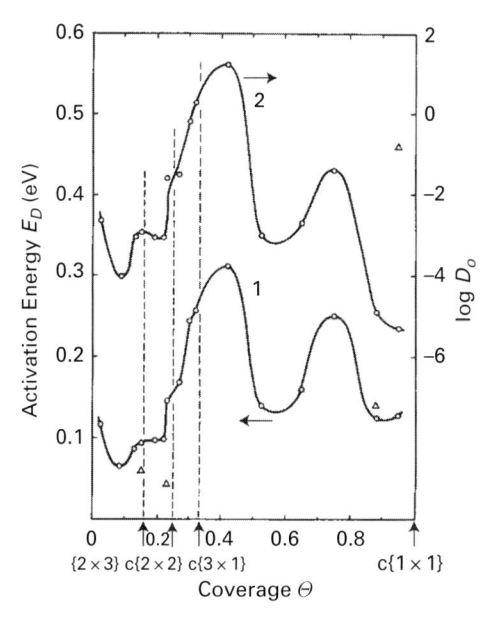

Fig. 5.61 Diffusion characteristics of Li on W(110), as a function of the surface coverage (after Loburets et al. [257]).

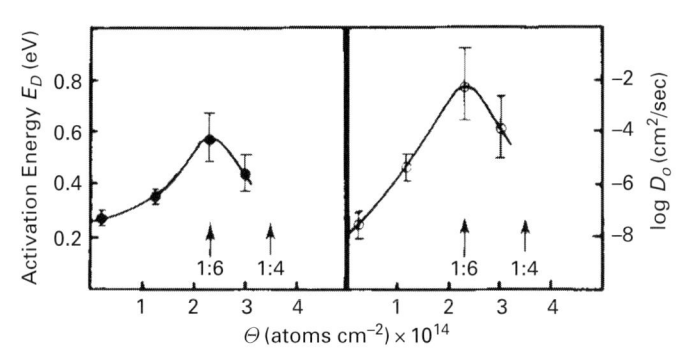

Fig. 5.62 Diffusion characteristics of Na on W(110) as a function of coverage, derived from field emission fluctuation studies (after Morin et al. [258]).

but we will only concentrate on the work by Dabrowski and Kleint [262]. They examined potassium on W(110) by using local field emission current fluctuations for coverages indicated in Fig. 5.63. At the lowest investigated coverage, around 0.1 monolayer, a value of 0.4 eV was obtained for the barrier, together with a prefactor in the range 10^{-3} cm^2/sec. However, this concentration may not be low enough to yield information about the behavior of single adatoms.

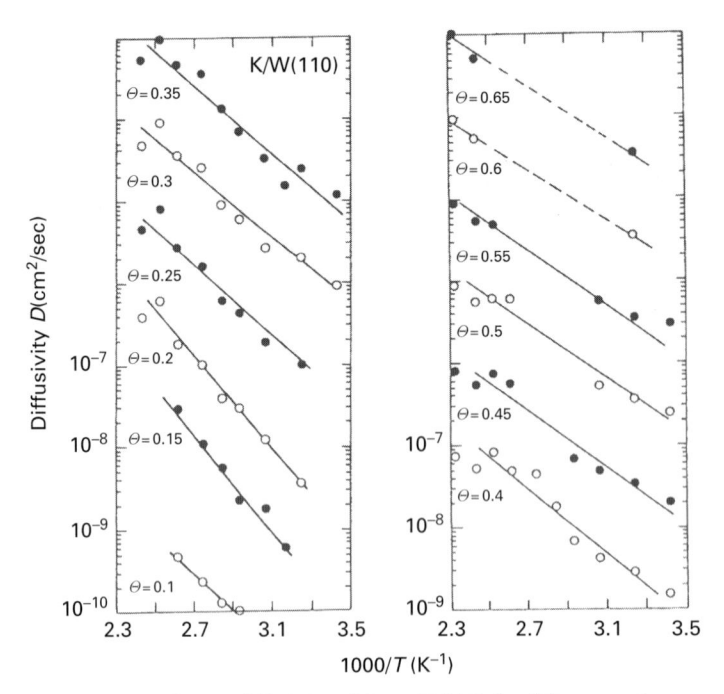

Fig. 5.63 Arrhenius plots for the diffusivity of K on W(110) for different surface concentrations. At a coverage of 0.1 ML, barrier is 0.4 eV (after Dabrowski and Kleint [262]).

5.17.4 Hetero-diffusion: manganese, iron, nickel

Dennler and Hafner [263] also did calculations for the diffusion of manganese atoms on the W(110) plane. In magnetic calculations, a barrier of 0.41 eV for hops along <111> was found; in non-magnetic estimates the barrier increased to 0.73 eV. Exchange processes again required much higher energy expenditures: in magnetic calculations the activation energies were 3.07 eV for exchange along the <100> direction and 4.22 eV for exchange along <110>. In non-magnetic work the activation energies for exchange were 3.30 eV along <100> and 4.54 eV along <110>. Exchange seems to be unlikely, but no experimental data are available for comparison.

Nahm and Gomer [264] examined the transformation of iron on W(110) from a low temperature to a high temperature form relying on LEED, AES as well as changes in the work function; they indirectly arrived at a diffusion barrier of 0.60 eV for a layer of one tenth of a monolayer. This was only a rough approximation, and the authors considered 0.43 eV a reasonable figure based on measurements at higher concentrations. Recently, Sladecek et al. [265] investigated the diffusion of 0.6 ML iron with nuclear resonant scattering. Their findings were similar to those of Nahm and Gomer [264] – they arrived at a barrier between 0.2 and 0.5 eV, but it is unlikely that they looked at atom movement without interactions.

Considerable work on tungsten surfaces has been done in Hafner's group in Vienna. Spisák and Hafner [50,51] have done density-functional calculations for the diffusion of iron on the (110) plane. They considered various possible transitions, including exchange, but arrived at an activation energy of 0.7 eV for hops along the <111> direction. Exchange along the <110> direction involved a much higher activation energy of 4.2 eV, exchange along the <001> direction required overcoming a barrier of 2.9 eV. This study showed that exchange was an unlikely mechanism for this system; unfortunately, there is no reliable experimental data for comparison.

Davydov [49,254] predicted for iron on W(110) a barrier of 0.85 eV from his cohesion approach, a value probably too high.

Bassett [266] looked at diffusion of nickel in 1978 with FIM, and from observations at two temperatures and the assumption of a prefactor of 10^{-3} cm^2/sec found an activation energy of 0.49 eV. It was a direct measurement of adatom displacements. Nine years later, Kellogg [267] also used FIM, and from measurements at four temperatures, obtained the same value of the barrier, 0.49 ± 0.02 eV, assuming a prefactor of 2×10^{-3} cm^2/sec. The data from the two studies are in excellent agreement, but both measurements relied on an assumed prefactor to arrive at an activation energy.

Davidov [49,254] with his cohesion approximation derived for Ni a value of 0.65 eV, clearly too high to describe this system.

5.17.5 Hetero-diffusion: copper, gallium, indium, tin

Early work on the spreading of copper over tungsten surfaces by field electron microscopy was done by Melmed [268,269]. He deposited copper atoms on one side of his emitter, maintained at 113 K, and then heated the surface and observed changes of the work function with time. Although he maintained good vacuum conditions and was able to deduce diffusion characteristics from the temperature dependence of the spreading, his observations were confined to vicinals of various surfaces, rather than detecting diffusion on individual planes.

Adsorption and diffusion of copper over tungsten was studied by Jones [270] in 1965. He deposited copper on one side of the emitter, and then followed diffusion out of the deposit on raising the temperature by examining field electron emission. No estimates of diffusion characteristics for individual planes were possible. Measurements were confined to vicinals of the (110) plane and to relatively high coverages. More reliable investigations of copper on W(110) unfortunately are not available.

With the cohesion approximation, Davidov [49,254] estimated for Cu a barrier of 0.53 eV, but this approximation was shown to be not reliable.

Nishikawa and Saadat [271] using field ion and field emission microscopy examined the behavior of gallium, indium, and tin on tungsten in an unusual fashion. They deposited these materials on an emitter tip by contact with the liquids, and then field evaporated the surface; this left the shank of the emitter covered. On heating, the adsorbed material would spread from the shank to the (110) plane at the center of the field emission image, and the spreading was followed by recording the field emission of electrons. Only measurements at a single temperature were made, and activation energies for diffusion on various vicinal planes were derived by assuming a value for the prefactor, but these are unlikely to be valid for single atom movement.

5.17.6 Hetero-diffusion: rhodium, palladium, cesium

In 1990, Dhanak and Bassett [160] looked at diffusion of rhodium on W(110) in the field ion microscope; with one atom deposited at the surface, they reported an activation energy of 0.72 ± 0.05 eV and a prefactor of 2.5×10^{-3} cm^2/sec. This is only a single study, but is a direct measurement with good statistics and should be quite reliable.

Bassett [266] in 1978 assumed a prefactor of 1×10^{-3} cm^2/sec and from observations in the field ion microscope on W(110) was able to derive a barrier for diffusion of palladium of 0.51 eV. In 2002, Fu et al. [272] again examined the diffusion of palladium on tungsten. On the (110) plane, they found a diffusion barrier of 0.51 ± 0.03 eV and a prefactor of $1.4 \times 10^{-3.0 \pm 1.9}$ cm^2/sec from measurements at five temperatures, shown in Fig. 4.40. This is in excellent agreement with Bassett's [266] earlier estimate. In the same year Oh et al. [273] looked at the influence of long jumps on the diffusion of palladium and also reported on FIM studies of palladium diffusivity on W(110), shown in Fig. 3.97. They reported an activation energy of 0.509 ± 0.009 eV, and a prefactor $D_o = 4.25(\times 1.7^{\pm 1}) \times 10^{-4}$ cm^2/sec, based on measurements at nine temperatures. In addition to single jumps they also found that double and vertical jumps participated at temperatures above 185 K. There appears to be general agreement achieved in three independent laboratories about the diffusion of palladium on W(110).

The cohesion approach of Davidov [49,254] yielded a barrier of 0.71 eV for palladium adatoms, clearly too high.

The first work on hetero-diffusion was carried out by Taylor and Langmuir [274,275] starting in 1932, who examined the diffusion of cesium on W(110). In their experiments, the sample was mounted inside cylindrical shields, as shown in Fig. 5.64a, and a small concentration of cesium, at a fractional coverage of 0.03, was allowed to accumulate on the surface. The material from the center of the wire was then removed by heating the sample, keeping the central cylinder negative and the others positive. Diffusion into the center was then measured by heating the sample, with all cylinders negative, but with only the current to the center one measured. Taylor and Langmuir deduced an activation energy for diffusion of 0.61 eV and a prefactor of 0.20 cm^2/sec from their measurements of the diffusivity at different temperatures, shown in Fig. 5.64b. Because of the low concentration in these experiments it is reasonable to believe that the results reflect the diffusion of single atoms. There is, however, a general concern about the validity of work looking at diffusion over macroscopic dimensions, because of the probable presence of steps that are likely to bind atoms. At the moment, there is no simple answer to this question.

Even though this may have been the only diffusivity measured on a clean metal surface before World War II, the results were confirmed by Love and Wiedrick [276] in 1968. They again studied diffusion of cesium on W(110), measuring the concentration by photoelectric emission in the apparatus shown in Fig. 5.65a. For the surface diffusivity at different temperatures they obtained the results in Fig 5.65b, which yielded an activation energy of 0.57 ± 0.02 eV and a prefactor of 0.23 cm^2/sec. These findings, obtained under

5.17 Tungsten: W(110)

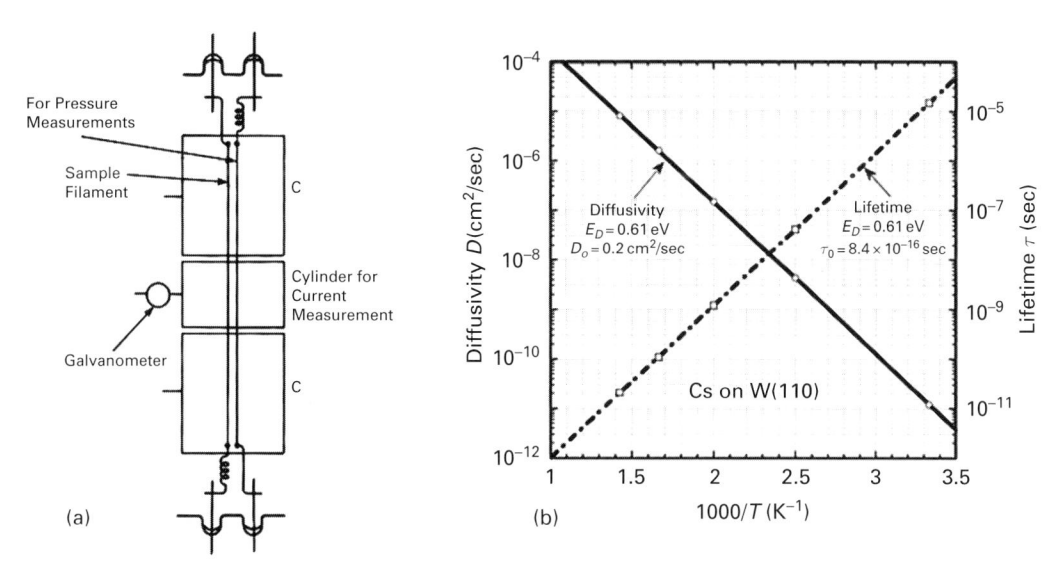

Fig. 5.64 Adsorption of Cs on W(110), as studied by Taylor and Langmuir [274,275]. (a) Apparatus to determine surface diffusion of Cs into central region of sample wire. (b) Arrhenius plot for diffusivity of Cs

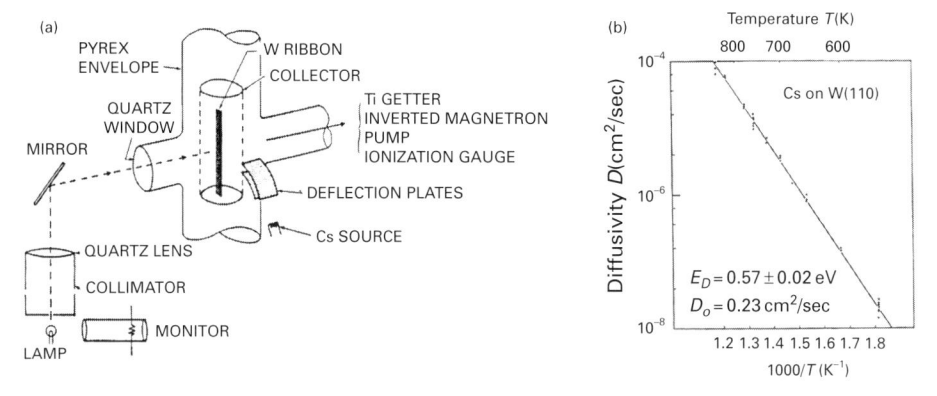

Fig. 5.65 Diffusivity of Cs on W(110) surface. (a) Schematic of photoelectric method for detecting concentration of Cs. (b) Diffusivity of Cs on W(110) in its dependence upon the reciprocal temperature (after Love and Wiederick [276]).

ultra-high vacuum conditions, are in excellent agreement with the previous work of Taylor and Langmuir [274,275]. It must be noted again that the (110) sample in this study was within two degrees of (110) and therefore had on it an unknown number of steps. The coverage of 5×10^{-4} monolayers was low enough so the barrier can be associated with single atom movement; however, the presence of steps may complicate matters. It appears that today we still do not have truly reliable information about the movement of Cs on an ideal W(110) surface.

5.17.7 Hetero-diffusion: barium, tantalum, rhenium

A number of studies has been carried out of the diffusion of barium on the W(110) plane. The first of these was done by Utsugi and Gomer [277], who observed the movement of Ba over the surface using the FEM, the lowest concentration being ~ 0.08 ML. In the absence of an applied field the activation energy turned out to be 0.41 eV, and the prefactor was 3.2×10^{-8} cm^2/sec. There are at least two problems with these measurements – uncertainty that the concentration was low enough to yield the diffusivity of single atoms unaffected by interactions, and then also the very low prefactor.

Measurements of barium were also reported by Naumovets et al. [278] who relied on their usual measurements of work function changes. The lowest barium concentration examined was 0.01 ML, and as appears from Fig. 5.66 the activation energy for diffusion was 0.078 eV with a prefactor of 1.3×10^{-4} cm^2/sec. These results appear more confidence inspiring than prior studies, but still control of the defect concentration on the surface is lacking.

Bassett and Parsley [243] examined the motion of tantalum by field ion microscopy, and obtained a diffusion barrier of 0.77 eV with a prefactor of 4.4×10^{-2} cm^2/sec. Measurements were done with several atoms present on the plane. Assuming a prefactor of 10^{-3} cm^2/sec, Tsong and Kellogg [279] reported a barrier of 0.7 eV for diffusion of tantalum adatoms observed in the field ion microscope at a single temperature and with one atom on the plane. The agreement of the data from the two groups is not bad, though

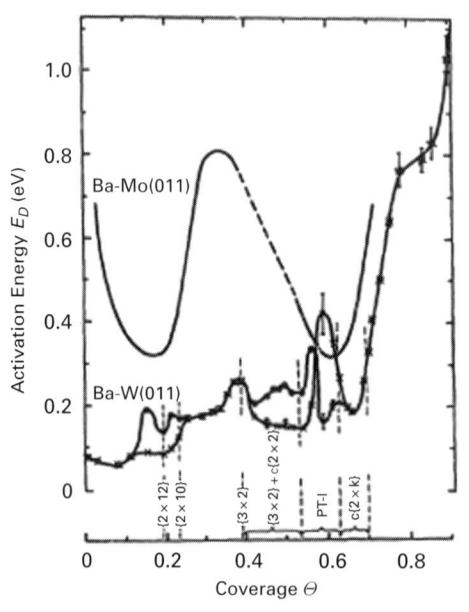

Fig. 5.66 Activation energy for diffusion of barium on W(110) plane in its dependence on the coverage. Crosses indicate high temperature results (after Naumovets et al. [278]).

5.17 Tungsten: W(110)

the first value may have been influenced by adatom interactions – an indication that interactions of Ta adatoms may be of only short range.

Davidov [49,254] arrived at a high value of 1.23 eV for the Ta barrier with cohesion scaling, a value not comparable with the experimental data.

In 1969, Bassett and Parsley [280] reported studies of rhenium diffusion on W(110) with several atoms on the plane, in which they found a diffusion barrier of 1.03 eV and a diffusion prefactor of $D_o = 3.0 \times 10^{-2}$ cm^2/sec. A year later, Bassett and Parsley [243] changed their previous prefactor for rhenium diffusion to 1.5×10^{-2} cm^2/sec. In their early work they also, for the first time, checked the effect of an electric field, of 0.9 V/Å, on the diffusion of individual Re atoms on W(110). For ordinary diffusion the activation energy was 1.034 eV; in the presence of the electric field it went to 1.081 eV and a prefactor 1.70×10^{-1} cm^2/sec; that is, hardly any change at all was detected in the diffusion barrier, although a significant effect on the prefactor was found.

Based only on 17 observations at 364 K, Tsong [281] in 1972 reported barriers of 0.88 eV and 0.95 eV for different diffusion paths of rhenium atoms on W(110). That such deductions were erroneous was pointed out by Johnson and White [282], and Tsong [283] later published a corrected value of 1.01 eV for the diffusion barrier. Two years later, from observations of one atom at a single temperature and an assumed prefactor of 10^{-3} cm^2/sec, Tsong and Kellogg [279] arrived at a barrier of 0.91 eV for diffusion of rhenium. The data for rhenium are not consistent and more extended investigations will be useful.

Davidov's [49,254] cohesion approach gave a not far away value of 1.12 eV for Re adatoms.

5.17.8 Hetero-diffusion: iridium, platinum, gold

Iridium atoms have been more frequently examined on W(110). In their first large FIM diffusion study, Bassett and Parsley [243] looked at iridium on W(110), and found an activation energy of 0.778 eV and a prefactor of 8.9×10^{-5} cm^2/sec, which seems a bit low, probably due to the presence of several atoms at the surface. Measurements by Tsong and Kellogg [279] in Fig. 5.67 gave a barrier of 0.70 eV and a prefactor of 1×10^{-5} cm^2/sec. The results again may be too low.

In a more extensive investigation in 1989 by Lovisa [284] with a total of 2200 observations spread over six temperatures, the barrier to iridium diffusion turned out to be much higher, 0.97 ± 0.02 eV, with a prefactor of $1.7(\times 2.3^{\pm 1}) \times 10^{-2}$ cm^2/sec, as appears from Fig. 5.68. Later, more extensive studies by Lovisa [285] gave an activation energy of 0.95 ± 0.03 eV and a prefactor of $8.48(\times 2.6^{\pm 1}) \times 10^{-3}$ cm^2/sec.

In the most recent work on iridium atoms, huge statistics of 1200 observations at each of nine temperatures, shown in Fig. 3.108, and detailed "zero-time" corrections were done by Antczak [286]; they gave quite similar results. For motion along <001>, a barrier height of 0.94 ± 0.02 eV and a prefactor $D_o = 3.0(\times 1.8^{\pm 1}) \times 10^{-3}$ cm^2/sec was found. Along <110>, the results were again 0.94 ± 0.02 eV for the barrier, and $D_o = 6.78(\times 1.8^{\pm 1}) \times 10^{-3}$ cm^2/sec, in good agreement with Lovisa [285]. They also found the presence of long jumps on the surface at temperatures higher than 350 K. Single jumps required an energy

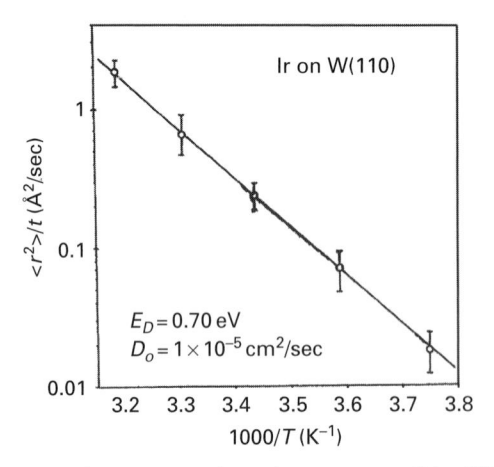

Fig. 5.67 Dependence upon reciprocal temperature of the diffusivity of Ir adatom on W(110) plane, as measured in FIM (after Tsong and Kellogg [279]).

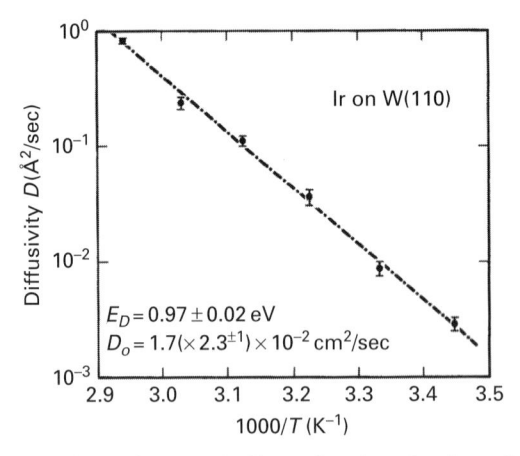

Fig. 5.68 Diffusivity of Ir on W(110) as a function of reciprocal temperature, derived from observations in FIM [284].

of 0.94 eV, double jumps 1.18 eV, horizontal transitions 1.25 eV, and vertical ones 1.27 eV. Details are described in Chapter 3.

For Ir the value Davidov [49,254] obtained with the cohesion approximation is not too far away, 0.86 eV compared with 0.94 obtained in experiments.

Platinum adatom movements have also been studied more than once on W(110). In their original work, Bassett and Parsley [243] in FIM studies obtained an approximate value for the activation energy, ~ 0.61 eV and a prefactor of ~ 10^{-4} cm^2/sec. The observations were done with several atoms present on the plane. Bassett [287] later reported a more carefully examined barrier of 0.67 ± 0.06 eV, with a prefactor of $3.1(\times 10^{\pm 1}) \times 10^{-3}$ cm^2/sec, shown in Fig. 5.69. Tsong and Kellogg [279] also looked at the diffusion of platinum adatoms on W(110) and found a barrier of 0.63 eV assuming a

5.18 Tungsten: W(100)

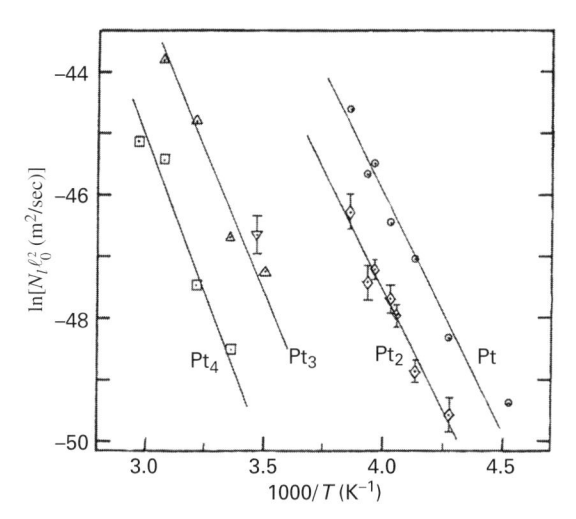

Fig. 5.69 Diffusivity of Pt atoms and clusters on W(110), obtained in FIM observations. For Pt adatom, diffusion barrier is 0.67 ± 0.06 eV (after Bassett [287]).

prefactor $D_o = 10^{-3}$ cm^2/sec. Their data seem to be in agreement with the work of Bassett [287] and it is worth noting that the results were obtained in two different laboratories. The cohesion approach of Davidov [49,254] for Pt gave a slightly higher barrier of 0.72 eV than the experimental value.

Diffusion of gold on W(110) was studied by Jones and Jones [288], who relied on field emission through a probe hole to examine the rate at which material arrived at the plane center. For a final deposit of half a monolayer they estimated an activation energy of 0.21 eV, but with a number of approximations to account for the influence of the applied field. Since spreading of material involves a correlated motion, the data are not appropriate for single atom diffusion. No information about single atom diffusion was attained. Davidov for Au [49,254] estimated a barrier of 0.54 eV with cohesion scaling. It is not clear how well this describes real movement.

5.18 Tungsten: W(100)

There is no information available about self-diffusion on this plane, which at room temperature and below is reconstructed into $W(100) - (\sqrt{2} \times \sqrt{2})R45°$ [289], shown in Fig. 5.70. There is a likelihood, however, that diffusion might occur by atom exchange.

Very little work has been done to examine diffusion of other atoms on this plane. One of the first studies was carried out by Jones and Jones [288], who did probe-hole field emission studies of the diffusion of gold on W(100). With the probe hole centered on the middle of the plane, the change in the recorded current as the gold swept past the probed area yielded, after some corrections, a very small activation energy of only 0.13 eV. There are a number of uncertainties about these measurements, and it is clear that what was measured was not single atom motion, inasmuch as the surface was eventually covered by a complete gold monolayer.

Table 5.6 Diffusion of potassium on flat and stepped W(100) [290].

Surface	Direction	E_D (eV)	$D_o(\times 10^{-2}$ cm^2/sec)
W(100)		0.43 ± 0.04	8.7
W(1019)	∥ and ⊥	0.38 ± 0.05	6.4
W(1011)	∥	0.42 ± 0.07	9.1
	⊥	0.62 ± 0.07	30.2
W(107)	∥	0.45 ± 0.07	6.9
	⊥	0.69 ± 0.05	57.4

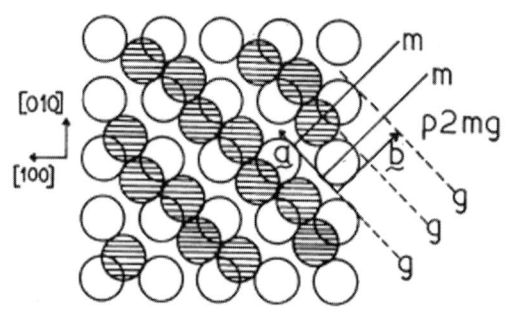

Fig. 5.70 Schematic of atomic arrangement of reconstructed W(100)-($\sqrt{2} \times \sqrt{2}$)R45° plane (after Debe and King [289]).

Bayat and Wassmuth [290] in 1984 examined diffusion of potassium on stepped (100) planes of tungsten using surface ionization microscopy. For the diffusion barrier on the (100) plane they obtained a value of 0.43 ± 0.04 eV with a prefactor $D_o = 8.7 \times 10^{-2}$ cm^2/sec. Their results for prefactors and activation energies on (100) vicinals are given in Table 5.6 and for W(100) are illustrated in Fig. 5.71. What is interesting here is that the prefactor for diffusion is some multiple of 10^{-2} cm^2/sec for all the easy directions of motion. Motion along the lattice steps takes place over a smaller barrier than at right angles to the steps. The barrier to motion parallel to the steps proved to be the same, ~0.42 eV, for most geometries examined, and the prefactors also have pretty much the same value. As the width of the terraces on the surface diminishes, the activation energy perpendicular to the steps increases from 0.62 to 0.69 eV, and the prefactor also rises. The value might reflect single atom movement since coverage was extremely low, below 10^{-5} ML.

Kellogg [291] looked at the migration of a nickel atom, making direct measurements by FIM at four temperatures. An activation energy of 1.01 ± 0.02 eV was determined by assuming a prefactor of 2×10^{-3} cm^2/sec. This barrier is twice a large as on the much smoother (110) plane.

Measurements of the diffusion of hafnium have been done relying on studies of fluctuations in the field emission current [292]. In determinations at temperatures ranging from 370 to 470 K, Beben and Gubernator arrived at the results in Fig. 5.72, with an activation energy of 0.53 and 0.54 eV in separate runs, and prefactors of $D_o = 2.7 \times 10^{-4}$ and 3.3×10^{-4} cm^2/sec. It must be noted that the surface concentration for these

5.18 Tungsten: W(100)

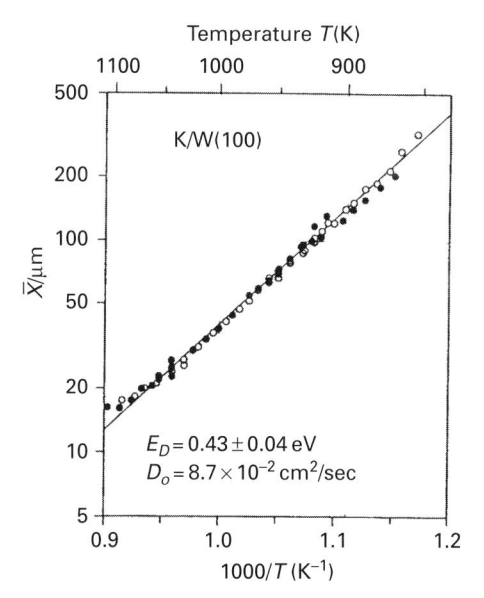

Fig. 5.71 Arrhenius plot for the mean diffusion length of K on W(100) determined by surface ionization microscopy (after Bayat and Wassmuth [290]).

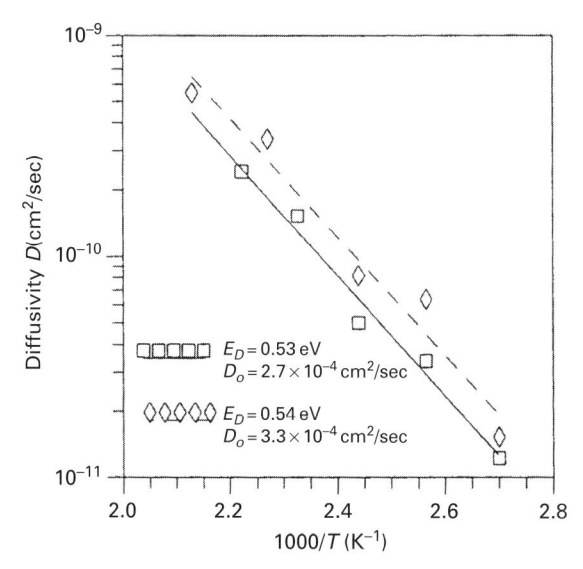

Fig. 5.72 Dependence of Hf diffusion on W(100) plane upon reciprocal temperature, determined from field emission current fluctuations (after Beben and Gubernator [292]).

measurements was not precisely determined; it was claimed to be less than one monolayer, but in all likelihood far from the infinitely dilute region necessary for our tabulation.

Diffusion of lead on tungsten surfaces was investigated by Morin and Drechsler [293] using field emission microscopy (FEM). They looked at spreading over an emitter tip and for the (100) region found a value of 1.3 eV for the activation energy. However, spreading was not controlled with a probe-hole, and it is not clear how steps on the (100) influenced the movement. Since a clear boundary was observed during spreading it is likely that the coverage was higher than 1 ML, making this study unsuitable for drawing conclusions about single atom movement.

Relatively little theoretical effort has gone to characterizing diffusion on the W(100) plane. Flahive and Graham [47] employed Morse potentials to evaluate the self-diffusion barrier on the (100) surface as 1.63 eV (see Fig. 4.9), but such estimates are not really appropriate due to the importance of many-body interactions. Recently Xu *et al.* [294] carried out calculations with the many-body Ackland potential [295] of the self-diffusion barrier on W(100) and found a value of 0.49 eV, much smaller than expected.

Spisák and Hafner [50,51] much more recently have reported density-functional calculations for diffusion of iron on W(100). Hopping on an unreconstructed surface required an energy of 1.6 eV, exchange 2.1 eV. This surface does undergo reconstruction, and hopping can therefore take place over a shorter and a longer bridge. Spisák and Hafner [50,51] found a barrier of 1.2 eV for hops across the shorter bridge and 1.3 eV along the longer, with atom exchange a much higher energy process at 2.3 eV. Diffusion of manganese on W(100) was also examined, using much the same techniques, by Dennler and Hafner [263], who found a diffusion barrier of 1.23 eV in magnetic estimates and a higher barrier of 1.75 eV for non-magnetic calculations on the non-reconstructed surface. On the reconstructed plane, the barrier for hopping over short and long bridges in magnetic calculations was 1.06 eV and 1.16 eV respectively. Exchange can again occur, but over a much higher activation barrier of 2.02 eV.

From this survey it is clear that relatively little is known about diffusion on W(100). What is needed, as a start, is definitive work on self-diffusion.

5.19   Tungsten: W(111)

5.19.1   Self-diffusion

In 1974, Graham and Ehrlich [296] identified two kinds of binding sites on the W(111) plane, the fault and lattice sites shown in Fig. 3.7. They made a rough estimate of 1.8 eV for the self-diffusion barrier between lattice sites and 1.3 eV for movement of an atom from a fault site to a lattice site.

As part of a larger study on thermal faceting of steps in 1988, Gong and Gomer [297] looked more intensely at diffusion on the (111) plane of tungsten by field emission. Their results, in Fig. 5.73, indicate that above 750 K the diffusivity first decreased and then increased again above 840 K, suggesting a roughening of the surface in this temperature range. They arrived at a diffusion barrier of 0.46 eV at the lower temperatures and a prefactor $D_o = 5.8 \times 10^{-8}$ cm^2/sec. The authors suspected that the applied field necessary for field emission affected the activation energy.

Recently Fu *et al.* [298] have examined tungsten behavior on the W(111) surface in greater detail. As shown in Fig. 5.74, a W adatom begins to move only above 650 K, once edge atoms have begun to dissociate from the (111) plane. Fu *et al.* give a diffusion

5.19 Tungsten: W(111)

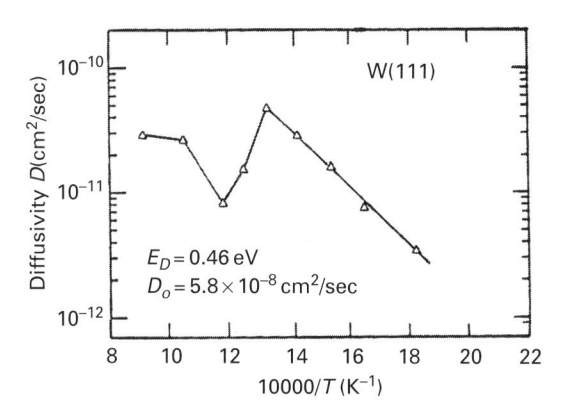

Fig. 5.73 Diffusivity of W on W(111) plane in its dependence upon reciprocal temperature, as measured by field emission current fluctuations (after Gong and Gomer [297]).

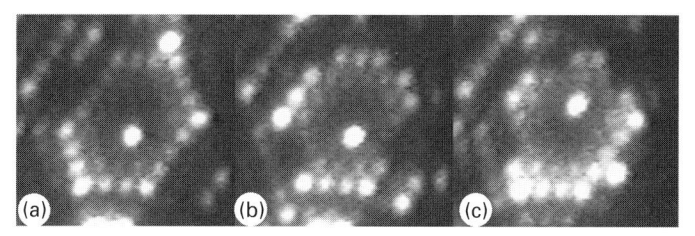

Fig. 5.74 FIM images of W adatom on W(111) surface. (a) Image after deposition. (b) After heating at 660 K, edge atoms start to detach, adatom stay in the same place. (c) After another heating at 660 K, more edge atom detached, adatom moved to new position (after Fu et al. [298]).

barrier of 1.9 eV, estimated assuming a prefactor $D_o = 10^{-3}$ cm^2/sec, a value not too far from the previous results of Graham and Ehrlich [296].

Theoretical estimates of diffusion on W(111) are rare. Only Flahive and Graham [47] have carried out calculations with Morse potentials to find a barrier of 2.92 eV for self-diffusion on this surface, but these ignore many-atom effects.

What is significant here is that in the field emission studies, prefactors five orders of magnitude below the usual values were obtained. It will be interesting to see if further determinations will verify this small prefactor. Measurements of self-diffusion would certainly be desirable, and some theoretical effort would be appropriate.

5.19.2 Hetero-diffusion

Flahive and Graham [299] in 1978 did FIM studies of single nickel atoms diffusing on small, well-defined (111) planes of tungsten. Their results, at temperatures from 390 to 420 K are shown in Fig. 5.75, and yield an activation energy of 0.873 ± 0.081 eV and a rather small prefactor of 1×10^{-6} cm^2/sec. Contrary to what was found with tungsten atoms, the nickel atoms occupy only the lattice sites.

Diffusion of lithium adsorbed on the (111) plane of tungsten has been examined in 1991 by Biernat et al. [300], who measured the fluctuations in the field emission

Diffusion on two-dimensional surfaces

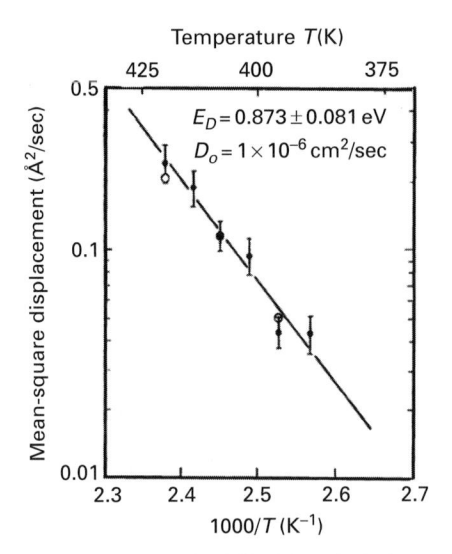

Fig. 5.75 Arrhenius plot for the diffusivity of Ni adatom on W(111) plane, measured by field ion microscopy (after Flahive and Graham [299]).

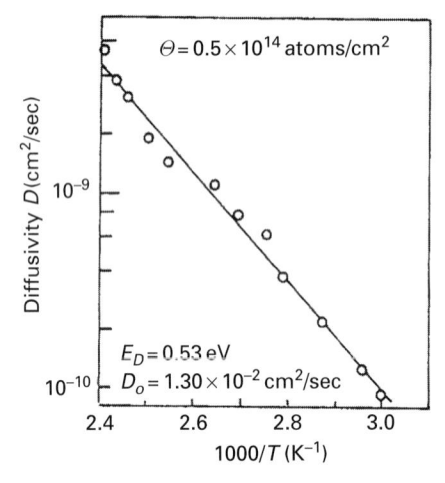

Fig. 5.76 Dependence of Li atom diffusivity on W(111) region upon inverse temperature, as examined by field emission fluctuations (after Biernat et al. [300]).

current at temperatures from 333 to 417 K. For studies at the lowest concentration, ~0.5 × 10^{14} atoms/cm^2, the activation energy for diffusion was found from the Arrhenius plot in Fig. 5.76 to be 0.53 eV and the prefactor was $D_o = 1.30 \times 10^{-2}$ cm^2/sec. It is not clear, however, that this movement is characteristic of single atom behavior.

Biernat and Dabrowski [301] used the same technique to examine the diffusion of titanium on the (111) plane of tungsten. At 0.2 monolayers, they found an activation energy of 1.77 ± 0.20 eV and a prefactor of ~3 × 10^{-2} cm^2/sec. The coverage seems too

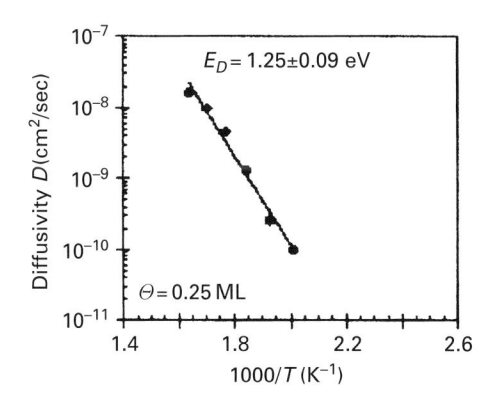

Fig. 5.77 Arrhenius plot of the diffusivity of Dy at 0.25 ML coverage on W(111) facet, obtain from fluctuations in FE current. Diffusion barrier of 1.25 eV (after Biernat and Blaszczyszyn [302]).

high to draw information about movement of atoms not influenced by the presence of other atoms on the surface.

Biernat and Blaszczyszyn [302] also looked at the diffusion of dysprosium on the same plane by the same current fluctuation method. At 0.25 of a monolayer they reported an activation energy of 1.25 ± 0.09 eV and a prefactor of almost 10^3 cm^2/sec, as in Fig. 5.77. The difficulty with these studies lies in the concentration of adsorbed material, which is not low enough to give information about single atom diffusion.

Diffusion of palladium has also been studied by Fu *et al.* [272] on W(111) using field ion microscopy. They found an activation energy for diffusion of 1.02 ± 0.06 eV, assuming a prefactor of 1×10^{-3} cm^2/sec. The data are shown in Fig. 4.40, but the plot is very uncertain. Other measurements with palladium have not been made on this plane.

5.20 Rhenium: Re(0001)

The structure of the hcp(0001) surface was shown in Fig. 3.27; it is similar to the fcc(111) in that both have two types of adsorption sites, fcc and hcp. Only little has been done to examine diffusion on rhenium. Goldstein [16] in 1999 studied diffusion of both rhenium and tungsten atoms on Re(0001), using FIM, with results shown in Fig. 5.78. In the diffusion of both W and Re atoms only one type of binding site was occupied, as indicated in Fig. 3.28, but it was impossible to identify which one. For rhenium atoms, the barrier to self-diffusion was found to be 0.48 ± 0.02 eV, with a prefactor $D_o = 6.13$ $(\times 2.6^{\pm 1}) \times 10^{-6}$ cm^2/sec. This prefactor seems to be suspiciously low. For tungsten diffusion, the activation energy turned out to be essentially the same, 0.48 ± 0.01 eV but the prefactor proved more normal, $2.17(\times 2.7^{\pm 1}) \times 10^{-3}$ cm^2/sec. Although these measurements did not occur in the same temperature range, the fact that tungsten diffusion seems to be normal suggests greater confidence in the results for rhenium. Also it is worth noting that this is a direct measurement of adatom movements. However, empirical comparisons with W(110) fail to correctly describe self-diffusion on Re(0001).

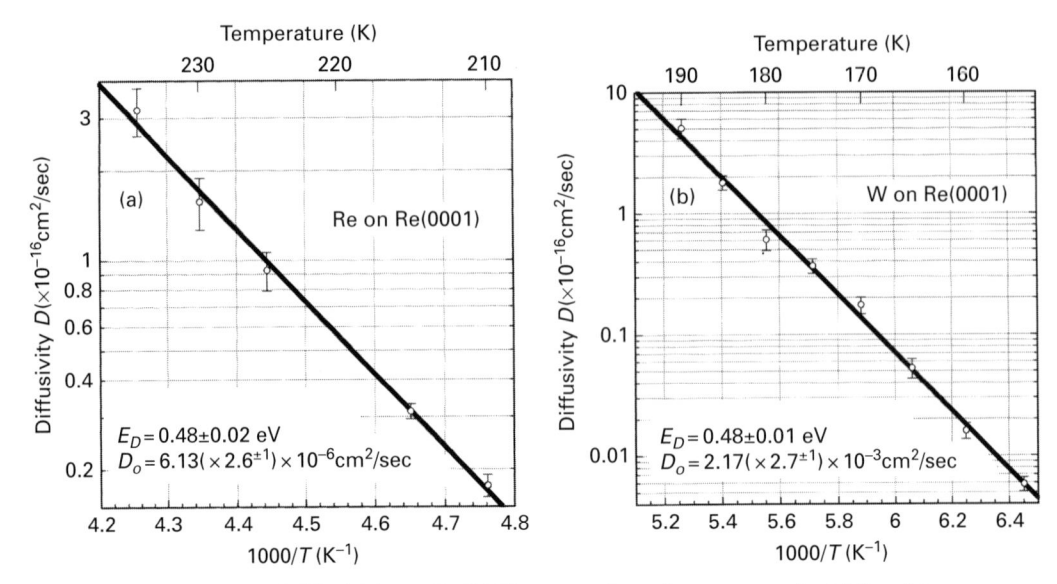

Fig. 5.78 Dependence of diffusivities on Re(0001) upon reciprocal temperature for (a) Re adatoms, (b) W adatoms [16].

For the Re(0001) plane only the rough estimates of Davydov [49] with his cohesion method are available, and yield a barrier of 1.01 eV for self-diffusion. This surface definitely needs more attention, both in experiments and theoretical calculations.

5.21 Iridium: Ir(111)

5.21.1 Self-diffusion

The first measurements of diffusion on Ir(111) were done by Wang [48] using field ion microscopy and working with only one atom on the plane; as shown in Fig. 5.79, he found an activation energy of 0.270 ± 0.004 eV and a prefactor of $1.0(\times 1.4^{\pm 1}) \times 10^{-4}$ cm^2/sec. The mapping of this plane, in Fig. 2.16, gives a clear indication of the existence of two adsorption sites, fcc and hcp, indicated in Fig. 3.3. However, jumps between the two kinds of adsorption sites were not separately noted. It must be appreciated that the identification of the different sites is easy since the shape of the atom image reveals the site directly, as is seen in Fig. 5.80. Unfortunately this is not true for other materials, as for example Re(0001).

Soon thereafter, Chen and Tsong [246] worked on the same system and reported a diffusion barrier of 0.022 ± 0.03 eV and a prefactor equal to $8.84(\times 8^{\pm 1}) \times 10^{-3}$ cm^2/sec, as shown in Fig. 5.81. This value for the activation energy is clearly much too low, and may have been a typing error. Chen and Tsong [303] published the material again, and gave a more sensible if low diffusion barrier of 0.22 ± 0.03 eV with the same prefactor.

Wang [304] in 1992 carried out extendent measurements and found much the same values as in his earlier work, $E_D = 0.27 \pm 0.03$ eV and $D_o = 9.0(\times 1.43^{\pm 1}) \times 10^{-5}$ cm^2/sec,

5.21 Iridium: Ir(111)

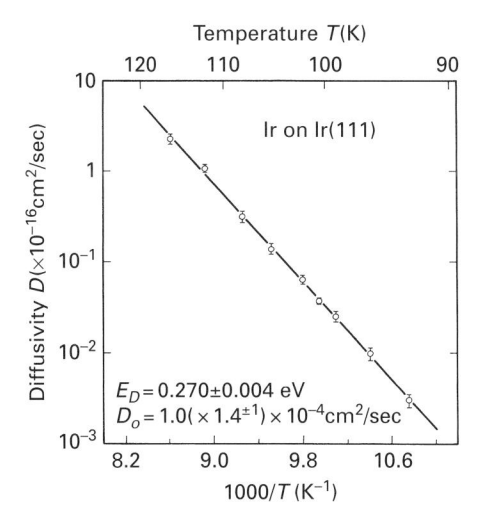

Fig. 5.79 Diffusivity as a function of the reciprocal temperature for Ir on Ir(111) [48].

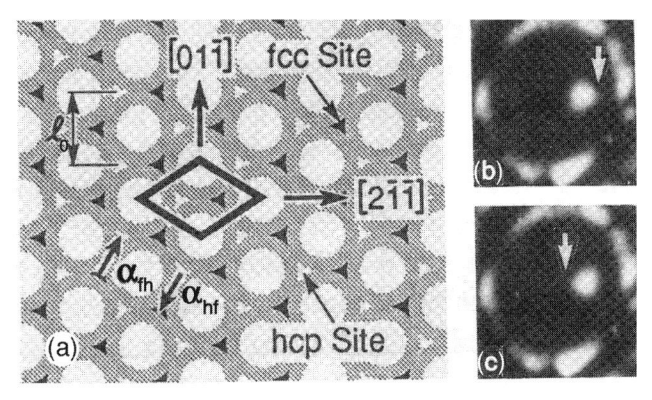

Fig. 5.80 (a) Diagram of fcc (111) plane, showing fcc and hcp binding sites. α_{fh} and α_{hf} indicate jump rates out of fcc and hcp sites. Field ion images of iridium adatom on Ir(111), shown in (b) with adatom at hcp site, apex pointed along $[2\bar{1}\bar{1}]$. In (c) iridium atom at fcc site, apex pointed in reverse direction [304].

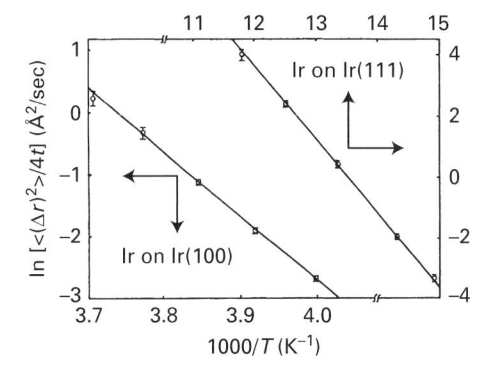

Fig. 5.81 Arrhenius plots for diffusivity of Ir adatoms on Ir(001) and Ir(111) planes obtained in FIM. Diffusion barriers: (001) – 0.93 ± 0.04 eV, (111) – 0.22 ± 0.03 eV (after Chen and Tsong [246]).

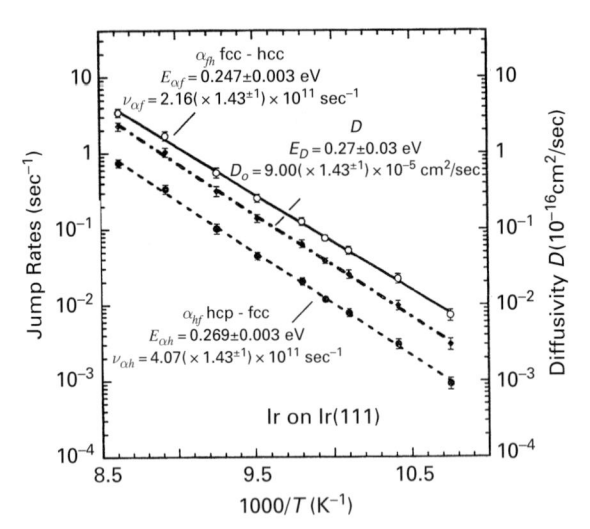

Fig. 5.82 Diffusivity and jump rates of Ir atoms on Ir(111) from FIM measurements as a function of the reciprocal temperature [304].

as appears in Fig. 5.82. He explored jumps between specific adsorption sites and for jumps from fcc to hcp sites obtained a barrier of 0.247 ± 0.003 eV and a frequency prefactor 2.16×10^{11} sec^{-1}; in the opposite direction, hcp to fcc, the activation energy proved to be 0.269 ± 0.003 eV and a prefactor of 4.07×10^{11} sec^{-1}. It appears that jumping from fcc to hcp site is slightly more favorable than the reverse process. These kinetics have been updated [305] with a more adequate resistance versus temperature calibration, giving a barrier of 0.289 ± 0.003 eV and $3.5(\times 1.4^{\pm 1}) \times 10^{-4}$ cm^2/sec, and Wang replotted his data in the same year [306] in Fig. 5.83. It should be noted however, that Fu et al. [307] still quote the early inappropriate value of 0.22 ± 0.03 eV for the diffusion barrier.

Busse et al. [34] using scanning tunneling microscopy looked at the saturation density of iridium islands on Ir(111) at different temperatures and at a coverage of 0.13 ± 0.01 monolayers; a plot is shown in Fig 5.5. They found an activation energy for diffusion of 0.30 ± 0.01 eV and a frequency prefactor of $5 \times 10^{11 \pm 0.5}$ sec^{-1}, in reasonable accord with the results of Wang [306].

Theoretical studies on the Ir(111) surface started with the work of Flahive and Graham [47] using Morse potentials; their findings are shown in Fig. 4.9 and they predicted a barrier of ~0.1 eV. Among the first extensive calculations of diffusion on Ir(111) was the work of Piveteau et al. [308], who evaluated energies in the tight-binding formalism. The activation energy for motion out of normal fcc sites proved to be 0.241 eV, and out of hcp sites 0.255 eV, reasonably close to experimental values. Molecular dynamics studies of self-diffusion on Ir(111) were carried out by Shiang et al. [309]. Using EAM potentials, the diffusion barrier turned out quite small, 0.11 eV. From an Arrhenius plot of molecular dynamics results for hopping, in Fig. 5.84, which relied on RGL estimates, they arrived at a higher barrier of 0.17 eV with a prefactor of 7.8×10^{-4} cm^2/sec; the static value proved to be 0.19 eV. They also found that hcp sites were slightly more favorable, iridium binding being ~0.03 eV stronger.

5.21 Iridium: Ir(111)

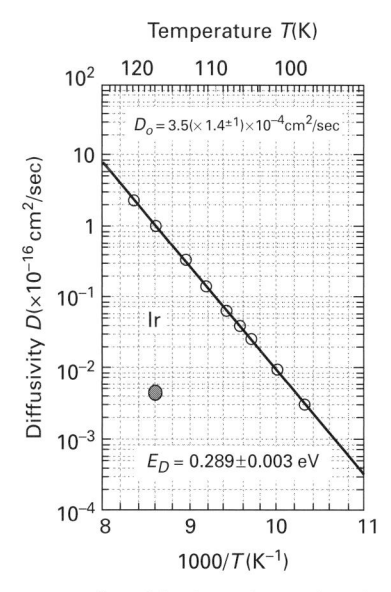

Fig. 5.83 FIM studies of the dependence of Ir adatom mobility on Ir(111) upon reciprocal temperature [306].

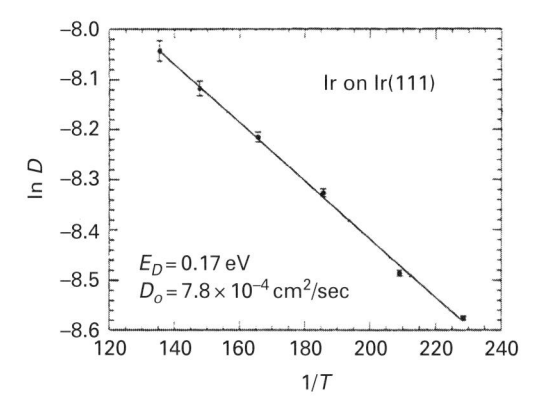

Fig. 5.84 Arrhenius plot for the diffusivity of Ir adatom on Ir(111), obtained in molecular dynamics simulations with RGL potentials (after Shiang et al. [309]).

Boisvert et al. [208] used the full-potential linear muffin-tin-orbital method to find a barrier of 0.24 ± 0.03 eV, one of the first theoretical reports close to experimental findings. With embedded atom potentials, Chang et al. [35,310] again reported a small diffusion barrier of 0.11 eV on Ir(111) using a 5×5 cell with 25 atoms in each of six layers. This value is in agreement with the findings of Shiang [309], but does not correspond with experiments [306]. Trushin et al. [311] worked with RGL potentials, and derived an activation energy of 0.20 eV; with Lennard-Jones potentials the result was higher, 0.24 eV. Any difference in the binding energy at hcp and fcc sites was lower than 0.005 eV. Two years later Davydov [49] reached a barrier of 0.25 eV using his cohesion approximation. Agrawal *et al.* [26] made estimates with Lennard-Jones potentials and found an activation barrier of 0.23 eV with a prefactor of 1.2×10^{-4} cm^2/sec. The barrier for self-diffusion on Ir(111) was estimated at 0.24 eV using RGL potentials by Kürpick [312], a result reasonably close to the value of 0.289 ± 0.003 eV from detailed observations [305].

Self-diffusion experiments on Ir(111) appear to be in a very good state. Some theoretical estimates for diffusion on Ir(111) are close to the experimental values, some not, but overall the situation for this material looks much better than for others.

5.21.2 Hetero-diffusion

In 1991, Wang and Ehrlich [313] carefully examined with FIM the sites on Ir(111) favored by different adatoms. They were able to do this by mapping out with atomic resolution where the adatoms were held on the surface. The maps so obtained were already shown in Fig. 3.19. Rhenium and tungsten adatoms were observed at hcp sites on Ir(111), but palladium adatoms favored fcc sites. Wang in 1992 [304] looked at the diffusion of rhenium atoms on Ir(111) and reported a barrier of 0.52 ± 0.01 eV to diffusion with a prefactor $D_o = 2.36(\times 1.51^{\pm 1}) \times 10^{-4}$ cm^2/sec. For jumps out of hcp sites he obtained an activation energy of 0.519 ± 0.007 eV and a frequency prefactor of $6.39(\times 1.57^{\pm 1}) \times 10^{11}$ sec^{-1}; for the return jump the activation energy was 0.378 ± 0.006 eV with a prefactor 2.07×10^{11} sec^{-1}. These results are illustrated in Fig. 5.85a.

For diffusion of tungsten Wang found an activation energy of 0.51 ± 0.01 eV with $D_o = 2.85(\times 1.89^{\pm 1}) \times 10^{-5}$ cm^2/sec. For jumps out of hcp sites he arrived at an activation energy of 0.506 ± 0.012 eV and a frequency prefactor 7.74×10^{10} sec^{-1}, and for the opposite jump the energy was 0.317 ± 0.008 eV with a prefactor of $3.56(\times 2^{\pm 1}) \times 10^{10}$ sec^{-1}. The results are shown in Fig. 5.85b. The prefactor for tungsten is quite low, but measurements were done in the same temperature range as for rhenium, with a tenfold higher prefactor. Wang and Ehrlich [314] observed that palladium atoms are mobile on Ir(111) at ~65 K. However, when palladium is deposited on an Ir(111) surface

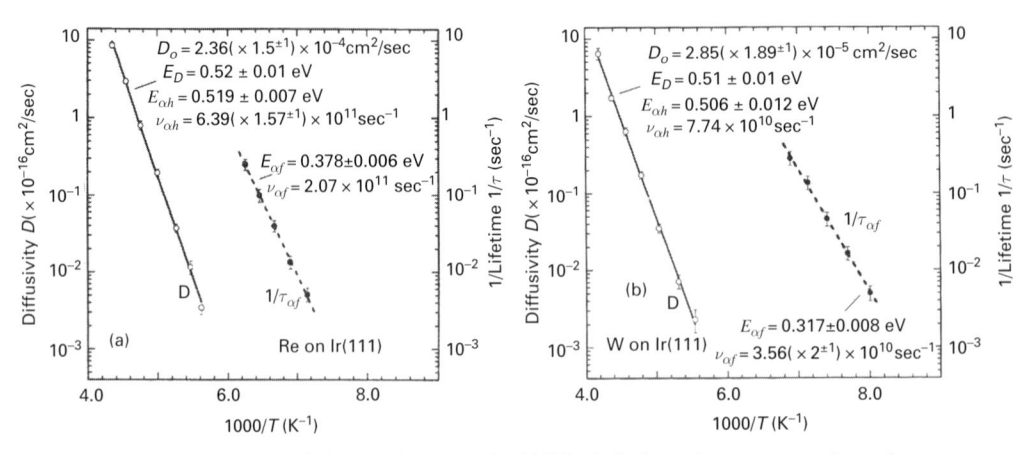

Fig. 5.85 Diffusion characteristics of adatoms on Ir(111) in their dependence upon reciprocal temperature derived from observations in FIM [304]. (a) For Re atoms. (b) For W atoms.

kept at ~20 K the distribution is random and atoms were immobile. The barrier to hopping from fcc to hcp sites was ~ 0.17 eV, and 0.13 eV for the jumps from hcp to fcc sites, clearly a significant difference.

Not enough results are available for hetero-diffusion to form any judgments about the reliability of these diffusion parameters, but the data were obtained by a direct technique and with good statistics, clearly an advantage.

5.22 Iridium: Ir(100)

5.22.1 Self-diffusion

Studies of iridium on Ir(100) have been reported by Chen and Tsong [246], who arrived at an activation energy of 0.93 ± 0.04 eV [315] and a prefactor of $1.4(\times 10^{\pm 1}) \times 10^{-2}$ cm^2/sec, shown in Fig. 5.81. Soon thereafter they published more extensive measurements [315], and as indicated in the Arrhenius plot in Fig. 3.55 they found a diffusion barrier of 0.84 ± 0.05 eV and a prefactor of $6.26(\times 11^{\pm 1}) \times 10^{-2}$ cm^2/sec. The sites occupied by the iridium atom after diffusion form a c(2 × 2) net illustrated in Fig. 5.86, and the final conclusion was that diffusion occurred by exchange of an adatom with an atom from the lattice.

Ir(100) was also carefully examined by Friedl et al. [316]. In self-diffusion of an Ir atom on a surface prepared by field evaporation at 30 K they again found a c(2 × 2) net of sites. Heating to 700 K and then allowing the atoms to diffuse, however, leads to a (1 × 1) net, and to a preference for jumps covering two spacings. Friedl et al. concluded that at low temperatures there exists a c(2 × 2) phase on the Ir(100) surface. However, if the low temperature phase were true, it should reveal itself regardless of the nature of the adatom. This was tested by Fu and Tsong [317] with rhodium as well as iridium adatoms, who found atom hopping for Rh and exchange for Ir, so this explanation did not withstand the test.

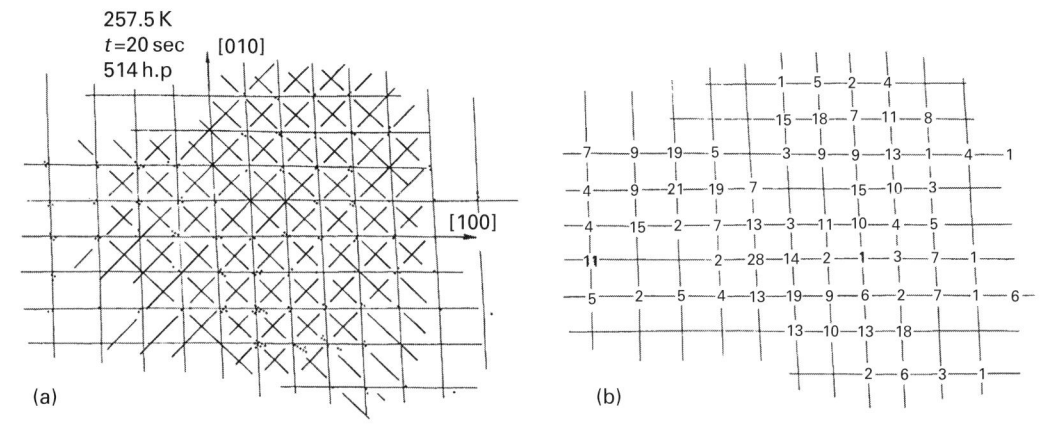

Fig. 5.86 Sites occupied by Ir atom after diffusion on Ir(100). (a) Map of sites indicates atom exchange mechanism in diffusion. (b) Frequency of site occupation (after Chen and Tsong [315]).

In 1998 Fu et al. [307] again studied self-diffusion on Ir(100) in the FIM and found an activation energy of 0.74 ± 0.02 eV and a prefactor of $7.30\times10^{-3\pm0.5}$ cm^2/sec. These values were reported again by Fu and Tsong [317] in 1999.

Theoretical work started with the investigation of Flahive and Graham [47] using Morse potentials; the data are presented in Fig. 4.9 and for self-diffusion on Ir(100) they found a barrier of ~0.97 eV. Molecular dynamics studies of self-diffusion were carried out by Shiang et al. [309] and diffusion by atom exchange occurred over a potential barrier of 0.79 eV in EAM calculations, while 0.77 eV using RGL potentials, in reasonable accord with experiments. Hopping on Ir(100) required a much higher activation energy of 1.58 eV using EAM and 1.57 eV from RGL. Boisvert et al. [208] used the full-potential linear muffin-tin-orbital method to find a barrier of 1.39 ± 0.04 eV for self-diffusion on the (100) plane, on the assumption that diffusion occurred by atoms hopping over the surface. In 1999, Feibelman and Stumpf [179] calculated diffusion barriers using density-functional theory. The activation energy for atom hops was found to be 1.13 eV, but only 0.51 eV for diffusion by atom exchange, not at all close to the experimental findings.

Davydov [49] using his cohesion approximation reached a barrier of 0.40 eV, much less than the experimental value and clearly much too low. No information about the diffusion mechanism was derived. Agrawal et al. [26] made estimates with Lennard-Jones potentials and found an activation barrier of 1.23 eV with a prefactor of 2.6×10^{-3} cm^2/sec for single hops. Exchange was not considered and the calculated results are too far from the experiments. The activation energy was evaluated by Chang and Wei [28] using VASP. They found a barrier of 1.17 eV, compared to 0.59 eV for atom exchange, which is rather lower than the value from experiments.

There is no real agreement between activation energies from experiments and from theory. Obviously the experimental barrier to self-diffusion on Ir(100) is not yet completely settled. What is interesting about this diffusion, however, is that the site distribution is c(2×2), suggesting that diffusion occurs by interchange of the adatom with a lattice atom.

5.22.2 Hetero-diffusion

Unfortunately hetero-diffusion on the Ir(100) surface has not been widely investigated. In addition to a lack of information about the energetics, it is also not clear what kind of atom will move on the surface, since both hopping and exchange are likely mechanisms. Diffusion of rhodium atoms on Ir(100) was investigated by Fu et al. [307], who reported a barrier of 0.80 ± 0.08 eV and a prefactor of $2.09\times10^{-3\pm1.5}$ cm^2/sec, as is shown in Fig. 5.87a. Rhodium turns out to diffuse on the Ir(100) surface by hopping not by atom exchange, as is clear from the site map in Fig. 5.87b. The same values were reported by Fu and Tsong in 1999 [317].

Tsong and Chen [318] also looked at rhenium atoms deposited on the Ir(100) surface. They found Re atoms not only exchange with the substrate, but also create a Re-Ir complex, making atom exchange in this system a two-part process, as in Fig. 3.58. More about this process is given in Chapter 3.

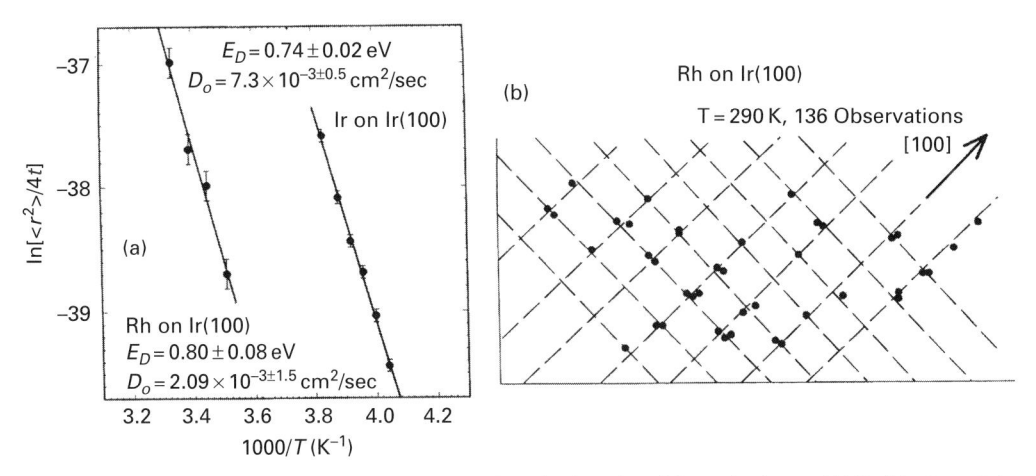

Fig. 5.87 Diffusion on Ir(100) in FIM. (a) Arrhenius plot for Rh and for Ir, both on Ir(100). (b) Map of sites occupied after diffusion of Rh on Ir(100) (after Fu et al. [307]).

5.23 Platinum: Pt(111)

The first observations of platinum diffusion on the platinum (111) plane were made with FIM by Bassett and Webber [319]. Platinum, iridium, and gold atoms were mobile on the surface at 77 K, the imaging temperature, and Bassett and Webber were unable to measure diffusivities. It was possible to investigate tungsten atoms, which are strongly bound, and they estimated an activation energy of 0.30 ± 0.1 eV assuming a value for the prefactor. They also used Morse potentials to calculate diffusion barriers and came up with a value of 0.07 eV for the barrier to self-diffusion on Pt(111). For Ir on Pt(111) they derived the same value.

5.23.1 Self-diffusion

In 1993 Henzler et al. [141] looked at the platinum on Pt(111) system in the sub-layer regime with SPA-LEED to derive the island density in nucleation. At a coverage of 0.5 ML they measured an activation energy of 0.12 eV for diffusion. It is not very likely that this value is valid for single atom motion. Feibelman et al. [320] made some simple observations on platinum in an FIM, cooling the sample to 65 K, and deduced a barrier of roughly 0.25 ± 0.02 eV for diffusion on Pt(111) with fcc sites preferred. This was based on measurements at a temperature of 100 K and assuming a standard prefactor. They also calculated a value of 0.38 eV for the diffusion barrier, using local-density-functional methods, and a preference of 0.18 eV for fcc against hcp sites.

The first STM experimental determination of the diffusion characteristics was carried out by Bott et al. [321]. They deposited a known number of platinum atoms on a Pt(111) surface in an STM and then heated the surface to different temperatures and allowed nucleation to occur. Assuming different values for the diffusivities they did Monte Carlo simulations to reproduce the experiments. From the temperature dependence the

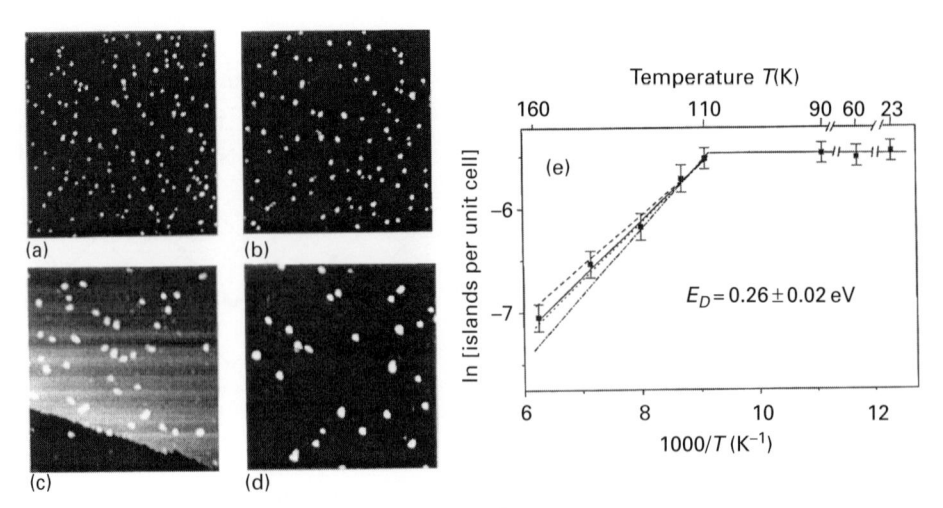

Fig. 5.88 STM images of Pt(111) after deposition of 4.2×10^{-3} ML of Pt at (a) 23 K (b) 115 K (c) 140 K and (d) 160 K. (e) Arrhenius plot for Pt island density. Black squares – experiments. Dotted line – simulation with $v_0 = 5 \times 10^{12}$ sec^{-1}, $E_D = 0.26$ eV. Other lines give simulations with barrier differing by ± 0.02 eV (after Bott et al. [321]).

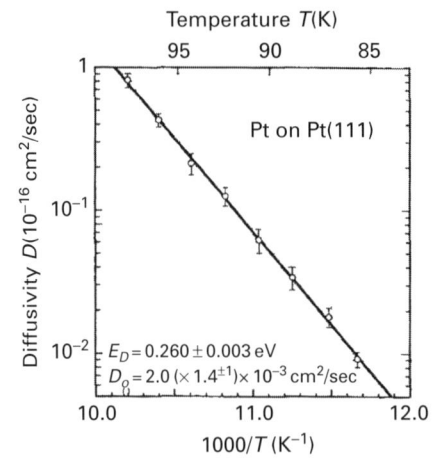

Fig. 5.89 Behavior of Pt atom diffusivity on Pt(111) surface obtain in FIM as a function of reciprocal temperature [305].

diffusion barrier of 0.26 ± 0.01 eV and a jump prefactor of $5 \times 10^{12 \pm 0.5}$ Hz were deduced. Measurements were done at two coverages, 4.2×10^{-3} and 1.0×10^{-3} monolayers. The same values for adatom motion were obtained from both measurements, as is illustrated in Fig. 5.88. Bott et al. claimed that direct observation of jump rates on Pt(111) with the STM was not possible due to the influence of the tip; this gave a low diffusion barrier of 0.17 eV and a prefactor of 1.3×10^{11} sec^{-1}. Measurements were done with a potential of 30 mV and a current of 0.2 nA.

Soon thereafter, Kyuno et al. [305] carried out detailed studies of diffusion in a standard field ion microscope cooled to ~20 K. The data, shown in Fig. 5.89, are in

5.23 Platinum: Pt(111)

excellent agreement with the previous STM study. They arrived at an activation energy of 0.260 ± 0.003 eV and a prefactor $D_o = 2.0(\times 1.4^{\pm 1}) \times 10^{-3}$ cm^2/sec. This good agreement is surprising, given the fact that quite different techniques were used in the measurements. Platinum atoms were observed to occupy fcc sites after heating; atoms were seen in hcp positions only after deposition on a cold surface, as was already clear in Fig. 3.24 [322]. Similarly to iridium, the type of binding site can be deduced from the shape of the atom image. Experimental data for Pt(111) are in exellent shape; there is agreement in the barrier obtained by two independent and different techniques, which is quite an achievement.

There has been a considerable number of theoretical estimates done for self-diffusion on Pt(111). First of course was the study carried out by Flahive and Graham [47] with Morse potentials, illustrated in Fig. 4.9. and they found a barrier around 0.05 eV. Among the first to employ semi-empirical potentials was Liu et al. [4], who used EAM potentials to evaluate an activation energy of 0.007 eV for hopping with AFW parameters, and 0.078 eV with VC; the prefactors were 1×10^{-4} cm^2/sec and 3.5×10^{-4} cm^2/sec. The energetics here are clearly much too low. Liu et al. also worked out the kinetics for Lennard-Jones and Morse potentials. With Lennard-Jones interactions, they reported a barrier of 0.19 eV and a prefactor 6.3×10^{-4} cm^2/sec; with Morse potentials the result, an activation energy of 0.07 eV, was less satisfactory.

Relying on EAM potentials, Villarba and Jónsson [323] in 1994 estimated a diffusion barrier for platinum atoms on Pt(111) of 0.08 eV. For Morse potentials the barrier proved to be 0.06 eV. In 1994 Liu et al. [324] used EMT to calculate a value of 0.13 eV for hopping. According to them the energy difference between fcc and hcp sites was only 0.01 eV. Wang and Fichthorn [325] did semi-empirical molecular dynamics and static potential energy calculations relying on CEM theory; on a relaxed Pt(111) surface they found a diffusion barrier of 0.038 eV; the value on a rigid surface was much higher, 0.18 eV. Stoltze [61] calculated the barrier using EMT potentials to be 0.159 eV. Diffusion of platinum atoms was examined by Jacobsen et al. [326], who found an activation energy of 0.16 eV using effective medium theory.

In 1996 Li and DePristo [62] used MD/MC CEM potentials to calculate a value of 0.048 eV for the barrier for hopping. Mortensen et al. [63] presented calculations of surface self-diffusion on several fcc(111) surfaces by the density-functional method. For diffusion on Pt(111) with a static substrate they found a barrier of 0.42 eV in the local-density approximation and 0.39 eV by the generalized gradient method. Feibelman [327] using the local-density approximation with the molecular dynamics VASP package [22–24] found a value of 0.29 eV, the first theoretical value close to the experimental results. Another density-functional theory estimate was made by Boisvert et al. [328] and arrived at a diffusion barrier of 0.33 eV from LDA as well as from GGA. For both they allowed the adatom and the top layer to relax. When the adatom and two top layers were relaxed the activation energy increased to 0.36 eV. Their work was done with (2×2) and (3×3) supercells, and they checked the influence of the number of layers in the system. For all cases the barrier was between 0.47–0.33 eV, somewhat on the high side. In the same year Ratsch and Scheffler [211], also using density-functional theory, found the value for the barrier to be equal to 0.15 eV.

Máca et al. [181,329] used RGL potentials for evaluating the activation energy of platinum self-diffusion as 0.17 eV. In 1999, Davydov [49] calculated diffusion barriers for a wide variety of metal surfaces by relying on the Wills and Harrison [83] theory of cohesion. For platinum jumps on the Pt(111) surface he arrived at a low value of 0.06 eV. Chang et al. [35] used density-functional calculations and found a barrier of 0.07 eV. Utilizing EAM interactions, Leonardelli et al. [330] came up with a very small activation energy of 0.046 eV for self-diffusion on Pt(111). They claimed that the Pt atom sank into the surface, creating a metastable 4-fold binding site, which has not been confirmed by experiment and makes estimation less reliable. Agrawal et al. [26] made calculations of the diffusion barrier for platinum atoms on Pt(111) using Lennard-Jones potentials and found a surprisingly good value of 0.20 eV and a prefactor of 1.2×10^{-4} cm^2/sec for atom jumps.

In 2005 Bulou and Massobrio [68] compared the exchange mechanism with atom hopping for Pt on Pt(111). They used a second-moment approximation of tight-binding theory and combined classical molecular dynamics with the nudged elastic band method to come up with an activation energy for hopping of 0.176 eV versus a much higher 2.105 eV for exchange. They also looked at the possibility of long-range exchange, but the energy for such an exchange was comparable with simple exchange, 2.111 eV versus 2.105 eV.

Recently Lee and Cho [331] presented an improved EAM simulation of non-bulk systems. For self-diffusion on the Pt(111) surface they used CY-EAM [332] and found a barrier of 0.096 eV. With CY-XEAM1 the value increased to 0.31 eV, and with CY-XEAM2 the value was even higher, 0.38 eV. Their results were either too low or too high and show why experiments are crucial to judge theoretical estimates of activation energies. For self-diffusion by atom jumps on Pt(111), Kim et al. [70] arrived at a barrier of 0.171 eV using molecular dynamics simulation with RGL interactions. For diffusion by atom exchange, the barrier was almost tenfold higher than expected – 1.627 eV. However, the estimate for atom jumps is much below the experimental value, as are many of the theoretical efforts. At the same time Yang et al. [333] relied on EAM potentials in their molecular dynamics simulations of a platinum atom diffusing on Pt(111). Their diffusion barrier turned out to be 0.19 eV with a prefactor of 5.1×10^{-3} cm^2/sec, in good agreement with earlier such estimates.

There are a huge number of theoretical estimates for this system and almost all of them are quite far off from the experimental barriers, which on the other hand are in excellent shape. The theoretical effort which stands out for agreeing with experiments is the work of Feibelman [327].

5.23.2 Hetero-diffusion

Blandin and Massobrio [334] resorted to molecular dynamics and embedded atom potentials to evaluate the mean-square displacement of silver atoms on Pt(111). From the Arrhenius plot in Fig. 5.90, they obtained a diffusion barrier of 0.058 ± 0.003 eV. The next year they did further calculations and came out with a value of 0.05 eV from static calculations. From an Arrhenius plot obtained using molecular dynamics simulations

5.23 Platinum: Pt(111)

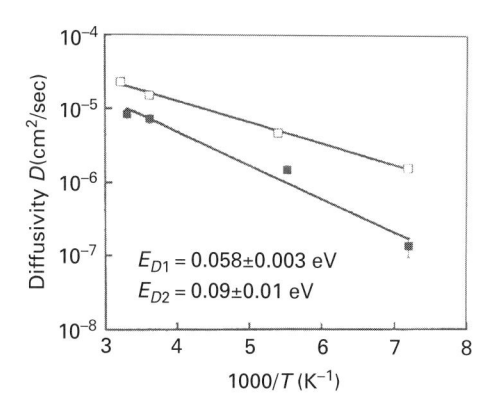

Fig. 5.90 Diffusivity of Ag on Pt(111) as a function of reciprocal temperature. Open squares – adatoms. Black squares – dimers. Data from molecular dynamics simulations using embedded atom potentials (after Blandin and Massobrio [334]).

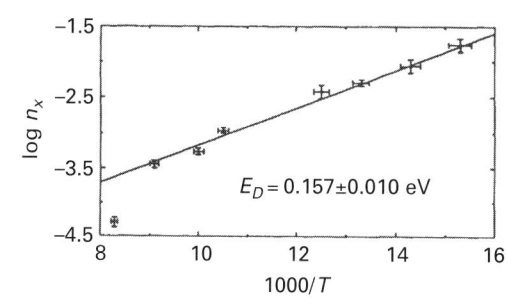

Fig. 5.91 Saturation Ag island density on Pt(111) measured with STM in its dependence upon the reciprocal temperature (after Brune et al. [337]).

they derived an activation energy of 0.060 ± 0.005 eV [335]. They also compared Arrhenius plots at low (140–300 K) and high temperatures (300–800K). From the former the activation energy amounted to 0.052 ± 0.005 eV, at higher temperatures the barrier increased to 0.08 ± 0.01 eV.

From scanning tunneling microscope studies of silver deposition on a Pt(111) surface, Röder et al. [336] were able to measure the density of islands at saturation for temperatures from 50 K to 140 K. This yielded a diffusion barrier of 0.14 ± 0.01 eV. Brune et al. [337] in 1994 probed the diffusion of silver on Pt(111) by relying on the kinetics of classical nucleation theory. Their measurements of the island density at temperatures from 65 to 120 K are given in Fig. 5.91, and lead to an activation energy of 0.157 ± 0.010 eV and a vibrational prefactor of $v_0 = 1 \times 10^{13 \pm 04}$ sec^{-1}. Diffusion of silver on Pt(111) was later examined by Brune et al. [204], who relied on effective medium theory to arrive at a barrier of 0.080 eV for hopping, rather different from the experimental value. Brune et al. [93] repeated these experiments, shown in Fig. 5.92, with a more careful analysis and obtained a diffusion barrier of 0.168 ± 0.005 eV and a frequency prefactor of $7 \times 10^{13 \pm 0.3}$ sec^{-1}.

The diffusion barrier of silver atoms on the (111) plane was also evaluated by Feibelman [338]; he used a local-density-functional calculation to find for the static

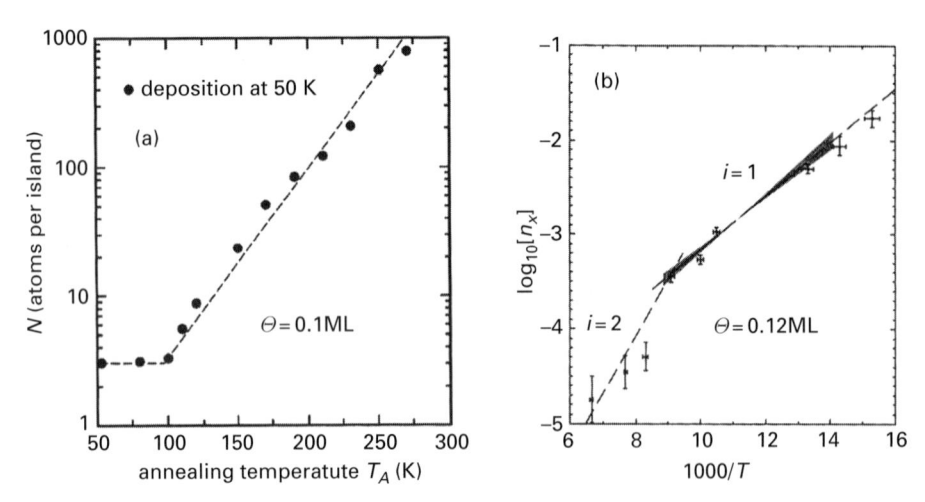

Fig. 5.92 Ag atoms on Pt(111) observed with STM. (a) Ag island size dependence upon annealing temperature. Increase, due to dimer dissociation, begins at $T>100$ K. (b) Saturation island density at 0.12 ML coverage as a function of reciprocal temperature. For critical island size $i=1$, diffusion barrier is 0.168 ± 0.005 eV (after Brune et al. [93]).

barrier a value of as 0.20 eV, reasonably close to the experiments. In his calculations, fcc sites are 0.03 eV more favorable than hcp sites. Ratsch et al. [210] also evaluated the activation energy for silver diffusion on Pt(111) using density-functional theory, and arrived at a value of 0.15 eV. There is quite good agreement of the experimental data with this study of Ratsch et al. [207].

Utilizing EAM interactions, Leonardelli et al. [330] looked at nickel on Pt(111), for which they found no metastable position similar to what they observed in self-diffusion. The activation energy for nickel diffusion was 0.178 eV. Nickel was also investigated by Kim et al. [70] with molecular dynamics simulations and RGL interactions. They derived a value of 0.114 eV. Unfortunately there is no experimental data for comparison.

In 2003, Sabiryanov et al. [339] looked at the influence of strain on the movement of Co on Pt(111) using *ab-initio* calculations [340] and the hopping barrier on the unstrained surface was estimated as 0.3 eV. Bulou and Massobrio [68] in 2005 used a second-moment approximation of tight-binding to calculate an activation energy for exchange of cobalt on Pt(111) as 2.065 eV. Long-range exchange required a higher energy of 2.248 eV, indicating that Co adatoms will be moving by jumps, not by exchange. Goyhenex [341] also employed the second-moment approximation to the tight-binding formalism to come up with an activation energy of 0.21 eV for cobalt atoms diffusing on Pt(111); the energy difference between fcc and hcp sites was negligible, less than 0.01 eV. From the calculated mean-square displacement at different temperatures and an Arrhenius plot, shown in Fig. 5.93, he derived a diffusion barrier of 0.23 ± 0.01 eV and a prefactor $D_o = 3.9 \times 10^{-5}$ cm^2/sec. It appears that jumping is the mechanism of movement for Co on the Pt(111) surface; exchange is very unlikely due to the huge energy required.

Graham and Toennies [342] looked at the diffusion of sodium on a Pt(111) surface with quasielastic helium atom scattering. They found that sodium at a coverage of

5.24 Platinum: Pt(100)

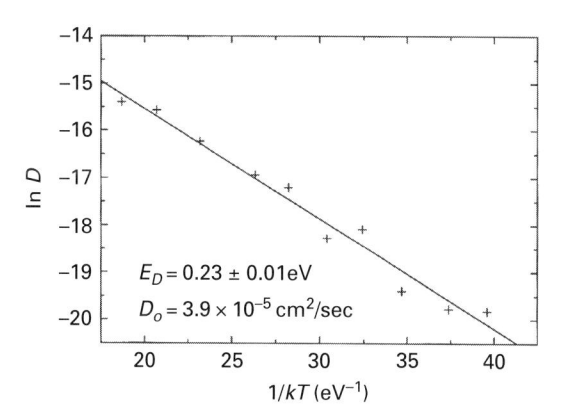

Fig. 5.93 Arrhenius plot for the diffusivity of Co atoms on Pt(111), leading to a diffusion barrier of 0.23 eV. Data obtained by molecular dynamics simulations (after Goyhenex [341]).

0.043 ML moved over the surface at a low temperature of 199 K, with an effective activation energy from an Arrhenius plot of 0.0215 ± 0.001 eV and a prefactor $(4.5 \pm 0.3) \times 10^{-4}$ cm^2/sec. The true diffusion barrier was estimated to be in the range of 0.030 to 0.044 eV, with $D_o = 6 \times 10^{-3}$ cm^2/sec. The coverage in this study should be suitable for an investigation of single atom movement.

Estimates for hetero-atom diffusion on Pt(111) were made by Kim et al. [70] from molecular dynamics simulations based on RGL interactions. The barriers were as follows: Ni 0.114 eV, Cu 0.111 eV, Pd 0.128 eV, Ag 0.090 eV, and Au 0.122 eV.

Only for Ag on Pt(111) is there enough data for comparisons. The experimental results look reasonable and are close to the density-functional estimates of Ratsch et al. [210].

5.24 Platinum: Pt(100)

The motivation for work on Pt(100) was provided by DeLorenzi and Jacucci [3] in 1985, who did molecular dynamics simulations of diffusion phenomena. They simulated atomic behavior using a potential derived by Price [343]. On a bcc(211) surface they observed primarily diffusion along the channels, with occasional exchange events involving a lattice atom from the <111> row. On (100), motion by single jumps was observed, but most interesting were more complicated transitions, indicated in Fig. 3.52. In these, an adatom enters the surface layer, replacing an atom from the lattice, which moves to the surface. These events, previously modeled in Fig. 3.53, provided the stimulus for closer examination of diffusion on (100) surfaces.

5.24.1 Self-diffusion

In 1990, Kellogg and Feibelman [344] finally examined self-diffusion on Pt(100) using the field ion microscope. Atoms were found to move diagonally, creating a c(2 × 2) net of sites such as shown in Fig. 3.54, which indicates atom exchange. From the mean-square displacement at 175 K, Kellogg and Feibelman estimated an activation energy of 0.47 eV for diffusion, assuming a prefactor of 1×10^{-3} cm^2/sec. Kellogg [345] has further looked at diffusion on the (100) plane of platinum. From an Arrhenius plot of the diffusivity, in Fig. 3.56, he arrived at an activation energy of 0.47 ± 0.01 eV and a prefactor $D_o = 1.3(\times 10^{\pm 1}) \times 10^{-3}$ cm^2/sec for self-diffusion.

A few years later, in 1994, Kellogg [346,347] looked at the effect of an electric field on the diffusion of Pt atoms on Pt(100), which occurs by atom exchange. These introductory studies showed an interesting effect. A negative field (that is with a negative voltage on the emitter) increased the diffusivity, a positive field diminished it. Quite different is the behavior on Pt(311), on which self-diffusion is in-channel and occurs by hopping. Here the field has really no effect. At higher positive fields on Pt(100), the activation energy for diffusion continued to rise and at 1.5 V/Å the map of sites at which atoms are seen, shown in Fig. 5.94, changed from c(2 × 2) to (1 × 1), suggesting a change from atom exchange to hopping.

The scanning tunneling microscope was used by Linderoth et al. [348] to measure the saturation density of platinum islands at a coverage of 0.07 ML on the reconstructed Pt(100)-hex surface in Fig. 5.95. Diffusion on this plane was highly anisotropic. They concluded that dimer mobility was negligible in the investigated range of temperatures

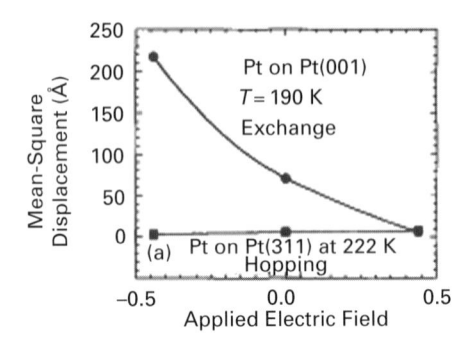

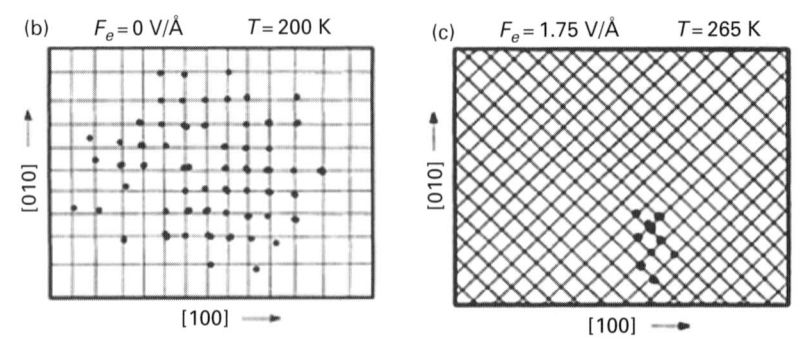

Fig. 5.94 Effect of electric field on diffusion of platinum on Pt(001) plane, observed in FIM. (a) Mean-square displacement of Pt in its dependence upon the applied electric field. (b) Sites filled after diffusion in zero field indicate movement by exchange with a lattice atom. (c) In electric field and at higher temperature, distribution suggests single atom hops (after Kellogg [346,347]).

5.24 Platinum: Pt(100)

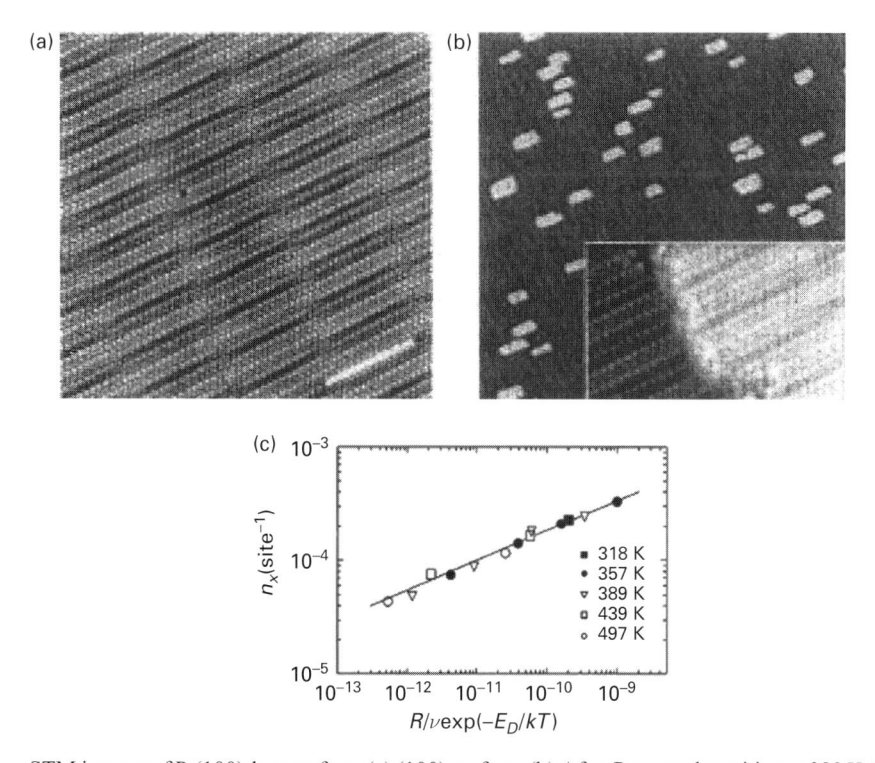

Fig. 5.95 STM images of Pt(100)-hex surface. (a) (100) surface. (b) After Pt atom deposition at 389 K. Inset: resolved edge of island. (c) Platinum island density measured at different temperatures (after Linderoth et al. [348]).

and used Monte Carlo simulations to explain their measurements, arriving at a diffusion barrier of 0.43 ± 0.03 eV. In 1998, Mortensen et al. [349] continued the study of Linderoth et al. [348] with an STM examination of platinum deposition on Pt(100)-hex. They found that diffusion along the six-atom wide channels dominated.

Only a small number of theoretical estimates has been made for diffusion on Pt(100), starting with the investigation by Flahive and Graham [47] using Morse potentials, shown in Fig. 4.9; they calculated a barrier of 0.8 eV. Early calculations were done by Liu et al. [4] with EAM interatomic potentials. With parametrization according to AFW they found an activation energy of 0.44 eV for atom jumps and a prefactor of 8×10^{-4} cm^2/sec. With VC parameters, they arrived at a much higher barrier of 1.25 eV, which also gave a higher prefactor of 5.0×10^{-3} cm^2/sec. Atom exchange occurred over a smaller activation energy of 0.31 eV with AFW parameters, and 0.64 with VC potentials, which gave a prefactor of 1×10^{-2} cm^2/sec. It is clear that exchange not hopping will dominate movement. Using Lennard-Jones potentials, hopping required overcoming a barrier of 1.05 eV and involved a prefactor of 4.9×10^{-3} cm^2/sec; with Morse interactions the barrier was lower, 0.82 eV. For atom exchange, Lennard-Jones interactions yielded a huge barrier of 3.97 eV and a prefactor of 4×10^{-2} cm^2/sec, showing that the L-J potential is not suitable for modeling processes on platinum surfaces.

Roughly at the same time Lynden-Bell [176] relied on molecular dynamics calculations for self-diffusion of platinum using the Sutton–Chen potential. She found a barrier of 0.611 ± 0.004 eV and a prefactor of $3.5(\times 0.2^{\pm 1}) \times 10^{-2}$ cm^2/sec for atom exchange. Sanders and DePristo [76] did transition-state theory estimates of atom hopping based on MD/MC-CEM interactions and reported an activation energy of 0.72 eV as well as a frequency prefactor of 2.5×10^{12} sec^{-1}. For hopping of platinum atoms, Stoltze [61], using effective medium interactions, estimated a barrier of 0.689 eV. In 1994 Villarba and Jónsson [323] used EAM potentials to obtain a barrier of 0.54 eV for atom exchange on the Pt(100) surface.

Calculations by density-functional theory were carried out by Feibelman and Stumpf [179] in 1999. For diffusion by atoms jumping over the surface the activation energy proved to be 1.04 eV. For diffusion by atom exchange with the substrate the barrier was only 0.38 eV, clearly the lower energy process, but lower than the experimental values. Davydov [49], with his cohesion based calculations, reported a barrier of only 0.07 eV for platinum jumps over this surface, definitely too low and out of range. Estimates were done by Zhuang and Liu [350] who derived an activation energy for exchange of 0.989 eV from conventional EAM and 0.865 eV using the potential of Oh and Johnson [58]. Zhuang and Liu [119] continued with the surface embedded atom method [234,351,352]. They arrived again at an activation energy of 0.98 ± 0.04 eV, in very good agreement with their static value of 0.99 eV to simple exchange, but much higher than the experimental value. For hopping they found a barrier of 1.84 eV. Further work by Zhuang and Liu [353] came up with a new mechanism for self-diffusion on Pt(100). In this process, labeled ad-dimer diffusion, an adatom moving into the lattice displaces two atoms from their sites, of which the second one remains on the surface as an adatom. Zhuang and Liu estimated a barrier 0.02 eV higher for conventional exchange than for ad-dimer diffusion, which should certainly participate in migration.

Estimates were made by Agrawal et al. [26] with Lennard-Jones potentials for atomic interactions, and once more show the importance of many-body effects in modeling platinum surfaces. They gave a barrier height of 1.02 eV and a prefactor of 2.1×10^{-3} cm^2/sec for atom jumps on this surface. In 2004, Xiao et al. [354] looked at a strained Pt(100) surface using the embedded atom method. They first examined exchange in a (5×5) cell and found an activation energy of 0.33 ± 0.05 eV. When they increased the cell to (10×20), the activation energy decreased to 0.27 ± 0.05 eV; a further increase of the cell size did not lower the energy. For hopping they found a barrier of ~ 0.56 eV± 0.05 eV for a (5×5) cell and 0.36 ± 0.05 eV for a 10×20 cell. Chang and Wei [28] in 2005 arrived at a very high energy of 1.2 eV for atom hopping on Pt(100) using the *ab initio* VASP package [22–24]. For diffusion by atom exchange the barrier was quite low, 0.31 eV. Kim et al. [70] found a barrier of 0.875 eV for Pt jumps using molecular dynamics estimates based on RGL potentials. For atom exchange, the barrier dropped to 0.388 eV, values which are again different from previous work.

The barriers calculated for self-diffusion over Pt(100) are all over the map, but the original work of Liu [4] with AFW-EAM potentials came quite close to the experiments.

5.24.2 Hetero-diffusion

For atoms other than platinum the experimental situation is not as good, since detailed measurements over a range of temperatures were not always possible. Kellogg et al. [355], however, from observations at one temperature of the diffusion of palladium deduced an activation energy of 0.70 ± 0.01 eV assuming a prefactor of 10^{-3} cm^2/sec. Note the much higher activation energy for diffusion of palladium compared to platinum, which is undoubtedly due to the difference in diffusion mechanisms – palladium diffused by hopping, as can be seen from the site map shown in Fig. 3.57. Kellogg looked briefly at nickel as well and found atom exchange occurring with the lattice.

Sanders and DePristo [76] calculated hetero-diffusion characteristics for atom hopping on Pt(100), and obtained the following results: nickel diffusion barrier 0.57 eV, prefactor 4.5×10^{12} sec^{-1}; copper 0.61 eV, 4.5×10^{12} sec^{-1}; rhodium 0.57 eV, 3.3×10^{12} sec^{-1}; palladium 0.49 eV, 3.2×10^{12} sec^{-1}; silver 0.41 eV, 9.3×10^{12} sec^{-1}; gold 0.58 eV, 2.1×10^{12} sec^{-1}. Hetero-atom diffusion barriers have also been given by Kim et al. [70], based on molecular dynamics simulations and RGL potentials. The barriers were 0.686 eV for Ni, 0.615 eV for Cu, 0.627 eV for Pd, 0.508 eV for Ag, and 0.507 eV for Au. No exchange was investigated in this study. The energy barrier for palladium lies somewhat closer to experiment than the findings of Sanders and DePristo [76].

The amount of information available about hetero-diffusion on Pt(100) is limited. The mechanism of Pd and Ni diffusion is clear, but other adatoms need further investigations. We note that the theoretical estimates did not consider the exchange mechanism at all.

5.25 Gold: Au(100)

5.25.1 Self-diffusion

The first studies of the self-diffusion for gold were done by Günther et al. [356] on the reconstructed Au(100)-hex surface. They relied on scanning tunneling microscopic examination of growth and determined the island density at saturation when 0.2 monolayers were deposited at different temperatures, as shown in Fig. 5.96. Interpretation of the measurements was not simple, as there were doubts about the critical island size, and diffusion was highly anisotropic. Their final decision was that the critical size at low temperatures was equal to three, yielding a diffusion barrier of ≈ 0.2 eV.

Liu et al. [357] offered a different set of assumptions to interpret the data of Günther et al.; using Monte Carlo simulations they arrived at a value of 0.35 eV for the barrier to atomic diffusion and 0.45 eV for the dimer, which gave reasonable agreement with experiments. Obviously the situation is not clear. Göbel and von Blanckenhagen [358] then did single surface scratch studies, measuring the width with an AFM, and came up with a diffusion barrier of 0.4 eV and a prefactor of 5×10^{-8} cm^2/sec. This work was done in air on a gold sphere and diffusion does not yield information about the movement of individual atoms. In any event, the value of the prefactor is so low that these results are in serious doubt.

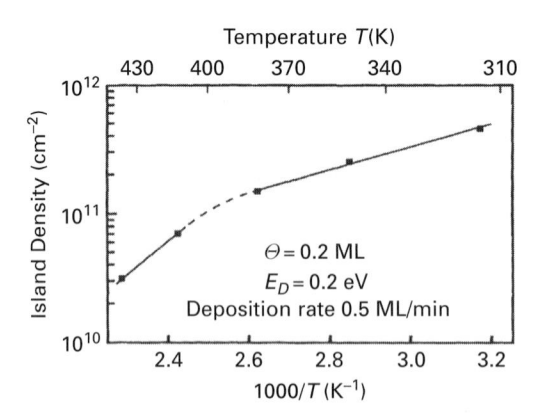

Fig. 5.96 Mean Au island density observed with STM on hex-reconstructed Au(100) surface as a function of the reciprocal temperature (after Günther et al. [356]).

It appears that at the moment there is no really reliable value from experiments for the barrier to diffusion of gold atoms on any gold surface.

In 1980 Flahive and Graham [47] used Morse potentials to estimate the barrier as 0.6 eV, as shown in Fig. 4.9. There have, however, been quite a number of more recent theoretical efforts to estimate diffusivities on gold surfaces. Liu et al. [4], with EAM atomic potentials, calculated a barrier to atom hops of 0.64 eV and a prefactor $D_o = 8 \times 10^{-4}$ cm^2/sec using AFW parameters; with VC potentials the barrier was even higher, 0.84 eV, and the prefactor amounted to 3.5×10^{-3} cm^2/sec. For diffusion by atom exchange the barrier with AFW potentials turned out to be 0.30 eV, and for VC parameters 0.32 eV, with a prefactor of 1×10^{-2} cm^2/sec. This suggests the exchange mechanism as the leading one, but this might change due to the reconstruction of this surface. The next calculations were done by Sanders and DePristo [76] using MD/MC-CEM for estimating interactions; they obtained a diffusion barrier for jumps of 0.67 eV and a frequency prefactor of 2.2×10^{12} sec^{-1} using transition-state theory. Stoltze [61], with effective medium theory potentials, looked at gold jumps and found activation energies of 0.49 eV.

Molecular dynamics simulations, based on embedded atom interactions, were done by Boisvert et al. [78]. For jumps the activation energy proved to be 0.43 eV with a prefactor of 2.11×10^{13} sec^{-1}, and for diffusion by atom exchange the numbers were 0.25 eV and 9.78×10^{12} sec^{-1}. In this instance, diffusion clearly occurred by atom exchange. This study was continued with the full-potential linear-muffin-tin-orbital method by Boisvert et al. [208] to find a barrier of 0.62 ± 0.04 eV. Only hopping was considered in this study. Further study in the same group with molecular dynamics simulations and EAM [359] derived for atom jumps an activation energy of 0.49 eV, and a prefactor 1.56×10^{-2} cm^2/sec; from TST the barrier was essentially the same, 0.50 eV. The barrier for atom exchange proved to be much smaller, 0.26 eV with a prefactor of 5.8×10^{-3} cm^2/sec, and 0.28 eV in TST estimates.

Diffusion of gold atoms on the hexagonally reconstructed Au(100) surface was probed by Bönig et al. [360], who resorted to EMT potentials for the interactions. The atom paths in both the x- and y-directions on this surface are indicated in Fig. 5.97a. Total energy

5.25 Gold: Au(100)

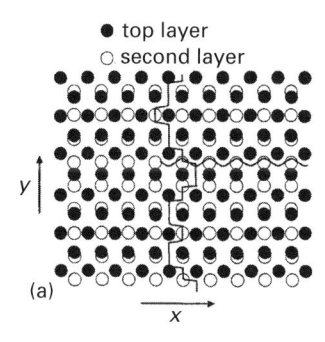

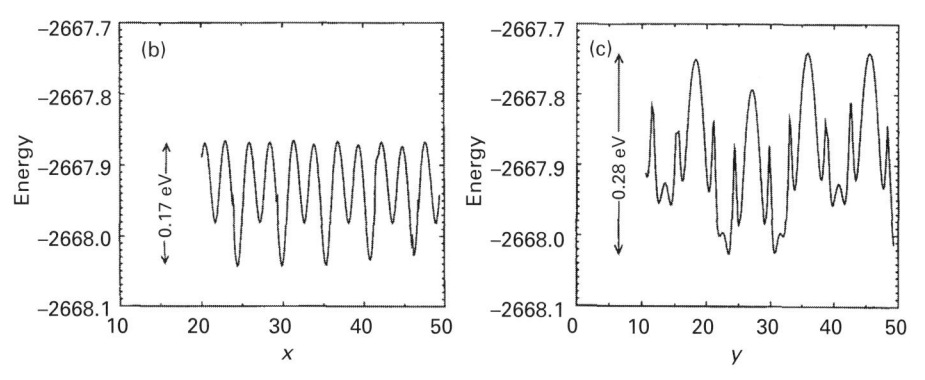

Fig. 5.97 Diffusion of Au on Au(100). (a) Paths for diffusion on reconstructed (100) along both x- and y-direction. Total energy obtained with EMT potentials in diffusion of adatom along (b) x-direction, (c) y-direction (after Bönig et al. [360]).

estimates as an adatom moves in the x-direction are plotted in Fig. 5.97b. Also shown, in Fig. 5.97c, is the variation of the energy along the y-axis. For diffusion along the x-axis the barrier was between 0.15 eV and 0.17 eV. The more complex diffusion along the y-axis has as its largest barrier the value of 0.28 eV.

In 1997 Yu and Scheffler [361] used density-functional theory to explore the possibility of exchange on fcc(100) planes. For self-diffusion on Au(100) they found that exchange is clearly favorable with an activation energy of 0.65 eV from the local-density approximation [25] and 0.40 eV from the generalized gradient approach (GGA), using a (3×3) cell. This compares with a barrier for hopping of 0.83 eV from LDA and 0.58 eV from GGA. Calculations were done by Mehl et al. [82] with EAM potentials using a 3×3 cell; they came up with a diffusion barrier of 0.70 eV for hopping on the surface. Davydov [49] also made estimates of gold diffusion in his cohesion approximation; it proved to be negative, an unrealistic −0.03 eV. In 2000, Baletto et al. [214] looked at the movement of single atoms of gold on gold Wulff polyhedra using RGL potentials. On a 201 atom Wulff polyhedron, consisting of 9 atoms for the (100) facet, the atom moved by exchange with an activation energy of 0.41 eV, whereas the hopping barrier was 0.53 eV. On a 1289 atom Wulff-cluster the activation energy for jumps was slightly higher, on the order of a few hundredths of an electron volt. Agrawal et al. [26], relying on Lennard-Jones estimates, evaluated a barrier of 0.64 eV with $D_o = 1.5 \times 10^{-3}$ cm^2/sec.

Chang and Wei [28] in 2005 resorted to density-functional theory and found a high barrier of 0.86 eV for atom hopping. For atom exchange the activation energy was only 0.32 eV. Müller and Ibach [127] did calculations using the local-density approximation to arrive at an activation energy of 0.64 eV. Estimates of the barrier to diffusion by atom exchange were also made, and yielded a similar result, 0.60 eV. Kim et al. [70] calculated a barrier of 0.531 eV, compared with 0.388 eV for atom exchange.

Right now, results for self-diffusion of gold are not in good shape – there is really no convincing experimental work, and theoretical estimates have yielded widely spread values. At the same time gold (100) surfaces are likely to reconstruct, making comparison of experiment and theory more chellanging. Most theoretical work predicts exchange rather than hopping as the mechanism of movement, although some do not consider exchange at all.

5.25.2 Hetero-diffusion

In 1993 He and Wang [362] examined the room temperature growth of a submonolayer of Fe on Au(100). From HRLEED and AES they claimed to see an atomic exchange of Fe with Au. No diffusion analyses were performed. Another indication of the exchange of iron atoms with Au(100) at room temperature was derivered by Hernán et al. [363] using STM and MC simulations, but again movement was not characterized energetically.

Sanders and DePristo [76] using MD/MC-CEM also examined hetero-diffusion on Au(100) and found the following characteristics: nickel barrier 0.61 eV, prefactor 4.6×10^{12} sec^{-1}; copper 0.66 eV, 5.0×10^{12} sec^{-1}; rhodium 0.62 eV, 3.6×10^{12} sec^{-1}; palladium 0.53 eV, 3.7×10^{12} sec^{-1}; silver 0.32 eV, 5.8×10^{12} sec^{-1}; platinum 0.82 eV, 2.5×10^{12} sec^{-1}. All of these estimates were for atom hopping on an unreconstructed surface, and without regard for the possibility of diffusion by atom exchange.

Estimates of hopping barriers for hetero-diffusion were also made by Kim et al. [70], who relied on molecular dynamics simulations using RGL interactions on Au(100). They found 0.591 eV for diffusion of Ni, 0.547 eV for Cu, 0.582 eV for Pd, 0.437 eV for Ag, and 0.994 eV for Pt, all for atom jumps.

Experiments, but also the many calculations, leave the activation energies for diffusion on gold in some doubt, and more definitive studies are clearly in order.

5.26 Gold: Au(111)

5.26.1 Self-diffusion

There is no experimental information about the diffusion of single atoms on the Au(111) surface. This surface reconstructs, making the analysis more difficult. There are, however, quite a few theoretical determinations available.

5.26 Gold: Au(111)

In 1980 Flahive and Graham [47] used Morse potentials, as shown in Fig. 4.9, and derived a value of ~0.05 eV for the diffusion barrier on Au(111). There have been quite a number of more recent theoretical efforts to estimate diffusivities as well. These started with the work of Liu et al. [4], using EAM atomic potentials. A diffusion barrier of 0.021 eV was estimated using AFW potentials, with a prefactor of 2×10^{-4} cm^2/sec; with VC potentials they found a barrier of 0.038 eV and a prefactor 3.7×10^{-4} cm^2/sec. Stoltze [61], with effective medium theory potentials, found an activation energy of 0.102 eV.

Ferrando and Tréglia [209] relied on RGL potentials and molecular dynamics simulations to obtain an activation barrier of 0.12 eV and a prefactor of 3.4×10^{-4} cm^2/sec, shown in Fig. 5.52; the binding energy difference between fcc and hcp positions was equal to 0.005 eV. Molecular dynamics simulations, based on embedded atom interactions, were done by Boisvert et al. [78] and for atom jumps they found a barrier of 0.013 eV with a prefactor of 0.698×10^{12} sec^{-1}. The full-potential linear-muffin-tin-orbital method was also carried out by Boisvert et al. [208] to find a self-diffusion barrier of 0.22 ± 0.03 eV. Boisvert and Lewis [359], resorting again to EAM potentials made molecular dynamics studies, as shown in Fig. 5.98. The activation energy for the atomic jumps proved to be 0.015 eV, with a prefactor of 7×10^{-5} cm^2/sec, in good agreement with the diffusivity as well as with transition-state theory, for which the barriers were 0.014 eV and 0.016 eV.

Density-functional calculations were made by Mortensen et al. [63] in the local-density approximation. They arrived at a barrier of 0.20 eV; in the generalized gradient method the barrier proved to be somewhat lower, 0.15 eV, but still much higher than in previous studies. In the same year Li and DePristo [62] used corrected medium theory to understand homoepitaxial growth on fcc(111) surfaces. For movement of a single gold atom on the gold (111) plane they found an activation energy of 0.029 eV.

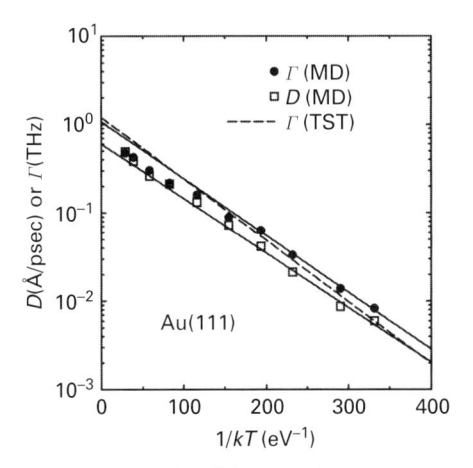

Fig. 5.98 Dependence of the diffusivity of Au adatoms on Au(111) upon the reciprocal temperature, calculated with EAM interactions. Diffusion barriers: from jump rate 0.015 eV, from diffusivity 0.014 eV, from TST 0.016 eV (after Boisvert and Lewis [359]).

Davydov [49] also made estimates of gold diffusion in his cohesion approximation and surprisingly found a barrier equal to zero. In 2000, Baletto et al. [214] looked at the movement of single atoms of gold on gold Wulff polyhedra using RGL potentials. The jump from hcp to fcc site required 0.11 eV. With embedded atom potentials Chang et al. [35] in the same year estimated a barrier of 0.04 eV. Agrawal et al. [26], relying on Lennard-Jones estimates, evaluated a barrier height of 0.12 eV and a prefactor of 8.5×10^{-5} cm^2/sec. In 2002 Liu [364] examined the strained $(23 \times \sqrt{3})$ Au(111) surface. Using the glue potential [365,366], he observed anisotropic movement on such a plane with the diffusion barrier changing from 0.12 to 0.8 eV for different positions.

Bulou and Massobrio [68] found a barrier to hopping of 0.112 eV and a much higher value, 0.878 eV for exchange, again using the second-moment approximation to tight-binding. Kim et al. [70] calculated a barrier of 0.117 eV for hops and 0.861 eV for exchange.

On Au(111), investigators have rarely seen a difference between fcc and hcp sites, and it is clear that movement should proceed by hopping. The estimated activation energies are rather spread out and movement on the reconstructed surface has rarely been investigated.

5.26.2 Hetero-diffusion

Diffusion of aluminum atoms on the (111)-$(22 \times \sqrt{3})$ plane of gold has been examined by Fischer et al. [367]; they determined the island density at a coverage of 0.1–0.15 ML at different temperatures and applied kinetic Monte Carlo simulations. Based on their Arrhenius plot they concluded that the critical size equaled one for temperatures lower than 230 K. In the temperature range from 100 to 200 K, shown in Fig. 5.99, they found an activation energy of 0.030 ± 0.005 eV for diffusion and a prefactor $v_0 = 7 \times 10^{3 \pm 1}$ sec^{-1}. The small value of the activation energy, but particularly the prefactor, casts doubt on this result, and raises concerns about the possible role of impurities and interactions on these findings. For temperatures higher than 245 K their distributions were bimodal and agreed with a critical nucleus of size $i = 0$, a sign of atomic exchange. Aluminum atoms sit on the Au(111) reconstruction–corrugation line, but there is no preferred place on such a line. Ovesson et al. [19] used density-functional theory calculations with GGA, and for Al on the Au(111) reconstructed surface derived a barrier of 0.12 eV and a frequency prefactor of 7×10^{12} sec^{-1}.

Meyer et al. [368] looked at the preferential adsorption sites for Ni atoms on a Au(111) herringbone-reconstructed surface, and explained the preferred sites by exchange of Ni atoms at the elbows. No energetics for this process were derived.

Adsorption of cobalt on Au(111)-$(22 \times \sqrt{3})$, also named as hex, was investigated by Goyhenex et al. [369] using tight-binding quenched molecular dynamics simulations. They derived the activation energy for diffusion which was measured as the difference between maximum and minimum in adsorption energies in adjacent places. For the discommensuration lines this difference was 0.18 eV, while in the fcc region 0.12 eV and in hcp 0.1 eV. This probably caused the slower diffusion across discommensuration lines with respect to other parts of the surface. Cobalt atom diffusion on unreconstructed

5.26 Gold: Au(111)

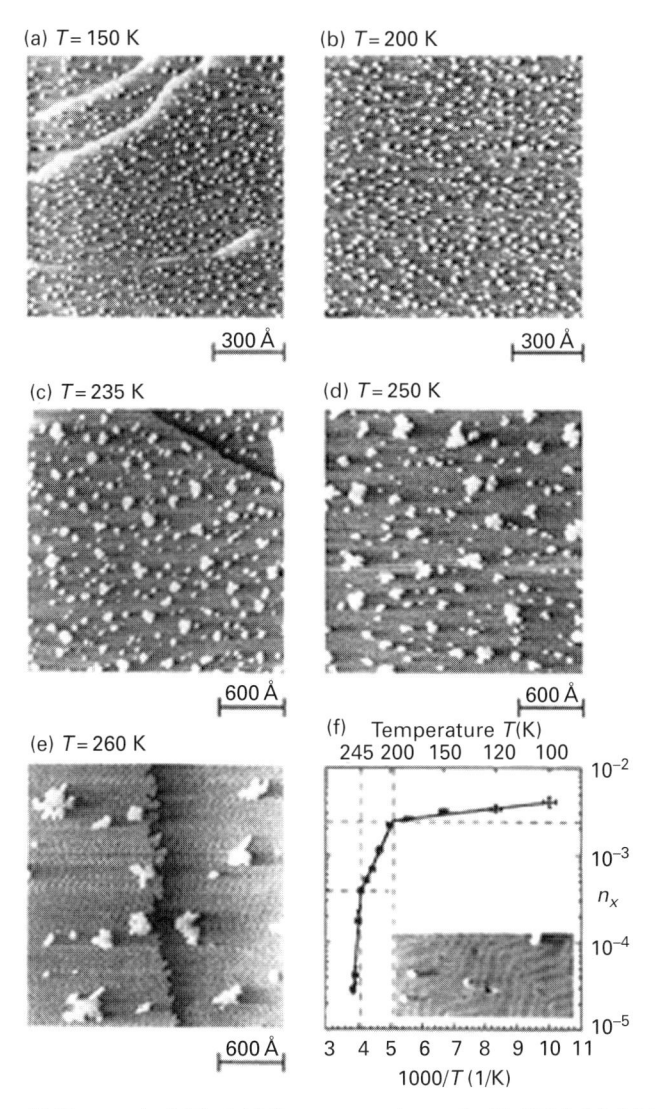

Fig. 5.99 Al films on Au(111), 0.1 ML coverage observed with STM. (a)–(e) Films deposited at different temperatures, at an impingement of 3.1×10^{-4} ML/sec. (f) Saturation Al island density as a function of reciprocal temperature. Diffusion barrier up to 200 K – 0.030 ± 0.005 eV (after Fischer et al. [367]).

Au(111) was tackled by Bulou et al. [370] in the second-moment approximation to tight-binding. From the mean-square displacement at different temperatures they were able to come up with an Arrhenius plot, shown in Fig. 5.100, which yielded an activation energy of 0.160 ± 0.003 eV and a prefactor $D_o = (5.8 \pm 0.9) \times 10^{-4}$ cm^2/sec. Static energy estimates gave a barrier of 0.155 eV for jumps from hcp to fcc sites and 0.137 eV for jumps in the opposite direction. Up from 810 K exchange started to be observed, with an activation energy of 1.1 eV. Two years later, Bulou and Massobrio [68] found an activation energy

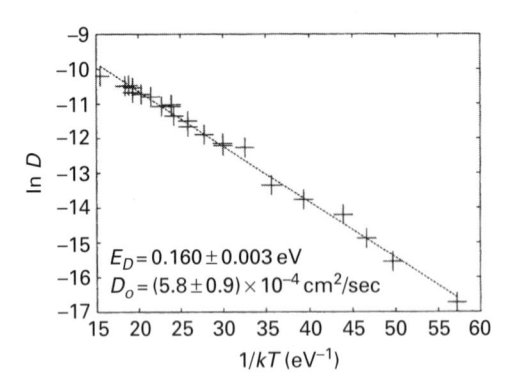

Fig. 5.100 Dependence of calculated Co diffusivity on Au(111) upon reciprocal temperature (after Bulou et al. [370]).

of 0.536 eV for atom exchange of cobalt on non-reconstructed Au(111) using a second-moment approximation to tight-binding theory. They [371] also looked at the behavior of Co atoms on the Au(111) herringbone-reconstructed surface, and realized that the longest displacements occurred for Co atoms which started on fcc or hcp areas and ended in discommensuration line areas (DL). Cobalt atoms in DL areas moved along lines with smaller displacements than in fcc and hcp areas. They also observed an exchange mechanism at 600 K, 47.5% on DL areas, 47.5 % in hcp areas and 5% in fcc areas.

Estimates of hetero-diffusion barriers were also made by Kim et al. [70], who did molecular dynamics simulations using RGL interactions. They estimated for atom jumps values for hopping as: Ni 0.078 eV, Cu 0.089 eV, Pd 0.111 eV, Ag 0.075 eV, and Pt 0.113 eV. These values are not in good agreement with previous estimates.

Experiments, but also the many calculations, leave the activation energies for diffusion on gold in some doubt, and more definitive studies are clearly in order.

5.27 Lead: Pb(111)

Regrettably there is hardly any information in the literature about diffusion on lead surfaces. However, Li et al. [372] did carry out calculations for surface self-diffusion on Pb(111) using EAM potentials and arrived at a barrier height of 0.045 eV.

5.28 Bismuth: Bi(111)

Recently homoepitaxial growth of submonolayers of Bi on a Bi(111) surface was investigated by Jnawali et al. [373] using SPA-LEED and STM at temperatures in the range 80–200 K. Based on nucleation theory for a coverage of 0.5 ML, they came out with a value of 0.135 eV for the diffusion barrier on the flat bismuth terrace. There are no other data available to compare this value.

5.29 Tables for 2D diffusion

Table 5.7 Fcc(100).

System	Mechanism	Experiment E_D (eV)	Experiment D_o (cm²/sec)	Theory E_D (eV)	Theory D_o (cm²/sec)	v_0 (sec⁻¹)	Technique	Ref.
Al/Al(100)	Jump			0.8			LDA	[1]
				0.65			LDA	[2]
				0.4	2×10^{-3}		EAM AFW	[4]
				0.46	1.5×10^{-2}		EAM VC	[4]
				0.6	1×10^{-2}		L-J	[4]
				0.45–0.58			EAM	[7]
				0.692			EMT	[9]
				0.74			DFT LDA	[11]
				0.30			EMT***	[12]
				0.51			DFT GGA	[13]
				0.57			DFT	[19]
				0.20	2.4×10^{-3}		RGL	[20]
				0.53			EAG	[18]
				0.52			LDA VASP^N	[21]
				0.5			GGA VASP^N	[21]
				0.60	6.1×10^{-3}		L-J	[26]
				0.46			VASP	[27]
				0.67			VASP	[28]
	Jump [100]			0.81			DFT LDA	[11]
	Ex			0.20			LDA	[2]
				0.25	4×10^{-2}		EAM VC	[4]
				0.69			EAM AFW	[4]
				2.28	9×10^{-2}		L-J	[4]
				0.1–0.24			EAM	[7]
				0.271			EMT	[9]
				0.35			DFT	[10]
				0.27			DFT LDA	[11]
				1.11	9.4×10^2		RGL	[20]
				0.56			EAG	[18]

Table 5.7 (cont.)

System	Mechanism	Experiment E_D (eV)	Experiment D_o (cm²/sec)	Theory E_D (eV)	Theory D_o (cm²/sec)	v_0 (sec⁻¹)	Technique	Ref.
				0.23			LDA VASP^N	[21]
				0.14			GGA VASP^N	[21]
				0.17			VASP	[27]
				0.13			DFT	[28]
	L-Ex			0.39–0.55			EAM	[7]
	A-Ex			0.85			EAG	[18]
Ni/Ni(100)		0.63$^£$				~10^{12}	FIM	[54]
		0.43				†10^{12}	STM	[74]
	Jump	0.60 ± 0.02					WF	[73]
				0.3			Morse	[47]
				0.63	1.6 × 10⁻³		EAM AFW	[4]
				0.68	5.4 × 10⁻³		EAM VC	[4]
				0.80	6.9 × 10⁻³		L-J	[4]
				0.61	3.2 × 10⁻³		CEM	[76]
				0.63			CEM$^¥$	[76]
				0.64	3.5 × 10⁻³		EAM FDB^S	[56]
				0.70	3.7 × 10⁻³		EAM VC^S	[56]
				0.45	3.5 × 10⁻³		EAM OJ^S	[56]
				0.93	6.4 × 10⁻³		EAM ATVF^S	[56]
				0.53	2.87 × 10⁻³		SC^M	[60]
				0.298	1.06 × 10⁻³		Morse	[60]
				0.67			CEM	[29]
				0.60			CEM$^¥$	[29]
				0.74			EAM	[7]
				0.558			EMT	[61]
				0.67			EAM	[78]
				0.632			EAM	[79]
				0.63	3.4 × 10⁻⁴	3.71 × 10^{12}	EAM FDB	[80]
				0.68	1.3 × 10⁻³		EAM VC	[80]
				0.631			EMT	[81]
				0.72^Å			DFT	[13]
				0.63			EAM	[82]
				0.20			Coh.	[49]
				0.61			EAM	[84,85]

System	Mechanism				Method	Ref
				0.82	DFT	[86]
				0.75	L-J	[26]
	Ex	0.59±0.03		0.376	RGL	[70]
					FIM	[55]
			$1 \times 10^{-5.2\pm0.6}$	0.93	EAM AFW	[4]
				1.15	EAM VC	[4]
				0.47	CEM	[29]
				1.0	CEM$^¥$	[29]
				0.85	EAM	[7]
			2.5×10^{-3}	1.29	EAM	[78]
				0.844	EMT	[81]
				0.78	DFT	[86]
				1.304	RGL	[70]
			4×10^{-2}	1.42	EAM	[7]
Cu/Ni(100)	L-Ex	0.35±0.02				
	Jump			0.62	STM	[87]
				0.45	CEM	[76]
				0.407	CEM$^¥$	[89]
			7.28×10^{12}	1.23	RGL	[70]
Rh/Ni(100)	Ex			0.67	CEM$^¥$	[89]
	Jump			0.65	CEM$^¥$	[76]
				0.85	CEM$^¥$	[89]
Pd/Ni(100)	Ex			0.55	CEM$^¥$	[89]
	Jump			0.64	CEM$^¥$	[76]
			$4 \times 10^{11\pm0.3}$	0.492	CEM$^¥$	[89]
			5.2×10^{12}	1.10	RGL	[70]
Ag/Ni(100)	Ex		4.15×10^{12}	0.42	CEM$^¥$	[89]
	Jump			0.44	CEM$^¥$	[76]
			3.76×10^{12}	0.347	CEM$^¥$	[89]
			4.5×10^{12}	1.47	RGL	[70]
In/Ni(100)	Ex	0.31±0.03	8.28×10^{12}		CEM$^¥$	[89]
Pt/Ni(100)	Ex	0.69	9.37×10^{12}		PACS	[88]
	Jump		4.2×10^{12}	0.55	FIM	[72]
			9.77×10^{12}	0.706	CEM$^¥$	[76]
Au/Ni(100)	Jump		$^†1 \times 10^{-3}$	0.48	RGL	[70]
				0.489	CEM$^¥$	[76]
			11.9×10^{12}		RGL	[70]
Cu/Cu(100)		0.40	10×10^{12}		He-Scat	[90]
		0.48	7.8×10^{12}		He-Scat	[91]
			4.45×10^{12}			
			1.4×10^{-4}			
			2.3×10^{-3}			
			3.7×10^{12}			
			2.9×10^{12}			

Table 5.7 (cont.)

System	Mechanism	Experiment E_D (eV)	Experiment D_o (cm²/sec)	Theory E_D (eV)	Theory D_o (cm²/sec)	v_0 (sec⁻¹)	Technique	Ref.
		0.34±0.07					STM	[93]
		0.28±0.06	~10⁻⁵			†1×10¹³	He-Scat	[92]
		0.39±0.06					LEIS	[94]
		0.36±0.03					HR LEED	[95]
		0.36					PACS	[96]
	Jump			0.248			Morse	[97]
				0.27			Morse	[47]
				0.38	1.2×10⁻³		EAM AFW	[4]
				0.53	5.2×10⁻³		EAM VC	[4]
				0.43			EMT	[98]
				0.66		5.5×10¹²	CEM$^¥$	[76]
				0.49			EAM	[100]
				0.398	4.7×10⁻⁴		EMT	[99]
				0.51			CEM$^¥$	[29]
				0.53			EAM VC	[77]
				0.46			CEM	[29]
				0.243	8.62×10⁻³		SC^M	[102]
				0.249	7.80×10⁻³		Morse	[102]
				0.425			EMT	[61]
				0.45			EAM	[103]
				0.69			DFT	[104]
				0.48			EAM	[105]
				0.399			EMT	[107]
				0.503			EAM	[79]
				0.52±0.04	1.6(±0.3)×10⁻³		CEM$^¥$	[108]
				0.43±0.02	3.4×10⁻³±0.2		RGL	[111]
				0.52±0.05			DFT GGA	[112]
				0.49±0.01		$20(\times e^{\pm 0.2})\times 10^{12}$	EAM^M	[112]
				0.51	3.2×10⁻⁴		EAM FDB	[80,114]
				0.53	3.3×10⁻⁴		EAM VC	[80,114]
				0.93			LDA	[111]
				0.49			EAM	[116]
				0.51±0.02		$18.2^{+4.03}_{-3.29}\times 10^{12}$	Thermo	[113]

0.01		Coh.	[49]
0.44		RGL	[119]
0.50		EAM	[121]
0.488		EAM	[374]
0.44	$(2.5 \pm 0.1) \times 10^{-3}$	RGL	[67]
0.67		DFT LDA	[21]
0.55		DFT GGA	[21]
0.59	2.3×10^{-3}	L-J	[26]
0.48		MS EAM	[122]
0.49		DFT	[123]
0.64		VASP	[28]
0.505	7.29×10^{-4}	EAM	[125]
0.67		LDA	[127]
0.477		RGL	[70]
0.505	5.3×10^{-4}	EAM	[126]
1.25		LDA	[11]
0.71 ± 0.05	$38 \times 10^{-3 \pm 0.5}$	RGL	[111]
0.72		EAM AFW	[4]
0.79	4×10^{-2}	EAM VC	[4]
−0.22		EMT	[98]
−0.20	8.5×10^{-4}	EMT	[99]
0.77		CEM*	[29]
0.79		EAM VC	[77]
0.18		CEM	[29]
0.97		DFT	[104]
0.80		EAM	[105]
0.996	$42 \times 10^{-3 \pm 0.3}$	EMT	[81]
0.7 ± 0.04		RGL	[111]
0.96 ± 0.1		DFT GGA	[112]
0.70 ± 0.04	$437(\times e^{\pm 0.7}) \times 10^{12}$	EAM^M	[112]
0.69		EAM	[116]
0.74 ± 0.02	$6.65^{+5.47}_{-3.00} \times 10^{14}$	Thermo	[113]
0.70 ± 0.04		RGL^M	[119]
0.85		RGL^S	[119]
0.75		EAM	[121]
0.94		DFT LDA	[21]
0.81		DFT GGA	[21]
0.69		EAM	[122]

Jump [100]
β
Ex

Table 5.7 (cont.)

System	Mechanism	Experiment E_D (eV)	Experiment D_o (cm²/sec)	Theory E_D (eV)	Theory D_o (cm²/sec)	v_0 (sec⁻¹)	Technique	Ref.
	Ex [110]			1.02			DFT	[123]
	D-Ex			0.94			VASP	[28]
	T-Ex			1.02			LDA	[127]
	Q-Ex			0.708			RGL	[70]
				2.81			LDA	[11]
				0.70 ± 0.05	$45 \times 10^{-3 \pm 0.6}$		RGL	[111]
				0.82 ± 0.08	$104 \times 10^{-3 \pm 1.0}$		RGL	[111]
				0.75 ± 0.01	$86 \times 10^{-3 \pm 0.9}$		RGL	[111]
Na/Cu(100)	Jump	0.051 ± 0.006				$(0.53 \pm 0.2) \times 10^{12}$	He-Scat	[133]
		0.0525 ± 0.0009					He-Scat	[132]
		0.075	$(6.3 \pm 0.6) \times 10^{-3}$				He-Scat	[134]
				0.077			LMD	[135]
		0.058 ± 0.007					³He-SE	[136]
				0.065 ± 0.006			LMD	[132]
Co/Cu(100)	Jump			0.61			VASP	[123]
				0.66			RGL	[139]
	Ex			1.00			VASP	[123]
				0.86			RGL	[139]
Co/Cu(100)^{NMG}	Jump			0.92			VASP	[123]
Ni/Cu(100)	Jump			0.64		5.2×10^{12}	CEM‡	[76]
	Jump			0.64		9.44×10^{12}	CEM‡	[89]
				0.439			RGL	[70]
Rh/Cu(100)	Ex			0.57		9.00×10^{12}	CEM‡	[89]
	Jump			0.71		4.6×10^{12}	CEM‡	[76]
				0.72		12.43×10^{12}	CEM‡	[89]
	Ex			0.42		10.61×10^{12}	CEM‡	[89]
Pd/Cu(100)	Jump			0.58		4.6×10^{12}	CEM‡	[76]
				0.73		17.57×10^{12}	CEM‡	[89]
				0.566			RGL	[70]
Ag/Cu(100)	Ex			0.59		15.58×10^{12}	CEM‡	[89]
	Jump			0.36		6.9×10^{12}	CEM‡	[76]
				0.48		9.07×10^{12}	CEM‡	[89]

System	Mech.			Method	Ref.
In/Cu(100)	Ex		0.393	RGL	[70]
	Jump		0.91	7.5×10^{12}	CEM$^¥$	[89]
		0.24 ± 0.03		LEIS	[128]
		0.28 ± 0.05		PACS	[88]
		0.20 ± 0.03		PACS	[129]
			0.32	EAM	[130]
			1.01	EAM	[130]
Cs/Cu(100)	Ex			^{3}He-SE	[137]
	Jump	0.31 ± 0.02		LMD	[137]
Pt/Cu(100)	Jump		0.24 ± 0.04	4.1×10^{12}	CEM$^¥$	[76]
			0.61	RGL	[70]
Au/Cu(100)	Jump		0.874		RGL	[70]
			0.56	3.1×10^{12}	CEM$^¥$	[76]
			0.554	RGL	[70]
			0.64	RGL^S	[138]
	Ex		0.58 ± 0.02	RGL	[138]
			0.57	RGL^S	[138]
Pb/Cu(100)	Jump	0.87 ± 0.07	0.68 ± 0.1		RBS	[131]
Rh/Rh(100)	Jump	0.83 ± 0.05	1×10^{-3}		FIM	[167]
		0.84	†10^{-3}		FIM	[169]
			†10^{-3}		FIM	[170]
			0.7		Morse	[47]
			0.87 ± 0.16	$4.06 \times (e^{\pm 0.85}) \times 10^{-3}$	L-J	[171]
			1.094		MCTST	[175]
			0.388 ± 0.002	$5.6(\times 2^{\pm 1}) \times 10^{-3}$	SC^M	[176]
			0.80	3×10^{-3}	TST CEM$^¥$	[76]
			0.832	1.36×10^{-3}	4.4×10^{12}	SC^M	[60,102]
			0.799	2.13×10^{-3}	Morse	[60,102]
			0.44		Coh.	[49]
			0.89		DFT	[179]
			1.02	3.0×10^{-3}	L-J	[26]
			0.99		DFT	[28]
	Jump^n		1.03 ± 0.01	$6.1(\times 1.5^{\pm 1}) \times 10^{-3}$	MCTST L-J	[173]
	Jump^m		0.92 ± 0.12	$3.7(\times 1.78^{\pm 1}) \times 10^{-3}$	MCTST L-J	[174]
	Ex		> 1.11		DFT	[179]
			1.20		DFT	[28]
Ni/Rh(100)	Jump		0.76	5.5×10^{12}	CEM$^¥$	[76]
			0.75	5.82×10^{12}	CEM$^¥$	[89]

Table 5.7 (cont.)

System	Mechanism	Experiment E_D (eV)	Experiment D_o (cm²/sec)	Theory E_D (eV)	Theory D_o (cm²/sec)	v_0 (sec^{-1})	Technique	Ref.
Cu/Rh(100)	Ex			1.21		6.5×10^{12}	CEM$^\Psi$	[89]
	Jump			0.74		5.4×10^{12}	CEM$^\Psi$	[76]
				0.53		2.46×10^{12}	CEM$^\Psi$	[89]
Pd/Rh(100)	Ex			1.44		3.62×10^{12}	CEM$^\Psi$	[89]
	Jump			0.66		4.1×10^{12}	CEM$^\Psi$	[76]
				0.73		4.07×10^{12}	CEM$^\Psi$	[89]
Ag/Rh(100)	Ex			1.46		8.64×10^{12}	CEM$^\Psi$	[89]
	Jump			0.50		4.9×10^{12}	CEM$^\Psi$	[76]
				0.51		3.31×10^{12}	CEM$^\Psi$	[89]
	Ex			1.91		4.14×10^{12}	CEM$^\Psi$	[89]
Pt/Rh(100)	Jump	0.92 ± 0.13	$2.0 \times 10^{-3 \pm 1.9}$				FIM	[182]
				0.66		3.7×10^{12}	CEM$^\Psi$	[76]
				0.55		2.7×10^{12}	CEM$^\Psi$	[76]
Au/Rh(100)	Jump	0.59–0.63				† 5×10^{12}	LEED	[185]
Pd/Pd(100)		0.51–0.62					RD	[186]
		~0.65				†5×10^{12}	Rean	[186]
	Jump			0.55			Morse	[47]
				0.71	1.2×10^{-3}		EAM AFW	[4]
				0.74	6.0×10^{-3}		EAM VC	[4]
				0.61		4.2×10^{12}	CEM$^\Psi$	[76]
				0.64			CEM	[77]
				0.75			CEM$^\Psi$	[77]
				0.503			EMT	[61]
				0.71			EAM	[82]
				0.71			DFT	[179]
				0.15			Coh.	[49]
				0.60	1.1×10^{-3}		L-J	[26]
				0.62 ± 0.04			EAM	[187]
				0.87			VASP	[28]
				0.621			RGL	[70]
				0.7		5.51×10^{13}	EAM	[78]
	Ex			0.61			EAM AFW	[4]
				0.59	3×10^{-2}		EAM VC	[4]

System	Mech.	Value	Prefactor	Method	Ref.
Ni/Pd(100)		0.70		CEM	[29]
		0.96		CEM*	[29]
		0.82		DFT	[179]
		0.85		VASP	[28]
		0.725		RGL	[70]
	Jump	0.51	3.41×10^{13}	EAM	[78]
		0.71	5.2×10^{12}	CEM*	[76]
		0.70	7.01×10^{12}	CEM*	[89]
Cu/Pd(100)		0.566		RGL	[70]
	Ex	0.78	9.74×10^{12}	CEM*	[89]
	Jump	0.72	5.3×10^{12}	CEM*	[76]
		0.52	2.63×10^{12}	CEM*	[89]
Rh/Pd(100)		0.558		RGL	[70]
	Ex	0.98	3.98×10^{12}	CEM*	[89]
	Jump	0.75	4.3×10^{12}	CEM*	[76]
		0.75	7.85×10^{12}	CEM*	[89]
	Ex	0.70	1.376×10^{13}	CEM*	[89]
Ag/Pd(100)		0.37 ± 0.03	8×10^9	He-Scat	[189]
	Jump	0.66		EAM	[189]
		0.42	8×10^{12}	CEM*	[76]
		0.50	4.96×10^{12}	CEM*	[89]
		0.60		EAM	[191]
		0.459		RGL	[70]
	Ex	1.34	6.38×10^{12}	CEM*	[89]
		0.75		EAM	[191]
In/Pd(100)		0.45 ± 0.04		PACS	[88]
Pt/Pd(100)		0.43	$^\dagger 10^{13}$	LEED	[188]
	Jump	0.64	4.0×10^{12}	CEM*	[76]
		0.985		RGL	[70]
Au/Pd(100)	Jump	0.54	2.8×10^{12}	CEM*	[76]
		0.59		RGL	[70]
Ag/Ag(100)		0.45	$^\dagger 1 \times 10^{12}$	SEM	[200]
		0.40 ± 0.05	$^\dagger 10^{12}$	LEIS	[223]
		0.330 ± 0.005	$^\dagger 10^{13}$	STM	[224]
		0.38	$^\dagger 3 \times 10^{13}$	STM	[225]
		0.40 ± 0.04	5×10^{12}	STM	[227]
		0.40	$^\dagger 10^{12 \pm 1}$	STM	[229]
	Jump	0.35 ± 0.03		PACS	[228]

Table 5.7 (cont.)

System	Mechanism	Experiment E_D (eV)	Experiment D_o (cm²/sec)	Theory E_D (eV)	Theory D_o (cm²/sec)	v_0 (sec^{-1})	Technique	Ref.
		0.37 ± 0.03					PACS	[96]
				0.35			Morse	[47]
				0.49			EAM	[230]
				0.48	1.2×10^{-3}		EAM AFW	[4]
				0.48	3.9×10^{-3}		EAM VC	[4]
				0.24		1.0×10^{12}	CEM$^\Psi$	[76]
				0.45			CEM$^\Psi$	[77]
				0.41			CEM	[77]
				0.479			EAM Univ.3	[207]
				0.365			EMT	[61]
				0.50 ± 0.03			Muffin tin	[208]
				0.48		1.53×10^{13}	EAM^M	[78]
				0.52			DFT LDA	[231]
				0.45			DFT GGA	[231]
				0.478			EAM	[79]
				0.367			EMT	[81]
				0.48	8.1×10^{-4}		EAM FDB	[233]
				0.48	2.3×10^{-3}		EAM VC	[233]
				0.43 ± 0.04	$(1.4 \pm 0.6) \times 10^{-3}$		RGL^M	[212]
				0.46			RGL^S	[212]
				0.48			EAM	[82]
				0.26 ± 0.04			SEAM	[119]
				0.06			Coh.	[49]
				0.43			RGL^N	[214]
				0.52	2.2×10^{-3}		L-J	[26]
				0.60			DFT	[28]
				0.62			LDA	[127]
				0.467			RGL	[70]
	Jump^z			0.49	3.4×10^{-4}		EAM FDB	[80]
				0.50	5.5×10^{-4}		EAM VC	[80]
	Ex	0.45 ± 0.03					PACS	[228]
		0.48 ± 0.03					PACS	[96]
				0.75			EAM AFW	[4]

		0.60			EAM VC	[4]
		0.70	2×10^{-2}		CEM$^\Psi$	[29]
		0.41			CEM	[77]
		0.78		3.92×10^{14}	EAM^M	[78]
		0.93			DFT LDA	[231]
		0.73			DFT GGA	[231]
		0.614			EMT	[81]
Ni/Ag(100)	D-Ex	0.59 ± 0.02	$(40 \pm 10) \times 10^{-3}$		RGL^M	[212]
	Jump	0.39			SEAM^S	[119]
		0.76			DFT	[28]
		1.00			LDA	[127]
		0.624			RGL	[70]
		0.74 ± 0.06	$(300 \pm 200) \times 10^{-3}$		RGL^M	[212]
		0.71		5.0×10^{12}	CEM$^\Psi$	[76]
		0.70		18.24×10^{12}	CEM$^\Psi$	[89]
		0.61			EAM	[239]
		0.529			RGL	[70]
Pd/Ag(100)	Ex	0.33		7.29×10^{12}	CEM$^\Psi$	[89]
		0.19			EAM	[239]
	Ex	0.43 ± 0.05		$^\dagger 5 \times 10^{12 \pm 0.5}$	Photo	[237]
		0.53			EAM	[237]
		0.38		4.01×10^{12}	CEM$^\Psi$	[89]
	Jump	0.58		5.0×10^{12}	CEM$^\Psi$	[76]
		0.76		16.08×10^{12}	CEM$^\Psi$	[89]
		0.674			RGL	[70]
Cu/Ag(100)	Jump	0.73		5.6×10^{12}	CEM$^\Psi$	[76]
		0.55		34.02×10^{12}	CEM$^\Psi$	[89]
		0.569			RGL	[70]
	Ex	0.52		14.44×10^{12}	CEM$^\Psi$	[89]
Rh/Ag(100)	Jump	0.74		4.5×10^{12}	CEM$^\Psi$	[76]
		0.73		9.69×10^{12}	CEM$^\Psi$	[89]
	Ex	0.26		3.43×10^{12}	CEM$^\Psi$	[89]
In/Ag(100)	Jump	0.31 ± 0.03		$^\dagger 10^{12}$	PACS	[235]
Pt/Ag(100)	Jump	0.59		3.7×10^{12}	CEM$^\Psi$	[76]
		1.189			RGL	[70]
Au/Ag(100)	Jump	0.54		3.2×10^{12}	CEM$^\Psi$	[76]
		0.662			RGL	[70]
Ir/Ir(100)	Ex	0.93 ± 0.04	$1.4 \times 10^{-2 \pm 1}$		FIM	[246]

Table 5.7 (cont.)

System	Mechanism	Experiment E_D (eV)	Experiment D_o (cm^2/sec)	Theory E_D (eV)	Theory D_o (cm^2/sec)	v_0 (sec^{-1})	Technique	Ref.
		0.84 ± 0.05	$6.26 \times 10^{-2 \pm 1}$				FIM	[315]
		0.74 ± 0.02	$7.30 \times 10^{-3 \pm 0.5}$				FIM	[307]
				0.79			EAM	[309]
				0.77			RGL	[309]
				0.51			DFT	[179]
				0.59			DFT	[28]
	Jump			~0.97			Morse	[47]
				1.58			EAM	[309]
				1.57			RGL	[309]
				1.39 ± 0.04			Muffin tin	[208]
				1.13			DFT	[179]
				0.40			Coh.	[49]
				1.23	2.6×10^{-3}		L-J	[26]
				1.17			DFT	[28]
Rh/Ir(100)	Jump	0.80 ± 0.08	$2.09 \times 10^{-3 \pm 1.5}$				FIM	[307]
Pt/Pt(100)	Ex	0.47 ± 0.01	$1.3 \times 10^{-3 \pm 1}$				FIM	[345]
				0.31			EAM AFW	[4]
				0.64	1×10^{-2}		EAM VC	[4]
				3.97	4×10^{-2}		L-J	[4]
				0.611 ± 0.004	$3.5(\times 0.2^{\pm 1}) \times 10^{-2}$		SC	[176]
				0.54			EAM	[323]
				0.38			DFT	[179]
				0.865			OJ	[350]
				0.98 ± 0.04			EAM	[119]
				0.99			EAM^S	[119]
				0.27 ± 0.05			EAM	[354]
				0.31			VASP	[28]
				0.388			RGL	[70]
	Jump			0.8			Morse	[47]
				0.44	8×10^{-4}		EAM AFW	[4]
				1.25	5×10^{-3}		EAM VC	[4]
				1.05	4.9×10^{-3}		L-J	[4]
				0.82			Morse	[4]

Pt/Pt(100) Hex	A-Ex	0.43±0.03				
	Jump		0.72		2.5×10¹²	CEM^¥ [76]
			0.689			EMT [61]
Ni/Pt(100)	Jump		1.04			DFT [179]
			0.07			Coh. [49]
			1.84			EAM [119]
			1.02	2.1×10⁻³		L-J [26]
			0.36±0.05			EAM [354]
			1.2			VASP [28]
			0.875			RGL [70]
			0.96			EAM [119]
Cu/Pt(100)	Jump					STM [348]
			0.57		4.5×10¹²	CEM^¥ [76]
			0.686			RGL [70]
Rh/Pt(100)	Jump		0.61		4.5×10¹²	CEM^¥ [76]
Pd/Pt(100)	Jump	0.70±0.01	† 1×10⁻³			RGL [70]
			0.615		3.3×10¹²	CEM^¥ [76]
			0.57			FIM [355]
			0.49		3.2×10¹²	CEM^¥ [76]
			0.627			RGL [70]
Ag/Pt(100)	Jump		0.41		9.3×10¹²	CEM^¥ [76]
			0.508			RGL [70]
Au/Pt(100)	Jump		0.58		2.1×10¹²	CEM^¥ [76]
			0.507			RGL [70]
Au/Au(100)	Jump		0.6			Morse [47]
			0.64	8×10⁻⁴		EAM AFW [4]
			0.84	3.5×10⁻³		EAM VC [4]
			0.67		2.2×10¹²	CEM^¥ [76]
			0.49			EMT [61]
			0.43		2.11×10¹³	EAM^M [78]
			0.62±0.04			Muffin-tin [208]
			0.49	1.56×10⁻²		EAM^M [359]
			0.50			EAM TST [359]
			0.83			DFT LDA [361]
			0.58			DFT GGA [361]
			0.70			EAM [82]
			−0.03			Coh. [49]
			0.53	1.5×10⁻³		RGL [214]
			0.64			L-J [26]

Table 5.7 (cont.)

System	Mechanism	Experiment E_D (eV)	Experiment D_o (cm²/sec)	Theory E_D (eV)	Theory D_o (cm²/sec)	v_0 (sec⁻¹)	Technique	Ref.
	Ex			0.86			DFT	[28]
				0.64			LDA	[127]
				0.531			RGL	[70]
				0.30			EAM AFW	[4]
				0.32	1×10^{-2}		EAM VC	[4]
				0.25			EAM^M	[78]
				0.26	5.8×10^{-3}	9.78×10^{12}	EAM^M	[359]
				0.28			EAM^S	[359]
				0.65			DFT LDA	[361]
				0.40			DFT GGA	[361]
				0.41			RGL	[214]
				0.32			DFT	[28]
				0.60			LDA	[127]
				0.388			RGL	[70]
Au/Au(100) Hex		≈ 0.2					STM	[356]
		0.35					Analysis	[357]
		0.4	5×10^{-8}				AFM	[358]
	x-Jump			0.15–0.17			EMT	[360]
	y-Jump			0.28			EMT	[360]
Ni/Au(100)				0.61		4.6×10^{12}	CEM*	[76]
				0.591			RGL	[70]
Cu/Au(100)	Jump			0.66		5.0×10^{12}	CEM*	[76]
				0.547			RGL	[70]
Rh/Au(100)	Jump			0.62		3.6×10^{12}	CEM*	[76]
Pd/Au(100)	Jump			0.53		3.7×10^{12}	CEM*	[76]
				0.582			RGL	[70]
Ag/Au(100)	Jump			0.32		5.8×10^{12}	CEM*	[76]
				0.437			RGL	[70]
Pt/Au(100)	Jump			0.82		2.5×10^{12}	CEM*	[76]
				0.994			RGL	[70]

*** on cluster; ^{M} molecular dynamics; ^{S} static; ^{z} 600 K; ^{m} Wide Temp.; ^{n} Low Temp.; ^{NMG} non-magnetic; † assumed value; $^£$ assumed $kT\ell^2/2h$; ^{N} NEB; ^{Å} paramagnetic; CEM* – MD/MC CEM

Table 5.8 Fcc(111).

System	Mechanism	Experiment E_D (eV)	Experiment D_o (cm²/sec)	Theory E_D (eV)	Theory D_o (cm²/sec)	v_0 (sec^{-1})	Technique	Ref.
Al/Al(111)	Jump			0.074	9×10^{-4}		EAM AFW	[4]
				0.054	1.6×10^{-3}		EAM VC	[4]
				0.11	1.3×10^{-3}		L-J	[4]
				0.04			DFT	[10]
				0.04			DFT	[31]
		0.042 ± 0.004	$2 \times 10^{-9 \pm 0.25}$				STM	[32]
		0.06				1×10^{11}	STM	[33]
		0.04				5×10^{11}	STM	[34]
				0.02			EMT***	[12]
				0.04			DFT	[35]
				0.03			EAM	[35]
				0.042			DFT	[19]
				0.067	1.4×10^{-3}	4×10^{12}	RGL	[20]
				0.13	3.2×10^{-4}		L-J	[26]
				0.04			VASP	[27]
	Ex	0.07 ± 0.01		0.63		$5 \times 10^{11 \pm 0.5}$	STM	[38]
							VASP	[27]
Ni/Ni(111)	Jump	$0.33^{£}$					FIM	[54]
		0.22 ± 0.02	$1 \times 10^{-5.2 \pm 0.7}$				FIM	[55]
				~0.01			Morse	[47]
				0.056	5×10^{-4}		EAM AFW	[4]
				0.063	6.2×10^{-4}		EAM VC	[4]
				0.15	9.0×10^{-4}		L-J	[4]
				0.06			EAM FBD	[56]
				0.07			EAM VC	[56]
				0.05			EAM OJ	[56]
				0.05			EAM ATVF	[56]
				0.086	8.74×10^{-4}		SC	[60]
				0.016	5.04×10^{-4}		Morse	[60]
				0.068			EMT	[61]
				0.036			CEM	[62]
				0.16 ± 0.02			DFT LDA	[63]

Table 5.8 (cont.)

System	Experiment			Theory				
	Mechanism	E_D (eV)	D_o (cm^2/sec)	E_D (eV)	D_o (cm^2/sec)	v_0 (sec^{-1})	Technique	Ref.
				0.11±0.02			DFT GGA	[63]
				0.06			EAM	[35]
				0.05			EAM	[64]
				0.063	(1.8±0.1)×10^{-4}		EAM	[67]
				0.14	1.7×10^{-4}		L-J	[26]
				0.049			2nd Moment	[68]
				0.061			RGL	[70]
	Ex			>2.0			2nd Moment	[68]
				2.15			EAM	[64]
				1.633			RGL	[70]
Cu/Ni(111)	Jump			0.050			RGL	[70]
Pd/Ni(111)	Jump			0.045			RGL	[70]
Ag/Ni(111)	Jump			0.038			RGL	[70]
Pt/Ni(111)	Jump	0.2–0.3					FIM	[72]
				0.018			RGL	[70]
				0.038			RGL	[70]
Au/Ni(111)	Jump	$0.03^{+0.01}_{-0.005}$					He-Scat	[140]
Cu/Cu(111)	Jump	0.064					SPA-LEED	[141]
				0.057			EMT	[142]
		0.037±0.005				5×10$^{13±1}$	STM*	[143]
		0.040±0.001				1×10$^{12±0.5}$	STM*	[144]
				0.05		1×10^{12}	DACAPO	[145]
				0.023–0.032			Morse	[97]
				0.026	3×10^{-4}		EAM AFW	[4]
				0.044	4.6×10^{-4}		EAM VC	[4]
				~0.12			EMT	[98]
				0.059	6.4×10^{-4}		SC	[102]
				0.064	5.24×10^{-4}		Morse	[102]
				0.14			Scaling	[15]
				0.053			EMT	[61]
				0.028			EAM	[105]
				0.039			CEM	[62]

	Jump^t	0.16±0.02		DFT LDA	[63]
	Jump^s	0.12±0.02		DFT GGA	[63]
		0.029		EAM	[116]
		0.05		Coh.	[49]
		0.05		DF	[149]
		0.05		EAM	[35]
		0.042	$(1.2±0.1) \times 10^{-4}$	RGL	[67]
		0.09		Emp.	[150]
		0.11	1.4×10^{-4}	L-J	[26]
		0.036±0.002	1.7×10^{-4}	EAM	[155]
		0.030±0.001	1.56×10^{-4}	Msd EAM	[155]
		0.043		2nd Moment	[68]
		0.026		EMT	[124]
		0.01		EAM	[117]
		0.043		RGL	[70]
		0.06	3.01×10^{-4}	EAM	[156]
	Jump^t [110]^t	0.041±0.002	$(2±0.1)\times 10^{-4}$	RGL	[146]
	Jump^s [110]^s	0.087±0.008	$(7±1)\times 10^{-4}$	RGL	[146]
	Jump $\perp$ [110]^t	0.043±0.001		RGL	[146]
	Jump $\perp$ [110]^s	0.082±0.007		RGL	[146]
	Jump $\perp$ [110]^t	0.040±0.001		RGL	[146]
	Jump $\perp$ [110]^s	0.10±0.01		RGL	[146]
	Ex	1.12		EAM	[105]
		0.5		EMT	[98]
		1.455		2nd Moment	[68]
		1.42		EAM	[117]
		1.352		RGL	[70]
	Ex^r	1.19		RGL	[146]
	α_{fh}	0.13		EAM	[148]
		0.03	1.84×10^{-4}	EAM	[126]
		0.041		EAM	[154]
		0.04		EAM	[153]
	α_{hf}	0.03		EAM	[122]
		0.11		EAM	[148]
		0.027	2.06×10^{-4}	EAM	[126]
Li/Cu(111)	α_{fh}	0.20	6.49×10^{-5}	LDA	[158]
	α_{hf}	0.054	4.54×10^{-5}	LDA	[158]
Na/Cu(111)	α_{fh}	0.30	5.32×10^{-5}	LDA	[158]

Table 5.8 (cont.)

System	Mechanism	Experiment E_D (eV)	Experiment D_o (cm²/sec)	Theory E_D (eV)	Theory D_o (cm²/sec)	v_0 (sec⁻¹)	Technique	Ref.
	a_{hf}			0.13	5.21×10^{-5}		LDA	[158]
K/Cu(111)	a_{fh}			0.20	2.02×10^{-5}		LDA	[158]
	a_{hf}			0.05	1.83×10^{-5}		LDA	[158]
Ni/Cu(111)	Jump			0.045			RGL	[70]
Pd/Cu(111)	Jump			0.041			RGL	[70]
Ag/Cu(111)	Jump			0.034			RGL	[70]
In/Cu(111)	Jump			0.02			EAM	[130]
	Jump			0.08			EAM	[157]
	Ex			1.26			EAM	[130]
Pt/Cu(111)	Jump			0.009			RGL	[70]
Au/Cu(111)	Jump			0.061			RGL^S	[138]
	Jump			0.039			RGL	[70]
	Jump^p			0.08 ± 0.01	3×10^{-4}		RGL^M	[138]
Pb/Cu(111)	Jump			0.03			VASP^N GGA	[159]
	Ex			1.17			VASP^N GGA	[159]
Rh/Rh(111)	Jump	0.16 ± 0.02	2×10^{-4}				FIM	[167]
		0.18 ± 0.06					STM	[168]
				0.05			Morse	[47]
				0.27 ± 0.026	$7.10 \times (e^{\pm 0.25}) \times 10^{-4}$		L-J	[171]
				0.29 ± 0.03	$7.1(\times 1.28^{\pm 1}) \times 10^{-4}$		MD	[173]
				0.035	6.2×10^{-4}		CEM	[76,177]
				0.092 ± 0.033	$(4.6 \pm 2.3) \times 10^{-4}$		CEM^M	[76,177]
				0.118 ± 0.043	$(7.1 \pm 4.5) \times 10^{-4}$		Morse^M	[76,177]
				0.234	4.5×10^{-4}		L-J	[76,177]
				0.106	6.09×10^{-4}		SC	[60,102]
				0.089	6.40×10^{-4}		Morse	[60,102]
				0.27 ± 0.01	$(1.1 \pm 0.2) \times 10^{-5}$		L-J^M	[108]
				0.26			Coh.	[49]
				0.15			RGL	[180]
				0.21	1.9×10^{-4}		L-J	[26]
	Jump^n			0.224 ± 0.003	$5.4(\times 1.37^{\pm 1}) \times 10^{-4}$		L-J MC	[173]
	Jump^m			0.20 ± 0.01	$2.91(\times 1.16^{\pm 1}) \times 10^{-4}$		MC TST	[173]
Pt/Rh(111)	Jump	0.22	†1×10^{-3}				FIM	[182]

System		Value 1	Value 2	Value 3	Method	Ref
Pd/Pd(111)	Jump	0.35 ± 0.04			STM	[192]
					Morse	[47]
			0.05		EAM AFW	[4]
			0.031		EAM VC	[4]
			0.059	5×10^{-4}	EMT	[61]
			0.104	4.5×10^{-4}	DFT LDA	[63]
			0.21		DFT GGA	[63]
			0.17		CEM	[62]
			0.034		Coh.	[49]
			0.10		L-J	[26]
			0.14	1.6×10^{-4}	EAM	[35]
			0.04		RGL	[70]
	Jump^l		0.109		RGL	[194]
	Jump^g		0.097	4.4×10^{-4}	RGL	[194]
	Ex		0.127	9.1×10^{-4}	RGL	[70]
K/Pd(111)	Jump	<0.130	1.135		RGL	[70]
		0.066 ± 0.011		$^\dagger 1.7 \times 10^{14}$	PEEM	[196]
					PEEM	[197]
		$1.3 \times 10^{-4 \pm 0.4}$				
Ni/Pd(111)	Jump		0.087		RGL	[70]
Cu/Pd(111)	Jump		0.091		RGL	[70]
Ag/Pd(111)	Jump		0.074		RGL	[70]
^{111}In/Pd(111)	Jump	0.23 ± 0.03		$^\dagger 10^{12}$	PACS	[195]
Pt/Pd(111)	Jump		0.127		RGL	[70]
Au/Pd(111)	Jump		0.11		RGL	[70]
Ag/Ag(111)	Jump	0.58			Resis.	[198]
		~0.1			Rean.	[200]
		0.15 ± 0.10			SEM	[199]
		0.12			SPA-LEED	[141]
		0.18			SPA-LEED	[202]
		0.051 ± 0.024		$2 \times 10^{11 \pm 0.5}$	He-Scat	[203]
		0.097 ± 0.010	0.068		STM	[204]
				10^{11}	EMT	[142]
		0.1		1.4×10^{12}	STM	[205]
			~0.01		Morse	[47]
			0.058		EAM	[206]
			0.12		EMT	[199]
			0.059	5×10^{-4}	EAM AFW	[4]
			0.044	4.1×10^{-4}	EAM VC	[4]
			0.023	6.2×10^{-4}	CEM$^\yen$ TST	[177]

Table 5.8 (cont.)

System	Mechanism	Experiment E_D (eV)	Experiment D_o (cm²/sec)	Theory E_D (eV)	Theory D_o (cm²/sec)	v_0 (sec⁻¹)	Technique	Ref.
				0.055	$(6.9 \pm 2.2) \times 10^{-4}$		CEM$^\Psi$ MD	[177]
				0.005	5.5×10^{-4}		Morse^S	[177]
				0.039 ± 0.009	$(4.7 \pm 1.4) \times 10^{-4}$		Morse^M	[177]
				0.06			EAM	[207]
				0.064			EMT	[61]
				0.14 ± 0.02			Muffin tin	[208]
				0.068	2.6×10^{-4}		RGL	[209]
				0.067			EMT	[204]
				0.055		2.29×10^{12}	EAM	[78]
				0.02			CEM$^\Psi$	[62]
				0.14			DFT LDA	[63]
				0.10			DFT GGA	[63]
				0.082		0.82×10^{12}	DFT	[211]
				0.04			Coh.	[49]
				0.04			EAM	[35]
				0.10	1.4×10^{-4}		L-J	[26]
				0.061			EAM	[215]
				0.064			RGL	[70]
	a_{fh}			0.07			RGL^M	[214]
	Jump^h			0.063			RGL^S	[212]
				0.069 ± 0.001			RGL^M	[212]
	Jump^g			0.098 ± 0.006			RGL^M	[212]
	Ex			1.126			RGL	[70]
Ag/Ag(111)ST	Jump	0.06 ± 0.01				$1 \times 10^{9 \pm 0.6}$	STM	[204]
				0.05			EMT	[204]
				0.06		5.9×10^{12}	DFT	[211]
Ni/Ag(111)	Jump			0.071			RGL	[70]
Cu/Ag(111)	Jump	0.065 ± 0.009				$1 \times 10^{12 \pm 0.5}$	STM	[219]
				0.08			EMT^M	[219]
				0.078			EGL	[70]
Pd/Ag(111)	Jump			0.087			RGL	[70]
Pt/Ag(111)	Jump			0.04			RGL	[70]
Ce/Ag(111)	Jump	0.011				$^\dagger 10^{12}$	STM	[221]

System							Method	Ref
Au/Ag(111)	Jump		0.102		(1.3–1.19)×10^{12}		EAM^M	[218]
			0.088				RGL	[70]
	Ex		0.421–0.805				EAM	[218]
	A-Ex		0.047–0.083				EAM	[218]
Ir/Ir(111)	Jump	0.27±0.004		1.0(×1.4$^{±1}$)×10$^{−4}$			FIM	[48]
		0.22±0.03		8.8×10$^{−3.0±0.9}$			FIM	[303]
		0.27±0.03		9.0(×1.43$^{±1}$)×10$^{−5}$			FIM	[304]
		0.289±0.003		3.5(×1.4$^{±1}$)×10$^{−4}$			FIM	[305]
		0.30±0.01			1.5×10^{12}		FIM	[34]
					5×10$^{11±0.5}$		STM	[47]
			~0.1				Morse	[309]
			0.11				EAM	[309]
			0.17	7.8×10$^{−4}$			RGL	[309]
			0.19				RGL^S	[309]
			0.24±0.03				Muffin-tin	[208]
			0.11				EAM	[310]
			0.20				RGL	[311]
			0.24				L-J	[311]
			0.25				Coh.	[49]
			0.23	1.2×10$^{−4}$			L-J	[26]
			0.24				RGL	[312]
			0.255				TB	[308]
	α_{hf}	0.269±0.003			4.07(×1.43$^{±1}$)×10^{11}		FIM	[304]
			0.241				TB	[308]
	α_{fh}	0.247±0.003			2.16(×1.43$^{±1}$)×10^{11}		FIM	[304]
Pd/Ir(111)	α_{fh}	−0.17					FIM	[314]
	α_{hf}	−013					FIM	[314]
W/Ir(111)	Jump	0.51±0.01		2.85(×1.89$^{±1}$)×10$^{−5}$			FIM	[304]
	α_{fh}	0.317±0.008			3.56(×2$^{±1}$)×10^{11}		FIM	[304]
	α_{hf}	0.506±0.012			7.74×10^{10}		FIM	[304]
Re/Ir(111)	Jump	0.52±0.01		2.36(×1.51$^{±1}$)×10$^{−4}$			FIM	[304]
	α_{fh}	0.378±0.006			2.07×10^{11}		FIM	[304]
	α_{hf}	0.519±0.007			6.39(×1.57$^{±1}$)×10^{11}		FIM	[304]
Pt/Pt(111)	Jump	0.12	0.07				Morse	[319]
		0.25±0.02		†1×10$^{−3}$			SPA-LEED	[141]
							FIM	[320]
		0.26±0.01	0.38				DFT LDA	[320]
		0.26±0.003		2.0(×1.4$^{±1}$)×10$^{−3}$	5×10$^{12±0.5}$		STM	[321]
							FIM	[305]

Table 5.8 (cont.)

System	Mechanism	Experiment E_D (eV)	Experiment D_o (cm²/sec)	Theory E_D (eV)	Theory D_o (cm²/sec)	v_0 (sec⁻¹)	Technique	Ref.
				0.05	1×10^{-4}		Morse	[47]
				0.007			EAM AFW	[4]
				0.078	3.5×10^{-4}		EAM VC	[4]
				0.19	6.3×10^{-4}		L-J	[4]
				0.07			Morse	[4]
				0.08			EAM VC	[323]
				0.06			Morse	[323]
				0.038			CEM^M	[325]
				0.13			EMT	[324]
				0.159			EMT	[61]
				0.16			EMT	[326]
				0.048			CEM*	[62]
				0.42			DFT LDA	[63]
				0.39			DFT GGA	[63]
				0.29			VASP	[327]
				0.36			DFT§	[328]
				0.15			DFT	[211]
				0.17			RGL	[181,329]
				0.06			Coh.	[49]
				0.07			DFT	[35]
				0.046			EAM	[330]
				0.20	5.1×10^{-3}		L-J	[26]
				0.176			2nd Moment	[68]
				0.19			EAM	[333]
				0.096			CY EAM	[331]
				0.31			CY-XEAM1	[331]
				0.38			CY-XEAM2	[331]
				0.171			RGL	[70]
	Ex			2.105			2nd Moment	[68]
	L-Ex			1.627			RGL	[70]
				2.111			2nd Moment	[68]
Na/Pt(111)	Jump	0.030–0.044	6×10^{-3}				He-Scat.	[342]
Co/Pt(111)	Jump			0.3			DFT	[339]

System					Method	Ref.
Ni/Pt(111)		0.21			2nd Moment	[341]
	Ex	0.23 ± 0.01	3.9×10^{-5}		Msd 2nd Moment	[341]
	L-Ex	2.065			2nd Moment	[68]
	Jump	2.248			2nd Moment	[68]
		0.178			EAM	[330]
Cu/Pt(111)	Jump	0.114			RGL	[70]
Pd/Pt(111)	Jump	0.111			RGL	[70]
Ag/Pt(111)	Jump	0.128		0.14 ± 0.01	RGL	[70]
		0.05			STM	[336]
		0.060 ± 0.005			EAM^S	[335]
		0.20			EAM^M	[335]
		0.08			DFT LDA	[338]
		0.15			EMT	[204]
			$7 \times 10^{13 \pm 0.3}$		DF	[210]
		0.09		0.168 ± 0.005	STM	[93]
		0.052 ± 0.005			RGL	[70]
		0.08 ± 0.01			EAM^M	[335]
					EAM^M	[335]
W/Pt(111)	Jump^j	0.07		0.30 ± 0.1	FIM	[319]
Ir/Pt(111)	Jump^k	0.122			Morse	[319]
Au/Pt(111)	Jump	−0.05			RGL	[70]
Au/Au(111)	Jump	0.021	2×10^{-4}		Morse	[47]
		0.038	3.7×10^{-4}		EAM AFW	[4]
		0.102			EAM VC	[4]
		0.12	3.4×10^{-4}		EMT	[61]
		0.013			RGL	[209]
		0.22 ± 0.03			EAM^M	[78]
		0.015	7×10^{-5}		Muffin-tin	[78]
		0.014	0.6×10^{-4}		EAM^M	[359]
		0.016			MD	[359]
		0.029	1.2×10^{12}		TST	[359]
		0.20			CEM	[62]
		0.15			DFT LDA	[63]
		0.04			DFT GGA	[63]
		0.117			EAM	[35]
		0.12	8.5×10^{-5}		RGL	[70]
		0.112			L-J	[26]
					2nd Moment	[68]

Table 5.8 (cont.)

System	Mechanism	Experiment E_D (eV)	Experiment D_o (cm^2/sec)	Theory E_D (eV)	Theory D_o (cm^2/sec)	ν_0 (sec^{-1})	Technique	Ref.
	Ex			0.878			2nd Moment	[68]
				0.861			RGL	[70]
	a_{fh}			0.11			RGL	[214]
Au/Au(111)-$(23 \times \sqrt{3})^{ST}$	Jump			0.12–0.8			GP	[364]
Al/Au(111)-$(22 \times \sqrt{3})$	Jump	0.30 ± 0.005				$7 \times 10^{3 \pm 1}$	STM	[367]
Co/Au(111)	Jump			0.12		7×10^{12}	DFT GGA	[19]
				0.160 ± 0.003	$(5.8 \pm 0.9) \times 10^{-4}$		2nd Moment	[370]
	a_{fh}			0.137			2nd Moment^S	[370]
	a_{fh}			0.155			2nd Moment^S	[370]
	Ex^l			1.1			2nd Moment^M	[370]
	Ex			0.536			2nd Moment	[68]
Co/Au(111)-$(22 \times \sqrt{3})$	Jump			0.18 (DL)			2nd Moment	[369]
				0.12 (FCC)			2nd Moment	[369]
				0.1 (HCP)			2nd Moment	[369]
Ni/Au(111)	Jump			0.078			RGL	[70]
Cu/Au(111)	Jump			0.089			RGL	[70]
Pd/Au(111)	Jump			0.111			RGL	[70]
Ag/Au(111)	Jump			0.075			RGL	[70]
Pt/Au(111)	Jump			0.113			RGL	[70]
Pb/Pb(111)	Jump			0.045			EAM	[372]
Bi/Bi(111)				0.135			LEED STM	[373]

g 500–900K; h 200–500K; i T>820K; j 140–300K; k 300–800K; l 150–500K; m Wide Temp.; n Low Temp.; ST strained surface; P T<700K; r T>100K; s 400–900K; t 80–300K; £ assumed $kTl^2/2h$; † assumed value; § two layers relaxed; *** on cluster; * direct measurement; N NEB; CEM$^¥$ – MD/MC CEM

Table 5.9 Bcc(100).

System	Mechanism	Experiment E_D (eV)	Experiment D_o (cm^2/sec)	Theory E_D (eV)	Theory D_o (cm^2/sec)	v_0 (sec^{-1})	Technique	Ref.
K/K	Jump	0.063 ± 0.015	(9.5 ± 0.9) × 10^{-4}				He-Scat.	[39]
Fe/Fe(100)		0.45 ± 0.08	7.2 × 10^{-4}				STM	[40,41]
		0.495 ± 0.050				$9 \times 10^{11 \pm 0.7}$	He-Scat	[42]
		0.37					RHEED STM	[43]
	Jump			0.6			Morse	[47]
				0.46			Scaling	[15]
				0.92	206 × 10^{-3}		EAM^M	[52]
				0.84			EAM^N	[52]
	Jump [110]			0.4°			VASP	[50]
	a			0.5°			VASP	[50]
	Ex^D			0.7°			VASP	[50]
				0.66	113 × 10^{-3}		EAM^M	[52]
				0.60			EAM^N	[52]
	ExND			0.99	306 × 10^{-3}		EAM^M	[52]
				0.97			EAM^N	[52]
Mo/Mo(100)	Jump			1.4			Morse	[47]
Ta/Ta(100)	Jump			1.0			Morse	[47]
W/W(100)	Jump			1.63			Morse	[47]
				0.49			Ackland	[294]
K/W(100)		0.43 ± 0.04	8.7 × 10^{-2}				SIM	[290]
Mn/W(100)^{MG}	Jump			1.23			VASP	[263]
Mn/W(100)^{NMG}	Jump			1.75			VASP	[263]
Mn/W(100)^R	Long a			1.06			VASP	[263]
	Short a			1.16			VASP	[263]
	Ex			2.02			VASP	[263]
Fe/W(100)^R	Short a			1.2			VASP	[50,51]
	Long a			1.3			VASP	[50,51]
	Ex			2.3			VASP	[50,51]
Fe/W(100)	Jump			1.6			VASP	[50,51]
	Ex			2.1			VASP	[50,51]
Ni/W(100)		1.01 ± 0.02	†2 × 10^{-3}				FIM	[291]
Hf/W(100)		0.53–0.54	2.7 × 10^{-4} – 3.3 × 10^{-4}				FEM	[292]
Au/W(100)		0.13					FEM	[288]
Pb/W(100)		1.3					FEM	[293]

^{N} NEB; ° on 1ML Fe on W(100); ^{R} on reconstructed surface; ^{MG} magnetic; ^{NMG} non-magnetic; ND non-diagonal; ^{D} diagonal; † assumed value; ^{M} molecular dynamics.

Table 5.10 Bcc(111) and Hcp(0001).

System	Mechanism	Experiment E_D (eV)	Experiment D_o (cm^2/sec)	Theory E_D (eV)	Theory D_o (cm^2/sec)	v_0 (sec^{-1})	Technique	Ref.
Fe/Fe(111)	Jump			1.4			Morse	[47]
				0.18			Scaled	[15]
Mo/Mo(111)	Jump			1.9			Morse	[47]
Ag/Ru(0001)	Jump			0.057			EAM	[166]
Ta/Ta(111)	Jump			2.2			Morse	[47]
W/W(111)	Jump	~1.8					FIM	[296]
		0.46	5.8 × 10^{-8}				FEM	[297]
				2.92			Morse	[47]
	a_{hf}	1.9	†10^{-3}				FIM	[298]
		1.3					FIM	[296]
Li/W(111)	Jump	0.53	1.30 × 10^{-2}				FEM	[300]
Ti/W(111)	Jump	1.77	3 × 10^{-2}				FEM	[301]
Ni/W(111)	Jump	0.873 ± 0.081	1 × 10^{-6}				FIM	[299]
Pd/W(111)	Jump	1.02 ± 0.06	†1 × 10^{-3}				FIM	[272]
Dy/W(111)	Jump	1.25 ± 0.09	10^3				FEM	[302]
Re/Re(0001)	Jump	0.48 ± 0.02	6.13($\times$2.6$^{\pm 1}$) × 10^{-6}				FIM	[16]
				1.01			Coh.	[49]
W/Re(0001)	Jump	0.48 ± 0.01	2.17($\times$2.7$^{\pm 1}$) × 10^{-3}				FIM	[16]

† assumed value.

Table 5.11 BCC(110).

System	Mechanism	Experiment E_D (eV)	Experiment D_0 (cm²/sec)	Theory E_D (eV)	Theory D_0 (cm²/sec)	v_0 (sec^{-1})	Technique	Ref.
Fe/Fe(110)	Jump			0.27			F-S	[44]
				−0.25			Morse	[47]
				0.65			Coh.	[49]
				0.65			DFT GGA	[53]
				0.40	1.27×10^{-2}		EAM^M	[46]
Mo/Mo(110)	Jump			0.40			Morse	[47]
				1.20			Coh.	[49]
Fe/Mo(110)	Jump	0.1 ± 0.05				$8 \times 10^{11 \pm 1}$	STM	[163]
Cu/Mo(110)	Jump	0.54	8.7×10^{-4}				Spreading	[161]
Ga/Mo(110)	Jump	0.25	2×10^{-5}				FIM	[162]
Rh/Mo(110)	Jump	0.62 ± 0.06	1×10^{-3}				FIM	[160]
In/Mo(110)	Jump	0.30	3×10^{-5}				FIM	[162]
Sn/Mo(110)	Jump	0.50	4×10^{-5}				FIM	[162]
Ba/Mo(110)	Jump	0.7	20				WF	[164]
Ta/Ta(110)	Jump			0.45			Morse	[47]
				1.19			Coh.	[49]
Pd/Ta(110)	Jump	0.49 ± 0.02	†2×10^{-3}				FIM	[240]
W/W(110)	Jump	0.92 ± 0.05	2.6×10^{-3}				FIM	[167,242]
		0.90 ± 0.07	$6.2(\times 13^{\pm 1}) \times 10^{-3}$				Rean	[245]
		0.91 ± 0.03	$1.2(\times 9^{\pm 1}) \times 10^{-2}$				FIM	[246]
				0.44			Morse	[248]
				0.54			Morse	[249]
				0.46			Morse	[47]
				0.52	1.3×10^{-3}		Morse	[250]
				0.70 ± 0.07	$1.1(\times 1.31^{\pm 1}) \times 10^{-3}$		L-J	[174]
				0.982	$3.6(\times e^{\pm 0.57}) \times 10^{-3}$		L-J^M	[251]
				1.14	1.11×10^{-3}		4th Moment	[252]
				1.22			Coh.	[49]
				0.89	6.92×10^{-3}		EAM^M	[253]
	x-eff Jump	0.92 ± 0.02	$1.61(\times 1.6^{\pm 1}) \times 10^{-3}$				FIM	[247]
	y-eff Jump	0.93 ± 0.01	$5.21(\times 1.4^{\pm 1}) \times 10^{-3}$				FIM	[247]
	α	0.94 ± 0.03				$5.92(\times 2.5^{\pm 1}) \times 10^{12}$	FIM	[247]
	β	1.24 ± 0.13				$8.06(\times 8.1^{\pm 1}) \times 10^{15}$	FIM	[247]

Table 5.11 (cont.)

System	Mechanism	Experiment E_D (eV)	Experiment D_0 (cm²/sec)	Theory E_D (eV)	Theory D_0 (cm²/sec)	v_0 (sec^{-1})	Technique	Ref.
	δ_x	1.28 ± 0.13				3.73(×8.6$^{\pm 1}$)×10^{16}	FIM	[247]
	δ_y	1.37 ± 0.13				1.00(×8.4$^{\pm 1}$)×10^{18}	FIM	[247]
	Ex			2.94			4th Moment	[252]
				3.11			EAM^M	[253]
				0.11	1.78 × 10^{-3}		WF	[257]
Li/W(110)	Jump						FEM	[258]
Na/W(110)	Jump	0.28	2 × 10^{-8}				FEM	[262]
K/W(110)	Jump	0.4	10^{-3}				Coh.	[254]
Sc/W(110)	Jump			0.87			Coh.	[254]
Ti/W(110)	Jump			1.03			Coh.	[254]
V/W(110)	Jump			1.08			Coh.	[254]
Cr/W(110)	Jump			1.01			Coh.	[254]
Mn/W(110)	Jump			0.95			Coh.	[254]
Mn/W(110)^{MG}	α			0.41			VASP	[263]
	Ex [110]			4.22			VASP	[263]
	Ex [100]			3.07			VASP	[263]
Mn/W(110)^{NMG}	α			0.73			VASP	[263]
	Ex [100]			3.30			VASP	[263]
	Ex [110]			4.54			VASP	[263]
Fe/W(110)	Jump	0.60					LEED WF	[264]
		0.2–0.5					NRS	[265]
				0.85			Coh.	[254]
	α			0.7			VASP	[50]
	Ex [110]			4.2			VASP	[51]
	Ex [001]			2.9			VASP	[51]
Co/W(110)	Jump			0.76			Coh.	[254]
Ni/W(110)	Jump	0.49	†10^{-3}				FIM	[266]
		0.49 ± 0.02	†2 × 10^{-3}				FIM	[267]
Cu/W(110)	Jump			0.65			Coh.	[254]
Y/W(110)	Jump			0.53			Coh.	[254]
Zr/W(110)	Jump			0.90			Coh.	[254]
Nb/W(110)	Jump			1.07			Coh.	[254]
	Jump			1.14			Coh.	[254]
Mo/W(110)	Jump			1.19			Coh.	[254]

System	Mechanism	Value	Extra	Method	Ref
Tc/W(110)	Jump		1.10		Coh.	[254]
Ru/W(110)	Jump		0.86		Coh.	[254]
Rh/W(110)	Jump	0.72 ± 0.05	2.5×10^{-3}		FIM	[160]
Pd/W(110)	Jump	0.51	$^{\dagger}1 \times 10^{-3}$	0.71	FIM	[266]
					Coh.	[254]
		0.51 ± 0.03	$1.4 \times 10^{-3 \pm 1.9}$		FIM	[272]
		0.509 ± 0.009	$4.25(\times 1.7^{\pm 1}) \times 10^{-4}$		FIM	[273]
Ag/W(110)	Jump		0.54		Coh.	[254]
Cs/W(110)	Jump	0.61	0.20		Desor	[274,275]
		0.57 ± 0.02	0.23		Photo	[276]
Ba/W(110)	Jump	0.41	3.2×10^{-8}		FEM	[277]
		0.078	1.3×10^{-4}		WF	[278]
Lu/W(110)	Jump		0.94		Coh.	[254]
Hf/W(110)	Jump		1.11		Coh.	[254]
Ta/W(110)	Jump	0.77	4.4×10^{-2}		FIM**	[243]
		0.70	$^{\dagger}10^{-3}$		FIM	[279]
			1.23		Coh.	[254]
Re/W(110)	Jump	1.03	3.0×10^{-2}		FIM**	[280]
		1.01			FIM	[283]
		0.91	$^{\dagger}10^{-3}$		FIM	[279]
			1.12		Coh.	[254]
Os/W(110)	Jump		1.01		Coh.	[254]
Ir/W(110)	Jump	0.97 ± 0.02	$1.7(\times 2.3^{\pm 1}) \times 10^{-2}$		FIM	[284]
		0.95 ± 0.03	$8.48(\times 2.6^{\pm 1}) \times 10^{-3}$		FIM	[285]
			0.86		Coh.	[254]
	x-eff Jump	0.94 ± 0.02	$3.0(\times 1.8^{\pm 1}) \times 10^{-3}$		FIM	[286]
	y-eff Jump	0.94 ± 0.02	$6.78(\times 1.8^{\pm 1}) \times 10^{-3}$		FIM	[286]
	α	0.94 ± 0.02		$5.45(\times 1.9^{\pm 1}) \times 10^{12}$	FIM	[286]
	β	1.18 ± 0.04		$1.17(\times 4.2^{\pm 1}) \times 10^{15}$	FIM	[286]
	δ_x	1.25 ± 0.08		$2.70(\times 1.3^{\pm 1}) \times 10^{16}$	FIM	[286]
	δ_y	1.27 ± 0.09		$4.61(\times 1.8^{\pm 1}) \times 10^{16}$	FIM	[286]
Pt/W(110)	Jump	0.67 ± 0.06	$3.1(\times 10^{\pm 1}) \times 10^{-3}$		FIM	[287]
		0.63	$^{\dagger}10^{-3}$		FIM	[279]
			0.72		Coh.	[254]
Au/W(110)	Jump		0.54		Coh.	[254]

** multi atom investigation; NMG non-magnetic; MG magnetic; † assumed value; M molecular dynamics

5.30 Predictions and comparisons

In the previous sections we have focused on presenting diffusivities obtained on various faces for different elements. Much is still unknown, however, and the question therefore immediately arises how one can predict diffusion properties not yet ascertained in experiments. The answer is, of course, obvious: we have to rely on theory as a guide. This is an activity which may not have been stressed sufficiently in these presentations. It is clear, however, that advances in quantum physics as well as in computational capabilities have brought about great progress. Predictions about the energetics of diffusion abound in the literature, but how reliable are they? To get a feeling for this question, we will compare calculations with experimentally determined diffusivities of materials for which the measured results are considered highly reliable.

To see how well the different theoretical approaches utilized in Chapters 4 and 5 work in giving us guidance about diffusivities, we will concentrate on self-diffusion for just a few systems: Cu(100) and Cu(111), Pt(111) as well as Ir(111), and finally W(110) and W(211). As is apparent from the previous sections, these surfaces have been thoroughly examined and there appears to be fairly uniform agreement about the experimental results. A number of different theoretical estimates are also available, so that some impression about the success of these efforts can be attained. The comparisons are made by presenting the results for the estimated activation energy, and sometimes also the prefactor for the diffusivity, which can then be contrasted with the experimental studies.

First to be examined will be copper surfaces. Self-diffusion on Cu(100) has been studied experimentally by many groups, and although the results are not in perfect agreement, a reasonable value for the diffusion characteristics can be deduced. The record of theoretical estimates is clear from Table 5.7; they generally give much the same results independent of the theoretical approach. Furthermore, they agree more or less with the experimental values. What is worrisome, however, is the not too good agreement about the mechanism of movement, with later studies favoring hopping over exchange.

On Cu(111) the experimental situation is much better, as recently two independent groups have arrived at essentially the same characteristics [143,144] directly measuring the movement of atoms. From Table 5.8 it appears that the number of attempts to calculate diffusion barriers is smaller than for Cu(100). For both (100) and (111) planes it is clear that theory and experiment are *generally* in reasonable agreement, independent of the approach used; this is especially true for some of the most recent estimates, but it must be remembered that in copper the d-shell is filled. That is why this agreement may not extend to other materials. It must also be emphasized that a 25% error in the diffusion barrier, not that uncommon in the calculated results, will have a very significant effect on the diffusivity.

The next surface to be examined is Pt(111), in Table 5.8, also an fcc metal, again with a filled d-shell. The diffusivity has been studied independently by two different groups using different experimental approaches, and the two results agree nicely [305,321]. Kyuno did direct measurements of single atom movement by FIM and Bott looked at

5.30 Predictions and comparisons

kinetics of island formation modeled with Monte Carlo simulations. Platinum is the only material for which two different approaches agree so well. Now, however, it is clear from Table 5.8 that the semi-empirical potentials employed in calculational estimates generally give approximations which are really in discord with experiments, either much too high or too low. The *ab initio* calculations are too high, but the latest estimate is only 12% away from the experimental findings.

The iridium (111) plane has been studied by several groups, but we adopt the work of Wang as representing reasonable experiments [304]. Only a few calculations have been made of iridium diffusivities, and it is clear from Table 5.8 that they vary a great deal and even the best differ from the experiments by 12%. Theoretical estimates have not come close enough to supplant experiments.

Finally to be examined are the (110) and (211) surfaces of tungsten. For these, excellent measurements by Antczak are available [247,375], and are in reasonable agreement with other studies as is apparent in Table 4.4 and 5.11. However, there is only one attempt to do calculations using a modern approach, the work of Xu and Adams [252]. On the flat (110) plane the activation energy for diffusion calculated by Xu and Adams is ~ 24% higher than the value derived from experiments. For the W(211) plane, experiment and theory agree within the limit of error. Recently, using EAM potentials and molecular dynamics, a value of 0.89 eV was derived by Chen et al. [253] for the diffusion barrier in good agreement with experimental data. The value for movement on the W(110) surface are in quite good shape.

The conclusions from these comparisons are clear. For copper, predictions agree well with experiments independent of the semi-empirical potential used. For the other systems, agreement can be achieved as well but to reach this goal careful choice of simulation settings is required and comparison with experiment is a key factor in this process. With fcc metals the RGL potential does reasonably well, but does not yield results of quantitative reliability. It seems that for quantitative information about diffusivities we are at the moment dependent upon experiments, which of late have not been intensively pursued. Calculated diffusivities are not usually reliable indicators of actual rates – they are, however, powerful for exploring novel mechanisms.

There is another approach to gaining insight into what diffusivities could be like, which was suggested some time ago by Wang and Ehrlich [14] and has also been used by Kief and Egelhoff [15] as well as Feibelman [13], and that is to compare the measured activation energy with the enthalpy of vaporization of the diffusing element. The enthalpies are known for all the elements [376], and one might expect some correlations between diffusion characteristics for similar materials. That idea is tested in Table 5.12, in which are listed the surface examined, the activation energy for diffusion E_D, the enthalpy of sublimation ΔH_s^o, and finally the ratio $R_s = E_D/\Delta H_s^o$; it is hoped that the last quantity will give us some help in making a guess about the diffusivity, inasmuch as the prefactor for the diffusivity can be expected to be around 10^{-3} cm^2/sec. In this listing we need to be particularly careful about fcc(100) and (110) surfaces; on these diffusion may occur either by an atom hopping or by exchange with an atom from the substrate, and these different mechanisms can be expected to give quite different ratios R_s. There is also some uncertainty in the measured results which can vitiate comparisons.

Diffusion on two-dimensional surfaces

Table 5.12 Comparison of activation energies and enthalpies of sublimation.

Surface	Rh(111)		Rh(100)		Rh(311)		Pd(111)	
Adatom	Rh	Pt	Rh	Pt	Rh	Pt	Pd	K
E_D (eV)	0.16	0.22	0.87	0.92	0.54	0.44	0.350	0.066
ΔH_s^o (eV)	5.754	5.841	5.754	5.841	5.754	5.841	3.908	0.920
R_s	0.0278	0.0377	0.151	0.158	0.0938	0.0753	0.0896	0.0717
Surface	Ta(110)					W(110)		
Adatom	Pd	W	W	Cs	Re	Ta	Ni	Ir	Pt
E_D (eV)	0.49	0.92	0.92	0.57	1.03	0.77	0.49	0.94	0.67
ΔH_s^o (eV)	3.908	8.776	8.776	0.790	7.955	8.080	4.440	6.874	5.841
R_s	0.125	0.105	0.105	0.722	0.129	0.953	0.110	0.137	0.115
Surface	W(110)			W(111)			Re(0001)	
Adatom	Pd	Rh		Ni	Li	Ti	Pd	Re	W
E_D (eV)	0.51	0.72		0.873	0.53	1.77	1.02	0.48	0.48
ΔH_s^o (eV)	3.908	5.754		4.440	1.646	4.887	3.908	7.955	8.776
R_s	0.131	0.125		0.197	0.322	0.362	0.389	0.0603	0.0547
Surface				W(211)				W(321)
Adatom	W	Rh	Re	Mo	Ir	Ni	Pd	W
E_D (eV)	0.81	0.536	0.83	0.71	0.67	0.46	0.32	0.81
ΔH_s^o (eV)	8.776	5.754	7.955	6.80	6.874	4.440	3.908	8.776
R_s	0.0923	0.0932	0.104	0.104	0.0975	0.104	0.0819	0.0923
Surface		Ir(111)			Ir(100)		Pt(111)	
Adatom	Ir	Re	W	Ir	Rh	Pt	Ag
E_D (eV)	0.289	0.52	0.51	0.74	0.80	0.260	0.168
ΔH_s^o (eV)	6.874	7.955	8.776	6.874	5.754	5.841	2.944
R_s	0.0420	0.0654	0.0581	0.108	0.139	0.0445	0.571

Correlations are certainly evident between the ratios $R_s = E_D/\Delta H_s^o$ for neighboring materials in Table 5.12. For example, on Ta(110), the ratio R_s is 0.125 for diffusion of palladium; on W(110), a neighbor in the periodic table, this ratio is 0.131 for palladium. Furthermore, for a given substrate, this ratio is fairly constant from one adatom to another, with the exception of alkali metal adatoms. This approach must, however, be used cautiously, as is clear comparing Re(0001) with W(110), which for self-diffusion have R_s values 0.0603 compared to 0.105. More experiments are clearly in order to obtain usable estimates of diffusivities.

The emphasis has so far been on diffusion barriers. Also necessary for predicting diffusivities are prefactors, and for these it is quite common in both experiment as well as theory to assume a standard prefactor for the diffusivity in the range $10^{-3} \text{cm}^2/\text{sec}$ or a standard frequency factor in the range $10^{12}/\text{sec}$. This assumption is probably appropriate if the atom is moving over the surface by hopping between nearest-neighbor places. However, it is still not clear how this value is influenced by the existence of long jumps, discussed in Chapter 3, or correlated motion on the surface. Such complicated mechanisms might change the values of the prefactor for the diffusivity, but experiments so far have not detected such a change. Obviously this problem needs further attention. The overall conclusion is: good experiments will still be needed to obtain a proper understanding of single atom diffusivities.

References

[1] P. J. Feibelman, Rebonding effects in separation and surface-diffusion barrier energies of an adatom pair, *Phys. Rev. Lett.* **58** (1987) 2766–2769.

[2] P. J. Feibelman, Diffusion path for an Al adatom on Al(001), *Phys. Rev. Lett.* **65** (1990) 729–732.

[3] G. DeLorenzi, G. Jacucci, The migration of point defects on bcc surfaces using a metallic pair potential, *Surf. Sci.* **164** (1985) 526–542.

[4] C. L. Liu, J. M. Cohen, J. B. Adams, A. F. Voter, EAM study of surface self-diffusion of single adatoms of fcc metals Ni, Cu, Al, Ag, Au, Pd, and Pt, *Surf. Sci.* **253** (1991) 334–344.

[5] J. B. Adams, S. M. Foiles, W. G. Wolfer, Self-diffusion and impurity diffusion of fcc metals using the five-frequency model and the Embedded Atom Method, *J. Mater. Res.* **4** (1989) 102–112.

[6] A. F. Voter, S. P. Chen, Accurate Interatomic Potentials for Ni, Al and Ni$_3$Al, *Mater. Res. Soc. Symp. Proc.* **82** (1987) 175–180.

[7] J. M. Cohen, Long range adatom diffusion mechanism on fcc (100) EAM modeled materials, *Surf. Sci. Lett.* **306** (1994) L545–549.

[8] J. E. Black, Z.-J. Tian, Complicated exchange-mediated diffusion mechanisms in and on a Cu(100) substrate at high temperatures, *Phys. Rev. Lett.* **71** (1993) 2445–2448.

[9] P. A. Gravil, S. Holloway, Exchange mechanisms for self-diffusion on aluminum surfaces, *Surf. Sci.* **310** (1994) 267–272.

[10] R. Stumpf, M. Scheffler, *Ab initio* calculations of energies and self-diffusion on flat and stepped surfaces of aluminum and their implications on crystal growth, *Phys. Rev. B* **53** (1996) 4958–4973.

[11] J.-M. Li, P.-H. Zhang, J.-L. Yang, L. Liu, Theoretical study of adatom self-diffusion on metallic fcc{100} surfaces, *Chin. Phys. Lett.* **14** (1997) 768–771.

[12] S. Valkealahti, M. Manninen, Diffusion on aluminum-cluster surfaces and the cluster growth, *Phys. Rev. B* **57** (1998) 15533–15540.

[13] P. J. Feibelman, Scaling of hopping self-diffusion barriers on fcc(100) surfaces with bulk bond energies, *Surf. Sci.* **423** (1999) 169–174.

[14] S. C. Wang, G. Ehrlich, Adatom diffusion on W(211): Re, W, Mo, Ir and Rh, *Surf. Sci.* **206** (1988) 451–474.

[15] M. T. Kief, W. F. Egelhoff, Growth and structure of Fe and Co thin films on Cu(111), Cu(100), and Cu(110): A comprehensive study of metastable film growth, *Phys. Rev. B* **47** (1993) 10785–10814.

[16] J. T. Goldstein, G. Ehrlich, Atom and cluster diffusion on Re(0001), *Surf. Sci.* **443** (1999) 105–115.

[17] F. Ercolessi, J. B. Adams, Interatomic potentials from first-principles calculations: the force-matching method, *Europhys. Lett.* **26** (1994) 583–588.

[18] O. S. Trushin, P. Salo, M. Alatalo, T. Ala-Nissila, Atomic mechanisms of cluster diffusion on metal fcc(100) surfaces, *Surf. Sci.* **482–485** (2001) 365–369.

[19] S. Ovesson, A. Bogicevic, G. Wahnstrom, B. I. Lundqvist, Neglected adsorbate interactions behind diffusion prefactor anomalies on metals, *Phys. Rev. B* **64** (2001) 125423 1–11.
[20] N. I. Papanicolaou, V. C. Papathanakos, D. G. Papageorgiou, Self-diffusion on Al(100) and Al(111) surfaces by molecular-dynamics simulation, *Physica B* **296** (2001) 259–263.
[21] T. Fordell, P. Salo, M. Alatalo, Self-diffusion on fcc (100) metal surfaces: Comparison of different approximations, *Phys. Rev. B* **65** (2002) 233408 1–4.
[22] G. Kresse, J. Hafner, *Ab initio* molecular dynamics for liquid metals, *Phys. Rev. B* **47** (1993) 558–561.
[23] G. Kresse, J. Hafner, *Ab initio* molecular dynamics simulation of the liquid-metal-amorphous-semiconductor transition in germanium, *Phys. Rev. B* **49** (1994) 14251–14269.
[24] G. Kresse, J. Furthmüller, Efficient iterative schemes for *ab initio* total-energy calculations using a plane-wave set, *Phys. Rev. B* **54** (1996) 11169–11186.
[25] L. Y. Chen, M. R. Baldan, S. C. Ying, Surface diffusion in the low-friction limit: Occurrence of long jumps, *Phys. Rev. B* **54** (1996) 8856–8861.
[26] P. M. Agrawal, B. M. Rice, D. L. Thompson, Predicting trends in rate parameters for self-diffusion on FCC metal surfaces, *Surf. Sci.* **515** (2002) 21–35.
[27] Y.-J. Sun, J.-M. Li, Self-diffusion mechanisms of adatom on Al(001), (011) and (111) surfaces, *Chin. Phys. Lett.* **20** (2003) 269–272.
[28] C. M. Chang, C. M. Wei, Self-diffusion of adatoms and dimers on fcc(100) surfaces, *Chin. J. Phys.* **43** (2005) 169–175.
[29] L. S. Perkins, A. E. DePristo, The influence of lattice distortion on atomic self-diffusion on fcc(001) surfaces: Ni, Cu, Pd, Ag, *Surf. Sci.* **325** (1995) 169–176.
[30] R. Stumpf, M. Scheffler, Theory of self-diffusion at and growth of Al(111), *Phys. Rev. Lett.* **72** (1994) 254–257.
[31] A. Bogicevic, J. Strömquist, B. I. Lundqvist, Low-symmetry diffusion barriers in homoepitaxial growth of Al(111), *Phys. Rev. Lett.* **81** (1998) 637–640.
[32] J. V. Barth, H. Brune, B. Fischer, J. Weckesser, K. Kern, Dynamics of surface migration in the weak corrugation regime, *Phys. Rev. Lett.* **84** (2000) 1732–1735.
[33] T. Michely, W. Langenkamp, H. Hansen, C. Busse, Comment on "Dynamics of Surface Migration in the Weak Corrugation Regime," *Phys. Rev. Lett.* **86** (2001) 2695.
[34] C. Busse, W. Langenkamp, C. Polop, A. Petersen, H. Hansen, U. Linke, P. J. Feibelman, T. Michely, Dimer binding energies on fcc(111) metal surfaces, *Surf. Sci.* **539** (2003) L560–566.
[35] C. M. Chang, C. M. Wei, S. P. Chen, Self-diffusion of small clusters on fcc metal (111) surfaces, *Phys. Rev. Lett.* **85** (2000) 1044–1047.
[36] S. Ovesson, Mean-field nucleation theory with nonlocal interactions, *Phys. Rev. Lett.* **88** (2002) 116102 1–4.
[37] C. Polop, H. Hansen, C. Busse, T. Michely, Relevance of nonlocal adatom-adatom interactions in homoepitaxial growth, *Phys. Rev. B* **67** (2003) 193405 1–4.
[38] C. Polop, H. Hansen, W. Langenkamp, Z. Zhong, C. Busse, U. Linke, M. Kotrla, P. J. Feibelman, T. Michely, Oscillatory interaction between O impurities and Al adatoms on Al(111) and its effect on nucleation and growth, *Surf. Sci.* **575** (2005) 89–102.
[39] D. Fuhrmann, E. Hulpke, Self-diffusion of potassium on ultra-thin epitaxial potassium layers, *J. Chem. Phys.* **106** (1997) 3407–3411.
[40] J. A. Stroscio, D. T. Pierce, R. A. Dragoset, Homoepitaxial growth of iron and a real space view of reflection-high-energy-electron diffraction, *Phys. Rev. Lett.* **70** (1993) 3615–3618.
[41] J. A. Stroscio, D. T. Pierce, Scaling of diffusion-mediated island growth in iron-on-iron homoepitaxy, *Phys. Rev. B* **49** (1994) 8522–8525.

References

[42] R. Pfandzelter, T. Igel, H. Winter, Real-time study of nucleation, growth, and ripening during Fe/Fe(100) homoepitaxy using ion scattering, *Phys. Rev. B* **62** (2000) R2299–2302.

[43] F. Dulot, B. Kierren, D. Malterre, Determination of kinetic parameters in layer-by-layer growth from RHEED profile analysis, *Thin Solid Films* **423** (2003) 64–69.

[44] U. Köhler, C. Jensen, A. C. Schindler, L. Brendel, D. E. Wolf, Scanning tunnelling microscopy and Monte Carlo studies of homoepitaxy on Fe(110), *Philos. Mag. B* **80** (2000) 283–292.

[45] M. W. Finnis, J. E. Sinclair, A simple empirical N-body potential for transition metals, *Philos. Mag. A* **50** (1984) 45–55.

[46] D. Chen, W. Hu, J. Yang, L. Sun, The dynamic diffusion behaviors of 2D small Fe clusters on Fe(110) surface, *J. Phys.: Condens. Matter* **19** (2007) 446009 1–8.

[47] P. G. Flahive, W. R. Graham, Pair potential calculations of single atom self-diffusion activation energies, *Surf. Sci.* **91** (1980) 449–462.

[48] S. C. Wang, G. Ehrlich, Self-adsorption sites on a close-packed surface: Ir on Ir(111), *Phys. Rev. Lett.* **62** (1989) 2297–2300.

[49] S. Y. Davydov, Calculation of the activation energy for surface self-diffusion of transition-metal atoms, *Phys. Solid State* **41** (1999) 8–10.

[50] D. Spisák, J. Hafner, Diffusion of Fe atoms on W surfaces and Fe/W films and along surface steps, *Phys. Rev. B* **70** (2004) 195426 1–13.

[51] D. Spisák, J. Hafner, Diffusion mechanisms for iron on tungsten, *Surf. Sci.* **584** (2005) 55–61.

[52] H. Chamati, N. I. Papanicolaou, Y. Mishin, D. A. Papaconstantopoulos, Embedded-atom potential for Fe and its application to self-diffusion on Fe(100), *Surf. Sci.* **600** (2006) 1793–1803.

[53] I. V. Shvets, S. Murphy, V. Kalinin, Nanowedge island formation on Mo(110), *Surf. Sci.* **601** (2007) 3169–3178.

[54] R. T. Tung, W. R. Graham, Single atom self-diffusion on nickel surfaces, *Surf. Sci.* **97** (1980) 73–87.

[55] T. Y. Fu, T. T. Tsong, Atomic processes in self-diffusion of Ni surfaces, *Surf. Sci.* **454–456** (2000) 571–574.

[56] B. M. Rice, C. S. Murthy, B. C. Garrett, Effects of surface structure and of embedded-atom pair functionals on adatom diffusion on fcc metallic surfaces, *Surf. Sci.* **276** (1992) 226–240.

[57] S. M. Foiles, M. I. Baskes, M. S. Daw, Embedded-atom-method functions for the fcc metals Cu, Ag, Ni, Pd, Pt, and their alloys, *Phys. Rev. B* **33** (1986) 7983–7991.

[58] D. J. Oh, R. A. Johnson, Simple embedded atom method model for fcc and hcp metals, *J. Mater. Res.* **3** (1988) 471–478.

[59] G. J. Ackland, G. Tichy, V. Vitek, M. W. Finnis, Simple *N*-body potentials for the noble metals and nickel, *Philos. Mag. A* **56** (1987) 735–756.

[60] K.-D. Shiang, Molecular dynamics simulation of adatom diffusion on metal surfaces, *J. Chem. Phys.* **99** (1993) 9994–10000.

[61] P. Stoltze, Simulation of surface defects, *J. Phys.: Condens. Matter* **6** (1994) 9495–9517.

[62] Y. Li, A. E. DePristo, Predicted growth mode for metal homoepitaxy on the fcc(111) surface, *Surf. Sci.* **351** (1996) 189–199.

[63] J. J. Mortensen, B. Hammer, O. H. Nielsen, K. W. Jacobsen, J. K. Nørskov, in: *Elementary Processes in Excitations and Reactions on Solid Surfaces*, A. Okiji, H. Kasai, K. Makoshi (eds.), Density functional theory study of self-diffusion on the (111) surfaces of Ni, Pd, Pt, Cu, Ag and Au, (Springer-Verlag, Berlin, 1996), p. 173–182.

[64] K. Haug, T. Jenkins, Effects of hydrogen on the three-dimensional epitaxial growth of Ni(100), (110), and (111), *J. Phys. Chem. B* **104** (2000) 10017–10023.

[65] M. S. Daw, M. I. Baskes, Embedded-atom method: Derivation and application to impurities, surfaces, and other defects in metals, *Phys. Rev. B* **29** (1984) 6443–6453.

[66] S. E. Wonchoba, W. H. Hu, D. G. Truhlar, Surface diffusion of H on Ni(100): Interpretation of the transition temperature, *Phys. Rev. B* **51** (1995) 9985–10002.

[67] U. Kürpick, Self-diffusion on (100), (110), and (111) surfaces of Ni and Cu: A detailed study of prefactors and activation energies, *Phys. Rev. B* **64** (2001) 075418 1–7.

[68] H. Bulou, C. Massobrio, Mechanisms of exchange diffusion on fcc(111) transition metal surfaces, *Phys. Rev. B* **72** (2005) 205427 1–6.

[69] L. T. Kong, L. J. Lewis, Transition state theory of the preexponential factors for self-diffusion on Cu, Ag, and Ni surfaces, *Phys. Rev. B* **74** (2006) 073412 1–4.

[70] S. Y. Kim, I.-H. Lee, S. Jun, Transition-pathway models of atomic diffusion on fcc metal surfaces. I. Flat surfaces, *Phys. Rev. B* **76** (2007) 245407 1–15.

[71] D. Passerone, M. Parrinello, Action-derived molecular dynamics in the study of rare events, *Phys. Rev. Lett.* **87** (2001) 108302 1–4.

[72] G. L. Kellogg, Surface diffusion of Pt adatoms on Ni surfaces, *Surf. Sci.* **266** (1992) 18–23.

[73] P. Schrammen, J. Hölzl, Investigation of surface self-diffusion of Ni atoms on the Ni(100) plane by means of work function measurements and Monte Carlo calculations, *Surf. Sci.* **130** (1983) 203–228.

[74] M. C. Bartelt, L. S. Perkins, J. W. Evans, Transitions in critical size for metal (100) homo-epitaxy, *Surf. Sci.* **344** (1995) L1193–1199.

[75] E. Kopatzki, S. Günther, W. Nichtl-Pecher, R. J. Behm, Homoepitaxial growth on Ni(100) and its modification by a preadsorbed oxygen adlayer, *Surf. Sci.* **284** (1993) 154–166.

[76] D. E. Sanders, A. E. DePristo, Predicted diffusion rates on fcc (001) metal surfaces for adsorbate/substrate combinations of Ni, Cu, Rh, Pd, Ag, Pt, Au, *Surf. Sci.* **260** (1992) 116–128.

[77] L. S. Perkins, A. E. DePristo, Self-diffusion mechanisms for adatoms on fcc(100) surfaces, *Surf. Sci.* **294** (1993) 67–77.

[78] G. Boisvert, L. J. Lewis, A. Yelon, Many-body nature of the Meyer–Neidel compensation law for diffusion, *Phys. Rev. Lett.* **75** (1995) 469–472.

[79] Z.-P. Shi, Z. Zhang, A. K. Swan, J. F. Wendelken, Dimer shearing as a novel mechanism for cluster diffusion and dissociation on metal (100) surfaces, *Phys. Rev. Lett.* **76** (1996) 4927–4930.

[80] U. Kürpick, T. S. Rahman, Vibrational free energy contribution to self-diffusion on Ni(100), Cu(100) and Ag(100), *Surf. Sci.* **383** (1997) 137–148.

[81] J. Merikoski, I. Vattulainen, J. Heinonen, T. Ala-Nissila, Effect of kinks and concerted diffusion mechanisms on mass transport and growth on stepped metal surfaces, *Surf. Sci.* **387** (1997) 167–182.

[82] H. Mehl, O. Biham, I. Furman, M. Karimi, Models for adatom diffusion on fcc (001) metal surfaces, *Phys. Rev. B* **60** (1999) 2106–2116.

[83] J. M. Wills, W. A. Harrison, Further studies on interionic interactions in simple metals and transition metals, *Phys. Rev. B* **29** (1984) 5486–5490.

[84] K. Haug, Z. Zhang, D. John, C. F. Walters, D. M. Zehner, W. E. Plummer, Effects of hydrogen in Ni(100) submonolayer homoepitaxy, *Phys. Rev. B* **55** (1997) R10233–10236.

[85] K. Haug, N.-K. N. Do, Kinetic Monte Carlo study of the effect of hydrogen on the two-dimensional epitaxial growth of Ni(100), *Phys. Rev. B* **60** (1999) 11095–11101.

[86] C. M. Chang, C. M. Wei, J. Hafner, Self-diffusion of adatoms on Ni(100) surfaces, *J. Phys.: Condens. Matter* **13** (2001) L321–328.

References

[87] B. Müller, L. Nedelmann, B. Fischer, H. Brune, K. Kern, Initial stages of Cu epitaxy on Ni(100): Postnucleation and a well-defined transition in critical island size, *Phys. Rev. B* **54** (1996) 17858–17865.

[88] G. Krausch, R. Fink, K. Jacobs, U. Kohl, J. Lohmüller, B. Luckscheiter, R. Platzer, B.-U. Runge, U. Wöhrmann, G. Schatz, Surface and interface studies with perturbed angular correlations, *Hyperfine Interactions* **78** (1993) 261–280.

[89] L. S. Perkins, A. E. DePristo, Heterogeneous adatom diffusion on fcc (100) surfaces: Ni, Cu, Rh, Pd, and Ag, *Surf. Sci.* **319** (1994) 225–231.

[90] J. J. de Miguel, A. Sanchez, A. Cebollada, J. M. Gallego, J. Perron, S. Ferrer, The surface morphology of a growing crystal studied by thermal energy atom scattering (TEAS), *Surf. Sci.* **189/190** (1987) 1062–1068.

[91] J. J. de Miguel, A. Cebollada, J. M. Gallego, J. Ferrón, S. Ferrer, Quantitative evaluation of the perfection of an epitaxial film grown by vapor deposition as determined by thermal energy atom scattering, *J. Cryst. Growth* **88** (1988) 442–454.

[92] H.-J. Ernst, F. Fabre, J. Lapujoulade, Nucleation and diffusion of Cu adatoms on Cu(100): A helium-atom-beam scattering study, *Phys. Rev. B* **46** (1992) 1929–1932.

[93] H. Brune, G. S. Bales, J. Jacobsen, C. Boragno, K. Kern, Measuring surface diffusion from nucleation island densities, *Phys. Rev. B* **60** (1999) 5991–6006.

[94] M. Breeman, D. O. Boerma, Migration of Cu adatoms on a Cu(100) surface, studied with low energy ion scattering (LEIS), *Surf. Sci.* **269/270** (1992) 224–228.

[95] H. Dürr, J. F. Wendelken, J.-K. Zuo, Island morphology and adatom energy barriers during homoepitaxy on Cu(001), *Surf. Sci.* **328** (1995) L527–532.

[96] C. R. Laurens, M. F. Rosu, F. Pleiter, L. Niesen, Soft-landing deposition of radioactive probe atoms on surfaces, *Hyperfine Interactions* **120/121** (1999) 59–68.

[97] P. Wynblatt, N. A. Gjostein, A calculation of relaxation, migration and formation energies for surface defects in copper, *Surf. Sci.* **12** (1968) 109–127.

[98] L. Hansen, P. Stoltze, K. W. Jacobsen, J. K. Nørskov, Self-diffusion on Cu surfaces, *Phys. Rev. B* **44** (1991) 6523–6526.

[99] L. B. Hansen, P. Stoltze, K. W. Jacobsen, J. K. Nørskov, Activation free energy and entropy for the normal and exchange selfdiffusion processes on Cu(100), *Surf. Sci.* **289** (1993) 68–74.

[100] Z.-J. Tian, T. S. Rahman, Energetics of stepped Cu surfaces, *Phys. Rev. B* **47** (1993) 9751–9759.

[101] A. P. Sutton, J. Chen, Long-range Finnis–Sinclair potentials, *Philos. Mag. Lett.* **61** (1990) 139–146.

[102] K.-D. Shiang, Theoretical studies of adatom diffusion on metal surfaces, *Phys. Lett. A* **180** (1993) 444–452.

[103] C.-L. Liu, Energetics of diffusion processes during nucleation and growth for the Cu/Cu(100) system, *Surf. Sci.* **316** (1994) 294–302.

[104] C. Lee, G. T. Barkema, M. Breeman, A. Pasquarello, R. Car, Diffusion mechanism of Cu adatoms on a Cu(001) surface, *Surf. Sci.* **306** (1994) L575–578.

[105] M. Karimi, T. Tomkowski, G. Vidali, O. Biham, Diffusion of Cu on Cu surface, *Phys. Rev. B* **52** (1995) 5364–5374.

[106] O. Biham, I. Furman, M. Karimi, G. Vidali, R. Kennett, H. Zeng, Models for diffusion and island growth in metal monolayers, *Surf. Sci.* **400** (1998) 29–43.

[107] J. Merikoski, T. Ala-Nissila, Diffusion processes and growth on stepped metal surfaces, *Phys. Rev. B* **52** (1995) R8715–8718.

[108] P. V. Kumar, J. S. Raul, S. J. Warakomski, K. A. Fichthorn, Smart Monte Carlo for accurate simulation of rare-event dynamics: Diffusion of adsorbed species on solid surfaces, *J. Chem. Phys.* **105** (1996) 686–695.

[109] V. Rosato, M. Guillope, B. Legrand, Thermodynamical and structural properties of f.c.c. transition metals using a simple tight-binding model, *Philos. Mag. A* **59** (1989) 321–336.

[110] F. Cleri, V. Rosato, Tight-binding potentials for transition metals and alloys, *Phys. Rev. B* **48** (1993) 22–33.

[111] G. A. Evangelakis, N. I. Papanicolaou, Adatom self-diffusion processes on (001) copper surfaces by molecular dynamics, *Surf. Sci.* **347** (1996) 376–386.

[112] G. Boisvert, L. J. Lewis, Self-diffusion of adatoms, dimers, and vacancies on Cu(100), *Phys. Rev. B* **56** (1997) 7643–7655.

[113] G. Boisvert, N. Mousseau, L. J. Lewis, Surface diffusion coefficients by thermodynamic integration: Cu on Cu(100), *Phys. Rev. B* **58** (1998) 12667–12670.

[114] U. Kürpick, A. Kara, T. S. Rahman, Role of lattice vibrations in adatom diffusion, *Phys. Rev. Lett.* **78** (1997) 1086–1089.

[115] U. Kürpick, T. S. Rahman, The role of vibrational entropy in surface diffusion: adatoms and vacancies on Ag(100), Cu(100), and Ni(100), *Surf. Sci.* **427–428** (1999) 15–21.

[116] O. S. Trushin, K. Kokko, P. T. Salo, W. Hergert, M. Kotrla, Step roughening effect on adatom diffusion, *Phys. Rev. B* **56** (1997) 12135–12138.

[117] S. Durukanoglu, O. S. Trushin, T. S. Rahman, Effect of step-step separation on surface diffusion processes, *Phys. Rev. B* **73** (2006) 125426 1–6.

[118] Q. Xie, Dynamics of adatom self-diffusion and island morphology evolution at a Cu(100) surface, *Phys. Status Solidi (b)* **207** (1998) 153–170.

[119] J. Zhuang, L. Liu, Global study of mechanisms for adatom diffusion on metal fcc(100) surfaces, *Phys. Rev. B* **59** (1999) 13278–13284.

[120] Z. Wang, Y. Li, J. B. Adams, Kinetic lattice Monte Carlo simulation of facet growth rate, *Surf. Sci.* **450** (2000) 51–63.

[121] J. B. Adams, Z. Wang, Y. Li, Modeling Cu thin film growth, *Thin Solid Films* **365** (2000) 201–210.

[122] J. Wang, H. Huang, T. S. Cale, Diffusion barriers on Cu surfaces and near steps, *Modeling Simul. Mater. Sci. Eng.* **12** (2004) 1209–1225.

[123] R. Pentcheva, Ab initio study of microscopic processes in the growth of Co on Cu(001), *Appl. Phys. A* **80** (2005) 971–975.

[124] M. O. Jahma, M. Rusanen, A. Karim, I. T. Koponen, T. Ala-Nissila, T. S. Rahman, Diffusion and submonolayer island growth during hyperthermal deposition on Cu(100) and Cu(111), *Surf. Sci.* **598** (2005) 246–252.

[125] H. Yildirim, A. Kara, S. Durukanoglu, T. S. Rahman, Calculated pre-exponential factors and energetics for adatom hopping on terraces and steps of Cu(100) and Cu(110), *Surf. Sci.* **600** (2006) 484–492.

[126] L. T. Kong, L. J. Lewis, Surface diffusion coefficient: Substrate dynamics matters, *Phys. Rev. B* **77** (2008) 165422 1–5.

[127] J. E. Müller, H. Ibach, Migration of point defects at charged Cu, Ag, and Au (100) surfaces, *Phys. Rev. B* **74** (2006) 085408 1–10.

[128] M. Breeman, D. O. Boerma, Sites, mobilities, and cluster formation of In atoms on a stepped Cu(100) surface, *Phys. Rev. B* **46** (1992) 1703–1709.

[129] M. F. Rosu, F. Pleiter, L. Niesen, Interaction between Cu atoms and isolated ^{111}In probe atoms on a Cu(100) surface, *Phys. Rev. B* **63** (2001) 165425 1–10.

[130] C. D. Van Siclen, Indium adatom diffusion and clustering on stepped copper surfaces, *Phys. Rev. B* **51** (1995) 7796–7804.

[131] C. Cohen, Y. Girard, P. Leroux-Hugon, A. L'Hoir, J. Moulin, D. Schmaus, Surface diffusion of Pb on (100) Cu: Coverage dependence and influence of ordered-phase formation, *Europhys. Lett.* **24** (1993) 767–772.

[132] A. P. Graham, F. Hofmann, J. P. Toennies, L. Y. Chen, S. C. Ying, Experimental and theoretical investigation of the microscopic vibrational and diffusional dynamics of sodium atoms on a Cu(001) surface, *Phys. Rev. B* **56** (1997) 10567–10578.

[133] J. Ellis, J. P. Toennies, Observation of jump diffusion of isolated sodium atoms on a Cu(001) surface by helium atom scattering, *Phys. Rev. Lett.* **70** (1993) 2118–2121.

[134] A. P. Graham, F. Hofmann, J. P. Toennies, L. Y. Chen, S. C. Ying, Determination of the Na/Cu(001) potential energy surface from helium scattering studies of the surface dynamics, *Phys. Rev. Lett.* **78** (1997) 3900–3903.

[135] A. Cucchetti, S. C. Ying, Diffusion of Na atoms on a Cu(001) surface, *Phys. Rev. B* **60** (1999) 11110–11117.

[136] G. Alexandrowicz, A. P. Jardine, H. Hedgeland, W. Allison, J. Ellis, Onset of 3D collective surface diffusion in the presence of lateral interactions: Na/Cu(001), *Phys. Rev. Lett.* **97** (2006) 156103 1–4.

[137] A. P. Jardine, G. Alexandrowicz, H. Hedgeland, R. D. Diehl, W. Allison, J. Ellis, Vibration and diffusion of Cs atoms on Cu(001), *J. Phys.: Condens. Matter* **19** (2007) 305010–305027.

[138] G. A. Evangelakis, G. C. Kallinteris, N. I. Papanicolaou, Molecular dynamics study of gold adatom diffusion on low-index copper surfaces, *Surf. Sci.* **394** (1997) 185–191.

[139] V. S. Stepanyuk, D. I. Bazhanov, A. N. Baranov, W. Hergert, A. A. Katsnelson, P. H. Dederichs, J. Kirschner, Atomistic processes and the strain distribution in the early stages of thin film growth, *Appl. Phys. A* **72** (2001) 443–446.

[140] W. Wulfhekel, N. N. Lipkin, J. Kliewer, G. Rosenfeld, L. C. Jorritsma, B. Poelsema, G. Comsa, Conventional and manipulated growth of Cu/Cu(111), *Surf. Sci.* **348** (1996) 227–242.

[141] M. Henzler, T. Schmidt, E. Z. Luo, in: *The Structure of Surfaces IV*, X. Xie, S. Y. Tong, M. A. V. Hove (eds.), Modes in homoepitaxial growth on Cu(111), Ag(111), and Pt(111), (World Scientific, Singapore, 1994), p. 619–264.

[142] D. C. Schlösser, K. Morgenstern, L. K. Verheij, G. Rosenfeld, F. Besenbacher, G. Comsa, Kinetics of island diffusion on Cu(111) and Ag(111) studied with variable-temperature STM, *Surf. Sci.* **465** (2000) 19–39.

[143] J. Repp, F. Moresco, G. Meyer, K.-H. Rieder, P. Hyldgard, M. Perrson, Substrate mediated long-range oscillatory interaction between adatoms, *Phys. Rev. Lett.* **85** (2000) 2981–2984.

[144] N. Knorr, H. Brune, M. Epple, A. Hirstein, M. A. Schneider, K. Kern, Long-range adsorbate interactions mediated by a two-dimensional electron gas, *Phys. Rev. B* **65** (2002) 115420 1–5.

[145] J. Repp, G. Meyer, K.-H. Rieder, P. Hyldgaard, Site determination and thermally assisted tunneling in homogeneous nucleation, *Phys. Rev. Lett.* **91** (2003) 206102 1–4.

[146] G. C. Kallinteris, G. A. Evangelakis, N. I. Papanicolaou, Molecular dynamics study of the vibrational and transport properties of copper adatoms on the (111) copper surface, compared with the (001) face, *Surf. Sci.* **369** (1996) 185–198.

[147] V. Rosato, M. Guillope, B. Legrand, Thermodynamical and structural properties of f.c.c. transition metals using simple tight-binding model, *Philos. Mag. A* **59** (1997) 321.

[148] M. Breeman, G. T. Barkema, M. H. Langelaar, D. O. Boerma, Computer simulation of metal-on-metal epitaxy, *Thin Solid Films* **272** (1996) 195–207.

[149] A. Bogicevic, S. Ovesson, P. Hyldgaard, B. I. Lundqvist, H. Brune, D. R. Jennison, Nature, strength, and consequences of indirect adsorbate interactions on metals, *Phys. Rev. Lett.* **85** (2000) 1910–1913.

[150] M. I. Larsson, Kinetic Monte Carlo simulations of adatom island decay on Cu(111), *Phys. Rev. B* **64** (2001) 115428 1–10.

[151] Y. X. Wang, Z. Y. Pan, Z. J. Li, Q. Wei, L. K. Zang, Z. X. Zhang, Effect of tensile strain on adatom diffusion on Cu(111) surface, *Surf. Sci.* **545** (2003) 137–142.

[152] Y. Mishin, M. J. Mehl, D. A. Papaconstantopoulos, A. F. Voter, J. D. Kress, Structural stability and lattice defects in copper: *Ab initio*, tight binding, and embedded-atom calculations, *Phys. Rev. B* **63** (2001) 224106 1–16.

[153] H. Huang, C. H. Woo, H. L. Wei, X. X. Zhang, Kinetics-limited surface structures at the nanoscale, *Appl. Phys. Lett.* **82** (2003) 1272–1274.

[154] M.-C. Marinica, C. Barreteau, M.-C. Desjonqueres, D. Spanjaard, Influence of short-range adatom-adatom interactions on the surface diffusion of Cu on Cu(111), *Phys. Rev. B* **70** (2004) 075415 1–14.

[155] J. Ferrón, L. Gómez, J. J. de Miguel, R. Miranda, Nonstochastic behavior of atomic surface diffusion on Cu(111) down to low temperatures, *Phys. Rev. Lett.* **93** (2004) 166107 1–4.

[156] J. Yang, W. Hu, J. Tang, M. Xu, Long-time scale dynamics study of diffusion dynamics of small Cu clusters on Cu(111) surface, *J. Phys. Chem. C* **112** (2008) 2074–2078.

[157] M. Breeman, G. T. Barkema, D. O. Boerma, Binding energies and stability of Cu-adatom clusters on Cu(100) and Cu(111), *Surf. Sci.* **323** (1995) 71–80.

[158] L. Padillo-Campos, A. Toro-Labbé, Theoretical study of the diffusion of alkali metals on a Cu(111) surface, *J. Mol. Struct. (Theochem)* **390** (1997) 183–192.

[159] M. L. Anderson, M. J. D'Amato, P. J. Feibelman, B. S. Swartzentruber, Vacancy-mediated and exchange diffusion in a Pb/Cu(111) surface alloy: Concurrent diffusion on two length scales, *Phys. Rev. Lett.* **90** (2003) 126102 1–4.

[160] V. R. Dhanak, D. W. Bassett, Field ion microscope studies of submonolayer rhodium films on (110) tungsten and molybdenum surfaces, *Surf. Sci.* **238** (1990) 289–292.

[161] A. D. Abramenkov, V. V. Slezov, L. V. Tanatarov, Y. M. Fogel, Investigation of surface diffusion of copper atoms on molybdenum by secondary ion-ion emission, *Sov. Phys.-Solid State* **12** (1971) 2365–2368.

[162] A. R. Saadat, Activation energies for surface diffusion and polarizabilities of gallium, indium and tin on a molybdenum surface, *J. Phys. D: Appl. Phys.* **27** (1994) 356–359.

[163] P.-O. Jubert, O. Fruchart, C. Meyer, Nucleation and surface diffusion in pulsed laser deposition of Fe on Mo(110), *Surf. Sci.* **522** (2003) 8–16.

[164] Y. S. Vedula, A. T. Loburets, A. G. Naumovets, Surface diffusion and interaction of adsorbed barium atoms on the (011) face of molybdenum, *Sov. Phys. JETP* **50** (1979) 391–396.

[165] A. G. Naumovets, Y. S. Vedula, Surface diffusion of adsorbates, *Surf. Sci. Rep.* **4** (1985) 365–434.

[166] J. C. Hamilton, Magic size effects for heteroepitaxial island diffusion, *Phys. Rev. Lett.* **77** (1996) 885–888.

[167] G. Ayrault, G. Ehrlich, Surface self-diffusion on an fcc crystal: An atomic view, *J. Chem. Phys.* **60** (1974) 281–294.

[168] F. Tsui, J. Wellman, C. Uher, R. Clarke, Morphology transition and layer-by-layer growth of Rh(111), *Phys. Rev. Lett.* **76** (1996) 3164–3167.

[169] G. L. Kellogg, Direct observation of substitutional-atom trapping on a metal surface, *Phys. Rev. Lett.* **72** (1994) 1662–1665.

[170] G. L. Kellogg, Oscillatory behavior in the size dependence of cluster mobility on metal surfaces: Rh on Rh(100), *Phys. Rev. Lett.* **73** (1994) 1833–1836.
[171] H. K. McDowell, J. D. Doll, Theoretical studies of surface diffusion: Self-diffusion in the fcc(100) system, *J. Chem.Phys.* **78** (1983) 3219–3222.
[172] H. K. McDowell, J. D. Doll, Theoretical studies of surface diffusion: Self-diffusion in the bcc (110) system, *Surf. Sci.* **121** (1982) L537–540.
[173] A. F. Voter, J. D. Doll, Surface self-diffusion constants at low temperature: Monte Carlo transition state theory with importance sampling, *J. Chem. Phys.* **80** (1984) 5814–5817.
[174] A. F. Voter, J. D. Doll, Transition state theory description of surface self-diffusion: Comparison with classical trajectory results, *J. Chem. Phys.* **80** (1984) 5832–5838.
[175] A. F. Voter, Classically exact overlayer dynamics: Diffusion of rhodium clusters on Rh(100), *Phys. Rev. B* **34** (1986) 6819–6829.
[176] R. M. Lynden-Bell, Migration of adatoms on the (100) surface of face-centered-cubic metals, *Surf. Sci.* **259** (1991) 129–138.
[177] D. E. Sanders, A. E. DePristo, A non-unique relationship between potential energy surface barrier and dynamical diffusion barrier: fcc(111) metal surface, *Surf. Sci.* **264** (1992) L169–176.
[178] M. R. Mruzik, G. M. Pound, A molecular dynamics study of surface diffusion, *J. Phys. F* **11** (1981) 1403–1422.
[179] P. J. Feibelman, R. Stumpf, Adsorption-induced lattice relaxation and diffusion by concerted substitution, *Phys. Rev. B* **59** (1999) 5892–5897.
[180] F. Máca, M. Kotrla, O. S. Trushin, Energy barriers for diffusion on stepped Rh(111) surfaces, *Surf. Sci.* **454–456** (2000) 579–583.
[181] F. Máca, M. Kotrla, O. S. Trushin, Energy barriers for interlayer diffusion in Pt/Pt(111) and Rh/Rh(111) homoepitaxy: Small islands, *Czech. J. Phys.* **49** (1999) 1591–1596.
[182] G. L. Kellogg, Diffusion of individual platinum atoms on single-crystal surfaces of rhodium, *Phys. Rev. B* **48** (1993) 11305–11312.
[183] G. L. Kellogg, Diffusion behavior of Pt adatoms and clusters on the Rh(100) surface, *Appl. Surf. Sci.* **67** (1993) 134–141.
[184] D. K. Flynn-Sanders, J. W. Evans, P. A. Thiel, Homoepitaxial growth on Pd(100), *Surf. Sci.* **289** (1993) 75–84.
[185] J. W. Evans, D. K. Flynn-Sanders, P. A. Thiel, Surface self-diffusion barrier of Pd(100) from low-energy electron diffraction, *Surf. Sci.* **298** (1993) 378–383.
[186] J. W. Evans, M. C. Bartelt, Nucleation and growth in metal-on-metal homoepitaxy: Rate equations, simulations and experiments, *J. Vac. Sci. Technol. A* **12** (1994) 1800–1808.
[187] A. V. Evteev, A. T. Kosilov, S. A. Solyanik, Atomic mechanisms and kinetics of self-diffusion on the Pd(001) surface, *Phys. Solid State* **46** (2004) 1781–1784.
[188] J. W. Evans, D. K. Flynn, P. A. Thiel, Influence of adsorption-site geometry and diffusion on thin-film growth: Pt/Pd(100), *Ultramicroscopy* **31** (1989) 80–86.
[189] C. Félix, G. Vandoni, W. Harbich, J. Buttet, R. Monot, Surface mobility of Ag on Pd(100) measured by specular helium scattering, *Phys. Rev. B* **54** (1996) 17039–17050.
[190] C. Félix, G. Vandoni, W. Harbich, J. Buttet, R. Monot, Surface diffusion of Ag on Pd(100) measured with specular helium scattering, *Surf. Sci.* **331–333** (1995) 925–929.
[191] C. Massobrio, B. Nacer, T. Bekkay, G. Vandoni, C. Felix, Low energy impact of silver atoms on Pd(100): comparison between helium scattering and microscopic scale simulation results, *Surf. Sci.* **385** (1997) 87–96.
[192] A. Steltenpohl, N. Memmel, Self-diffusion on Pd(111), *Surf. Sci.* **454–456** (2000) 558–561.

[193] A. Steltenpohl, N. Memmel, Energetic and entropic contributions to surface diffusion and epitaxial growth, *Phys. Rev. Lett.* **84** (2000) 1728–1731.
[194] N. I. Papanicolaou, Self-diffusion on Pd(111) by molecular dynamics simulation, *Comput. Mater. Sci.* **24** (2002) 117–121.
[195] E. Hunger, H. Haas, Adsorption sites and diffusion steps of In and Cd on Pd(111) surfaces, *Surf. Sci.* **234** (1990) 273–286.
[196] M. Ondrejcek, W. Hencel, H. Conrad, V. Cháb, Z. Chvoj, W. Engel, A. M. Bradshaw, Surface diffusion of K on Pd 111 studied by photoemission electron microscopy, *Chem. Phys. Lett.* **215** (1993) 528–534.
[197] M. Snábl, M. Ondrejcek, V. Cháb, W. Stenzel, H. Conrad, A. M. Bradshaw, Temperature dependence of the surface diffusion coefficient of K atoms on Pd{111} measured with PEEM, *Surf. Sci.* **352–354** (1996) 546–551.
[198] D. Stark, Diffusion processes on stepped surfaces of thin metal films: Migration of silver adatoms on silver (111) terraces, *Surf. Sci.* **189/190** (1987) 1111–1116.
[199] G. W. Jones, J. M. Marcano, J. K. Nørskov, J. A. Venables, Energies controlling nucleation and growth processes: The case of Ag/W(110), *Phys. Rev. Lett.* **65** (1990) 3317–3320.
[200] J. A. Venables, Nucleation calculations in a pair-binding model, *Phys. Rev. B* **36** (1987) 4153–4162.
[201] D. T. Spiller, P. Akhter, J. A. Venables, UHV-SEM study of the nucleation and growth of Ag/W(110), *Surf. Sci.* **131** (1983) 517–533.
[202] E. Z. Luo, J. Wollschläger, F. Wegner, M. Henzler, SPA-LEED studies of growth of Ag on Ag(111) at low temperatures, *Appl. Phys. A* **60** (1995) 19–25.
[203] G. Rosenfeld, N. N. Lipkin, W. Wulfhekel, J. Kliewer, K. Morgenstern, B. Poelsema, G. Comsa, New concepts for controlled homoepitaxy, *Appl. Phys. A* **61** (1995) 455–466.
[204] H. Brune, K. Bromann, H. Röder, K. Kern, J. Jacobsen, P. Stoltze, K. Jacobsen, J. Nørskov, Effect of strain on surface diffusion and nucleation, *Phys. Rev. B* **52** (1995) R14380–14383.
[205] E. Cox, M. Li, P.-W. Chung, C. Ghosh, T. S. Rahman, C. J. Jenks, J. W. Evans, P. A. Thiel, Temperature dependence of island growth shapes during submonolayer deposition of Ag on Ag(111), *Phys. Rev. B* **71** (2005) 115414 1–9.
[206] W. K. Rilling, C. M. Gilmore, T. D. Andreadis, J. A. Sprague, An embedded-atom-method study of diffusion of an Ag adatom on (111) Ag, *Can. J. Phys.* **68** (1990) 1035–1040.
[207] R. C. Nelson, T. L. Einstein, S. V. Khare, P. J. Reus, Energetics of steps, kinks, and defects on Ag{100} and Ag{111} using the embedded atom method, and some consequences, *Surf. Sci.* **295** (1993) 462–484.
[208] G. Boisvert, L. J. Lewis, M. J. Puska, R. M. Nieminen, Energetics of diffusion on the (100) and (111) surfaces of Ag, Au, and Ir from first principles, *Phys. Rev. B* **52** (1995) 9078–9085.
[209] R. Ferrando, G. Tréglia, Tight-binding molecular dynamics study of diffusion on Au and Ag(111), *Surf. Sci.* **331–333** (1995) 920–924.
[210] C. Ratsch, A. P. Seitsonen, M. Scheffler, Strain dependence of surface diffusion: Ag on Ag(111) and Pt(111), *Phys. Rev. B* **55** (1997) 6750–6753.
[211] C. Ratsch, M. Scheffler, Density-functional theory calculations of hopping rates of surface diffusion, *Phys. Rev. B* **58** (1998) 13163–13166.
[212] N. I. Papanicolaou, G. A. Evangelakis, G. C. Kallinteris, Molecular dynamics description of silver adatom diffusion on Ag(100) and Ag(111) surfaces, *Comput. Mater. Sci.* **10** (1998) 105–110.
[213] G. Antczak, G. Ehrlich, Jump processes in surface diffusion, *Surf. Sci. Rep.* **62** (2007) 39–61.

References

[214] F. Baletto, C. Mottet, R. Ferrando, Molecular dynamics simulation of surface diffusion and growth on silver and gold clusters, *Surf. Sci.* **446** (2000) 31–45.

[215] Z. Chvoj, C. Ghosh, T. S. Rahman, M. C. Tringides, Prefactors for interlayer diffusion on Ag/Ag(111), *J. Phys.: Condens. Matter* **15** (2003) 5223–5230.

[216] J. Vrijmoeth, H. A. van der Vegt, J. A. Meyer, E. Vlieg, R. J. Behm, Surfactant-induced layer-by-layer growth of Ag on Ag(111): Origin and side effects, *Phys. Rev. Lett.* **72** (1994) 3843–3846.

[217] H. A. van der Vegt, J. Vrijmoeth, B. J. Behm, E. Vlieg, Sb-enhanced nucleation in the homoepitaxial growth of Ag(111), *Phys. Rev. B* **57** (1998) 4127–4131.

[218] M. I. Haftel, M. Rosen, New ballistically and thermally activated exchange processes in the vapor deposition of Au on Ag(111): a molecular dynamics study, *Surf. Sci.* **407** (1998) 16–26.

[219] K. Morgenstern, K.-F. Braun, K.-H. Rieder, Direct imaging of Cu dimer formation, motion, and interaction with Cu atoms on Ag(111), *Phys. Rev. Lett.* **93** (2004) 056102 1–4.

[220] K. Morgenstern, Fast scanning tunneling microscopy as a tool to understand changes on metal surfaces: from nanostructures to single atoms, *Phys. Stat. Sol.* **242** (2005) 773–796.

[221] F. Silly, M. Pivetta, M. Ternes, F. Patthey, J. P. Pelz, W.-D. Schneider, Creation of an atomic superlattice by immersing metallic adatoms in a two-dimensional electron sea, *Phys. Rev. Lett.* **92** (2004) 016101 1–4.

[222] K. Hartig, A. P. Janssen, J. A. Venables, Nucleation and growth in the system Ag/Mo(100): A comparison of UHV-SEM and AES/LEED observations, *Surf. Sci.* **74** (1978) 69–78.

[223] M. H. Langelaar, M. Breeman, D. O. Boerma, Mobility of Ag adatoms on Ag(100), *Surf. Sci.* **352–354** (1996) 597–601.

[224] C.-M. Zhang, M. C. Bartelt, J.-M. Wen, C. J. Jenks, J. W. Evans, P. A. Thiel, The initial stages of Ag/Ag(100) homoepitaxy: scanning tunneling microscopy experiements and Monte Carlo simulations, *J. Cryst. Growth* **174** (1997) 851–857.

[225] C.-M. Zhang, M. C. Bartelt, J.-M. Wen, C. J. Jenks, J. W. Evans, P. A. Thiel, Submonolayer island formation and the onset of multilayer growth during Ag/Ag(100) homoepitaxy, *Surf. Sci.* **406** (1998) 178–193.

[226] P. A. Thiel, J. W. Evans, Nucleation, growth, and relaxation of thin films: Metal(100) homoepitaxial systems, *J. Phys. Chem. B* **104** (2000) 1663–1676.

[227] L. Bardotti, C. R. Stoldt, C. J. Jenks, M. C. Bartelt, J. W. Evans, P. A. Thiel, High-resolution LEED profile analysis and diffusion barrier estimation for submonolayer homoepitaxy of Ag/Ag(100), *Phys. Rev. B* **57** (1998) 12544–12549.

[228] M. F. Rosu, C. R. Laurens, A. Falepin, M. A. James, M. H. Langelaar, F. Pleiter, O. C. Rogojanu, L. Niesen, Direct observation of self-diffusion mechanism on the Ag(100) surface, *Phys. Rev. Lett.* **81** (1998) 4680–4683.

[229] S. Frank, H. Wedler, R. J. Behm, J. Rottler, P. Maass, K. J. Caspersen, C. R. Stoldt, P. A. Thiel, J. W. Evans, Approaching the low-temperature limit in nucleation and two-dimensional growth of fcc (100) metal films Ag/Ag(100), *Phys. Rev. B* **66** (2002) 155435 1–7.

[230] A. F. Voter, Simulation of the layer-growth dynamics in silver films: Dynamics of adatom and vacancy clusters on Ag(100), *Proc. SPIE* **821** (1987) 214–226.

[231] B. D. Yu, M. Scheffler, Anisotropy of growth of the close-packed surfaces of silver, *Phys. Rev. Lett.* **77** (1996) 1095–1098.

[232] B. D. Yu, M. Scheffler, *Ab initio* study of step formation and self-diffusion on Ag(100), *Phys. Rev. B* **55** (1997) 13916–13924.

[233] U. Kürpick, T. S. Rahman, Diffusion processes relevant to homoepitaxial growth on Ag(100), *Phys. Rev. B* **57** (1998) 2482–2492.

[234] M. I. Haftel, Surface reconstruction of platinum and gold and the embedded-atom method, *Phys. Rev. B* **48** (1993) 2611–2622.
[235] R. Fink, R. Wesche, T. Klas, G. Krausch, R. Platzer, J. Voigt, U. Wöhrmann, G. Schatz, Step-correlated diffusion of In atoms on Ag(100) and Ag(111) surfaces, *Surf. Sci.* **225** (1990) 331–340.
[236] R. Fink, G. Krausch, B. Luckscheiter, R. Platzer, J. Voigt, R. Wesche, U. Wöhrmann, X. L. Ding, G. Schatz, Diffusion of isolated In atoms on Ag and Cu surfaces, *Vacuum* **41** (1990) 1643–1645.
[237] F. Patthey, C. Massobrio, W.-D. Schneider, Dynamics of surface alloying: Determination of diffusion barriers from photoelectron spectra, *Phys. Rev. B* **53** (1996) 13146–13149.
[238] M. Canepa, E. Magnano, A. Campora, P. Cantini, M. Salvietti, L. Mattera, Diffusion by atomic place exchange in ultrathin iron films on Ag(100): an ion scattering spectroscopy study, *Surf. Sci.* **352–354** (1996) 36–40.
[239] M. Caffio, A. Atrei, U. Bardi, G. Rovida, Growth mechanism and structure of nickel deposited on Ag(001), *Surf. Sci.* **588** (2005) 135–148.
[240] P. R. Schwoebel, G. L. Kellogg, Palladium diffusion and cluster nucleation on Ta(110), *Phys. Rev. B* **38** (1988) 5326–5331.
[241] G. Antczak, R. Blaszczyszyn, Diffusion of palladium on tantalum microcrystal, *Acta Phys. Pol.* **103** (2003) 57–65.
[242] G. Ehrlich, F. G. Hudda, Atomic view of surface self-diffusion: tungsten on tungsten, *J. Chem. Phys.* **44** (1966) 1039–1049.
[243] D. W. Bassett, M. J. Parsley, Field ion microscope studies of transition metal adatom diffusion on (110), (211) and (321) tungsten surfaces, *J. Phys. D* **3** (1970) 707–716.
[244] P. Cowan, T. T. Tsong, Diffusion behavior of tungsten clusters on the tungsten (110) planes, *Phys. Lett.* **53A** (1975) 383–383.
[245] G. L. Kellogg, T. T. Tsong, P. Cowan, Direct observation of surface diffusion and atomic interactions on metal surfaces, *Surf. Sci.* **70** (1978) 485–519.
[246] C. Chen, T. T. Tsong, Behavior of Ir atoms and clusters on Ir surfaces, *Phys. Rev. B* **41** (1990) 12403–12412.
[247] G. Antczak, G. Ehrlich, Long jump rates in surface diffusion: W on W(110), *Phys. Rev. Lett.* **92** (2004) 166105 1–4.
[248] G. Ehrlich, C. F. Kirk, Binding and field desorption of individual tungsten atoms, *J. Chem. Phys.* **48** (1968) 1465–1480.
[249] P. Wynblatt, N. A. Gjostein, A calculation of migration energies and binding energies for tungsten adatoms on tungsten surfaces, *Surf. Sci.* **22** (1970) 125–136.
[250] J. R. Banavar, M. H. Cohen, R. Gomer, A calculation of surface diffusion coefficients of adsorbates on the (110) plane of tungsten, *Surf. Sci.* **107** (1981) 113–126.
[251] J. D. Doll, H. K. McDowell, Theoretical studies of surface diffusion: Self-diffusion in the bcc(211) System, *Surf. Sci.* **123** (1982) 99–105.
[252] W. Xu, J. B. Adams, W single adatom diffusion on W surfaces, *Surf. Sci.* **319** (1994) 58–67.
[253] D. Chen, W. Hu, H. Deng, L. Sun, F. Gao, Diffusion of tungsten clusters on tungsten (110) surface, *Eur. Phys. J. B* **68** (2009) 479–485.
[254] S. Y. Davydov, S. K. Tikhonov, On the cohesive approach to the calculation of d-metal on d-metal adsorption properties, *Surf. Sci.* **371** (1997) 157–167.
[255] W. A. Harrison, J. M. Wills, Interionic interactions in simple metals, *Phys. Rev. B* **25** (1982) 5007–5017.
[256] J. M. Wills, W. A. Harrison, Interionic interactions in transition metals, *Phys. Rev. B* **28** (1983) 4363–4373.

References

[257] A. T. Loburets, A. G. Naumovets, Y. S. Vedula, Surface diffusion of lithium on (011) face of tungsten, *Surf. Sci.* **120** (1982) 347–366.

[258] R. Morin, Diffusion and compressibility of sodium on the (110) plane of tungsten, *Surf. Sci.* **155** (1985) 187–202.

[259] L. Schmidt, R. Gomer, Adsorption of potassium on tungsten, *J. Chem. Phys.* **42** (1965) 3573–3598.

[260] A. G. Naumovets, Desorption of potassium from tungsten in an electric field, *Soviet Phys. Sol. State* **5** (1964) 1668–1674.

[261] R. Meclewski, Measurement of the surface-diffusion activation energy of potassium on tungsten, *Acta Phys. Pol. A* **37** (1970) 41–47.

[262] A. M. Dabrowski, C. Kleint, Cross-correlation function of field emission flicker noise from K/W(110): interpretation by adparticle surface diffusion, *Surf. Sci.* **172** (1986) 372–384.

[263] S. Dennler, J. Hafner, First-principles study of ultrathin Mn films on W surfaces. II. Surface diffusion, *Phys. Rev. B* **72** (2005) 214414 1–9.

[264] T.-U. Nahm, R. Gomer, The conversion of Fe on W(110) from the low temperature to the high temperature form, *Surf. Sci.* **380** (1997) 52–60.

[265] M. Sladecek, B. Sepiol, J. Korecki, T. Slezak, R. Rüffer, D. Kmiec, G. Vogl, Dynamics in submonolayer Fe-films, *Surf. Sci.* **566–568** (2004) 372–376.

[266] D. W. Bassett, Field ion microscopic studies of submonolayer films of nickel, palladium and platinum on (110) tungsten surfaces, *Thin Solid Films* **48** (1978) 237–246.

[267] G. L. Kellogg, Surface diffusion and clustering of nickel atoms on the (110) plane of tungsten, *Surf. Sci.* **187** (1987) 153–164.

[268] A. J. Melmed, Adsorption and surface diffusion of copper on tungsten, *J. Chem. Phys.* **43** (1965) 3057–3062.

[269] A. J. Melmed, Influence of adsorbed gas on surface diffusion and nucleation, *J. Appl. Phys.* **37** (1966) 275–279.

[270] J. P. Jones, The adsorption of copper on tungsten, *Proc. Roy. Soc. (London) A* **284** (1965) 469–487.

[271] O. Nishikawa, A. R. Saadat, Field emission and field ion microscope study of Ga, In, and Sn on W: Structure, work function, diffusion and binding energy, *Surf. Sci.* **60** (1976) 301–324.

[272] T.-Y. Fu, L.-C. Cheng, Y.-J. Hwang, T. T. Tseng, Diffusion of Pd adatoms on W surfaces and their interactions with steps, *Surf. Sci.* **507–510** (2002) 103–107.

[273] S.-M. Oh, S. J. Koh, K. Kyuno, G. Ehrlich, Non-nearest-neighbor jumps in 2D diffusion: Pd on W(110), *Phys. Rev. Lett.* **88** (2002) 236102 1–4.

[274] I. Langmuir, J. B. Taylor, The mobility of cesium atoms adsorbed on tungsten, *Phys. Rev.* **40** (1932) 463–464.

[275] J. B. Taylor, I. Langmuir, The evaporation of atoms, ions and electrons from cesium films on tungsten, *Phys. Rev.* **44** (1933) 423–458.

[276] H. M. Love, H. D. Wiederick, Cesium diffusion at a tungsten surface, *Can. J. Phys.* **47** (1969) 657–663.

[277] H. Utsugi, R. Gomer, Field desorption of barium from tungsten, *J. Chem. Phys.* **37** (1962) 1706–1719.

[278] A. G. Naumovets, V. V. Poplavsky, Y. S. Vedula, Diffusion and phase transitions in barium monolayers on the (011) plane of tungsten, *Surf. Sci.* **200** (1988) 321–334.

[279] T. T. Tsong, G. Kellogg, Direct observation of the directional walk of single adatoms and the adatom polarizability, *Phys. Rev. B* **12** (1975) 1343–1353.

[280] D. W. Bassett, M. J. Parsley, The effect of an electric field on the surface diffusion of rhenium adsorbed on tungsten, *Brit. J. Appl. Phys. (J. Phys. D)* **2** (1969) 13–16.
[281] T. T. Tsong, Direct observation of interactions between individual atoms on tungsten surfaces, *Phys. Rev. B* **6** (1972) 417–426.
[282] R. A. Johnson, P. J. White, Interpretation of field-ion microscopy data for surface diffusion and clustering, *Phys. Rev. B* **7** (1973) 4016–4018.
[283] T. T. Tsong, Surface diffusion and cluster binding energy of individual atoms on tungsten surfaces, *Phys. Rev. B* **7** (1973) 4018–4020.
[284] M. Lovisa, G. Ehrlich, Adatom diffusion on metals: Ir on W(110), *J. Phys. (Paris)* **50** (1989) C8-279–284.
[285] M. Lovisa, Diffusivity of Ir atoms on W(110), Private communication (1993).
[286] G. Antczak, G. Ehrlich, Long jumps in diffusion of iridium on W(110), *Phys. Rev. B* **71** (2005) 115422 1–9.
[287] D. W. Bassett, Migration of platinum adatom clusters on tungsten (110) surfaces, *J. Phys. C* **9** (1976) 2491–2503.
[288] J. P. Jones, N. T. Jones, Field emission microscopy of gold on single-crystal planes of tungsten, *Thin Solid Films* **35** (1976) 83–97.
[289] M. K. Debe, D. A. King, Space-group determination of the low-temperature W{100}(2 × 2) R45 surface structure by low-energy-electron diffraction, *Phys. Rev. Lett.* **39** (1977) 708–711.
[290] B. Bayat, H.-W. Wassmuth, Desorption kinetics and directional dependence of the surface diffusion of potassium on stepped W(100) surfaces, *Surf. Sci.* **140** (1984) 511–520.
[291] G. L. Kellogg, The mobility and structure of nickel atoms on the (100) plane of tungsten, *Surf. Sci.* **192** (1987) L879–886.
[292] J. Beben, W. Gubernator, Investigation of surface diffusion of Hf on W(100) by the density fluctuation method, *Surf. Sci.* **304** (1994) 59–64.
[293] R. Morin, M. Drechsler, A study of a coadsorbate surface diffusion (Pb on C/W), *Surf. Sci.* **111** (1981) 140–148.
[294] Z. Xu, L. G. Zhou, J. Wang, T. S. Cale, H. Huang, Three-dimensional Ehrlich-Schwoebel barriers of W, *Computers, Materials, and Continua* **3** (2006) 43–47.
[295] G. J. Ackland, R. Thetford, An improved n-body semiempirical model for body-centered cubic transition-metals, *Philos. Mag. A* **56** (1987) 15–30.
[296] W. R. Graham, G. Ehrlich, Direct identification of atomic binding sites on a crystal, *Surf. Sci.* **45** (1974) 530–552.
[297] Y. M. Gong, R. Gomer, Thermal roughening on stepped tungsten surfaces I. The zone (011) – (112), *J. Chem. Phys* **88** (1988) 1359–1369.
[298] T. Y. Fu, W. J. Weng, T. T. Tsong, Dynamic study of W atoms and clusters on W(111) surfaces, *Appl. Surf. Sci.* **254** (2008) 7831–7834.
[299] P. G. Flahive, W. R. Graham, Surface site geometry and diffusion characteristics of single Ni atoms on W(111), *Thin Solid Films* **51** (1978) 175–184.
[300] T. Biernat, C. Kleint, R. Meclewski, Field emission current fluctuations due to lithium adsorbed on the W(111) region, *Surf. Sci.* **246** (1991) 54–59.
[301] T. Biernat, A. M. Dabrowski, Two-process diffusion of titanium on the (111) tungsten plane by the current fluctuation method in FEM, *Vacuum* **63** (2001) 113–118.
[302] T. Biernat, R. Blaszczyszyn, Surface diffusion of dysprosium on the W(111) facet, *Appl. Surf. Sci.* **230** (2004) 81–87.
[303] C. L. Chen, T. T. Tsong, Self-diffusion on reconstructed and nonreconstructed Ir surfaces, *J. Vac. Sci. Technol. A* **10** (1992) 2178–2184.

[304] S. C. Wang, G. Ehrlich, Atomic behavior at individual binding sites: Ir, Re, and W on Ir(111), *Phys. Rev. Lett.* **68** (1992) 1160–1163.

[305] K. Kyuno, A. Gölzhäuser, G. Ehrlich, Growth and the diffusion of platinum atoms and dimers on Pt(111), *Surf. Sci.* **397** (1998) 191–196.

[306] S. C. Wang, U. Kürpick, G. Ehrlich, Surface diffusion of compact and other clusters: Ir$_x$ on Ir(111), *Phys. Rev. Lett.* **81** (1998) 4923–4926.

[307] T.-Y. Fu, H.-T. Wu, T. T. Tsong, Energetics of surface atomic processes near a lattice step, *Phys. Rev. B* **58** (1998) 2340–2346.

[308] B. Piveteau, D. Spanjaard, M. C. Desjonquères, Inversion of the stability between normal and fault sites for transition-metal adatoms on (111)fcc and (0001)hcp transition metal surfaces, *Phys. Rev. B* **46** (1992) 7121–7126.

[309] K.-D. Shiang, C. M. Wei, T. T. Tsong, A molecular dynamics study of self-diffusion on metal surfaces, *Surf. Sci.* **301** (1994) 136–150.

[310] C. M. Chang, C. M. Wei, S. P. Chen, Modeling of Ir adatoms on Ir surfaces, *Phys. Rev. B* **54** (1996) 17083–17096.

[311] O. S. Trushin, M. Kotrla, F. Máca, Energy barriers on stepped Ir/Ir(111) surfaces: a molecular statics calculation, *Surf. Sci.* **389** (1997) 55–65.

[312] U. Kürpick, Self-diffusion on stepped Ir(111) surfaces, *Phys. Rev. B* **69** (2004) 205410 1–6.

[313] S. C. Wang, G. Ehrlich, Determination of atomic binding sites on the fcc(111) plane, *Surf. Sci.* **246** (1991) 37–42.

[314] S. C. Wang, G. Ehrlich, Atom condensation on an atomically smooth surface: Ir, Re, W, and Pd on Ir(111), *J. Chem. Phys.* **94** (1991) 4071–4074.

[315] C. Chen, T. T. Tsong, Displacement distribution and atomic jump direction in diffusion of Ir atoms on the Ir(001) surface, *Phys. Rev. Lett.* **64** (1990) 3147–3150.

[316] A. Friedl, O. Schütz, K. Müller, Self-diffusion on iridium (100). A structure investigation by field-ion microscopy, *Surf. Sci.* **266** (1992) 24–29.

[317] T.-Y. Fu, T. T. Tsong, Structure and diffusion of small Ir and Rh clusters on Ir(001) surfaces, *Surf. Sci.* **421** (1999) 157–166.

[318] T. T. Tsong, C. L. Chen, Atomic replacement and vacancy formation and annihilation on iridium surfaces, *Nature* **355** (1992) 328–331.

[319] D. W. Bassett, P. R. Webber, Diffusion of single adatoms of platinum, iridium and gold on platinum surfaces, *Surf. Sci.* **70** (1978) 520–531.

[320] P. J. Feibelman, J. S. Nelson, G. L. Kellogg, Energetics of Pt adsorption on Pt(111), *Phys. Rev. B* **49** (1994) 10548–10556.

[321] M. Bott, M. Hohage, M. Morgenstern, T. Michely, G. Comsa, New approach for determination of diffusion parameters of adatoms, *Phys. Rev. Lett.* **76** (1996) 1304–1307.

[322] A. Gölzhäuser, G. Ehrlich, Atom movement and binding on surface clusters: Pt on Pt(111) clusters, *Phys. Rev. Lett.* **77** (1996) 1334–1337.

[323] M. Villarba, H. Jónsson, Diffusion mechanisms relevant to metal crystal growth: Pt/Pt(111), *Surf. Sci.* **317** (1994) 15–36.

[324] S. Liu, Z. Zhang, J. Nørskov, H. Metiu, The mobility of Pt atoms and small Pt clusters on Pt(111) and its implications for the early stages of epitaxial growth, *Surf. Sci.* **321** (1994) 161–171.

[325] R. Wang, K. A. Fichthorn, An investigation of the energetics and dynamics of adatom motion to descending step edges in Pt/Pt(111) homoepitaxy, *Surf. Sci.* **301** (1994) 253–259.

[326] J. Jacobsen, K. W. Jacobsen, P. Stoltze, J. K. Nørskov, Island shape-induced transition from 2D to 3D growth for Pt/Pt(111), *Phys. Rev. Lett.* **74** (1995) 2295–2298.

[327] P. J. Feibelman, Interlayer self-diffusion on stepped Pt(111), *Phys. Rev. Lett.* **81** (1998) 168–171.
[328] G. Boisvert, L. J. Lewis, M. Scheffler, Island morphology and adatom self-diffusion on Pt(111), *Phys. Rev. B* **57** (1998) 1881–1889.
[329] F. Máca, M. Kotrla, O. S. Trushin, Energy barriers for diffusion on stepped Pt(111) surface, *Vacuum* **54** (1999) 113–117.
[330] G. Leonardelli, E. Lundgren, M. Schmid, Adatom interlayer diffusion on Pt(111): an embedded atom method study, *Surf. Sci.* **490** (2001) 29–42.
[331] B. Lee, K. Cho, Extended embedded-atom method for platinum nanoparticles, *Surf. Sci.* **600** (2006) 1982–1990.
[332] J. Cai, Y. Y. Ye, Simple analytical embedded-atom-potential model including a long-range force for fcc metals and their alloys, *Phys. Rev. B* **54** (1996) 8398–8410.
[333] J. Yang, W. Hu, G. Yi, J. Tang, Atomistic simulation of Pt trimer on Pt(111) surface, *Appl. Surf. Sci.* **253** (2007) 8825–8829.
[334] P. Blandin, C. Massobrio, Diffusion properties and collisional dynamics of Ag adatoms and dimers on Pt(111), *Surf. Sci.* **279** (1992) L219–224.
[335] C. Massobrio, P. Blandin, Structure and dynamics of Ag clusters on Pt(111), *Phys. Rev. B* **47** (1993) 13687–13694.
[336] H. Röder, H. Brune, J.-P. Bucher, K. Kern, Changing morphology of metallic monolayers via temperature controlled heteroepitaxial growth, *Surf. Sci.* **298** (1993) 121–126.
[337] H. Brune, H. Röder, C. Boragno, K. Kern, Microscopic view of nucleation on surfaces, *Phys. Rev. Lett.* **73** (1994) 1955–1958.
[338] P. J. Feibelman, Diffusion barrier for a Ag adatom on Pt(111), *Surf. Sci.* **313** (1994) L801–805.
[339] R. F. Sabiryanov, M. I. Larsson, K. J. Cho, W. D. Nix, B. M. Clemens, Surface diffusion and growth of patterned nanostructures on strained surfaces, *Phys. Rev. B* **67** (2003) 125412 1–8.
[340] G. Kresse, J. Joubert, From ultrasoft pseudopotentials to the projector augmented-wave method, *Phys. Rev. B* **59** (1999) 1758–1775.
[341] C. Goyhenex, Adatom and dimer migration in heteroepitaxy: Co/Pt(111), *Surf. Sci.* **600** (2006) 15–22.
[342] A. P. Graham, J. P. Toennies, Macroscopic diffusion and low-frequency vibration of sodium on Pt(111) investigated with helium atom scattering, *J. Phys. Chem. B* **105** (2001) 4003–4009.
[343] D. L. Price, Effects of a volume-dependent potential on equilibrium properties of liquid sodium, *Phys. Rev. A* **4** (1971) 358–363.
[344] G. L. Kellogg, P. J. Feibelman, Surface self-diffusion on Pt(001) by an atomic exchange mechanism, *Phys. Rev. Lett.* **64** (1990) 3143–3146.
[345] G. L. Kellogg, Temperature dependence of surface self-diffusion on Pt(001), *Surf. Sci.* **246** (1991) 31–36.
[346] G. L. Kellogg, The effect of an externally applied electric field on the diffusion of Pt adatoms, dimers, and trimers on Pt(001), *Appl. Surf. Sci.* **76/77** (1994) 115–121.
[347] G. L. Kellogg, Electric field inhibition and promotion of exchange diffusion on Pt(001), *Phys. Rev. Lett.* **70** (1993) 1631–1634.
[348] T. R. Linderoth, J. J. Mortensen, K. W. Jacobsen, E. Laegsgaard, I. Stensgaard, F. Besenbacher, Homoepitaxial growth of Pt on Pt(100)-hex: Effects of strongly anisotropic diffusion and finite island sizes, *Phys. Rev. Lett.* **77** (1996) 87–90.
[349] J. J. Mortensen, T. R. Linderoth, K. W. Jacobsen, E. Laegsgaard, I. Stensgaard, F. Besenbacher, Effects of anisotropic diffusion and finite island sizes in homoepitaxial growth Pt on Pt(100)-hex, *Surf. Sci.* **400** (1998) 290–313.

[350] J. Zhuang, L. Liu, Exchange mechanism for adatom diffusion on metal fcc(100) surfaces, *Phys. Rev. B* **58** (1998) 1173–1176.

[351] M. I. Haftel, M. Rosen, T. Franklin, M. Hettermann, Molecular dynamics observations of the interdiffusion and Stranski-Krastanov growth in the early film deposition of Au on Ag(100), *Phys. Rev. Lett.* **72** (1994) 1858–1861.

[352] M. I. Haftel, M. Rosen, Molecular-dynamics description of early film deposition of Au on Ag(110), *Phys. Rev. B* **51** (1995) 4426–4434.

[353] J. Zhuang, L. Liu, Adatom self-diffusion on Pt(100) surface by an ad-dimer migrating, *Science in China A* **43** (2000) 1108–1113.

[354] W. Xiao, P. A. Greaney, D. C. Chrzan, Pt adatom diffusion on strained Pt(001), *Phys. Rev. B* **70** (2004) 033402 1–4.

[355] G. L. Kellogg, A. F. Wright, M. S. Daw, Surface diffusion and adatom-induced substrate relaxations of Pt, Pd, and Ni atoms on Pt(001), *J. Vac. Sci. Technol. A* **9** (1991) 1757–1760.

[356] S. Günther, E. Kopatzki, M. C. Bartelt, J. W. Evans, R. J. Behm, Anisotropy in nucleation and growth of two-dimensional islands during homoepitaxy on 'hex' reconstructed Au(100), *Phys. Rev. Lett.* **73** (1994) 553–556.

[357] S. Liu, L. Bönig, H. Metiu, Effect of small-cluster mobility and dissociation on the island density in epitaxial growth, *Phys. Rev. B* **52** (1995) 2907–2913.

[358] H. Göbel, P. von Blanckenhagen, A study of surface diffusion on gold with an atomic force microscope, *Surf. Sci.* **331–333** (1995) 885–890.

[359] G. Boisvert, L. J. Lewis, Self-diffusion on low-index metallic surfaces: Ag and Au(100) and (111), *Phys. Rev. B* **54** (1996) 2880–2889.

[360] L. Bönig, S. Liu, H. Metiu, An effective medium theory study of Au islands on the Au(100) surface: reconstruction, adatom diffusion, and island formation, *Surf. Sci.* **365** (1996) 87–95.

[361] B. D. Yu, M. Scheffler, Physical origin of exchange diffusion on fcc(100) metal surfaces, *Phys. Rev. B* **56** (1997) R15569–15572.

[362] Y.-L. He, G.-C. Wang, Observation of atomic place exchange in submonolayer heteroepitaxial Fe/Au(001) films, *Phys. Rev. Lett.* **71** (1993) 3834–3837.

[363] O. S. Hernán, A. L. Vázquez de Parga, J. M. Gallego, R. Miranda, Self-surfactant effect on Fe/Au(100): place exchange plus Au self-diffusion, *Surf. Sci.* **415** (1998) 106–121.

[364] Y. B. Liu, D. Y. Sun, X. G. Gong, Local strain induced anisotropic diffusion on $(23\times\sqrt{3})$-Au(111) surface, *Surf. Sci.* **498** (2002) 337–342.

[365] F. Ercolessi, M. Parinello, E. Tosatti, Au(100) reconstruction in the glue model, *Surf. Sci.* **177** (1986) 314–328.

[366] F. Ercolessi, M. Parrinello, E. Tosatti, Simulation of gold in the glue model, *Philos. Mag. A* **58** (1988) 213–226.

[367] B. Fischer, H. Brune, J. V. Barth, A. Fricke, K. Kern, Nucleation kinetics on inhomogeneous substrates: Al/Au(111), *Phys. Rev. Lett.* **82** (1999) 1732–1735.

[368] J. A. Meyer, I. D. Baikie, E. Kopatzki, R. J. Behm, Preferential island nucleation at the elbows of the Au(111) herringbone reconstruction through place exchange, *Surf. Sci.* **365** (1996) L647–651.

[369] C. Goyhenex, H. Bulou, J.-P. Deville, G. Tréglia, Atomistic simulations of relaxation and reconstruction phenomena in heteroepitaxy: Co/Au(111), *Appl. Surf. Sci.* **188** (2002) 134–139.

[370] H. Bulou, O. Lucas, M. Kibaly, C. Goyhenex, Long-time scale molecular dynamics study of Co diffusion on the Au(111) surface, *Comput. Mater. Sci.* **27** (2003) 181–185.

[371] H. Bulou, C. Massobrio, Dynamical behavior of Co adatoms on the herringbone reconstructed surface of Au(111), *Superlattices Microstruct.* **36** (2004) 305–313.

[372] S.-C. Li, Y. Han, J.-F. Jia, Q.-K. Xue, F. Liu, Determination of the Ehrlich-Schwoebel barrier in epitaxial growth of thin films, *Phys. Rev. B* **74** (2006) 195428 1–5.

[373] G. Jnawali, H. Hattab, C. A. Bobisch, A. Bernhart, E. Zubkov, R. Möller, M. Horn-von Hoegen, Homoepitaxial growth of Bi(111), *Phys. Rev. B* **78** (2008) 035321 1–9.

[374] O. S. Trushin, P. Salo, T. Ala-Nissila, Energetics and many-particle mechanisms of two-dimensional cluster diffusion on Cu(100) surfaces, *Phys. Rev. B* **62** (2000) 1611–1614.

[375] G. Antczak, Long jumps in one-dimensional surface self-diffusion: Rebound transitions, *Phys. Rev. B* **73** (2006) 033406 1–4.

[376] D. R. Lide, *CRC Handbook of Chemistry and Physics, 86th Edition* (CRC Press, Taylor & Francis, Boca Raton, 2008).

6 Diffusion in special environments

Single adatom diffusion on a variety of one- and two-dimensional surfaces has now been surveyed. However, events occurring at plane edges and other types of defects play an important role in affecting diffusion over the entire surface. There are also a huge number of indirect indications that impurities influence atom movement as well as other processes. The amount of direct information available for this field is not large, but we will examine it to gain at least some insight into such phenomena.

6.1 Near impurities

Some effort has been made to uncover the way in which atomic diffusion is affected by impurities in the substrate. A start was made by Cowan and Tsong [1] in 1977, who explored the effect of rhenium atoms dissolved in tungsten on the surface diffusion of tungsten atoms on the (110) plane. This surface was prepared from W-3% Re alloy, and so had a number of rhenium substitutional atoms on the (110) plane. Tungsten atoms were then deposited on the surface, and the spatial distribution of places visited by atoms during their thermal movements was measured and compared with that for tungsten moving on the clean W(110), as shown in Fig. 6.1. Tungsten adatoms diffusing over a W-3% Re(110) surface were found to spend more time at a few sites on the surface, which the authors associated as most likely being next to interstitial rhenium atoms embedded in the surface. They estimated an increase in the binding energy of 0.090 ± 0.007 eV for a tungsten adatom in the nearest-neighbor position along the [001] direction from a rhenium atom embedded in the substrate. In the next-nearest-neighbor position, along the [110] direction, repulsion was equal to or larger than 0.080 eV. Cowan and Tsong also examined the kinetics of atom motion close to rhenium atoms, and found a diffusion barrier 0.041 ± 0.006 eV higher than on a clean part of the surface.

The difficulty with this type of study is lack of exact knowledge of the location of the impurity atoms. This was remedied in 1994 by Kellogg [2], who embedded an iridium atom on a rhodium(100) surface by the exchange mechanism, as in Fig. 6.2. In this process, a rhodium adatom is produced on the surface and since the position of the embedded Ir atom is exactly known, details of the association of rhodium with the bound iridium, as well as changes in diffusion characteristics associated with the isolated impurity atom, can be studied. It must also be noted that iridium atoms on Rh(100) can be distinguished from

424   Diffusion in special environments

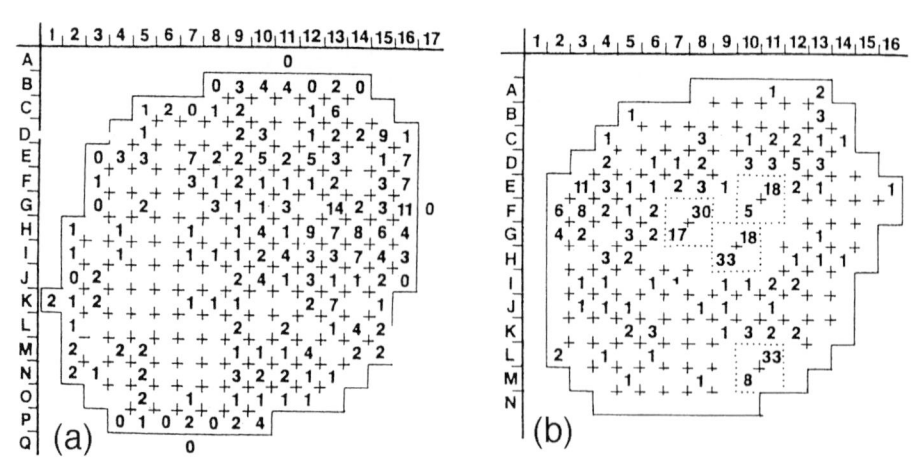

Fig. 6.1   Distribution of tungsten atoms over W(110) surface. (a) On clean (110) plane. (b) On (110) plane of W-3%Re after diffusion at 340 K. Concentration at specific sites is evident (after Cowan and Tsong [1]).

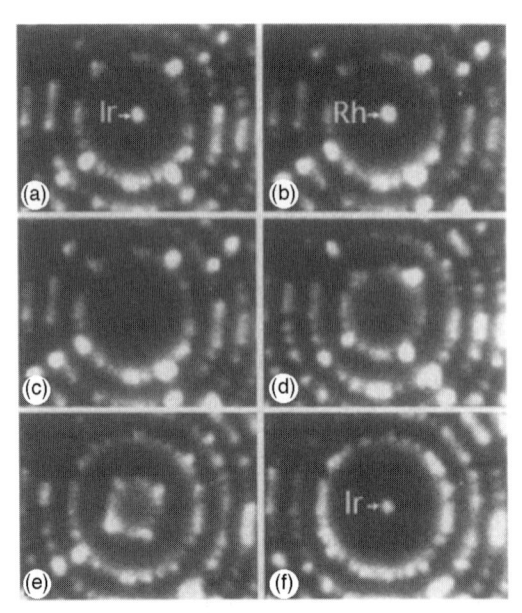

Fig. 6.2   FIM Ne images at 77 K show embedding of Ir atom on Rh(100). (a) – (b) Ir adatom exchanges with Rh substrate. (c)–(f) Field evaporation of outermost Rh plane reveals Ir atom embedded in lattice (after Kellogg [2]).

rhodium atoms as the voltage for their field evaporation is considerably lower than for rhodium; rhodium at a site adjacent to a substitutional iridium desorbs at an intermediate value. It turned out that a rhodium adatom was bound at four sites around iridium, as illustrated by the field ion microscope images in Fig. 6.3. From studies at temperatures 248 K to 265 K, Kellogg estimated a barrier for diffusion around the

6.1 Near impurities

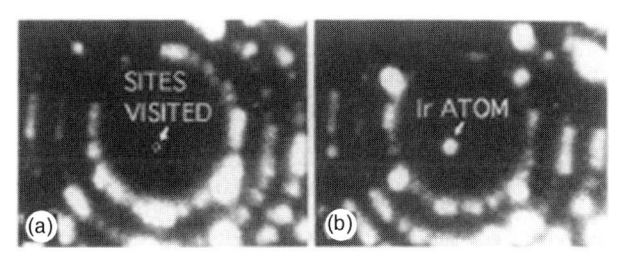

Fig. 6.3 FIM images reveal the trapping of a Rh atom by an Ir atom substituted into Rh(100) plane. (a) White dots indicate location of Rh atom after > 240 cycles at ~255 K. (b) After field evaporation of one layer, Ir atom is revealed below the atom sites shown in (a) (after Kellogg [2]).

inserted iridium of 0.76 eV, assuming $D_o = 1 \times 10^{-3}$ cm^2/sec. At higher temperatures, the rhodium detached from the impurity and self-diffused over the surface with an activation energy of 0.83 ± 0.05 eV, reasonably close to Ayrault's [3] value. Kellogg also determined the energy for a rhodium atom to dissociate from a substitutional iridium as 0.95 ± 0.02 eV, assuming the same prefactor as for diffusion around the iridium trap.

Exchange of iron atoms with gold on the Au(100)-(26 × 5) surface, described as hex, was observed by Hernán et al. [4] in 1998 using STM. They were able to directly image the Fe atoms embedded into the first Au layer at room temperature and established that the distribution was random, probably due to Fe-Fe repulsive interactions, given that Fe atoms are immobile at room temperature. Based on their STM observations and on Monte Carlo simulations, Hernán et al. concluded that the Au atoms expelled during the exchange process moved almost in the same way as on a clean Au(100) surface. However, if Fe atoms were embedded into the corners of a reconstructed cell, they acted as preferential sites for nucleation.

The effect of gases, specifically adsorbed hydrogen, on the diffusivity of individual adatoms has been examined a number of times. In 1980 Casanova and Tsong [5] looked at how hydrogen affects the self-diffusion of tungsten atoms on the (321) plane of tungsten. Their results are shown in Fig. 6.4 – hydrogen, presumably adsorbed as atoms, inhibited the diffusion. The diffusion barrier was increased by 0.05 eV, but this is probably within the limit of error for the measurements. Nevertheless, based on Fig. 6.4, there is little doubt that the tungsten atom mobility was decreased.

A more detailed study of the effect of hydrogen on self-diffusion of nickel atoms on a nickel crystal was carried out at essentially the same time by Tung and Graham [6]. Among other things they compared a nickel tip prepared by thermal annealing with one that had been field evaporated in hydrogen. The differences in the diffusivity on Ni(110) are shown in Fig. 3.46. After hydrogen treatment of the tip, atom diffusion set in at a temperature ~60 K below that on a surface not exposed to hydrogen. All these results were obtained after heating and field evaporation of the tip, which eliminated atom diffusion at cryogenic temperatures at and below 30 K, observed on (110), (331), and (311) planes when hydrogen was still present. Although an understanding of how hydrogen promotes diffusion of nickel atoms, especially on Ni(110), does not emerge from this work, it is clear that there is a very major effect.

Diffusion in special environments

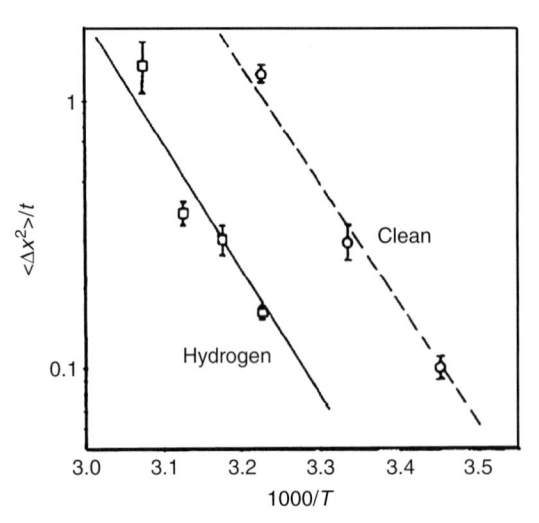

Fig. 6.4 Arrhenius plots for self-diffusion of W adatom on W(321) plane. Diffusion on clean surface indicated by dashed line, on hydrogen saturated surface by solid line (after Casanova and Tsong [5]).

The influence of low coverages of hydrogen impurities on the movement of Ag atoms on the Pt(111) surface was investigated in 1995 by Blandin and Ballone [7] using molecular dynamics with the embedded atom method. They deposited Ag and hydrogen at random positions on the surface, and showed that the presence of hydrogen increased the barrier for Ag movement by around 0.1 eV. Ag and H "dynamically" form a dimer on the surface with a characteristic lifetime around 0.4 psec. However, H embedded between the first and second layer of Pt did not influence the movement of Ag atoms.

A year later, a theoretical discussion of the effect of hydrogen on surface diffusion has been given by Stumpf [8], who relied on density-functional calculations for self-diffusion of Be(0001). He established that hydrogen atoms sat on top of beryllium adatoms and formed a strong bond which weakened the binding of the adatom to the substrate, known as the sky-hook mechanism. This lowered the barrier for surface diffusion from 0.06 eV on the bare surface to 0.02 eV when interacting with hydrogen.

The effect of hydrogen on self-diffusion on Rh(100), and also on Rh(311), has been looked at the next year in field ion microscopic observations by Kellogg [9]. The effect, as shown in Fig. 6.5, is an apparent reduction in the diffusion barrier by the hydrogen. This is only apparent, however, as raising the temperature now not only changed the rate but also the hydrogen coverage on the surface. Nevertheless, it is clear that rhodium self-diffusion occurs more rapidly and at lower temperatures. The effect of hydrogen on Rh(311) is similar, increasing the rate of diffusion. A clear indication of changes in the mean-square displacement for different hydrogen coverages was presented by Kellogg [9] but he claimed that the sky-hook mechanism, a decrease of metal-to-metal bonding through formation of strong metal-hydrogen bonds, was not adequate to explain his finding. The same year [10] Kellogg looked at the effect of hydrogen on self-diffusion on

6.1 Near impurities

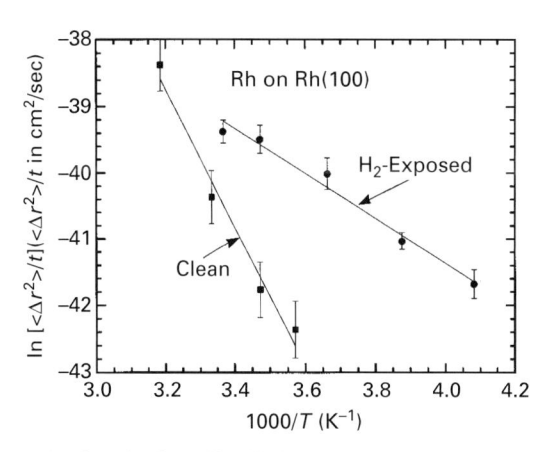

Fig. 6.5 Arrhenius plot for self-diffusion on Rh(100), for clean and H$_2$ exposed surface (after Kellogg [9]).

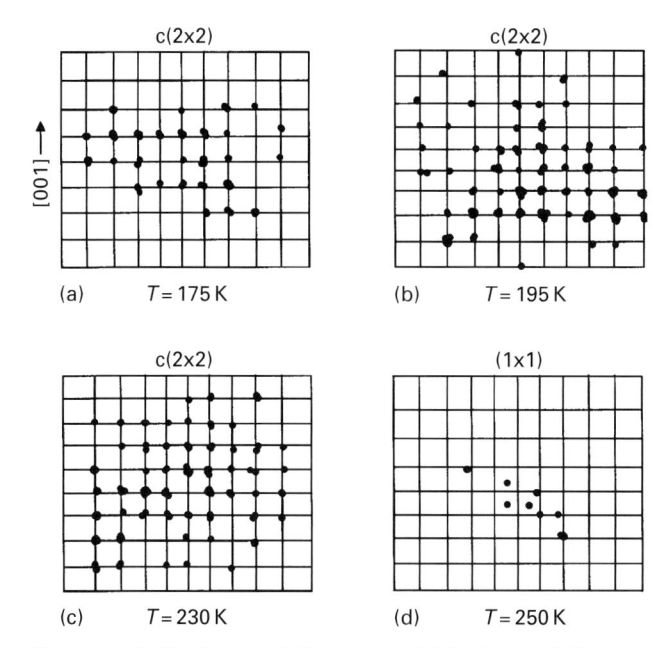

Fig. 6.6 Sites occupied by Pt atom diffusing on Pt(100) plane at different temperatures and H$_2$ pressures. (a)–(c) c(2 × 2) maps indicate exchange as the diffusion mechanism. (d) (1 × 1) map shows that Pt atom hopping is now dominant. H$_2$ pressures in Torr: (a) 0 (b) 1 × 10^{-10} (c) 3 × 10^{-10} (d) 1 × 10^{-9} (after Kellogg [10]).

the Pt(100) surface. On the clean surface, movement proceeded by exchange, but with a hydrogen partial pressure of ~1 × 10^{-10} Torr the diffusion rate decreased by three orders of magnitude over 2–3 hours. Exposing the surface to hydrogen in the range 1 × 10^{-9} Torr at 200 K completely stopped diffusion of the Pt atom on the surface. Kellogg additionally investigated the map of sites occupied by a Pt atom and noticed a change in the mechanism of movement from exchange to hopping, shown in Fig. 6.6. The activation energy for hopping, assuming a standard prefactor for the diffusivity, was estimated as 0.63–0.66 eV. This study showed that hydrogen does not always increase the mobility of adatoms: slowing down movement seems to be correlated with the mechanism involved in adatom diffusion.

Calculations of the effect of hydrogen on self-diffusion on a Ni(100) surface were carried out in the same year with EAM interactions by Haug et al. [11]. They arrived at a diffusion barrier of 0.61 eV on the bare surface. Hydrogen added to the surface interacted attractively (~ 0.02 eV) with a nickel adatom in a nearest-neighbor position, but this attractive force did not change much in the movement of hydrogen adatoms. The presence of hydrogen did, however, change the mechanism of movement of the Ni adatom on the surface. The nickel jumped on top of the hydrogen atom, over a barrier of 0.47 eV, and then to the other side, with an energy barrier of 0.067 eV. Hydrogen was also able to move as a pair together with a nickel atom over a barrier of 0.47 eV. The net effect of hydrogen addition to Ni(100) was to enhance the diffusivity of an adatom. A hydrogen adatom adsorbed close to a Ni island on Ni(100) was also able to destabilize the island by reducing the activation energy for detaching a corner atom. Haug and Jenkins [12] in 2000 continued these investigations, examining Ni(110) and Ni(111) planes. On the former, hydrogen had little influence on movement; the barrier for Ni hopping on the clean surface was 0.39 eV, while in the presence of H it was 0.38 eV. However, the barrier for atom exchange on Ni (110) was lowered from 0.42 eV to 0.38 eV, and now was comparable to the barrier for regular hopping on this plane, raising the possibility that both mechanisms coexist on the surface. On the Ni(111) surface, the movement of atoms between fcc sites required 0.05 eV, while in the presence of hydrogen the barrier was only 0.02 eV. What is interesting is that Ni and H create a dimer when the Ni atom sits in an fcc site while H is in an adjacent hcp site, and the movement of the Ni atom towards (or away from) the H atom is very expensive − 0.07 eV.

How hydrogen affects self-diffusion on a Pt(110)-(1 × 2) surface has been probed with an STM by Horch et al. [13]. Addition of hydrogen at a pressure of 3×10^{-8} Torr caused bright spots to appear on the surface at 302 K and platinum adatoms moved more rapidly. At $\sim 5 \times 10^{-7}$ Torr hydrogen the mean-square displacement was roughly 500 times larger than without hydrogen, and the diffusion barrier was lowered by 0.16 ± 0.02 eV. In separate density-functional calculations it was found that the most stable arrangement was for the hydrogen to be on top of a platinum adatom (the sky-hook mechanism). The situation for an adatom on Pt(110)-(1 × 2) is somewhat complicated, however. The adatom can move along the bottom of the channel, or else it can jump to the (111) walls and continue there, as suggested in Fig. 6.7. On the clean platinum surface the diffusion barrier was estimated at 0.94 eV (somewhat above the experimentally measured value) and movement on the {111} walls was preferred. In the presence of hydrogen the path up the sidewalls becomes more demanding energetically, but along the bottom the barrier is reduced by 0.09 eV, as indicated in Fig. 6.8; this lowering is close to the experimentally found reduction of 0.16 eV. The presence of hydrogen changes the preferred diffusion path from a metastable walk on the sidewalls to a straight movement on the bottom of the channel.

6.1 Near impurities

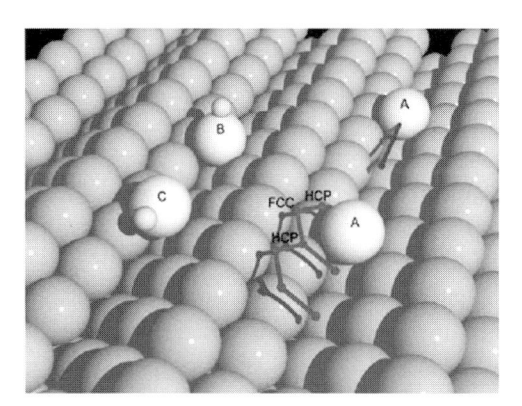

Fig. 6.7 Possible diffusion paths of Pt atoms on Pt(110)-(1 × 2) surface. *A* atom moves along channel. Alternative paths on the channel walls are possible at fcc and hcp sites. *B* indicates equilibrium position of Pt-H, with transition state at *C* (after Horch *et al.* [13]).

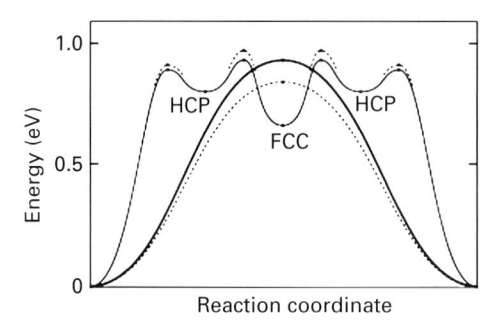

Fig. 6.8 Calculated energy changes for different diffusion paths of Pt on Pt(110)-(1 × 2). Solid curves indicate motion on clean surface, dashed curves are for Pt-H complex. Thick curves give change along direct path (after Horch *et al.* [13]).

Recently, in 2007, Prévot *et al.* [14] used SPA-LEED to look at the movement of Au atoms on a Cu(100) surface on which nitrogen atoms had been embedded. They concluded that the embedded nitrogen worked as a trap for moving Au atoms. They fitted the transition temperature between different nucleation regimes obtained from mean-field nucleation theory with kinetic Monte Carlo simulations. Using MC simulations with an embedded array of traps at the surface to model their finding they came out with an energy for migrating Au atoms of 0.4 eV and an energy of traps between 0.2–0.3 eV; these energies did not depend on the size of the traps.

In many of the more recent examples, addition of hydrogen has increased the surface diffusivity of adatoms. Only self-diffusion on W(321), Pt(100) and movement of Ag on Pt(111) have shown the opposite behavior, and it will be interesting to test this further. At the moment, this is all the direct information available about diffusion around impurity atoms in the surface, but this is an area interesting enough for additional studies. There are a number of indirect indications of the effect of impurities on diffusion, some of them from island fluctuation measurements others from the creation of "wedding cake" structures, but we will not describe these in detail.

6.2 To descending step edges

Even low index surfaces will have occasional lattice steps on them, as it clear from Fig. 6.9 [15], and the question therefore arises how atoms diffuse toward and over such steps. We will first look at the approach of atoms toward descending steps, to find out if adatoms during their descending motion experience only an additional edge barrier E_a exactly at the step edge, as shown in Fig. 6.10, or if the influence of steps has a longer range. We also will explore the mechanism of movement and its dependence upon the chemical identity of the adatoms. In this book we will describe steps by the direction in which they are oriented. For example a <110> step on the fcc(100) surface means a step along the close packed direction <110>. For fcc(111) steps we will use a notation A and B, where both mean steps in the <110> direction; however, the first creates a {100} plane while the second a {111} plane. On an fcc(110) step <110> means parallel to the channels of the plane, while the <100> step is perpendicular.

Investigation of adatom movement over descending steps started to be explored by Ehrlich and Hudda [16] in 1966. Using the FIM they established that (110), (211), and (321) tungsten surfaces were bounded by barriers that reflected tungsten adatoms which in diffusion collided with them. From measurements in the vicinity of 300 K they concluded that an additional barrier E_a just 0.065 eV higher than that surmounted in diffusion in the center of the plane would suffice to account for reflection. This study, however, did not provide a truly quantitative value for the height of the barrier, and showed only that interlayer transport of material was limited by the step edge. The

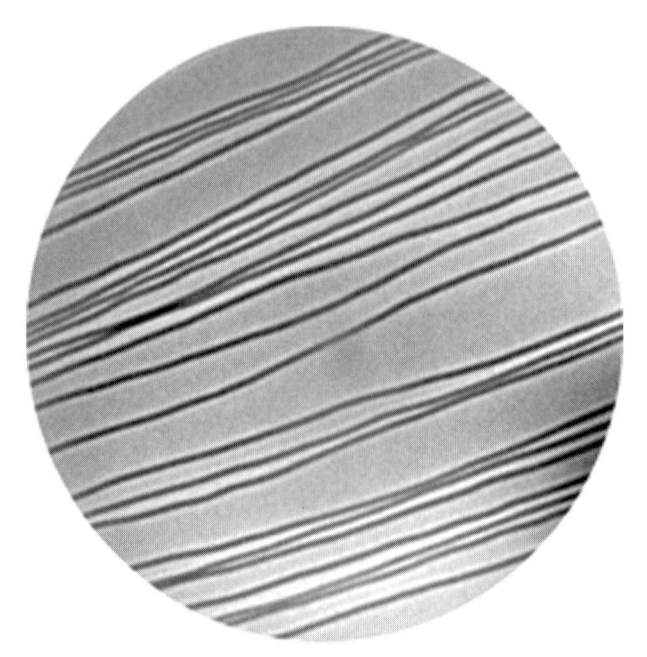

Fig. 6.9 Image of monoatomic steps on a Pt(111) surface obtained in low energy electron microscope LEEM (after Swiech [15]).

6.2 To descending step edges

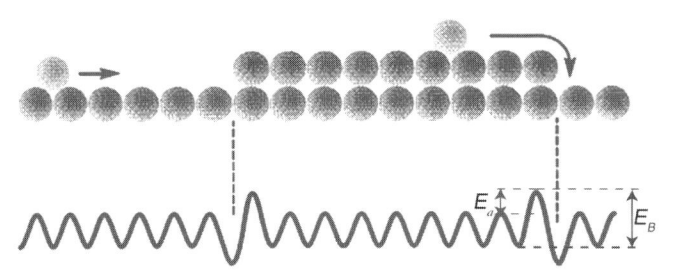

Fig. 6.10 Atom incorporation into ascending and descending steps. Top: side view of adatoms on a stepped surface. Bottom: Potential energy of adatom, showing additional step-edge barrier E_a.

presence of these reflecting barriers is important, as it impedes incorporation of atoms into a descending step, which in crystal growth had previously been assumed to occur readily. The rate of incorporation r_i will be given by

$$r_i = v_{0B} \exp(-E_B/kT), \tag{6.1}$$

where v_{0B} is the kinetic prefactor and E_B the height of the step-edge barrier. The origin of this additional step-edge barrier E_a, where $E_a = E_B - E_D$, is easy to understand. When an adsorbed atom attempts to cross a descending step, the number of interactions with lattice atoms is decreased. The same year as the work of Hudda and Ehrlich, Schwoebel [17,18] invoked step-edge barriers in his theoretical investigations on transport on stepped surfaces. In the literature the additional barrier E_a for atom descent has quite frequently been named the Ehrlich–Schwoebel barrier.

Some years later, Tsong [19] examined diffusion of rhenium on W(321) at one temperature, 332 K, and gave a diffusion barrier of 0.88 eV, in agreement with earlier, more extensive studies by Bassett and Parseley [20]; for escape from the plane he estimated a total activation energy E_B of 0.932 eV, just 0.05 eV higher than the activation energy for normal diffusion E_D.

Attempts to define the step-edge barrier E_B more quantitatively were made by Bassett [21,22] and his group in 1975. They examined the rate of escape of Ta, W, Re, Ir, and Pt atoms from a W(110) surface at different temperatures. To their surprise, the activation energy they found for escape from the temperature dependence agreed with that for diffusion within their limit of error; in their analysis, they could not detect any additional barrier to leaping over the step edge, yet they did not lose atoms from the surface. The measured values for the step-edge barrier E_B were 0.72 ± 0.10 eV for Ta, 0.89 ± 0.10 eV for W, 0.89 ± 0.10 eV for Re atoms, 0.77 ± 0.10 eV for Ir and less than 0.60 ± 0.05 eV for Pt atoms, while diffusion examined in the same study required 0.77 eV for Ta, 0.86 eV for W, 0.99 eV for Re, 0.77 eV for Ir, and 0.62 eV for Pt. This yielded the additional step-edge barrier E_a of -0.05 eV for Ta, 0.03 for W, -0.1 eV for Re, 0.0 for Ir, and -0.02 eV for Pt, all negligibly small quantities.[1]

---

[1] Bassett [21,22] also calculated step-edge barriers assuming a standard frequency prefactor and taking the mean time spent by an atom in the site adjacent to the step; this yielded for the step-edge barrier E_B values 1.37 eV for W, 1.26 eV for Re, 1.14 eV for Pt.

Diffusion in special environments

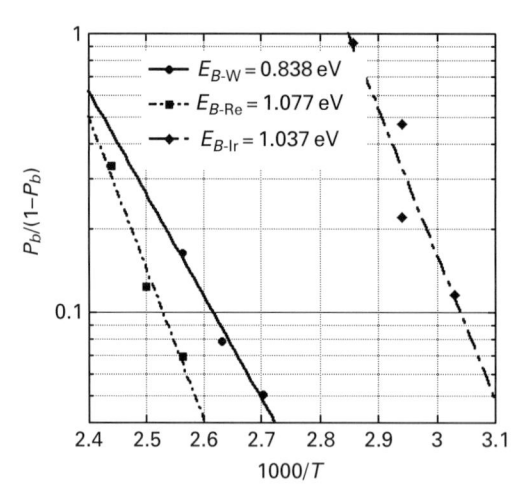

Fig. 6.11 Arrhenius plots for the escape probability of W, Re, and Ir atoms from W(110) plane, as measured by Wang and Tsong [23].

Wang and Tsong [23] took another look at the step-edge barrier on the W(110) plane, and examined the rate of loss for W, Re, and Ir adatoms. The (110) plane of tungsten is not ideal for such studies, as the atoms incorporating into a step are not resolved in the field ion microscope. For tungsten atoms on W(110) they estimated an activation energy in two ways. One, by assuming a prefactor $D_o = 3.6 \times 10^{-3}$ cm^2/sec for the diffusivity, giving an increase to the barrier of $E_a = 0.228 \pm 0.009$ eV, which ends up as an overall barrier E_B of 1.116 eV. The other, by assuming a frequency factor of $v_{OB} = kT/h$; in this way they obtained an increase of $E_a = 0.199 \pm 0.009$ eV, which yields an overall activation energy E_B of 1.087 eV. For Re and Ir atoms on W(110) they estimated an activation energy only using a frequency factor $v_{OB} = kT/h$. For the temperature range 390–410 K they found for Re an increase $E_a = 0.145 \pm 0.006$ eV, which gives a total E_B of 1.135 eV. For Ir atoms in the temperature range 330–350 K, they arrived at an additional barrier $E_a = 0.195 \pm 0.012$ eV, or a total escape barrier E_B of 0.935 eV. These values were obtained from the formula

$$P_b/(1 - P_b) = (n_{out}/n_{in}) g\tau v_{0B} \exp[-(E_D + E_a)/kT]; \qquad (6.2)$$

here n_{out} and n_{in} give the number of paths for an adatom going over the boundary or else inward, and g is a geometrical factor. The authors claimed to obtain erratic results for the activation energy and the prefactor from an Arrhenius plot. However, such a plot for W as well as Re and Ir atoms is shown in Fig. 6.11, and seems quite reasonable. The total activation energy for tungsten is just a little lower than for self-diffusion, for Re we obtain a slightly higher value ($E_a = 0.087$ eV) than arrived at in this study, and for Ir a barrier still larger ($E_a = 0.297$ eV). This is similar to what has been reported by Bassett [21,22], and leaves the subject in uncertainty.

Fink and Ehrlich [24] in 1984 carried out an interesting exploration of the free energy of the sites on a (211) plane of tungsten. On this surface, shown in Fig. 3.1, self-diffusion occurs entirely along the <111> channels, and by determining the number of times an

6.2 To descending step edges

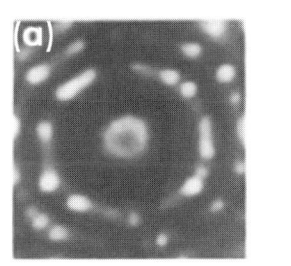

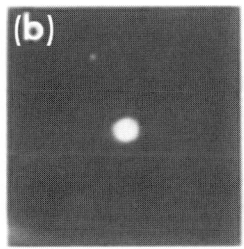

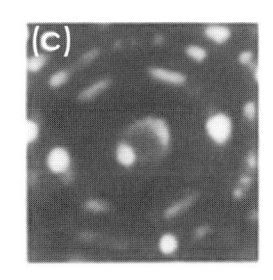

Fig. 6.12 FIM images of Ir atom incorporation into descending step of Ir$_{12}$ cluster [25]. (a) Ir cluster on Ir(111). (b) Ir atom deposited on top of cluster dominates image. (c) After heating to 120 K, iridium atom is incorporated into cluster.

atom is observed at the different sites their relative free energy can be ascertained. The results for tungsten adatoms were shown in Fig. 3.29. Atoms are fairly uniformly distributed over sites in the center, but the occupation increases dramatically close to the ends of the channel. The free energy of the sites, in Fig. 3.30, shows the expected effects – a significant diminution at the ends, with a small high energy hump on the side toward the (111) plane. For rhenium atoms the binding is somewhat different. At the last position on the plane the free energy is high, with a sharp drop for the next site closer in. Strong binding close to the plane ends is obviously important when an atom attempts to diffuse away from a surface over the step, but in these experiments no attempts to study atom escape over the step-edge barrier were made. However, based on this study, it can be expected that a step on the W(211) plane will influence the descending motion of an adatom not only at the step edge, but also 2–3 spaces away from the edge.

A closer examination of atom behavior on top of clusters was done by Wang and Ehrlich [25] in 1991. They found that on a small cluster, made up of 12 iridium atoms and shown in Fig. 6.12, an iridium adatom diffuses at a temperature ~ 20 K higher than on a large plane. However, when the atom reached the cluster edge, it incorporated into the edge at the temperature of migration. This observation suggested no extra barrier for an atom descending from a cluster of size 12. Wang, however, continued his investigations with bigger clusters and found that on larger clusters of 18 or more atoms, self-diffusion occurred as on a large (111) surface, at ~ 100 K. When an atom comes to the edge of the cluster, however, it was trapped there. On warming to $T \geq 150$ K the atom escaped over the edge in preference to returning to the center. A crude diagram of the potential energy close to the step deduced from these data is given in Fig. 6.13.

It has already been noted that iridium clusters on Ir(111) are bounded by facets of two different structures. As is apparent from Fig. 6.13a, edges of type A are made up of {100} facets, type B edges have a {111} structure, and their properties are different. Incorporation of an iridium adatom occurs at B-type edges by heating to ~ 160 K, as in Fig 6.14c; to incorporate an atom at an A-type edge requires ~175 K, as in Fig. 6.14b. The mechanism of escape is also different at the two edges. That has been tested with the incorporation of a tungsten atom. Incorporation may occur as the result of an atom jumping over the edge and attaching itself to the lattice step, as illustrated in Fig. 6.15a. An alternative, however, is for the adatom to exchange with a lattice atom, as in Fig. 6.15b, leaving the adatom buried in the cluster and pushing out a lattice atom.

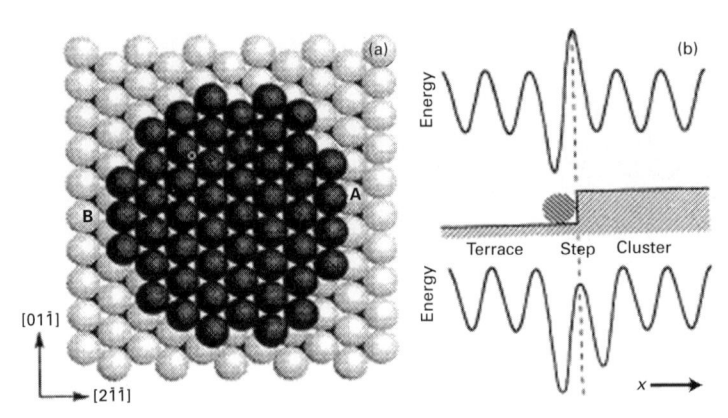

Fig. 6.13 Ir cluster on Ir(111) plane [25]. (a) Schematic of Ir$_{50}$, showing A- and B-type edges. (b) Schematic of potential energy at step edge. Top: traditional view. Bottom: Results consistent with experiments.

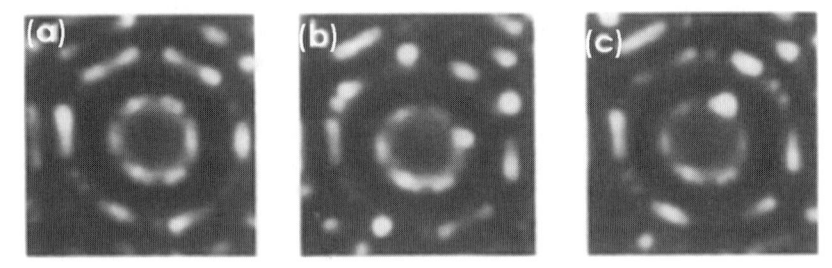

Fig. 6.14 FIM images of incorporation of Ir atom into Ir$_{50}$ cluster [25]. (a) Initially bare cluster. (b) FIM image after heating cluster with adatom on it to ~175 K, showing atom incorporation at A-type step. (c) Atom incorporation at B-type step after warming at ~160 K.

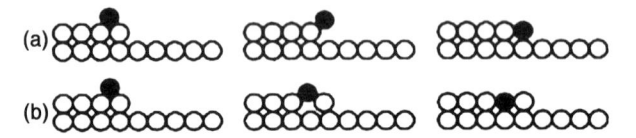

Fig. 6.15 Schematic of possible processes for atom incorporation into lattice steps [25]. (a) By atom jump into descending step. (b) By exchange of adatom with cluster atom.

Incorporation of a foreign atom such as tungsten makes it possible to distinguish between the two processes. If a simple jump gives incorporation, a tungsten atom can be detected attached to the cluster. On the other hand, atom exchange will leave an iridium atom attached, and the two alternatives can be easily discerned using FIM since the field to remove an iridium atom attached to a cluster is ~5% higher than for a tungsten atom.

Incorporation has been tested, as shown in Fig. 6.16, by depositing a tungsten atom on a cluster of 33 iridium atoms formed on the Ir(111) surface. On warming to 220 K, the atom attaches to a site of type A, and is removed by a low voltage of 9.2 kV. Incorporation into a B-type step, shown in Fig. 6.17, is examined in the same way. To remove the atom

6.2 To descending step edges

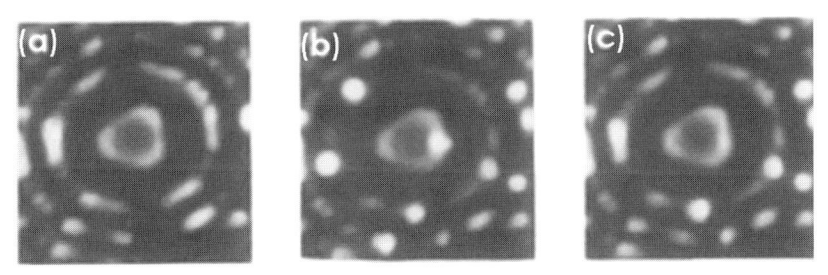

Fig. 6.16 FIM images of incorporation of tungsten atom into A-type step of Ir$_{33}$ [25]. (a) Ir cluster, on top of which Ir atom will be deposited. (b) After heating to 220 K, atom incorporates into A-type step. (c) Protruding atom has been field evaporated at 9.2 kV, typical of tungsten.

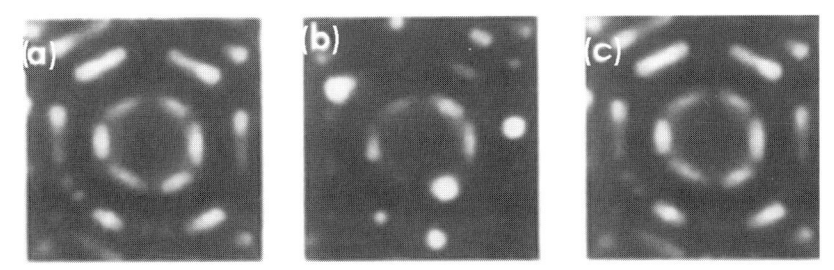

Fig. 6.17 FIM images of tungsten atom incorporation into step of Ir$_{69}$ cluster [25]. (a) Field evaporation has created cluster on Ir(111), and W atom was then deposited on it. (b) After heating to 220 K, atom has incorporated into B-type step. (c) Field evaporation removes atom at 9.8 kV, typical of Ir atom.

attached to the step now requires a higher voltage of 9.8 kV, indicating an iridium atom, *not* tungsten, has been removed.

Such observations are a direct proof that at a B-type step incorporation occurs by atom exchange, while at an A-type step a simple hop is involved. The mechanism seems to be nicely correlated with the atomic structure of edges since step A has an atom lying directly in front of every edge atom, which might block exchange, while step B provides a channel, which probably constitutes an easy path for movement of the edge atom. A similar situation was observed for Re on the Ir(111) surface [26]. Descent at a B-type step proceeded here also by exchange; when incorporation took place at a step of type A it occurred by simple hopping. The work of Wang and Ehrlich showed that the descending movement of an adatom is affected not only by the additional energy barrier at the step edge, but also by the mechanism of movement.

Up to this point, experimental information about step-edge barriers had been derived from observations on fairly high melting materials. Interest in crystal growth had constantly grown, however, and there clearly was a need for data on step-edge barriers of metals like copper or silver, not readily susceptible to FIM analysis. Models therefore began to be developed to derive this information directly from growth experiments, as second layer nucleation on islands is affected by the size of the step-edge barrier. Theoretical and experimental investigations started blooming and for clarity of the presentation we therefore divided further findings according to the material, in the order they appear in the periodic table.

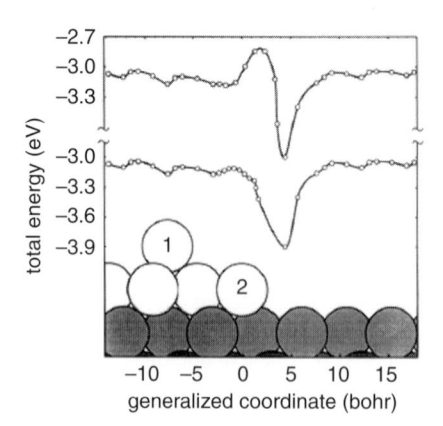

Fig. 6.18 Energy along diffusion path of Al atom at a {111} step on Al(111). Upper curve for hopping, lower one for atom exchange (after Stumpf and Scheffler [28]).

6.2.1 Aluminum

The energetics of atom incorporation from the top of an aluminum cluster were investigated by Stumpf and Scheffler [27–29] in 1994, using density-functional theory and the local-density approximation. Their estimates for the total energy of an aluminum adatom on the Al(111) plane with B-type {111} edges are shown in Fig. 6.18. For hopping over the edge the barrier E_B was found as 0.33 eV, but exchange required only 0.06 eV and was clearly favored for incorporation. For step A the energy E_B required for hopping amounted to 0.45 eV, compared with an energy of 0.08 eV for exchange. Both steps involve an additional barrier E_a for descent, since diffusion on the terrace requires only 0.04 eV. It is interesting to note that the adatom energy at the edge is 0.15 eV lower than in the plane center, which is similar to effects found earlier in experiments on Ir(111). The authors ruled out electrostatic attraction of adatoms and step dipoles as well as elastic interactions of adsorbate-induced and step-induced relaxation as the explanation. They speculated that it can be caused by interaction of adatoms and step-induced surface states.

Diffusion of dimers and trimers down an Al(111) surface was examined by Bockstedte et al. [30] using the EAM potentials of Ercolessi and Adams [31]. Both dimers as well as trimers dissociate during descent, and for both the descent proceeds by exchange. The descent of a dimer down {111} steps (B steps) can be divided into two processes: first the whole dimer moves down by exchange of one atom, the dimer bond remaining unchanged. Then the dimer dissociates as one atom is incorporated into the step, while the second atom stays on the island but may descend later by exchange, not necessarily at the same place. The energies associated with these two processes are 0.1 eV and 0.08 eV respectively, as illustrated in Fig. 6.19. If the remaining atom diffuses down the step following the first one such a transition requires an energy of 0.11 eV.

Diffusion of trimers down a B step proceeds in analogous fashion. After the trimer approaches the step, it starts moving down by exchange. Dissociation occurs and the remaining dimer now moves over the surface. The process is illustrated in Fig. 6.20a. The energy barrier associated with the movement of the trimer by exchange involves

6.2 To descending step edges

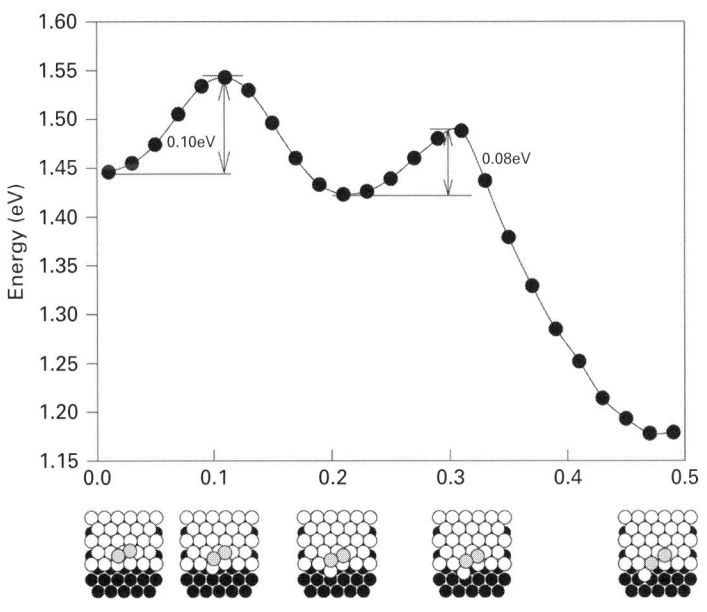

Fig. 6.19 Potential energy calculated with Ercolessi and Adams potential [31] in diffusion of Al dimers down {111} faceted step of Al(111) surface (after Bockstedte et al. [30]).

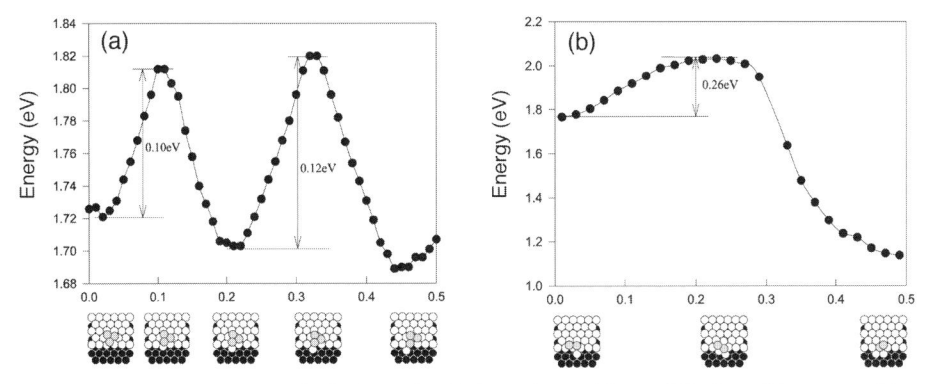

Fig. 6.20 Potential energy in diffusion of Al$_3$ down {111} faceted step. (a) For diffusion of first atom. (b) Diffusion of second atom (after Bockstedte et al. [30]).

overcoming two barriers, 0.10 eV and then 0.12 eV for dissociation. The descent of the remaining dimer next to a newly created kink costs 0.26 eV and is shown in Fig. 6.20b. Migration of the last trimer atom away from the kink occurs over a barrier of 0.16 eV. Movement down {100} A-type steps is much more energetically demanding. For a dimer the descent over this step costs 0.39 eV, for a trimer 0.38 eV is needed for the first part and 0.13 eV to continue the process. Descent of the last atom in the dimer requires 0.15 eV, and in the trimer 0.16 eV. Activation energies at B-type steps as well as at A-type {100} facets are both listed for dimers and trimers in Table 6.1. It is clear that the initial step of going over the edge occurs much more easily at B-type edges for both dimers and trimers.

Table 6.1 Energetics of an aluminium cluster diffusing down a (111) island [30].

Facet	Diffusion barrier E_B (eV)	
	A-type	B-type
Dimer	0.39	0.10/0.08
Trimer (process 1)	0.38	0.10/0.12
Trimer (process 2)	0.13	0.26
Last atom in the dimer	0.15	0.11
Last atom in the trimer	0.16	0.16

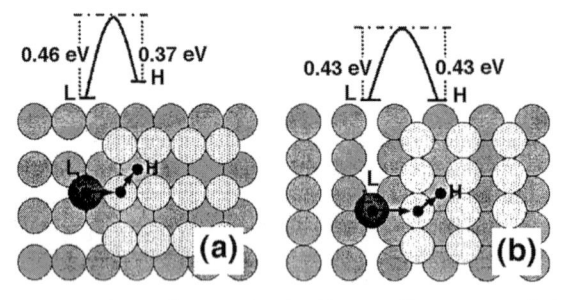

Fig. 6.21 Atom motion and energetics for Pb ascending monatomic height step on Al(110). (a) In-channel jump. (b) Cross-channel jump (after Zhu et al. [33]).

In 2002 Liu et al.[32] used molecular statics to investigate descending motion over multilayer steps, a subject discussed in Section 6.5. For comparison they also derived energetics for descending a monolayer step. Descending an Al(111) surface step along the <110> direction an energy E_B of 0.06 eV for step type B and 0.30 for step type A were required. While descending from Al(111) over a <110>/{110} step[2] required the same energy as descending down step B. Descending from Al(100) over a <100>/{100} step required an energy E_B of 0.35 eV, while over a <100>/{110} step required an energy E_B of 0.34 eV; stepping down a <110>/{111} step is more demanding – $E_B = 0.45$ eV. On Al(110) descent over a <111>/{110} step needed $E_B = 0.33$ eV, while descents over a <100>/{100} step perpendicular to the channels of the plane and a <110>/{111} step parallel to the channels of the plane were more demanding, with E_B of 0.63 and 0.7 eV respectively. All these transitions were accomplished by atom exchange. Liu et al. did not derive energetics for terrace movement in this study.

Aspects of diffusion on Al(110) have also been examined in density-functional calculations by Zhu et al.[33], who looked at atom ascent and descent at steps. Two different geometries, shown in Fig. 6.21, were considered, in which the adatom exchanges with a lattice atom. For ascent in an aluminum channel, an aluminum atom was found to require 0.60 eV; the downward step involved a barrier of only 0.43 eV. An

---

[2] Step <110>/{110} means the step position is along the <110> direction with the structure of the {110} plane.

upward move in the cross-channel direction entailed a barrier of 0.67 eV, while 0.71 eV for going down.

No experimental data are available for descent down aluminum steps, and the theoretical information available is unfortunately also limited. However, from existing data the most likely mechanism of descent will be exchange, not hopping, for all types of edges.

6.2.2 Iron

Information about the step-edge barrier on the iron(100) surface is provided by the STM experimental observations of island fluctuations by Stroscio et al. [34]. The first analytical analysis based on these data was done by Smilauer and Harris [35] who were able to deduce an additional step-edge barrier E_a in the range 0.0 eV to 0.10 eV. Estimates were continued by Amar and Family [36], who based their work on the experimentally measured diffusion barrier E_D of 0.454 eV from Stroscio et al. [34]. They then looked at the RHEED intensity and compared it with Monte Carlo simulations using different barriers for incorporation into an edge site. With the STM findings of Stroscio et al. at room temperature they found an additional step-edge barrier E_a between 0.03 eV and 0.06 eV, as indicated in Fig. 6.22. The mechanism of atom descent was not determined. In 1999, Bartelt and Evans [37] simulated the epitaxial growth of Fe on Fe(100) and compared these with measurements carried out with the STM, again by Stroscio et al. [34]. In this way they came up with a step-edge barrier E_a of 0.03 eV to 0.04 eV, values in agreement with the earlier findings of Amar and Family [36].

The epitaxial growth of iron on Fe(110) was examined by Köhler et al. [38] with a scanning tunneling microscope, and the atomistic results were then compared with Monte Carlo simulations to give the following additional step barriers E_a for transitions in the $[1\bar{1}1]$ direction: <111> edge, 0.41 eV; <110> edge, 0.34 eV and <001> edge,

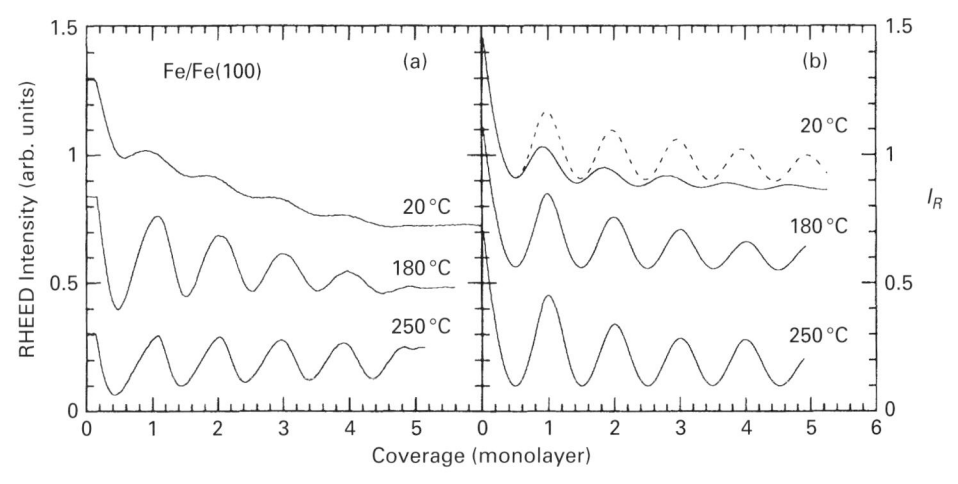

Fig. 6.22 Diffusion of Fe atoms on Fe(100) surface. (a) RHEED intensity data at different temperatures. (b) Simulation with step barrier of 0.03 eV (solid line) and without step barrier (dashed line) (after Amar and Family [36]).

0.36 eV. The anisotropy of the step-edge barriers agrees very nicely with the elongated shape of the islands observed with STM.

There are no theoretical insights into the step-edge barriers of iron surfaces, and experimental findings are also limited to one measurement on Fe(100) and one on Fe(110).

6.2.3 Nickel

Adatom descent over nickel as well as copper steps have received quite wide theoretical attention, unfortunately with a complete lack of experimental findings for comparison. Calculations started in 1992 with the work of Liu and Adams [39], who used the embedded atom method to calculate activation energies for a Ni atom descending at Ni steps. They did not observe any exchange; however they did not clearly identify the kind of step for the Ni(111) surface. The highest activation energy for descent was observed on Ni(110), $E_B = 1.34$ eV, where they looked at the steps along the channels of the plane. This value should be compared with the movement along the channels on the terrace, which required only $E_D = 0.44$ eV. The second highest energy for descent was for Ni(100) over a <110> step, where Liu and Adams found a barrier E_B of 0.91 eV, while diffusion on the terrace cost only 0.63 eV, and the lowest, on Ni(111), was $E_B = 0.55$ eV; diffusion on the (111) terrace required only 0.056 eV. The fact that a jump instead of exchange occurred suggests a step of type A rather than type B. We can also see that the order of the energies needed for descent is not the same as for movement on the terraces.

Liu and Adams [40] continued their work using EAM and molecular statics. They found a forbidden region next to both ascending and descending steps, with an interior barrier 2 to 3 spaces away from the step on Ni(111). The diffusion barrier had a constant value of 0.056 eV, as long as the atom is a distance of > 5 spacings away from the step. Approaching the descending step the barrier rose to 0.064 eV, and then the adatom experienced lower barriers than for movement on the terrace. When the adatom reached the step the barrier rose again. It was also shown that on a B-type step descending by exchange is more likely than by a jump over the step, the barrier E_B amounting to 0.169 eV compared with 0.505 eV for jumps. For a type A step, the energetics for exchange and jumps were comparable, E_B being in the range of 0.55 eV compared with 0.585 eV. Descending from Ni(110) at a <001> step and from Ni(100) by a <110> step proceeded by exchange with an energy E_B of 1.28 and 0.85 eV. For jumping over steps they found the same values as in their previous study. They did not observe any forbidden regions on the latter planes. The movement on the terrace occurs over a barrier of 0.41 eV for in-channel diffusion and 0.49 for cross-channel diffusion. On Ni(100), the descending barrier by exchange was 0.22 eV higher than for movement on the terrace.

The next investigations of Ni steps were done in 1994 by Stoltze [41], who looked briefly on atom movement over step edges using effective medium theory. For Ni(100) over <110> steps he found a value E_B of 0.787 eV for hopping compared with 0.890 eV for exchange. In his investigation, movement on the terrace required 0.558 eV. For Ni(110) he looked at movement over <110> steps (across channels), for which jumping cost $E_B = 1.178$ eV while exchange required $E_B = 0.917$ eV. An adatom descending along channels of Ni(110) cost $E_B = 0.68$ eV, while exchange was more demanding, needing

6.2 To descending step edges

Table 6.2 Comparison of energy barriers (eV) for diffusion of fcc metals from fcc(111) calculated with CEM [42].

	Ag	Cu	Au	Pd	Ni	Pt
E_D (Jump)	0.020	0.039	0.029	0.034	0.036	0.048
E_B (Ex)	0.064	0.176	0.090	0.030	0.094	0.044
E_K (Ex)	0.030	0.122	0.047	0.005	0.067	0.012

$E_B = 0.81$ eV. Terrace movement on Ni(110) involved a barrier of 0.407 eV for the in-channel direction and 1.157 eV across channels. On Ni(111) jumps over the edge cost $E_B = 0.529$ eV, exchange $E_B = 0.669$ eV. A- and B-type steps were not recognized, and movement on the (111) terrace cost only 0.068 eV.

Li and DePristo [42,43] undertook an effort to better understand homoepitaxy on fcc(111) surfaces by calculating barriers to surface diffusion compared to atom incorporation into lattice steps by exchange. For this they employed the corrected effective medium theory, CEM, and obtained the barriers to surface diffusion over B-type step edges listed in Table 6.2. For Ni(111) the barrier for descending was $E_B = 0.094$ eV, while descending at a kink was easier, requiring $E_K = 0.067$ eV. To overcome the step-edge barrier, an adatom needs a higher energy than for diffusion on the flat surface, for which the possibility of exchange was not taken into account. It is clear after examining the data in Chapter 5 that some of these estimates are way off. In fact, the authors pointed out that theory was not designed for quantitative data and that only relative information was valid. Also examined were the interlayer diffusion processes. For small clusters of seven atoms or less the barrier to incorporation was very high, but decreased by an order of magnitude once the cluster size exceeded ten atoms and then was almost size independent. This is not what has been found in the few experiments done, however. Although the results of Li and DePristo's work should not be taken too seriously, they do show that for palladium, nickel, and platinum the incorporation barrier is smaller for nineteen atom clusters than at kinks in the steps.

In 1997 Merikoski et al. [44] looked at step characteristics on fcc(100) planes. For the Ni(100) surface descending by exchange over a <110> step seemed to be more probable, with a barrier E_B of 0.744 eV compared to 0.920 eV for jumps. Movement on the terrace proceeded by jumping which required only 0.631 eV compared to 0.844 eV for exchange. More interesting are the findings of Merikoski et al. for kinks. For Ni the value of E_K for exchange at kink sites was a bit lower than for hopping, 0.522 eV for exchange and 0.695 eV for jumps. The barrier for exchange at the kink was lower than for hopping on the flat plane. For all the three materials they investigated, the activation energy for descending into kinks is lower than the energy of descending at a straight step.

The influence of hydrogen on atom descent on nickel surfaces was investigated by Haug and Jenkins [12] using EAM interactions. For descending a straight <110> step by jumps on Ni(100) without any hydrogen, the energy barrier E_B was 0.95 eV, leaving an additional step-edge barrier E_a in the range of 0.34 eV; descending at a kink was a bit easier, requiring $E_K=0.7$ eV or an additional barrier of 0.09 eV. Hydrogen promoted the descending motion at both straight and kinked steps. On the former the barrier for descending with H present was 0.28 eV lower than without, while at the kink the barrier was 0.09 eV lower. However, the movement on the flat surface in the presence of H was also easier, 0.47 eV compared with 0.61 eV in the absence of H. For Ni(110) surfaces, descent by adatom exchange along the short axis, along the channels, cost $E_B=0.69$ eV on the clean surface, and almost the same, 0.68 eV, in the presence of H. Stepping down by jumps on the clean surface cost 1.33 eV along the long axis, across the channels, and 0.75 eV along the short. The barrier for jumping was lowered by the presence of hydrogen to 1.05 eV across the channels and 0.73 eV along them, but jumps were still more demanding than exchange. On this plane, hydrogen does not seem to play a big role in interlayer transport. Terrace diffusion also does not feel the influence of hydrogen: diffusion occurs over a barrier of 0.39 eV on the clean surface compared with 0.38 eV on the surface with hydrogen. Hydrogen, however, does seem to influence movement on Ni(111). At an A-type step, descent by exchange without H cost $E_B=0.23$ eV, while with H only 0.18 eV. These values can be compared with movement on the terrace, which occurs over a barrier of 0.05 eV without H and 0.02 eV with H present. For type B steps the barriers were in the same range as for A-type steps.

For Ni(111) steps the mechanism of descent was not clearly deduced. The study of Liu and Adams [40] slightly prefers exchange over atom jumps for both types of edges, but their barrier is much higher than obtained later for jumping by Haug and Jenkins [12], leaving the situation confused. On Ni(100), two studies favor exchange over jumps while one indicates jumping over exchange, the rest do not consider both mechanisms for comparison. However, kink sites seem to provide an easier descending path. There is also disagreement between the values derived for the Ni(110) surface.

6.2.4 Copper(111): experiment

The first attempt to derive the barrier for interlayer movement on Cu(111) based on experimental findings was done by Markov [45] who arrived at the critical temperature for the transition from crystal growth by step flow to two-dimensional nucleation. Using experimental data from diffraction studies, he was able to give the additional step-edge barriers. He relied on experimental findings of van der Vegt et al. [46] to arrive at a value of $E_a=0.113$ eV. From experiments of Wulfhekel et al. [47] a slightly different value of $E_a=0.080$ eV to 0.055 eV was derived.

Icking-Konert et al. [48] looked at the decay of Cu adatom islands on Cu(111) in STM experiments at varying temperatures. They arrived at a value of $E_a=0.116\pm0.002$ eV for the additional step-edge barrier by comparing experimental findings and simulations. A year later, the same group [49,50] analyzed the decay of a 40 multilayer Cu island on Cu(111) and found a step-edge barrier E_a of 0.224 ± 0.009 eV, as shown in Fig. 6.23. The pre-exponential factor for hopping over the step was in the range $3.5\times10^{12\pm1}$ sec^{-1}. This is the weighted average over all possible crossing processes at A- and B-type steps. They also found that the value of the step-edge barrier was independent of terrace width, as long as this was larger than 14 ± 2Å. The authors reanalyzed results from a previous paper

6.2 To descending step edges

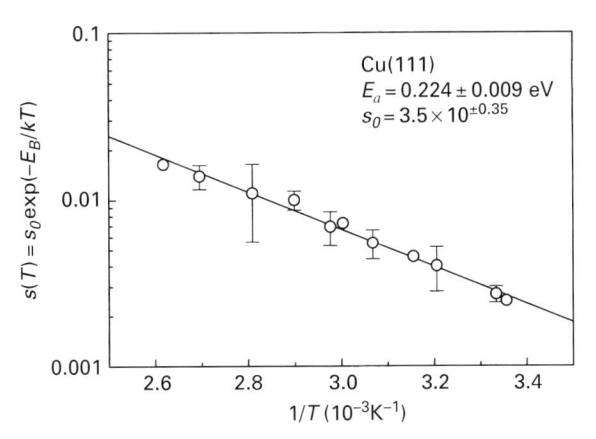

Fig. 6.23 Arrhenius plot for decay of multilayer Cu(111) island obtained by fitting STM data to rate equations, yielding a step-edge barrier of 0.224 ± 0.009 eV (after Giesen and Ibach [50]).

[48] with a "sticking coefficient" prefactor $s_0 = 3.5 \times 10^{\pm 0.35}$ and came out with a value of $E_a = 0.19 \pm 0.02$ for the additional step-edge barrier [50].

Diffusion of copper atoms on Cu(111) and Cu(100) was looked at by Ferrón et al. [51], who relied on a second-moment approximation to tight-binding theory [52]. From He scattering data they concluded that Cu(111) had poor interlayer transport; for Cu(100), however, layer-by-layer growth was observed, suggesting that on this surface good interlayer mass transport prevailed, as indicated in Fig. 6.24. From Monte Carlo simulations Ferrón et al. established that diffusion over the flat Cu(111) plane occurred by hopping. At both A- and B-type edges, however, the atoms go to the lower level by exchange with an atom from the lattice, as shown in Fig. 6.25. Exchange descent was also suggested for the Cu(100) plane. The additional step barrier E_a was calculated as 0.03 eV on Cu(100), and 0.057 eV for Cu(111). Differences between descent at the two kinds of steps were not investigated. They also looked at the presence of 1ML of Pb on Cu(111) and found that this surfactant changed the mechanism of atom movement on the surface from hopping to exchange, and suppressed the step-edge barrier, facilitating layer-by-layer growth. What is more surprising was that movement of Cu proceeded *below* the Pb layer.

6.2.5 Copper(111): theory

The investigations of interlayer movement for copper surfaces started with the study of Stoltze and Nørskov [53]. They carried out effective medium theory calculations for diffusion and incorporation into a lattice step from above for copper adatoms on copper clusters of three and seven atoms with the Cu(111) structure. Estimates were always made for B-type edges. For adatom jumping over the edge on the trimer they found an activation energy E_B of 0.284 eV, compared to 0.230 eV for exchange, and on the heptamer these figures were 0.418 eV and 0.267 eV for exchange. Diffusion on the cluster required overcoming a barrier of 0.093 eV, revealing a considerable energy requirement

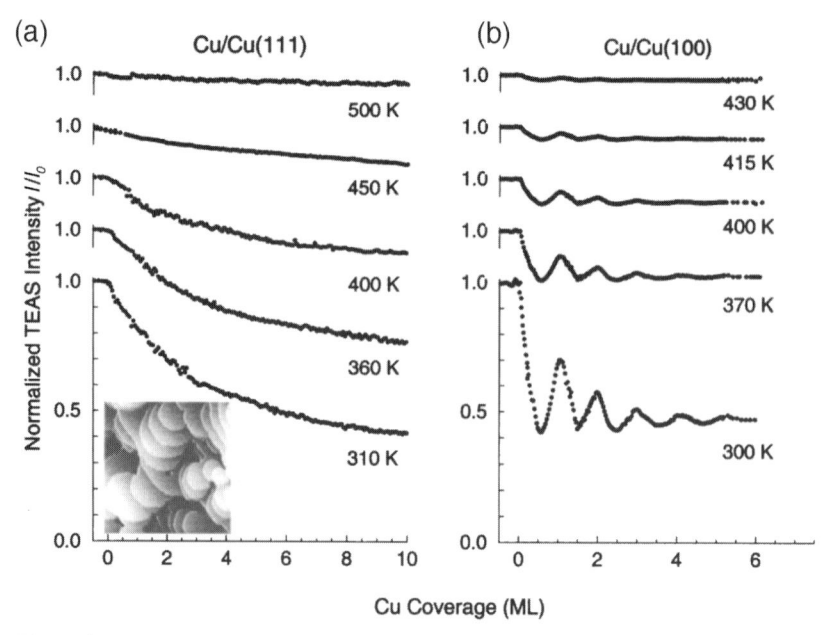

Fig. 6.24 Thermal energy scattering showing homoepitaxial Cu growth. (a) Interlayer diffusion is suppressed on Cu(111), giving multilayer growth. This is shown by monotonic decrease of specular intensity and by STM image. (b) Film grows easily on Cu(100) surface (after Ferrón et al. [51]).

for incorporation. This work was presented at a meeting in 1991, and stimulated the experiments by Wang and Ehrlich [25]. Based on their own findings, Stoltze and Nørskov also postulated that the barrier for hopping over the edge was very size dependent, while exchange did not seem to be so dependent, findings opposite to what was seen in theoretical investigations on flat surfaces. Stoltze [41] continued working with EMT and for Cu(111) steps, an adatom jump cost only $E_B = 0.365$ eV and is more probable than exchange, which required 0.45 eV. Again going over the step was much more demanding than movement on the flat plane, which occurred over 0.053 eV.

Karimi et al. [54] utilized EAM potentials to derive the activation energy of diffusion E_D for an adatom as 0.028 eV; movement from an upper to a lower level occurred over a barrier E_B of 0.085 eV by exchange with a lattice atom, compared to 0.49 eV for a jump. Only steps of type B were investigated. For all steps investigated, atom descent by exchange was energetically favorable. They also examined diffusion of atoms near the steps. Movement along steps required similar energies of 0.29 eV, movement perpendicular to the ledge needed 0.67 eV for Cu(111).

Breeman [55] used the Finnis–Sinclair potential [56] and found that the movement from fcc to hcp sites required 0.13 eV, and for the reverse the barrier was 0.11 eV. Interlayer step diffusion at a {111} step (B-type) cost $E_B = 0.27$ eV for exchange and 0.53 eV for hopping. Descent from {100} steps (A-type) by a jump was comparable to exchange, $E_B = 0.53$ eV for jump and 0.56 eV for exchange. Li and DePristo [42,43] calculated barriers to surface diffusion and compared them to atom incorporation by exchange into lattice steps using the corrected effective medium theory, CEM; the

6.2 To descending step edges

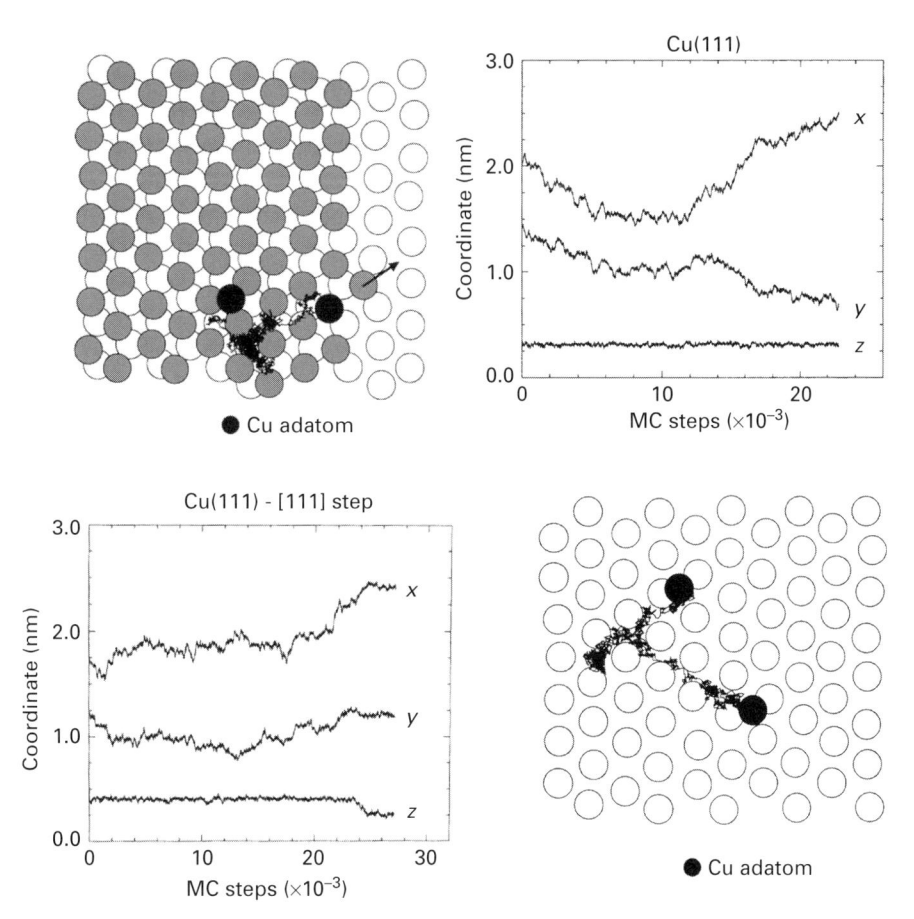

Fig. 6.25 Monte Carlo simulations of Cu diffusion on Cu(111). Left top: adatom approaches step and then exchanges with edge atom. Left bottom: time evolution of adatom coordinate. Right top: coordinate of Cu atom diffusing on Cu(111) terrace. Right bottom: adatom trajectory (after Ferrón et al. [51]).

barriers to surface diffusion over B-type step edges are listed in Table 6.2. For Cu(111) they found a barrier equal to $E_B = 0.176$ eV, at the kink $E_K = 0.122$ eV, and the barrier for interlayer movement was much higher than for diffusion on the flat surface. It should be noted that their values are way off from other investigations. They also examined interlayer processes for small clusters of seven atoms or less and found the barrier to incorporation was very high; the barrier decreased by an order of magnitude once the cluster size exceeded ten atoms and then was almost size independent. This is not what has been found in the few experiments done, however.

Trushin et al.[57] used EAM to look at the Cu(111) plane, and found that jumps over a {100} faceted step of type A required $E_B = 0.51$ eV, but only 0.28 eV for exchange; at type B steps, {111} faceted, $E_B = 0.5$ eV was needed for jumps and 0.085 eV for exchange. Movement on the flat surface required an energy E_D of only 0.029 eV. As we can see, descent on Cu steps seems to proceed by exchange rather than by jumping; additionally descent through step B is much more likely than over step A. On the Cu(111) plane, descending by exchange close to a kink in an A-type step required $E_K = 0.2$ eV, by a jump 0.4 eV. For a kink in the B-type step, exchange was also more favorable than a jump, with a barrier of $E_K = 0.31$ eV compared with 0.39 eV for the jump. However for B-type steps, kinks did not provided an easy descent, compared to straight steps.

The fast island decay observed by Giesen et al. [58] for Cu(111) mounds as well as by Morgenstern et al. [59,60] for Ag(111) triggered theoretical investigations by Larsson [61], who used a Monte Carlo bond-counting model to show that such fast decay of Cu(111) islands can be explained by adatoms hopping over the edge, not by exchange. For crossing an A-type step he derived a barrier E_B of 0.24 eV, while for a B-type step it was 0.47 eV. The activation energy for diffusion E_D on the terrace was estimated as 0.09 eV. He claimed that enhanced intralayer diffusion at the corners as well as step–step interactions can be responsible for the rapid decay of small islands.

Feibelman [62] used *ab initio* barriers for exchange at the vicinal steps of the Cu(111) plane and found them to be in the range $E_K = 0.66$–0.85 eV when kinks on the terraces were close. In 2003 Huang et al. [63] used EAM potentials in molecular dynamics/statics simulations to look at the multilayer descent from the Cu(111) surface. Monolayer descent to another (111) plane required E_B of 0.1 eV over a B-type step and 0.34 eV over a step of type A, but terrace movement on Cu(111) needed only $E_D = 0.04$ eV.

In 2004 Wang et al. [64] investigated with EAM the movement of adatoms between facets and for monolayer descent from Cu(111) over <110> steps. For B steps they showed that descent required E_B 0.08 eV by exchange and 0.5 eV by hopping. A step exchange required a barrier E_B of 0.28 eV, while hopping had the same energy as at the B step – 0.5 eV. Compared with an energy E_D of 0.04 eV for hopping on the flat plane, there is clear existence of an additional barrier E_a at both types of steps.

Recently Smirnov et al. [65] used *ab initio* and MC simulations to look at the interactions of a Cu(111) island with adatoms deposited on this island. They found a repulsive interaction of 30 meV at the first row of hollow sites from descending steps A and B. In the second row of hollow sites the interaction was also repulsive, ~10 meV. Attractive minima of -19 meV (A-type of step) and -20 meV (B-type of step) were observed at distances 1.5 Å from step A and 2.9 Å from step B. They associated this effect with redistribution of the electron charge-density at the step. Such interactions may cause empty zones (zones where adsorption of adatoms was not noticed) to be observed experimentally on the cluster.

Recently the descending motion of iron was investigated by Mo et al. [66] and for movement down a step by exchange they derived a barrier of 0.07 eV; diffusion on the terrace required 0.025 eV. The same system was also investigated by Ding et al. [67]. An oscillatory dependence was observed with the barrier at a descending edge of 0.173 eV; this was for a jump, not by exchange, as in the previous study.

Descending motion from B-type of steps on Cu(111) seems to be preferably by exchange and theoretical values for the additional barrier varied from 0.038 to 0.14 eV. The situation for descending over step A is more confusing – half the studies suggest exchange, the other half jumps. Theory however is in agreement that descent over step A is more demanding than descent over step B. Unfortunately experiments so far have derived only indirect measurements averaged over all types of steps.

6.2.6 Copper(100)

There is no experimental investigation of adatom descent from Cu(100) surfaces. Most theoretical studies tackled descending motion over close-packed <110> steps. Investigations where done by Stoltze [41] with effective medium theory. Adatom descent over a <110> step needed an energy E_B of 0.572 eV by jumps and 0.626 eV by exchange. This should be compared with 0.425 eV for movement by jumps on the terrace. Liu [68] also used EMT interactions for determining the energy necessary for jumping and exchange over <110> steps. For jumps $E_B = 0.81$ eV was required, while exchange only needed 0.65 eV, and diffusion on the terrace was over a 0.45 eV barrier. Calculational efforts to determine barriers continued, and Karimi et al. [54] made detailed calculations using EAM potentials for self-diffusion of copper atoms on copper. At <110> steps diffusion by jumps required overcoming a barrier of $E_B = 0.77$ eV, but moving over an edge by atom exchange took an energy E_B of 0.51 eV. Movement on the flat surface cost 0.48 eV. Movement along steps by hops required 0.25 eV while perpendicular to the ledge 0.84 eV.

In 1996 Breeman et al. [55] looked at diffusion on the Cu(100) plane with the Finnis–Sinclair potential [56] with and without indium present. A copper adatom jumping over the <110> step edge required $E_B = 0.51$ eV, jumping next to a kink required $E_K = 0.4$ eV. Exchange at the kink was comparable with jumping over a straight edge, costing 0.52 eV. Exchange with an indium atom embedded into a straight step cost only 0.26 eV, exchange with an In atom embedded in the kink site cost even less, 0.16 eV. Breeman did not derive the energetics for movement of Cu on the flat (100) surface, that is why it is difficult to know if there was an additional barrier at the step edge.

Trushin et al. [57] also looked at copper surfaces with EAM. On a flat Cu(100), atom jumps occurred over an activation energy E_D of 0.49 eV and of 0.69 eV for exchange. Descent over a <100> step required $E_B = 0.33$ eV for exchange and 0.57 eV for jumps, so the barrier for exchange over this step is lower than the barrier for motion on the flat terrace. For the close-packed <110> step of this plane the barriers are a bit higher, $E_B = 0.54$ eV for exchange and 0.77 eV for jumps. The descent in the presence of a kink in the <110> step was also investigated. Exchange with a corner atom of the kink required $E_K = 0.34$ eV, while a jump was more energetic and cost 0.57 eV. Kinks seem to provide the easier way for interlayer motion. However, what is surprising in these investigations is that the descending motion of an atom on Cu(100) by exchange has a lower energy barrier than movement on the flat surface. Unfortunately there are no direct measurements available to support this claim. Additionally, descending over <100> steps (if they exist on the surface) is easier than over <110> steps.

In 1997 Merikoski et al. [44] looked at step characteristics on Cu(100) planes. Movement on the flat surface proceeded by hopping over a barrier E_D of 0.399 eV. Based on their findings, jumping over the <110> edge was more favorable than exchange, with an activation energy of $E_B = 0.578$ eV compared with 0.631 eV. The favored mechanism of descent for copper does not agree with the previous investigation of Trushin et al. [57] which suggested exchange with a very low energy as the mechanism on this plane. More interesting are the findings of Merikoski et al. for kinks. For both Cu and Ag, hopping and exchange at kinks seem to have almost the same energy. For Cu, $E_K = 0.442$ eV for jumps compared with 0.422 eV for exchange. For all three materials, including Ni, investigated by Merikoski et al., the activation energy for descending into kinks is lower than the energy of descending at a straight step, and for all three of them the barrier for descending motion was higher than the barrier for movement on the flat surface.

Montalenti and Ferrando [69] used RGL potentials to calculate the step-edge barriers on Cu(100). They did not describe the step orientation but most likely they were <110> steps. Descent via jumps required $E_B = 0.71$ eV, while exchange needed only 0.52 eV. Both barriers are higher than for movement over the plane, which required $E_D = 0.39$ eV. Ferrón et al. [51] relied on a second-moment approximation to tight-binding theory [52] and He scattering data. For Cu(100) they observed layer-by-layer growth, suggesting that on this surface good interlayer mass transport prevailed, as indicated in Fig. 6.24. Exchange descent was suggested with the additional step barriers E_a as 0.03 eV.

Wang et al. [64] investigated descent from a Cu(100) surface over <100> steps with EAM and found that an energy E_B of 0.32 eV was needed for exchange, while hopping required 0.55 eV. Descent over <110> steps required an energy of 0.57 for exchange, 0.81 for hopping. Movement on the flat Cu(100) was shown to proceed by hopping with an energy E_D of 0.48 eV. Surprisingly, the value for descent by exchange was lower than for movement on the terrace. Again, descent over a <100> step was easier than over a <110> step, findings in excellent agreement with the results of Trushin et al. [57]. In 2006 Yildirim et al. [70] did calculations using EAM semi-empirical potentials. For jumping over a <110> step they arrived at a barrier E_B of 0.79 eV, which indicated an additional step-edge barrier E_a of 0.185 eV.

Most of the calculations prefer exchange over hopping as the mechanism of descent from the Cu(100) surface. Movement seems to be easier through <100> rather than <110> steps; however, the latter is more frequently observed on the surface. Studies also suggest easier descent at kinks.

6.2.7 Copper(110)

Descent motion from the channeled Cu(110) surface was studied by Stoltze [41] with EMT. Two directions of descent, $[1\bar{1}0]$ and $[001]$, from the copper (110) surface were explored. For jumping along the $[1\bar{1}0]$ direction down a <001> step, an adatom needed $E_B = 0.48$ eV, while in direction [001] 0.84 eV was required. Exchange is more favorable for the [001] direction, requiring $E_B = 0.69$ eV, and more demanding in the $[1\bar{1}0]$ direction, at 0.574 eV. Overcoming both steps is associated with an additional step-edge barrier, since jumps on the terrace cost 0.292 eV in the $[1\bar{1}0]$ direction and 0.826 eV along [100].

Interlayer diffusion on this plane continued to be studied by Karimi et al. [54] using EAM potentials. Diffusion on the flat Cu(110) involved a barrier E_D of 0.24 eV, and diffusion down a <011> step by atom exchange required an activation energy E_B of 0.72 eV; jumps are unlikely, as they cost 1.17 eV. Movement along a <011> step by jumps required 0.28 eV, perpendicular to the ledge 1.17 eV.

6.2 To descending step edges

Diffusion of copper over steps on Cu(110) was probed by Zhu et al. [33] in density-functional calculations. An in-channel upward exchange involved a barrier of 0.84 eV, and 0.54 eV downward. To move up in the cross-channel direction, previously shown in Fig. 6.21, required 0.75 eV, but 0.82 eV down. Yildirim et al. [70], using EAM, found a barrier $E_B = 0.644$ eV for jumps over Cu(110) steps perpendicular to the channel, so the additional step-edge barrier E_a was 0.414 eV. Prefactors were all on the order of 10^{-3} cm^2/sec.

Recently movement of Co adatoms on the Cu(110) was investigated with molecular statics. For descent via exchange down <110> steps a barrier E_B of 0.66 cV was derived, while descent over <100> cost 0.60 eV. Movement of Co adatoms on the terrace by exchange required 0.32 eV.

The number of theoretical investigations on Cu surfaces is huge compared with a very limited experimental picture. It will be important to have more direct measurements of adatoms overcoming steps, even though for copper, calculations have yielded good results.

6.2.8 Rhodium

Very limited information is available in the literature about rhodium steps. The only data come from an investigation of self-diffusion on small islands on Rh(111), examined by Máca et al. [71,72] using RGL interactions between atoms. The activation energy for rhodium self-diffusion on defect free (111) surfaces was found to be 0.15 eV, in very good agreement with the experimental value of 0.16 ± 0.02 eV [3]. For jumps over the straight edges the activation energy E_B was found to be similar for the two types: 0.73 eV for Rh at A-type steps and 0.74 at B-type. For exchange at straight steps the barriers E_B were considerably smaller, 0.39 eV for B-type steps and 0.47 eV for A-type steps. The presence of kinks seemed to influence descent on Rh; the activation energy E_K for exchange at B-type steps reached a low of 0.24 eV. For steps of type A, descent over a kink by exchange was comparable with descent over a straight step. The authors also looked at descent from 3×3 islands. For small Rh islands the energy for descending by exchange was much lower for type B steps, 0.24 eV, but a bit higher at 0.43 eV for type A.

Rhodium steps definitely require much more scientific attention from both experiment and theory.

6.2.9 Palladium

The first investigations of palladium steps were done by Stoltze [41] using effective medium theory. For Pd on Pd(100) overcoming a probable <110> step required an energy E_B of 0.573 eV for jumps and 0.698 eV for exchange. These values can be compared with the energy for diffusion on the terrace, in which a barrier E_D of 0.503 eV had to be overcome in jumps. On Pd(110), interlayer movement of an adatom along the $[1\bar{1}0]$ direction needed an energy E_B of 0.42 eV for jumps, and 0.658 eV for exchange. To get over the <110> step along the [001] direction on the same plane, a palladium atom needed an energy E_B of 0.779 eV for jumps and almost the same, 0.777 eV, for exchange. Movement on the terrace was estimated as 0.366 eV for the $[1\bar{1}0]$ direction and 0.776 eV for cross-channel jumps in the [001] direction. On the Pd(111) plane, jumping over steps was more favorable than exchange, $E_B = 0.306$ eV compared to 0.478 eV, but movement on the terrace had a much lower activation energy of 0.104 eV.

Li and DePristo [42,43] utilized the corrected effective medium theory, CEM, to investigate surface diffusion on Pd(111) and compared this to atom incorporation into lattice steps. The barriers obtained for diffusion over B-type step edges, listed in Table 6.2, amounted to $E_B = 0.03$ eV for a descending step, and 0.04 eV at a kink. What is interesting is that for palladium, the barrier to interlayer movement is lower than for diffusion on the flat surface. They also looked at small clusters of seven atoms or less and found the barrier to incorporation to be very high. However, the barrier decreased by an order of magnitude once the cluster size exceeded ten atoms and then started to be almost size independent. Although there is a bit of doubt about these findings, they do show that for palladium, as well as for nickel and platinum, the incorporation barrier is smaller for nineteen atom clusters than at kinks in the steps.

6.2.10 Silver(111): experiment

Descending motion of adatoms from silver (111) was investigated in 1991 with STM by Vrijmoeth et al. [73] and from observations of the growth under different conditions, they were able to arrive at an additional step-edge barrier E_a of 0.150 ± 0.030 eV. They also looked at the effect of Sb on the step-edge barrier, and at a coverage of 0.08 ML found that the additional barrier for descending motion decreased by 0.03 eV. Smilauer and Harris [35] in 1995 came up with a quantitative relation for the onset of nucleation on islands, and arrived at values of $E_a = 0.13$ eV to 0.23 eV for the additional step-edge barrier of Ag on Ag(111). At the same time, Meyer et al. [74] derived a different model to analyze growth and found a step-edge barrier E_a of $0.150^{+0.040}_{-0.020}$ eV for the same material. Using the rate of nucleation on islands measured in STM experiments, Bromann et al. [75], the same year, found an effect of strain on the step-edge barrier. Based on the temperature dependence of the step-down diffusion of Ag adatoms deposited on different substrates they arrived at the results in Fig. 6.26; an additional step-edge barrier of 0.120 ± 0.015 eV for Ag on Ag(111), and 0.030 ± 0.005 for Ag on Ag islands on Pt(111). The difference between the step-edge barriers can be associated with differences

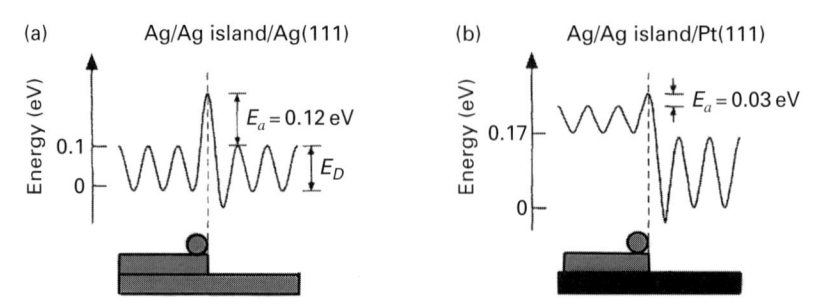

Fig. 6.26 Potential energy of Ag adatom diffusion on (111) plane [75]. (a) Ag deposited on Ag(111). (b) Ag deposited on Ag island on Pt(111).

6.2 To descending step edges

in binding of Ag atoms on Ag(111) and on a monolayer of silver, or with strain relief close to the island edges. The authors suggested that strain relief rather than electronic coupling was responsible for the difference observed in the step-edge barrier.

Markov [45] in 1997 derived an expression for the critical temperature for the transition from crystal growth by step flow to two-dimensional nucleation. Using experimental data from diffraction studies he was able to give the additional step-edge barriers E_a for Ag/Ag(111); he obtained a value $E_a=0.15$ eV based on the experimental findings of Elliot et al. [76], Vrijmoeth et al. [73], and Rosenfeld et al. [77]. Based on findings of Brune et al. [78] and Broman et al. [75], he arrived at a value of $E_a=0.12$ eV for the additional step-edge barrier. A different approach was followed by Roos et al. [79] who studied the dependence of interlayer diffusion on the ratio of the diffusivity to the flux, D/F_f. For Ag on Ag(111), the step-edge barrier E_a was estimated at 0.040 eV, rather lower than previous values. In the year 2000, Roos and Tringides [80] investigated self-diffusion on the (111) plane of silver. From their RHEED measurements, which will not be described here, and Monte Carlo simulations, they concluded that the frequency factor for interlayer diffusion was much larger than for diffusion over the surface, by something like a factor of 100. This conclusion has been challenged a number of times, by Krug [81], by Morgenstern and Besenbacher [82], as well as by Heinrichs and Maass [83]. We forgo any further comments here but note that subsequent additional work by Chvoj et al. [84] called for further studies.

Interesting STM experiments were carried out by Morgenstern et al. [85], who compared the filling of a vacancy island surrounded by an ascending step with the decay rate of an adatom island, sketched in Fig. 6.27. They found an additional step-edge barrier E_a of 0.13 ± 0.04 eV, with $v_s/v_0 = 10^{-0.6\pm0.5}$, where v_0 is the frequency factor for atomic diffusion and v_s for motion across the step. The measurements directly showed the energy required for descending motion, but the value obtained was an average over all types of steps, as well as for descent at kinks and corners.

Recently Li et al. [86] looked at "wedding cakes" morphologies with STM. To account for such morphologies created in the Ag/Ag(111) system with KMC simulations they needed to introduce a non-uniform step-edge barrier. For a step of type B the value of the

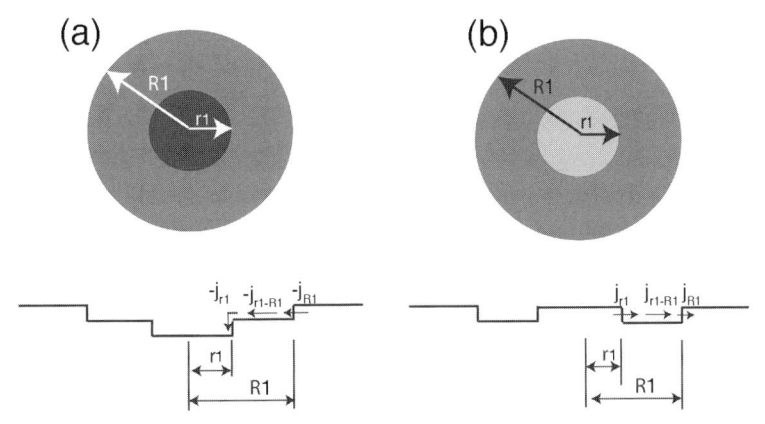

Fig. 6.27 Schematics of geometry for measuring step-edge barrier [85]. (a) Vacancy island. (b) Adatom island, both surrounded by ascending steps.

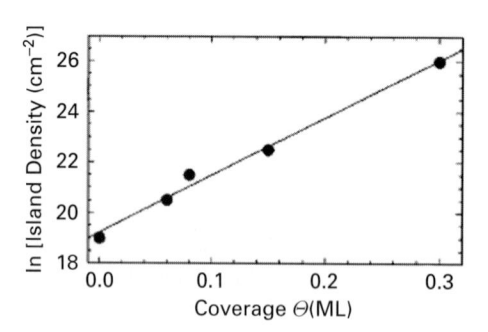

Fig. 6.28 Density of silver islands on Ag(111) as a function of the amount of deposited antimony (after van der Vegt et al. [87]).

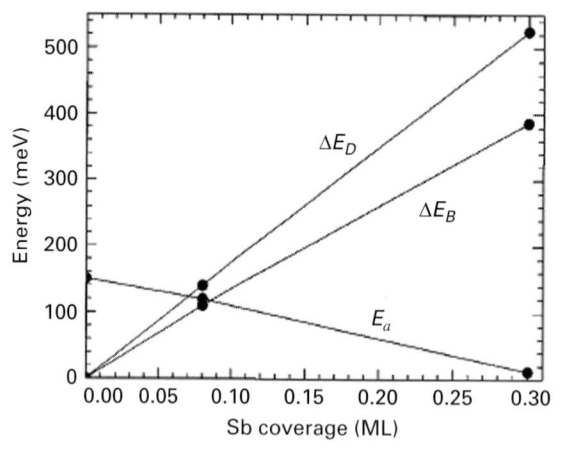

Fig. 6.29 Changes in diffusion barriers of Ag on Ag(111) due to the presence of antimony. E_a gives additional step-edge barrier, ΔE_D is the additional barrier to surface diffusion. ΔE_B is the change in the total step-edge barrier (after van der Vegt et al. [87]).

additional step-edge barrier E_a was 0.08 eV and for step A, E_a was twice bigger to explain experimental observations at 150 and 180 K.

Van der Vegt et al. [87] examined the effect of antimony, amounting to ≤ 0.2 monolayers, on the homoepitaxial growth of Ag(111). The saturation density of islands, n_x, at a given coverage is found according to nucleation theory as in Eq. (2.15): $n_x \propto \left(\frac{F_f}{D}\right)^{i/(i+2)}$, where F_f is the rate of deposition and i the critical size of the nucleus. The deposition rate was fixed and, assuming that the critical nuclear size does not change, an increase in the island density must be due to a change in the diffusivity D. Antimony increased the silver island density exponentially, as in Fig. 6.28, and this was interpreted to be caused by an increase in the activation energy for diffusion, given by

$$E_D = E_o + 1.7\Theta, \tag{6.3}$$

where Θ is the amount of antimony deposited on the surface. Also estimated was the step-edge barrier, and the effect on it of the antimony on the surface. The changes in these quantities are shown in Fig. 6.29. The total energy barrier to escape, E_B, increases. But it

6.2.11 Silver(111): theory

Stoltze [41] used effective medium theory in 1994 for investigations of the Ag(111) plane and found jumping over steps easier than exchange, $E_B = 0.359$ eV against 0.455 eV. Both values are higher than for movement on the plane, which proceeded with an activation energy E_D of 0.064 eV. No information about the type of step was given. Li and DePristo [42,43] calculated barriers with the corrected effective medium theory, CEM; for crossing B-type steps by exchange they obtained the barriers listed in Table 6.2. All barriers for crossing the steps are higher than for diffusion over the flat terrace. For descending a perfectly straight A-type step by hopping an energy E_B of 0.44 eV was required, while exchange on the same step was slightly lower at 0.31 eV. Descending a perfect type B step by hopping cost $E_B = 0.43$ eV; however, exchange required only 0.06 eV. The energies were slightly lower on a 19 atom hexagon; hopping cost $E_B = 0.39$ eV over an A-type step and 0.40 eV for a B-type, exchange was more likely on an A-type step with a barrier of 0.22 eV but only 0.02 eV on a B-type step. Descent on a zigzag step cost $E_B = 0.12$ eV by exchange. Descending next to a kink on an A-type step required $E_K = 0.19$ eV, while at a step of type B it was only 0.05 eV. These data are a bit uncertain since there is not perfect agreement between the two papers published by Li and DePristo.

In 2002 Heinrichs and Maass [88] looked at the influence of interactions on epitaxial growth and found that the height deduced for the step-edge barrier was very sensitive to interactions. Based on the experimental findings of Brune et al. [78] for Ag on 1ML of silver deposited on Pt(111) they found that the additional step-edge barrier E_a can change from 0.068 ± 0.001 eV without interactions to 0.074 ± 0.002 eV when ring barriers between two adatoms are allowed. This paper showed clearly that adatom interactions need to be investigated in detail to arrive at proper values for step-edge barriers.

Hong et al. [89,90] studied the effect of strain on the step-edge barrier on Ag(111) using the EAM potentials of Cai and Ye [91]. Their data are summarized in Fig. 6.30a. They found that strain had the opposite influence on the step-edge barrier compared to movement on the terrace. As already shown in Section 3.2.4, the surface diffusion barrier E_D increases with lattice stress, but the step-edge barrier E_a monotonically decreased. Tensile stress makes descent easier while compressive stress makes it harder, probably due to the open structure of the steps.

Haftel and Rosen [92] in 1998 looked at the descending motion of an Au adatom over steps of the Ag(111) surface with molecular dynamics and EAM potentials. The Au adatom moved over the terrace by hopping, with an energy E_D of 0.102 eV. Haftel and Rosen were unable to detect any adatom hopping over A-type steps, hopping over B-type steps required an energy $E_B = 0.716$ eV. Exchange over steps of both A- and B-type happened over a very low barrier; 0.051–0.061 eV for steps of type A and 0.047 eV for steps of type B.

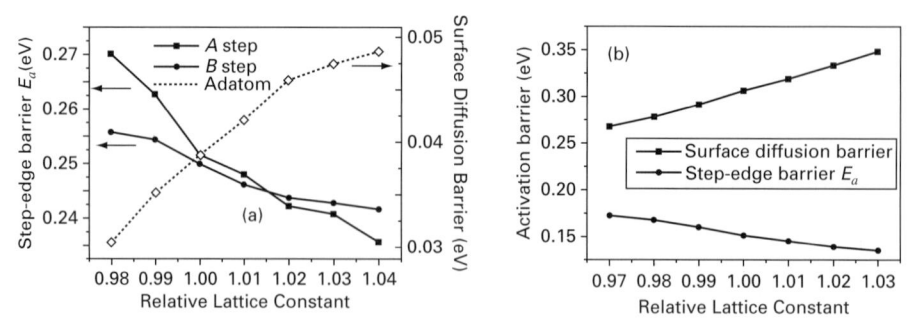

Fig. 6.30 Dependence of Ehrlich–Schwoebel barrier on lattice strain calculated with EAM for (a) two types of steps on Ag(111), (b) step on Ag(100) (after Hong et al. [89]).

Experimental values for descending motion from the Ag(111) surface seem to be in quite good agreement, unfortunately the mechanism of descent has not been uncovered. Theoretical values are confusing – one prefers hopping, the other exchange, and barriers are also not in agreement.

6.2.12 Silver(100)

Studies of descending motion from Ag(100) started with the work of Stoltze [41] using effective medium theory. The jump down the probably <110> step cost $E_B = 0.45$ eV, and exchange required an even higher energy of 0.559 eV. Terrace movement required only 0.365 eV. In 1997 Merikoski et al. [44] looked at step characteristics on fcc(100) planes using EMT. The <110> step-edge barriers E_B were 0.481 eV compared with 0.516 for exchange, and higher than for movement on a perfect (100) plane, which cost 0.367 eV; these values are in quite good agreement with previous findings by the same method. For descending at a kink by jumps they found a value $E_K = 0.401$ eV, which is comparable with 0.404 eV derived for exchange. Descending into kinks has lower energy requirements than descending at a straight step, but the barrier for interlayer motion was higher than the barrier for intralayer transport. The same year Yu and Scheffler [93] did density-functional theory calculations using both the local-density and the generalized gradient approximations. For an atom descending a <110> step by jumps, they got an activation energy $E_a = 0.18$ eV higher than for diffusion on the terrace using LDA and $E_a = 0.1$ eV from GGA. But more interesting were values for exchange, which turned out to be comparable with terrace diffusion on this plane, 0.52 eV from LDA and 0.45 eV from GGA. Yu and Scheffler claimed that the lack of a barrier at the step explained the smooth 2D growth on the silver (100) plane.

Kürpick and Rahman [94] using the embedded atom method with FBD and VC potentials calculated diffusion on the Ag(100) plane. They considered two kinds of steps for this plane, <110> and <100>. For the <110> step using FBD potentials, hopping required $E_B = 0.59$ eV against 0.64 eV for exchange, while VC potentials showed the opposite tendency: hopping needed $E_B = 0.70$ eV and exchange 0.51 eV, so it is difficult to draw conclusions about the mechanism of descent based on these results. Even more interesting were the <100> steps. Here using both potentials the energetically lower one was the exchange mechanism. From FBD, hopping had to overcome a barrier E_B of 0.51 eV and exchange 0.38 eV; from VC, hopping needed 0.55 eV versus 0.31 eV for exchange. But more surprising is that the exchange over the edge required less energy than diffusion on the terrace, which faced a barrier E_D of 0.48 eV, suggesting a negative barrier at the step edge, seen previously by Trushin et al. [57] for descent by exchange on a <100> type step of Cu(100). This negative barrier would explain the rare existence of this type of step on the fcc(100) surface.

Montalenti and Ferrando [69] calculated step-edge barriers with RGL potentials and for the descent of an Ag adatom on Ag(100) they found a barrier E_B of 0.59 eV for jumps; for descent by exchange, most likely over a <110> type edge, the barrier was very close, 0.55 eV. Both values are higher than the barrier to diffusion on the terrace, E_D, which according to their estimate amounted to 0.43 eV.

Hong et al. [89,90] studied the effect of strain on the step-edge barrier on Ag(100) using EAM potentials; their data are summarized in Fig. 6.30b. Similarly to Ag(111), they found that strain had the opposite influence on the step-edge barrier compared to movement on the terrace.

From an analysis of Ag film roughness observed with a STM at 230 K, Stoldt et. al. [95] deduced the additional step edge barrier of 0.03 eV. Two years later the same group [96] again arrived at a much higher value of 0.07 eV from film roughness and the mean mound diameter.

The situation of descending motion on Ag(100) is not certain, since half of the theoretical studies favor exchange, the other half jumping, and there is no experimental data available for verification. One thing seems to be clear – <100> steps will not exist on the surface at temperatures where diffusion is not frozen, since as soon as an adatom reaches them incorporation will be immediate.

6.2.13 Silver(110)

Just as for the silver(100) surface, there are no experimental data for descending motion from the Ag(110) surface. There are, however, a few theoretical results. These started with the work of Stoltze [41], who using EMT investigated two kinds of steps parallel and perpendicular to the channels of the anisotropic Ag(110) plane. Jumping down along the $[1\bar{1}0]$ direction required an energy E_B of 0.36 eV, while over <001> steps the barrier E_B was 0.646 eV. The anisotropic movement on the flat plane required 0.291 eV for the $[1\bar{1}0]$ direction and 0.639 eV in the [001]. Exchange over <001> steps was more demanding, $E_B = 0.537$ eV, while for the [001] direction it was less favorable – 0.605 eV.

Work was done also by Hontinfinde et al. [97] using RGL interactions that generally give reasonable results. For in-channel terrace motion their diffusion barrier E_D was 0.28 eV, the activation energy for cross-channel motion by atom exchange amounted to 0.38 eV. This is to be compared with a barrier of $E_B = 0.56$ eV for incorporation at an A-type step of {100} facets (<110> steps); incorporation proceeded by atom exchange since jumps required a much higher activation energy of 0.8 eV. For B-type edges of {111} facets (<001> steps), incorporation occurred by hopping over a barrier E_B of

0.35 eV. The barrier is comparable with the energy of terrace exchange and only 0.07 eV higher than the activation energy for in-channel hopping. The barrier to descent over a step of type B increased strongly, to $E_B = 0.53$ eV, if two atoms formed a dimer. Descending by exchange at a kink site at an A-type step required a barrier E_K for incorporation of 0.38 eV on a large island and 0.29 eV on a small one. These estimates clearly indicate that kinks favor incorporation.

Density-functional calculations for silver atoms diffusing over ascending lattice steps have also been done by Zhu et al. [33]. For in-channel motion, already illustrated in Fig. 6.21, the silver atom had to overcome a barrier of 0.68 eV to go up by exchanging with a lattice atom; the barrier downward was 0.54 eV. For cross-channel ascent by exchange the barrier was 0.62 eV and again 0.62 eV going down.

Descending motion from Ag(110) seems to preferably proceed by jumping down along channels of the plane. Existing theoretical estimates seem to be in agreement that such movement would require ~0.07 eV, while cross-channel descending motion required a four times higher energy.

6.2.14 Tungsten

Tungsten was the first material on which step crossing was investigated and the additional step-edge barrier was discovered. The reason is clear – tungsten is a very suitable material for FIM, and FIM has been the most useful technique for direct observations of single atom movement. The early findings have already been described in which a small additional barrier E_a~0.065 eV was discovered for self-diffusion on W(110) [16] and ~0.05 for W(321) [19]. For hetero-atom descent on W(110), a reflective barrier was also shown to exist but to be very small [21,22]. Here we will continue with later results. In 2002, Fu et al. [98], using the FIM, determined the rate of escape of palladium atoms from the (111) plane of tungsten at different temperatures, and arrived at an escape barrier of 1.63 ± 0.06 eV and a prefactor of $3.9 \times 10^{13 \pm 0.7}$ sec^{-1}. Assuming a prefactor for diffusion of 10^{-3} cm^2/sec, they came up with a diffusion barrier of 1.02 ± 0.06 eV for movement on the terrace, yielding an additional step-edge barrier of 0.61 eV. This large value should not be surprising, as the bcc(111) plane is bounded by quite rough steps.

In the same year, Fu et al. [98,99] continued to look at the escape of palladium atoms from the W(110), the W(111), and from the W(211) plane, as shown in Fig. 6.31. For escape from W(110) they give an extra escape barrier of 0.27 ± 0.02 eV and a frequency prefactor of 6.3×10^{12} sec^{-1}. For W(211), the additional activation energy for escape was given as 0.29 ± 0.08 eV, always assuming a prefactor of 3.7×10^{12} sec^{-1}. These results should not be taken too seriously, as the fit to the data points in Fig. 6.31 for these characteristics is surprisingly poor. Additionally it was shown by Fink and Ehrlich [24] that on W(211) the barrier for W, Re and Si was not localized at the step. Such a possibility should also be taken into account in an analysis of the data for Pd.

There are no theoretical efforts for deriving the step-edge barrier on tungsten surfaces.

6.2 To descending step edges

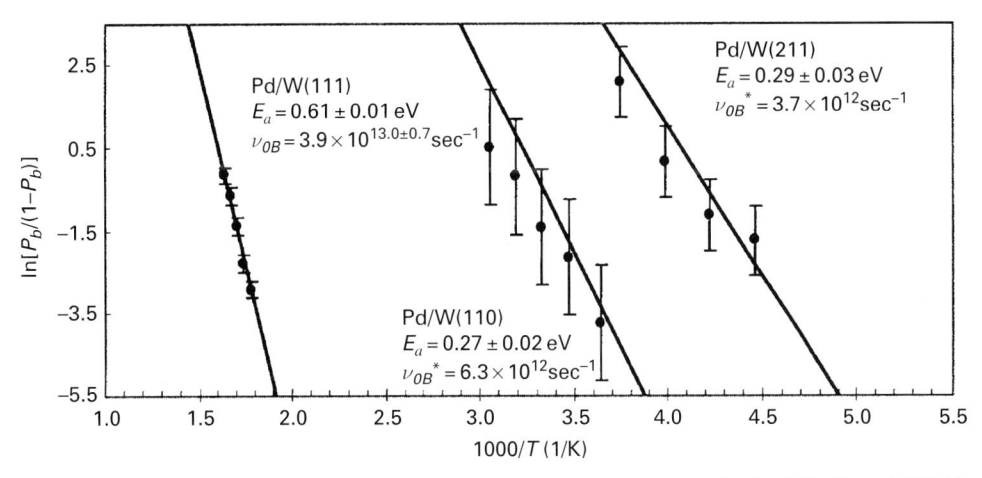

Fig. 6.31 Arrhenius plots for rate of descent of Pd over tungsten steps on W(111), W(110), and W(211). Listed are the additional step-edge barriers, and assumed prefactors indicated by * (after Fu et al. [99]).

6.2.15 Rhenium

Not much information is available about atom behavior on rhenium, but an interesting FIM study was reported by Goldstein [100] in 1999. He was primarily concerned with surface diffusion on Re(0001), the only hcp metal studied, and in his work determined an activation energy for self-diffusion E_D of 0.481 ± 0.019 eV and a prefactor of $6.13(\times 2.6^{\pm 1}) \times 10^{-6}$ cm^2/sec. For a tungsten adatom on the same surface, diffusion occurred over a similar barrier of 0.478 ± 0.005 eV, with a more normal prefactor of $2.17(\times 2.7^{\pm 1}) \times 10^{-3}$ cm^2/sec. More interesting in the present context, Goldstein found that at temperatures below 250 K, rhenium atoms diffusing over the surface were trapped on reaching the edge. For tungsten atoms, trapping occurred below 210 K. When the temperature was raised above these limits, the adatom detached from the edge and moved back to the center, an event observed many times. Under these conditions, collisions with the step edge were frequent, but the atom was never lost from the plane, suggesting the potential energy curve shown in Fig. 6.32. The plane is bounded by a substantial barrier E_B, but adjacent to it is a trap of significant depth bounded on the other side by a lower barrier E_R that obstructs the return to the center.

Goldstein assumed that the prefactor for diffusion was constant over the entire plane up to the step edge, and in this way arrived at a return barrier E_R of 0.57 ± 0.02 eV for Re and 0.58 ± 0.02 eV for W atoms, almost 0.1 eV higher than for diffusion in the central region. The step-edge barriers E_B arrived at under the same assumptions were given as 0.68 ± 0.02 eV for Re and 0.83 ± 0.02 eV for W atoms. The qualitative behavior of atoms at the step edge of rhenium therefore somewhat resembles what has been found for the edge of the Pt(111) plane, and it is clear that direct examination of the behavior of single adatoms provides the most valuable information. Here it should be noted that Goldstein also roughly tested the mechanism by which an atom crosses the step edge. He deposited a tungsten atom on the surface, and then identified by field evaporation the atom attached

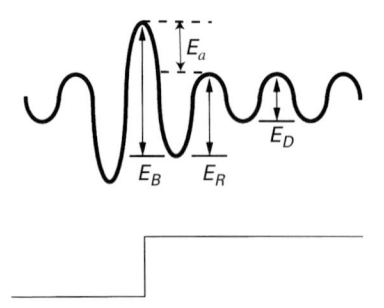

Fig. 6.32 Schematic energy of atom diffusing near edge of Re(0001) plane, deduced from observations of single adatoms [100].

to the edge after incorporation. In his only two trials, tungsten was found at the edge. This suggested that a simple jump had occurred at least on one kind of the steps bounding this plane, but more experiments are needed to demonstrate this unequivocally. Unfortunately, no reasonable attempts have been made at calculating the step-edge barriers.

6.2.16 Iridium: experiment

Experimental investigations by Wang and Ehrlich [25,26] of W, Re, and Ir atoms overcoming A- and B-type steps on Ir(111) for the first time demonstrated descent by exchange; these were described earlier, in Section 6.2. It is clear from this study that descent from Ir(111) is associated with a reflective barrier, since higher temperatures are required to initiate descent. Unfortunately, Wang did not carry out investigations of the energy for this process. Here we will continue our survey with subsequent findings for iridium surfaces.

In 1998, Fu et al. [101] examined atom escape for three systems: rhodium and iridium atoms from Ir(100), and iridium atoms from the iridium (111) plane. FIM measurements were done at a number of different temperatures, yielding Arrhenius plots from which detailed kinetics were derived, as shown in Fig. 6.33. For rhodium leaving the Ir(100) surface, the activation energy E_B was 0.84 ± 0.08 eV, with a frequency prefactor of $2.4 \times 10^{13 \pm 1.4}$ sec^{-1}. The diffusion characteristics of rhodium on the same surface were quite close to these values, $E_D = 0.80 \pm 0.08$ eV for the diffusion barrier and a prefactor of $2.09 \times 10^{-3 \pm 1.5}$ cm^2/sec, which makes it quite uncertain that for this system there really is a step-edge barrier E_a. For iridium atoms on the same plane, the barrier for escape E_B was 0.72 ± 0.08 eV, compared with an activation energy for diffusion E_D of 0.74 ± 0.02 eV; the prefactors were given as $5.3 \times 10^{13 \pm 1.6}$ sec^{-1} and $7.30 \times 10^{-3 \pm 0.5}$ cm^2/sec for diffusion. Here again there is doubt about the existence of a step-edge barrier. For iridium leaving Ir(111), Fu et al. found a barrier E_B of 0.41 ± 0.03 eV and a frequency prefactor of $1.8 \times 10^{13 \pm 1.1}$ sec^{-1}. To arrive at a step-edge barrier E_a, they picked a value of 0.21 ± 0.03 eV for the diffusion barrier given by Kellogg [102], but this is clearly too low. The self-diffusion characteristics on Ir(111) are an activation energy of 0.289 ± 0.003 eV, and a prefactor $D_o = 3.5(\mathrm{x}1.4^{\pm 1}) \times 10^{-4}$ cm^2/sec [103], so that the

6.2 To descending step edges

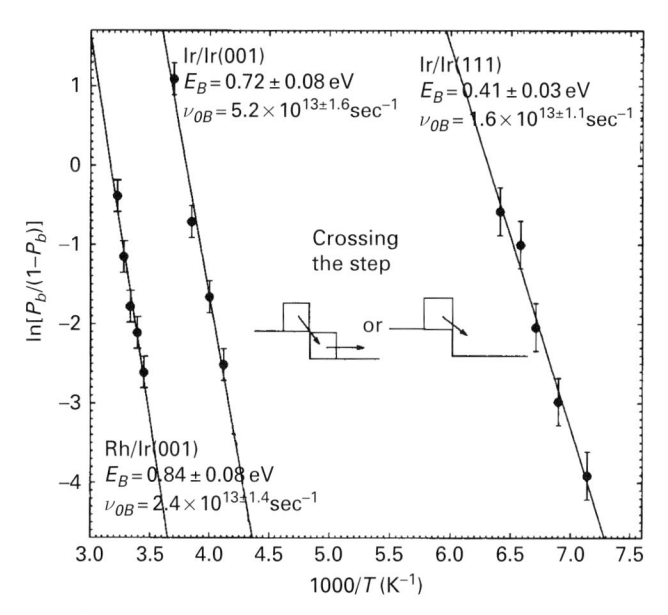

Fig. 6.33 Arrhenius plots for the descent of atoms from Ir(100) and Ir(111) planes (after Fu et al. [101]).

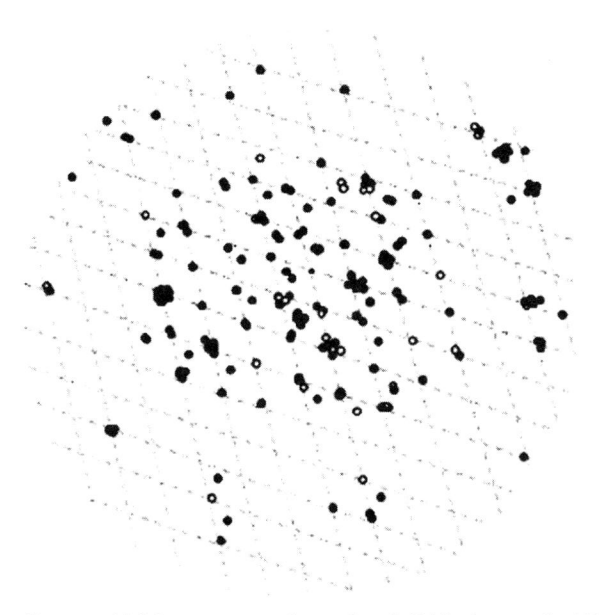

Fig. 6.34 Sites at which Ir atoms were located on Ir(111) cluster of ~ 150 atoms. Open circles indicate location after deposition at ~ 30 K, black circles after diffusion (after Fu et al. [101]).

additional step-edge barrier E_a amounted to 0.12 eV. Here it should be noted that the (111) plane of iridium is similar to that of platinum: both have a central region populated by adatoms but surrounded by an empty zone. Sites at the very edges of the plane bind atoms, as shown in Fig. 6.34, but no estimates were made of how this affected the escape process [101,104].

Diffusion in special environments

The mechanism of descent was not examined, but based on the exchange of Ir on Ir(100), the authors suggested that descent would be by exchange as well. For Rh on Ir(100) they mentioned that jumps might be possible, but this idea was not investigated separately. To determine the difference in the barrier for an Ir atom to descend over steps of type A and B on Ir(111), the authors looked at 58 descending runs in which atoms crossed step A 18 times and step B 40 times. Based on this observation the difference in activation energy was estimated as 0.012 eV. The statistics are rather poor, and the mechanism of descent was not probed, but the results are in agreement with the earlier work of Wang and Ehrlich [25].

A number of theoretical studies suggest that vacancies, corners, and kink sites accelerate the descent of atoms. This proposition has been studied directly by Wang [105], who recently tested the effect of edge imperfections on the rate of incorporation. In the first experiments, an Ir atom was deposited on an Ir$_{19}$ cluster sitting on an Ir(111) plane. The diffusion path of the atom at 160 K was observed, and as shown in Fig. 6.35, incorporation at an A-type step occurred after ten 5 sec heating periods. The experiment was repeated 28 times, and 13 times the atom incorporated at a type A step, showing equal probability of incorporation to step A and B for this cluster. Here it is worth mentioning that the FIM should be able to detect changes in the incorporation barrier bigger than 5%. Next an atom was attached to the side of an Ir$_{19}$ cluster formed on Ir(111), shown in Fig. 6.36, to introduce a kink. The atom in the lower left-hand side should play the role of catalyst if incorporation is preferred at this defect. An adatom was deposited on the surface, and its path in diffusion after heating to 160 K is shown in Fig. 6.36, but even though the adatom twice was close to the "kinked edge," incorporation eventually

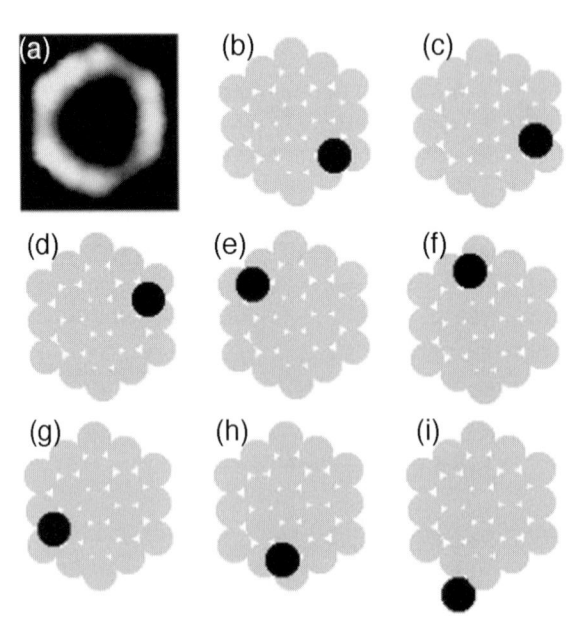

Fig. 6.35 Incorporation of Ir atom deposited on top of Ir$_{19}$ cluster, showing the track of the atom in diffusing at 160 K [105]. Incorporation finally takes place at lower left step edge.

6.2 To descending step edges

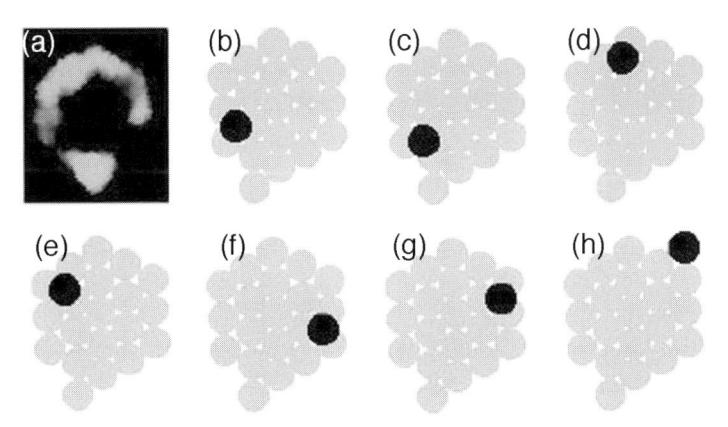

Fig. 6.36 Ir atom incorporation into Ir$_{20}$ cluster, with one atom attached to central Ir$_{19}$ core [105]. (a) FIM image of Ir$_{20}$ cluster. (b) Shows atom after deposition. (c)–(g) Atom movement around edge after heating to 160 K. (h) Atom has incorporated into straight edge after 13 heating intervals.

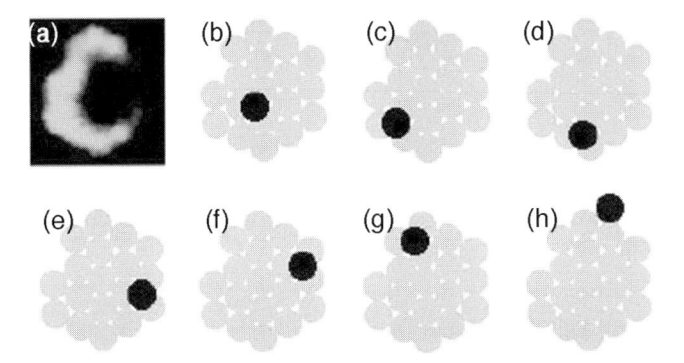

Fig. 6.37 Ir atom incorporation into Ir$_{18}$ cluster [105]. (a) FIM image of Ir$_{18}$ cluster. Ir atom has been placed on top of cluster in (b), and was then heated for 5 sec at 120 K, moving atom close to edge in (c). (d)–(g) Motion along edge after heating to 160 K. (h) Atom incorporates into straight edge after nine heating intervals.

occurred into a straight edge opposite to the defect. Out of a total of 23 repetitions of this experiment, the adatom on the cluster incorporated 21 times into straight edges, 8 of them of type A. If the defect had any effect on atom behavior it proved to be small, with less than a one percent reduction of the incorporation barrier. It appears that the atom added to the straight step of an Ir$_{19}$ cluster did not provide an especially low-energy path for incorporation.

Also examined was the activity of different types of kink sites which calculations have shown to be very active in promoting atom attachment into a step. The kink was created by removing an edge atom from an Ir$_{19}$ cluster. Measurements of atom diffusion on an Ir$_{18}$ cluster, again on Ir(111), are shown in Fig. 6.37. The atom again migrated around on the cluster, passing near the kink site, but incorporation once again occurred into a straight edge rather than into the kink. This type of incorporation was repeated 27 times. In these trials, only one ended with the atom incorporating at the defect site,

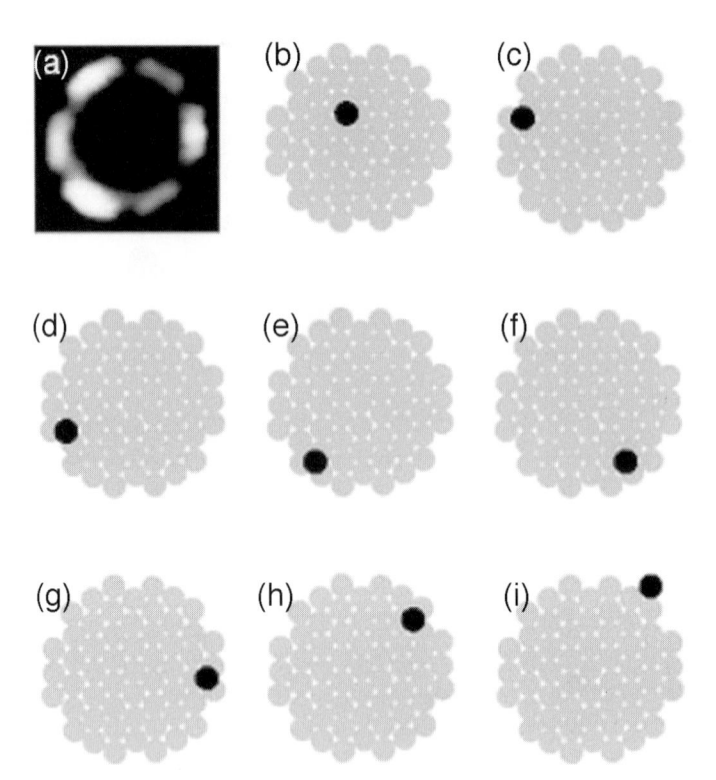

Fig. 6.38 Iridium atom incorporation into Ir$_{55}$ cluster, showing the track of the atom diffusing over the cluster [105]. Incorporation takes place at upper right edge.

suggesting a rather higher barrier to incorporation there. Finally an Ir$_{55}$ cluster was created by evaporating all 6 corner atoms from a cluster of 61 Ir atoms. On this cluster the same procedure was followed, as indicated in Fig. 6.38. In 51 attempts, atoms incorporated in the empty corner only 4 times. This observation suggested a barrier at the kink rather higher than at a straight edge. The last test was done on a cluster of 63 atoms created by adding two atoms to Ir$_{61}$ to create a kink. In 29 repetitions, atoms incorporated at kink sites three times, once in a kink created at a *B*-type edge and twice into a kink on an *A*-type edge. Here kinks might be slightly more favorable than at ordinary edges. Some other trials of incorporation were also done, but overall there was no indication of defects in the cluster edge significantly promoting incorporation. This, of course, raises serious questions of how to understand the disagreement with calculations showing preferred incorporation into defect sites, which may again have something to do with the different behavior of straight steps and steps at small islands, but this is clearly a matter that will have to be examined.

6.2.17 Iridium: theory

RGL interactions were used by Trushin *et al.* [106] in estimates of self-diffusion on Ir(111). On a terrace the activation energy E_D was 0.20 eV, but to go over the step and

6.2 To descending step edges

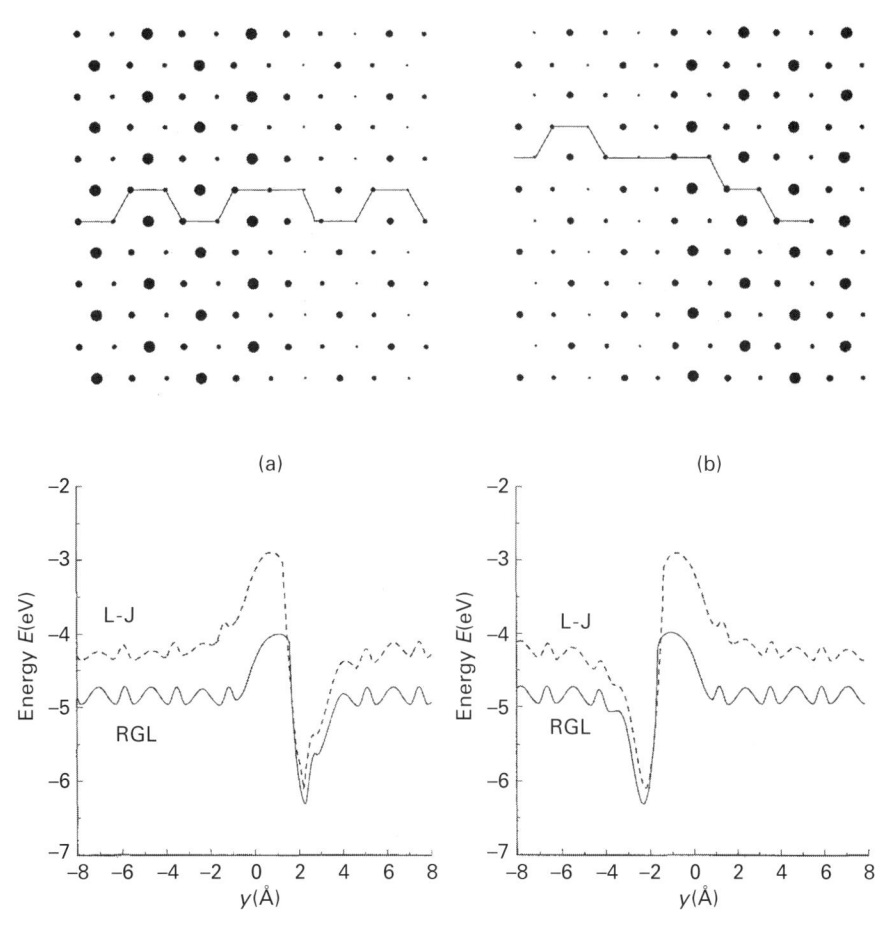

Fig. 6.39 Diffusion of Ir atom on Ir(111) plane. Top: diffusion path over step. Bottom: energy along diffusion path. (a) *A*-type steps. (b) *B*-type steps. Dashed line: Lennard-Jones calculations. Solid line: RGL estimates (after Trushin *et al.* [106]).

incorporate required more energy, as is apparent in Fig. 6.39. For *A*-type edges, made up of (100) facets, atoms had to jump over a barrier E_B of 0.97 eV to incorporate; exchange required a much higher energy of 1.75 eV. Incorporation at a kink site occurred somewhat more easily, over a barrier E_K of 0.83 eV for atom jumps and 1.50 eV for atom exchange, as shown in Fig. 6.40. The situation at *B*-type edges, of (111) facets, is different. Jumps over the step required an energy E_B of 0.98 eV, but exchange with a lattice atom took place more easily over a barrier of 0.74 eV. Incorporation at a kink site occurred over an energy E_K of 0.83 eV for a jump, and 0.90 eV for atom exchange. Here we have to note that the estimate for the self-diffusion barrier E_D made with RGL potentials is below the experimental value of 0.289 eV [103].

Studies of iridium atoms incorporating into a step edge on Ir(111) have also been done by Kürpick [107], who used RGL potentials to explore atom behavior at straight edges as well as at steps with vacancies. The expected behavior at both *A*- and *B*-type steps is

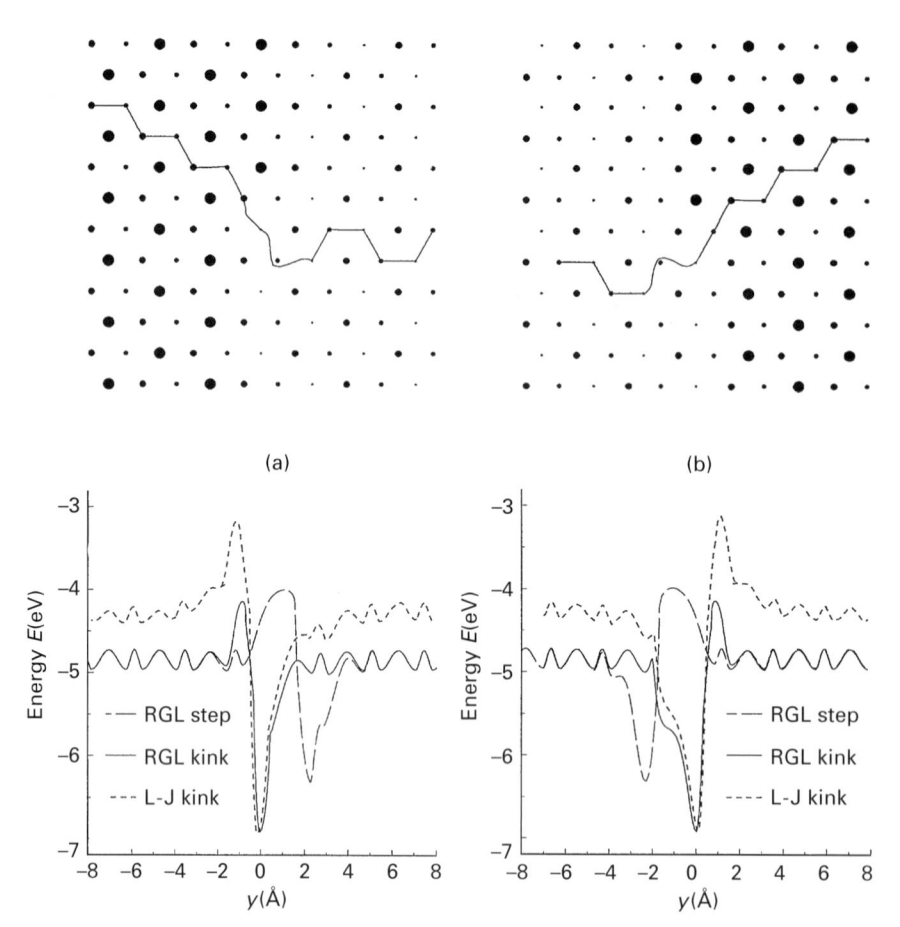

Fig. 6.40 Diffusion of Ir atom on Ir(111) with descent over kinked step. Top: diffusion path over step. Bottom: energy along diffusion path. (a) *A*-type step. (b) *B*-type step (after Trushin *et al.* [106]).

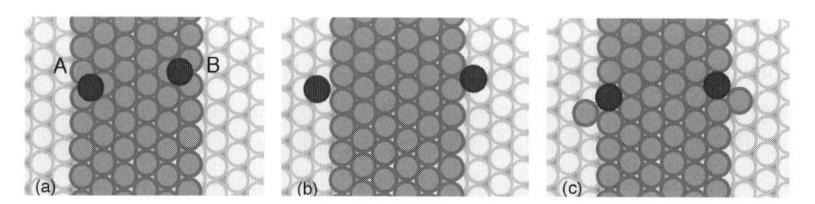

Fig. 6.41 Ir atom descent from Ir(111) surface [107]. (a) Initial position of adatoms at *A*- and *B*-type steps. (b) Final position after atom hop. (c) Final position after atom exchange.

shown in Fig. 6.41. However, in molecular dynamics studies she did not obtain any incorporation events at straight steps at 900 K. Using transition-state theory she found the energy involved in such transitions was quite high, as is apparent in Table 6.3. At (100) or *A*-type steps, jumps to incorporation required an activation energy E_B of 0.90 eV and a prefactor of $(7.4 \pm 0.1) \times 10^{-4}$ cm^2/sec, whereas exchange occurred over a much higher

6.2 To descending step edges

Table 6.3 Kinetic parameters for descent of Ir atom at Ir(111) step edges [107].

	(100) facet (A-type):					
	Straight step		Kink		Vacancy	
	Jump	Ex	Jump	Ex	Jump	Ex
E_B (eV)	0.90	1.58	0.77	0.85	0.75	0.82
D_{oB}(cm²/sec)	$(7.4\pm0.1)\times10^{-4}$	$(2.7\pm0.1)\times10^{-3}$			$(7.5\pm0.1)\times10^{-4}$	$(2.1\pm0.1)\times10^{-3}$

	(111) facet (B-type):					
	Straight step		Kink		Vacancy	
	Jump	Ex	Jump	Ex	Jump	Ex
E_B (eV)	0.91	1.03	0.77	1.45	0.75	1.37
D_{oB}(cm²/sec)	$(7.0\pm0.1)\times10^{-4}$	$(5.6\pm0.1)\times10^{-3}$			$(7.4\pm0.1)\times10^{-4}$	$(2.2\pm0.1)\times10^{-3}$

barrier of 1.58 eV and a prefactor of $(2.7\pm0.1)\times10^{-3}$ cm²/sec. At (111) facets, B-type steps, jumps occurred over a barrier of 0.91 eV with a prefactor of $(7.0\pm0.1)\times10^{-4}$ cm²/sec, while exchange involved an activation energy of 1.03 eV and a prefactor of $(5.6\pm0.1)\times10^{-3}$ cm²/sec. These values are to be compared with a barrier of 0.24 eV to diffusion calculated for the (111) plane.

The situation was different when vacancies at the step were present, as suggested in Fig. 6.42. At an A-type step, descent by an atom jumping at a vacancy occurred over a barrier E_V of 0.75 eV, with a prefactor of $(7.5\pm0.1)\times10^{-4}$ cm²/sec. Exchange involved a higher barrier of 0.82 eV and a prefactor of $(2.1\pm0.1)\times10^{-3}$ cm²/sec. Descent at a B-type step again occurred over a barrier E_V of 0.75 eV, with a prefactor of $(7.4\pm0.1)\times10^{-4}$ cm²/sec, compared to a barrier for exchange of 1.37 eV and a prefactor of $(2.2\pm0.1)\times10^{-3}$ cm²/sec. Descent at a step with several adjacent vacancies took place over a somewhat higher barrier than when only a single vacancy was present. For exchange, the prefactor was 3 to 8 times higher than for hopping and in molecular dynamics simulations exchange on a type A step with a single vacancy was observed at 1100 K. This was explained by the Arrhenius plots in Fig. 6.43 for hopping and exchange, which intersect at 788 K so that exchange might get favorable at higher temperatures. These results are interesting, as they differ from experiments on clusters, for which descent, presumably by exchange, occurred preferentially at B-type edges. This may indicate a difference in atom descent at straight steps and at steps of small islands.

On Ir(111) steps direct observations of step-edge barriers have been made and the data are in good standing, but diffusion over iridium needs much more attention. Theory for movement on the flat surface predicts values comparable with experiment. For descending motion, however, the predicted additional step-edge barrier E_a is huge, around twice the value for surface movement. The existing reports also are not in agreement about the mechanism of descent.

466 Diffusion in special environments

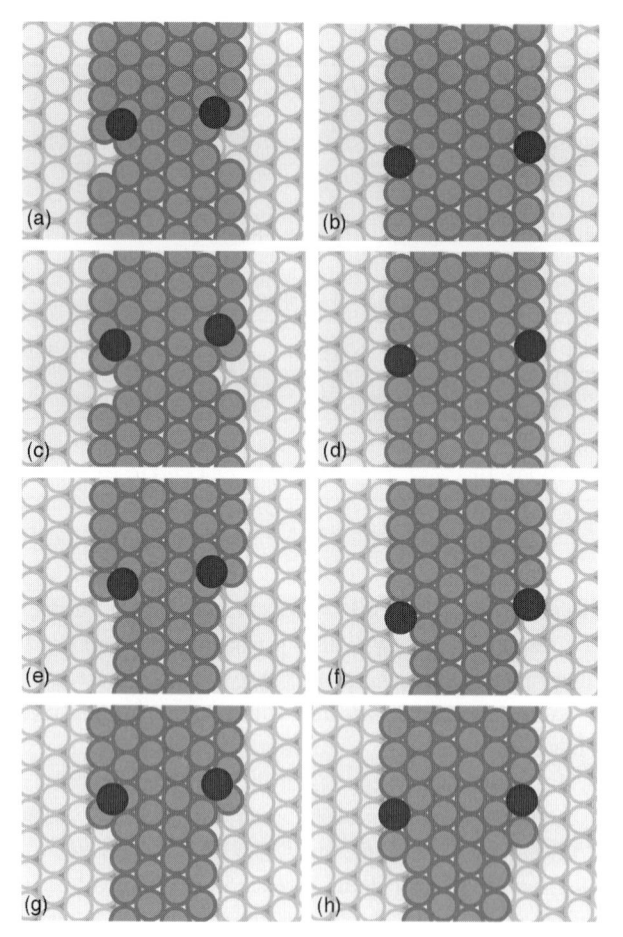

Fig. 6.42 Ir atom descent from Ir(111) plane with edge vacancies [107] (a) and (b) for atom hopping, (c) and (d) for atom exchange transition. (e)–(h) Atom descent when edge atoms are missing.

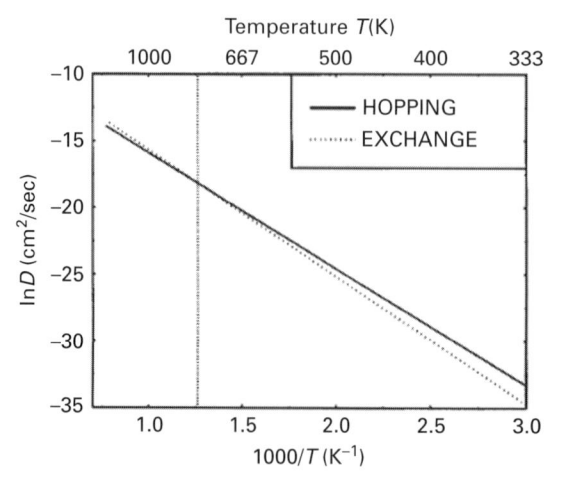

Fig. 6.43 Arrhenius plot for descent from Ir(111) plane with {100} faceted step edge with one vacancy, for hopping (solid line) and exchange (dotted line). Intersection at 788 K (after Kürpick [107]).

6.2.18 Platinum: experiment

In 1995 Smilauer and Harris [35] came up with a quantitative relation for the onset of nucleation on islands, and arrived at values for the additional step-edge barrier E_a of 0.15 eV to 0.27 eV for Pt on Pt(111). At the same time, Meyer et al. [74] derived a different model to analyze growth and found a step-edge barrier E_a $0.165^{+0.050}_{-0.020}$ eV for Pt on Pt(111).

Markov [108] in the same year derived another relation for second layer nucleation in homoepitaxy and arrived at two values for the step-edge barrier on Pt(111) from the experiments of Bott et al. [109]: $E_a = 0.31$ eV and 0.44 eV. The lower value was attributed to A-type steps, (100) facetted, and the higher one to passage over B-type steps. The following year, Markov [45] looked at the transition from crystal growth by step flow to two-dimensional nucleation, and using experimental data from diffraction studies done by Poelsema et al. [110] he arrived at a step-edge barrier of $E_a \approx 0.21$ eV.

A detailed FIM probe of how atoms escape from a plane was presented by Kyuno and Ehrlich [111] for platinum atoms on Pt(111). The Pt(111) surface is like Ir(111) in that only the center is occupied by adatoms; this center is surrounded by an empty zone of higher energy, which extends almost to the boundary of the plane. Atoms can again be held at the plane edge, as illustrated in Fig. 6.44. Kyuno and Ehrlich measured the rate at which platinum atoms escaped from this plane at temperatures from 120 to 140 K. The results are shown in the Arrhenius plot in Fig. 6.45, which yields an activation energy of 0.32 ± 0.01 eV. To properly interpret this figure it is necessary to know something about the potential energy variations on going across the plane. A rough diagram of the potentials close to the edge is shown in Fig. 6.46. The rate of escape r_e can be written as

$$r_e = P_E v_{0B} \exp(-E_{B_2}/kT), \tag{6.4}$$

where P_E is the probability that an atom occupies an edge site, v_{0B} is the frequency factor, and E_{B_2} is the barrier to escape and incorporate. From observations of atom transitions, Kyuno was able to arrive at a barrier E_{B_2} (which prevents atoms from incorporating from sites next to the step) equal to 0.36 eV, and for the barrier E_{B_1} in Fig. 6.46 which confines atoms to the interior a value of ~0.33 eV. The activation energy for diffusion E_D at the center of the Pt(111) plane has been found as 0.26 eV, so that the barrier at the step edge

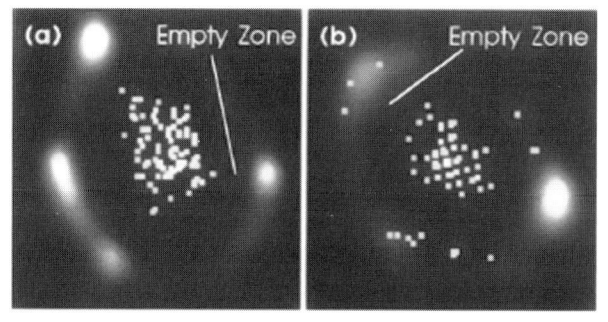

Fig. 6.44 Location of Pt on Pt(111) plane [111]. (a) Positions at which Pt$_5$ cluster was observed, all located in inner region. (b) Single platinum atoms were found either in center or at edges.

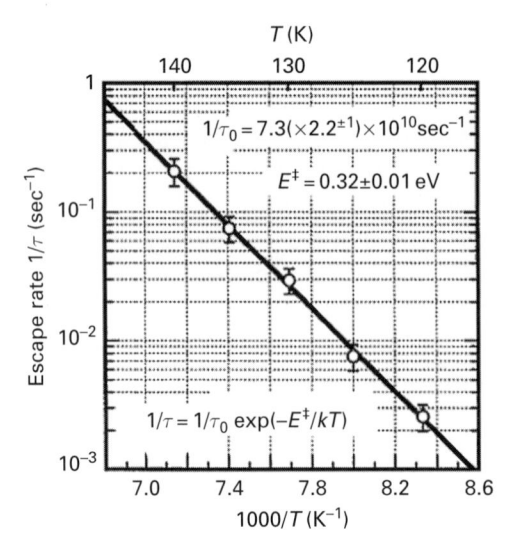

Fig. 6.45 Arrhenius plot for the rate of Pt atom escape from Pt(111) island with radius of ~ 20 nearest-neighbor spacings [111].

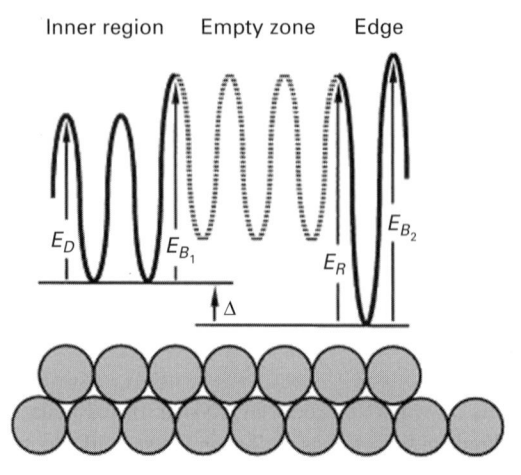

Fig. 6.46 Schematic of energy barriers to Pt atom jumps on Pt(111) cluster, sketched below [111]. Dotted curve indicates values were not well known.

E_B is 0.1 eV higher. Direct measurement of the rate of escape delivered an activation energy E_a of 0.06 ± 0.01 eV above that for diffusion. This appears to be one of the first more reliable determinations of the potential opposing escape.

A year later, in 2000, Krug et al. [112] developed a new theory for nucleation on two-dimensional islands based on the residence time of an adatom on that island. For CO decorated Pt(111) the step-edge barrier E_a, derived from work by Kalff et al. [113], was 0.36 eV, as indicated in Fig. 6.47, and dropped significantly, to 0.12 eV, as the contamination was removed. The value predicted from the procedure described by Tersoff et al.

6.2 To descending step edges

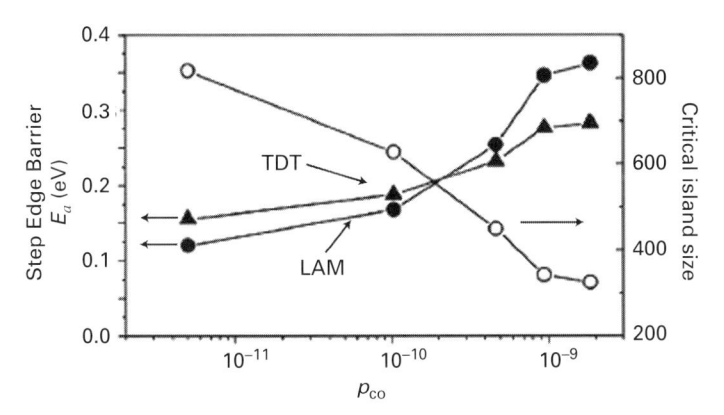

Fig. 6.47 Step-edge barrier (full symbols) and critical island size in their dependence on CO. Full circles from formula of Krug [112], triangles from Tersoff et al.[114] (after Krug et al. [112]).

[114], labeled TDT, differed significantly in the higher effect of CO, 0.28 eV compared with 0.36 eV. With a decreasing influence of impurities, the two analyzing techniques started to give comparable values, 0.15 eV from TDT compared with 0.12 eV from the lonely atom method (LAM) of Krug et al. [112]. From an analysis of multilayer growth on Pt(111) in the presence of a partial CO pressure of 1.9×10^{-9} mbar at 440 K they found a comparable barrier E_a of 0.33 eV. According to Krug et al. [112], the lonely atom method can be trusted only in the presence of a strong step-edge barrier, which is the case for a high CO pressure; in this region Tersoff et al. underestimated the value of the step-edge barrier. Krug et al. also claimed that for small pressures the LAM method was unreliable, as it underestimated the nucleation rates. However, surprisingly enough, the value obtained in the limit of small pressures agreed reasonably with the barrier measured directly by FIM.

Very interesting examinations of cobalt deposits on a Pt(111) surface, on which vacancy islands had been created, were done by Lundgren et al. [115] using atomic resolution STM. The image of platinum and cobalt atoms can be distinguished since Co appear to be darker than Pt. Lundgren et al. deposited 0.04 ML of cobalt and found cobalt atoms inside the vacancy islands at the edge as well as in the second row from the edge, shown in Fig. 6.48. The presence of atoms in the second row suggested exchange as the mechanism of step descent. To get more information about the mechanism, the amount of Co at the edge and in the second row from the edge was investigated as a function of the vacancy size, as in Fig. 6.49.

The number of Co atoms at the edge agreed with that expected from the rate of deposition and this amount depended on the size of the vacancy island. The number of atoms situated behind the first row of platinum atoms did not depend on the size of the vacancy island, and was associated with the atoms deposited on top. However, the number of descending atoms did depend on the step length. Since the number of atoms embedded in the second row was independent of island size, Lundgren et al. concluded that these atoms had diffused over the top and had exchanged with kink or corner platinum atoms from the lattice. Additionally, Co atoms liked to assemble in rows,

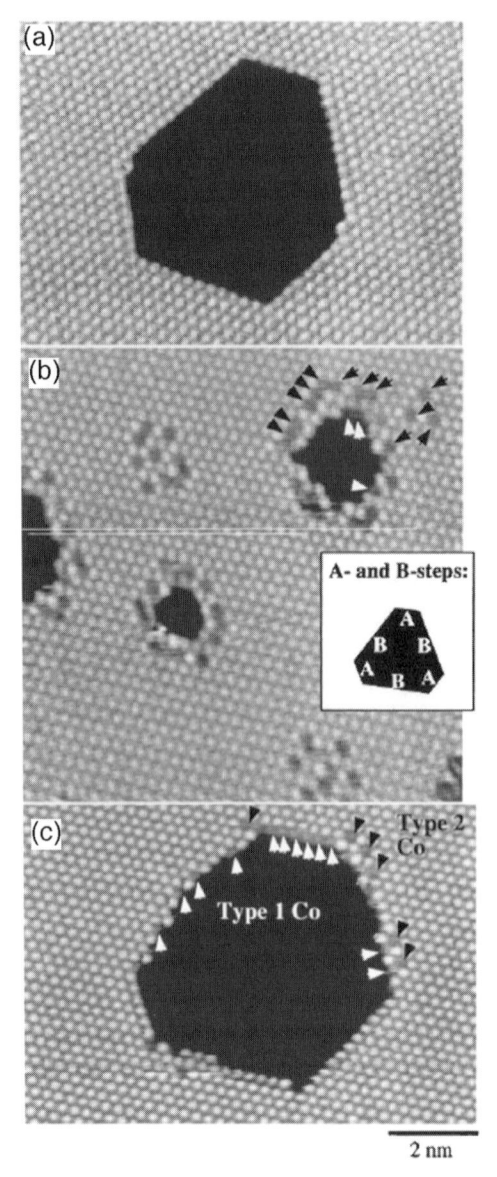

Fig. 6.48 STM image of vacancy island on Pt(111) surface. (a) Island formed in sputtering. (b) 0.04 ML of Co deposited on Pt(111) surface. (c) As in (b), but a different part of the surface. White arrows indicate atoms at the edge (type 1), black arrows show atoms behind the edge (type 2) (after Lundgren *et al.* [115]).

which also supported descents over kinks, since such exchange creates shifts, shown in Fig. 6.50. The authors never observed Co atoms in the second row, next to long straight steps without kinks. The Co atoms were not found uniformly all around the vacancy cluster – they appeared to be close to the corners of the island. The conclusion was that Co did not jump over the step-edge barrier. Instead, exchange occurred and was

6.2 To descending step edges

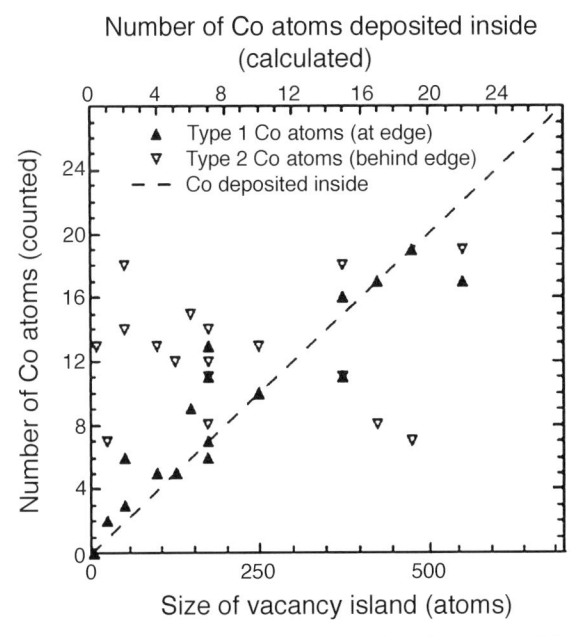

Fig. 6.49 Plot of type-1 and type-2 Co atoms against the vacancy island size. Dashed line gives number of Co atoms calculated to be in the vacancy island (after Lundgren et al. [115]).

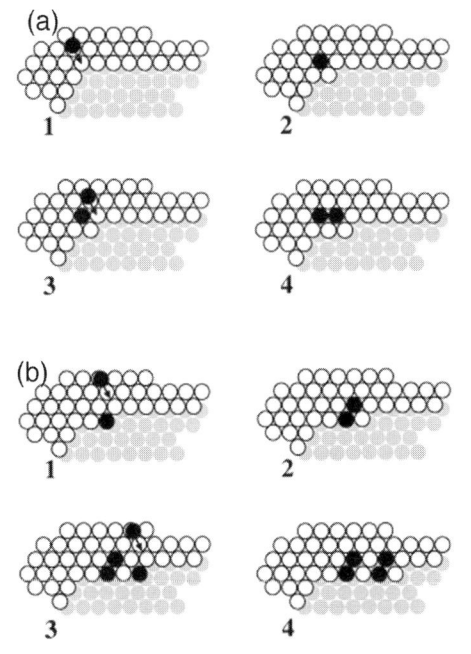

Fig. 6.50 Models for exchange processes at corners, and kink creation. (a) Creation of kink by exchange. (b) Kink arises from Pt atom in corner (after Lundgren et al. [115]).

Diffusion in special environments

accelerated at the corners, whereas at straight steps atom descent took place much more slowly with a mechanism not uncovered by this investigation.

6.2.19 Platinum: theory

The study of platinum steps started with calculations by Villarba and Jónsson [116] who made extensive estimates, using the embedded atom method, of the energy barriers important in the incorporation of platinum atoms into steps on Pt(111). On the Pt(111) plane itself the activation energy for diffusion E_D amounted to 0.08 eV. They found that hopping over a step is highly unlikely, due to the high energy barrier E_B of 0.7–0.8 eV. Of special interest are the barriers for an atom approaching A- or B-type straight, long steps shown in Fig. 6.51. Close to either step edge there is a very considerable alteration of the potentials. Exchange descent at the straight step required an energy E_B of 0.30 eV from an fcc site and 0.25 eV from an hcp site at the A-type edge; at a B-type edge the barrier from an hcp site was 0.18 eV. Descent at the A-type edge was more demanding, due to the bigger number of atoms influenced by such a movement. In the exchange of an adatom at

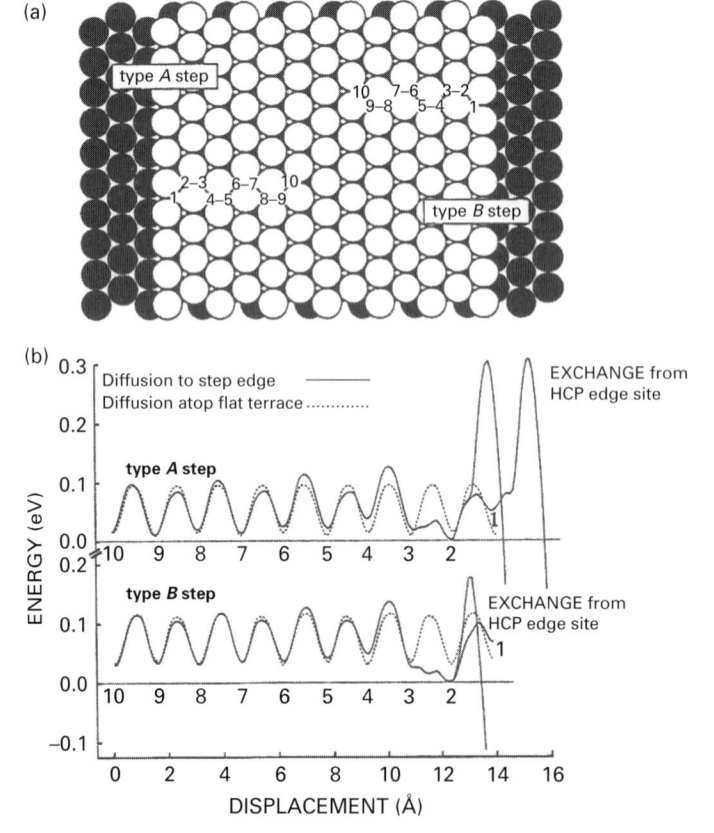

Fig. 6.51 Platinum atom behavior on Pt(111) surface. (a) Model of A- and B-type steps. (b) EAM calculated energy of platinum atom diffusing on flat (111) surface (dashed) and over steps (solid) by atom exchange (after Villarba and Jónsson [116]).

6.2 To descending step edges

Table 6.4 Barriers for exchange descent of Pt from compact Pt islands [116].

Cluster	Barrier E_B (eV)	
Trimer	0.49	
Heptamer	0.48 (fcc)	0.29 (hcp)
	A-type	B-type
19-atoms	0.38	0.03
37-atoms	0.38	0.06
61-atoms	0.33	0.14
91-atoms	0.32	0.16
127-atoms	0.31	0.17
169-atoms	0.31	0.17
Straight Edge	0.30	0.18
Near Kink	0.08	0.06

an *A*-type edge, the adatom must go around an underlying atom, displacing the nearest and next-nearest edge atoms. For descent by atom exchange at kink sites, they estimated a much lower barrier E_K of 0.08 eV at an edge of type *A* and 0.06 eV at a *B*-type edge. For a *B*-type kink the activation energy was lower than for movement on the terrace. They also estimated how the barrier to incorporation varied with cluster size. As is apparent from Table 6.4, for *B*-type edges the barrier E_B increases from 0.03 eV for a 19 atoms cluster to 0.17 eV at a 169 atoms cluster. Over the same range of cluster sizes the barrier E_B at *A*-type edges decreases somewhat, from a high of 0.38 eV to 0.31 eV.

Villarba and Jónsson also compared incorporation at a straight *B*-type step with what happens on a 37 atoms cluster, as shown in Fig. 6.52. What is apparent here is that for an atom on the island that has reached the edge (site 2), the descent is easier than a return to the interior. For displacing a corner atom of a 19 atom island by exchange the barrier was only 0.19 eV. The barrier for an adatom descending from small clusters, such as trimers and heptamers, was higher than at the straight edges, 0.49 eV for descending from a trimer and 0.29 eV from an heptamer starting at an hcp site. Although the embedded atom method does not give good diffusion barriers on Pt(111), these estimates are certainly the most extensive and seem to show effects revealed in experiments. It should be noted that Siswanto and Jónsson [117] more recently have done density-functional calculations which show behavior quite similar to the embedded atom work, as seen in Fig. 6.53.

At roughly the same time, Wang and Fichthorn [118] performed molecular dynamics simulations with the corrected effective medium theory (CEM) for incorporation of platinum atoms into a descending step on Pt(111). On a 3 × 3 cluster with *A*-type edges, incorporation always occurred by exchange with an energy E_B of 0.03 eV, which is a slightly lower value than the $E_D = 0.038$ eV calculated by the authors for movement on the clean relaxed surface. They also did simulations on 6 × 6 clusters, and again exchange was the significant mechanism. Incorporation by exchange was seen on both types of edges, but exchange at *B*-type steps was more frequent. On the larger 6 × 6 clusters a sizeable number of atoms did not incorporate, and they could not exclude the

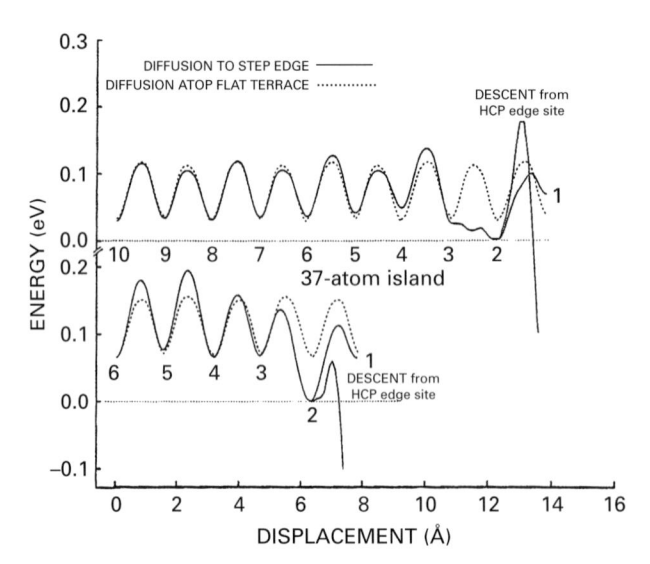

Fig. 6.52 Energy of Pt atom diffusing on Pt(111) surfaces. Comparison of behavior at straight *B*-type step and at cluster of 37 atoms. Dashed curve indicates diffusion on flat (111) plane (after Villarba and Jónsson [116]).

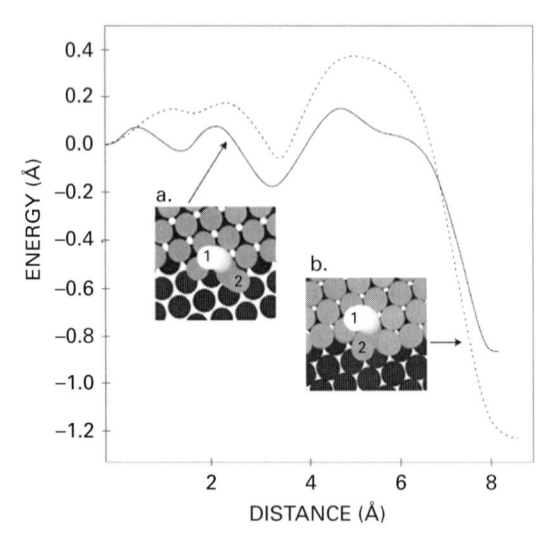

Fig. 6.53 Density-functional calculations of Pt atom diffusing over *B*-type step on Pt(111) surface (after Jónsson [117]).

possibility of jumps occurring. From static potential energy estimates for atoms diffusing toward a step edge they found the energy at a site two steps from the edge was considerably higher than in the interior for both 7×7 and 10×10 clusters, as in Fig. 6.54. A year earlier Wang and Fichthorn [119] looked at descent by hopping from 3×3 and 4×4 islands; the barriers E_B were much higher, 0.50 eV for a 3×3 island and

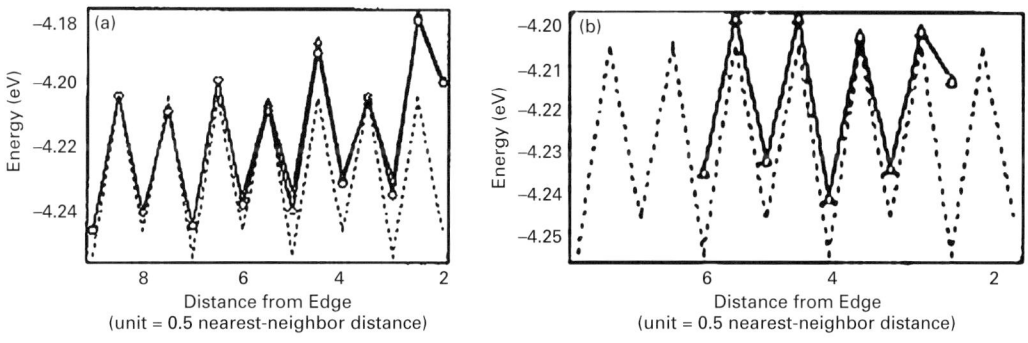

Fig. 6.54 Energy profile near edge for Pt atom on Pt(111). (a) At B-type edge of 7×7 (triangles) and 10×10 cluster. (b) At A-type edge of 7×7 island. Dashed curves indicate energy on a flat surface (after Wang and Fichthorn [118]).

0.51 eV for a 4×4. They found exchange to be a much more probable mechanism for descent, but also that exchange might be enhanced by momentum transfer between an impinging atom and the island.

In the same year Stoltze [41] used effective medium theory to look at platinum steps. He calculated the energy required for moving over different kinds of steps. On Pt(100), an energy E_B of 0.755 eV was needed to jump over the <110> step, 0.066 eV higher than for jumps on the flat surface. Exchange over this step was slightly more demanding at $E_B = 0.797$ eV. Jumps in the [001] direction on Pt(110) required $E_B = 0.944$ eV, while exchange occurred over a barrier of 1.294 eV. Exchange along the $[1\bar{1}0]$ direction cost a bit less, $E_B = 0.782$ eV. Movement on the plane in the $[1\bar{1}0]$ direction proceeded over an energy barrier E_D of 0.42 eV while in the [001] direction diffusion took place over a barrier of 0.945 eV; the latter value was the same as for jumping down to a lower level in the [001] direction. According to Stoltze, jumping over a step edge was more favorable on Pt(111) than exchange, $E_B = 0.363$ eV compared with 0.470 eV for exchange. Both values were much higher than for movement on the terrace, which needed only $E_D = 0.159$ eV. It should be noted, however, that this value was much smaller than what has been obtained in experiments.

Monte Carlo estimates of the diffusion of platinum atoms on Pt(111) were made by Jacobsen et al. [120] in 1995 also using effective medium theory. Their estimates of the energy variations as an atom approaches and then crosses a lattice step are shown in Fig. 6.55. The change in potential two spaces away, as an adatom approaches a descending step, is clear from this figure. Incorporation proceeded by an atom jumping over a step of type A and by exchange over B-type steps. For the self-diffusion barrier E_D on Pt(111) they estimated a value of 0.16 eV, which they recognized as too low. For motion over the edge at an A-type step they found a barrier E_B of 0.41 eV, compared to an exchange barrier $E_B = 0.37$ eV over a B-type step. The lowest activation energy for incorporation, of $E_K = 0.26$ eV, was found at a kink site in a B-type step.

Li and DePristo [42,43] using the corrected effective medium theory, CEM, obtained the barriers to self-diffusion on Pt(111) surfaces. Movement over B-type step edges are

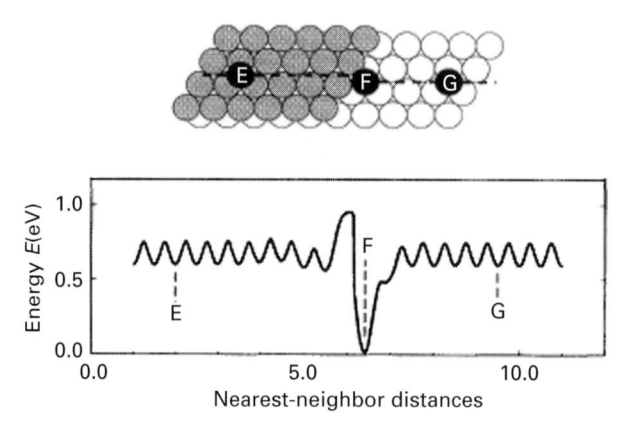

Fig. 6.55 Platinum atom diffusion over Pt(111) clusters. Top: Model for diffusive atomic motion over step. Bottom: EMT calculations of energy changes accompanying above movement (after Jacobsen et al. [120]).

listed in Table 6.2. For descent of platinum by exchange they obtained a value for overcoming a straight step ($E_B = 0.044$ eV) and kink ($E_K = 0.012$ eV) lower than for movement on the terrace. These findings have to be treated carefully since most of them do not agree with any other results. They also showed that for platinum the incorporation barrier was smaller for nineteen atom clusters than at kinks in the steps.

Ab initio simulations of self-diffusion of platinum atoms over step edges on Pt(111) were made by Feibelman [121] using the Vienna simulation package (VASP) [122–124]. The atoms preferred sitting in fcc sites on the surface by 0.21 eV, rather larger than the >0.06 eV found in experiments [125]. Diffusion on a (111) terrace in this study involved overcoming a barrier E_D of 0.29 eV. For jumping over the edges the energy E_B required was 0.53 eV at an *A*-type step and 0.80 eV at a *B*-type step. Exchange down an *A*-type step involved overcoming a barrier E_B of 0.31 eV, whereas down a *B*-type step the barrier proved to be higher, 0.64 eV. The step edge barrier E_a, according to Feibelman, was therefore 0.02 eV for an *A*-type step and interlayer movement proceeded preferentially via step *A*. The mechanism for descending *A*- and *B*-type edges is, of course, the reverse of what had previously been found in experiments for hetero-descent on Ir(111).

Self-diffusion on small islands on Pt(111) has been the topic examined by Máca et al. [71,72,126] using RGL interactions between atoms. For platinum on Pt(111) the diffusion barrier E_D proved to be only 0.17 eV, much lower than the value of 0.26 eV from experiments. For jumps over the straight edges the activation energy E_B was found to be similar for the two types: 0.53 eV for an *A*-type of step and 0.52 eV for *B*-type. For exchange at straight *B*-type steps the barrier E_B was considerably smaller, 0.40 eV. The activation energy E_B for exchange on *A*-type steps was also lower than for jumps: 0.50 eV. The presence of kinks seemed to influence interlayer transport. The energy E_K for descent by exchange over type *B* steps with kinks was as low as 0.26 eV, but only 0.42 eV on type *A*.

Máca et al. also looked at descent from 3 × 3 and 6 × 6 islands on Pt(111). For small Pt islands no change was found in the barrier for jumping, but there were huge differences

6.2 To descending step edges

for exchange. In the case of 3 × 3 islands exchange required E_B = 0.26 eV for type A steps and 0.13 eV for type B. The values for 36 atom islands were between those for 3 × 3 island and long straight edges: E_B = 0.30 eV for A-type steps and 0.16 eV for B-type. This reduction as the size becomes smaller was stronger at B-type edges than at edges of type A. What is of interest here is that transitions down steps are likelier at B-type steps than at steps of type A, the opposite of what has been found by Feibelman [121]. It must be noted, however, that the results of Máca et al. differ significantly from the experimental work of Kyuno [111], which yielded rather larger values.

Leonardelli et al. [127] made calculations with EAM potentials for platinum diffusion over steps on Pt(111). They observed that in considering the transition state, Pt sank deeper into the surface, creating four fold metastable adsorption places which lowered the energy barrier for diffusion E_D to 0.046 eV. The metastable four fold adsorption places were not observed in direct experiments by Gölzhäuser and Ehrlich [125] where two kinds of adsorption sites (fcc and hcp) were clearly distinguished. According to Leonardelli nickel atoms, however, deposited on the surface did not create such metastable sites and the activation energy E_D was around 0.178 eV. For descent over A-type steps by a platinum adatom at an fcc site, the activation energy for exchange E_B was 0.29 eV; it was 0.28 eV if the atom at first was in an hcp site. For movement over B-type edges the barriers were smaller, E_B = 0.17 eV if starting from an fcc site and 0.19 eV from hcp. Leonardelli arrived at a step edge barrier of 0.08 eV for B-type edges and 0.20 eV for A-type steps. All these estimates were for (4,3,3) and (4,4,3) vicinal surfaces. The barriers for jumps rather than atom exchange proved much higher and are therefore not given. At straight steps of a four row stripe-shaped island the barriers were also higher. From an fcc site at an A-type edge the activation energy E_B proved to be 0.36 eV, and from an hcp site 0.34 eV. For a B-type edge from an fcc site a barrier of 0.19 eV had to be overcome, and 0.22 eV from an hcp site. It is interesting that the barrier for exchange was higher for stripe islands but the barrier for hopping was exactly the same.

Leonardelli et al. [127] also did estimates for the behavior of nickel atoms on Pt(111). For vicinal surfaces, for an atom coming from an fcc site and encountering an A-type edge, the barrier E_B was 0.37 eV, versus 0.39 eV when originally at an hcp site. For B-type edges the barriers E_B were comparable, 0.32 eV and 0.30 eV, and the step-edge barrier E_a proved to be 0.09 eV for B-type steps and 0.17 for A-type steps. At straight edges of four row-striped islands the activation energy was somewhat higher, E_B = 0.41 eV and 0.45 eV at A-type steps and 0.35 eV versus 0.33 eV at B-type steps. Again exchange was more favorable than hopping. Comparisons with experiments are not available here. Descent at a kink site was less demanding than descent at straight steps. For both atoms, Pt and Ni, the descent at a kink in step B had the lowest barrier, lower than for movement of the adatom on the flat surface. However the strange observations of Pt atoms in four fold sites suggest that the data may not be reliable.

Information about descending platinum steps is quite wide, but disagreement between direct measurements and theoretical calculations should be clarified. Also quite surprising is that theory predicted the exchange mechanism on both types of steps.

6.2.20　Gold

There are only a few investigations about descending motion over gold steps, started by Stoltze [41] with effective medium theory. For jumps over <110> steps at the gold (100) plane, an adatom needed $E_B=0.546$ eV, while the value for exchange was comparable at 0.528 eV. Interlayer motion was slightly more demanding than movement on the terrace, which required $E_D=0.490$ eV.

For Au(110), Stoltze looked at atom interlayer jumping in the [001] direction and found a value E_B of 0.675 eV; for exchange at this step the barrier is slightly higher, 0.691 eV. Interlayer exchange in the [1$\bar{1}$0] direction required $E_B=0.582$ eV. Jumping on the Au(110) terrace in the [1$\bar{1}$0] direction was much less demanding at $E_D=0.265$ eV, but jumps along [100] required 0.67 eV. Exchange on the terrace in the [100] direction is more probable, occurring over 0.554 eV.

On the Au(111) only one type of step was investigated and it turned out that jumping was more favorable than exchange, with a barrier E_B of 0.260 eV compared to 0.326 eV. Movement on the flat (111) surface also proceeded by hopping with an energy E_D of 0.102 eV.

Li and DePristo [42,43] calculated barriers to surface diffusion with the corrected effective medium theory, CEM, in order to explain homoepitaxial growth. The barriers to surface diffusion over B-type step edges are listed in Table 6.2. Movement on the flat surface required $E_D=0.029$ eV, overcoming a step cost more, $E_B=0.090$ eV, but stepping down at the kink involved a barrier E_K of 0.047 eV, while descending from a 19 atom island took 0.034 eV. These findings have to be treated with caution.

The interlayer movement of adatoms over gold steps needs extensive investigation – it is impossible to draw any conclusions based on the data available.

6.2.21　Lead

Zhu et al. [33] carried out density-functional estimates of lead atoms moving over lattice steps on Pb(110). The adatom undergoes exchange with the lattice, but this involves an activation energy of 0.46 eV for an in-channel ascent, and 0.37 eV for the reverse process. For a cross-channel ascent, already shown in Fig. 6.21, the barrier was 0.43 eV, with a barrier of the same height for descent.

An investigation of lead steps was also done by Li et al. [128] in 2006. They compared the growth of a cluster with the decay of a vacancy island and found a new method to determine the step-edge barrier. From STM observations of Pb on Pb(111) they deduced an additional step-edge barrier E_a of 0.083 ± 0.060 eV. They named this an effective barrier, since it was an average for descending A-type steps, B-type steps, as well as at kinks. Based on EAM calculation they derived the barrier E_a for hopping over an A-type step as 0.1 eV via a rolling-over motion (jump), and the barrier over B steps as 0.03 eV via concerted motion, probably exchange. The effective barrier derived from experiments lies between the EAM values. The mechanism of descent is also in agreement with what was found for Ir(111) – A-type hopping and B-type exchange.

6.3 Atom lifetime versus the step-edge barrier

Having examined the energetics of atom incorporation at steps it is now worthwhile to briefly look at the detailed kinetics of adatom incorporation and its dependence on the potential landscape near the descending step. The effects of step-edge barriers on diffusion were discussed in 1997 by Kyuno and Ehrlich [129], who envisioned the edge potentials shown in Fig. 6.56. The lifetime τ_i of an atom at site i before escaping is described by

$$\tau_i = 1/(\lambda_i + \mu_i) + \lambda_i/(\lambda_i + \mu_i)\tau_{i+1} + \mu_i/(\lambda_i + \mu_i)\tau_{i-1}; \quad l < u. \tag{6.5}$$

Here λ_i represents the rate of jumping to the right and μ_i the rate to the left for an atom at position i. The rate of escape, either from site l or u, will depend upon the rate $\nu_{0B}\exp(-E_B/kT)$ at which the final barrier E_B is surmounted. However, atoms can also move back into the plane interior, giving for the rate of escape from an interior site

$$1/\tau_x \sim \nu_{0B}\exp(-E_B/kT)/\{1 + b - 1/2) \times \exp[-(E_R - E_D)/kT]\}, \tag{6.6}$$

where b indicates the width of the plane and the barrier E_R preventing an adatom from returning to the center is indicated in Fig. 6.56. When next to the edge there is a trap which keeps atoms in a deep asymmetric well, then the rate of return j_R will be negligible and the rate of escape can be described by

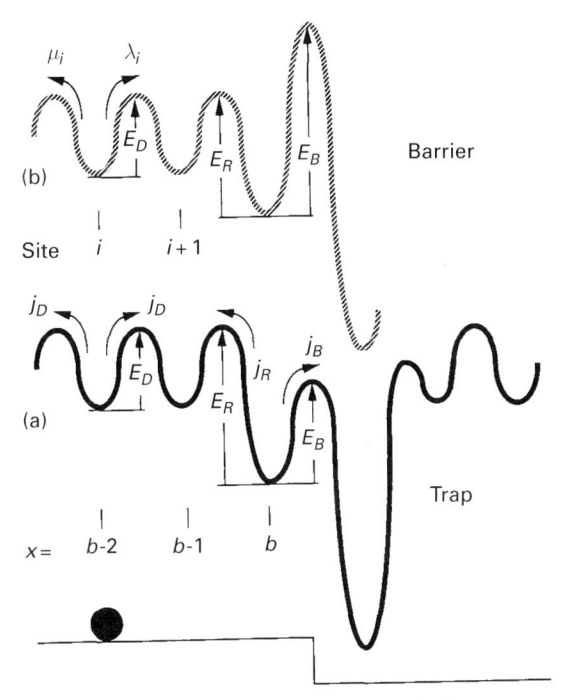

$$1/\tau_x \sim j_B = \nu_{0B}\exp(-E_B/kT). \tag{6.7}$$

Fig. 6.56 Energy barrier E as well as jump rates j in diffusion over a descending step [129]. (a) Potential for trap at end. (b) For repulsive barrier with adjacent deep well.

Another extreme case occurs when there is a repulsive barrier at the edge but the rate of return jumps exceeds the jump rate in diffusion, $j_R \gg j_D$. Atoms at the edge equilibrate with the interior and the rate of escape can be written as

$$1/\tau_x \sim [v_{0B}/(b-1/2)] \exp[-(E_B - E_R + E_D)/kT]. \tag{6.8}$$

It is clear that the lifetime of atoms before incorporation depends upon the detailed energetics close to the edge. In Fig. 6.57 we present a comparison of the effect of a conventional reflective barrier, a trap as in Fig. 6.56b and a barrier with a trap shown in Fig. 6.56a on the rate of adatom escape from a plane with a radius of seven spacings. The plots are from Monte Carlo simulations with an input energy for terrace diffusion of $E_D = 0.3$ eV, a step-edge barrier of $E_B = 0.52$ eV, and a standard frequency factor 10^{12} sec^{-1}. From the plot it is clear that the presence of a trap next to the step influences the rate of adatom escape, as well as the energy and prefactor derived from an Arrhenius plot.

Up till now the barrier to diffusion over the step edge has been emphasized. However, barriers are also possible away from the edge, for example on Pt(111) and Ir(111), and such barriers influence the rate of incorporation on those planes. The influence of such interior barriers on the lifetime of an atom was investigated by Oh et al. [104]. On a one-dimensional plane the lifetime of an atom in position i has already been given by Eq. (6.5). For an atom deposited at the center of a 1D cluster without additional traps and barriers it was found that the lifetime τ_c was

$$\tau_c = (b_2 + 1)^2/(2j_D), \tag{6.9}$$

where b_2 gives the size of the cluster. For a traditional reflective barrier at the edge, shown in Fig. 6.58a, the lifetime was given by two terms,

$$\tau_c = b_2^2/(2j_D) + \left(b_2 + \frac{1}{2}\right)/j_{B_2}, \tag{6.10}$$

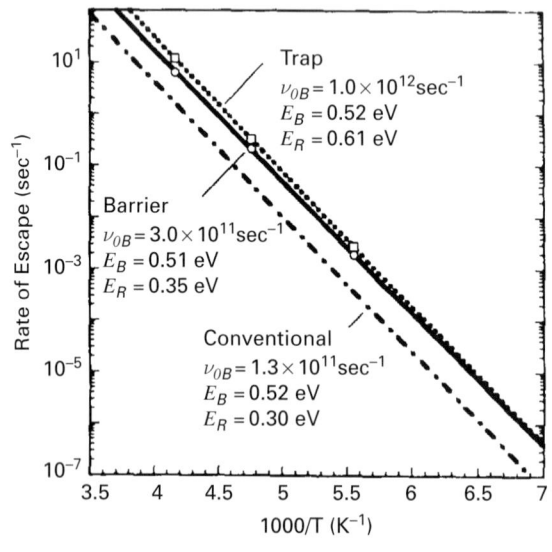

Fig. 6.57   Dependence of atom incorporation rate at descending step upon the reciprocal temperature [129]. Trap corresponds to potential curve in Fig. 6.56a, Barrier to Fig. 6.56b, Conventional to Fig. 6.58a.

6.3 Atom lifetime versus the step-edge barrier

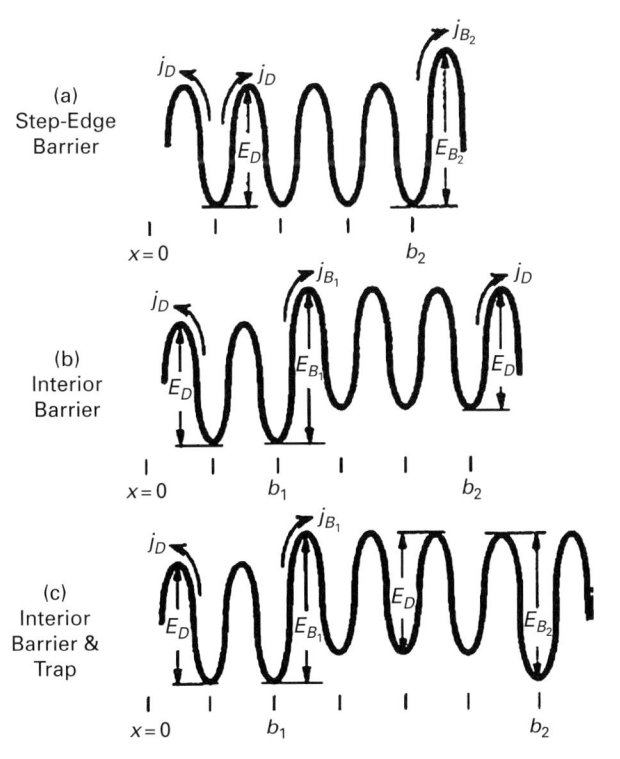

Fig. 6.58 Possible potentials at edge of 1D cluster [104]. Only right half of cluster is shown.

where the first term described the time to reach the edge site, the second the lifetime for incorporation starting at that point. The influence of the size of the cluster on the lifetime is shown in Fig. 6.59, where the lifetime to incorporation increases by a factor of ~ 30 due to the presence of the step-edge barrier.

Barriers not at the step edge but in the interior have a more dramatic effect. What has to be taken into account is that an atom, after overcoming the interior barrier, still has a chance to return and continue movement in the interior. The lifetime is now described by

$$\tau_c = b_1^2/(2j_D) + \left(b_1 + \frac{1}{2}\right)/j_{B_1} + \tau_{b_1+1}, \tag{6.11}$$

where the additional term τ_{b_1+1} gives the lifetime starting just beyond the interior barrier, which was given by

$$\tau_{b_1+1} = (b_2 - b_1)[(b_2 - b_1 + 1)/(2j_D) + (b_1 + 1/2)/j_{B_1}]. \tag{6.12}$$

Investigated was the influence of an interior barrier at different positions on a cluster of 20 atomic spaces. For starts at the center, the lifetime increased more than 5 times depending on the location of the interior barrier, as shown in Fig. 6.60. The presence of an outermost trap in the cluster (Fig 6.58c) in addition to the interior barrier is not felt strongly by an atom and the lifetime overlaps with the lifetime for a single interior barrier, shown in Fig 6.60 by solid circles. However, the additional step-edge barrier has

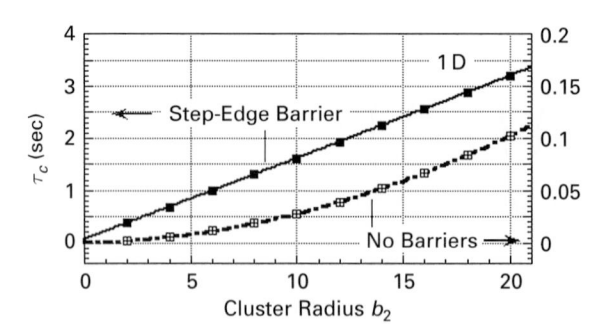

Fig. 6.59 Lifetime τ_c for atom starting at center to incorporate into step of 1D cluster of atoms [104]. Diffusion barrier 0.26 eV, step-edge barrier 0.33 eV, $T = 140$ K.

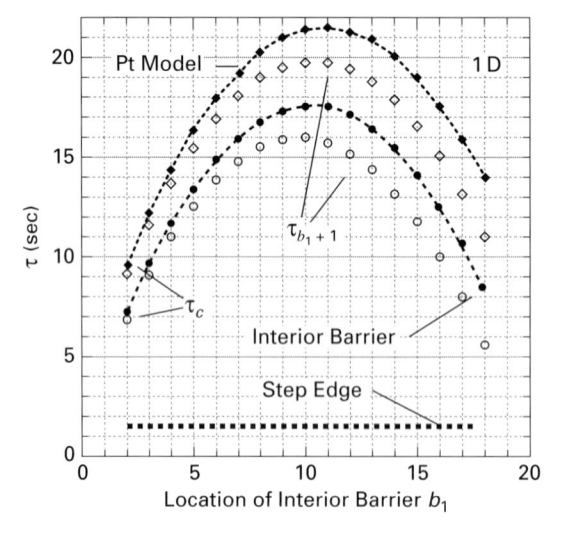

Fig. 6.60 Effect of interior barrier at b_1 on lifetime for incorporation on a 1D cluster with radius of 20 nearest-neighbor spacings. [104]. Solid points give lifetime starting at center. Dashed horizontal indicates lifetime just in presence of step-edge barriers.

significant influence on the lifetime of the atom, indicated by solid diamonds (or by Pt model) in Fig 6.60; the lifetime is around 20% higher. For atoms deposited at random positions, which is a more realistic picture for vapor deposition, the lifetime is given in Fig. 6.61, with the position of the interior barrier shifting in the direction of the step edges, since atoms deposited close to the edge have a shorter lifetime than when deposited at the center. For interior barrier positions between 12 and 2 the random start can lower the lifetime by one fifth or more. The effect is the strongest for interior barrier positions midway between the center and edges of the cluster.

The lifetime was also investigated for a 2D cluster. On such a cluster with a reflective step-edge barrier the lifetime, averaged over all positions, is described by

$$<\tau> = (b_2 + j_D/j_{B_2})b_2/(2j_D). \qquad (6.13)$$

6.3 Atom lifetime versus the step-edge barrier

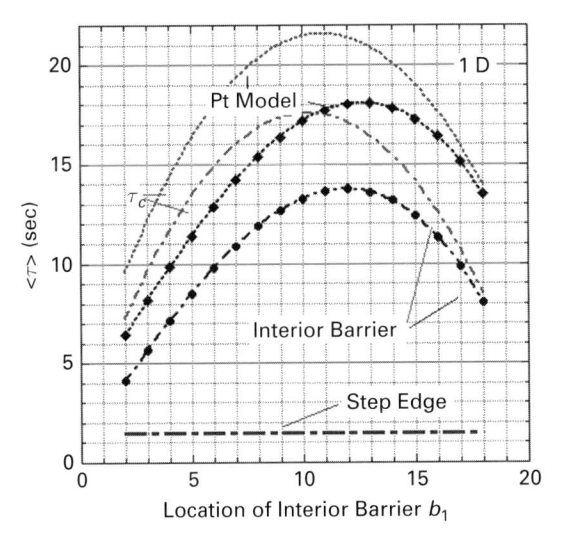

Fig. 6.61 Lifetimes for atoms starting at random on 1D cluster with radius of 20 nearest-neighbor spacings compared with lifetime starting in center (in grey) [104].

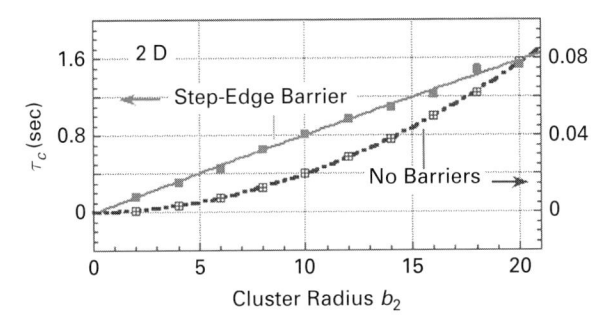

Fig. 6.62 Comparison of lifetime of atom starting at center on a 2D circular cluster with and without a step-edge barrier [104]. $E_{B_2} = 0.33$ eV, $E_D = 0.26$ eV, $T = 140$ K.

The dependence of the lifetime, with and without a step-edge barrier, on the cluster size is given in Fig. 6.62. On the circular cluster the lifetime is roughly half that on the 1D cluster and follows the same trend. It is clear the lifetime is influenced by both the step-edge barrier and the cluster size. For a cluster of 20 atomic spacings an interior barrier significantly influences the lifetime of atoms deposited at the center, as in Fig. 6.63 An interior barrier between 4 and 18 increases the adatom lifetime at least three fold compared to the lifetime with only a step-edge barrier. For random starts of the atom the lifetime is cut by more than 25% when the interior barrier is located at < 14, compared with central starting, as is clear in Fig. 6.64. However, the interior barrier still increases the overall lifetime of atoms. On 2D clusters lifetimes are smaller than on 1D clusters, but the qualitative trends are the same.

The presence of interior barriers has an effect not just on atom lifetime for incorporation. Oh et al. [130] also showed such barriers increase the rate of dimer nucleation.

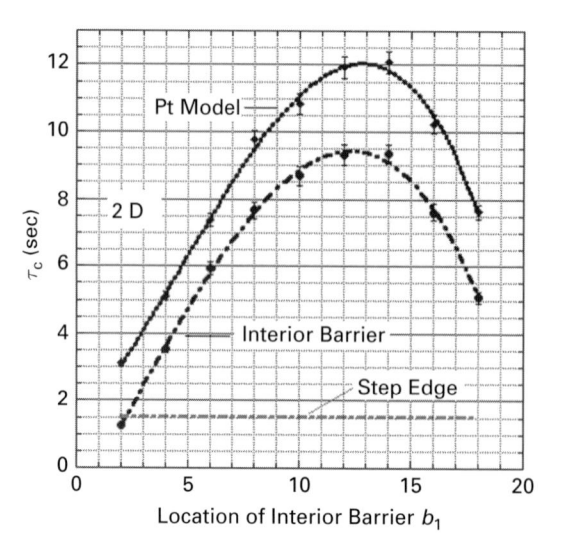

Fig. 6.63 Effect of interior barrier on lifetime to incorporation starting from the center of a 2D cluster with a radius of 20 nearest-neighbor spacings [104].

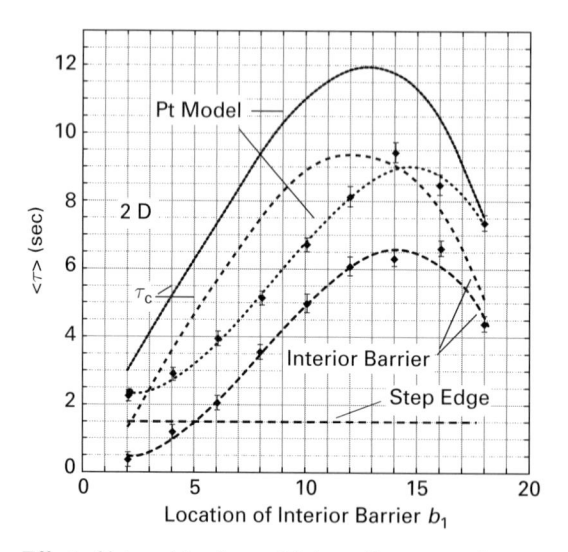

Fig. 6.64 Effect of internal barrier on lifetime of atoms starting at random positions on a 2D cluster [104].

Fig. 6.65 gives the fraction nucleated on a square of half-width 20, with an internal barrier of 0.33 eV and a diffusion barrier of 0.26 eV. As the internal barrier is moved toward the plane edges, there is a very significant increase in nucleation, as collisions between adatoms increase. The interior barrier together with the step-edge barrier cause a faster decrease in the rate of nucleation at increasing temperature. What emerges from these studies is that the detailed energetics over a cluster have a really noteworthy effect on diffusion and also on surface reactions.

6.4 Comparisons

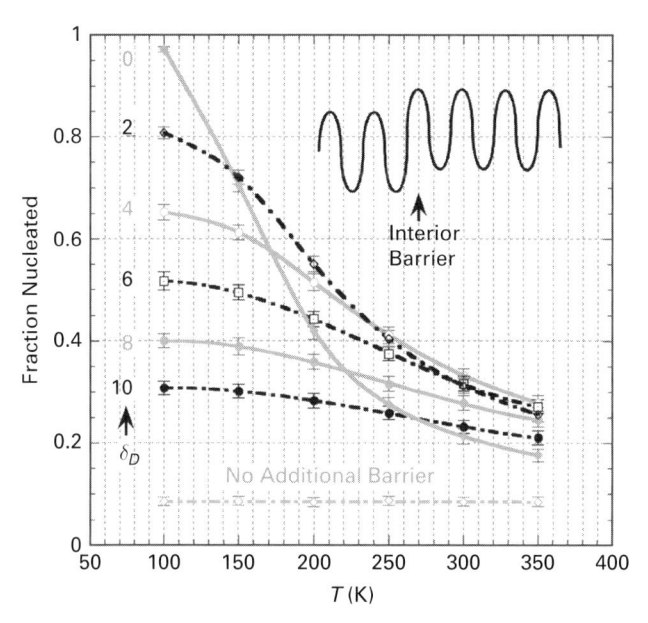

Fig. 6.65 Effect of interior barrier on nucleation of dimers on square (100) plane with half-width 20 [130]. Interior barrier height 0.33 eV, diffusion barrier 0.26 eV, no additional barrier at the step edge; δ_D gives distance between interior barrier and step edge.

6.4 Comparisons

The amount of experimental data available for step-edge barriers is not small, but not all of it is consistent. It is therefore useful to summarize some of the results in Tables 6.5 through 6.7. What is clear from the interpretation of growth experiments for Ag(111) in Table 6.5 is that the results for the additional step-edge barrier E_a are all reasonably close together, regardless of the particular analysis, except for the work of Roos et al. [79], which is drastically lower. Unfortunately, there are no direct studies of atom loss over step edges available for comparison.

The situation is different for studies of the step-edge barrier on Pt(111), given in Table 6.6, where deductions are listed from experiments. These are interesting for their considerable spread. Markov's [108] early findings are unusual as they are so large compared to other results. It is not clear that such a spread in the additional step-edge barrier is connected with the different experiments analyzed or just due to a different method of analysis. Noteworthy is the small barrier found in direct observations of atom loss from a Pt(111) plane [111]. It is worth noting that the direct observations by Kyuno and Ehrlich took into account the existence of an interior barrier on this plane. The other findings considered only a traditional step-edge barrier. As shown by Oh et al. [104], an interior barrier can have significant influence on the lifetime of an atom on the cluster and will influence the growth characteristics. However, it is still not clear if interior barriers will be present for straight steps. Interior barriers might be responsible for obtaining different values of the additional step-edge barrier for the

Table 6.5 Summary of step-edge barriers E_a for self-diffusion on Ag(111) derived from experiments.

Reference	Step-edge barrier E_a(eV)
Vrijmoeth et al. [73]	0.150 ± 0.030
Smilauer and Harris [35]	0.13 – 0.23
Meyer et al. [74]	$0.150^{+0.040}_{-0.020}$
Bromann et al. [75]	0.120 ± 0.015
Markov [45]	0.15
Roos et al. [79]	0.040
Morgenstern et al. [85]	0.13 ± 0.04

Table 6.6 Step-edge barriers derived in experiments for self-diffusion on Pt(111).

Reference	Step-edge barrier E_a (eV)
Smilauer and Harris [35]	0.15–0.27
Meyer et al. [74]	$0.165^{+0.050}_{-0.020}$
Markov [108]	0.31, 0.44
Markov [45]	≈ 0.21
Kyuno and Ehrlich [111]	0.06 ± 0.01
Krug et al. [112]	0.12

Table 6.7 Total calculated barrier for escape of Pt from Pt(111).

Reference	Step A	Step B	Kink A	Kink B
Villarba and Jónsson [116]	0.30 (Ex)	0.18 (Ex)	0.08 (Ex)	0.06 (Ex)
Jacobsen et al. [120]	0.41(Jump)	0.37 (Ex)		0.26 (Ex)
Li and DePristo [42]		0.044 (Ex)		0.012(Ex)
Feibelman [121]	0.31 (Ex)	0.64(Ex)		
Máca et al. [126]	0.50 (Ex)	0.40(Ex)	0.42(Ex)	0.26(Ex)
Leonardelli et al. [127]	0.28(Ex)	0.17(Ex)		

Total escape barrier E_B (eV)

same material on analyzing different experiments. That is why the detailed analysis of experiments is so important. There is as large a difference in the various calculated barriers shown in Table 6.7 for crossing a step edge on Pt(111), but some of the simpler efforts are in excellent agreement with experiments. Theory clearly distinguishes between types of steps and suggests an acceleration of interlayer movement at kinks, which has not been observed experimentally so far. The mechanism of descent is still not uncovered; however, exchange seems to be the most likely.

Theoretical results for the barrier to atom incorporation on Ir(111) are in reasonable agreement, and the experimental work is in good agreement with the results for Pt(111). Unfortunately, there is a factor of two difference between the barriers from theory and experiment. Surprisingly, theory is quite good at predicting the value for the diffusion barrier on a flat surface for this system and clearly predicts easier descent at kinks, a fact not yet confirmed by experiment. There still stays open a question about the mechanism of descent on Ir(111). Exchange was shown experimentally to be involved in descent of a hetero-atom from steps of type B. Unfortunately, experiments cannot distinguish between a jump and exchange for self-diffusion, while theory offers a confusing image for B-type steps, since one study suggests exchange as the leading mechanism [106] while another favors a simple jump [107].

There is a wealth of theoretical investigations for Cu(111) but no reliable experimental values have been derived so far for additional step-edge barriers, they are scattered from 0.113 to 0.224 eV. Scatter in theoretical values is in the same range from 0.04 to 0.14 eV for steps of type B and from 0.15 to 0.4 for A-type steps. For B steps, theory seems to agree that descent will proceed by exchange, while for steps A the situation is not resolved.

Steps for fcc(100) surfaces have received less attention from both experiment and theory. There is no clear understanding if descent would proceed by hopping or exchange. Half of the investigations prefer the latter, the other half the former. Most investigated is the densely packed <110> step. Descent over the <100> step is usually associated with a negative barrier; in practice this means that if the temperature is high enough to promote movement, such steps will not exist at the surface.

On fcc(110) surfaces the easier descent is usually in the direction along the channels of this plane, in most cases by atom jumps.

The calculational efforts have been useful for revealing interesting features, but it is clear that detailed experimental information is really needed to put diffusion over step edges on a sound footing. The mechanism of descent has not always been clear. Quite interesting is the prediction of a negative step-edge barrier; this finding will definitely have to be verified by experiment. A survey of the available data on step-edge barriers is provided in Table 6.8.

6.5 Atom descent over many-layered steps and between facets

Up to this point, atoms have been assumed to descend over monoatomic steps. However, this is not always the case. It is possible that the descending movement over multilayer steps or over whole facets has to be taken into account as well in the detailed description of interlayer movement. Here we will consider a few investigations tackling this problem. Unfortunately study in this area is limited to

Table 6.8 Additional step-edge barriers.

System	Type of step	Mechanism	Experiment E_a (eV)	Theory E_a (eV)	Method	Ref.
Al/Al(111)	$A^§$	Ex/Jump		0.04/0.41	DFT LDA	[27–29]
		Ex		$0.26^\&$	MS	[32]
Al/Al(111)	$B^§$	Ex/Jump		0.02/0.29	DFT LDA	[27–29]
		Ex		$0.02^\&$	MS	[32]
	<110>/{110}	Ex		$0.02^\&$	MS	[32]
Al/Al(100)	<100>/{100}	Ex		$\sim 0.10^\&$	MS	[32]
	<100>/{110}	Ex		$\sim 0.09^\&$	MS	[32]
	<100>/{111}	Ex		$\sim 0.20^\&$	MS	[32]
Al/Al(110)	<001>	Ex		$\sim 0.13^\&$	DFT	[33]
	<100>/{100}	Ex		$\sim 0.33^\&$	MS	[32]
	<110>	Ex		$\sim 0.41^\&$	DFT	[33]
	<110>/{111}	Ex		$\sim 0.4^\&$	MS	[32]
	<111>/{110}	Ex		$\sim 0.03^\&$	MS	[32]
Fe/Fe(100)			0–0.1		Nucl	[35]
			0.03–0.06		MC RHEED	[36]
			0.03–0.04		Sim	[37]
Fe/Fe(110)	<111>		0.41		STM MC	[38]
	<110>		0.34		STM MC	[38]
	<001>		0.36		STM MC	[38]
Ni/Ni(111)		Jump/Ex		0.461/0.601	EMT	[41]
	A	Ex/Jump		0.494/0.529	EAM	[40]
		Ex		0.18	EAM	[12]
	B	Ex/Jump		0.113/0.449	EAM	[40]
		Ex		0.058	CEM	[42,43]
	BK	Ex		0.031	CEM	[42,43]
Ni/Ni(100)	<110>	Ex/Jump		0.22/0.28	EAM	[40]
		Jump/Ex		0.229/0.332	EMT	[41]
		Ex/Jump		0.113/0.289	EMT	[44]
		Jump		0.34	EAM	[12]
	<110> K	Jump		0.09	EAM	[12]
		Ex/Jump		−0.109/0.064	EMT	[44]
Ni/Ni(110)	<001>	Ex/Jump		0.3/0.36	EAM	[12]
		Jump/Ex		0.273/0.403	EMT	[41]
		Ex/Jump		0.87/0.93	EAM	[40]
	<110>	Ex/Jump		0.51/0.771	EMT	[41]
		Jump		0.94	EAM	[12]
Cu/Cu(111)			0.113		Nucl	[45]
			0.080–0.055		Nucl	[45]
			0.116 ± 0.002		STM	[48]
			0.224 ± 0.009		STM	[49,50]
			0.19 ± 0.002		Rean STM	[48]
		Jump/Ex		0.312/0.397	EMT	[41]
		Ex		0.057	2nd Moment	[51]
	K	Ex		~ 0.62–$0.81^\&$	DFT	[62]
	A	Jump/Ex		0.4/0.43	F-S	[55]
		Ex/Jump		0.251/0.481	EAM	[57]

6.5 Atom descent over many-layered steps and between facets

Table 6.8 (cont.)

System	Type of step	Mechanism	Experiment E_a (eV)	Theory E_a (eV)	Method	Ref.
		Jump		0.15	MC	[61]
		Ex/Jump		0.24/0.46	EAM	[64]
				0.3	MD EAM	[63]
	AK	Ex/Jump		0.171/0.371	EAM	[57]
	B	Ex		0.137	CEM	[42,43]
		Ex/Jump		0.057/0.462	EAM	[54]
		Ex/Jump		0.14/0.4	F-S	[55]
		Ex/Jump		0.056/0.471	EAM	[57]
		Jump		0.38	MC	[61]
		Ex/Jump		0.04/0.46	EAM	[64]
				0.06	MD EAM	[63]
	BK	Ex/Jump		0.281/0.361	EAM	[57]
		Ex		0.083	CEM	[42,43]
Fe/Cu(111)	B	Jump		0.148	DFT	[67]
		Ex		0.045	DFT	[66]
Cu/Cu(100)	<110>	Jump/Ex		0.147/0.201	EMT	[41]
		Ex/Jump		0.20/0.36	EMT	[68]
		Jump		~0.11[&]	F-S	[55]
		Jump/Ex		0.179/0.232	EMT	[44]
		Ex/Jump		0.13/0.32	RGL	[69]
		Jump		0.185	EAM	[70]
		Ex/Jump		0.05/0.28	EAM	[57]
		Ex/Jump		0.03/0.29	EAM	[54]
		Ex/Jump		0.09/0.33	EAM	[64]
		Ex		0.03	2ndMoment	[51]
	<110> K	Ex/Jump		−0.15/0.08	EAM	[57]
		Jump/Ex		~0[&]/~0.12[&]	F-S	[55]
		Ex/Jump		0.023/0.043	EMT	[44]
	<001>	Ex/Jump		−0.16/0.08	EAM	[57]
		Ex/Jump		−0.16/0.07	EAM	[64]
Cu/Cu(110)	<110>	Ex/Jump		0.48/0.93	EAM	[54]
		Jump		0.414	EAM	[70]
		Ex/Jump		0.398/0.548	EMT	[41]
		Ex		~0.57[&]	DFT	[33]
	<001>	Ex		~0.29[&]	DFT	[33]
		Jump/Ex		0.188/0.282	EMT	[41]
Co/Cu(110)	<110>	Ex		0.34	MS	[131]
	<001>			0.28	MS	[131]
Rh/Rh(111)	A	Ex/Jump		0.32/0.58	RGL	[71,72]
	AK	Ex		0.32	RGL	[71,72]
	B	Ex/Jump		0.24/0.59	RGL	[71,72]
	BK	Ex		0.09	RGL	[71,72]
Pd/Pd(111)		Jump/Ex		0.202/0.374	EMT	[41]
	B	Ex		−0.004	CEM	[42,43]
	BK	Ex		−0.029	CEM	[42,43]
Pd/Pd(100)	<110>	Jump/Ex		0.070/0.195	EMT	[41]

Table 6.8 (cont.)

System	Type of step	Mechanism	Experiment E_a (eV)	Theory E_a (eV)	Method	Ref.
Pd/Pd(110)	<001>	Jump/Ex		0.054/0.292	EMT	[41]
	<110>	Ex/Jump		0.411/0.413	EMT	[41]
Ag/Ag(111)			0.150 ± 0.030		STM	[73]
			0.13–0.23		Nucl	[35]
			$0.15^{+0.040}_{-0.020}$		STM	[74]
			0.120 ± 0.015		STM	[75]
			0.15		Nucl	[45]
			0.12		Nucl	[45]
			0.04		Nucl	[79]
			0.13 ± 0.04		STM	[85]
		Jump/Ex		0.295/0.391	EMT	[41]
	A	Ex/Jump		0.29/0.42	CEM	[42]
			0.16		STM	[86]
	AK	Ex		0.17	CEM	[42]
	B	Ex/Jump		0.044/0.41	CEM	[42,43]
			0.08		STM	[86]
	BK	Ex		0.03	CEM	[42]
Ag/Ag island / Pt(111)			0.030 ± 0.005		STM	[75]
		No Inter		0.068 ± 0.001	MC	[88]
		Inter		0.074 ± 0.002	MC	[88]
Au/Ag(111)	A	Ex		−0.051 to −0.041	MD EAM	[92]
	B	Ex/Jump		−0.055/0.614	MD EAM	[92]
Ag/Ag(100)			0.03		Sim, STM	[95]
			0.07		Sim, STM	[96]
	<110>	Jump/Ex		0.085/0.194	EMT	[41]
		Jump/Ex		0.114/0.149	EMT	[44]
		Ex/Jump		~0/0.10	DFT GGA	[93]
		Ex/Jump		~0/0.18	DFT LDA	[93]
		Ex/Jump		0.12/0.16	RGL	[69]
		Jump/Ex		0.11/0.16	EAM FDB	[94]
		Ex/Jump		0.03/0.22	EAM VC	[94]
	<001>	Ex/Jump		−0.1/0.03	EAM FDB	[94]
		Ex/Jump		−0.17/0.07	EAM VC	[94]
	<110> K	Jump/Ex		0.034/0.037	EMT	[44]
Ag/Ag(110)	<001>	Jump/Ex		0.069/0.246	EMT	[41]
		Ex		~0.24$^{\&}$	DFT	[33]
		Jump		0.07	RGL	[97]
	<110>	Ex		~032$^{\&}$	DFT	[33]
		Ex/Jump		0.28/0.52	RGL	[97]
		Ex/Jump		0.314/0.355	EMT	[41]
	<110> K	Ex		0.10	RGL	[97]
Pd/W(110)			0.27 ± 0.02		FIM	[98]
Ta/W(110)			−0.05		FIM	[21,22]
			0.36^{Æ}		FIM	[21,22]
W/W(110)			~0.065		FIM	[16]

6.5 Atom descent over many-layered steps and between facets

Table 6.8 (cont.)

			Experiment	Theory		
System	Type of step	Mechanism	E_a (eV)	E_a (eV)	Method	Ref.
			0.03		FIM	[21,22]
			$0.51^{Æ}$		FIM	[21,22]
			−0.05		FIM	[23]
			0.228 ± 0.009^{æ}		FIM	[23]
			0.199 ± 0.009^{Æ}		FIM	[23]
Re/W(110)			−0.10		FIM	[21,22]
			$0.27^{Æ}$		FIM	[21,22]
			0.087		FIM	[23]
			0.145 ± 0.006^{Æ}		FIM	[23]
Ir/W(110)			0		FIM	[21,22]
			$\sim 0.54^{Æ}$		FIM	[21,22]
			0.297		FIM	[23]
			0.195 ± 0.012^{Æ}		FIM	[23]
Pt/W(110)			−0.02		FIM	[21,22]
			$0.52^{Æ}$		FIM	[21,22]
Pd/W(111)			0.61 ± 0.06		FIM	[98]
Re/W(321)			0.05		FIM	[19]
Pd/W(211)			0.29 ± 0.08		FIM	[98]
Re/Re(0001)		Jump	0.20 ± 0.028		FIM	[100]
W/Re(0001)		Jump	0.35 ± 0.021		FIM	[100]
Ir/Ir(111)			0.20 ± 0.03		FIM	[101]
	A	Jump/Ex		0.66/1.34	RGL	[107]
		Jump/Ex		0.77/1.55	RGL	[106]
	AK	Jump/Ex		0.63/1.30	RGL	[106]
		Jump/Ex		0.53/1.21	RGL	[107]
	B	Ex/Jump		0.54/0.78	RGL	[106]
		Jump/Ex		0.67/0.79	RGL	[107]
	BK	Jump/Ex		0.63/0.70	RGL	[106]
		Jump/Ex		0.53/0.61	RGL	[107]
	AV	Jump/Ex		0.51/0.58	RGL	[107]
	BV	Jump/Ex		0.51/1.13	RGL	[107]
Ir/Ir(100)		Ex	-0.02 ± 0.08		FIM	[101]
Rh/Ir(100)		Jump	0.04 ± 0.08		FIM	[101]
Ni/Pt(111)	A	Ex/Jump		0.17/0.44	EAM	[127]
	B	Ex/Jump		0.09/0.44	EAM	[127]
Pt/Pt(111)		Jump/Ex		0.204/0.311	EMT	[41]
			0.15–0.27		Nucl	[35]
			$0.165^{+0.05}_{-0.020}$		Nucl	[74]
			~0.21		Nucl	[108]
			0.06 ± 0.01		FIM	[111]
			0.10		FIM	[111]
				0.12	LAM	[112]
				0.15	TDT	[112,114]
	A	Jump		0.25	EMT	[120]
		Ex/Jump		$0.17^{hcp}0.22^{fcc}/$	EAM	[116]
				0.62–0.72		
			0.31		Nucl	[108]

Table 6.8 (cont.)

System	Type of step	Mechanism	Experiment E_a (eV)	Theory E_a (eV)	Method	Ref.
		Ex/Jump		0.02/0.24	VASP	[121]
		Ex/Jump		0.33/0.35	RGL	[71,72,126]
		Ex/Jump		0.20/0.74	EAM	[127]
	AK	Ex		0.00	EAM	[116]
		Ex		0.25	RGL	[71,72,126]
	B	Ex		0.10^{hcp}	EAM	[116]
			0.44		Nucl	[108]
		Ex		0.21	EMT	[120]
		Ex		−0.004	CEM	[42,43]
		Ex/Jump		0.35/0.51	VASP	[121]
		Ex/Jump		0.23/0.36	RGL	[71,72,126]
		Ex/Jump		0.08/0.73	EAM	[127]
	BK	Ex		−0.02	EAM	[116]
		Ex		0.10	EMT	[120]
		Ex		−0.036	CEM	[42,43]
		Ex		0.09	RGL	[71,72,126]
Pt/Pt(100)	<110>	Jump/Ex		0.066/0.108	EMT	[41]
Pt/Pt(110)	<001>	Ex		0.362	EMT	[41]
	<110>	Jump/Ex		0.524/0.874	EMT	[41]
Au/Au(111)		Jump/Ex		0.158/0.224	EMT	[41]
	B	Ex		0.061	CEM	[42,43]
	BK	Ex		0.018	CEM	[42,43]
Au/Au(100)	<110>	Ex/Jump		0.038/0.056	EMT	[41]
Au/Au(110)	<110>	Ex/Jump		0.41/0.426	EMT	[41]
	<001>	Ex		0.317	EMT	[41]
Pb/Pb(111)			0.083 ± 0.060		STM	[128]
	A	Jump		0.1	EAM	[128]
	B	Ex		0.03	EAM	[128]
Pb/Pb(110)	<001>	Ex		~0.17$^\&$	DFT	[33]
	<110>	Ex		~0.23$^\&$	DFT	[33]

$^\&$ The activation energy for terrace diffusion E_D used was based on values in Chapters 4 and 5; These are Al(100) 0.25 eV, Al(110) 0.3 eV, Al(111) 0.04 eV, Cu(100) 0.4 eV, Cu(111) 0.04 eV, Cu(110) 0.25 eV, Ag(110) 0.3 eV, Pb(110) 0.2 eV; ^{Æ} assumed kT/h; § A-type steps on the fcc(111) plane are sometimes designated by <110>/{100}, <110> being the orientation of the step and {100} its faceting; similarly B-type steps are sometimes shown as <110>/{111}; ^{œ} assumed $D = 3.6 \times 10^{-3}$ cm^2/sec

theoretical effort. As first shown by Valkealahti and Manninen [132] using effective medium theory, diffusion between facets of aluminum polyhedra occurred from (111) and (100) planes. For jumping between (100) and (111) facets, the static value of the energy was 0.47 eV while the dynamic one amounted to 0.48 eV. The opposite jumps, from (111) to (100), cost a bit less, only 0.28 eV for the static value and 0.30 eV for the dynamic estimate. Jumps between (111) planes had a comparable value, 0.28 eV for the static and 0.29 eV for the dynamic barrier. Exchange between two (111) facets

6.5 Atom descent over many-layered steps and between facets

turned out to be less demanding, requiring only 0.17 eV. They also looked at jumps with "pull," which means hopping across the edge with the help of a nearby facet. Such jumps with "pull" turn out to have quite a low energy. For moving between (111) facets only 0.11 eV, and between (111) and (100) facets with "pull" cost 0.31 eV. Chain exchange, between (111) facets, involving a row of atoms from the (100) plane, also had a very low energy, which increased with the number of atoms involved in a process. For three atoms, the energy required was only 0.08 eV, with four atoms 0.17 eV, and with a chain of five atoms 0.28 eV.

The diffusion mechanism and energetics between (111) and (100) facets of gold and silver were examined with RGL potentials using the nudged elastic band method by Baletto et al. [133]. They worked with two polyhedras with 1289 atoms (25 atoms on the (100) plane) and 201 atoms (9 atoms on the (100) plane). Adatom jumps from the (111) plane to the (100) cost 0.36 eV on Ag clusters and 0.34 eV on Au clusters. Exchange between the same planes required slightly less energy, 0.32 eV for silver and 0.24 for gold. Jumping between two (111) planes was comparable and cost 0.34 eV for silver and 0.29 eV for gold; exchange was more favorable, with 0.23 eV for silver and 0.15 eV for gold. They also observed a chain process which covered movement to (111) planes through the (100) plane, a process illustrated in Fig. 6.66a. This cost only 0.19 eV for silver and 0.21 eV for gold. Multiple exchange, named a "half-chain" process, shown in Fig. 6.66b where the adatom pushes some atoms in a row and an atom is emitted on the (100) plane a few atomic spaces apart from the original entrance of the adatom. Such a process cost 0.38 eV for silver and less, only 0.29 eV, for gold. Based on their finding they concluded that diffusion on and from (100) facets took place only at high temperatures due to the presence of low-energy long-range mechanisms which bypass this plane.

Fig. 6.66 Mechanisms for atom motion over Ag and Au polyhedra. (a) Chain process between opposite (111) facets. Atom pushes a row until end atom falls onto (111). (b) Half-chain for transfer from (111) to (100) facet (after Baletto et al. [133]).

Table 6.9 Step-edge barriers by exchange (in eV) for aluminum [32].

| | | Number of Layers ||
Initial Facet	Step Orientation/Final Facet	1	Multi
{111}	<110>/{100}	0.30	0.30
	<110>/{110}	0.06	0.04
	<110>/{111}	0.06	0.21
{100}	<100>/{100}	0.35	0.25
	<100>/{110}	0.34	0.35
	<110>/{111}	0.45	0.68
{110}	<100>/{100}	0.63	0.72
	<111>/{110}	0.33	0.47
	<110>/{111}	0.70	0.83

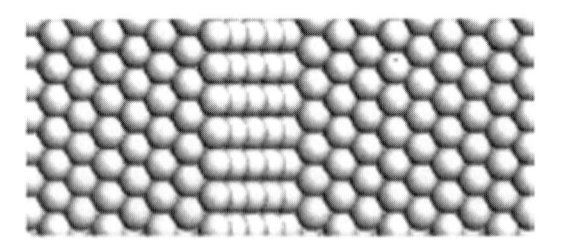

Fig. 6.67 Multilayer step, a {100} surface, between two {111} planes of aluminum (after Liu et al. [32]).

In 2001 Liu et al. [134] examined the movement of adatoms between two facets of aluminum. For jumps between two {111} planes an energy of 0.45 eV was required, while exchange between these two faces needed only 0.22 eV. For jumps from the {111} to the {100} plane, an atom had to overcome a barrier of 0.48 eV, while exchange again was less demanding, requiring only 0.30 eV. Jumps between two {100} planes cost 0.60 eV, exchange only 0.25 eV. Jumps from a {100} to a {111} plane needed 0.78 eV, exchange 0.68 eV. What is interesting in this investigation is that exchange seemed to be generally less demanding than hopping.

Liu et al. [32] in 2002 used molecular statics to discuss a new topic – atom descent occurring over a step made up of many layers, as shown in Fig. 6.67 for diffusion from the (111) plane of an fcc crystal. For aluminum they made estimates, in Table 6.9, giving the diffusion barriers over monoatomic steps compared with multilayers for diffusion by atom exchange.[3] From this study it can be concluded that diffusion across a 4-layer step can be treated as the movement from one type of facet to another, since the energy of crossing saturates at this thickness. It appears that multilevel transitions are sometimes more favorable than descent over a single layer step. The same year, Huang [135] introduced the step-facet barrier for the intersection of two <110>/{111}

---

[3] The orientation of the steps appears first, followed by a description of the steps faceting.

6.5 Atom descent over many-layered steps and between facets

Table 6.10 Step-edge diffusion barriers (in eV) on Cu(111) [63].

Multilayer	{111} → {111}	0.33
	{111} → {100}	0.40
	{100} → {111}	0.67
	{100} → {100}	0.20
Monolayer	<110>/{111}	0.10
	<110>/{100}	0.34

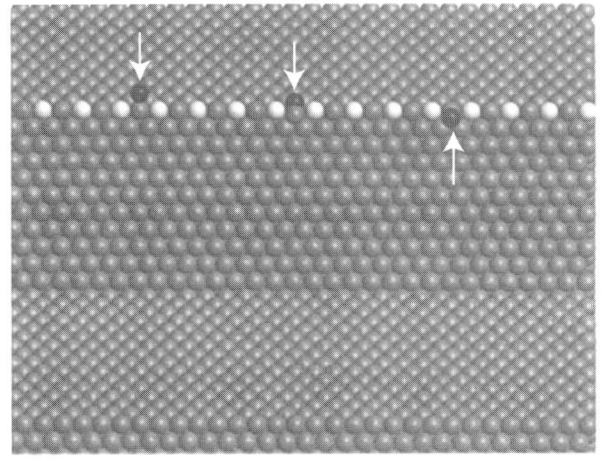

Fig. 6.68 Two Cu(111) surfaces separated by multilayer (111) step. Dark spheres (indicated by arrow) represent different stages of adatom migration: initial, transition, and final state (after Huang et al. [63]).

steps on a Cu(100) surface. The energy for such crossing, however, was quite high, 0.98 eV, making such a process insignificant.

This work was continued by Huang et al. [63] with embedded atom calculations [136] for multilayer steps from copper (111), as in Fig. 6.68. Their results, in Table 6.10, reveal that monolayer steps allow easier passage of copper atoms than do multilayer structures. However, multilayer descents are associated with energetics which should be achievable at higher temperatures. Diffusion on Cu(111) in their estimates required overcoming a barrier of 0.04 eV, in good agreement with experiments. They also introduced indium as a surfactant and, based on the difference in sublimation energies, estimated that the bond for indium was about 71% of the copper-copper bond. Huang et al. postulated an 84% reduction of the step-edge barrier due to the presence of indium on the surface, that is a lowering of the step-edge barrier from 0.33 eV to 0.28 eV. Carrying out growth experiments of copper as well as of copper with indium, they showed that the copper layer grew much smoother when indium was present, supporting their claim that indium lowered the step-edge barrier.

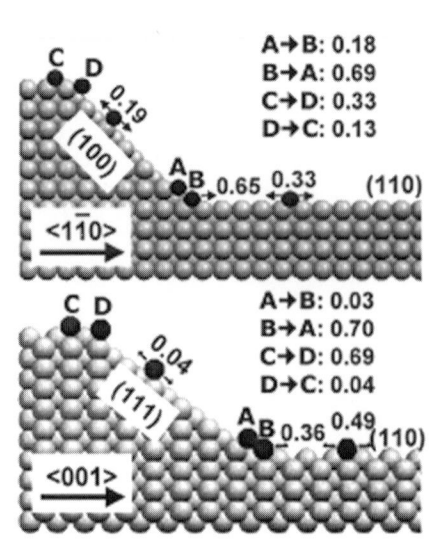

Fig. 6.69 Self-diffusion events over multilayer steps between Al(100) planes. Energy barriers for diffusion calculated with VASP program are indicated in the figure (after Buatier de Mongeot et al. [137]).

In an attempt to explain their atomic force microscope observations of huts, Buatier de Mongeot et al. [137] in 2003 did density-functional calculations in the generalized gradient approximation of VASP for aluminum atoms diffusing from Al(110) over multistep facets, as shown in Fig. 6.69. The energetics of diffusion over steps were comparable to what had been found for cross-channel transitions on flat Al(110), namely 0.49 eV. What was surprising was that the outer-corner crossing required more energy than inner-corner crossing, and movement from (111) to (100) facets was preferred over the opposite, making the multilayer barrier asymmetric.

In 2004, calculations with the VASP package were done by Zhu et al. [33] to examine atom diffusion at edges of fcc(110) planes. They also considered transitions from fcc(110) over facets or ridges, displayed in Fig. 6.70. The energetics for these steps are listed in Table 6.11, and it is clear that these processes do not involve overcoming very high barriers and will influence interlayer motion. What is interesting is that an in-channel climb onto a (100) facet was energetically favorable by exchange, but for a cross-channel climb onto a (111) facet atomic hopping was preferred.

Calculations with EAM potentials were carried out by Wang et al. [64] to characterize transitions down steps from (100) and (111) copper surfaces. Their findings for monoatomic steps are given in Table 6.12, and just as seen previously for other materials, some of the barriers are not too high. The descent from the (111) plane via a <110>/{111} step cost much less than descending over a <110>/{100} step, 0.08eV compared to 0.28 eV. On a flat (111) surface a moving atom overcame a barrier of 0.04 eV. Both steps were descended by exchange rather than hopping. However, on a Cu(100) surface, an atom moved by hopping according to the authors, with an energy of 0.48 eV. Descent proceeds by exchange, overcoming a barrier of 0.57 eV on the <110>/{111} step and

6.5 Atom descent over many-layered steps and between facets

Table 6.11 Self-diffusion barriers (in eV) calculated with VASP program for ridge crossing on (110) surface in Fig. 6.70 [33].

	Al		Cu	
	In-channel	Cross-channel	In-channel	Cross-channel
A → B Ex	0.18	0.03	0.19	0.03
B → A Ex	0.69	0.70	0.47	0.55
B → E Jump	0.65	0.36	0.33	0.42
E → B Jump	0.40	0.38	0.24	0.37
C → D Ex	0.33	0.69	0.34	0.43
D → C Ex	0.13	0.04	0.11	0.03

Table 6.12 Diffusion barriers calculated with EAM potentials for Cu adatoms on copper [64].

On (111) surface	Barrier	On (100) surface	Barrier
Flat surface	0.04 Jump	Flat surface	0.48 Jump
Down step <110>/{111}	0.08 Ex	Down step <110>/{111}	0.57 Ex
Down step <110>/{100}	0.28 Ex	Down step <100>/{100}	0.32 Ex
Step			
<110>/{100}→<110>/{111}	0.40 Ex	<110>/{111}→<100>/{100}	0.55 Jump
<100>/{111} → <110>/{100}	0.42 Ex	<100>/{100} → <110>/{111}	0.85 Jump
		<110>/{111} → <110>/{111}	0.57 Jump
		<100>/{100} → <100>/{111}	0.56 Ex

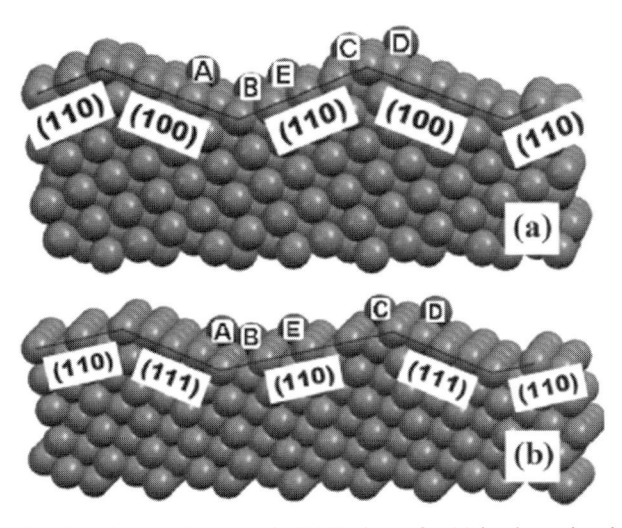

Fig. 6.70 Facets and ridges between fcc(110) planes for (a) in-channel and (b) cross-channel directions (after Zhu et al. [33]).

Table 6.13 Facet–facet diffusion barriers (in ev) calculated for Cu adatoms with EAM potentials for different number of layers N [64].

	$N=1$		$N \geq 4$	
Process	Ex	Jump	Ex	Jump
$\{111\} \rightarrow <110>/\{111\}$	0.08	0.50	0.30	0.57
$\{111\} \rightarrow <110>/\{100\}$	0.28	0.50	0.40	0.57
$\{100\} \rightarrow <110>/\{111\}$	0.57	0.81	0.75	0.82
$\{100\} \rightarrow <100>/\{100\}$	0.32	0.55	0.17	0.55

0.32 eV on the <100>/<100> step. The barrier on the latter step was lower than for movement on the terrace. Barriers for transitions from one facet to an adjacent one are listed in Table 6.13; some of these processes again involve crossing reasonably small barriers. The authors also considered multilayer transitions, but these are generally of a much higher energy.

Transitions over many-layered steps have so far been explored in calculational studies, and it should be interesting to compare these with experimental results once these become available.

6.6 To ascending step edges

In the past, the standard view was that atoms diffuse randomly over a surface until they strike an ascending lattice step and incorporate there. A direct look at this process was taken by Wang [138] in 1993. He created a small iridium cluster of 12 atoms on top of the Ir(111) plane, and deposited an iridium adatom on the terrace nearby the cluster, then observed the movement of the adatom, which is shown in Fig. 6.71. The atom diffused on the plane for a while and eventually incorporated into the central Ir_{12} cluster, but without appearing at sites in the immediate vicinity of the cluster. The combination of several deposition experiments is given in Fig. 6.72, which clearly reveals an empty zone around the Ir_{12} in the center. This effect was also noted for larger clusters, containing up to twenty-one atoms. One possible way of accounting for this behavior is to invoke a small potential barrier around the cluster which keeps atoms from coming closer. This idea has been tested by depositing an atom on a terrace with a cluster on it, and then comparing the probability of incorporating into the cluster with the probability of being trapped close to the edge of the (111) plane. Results at different temperatures are given in Fig. 6.73, and show no temperature dependence for the ratio of incorporated atoms to atoms trapped at the edge. This suggests the absence of an additional barrier around the cluster. Instead, the activation energy for atom motion in the vicinity of the island has to diminish, so that atoms are sucked into the cluster. From these experiments it was estimated that a diminution of the normal activation energy for diffusion by ~10% could account for these results.

6.6 To ascending step edges 499

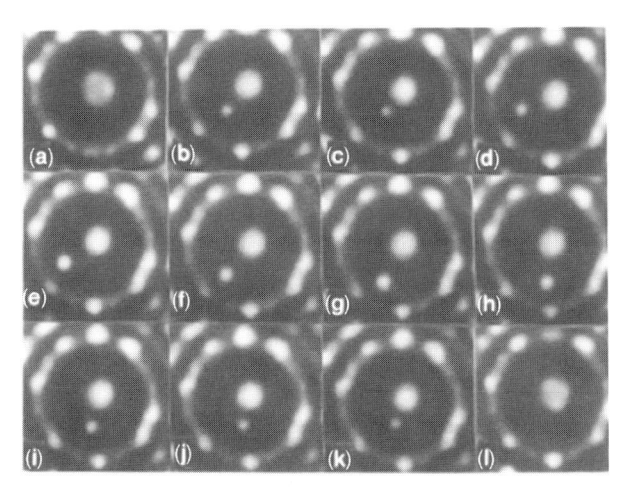

Fig. 6.71 Ir atom incorporation into ascending step of Ir cluster on Ir(111) [138]. (a) Cluster formed by field evaporation. (b) Ir atom has been deposited on (111) plane. (c)–(k) Atom moves over (111) surface after heating. (l) Atom has incorporated into cluster after 118 diffusion intervals.

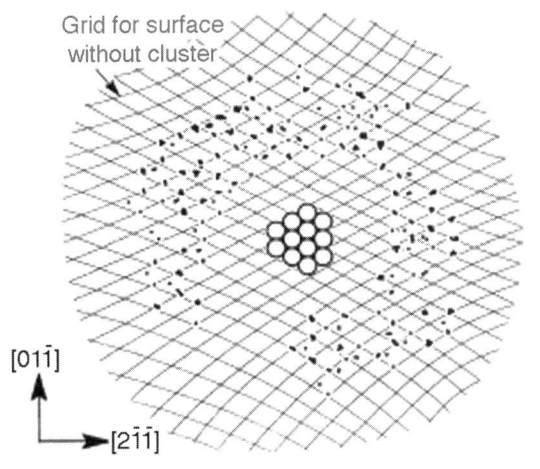

Fig. 6.72 Location of Ir atom after diffusion on Ir(111) surface with Ir$_{12}$ cluster at the center, showing empty region around cluster [138].

Additional insight into this phenomenon was obtained in further experiments, in which Wang [139] deposited iridium atoms onto an Ir(111) plane kept at ≈ 20 K with a cluster in the center. At this temperature deposited adatoms were immobile on a flat cluster free surface; however, as shown in Fig. 6.74, for a surface with a cluster, atoms are found all over the surface, except in a ring around the cluster, with the ring of roughly the same size as in the diffusion experiments. In this example, the experiment was done with a central cluster of 12 iridium atoms, but the same sort of empty zone was found when the cluster size was increased to 59 atoms, as in Fig. 3.136c. The one way of accounting for this behavior at low temperatures, where thermal diffusion does not occur, is to assume the potentials at sites around the cluster are seriously perturbed, as indicated in Fig. 3.137,

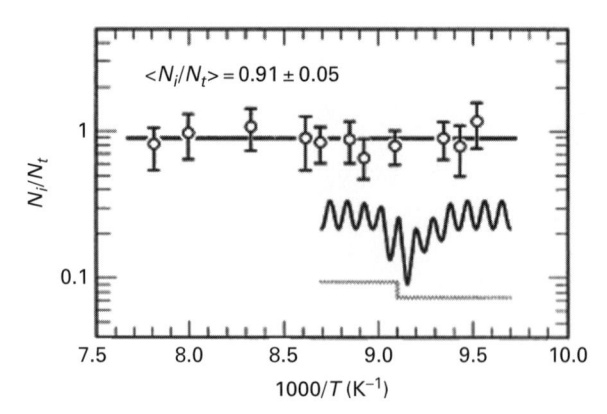

Fig. 6.73 Ratio of Ir atoms incorporated into central Ir$_{12}$ cluster, N_i, compared to N_t, the number trapped at the edge of the plane [138]. Lower right: sketch of potential in edge vicinity consistent with experiments.

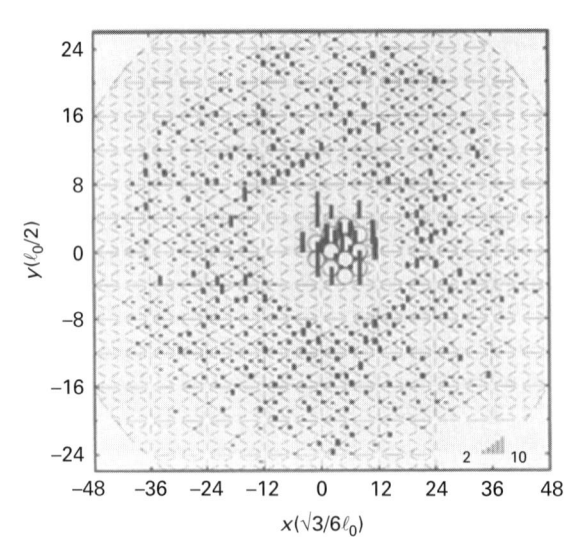

Fig. 6.74 Distribution of Ir atoms deposited on Ir(111) plane at ≈ 20 K, with Ir$_{12}$ cluster at the center [139]. Empty zone surrounds cluster.

much more than was necessary to understand the experiments exploring diffusion to the cluster. Theoretical estimates have now been made showing such deformations in the immediate vicinity of a step. For example, Villarba and Jónsson [116], working with EAM interactions, plotted the energy of a platinum atom approaching an ascending *A*- as well as a *B*-type step on Pt(111). As is clear in Fig. 6.75, the well adjacent to the step disappears due to interactions. Villarba and Jónsson in Fig. 6.75 also found decreased barriers for an atom going from site 5 to 4 at the type *A* step and between site 4 and 3 for a *B*-type step. Their results, like the results of Wang [139], show that once an atom reaches site 3 it can easily hop towards the island and incorporate there. This view of the

6.6 To ascending step edges

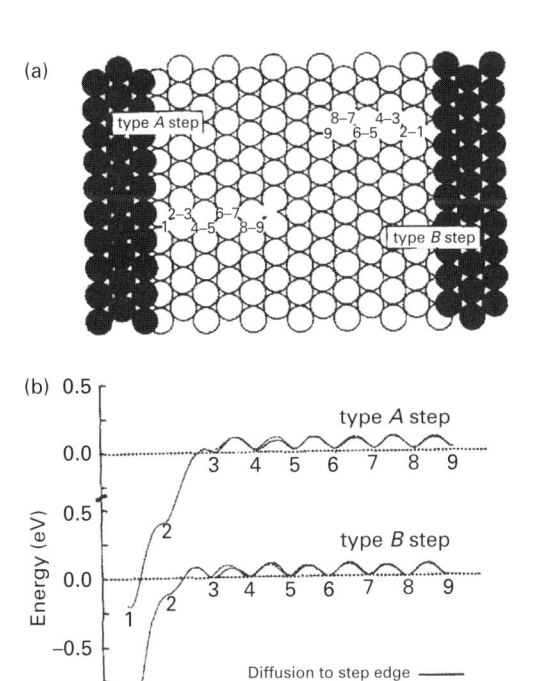

Fig. 6.75 Deformation of normal surface potentials at ascending A- and B-type steps on Pt(111), estimated using EAM interactions. (a) Hard-sphere model of steps on fcc(111). (b) Potential energy plots at two types of steps (after Villarba and Jónsson [116]).

incorporation of atoms into ascending steps has been criticized by Kellogg [140], who labeled the proposed energetics around an island as unphysical. However, at the moment all the evidence points to this view as being correct.

Atom approach toward ascending Ni steps was investigated by Liu and Adams [40] using EAM potentials and molecular statics. They found forbidden regions next to both descending and ascending steps on the Ni(111) plane. On approaching an ascending step of type B, the energy barrier rose to 0.06 eV three spacings away from the step. Close to the step, the diffusion barrier dropped suddenly to 0.04 eV. When one space apart from the ascending step the adatom is very unstable and usually reaches the step by a double jump and incorporates. A movement of clusters with up to five atoms on the Ni(111) plane showed similar effects. Approaching an A-type step, adatoms experienced the same situation, only the energy barriers were different. Jumps to the forbidden region cost the same energy 0.06 eV, while near the step, the decrease of energy was around 0.022 eV. The forbidden region was probably caused by stress around the step. Hontinfinde et al. [97], who did molecular dynamics using RGL potentials on a Ag(110) surface, noticed slightly lower barriers approaching a step from the lower terrace than for movement on the terrace.

Recently Smirnov et al. [65] using *ab initio* and MC simulations examined the interactions of Cu adatoms with Cu(111) islands. They found concentric repulsive

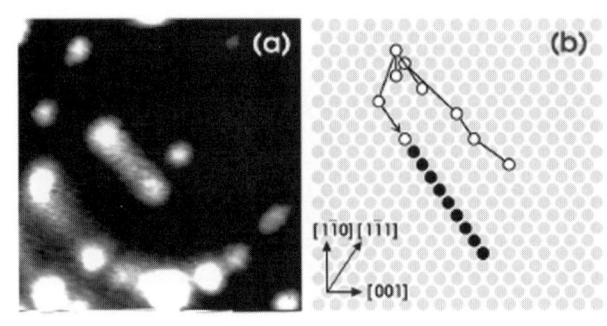

Fig. 6.76 Incorporation of Ir atom into Ir chain on W(110) plane [141]. (a) FIM image of chain with one Ir atom nearby. (b) Trajectory of Ir atom after diffusion at 360 K for 5 sec intervals. Atom never approaches side of chain.

rings of 25 meV around islands at 4–7 Å, an attractive ring of −5 meV at 8–14 Å and repulsive ring of 2 meV at 15–20 Å. Based on their MC simulation they claim that at low temperatures adatoms prefers to stay around 9 Å away from the island. They associate this effect with quantum confinement of electrons at the surface close to the step. However overcoming the repulsive barrier should be temperature dependent and that is not what was observed by Wang for Ir(111) indicating a different origin of the empty zone for this system.

Somewhat related to the approach of diffusing atoms toward ascending steps is their behavior in the vicinity of one-dimensional iridium and palladium chains on W(110), which has been examined by Koh and Ehrlich [141]. When an iridium atom is deposited on a surface on which there is such a chain, and the surface is heated, the atom does not incorporate into the long side of the chain. Instead, as shown in Fig. 6.76, the atom diffuses parallel to the chain but at some distance from it until it finally attaches itself to the chain end. When there is a second chain present on the plane, an iridium adatom again avoids the chain sides, as in Fig. 6.77, and on incorporation goes to the end of the chain. Enough is known about interactions between iridium adatoms to make an estimate of how the potential energy looks around the chains. That is shown in Fig. 6.78, and it is clear that going to the chain end is the lowest energy path. Palladium adatoms deposited between palladium chains behave quite similarly despite a large difference in binding energy of Pd and Ir. However, due to differences in second and third nearest-neighbor interactions, as well as trio interactions, palladium is expected to attach to the side of a chain when it reaches it. Once a Pd adatom reaches the side of the chain it than moves very rapidly along this chain to attach at the end (something never observed for Ir adatoms). This is an example of how long-range interactions alter the normal diffusion behavior; in the absence of chains, atom movement on bcc(110) occurs equally along all <111> directions. The problem of long-range interactions will be discussed in detail in Chapter 10.

Similar effects have more recently also been discovered by Stepanyuk *et al.* [142], who looked at the diffusion of Co atoms on Cu(111) by scanning tunneling microscopy. What they found is shown in Fig. 6.79 – the formation of metastable one-dimensional chains of Co atoms. Interactions between Co atoms had been examined in both theory and

6.6 To ascending step edges

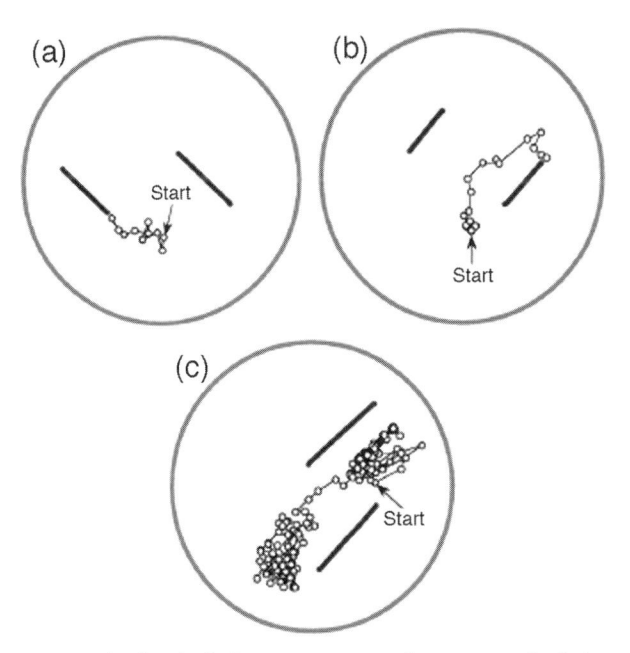

Fig. 6.77 Trajectories for single Ir atom movement between two Ir chains on W(110) after 10 sec heating interval at 340 K [141].

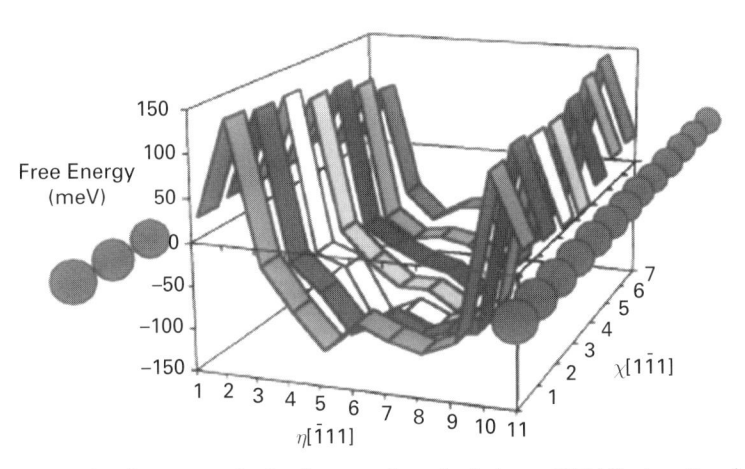

Fig. 6.78 Interaction free energy for Ir adatom and two Ir chains on W(110) at $\eta=0$ and $\eta=12$ [141]. Estimates based on previously measured pair- and trio-interactions.

experiments, as shown in Fig. 6.80. This allowed Stepanyuk *et al.* to calculate interactions between a Co atom and a Co chain, as illustrated in Fig. 6.80. Approaching the chain from the side requires overcoming a higher diffusion barrier than does moving parallel to the chain, so that atoms incorporate into the chain ends. These profound morphological effects are brought about by interactions of only a couple of meV, illustrating the importance of long-range effects in growth phenomena.

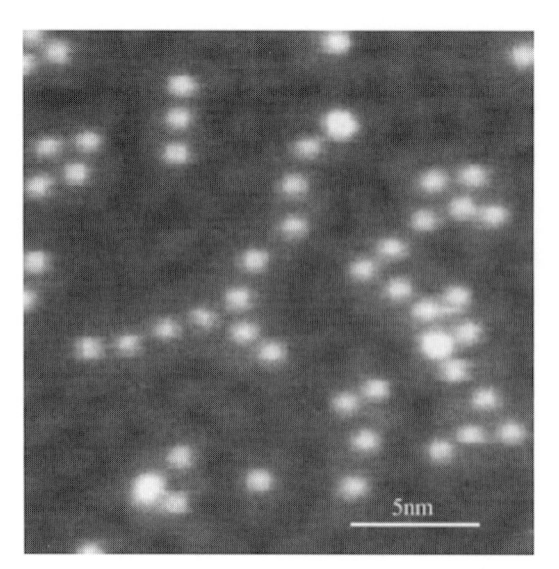

Fig. 6.79 STM image of Co atoms deposited on Cu(111) surface at 11 K at a coverage of 0.006 ML. Long one-dimensional chains are formed, due to long-ranged interactions (after Stepanyuk et al.[142]).

Most recently, Negulyaev et al. [143] have carried out STM observations of the diffusion of copper adatoms near copper chains on the Cu(111) plane. This work is analogous to the earlier FIM observations by Koh and Ehrlich [141] on iridium adatom motion near Ir chains and Pd atoms migrating near Pd chains on W(110). In the latter study the interactions were obtained from experimental determinations of the interaction between adatoms, whereas Negulyaev et al. resorted to DFT to evaluate them. Next to a single chain of Cu atoms, the copper adatom diffused parallel to the chain at a separation of ~25 Å. For a copper adatom between two copper chains separated by 55 Å diffusion again occurred parallel to the chains. Estimates of the linear density of states from calculations and from experiment are given in Fig. 6.81, and reveal small oscillations. Interactions between chain and adatom are strongly attractive at a nearest-neighbor distance from the chain followed by repulsion at a larger separation preventing incorporation. Negulyaev et al. operated at 12 K, at which the copper jump rate was ~1/200 that of the iridium on W(110) [141], which could eliminate the random excursions of the Ir adatom. In any event, the effect of long-range interactions near steps and chains is now quite well established.

In 2005, Mo et al. [66] undertook density-functional calculations of iron diffusion on stepped copper(111) surfaces. At a descending step, the barrier to diffusion and incorporation of the iron adatom by exchange with the substrate amounted to 0.070 eV. For incorporation at an ascending step the barrier to exchange was much higher, and amounted to 0.66 eV. The displaced Cu atom in an ascending exchange process jumps down over a barrier of 0.34 eV. The second Fe atom approaching the ascending step joins the first Fe atom embedded in the step. The authors predict that a similar mechanism will also be observed for W atoms at a Cu(111) surface.

Ding et al. [67] also made DFT estimates for the same system, and for iron incorporating at a descending step the barrier amounted to ≈ 0.173 eV, as is indicated in Fig. 6.82. At

6.6 To ascending step edges

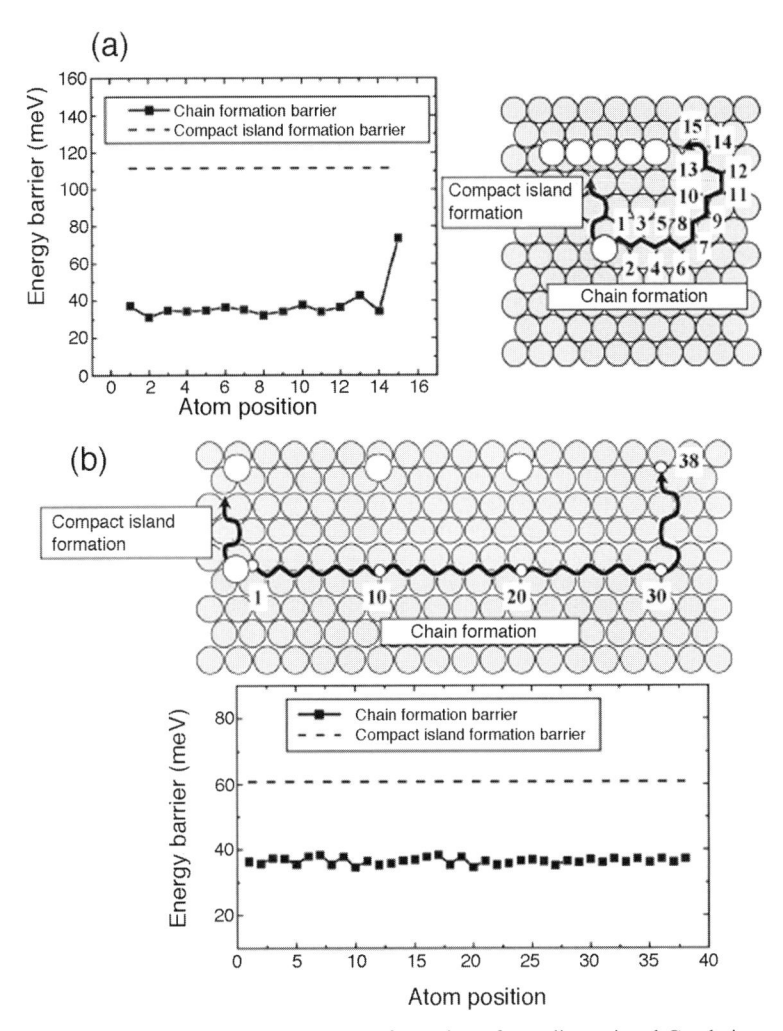

Fig. 6.80 Effects of long-ranged interactions on formation of one-dimensional Co chains, calculated for Co adatom and chain of Co atoms. Shown is the energy barrier, and separately the path of an incorporating atom. (a) Co atoms in the chain positioned at first nearest-neighbor distance. (b) Co atoms positioned at first minimum in long-range potential. In both, incorporation only occurs at end of chain (after Stepanyuk et al.[142]).

an ascending step, the barrier to incorporation was much smaller, and was only 0.020 eV. They observed an oscillatory dependence of adatom–step interactions for both descending and ascending steps with a period of about 1.5 nm. They associated this effect with the electron-charge distribution at the step. It should be noted here, however, that the possibility of atom exchange occurring was not mentioned, and this could account for these high values compared to what was found by Mo et al.

Only little has so far been done to characterize the approach of atoms to ascending step edges, and it is clear that more work is needed to arrive at a clear view of these phenomena. Right now it appears, however, that the standard picture of randomly

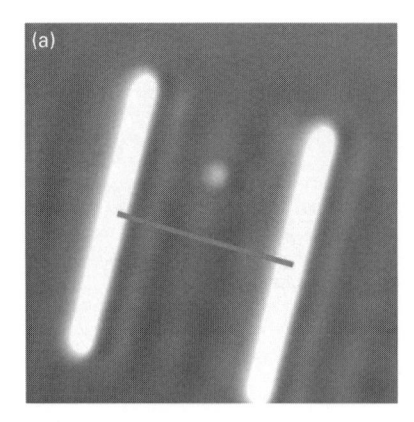

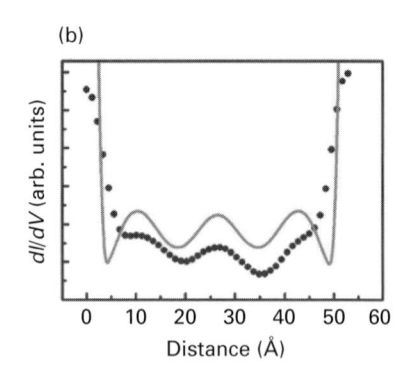

Fig. 6.81 Diffusion of single Cu adatom between two chains of Cu atoms on Cu(111). (a) STM image of adatom and chains, separated by 55Å. Temperature 12 K. (b) Calculated and measured LDOS at right angle to chains (after Negulyaev et al. [143]).

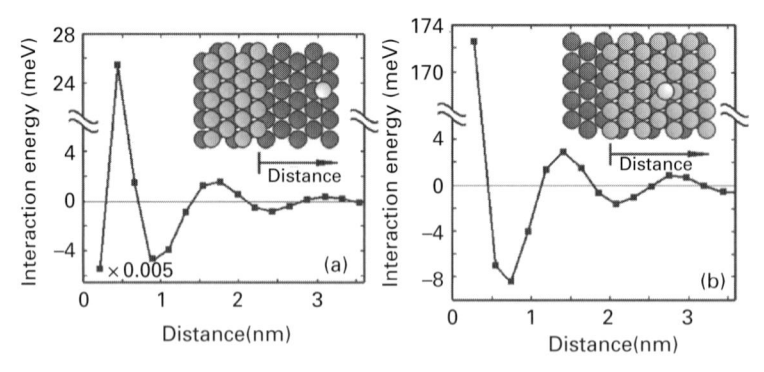

Fig. 6.82 Energy of interaction between Fe atom on copper steps on Cu(111) plane, derived from DFT calculations. (a) At ascending step. (b) At descending step. Energy change for incorporation is much higher for ascent (after Ding et al. [67]).

diffusing atoms striking islands will have to be modified to allow more generally for the interactions exerted on the adatoms.

6.7 Diffusion near dislocations

In the theory of crystal growth a breakthrough occurred in 1951 when Burton, Cabrera, and Frank [144] showed that a screw dislocation emanating from a surface, as depicted in Fig. 6.83, could rapidly bring about incorporation of atoms deposited on the surface, forming a growth spiral. Much work at atomic resolution was carried out in the next decades [145], but nothing at all was done to characterize the diffusion of atoms over the surface as influenced by such dislocations. From Fig. 6.83 it seems clear that a screw dislocation might have effects similar to those of a lattice step, and in the literature, growth has been represented as occurring via atom diffusion just as on a normal surface, with atoms eventually incorporating at the ascending lattice step [146].

6.7 Diffusion near dislocations

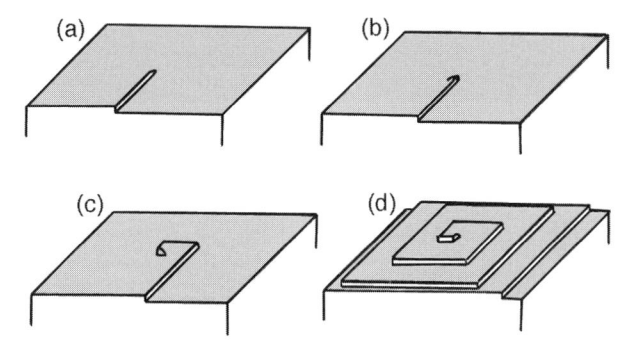

Fig. 6.83 Schematic showing formation of growth spiral by addition of atoms to a screw dislocation emerging in (a) from the surface.

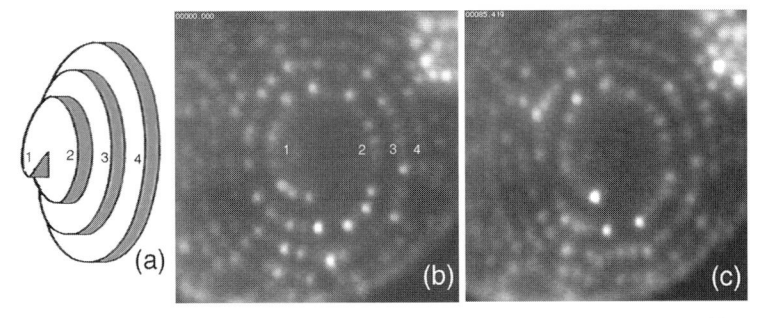

Fig. 6.84 End of tip intercepted by screw dislocation [147]. (a) Schematic of surface, with numbers indicating the height. (b) He field ionization image of screw dislocation, with numbers corresponding to schematic. (c) Image after field evaporation of four layers. Edge atoms in (b) and (c) can be superposed, indicating core is perpendicular to the surface.

Recently, however, Antczak and Józwik [147] carried out the first atomic scale observations of atom behavior on a dislocated surface, using the field ion microscope to examine a tungsten atom on a W(110) plane intersected by a screw dislocation. An image of the well-annealed surface is shown in Fig. 6.84 before and after field evaporation to remove four atomic layers. The image obtained after evaporation was directly superposed on the original image, proving that the defect was indeed primarily a screw dislocation. When a tungsten atom was evaporated onto this surface and its locations after two hundred ninety 10 sec diffusion intervals at 340 K were recorded, the results, in Fig. 6.85a were surprising. The $[1\bar{1}1]$ grid through the atom positions near the dislocation core appeared tilted to the right at the lower ramp of the dislocation with respect to similar lines at the left on the upper ramp of the dislocation, suggesting a possible change in interatomic spacings arising from stress. However, the $[\bar{1}11]$ grids from the lower and upper ramps match each other quite nicely in Fig. 6.85b.

Furthermore, the mean-square displacement in the $[1\bar{1}1]$ direction amounted to 4.00 ± 0.49 $(a_l\sqrt{3}/2)^2$, while in the $[\bar{1}11]$ direction it was considerably smaller, only 0.93 ± 0.12 $(a_l\sqrt{3}/2)^2$. Distortion in the $[1\bar{1}1]$ direction created an easy path for the adatom to move over the surface, while on the undefected surface the mean-square displacement

508 **Diffusion in special environments**

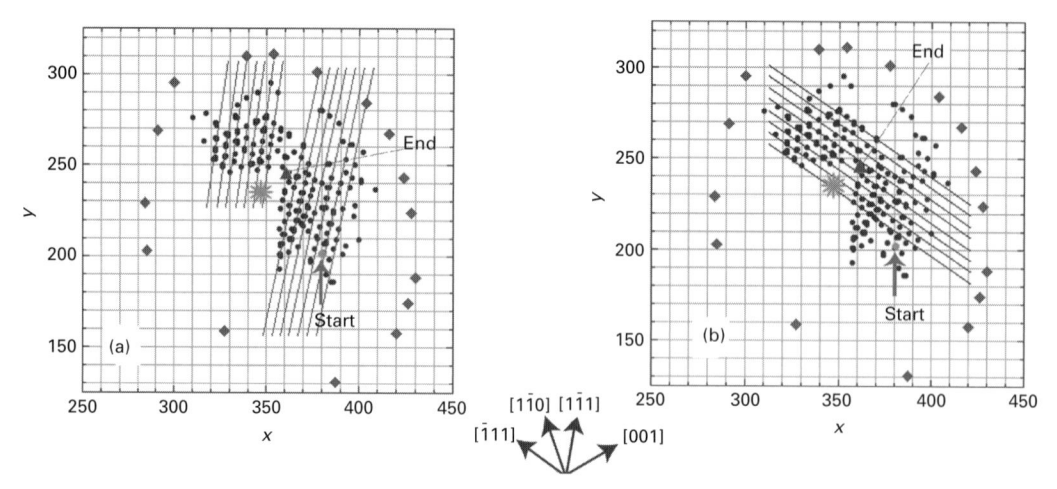

Fig. 6.85 Location of surface sites on W(110) at which tungsten adatom was observed after diffusion at 340 K [147]. (a) Grid, associated with the $[1\bar{1}1]$ direction is tilted at the right with respect to the left. (b) Grid associated with $[\bar{1}11]$ direction does not reveal any tilt.

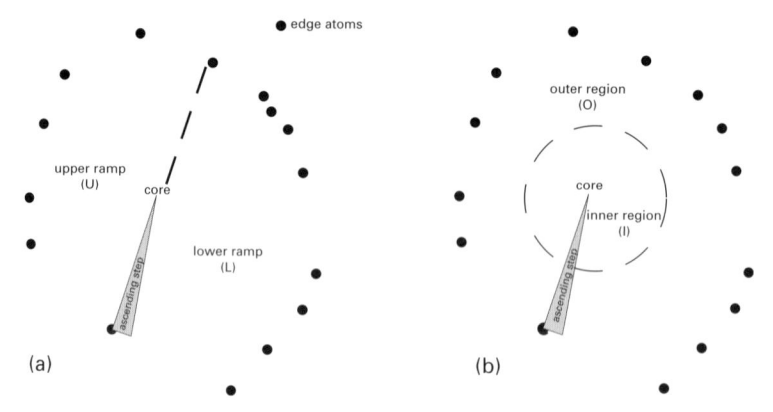

Fig. 6.86 Schematic of different regions of the dislocation loop [147]. (a) Upper and lower ramps. (b) Inner and outer regions, the former out to ~ six atom spacings from dislocation core.

for both the <111> directions were equal. Most interesting, however, are the details of the atomic displacements in diffusion in different locations. On an ordinary flat W(110), diffusivities were measured for three separate sets of 100 observations of a tungsten atom, and were found to be close to each other: 0.206 ± 0.030, 0.175 ± 0.030, and 0.168 ± 0.021, all in units of $(a_l/2)^2$ per second. On the dislocated surface, atomic behavior varied depending upon the location on the surface, and it proved useful to subdivide the atomic transitions, as indicated in Fig. 6.86, between lower (L) and upper ramps (U), as well as between regions closer (I) and farther from the core (O). As is clear from the diffusivities in Table 6.14, diffusion on the outer region on the lower ramp (LO) of the loop is roughly five times that on the upper loop (U), where it amounts to only 64% the normal value. In the inner region (I), close to the core of the dislocation, the adatom moved approximately 2.5 times slower than in the outer region. Finally, the adatom does

Table 6.14 Diffusivity of adatom at 340 K on W(110) intersected by screw dislocation and on undefected W(110) plane [147].

Type of surface	Region	N	t (sec)	$\langle\Delta x^2\rangle$ $(a_l/2)^2$	$\langle\Delta y^2\rangle$ $(a_l\sqrt{2}/2)^2$	D_x $(a_l/2)^2 \sec^{-1}$	D_y $(a_l\sqrt{2}/2)^2 \sec^{-1}$
Defect-free		1200	4	1.353 ± 0.065	1.503 ± 0.079	0.169 ± 0.008	0.188 ± 0.010
Screw	All data	290	10	4.583 ± 0.531	4.459 ± 0.483	0.229 ± 0.027	0.223 ± 0.024
Dislocation	(O)	140	10	6.500 ± 1.003	6.071 ± 0.864	0.325 ± 0.051	0.304 ± 0.044
	(I)	135	10	2.556 ± 0.357	2.763 ± 0.476	0.128 ± 0.018	0.138 ± 0.024
	(U)	123	10	2.081 ± 0.378	1.659 ± 0.236	0.104 ± 0.019	0.083 ± 0.012
	(L)	167	10	6.425 ± 0.852	6.521 ± 0.785	0.321 ± 0.043	0.326 ± 0.039
	(OU)	68	10	2.162 ± 0.531	1.397 ± 0.235	0.108 ± 0.027	0.070 ± 0.012
	(OL)	72	10	10.597 ± 1.758	10.486 ± 1.492	0.530 ± 0.088	0.524 ± 0.075
	(IU)	42	10	1.048 ± 0.202	0.952 ± 0.190	0.052 ± 0.010	0.048 ± 0.010
	(IL)	91	10	3.220 ± 0.507	3.440 ± 0.670	0.161 ± 0.026	0.172 ± 0.034

not appear to find incorporation into the outer descending lattice step advantageous – it stays two spacings away.

A full explanation for the various observations is not yet available. What appears to be happening, however, is that the dislocation introduces strain into the surface, above and beyond that predicted by elastic theory, and this affected the diffusion in entirely unexpected ways. From this study strain parallel to the surface can be expected as indicated by the lateral changes in atomic distances visible in the $[1\bar{1}1]$ grids. There are also differences in movement between upper and lower ramps probably associated with strain perpendicular to the surface. The strain also appears to change from tensile to compressive with distance from the core. Clearly much more will have to be done to establish possible contributions from edge components of the dislocation and to determine if the distribution observed here is general for other materials as well. What is clear already, however, is that adatom diffusion is very strongly affected by the presence of the dislocation and cannot be approximated as being similar to atom behavior on an ordinary surface. It is also clear that novel, unexpected phenomena are still being discovered in the field of surface diffusion.

References

[1] P. L. Cowan, T. T. Tsong, Direct observation of adatom-substitutional atom interaction on dilute metal alloy surfaces, *Surf. Sci.* **67** (1977) 158–179.
[2] G. L. Kellogg, Direct observation of substitutional-atom trapping on a metal surface, *Phys. Rev. Lett.* **72** (1994) 1662–1665.
[3] G. Ayrault, G. Ehrlich, Surface self-diffusion on an fcc crystal: An atomic view, *J. Chem. Phys.* **60** (1974) 281–294.
[4] O. S. Hernán, A. L. Vázquez de Parga, J. M. Gallego, R. Miranda, Self-surfactant effect on Fe/Au(100): Place exchange plus Au self-diffusion, *Surf. Sci.* **415** (1998) 106–121.
[5] R. Casanova, T. T. Tsong, Surface diffusion of single W atoms on hydrogen saturated W{123} plane, *Surf. Sci.* **94** (1980) L179–183.

[6] R. T. Tung, W. R. Graham, Single atom self-diffusion on nickel surfaces, *Surf. Sci.* **97** (1980) 73–87.

[7] P. Blandin, P. Ballone, Diffusion of metal adatom on compact metal surfaces in the presence of defects and impurities, *Surf. Sci.* **331–333** (1995) 891–895.

[8] R. Stumpf, H-enhanced mobility and defect formation at surfaces: H on Be(0001), *Phys. Rev. B* **53** (1996) R4253–4256.

[9] G. L. Kellogg, Hydrogen promotion of surface self-diffusion on Rh(100) and Rh(311), *Phys. Rev. B* **55** (1997) 7206–7212.

[10] G. L. Kellogg, Hydrogen inhibition of exchange diffusion on Pt(100), *Phys. Rev. Lett.* **79** (1997) 4417–4420.

[11] K. Haug, Z. Zhang, D. John, C. F. Walters, D. M. Zehner, W. E. Plummer, Effects of hydrogen in Ni(100) submonolayer homoepitaxy, *Phys. Rev. B* **55** (1997) R10233–10236.

[12] K. Haug, T. Jenkins, Effects of hydrogen on the three-dimensional epitaxial growth of Ni(100), (110), and (111), *J. Phys. Chem. B* **104** (2000) 10017–10023.

[13] S. Horch, H. T. Lorensen, S. Helveg, E. Laegsgaard, I. Stensgaard, K. W. Jacobsen, J. K. Nørskov, F. Besenbacher, Enhancement of surface self-diffusion of platinum atoms by adsorbed hydrogen, *Nature* **398** (1999) 134–136.

[14] G. Prévot, H. Guesmi, B. Croset, Ordered growth of nanodots on a pre-structured metallic template Au/N/Cu(001), *Surf. Sci.* **601** (2007) 2017–2025.

[15] W. Swiech, Stepped Pt(111), Personal communication, 2008.

[16] G. Ehrlich, F. G. Hudda, Atomic view of surface self-diffusion: Tungsten on tungsten, *J. Chem. Phys.* **44** (1966) 1039–1049.

[17] R. L. Schwoebel, E. J. Shipsey, Step motion on crystal surfaces, *J. Appl. Phys.* **37** (1966) 3682–3686.

[18] R. L. Schwoebel, Step motion on crystal surfaces. II, *J. Appl. Phys.* **40** (1969) 614–618.

[19] T. T. Tsong, Direct observation of interactions between individual atoms on tungsten surfaces, *Phys. Rev. B* **6** (1972) 417–426.

[20] D. W. Bassett, M. J. Parsley, Field ion microscopic studies of transition metal adatom diffusion on (110), (211) and (321) tungsten surfaces, *J. Phys. D* **3** (1970) 707–716.

[21] D. W. Bassett, Surface atom displacement processes, *Surf. Sci.* **53** (1975) 74–86.

[22] D. W. Bassett, C. K. Chung, D. Tice, Field ion microscope studies of atomic displacement processes on metal surfaces, *La Vide* **176** (1975) 39–43.

[23] S.-C. Wang, T. T. Tsong, Measurement of the barrier height of the reflective W(110) plane boundaries in surface diffusion of single atoms, *Surf. Sci.* **121** (1982) 85–97.

[24] H.-W. Fink, G. Ehrlich, Lattice steps and adatom binding on W(211), *Surf. Sci.* **143** (1984) 125–144.

[25] S. C. Wang, G. Ehrlich, Atom incorporation at surface clusters: An atomic view, *Phys. Rev. Lett.* **67** (1991) 2509–2512.

[26] S. C. Wang, G. Ehrlich, Atom condensation on surface clusters: Adsorption or incorporation? *Phys. Rev. Lett.* **75** (1995) 2964–2967.

[27] R. Stumpf, M. Scheffler, Mechanisms of self-diffusion on flat and stepped Al surfaces, *Surf. Sci.* **307–309** (1994) 501–506.

[28] R. Stumpf, M. Scheffler, Theory of self-diffusion at and growth of Al(111), *Phys. Rev. Lett.* **72** (1994) 254–257.

[29] R. Stumpf, M. Scheffler, *Ab initio* calculations of energies and self-diffusion on flat and stepped surfaces of aluminum and their implications on crystal growth, *Phys. Rev. B* **53** (1996) 4958–4973.

[30] M. Bockstedte, S. J. Liu, O. Pankratov, C. H. Woo, H. Huang, Diffusion of clusters down (111) aluminum islands, *Comput. Mater. Sci.* **23** (2002) 85–94.

[31] F. Ercolessi, J. B. Adams, Interatomic potentials from first-principles calculations: The force-matching method, *Europhys. Lett.* **26** (1994) 583–588.

[32] S. J. Liu, H. Huang, C. H. Woo, Schwoebel-Ehrlich barrier: From two to three dimensions, *Appl. Phys. Lett.* **80** (2002) 3295–3297.

[33] W. Zhu, F. Buatier de Mongeot, U. Valbusa, E. G. Wang, Z. Zhang, Adatom ascending at step edges and faceting on fcc metal (110) surfaces, *Phys. Rev. Lett.* **92** (2004) 106102 1–4.

[34] J. A. Stroscio, D. T. Pierce, R. A. Dragoset, Homoepitaxial growth of iron and a real space view of reflection-high-energy-electron diffraction, *Phys. Rev. Lett.* **70** (1993) 3615–3618.

[35] P. Smilauer, S. Harris, Determination of step-edge barriers to interlayer transport from surface morphology during the initial stages of homoepitaxial growth, *Phys. Rev. B* **51** (1995) 14798–14801.

[36] J. G. Amar, F. Family, Step barrier for interlayer-diffusion in Fe/Fe(100) epitaxial growth, *Phys. Rev. B* **52** (1995) 13801–13804.

[37] M. C. Bartelt, J. W. Evans, Temperature dependence of kinetic roughening during metal (100) homoepitaxy: Kinetic phase transition from "mounding" to smooth growth, *Surf. Sci.* **423** (1999) 189–207.

[38] U. Köhler, C. Jensen, A. C. Schindler, L. Brendel, D. E. Wolf, Scanning tunnelling microscopy and Monte Carlo studies of homoepitaxy on Fe(110), *Philos. Mag. B* **80** (2000) 283–292.

[39] C.-L. Liu, J. B. Adams, Diffusion mechanisms on Ni surfaces, *Surf. Sci.* **265** (1992) 262–272.

[40] C.-L. Liu, J. B. Adams, Diffusion behavior of single adatoms near and at steps during growth of metallic thin films on Ni surfaces, *Surf. Sci.* **294** (1993) 197–210.

[41] P. Stoltze, Simulation of surface defects, *J. Phys.: Condens. Matter* **6** (1994) 9495–9517.

[42] Y. Li, A. E. DePristo, Predicted growth mode for metal homoepitaxy on the fcc(111) surface, *Surf. Sci.* **351** (1996) 189–199.

[43] Y. Li, A. E. DePristo, Potential energy barriers for interlayer mass transport in homoepitaxial growth on fcc(111) surfaces: Pt and Ag, *Surf. Sci.* **319** (1994) 141–148.

[44] J. Merikoski, I. Vattulainen, J. Heinonen, T. Ala-Nissila, Effect of kinks and concerted diffusion mechanisms on mass transport and growth on stepped metal surfaces, *Surf. Sci.* **387** (1997) 167–182.

[45] I. Markov, Surface energetics from the transition from step-flow growth to two-dimensional nucleation in metal homoepitaxy, *Phys. Rev. B* **56** (1997) 12544–12552.

[46] H. A. van der Vegt, H. M. van Pinxteren, M. Lohmeier, E. Vlieg, J. M. C. Thornton, Surfactant-induced layer-by-layer growth of Ag on Ag(111), *Phys. Rev. Lett.* **68** (1992) 3335–3338.

[47] W. Wulfhekel, N. N. Lipkin, J. Kliewer, G. Rosenfeld, L. C. Jorritsma, B. Poelsema, G. Comsa, Conventional and manipulated growth of Cu/Cu(111), *Surf. Sci.* **348** (1996) 227–242.

[48] G. S. Icking-Konert, M. Giesen, H. Ibach, Decay of Cu adatom islands on Cu(111), *Surf. Sci.* **398** (1998) 37–48.

[49] M. Giesen, G. S. Icking-Konert, H. Ibach, Interlayer mass transport and quantum confinement of electronic states, *Phys. Rev. Lett.* **82** (1999) 3101–3104.

[50] M. Giesen, H. Ibach, Step edge barrier controlled decay of multilayer islands on Cu(111), *Surf. Sci.* **431** (1999) 109–115.

[51] J. Ferrón, L. Gómez, J. Camarero, J. E. Prieto, V. Cros, A. L. Vázquez de Parga, J. J. de Miguel, R. Miranda, Influence of surfactants on atomic diffusion, *Surf. Sci.* **459** (2000) 135–148.

[52] D. Tománek, S. Mukherjee, K. H. Bennemann, Simple theory for the electronic and atomic structure of small clusters, *Phys. Rev. B* **28** (1983) 665–673.

[53] P. Stoltze, J. K. Nørskov, Accommodation and diffusion of Cu deposited on flat and stepped Cu(111) surfaces, *Phys. Rev. B* **48** (1993) 5607–5611.

[54] M. Karimi, T. Tomkowski, G. Vidali, O. Biham, Diffusion of Cu on Cu surface, *Phys. Rev. B* **52** (1995) 5364–5374.

[55] M. Breeman, G. T. Barkema, M. H. Langelaar, D. O. Boerma, Computer simulation of metal-on-metal epitaxy, *Thin Solid Films* **272** (1996) 195–207.

[56] M. W. Finnis, J. E. Sinclair, A simple empirical N-body potential for transition metals, *Philos. Mag. A* **50** (1984) 45–55.

[57] O. S. Trushin, K. Kokko, P. T. Salo, W. Hergert, M. Kotrla, Step roughening effect on adatom diffusion, *Phys. Rev. B* **56** (1997) 12135–12138.

[58] M. Giesen, G. S. Icking-Konert, H. Ibach, Fast decay of adatom islands and mounds on Cu(111): A new effective channel for interlayer mass transport, *Phys. Rev. Lett.* **80** (1998) 552–555.

[59] K. Morgenstern, G. Rosenfeld, G. Comsa, E. Laegsgaard, F. Besenbacher, Comment on "Interlayer mass transport and quantum confinement of electronic states", *Phys. Rev. Lett.* **85** (2000) 468.

[60] K. Morgenstern, G. Rosenfeld, G. Comsa, M. R. Sorensen, B. Hammer, E. Laegsgaard, F. Besenbacher, Kinetics of fast island decay on Ag(111), *Phys. Rev. B* **63** (2001) 045412 1–5.

[61] M. I. Larsson, Kinetic Monte Carlo simulations of adatom island decay on Cu(111), *Phys. Rev. B* **64** (2001) 115428 1–10.

[62] P. J. Feibelman, Accelerated mound decay at adjacent kinks on Cu(111), *Surf. Sci.* **478** (2001) L349–354.

[63] H. Huang, C. H. Woo, H. L. Wei, X. X. Zhang, Kinetics-limited surface structures at the nanoscale, *Appl. Phys. Lett.* **82** (2003) 1272–1274.

[64] J. Wang, H. Huang, T. S. Cale, Diffusion barriers on Cu surfaces and near steps, *Modeling Simul. Mater. Sci. Eng.* **12** (2004) 1209–1225.

[65] A. S. Smirnov, N. N. Negulyaev, L. Niebergall, W. Hergert, A. M. Saletsky, V. S. Stepanyuk, Effect of quantum confinement of surface electrons on an atomic motion on nanoislands: *Ab initio* calculation and Kinetic Monte Carlo simulations, *Phys. Rev. B* **78** (2008) 041405(R) 1–4.

[66] Y. Mo, K. Varga, K. Kaxiras, Z. Zhang, Kinetic pathway for the formation of Fe nanowires on stepped Cu(111) surfaces, *Phys. Rev. Lett.* **94** (2005) 155503 1–4.

[67] H. F. Ding, V. S. Stepanyuk, P. A. Ignatiev, N. N. Negulyaev, L. Niebergall, M. Wasniowska, C. L. Gao, P. Bruno, J. Kirschner, Self-organized long-period adatom strings on stepped metal surfaces: Scanning tunneling microscopy, *ab initio* calculations, and kinetic Monte Carlo simulations, *Phys. Rev. B* **76** (2007) 033409 1–4.

[68] C.-L. Liu, Energetics of diffusion processes during nucleation and growth for the Cu/Cu(100) system, *Surf. Sci.* **316** (1994) 294–302.

[69] F. Montalenti, R. Ferrando, Jumps and concerted moves in Cu, Ag, and Au(110) adatom self-diffusion, *Phys. Rev. B* **59** (1999) 5881–5891.

[70] H. Yildirim, A. Kara, S. Durukanoglu, T. S. Rahman, Calculated pre-exponential factors and energetics for adatom hopping on terraces and steps of Cu(100) and Cu(110), *Surf. Sci.* **600** (2006) 484–492.

References

[71] F. Máca, M. Kotrla, O. S. Trushin, Energy barriers for interlayer diffusion in Pt/Pt(111) and Rh/Rh(111) homoepitaxy: Small islands, *Czech. J. Phys.* **49** (1999) 1591–1596.

[72] F. Máca, M. Kotrla, O. S. Trushin, Energy barriers for diffusion on stepped Rh(111) surfaces, *Surf. Sci.* **454–456** (2000) 579–583.

[73] J. Vrijmoeth, H. A. van der Vegt, J. A. Meyer, E. Vlieg, R. J. Behm, Surfactant-induced layer-by-layer growth of Ag on Ag(111): Origin and side effects, *Phys. Rev. Lett.* **72** (1994) 3843–3846.

[74] J. A. Meyer, J. Vrijmoeth, H. A. van der Vegt, E. Vlieg, R. J. Behm, Importance of the additional step-edge barrier in determining film morphology during epitaxial growth, *Phys. Rev. B* **51** (1995) 14790–14793.

[75] K. Bromann, H. Brune, H. Röder, K. Kern, Interlayer mass transport in homoepitaxial and heteroepitaxial metal growth, *Phys. Rev. Lett.* **75** (1995) 677–680.

[76] W. C. Elliott, P. F. Miceli, T. Tse, P. W. Stephens, Temperature and orientation dependence of kinetic roughening during homoepitaxy: A quantitative x-ray-scattering study of Ag, *Phys. Rev. B* **54** (1996) 17938–17942.

[77] G. Rosenfeld, K. Morgenstern, I. Beckmann, W. Wulfhekel, E. Laegsgaard, F. Besenbacher, G. Comsa, Stability of two-dimensional clusters on crystal surfaces: From Ostwald ripening to single-cluster decay, *Surf. Sci.* **402–404** (1998) 401–408.

[78] H. Brune, K. Bromann, H. Röder, K. Kern, J. Jacobsen, P. Stoltze, K. Jacobsen, J. Nørskov, Effect of strain on surface diffusion and nucleation, *Phys. Rev. B* **52** (1995) R14380–14383.

[79] K. R. Roos, R. Bhutani, M. C. Tringides, Inter-layer mass transport in a low-coverage, low island-density regime, *Surf. Sci.* **384** (1997) 62–69.

[80] K. R. Roos, M. C. Tringides, Determination of interlayer diffusion parameters for Ag/Ag(111), *Phys. Rev. Lett.* **85** (2000) 1480–1483.

[81] J. Krug, Comment on "Determination of interlayer diffusion parameters for Ag/Ag(111), *Phys. Rev. Lett.* **87** (2001) 149601 1.

[82] K. Morgenstern, F. Besenbacher, Comment on "Determination of interlayer diffusion parameters for Ag/Ag(111), *Phys. Rev. Lett.* **87** (2001) 149603 1.

[83] S. Heinrichs, P. Maass, Comment on "Determination of interlayer diffusion parameters for Ag/Ag(111), *Phys. Rev. Lett.* **87** (2001) 149605 1.

[84] Z. Chvoj, C. Ghosh, T. S. Rahman, M. C. Tringides, Prefactors for interlayer diffusion on Ag/Ag(111), *J. Phys.: Condens. Matter* **15** (2003) 5223–5230.

[85] K. Morgenstern, G. Rosenfeld, E. Laegsgaard, F. Besenbacher, G. Comsa, Measurement of energies controlling ripening and annealing on metal surfaces, *Phys. Rev. Lett.* **80** (1998) 556–559.

[86] M. Li, Y. Han, P. A. Thiel, J. W. Evans, Formation of complex wedding-cake morphologies during homoepitaxial film growth of Ag on Ag(111): Atomistic, step-dynamics, and continuum modeling, *J. Phys.: Condens. Matter* **21** (2009) 084216 1–12.

[87] H. A. van der Vegt, J. Vrijmoeth, B. J. Behm, E. Vlieg, Sb-enhanced nucleation in the homoepitaxial growth of Ag(111), *Phys. Rev. B* **57** (1998) 4127–4131.

[88] S. Heinrichs, P. Maass, Influence of adatom interactions on second-layer nucleation, *Phys. Rev. B* **66** (2002) 073402 1–4.

[89] K.-H. Hong, P.-R. Cha, J.-K. Yoon, The effect of lattice strain on the step edge diffusion and morphological development during epitaxial growth, *Mater. Sci. Forum* **426–432** (2003) 3463–3468.

[90] K.-H. Hong, P.-R. Cha, H.-S. Nam, J.-K. Yoon, The effect of lattice strain on step edge diffusion, *Met. and Mat. Int.* **9** (2003) 129–134.

[91] J. Cai, Y. Y. Ye, Simple analytical embedded-atom-potential model including a long-range force for fcc metals and their alloys, *Phys. Rev. B* **54** (1996) 8398–8410.

[92] M. I. Haftel, M. Rosen, New ballistically and thermally activated exchange processes in the vapor deposition of Au on Ag(111): a molecular dynamics study, *Surf. Sci.* **407** (1998) 16–26.

[93] B. D. Yu, M. Scheffler, *Ab initio* study of step formation and self-diffusion on Ag(100), *Phys. Rev. B* **55** (1997) 13916–13924.

[94] U. Kürpick, T. S. Rahman, Diffusion processes relevant to homoepitaxial growth on Ag(100), *Phys. Rev. B* **57** (1998) 2482–2492.

[95] C. R. Stoldt, K. J. Caspersen, M. C. Bartelt, C. J. Jenks, J. W. Evans, P. A. Thiel, Using temperature to tune film roughness: Nonintuitive behavior in a simple system, *Phys. Rev. Lett.* **85** (2000) 800–803.

[96] K. J. Caspersen, A. R. Layson, C. R. Stoldt, V. Fournee, P. A. Thiel, J. W. Evans, Development and ordering mounds during metal (100) homoepitaxy, *Phys. Rev. B* **65** (2002) 193407 1–4.

[97] F. Hontinfinde, R. Ferrando, A. C. Levi, Diffusion processes relevant to the epitaxial growth of Ag on Ag(110), *Surf. Sci.* **366** (1996) 306–316.

[98] T.-Y. Fu, L.-C. Cheng, Y.-J. Hwang, T. T. Tseng, Diffusion of Pd adatoms on W surfaces and their interactions with steps, *Surf. Sci.* **507–510** (2002) 103–107.

[99] T. Y. Fu, L. C. Cheng, T. T. Tsong, Determination of atomic potential energy for Pd adatom diffusion across W(111) islands and surfaces, *J. Vac. Sci. Technol. A* **20** (2002) 897–899.

[100] J. T. Goldstein, G. Ehrlich, Atom and cluster diffusion on Re(0001), *Surf. Sci.* **443** (1999) 105–115.

[101] T.-Y. Fu, H.-T. Wu, T. T. Tsong, Energetics of surface atomic processes near a lattice step, *Phys. Rev. B* **58** (1998) 2340–2346.

[102] G. L. Kellogg, Field ion microscope studies of single-atom surface diffusion and cluster nucleation on metal surfaces, *Surf. Sci. Rep.* **21** (1994) 1–88.

[103] K. Kyuno, A. Gölzhäuser, G. Ehrlich, Growth and the diffusion of platinum atoms and dimers on Pt(111), *Surf. Sci.* **397** (1998) 191–196.

[104] S.-M. Oh, K. Kyuno, S. C. Wang, G. Ehrlich, Step-edge versus interior barriers to atom incorporation at lattice steps, *Phys. Rev. B* **67** (2003) 075413 1–7.

[105] S. C. Wang, G. Ehrlich, Atom incorporation at edge defects in clusters, *Phys. Rev. Lett.* **93** (2004) 176101 1–4.

[106] O. S. Trushin, M. Kotrla, F. Máca, Energy barriers on stepped Ir/Ir(111) surfaces: A molecular statics calculation, *Surf. Sci.* **389** (1997) 55–65.

[107] U. Kürpick, Self-diffusion on stepped Ir(111) surfaces, *Phys. Rev. B* **69** (2004) 205410 1–6.

[108] I. Markov, Method for evaluation of the Ehrlich-Schwoebel barrier to interlayer transport in metal homoepitaxy, *Phys. Rev. B* **54** (1996) 17930–17937.

[109] M. Bott, T. Michely, G. Comsa, The homoepitaxial growth of Pt on Pt(111) studied with STM, *Surf. Sci.* **272** (1992) 161–166.

[110] B. Poelsema, A. F. Becker, R. Kunkel, G. Rosenfeld, L. K. Verheij, G. Comsa, in: *Surface Science: Principles and Applications, Springer Proceedings in Physics Vol. 73*, R. F. Howe, R. N. Lamb, K. Wandelt (eds.), The role of kinetic effects in the growth of Pt on Pt(111), (Springer-Verlag, Berlin, 1993), p. 95–104.

[111] K. Kyuno, G. Ehrlich, Step-edge barriers on Pt(111): An atomistic view, *Phys. Rev. Lett.* **81** (1998) 5592–5595.

[112] J. Krug, P. Politi, T. Michely, Island nucleation in the presence of step-edge barriers: Theory and applications, *Phys. Rev. B* **61** (2000) 14037–14046.

[113] M. Kalff, G. Comsa, T. Michely, How sensitive is epitaxial growth to adsorbates? *Phys. Rev. Lett.* **81** (1998) 1255–1258.

[114] J. Tersoff, A. W. Denier van der Gon, R. M. Tromp, Critical island size for layer-by-layer growth, *Phys. Rev. Lett.* **72** (1994) 266–269.

[115] E. Lundgren, B. Stanka, G. Leonardelli, M. Schmid, P. Varga, Interlayer diffusion of adatoms: A scanning-tunneling microscopy study, *Phys. Rev. Lett.* **82** (1999) 5068–5071.

[116] M. Villarba, H. Jónsson, Diffusion mechanisms relevant to metal crystal growth: Pt/Pt(111), *Surf. Sci.* **317** (1994) 15–36.

[117] H. Jónsson, Theoretical studies of atomic-scale processes relevant to crystal growth, *Annu. Rev. Phys. Chem.* **51** (2000) 623–653.

[118] R. Wang, K. A. Fichthorn, An investigation of the energetics and dynamics of adatom motion to descending step edges in Pt/Pt(111) homoepitaxy, *Surf. Sci.* **301** (1994) 253–259.

[119] R. Wang, K. A. Fichthorn, An investigation of adsorption-induced smoothing mechanisms in Pt/Pt(111) homoepitaxy, *Molec. Simul.* **11** (1993) 105–120.

[120] J. Jacobsen, K. W. Jacobsen, P. Stoltze, J. K. Nørskov, Island shape-induced transition from 2D to 3D growth for Pt/Pt(111), *Phys. Rev. Lett.* **74** (1995) 2295–2298.

[121] P. J. Feibelman, Interlayer self-diffusion on stepped Pt(111), *Phys. Rev. Lett.* **81** (1998) 168–171.

[122] G. Kresse, J. Hafner, *Ab initio* molecular dynamics for liquid metals, *Phys. Rev. B* **47** (1993) 558–561.

[123] G. Kresse, J. Hafner, *Ab initio* molecular dynamics simulation of the liquid-metal-amorphous-semiconductor transition in germanium, *Phys. Rev. B* **49** (1994) 14251–14269.

[124] G. Kresse, J. Furthmüller, Efficient iterative schemes for *ab initio* total-energy calculations using a plane-wave set, *Phys. Rev. B* **54** (1996) 11169–11186.

[125] A. Gölzhäuser, G. Ehrlich, Atom movement and binding on surface clusters: Pt on Pt(111) Clusters, *Phys. Rev. Lett.* **77** (1996) 1334–1337.

[126] F. Máca, M. Kotrla, O. S. Trushin, Energy barriers for diffusion on stepped Pt(111) surface, *Vacuum* **54** (1999) 113–117.

[127] G. Leonardelli, E. Lundgren, M. Schmid, Adatom interlayer diffusion on Pt(111): An embedded atom method study, *Surf. Sci.* **490** (2001) 29–42.

[128] S.-C. Li, Y. Han, J.-F. Jia, Q.-K. Xue, F. Liu, Determination of the Ehrlich-Schwoebel barrier in epitaxial growth of thin films, *Phys. Rev. B* **74** (2006) 195428 1–5.

[129] K. Kyuno, G. Ehrlich, Step-edge barriers: thruths and kinetic consequences, *Surf. Sci.* **394** (1997) L179–187.

[130] S.-M. Oh, K. Kyuno, G. Ehrlich, Interior barriers and dimer nucleation on islands, *Surf. Sci.* **540** (2003) L583–586.

[131] V. S. Stepanyuk, N. N. Negulyaev, A. M. Saletsky, W. Hergert, Growth of Co nanostructures on Cu(110): Atomistic scale simulations, *Phys. Rev. B* **78** (2008) 113406 1–4.

[132] S. Valkealahti, M. Manninen, Diffusion on aluminum-cluster surfaces and the cluster growth, *Phys. Rev. B* **57** (1998) 15533–15540.

[133] F. Baletto, C. Mottet, R. Ferrando, Molecular dynamics simulation of surface diffusion and growth on silver and gold clusters, *Surf. Sci.* **446** (2000) 31–45.

[134] S. J. Liu, E. G. Wang, C. H. Woo, H. Huang, Three-dimensional Schwoebel-Ehrlich barrier, *J. Comp. Aid. Mat. Des.* **7** (2001) 195–201.

[135] H. Huang, Adatom diffusion along and down island steps, *J. Comp. Aid. Mat. Des.* **9** (2002) 75–80.

[136] Y. Mishin, M. J. Mehl, D. A. Papaconstantopoulos, A. F. Voter, J. D. Kress, Structural stability and lattice defects in copper: *Ab initio*, tight binding, and embedded-atom calculations, *Phys. Rev. B* **63** (2001) 224106 1–16.
[137] F. Buatier de Mongeot, W. Zhu, A. Molle, R. Buzio, C. Boragno, U. Valbusa, E. G. Wang, Z. Zhuang, Nanocrystal formation and faceting instability in Al(110) homoepitaxy: *True* upward adatom diffusion at step edges and island corners, *Phys. Rev. Lett.* **91** (2003) 016102 1–4.
[138] S. C. Wang, G. Ehrlich, Adatom motion to lattice steps: A direct view, *Phys. Rev. Lett.* **70** (1993) 41–44.
[139] S. C. Wang, G. Ehrlich, Atom condensation at lattice steps and clusters, *Phys. Rev. Lett.* **71** (1993) 4174–4177.
[140] G. L. Kellogg, Experimental observation of ballistic atom exchange on metal surfaces, *Phys. Rev. Lett.* **76** (1996) 98–101.
[141] S. J. Koh, G. Ehrlich, Self-assembly of one-dimensional surface structures: Long-range interactions in the growth of Ir and Pd on W(110), *Phys. Rev. Lett.* **87** (2001) 106103 1–4.
[142] V. S. Stepanyuk, A. N. Baranov, D. V. Tsivlin, W. Hergert, P. Bruno, N. Knorr, M. A. Schneider, K. Kern, Quantum interference and long-range adsorbate-adsorbate interactions, *Phys. Rev. B* **68** (2003) 205410 1–5.
[143] N. N. Negulyaev, V. S. Stepanyuk, L. Niebergall, P. Bruno, W. Hergert, J. Repp, K. H. Rieder, G. Meyer, Direct evidence for the effect of quantum confinement of surface-state electrons on atomic diffusion, *Phys. Rev. Lett.* **101** (2008) 226601 1–4.
[144] W. K. Burton, N. Cabrera, F. C. Frank, The growth of crystals and the equilibrium structure of their surfaces, *Phil. Trans. Roy. Soc. A* **243** (1951) 299–358.
[145] K. M. Bowkett, D. A. Smith, *Field Ion Microscopy* (North-Holland, Amsterdam, 1970), Chapter 5.
[146] B. Mutaftschief, *The Atomistic Nature of Crystal Growth* (Springer-Verlag, Berlin, 2001), Section 17.3.
[147] G. Antczak, P. Jóźwik, Atom movement on a dislocated surface, *Langmuir* **24** (2008) 9970–9973.

7 Mechanism of cluster diffusion

So far we have concentrated on the behavior of single atoms. However, when several atoms are present on a plane and they diffuse, atoms may collide with each other and form a cluster. Such events are illustrated in Figs. 7.1, 7.2, and 7.3 [1–3], where coalescence of two as well as three atoms is observed directly using the field ion microscope. These clusters are of considerable interest for the roles they play in the growth and dissolution of a crystal, as well as their effects on surface chemical reactions. Of primary concern here is the ability of atom clusters to diffuse over a crystal surface, and it is this aspect of cluster properties that we will emphasize. We will look at the conditions under which a cluster moves as a whole and also when parts of a cluster start moving independently. Different mechanisms of diffusion will be discussed in some detail.

In probing the diffusion of clusters, it is worthwhile to distinguish two different types of mechanisms – movement by single atom jumps, and by concerted atom displacements. In the first category, five types of movement have so far been identified, which are: 1. diffusion by sequential atom jumps (Fig. 7.4a), 2. peripheral displacements (Fig. 7.4b), 3. by the leapfrog mechanism (Fig. 7.4c), 4. by the correlated evaporation–condensation mechanism also known as detachment–attachment (Fig. 7.4d) or terrace limited diffusion, and 5. by the evaporation–condensation mechanism (Fig. 7.4e), in which one atom leaves a cluster and then a different atom from the terrace attaches to the cluster.

Concerted displacements recognized so far are illustrated in Fig. 7.5. They may occur by sliding of the cluster as a whole (Fig 7.5a), by dimer or trimer shearing shown in Fig 7.5b or by reptation in Fig 7.5c, and finally by motion of a dislocation (Fig. 7.5d). Recently a new concerted motion was discovered. Trushin *et al.* [4], during investigations of Cu clusters on a Cu(100) surface using EAM, noticed an interesting mechanism contributing to movements of large clusters – internal dimer rotation, illustrated in Fig. 7.6. Internal dimer rotation differs from isolated dimer rotation since it involves concerted motion of all cluster atoms at the moment of transition.

We will start with a rather simple example, the diffusion of a dimer, made up of two atoms, moving in one dimension. This will provide a formal but simple introduction to brief presentations of the motion of larger clusters. Only after this outline of the formalism will we discuss the investigations, both experimental as well as theoretical, carried out to examine processes involved in cluster diffusion.

518   **Mechanism of cluster diffusion**

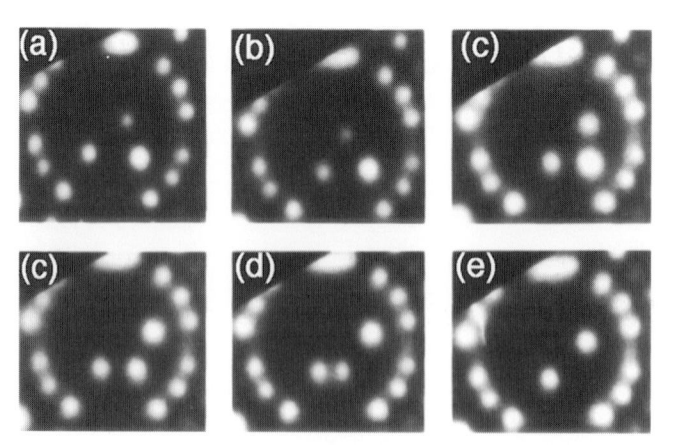

Fig. 7.1   Formation of Re$_2$ on W(211) plane [1]. In (a), three rhenium atoms have been deposited on the plane, with two in one channel. Warming for 30 sec to 286 K induces motion in (b)–(e), and ends in creating an in-channel dimer.

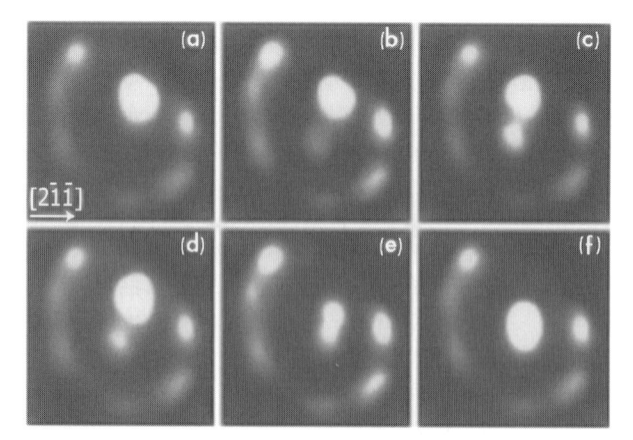

Fig. 7.2   Creation of Pt$_2$ cluster on Pt(111) plane [2]. In (a) a single Pt atom has been deposited on the surface maintained at ~ 30 K. Another atom has been deposited in the center of the plane in (b). Five second intervals of heating to ~ 100 K bring about atom motion in (c)–(e). Dimer, which starts to diffuse at ≈ 150 K, has been formed in (f).

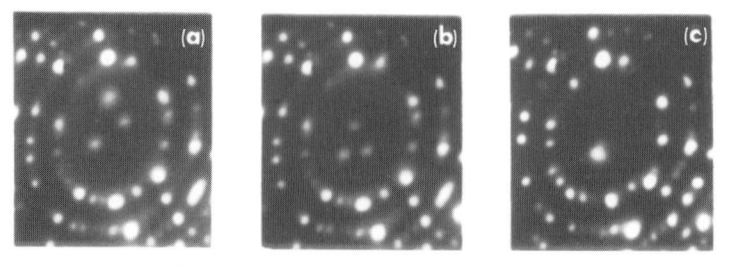

Fig. 7.3   Formation of rhenium trimer on W(110) plane [3]. In (a) three Re atoms have been deposited on a cold (~ 20 K) surface. (b) After 10 sec at 420 K, atoms have moved, and after another such heating interval have combined into a triangular trimer in (c).

Mechanism of cluster diffusion

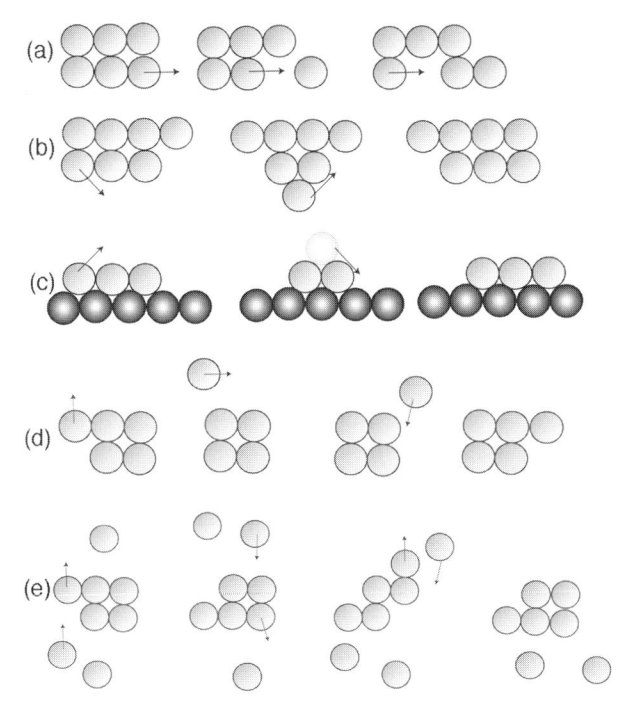

519

Fig. 7.4 Schematics of single atom events in cluster diffusion. (a) Sequential displacements, top view. (b) Peripheral diffusion, top view. (c) Leapfrog movement, side view. (d) Detachment–attachment mechanism, top view. (e) evaporation–condensation.

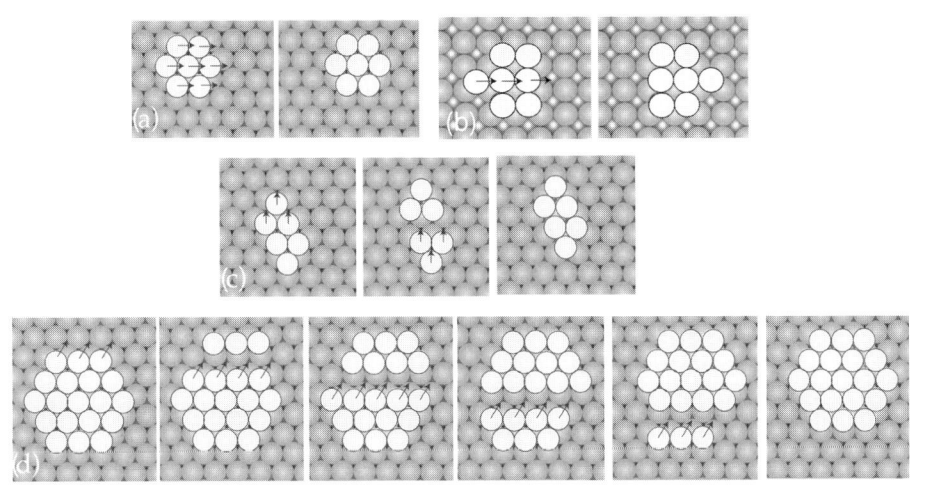

Fig. 7.5 Schematics showing concerted mechanisms of cluster motion. (a) Gliding. (b) Shearing. (c) Reptation. (d) Via Dislocation.

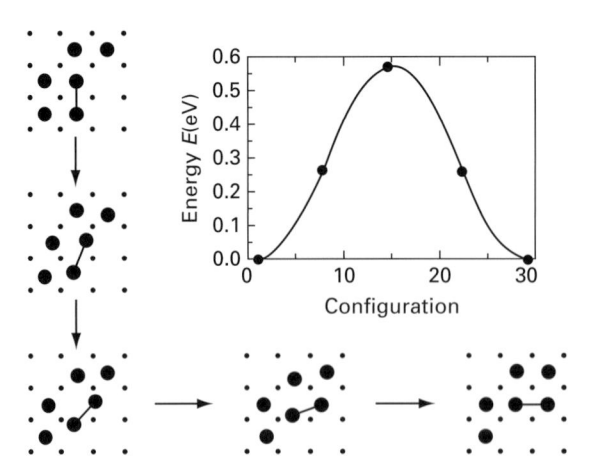

Fig. 7.6 Potential energy calculated for internal Cu dimer rotation in hexamer on Cu(100) using EAM interactions. Schematics at the border show intermediate states of dimer rotating in hexamer (after Trushin et al. [4]).

7.1 Via single atom jumps

7.1.1 One-dimensional movement of dimers

At the start we confine our attention to one-dimensional motion of a dimer with one atom in each of two adjacent diffusion channels, as on the (211) plane of a bcc crystal, and will focus on the displacement of the center of mass (COM). What we present is based largely on work by Reed [5], but an alternative presentation, based on analogy with electrical networks, has been given by Titulaer and Deutch [6]. Dimer movement is sketched schematically in Fig. 7.7. At the top of the diagram are shown the two configurations of the dimer, designated by 0 when the two atoms are in line, so that the dimer is straight, and by 1 when the dimer is slanted. The spacing of sites in the channel is taken as ℓ_0, the nearest-neighbor spacing, so that when one atom in the dimer jumps, the center of mass moves a distance $\ell_0/2$, which will be the unit of length for now. In state 0, the straight state, an atom can jump to the right at the rate a, and to the left at the rate d. In state 1, an atom jump to the right occurs at the rate b, and to the left at the rate c. Rates of jumps to the right at position x can more generally be written as λ_x, and the rate to the left, again from position x, as μ_x.

Just as we did for the motion of single atoms, we can write out the Kolmogorov relation for dimers as

$$dp_x/dt = \lambda_{x-1} p_{x-1} - (\lambda_x + \mu_x) p_x + \mu_{x+1} p_{x+1} \qquad x = 0, \pm 1, \pm 2, \ldots \qquad (7.1)$$

where p_x is the probability of having the center of mass at position x. On multiplying this equation by x and summing over all positions, we obtain the differential equation governing the average of x. The average is given by

$$\langle x \rangle = \sum_x x p_x, \qquad (7.2)$$

7.1 Via single atom jumps

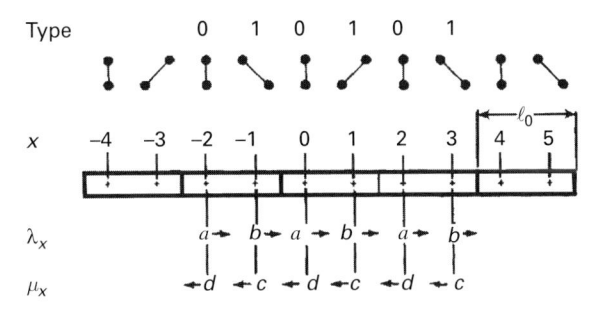

Fig. 7.7 Diffusion of cross-channel dimer represented as a random walk [5]. Dimer configurations are indicated for center of mass position x. Heavy lines show unit cell. λ_x gives rate of center of mass motion to the right, μ_x to the left.

so that

$$d\langle x\rangle/dt = \sum_x \lambda_{x-1} p_{x-1} x - \sum_x (\lambda_x + \mu_x) p_x x + \sum_x \mu_{x+1} p_{x+1} x \qquad (7.3)$$

$$= \langle (x+1)\lambda_x\rangle - \langle x\lambda_x\rangle + \langle (x-1)\mu_x\rangle - \langle x\mu_x\rangle \qquad (7.4)$$

and finally

$$d\langle x\rangle/dt = \langle \lambda_x\rangle - \langle \mu_x\rangle. \qquad (7.5)$$

Integrating over time we find that

$$\langle x\rangle = (\langle \lambda_x\rangle - \langle \mu_x\rangle)t, \qquad (7.6)$$

as at time $t=0$, the mean displacement is equal to zero. Henceforth it will be useful to indicate the state of the dimer at the starting site by the superscript z, where z can be either 0 or 1. The probability of the center of mass being at a site of type 0 is given by

$$P_0^{(z)} = \sum_k p_k \qquad k = 0, \pm 2, \pm 4, \ldots \qquad (7.7)$$

Similarly, for being at a site of type 1, the probability is

$$P_1^{(z)} = \sum_k p_k \qquad k = 0, \pm 1, \pm 3, \pm 5, \ldots \qquad (7.8)$$

furthermore

$$P_0^{(z)} + P_1^{(z)} = 1, \qquad (7.9)$$

as there are only two types of sites present, type 0 and 1.
The rate constants averaged over all positions may be written as

$$\langle \lambda_x\rangle = \sum_x p_x \lambda_x = a P_0^{(z)} + b P_1^{(z)} \qquad (7.10)$$

$$\langle \mu_x \rangle = \sum_x p_x \mu_x = dP_0^{(z)} + cP_1^{(z)}. \tag{7.11}$$

The mean displacement for our dimer from Eq. (7.6) therefore becomes

$$\langle x \rangle = \left[(a-d)P_0^{(z)} + (b-c)P_1^{(z)} \right] t, \tag{7.12}$$

and the mean displacement will in general have a finite value as time increases. It is now a straightforward matter to work out the variance of the displacement, that is the displacement fluctuation $\langle \Delta x^2 \rangle = \langle x^2 \rangle - \langle x \rangle^2$, since we already know the value of the average of x, $\langle x \rangle$. Following the procedures used to obtain $\langle x \rangle$, we multiply the probability p_x by x^2 and sum over all values of x to get

$$\langle x^2 \rangle = \sum_x x^2 p_x, \tag{7.13}$$

so that

$$d\langle x^2 \rangle / dt = \sum_x \lambda_{x-1} p_{x-1} x^2 - \sum_x (\lambda_x + \mu_x) p_x x^2 + \sum_x \mu_{x+1} p_{x+1} x^2 \tag{7.14}$$

$$= \left\langle (x+1)^2 \lambda_x \right\rangle - \left\langle x^2 \lambda_x \right\rangle + \left\langle (x-1)^2 \mu_x \right\rangle - \left\langle x^2 \mu_x \right\rangle \tag{7.15}$$

$$= \langle \lambda_x (2x+1) \rangle - \langle \mu_x (2x-1) \rangle. \tag{7.16}$$

For the time derivative of the displacement fluctuation we have

$$d\langle \Delta x^2 \rangle / dt = d\langle x^2 \rangle / dt - 2\langle x \rangle d\langle x \rangle / dt \tag{7.17}$$

and therefore

$$\begin{aligned} d\langle \Delta x^2 \rangle / dt &= \langle (2x+1)\lambda_x \rangle - \langle (2x-1)\mu_x \rangle - 2\langle x \rangle \langle \lambda_x \rangle + 2\langle x \rangle \langle \mu_x \rangle \\ &= 2(\langle x\lambda_x \rangle - \langle x\mu_x \rangle) + \langle \lambda_x \rangle (1 - 2\langle x \rangle) + \langle \mu_x \rangle (1 + 2\langle x \rangle). \end{aligned} \tag{7.18}$$

In evaluating the displacement fluctuation, we will limit ourselves here to symmetrical dimer diffusion, in which the jump rate d to the left is equal to the rate a to the right, and the rate c to the left equals the rate b, so that in accord with Eq. (7.12) the mean displacement vanishes.

Under these conditions,

$$\langle x\lambda_x \rangle = \sum_x x\lambda_x p_x = a\langle x \rangle_0 + b\langle x \rangle_1 \tag{7.19}$$

$$\langle x\mu_x \rangle = \sum_x x\mu_x p_x = a\langle x \rangle_0 + b\langle x \rangle_1 \tag{7.20}$$

Here $\langle x \rangle_i$ denotes the value of the position x averaged over all sites of type i. The differential equation for the displacement fluctuation now appears as

7.1 Via single atom jumps

$$d\langle\Delta x^2\rangle/dt = 2[a\langle x_0\rangle + b\langle x_1\rangle - (a\langle x_0\rangle + b\langle x_1\rangle)]$$
$$+ (aP_0 + bP_1)(1 - 2\langle x\rangle) + (aP_0 + bP_1)(1 + 2\langle x\rangle)$$
$$= 2(aP_0 + bP_1). \tag{7.21}$$

The probability $P_0^{(z)}$ is readily accessible by summing the probability p_k over all sites of type 0, so it follows from Eq. (7.1) that

$$dP_0/dt = \sum_x \lambda_{x-1} p_{x-1} - \sum_x (\lambda_x + \mu_x) p_x + \sum_x \mu_{x+1} p_{x+1} \quad x = 0, \pm 2, \pm 4, \cdots \tag{7.22}$$

For the general case of unsymmetric motion we find

$$dP_0/dt = bP_1 - (a+d)P_0 + cP_1 = (b+c)P_1 - (a+d)P_0. \tag{7.23}$$

Since $P_1 = 1 - P_0$

$$dP_0/dt + (a+b+c+d)P_0 = b+c. \tag{7.24}$$

Solving this differential equation we obtain

$$P_0^0 = \frac{1}{a+b+c+d}[b+c+(a+d)\exp-(a+b+c+d)t], \tag{7.25}$$

inasmuch as at $t=0$, $P_0^0 = 1$. If now we again restrict ourselves to symmetrical walks with $a = d$ and $b = c$, then

$$P_0^0 = \frac{1}{a+b}[b + a\exp-2(a+b)t] \tag{7.26}$$

$$P_1^0 = \frac{a}{a+b}[1 - \exp-2(a+b)t]. \tag{7.27}$$

In a similar series of steps we find

$$P_0^1 = \frac{b}{a+b}[1 - \exp-2(a+b)t] \tag{7.28}$$

$$P_1^1 = \frac{1}{a+b}[a + b\exp-2(a+b)t]. \tag{7.29}$$

Inserting these two results for starting at position 0 in the differential equation for the displacement fluctuation, Eq. (7.21), we see that

$$d\langle\Delta x^2\rangle^0/dt = \frac{2a}{a+b}[2b + (a-b)\exp-2(a+b)t]. \tag{7.30}$$

After integration, this transforms to

$$\langle\Delta x^2\rangle^0 = \frac{2a}{a+b}\left[2bt + \frac{(a-b)}{2(a+b)}(1 - \exp-2(a+b)t)\right]. \tag{7.31}$$

If instead we insert the expressions appropriate for starting at position 1 we obtain

$$d\langle \Delta x^2 \rangle^1 / dt = \frac{2b}{a+b}[2a - (a-b)\exp{-2(a+b)t}], \quad (7.32)$$

and on integrating get

$$\langle \Delta x^2 \rangle^1 = \frac{2b}{a+b}\left[2at - \frac{(a-b)}{2(a+b)}(1 - \exp{-2(a+b)t})\right], \quad (7.33)$$

all of which is in agreement with what was found by Reed [5].

Our concern is primarily with longer distance diffusion, that is with long diffusion times. Under these conditions, the second term in Eqs. (7.31) and (7.33) becomes vanishingly small, and the displacement fluctuation simplifies to

$$\langle \Delta x^2 \rangle = \frac{4abt}{a+b}, \quad (7.34)$$

and is independent of the starting point. For long diffusion times the probability of being at a particular type of site also simplifies, and we have

$$P_0 = \frac{b}{a+b} \qquad P_1 = \frac{a}{a+b}. \quad (7.35)$$

Under these circumstances, local equilibrium prevails for the dimer configurations, so that

$$P_0 a = P_1 b, \quad (7.36)$$

and from the ratio of the occupation probabilities we can obtain the ratio of jump rates

$$\frac{P_1}{P_0} = \frac{a}{b}. \quad (7.37)$$

So far we have used the displacement $\ell_0/2$ as the unit of length. If, as is more usual, we resort to standard units then

$$\langle \Delta x^2 \rangle = \frac{ab}{a+b}\ell_0^2 t = \frac{a}{1+a/b}\ell_0^2 t. \quad (7.38)$$

It is obvious that from diffusion measurements of the center of mass we do not find out the jump rates a and b. However, by combining determinations of the displacement fluctuation $\langle \Delta x^2 \rangle$ with measurements of the ratio P_1/P_0 it is simple to ascertain the individual jump rates. Here we note that in position 1 two configurations are possible, whereas at 0 positions only one can exist. For the rate a we therefore have

$$a = 2\nu_a \exp(-E_a/kT), \quad (7.39)$$

and for b

$$b = \nu_b \exp(-E_b/kT). \quad (7.40)$$

7.1 Via single atom jumps

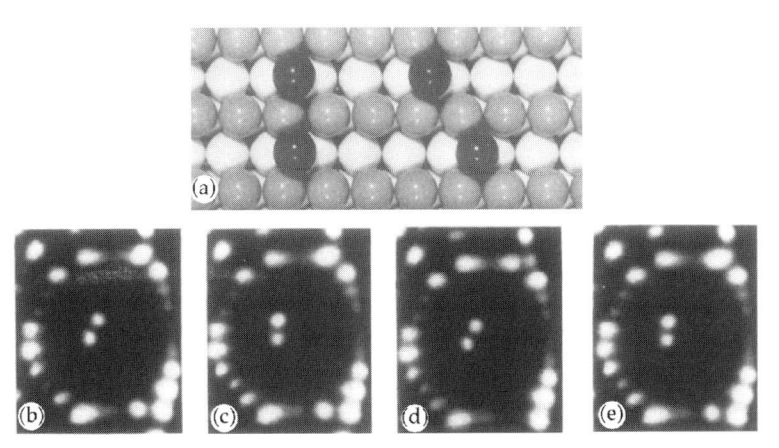

Fig. 7.8 Tungsten dimer diffusion on W(211) [8]. (a) Hard-sphere model of W(211) plane with cross-channel dimer in straight and staggered configuration. (b)–(e) FIM images of tungsten dimer diffusing on W(211) plane at 255 K.

It is worth noting that expressions for the displacement fluctuation could alternatively have been obtained from the moment generating function, as was done for the diffusion of single atoms. Derivations of this type have been given by Wrigley et al. [7], but do not yield much more than has been found here by direct calculations. We also want to emphasize that these derivations have been important for giving us Eqs. (7.37) and (7.38), which have been instrumental in analyzing data on dimer diffusion.

Movement of cross-channel dimers can be easily observed in the FIM, where both dimer atoms are recognizable and the configuration is readily observed. Diffusion of a tungsten cross-channel dimer on the W(211) plane was reported in 1974 by Graham and Ehrlich [8]. Their data, illustrated in Fig. 7.8, clearly showed that diffusion occurred by single atom transitions.

Let us now look at the diffusion of an in-channel dimer, as was first done by Reed [1]. Such a dimer may be able to move as a whole, in a concerted jump, or else by two jumps of single atoms. The latter mechanism is illustrated in Fig. 7.9, and the displacement fluctuation is given by Eq. (7.38), derived for the diffusion of cross-channel dimers. There is, however, a difference in the rate constant a of the in-channel dimer. This lacks the factor of two present for cross-channel dimers, shown in Eq (7.39); the latter can exist in two energetically equal states, which is not the case for the in-channel geometry. If motion occurs by concerted jumps, diffusion can be described as a random walk of the center of mass, and follows the rules already laid out in Chapter 1.

When dealing with the diffusion of single atoms, we derived the distribution of displacements in addition to the displacement fluctuation, as that provided information about the kinds of jumps participating in the diffusion. The displacement distribution has also been obtained for the diffusion of dimers on the same assumptions used here. From the distribution it is possible to arrive at the jump rates a and b, but knowledge of these is more readily attained from the displacement fluctuation combined with the ratio of occupation probabilities. For this reason we eschew discussion of the distribution of displacements and just refer to the literature [7].

Fig. 7.9 Comparison of atom motion in cross-channel and in-channel dimers, showing configurations and transition rates for motion by jumps of individual atoms [1].

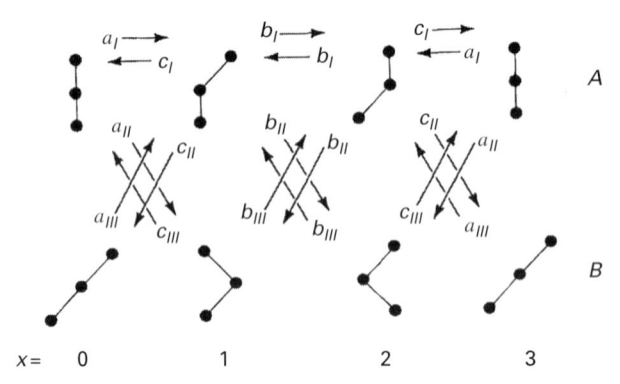

Fig. 7.10 Cross-channel trimer configurations and rate constants in diffusion by transitions of individual atoms [7]. Configurations of different energies at the same center of mass site x are distinguished by A and B.

7.1.2 One-dimensional movement of trimers

The same general procedures can also be applied to the examination of larger clusters, such as one-dimensional trimer diffusion, but the schematic of trimer jump rates and configurations in Fig. 7.10 makes it clear that this is a more complicated task. This problem has been worked out by Wrigley et al. [7]. The details will not be considered here. The trimer can have a center of mass in the same place for two energetically different configurations, which are marked in Fig. 7.10 by A and B. The smallest displacement of the center of mass made by a trimer moving by individual jumps is $\ell_0/3$. The displacement fluctuation (in units of $\ell_0/3$) of the trimer center in the limit of long times is given by

7.1 Via single atom jumps

$$\langle(\Delta x^2)\rangle = 4tP_{0A}[(b_I - c_I + b_{II} - c_{II})\Delta_A/\Delta + (b_{III} - c_{III})\Delta_B/\Delta \\ + c_I(1 + a_{II}/a_I) + (b_I + 2b_{II} + 2c_{II})/2]a_I/c_I \quad (7.41)$$

where

$$\Delta_A = [b_{III}(b_{II} - a_{II}c_I/a_I) - (b_I - c_I + b_{II} - c_{II})(b_{III} + c_{III})] \quad (7.42)$$

$$\Delta_B = [b_{II}(b_I - c_I + b_{II} - c_{II}) - (b_{II} - a_{II}c_I/a_I)(2b_I + c_I + b_{II} + c_{II})] \quad (7.43)$$

$$\Delta = (2b_I + c_I + b_{II} + c_{II})(b_{III} + c_{III}) - b_{II}b_{III} \quad (7.44)$$

and P_{0A}, the probability of finding a trimer in configuration $0A$, regardless of the position x, is

$$P_{0A} = [1 + 2a_I/c_I + 2a_{II}/c_{III} + a_Ic_{II}/(c_Ia_{III})]^{-1}. \quad (7.45)$$

These tedious expressions simplify greatly under some obvious conditions. Should diffusion occur only through configurations A, as would be the case if $a_{II} = b_{II} = c_{II} = 0$, then

$$<(\Delta x)^2> = \frac{18a_Ib_Ic_It}{(2a_I + c_I)(2b_I + c_I)}. \quad (7.46)$$

If, on the other hand diffusion takes place through $0B$, $1A$, and $2A$, as it would when $b_{II} = c_I = 0$, then the displacement fluctuation is given by

$$<(\Delta x)^2> = \frac{18a_{III}b_Ic_{II}t}{(2a_{III} + c_{II})(2b_I + c_{II})}. \quad (7.47)$$

Here it must be remembered that the unit displacement is always $\ell_0/3$.[1]

Cross-channel tungsten trimers were observed in the FIM by Graham and Ehrlich on a W(211) surface [8] and are shown in Fig. 7.11. All atoms in the trimer are resolved and it

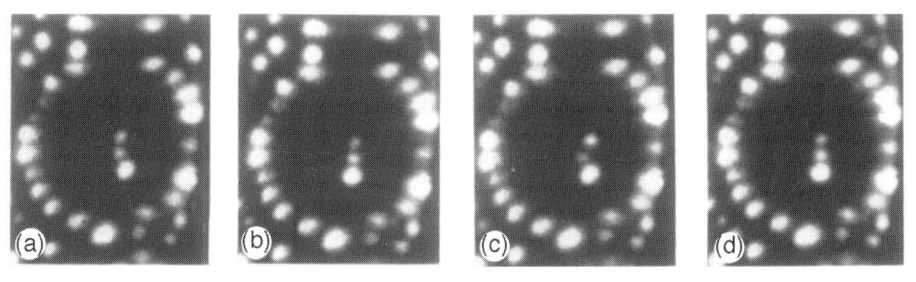

Fig. 7.11 Tungsten cross-channel trimer diffusion on W(211) plane observed in FIM after diffusion at 277 K [8].

---

[1] Also worked out has been the displacement fluctuation of dimers in one-dimensional diffusion undergoing dissociation, a rather more complicated problem [7].

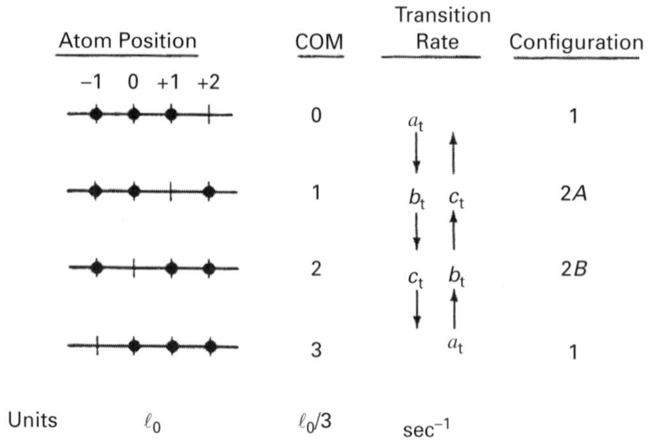

Mechanism of cluster diffusion

Atom Position −1 0 +1 +2	COM	Transition Rate	Configuration
—•—•—•—⊢—	0	a_t ↓ ↑	1
—•—•—⊢—•—	1	b_t ↓ ↑ c_t	2A
—•—⊢—•—•—	2	c_t ↓ ↑ b_t	2B
—⊢—•—•—•—	3	↓ ↑ a_t	1
Units	ℓ_0	$\ell_0/3$	sec^{-1}

Fig. 7.12 Schematic of in-channel trimer motion, showing jump rates and spatial configurations assuming single atom transitions [1].

appears that, just as with cross-channel dimers, diffusion takes place by single atom transitions.

Diffusion of in-channel trimers is easy to describe if it occurs by jumps of individual atoms. A schematic for such motion is given in Fig. 7.12, and by analogy with previous discussions of cross-channel trimers, the displacement fluctuation for the in-channel trimer is found as

$$\langle \Delta x^2 \rangle = \frac{2 a_t b_t c_t \ell_0{}^2 t}{(2a_t + c_t)(2b_t + c_t)}. \tag{7.48}$$

It must be emphasized again that everything has been based on the assumption that dimer and trimer atoms make individual jumps. This certainly seems to be the case for cross-channel clusters, but is probably not generally valid when all the cluster atoms lie in one channel, as in Fig. 7.9 or 7.12. It should be noted that we have furthermore assumed that clusters move by atoms making simple jumps, with no exchanges with the lattice. The latter jump mechanism would be readily discerned, as it would result in cross-channel transitions that are immediately apparent.

7.1.3 Leapfrog cluster diffusion

So far we have considered diffusion in which a single cluster atom advances in some way; the rest of the cluster atoms follow one by one, and in this way the cluster moves over the crystal surface. An alternative mechanism to such diffusion was discovered by Linderoth et al. [9], who studied motion of platinum clusters on Pt(110)-(1 × 2). What they found is demonstrated by STM images as well as schematics in Fig. 7.13. Instead of the front atom of the in-channel tetramer displacing, the atom in the back jumps onto the top of the cluster, advances over it, and eventually returns to a position in front of what

7.1 Via single atom jumps

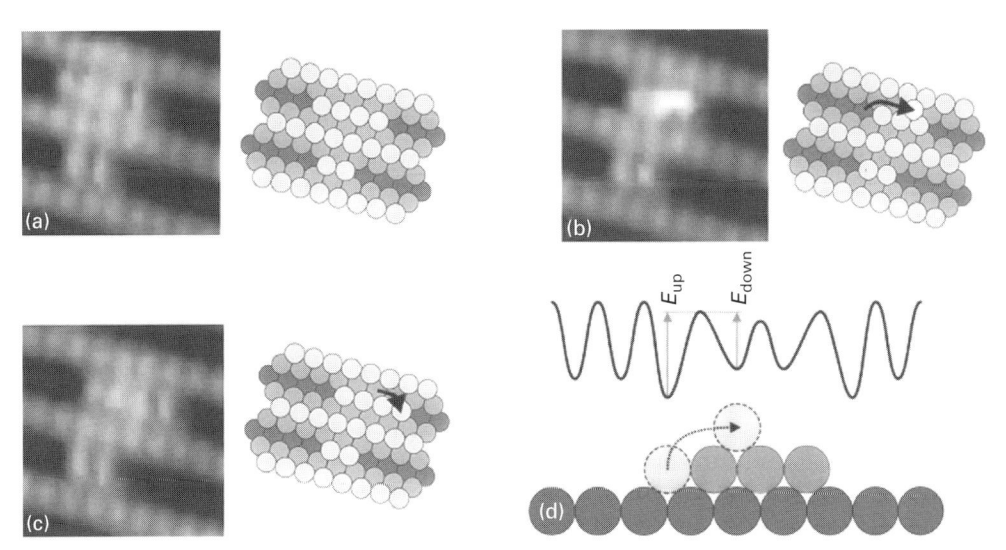

Fig. 7.13 Leapfrog movement of in-channel Pt tetramer on Pt(110)-(1 × 2) surface at 313 K. STM images at the left, schematics at the right. In (b) atom from end of cluster has transitioned to the top, and incorporated at the front in (c). (d) Schematic of leapfrog movement and potential for Pt end-atom diffusing over the cluster on Pt(110)-(1 × 2) (after Linderoth et al. [9]).

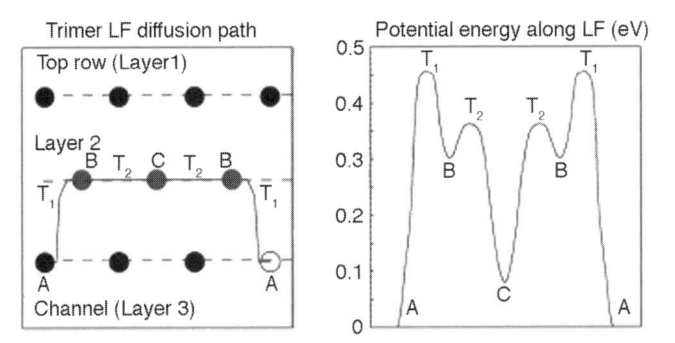

Fig. 7.14 Leapfrog movement of trimer. Left: Diffusion path of trimer. Right: Potential energy changes in leapfrog cluster motion (after Montalenti and Ferrando [10]).

had been the first atom. In this way the center of mass of the whole cluster changes its position. A schematic potential for this transition is given in Fig. 7.13d. It should be noted that such transitions are peculiar of the reconstructed fcc(110) surface, and are analogous to the metastable walks observed in single atom diffusion on Pt(110)-(1 × 2) and discussed in Section 3.3.3, in which the adatom can diffuse speedily by jumping up onto the sidewalls of the channel. Leapfrog diffusion over a cluster has been shown energetically advantageous in calculations with RGL potentials by Montalenti and Ferrando [10], whose results for an in-channel Pt$_3$ cluster are shown in Fig. 7.14. In the absence of reconstruction of the fcc(110) surface, this process has so far never been observed and is unlikely to occur.

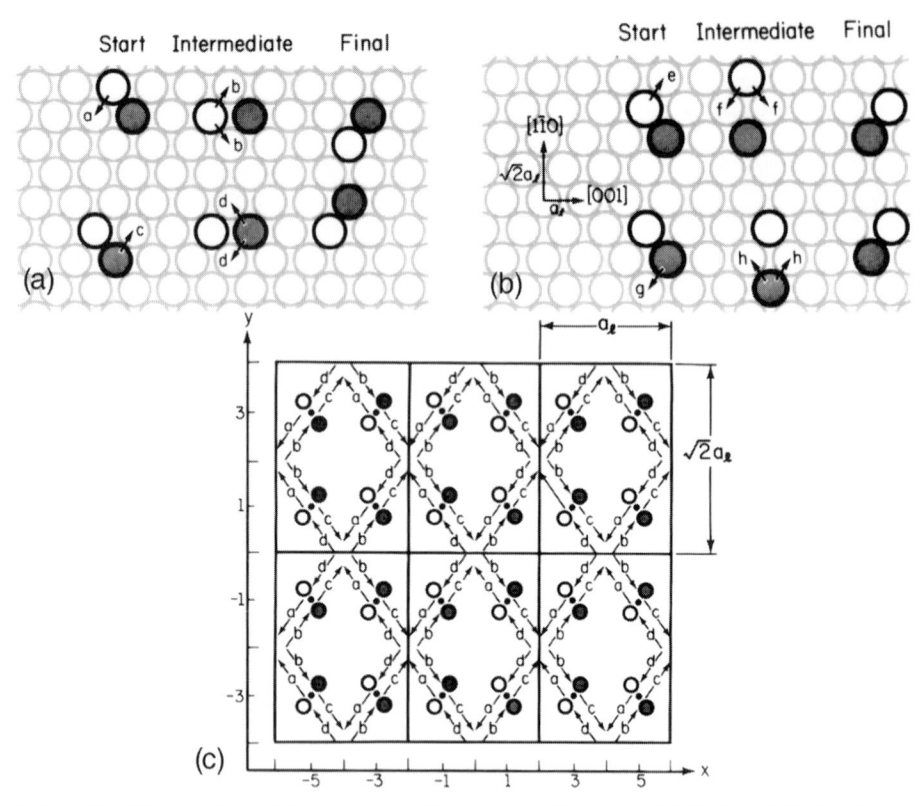

Fig. 7.15 Schematic of hetero-dimer states and rates of jumps occurring on bcc(110) plane [11]. (a) Via horizontal. (b) Vertical intermediates. (c) Grid for center of position COP, jump rates, and cluster configurations in diffusion via horizontal transitions on bcc(110) plane, assuming atoms make individual transitions. Solid lines indicate unit cell for COP grid.

7.1.4 Two-dimensional dimer movement

Analysis has not been confined to just one-dimensional motion of clusters. The two-dimensional movement of hetero-dimers has also been carried through [11], assuming that diffusion occurs by single atom jumps. On a bcc(110) surface, diffusion can occur via horizontal intermediates, shown in Fig. 7.15a, or through vertical intermediates, in Fig. 7.15b. Expressions have been worked out for the displacement fluctuation, again on the assumption that jumps of the two atoms take place independently. A grid of the positions of the dimer center, jump rates, and configurations is shown in Fig. 7.15c for motion via horizontal transitions, and the displacement fluctuation along the x-axis in the limit of long time intervals and in units of $a_\ell/4$, where a_ℓ is the lattice spacing, turns out to be

$$\langle(\Delta x)^2\rangle = \frac{16bct}{[a+c+2(b+d)]}. \tag{7.49}$$

The mean-square displacement along the y-axis, again at long times and in units of $\sqrt{2}a_\ell/4$, is given by

7.1 Via single atom jumps

$$<(\Delta y)^2> = \frac{8act}{[(a+c)(1+a/2b)]} \approx 8act/(a+c). \tag{7.50}$$

When we take into account horizontal as well as vertical intermediate states, the displacement fluctuation along the x-axis changes to

$$<(\Delta x)^2> = 8P_1 t\left[\frac{ac}{a+c} + \frac{eg}{e+g}\right], \tag{7.51}$$

with the new rate constants defined in Fig. 7.15b. The displacement fluctuation depends on the time t, on the probability P_1 of dimers being in an odd state, as well as on the effective rates of jumping out of that state. The probability P_1 of dimers is given by

$$P_1 = 1/(1 + \frac{a}{2b} + \frac{e}{2f}). \tag{7.52}$$

For a simple dimer, with both atoms of the same material, $c = a$ and $d = b$.

For such a dimer the displacements along x and y are equal except for the length of the unit step: $a_\ell/4$ along x and $\sqrt{2}a_\ell/4$ along y. The displacement fluctuation emerges as

$$16\langle\Delta x^2\rangle = 8\langle\Delta y^2\rangle = \frac{a_\ell^2}{a+2b}\left\{8abt + \frac{a(a-2b)}{a+2b} \times [1 - \exp{-2(a+2b)t}]\right\}. \tag{7.53}$$

Atom jumps leading to dimer diffusion on the (100) plane are shown in Figs. 7.16a and b when simple hops take place. The same final configurations can be obtained in translation and rotation of dimers when atom exchange with the lattice occurs, as is described in the next section. From Fig. 7.16a and b it is clear that a COM displacement of $a_\ell/2$ will occur along both coordinates when an atom moves by hopping. Rotation of the dimer as well as translation can be achieved by single atom hopping, however rotation requires only one jump while translation involves at least two. Based on the previous presentation it is possible to work out an expression for the mean-square displacement of a simple dimer on the (100) plane. After changing to ψ and ξ axes at 45° to the Cartesian coordinates, as in Fig. 7.16c (which also defines the various symbols), this immediately gives the displacement fluctuation in Eq. (7.54)

$$\langle\Delta\psi^2\rangle = \langle\Delta\xi^2\rangle = \frac{a_\ell^2}{a+2b}\left\{8abt + \frac{a(a-2b)}{a+2b} \times [1 - \exp{-2(a+2b)t}]\right\}. \tag{7.54}$$

For long diffusion times this reduces to

$$\langle\Delta\psi^2\rangle = \langle\Delta\xi^2\rangle = \frac{8abta_\ell^2}{a+2b}. \tag{7.55}$$

The mechanism by which rotation of the dimer takes place is not easy to establish, however. We have so far assumed that ordinary atom hops are involved in diffusion. An alternative is possible – diffusion by one of the atoms undergoing exchange with a lattice atom, as was illustrated in Fig. 3.53. This yields exactly the same displacement as ordinary atom jumps, and it may therefore not be simple to distinguish the two, even by measuring the distribution of displacements.

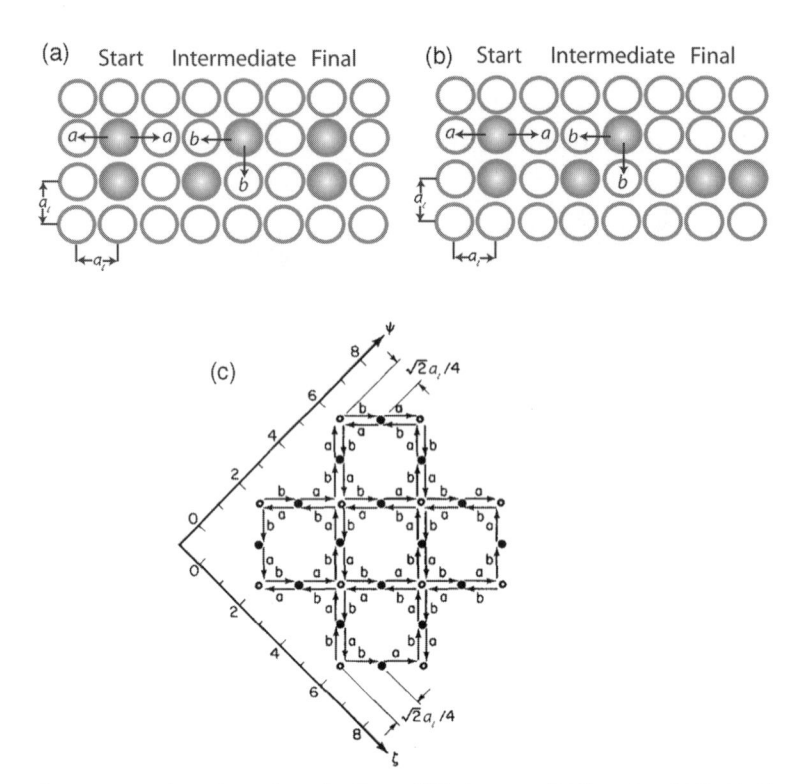

Fig. 7.16 Jump rates and configurations for dimer diffusion on a (100) surface [11]. (a) Simple dimer diffusion which results in translation of dimer. (b) Diffusion with rotation. (c) Schematic showing grid for center of mass as well as transitions for simple dimer on (100) referenced to ψ and ζ coordinates, along which dimer states fall into either even or odd.

7.1.5 Cluster diffusion by atom exchange

So far, the emphasis has generally been on simple atom jumps in the mobility of atom clusters over a surface. However, as was already pointed out in Chapter 3, exchange of an adatom with an atom from the substrate is known to play an important role on fcc planes. The possibility therefore also exists that such exchange processes may participate in the diffusion of clusters. For the sake of simplicity, we shall limit ourselves to the diffusion of dimers. A larger examination of the mechanisms of cluster diffusion has been given by Chang et al. [12], on which we have relied in this brief summary.

The most obvious possibility of exchange holds for dimer motion on the fcc(110) plane. The usual picture for an in-channel dimer, shown in Fig. 7.17a, is for one atom to jump along the channel, with the second one following subsequently. Also possible, however, is a concerted transition of both atoms, as in Fig. 7.17b. Cross-channel motion of the dimer can also occur by single atom hopping from one channel to the next, but from what is known about single atom transitions this is probably a high energy process. A more likely possibility for a cross-channel transition is sketched in Fig. 7.18; one atom can carry out an exchange with an atom from the lattice, followed by a similar transition

7.1 Via single atom jumps

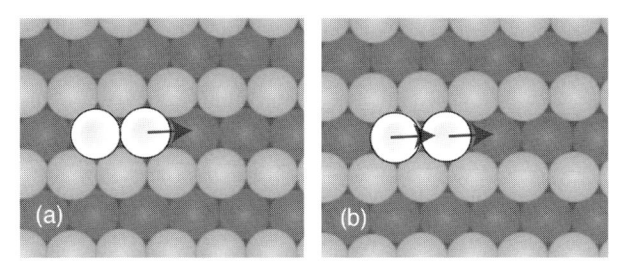

Fig. 7.17 Schematics of in-channel motion of dimer. (a) One dimer atom jumps, the second follows subsequently. (b) Two-atom movement of dimer.

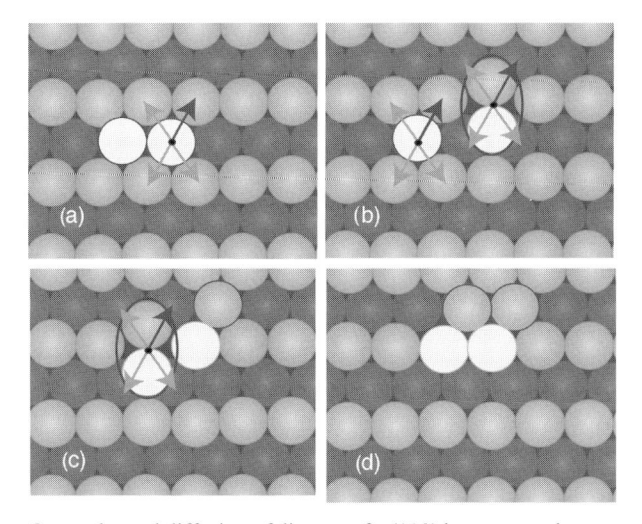

Fig. 7.18 Cross-channel diffusion of dimer on fcc(110) by atom exchange, one atom at a time. (a) First dimer atom starting exchange with the wall atom. (b) First dimer atom creates dumbbell with wall atom, second atom starting exchange. (c) First atom finished exchange, while second atom creates dumbbell with the wall atom. (d) Both dimer atoms have finished exchange, creating dimer in next channel. Arrows show all possible directions for exchange, black arrows indicate direction chosen for illustration. Dumbbell transition state is marked by ellipse.

by the second atom. We illustrate one possible way for the dimer to exchange in a single-atom event. Arrows in Fig. 7.18a indicate possible directions of exchange, the black arrow shows the direction chosen for the illustration. The intermediate stage of the exchange mechanism is creation of a dumbbell, as already discussed in Chapter 3 and shown in Fig. 3.42. Dissociation of the dumbbell can proceed in four directions on an fcc(110) surface. For single atom exchange all these directions are supposed to be equal; however, experiments indicate otherwise. In the case of a dimer after creation of the dumbbell, shown in Fig. 7.18b, it is clear that dissociation of the dumbbell will *not* occur with the same probability in all indicated directions, since dissociation will be influenced by the presence of the second adatom on the surface. Exchange of only one dimer atom will result in the creation of a cross-channel dimer. As an alternative, the two dimer atoms

534   **Mechanism of cluster diffusion**

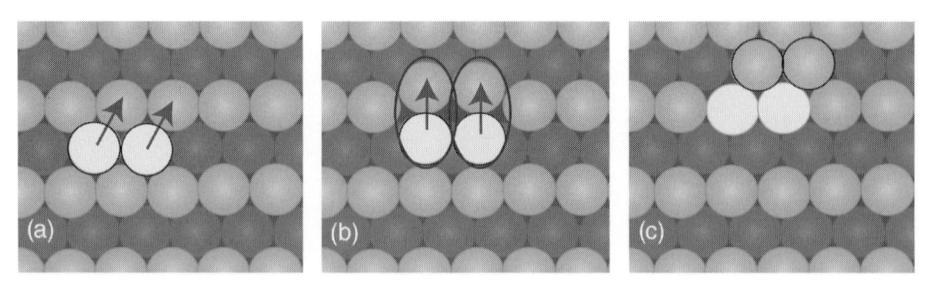

Fig. 7.19   Cross-channel diffusion of dimer on fcc(110) by a two atom exchange process. Dumbbell transition state is marked by ellipse.

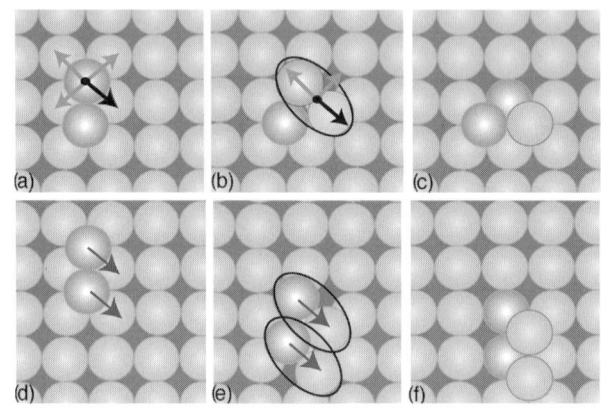

Fig. 7.20   Dimer diffusion on (100) plane. (a)–(c) By single atom exchange with substrate atom. (d)–(f) By two atom exchange with surface atom. Arrows indicate all possible directions for exchange, black arrow indicate direction chosen for illustration. Dumbbell transition state is marked by ellipse.

may also undergo simultaneous exchange, for which one possible path is illustrated in Fig. 7.19. We suspect that simultaneous exchange with atoms moving in two different directions and condensing after exchange is rather unlikely. Distinguishing between atom-by-atom and simultaneous exchange processes experimentally is not straightforward, but calculations of the activation energies for the two processes should give us some insight into what actually happens.

Single atom exchange processes have not been confined to fcc(110) planes – these have also been detected on (100) surfaces, so it is likely that diffusion of dimers may occur in this way. A single atom exchange process for a dimer is depicted in Fig. 7.20a–c, resulting in dimer rotation. Exchange proceeds in analogy to the process on fcc(110): the first dimer atom can exchange with a surface atom in four possible directions, indicated in Fig. 7.20a by arrows; the black arrow shows the direction we chose for the illustration. Not all directions will be equal due to the presence of the second atom on the surface. The directions away from the dimer are likely to be less probable. The intermediate stage of the exchange will also be a dumbbell; however, on this surface there are only three possible directions for dumbbell dissociation, as shown in Fig. 7.20b, and one

7.1 Via single atom jumps

of them (out from the other atom) is less likely than others. A translation of the dimer can be obtained by double rotation or without rotation. As shown in Fig. 7.20d–f, both atoms may execute a simultaneous exchange. On the (100) plane, the occurrence of dimer rotation provides a simple criterion to distinguish exchange, atom-by-atom, from simultaneous exchange of both atoms, provided it is known in an experiment that only exchange occurs. However, the same configurations can be achieved by a sequence of hops on the surface.

The coexistence of movement both by hopping and exchange has been detected in experiments of atom diffusion. Such coexistence might be also occur in dimer movement, complicating the interpretation of cluster diffusion.

7.1.6 Peripheral diffusion and the evaporation–condensation mechanism

In diffusion of larger clusters, movement by single atoms is usually associated with a sequence of peripheral displacements. Here it must be noted that the rate of such perimeter movement depends on the kind of cluster edge at which the atom is moving. Rates are also different for crossing corners (known as corner rounding) or for attaching–detaching from kinks; the last one depends on the number of kinks present. As an example, for the cluster on an fcc(111) plane shown in Fig 7.21a, movement along steps A proceeds at a rate a_s, while along steps B it occurs at a rate b_s; movement around corners takes place at the rate h_r. Corner breaking at the rate h_{re}, in which an atom detaches from the corner and starts movement along the cluster edges, might also play a role. When in addition kinks are present, the rates of detachment h_k from a kink to the straight edge and attachment k_k from a straight edge to kinks, shown in Fig. 7.21b, also

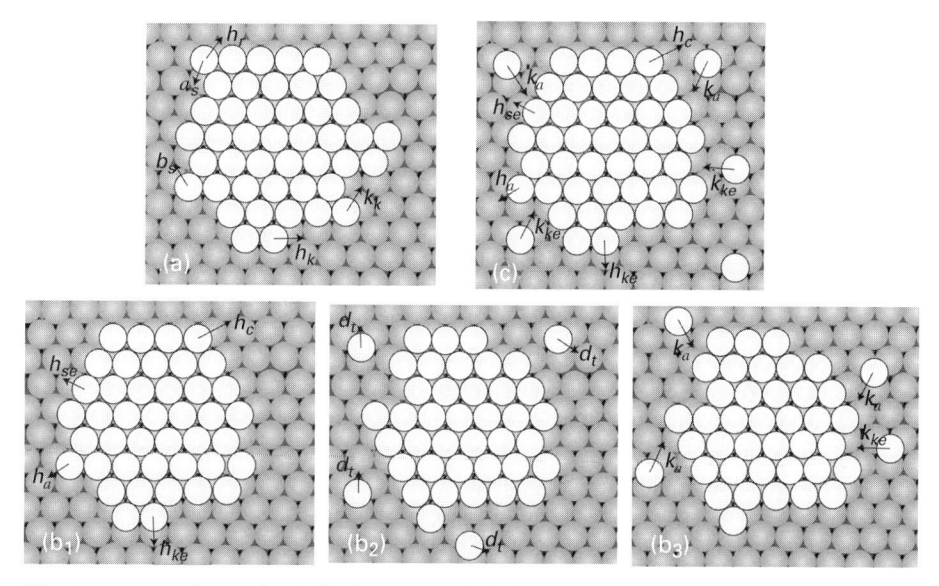

Fig. 7.21 Kinetic processes in peripheral displacements. (a) Perimeter movement. (b) Correlated evaporation–condensation. (c) Evaporation–condensation mechanism.

have to be taken into account. For some structures, movement along the edges might proceed by the exchange mechanism; for example, for Al(111), theory predicts the exchange mechanism at *B*-type steps as a leading contender [13]. Movement of the center of mass will be described by a combination of all these rates together, and the problem is further described in Section 7.5. At higher temperatures, diffusion along the steps can proceed by longer jumps as well, which will proceed at different rates.

In the movement of clusters by correlated evaporation–condensation (CEC), known also as the terrace limited mechanism, cluster atoms separate from the cluster and after diffusion on the terrace come back to the same cluster and combine again. In such mechanism at least four rates are involved for a compact cluster. The first one indicated in Fig. 7.21b as h_{se} is the rate of detachment of an atom from the straight step or at the rate h_c from the corner (core breakup) to the terrace; after detachment the adatom moves on the terrace with a rate typical for terrace diffusion, indicated in the picture as d_t. Finally, when the atom meets the edge of the cluster it can again attach, at the rate of k_a if the adatom meets a straight edge. When kinks are present, different rates h_{ke} will be associated with detachment of atoms from the kink to the terrace and k_{ke} for attachment of adatoms to the kink from the terrace. When a compact cluster has an atom attached to a straight step, this atom usually detaches first at the rate of h_a. In carrying out simulations of diffusion by these mechanisms, all the different rate constants must be included to attain a proper description – a single rate will not do justice to this problem. However, diffusion around cluster edges is expected to take place over smaller barriers than atom detachment, so that at low temperatures the latter process should not be important. One can also expect that attachment processes are very rapid and influence movement less than detachment.

In the last named mechanism, evaporation–condensation (EC), shown in Fig. 7.21c, the cluster atom leaves the cluster for the terrace, but a different adatom from the terrace attaches to the cluster, maintaining constant average cluster size. The rates for such a mechanism are the same as in the correlated evaporation–condensation mechanism, but the existence of such a mechanism will depend on the presence of adatoms on the terrace at some distance from the cluster. Both EC type mechanisms are rather unlikely at low temperatures, due to the high energy of detachment.

7.2 Concerted displacements

The examples of cluster motion considered up to this point have all depended upon jumps of individual atoms. This may not always be the case, however, and in this section we will present examples of cluster diffusion by concerted motions.

7.2.1 Gliding

A good demonstration of concerted atom diffusion is provided by the diffusion of hexagonal clusters made up of 7 or 19 atoms on an fcc(111) surface, in the literature sometimes described as magic clusters. For Ir(111) such diffusion has been examined

7.2 Concerted displacements

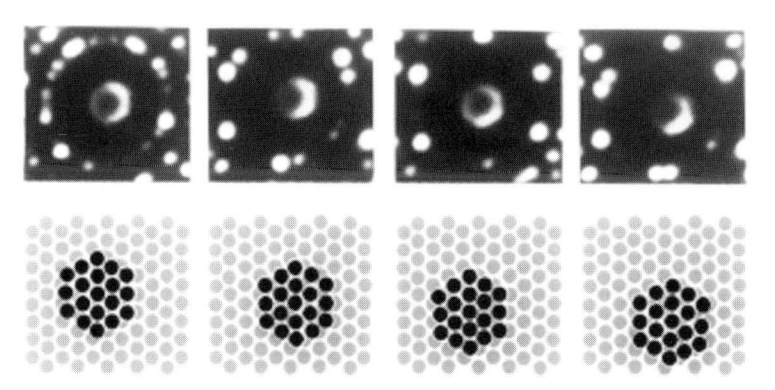

Fig. 7.22   Motion of compact hexagonal Ir$_{19}$ cluster over Ir(111) plane after heating to ~ 690 K [14]. FIM images on top, schematics below. No individual atom transitions were detected.

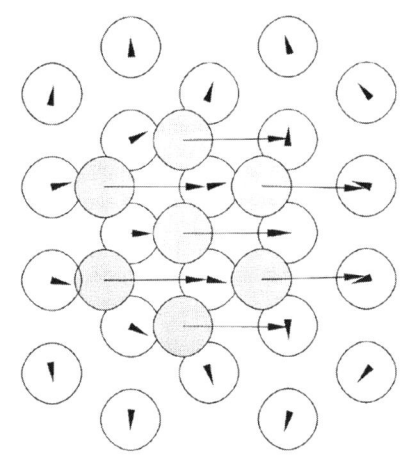

Fig. 7.23   Pattern of Ir$_7$ displacements on Ir(111) obtained from Sutton–Chen potential, for vibrational mode of 0.75 THz [16].

by Wang [14,15], and is shown for an hexagonal Ir$_{19}$ cluster in Fig. 7.22. What is clear here is that the center of the cluster moves over the surface, but there is no evidence in the FIM images of any shape changes, suggesting that the cluster glides over the surface as a whole. Atomic simulations to explain details of this diffusion for Ir$_7$ have been carried out by Kürpick *et al.* [16] using Sutton–Chen potentials. Shown in Fig. 7.23 is a view of the pattern of surface atom displacements in a vibrational mode at 0.75 THz during movement of an Ir$_7$ cluster. What is very surprising in this study is that the surface is not rigid and seems to create an easy path for moving the cluster.

A more detailed picture was obtained in molecular dynamics simulations at the elevated temperature of 1350 K. In Fig. 7.24 are given three separate sequences, indicating distortions of the atomic arrangement as the cluster moves from one equilibrium site to a neighboring one. However, on reaching equilibrium, the hexagonal shape

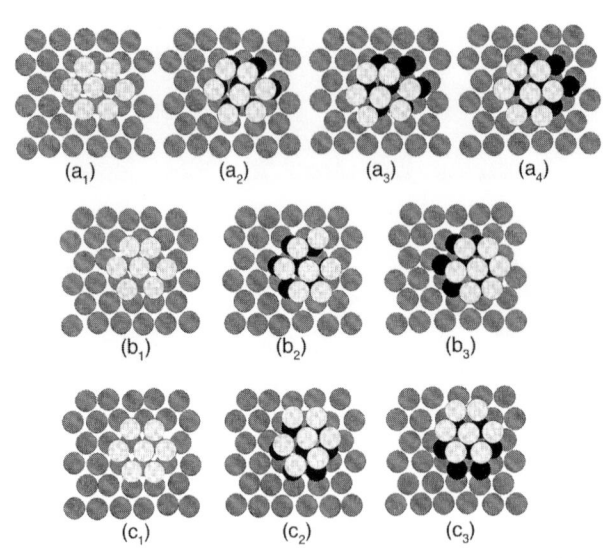

Fig. 7.24 Molecular dynamics simulations with Sutton–Chen potential for Ir$_7$ diffusing on Ir(111) at 1350 K [16]. Black circles show initial sites of minimum energy. Cluster undergoes distortions while maintaining overall shape.

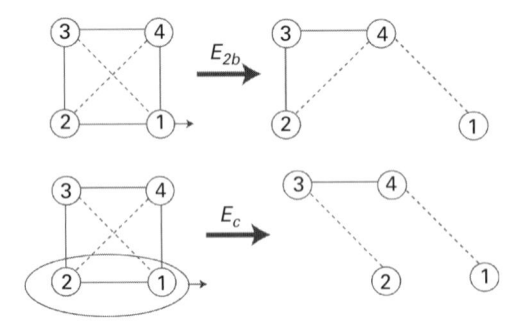

Fig. 7.25 Competing mechanisms for tetramer motion on (100) plane [17]. Top: by severing nearest-neighbor links. Bottom: by shearing of dimer.

is always recovered. It is evident from this study that this hexagonal cluster does not diffuse over the surface by jumps of individual atoms, and that concerted motion of the whole cluster is involved.

7.2.2 Cluster shearing and reptation

In 1996, Shi et al. [17] came out with a novel way for a cluster to diffuse on a (100) surface – by shearing rather than by simple atom jumps. The two types of processes are illustrated in Fig. 7.25 for a tetramer: movement by a single atom, and by two atoms together in a shearing motion. Is shearing an energetically favorable process? To examine this question, Shi et al. did EAM calculations for the two events in three systems, as indicated

7.2 Concerted displacements

Table 7.1 Self-diffusion of clusters on (100) surfaces by shearing, with barriers in eV [17]. Second barrier is for cluster atom detachment.

Cluster Geometry	Cu(100)	Ag(100)	Ni(100)
	0.503	0.478	0.632
	0.494	0.480	0.611
	0.501/0.552	0.491/0.503	0.621/0.673
	0.688/0.815	0.637/0.723	0.842/0.989
	0.551	0.510	0.675
	0.713/0.835	0.658/0.737	0.870/1.008
	0.552	0.512	0.677
	0.758/0.838	0.676/0.740	0.918/1.011
	0.857/0.975	0.763/0.903	1.051/1.215
	0.554	0.516	0.678

in Table 7.1; for clusters of 4, 6, or 8 atoms they found dimer shearing rather than single jumps to be the low-energy process. The shearing process also can explain the oscillatory behavior of cluster motion observed in experiments by Kellogg [18] for self-diffusion on the Rh(100) surface. Depending on the size of the cluster and the surface geometry, dimer or trimer shearing may be energetically favorable. Three years later, a related mechanism on fcc (111) planes, reptation, was put forth by Chirita et al. [19]. In this process, part of a cluster glides over the surface, and is then followed by the remaining atoms of the cluster, as illustrated in Fig. 7.26. The two parts need not be of the same size. Movement proceeds by a high-energy transient state, in which part of the cluster is in fcc sites and a second part in hcp sites. The cluster can relax back from this state or move forward to a new position. Estimates of the diffusion barrier were made with EAM potentials and for clusters with more than seven atoms, reptation was found to be the low-energy process.

7.2.3 Dislocation mechanism

Hamilton et al. [20] in 1995 made an interesting suggestion about diffusion of clusters on fcc(111) surfaces – it could occur by nucleation of a dislocation. For motion of nickel

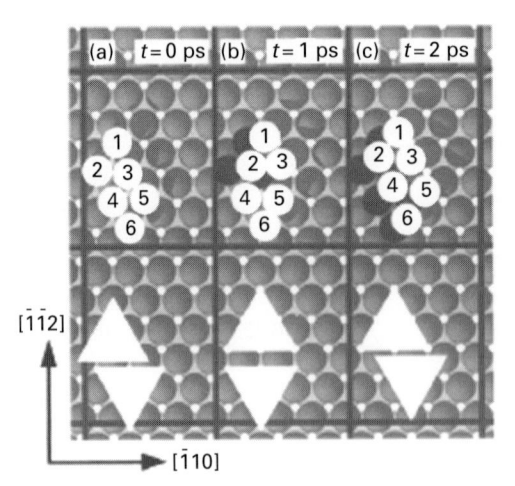

Fig. 7.26 Schematic for diffusion of hexamer on fcc(111) surface by reptation. Initial positions indicated by black circles. Bottom shows subcluster units (after Chirita et al. [19]).

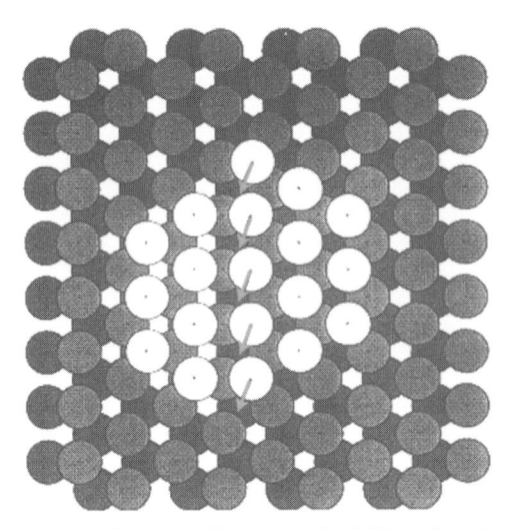

Fig. 7.27 Motion of 19 atom cluster over fcc(111) surface via a misfit dislocation. Arrows show motion toward bridge positions, allowing cluster to diffuse (after Hamilton et al. [20]).

clusters on the Ni(111) plane they made calculations for islands of 19, 27, 37, and 43 atoms using EAM potentials. Motion for a cluster of 19 atoms via a dislocation is illustrated in Fig. 7.27. They came to the conclusion that for islands in the range of 20 to 100 atoms, dislocation motion by gliding of cluster rows should be the process of the lowest activation energy. The activation energy to create a misfit on a Ni(111) surface was calculated as 0.268 eV per atom. However, for clusters bigger than 100, they estimated that kink motion has the lowest energy requirement, making it the more likely mechanism of movement.

7.2 Concerted displacements

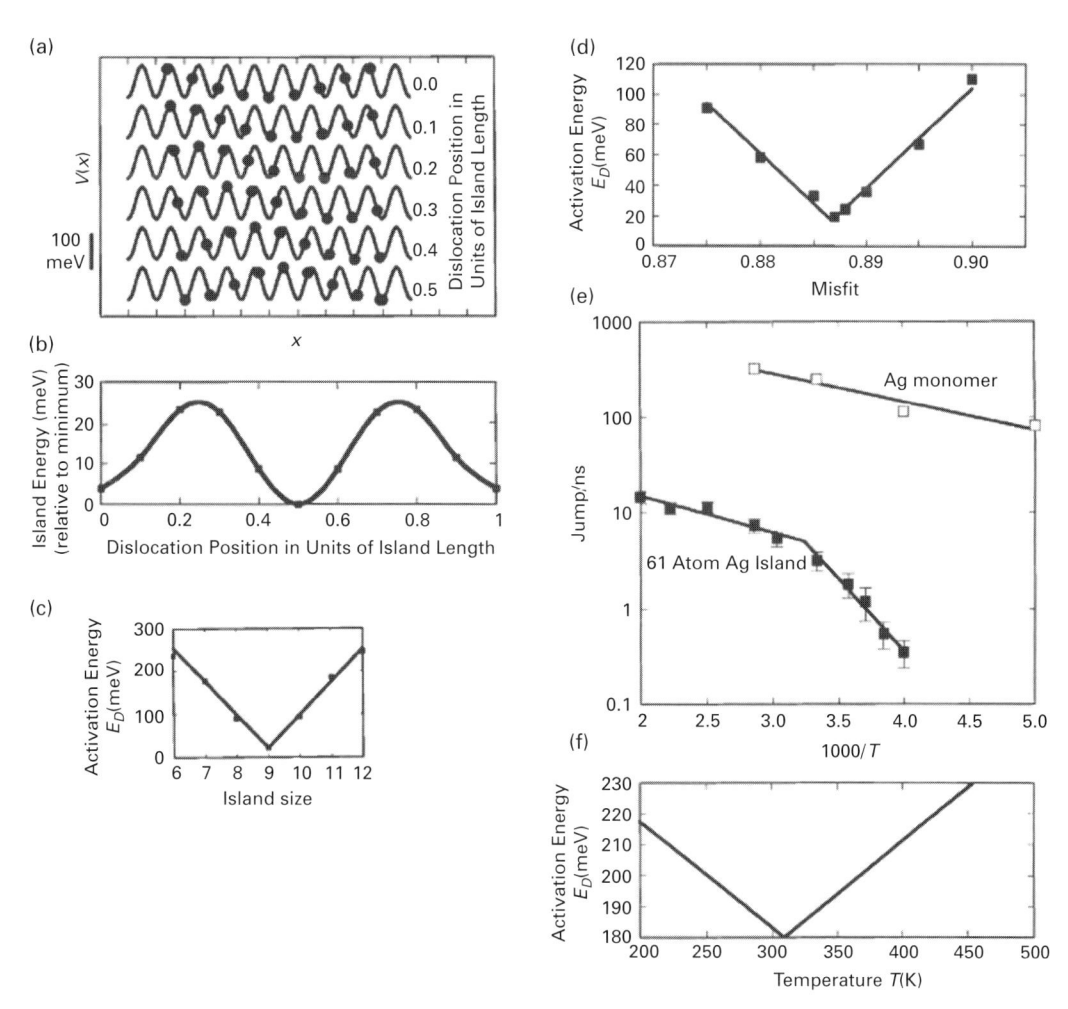

Fig. 7.28 Cluster motion via nucleation of a dislocation. (a) Position of atoms in cluster of nine in one-dimensional diffusion over substrate potential is shown on top. (b) Energy of island relative to minimum in its dependence on position of dislocation. (c) Activation energy for diffusion via dislocation mechanism as a function of island size. (d) Activation energy for diffusion via dislocation mechanism as a function of misfit. (e)–(f) Diffusion of silver islands calculated for Ru(0001) plane. (e) Plot for single silver adatom and for island of 61 atoms, the latter fit with activation energy for diffusion of 61 atom Ag cluster varying with temperature (after Hamilton [21]).

Hamilton [21] also examined diffusion of heteroepitaxial islands, in which the lattice constant of the island differed from that of the underlying crystal. Calculations were made on the Frenkel–Kontorova model [22] using EAM techniques for a cluster with a spacing 11% smaller than that of the substrate. In Fig. 7.28a is shown a one-dimensional island of nine atoms in the potential of the substrate. A dislocation was nucleated at the 0.1 position on the left side of the island; the 0.0 schematic shows the dislocation-free island. In schematic 0.2, a dislocation starts to move and at 0.5 is in the center of the island. Introduction of a dislocation into the cluster changed the energy as shown in Fig. 7.28b, resulting in an activation energy of only 0.025 eV, compared to 0.10 eV for diffusion of a single adatom in the same potential. As the size of the island changed so did the activation energy, as indicated in Fig. 7.28c, reaching a minimum value for an island of nine atoms. The diffusion barrier is of course a function of the misfit between the island and the crystal, shown in Fig. 7.28d, which is dependent upon the temperature. Hamilton also calculated the energetics of two-dimensional islands of silver on Ru(0001); the activation energy for diffusion was found to be a sensitive function of the temperature. This is indicated in Fig. 7.28e–f for an island of 61 silver atoms. What is clear is that the movement of the 61 atom island cannot be fitted by single-atom diffusion; the authors attribute this to the temperature dependence of the misfit between adlayer and substrate. The paper showed the dependence of the activation energy for diffusion on the misfit as well as the size of the clusters.

These are very interesting ideas, but right now only little information is available about the behavior of hetero-epitaxial islands, so the importance of these notions is difficult to judge.

7.2.4 Concerted translation and rotation on FCC(111)

Hamilton et al. [23] have also come up with another intriguing mechanism for cluster diffusion. On putting a layer of iridium on an Ir(111) plane, they discovered in their first-principles calculations that such a pseudomorphic layer had a minimum energy configuration when the cluster atoms were on top of the substrate atoms slightly above a stable fcc configuration. With this as a start, they conjured up the four mechanisms in Fig. 7.29 in which concerted translation or translation plus rotation is accomplished for hexagonal clusters. The activation energies found for the different mechanisms are shown in Table 7.2, and it is clear that bridge glide offers the lowest energy path. However, cartwheel shuffle and cartwheel glide have a not much higher energy for 7 atom clusters and might participate at higher temperature. For 19 atom clusters the energy of a cartwheel shuffle is even closer, making coexistence possible. Whether such a pathway exists for other clusters as well still remains to be explored.

7.2.5 Concerted exchange mechanisms

Molecular dynamics simulations of dimer diffusion have been carried out by Zhuang et al. [24] on the (100) surfaces of platinum, silver, aluminum, and gold using EAM potentials. For platinum and silver, interactions were from the EAM potentials of Haftel et al. [25–27], for aluminum and gold from Johnson [28]. The mechanism for dimer diffusion preferred on all these surfaces is the conventional exchange process, shown in Fig. 7.20, or simple hopping. It should be noted, however, that on Pt(100) the hopping event does not occur. However, other processes, which take place less often, were also observed in this study. These are interesting nevertheless and are shown in Fig. 7.30 for a temperature of 700 K. In the 180° exchange rotation, adatom 1 interacts with lattice atom 2, lifting it out of the surface. This atom then interacts with atom 3, also raising it. Atom 1

7.2 Concerted displacements

Table 7.2 Energetics (in eV) of concerted motion of iridium clusters on Ir(111) [23].

Cluster size	Bridge glide	Top glide	Cartwheel shuffle	Cartwheel glide
7 atoms	1.54	3.02	1.92	1.92
19 atoms	3.2	5.2	3.6	5.1

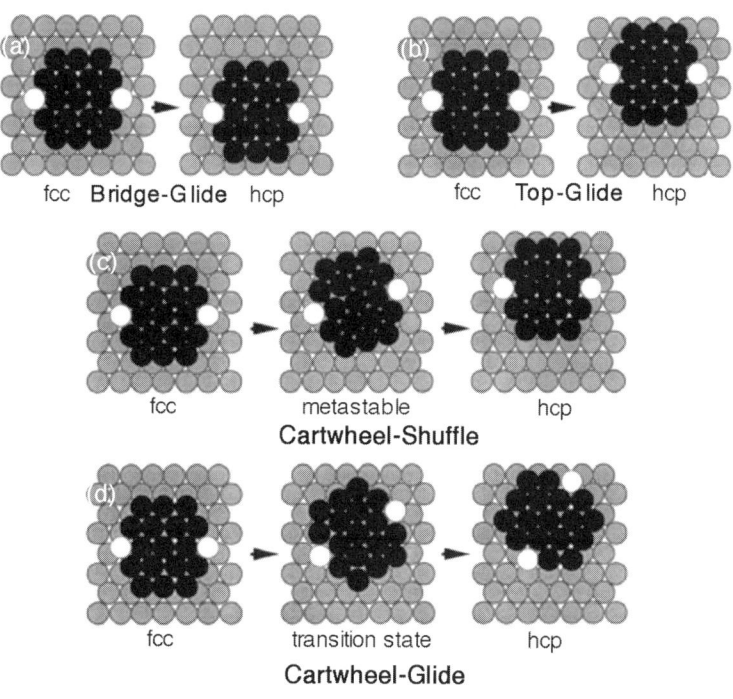

Fig. 7.29   Possible movements of 19 atom Ir cluster on Ir(111) surface by concerted atom displacements. White atoms serve to indicate rotation. Energies are indicated in Table 8.2 (after Hamilton et al. [23]).

then sinks back, as does atom 2, leaving a dimer rotated 180° with respect to the original position. The trajectory in Fig. 7.30f illustrates this process more clearly. Also observed on Pt(100) was a 270° exchange rotation, in which the final configuration of the dimer is 270° rotated compared to the original one. For Al(100), simulations at 450 K revealed both 180° and 270° exchange rotations. Also possible is a 360° exchange rotation, rarely observed for Pt(100) and Al(100). The authors labeled these for Pt(100) and Ag(100) as concerted complicated exchange events, and for Al(100) as strain-induced events. For platinum at 950 K exchange rotation events amounted to roughly 10% of conventional exchange, for aluminum at 550 K to only 6%. A diagram of the energy changes in 180° exchange rotation, which is observed most frequently, is given in Fig. 7.31, keyed to the pictures in Fig. 7.30. Other processes, such as simultaneous jumps of both atoms, and others, were also identified in the simulations, but only quite rarely.

Liu et al. [29] extended calculations by considering in addition Cu(100) as well as Ni(100), and relying on three different potentials – EAM interactions [28],

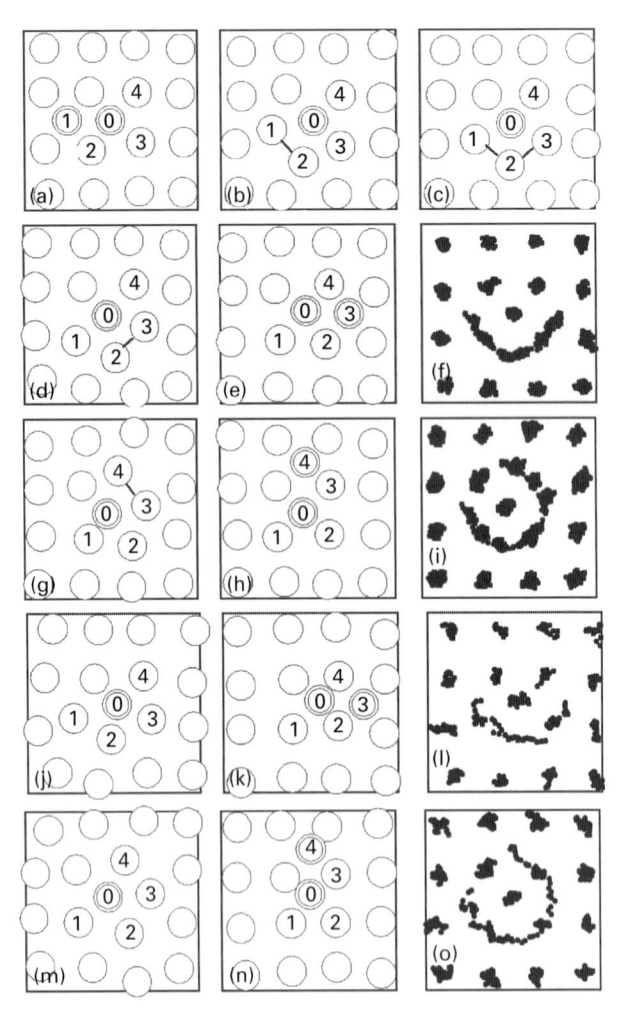

Fig. 7.30 Exchange events during dimer diffusion on fcc(100) plane. (a)→(b)→(c)→(d)→(e) 180° exchange rotation. (a)→(b)→(c)→(d)→(g)→(h) 270° exchange rotation on Pt(100) at 700 K. (f), (i) Molecular dynamics trajectories. (a)→(j)→(k) and (a)→(m)→(n) depict 180° and 270° exchange rotation on Al(100) at 450 K. (l) and (o) give molecular dynamics trajectories (after Zhuang et al. [24]).

surface-embedded-atom potentials (SEAM) [25–27], and RGL interactions [30,31]. The results for static activation energies are given in Table 7.3 for dimer hopping as well as simple exchange on (100) surfaces. For nickel and copper, hopping occurs over the lower barrier, but for aluminum, gold, and platinum, the situation is reversed and hopping has to proceed over a higher barrier. The first group, for which $E_e > E_h$, is classified as hard, the second, where $E_h > E_e$, as soft. For silver, different potentials give inconsistent results, as shown in Table 7.3, making classification of this material difficult. For platinum the dimer in the transition state is actually locally stable, and this is therefore referred to as a soft surface.

7.2 Concerted displacements

Table 7.3 Activation energy of hopping and simple exchange mechanism on metal(100) surfaces [29].

Surfaces	Hopping (eV)	Simple exchange (eV)
Ag(E)	0.4836	0.5892
Ag(R)	0.5001	1.0112
Ni(E)	0.5607	1.349
Cu(R)	0.4307	0.7853
Al(E)	0.4614	0.3545
Ag(S)	0.5818	0.3615
Au(E)	0.8151	0.2647
Pt(E)	1.1976	0.6437
Au(R)	0.5962	0.3715
Pt(S)	1.2319	0.2923

E-EAM, R-RGL, S-SEAM potential

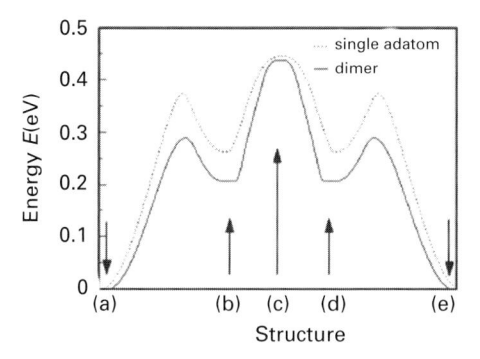

Fig. 7.31 Variation of the energy during exchange rotation of the dimer on Pt(100). Solid curve gives changes for 180° exchange rotation. Letters refer to the structures in Fig. 7.30 (after Zhuang et al. [24]).

In addition to simple exchange and exchange rotation, already discussed above, other diffusion mechanisms were also found in molecular dynamics simulations. These were cooperative hopping, hopping rotation, and cooperative exchange. In cooperative hopping, illustrated in Fig. 7.32, atom A of the dimer hops over the potential barrier, at the same time this movement triggers an instantaneous jump of atom B to a nearest-neighbor site. The directions for the jump of atom B are fixed by the movement of atom A; atom B can jump only towards atom A horizontally (c_2) or vertically (c_1). Hopping rotation is made up of two consistent hopping events which rotate the dimer, shown in Fig. 7.16b; cooperative exchange occurs via a two-atom dimer exchange, already shown in Fig. 7.20d–f. These three movements occur not too frequently by comparison with simple hopping and exchange. It must be emphasized that on middle surfaces, exchange of the strain-induced kind predominates; on soft surfaces, concerted motion exchange is the important process.

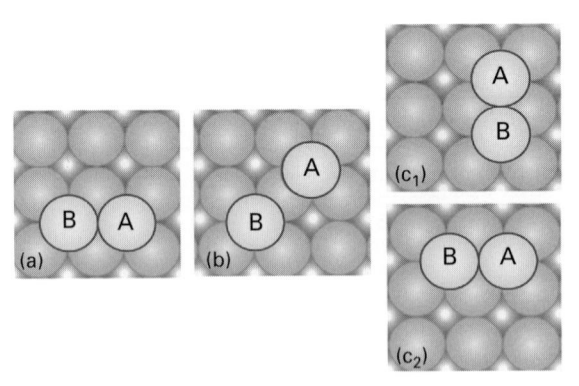

Fig. 7.32 Cooperative hopping of dimer on Cu(100) surface, obtained in RGL calculations based on the work of Liu et al. [29].

7.3 Mechanism of dimer diffusion versus bond length

A simple model of dimer diffusion in a two-dimensional periodic potential has been given by Pijper and Fasolino [32]. The two atoms in the dimer at a separation r interact via a Morse potential, and the resulting potential $V(s)$ for diffusive motion is shown in Fig. 7.33a, where a_ℓ is the substrate lattice constant. Movement of a rigid dimer from position A_1 to the neighboring site A_2 will occur via saddle point T_3 when the dimer moves as a unit, with $r = r_{eq}$. Piecewise diffusion, in which stretching of the dimer is possible, occurs over the path $A_1 \rightarrow T_1 \rightarrow B \rightarrow T_2 \rightarrow A_2$, where B is a local minimum. During such a movement the dimer bond length almost doubles after the first atom jump and regains its length after diffusion. This piecewise rotation may be favorable for a larger width w of the Morse potential. Pijper and Fasolino looked at dimer movement as a function of stiffness. For larger width w (smaller force constant) piecewise motion became energetically more favorable, as is apparent in Fig 7.33a. A similar effect occurs changing the effectiveness of D_M/V_0, where D_M is the depth of the Morse potential and V_0 that for the atom-substrate energy, as is shown in Fig. 7.33b. The broken lines indicate piecewise movement, solid lines rigid movement and the ratio D_M/V_0 is proportional to the force constant k_{force}. An increase in D_M/V_0 increases the barrier and makes piecewise diffusion less favorable than rigid diffusion.

Additionally to stretching of dimers the authors also looked at movement of dimers with a contracted bond ($r < r_{eq}$). They found that movement of such a dimer should proceed over a lower energy barrier than for a rigid or a stretched dimer, as indicated in Fig. 7.33c. The model suggests that movement of dimers over a surface is the result of gaining energy due to interactions between the dimer atoms and losing of energy due to stretching the dimer bond during movement. Further investigations will be required to see if this model describes the nature of dimer movement correctly on real surfaces. It was already shown that using Morse potentials was rather unsuccessful in predicting the correct values of activation energies for single atom movement; however, the trends might be better described by this potential.

7.4 Kinetic mechanisms of larger clusters

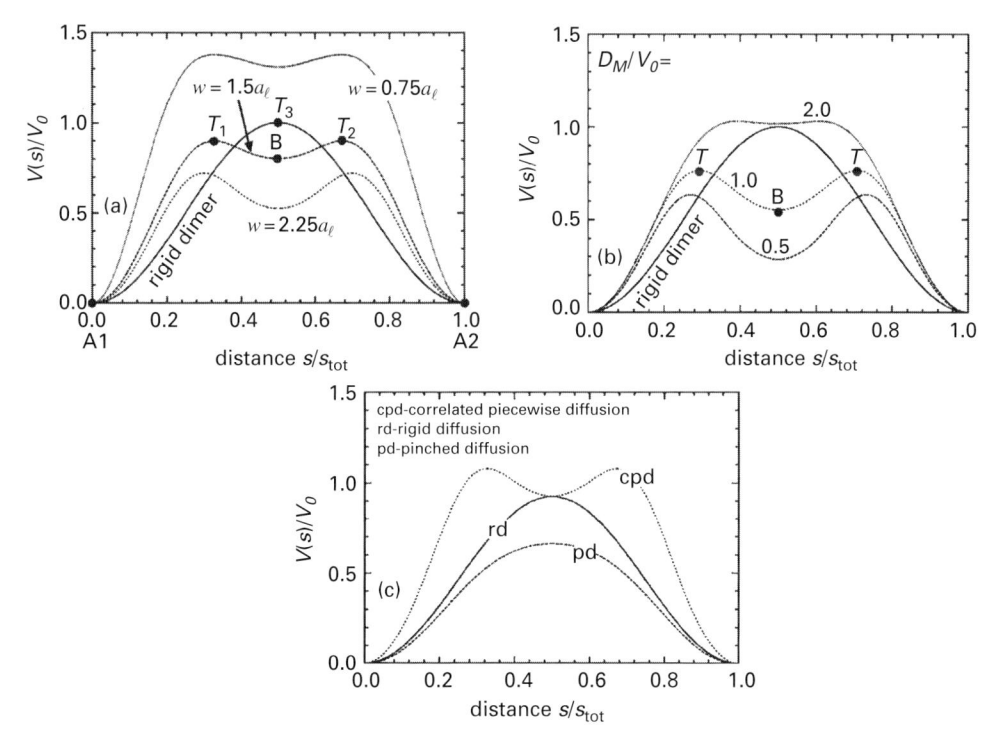

Fig. 7.33 Energy changes for movement of dimer between adsorption sites. (a) For different values of the width w of the Morse potential. (b) For different values of D_M/V_0 where V_0 is the effective barrier for non-interacting atoms and D_M is the well depth between two atoms. (c) Comparison of extended, rigid and contracted dimer bond (after Pijper and Fasolino [32]).

7.4 Kinetic mechanisms of larger clusters

The movement of large clusters has been found to have a significant influence on diffusion over surfaces, so a number of investigations has been carried out to get insight into such diffusivity. We will explore this problem in somewhat more detail in the next section. It is interesting to note that in 1994 the center of mass of large clusters on an Ag(100) surface was found to proceed by a random walk on the surface [33], as shown in Fig. 7.34. However, the mechanism of diffusion constitutes a much more complicated problem than the movement of the center of mass. For big clusters the most probable mechanism is peripheral movement of cluster atoms, shown in Fig. 7.21a. Needless to say, the expressions for the diffusivity become complicated as cluster size increases, and have so far not been applied to real systems. For this reason we will not examine the derivations in any detail.

Investigations of diffusion for bigger clusters started with the work of Voter [34] in 1986, who looked at Rh clusters on the Rh(100) surface with molecular dynamic and lattice gas simulations, relying on Lennard-Jones interactions. He concentrated on movement by sequential jumps and found that jumps away from a nearest-neighbor atom, as in

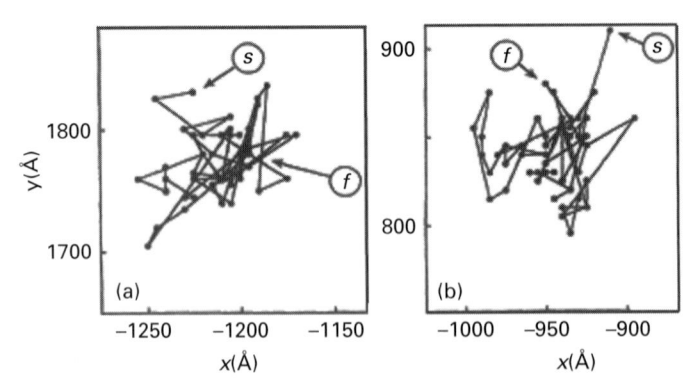

Fig. 7.34 Trajectories for two Ag clusters of different size on Ag(100) measured with STM. Start is indicated by *s*, and final location by *f*. (a) Cluster of 100 atoms. (b) 290 atoms (after Wen et al. [33]).

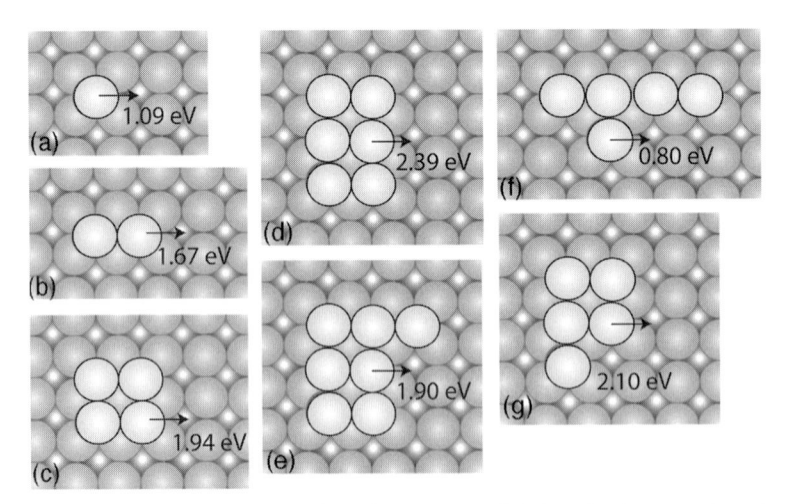

Fig. 7.35 Schematics for atomic jumps of rhodium atoms and their activation energies calculated with Lennard-Jones potentials relevant for diffusion of rhodium cluster on Rh(100) [34].

Fig. 7.35b, required more energy than for a single atom jump on the surface, as in Fig. 7.35a, demonstrating that cluster interactions are attractive. Moving along an edge, in Fig. 7.35f, was easier than jumping out of a block of atoms, shown in Fig 7.35d.

Considerable progress has been made recently by Sanchez and Evans [35] who looked at self-diffusion of clusters, taking into account different hopping rates in movement on the square fcc(100) surface. Their investigation did not allow for the possibility of an exchange mechanism on this surface, or for atom detachment and attachment. However, they considered a number of different perimeter events, such as straight edge hopping, which proceeded at the rate h_e, corner rounding at a rate h_r, kink escape at rate h_k, and core breakup at the rate h_c. The first two transitions did not require changes in the bonding or in the energy of the system, while the second two required breaking of one bond. They also considered the edge "breakup," which involves breaking two bonds. For small clusters, such transitions are in the negligible range, but play a role for bigger islands.

7.4 Kinetic mechanisms of larger clusters

A number of other transitions should probably be considered in explaining the movement of real systems, for example diffusion of dimers around the perimeter, movement by longer jumps, as well as kink movement; however, as a starting point, the rates presented are of interest. Sanchez and Evans carried out detailed investigations for smaller clusters first; then they turned their attention to bigger entities. We will not present here their work for small clusters, as this was described in previous sections; however, we will present a few comments.

For dimers, the authors did not consider the possibility of clusters being spread out over a distance longer than a nearest-neighbor separation. With this assumption, the movement of a dimer can only take place by transition between vertical and horizontal states, and the possibility of a dimer being in an intermediate state was ignored. This assumption greatly simplifies their calculations, since instead of the two rates of jumping a and b, that describe the conversion of a vertical into a horizontal dimer, or the reverse, shown in Fig. 7.16, they have only one, h_r. The same assumptions were applied in the investigation of trimers, which led them to the diffusivity

$$D = \frac{h_r h_e}{3(h_r + h_e)}, \tag{7.56}$$

where h_e indicates the rate of straight edge hopping. For tetramers the diffusivity was described by

$$D = \frac{6 h_r h_c}{(h_r + 18 h_c)}, \tag{7.57}$$

with h_c being the rate of core breakup.

The assumption that cluster atoms are always in nearest-neighbor positions and that the detachment–attachment mechanism is not active may describe the movement of big clusters quite correctly. Sanchez and Evans [35] arrived at relatively simple expressions for the mean-square displacement and diffusivities of larger clusters on (100) surfaces by refining the work of Titulaer and Deutch [6]. We will here illustrate an example for pentamers. A low energy path for diffusion of such a cluster is illustrated in Fig. 7.36.

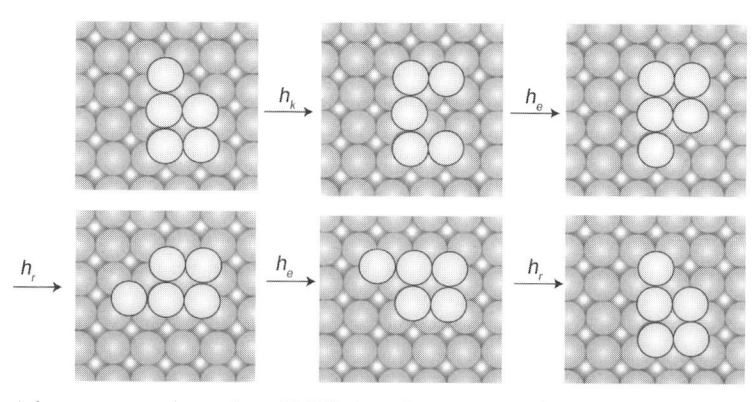

Fig. 7.36 A low-energy pathway for self-diffusion of pentamer on fcc(100) plane [35].

Mechanism of cluster diffusion

Such a path is chosen to avoid direct core breakup, and movement is mediated by corner rounding or kink escape, whichever is more efficient.

The assumption is made that the rates h_r and h_k remain finite, and Sanchez and Evans considered the limit $h_e \to \infty$, which, from detailed balance, requires that $h_c \to 0$. If unfolding of the configuration shaped like a C is ignored this leaves only the rates h_k and h_r. The diffusivity D for the pentamer ends up quite simply as

$$D_5 = \frac{h_r h_k}{8(h_r + h_k)}. \tag{7.58}$$

Sanchez and Evans also briefly looked at tetramer diffusion with dimer shearing active, and later gave expressions for other clusters, which, however, will be downplayed here to concentrate on the work of Salo et al. [36] that expanded considerations of the diffusivity of clusters on (100) surfaces and considered both dimer and trimer shearing. Salo et al. looked at cluster made up of 5 to 9 atoms and followed the same nomenclature as that of Sanchez and Evans [35].

Concentrating first on pentamers as our primary example, we display in Fig. 7.37 the different rate processes assumed as being involved in diffusion of the clusters. The activation energies for these events have been evaluated by Trushin et al. [4] for copper on Cu(100) using effective medium theory, and this makes it possible to sort out transitions, indicated by italics in Fig. 7.37, that occur either very rapidly or slowly and can therefore be neglected; the rate limiting step is shown bold-faced. The remaining configurations for pentamer diffusion are indicated in Fig. 7.38 and yield the diffusivity

$$D_5 = \frac{2[h'_{si} + 2(h'_s + h_r)]}{[2(h_{s1} + h_{si}) + h_r]} \frac{[2h_{si}(h_{s1} + h_{si}) + h_r(h_{s1} + 2h_{si})]}{(8h'_{si} + 16h'_s + 21h_{si} + 16h_r)}, \tag{7.59}$$

where h_s indicates the rate of shearing and h'_s the reverse process. If $h_{s1} = h_k/2$ and other rates for dimer shearing are set to zero, Eq. (7.58), the result previously obtained by Sanchez and Evans [35], is found. Expressions for other clusters, containing from 6 to 9 atoms, were also derived. Using the rate constants for jumps of copper on Cu(100) from Trushin et al. [4], the diffusivities at different temperatures were calculated and are shown in Fig. 7.39. What is clear are the oscillations with size, with local minima for clusters of 4, 6, and 9 atoms, already seen in prior studies. New processes of dimer and trimer shearing (described earlier) increase the diffusion rate and affect island diffusion. What is evident is that knowing the individual jump rates it is now possible to write out the cluster diffusivities, at least on (100) surfaces. Finding these rates, however, still remains a task to be addressed in experiments.

7.5 Derivation of the mechanism of large cluster movements

Observations of the movement of large clusters triggered a number of discussions of how the cluster diffusivity D was affected by the size and the specific mechanism of diffusion. Soler [37], in 1994, did Monte Carlo calculations with a triangular lattice model. In his

7.5 Derivation of the mechanism of large cluster movements

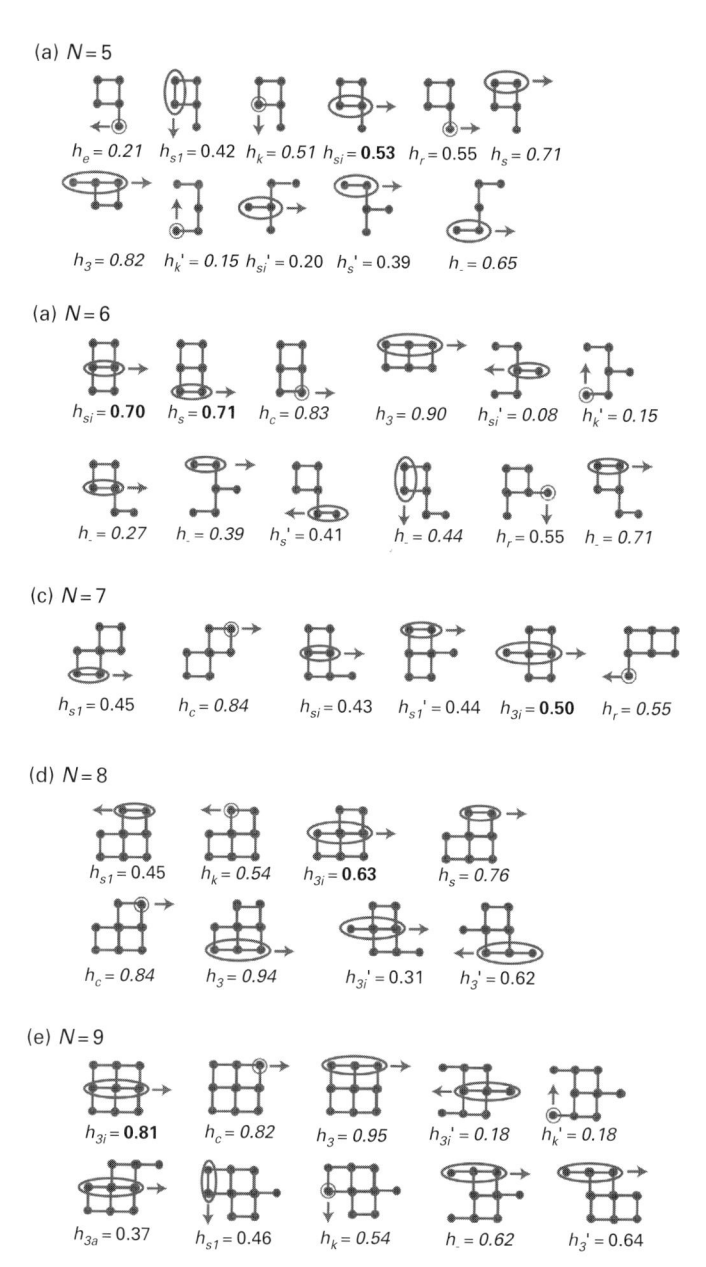

(a) N = 5

$h_e = 0.21$ $h_{s1} = 0.42$ $h_k = 0.51$ $h_{si} = \mathbf{0.53}$ $h_r = 0.55$ $h_s = 0.71$

$h_3 = 0.82$ $h_k' = 0.15$ $h_{si}' = 0.20$ $h_s' = 0.39$ $h_- = 0.65$

(a) N = 6

$h_{si} = \mathbf{0.70}$ $h_s = \mathbf{0.71}$ $h_c = 0.83$ $h_3 = 0.90$ $h_{si}' = 0.08$ $h_k' = 0.15$

$h_- = 0.27$ $h_- = 0.39$ $h_s' = 0.41$ $h_- = 0.44$ $h_r = 0.55$ $h_- = 0.71$

(c) N = 7

$h_{s1} = 0.45$ $h_c = 0.84$ $h_{si} = 0.43$ $h_{s1}' = 0.44$ $h_{3i} = \mathbf{0.50}$ $h_r = 0.55$

(d) N = 8

$h_{s1} = 0.45$ $h_k = 0.54$ $h_{3i} = \mathbf{0.63}$ $h_s = 0.76$

$h_c = 0.84$ $h_3 = 0.94$ $h_{3i}' = 0.31$ $h_3' = 0.62$

(e) N = 9

$h_{3i} = \mathbf{0.81}$ $h_c = 0.82$ $h_3 = 0.95$ $h_{3i}' = 0.18$ $h_k' = 0.18$

$h_{3a} = 0.37$ $h_{s1} = 0.46$ $h_k = 0.54$ $h_- = 0.62$ $h_3' = 0.64$

Fig. 7.37 Jump processes for islands in which the number of atoms ranges from five to nine atoms on fcc (100) plane (after Salo et al. [36]). Activation energies for jump processes of Cu islands on Cu (100), evaluated by EAM [4]. Italics show processes not included in transition matrix as either too fast or slow. Rate limiting step is shown in boldface; h_s describes a shearing process, h_s' reverse shearing process.

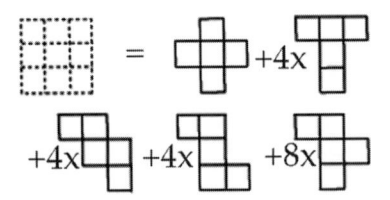

Fig. 7.38 Quasi-configurations for pentamer diffusion on fcc(100) plane, with multipliers giving the degeneracy (after Salo et al. [36]).

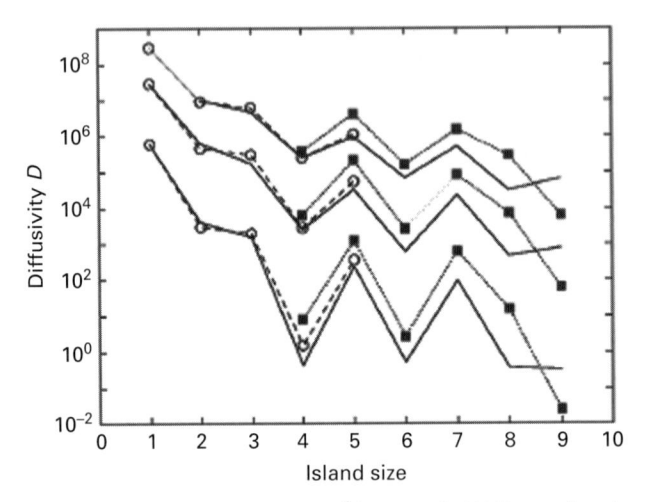

Fig. 7.39 Copper cluster diffusivity (units a_ℓ^2/sec) on Cu(100) as a function of size at temperatures of 300 (bottom), 400, and 500 K. Solid line –Monte Carlo simulations, open circle –analytical results with single atom processes, filled square – many-particle processes included (after Salo et al. [36]).

investigations he considered motion of large clusters by (a) movement along the perimeter, (b) movement via an evaporation–condensation mechanism, and (c) via vacancy movement. In Fig. 7.40a, the Arrhenius plot for clusters consisting of 300 atoms is presented. We can see that movement along the perimeter has a lower energy than an evaporation–condensation process or vacancy movement; the two latter mechanisms required comparable energy. The dependence of diffusivity on the clusters size, for all three mechanisms, is shown in Fig. 7.40b. Soler suggested that perimeter diffusion of atoms was the significant mechanism, evaporation–condensation being negligible at low temperatures.

Van Siclen [38] then considered the role of correlations in the evaporation–condensation mechanism of diffusion. For uncorrelated events, the diffusivity D varied as $1/R_r$, where R_r denotes the cluster radius. However, an evaporating atom is very likely to redeposit close to the point of emission, and this changed the dependence to $D \propto 1/R_r^2$. In case of cluster diffusion caused by an atom running along the perimeter, the dependence of the diffusivity was given by $D \propto 1/R_r^3$. His investigations viewed the cluster as circular and the structure of the substrate was not taken into account. He also investigated facetted islands and concluded that such islands could only exist above a certain size; for

7.5 Derivation of the mechanism of large cluster movements

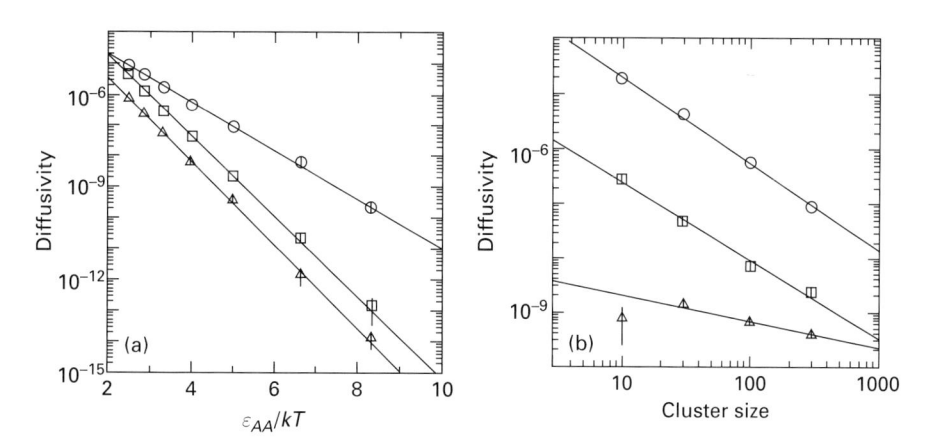

Fig. 7.40 Cluster diffusivities calculated for a triangular lattice from MC simulations. (a) Arrhenius plots of 300 atom cluster (circles). Contribution of evaporation–condensation processes is indicated by squares, of vacancy movement inside cluster by triangles. (b) Size dependence of cluster diffusion, separated as in (a) (after Soler [37]).

such clusters the diffusivity would be independent of the size. A year later, Soler [39] examined diffusion of clusters by evaporation and condensation of atoms as affected by correlations, and came to the conclusion that correlation depended on the diffusivity as well as on cluster size. He claimed that the dependence of diffusivity on $1/R_r$ was valid only for short times due to the increasing importance of the correlated evaporation–condensation mechanism. Under experimental conditions the diffusivity varied with size according to

$$D \sim R_r^{-(2+\Delta b)}, \qquad (7.60)$$

where Δb describes the non-negligible perimeter energy, which was positive for atom clusters and negative for vacancy clusters.

Diffusion of large clusters was analyzed in a rather different way by Khare et al. [40,41], who pointed out the analogy with equilibrium fluctuations on vicinal surfaces, which had been extensively examined in the continuous limit by resorting to Langevin dynamics. The diffusivity of a cluster of radius R_r was found to be

$$D = D^* R_r^{-\alpha_L}, \qquad (7.61)$$

where the exponent α_L, derived for different mechanisms, agrees with the findings of Van Siclen [38]; it equaled three for atoms diffusing around the cluster edges, two if diffusion across the crystal terrace was limiting, known also as the correlated evaporation–condensation mechanism, and one for evaporation–condensation defining cluster motion. Expressions for the prefactor D^* in the different cases were also given. To check if this approach was valid for islands of 100 or more atoms, Monte Carlo simulations were done, which proved to be in good agreement with predictions from the continuum approach. In these investigations, the clusters were circular in shape and the atomic structure of the surface was not considered.

In the derivation of large cluster movement the structure of the cluster proved to be important, so the above derivations only partly describe the situation; the experimental data will be presented at length in Chapter 9.

With some of the basics established, we will now examine the empirical data and theoretical speculations about diffusion of clusters on various surfaces, which still leave uncertainties about this subject.

References

[1] D. A. Reed, G. Ehrlich, In-channel clusters: Rhenium on W(211), *Surf. Sci.* **151** (1985) 143–165.

[2] K. Kyuno, G. Ehrlich, Diffusion and dissociation of platinum clusters on Pt(111), *Surf. Sci.* **437** (1999) 29–37.

[3] H.-W. Fink, G. Ehrlich, Pair and trio interactions between adatoms: Re on W(110), *J. Chem. Phys.* **81** (1984) 4657–4665.

[4] O. S. Trushin, P. Salo, T. Ala-Nissila, Energetics and many-particle mechanisms of two-dimensional cluster diffusion on Cu(100) surfaces, *Phys. Rev. B* **62** (2000) 1611–1614.

[5] D. A. Reed, G. Ehrlich, One-dimensional random walks of linear clusters, *J. Chem. Phys.* **64** (1976) 4616–4624.

[6] U. M. Titulaer, J. M. Deutch, Some aspects of cluster diffusion on surfaces, *J. Chem. Phys.* **77** (1982) 472–478.

[7] J. D. Wrigley, D. A. Reed, G. Ehrlich, Statistics of one-dimensional cluster motion, *J. Chem. Phys.* **67** (1977) 781–792.

[8] W. R. Graham, G. Ehrlich, Surface diffusion of atom clusters, *J. Phys. F* **4** (1974) L212–214.

[9] T. R. Linderoth, S. Horch, L. Petersen, S. Helveg, E. Laegsgaard, I. Stensgaard, F. Besenbacher, Novel mechanism for diffusion of one-dimensional clusters; Pt/Pt(110)-(1×2), *Phys. Rev. Lett.* **82** (1999) 1494–1497.

[10] F. Montalenti, R. Ferrando, Leapfrog diffusion mechanism for one-dimensional chains on missing-row reconstructed surfaces, *Phys. Rev. Lett.* **82** (1999) 1498–1501.

[11] J. D. Wrigley, G. Ehrlich, Two-dimensional random walks of diatomic clusters, *J. Chem. Phys.* **84** (1986) 5936–5954.

[12] C. M. Chang, C. M. Wei, S. P. Chen, Modeling of Ir adatoms on Ir surfaces, *Phys. Rev. B* **54** (1996) 17083–17096.

[13] A. Bogicevic, J. Strömquist, B. I. Lundqvist, Low-symmetry diffusion barriers in homoepitaxial growth of Al(111), *Phys. Rev. Lett.* **81** (1998) 637–640.

[14] S. C. Wang, G. Ehrlich, Diffusion of large surface clusters: Direct observations on Ir(111), *Phys. Rev. Lett.* **79** (1997) 4234–4237.

[15] S. C. Wang, U. Kürpick, G. Ehrlich, Surface diffusion of compact and other clusters: Ir$_x$ on Ir(111), *Phys. Rev. Lett.* **81** (1998) 4923–4926.

[16] U. Kürpick, B. Fricke, G. Ehrlich, Diffusion mechanisms of compact surface clusters: Ir$_7$ on Ir(111), *Surf. Sci.* **470** (2000) L45–51.

[17] Z.-P. Shi, Z. Zhang, A. K. Swan, J. F. Wendelken, Dimer shearing as a novel mechanism for cluster diffusion and dissociation on metal (100) surfaces, *Phys. Rev. Lett.* **76** (1996) 4927–4930.

[18] G. L. Kellogg, Oscillatory behavior in the size dependence of cluster mobility on metal surfaces: Rh on Rh(100), *Phys. Rev. Lett.* **73** (1994) 1833–1836.

References

[19] V. Chirita, E. P. Münger, J. E. Greene, J.-E. Sundgren, Reptation: A mechanism for cluster migration on (111) face-centered-cubic metal surfaces, *Surf. Sci.* **436** (1999) L641–647.

[20] J. C. Hamilton, M. S. Daw, S. M. Foiles, Dislocation mechanism for island diffusion on fcc (111) surfaces, *Phys. Rev. Lett.* **74** (1995) 2760–2763.

[21] J. C. Hamilton, Magic size effects for heteroepitaxial island diffusion, *Phys. Rev. Lett.* **77** (1996) 885–888.

[22] S. Aubry, P. Y. Le Daeron, The discrete Frenkel-Kontorova model and its extensions: I. Exact results for the ground states, *Physica (Amsterdam) D* **8** (1983) 381–422.

[23] J. C. Hamilton, M. R. Sorensen, A. F. Voter, Compact surface-cluster diffusion by concerted rotation and translation, *Phys. Rev. B* **61** (2000) R5125–5128.

[24] J. Zhuang, Q. Liu, M. Zhuang, L. Liu, L. Zhao, Y. Li, Exchange rotation mechanism for dimer diffusion on metal fcc(001) surfaces, *Phys. Rev. B* **68** (2003) 113401 1–4.

[25] M. I. Haftel, Surface reconstruction of platinum and gold and the embedded-atom method, *Phys. Rev. B* **48** (1993) 2611–2622.

[26] M. I. Haftel, M. Rosen, T. Franklin, M. Hettermann, Molecular dynamics observations of the interdiffusion and Stranski-Krastanov growth in the early film deposition of Au on Ag(100), *Phys. Rev. Lett.* **72** (1994) 1858–1861.

[27] M. I. Haftel, M. Rosen, Molecular-dynamics description of early film deposition of Au on Ag (110), *Phys. Rev. B* **51** (1995) 4426–4434.

[28] R. A. Johnson, Analytical nearest-neighbor model for fcc metals, *Phys. Rev. B* **37** (1988) 3924–3931.

[29] Q. Liu, Z. Sun, X. Ning, Y. Li, L. Liu, J. Zhuang, Systematical study of dimer diffusion on metal fcc(001) surfaces, *Surf. Sci.* **554** (2004) 25–32.

[30] V. Rosato, M. Guillope, B. Legrand, Thermodynamical and structural properties of f.c.c. transition metals using a simple tight-binding model, *Philos. Mag. A* **59** (1989) 321–336.

[31] M. Guillope, B. Legrand, (110) surface stability in noble metals, *Surf. Sci.* **215**(1989) 577–595.

[32] E. Pijper, A. Fasolino, Mechanism for correlated surface diffusion of weakly bonded dimers, *Phys. Rev. B* **72** (2005) 165328 1–5.

[33] J.-M. Wen, S.-L. Chang, J. W. Burnett, J. W. Evans, P. A. Thiel, Diffusion of large two-dimensional Ag clusters on Ag(100), *Phys. Rev. Lett.* **73** (1994) 2591–2594.

[34] A. F. Voter, Classically exact overlayer dynamics: Diffusion of rhodium clusters on Rh(100), *Phys. Rev. B* **34** (1986) 6819–6829.

[35] J. R. Sanchez, J. W. Evans, Diffusion of small clusters on metal (100) surfaces: Exact master-equation analysis for lattice-gas models, *Phys. Rev. B* **59** (1999) 3224–3233.

[36] P. Salo, J. Hirvonen, I. T. Koponen, O. S. Trushin, J. Heinonen, T. Ala-Nissila, Role of concerted atomic movements on the diffusion of small islands on fcc(100) metal surfaces, *Phys. Rev. B* **64** (2001) 161405 1–4.

[37] J. M. Soler, Monte Carlo simulation of cluster diffusion in a triangular lattice, *Phys. Rev. B* **50** (1994) 5578–5581.

[38] C. D. Van Siclen, Single jump mechanisms for large cluster diffusion on metal surfaces, *Phys. Rev. Lett.* **75** (1995) 1574–1577.

[39] J. M. Soler, Cluster diffusion by evaporation-condensation, *Phys. Rev. B* **53** (1996) R10540–10543.

[40] S. V. Khare, N. C. Bartelt, T. L. Einstein, Diffusion of monolayer adatom and vacancy clusters: Langevin analysis and Monte Carlo simulations of their Brownian motion, *Phys. Rev. Lett.* **75** (1995) 2148–2151.

[41] S. V. Khare, T. L. Einstein, Brownian motion and shape fluctuations of single-layer adatom and vacancy clusters on surfaces: theory and simulations, *Phys. Rev. B* **54** (1996) 11752–11761.

8 Diffusivities of small clusters

In the previous chapter we presented possible mechanisms which can contribute to cluster diffusion; in this chapter, we will concentrate on the energetics of cluster movement, mostly the movement of the center of mass. Early studies of cluster diffusion were all done on tungsten surfaces using FIM, but as techniques other than field ion microscopy were applied to learning more about this subject, other surfaces came under scrutiny. The biggest change in the level of activity, however, was made by theoretical calculations. These now dominate the field and have usually covered several surfaces of different materials in one examination. The number of experimental studies of cluster behavior decreased markedly as computational efforts reached new intensities. Unfortunately, theoretical investigations still are quite uncertain and experiments are urgently needed for comparison and verification. Nevertheless we will try to arrange our comments chronologically in the description of each material, but with experiments and theoretical calculations separated.

8.1 Early investigations

8.1.1 Experiments

Work on the diffusion of single adatoms on a metal surface had been going on for just a few years when Bassett began to look at clusters formed by association of several atoms [1]. In 1969 he noted that after depositing several atoms on the (211) and (321) planes of tungsten, clusters formed, with a mobility smaller than that of single atoms, provided that deposition took place with the atoms in the same channel. On these planes, clusters moved in only one dimension, along the channels of the planes. Studied were tantalum, tungsten, rhenium, iridium, and platinum atoms. On the (110) plane of tungsten behavior was different – as one can expect from the structure of this plane, movement was two-dimensional. Tantalum formed stable two-dimensional clusters of several atoms, with dimers having a mobility roughly one tenth that of singles. As shown in Fig. 8.1, atoms in platinum and iridium dimers were resolved in the FIM, and were oriented along the close-packed <111> directions of the surface. They diffused only slightly less rapidly than individual adatoms. Easily formed dimers turned out not to be typical for all metallic adatoms: rhenium adatoms did not form dimers at all, and only bigger Re clusters were observed. This study showed, however, that larger clusters could be built up one atom at a time with very precise control.

8.1 Early investigations

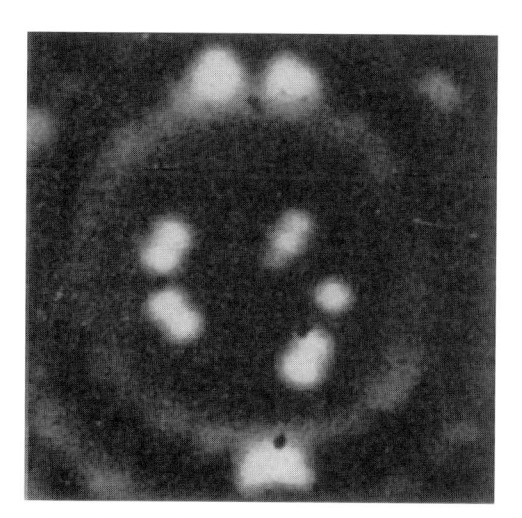

Fig. 8.1 Three dimers and one trimer (nearest center) are seen in FIM on W(110) plane after allowing Pt atoms deposited on the surface to diffuse at 240 K (after Bassett and Parsley [1]).

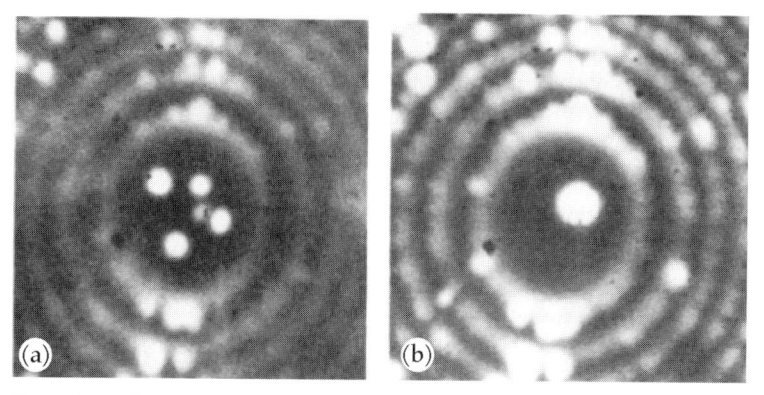

Fig. 8.2 Formation of Ta$_5$ cluster on W(110) surface observed in FIM. Five Ta atoms were deposited on (110) plane at 77 K in (a). After 2 min at 291 K, pentamer has formed in (b) (after Bassett [2]).

At the beginning of 1970 Bassett [2] followed his first investigation with a number of small descriptions of various clusters. A two-dimensional tantalum pentamer on W(110), shown in Fig. 8.2, was created as well as a rhenium and a linear iridium heptamer [3]. A much more detailed description of the technique for making and observing clusters followed, showing the creation of a hexamer of iridium on W(110), in Fig. 8.3, as well as rhenium and tantalum clusters [4]. The diffusion of clusters was mentioned only briefly, however. Dimer mobility was noted as being smaller than that of adatoms by a factor ranging from two to ten, but tantalum dimers had a mobility comparable with that of tantalum adatoms.

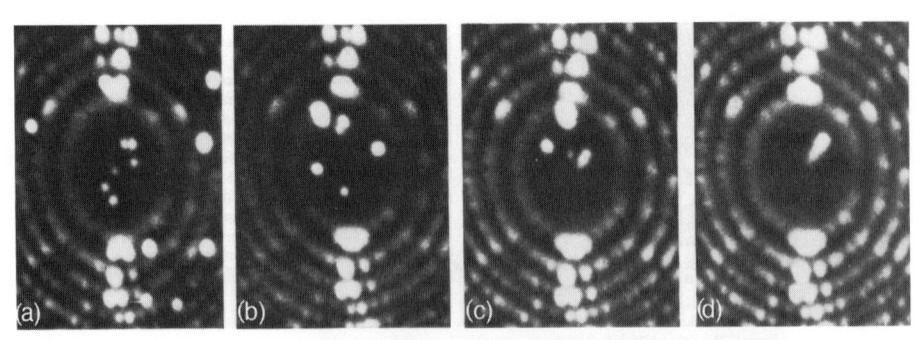

Fig. 8.3 Steps in the formation of a linear Ir$_6$ cluster on W(110) examined in FIM, starting with an initial deposition in (a) on surface cooled to 78 K. (b) After heating to 380 K for 1 min. (c) After an additional 3 min heating. (d) After a further 4 min heating a cluster has formed (after Bassett [4]).

A year after Bassett, Tsong [5,6] looked at W-Re, as well as W$_2$ dimers on W(110) and Re$_3$ as well as ReW$_2$ trimers, and found that for a W-Re hetero-dimer at temperatures between 300 and 340 K the tungsten atom can migrate away, while the Re atom stays stationary. A stable Re-W cluster, as well as a Re$_2$ cluster with atoms in nearest-neighbor positions was not observed below 332 K. However, a rhenium atom was found to easily combine with a cluster of two tungsten atoms, which is stable up to 460 K, and starts migration above 380 K. Clusters made up of three Re atoms were also stable on this plane. The situation differs for tungsten clusters. W dimers were easily created and stable up to 390 K, with dimer atoms always observed at the nearest-neighbor distance. Tungsten clusters with up to six atoms dissociated below 490 K. The energetics of cluster movement was not investigated in this study.

Tsong [5,6] also examined a chain of four tungsten atoms in adjacent channels of the W(211) plane, which was observed to migrate as a unit. Somewhat later there followed a further description of clusters on tungsten [6]. A tungsten dimer with atoms in adjacent rows was observed to migrate as a unit on W(211) at up to 330 K and a tetramer with atoms in adjacent rows was seen to diffuse at 300 K. Some of these clusters are shown in Fig. 8.4. When a rhenium atom was present in a channel next to tungsten, the two migrated together but dissociated frequently even below a temperature of 320 K. When both atoms were present in the same channel they formed a stable Re-W hetero-dimer, with much lower mobility. Two Re adatoms deposited in the same channel also readily combined to form a Re$_2$ dimer, which behaves very similarly to a dimer of tungsten. Also examined were the structures of various clusters. Two configurations were identified for W$_2$ on W(211), and possible structures were proposed for W$_2$, W$_3$, W$_4$, and W$_6$, as well as for rhenium clusters on the W(110) plane; the latter have to be looked at as speculative. What is clear from this study is that creating clusters depends on the structure of the surface.

These early studies were quite qualitative; they provided only little information about the migration of clusters, and were more concerned with cluster structure. A break in this trend was made by Bassett and Tice [7], who did quantitative studies of cluster stability, but that is not our primary concern here.

8.1 Early investigations

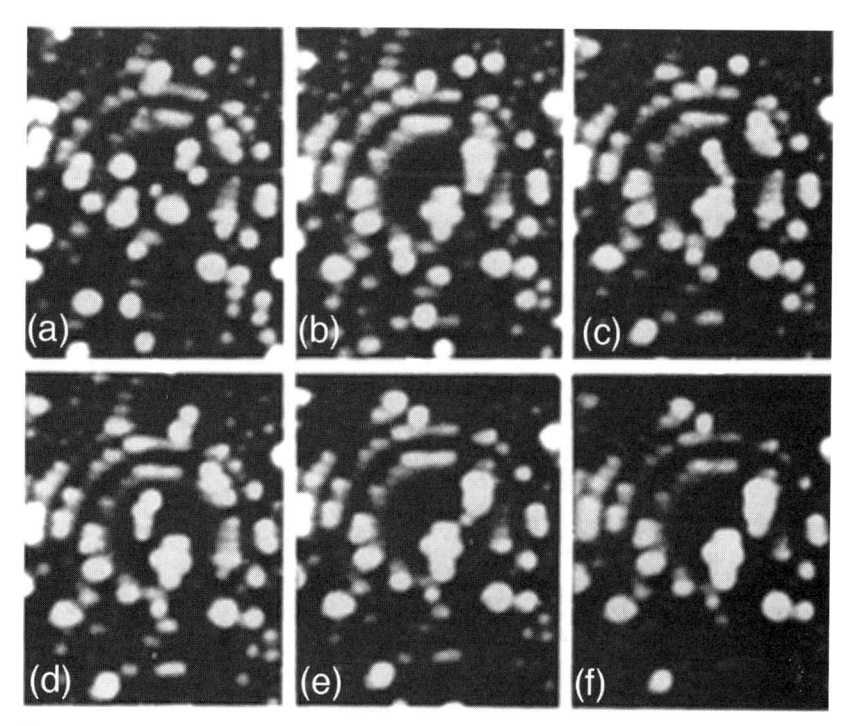

Fig. 8.4 Tungsten clusters on W(211) plane observed in FIM. (a) Ten tungsten atoms were deposited on W(211) plane. On heating to 300 K, six combined to form W_6, the rest combined into linear W_4 in (b). (c)–(f) Displacements of clusters on heating to 300 K (after Tsong [6]).

8.1.2 Theoretical investigations

Calculational efforts began with high intensity in the early 1990s. As will quickly become apparent, these studies nicely supplement experimental work in providing detailed information not otherwise accessible. The first molecular dynamics simulations of clusters were done by Tully *et al.* [8] on an fcc(100) surface using a ghost-particle theory to avoid dealing with the many surface and bulk atoms. Adatoms interacted with each other via Lennard-Jones potentials, and the crystal was not allowed to relax. From an Arrhenius plot of the diffusivities in Fig. 8.5 they were able to arrive at the diffusion parameters shown in Table 8.1. From dimers upward there was a gradual increase in the activation energy, and generally a decrease in the

Table 8.1 Rate parameters for diffusion of Lennard-Jones Pt clusters on fcc(100) [8].

Cluster size N	Prefactor $V_0 \times 10^{12}$	Activation Energy E_D (eV)
1	7.62	0.390 ± 0.005
2	7.34	0.338 ± 0.007
3	6.58	0.360 ± 0.010
4	6.84	0.415 ± 0.013
5	5.57	0.418 ± 0.013
6	4.30	0.425 ± 0.013

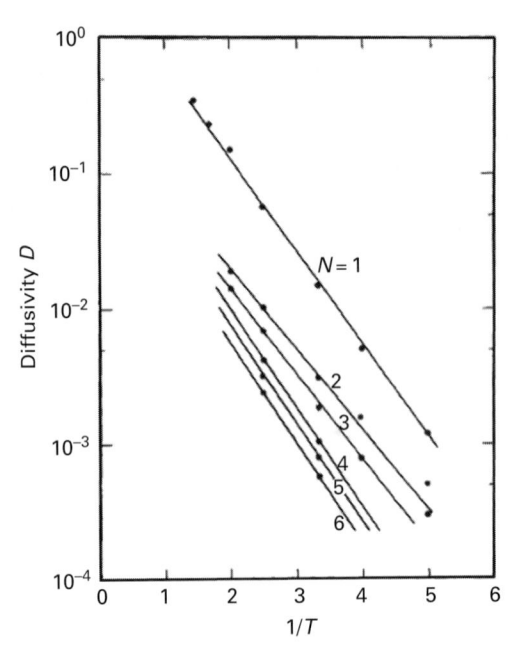

Fig. 8.5 Arrhenius plots for the diffusion of differently sized clusters (after Tully et al. [8]) interacting via Lennard-Jones potentials.

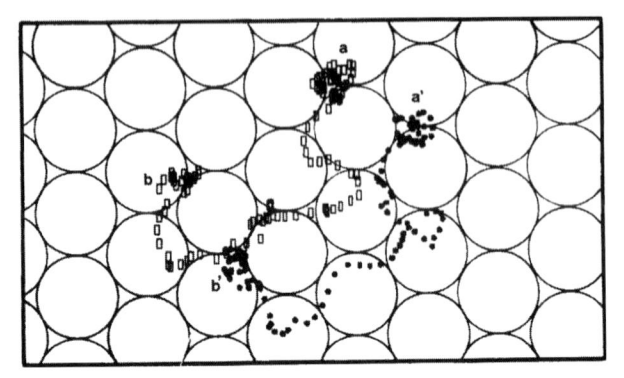

Fig. 8.6 Trajectory of dimer at $T=0.34\ \varepsilon_{LJ}/k$ during multiple jumps on fcc(111) surface obtained in molecular dynamics simulations with Lennard-Jones potentials (after Ghaleb [9]).

prefactor, as the size of the cluster increased from two up to six atoms. The energy unit employed in the molecular dynamics simulations here was 0.25 eV; distances were measured in terms of the spacing of the Lennard-Jones minimum, 1.39 Å. It must be noted that for platinum the unit of temperature was 2910 K, so that diffusion parameters were obtained over a very wide energy range.

In 1984, Ghaleb [9] did computer simulations of dimer diffusion on an fcc(111) crystal also relying on Lennard-Jones potentials and came up with very interesting results. At an elevated temperature of 0.34 ε_{LJ}/k, where ε_{LJ} is the depth of the Lennard-Jones potential, both atoms were found to jump simultaneously and to execute long transitions, indicated in Fig. 8.6. This was a surprising and most important result, which changed how cluster diffusion was interpreted. At a lower temperature of 0.22 ε_{LJ}/k transitions were more usual; they occurred between nearest-neighbor adsorption sites, through jumps of one atom to an adjacent site, and the mobility of the center of mass of dimers was very close to the mobility of single atoms.

With the early cluster studies out of the way, we will now begin listing work according to the surface examined.

8.2 Clusters on aluminum surfaces

8.2.1 Aluminum (100)

There is no experimental information available about the energetics of cluster movement on aluminum surfaces. Theoretical studies started with local-density-functional calculations for the diffusion of a pair of aluminum atoms on Al(100), done by Feibelman [10] in 1987. For single Al adatoms he arrived at a barrier to surface diffusion of 0.80 eV. The activation energy for the dimer was lower, however, amounting to only 0.66 eV. In the dimer, the two adatoms were found to be sitting higher from the surface than does a single atom, and the bonding to the surface was somewhat diminished as the two adatoms interacted. Diffusion occurred by the jumping of one atom at a time, so the distance between the two adatoms increased, and the atom not jumping could strengthen its interactions with the lattice. The energy gained was transferred to the moving adatom, lowering the diffusion barrier. This reduction in the activation energy for dimer movement compared to that of a single adatom was also found on Ni(100) and Pt(100) [11].

Further estimates for diffusion of aluminum clusters, illustrated in Fig. 8.7, were done by Valkealahti and Manninen [12] with effective medium theory. Diffusion of dimers on the (100) face was slower than on the (111) plane, and close to 600 K exchange events occurred. Dimers were not dissociated below 600 K, and trimers behaved similarly. No energetics for dimers were derived, however.

Bogicevic et al. [13] used the VASP simulation package, as well as DACAPO [14] to look at self-diffusion of Al, Au, and Rh dimers on (100) planes. In this study, contrary to the findings of Feibelman [10], dimer atoms occupy neighboring hollow sites. A number

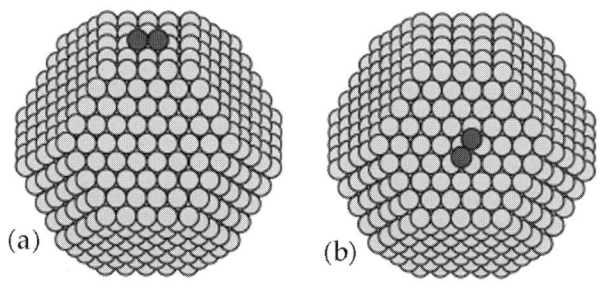

Fig. 8.7   Aluminum dimer on aluminum clusters, calculated with EMT. (a) On Al(100). (b) On Al(111) (after Valkealahti and Manninen [12]).

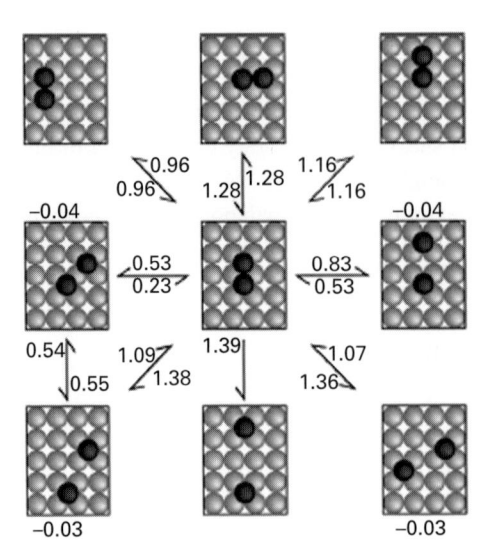

Fig. 8.8 Activation energies (in eV) for diffusion and dissociation of Al$_2$ on Al(100) plane obtained from density-functional calculations. Binding energies are given at top or bottom of each configuration (after Bogicevic et al. [13]).

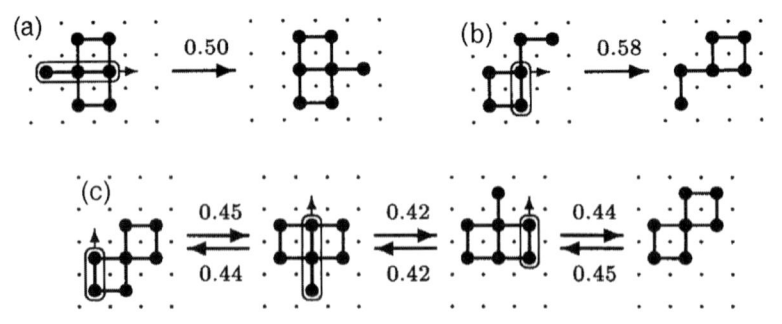

Fig. 8.9 Diffusion of copper pentamer on Cu(100) plane. (a) Internal three atom shearing. Activation energies (in eV) from EAM calculations. (b) Dimer rotation. (c) Schematic for diffusion of Al$_7$ cluster on Al(100) plane (after Trushin et al. [15]).

of processes was considered; two of them seemed to be more important than the others: shearing (middle of the second row in Fig. 8.8 and to the left) as well as stretching (the middle row, center to right). The process described by Bogicevic et al. [13] as shearing is in fact that of a dimer atom moving perpendicular to the dimer axis, which can be achieved by an atom jump to the side. For aluminum dimers the barrier for such described shearing (0.53 eV) was 0.3 eV lower than for stretching (0.83 eV). It appears that for these systems this process is the low-energy path for the dimers. The authors argue that this is likely to be a general effect for other (100) fcc metals.

Slightly bigger clusters than dimers were investigated by Trushin et al. [15] on Al(100) with the Ercolessi–Adams [16] glue potential. The new processes considered, internal three-atom shearing, row shearing at the edges, and dimer rotation, are all illustrated in Fig. 8.9. Here by shearing the authors mean the movement of two (dimer shearing) or three atoms (trimer shearing) together in concerted motion. With Al$_7$ clusters they found

8.2 Clusters on aluminum surfaces

Table 8.2 Activation energies (in eV) for dimer diffusion evaluated by density-functional calculations on fcc(100) surfaces [20].

		Al	Rh	Ir	Ni	Pd	Pt	Cu	Ag	Au
Dimer	E_D(Jump)	0.54	1.33	1.87	0.88	0.86	1.30	0.66	0.55	0.82
	E_D(Ex)	0.40	1.61	1.30	0.69	0.75	0.18	0.77	0.69	0.24
	E_D(expt)			0.88			0.41			

that dimer shearing at the edges occurred over a barrier of 0.44–0.45 eV, and internal three-atom row shearing occurred over 0.42 eV, small compared to the activation energy for atom jumps of 0.53 eV and 0.56 eV for exchange. The energetics of Al$_7$ diffusion are illustrated in Fig. 8.9; exchange events of low energy were not found.

Molecular dynamics simulations of dimer self-diffusion have been carried out by Zhuang et al. [17] on the (100) surface of aluminum using the EAM potentials of Johnson [18]. The mechanism for dimer diffusion preferred was the conventional exchange process with a value of 0.342 eV for the barrier compared to 0.425 for conventional hopping. More complicated mechanisms of movement, such as 180° and 270° exchange processes were also observed for aluminum dimers, however, less frequently and their energetics were not revealed. Very similar findings were reported by Liu et al. [19]. The results for static activation energies are given in Table 7.3 and hopping has to proceed over a higher barrier (0.4614 eV) than exchange (0.3545 eV). More complicated mechanisms on this surface were also observed.

In 2005 Chang and Wei [20] used a different theoretical approach and carried out *ab initio* density-functional calculations using the VASP [21,22] package to establish the activation energy for self-diffusion of dimers on different fcc(100) planes. Their results are given in Table 8.2, calculated with a (5 × 5) surface cell using the LDA approximation and the Ceperley–Alder potential [23,24]. Diffusion by atom exchange was predicted for aluminum with an activation energy of 0.40 eV compared to 0.54 eV for simple hopping. More complicated mechanisms were not taken into account.

Unfortunately there is no experimental information for movement of clusters on aluminum and the theoretical investigations are not conclusive. The activation energies for movement of dimers by hopping are quite spread out, and range from 0.425 eV to 0.66 eV. However, most investigations indicate exchange as a leading mechanism and three values obtained so far are not too far from each other: 0.34 eV, 0.35 eV, and 0.4 eV [17,19,20]. Only Trushin [15] looked at diffusion of clusters bigger than a dimer on Al(100) and concluded that shearing played an important role.

8.2.2 Aluminum (111)

Investigations of the motion of aluminum dimers on Al(111) started with the study of Bogicevic et al. [25] using density-functional theory. Both intra- and inter-cell movements were considered. In the former, the dimer is confined to an hexagonal cell containing six sites around one surface atom and long-range diffusion does not take place, whereas in the latter case movement over long distances is accessible. These steps of

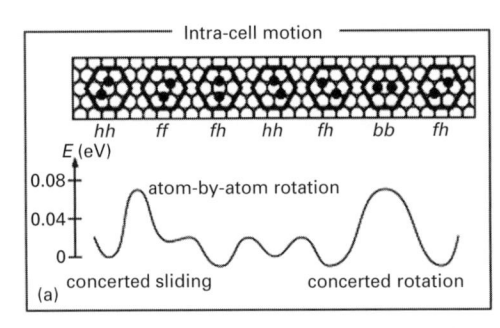

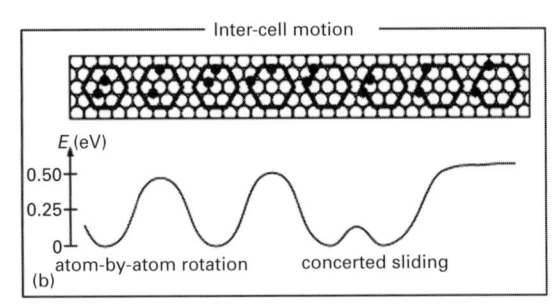

Fig. 8.10 Motion of aluminum dimer on Al(111) derived from density-functional calculations. (a) Confined (non-diffusive) motion. (b) Inter-cell diffusion (after Bogicevic et al. [25]).

dimer movement are illustrated in Fig. 8.10. Dimer motion for aluminum is different from what was reported for Ir$_2$ on Ir(111) [26], where single atom jumps were assumed. With aluminum there is a strong attraction between the two atoms, and rotation out of the cell requires overcoming a barrier of 0.50 eV. Once done, the second step occurs without any barrier, and the dimer moves by concerted sliding. Concerted inter-cell sliding in the <112> direction required only 0.13 eV. In intra-cell rotation by individual atom jumps over the bridge, the highest barrier found was 0.03 eV. Rotational motion of the dimer already began at 8 K, but leads only to local movement. Concerted intra-cell rotation of the dimer required three times more energy, 0.09 eV, as estimated in the GGA approximation. Concerted intra-cell sliding occurred over a barrier of 0.08 eV. The effect of compressive strain, a reduction in the lattice spacing, is to significantly interfere with dimer diffusion, as the energy of dimers on a pair of adjacent fcc or hcp sites is increased. As is clear from the energetics, intra-cell movement will proceed mostly through atom by atom transitions. However, for longer-range diffusion, the dimer must overcome a bigger barrier than for local movement.

Valkealahti and Manninen [12] investigated diffusion of aluminum dimers on a polyhedron, illustrated in Fig. 8.7, with effective medium theory. On the (111) plane, dimers were quite mobile starting at 100 K, and did not dissociate below 600 K. At higher temperatures exchange events occurred with cluster atoms.

Self-diffusion of metal clusters with up to five atoms was examined by Chang et al. [27] with density-functional theory in the local-density approximation (VASP). For the aluminum dimer Al$_2$ on Al(111), the lowest energy was obtained for a "mixed fcc-hcp" dimer with one atom in an hcp site and the other in an fcc position. Movement with the lowest activation energy occurred via zigzag motion illustrated by the sequence $a_1 \leftrightarrow a_2 \leftrightarrow a_3$, a_4, or a_5, in Fig 8.11a, with only one atom moving at a time. Zigzag transitions result in local movement which can occurr by the translation $a_1 \leftrightarrow a_2 \leftrightarrow a_3$, or by dimer rotation $a_1 \leftrightarrow a_2 \leftrightarrow a_4$ or a_5. Local motion had a small activation energy of 0.05 eV, local movement of two atoms together required 0.09 eV, making it rather unlikely. Long-range diffusion (movement out of a cell) meant overcoming a barrier of 0.08 eV.

Details of Al$_3$ diffusion over the Al(111) surface have also been studied by Chang et al. [28]. The cluster can exist in a number of different configurations on the surface, of which

8.2 Clusters on aluminum surfaces

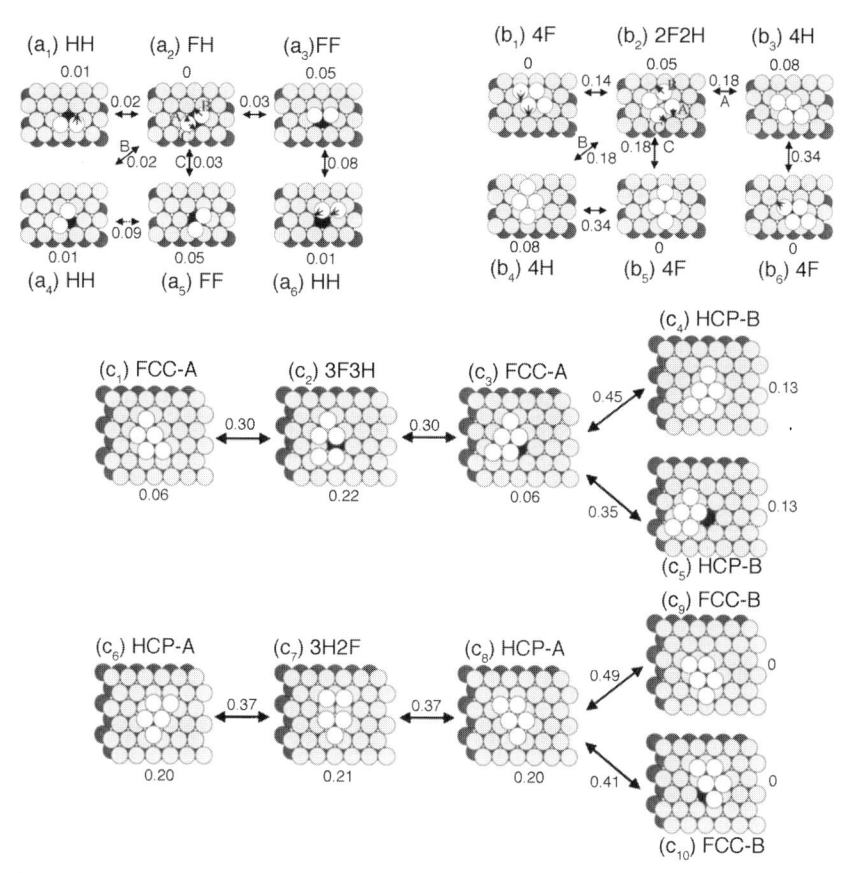

Fig. 8.11 Structure and diffusion path for Al clusters on Al(111) surface. Energetics from EAM calculations for dimer on Al(111). Numbers indicate structural energies in eV (after Chang et al. [27]).

the configuration of the lowest energy is a triangular trimer in fcc sites with the center over an empty site, designated FCC-H; a final letter T indicates that the cluster center is on top of an surface atom. It should be noted that this was also the preferred arrangement in experiments with iridium trimers on Ir(111) [26]. The possible arrangements are shown in Fig. 8.12, L indicating linear, and it is clear that the linear clusters have a higher energy, as do non-compact triangular trimers, shown in Fig. 8.12c. Three different non-local diffusion steps were identified: concerted translation between FCC-H to HCP-T with an energy of 0.24 eV, translation between FCC-T to HCP-H, with energy of 0.22 eV, and transformations between compact triangular clusters and linear trimers with an activation energy of 0.21 eV. Translation of linear trimers proceeds over a higher barrier of 0.28 eV. The authors conclude that the most likely diffusion path for aluminum trimers is that shown in Fig. 8.12a, since it involves the largest displacements with the lowest number of steps. This path involves transitions of a compact trimer via a jump to an HCP site over a lattice atom, then rotation to an FCC site, and finally a jump to an HCP-H site. A variety of other migrations were considered with this as the most probable one, which takes place by concerted

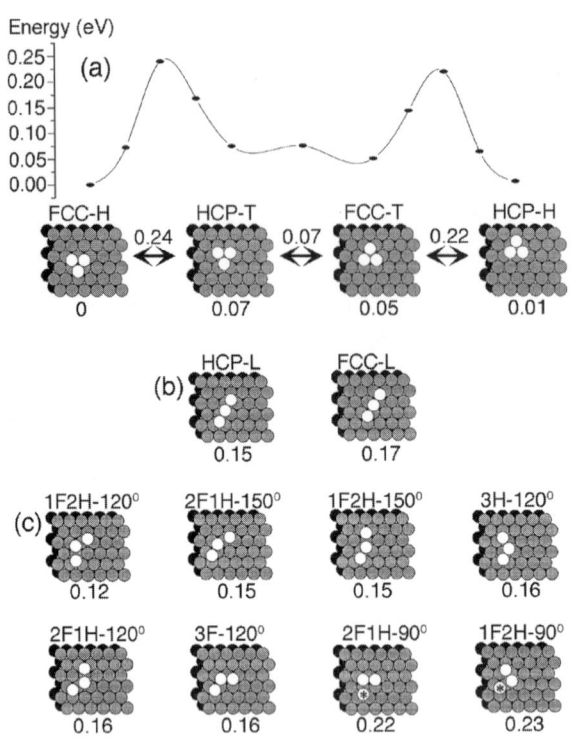

Fig. 8.12 Al$_3$ configurations on Al(111) surface studied with DFT. (a) Structure and diffusion path for concerted movement of triangular Al$_3$ on Al(111). Number over arrow gives transition state energy in eV. (b) Linear trimer. Number under structure gives calculated total energy (in eV) compared to most stable. (c) Non-compact triangular trimers (after Chang et al. [28]).

jumps and rotation of the whole trimer. This was confirmed by molecular dynamics simulations based on the embedded atom method, which also detected the trimer rotation HCP-T ↔ FCC-T. The study suggests that trimers exist mostly in the compact triangular form, and it would be crucial to experimentally check the presence of linear trimers on the Al(111) surface.

Diffusion of larger clusters was also considered, the most interesting being the tetramer, in Fig. 8.11b. There the lowest energy path was observed by movement between compact configurations. Translation took place by the transition $b_1 ↔ b_2 ↔ b_3$, and rotation via $b_1 ↔ b_2 ↔ b_4$ or b_5. The concerted intra- and inter-cell motion of tetramers is more demanding than zigzag motion with two atoms moving together. The activation energy for long-range Al$_4$ diffusion was only 0.18 eV, but increased again for pentamers, where it was 0.30 eV for movement of two pentamer atoms from fcc sites to adjacent hcp sites or 0.37 eV for motion from hcp to fcc sites. For pentamers the most favorable process was rotation with only two atoms moving (dimer shearing), $c_1 → c_2 → c_3$ in Fig 8.11c. Long-range concerted motion required more energy of 0.45 eV for movement between fcc to hcp sites or 0.49 eV for the opposite direction. Self-diffusion barriers were also calculated for other metals with EAM potentials and are shown in Fig. 8.13. What is

8.3 Clusters on iron surfaces

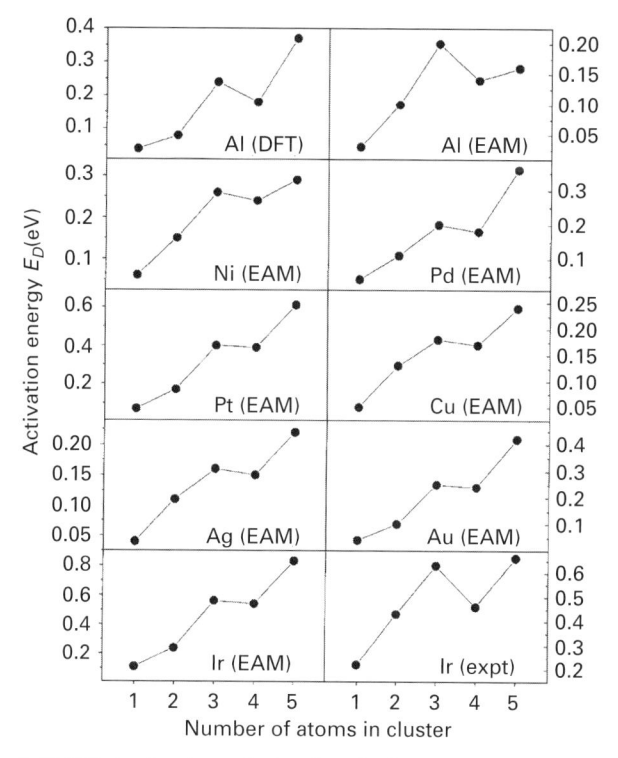

Fig. 8.13 Self-diffusion barriers calculated on fcc(111) planes of various metals with EAM potentials (after Chang et al. [27]).

important here is that all show a decrease in the diffusion barrier in going from trimers to tetramers, and then increase again for pentamers, as found a little earlier for Ir(111) by Wang [29].

Unfortunately we have no experimental information for clusters on Al(111). The theoretical predictions agree that there is no exchange involved in the movement of clusters, and a number of correlated and uncorrelated motions were studied. The low energy for rotation of dimers seems to be established. Longer movement of dimers is more demanding. Only Chang et al. [27] looked at the change in the energetics with the size of clusters and noticed a minimum for the energy of tetramer diffusion. Further experimental investigations are definitely in order to uncover the true nature of cluster movement on this plane.

8.3 Clusters on iron surfaces

In 2007, Chen et al. [30] calculated the diffusivity of iron clusters on Fe(110), using modified analytical EAM potentials [31,32]. Binding of the cluster is greater for islands than chains, as is clear from Fig. 8.14. They evaluated the mean-square displacement at different temperatures, and from Arrhenius plots of the diffusivity were able to arrive at

Table 8.3 Diffusion characteristics for iron clusters on Fe(110) from EAM calculations [30].

Cluster size (atoms)	Activation energy E_D (eV)	Prefactor for diffusivity D_o (cm²/sec)
1	0.40	1.27×10^{-2}
2	0.36	1.59×10^{-3}
3	0.49	2.81×10^{-3}
4	0.84	1.89×10^{-1}
5	0.71	7.71×10^{-3}
6	0.75	2.98×10^{-2}
7	1.16	1.01×10^{-1}
8	0.89	1.18×10^{-2}
9	0.65	6.60×10^{-4}

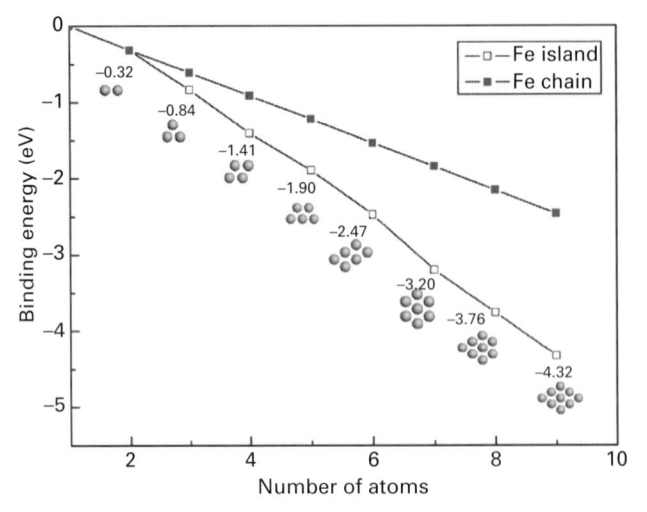

Fig. 8.14 Energy of iron chain compared with that of islands obtained in EAM calculations on Fe(110) (after Chen et al. [30]).

the activation energies and prefactors, listed in Table 8.3 for clusters made up of up to nine atoms. It is of interest that the tetramer and heptamer have unusually high activation energies as well as prefactors. Movement of dimers required a lower energy than monomers. Chen et al. concluded that dimer movement proceeds by sequential jumping of dimer atoms and that the lower energy is due to the higher coordinational number. For bigger clusters, the sequential displacement of atoms along cluster edges was suggested and the increasing mobility of non-compact clusters was attributed to diffusion of the extra atom along the periphery. Regrettably there are no measurements available to validate these estimates.

8.4 Clusters on nickel surfaces

8.4.1 Nickel (100)

For nickel just as for aluminum surfaces we have no experimental information about cluster movement. Investigations on Ni(100) started with calculations using embedded atom potentials by Liu and Adams [11], done for clusters both on the (100) and (111) planes of nickel. The EAM potentials have not always given good agreement with experiments but show two-dimensional islands for all cluster sizes on Ni(100), and dissociation energies oscillating with size. However, as can be seen in Table 8.4, on the (100) plane Liu and Adams observed an oscillation in the diffusion barrier with size different from that for self-diffusion on the (111) plane. Surprisingly, the migration energy for Ni and Pt dimers is slightly lower than for single atom movement.

Table 8.4 Cluster diffusion parameters calculated with EAM potentials [11].

	\multicolumn{5}{c}{Activation energy E_D(eV)}
Cluster	Ni/Ni(100)	Pt/Ni(100)	Pt/Pt(100)	Ni/Ni(111)	Ir/Ir(111) [26]
Single	0.63	0.76	0.47(0.64)	0.056	0.27
Dimer	0.61	0.69	0.41(0.52)	0.12	0.43
Trimer	0.68	0.81	0.49(0.65)	0.20	0.63
Tetramer	1.00	1.00		0.21	0.46
Pentamer	0.68	0.78		0.26	0.66

Energetics in parenthesis from Kellog [33].

The authors looked at the details of Ni and Pt dimer movement on the Ni(100) surface, the energetics of which are illustrated in Fig. 8.15a$_2$–c$_2$. In the initial movement of the dimer there are two options for the atomic jump. Energetically less demanding is the jump in the direction perpendicular to the dimer axis, which costs 0.61 eV for Ni and 0.69 eV for a Pt atom. A more demanding jump is in the direction of the dimer axis, which required 1.01 eV for Ni and 1.12 eV for Pt. Subsequent jumps of the second dimer atom occurs with a much lower energy of 0.25 eV for Ni and 0.43 eV for Pt.

For trimers the lowest energy transition is rotation, 0.26 eV for nickel and 0.38 eV for platinum, which involves changes between the triangular forms. Since such transitions do not shift the center of mass significantly, long-range movement proceeds through the linear trimers. The optimal path for such a transition is presented in Fig. 8.15a$_3$–d$_3$. The most demanding step of this path is the initial one of breaking the triangular trimer, costing 0.68 eV for nickel and 0.81 eV for platinum. The initial step of breaking a compact shape is also most demanding in tetramer movement, 1.00 eV for both nickel and platinum, shown in Fig. 8.15a$_4$–d$_4$. For the pentamer, the energy of moving atoms is lower than for the tetramer and comparable with trimer motion, as is clear from the optimal path illustrated in Fig 8.15a$_5$–d$_5$. The investigations did not consider dimer shearing as a possible mechanism, which might play quite an important role, especially

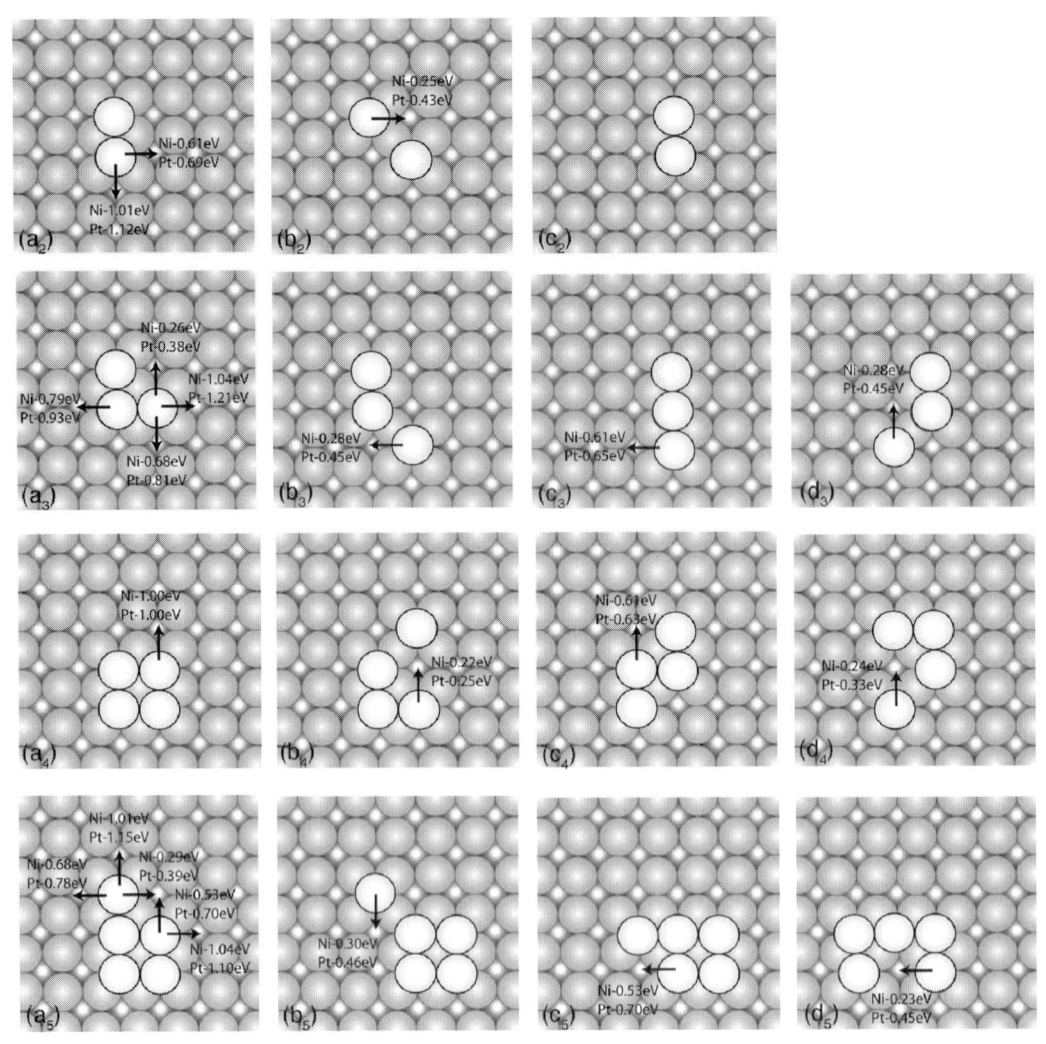

Fig. 8.15 Schematic of optimal diffusion paths and energetics for Ni and Pt clusters on Ni(100) plane, based on the work of Liu and Adams [11] using EAM potentials.

in the movement of tetramers or pentamers. What is surprising is that for Pt clusters on Ni(100), Liu and Adams did not observe exchange, since Kellogg [34] showed that for single atom movement of Pt on Ni(100) exchange was the basic mechanism of transport.

A new mechanism for movement on Ni(100) was discovered by Shi et al. [35] using EAM interactions and bond counting. They found that displacements by concerted motion of two or three atoms can be more favorable than movement by single atoms. The mechanism is illustrated in Fig. 7.25, and the energetics of the different steps in the movement of a Ni cluster are summarized in Table 7.1. For nickel tetramers the initial single jump required 0.989 eV, while the initial step by dimer shearing needed only 0.842 eV. For a nine-atom cluster, trimer shearing seems to be more favorable than movement by single atoms, at 1.051 eV compared to 1.215 eV.

8.4 Clusters on nickel surfaces

Further theoretical investigations of dimers on Ni(100) were done by Liu *et al.* [19] who relied on EAM interactions [18] in their simulations. They found that movement of nickel dimers proceeded by hopping rather than by exchange. The energy for hopping of the dimer was 0.561 eV, and for exchange 1.349 eV. A year later Chang and Wei [20] used the VASP [21,22] simulation package to also find the diffusion barrier and mechanism of single atoms and dimers. Their results for dimers are presented in Table 8.2 and for nickel dimers they suggest exchange as a leading mechanism, with an activation energy 0.69 eV, rather than hopping over a higher barrier of 0.88 eV. This investigation seems to contradict the previous findings of Liu *et al.* [19].

Recently Kim *et al.* [36] used action-derived molecular dynamics, with interactions described by a tight-binding second-moment approximation, to look at cluster movement for a range of materials. For Ni dimers on a Ni(100) surface, they found hopping in the direction of the dimer axis needed 0.747 eV, hopping perpendicular to the dimer axis required 0.376 eV; exchange of the dimer with the substrate was unlikely and required 1.351 eV. What is very surprising is that the jump of an atom in the direction perpendicular to the dimer axis has exactly the same energy as calculated by them for the jump of a single isolated adatom, making this investigation a bit suspicious. They also looked at the energetics of atom movement for bigger clusters. In their study they only considered a movement of trimer atoms from the linear state along the [110] direction. Jumps in the direction [110], the direction of the trimer axis, were contemplated and required 1.109 eV. It is clear that it is not the direction of lowest energy. In contrast, exchange in the [100] direction required 1.371 eV. A similar situation holds for clusters of four atoms. They again started only as linear clusters, even though the close-packed tetragon is usually observed on fcc(100) planes. They looked only on the energetics of jumps in the [110] direction, which is the direction of the linear tetramer. Such jumps require 0.748 eV. Exchange in the [001] direction was found to need 1.347 eV.

The investigations of clusters on the Ni(100) surface are far from conclusive. Two investigations suggest hopping as the primary mechanism, while one favors exchange. Here it is worth noting that for single atoms the movement by atom exchange was suggested in experiments by Fu and Tsong [37]. The activation energies for hopping of dimers are spread out from 0.56eV to 0.88 eV, while for exchange they range from 0.69 eV to 1.35 eV. Bigger clusters were only investigated by Liu *et al.* [19], Shi *et al.* [35], and Kim *et al.* [36]. Shi *et al.* showed that dimer and trimer shearing might play a very important role in such movement; studies of Liu *et al.* and Kim *et al.*, even though done later, did not take shearing into account.

8.4.2 Nickel (111)

Theoretical work on the Ni (111) surface started together with the study on Ni(100) by Liu and Adams [11] using EAM potentials. On Ni(111), just as on Ni(100), the dissociation energies oscillated with cluster size. Oscillations were also observed in the diffusion energies, as is apparent from Table 8.4. The activation energy for movement of dimers increased to 0.12 eV, and a further rise to 0.20 eV was observed for trimers. For tetramers the energy was almost the same, 0.21 eV, but increased further to 0.26 eV for pentamers.

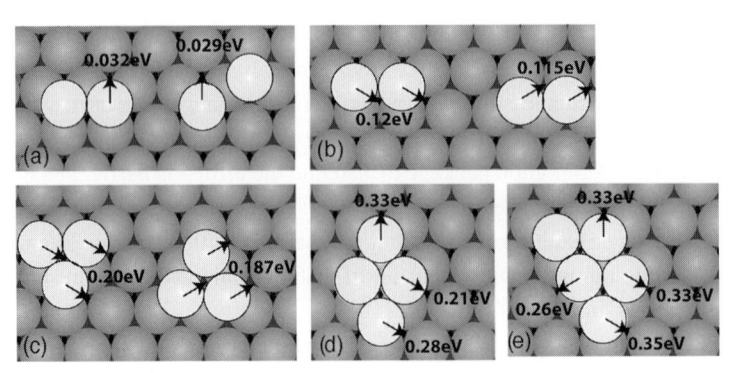

Fig. 8.16 Schematics of optimal diffusion path of nickel clusters on Ni(111) plane, together with EAM energetics of atom jumps, based on studies by Liu and Adams [11].

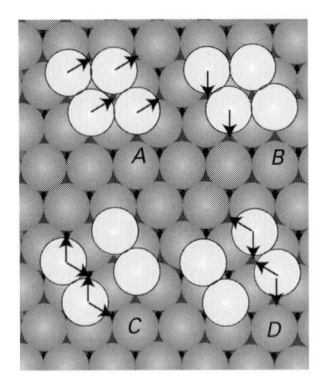

Fig. 8.17 Mechanisms of diffusion for Ni$_4$ cluster on Ni(111) surface, based on Hamilton et al. [38] using EAM potentials. *A* shows Ni atoms moving almost simultaneously through bridge sites. *B* and *D* illustrate dislocation process. In *B* two atoms move to produce the dislocation in either *C* or *D*; this can relax following arrows.

Optimal migration paths for clusters are shown in Fig. 8.16. For dimers the initial atom jump out of the cluster required 0.032 eV, while the following transition of the second atom needed only 0.029 eV. Collective jumping from fcc to hcp sites and vice versa required 0.12 eV and 0.115 eV, making collective jumps unlikely. The authors found that the collective steps between fcc and hcp sites are the most likely mechanism of movement for bigger clusters. For collective movement of trimers an energy of 0.20 eV from fcc, or 0.187 eV from hcp, is required. The values in Fig. 8.16d–e show the energetics of possible initial jumps in atom-by-atom movement for tetramers and pentamers, the energy for the collective motion of the tetramer from fcc to hcp sites is in a comparable range and amounts to 0.21 eV, while for the pentamer 0.26 eV was required.

Hamilton et al. [38] in 1995 came up with a new mechanism for cluster diffusion on fcc(111) surfaces – the nucleation of misfit dislocations. Working with nickel islands on Ni(111) surfaces and EAM potentials, they did molecular dynamics at temperatures from 400 K to 1200 K. For a tetramer they came up with the configurations shown in Fig. 8.17. All the atoms can move, as in *A*, to give a gliding transition. However, an alternative is also possible, and is suggested by *B*, where two atoms translate, and create a dislocated structure that can then continue movement in four different ways. For the gliding of a tetramer they found an energy of 0.30 eV, while for dislocation movement it was 0.36 eV, making gliding of the cluster as a whole slightly more likely. Simulations were also done for larger clusters with up to 43 atoms, and gliding as well as dislocation transitions were observed. In Fig. 7.27 is shown a dislocated cluster of 19 atoms, in which the rows of atoms can move one at a time. Hamilton et al. suggested that gliding and dislocation movement are favored over atom motion along the edges. It must be noted, however, that this mechanism does not account for the long jumps that have been observed in experiments with Ir_{19} clusters diffusing on Ir(111) [39].

Chang et al. [27] using EAM potentials investigated clusters up to pentamers, as shown in Fig. 8.13, deriving the activation energies for movement of the center of mass. For dimers on Ni(111), the activation energy was 0.15 eV and for trimers 0.26 eV. Movement of tetramers required a slightly lower energy of 0.24 eV than for trimers. The energy again increased to 0.29 eV for pentamers. Detailed information about the transition state for movement was provided only for aluminum clusters.

Calculations of seven-atom hexagonal clusters of nickel, palladium, platinum, copper, silver, and gold on the (111) plane of the same metal were done by Longo et al. [40] using EAM potentials parameterized according to Voter and Chen [41]. They evaluated the activation energies for a cluster gliding over its own crystal, and for nickel they came up with a value of 0.58 eV. For Ni_7, Pt_7, and Cu_7 translational motion was only observed on heating the cluster to the melting temperature. For Ni_7 this was expected, as the barrier for peripheral movement was anticipated to be higher than for gliding.

Recently Kim et al. [36] published their findings for clusters on Ni(111). In their investigations it is not clearly stated where the cluster is sitting in the initial stage – in fcc or hcp sites; from the choice of their hopping direction for dimers it appears that clusters are actually sitting in hcp sites. The dimer is oriented along the $[1\bar{1}0]$ direction and the jump occurs along $[1\bar{2}1]$, which is a jump 30° to the axis of the dimer and requires 0.128 eV. The jump at 90° to the dimer axis requires less energy, 0.087 eV. Exchange is unlikely and has a huge barrier of 1.635 eV. For trimers they considered only a triangular form. It was claimed that hopping required 0.172 eV, while exchange was unlikely at 1.873 eV.

The information available for cluster movement on the Ni(111) surface is very uncertain. The findings of Liu and Adams [11] are quite similar to findings of Chang et al [27], but there is no experimental data for comparison. However, it appears that gliding as well as dislocation movement play a role in the diffusion of bigger clusters on this plane.

8.4.3 Nickel (110)

One-dimensional diffusion of linear clusters of three to five atoms along the channels of fcc(110)-(2 × 1) surfaces was examined on Pt, Ag, Cu, and Ni surfaces with EAM potentials by Kürpick [42]. Four different diffusion mechanisms were considered and are indicated in Fig. 8.18. In row A, the cluster atoms are all in minimum energy positions, in row B they are at bridge sites, the transition state for concerted cluster

Diffusivities of small clusters

Table 8.5 Self-diffusion barriers (in eV) for clusters on (110)-(2 × 1) plane calculated with EAM for concerted motion CM, hopping onto the facet HOP$_{up}$, HOP$_{down}$, and exchange Ex [42]. Values in parentheses are for unreconstructed surfaces.

	CM	HOP$_{up}$	HOP$_{down}$	Ex
Pt$_3$	1.55(1.48)	0.98	0.70	1.30
Pt$_4$	2.02(1.93)	0.98	0.67	1.36
Pt$_5$	2.48(2.39)	0.98	0.67	1.36
Ag$_3$	0.87(0.72)	0.62	0.38	0.74
Ag$_4$	1.15(0.96)	0.62	0.37	0.75
Ag$_5$	1.43(1.19)	0.62	0.37	0.76
Cu$_3$	1.03(0.82)	0.78	0.46	0.87
Cu$_4$	1.37(1.09)	0.77	0.43	0.89
Cu$_5$	1.71(1.37)	0.77	0.43	0.89
Ni$_3$	1.44(1.16)	1.06	0.65	1.19
Ni$_4$	1.93(1.54)	1.05	0.62	1.21
Ni$_5$	2.41(1.92)	1.04	0.60	1.22

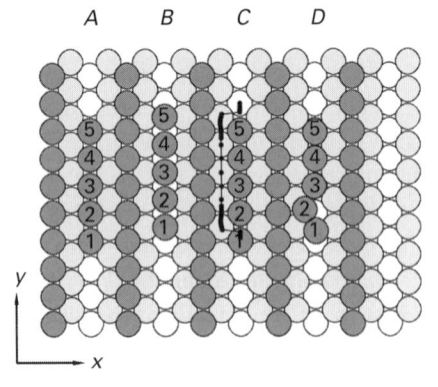

Fig. 8.18 A–D indicate different positions calculated for pentamer with EAM in diffusion on fcc(110)-(2 × 1) plane, starting in minimum energy. B gives transition state for concerted atom motion, C shows minimum energy path for leapfrog transition. D indicates transition state for exchange between atom 2 and 1 (after Kürpick [42]).

motion. In row C, atom 1 climbed out of its equilibrium position and over the other cluster atoms to carry out a leapfrog transition, indicated by the black dots at the left. Finally, in row D is shown the transition state for atom exchange. Atom 2 initiates the process, moving into the next layer, and atom 1 then hops to the initial position of atom 2, which continues movement by the leapfrog mechanism and descends after atom 5. The activation energies on the different (110)-(2 × 1) surfaces are given in Table 8.5 for the various mechanisms, the figures listed in parenthesis being the barrier to concerted jumps on the unreconstructed surface. It is clear that the leapfrog mechanism requires the smallest energy, making it the leading candidate. Jump-up in the leapfrog movement is the limiting step for all the systems treated here; both exchange as well as concerted motion require a much higher energy.

Recently the movement of in-channel clusters on the unreconstructed Ni(110) plane was investigated by Kim et al. [36] who resorted to action-derived molecular dynamics. All the clusters they investigated had an in-channel linear arrangement. In-channel movement of Ni dimers required 0.572 eV, exchange with a wall atom was more energetically demanding at 0.730 eV, and jumps to the next channel occurred over a huge barrier of 1.954 eV. What is more surprising, for an in-channel trimer, exchange has an energy comparable to in-channel hopping, 0.609 eV compared to 0.619 eV. A similar situation was observed for an in-channel Ni tetramer; the energy for exchange amounted to 0.626 eV, while for hopping it was 0.630 eV. Jumping of clusters over a channel is unlikely for all investigated systems.

8.5 Clusters on copper surfaces

8.5.1 Experiments

In 2000, Schlösser et al. [43] carried out both extensive STM measurements and calculations to better understand the diffusion of islands on Cu(111) as well as Ag(111). They investigated diffusivity as a function of island radius and discovered that it was dependent on the size of the island. For vacancy islands with a radius larger than 15 Å on the Cu(111) surface they measured an activation energy for diffusion of 0.49 ± 0.01 eV. Schlösser et al. investigated a number of different basic processes using the EMT potential of Stoltze to account for their findings and ended their work with the conclusion that "a simple interpretation of island diffusion in terms of a single rate-limiting process is not possible."

Repp et al. [44] carried out low-temperature STM studies of Cu dimers on Cu(111), addressing the question which sites were preferentially occupied. They found that there was a small energy difference of $(1.3\pm0.5)\times 10^{-3}$ eV that stabilized occupation of fcc-fcc sites compared to fcc-hcp. The hcp-hcp configuration was not observed. In the temperatures accessible to their instrument below 21 K, they detected only local dimer movement, not long-range translation. They were able to trace single dimer jumps at 5 K. As indicated in Fig. 3.25, they measured the jump rates at different temperatures, and obtained a curved Arrhenius plot. At temperatures $T\leq 5$ K they found that thermally assisted tunneling played a role; at higher temperatures, transitions occurred mostly by thermal diffusion over a barrier of 0.018 ± 0.003 eV with a frequency prefactor of $8\times 10^{11\pm 0.5}$ sec^{-1} for movement of the dimer from the fcc-fcc to the fcc-hcp configuration. Density-functional theory calculations gave a higher value of 0.028 eV for local diffusion, and also a much higher energy difference of 0.021 eV between an fcc-hcp dimer and one in the lowest energy configuration. It should be noted that the dimer as well as trimer preferred fcc sites.

8.5.2 Theoretical investigations: Cu(100)

Theoretical studies started in 1994 with the investigation of Liu [45], who made estimates of the diffusion behavior of copper clusters on Cu(100) relying on EAM interactions.

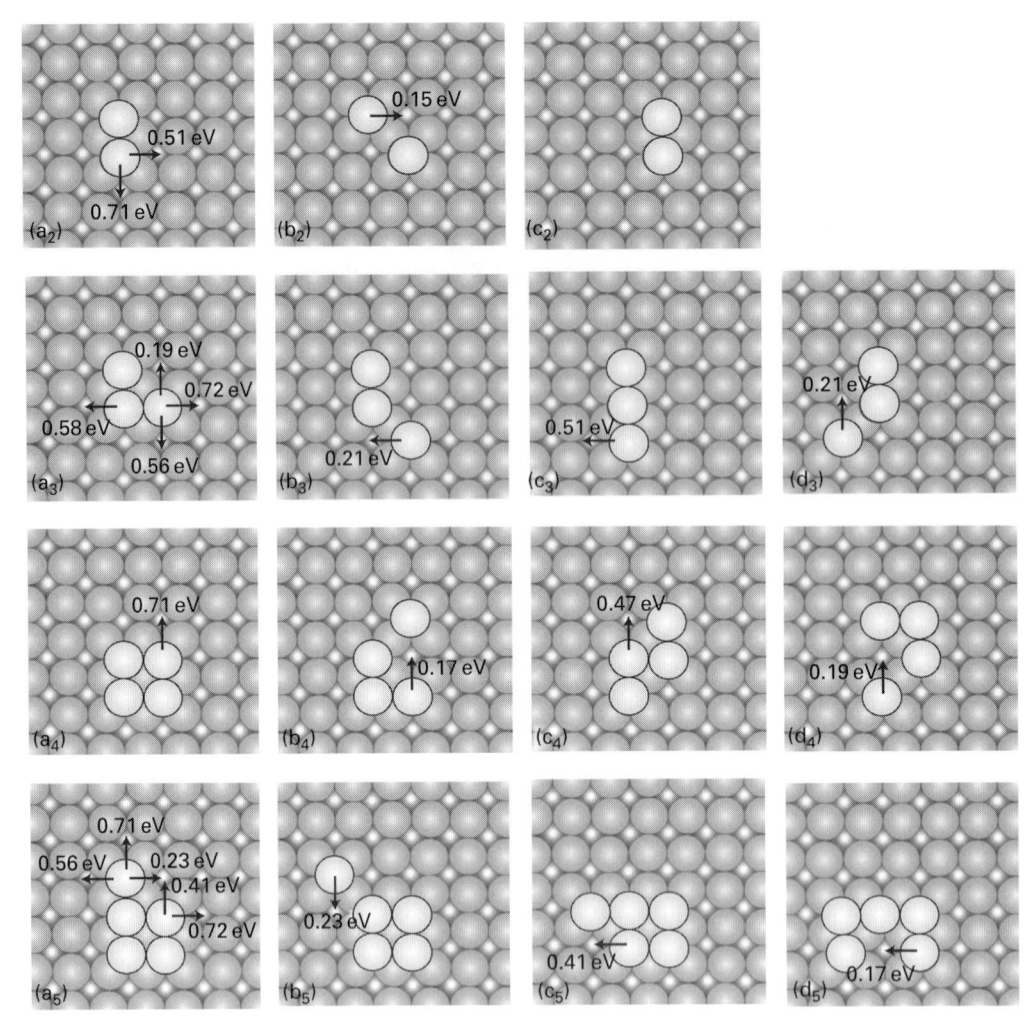

Fig. 8.19 Low-energy configurations of small copper clusters on Cu(100) and optimal diffusion path as evaluated with embedded atom method by Liu [45].

The most stable cluster structures are indicated in Fig. 8.19a, and foreshadow how diffusion occurs. The optimal diffusion paths for dimers to pentamers are shown in Fig. 8.19a–d. Shearing was not considered, nor was the possibility of exchange. For dimers the initial step cost 0.51 eV, while the subsequent step required only 0.15 eV. For trimers the low-energy transformation is rotation of the triangular trimer at 0.19 eV, but this will not lead to long-range diffusion. The optimal long path is shown in Fig. 8.19a$_3$–d$_3$ and involves transitions between triangular and linear shapes of the trimer. For tetramer movement the limiting step is breaking the tight initial shape, which required 0.71 eV; the following steps are less energetically demanding. Pentamer movement involved lower energies than for tetramers. Activation energies for cluster movement were calculated and are given

8.5 Clusters on copper surfaces

Table 8.6 EAM diffusion barriers for cluster migration on Cu(100) [45].

Cluster size	Diffusion barrier (eV)
Monomer	0.45
Dimer	0.51
Trimer	0.56
Tetramer	0.71
Pentamer	0.56

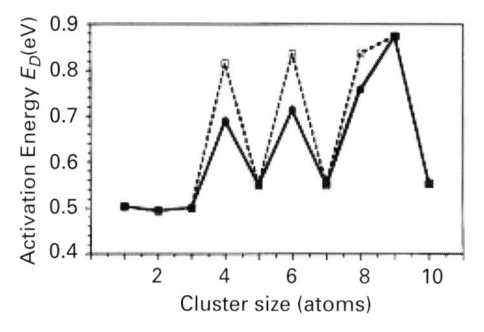

Fig. 8.20 Activation energy for diffusion of Cu islands over Cu(100) calculated using embedded atom method. Solid curve: dimer shearing. Dashed curve: sequential movement of atoms (after Shi et al. [35]).

in Table 8.6. What is striking here is that there is an oscillation of the barrier with size from trimer to pentamer, as was seen in the work of Kellogg [46]. This is easy to understand. For a cluster with an odd number of atoms there is a central core to which an extra atom is attached, but only weakly, so it can easily diffuse around the center. This, however, does not allow long-range diffusion, which would require higher energy detachment of an atom from the central part of the cluster, unless shearing of the cluster can occur. The clusters up to pentamers are quite mobile, even at room temperature.

In the diffusion of clusters on surfaces most of the emphasis has been on the jumping of single atoms. However, Shi et al. [35] have suggested that on (100) surfaces shearing of two atoms, illustrated in Fig. 7.25, could make a contribution as the easiest path for clusters with an even number of atoms. Here it should be noted that similar structures were actually observed for iridium tetramers on Ir(111) [26], which diffuse unusually rapidly. The energetics of different steps in cluster movement are summarized in Table 7.1. For example, for tetramers, it was shown that the initial step by movement of a single atom required 0.815 eV, while by dimer shearing it took only 0.688 eV. For nine atom clusters, the initial step, a single atom jump, required 0.975 eV against 0.857 eV by trimer shearing. Shearing also limited the number of intermediate steps involved in diffusion. For clusters of five, seven, and ten atoms, with a peripheral atom outside the core, the initial step can proceed in the traditional way, by single atom movement, but the next steps may involve shearing. Even in this case, movement by shearing only is possible. Quantitative estimates using bond counting and embedded atom calculations indicated an activation energy for diffusion of copper clusters on Cu(100) that increases with size, as shown in Fig. 8.20. The oscillation seen there agrees nicely with the results for rhodium clusters on Rh(100) found by Kellogg [46].

Various estimates for the diffusion of copper dimers have been made by Boisvert and Lewis [47]. Two possibilities, atom jumps and atom exchange with the lattice were considered, both with EAM potentials and by *ab initio* calculations. For dimer jumps they found an activation energy of 0.48 ± 0.03 eV and a frequency prefactor of $13(\times e^{\pm 0.5}) \times 10^{12}$ sec^{-1}; for diffusion by atom exchange the barrier was higher, 0.73 ± 0.05 eV, as well as the frequency prefactor, which amounted to $320(\times e^{\pm 0.8}) \times 10^{12}$ sec^{-1}, both obtained from molecular dynamics simulations in the EAM approximation. With *ab initio* methods the activation energy turned out to be 0.57 ± 0.06 eV for jumps and for atom exchange 0.79 ± 0.15 eV. At low temperatures dimer diffusion should therefore proceed by atoms jumping. It should be noted here that the barriers calculated for jumps of single atoms, 0.49 eV by EAM and 0.52 eV by *ab initio* techniques, are somewhat higher than the experimental results, but comparable with the energy calculated for dimer movement. Based on their data, the authors concluded that on a Cu(100) surface, dimers contribute to mass transport as much as do single copper atoms.

Monte Carlo simulations with EAM potentials were reported by Biham et al. [48] for copper clusters. The number of jumps per second at 250 K was as follows: monomers 172 ± 5, dimers 237 ± 8, trimers 5.6 ± 0.2, pentamers 1.12 ± 0.04, and heptamers 0.28 ± 0.02. Only single atom jumps, and no correlated movements were considered. Islands with 4, 6, and 8 atoms did not move at a temperature of 250 K. What is interesting here is the high rate of dimer jumps compared with single atoms and also trimers.

Trushin et al. [49] have more closely examined the concerted dimer shearing mechanism for cluster diffusion proposed by Shi et al. [35]. The energetics for a number of cluster sizes, for the initial single atom movement and for dimer or trimer shearing, are shown in Fig. 8.21. For dimers they found that movement by individual atom jumps is easier than by the whole dimer gliding, requiring 0.484 eV compared to 0.820 eV. For trimers single atom jumps again seem to be the more likely initial step

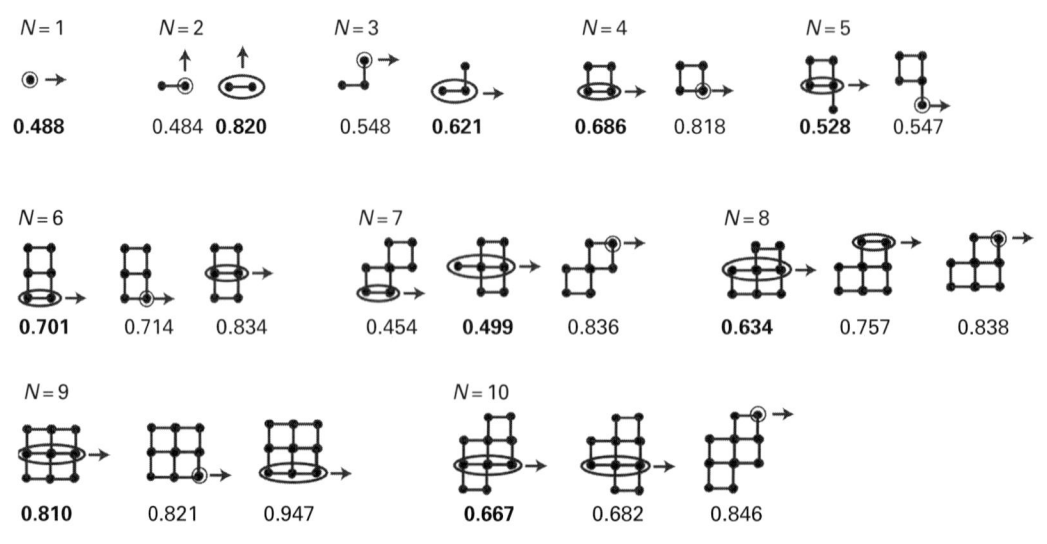

Fig. 8.21 Equilibrium shape of Cu clusters on Cu(100), together with activation energy for center of mass diffusion (bold) and dissociation calculated with EAM (after Trushin et al. [49]).

8.5 Clusters on copper surfaces

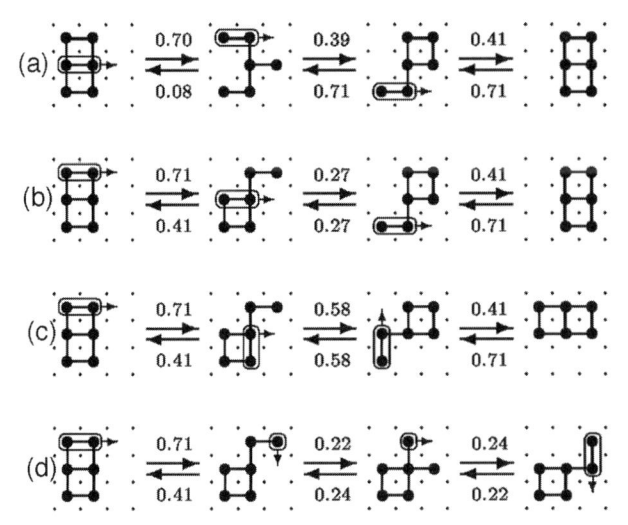

Fig. 8.22 Center-of-mass motion calculated using EAM for hexamer on fcc(100). (a) and (b) Translation. (c) and (d) Rotation (after Trushin et al. [49]).

than dimer shearing jumps. The situation changes for tetramers, where dimer shearing provided an easier path. For pentamers both initial types of steps are equally probable. Six atoms clusters again seem to prefer dimer shearing over atom-by-atom movement and for bigger clusters trimer shearing is a likely mechanism. The authors also demonstrated how it is possible to translate and rotate a cluster via dimer shearing, as shown for a hexamer in Fig. 8.22. The possibility of internal dimer rotation is illustrated in Fig. 7.6.

Salo et al. [50] now used the formalism derived by Sanchez and Evans [51] to evaluate diffusivities for copper islands on Cu(100). The diffusivity for pentamers was given by Eq. (7.58), and the energetics of different initial steps were presented in Fig. 7.37; Salo et al. also derived the equation for the diffusivity of clusters made up of five to nine atoms. The dependence of diffusivity on the size of the cluster, taking into account the shearing mechanism, was shown in Fig. 7.39. Given in this figure are the Monte Carlo simulations for single particle events based on EMT energetics, as well as the results for single atom events from the analytical theory (again relying on EMT) and also for many-atom processes (from EAM potentials). What is clear here are the oscillations in the diffusivity with the size of the cluster for all the estimates, and that the new many-atom events have not disturbed this oscillation. However, the new events considered here quantitatively increase the diffusivities.

Trushin et al. [15] by taking advantage of advances in identifying saddle points for clusters have been able to establish additional opportunities for diffusion. As an example, for Cu_{10} in Fig. 8.23 are shown different paths, including new internal-row shearing, which have reasonably low activation energies. These constitute further opportunities for diffusion, which can take place prior to cluster dissociation. In addition, Trushin et al. also observed an interesting event for a copper adatom on top of a cluster of 17 Cu atoms on Cu(100). By internal dimer shearing near a kink site, illustrated in Fig. 8.24, the

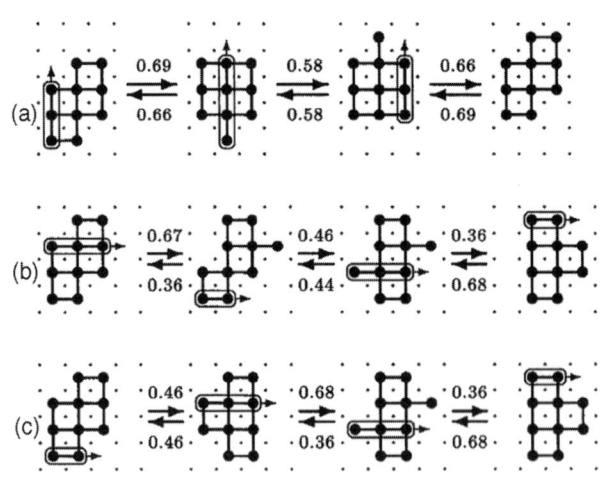

Fig. 8.23 Diffusion paths, with trimer shearing, calculated with EAM for center of mass diffusion of Cu_{10} on Cu(100) (after Trushin et al. [15]).

Fig. 8.24 Interlayer diffusion of copper atom inside Cu_{17} cluster on Cu(100), accomplished by concerted dimer shearing next to kink site (after Trushin et al. [15]).

adatom incorporates into the cluster over an activation energy of 0.43 eV, which is lower than the barrier of 0.49 eV for atom jumps on the Cu(100) surface.

Liu et al. [19] looked at the movement of dimers on fcc(100) with RGL interactions [52,53]. Their results for static activation energies are given in Table 7.3 for dimer hopping as well as simple exchange. For dimers on the copper(100) surface, they found that simple hopping with an activation energy 0.4307 eV was more likely than exchange, which proceeded with an energy of 0.7853 eV. They also did molecular dynamics simulations and in a total time of 2 nsec at 650 K observed 38 hops but only one exchange and no exchange rotations. Additionally they observed cooperative hopping (Fig. 7.32), where the jump of one dimer atom triggers the jump of a second atom, and rotation hopping, which is simple hopping leading to rotation of the dimer. A year later, Chang and Wei [20] carried out *ab initio* density-functional calculations using the VASP [21,22] package also for dimers. Activation energies calculated with a (5 × 5)

8.5 Clusters on copper surfaces

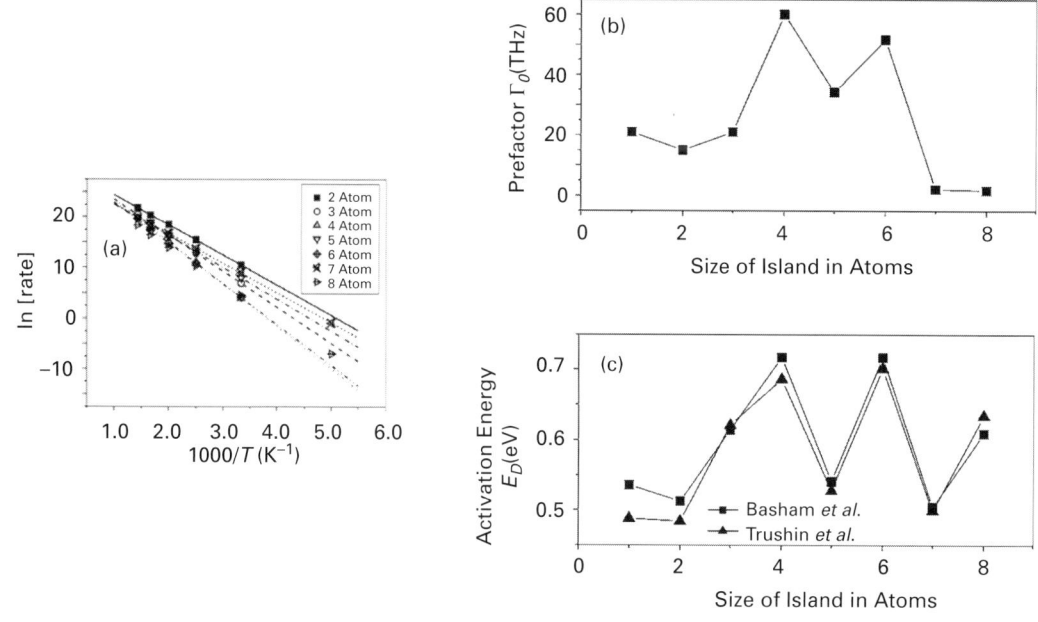

Fig. 8.25 Kinetics of Cu clusters diffusing on Cu(100) plane as a function of size. (a) Arrhenius dependence for movement of center of mass of clusters. (b) Prefactor. (c) Activation energy, obtained from EAM calculations (after Basham et al. [54]).

surface cell are given in Table 8.2. No information about the mechanism was derived but for Cu dimers, diffusion by hopping was found to be more likely than by exchange, over a barrier of 0.66 eV compared with 0.77 eV.

Basham et al. [54] have done self-learning kinetic Monte Carlo simulations based on optimized EAM potentials to examine diffusion of copper islands and showed that movement of the center of mass of small islands followed an Arrhenius dependence, as in Fig. 8.25a. The results for activation energies and prefactors are shown in Fig. 8.25b,c. Several things are of interest. The activation energies for diffusion in this study are in quite good agreement with the earlier work of Trushin et al. [49,50] as well as with the findings of Shi et al. [35], which took into account dimer and trimer shearing. What is noteworthy again is the oscillation of the activation energy with size, with tetramers and hexamers having large barriers, as found some time ago in experiments with rhodium clusters on Rh(100) by Kellogg [46].

Recently, an investigation of copper clusters using action-driven molecular dynamics was done by Kim et al. [36]. For dimers they looked at the energetics of the jump of a dimer in the [110] direction and found an activation energy of 0.871 eV. Perpendicular jumps are more likely at 0.477eV. Exchange required more energy at 0.686 eV. They also examined jumps of linear trimers, elongated along the [110] direction, but only considered jumps in the direction of the trimer axis, which required 1.256 eV. Exchange in the [100] direction has a lower energy of 0.712 eV, but to compare which mechanism is favorable we should know the energetics of jumps perpendicular to the trimer axis, which is not given by the authors. For linear tetramers aligned in the [110] direction, the situation is similar to that described for trimers. Jumps in the [110] direction cost 1.615 eV, exchange in the [100] direction 0.718 eV. Again values are not suitable for comparison or conclusions about the mechanism of movement.

There are quite a number of theoretical investigations for the movement of Cu clusters on Cu(100), with no experimental data for comparison. Most of the calculations concentrate on the energetics of certain steps during diffusion rather than on the movement of the center of mass. From all of the work it is clear that exchange is rather unlikely on this plane and that shearing definitely plays a role in diffusion. The dimer is estimated to move over the surface by simple jumps with an activation energy lying between 0.43 and 0.66 eV, not a bad agreement. The energy needed for long-range movement of trimers is in the range of 0.5–0.6 eV; for tetramers the energy for center-of-mass movement is 0.68–0.7eV, and for pentamers 0.53–0.56 eV; the energy required for diffusion of Cu_6 ~0.7 eV, for Cu_7 ~0.45–0.55eV, for Cu_8 ~0.63–0.76 eV, for Cu_9 ~0.81–0.85 eV, and for Cu_{10} ~0.55–0.67 eV. Almost all these data were derived in calculations based on the EAM potential, and that is worrisome. Only for dimers were *ab initio* methods as well as RGL potentials employed, and the results from these methods differ slightly from the rest. For dimer diffusion an energy of 0.43 eV was derived using RGL potentials, compared with 0.66 eV using VASP. Since there are no experimental values for comparison it is difficult to judge if EAM well describes diffusion on Cu(100).

8.5.3 Theoretical investigations: Cu(111)

Investigations of the movement of clusters on the Cu(111) surface started with the work of Chang et al. [27] who used EAM potentials. Self-diffusion barriers are shown in Fig. 8.13. The movement of copper dimers required 0.13 eV, trimers 0.18 eV, tetramers 0.17 eV, and pentamers 0.24 eV. The detailed path of movement was investigated only for aluminum clusters. What is worth noting is essentially the same energy for movement of trimers and tetramers.

The same year, Longo et al. [40] looked at the movement of a seven atom Cu cluster on the Cu(111) surface, using EAM potentials parameterized according to Voter and Chen [41]. The Cu_7 cluster glided with an energy of 0.36 eV; small deformations of the cluster by translational motion from fcc to hcp sites and vice versa were observed on heating the cluster to the melting temperature; the process started at 430 K.

Marinica et al. [55] in 2004 carried out extensive calculations of energy barriers for cluster diffusion on Cu(111) using embedded atom potentials devised by Mishin et al. [56]. They addressed the various configurations shown in Fig. 8.26. It should be noted that the authors have designated fcc sites by *f* and hcp ones by *h*. Their results for dimers are shown in Fig. 8.27a for intra-cell movement and in Fig. 8.27b for inter-cell diffusion. In intra-cell movement the smallest barrier of 0.016 eV occurs in atom-by-atom rotation from *ff* to *fh*, in excellent agreement with the experimental findings of Repp et al. [44] for the same kind of movement. A much higher barrier of 0.026 eV intervenes for atom-by-atom rotation between *fh* and *hh*, which implies that movement backward and forward between *ff* and *fh* places is more likely than a full rotation. In inter-cell movement, by far

8.5 Clusters on copper surfaces

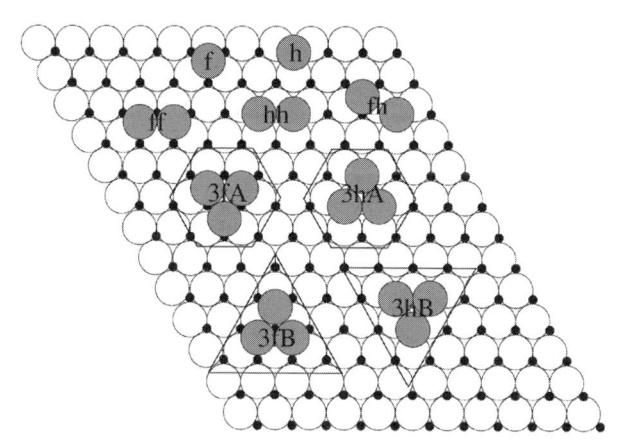

Fig. 8.26 Cluster configurations found in EAM calculations on fcc(111) plane. Black dots indicate hcp sites (after Marinica et al. [55]).

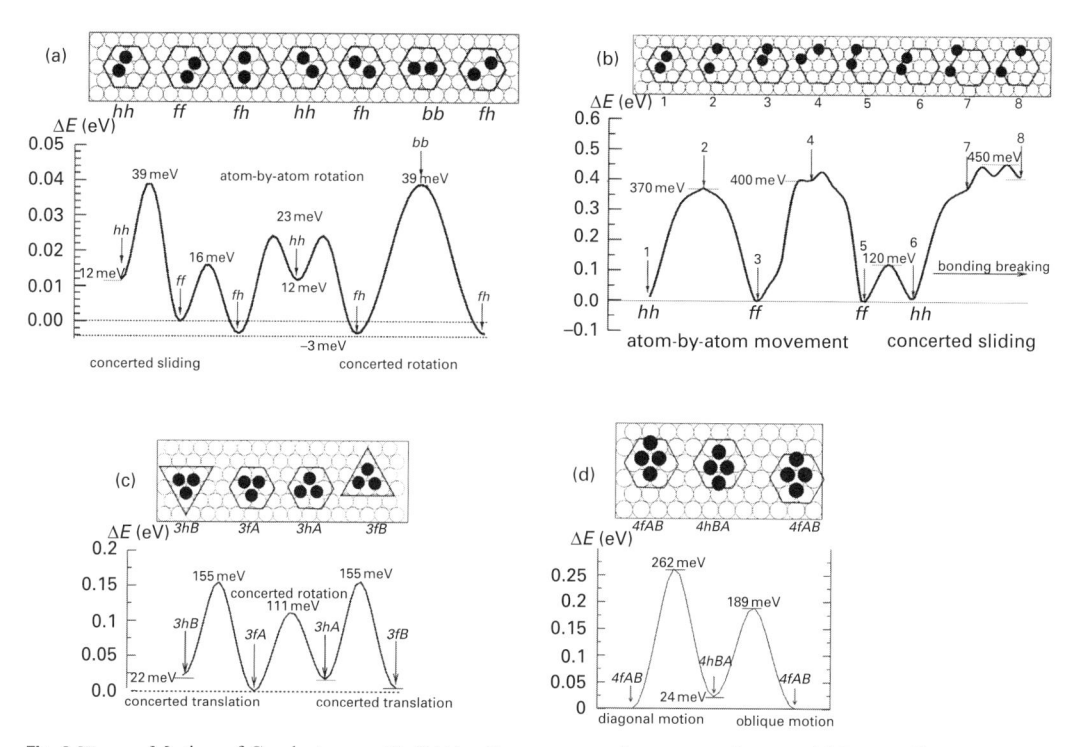

Fig. 8.27 Motion of Cu clusters on Cu(111) with accompanying energy changes. (a) Intra-cell movement of dimer. (b) Inter-cell movement of dimer. (c) Concerted motion of Cu trimer. (d) Concerted diffusion of Cu tetramer (after Marinica et al. [55]).

the lowest barrier exists for concerted sliding between *ff* and *hh*, and amounts to 0.120 eV; the same path covered in the opposite direction required only 0.039 eV. Longer atom-by-atom motion requires overcoming a much higher barrier of 0.40 eV.

Trimer energetics are shown in Fig. 8.27c and have a barrier of 0.155 eV for concerted translation compared with 0.111 eV for concerted rotation. It should be noted here that

Diffusivities of small clusters

Table 8.7 EAM diffusion kinetics for concerted motion of clusters on Cu(111) [55].

Adisland	Cu	Cu$_2$	Cu$_3$	Cu$_4$(D)	Cu$_4$(O)	Cu$_7$
Barrier (eV)						
f→h	0.041	0.132	0.155	0.262	0.189	0.388
h→f	0.036	0.120	0.133	0.238	0.165	0.348
attempt frequency (10^{12} sec^{-1})						
f→h	1.14	1.32	4.51	4.12	4.65	13.17
h→f	1.16	1.40	6.47	5.05	5.70	18.00

D – diagonal motion, O – oblique motion

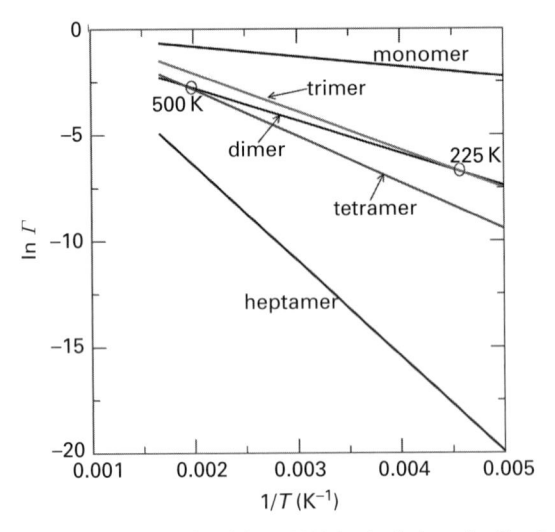

Fig. 8.28 Arrhenius plots derived from EAM calculations for Cu clusters hopping from fcc to hcp sites on Cu(111) plane. Tetramer curve is for oblique transition (after Marinica et al. [55]).

atom-by-atom jumps have been disregarded as requiring too much energy; linear trimers were not considered. Energetics for tetramer motion are given in Fig. 8.27d, and reveal that what is called oblique motion as energetically preferred, with an energy of 0.189 eV from f to h and 0.165 eV from h to f. Diagonal motion, along the long diagonal of the diamond, required 0.262 eV from f to h, and 0.238 eV for the reverse process. For hexagonal heptamers the maximum barrier was found to be 0.388 eV, for movement from f to h and 0.346 eV in the opposite direction. An Arrhenius plot for the jump rates of all these clusters is given in Fig. 8.28, and shows how differently the heptamer behaves. The values of diffusion barriers and attempt frequencies for concerted motion of clusters are summarized in Table 8.7, from which the increase in attempt frequency for three atom clusters and higher is clear.

Morgenstern et al. [57] did molecular dynamics simulations of the movement of Cu dimers on Cu(111) and Ag(111), shown in Fig. 8.29. The energy for self-diffusion on the

8.5 Clusters on copper surfaces

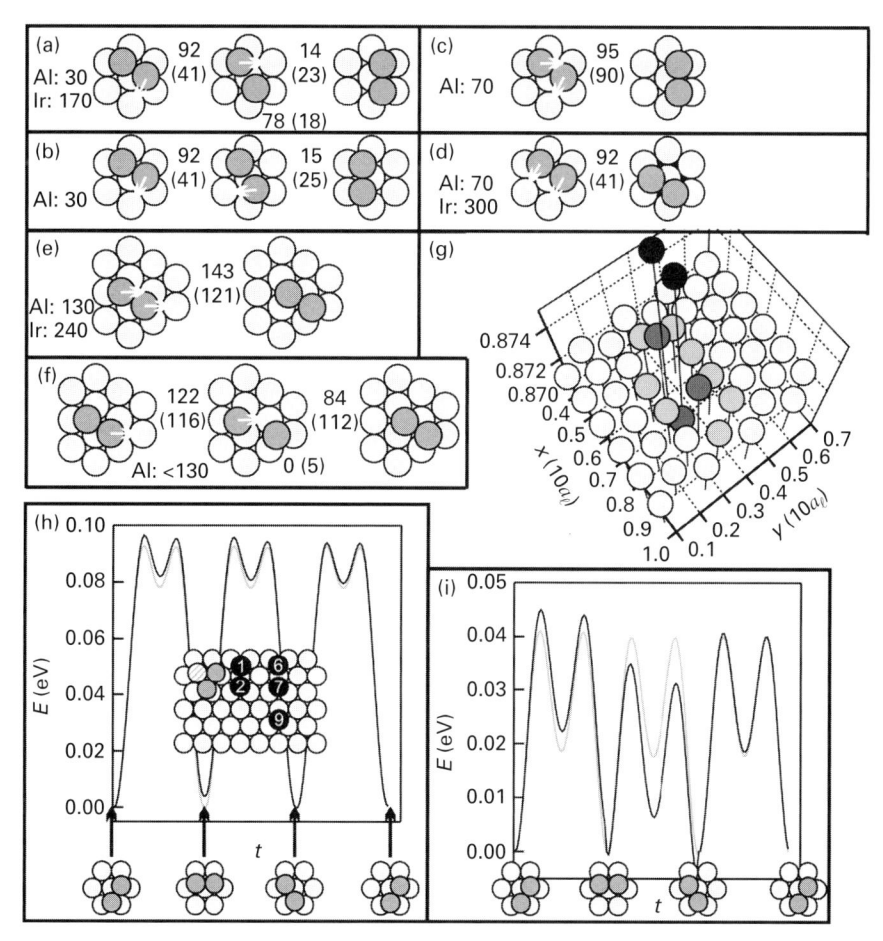

Fig. 8.29 Molecular dynamics simulation of Cu dimer on Ag(111) and Cu(111) using EMT potentials. (a), (b) Sequential rotation, (c) collective rotation, (d) collective intra-cell hopping over bridge site, (e) collective inter-cell hopping over bridge site, (f) sequential inter-cell motion. (g) Out-of-place displacement of an adatom around a dimer. Potential energy changes for (h) Cu on Ag(111) and (i) Cu on Cu(111). Values for diffusion on Cu(111) are given in the parenthesis. Calculated values for movement of Ir dimers on Ir(111) [115] and Al dimers on Al(111) [25] also determined by EMT are given for comparison (after Morgenstern et al. [57]).

copper surface is indicated in parenthesis in the figure, the first jump requiring 0.041 eV, while the subsequent transition needed only 0.023 eV. Concerted motions are also illustrated, with the lowest energy for the concerted two-atom jump, requiring 0.041 eV, which is comparable with the jump of the first atom. The lowest energy of 0.116 eV for inter-cell motion was obtained for sequential dimer translation.

The diffusion of small copper islands on Cu(111) has recently been examined by Karim et al. [58], using EAM interactions [59]. They relied on a smart kinetic Monte Carlo scheme, and at the start only occupation of fcc sites was allowed. Karim et al. also carried out molecular dynamics simulations at 500 K to learn about processes involved in the movement of clusters, and found a number of concerted movements as well as

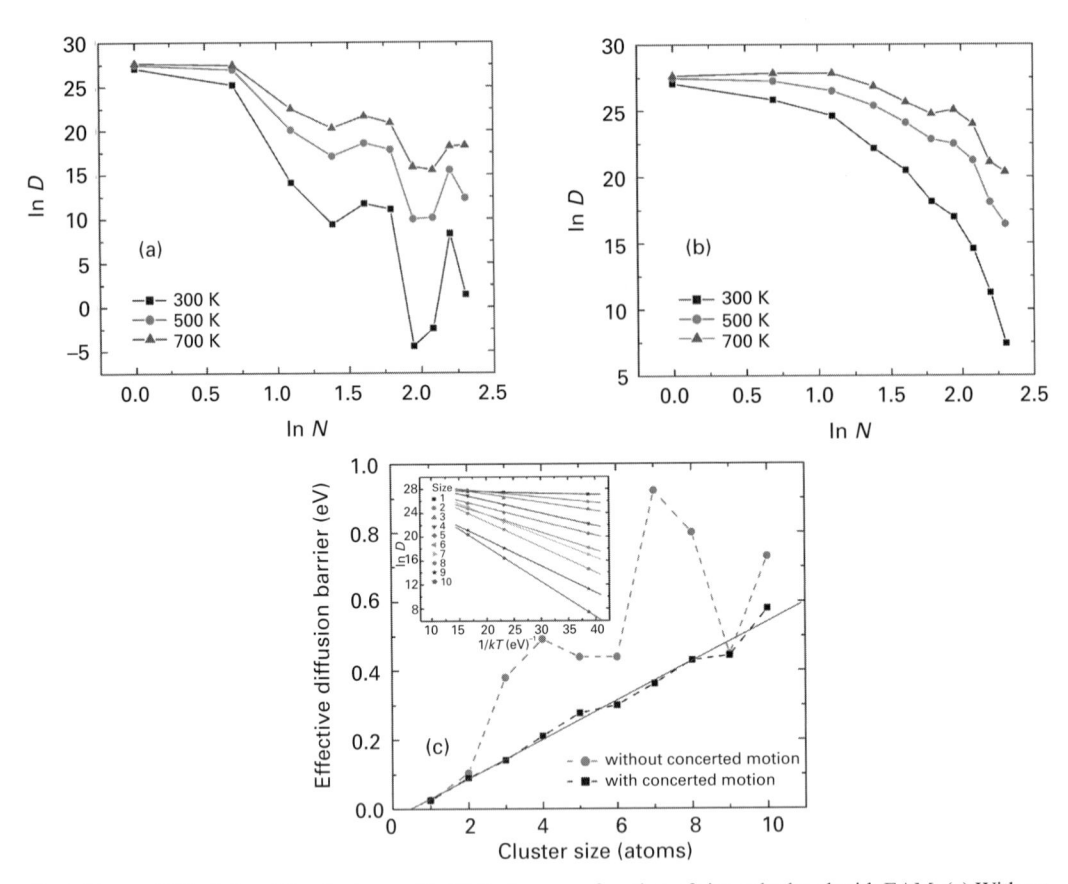

Fig. 8.30 Diffusivity of Cu clusters on Cu(111) plane as a function of size calculated with EAM. (a) Without inclusion of concerted motion. (b) With inclusion of concerted motion. (c) Diffusion barrier as a function of cluster size (after Karim et al. [58]).

transitions involving hcp sites with not too high a barrier. Results obtained with no concerted motion and with no possibility of hcp site occupation for clusters made of two to ten atoms are shown by circles in Fig. 8.30a. These clearly reveal oscillations in the logarithm of diffusivity with the size of the cluster. After including concerted motion in MD simulations the diffusivity no longer shows an oscillatory behavior and the effective activation energy has a linear dependence on cluster size. This is quite surprising, since an oscillation in the activation energy was observed experimentally for clusters on Ir(111); however, Cu(111) may have different characteristics.

Karim et al. also provided a detailed energy characteristic of transport. For example, there are two ways for a dimer deposited at fcc sites to move to another fcc site. One required an energy of 0.101 eV, while the other needed only 0.015 eV, as shown in Fig. 8.31a. The effective barrier for atom movement amounts to 0.104 eV. Molecular dynamics simulations revealed 13 different mechanisms for dimer motion, which includes sliding and rotation through hcp sites, shown in Fig 8.32a. Energetically the lowest process discovered was jumping of one atom from a dimer deposited in an hcp site

8.5 Clusters on copper surfaces

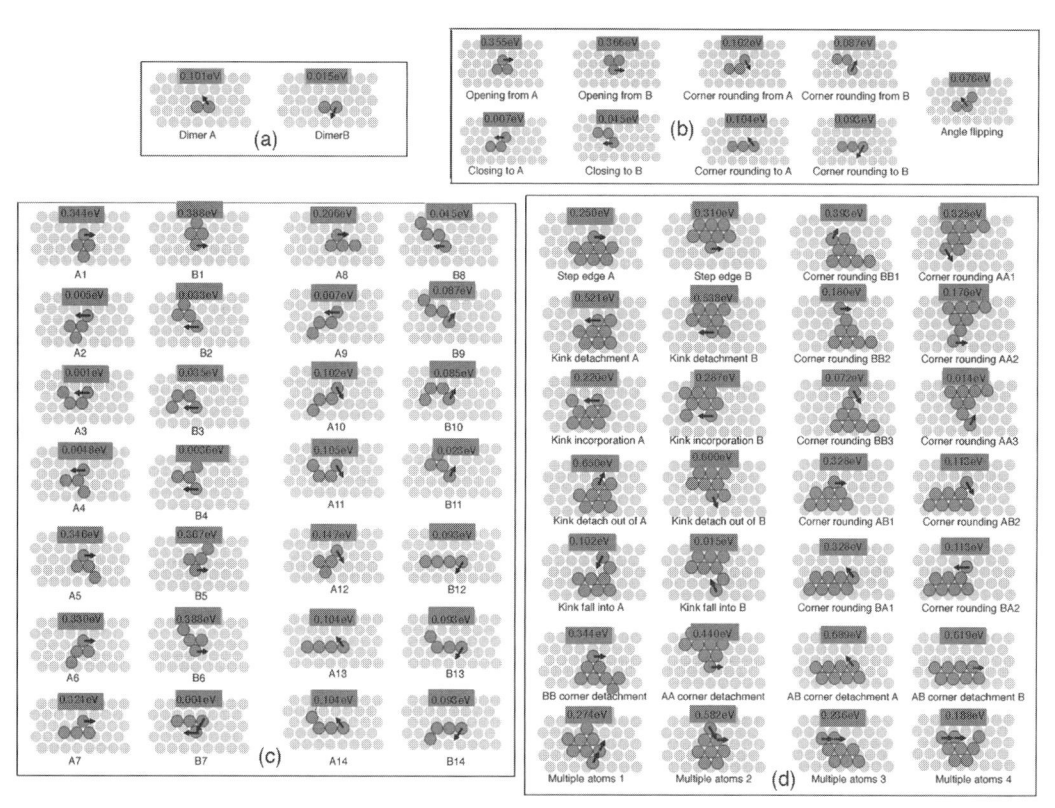

Fig. 8.31 Schematics of Cu cluster diffusion on Cu(111) plane, together with diffusion barriers estimated using EAM potentials. (a) Mechanisms for dimer jumps between fcc sites. (b) Trimer diffusion from fcc to fcc sites. (c) Diffusion of tetramer between fcc sites. (d) Cluster of eight atoms diffusing between fcc sites (after Karim et al. [58]).

to an fcc site, creating a mixed hcp-fcc dimer. Such movement required only 0.005eV, in agreement with observations of Repp et al. [44] who did not observe hcp-hcp dimers. Jumps from an fcc-fcc dimer to an fcc-hcp configuration needed only 0.009 eV, which is only half the value measured by Repp et al. Additionally, this low-energy mechanism does not shift the center of mass much; more complicated mechanisms like long jumps or concerted motion greatly influence the movement of the center of mass and have a significant effect on the effective barrier. With concerted motion included in simulations the effective barrier for dimer movement was lowered to 0.092 eV.

Diffusion of trimers via atom-by-atom movement required 0.380 eV. Single atom movement, shown in Fig 8.31b together with the associated energetics, does not facilitate long-range diffusion. To achieve long-range movement concerted gliding and rotation, as in Fig. 8.32b, should be included. The lowest energy process for long-range movement is a rotation between fcc and hcp sites, which required an energy of only 0.038 eV from fcc to hcp and 0.062 eV in the opposite direction. Concerted sliding from fcc to hcp required 0.125 eV, and 0.015 eV in the reverse direction. With all this concerted motion in the picture the effective barrier was lowered from 0.380 eV to 0.141 eV.

588 Diffusivities of small clusters

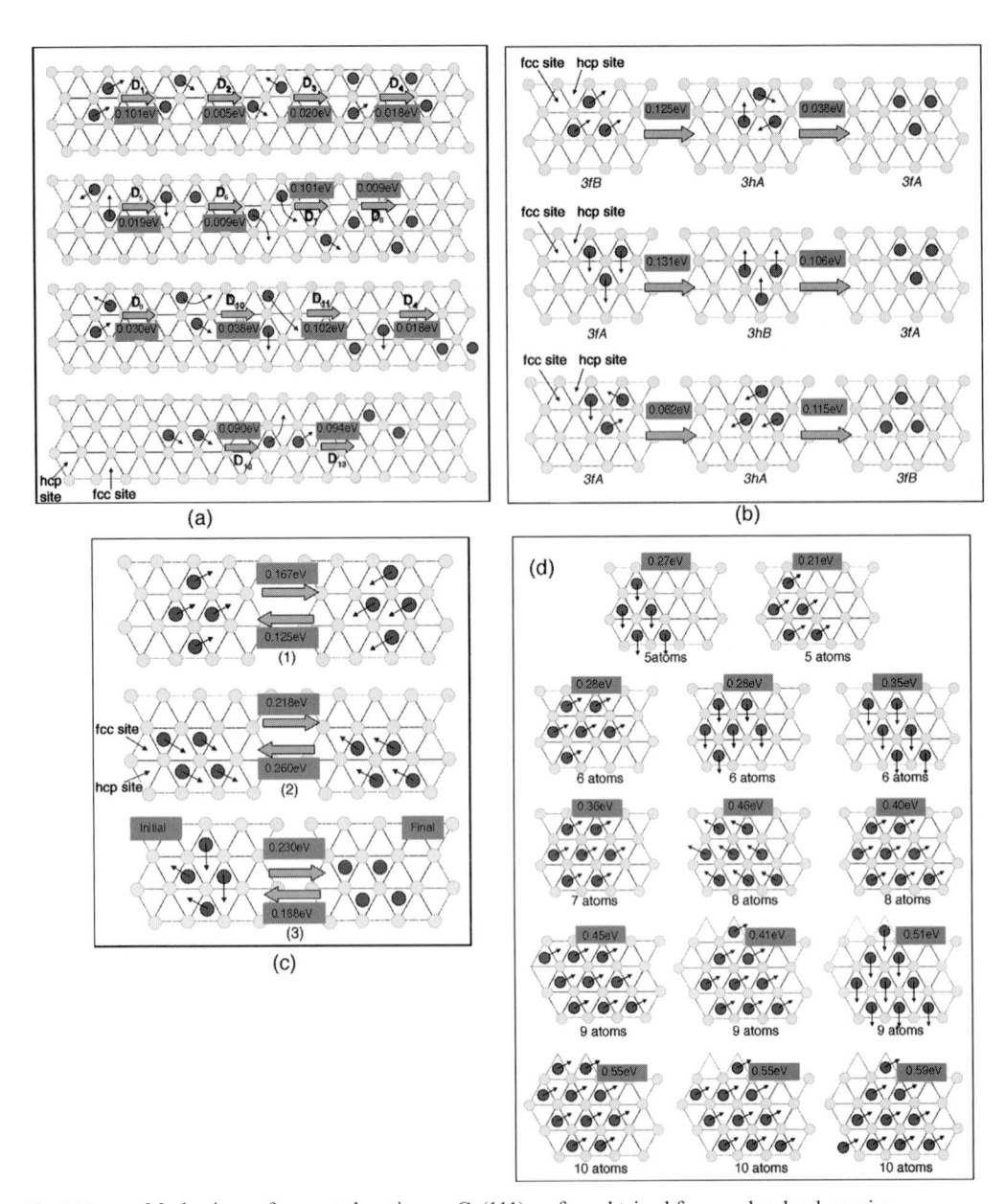

Fig. 8.32 Mechanisms of concerted motion on Cu(111) surface obtained from molecular dynamics simulations. (a) For dimer movement. (b) Trimer. (c) Tetramer diffusion via (1) diagonal glide, (2) vertical glide, (3) shearing. (d) Diffusion of clusters made up of 5 to 10 atoms (after Karim et al. [58]).

For tetramers, all the different events for atom-by-atom movement are illustrated in Fig. 8.31c. There are 28 jump processes included, which lead to an effective barrier of 0.492 eV. Molecular dynamics simulations revealed concerted motion and shearing of diamond-shaped tetramers, which takes place by sliding between fcc and hcp sites along the diagonals shown in Fig. 8.32c. After taking into account concerted motions, the

8.5 Clusters on copper surfaces

effective barrier for movement of this type of tetramer was lowered to 0.212 eV. For movement of 5 to 10 atom clusters, the corner detachment and step edge movement, illustrated in Fig. 8.31d with corresponding energetics, have to be considered for atom-by-atom movement. Molecular dynamics simulations also reveal a number of concerted motions shown in Fig. 8.32d; the barriers for concerted motion are lower than for atom-by-atom movement. Including these additional processes causes reduction of the logarithm of the diffusivity with increasing size of the cluster, as shown in Fig. 8.30b by the squares. The oscillations in effective energy disappeared when concerted motion was taken into account, as in Fig. 8.30c. It appears that the diffusion barrier on Cu(111), after allowing for all events, rises linearly with size. It should be noted that the low diffusion barrier found for iridium tetramers on Ir(111) [26] is not revealed for copper islands on Cu(111), suggesting a dependence on substrate chemistry.

Kim et al. [36] recently looked at the movement of atoms in clusters using action-driven molecular dynamics. The most worrying part of this study is that they do not indicate where the cluster is sitting, in fcc or hcp sites, so it is not clear between which sites atoms are jumping. For dimers they consider 30 degree dimer atom hopping along $[1\bar{2}1]$ and a 90 degree transition along $[11\bar{2}]$; for the first the energy was 0.128 eV, for the second 0.087 eV. Exchange was unlikely, at an energy of 1.635 eV. For triangular trimers jumps involved an energy of 0.172 eV, while exchange was of course more demanding at 1.873 eV.

In 2008, Yang et al. [60] used modified analytical EAM potentials [31,32] to do molecular dynamics simulations of clusters made up of one to eight Cu atoms on Cu(111). From the temperature dependence they were able to draw Arrhenius plots, and deduce the prefactors and activation energies of clusters shown in Fig. 8.33. They observed an increase in activation energy up to trimers, a drop for tetramers, and the another rise again. What is more interesting is that static estimates [27] seem not to reproduce such a drop. Such progression with size is quite different from what had previously been found by Karim et al. [55], but similar to what was observed by Chang et al. [27]. Yang et al. suggest that concerted motion influenced the octamer movement. The prefactor for Cu_7 is only an order of magnitude larger than that for single adatoms, and they claim that the hard character of the surface is responsible for this small difference.

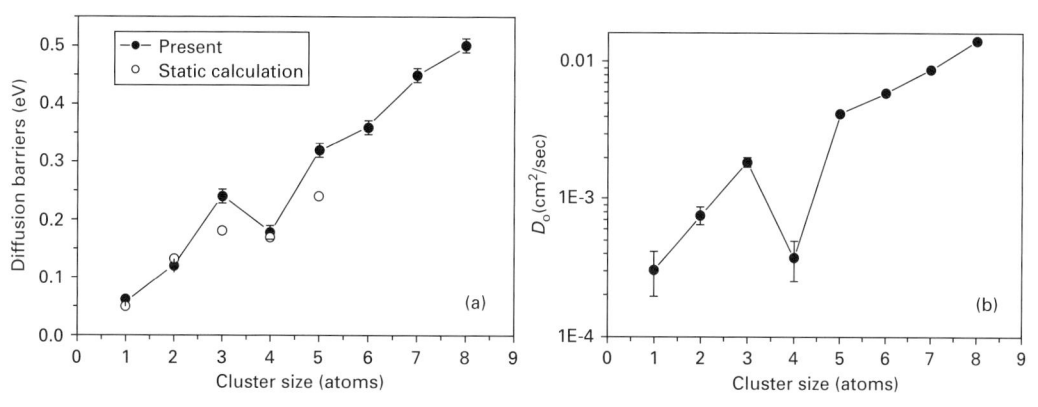

Fig. 8.33 Diffusion characteristics calculated using MAEAM for copper clusters on Cu(111). (a) Barrier. (b) Prefactors. Both as a function of size (after Yang et al. [60]).

Papathanakos and Evangelakis [61] have carried out molecular dynamics simulations, based on RGL interactions for a hetero-system of small gold clusters diffusing on Cu(111). They found that cluster atoms did not always occupy equilibrium positions and spent noticeable time in perturbed configurations. They frequently saw clusters at the hcp-fcc bridge. In some instances they found concerted movement of the cluster as a whole. For a trimer, diffusion involved rotation of the triangular atom arrangement, usually a clockwise 60°, but a pentamer translated without rotation. Most frequently, however, diffusion involved concerted movement of parts of the cluster. One such example is shown in Fig. 8.34; there, a heptamer diffused by atoms A, B, C, and D rotating anticlockwise while atoms E, F, and G rotated clockwise, the type of behavior already noted by Chirita et al. [62]. They also observed heptamer evolution from the compact to the linear form and back to compact, illustrated in Fig. 8.35. The process is possible but not favorable due to the straining of interatomic bonds. From observations of the individual cluster atoms as well as of the center of mass for the cluster at different temperatures the diffusion characteristics were derived and are given in Table 8.8. The diffusion coefficients obtained from the movement of the center of mass follow an Arrhenius dependence on temperature. A clear oscillation of the diffusion barrier of the cluster with size emerges, with tetramers and hexamers showing low values. Prefactors for center-of-mass motion (D_{COM}) proved considerably higher than for individual cluster atoms (D_a), which tended to balance out the higher activation energies for cluster motion compared to that of cluster atoms. It is also important to remember that there is quite a big mismatch between Cu(111) and Au(111) clusters, which might in part be responsible for finding so many metastable configurations, since sub-processes were observed releasing stress locally.

Raeker and DePristo [63] looked briefly at movement of Fe clusters on a Cu(111) surface with MD/MC-CEM. They found that the mobility of clusters of 2 to 4 atoms are 10 times lower than the mobility of single atoms. The mobility of clusters of 5 to 7 atoms were, however, almost one order of magnitude lower than the movement of clusters made up of 2 to 4 atoms.

Repp et al. [44] derived experimental values for the movement of dimers on the Cu(111) surface using the STM. Their 0.018 eV barrier agreed nicely with the estimate of 0.016 eV by Marinica et al. [55] for the energetics of jumps between fcc and hcp sites

Table 8.8 Kinetic parameters for atomic D_a and COM diffusion of Au clusters on Cu(111), calculated with RGL interactions [61].

Cluster size (atoms)	$D_o(10^{-5}$ cm^2/sec) D_a	D_{COM}	E_D(eV) D_a	D_{COM}
3	3.4	22.3	0.058	0.108
4	2.5	12.4	0.057	0.093
5	5.8	25.4	0.118	0.157
6	3.4	11.8	0.091	0.120
7	7.2	22.0	0.137	0.166

8.5 Clusters on copper surfaces

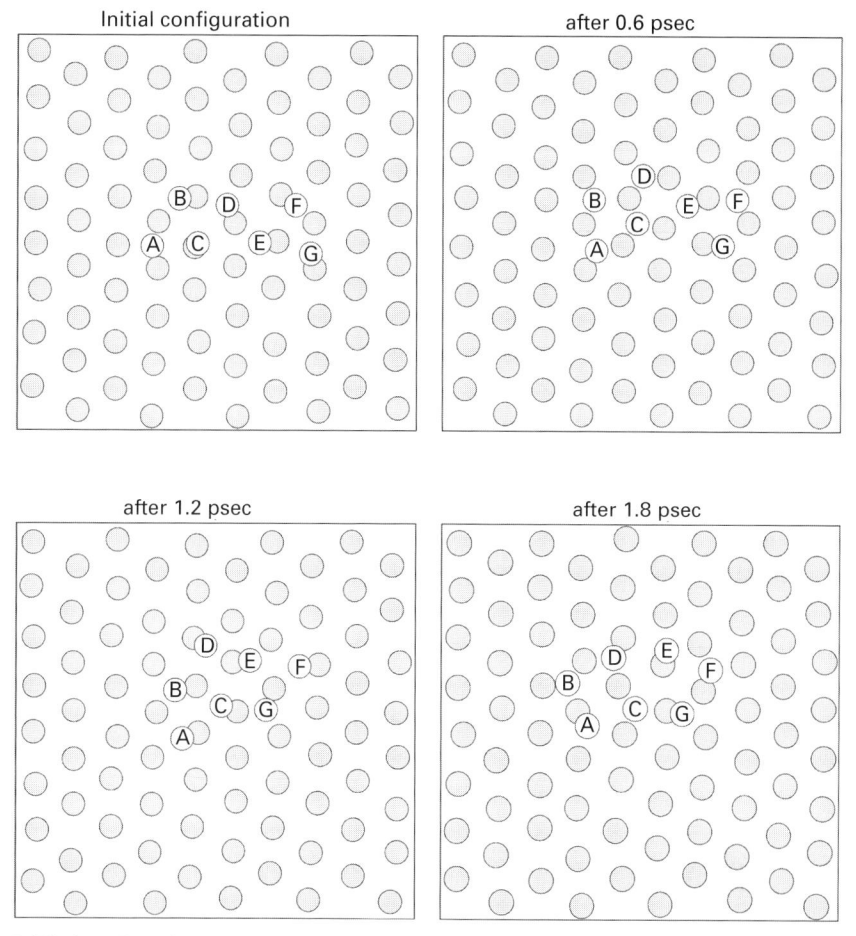

Fig. 8.34 Diffusion of Au$_7$ by subprocesses calculated with RGL on Cu(111). Tetramer A, B, C, and D rotates anti-clockwise, while trimer E, F, G moves clockwise and cluster diffuses to the right (after Papathanakos and Evangelakis [61]).

but exceeded the value of 0.009 eV obtained by Karim *et al.* [58]. Unfortunately Repp did not look at longer non-local transitions of dimers; he only stated that they would occur at temperatures higher than 21 K. This agrees with calculations that indicate the movement of the dimer center of mass has a much higher energy, 0.13 eV from Chang *et al.* [27], and 0.092 by Karim *et al.* [58]. The movement of the center of mass of the trimer deduced in these groups are also not too far away, 0.141 by Karim *et al.* [58] and 0.18 eV by Chang *et al.* [27]. For tetramers the barriers were 0.17 by Chang *et al.* [27] and 0.212 eV by Karim *et al.* [58]. Quite good agreement was also obtained for compact 7 atom clusters: 0.36 eV by Karim *et al.* [58] and Longo *et al.* [40]. However, Papathanakos and Evangelakis [61] opened another important question about hetero-clusters on Cu(111): how does the mismatch of substrate and cluster influence diffusion? Further investigations are necessary to answer this question.

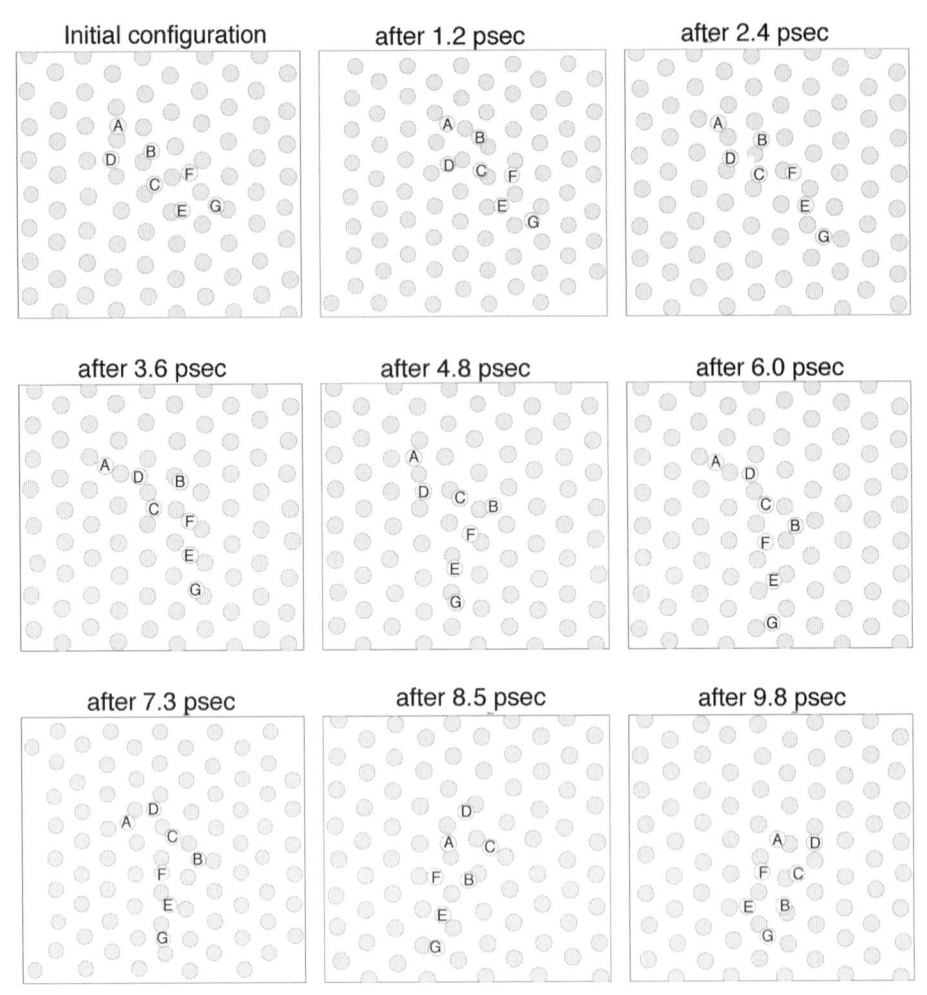

Fig. 8.35 Diffusion of gold heptamer on Cu(111) plane. Heptamer assumes linear form, but eventually returns to compact configuration (after Papathanakos and Evangelakis [61]).

8.5.4 Theoretical investigations: Cu(110)

The movement of linear clusters on the one-dimensional Cu(110)-(2 × 1) surface was examined with EAM potentials by Kürpick [42]. Four different diffusion mechanisms were considered, shown in Fig. 8.18. The details of the mechanisms were already described in Section 8.3.3 and the activation energies for the different processes are summarized in Table 8.5. Concerted motion of Cu islands is energetically not favorable and required 1.03 eV for trimers, 1.37 eV for tetramers and 1.71 eV for pentamers. Values are slightly lower for the unreconstructed surface: 0.82 eV for trimers, 1.09 eV for tetramers, and 1.37 eV for pentamers. The jump of an atom up the {111} type wall required only 0.78 eV for a trimer atom, and 0.77 eV for a tetramer or pentamer atom. Stepping down from a {111} wall was even easier, requiring only 0.46 eV for a trimer atom and 0.43 eV for tetramer and pentamer atoms. The author also considered movement by exchange, but this did not provide an easy path. For a trimer, atom exchange needed 0.87 eV, while for a tetramer or pentamer 0.89 eV. A leapfrog jump seems to provide the energetically easiest way for movement.

Kim et al. [36] used molecular dynamics simulations to look at the movement of dimers, trimers, and tetramers on the unreconstructed surface of Cu(110). A jump of a dimer in-channel required 0.482 eV, jumping across a channel as well as exchange were unlikely, 1.954 eV for the first and 0.73 eV for the second. The situation is not so clear for trimers, since the energy for in-channel hopping and for exchange are very close, 0.718 eV for hopping and 0.765 eV for exchange. Jumping across a channel can be easily ruled out at 2.693 eV. Surprisingly, for an in-channel tetramer exchange started to be more favorable than hopping, requiring 0.777 eV for exchange against 0.985 eV for hopping.

Unfortunately there is no experimental information to confirm any of these findings.

8.6 Clusters on rhodium surfaces

8.6.1 Experiments

Ayrault [64] made brief observations of two rhodium atoms diffusing in [110] channels on the rhodium (110) plane. With two atoms in adjacent channels, they found strongly correlated motion; when an empty channel intervened, correlation weakened. No detailed observation of dimer diffusion was reported, however.

The diffusion of platinum on rhodium(100) was examined by Kellogg [65]. Platinum clusters with two to five atoms were generally in a chain configuration, although pentamers were balanced by two-dimensional islands. Clusters of six platinum atoms preferred to be in a two-dimensional form. Dimers were observed mostly in a close-packed configuration. Rough estimates of the diffusion barrier as a function of the size of the cluster are shown in Fig. 8.36. The barrier for single atoms and dimers appear to be the same with an activation energy of 0.91 eV, and that also holds for trimers, tetramers, and pentamers at an energy of 1.03 eV. All of these clusters diffuse not by exchange with substrate atoms but rather by atom hops. The similarity of the barrier for clusters of three through five atoms suggests it is the end atom of the chain that makes the rate limiting step, and the increase in going to six atoms with an energy of 1.16 eV is consistent with the change to a two-dimensional form. The existence of the linear configuration up to pentamers is an indication of the importance of long-range interactions for this system.

Kellogg [46] also studied motion of rhodium clusters on Rh(100) in FIM observations. Examined were entities made up of two to ten atoms and also one with twelve atoms. All existed as two-dimensional islands, with the exception of the trimer, which was usually in the shape of a chain. Rough estimates of the diffusion barrier, based on measurement at one temperature and assuming a standard prefactor for the diffusivity in the range of 10^{-3} cm^2/sec, are plotted as a function of the number of constituent atoms in Fig. 8.37.

Diffusivities of small clusters

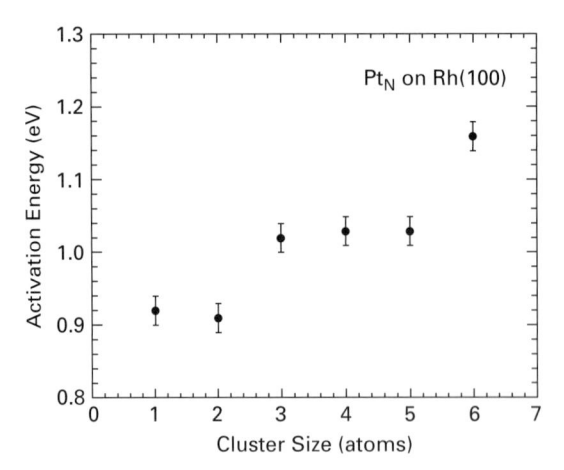

Fig. 8.36 Activation energies for diffusion of platinum clusters on Rh(100) measured in FIM (after Kellogg [65]).

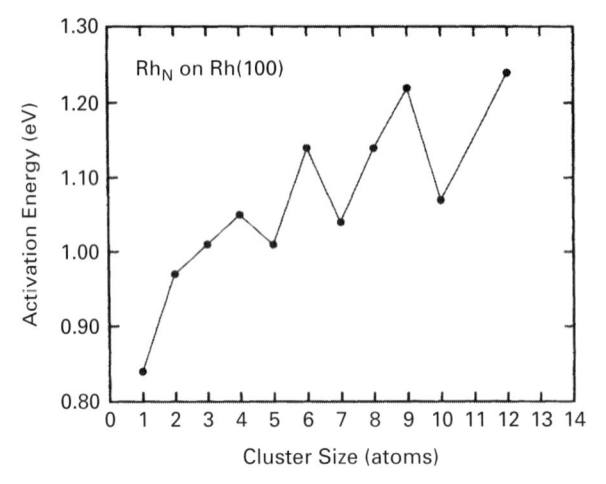

Fig. 8.37 Activation energy estimated from FIM observations for diffusion of rhodium clusters on Rh(100) plane (after Kellogg [46]).

What is seen here again is a clear oscillation, with an even number of atoms giving higher barriers. This effect is easy to understand. With an even number of atoms, a corner atom must move away from the others to initiate a jump, which is a higher energy process. In clusters with an odd number of atoms, one atom is attached to a central core and is easier to move. There is, however, one concern. Some of the mean-square displacements measured are quite small, and could possibly have arisen from slight oscillations of the center of mass. This could come about from an odd atom moving around the periphery, without any long-range diffusion of the cluster. The existence of linear Pt trimers and compact Rh trimers on the same surface indicates a larger importance of long-range interactions for Pt compared to Rh.

8.6 Clusters on rhodium surfaces

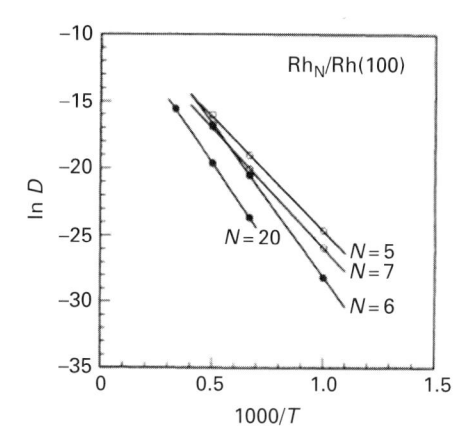

Fig. 8.38 Arrhenius plots for rhodium clusters on Rh(100) evaluated from transition-state theory and Lennard-Jones potentials (after Voter [66]).

8.6.2 Theoretical investigations

Estimates of the diffusivity of rhodium clusters on Rh(100) were made by Voter [66] using transition-state theory and Lennard-Jones interactions between atoms, without concern for many-atom effects. Activation barriers for atom jumps in small clusters were evaluated and are shown in Fig. 7.35. What is important here is that atom motion along a straight edge is the lowest energy step in cluster diffusion. To continue cluster movement, atoms will also occasionally have to emerge from kink sites in a cluster edge, a process more energy demanding. This can occur over a barrier of 2.1 eV as shown in Fig. 7.35g, and was used in examining the mobility of a much larger cluster, to be discussed later. The diffusivity of clusters followed an Arrhenius dependence, shown in Fig 8.38. The activation energy for pentamers amounted to 1.45 ± 0.02 eV, for hexamers 1.96 ± 0.02 eV and for heptamers the energy was lowered to 1.52 ± 0.01 eV. In this study, the possibility of diffusion by dimer or trimer shearing was not investigated.

Bogicevic et al. [13] relied on VASP simulation as well as DACAPO [14] to learn about self-diffusion of Rh dimers on the Rh(100) plane. Two important processes were taken into account. First were jumps perpendicular to the dimer axis, named shearing by the authors (middle of second row in Fig. 8.8 and to the left), second was jumping in the direction of the dimer axis called stretching (middle row in Fig. 8.8, center to right). It appears that a jump perpendicular to the dimer axis is the low-energy path for dimers with an energy of 1.07 eV compared with 1.27 eV for stretching. Concerted motion was found to be unfavorable.

Chang and Wei [20] resorted to *ab initio* density-functional calculations using the VASP [21,22] package with the LDA aproximation and a (5 × 5) surface cell to establish the activation energy for self-diffusion of dimers on Rh(100) planes. Their findings are summarized in Table 8.2. Movement by hopping required 1.33 eV while by exchange 1.61 eV; the value for dimer hopping is much higher than the 0.97 eV measured in experiments by Kellogg [46]. The study of Bogicevic et al. [13] is closer to the experimental value, assuming that the first step is limiting for the movement of the dimer.

No information is available, from experiment or theory, about movement of clusters on Rh(111) or Rh(110). The only data are for Rh(100), and these calculational estimates are quite spread out.

8.7 Clusters on palladium surfaces

No experimental information is available about movement of palladium nor of heteroclusters, but nevertheless theoretical investigations were carried out. Fernandez et al. [67] looked at the stability of copper clusters on Pd(110) surfaces using EAM interactions, and for dimers came up with a barrier of 0.54 eV for jumps along the channel and 1.40 eV for jumps across. Next year, Massobrio and Fernandez [68] again examined Cu as well as Pd clusters with EAM interactions. For clusters one to five atoms long, in-channel hopping was always more favorable than exchange. For Cu dimers, hopping required an energy of 0.50 eV compared to 0.75 eV for exchange; the value for hopping is slightly lower than in their previous investigation. For Cu trimers they found the hopping energy to be 0.48 eV, while exchange took 0.76 eV; for Cu tetramers and pentamers hopping required 0.47 eV and exchange 0.81 eV. Jumping across a channel was much more demanding. For Pd clusters of two to five atoms, exchange required 0.71 eV. Hopping involved 0.56 eV for Pd dimers and trimers, and 0.55 eV for Pd tetramers and pentamers. It is surprising that for both mechanisms, hopping and exchange, the energy did not change much with cluster size.

The theoretical investigations focus on two-dimensional surfaces. Started with the work of Chang et al. [27], who used EAM potentials for investigation of the Pd(111) surface; their results are illustrated in Fig. 8.13. Detailed investigations were done for aluminum clusters; for palladium clusters only energies for the movement of the center of mass were listed. Motion of palladium dimers required an energy of 0.11 eV, trimers 0.20 eV, tetramers 0.18 eV, and pentamers 0.36 eV.

Using EAM potentials parameterized according to Voter and Chen [41], Longo et al. [40] looked at the seven-atom Pd cluster on the Pd(111) surface. For cluster gliding they found an activation energy of 0.65 eV, and translational motion was not observed on heating the cluster to the melting temperature. For Pd_7, the barrier for edge diffusion was calculated as 0.49 eV, again smaller than for gliding, and edge running was observed in melting simulations at temperatures higher than 800 K.

The same year da Silva and Antonelli [69] examined the movement of a Pd_7 cluster on a Pd(111) surface with canonical ensemble molecular dynamics and RGL potentials at 500 K and 800 K. The diffusion coefficient at 500 K was $3.7 \times 10^{-6} cm^2$/sec and $1.5 \times 10^{-5} cm^2$/sec at 800 K. Most of the time the cluster moved via edge gliding, changing position from fcc to hcp sites and vice versa, with a compact shape maintained, as shown in Fig. 8.39e. However, distortions of the cluster were also observed, as in Fig 8.39e–k; the cluster finally came back to a close-packed structure and continued movement in Fig. 8.39l–o. Distortion of the cluster involved a number of single atom as well as concerted movements. The cluster diffuses at a more rapid rate when it is distorted, and is stable in its compact form. Note that the time line of the snapshots is

8.7 Clusters on palladium surfaces

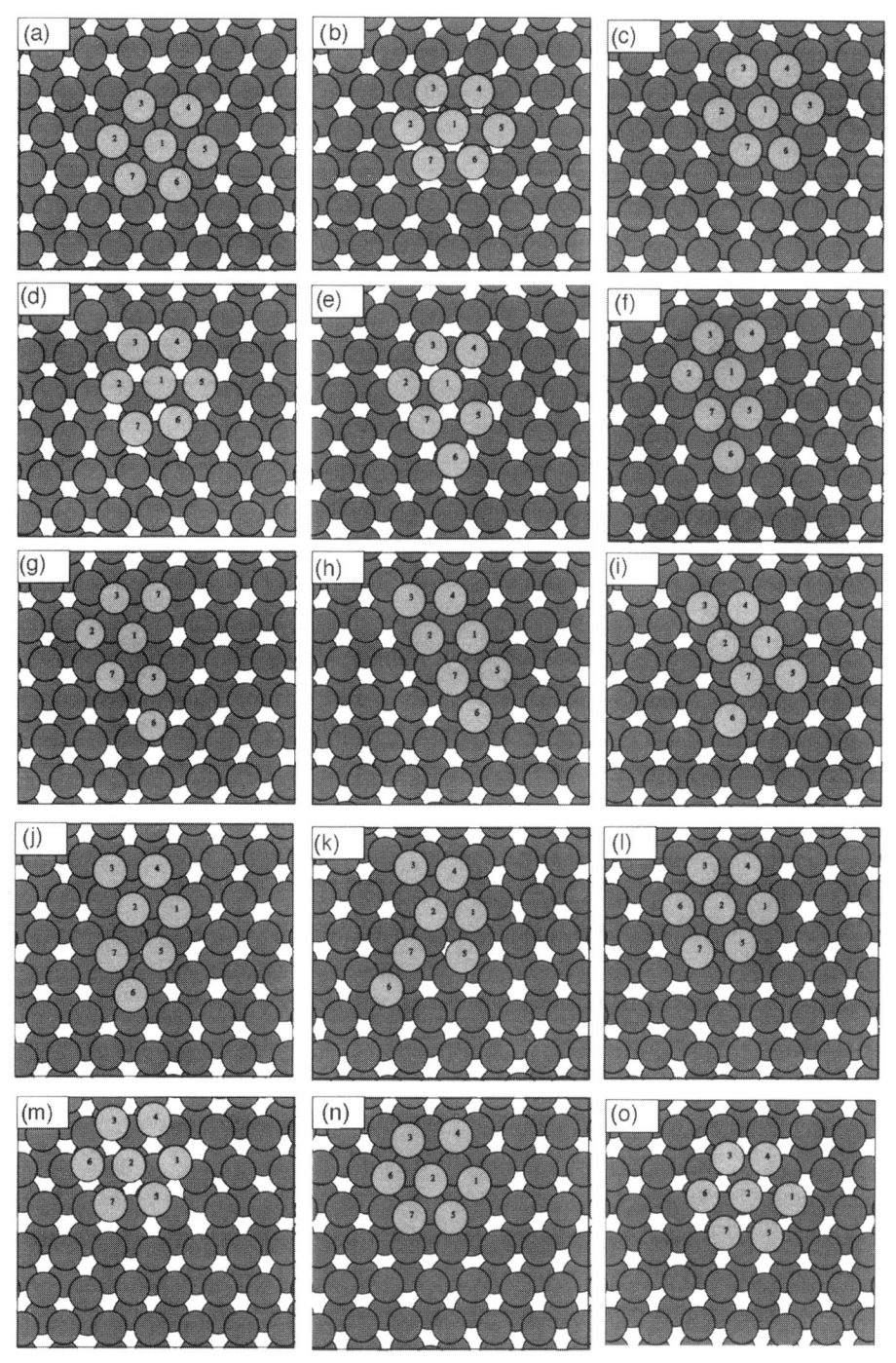

Fig. 8.39 Molecular dynamics simulations of Pd heptamer diffusing on Pd(111) obtained using RGL interactions and showing changes in overall configuration (after da Silva and Antonelli [69]).

not linear. Concerted row gliding was observed in these simulations by da Silva and Antonelli [69], but these findings seem to contradict those of Longo et al. [40] who did not observe gliding for Pd$_7$, and instead saw edge running.

In 2005 Chang and Wei [20] using the VASP [21,22] package established the activation energy for diffusion of dimers on the Pd(100) surface. They found that exchange was more likely than hopping on this surface; exchange required 0.75 eV against 0.85 eV for hopping. The molecular dynamics study of Kim et al. [36] looked at the movement of clusters of two to four atoms on different palladium surfaces. For dimers on Pd(100), jumps perpendicular to the dimer axis were most likely and required 0.621 eV, jumps in the direction of the axis cost 0.963 eV, against 0.721 eV for exchange. The value of the activation energy derived for exchange was quite similar to one derived by Chang and Wei [20], but Kim et al. claimed that hopping, not exchange, was the most likely mechanism. On Pd(110) in-channel jumps occurred over 0.671 eV, exchange took 1.36 eV, and cross-channel jumps 1.670 eV. Values are higher than the findings of Fernandez et al. [67] on the same plane. On Pd(111) it is not clear if dimers sit in fcc or hcp sites, so it is not certain between which places the dimer jumped. However, jumps 30 degrees to the dimer axis required 0.15 eV; 90-degree jumps occurred over 0.119 eV, and exchange was unlikely at 1.342 eV. For linear trimers aligned along the [110] direction on Pd(100) exchange required an energy of 0.766 eV, while hopping in the [110] direction took 1.331 eV. To distinguish which mechanism is more likely, hopping in the direction perpendicular to [110] should be considered. For linear trimers along Pd(110) channels, in-channel hopping required 0.96 eV while exchange occurred over 1.071 eV, again higher than the results of Fernandez et al. [67]. On Pd(111) triangular trimers, without any indications of the binding site, were investigated and jumps cost 0.266 eV while exchange took 1.372 eV. Linear tetramers were examined on the Pd(110) and (100) surfaces. On the (100) surface, the tetramers were aligned in the [110] direction and hopping in the same direction was very demanding at 1.661 eV. Exchange required a smaller energy of 0.786 eV; however, the most probable hopping in the direction perpendicular to the tetramer axis was again not given for comparison. On Pd(110) exchange is more probable than in-channel hopping at 1.07 eV compared to 1.275 eV, again in disagreement with the findings of Fernandez et al. [67], where hopping not exchange was more likely. Unfortunately no experiments have been done on surface diffusion of palladium, so it is difficult to evaluate the theoretical efforts which appear to be reasonably different.

8.8 Clusters on silver surfaces

8.8.1 Experiments

Experimental investigations of atomic diffusion on the Ag(111) surface started with the work of Schlösser et al. [43] who from their STM observations were able to arrive at a diffusion barrier of 0.51 ± 0.05 eV for vacancy islands with a radius larger than 18 Å. For silver islands prepared on Ag(111), diffusion proceeded with similar energetics of 0.53 ± 0.05 eV.

8.8 Clusters on silver surfaces

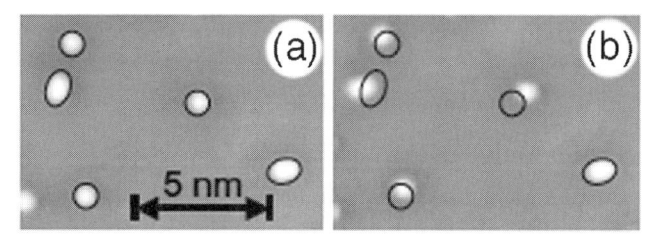

Fig. 8.40 Copper dimer inter-cell motion on Ag(111) surface for 80 sec at $T = 24$ K, observed with STM. Circles and ellipse in identical position (after Morgenstern et al. [57]).

Scanning tunneling microscope observations of copper dimers diffusing on Ag(111) have been made by Morgenstern et al. [57], and show that dimer rotation begins at 16 K, while inter-cell diffusion starts above 24 K, as illustrated in Fig. 8.40. An activation energy of ≈ 0.073 eV was estimated for the latter assuming a prefactor of 10^{12} sec^{-1}, compared with a barrier of 0.065 ± 0.009 eV for diffusion of copper adatoms, which occurs primarily between fcc sites. Dimers were observed in three positions – right inclined, left inclined, and horizontal, and occupied these with equal frequency in the absence of other atoms or dimers. Molecular dynamics simulations were also carried out for the movement of Cu dimers on Ag(111). These are shown in Fig. 8.29; the energy required for the first jump of a dimer atom amounted to 0.092 eV, while the subsequent jump needed only 0.014 eV. Concerted motion is also illustrated in Fig. 8.29; the lowest energy was required for concerted two-atom jumps which occurred over 0.092 eV, the same as for the first jump of a dimer atom. Dimer zigzag motion required 0.122 eV, and is the lowest energy path for inter-cell diffusion. The values obtained from MD calculations are noticeable higher than experimental determinations.

8.8.2 Theoretical investigations: Ag(100)

Theoretical efforts began with the work of Voter [70] who relied on EAM potentials for his estimates of the activation energies of silver clusters on Ag(100). For a single Ag adatom, he estimated a barrier of 0.489 eV. For a dimer atom jump in the direction of the molecular axis the energy rose to 0.770 eV, and transverse to the axis to 0.502. The barriers for the rate limiting step in diffusion of larger clusters were as follows: tetramer 0.809 eV, pentamer 0.555 eV, and block hexamer 0.965 eV. For a hexamer made up of a tetramer with a dimer attached at one site the diffusion energy was 0.502 eV. This was, of course, an introductory study to demonstrate the possibility of achieving quantitative barriers for clusters involved in crystal growth.

Investigation continued with the work of Shi et al. [35] who again used embedded atom calculations to check a new mechanism of movement – dimer shearing – described in Chapter 7. Their findings are summarized in Table 7.1. For movement of dimers, atom-by-atom jumps are still favorable, with the first jump requiring 0.48 eV, but for tetramers shearing is energetically advantageous, requiring 0.637 eV compared with 0.723 eV for atom-by-atom movement. Similarly for hexamers, 0.658 eV was needed for shearing compared with 0.737 eV for atom-by-atom movement. Compact clusters of nine atoms prefer trimer shearing over atom-by-atom jumps, at 0.763 eV compared to 0.903 eV. The message from this paper is clear – dimer shearing for bigger compact clusters is more favorable than atom-by-atom jumping, but it is also worth noting that exchange was not considered in this study.

Molecular dynamics simulations of dimer diffusion have been carried out by Zhuang *et al.* [17] using the EAM potentials of Haftel *et al.* [71–73]. For silver, exchange with an energy of 0.348 eV was more favorable then hopping, which required overcoming a barrier of 0.585 eV. In a molecular dynamics study not only conventional but also rotationally complicated exchanges were observed on this surface. A similar study was pursued by Liu *et al.* [19] with EAM interactions [18], surface-embedded-atom potentials (SEAM) [71–73], and RGL interactions [52,53]. Their findings for silver are not conclusive, since from EAM and RGL calculations exchange is less favorable than hopping, with an energy of 0.5892 eV (EAM) and 1.0112 eV (RGL), while hopping required 0.4836 eV (EAM) and 0.5001 eV (RGL). From SEAM, hopping cost 0.5818 eV, but exchange only 0.3615 eV. Molecular dynamics studies with SEAM potentials at 550 K for 2 nsec time showed 3 hopping events, 55 simple exchanges and 6 exchange rotations. However, they did not show molecular dynamics simulations with different potentials. *Ab initio* density-functional calculations in the VASP [21,22] package were resorted to by Chang and Wei [20] also to calculate barriers for diffusion of dimers. For silver dimers they found hopping more probable than exchange, with an energy barrier of 0.55 eV for hopping and 0.69 eV for exchange.

Kim *et al.* [36] surveyed the energetics of jumping and exchange in clusters of two to four atoms with molecular dynamics. For dimers they found that perpendicular jumping is more likely, with an energy of 0.467 eV, whereas exchange required 0.615 eV and tangential hopping 0.775 eV. For a linear trimer aligned along the [110] direction, exchange cost 0.652 eV. Only jumps in the [110] direction, over a 1.09 eV barrier were examined, but they did not consider jumping in a direction perpendicular to the trimer axis, which is needed for comparison with exchange. For linear tetramers aligned in the same direction, the situation is similar to trimers: hopping in the [110] direction required 1.376 eV while exchange needed only 0.675 eV.

From these theoretical findings it is not clear which mechanism leads in the diffusion of dimers on Ag(100), hopping or exchange, or do both mechanisms coexist on this surface. It is not surprising that values of the energetics are very spread out. Unfortunately there is no experimental study for comparison. For bigger clusters, Shi *et al.* [31] suggested that the possibility of dimer or trimer shearing should be taken into account; however, they did not consider movement by exchange, which might change the energetics of the transition.

8.8.3 Theoretical investigations: Ag(111)

Theoretical research on self-diffusion on the silver (111) surface started with the EAM calculations of Nelson *et al.* [74] who for the initial step in dimer motion on this surface estimated a value of 0.121 eV. Bigger clusters were not investigated. EAM potentials were also used by Chang *et al.* [27] to briefly look at the energetics of cluster movement, as illustrated in Fig 8.13. For movement of a dimer an energy of 0.11 eV was derived, trimers needed 0.16 eV, tetramers only 0.15 eV, and pentamers 0.22 eV.

Using EAM potentials parameterized according to Voter and Chen [41], Longo et al. [40] evaluated the activation energies for a heptamer gliding over its own crystal. For silver the energy of gliding was 0.66 eV. However they failed to observe gliding for Ag$_7$ in MD, which was attributed to the vertical mobility of the substrate at elevated temperatures.

Kim et al. [36] looked at the movement of atoms in clusters on Ag(111); however, they did not specify at which adsorption sites clusters were positioned, fcc or hcp. As expected for fcc(111) surfaces, exchange was always associated with a high energy, so it is not listed here. For dimers, 90-degree hopping required a low energy of 0.108 eV, 30-degree hopping cost 0.114 eV, values quite close to the findings of Chang et al. [27] and Nelson et al. [74]. For triangular trimers, jumps required 0.185 eV. Regrettably there are no experiments available to test the calculations and diffusion for bigger clusters was not explored.

8.8.4   Theoretical investigation: Ag(110)

One-dimensional cluster diffusion on Ag (110)-(2 × 1) surfaces was examined with EAM potentials by Kürpick [42]. Four different diffusion mechanisms were considered, described in the section for Ni(110), and are indicated in Fig. 8.18. They took into account movement by the leapfrog mechanism versus sequential hopping and exchange. For silver, similar to other fcc materials investigated, leapfrog motion seemed to be the most likely mechanism. Concerted hopping of a trimer cost 0.87 eV, for a tetramer 1.15 eV, and a pentamer 1.43 eV. Exchange of a trimer atom required 0.74 eV, a tetramer 0.75 eV, and a pentamer 0.76 eV. While hopping up a {111} wall took only 0.62 eV for clusters of three to five atoms, hopping down involved a barrier 0.38 eV for trimers and 0.37 eV for tetramers and pentamers. Concerted hopping on an unreconstructed surface required 0.72 eV for trimers, 0.96 eV for tetramers, and 1.19 eV for pentamers.

Kim et al. [36] examined the movement of atoms in linear clusters on an unreconstructed Ag(110) plane; cluster jumping out of a channel was very energetically demanding. In-channel jumps for a dimer required 0.511 eV, exchange 0.834 eV. In-channel jumps for trimers cost 0.740 eV, while the barrier to exchange was higher at 0.779 eV. For a linear tetramer, in-channel jumps cost 0.985, while exchange required 1.002 eV. The values they obtained are quite close to the energetics derived by Kürpick [42] for concerted motion of clusters on an unreconstructed surface. Again there are no experiments for comparison with these findings.

8.9   Clusters on tantalum surfaces

FIM observations of cluster structures of palladium were made on Ta(110) by Schwoebel and Kellogg [75] in 1988. Palladium was deposited and allowed to diffuse at ~200 K to form clusters. Clusters made up of two to eight palladium atoms were in the form of

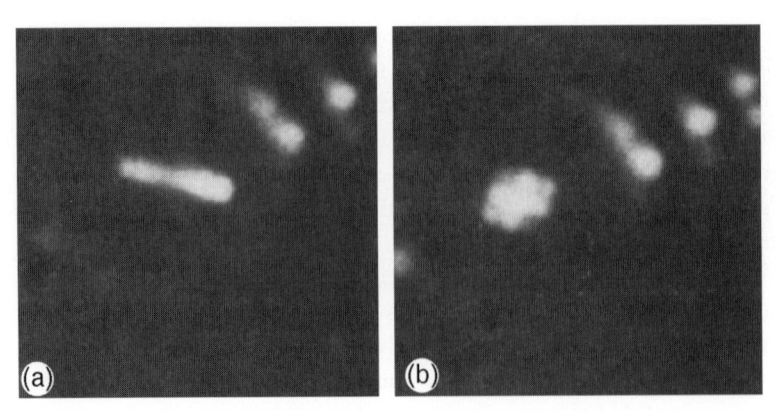

Fig. 8.41 Palladium clusters on Ta(110) plane observed in FIM. (a) Linear chain of nine Pd atoms. (b) Two-dimensional nanomer formed from chain by equilibration at 225 K (after Schwoebel and Kellogg [75]).

chains oriented along [111], as shown in Fig. 8.41. The chains became mobile at a temperature of ~250 K. With nine or more palladium atoms two-dimensional islands formed, mobile from ~250 K to ~325 K. Similar behavior was reported earlier on W(110) by Bassett [76].

8.10 Clusters on tungsten surfaces

8.10.1 Experiments: tungsten(211) and (321)

Quantitative kinetic studies were begun by Graham [77], who compared W_2 clusters made of atoms in adjacent channels of W(211), with the diffusion of single atoms at a series of temperatures. The measurements for the latter were made with just one atom on the plane, eliminating the correlations that had earlier led to quite incorrect kinetics. In analogy with the analysis of single atoms, movement of the dimer center of mass was studied. However, the results for dimers gave quite surprising results, with an activation energy of only 0.30 eV and a tiny prefactor of 10^{-12} cm^2/sec. One thing was clear, dimer atoms do not move independently; however, the exact mechanism was not determined. The study did not take into account the trapping characteristics of the edges, but this is probably not the only reason for such a low prefactor.

In a more detailed presentation, Graham [78,79] established the different configurations for tungsten dimers with atoms in adjacent rows on W(211), as given by the models in Fig. 7.8a and in FIM images of the dimers in Fig. 7.8b. The kinetics described for the center of mass of the dimers, again without taking into account the influence of edges, were much the same as the previously derived values, an activation energy of 0.37 ± 0.04 eV and a low prefactor of 10^{-11} cm^2/sec, but the important result of the observations here was the realization that dimer motion occurred by single jumps of the individual atoms, which is of course the assumption of the derivations in Section 7.1. The same seemed to be true for linear trimers, for which the measured activation energy

8.10 Clusters on tungsten surfaces

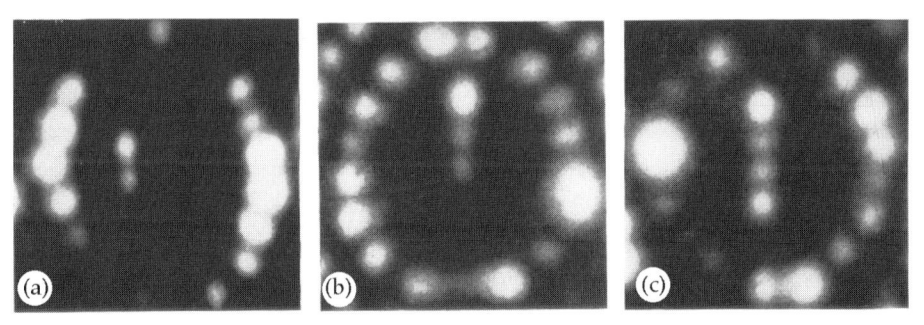

Fig. 8.42 Iridium clusters on W(211) plane [80]. (a) Ir$_2$. (b) Ir$_3$. (c) Ir$_4$. All these clusters are always seen in the straight configuration in the FIM.

was 0.36 ± 0.06 eV with $D_o \sim 10^{-11}$ cm^2/sec, with each atom in an adjacent channel. Later on some possible explanations of such a low value of the prefactor will be presented. More detailed field ion micrographs of dimer as well as trimer movement appeared soon thereafter; these pictures, shown in Fig. 7.11, revealed more clearly the individual atomic jumps in diffusion. Valuable information that came out of this study was that two W atoms separated by an empty channel, that is separated by a distance of 8.95 Å, do not feel each others presence, while for atoms in nearest-neighbor channels the interactions are very strong. On W(321), adatoms in nearest-neighbor channels (7.09 Å distance) moved independently.

Motion of iridium clusters, always with atoms in adjacent channels of W(211), was explored by Reed [80]. This diffusion proceeds differently than the motion of clusters made up of tungsten atoms. Only clusters with atoms lined up straight are seen, as shown in Fig. 8.42 for dimers, trimers and tetramers. With tungsten clusters, the free energies of the different forms are comparable, so that both straight and staggered configurations are seen; for iridium, the free energy of the straight configuration is considerably lower. From Arrhenius plots both for single atoms and for clusters it is clear that dimers have a higher barrier to overcome for diffusion. However, the quantitative values for the diffusion parameters derived in this study (for dimers 0.67 ± 0.06 eV and $D_o = 9 \times 10^{-6}$ cm^2/sec) both for atoms and dimers, were not correct due to the low prefactor measured. Data from positions close to the edges, which had a limiting effect on movement, were not withdrawn.

Molybdenum dimers, with atoms in adjacent channels of the W(211) plane, together with single atom motion both on W(211) and (321), were examined by Sakata and Nakamura [81]. Their results were already shown in Fig. 4.37. For cross-channel molybdenum dimers they arrived at a diffusion barrier of 0.26 eV and a prefactor of 2.3×10^{-12} cm^2/sec. They again identified three different configurations for the molybdenum dimers; just as found previously for W$_2$, the free energies were not too different. The quantitative values for the diffusion parameters of the different entities appear to be inaccurate, with prefactors and activation energies too low, and there was no indication of any correction for edge effects.

The beginnings to truly quantitative work on dimers were made by Stolt *et al.* [82], who studied rhenium dimers with atoms in adjacent rows of the (211) plane of tungsten.

Stolt *et al.* first recognized the importance of interactions of dimer atoms with lattice edges. In his study, he always started measurements with a dimer in the center of a channel roughly twenty-five spacings long, and terminated measurements when the dimer reached the edges. They also monitored the influence of edges by running Monte Carlo simulations, and furthermore noted that the two edges on the W(211) plane were different; on the one towards the [111] direction dimers are trapped very efficiently, while on the other, towards [100], less frequently.

For energetic studies dimers close to both edges were withdrawn from the data set. Rhenium dimers appear in both straight and staggered configurations, similar to tungsten dimers, and are stable up to 350 K; their energetics are easy to establish. A pair of atoms in two rows, each with L sites, can be arranged in $(2 - \delta_{xo})(L - x)$ ways, where x is the distance apart and δ_{xo} is the Kronecker delta. The probability of finding a pair of atoms at the separation x, P_x, is therefore

$$P_x = C(2 - \delta_{xo})(L - X) \exp(-F_x/kT), \tag{8.1}$$

where F_x is the free energy for the system with the atoms X units apart; C is just a normalization constant. The frequency with which the staggered configuration 1 will be found compared to that of the straight configuration 0 is just

$$(P_1/P_0)_L = 2[(L - 1)/L] \exp[(S_1 - S_0)/k] \exp[-(E_1 - E_0)/kT]. \tag{8.2}$$

This ratio was measured at equilibrium at temperatures from 270 to 327 K. The difference in energy, $E_1 - E_0$, was -0.055 ± 0.011 eV and $S_1 - S_0$ was given by $-(1.5 \pm 0.04) \times 10^{-3}$ eV/K, that is, the staggered configuration was slightly more stable. The values in these measurements were, however, affected by redistribution while the sample cooled. Corrections for changes during quenching of the sample were possible [83] and the Arrhenius plot in Fig. 8.43a yielded -0.061 ± 0.011 eV for $E_1 - E_0$ and $-(1.64 \pm 0.39) \times 10^{-4}$ eV/K for $S_1 - S_0$; there is a bigger difference in the thermodynamics of the staggered and straight state than previously estimated. Stolt *et al.* took into account transient cooling; however, what was still not allowed for was transient heating. However, transient heating has much less effect on the data than cooling – it is possible to lower its influence by increasing the starting current.

As is indicated in Fig. 7.7, the jump rate of a dimer atom out of the straight into the staggered configuration is given by a, and the return rate from staggered to straight by b. As already shown in the previous section, in Eq. (7.35), the probabilities of the two configurations are: $P_0 = b/(a + b)$ and $P_1 = a/(a + b)$.

The displacement fluctuation of the center of mass of the dimer therefore appears as described by Eq. (1.36), $<\Delta x^2> = 2Dt$, with

$$D = \frac{a}{1 + (P_1/P_0)_L [L/(L-1)]} \frac{\ell_0^2}{2}. \tag{8.3}$$

From measurements at temperatures from 263 to 351 K, Stolt *et al.* [82] obtained a diffusion barrier of 0.78 ± 0.013 eV with a prefactor of $4.5(\times 1.7^{\pm 1}) \times 10^{-4}$ cm^2/sec for the dimer, compared to a barrier of 0.86 ± 0.03 eV and a prefactor $D_o = 2.2(\times 2.8^{\pm 1}) \times 10^{-3}$ cm^2/sec for diffusion of a single rhenium atom. The Arrhenius plots for these results are

8.10 Clusters on tungsten surfaces

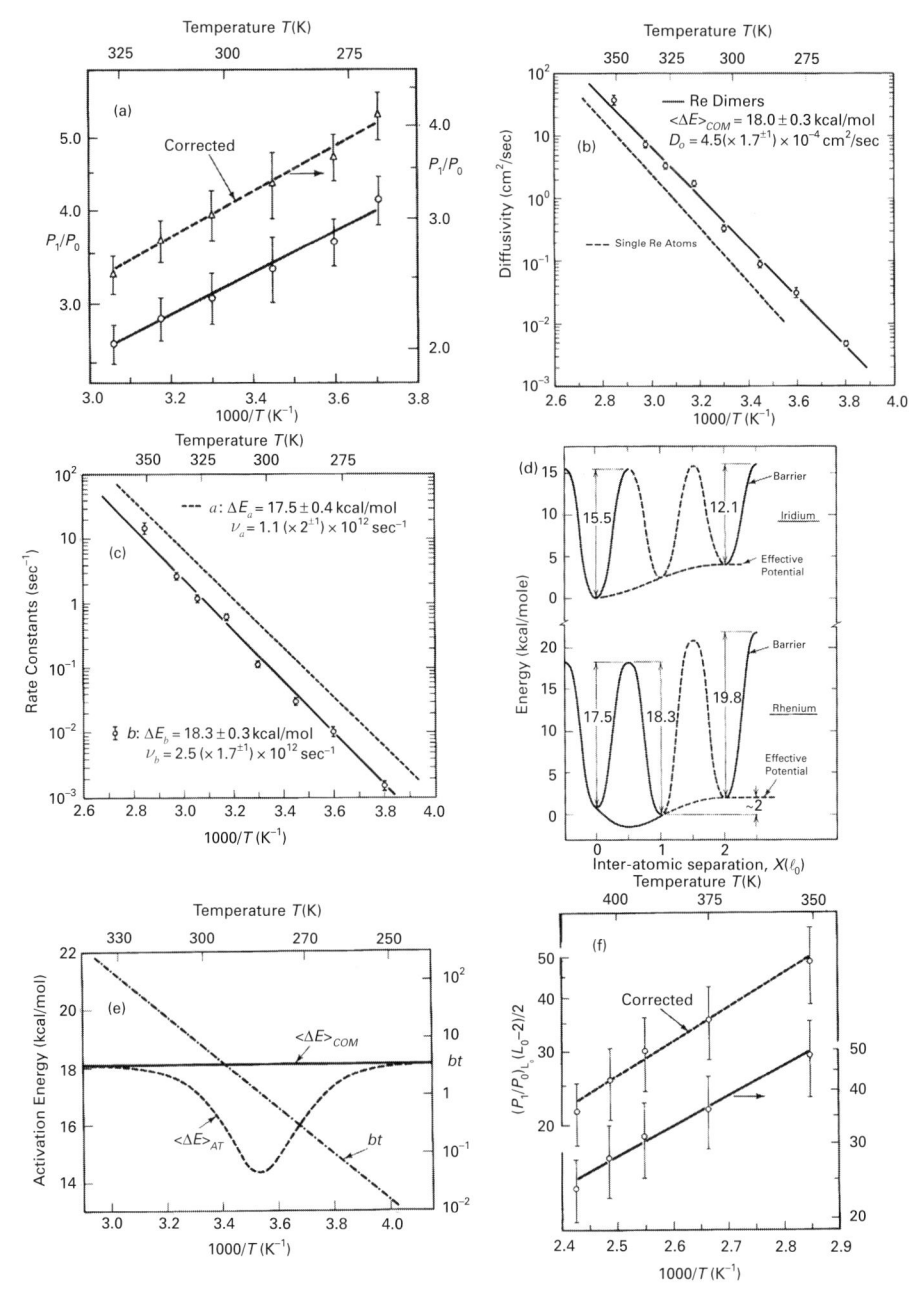

Fig. 8.43 Behavior of Re$_2$ dimer on W(211) plane [83] studied in FIM. (a) Ratio of Re$_2$ in state 1 compared to state 0 as a function of reciprocal temperature, before and after correction for changes during quench. (b) Arrhenius plot for center of mass of Re$_2$ diffusing on W(211) plane [82]. (c) Temperature dependence of cross-channel dimers on W(211) plane for Re$_2$. Dashed lines indicate lack of quantitative information. (d) Dependence of dimer energy on interatomic separation, for iridium (top) and rhenium (bottom) cross-channel dimer. (e) Dimer diffusion analyzed in terms of individual atom displacement as a function of reciprocal temparature. (f) Arrhenius plot for dissociation of Re$_2$ on W(211).

shown in Fig. 8.43b. Diffusion of the dimer is clearly more rapid than that of individual adatoms. The center of mass moves due to the two jump processes *a* and *b*, and their energetics are given in Fig. 8.43c.

The energetics of rhenium and iridium dimers diffusion on W(211) are shown in the potential diagram of Fig. 8.43d. For iridium the potential curve is based on a previous study by Reed [80] and the fact that dimers were not observed in a staggered configuration. Interactions between the two atoms of the dimer in the staggered configuration lower the energy compared to the straight configuration, and make the transition from the staggered to the straight configuration the limiting step in diffusion. Only rhenium dimers were investigated in detail, but from the data available on W(211) for dimer motion around 280 K, it is possible to deduce how the relative population of staggered to straight configurations depends on the chemical specificity of interactions. For rhenium dimers, the ratio of P_1/P_0 is 4, for molybdenum 2, for tungsten 0.4 and for iridium 0. Two additional things are worth noting. Previously, low values for the diffusion barrier and prefactor have been found for dimers. However, these were based on measurements of the displacement of individual atoms rather than for the center of mass. For studies over a limited temperature range, this can give anomalously low values, both for the barrier and the prefactor. The effective activation energy $<\Delta E>_{AT}$ measured from the logarithmic derivative with respect to $1/T$, can be described by

$$<\Delta E>_{AT} = <\Delta E>_{COM} + \Delta E_b \left(\frac{(2bt+1)\exp(-2bt) - 1}{2bt + 1 - \exp{-2bt}} \right); \qquad (8.4)$$

it has a local minimum for $bt \sim 1$, shown in Fig. 8.43e. Making measurement close to the minimum can lead to very low values of the diffusion energy and prefactor.

It is also interesting to note that the dissociation energy for rhenium dimers was examined a bit later. The equilibrium between dimer and dissociation products at 392 K is illustrated in Fig. 8.44, and the ratio of staggered to dissociated dimers, from

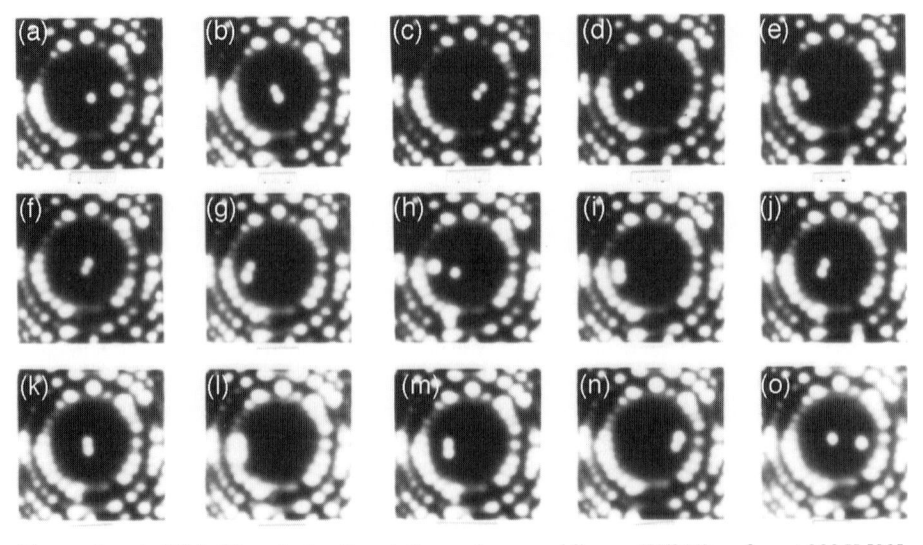

Fig. 8.44 Observations in FIM of Re$_2$ cluster dissociating and reassembling on W(211) surface at 392 K [83].

8.10 Clusters on tungsten surfaces

351 to 412 K, is plotted in Fig. 8.43f. This yields a value for E_1 of -0.16 ± 0.048 eV and an entropy $S_1 = (1.21 \pm 1.25) \times 10^{-4}$ eV/K. Ehrlich and Stolt [84] looked at the movement of cross-channel trimers and surprisingly, the energetics for movement of the center of mass were very similar to dimers, despite the availability of different diffusion paths. The prefactor for diffusivity was $D_o = 5.2(2.3^{\pm 1}) \times 10^{-4}$ cm^2/sec, and the activation energy 0.79 ± 0.02 eV.

The behavior of rhenium clusters on W(211) is quite different when two atoms are deposited into the same channel; when the surface is warmed till diffusion starts at 296 K, the two atoms eventually associate [85]. This happens after a distance of three spacings is achieved between the two atoms. Ordinarily it was not possible to resolve the dimer atoms, but the image had an elongated shape, and atoms were distinguished at temperatures below 15 K. Occasionally the dimer dissociated, usually near the plane edges. The dimer is immobile at this temperature, but on heating to 381 K it diffuses rapidly, always in a single channel. Measurements over a range of 48 K yielded the Arrhenius plot, which is compared to that for cross-channel dimers and also for single atoms in Fig. 8.45a. The diffusion barrier for the in-channel dimer amounted to 1.01 ± 0.051 eV, with a prefactor of $1.8(\times 5.1^{\pm 1}) \times 10^{-3}$ cm^2/sec. The barrier is clearly significantly higher than for the cross-channel dimer, but the prefactors are roughly the same. Surprisingly, removal of one atom from the dimer by field evaporation proceeds at 88% of the field for evaporation of a single atom.

The in-channel trimers were formed and have been found to diffuse at 435 K, with no indication of any dissociation at this temperature. The kinetics of diffusion for an in-channel dimer, illustrated in Fig. 7.9, are quite similar to those of cross-channel dimers, discussed in the previous section, assuming diffusion occurs by single jumps of individual atoms. The mean-square displacement is given in terms of the two jump rates a and b, a for the displacement of one atom against the other, and b for rejoining the two. The displacement fluctuation can therefore be written as Eq. (7.38),

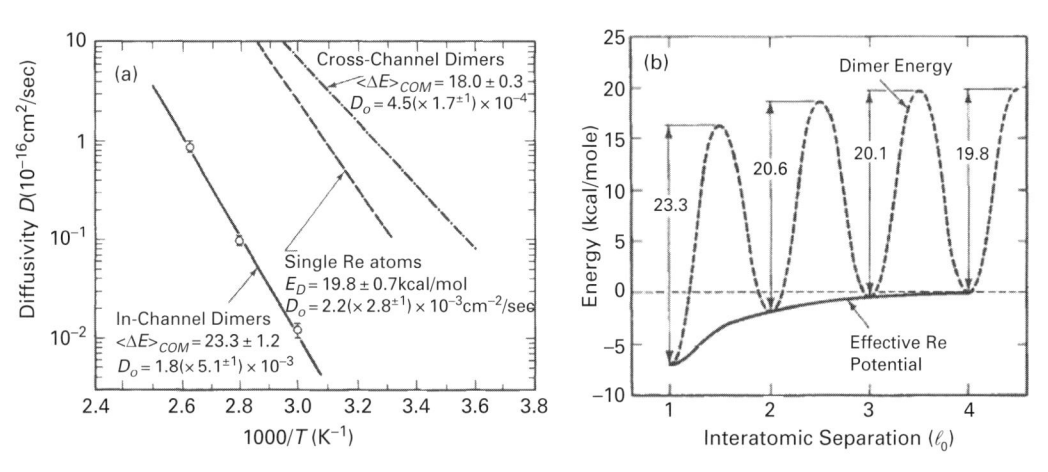

Fig. 8.45 Re dimers on the W(211) plane [85] observed in FIM. (a) Comparative Arrhenius plots for the diffusivities of rhenium atoms, cross-channel dimers, and in-channel Re dimers. (b) Potential barriers for in-channel rhenium dimer.

$\langle \Delta x^2 \rangle = \frac{ab}{a+b} t \ell_0^2$, where ℓ_0 is the nearest-neighbor distance and t the length of the diffusion time interval. The ratio of dimers in state 2 (dimer atoms are separate by a distance $2\ell_0$) to those in state 1 (when dimer atoms are at the nearest-neigbour distance) is according to Eq. (7.37) $P_2/P_1 = a/b$, so that the displacement fluctuation can also be written as

$$\langle \Delta x^2 \rangle = P_1 a \ell_0^2 t. \qquad (8.5)$$

The dimer is seen only in state 1 with atoms adjacent to each other, so the parameter given above refers to the jump rate a. What is important to note, however, is that dimers have not been observed to dissociate. That means that when in state 2, there is still a barrier preventing dissociation, that is, interactions are not just limited to nearest neighbors. A rough potential diagram for rhenium in-channel dimers was given in Fig. 8.45b, but it must be emphasized that only the rate a is based on measurements; the other values are reasonable guesses based on information about the energy difference between the bound cross-channel dimer and the dissociated pair. Furthermore, single jumps have been assumed in diffusion, but the possibility is strong that two atoms can carry out a concerted move. It is clear that the potential for in-channel dimers differ significantly from that for cross-channel dimers.

The triangular trimer with a dimer in one channel and one atom in a nearest-neighbor channel was also investigated and movement of such clusters occurs in the same temperature range as for isolated in-channel dimers, indicating the limiting step in movement. Pentamers with two atoms in the first channel, two in the second and one atom in the third also move in the same temperature range as in-channel dimers. The conclusion from this study was that cross-channel interactions were much weaker than in-channel ones, which dominate movement of bigger clusters.

8.10.2 Experiments: tungsten (110)

Tungsten clusters were studied on the (110) plane of tungsten by Cowan and Tsong [86] with a standard FIM. The dimer appeared in only a single configuration, with atoms in nearest-neighbor sites. The quantitative diffusion kinetics for dimers on W(110) showed a rather slower motion than for adatoms. From an Arrhenius plot of the diffusivity, in Fig. 8.46, an activation energy of 0.77 ± 0.06 eV and a prefactor of 4.5×10^{-4} cm^2/sec were obtained for single tungsten atoms; for dimers, the activation energy was given as 0.80 ± 0.080 eV with a prefactor of 1.6×10^{-4} cm^2/sec. However, the values here are again no longer valid. In 1978 the temperature calibration was corrected and yielded an activation energy of 0.90 ± 0.07 eV for tungsten atoms with a prefactor $D_o = 6.2(\times 13^{\pm 1}) \times 10^{-3}$ cm^2/sec; for tungsten dimers the activation energy was 0.92 ± 0.14 eV and $D_o = 1.4(\times 160^{\pm 1}) \times 10^{-3}$ cm^2/sec [87]. Diffusion of trimers started at 340 K, but at this temperature frequent dissociation was observed.

In 1975 Bassett et al. [88] studied the lifetime of dimers. For tantalum dimer diffusion, they obtained an activation energy of 0.77 eV, for rhenium dimers 0.99 eV, iridium dimers 0.77 eV, platinum dimers 0.62 eV, palladium dimers 0.46 eV, and nickel dimers 0.41 eV. Unfortunately, measurements were done with more then one dimer on

8.10 Clusters on tungsten surfaces

Table 8.9 Activation energy and prefactor for platinum diffusion on W(110) [90].

Cluster	E_D (eV)	Prefactor (cm²/sec)
Pt	0.67 ± 0.06	$30.6(\times 20.1^{\pm 1}) \times 10^{-8}$
Pt$_2$	0.67 ± 0.06	$9.21(\times 20.1^{\pm 1}) \times 10^{-8}$
Pt$_3$	0.79 ± 0.16	$15.2(\times 403^{\pm 1}) \times 10^{-8}$
Pt$_4$	0.87 ± 0.16	$45.6(\times 403^{\pm 1}) \times 10^{-8}$

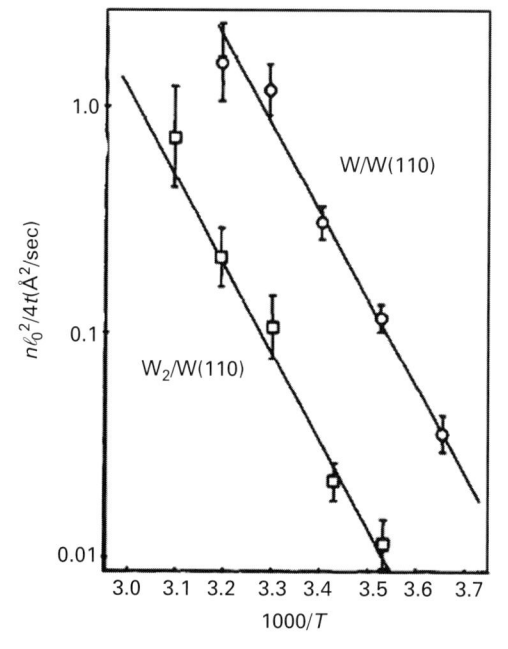

Fig. 8.46 Arrhenius plots for the diffusivity of W and W$_2$ on W(110) studied in FIM. A diffusion barrier of 0.80 ± 0.080 eV with a prefactor of 1.6×10^{-4} cm²/sec was derived. These were corrected in 1978 to 0.92 ± 0.14 eV and $1.4(\times 160^{\pm 1}) \times 10^{-3}$ cm²/sec [87] (after Cowan and Tsong [86]).

the plane, which could influence the results. The influence of other atoms on the surface is obvious for Re$_2$, since from later studies it is clear that the rhenium dimer is unstable on the W(110) surface [89].

More concentrated work on cluster motion over a two-dimensional surface, W(110), was also done in this period. Bassett [90] looked at platinum clusters from dimers through tetramers. From measurements at different temperatures he arrived at the Arrhenius plot already shown in Fig. 5.69. The diffusion parameters derived in that way are listed in Table 8.9. What must be noted here is that these values are undoubtedly not quite correct. From present knowledge it is clear that the prefactors are much too low. Nevertheless, the relation between the energetics is interesting. For dimers the diffusion barrier is the same as for single atoms, but after that the activation energy goes up with size. The clusters generally are observed in the straight configuration, oriented along the close-packed <100> direction of the substrate. The diffusion barrier increases with size and observation

Table 8.10 Temperature for observable migration (T_D), rearrangement (T_r) and dissociation (T_d) of adatom clusters A_N and islands of more than ten adatoms on a (110) tungsten surface [76].

Adsorbate	Nickel	Palladium	Platinum
T_D (A$_2$) (K)	225	190	245
T_d (A$_2$) (K)	250	220	260
T_d (A$_4$) (K)	295	>220	340
T_r (island) (K)	340	290	<410
T_d (island) (K)	370	340	410

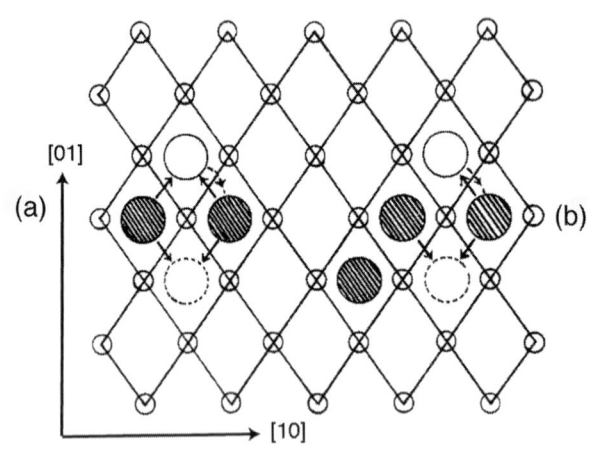

Fig. 8.47 Location of Pt atoms in mechanism of Pt$_2$ and Pt$_3$ diffusion by adatom jumps on W(110), examined in FIM (after Bassett [90]).

of linear, not compact clusters, suggests that interatomic forces are longer than nearest-neighbor, and that interactions in a longer chain make a significant contribution.

For dimers the location of the binding sites was carefully examined, with the conclusion that dimers were sitting in lattice sites, at least when observed by field ion microscopy. Just as for one-dimensional diffusion of dimers on W(211), Bassett assumed that diffusion on (110) occurred by individual single atom jumps, sketched in Fig. 8.47. Such jumps lead not only to translation but to rotation of the cluster as well, with the probability of rotation predicted in agreement with observations. Sequential adatom motion accounted for observations on dimers, but also on trimers, with the assumption that the barrier to formation of the close-packed trimer was high. For tetramers the reconfiguration probability is again small.

Bassett [76] looked briefly at clusters of palladium, nickel, and platinum. All these clusters appeared as linear chains oriented along the <111> directions with atoms in nearest-neighbor positions. The temperatures for significant mobility of dimers are listed in Table 8.10, together with the temperature for dissociation of dimers and tetramers. Diffusion sets in at relatively low temperatures, but for Pt dissociation begins at temperatures just 15 degrees higher. Chains were the stable form for nickel clusters with

8.10 Clusters on tungsten surfaces

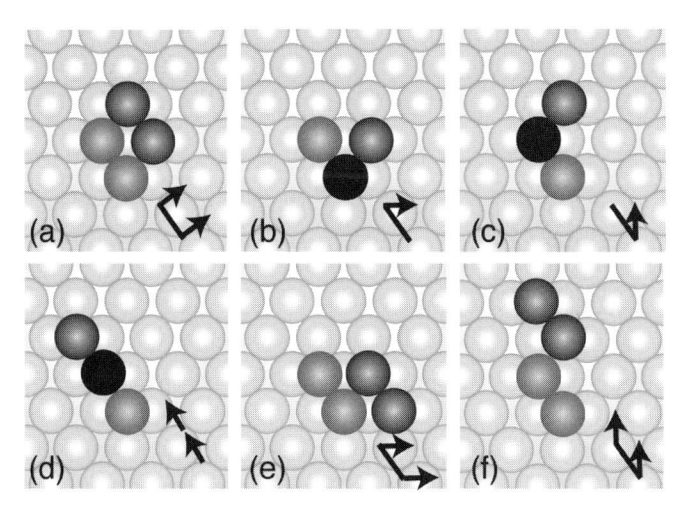

Fig. 8.48 Possible steps for dimer diffusion on W(110) plane identified in FIM observations [91]. Very light grey – surface atoms, light grey – position of dimer before diffusion, dark grey – position of atom after diffusion, black – atom at the same place before and after diffusion, arrows indicate change in position of atom during diffusion.

up to 6 atoms, for palladium up to 10 atoms, and for platinum up to 30 atoms, a clear indication of the importance of long-range interactions for these systems.

Tsong and Casanova [91,92] also examined diffusion on W(110), but this time with tungsten dimers. They interpreted their observations on the assumption that dimer atoms sit in adjacent lattice sites, and came to the conclusion that there were six elementary steps in the diffusion, shown in the diagram in Fig. 8.48. This was based on studies of the distribution of observed displacements and identification of the elementary steps. Some of these elementary processes required two jumps along the <111> direction with an intermediate bound in the <110> or <100> direction. These two jumps are correlated with a correlation factor of 0.1 at 299 K and 0.23 at 309 K.

Rhenium clusters have been examined on W(110) and have been shown to diffuse in an interesting fashion [93]. Dimers were not stable, but trimers were and existed in either a linear or triangular form, with the former configuration slightly less stable than the triangular one. Rhenium tetramers as well as pentamers, on the other hand, appeared to have a very open structure, with an interatomic separation of three spacings along [001] and two along [$\bar{1}$10], shown in Fig. 8.49a. This is quite different from what was proposed earlier by Tsong [6]. The pentamer looks just like a tetramer, but has an atom stuck in the center; it can also exist in a linear form, as shown in Fig. 8.49a, after heating to a temperature higher than 400 K. Movement of the trimer over W(110) is illustrated in Fig. 8.50, and of the pentamer in Fig. 8.51. The trimer clearly undergoes a change in structure during diffusion, the pentamer does not. The temperature for mobility of the clusters is shown in Fig. 8.49b. The open tetramer and the pentamer are comparable in diffusivity to single adatoms, which is much higher than for trimers. What is curious is the distribution of displacements. For single atoms this looks normal, with a small likelihood of larger displacements. With the pentamer, quite a long displacement is favored, suggesting a mechanism other than single atom jumps.

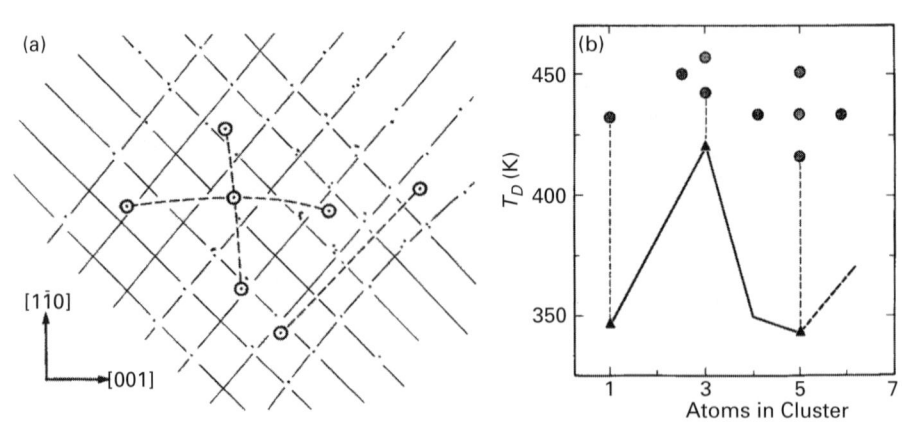

Fig. 8.49 Rhenium clusters studied with FIM on W(110) plane [93]. (a) Location of adatoms for compact as well as linear Re$_5$ on W(110). For the latter, only end atoms are resolved. (b) Dependence of temperature for diffusion (at which msd is ~ 0.5 Å^2/sec) upon size for rhenium clusters on W(110) plane.

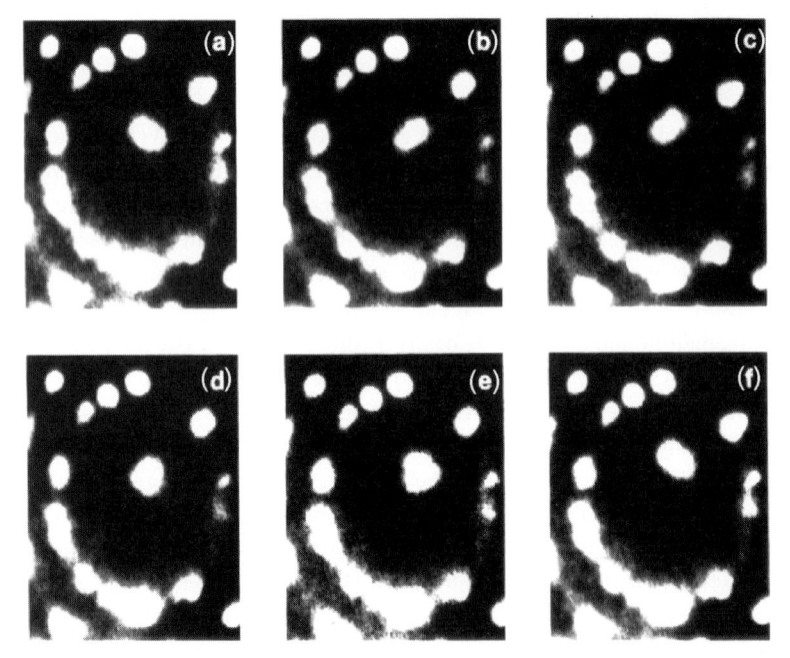

Fig. 8.50 Diffusion of rhenium trimer observed with FIM during 20 sec intervals on W(110) surface at 403 K [93].

Interesting observations of Re-Pd dimers were reported on W(110) by Watanabe [94]. In this dimer, the rhenium atom stays anchored at one site, but the palladium, in the temperature range 148–168 K, can rotate between nearest neighbors such as (1,1) and (1,−1), depicted in Fig. 8.52a. During 100 observations of equilibration at 160 K, the Pd adatom was never seen in the (2,0) position relative to the Re adatom. The potential that must be confronted in such a rotation is sketched in Fig. 8.52b, and it is clear that the

8.10 Clusters on tungsten surfaces

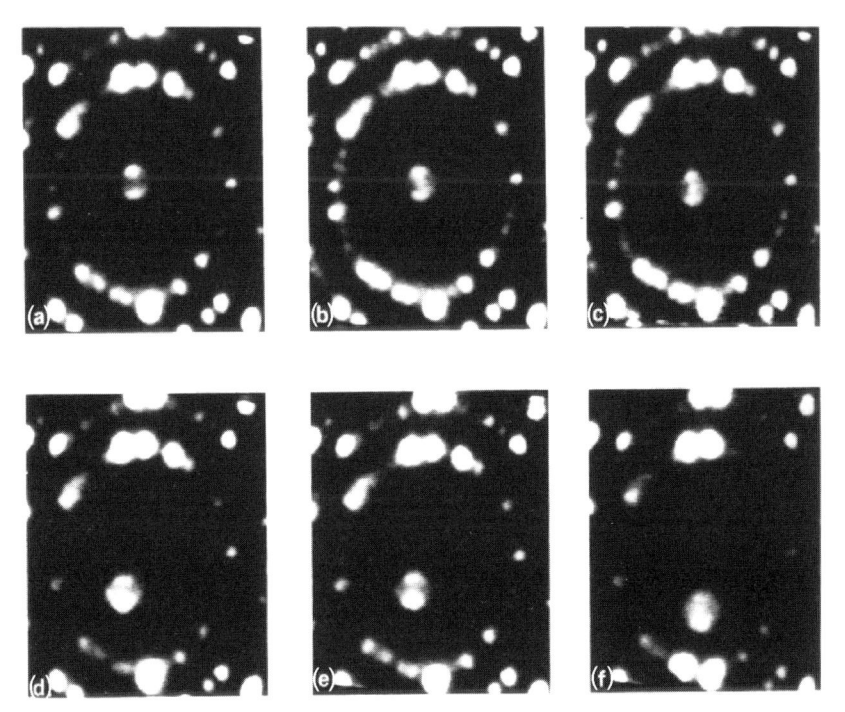

Fig. 8.51 Rhenium pentamer studied with FIM diffusing on W(110) plane during 60 sec interval at 413 K [93].

barrier $E_A(1,1)$ can be derived by measuring the temperature dependence of the waiting time. As shown in Fig. 8.52c, this was actually done at five temperatures, and yielded the barrier $E_A(1,1) = 0.48 \pm 0.02$ eV and a prefactor for the waiting time of 7.9×10^{-13} sec, quite a surprising result. The barrier for rotation is within the limits of error of that found in measurements for a single Pd atom diffusing on W(110), 0.509 ± 0.009 eV with a prefactor $D_o = 4.25(\times 1.7^{\pm 1}) \times 10^{-4}$ cm^2/sec [76,95,96]. Interactions with the neighboring rhenium atom have had only quite a minor effect on the activation energy of movement, but the hetero-dimer does not dissociate at this temperature range.

Careful FIM measurements of the spacing of Ir dimers on W(110) were made by Chambers [97] in 1991. By a detailed site mapping technique he was able to establish something quite important: the dimer atoms occupied nearest-neighbor sites. Changes in the shape of palladium and iridium clusters on W(110) were investigated by Koh [98] in terms of interactions rather than diffusion, and will be described in detail in Chapter 10. Here we only mention that chains of 9 Pd atoms are metastable, clusters made from 8 atoms change from a linear to a compact shape at 242 K. Cluster of 7 and fewer atoms are always seen in the linear configuration at 242 K, but for 9 and more atoms only a compact, two-dimensional shape is observed. The situations changes a bit with Ir, for which linear clusters of 15 atoms and longer are thermodynamically stable.

Clusters of palladium have also been examined by Fu *et al.* [99] on W(110). In keeping with previous work they again find that linear chains are preferred from dimers to Pd$_8$, where two-dimensional clusters become dominant, as shown in Fig. 8.53. Estimates of

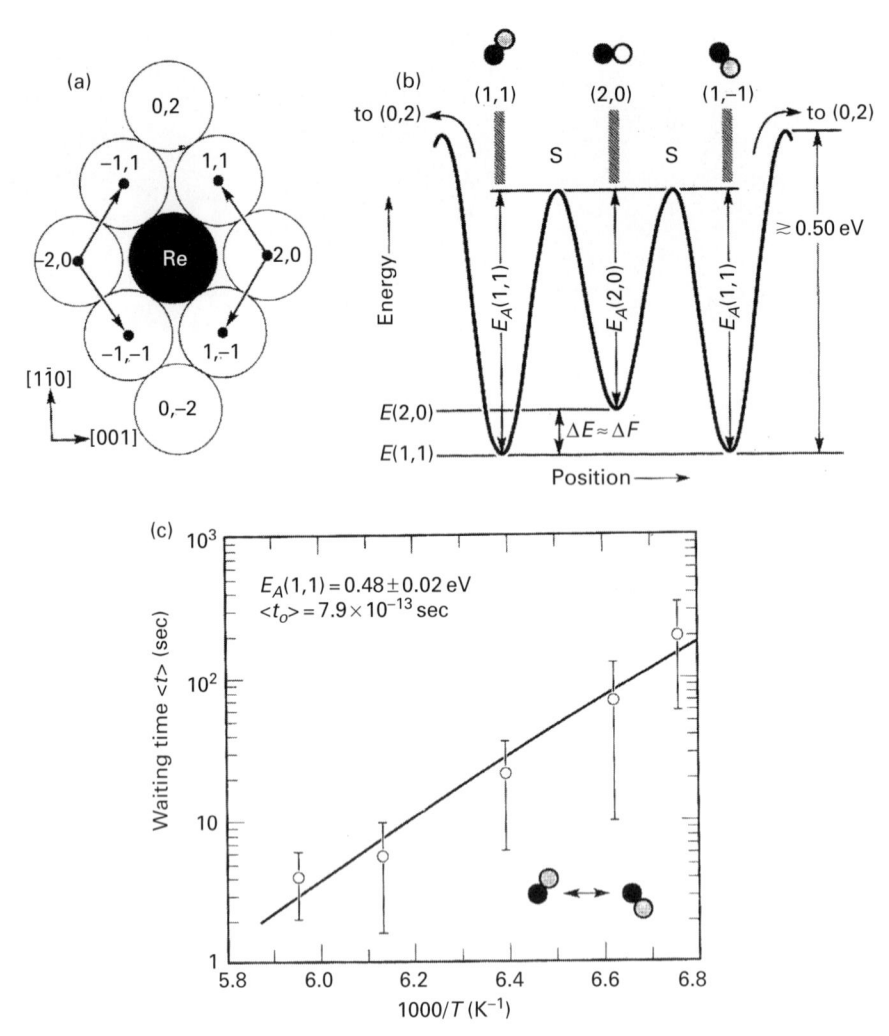

Fig. 8.52 (a) Path of palladium atom rotating around rhenium nearest neighbor on W(110) plane studied in FIM at a low temperature, ~160 K. (b) Potential energy schematic for Pd atom rotating around Re, as indicated in (a). (c) Arrhenius plot for Pd atom rotation around rhenium neighbor on W(110) [94].

the diffusion barrier were made, assuming a prefactor for the diffusivity of 10^{-3} cm^2/sec, and are plotted as a function of size in Fig. 8.54. The height of the barrier increased with size, as usually found. Movement appears to occur by a cluster atom hopping to a nearest-neighbor site along the <111> directions, followed subsequently by other atoms.

8.10.3 Experiments: tungsten(111)

Diffusion of tungsten clusters have recently been examined in the FIM on the W(111) plane by Fu *et al.* [100]. Results for different barriers, estimated assuming a prefactor

8.10 Clusters on tungsten surfaces

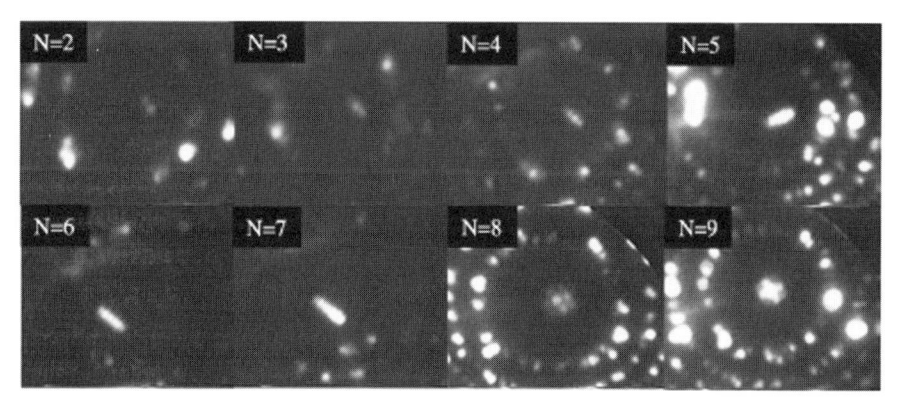

Fig. 8.53 Stable configurations of palladium clusters observed in FIM on W(110) plane (after Fu *et al.* [99]).

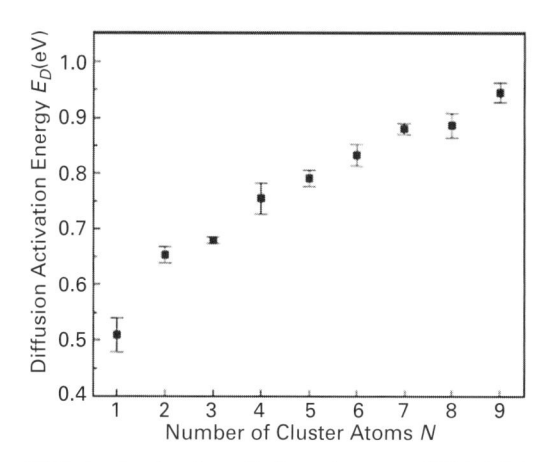

Fig. 8.54 Diffusion barriers for palladium clusters on W(110) (after Fu *et al.* [99]).

$D_o = 10^{-3}$ cm^2/sec, are plotted in Fig. 8.55. The barrier drops significantly from adatoms to dimers, than increases till tetramers, but than again undergoes a slight dip. For clusters of nine or more the temperature to induce diffusion is high enough to bring about dissociation of atoms from the edges of the underlying (111) plane. The barrier for nanomers is about the same as for single atoms. It must be noted, however, that in diffusion the clusters come close to the edges of the plane, which may introduce some uncertainty into these estimates.

8.10.4 Theoretical investigations

It is of interest to note that theoretical estimates of the surface diffusivity of tungsten dimers have been made by Xu and Adams [101] for the (110) and (211) planes of tungsten, using a fourth-moment approximation to tight-binding theory. On W(110),

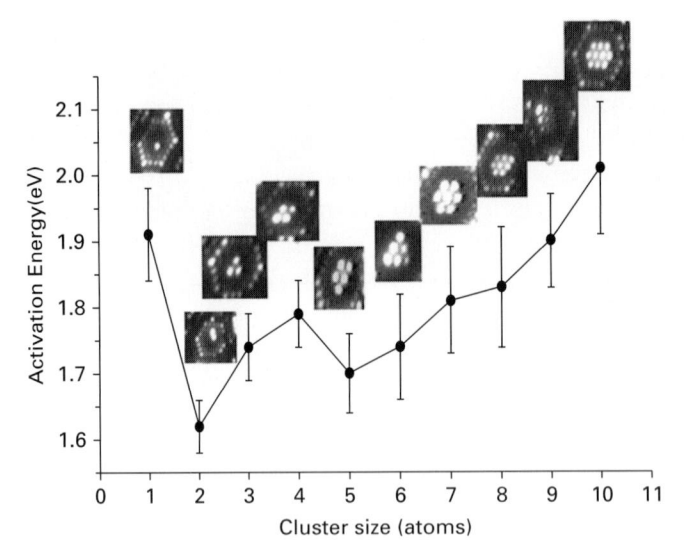

Fig. 8.55 Dependence of diffusion barrier for tungsten clusters observed in FIM on the cluster size on W(111) (after Fu et al. [100]).

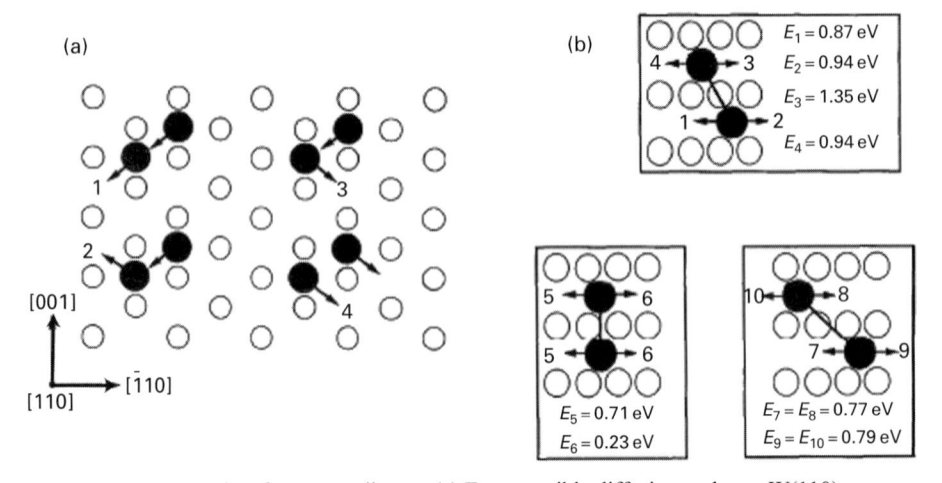

Fig. 8.56 Possible diffusion paths of tungsten dimers. (a) Four possible diffusion paths on W(110). (b) Energy barriers for movement of cross-channel dimers on W(211) (after Xu and Adams [101]).

four different paths for diffusion, shown in Fig. 8.56a, were considered. Motion along path 1, along the <111> direction, occurred most rapidly, with an activation energy of 1.38 eV and a prefactor $D_o = 9.83 \times 10^{-4}$ cm^2/sec. Path 2 required overcoming a barrier of 1.76 eV, with $D_o = 1.06 \times 10^{-3}$ cm^2/sec, while path 3 required 1.89 eV and $D_o = 1.10 \times 10^{-3}$ cm^2/sec. They did not observe Path 4 in the simulations. By forcing such a movement, they obtained a very high energy of 3.63 eV. Here, when one atom moves, the second one follows; that is, diffusion occurs by concerted jumps in the close-packed <111> direction. This is also what happens for a dimer with both atoms in the same

8.10 Clusters on tungsten surfaces

channel of the (211) plane of tungsten; both atoms move, maintaining a constant bond length, but the calculated activation energy is very high, 1.98 eV and the prefactor for the diffusivity was $D_o = 1.84 \times 10^{-3}$ cm²/sec. In adjacent channels, the two atoms of the dimer jumped with energy barriers indicated in Fig. 8.56b; these data indicate that the staggered configuration is most stable, contrary to what was seen in experiments. They also observed a huge influence of the anisotropy of the surface on the energetics of movement, not observed in experiments. An atom jumping out from the straight configuration to the right is much less demanding than to the left. Xu and Adams examined a number of paths for movement between stable configurations. The path with the lowest energy included individual jumps 1 and 6 in Fig. 8.56b and required 0.87 eV, with a prefactor for the diffusivity of 1.34×10^{-3} cm²/sec. However, this path will result only in local motion of the dimer; for longer motion, a slightly higher energy is needed. What is very surprising is that the energy barrier to diffusion is higher than that for dissociation, which they assumed happened at a bond length of 7.07 Å and is illustrated in Fig. 8.46b. These results are independent of the estimates of the diffusion barriers, which are readily obtained from experiments, and are just slightly lower. Further studies are clearly needed.

The diffusivity parameters of tungsten clusters, from dimers to nanomers, on W(110) have been evaluated by Chen et al. [102] using a modified analytical embedded atom method [31,32] to calculate the mean-square displacement. As shown in Fig. 8.57, clusters are all in the island configuration with a higher binding energy than chains. From Arhhenius plots of the diffusivity, Chen et al. derived the prefactors and activation energies listed in Table 8.11. What is of interest here is the oscillation of the activation energy as the size of the cluster increases, with local maxima for four, six and eight atoms; the prefactor shows similarly values of 10^{-1} cm²/sec. The authors point out that the dimer shearing mechanism participates in the diffusion of both hexamer and heptamer, but involves less energy for the heptamer as it occurs at the periphery. For clusters of up to five atoms, dimer shearing is not present and movement proceeded by successive hopping along the edges. Clusters of nine atoms frequently change their shape due to peripheral movement of two atoms, while for eight atom clusters the change of shape is not frequent.

Table 8.11 Diffusion characteristics of tungsten clusters calculated with MAEAM on W(110) [102].

Cluster size (atoms)	Activation energy E_D(eV)	Prefactor for diffusivity D_o (cm²/sec)
1	0.89	6.92×10^{-3}
2	0.81	1.18×10^{-3}
3	0.98	6.32×10^{-4}
4	1.87	2.06×10^{-1}
5	1.21	3.64×10^{-4}
6	2.06	1.83×10^{-1}
7	1.69	9.41×10^{-4}
8	2.51	7.64×10^{-1}
9	1.32	2.40×10^{-4}

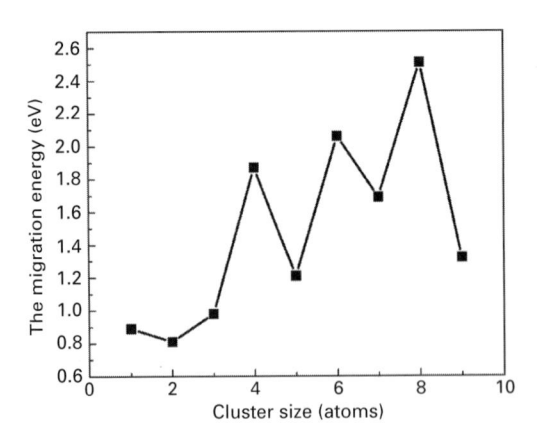

Fig. 8.57 Diffusion barrier of tungsten clusters on W(110) as a function of size calculated using MAEAM (after Chen et al. [102]).

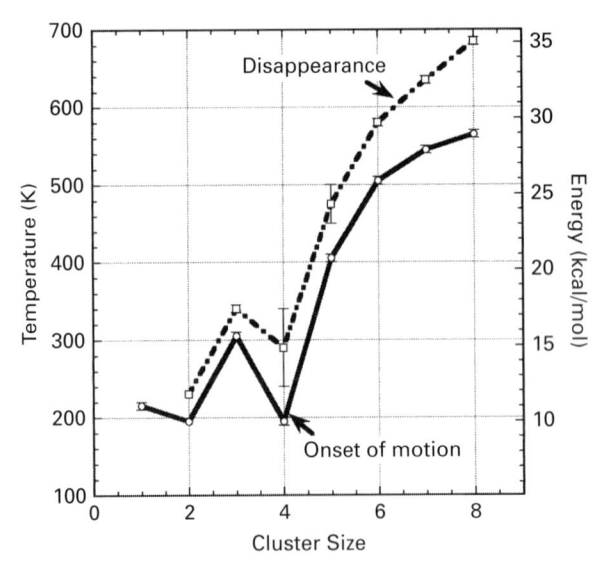

Fig. 8.58 FIM observations of onset temperature and activation energy for cluster diffusion and disappearance on Re(0001) [103].

8.11 Clusters on rhenium surfaces

Cluster diffusion on quite a different metal surface, Re(0001), was looked at briefly, using field ion microscopy, by Goldstein [103]. The onset of rhenium cluster diffusion was measured quantitatively, and an activation energy was estimated by assuming a prefactor like that for adatom diffusion on the same surface. The movement of dimers was faster than of monomers, and their dissociation proceeded at temperatures comparable with that of single atom motion. The results are shown in Fig. 8.58; they are quite similar to what has been observed for iridium clusters on Ir(111); after rhenium trimers, the diffusion barrier drops suddenly for tetramers and is equal to that for dimers. Thereafter, as the cluster size increases, so does the activation energy. After cluster dissociation, the products are rarely seen on the Re(0001) plane, except for dimers, due to their much higher mobility.

8.12   Clusters on iridium surfaces

8.12.1   Experiments: Ir(100)

Schwoebel and Kellogg [104] found that on Ir(100), clusters of iridium form chains along the [110] direction if the number of atoms was five or less. With six atoms in a cluster, a two-dimensional arrangement turned out to be stable, as illustrated in Fig. 8.59. Mobility of clusters, however, was not investigated. Further investigations on clusters on the (100) plane of iridium were done by Chen and Tsong [105]. At low temperatures, iridium trimers and tetramers were linear, oriented along [110], but were able to transform to a two-dimensional structure at higher temperatures. Hexamers existed in both forms, and a cluster of nine atoms was observed one-dimensional at an elevated temperature. The ratio of linear to two-dimensional structures for iridium trimers was measured over a range of temperatures shown in Fig. 8.60, and the logarithm of their ratio was found to be made up of two intersecting straight lines, with quite different energetics; the significance of this is not clear. Some caution in interpreting these results is in order, however. For diffusion of single iridium adatoms on Ir(100), Chen and Tsong report a diffusion barrier of 0.93 ± 0.04 eV and a prefactor $D_o = 1.4(\times 10^{\pm 1}) \times 10^{-2}$ cm^2/sec. A few months later they found a significantly lower value, a barrier of 0.84 ± 0.05 eV and a higher prefactor, $D_o = 6.26(\times 11^{\pm 1}) \times 10^{-2}$ cm^2/sec [106].

A surprising complex forms when a single rhenium adatom is deposited on a (100) plane of iridium [107], as in Fig. 8.61. After depositing a rhenium atom and warming to 240 K, the atom pulls one of the iridium atoms out of the lattice, and forms a Re-Ir dimer on the surface, with the center above an empty lattice site. Here it should be noted that Re atoms deposited on the surface had a circular image, while a Re-Ir complex showed an elongated shape, making direct recognition of the type of dimer possible. This dimer can

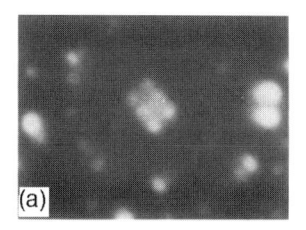

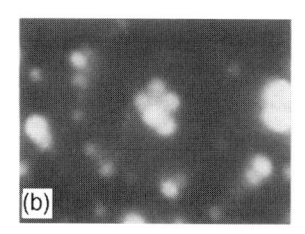

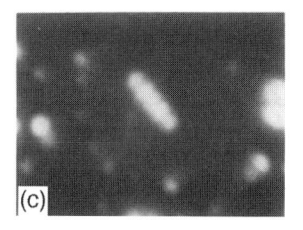

Fig. 8.59   FIM images of iridium clusters on Ir(100). (a) Iridium hexamer equilibrated at 460 K. (b) One atom has been field evaporated from hexamer at 77 K. (c) After equilibration at 450 K, linear pentamer has formed (after Schwoebel and Kellogg [104]).

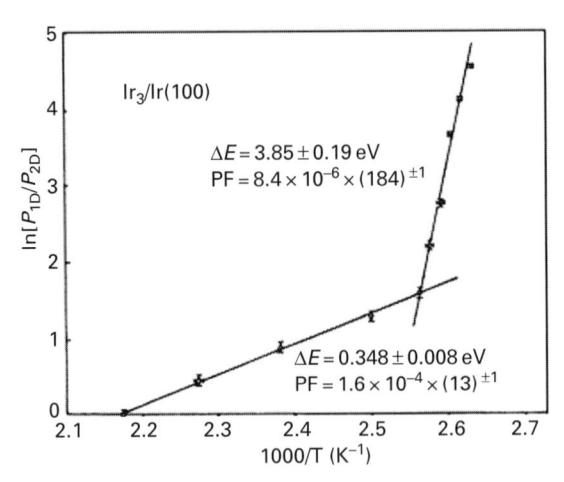

Fig. 8.60 Dependence of the ratio of one- to two-dimensional arrangements of Ir$_3$ observed on Ir(100) in FIM upon reciprocal temperature (after Chen and Tsong [105]). PF = pretactor.

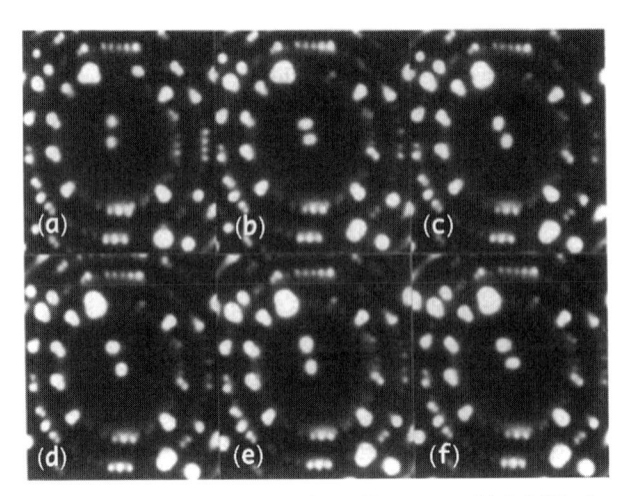

Fig. 8.61 Observations in FIM of interaction of Re atoms with Ir(100) plane. (a) Two Re atoms deposited on Ir(100) surface. On heating to 240 K, two Re-Ir complexes form and diffuse in (c)–(f) (after Chen and Tsong [107]).

move over the surface in two stages. Above 210 K, the iridium atom of the complex can exchange with an iridium atom from the lattice, allowing for local movement over four nearest-neighbor sites, shown in Fig. 8.62a. In this way the hetero-dimer orientation changes 90 degrees during movement. Above ~235 K, the Re-Ir dimer can move further away leaving behind a vacancy at the surface, as in Fig. 8.62b. The two different paths, labeled α_{Re} for short-range (local) movement and β_{Re} for long-range movement, have different characteristics, as indicated in Fig. 8.62c. In the α_{Re} mode, the activation energy amounted to 0.60 ± 0.07 eV with a prefactor $D_o = 4 \times 10^{-4.0 \pm 1.5}$ cm^2/sec; for β_{Re} transitions

8.12 Clusters on iridium surfaces

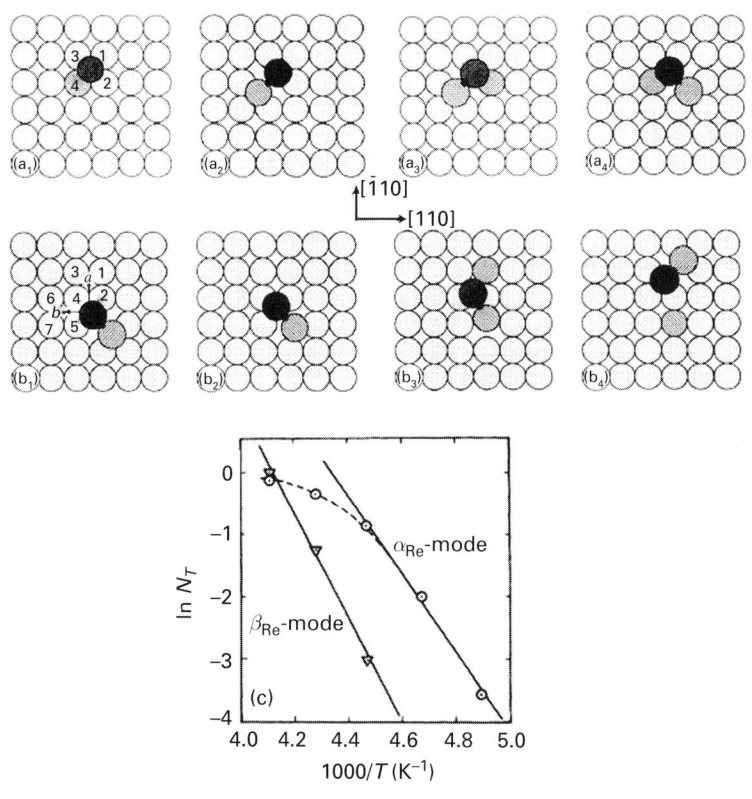

Fig. 8.62 Re-Ir dimers on Ir(100) studied in FIM. (a) Mechanism of α_{Re} diffusion of Re-Ir complex on Ir(100). (a$_1$) Re atom in position to displace Ir atoms 1–4. (a$_2$) Ir atom 4 is moved out of its site. (a$_3$)–(a$_4$) Center of complex moves to right through exchange with atom 2. In this mode, only diffusion in one unit cell of substrate can take place. (b) Mechanism of β_{Re} diffusion of Re-Ir on Ir(100). (b$_1$) Re atom can move to site a or b. (b$_2$)–(b$_4$) Example of motion when Re atom moved to a. Atom 1 is displaced from substrate, initiating β_{Re} step. (c) Arrhenius plots for α_{Re} and β_{Re} diffusion modes of Re-Ir complex on Ir(100). For α_{Re} the activation energy was 0.60 ± 0.07 eV, for β_{Re} 0.73 ± 0.07 eV (after Chen and Tsong [107]).

the barrier is 0.73 ± 0.07 eV and the prefactor D_o amounted to $2 \times 10^{-2.0 \pm 1.5}$ cm^2/sec. At temperatures above 280 K, the complex may dissociate, and the rhenium atom combines with the vacant site, leaving the iridium atom to diffuse by an exchange process.

Not much work has been done in the past on cluster diffusion on Ir(100) plane. However, this lack was rectified by Fu and Tsong [108] who studied iridium and rhodium dimers and trimers. Diffusion characteristics were derived from Arrhenius plots in Fig. 8.63a. For self-diffusion of iridium dimers, the diffusion barrier was 0.88 ± 0.05 eV with a prefactor of $1.42 \times 10^{-3 \pm 0.8}$ cm^2/sec; for the rhodium dimer these values were similar, $E_D = 0.83 \pm 0.06$ eV and $D_o = 3.53 \times 10^{-4 \pm 1.1}$ cm^2/sec. Rhodium dimers diffuse more rapidly than iridium dimers, which is opposite to the behavior of single adatoms. Rhodium dimers also show fewer orientation changes than Ir dimers. Diffusion by translation of Ir dimers was rarely observed below 300 K, but became more frequent above this temperature.

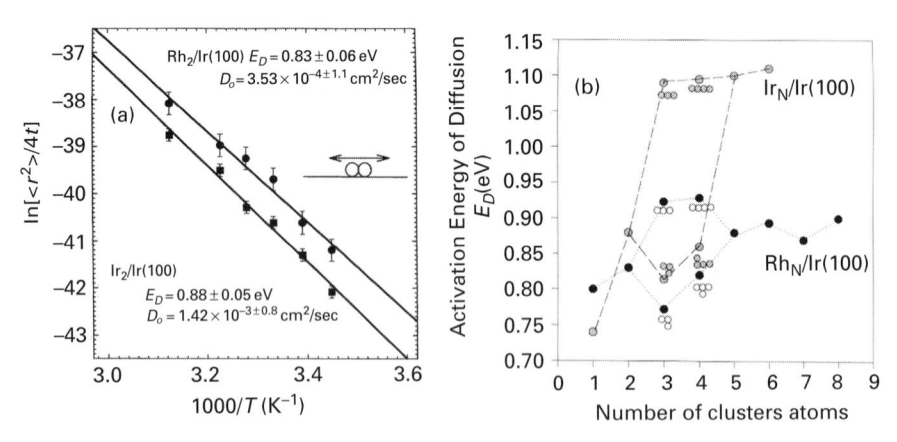

Fig. 8.63 Rh and Ir clusters examined in FIM on Ir(100). (a) Dependence of Rh$_2$ and Ir$_2$ diffusivity upon reciprocal temperature. (b) Dependence of activation energy on the size of clusters (after Fu and Tsong [108]).

Here it must be kept in mind that on Ir(100), diffusion of single iridium atoms occurs by atom exchange, a generally low-energy process, and single rhodium atoms diffuse by atom jumps. An unequivocal decision about the jump mechanism for diffusion of iridium dimers has not been achieved. However, atom exchange is probable in diffusion of the iridium dimer. The displacement distribution has been fitted on the assumption that two elemental processes can contribute: dimer atom exchange, as well as an atom making two jumps to change orientation. Good agreement was found up to 305 K. At higher temperatures, the fit worsens, and other mechanisms must also participate. For rhodium dimers the situation is simple – atom exchange cannot occur. After the first exchange, iridium atoms drawn from the lattice would continue diffusion and this would cause a change in the diffusion characteristics of the dimer, which is *not* observed. Two-dimensional iridium trimers diffuse much the same as two-dimensional rhodium trimers, and the conclusion therefore is that the motion in both arises from atom jumps without exchange. For the rhodium trimer the activation energy is lower than for single atom movement, which might be explained by the saddle point being coordinated by two other atoms, which cause that atom to hop over the two. Rough estimates of the diffusion barrier, assuming a standard prefactor for the diffusivity, for Rh$_4$ through Rh$_8$ and Ir$_4$ through Ir$_6$, were also made and show the expected increase with size, given in Fig. 8.63b. Fu and Tsong analyzed the movement of differently shaped clusters separately, but this separation may lead to confusion.

Iridium and rhodium tetramers were separately examined on Ir(100) but only the atomic transformations were considered [109]. The iridium tetramer usually exists as a linear chain. At temperatures around 290 K, movement involves a change in shape, during which one atom diffuses along a three atom chain. At a temperature of ~330 K, a ledge atom diffuses around a corner and attaches itself to the end. Diffusion at temperature higher than 370 K is presumed to involve the displacement of an end atom, followed by subsequent displacements of the others. For Rh$_4$ diffusion occurs in a much more complicated fashion. The most stable structure is also a chain of atoms, as in

8.12 Clusters on iridium surfaces

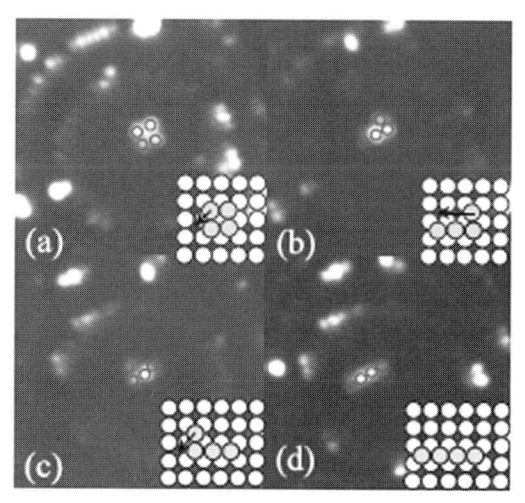

Fig. 8.64 FIM images of Rh$_4$ diffusing on Ir(100) at 350 K. Possible atom movement indicated by arrow (after Fu and Tsong [109]).

Fig. 8.64d, but other arrangements do not differ much in energy. For long-range diffusion, the structures shown in Fig. 8.64 are envisioned, which allow translation as well as rotation, in fact observed. The activation energy for Ir tetramers was 1.09 eV, and 0.93 eV for rhodium tetramers. No observations were made of tetramers shearing.

8.12.2 Experiments: Ir(111)

The first study of cluster movement on a close-packed surface, the (111) plane of iridium, was made by Wang [26,110]. The atomic arrangement of the clusters is quite interesting. With the field ion microscope it was possible to map the binding sites for atoms on the (111) surface, which fall into two categories: fcc sites, which are the locations at which atoms in the crystal are bound, and hcp sites, both shown in Fig. 3.3. Single iridium atoms prefer to sit in hcp sites, and the same sites seem to be favored for dimers, in which the atoms are located in sites adjacent to each other, as in Fig. 8.65. Iridium trimers have been observed either in linear or triangular configurations, with the latter just slightly favored. The atoms can be either in fcc or hcp locations, and the different positions of triangular trimers are sketched in Fig. 8.66. Type *A* configurations, in which the center is above a lattice atom are seen in only ~20% of observations. Linear trimers have a free energy less than 0.004 eV higher than the triangular one. Tetramers are always observed in the tetragonal form and can exist in six different configurations, with fcc sites strongly favored compared to hcp sites. With larger clusters, atoms are usually in fcc sites, marking a significant change. Schematics of the bigger islands are given in Fig. 8.67. It is clear that as the size increases, clusters tend to occupy fcc sites and continue normal crystal growth. Linear clusters are observed only for trimers.

624   Diffusivities of small clusters

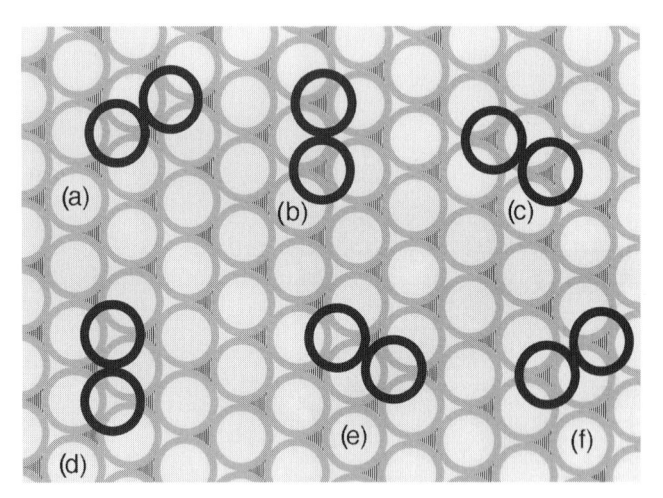

Fig. 8.65   Ir dimer positions on Ir(111) determined in FIM after diffusion at 180 K [26].

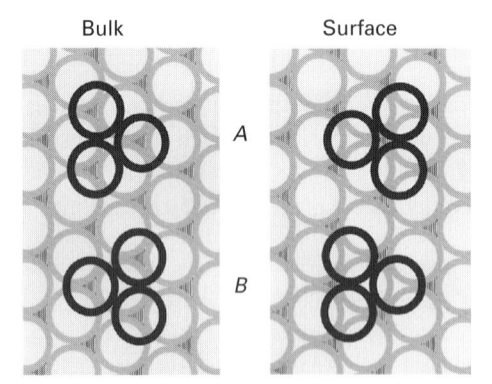

Fig. 8.66   Configurations of triangular Ir clusters observed on Ir(111) plane in FIM [26]. *A*-type clusters have center above binding site, *B*-type center is above surface atom.

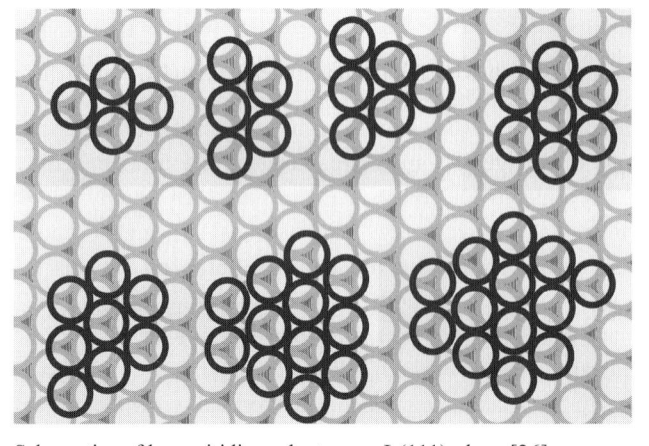

Fig. 8.67   Schematics of larger iridium clusters on Ir(111) plane [26].

8.12 Clusters on iridium surfaces

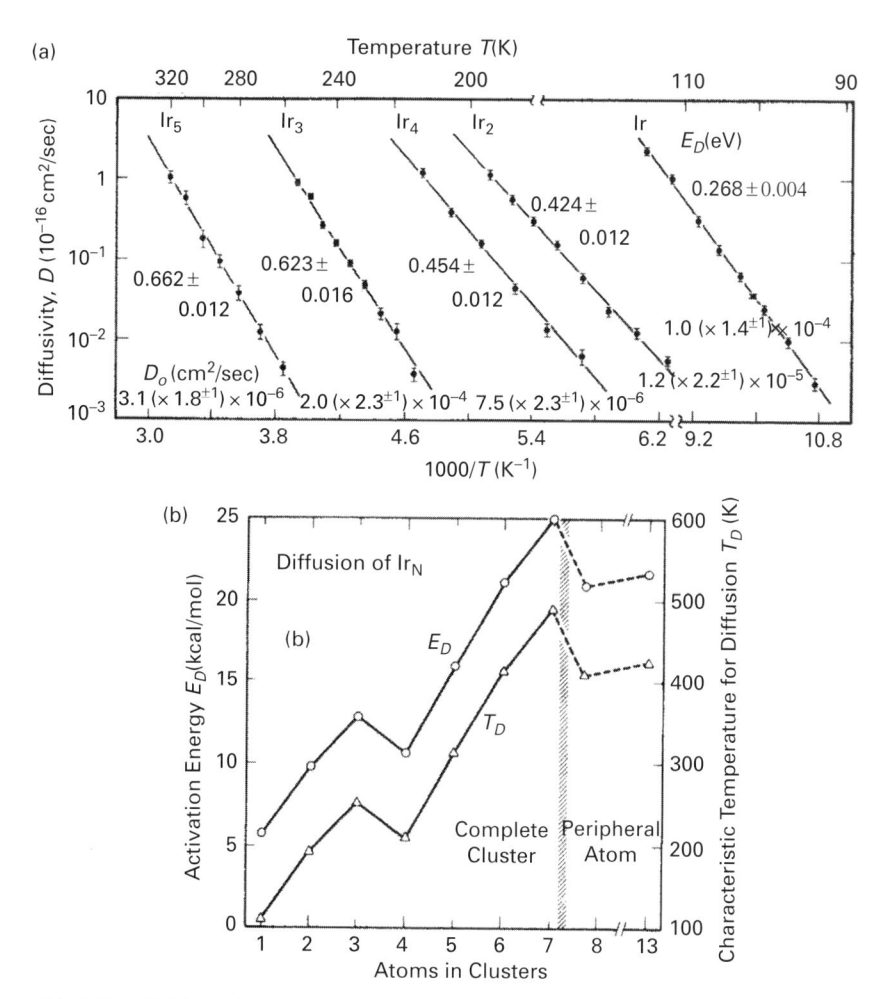

Fig. 8.68 Diffusivity of iridium clusters on Ir(111) surface studied in FIM [26]. (a) Arrhenius plots for the diffusion of iridium clusters on Ir(111). (b) Characteristic temperature, at which diffusivity is 0.7×10^{-16} cm^2/sec for iridium clusters on Ir(111) plane.

Diffusion of these clusters has been looked at in detail up to Ir$_5$, and Arrhenius plots are given in Fig. 8.68a. What is especially interesting is that there is a large increase in the diffusion barrier for the center of mass in going from Ir$_2$ to Ir$_3$, but this drops for the tetramer, and then increases again strongly for Ir$_5$. The prefactors are lower than usual for single atom diffusion, but not low enough to cause concern. For larger clusters, only limited observations were made of diffusion, but the temperature at which the mean-square displacement is 0.7 Å^2/sec is plotted in Fig. 8.68b, together with a crude estimate of the diffusion barrier. The points for clusters with 8 and 13 atoms refer to motion of the added atom around the core of the central cluster.

Changes in the atomic arrangement have been observed for many of the clusters, but only a few are shown. The motion of the dimer, illustrated by schematics in Fig. 8.69, can be divided into two categories: it can move in a cell of six sites around a given atom of the

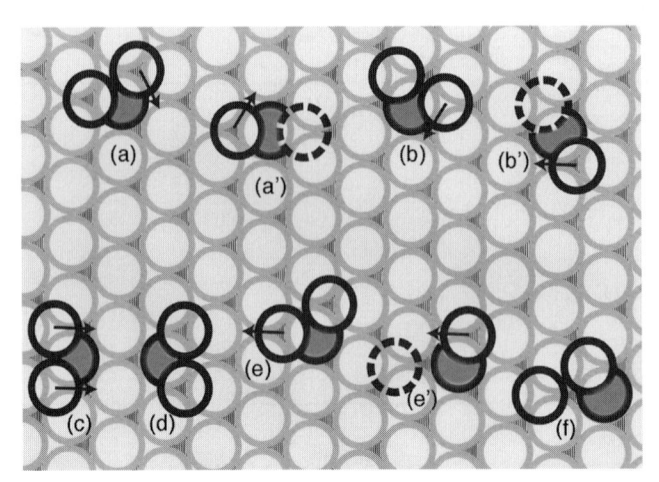

Fig. 8.69 Schematic showing intra- and inter-cell jump of Ir dimer examined in FIM on Ir(111) [26]. Intra-cell movement: (a)→(d). Inter-cell movement: (e)→(f). Primes indicate transition state, not observed.

lattice, but an inter-cell jump is also possible. In 372 observations after 15 sec heating periods at 168 K, Wang found 273 in which the position of the dimer did not change, 22 translations as well as 66 rotations localized in one cell, and 11 inter-cell movements. If movement proceeded by concerted dimer motion, then translation should be more frequent than rotations; that is why atom-by-atom movement was suggested.

The different transition possibilities for trimers, assuming motion by single nearest-neighbor jumps, are shown in Fig. 8.70. Movement of trimers involves transformations between triangular and linear configurations. In 420 observations after heating to 235 K for 10 sec, changes were found 144 times. Translation was observed 58 times while rotation 14; inter-conversion between triangular and linear forms was seen 72 times. During observations at this temperature, the trimer did not undergo long-range diffusion. At higher temperatures movement was observed over longer distances, with interchanges between linear and triangular configurations. Complicated trimer movements suggest that individual jumps were involved in diffusion.

Tetramers are most interesting and changes in the arrangement of the cluster atoms observed in diffusion are indicated in Fig. 8.71. In the shear-type transition shown there, motion may of course occur for both atoms at the same time, making movement easier, but that is not revealed directly by the observations. The tetramer in the metastable configuration was observed experimentally and is shown in Fig. 8.71f. Since tetramers were seen in surface sites in only 15% of the observations, one can expect that translation will be less likely than rotation. In fact the rate of rotation compared to translation is 3 to 1. The higher ratio of rotation to translation is also expected for the shearing motion. For pentamers only two atoms have to move to accomplish rotation, while translation requires movement of all atoms and additionally the cluster is in an unfavorable hcp state, from which it has to move again to an fcc site. Translation is expected only at high temperatures. In 255 observations at 260 K, rotation was observed 109 times, while translations never.

For heptamers, a change in the shape was never observed at 460 K. Movement of an atom around the heptamer core was observed for clusters made of eight atoms, with

8.12 Clusters on iridium surfaces

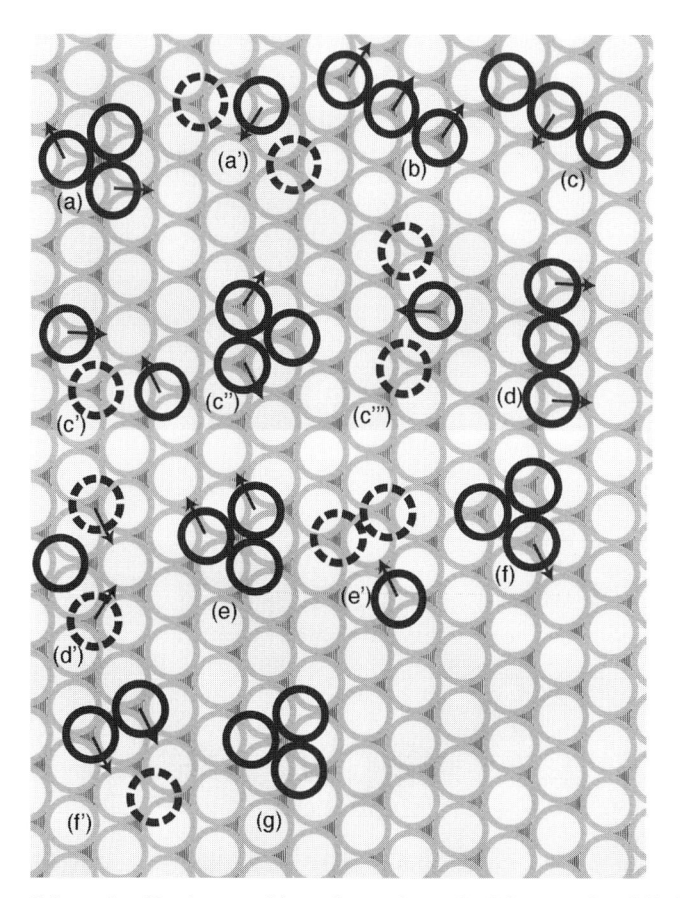

Fig. 8.70 Schematic of Ir trimer positions observed on Ir(111) in FIM after diffusion for 10 sec at 235 K [26]. Primes indicate transition state which was not observed. (e)→(g) necessary for long-range diffusion.

detachment of the eighth atom at 480 K, while dissociation of the heptamer started at 510 K. A similar situation was observed for Ir_{13}; the peripheral atom moved around the stationary Ir_{12} core at a temperature of 420 K and detached at 500 K, leaving the core unchanged. Possible atom displacements have been given for clusters up to Ir_{13}, and it must be remembered that the indicated transitions are mostly based on the assumption of single atom jumps, but shearing motions were also detected for tetramers and pentamers.

At roughly the same time, the diffusion of clusters on Ir(111) was also investigated, but in less detail, by Chen and Tsong [105]. No attempt to relate the atomic arrangement of the cluster to the location of the atoms of the surface layer was made, but temperatures for the onset of diffusion were measured and are shown in Fig. 8.72. The shape of this curve is much the same as measured by Wang in a previous study, but the diffusion temperatures are significantly lower. They also looked at the ratio of linear to compact two-dimensional trimers and found a linear dependence of the logarithm of this ratio plotted against reciprocal temperature, with a slope of -0.098 ± 0.004 eV and a prefactor equal to $1.1(\times 13^{\pm 1}) \times 10^3$.

Table 8.12 Diffusion characteristics of iridium clusters on Ir(111) [29].

Cluster	D_o (cm^2/sec)	E_D (eV)
Ir	$3.8(\times 1.4^{\pm 1}) \times 10^{-4}$	0.290 ± 0.003
Ir$_2$	$2.6(\times 2.2^{\pm 1}) \times 10^{-5}$	0.45 ± 0.01
Ir$_3$	$4.2(\times 2.4^{\pm 1}) \times 10^{-4}$	0.65 ± 0.02
Ir$_4$	$1.5(\times 2.3^{\pm 1}) \times 10^{-5}$	0.48 ± 0.01
Ir$_5$	$6.0(\times 1.9^{\pm 1}) \times 10^{-6}$	0.69 ± 0.02
Ir$_6$	$\sim 1 \times 10^{-5}$	~ 0.92
Ir$_7$	$1.4(\times 2.4^{\pm 1})$	1.49 ± 0.03

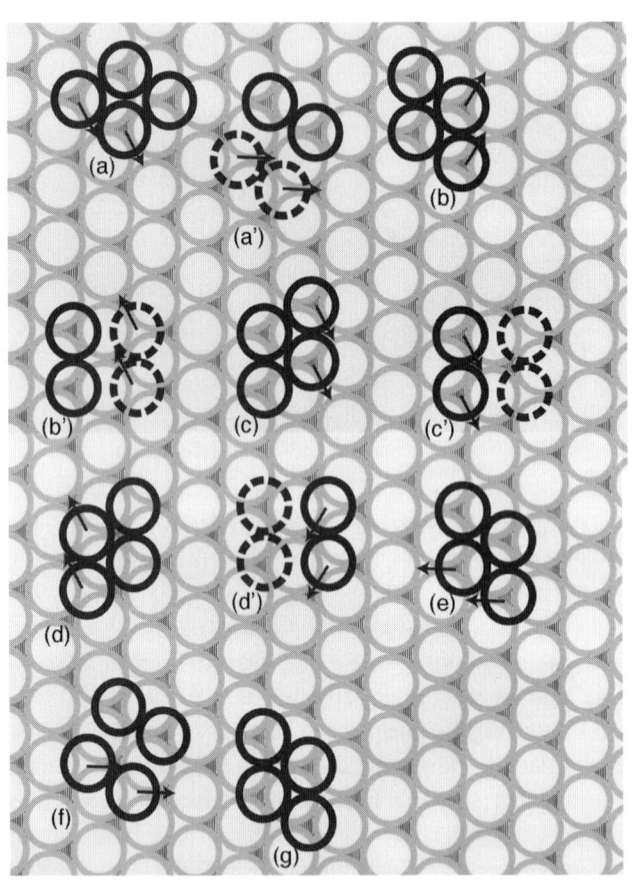

Fig. 8.71 Configurations of Ir$_4$ after 10 sec diffusion on Ir(111) at 184 K [26]. Intermediate configurations shown in primes were not actually observed, except for configuration (f). (c)–(d) shows pure translation.

At roughly this time, Wang [29] modified the resistance versus temperature calibration appropriate for his experiments on the diffusion of iridium clusters on Ir(111) and came up with more reliable values for the diffusion characteristics. These are listed in Table 8.12, and are also included in Fig. 8.73. It is worth noting that the barrier to diffusion of Ir$_4$ still has a significant local minimum probably associated with shearing motion, and not seen for Pt$_4$.

8.12 Clusters on iridium surfaces

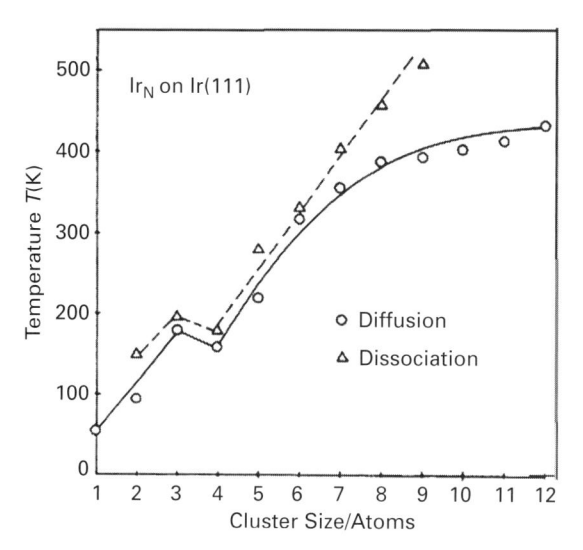

Fig. 8.72 Temperature for the onset of Ir cluster diffusion studied in FIM on Ir(111) plane (after Chen and Tsong [105]).

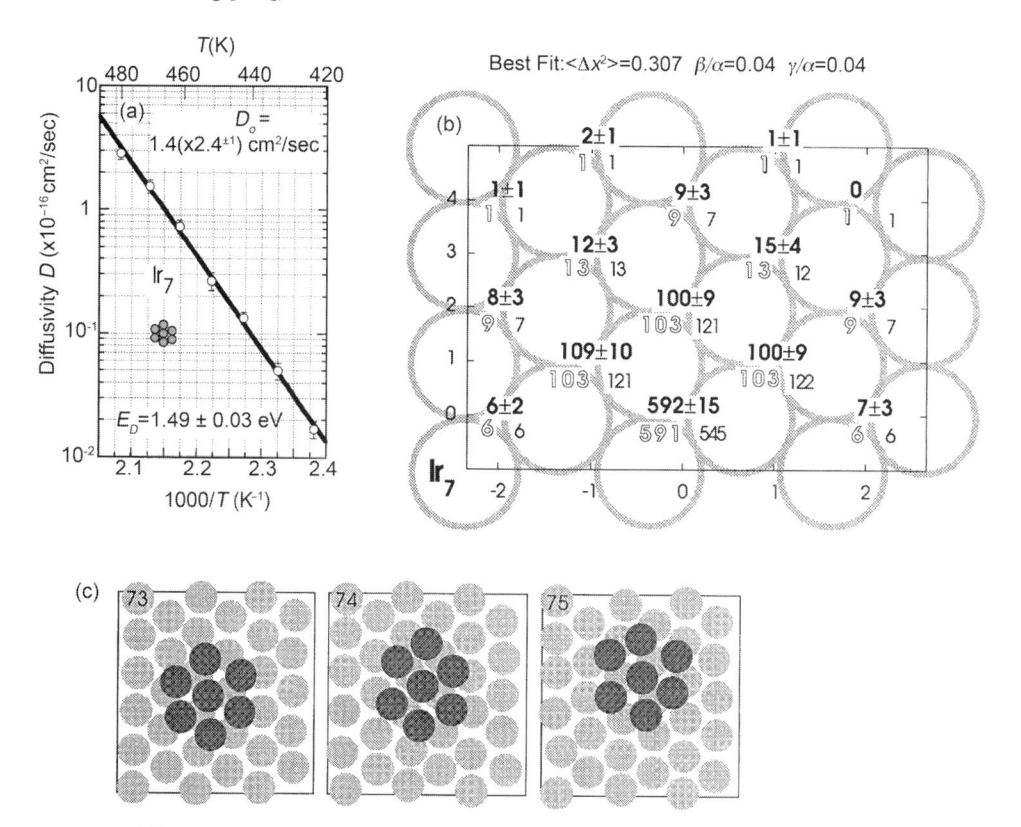

Fig. 8.73 Diffusion of iridium cluster on Ir(111) [111] examined in FIM. (a) Arrhenius plot for Ir$_7$. (b) Distribution of displacements. Experimental results in bold letters, best fit in outlined letters, normal lettering gives random distribution. (c) Molecular dynamics simulations show gliding to be the main diffusion mechanism.

Wang et al. [111] in 1998 looked at the diffusion of compact clusters on the Ir(111) surface. The heptamer moved, preserving its shape, with an activation energy of 1.49 ± 0.03 eV, and with a huge prefactor of 1.4(×2.4$^{\pm 1}$)cm^2/sec, as shown in Fig 8.73a. Such huge prefactors seem to be associated with diffusion of compact clusters on Ir(111). The distribution of displacements was also investigated, as indicated in Fig. 8.73b, and did not reveal a very significant contribution of long jumps in heptamer movement. Molecular dynamics studies indicated in Fig. 8.73c revealed gliding of the cluster as the main mechanism of movement. In 2000 Kürpick et al. [112] continued investigating the movement of Ir$_7$ clusters on Ir(111) with molecular dynamics simulations, applying the SC and RGL potentials to derive the activation energy for motion. For the static barrier for concerted displacements of cluster atoms from fcc to bridge sites a barrier of 1.23 eV and a prefactor $D_o = 3.9 \times 10^{-4}$ cm^2/sec were derived from SC and 1.51 eV and $D_o = 4.3 \times 10^{-4}$ cm^2/sec from RGL potentials. Push of two atoms of an A-type ((100) microfacet) edge required 1.27 eV and $D_o = 4.6 \times 10^{-4}$ cm^2/sec as derived from SC, and 1.47 eV and $D_o = 4.9 \times 10^{-4}$ cm^2/sec from RGL. When pulling two atoms of B-type edges the barrier was 1.47 eV and $D_o = 4.9 \times 10^{-4}$ cm^2/sec with RGL interactions. However, both potentials predicted a much lower prefactor for the diffusivity than observed, and Kürpick speculated that this might come from the large number of non-equivalent processes involved in movement.

8.12.3 Experiments: Ir(110)

Field ion microscopic measurements of the diffusivity of iridium dimers, Ir$_2$, were done on the Ir(110) surface by Tsong and Chen [113]. The surface was formed in the (1 × 1) configuration by field evaporation. Iridium atoms deposited on this surface carried out cross-channel transitions on warming and combined to form a cluster in one channel. It is interesting that cross-channel dimers are not stable. Adatoms probably move by exchange to the next channel, creating an in-channel dimer. Such a dimer can make three different types of transitions. It can diffuse in the channel, it can make a cross-channel movement along the <211> directions, or along <100>. The map of sites at which the center of the dimer was observed after diffusion was much the same as that for single iridium atom motion. The distribution of displacements was also measured, but with some difficulty, as the temperature required for diffusion was close to that for reconstruction of the (1 × 2) lattice. However, it was possible to obtain the characteristics for both in-channel and cross-channel dimer motion, as shown in Fig. 8.74. The barrier to cross-channel diffusion of the dimer was 1.18 ± 0.12 eV, with a prefactor of $2.6 \times 10^{-4.0 \pm 0.7}$ cm^2/sec; in-channel movement was faster, with an activation energy of 1.05 ± 0.14 eV and a prefactor of $3.7 \times 10^{-5.0 \pm 1.3}$ cm^2/sec. Tsong and Chen also speculated whether single atom exchange was responsible for dimer diffusion or if it occurred through concerted exchange of two atoms, but this analysis did not lead to a certain answer.

8.12 Clusters on iridium surfaces

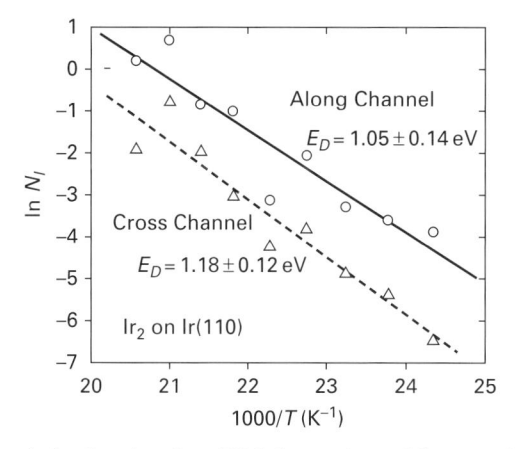

Fig. 8.74 Arrhenius plots from FIM observations of the rate of Ir$_2$ jumps along and across channels of the Ir(110) plane (after Tsong and Chen [113]).

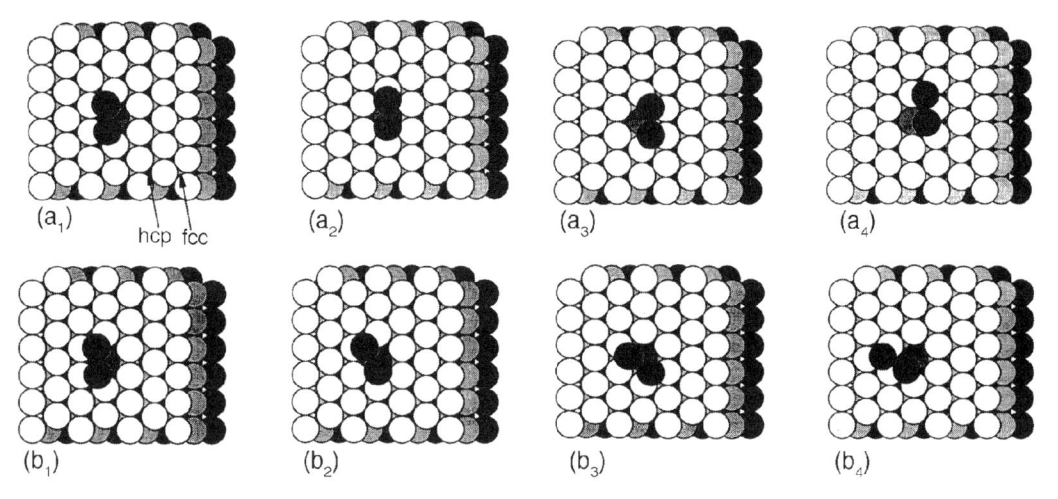

Fig. 8.75 RGL calculations of the diffusion of iridium dimer on Ir(111) surface. (a$_1$) → (a$_2$) → (a$_3$) constitutes intra-cell transition, (a$_3$) → (a$_4$) an inter-cell translation. (b$_1$) → (b$_2$) → (b$_3$) describes intra-cell rotation, (b$_3$) → (b$_4$) inter-cell rotation (after Shiang and Tsong [114]).

8.12.4 Theoretical investigations

Atom jumps made in diffusion of clusters are not always easy to ascertain, and molecular dynamics simulations have therefore been made by Shiang and Tsong [114] using the RGL potential, which gives generally good predictions for fcc metals. On the close-packed Ir(111) plane, dimer motion can take place as diagrammed in Fig. 8.75, by intra-cell movement a$_1$ to a$_2$ to a$_3$, or by inter-cell diffusion, a$_3$ to a$_4$. Alternatively, the dimer can rotate, where the change from b$_1$ to b$_2$ occurs intra-cell and b$_3$ to b$_4$ inter-cell. Here inter-cell movement means motion within a cell of six adsorption sites around a given substrate atom. For intra-cell translation the barrier was 0.341 eV, for intra-cell rotation 0.295 eV, and for inter-cell rotation 0.467 eV; the second mechanism should occur most frequently. The value for inter-cell motion agreed with the experiments by Wang [29], which yielded 0.45 eV.

On the (100) surface an adatom can either jump from one four fold site to an adjacent one over the intervening two-atom bridge, or else an exchange between one of the dimer's adatoms and an adjacent lattice atom can take place, with the latter moving to an adjacent four fold site. The latter dimer process involves overcoming a barrier of 1.078 eV, considerably smaller than other alternatives, and is most frequently observed, 39 times out of 40 diffusion events. The value of the activation energy is higher than that determined in experiments on this plane. On the (110) plane, dimer diffusion can also occur in two different ways, by motion in the channels, or by exchange with lattice atoms. The activation energy to in-channel hopping was estimated as 1.192 eV, and for exchange 1.268 eV. Both processes are therefore likely to coexist and in 40 diffusion events 12 were by exchange of both dimer atoms, 28 by in-channel hopping. The barriers are quite close to the experimental value of Tsong and Chen [113]. Somewhat similar results were found for iridium trimers: a barrier of 1.303 eV for in-channel diffusion and 1.371 eV for cross-channel motion by atom exchange.

Extensive calculations with EAM potentials were carried out by Chang *et al.* [27,115] for Ir clusters on Ir(100), (110), and (111). Dimer atoms were found to preferentially bond in nearest-neighbor positions on all three planes. Trimers proved more complicated. On (100) and (110) planes, trimers were found to prefer a one-dimensional linear arrangement; on (111), however, close-packed triangles had the lowest energy, in reasonable accord with experiments. A variety of possible diffusion paths were considered and are shown in Fig. 8.76a. On the (111) plane, the motion depicted has the lowest energy – one

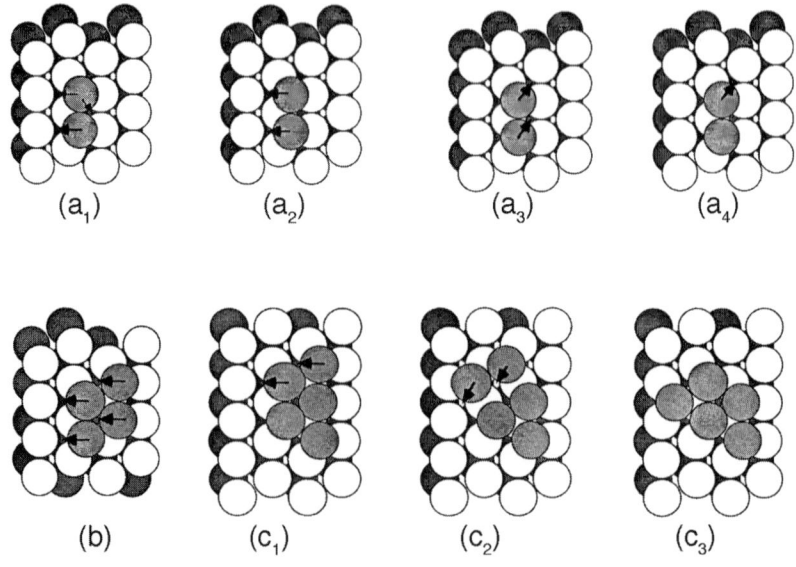

Fig. 8.76 Diffusion of Ir clusters on Ir(111) surface examined using EAM potentials. (a_1) Large arrow shows first transition, small arrow the second of dimer. (a_2) Local translation. (a_3) Inter-cell motion. (a_4) Forbidden transition. (b) Most favorable path for diffusion of Ir$_4$ on Ir(111). (c) Rotation of Ir$_5$ on Ir(111) (after Chang *et al.* [115]).

8.12 Clusters on iridium surfaces

atom of the dimer moves into an adjacent site over a barrier of 0.17 eV, and the other one follows with very little (0.02 eV) expenditure of energy. Movement of both atoms of the dimer as shown in Fig. 8.76a$_2$ requires 0.30 eV, hopping of both dimer atoms as in Fig. 8.76a$_3$ needed only 0.24 eV. Chang *et al.* did not observe any difference in energy of the dimers in fcc and hcp sites, so the barrier was not influenced by the nature of the sites. This observation, however, gives a much lower value of the activation energy than the 0.45 eV from experiment. One possibility to explain such a low value is that only local motion was taken into account.

On the (110) plane, a number of mechanisms were considered: in-channel jumps of both dimer atoms (1.25 eV), in-channel jumps of two individual dimer atoms (1.31 eV), exchange of individual dimer atoms (1.52 eV) and exchange of both dimer atoms (1.94 eV). The lowest energy mechanism turned out to be hopping along the channels, with both atoms jumping instead of only one at a time, over a barrier of 1.25 eV. These values are only slightly higher than the experimental determinations. Experiments also suggested the presence of both mechanisms, with hopping a bit more favorable than exchange. On the Ir(100) plane, Chang *et al.* considered a number of hopping and exchange processes, including hopping over a bridge by individual atoms and by concerted motion, as well as simple exchange by single atoms and concerted exchange by the dimer. Simple atom exchange by individual atoms which leads to rotation of the dimer was the preferred mechanism of diffusion on Ir(100), with an activation energy of 0.65 eV; all other investigated mechanisms had a much higher energy. The value of the activation energy for simple exchange of a dimer is lower than for single atom exchange, as well as the experimentally measured value for dimer movement, 0.88 eV, reported by Fu and Tsong [108].

Larger clusters were also considered on the Ir(111) surface. For the triangular trimer, the atoms essentially move together, with an energy barrier of 0.54 eV for movement of the front atom, which drags the two remaining atoms along, or a slightly higher value of 0.56 eV for movement involving the back atom pushing the remaining atoms. The values are in good agreement with the experimental barrier of 0.65 eV [29]. The experimental findings are of course somewhat more complicated, as the trimers, for example, exist equally likely in a linear as well as triangular form, but these estimates were important for pointing to the significance of coordinated jumps of the individual cluster atoms. For the tetramer, motion also occurs by concerted translation of all the atoms as in Fig. 8.76b over a barrier of 0.54 eV. Rotation of trimers and tetramers was not observed, opposite to the experimental findings. In experiments with tetramers, rotation was three times more frequent than translation. Similar to experimental rotation, illustrated in Fig. 8.76c, was the lowest energy path for pentamers; this required 0.83 eV, and involved concerted motion of only two atoms. Translation of the whole cluster was more energetically demanding, requiring 0.92 eV or 1.18 eV depending on whether the initial move was by one or two atoms.

Hamilton *et al.* [116] used first-principles calculations to look at diffusion of compact iridium clusters on Ir(111) and investigated new mechanisms, such as cartwheel shuffle and cartwheel glide. For Ir$_7$ clusters they showed that bridge gliding is the most likely mechanism of movement, with an energy of 1.54 eV, in very good agreement with the experimentally measured value of 1.49 eV.

In 2005 Chang and Wei [20] used *ab initio* density-functional calculations and the VASP [21,22] package with a (5 × 5) surface cell to investigate the movements of dimers. For iridium dimers they came out with a value for exchange lower than for hopping, a barrier of 1.30 eV compared with 1.87 eV for hopping. It should be noted that the barriers calculated for Ir_2 and Pt_2 are not in agreement with the experimental results, but the occurrence of exchange in dimer diffusion agrees with that in the diffusion of the adatoms.

Movement of Ir dimers has been widely investigated and all calculations are in agreement about the mechanism of dimer diffusion. On Ir(100), dimers move by exchange, on Ir(111) by hopping and on Ir(110) both by hopping and exchange. Calculations are less certain if movement of dimers is sequential or concerted. There is also quite a big scatter in the derived energetics. The activation energy for dimer movement on Ir(100) ranges from 0.65 to 1.3 eV, with the experimental value in the middle of this range, on Ir(111) from 0.24 to 0.47 eV, with the first figure close to the experimental value. On Ir(110) theoretically obtained values are slightly higher, but the relation between hopping and exchange seems to be similar to experiment. Only Chang *et al.* [27] looked at the energetics of bigger clusters on Ir(111), and came out with the same change in energetics with size as found in experiments.

8.13 Clusters on platinum surfaces

8.13.1 Experiments

Schwoebel *et al.* [117] using an FIM showed that on Pt(100), chains were the low-energy configuration for platinum trimers and also pentamers. For tetramers and hexamers, the two-dimensional islands were more stable thermodynamically. When a single atom was added to the hexamer, it diffused around the central cluster. Using embedded atom potentials the authors were able to get the same cluster configurations as in observations, provided the lattice was appropriately relaxed. Regrettably no indications were given about the diffusivity of these clusters

Estimates relying on EAM interactions were made by Kellogg and Voter [33] for the diffusion of Pt_2 and Pt_3 clusters over the Pt(100) surface, and were combined with results of observations in the field ion microscope. For single Pt atoms it was already known that diffusion occurred by atom exchange with the lattice. For dimers, either atom jumps or exchange are possible. From the displacement of the center of mass of dimers at 175 K an activation energy of 0.41 eV was deduced assuming a standard prefactor for the diffusivity. This value is lower than the activation energy for movement of monomers, 0.47 eV. Unfortunately the map of sites occupied does not provide answers about the mechanism of dimer movement. The authors relied on EAM investigation to find which mechanism is operating in diffusion. From EAM it is clear that the activation energy for motion by atom exchange, 0.52 eV, is much lower than for atom jumping, and is not too far away from the experimentally determined value. Regular jumps required a huge energy of 1.3 eV, making this an unlikely process.

8.13 Clusters on platinum surfaces

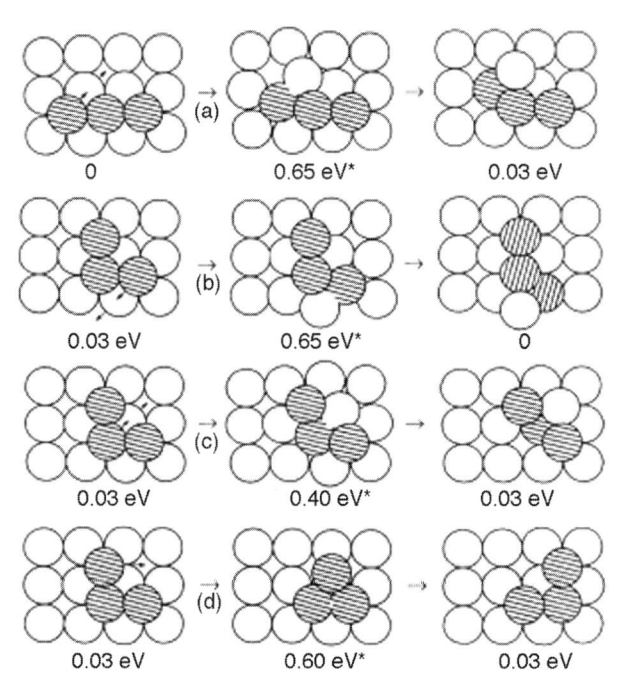

Fig. 8.77 Schematics of the mechanisms for Pt$_3$ diffusion on Pt(100) plane (after Kellogg and Voter [33]) obtained with EAM. * indicate transition state.

Trimers are always seen in the linear configuration, as chains oriented along [110] or [1$\bar{1}$0], and they diffuse with changes in orientation over a barrier of 0.49 eV, somewhat above that for single platinum adatoms. A triangular shape was never observed, but it could exist as a metastable configuration. Motion by atom exchange, as in Fig. 8.77a and b, does not yield any movement of the center of mass; combined with step c it does, however. The (1 × 1) pattern of sites at which the trimer center is observed (Fig. 8.78) can be regained if process 8.77d can take place. Diffusion by the conventional jump of monomers, dimers, or trimers all would have an activation energy larger than 1.2 eV. From molecular statics it turned out that coordination is higher at the saddle point for adatom hopping in trimers, furthering motion through the metastable triangular trimer configuration shown in Fig. 8.77d. This has a much lower barrier, estimated at only 0.60 eV, of the same order as the exchange process, and may therefore occur in the experiments together with hopping.

Kellogg [118] also checked the effect of an electric field on the activation energy for cluster diffusion. As shown in Fig. 8.79, changing from a negative to zero field raises the activation energy for dimers, but not trimers; a further increase, however, increases the activation energy, estimated assuming a prefactor of 1×10^{-3} cm^2/sec. This dependence of the activation energy on the field confirmed earlier determinations of the mechanism of movement for dimers and trimers. Since single atoms move over the surface by exchange and the changes with electric field look the same for monomers as for dimers and trimers, it probably means that the same mechanism is involved in their

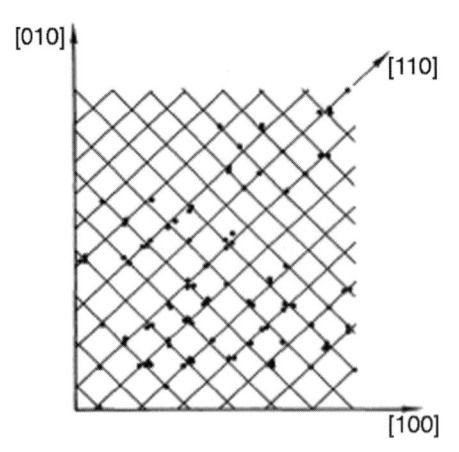

Fig. 8.78 Map of sites visited by platinum trimer center of mass on Pt(100) (after Kellogg and Voter [33]).

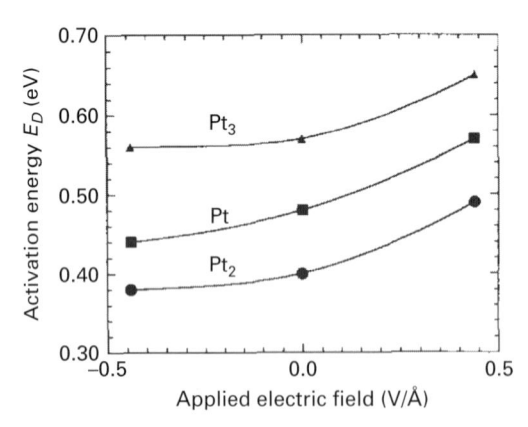

Fig. 8.79 Effect of electric field on diffusion barrier of platinum clusters on Pt(100) observed in FIM (after Kellogg [118]).

diffusion. Although there is yet no rigorous explanation, it is clear that the field has a more powerful effect than surmised by Basset and Parsley [119].

In this general time interval a number of field ion microscopic studies of cluster mobility appeared. Platinum clusters, ranging from dimers to heptamers were examined by Kyuno [120,121] on the (111) surface of platinum. The procedure for cluster formation is illustrated in Fig. 7.2. In Fig. 8.80a the grid of sites visited by a trimer is illustrated by comparison with the movement of a single atom, and it was possible to deduce that trimers visited fcc sites on the surface. As shown in Fig. 8.80b, linear trimers occupy fcc sites, but the trimer can also exist in a triangular arrangement, with the atoms in hcp sites. The atoms in the heptamer occupy fcc sites. For most of the clusters, relating the binding sites to the atomic arrangement of the substrate did not prove possible; only for trimers and heptamers was this feasible.

8.13 Clusters on platinum surfaces

Table 8.13 Diffusion parameters for Pt clusters on Pt(111) [121].

Cluster	D_o(cm^2/sec)	E_D(eV)	T_D(K)*
Pt	$2.0(\times 1.4^{\pm 1}) \times 10^{-3}$	0.260 ± 0.003	103
Pt$_2$	$1.9(\times 4.5^{\pm 1}) \times 10^{-4}$	0.37 ± 0.02	160
Pt$_3$	$1.1(\times 2.1^{\pm 1}) \times 10^{-3}$	0.52 ± 0.01	211
Pt$_4$	$6.6(\times 6.1^{\pm 1}) \times 10^{-5}$	0.57 ± 0.04	254
Pt$_5$	$1.8(\times 2.3^{\pm 1}) \times 10^{-2}$	0.78 ± 0.02	289
Pt$_6$	$4.9(\times 5.1^{\pm 1}) \times 10^{-3}$	0.89 ± 0.04	342
Pt$_7$	$5.1(\times 3.8^{\pm 1}) \times 10^{-1}$	1.17 ± 0.04	390

* T_D – characteristic temperature for diffusion

The results for the diffusivities at different temperatures are shown in Fig. 8.80c, and the diffusion parameters are listed in Table 8.13. What should be noted here is that the diffusion barrier increases regularly as the cluster size rises, reaching 1.17 ± 0.04 eV with heptamers, for which the prefactor D_o is unusually high, $5.1(\times 3.8^{\pm 1}) \times 10^{-1}$ cm^2/sec. For heptamers gliding is also very likely. The change in activation energy with size is shown in Fig. 8.80d and the noticeable difference between diffusion of Pt clusters and Ir clusters is obvious – for Pt clusters there is no minimum for tetramers. Experimental results for platinum clusters do not coincide with theoretical estimates by Liu *et al.* [122] described later using effective medium theory, but the trend of the diffusion barriers with size is roughly the same.

Linderoth *et al.* [123] looked at the movement of linear clusters made up of three to six atoms on the Pt(110)-(1 × 2) reconstructed surface using a fast STM at temperatures over the range 334–382 K. They succeeded in directly observing atoms climbing on the linear cluster and moving by the leapfrog mechanism, and assumed that the influence of the cluster length was negligible. Looking at the clusters displacement rate, they came up with a value for the on-top promotion step of $E_{up} = 0.91 \pm 0.05$ eV. The barrier for movement down was estimated from the lifetime of the metastable configuration of the cluster with an atom on top. However, since the metastable stage was not resolved, for most displacements of the cluster a simulation was carried out to evaluate the probability of observing the promoted atom under their scanning conditions and they arrived at a value of $E_{down} = 0.70$ eV.

8.13.2 Theoretical investigations: Pt(100)

Molecular dynamics simulations of dimer diffusion have been carried out by Zhuang *et al.* [17] using EAM potentials of Haftel *et al.* [71–73]. The mechanism for dimer diffusion preferred is the conventional exchange process; it should be noted that on Pt(100) the hopping event does not occur. Other processes, described in previous chapters like the 180° exchange rotation, 270° exchange rotation and even 360° exchange rotation, were observed at 700 K. For platinum at 950 K exchange rotation events amount to roughly 10% of conventional exchange. The energy of dimer exchange on Pt(100) surfaces amounted to 0.372 eV, while for hopping it was 2.029 eV.

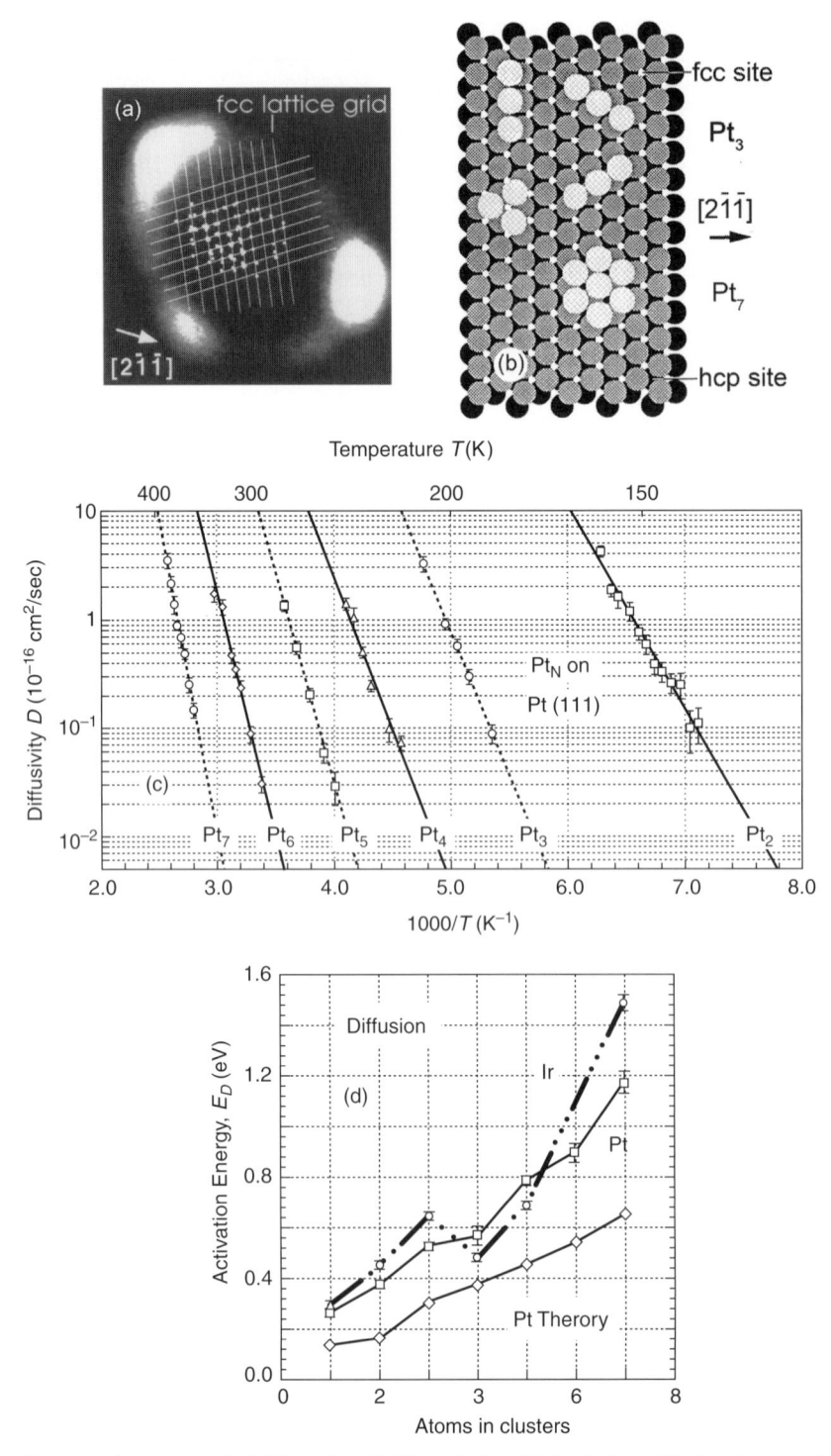

Fig. 8.80 Platinum clusters on a Pt(111) surface [121] studied in FIM. (a) Fcc grids for trimer positions. (b) Models showing configurations of platinum trimers and heptamers on Pt(111) surface. Structures agree with experimental observations. (c) Arrhenius plots for the diffusivity of platinum clusters on Pt(111). (d) Dependence of activation energy for diffusion on cluster size.

Liu et al. [19], relying on two different potentials, EAM interactions [18] and surface-embedded-atom potentials (SEAM) [71–73], also looked at dimer movement. With both EAM and SEAM, exchange was found more probable than hopping, but values of the energy for exchange were quite far apart – 0.6437 eV from EAM and 0.2923 eV from SEAM. Values for hopping are close together, 1.1976 eV (EAM) and 1.2319 eV (SEAM). They also frequently saw simple exchange during molecular dynamics simulations at 650 K, as well as less frequent rotation-exchange and two-atom exchange.

Chang and Wei [20] carried out *ab initio* density-functional calculations using the VASP [21,22] package and found that the energy of exchange required for dimers was much lower than for hopping, 0.18 eV compared to 1.30 eV. It should be noted that the barriers calculated for Pt_2 in this study are not in agreement with the experimental results or with other theoretical estimates, but the occurrence of exchange in dimer diffusion agrees with that in the diffusion of the adatoms.

Kim et al. [36] examined the energy of movement for linear clusters along the [110] direction. Dimer movement by exchange had the lowest energy of 0.799 eV, hopping perpendicular to the dimer axis required 0.875 eV, while jumping in the [110] direction involved a barrier of 1.278 eV. For linear trimers only jumps in the direction of the trimer axis were considered and the energy was 1.740 eV. Exchange required 0.898 eV, but for comparison the energy of a perpendicular jump would be needed. For linear tetramers the situation looked similar – jumps in the direction of the tetramer required 2.132 eV, while perpendicular exchange was over a barrier of 0.945 eV.

The barriers for dimer diffusion on Pt(100) derived from theory are spread from 0.799 eV to 0.18 eV, quite a big spread compared with the experimentally determined activation energy of 0.41 eV. However all of the calculations predict the same mechanism of movement – exchange. Movement of trimers turned out to be by exchange as well; however, experiments also indicated some hopping.

8.13.3 Theoretical investigations: self-diffusion on Pt(111)

The first theoretical study for platinum clusters on Pt(111) was done by Liu et al. [122] with the effective medium approximation. The energy difference for small clusters sitting in fcc sites compared to ones sitting in hcp sites was within the error of estimates; they therefore considered only clusters in fcc sites. Dimer spacings were slightly lower than the lattice distance of 2.56 Å, which means that the dimer was slightly elevated above the surface. Dimers were able to translate as well as rotate, both processes requiring the same energy of 0.16 eV. At 300 K, the movement of dimers was not site-to-site hopping any more. The barrier determined in this study was, however, much lower than the experimentally determined value of 0.37 eV.

Trimers were observed in both linear and triangular shapes, with the triangular one more stable. Translation of linear trimers required 0.20 eV, other processes preserving linear shape were more energetically demanding. The transformation from linear to triangular trimer was estimated as 0.16 eV, the opposite transition involved a much higher energy of 0.56 eV. The authors suggested that the linear trimers will first change shape to a triangular form and then move over the surface. This, however, is not in agreement with experiments in which linear trimers were observed during diffusion. For triangular clusters in fcc sites rotation as well as translation of triangular trimers without shape change required 0.30 eV. For triangular clusters in hcp sites only translation is allowed at 0.29 eV, rotation is blocked by surface atoms. The activation energy for trimer diffusion was much lower than the 0.52 eV measured in experiments [120,121]. Tetramers were found to move as a whole, with an energy of 0.37 eV. Detaching an atom from a tetramer cost 0.5 eV. A more surprising finding was that the opposite process of attaching required 0.8 eV. For movement of the pentamer, motion of the cluster as a whole was most likely, since the collective energy for movement amounted to only 0.46 eV. Additionally, for detachment of an atom 0.83 eV was needed. Concerted motion of hexamers required 0.56 eV, while for heptamers it amounted to 0.65 eV. All activation energies for diffusion were lower than those experimentally measured by Kyuno [120,121], but they have the same tendency to increase with size.

Molecular dynamics simulations were made by Münger et al. [124] of Pt_7 clusters on Pt(111) at 1000 K. For interactions they used the embedded atom potentials of Johnson and Oh [18,125]. At this high temperature, movement or restructuring of the hexagonal cluster took place only rarely. However, when a single platinum atom was deposited on the cluster its configuration changed and enhanced cluster motion was observed. Actual experiments at this high temperature have not been reported, but Ir and Re atoms have been deposited on iridium clusters on Ir(111) [126]. Heating these to 200 K did not reveal any change in motion or structural changes. The mobility of Pt_6, Pt_7, and Pt_8 clusters on the Pt(111) plane has been probed in molecular dynamics simulations by Münger et al. [127,128], using embedded atom potentials [18]. Simulations were done again at 1000 K, a temperature at which Pt_7 hardly moved, but removal of the central atom brought about notable changes. The cluster underwent a variety of shape changes by concerted translation and by double-shear glide. Addition of another platinum atom on top of Pt_7 induced bond breaking, reshaping, and migration. The latter events have been reported before [124]. Once the atom added to the cluster top incorporated into the side, diffusion again slowed down. These processes have so far not been seen in experiments. What is surprising is that at 1000 K, the authors observed no motion of the Pt_7 cluster while in experiments motion of this cluster was already seen at 400 K [120,121].

Chirita et al. [129] more intensively pursued their calculations of platinum cluster diffusion on Pt(111). The mechanism they conclude as being important is double-shearing, named reptation by them, and illustrated in Fig. 7.26, in which one subunit moves as a whole against another. This mechanism was also found for larger clusters, and at 1000 K accounted for 40% of the observed diffusion processes. For Pt_6, the activation energy amounted to ~1.55 eV, and as cluster size increased, double-shearing was expected to become a more important contributor. Double-shearing has been suggested in the diffusion of iridium tetramers on Ir(111) [26], but so far there has been no direct indication of its importance in other experimental observations of cluster diffusion. It should be noted that the value of the activation energy for diffusion of Pt_6 clusters seems to be much too high compared with the experimental value of 0.89 eV [120,121].

Boisvert and Lewis [130] have done density-functional calculations using the local-density approximation in order to better understand the diffusion of platinum dimers on

8.13 Clusters on platinum surfaces

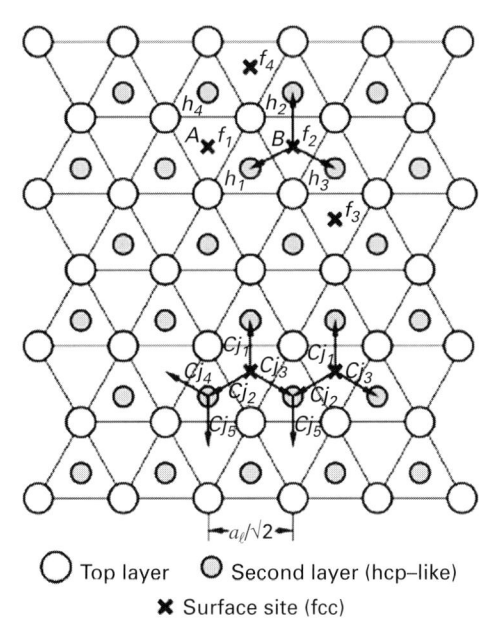

Fig. 8.81 Schematic of platinum dimer diffusion processes on Pt(111) obtained in DFT calculations (after Boisvert et al. [130]).

Pt(111). According to their estimates, the most stable configuration of a dimer is with the atoms in neighboring fcc sites. With atoms in adjacent hcp sites the energy is 0.30 eV higher, and other configurations are still higher in energy. The dissociation energy of the dimer was estimated at 0.81 eV. For diffusion, at least one of the platinum atoms will have to jump to an adjacent hcp site. The lowest energy arrangement is with atom A at site f_1 and atom B at h_2, as in Fig. 8.81, has an energy 0.34 eV above the equilibrium arrangement. Other configurations have a considerably higher energy. The activation energy of such a jump to an adjacent hcp site was calculated to be 0.35 eV; the subsequent jump back to restore both atoms to neighboring fcc sites takes place over a very small barrier. However, a jump further would lead to a rather higher energy arrangement, so that single atom jumps would limit diffusion to local movement of the dimer. Boisvert and Lewis therefore also considered movement by concerted jumps of both atoms, also illustrated in Fig. 8.81. Here there are three possibilities. Concerted jump *1*, designated by cj_1, has the highest activation energy, ~ 0.8 eV, as both atoms must pass on either side of a lattice atom. Concerted jumps cj_2 and cj_3, on the other hand, can proceed over a barrier of only 0.37 eV, and the dimer can then easily transform to an adjacent equilibrium state. Using cj_4 or cj_5, the dimer can come back to an fcc-fcc configuration. Combined with single atom jumps, the dimer would thus be able to carry out diffusion over the entire plane. The obtained barrier is in very good agreement with the experimentally measured value of 0.37 eV. Experiments to track the steps involved in the motion of platinum dimers have not been done on Pt(111), but have been reported for Ir_2 on Ir(111) [26]. There, both intra-cell motion as well as inter-cell diffusion were noted, but it was assumed that motion occurred via jumps of individual atoms.

Diffusivities of small clusters

Chang et al. [27] using EAM potentials calculated barriers for self-diffusion on fcc(111) surfaces. For platinum clusters the energy was 0.17 eV for dimers, 0.40 eV for trimers, 0.39 eV for tetramers, and 0.61 eV for pentamers. What is important here is the decrease in the diffusion barrier in going from trimers to tetramers, followed by an increase again with pentamers. It must be noted that such a pattern was not observed experimentally for clusters of platinum self-diffusing on Pt(111) [121], where there was no distinct minimum at tetramers. Also, the experimentally derived values were higher than what has been presented here.

Calculations of diffusion of seven-atom platinum clusters were done by Longo et al. [40] using EAM potentials parameterized according to Voter and Chen [41]. They evaluated the activation energy for a cluster gliding over its own crystal, which for Pt was only 0.91 eV. For Pt_7, translational motion was also observed on heating the cluster to the melting temperature; gliding started at a temperature of 1200 K, quite high compared with the experimentally observed diffusion at 400 K. Surprisingly, the activation energy for gliding was lower than the experimental value of 1.17 eV.

Motion of platinum dimers was briefly described by Albe and Müller [131], who carried out molecular dynamics simulations relying on a bond-order potential [132]. The dimer was deposited at random on a Pt(111) plane at 600 K, and landed on fcc sites. As shown in the images in Fig. 8.82, diffusion occurred by concerted jumps towards a neighboring fcc location. Regrettably no energetics were explored, and only one temperature was studied.

Recently the stability and diffusion of Pt trimers on Pt(111) were investigated by Yang et al. [133] using molecular dynamics simulations and EAM potentials. For single atom movement they derived a value for the activation energy of 0.19 eV, lower than in the experiments. The triangular clusters were more stable than the linear configurations. The most energetically favorable path was concerted translation, which required 0.40 eV. The Arrhenius analysis of migration is shown in Fig. 8.83 and yields an activation energy

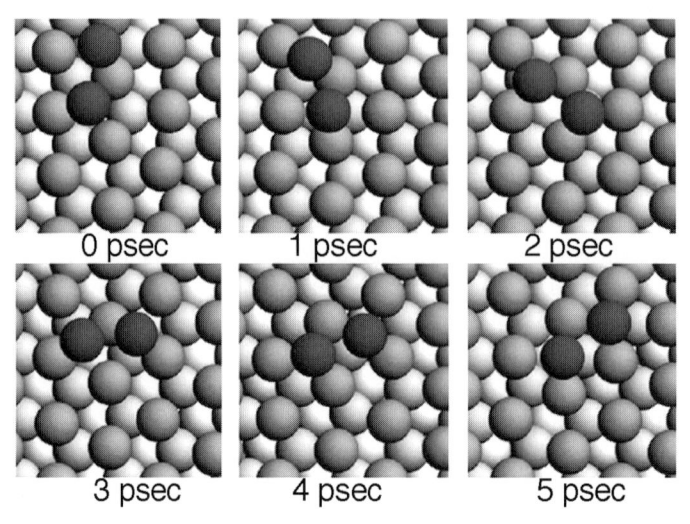

Fig. 8.82 Cooperative jumps of platinum dimer on Pt(111) plane at 600 K, derived in molecular dynamics simulations with bond-order potential (after Albe and Müller [131]).

8.13 Clusters on platinum surfaces

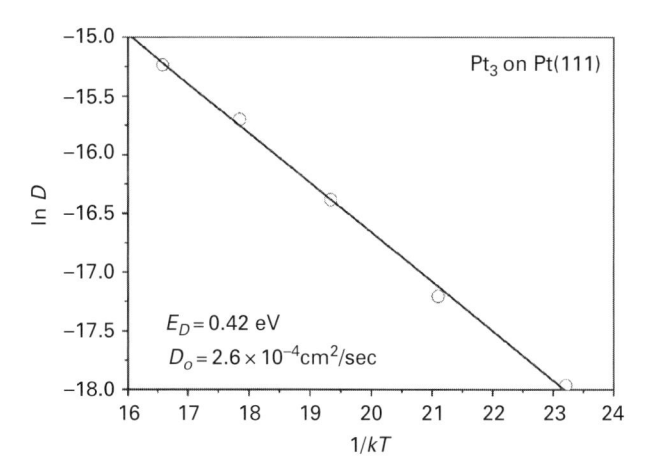

Fig. 8.83 Arrhenius plot for diffusion of platinum trimer calculated with EAM. Solid line indicates best fit to molecular dynamics simulations, open circles give calculated diffusivities (after Yang et al. [133]).

of 0.42 eV and a prefactor of 2.6×10^{-4} cm^2/sec, which agreed well with their static estimation for translation of triangular clusters. Transitions from triangular to linear form were not considered in this paper.

Kim et al. [36] used molecular dynamics to look at the diffusion of clusters with two and three atoms. They did not specify whether the adsorption sites were fcc or hcp. For dimers the lowest energy, only 0.186 eV, was for a jump of 30 degrees from the dimer axis in the direction $[1\bar{2}1]$. Jumping perpendicular to the dimer axis required 0.324 eV, exchange was unlikely at 2.734 eV. For triangular clusters a hopping energy of 0.369 eV was derived, and exchange was again unlikely.

In 2008, Yang et al. [60] resorted to modified analytical EAM potentials to do molecular dynamics simulations of the diffusion of Pt and Pt$_7$ on the Pt(111) surface. They found an activation energy for the movement of the center of mass of 1.08 eV and a prefactor 6.1×10^{-1} cm^2/sec, values in good agreement with the experimental results [121].

The diffusion barriers calculated for platinum clusters on Pt(111) cover a wide range of values but, with the exception of the work of Boisvert and Lewis as well as of Yang et al., are generally far from the experiments. There is also a huge disagreement between the onset temperature for cluster movement obtained from theory and experiment.

8.13.4 Theoretical investigations: hetero-diffusion on Pt(111)

Hetero-diffusion studies on the Pt(111) surface started in 1992 with the work of Blandin and Massobrio [134], who did molecular simulations of Ag$_2$ diffusing on Pt(111) as calculated with embedded atom potentials. From observations between 130 K and 320 K they were able to come up with an Arrhenius plot which yielded an activation energy of 0.09 ± 0.01 eV for the dimer, compared with 0.058 ± 0.003 eV for single Ag atoms.

Massobrio and Blandin [135] further examined silver dimers and trimers on Pt(111) using slightly modified EAM potentials. They calculated the lowest energy cluster conformations on the surface, and as indicated in Fig. 8.84, found that linear chains are

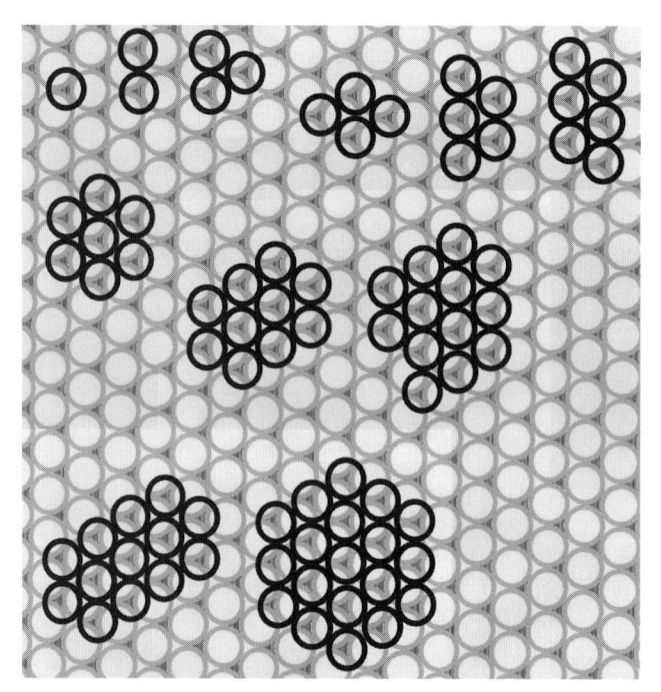

Fig. 8.84 Stable structures of Ag clusters on Pt(111) derived in EAM calculations, based on the work of Massobrio and Blandin [135].

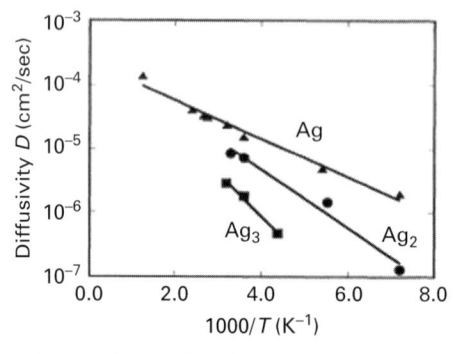

Fig. 8.85 Arrhenius plots for diffusion of silver adatoms (triangles), dimers (circles), and trimers (squares) on Pt(111) obtained with EAM interactions. Diffusion barriers: adatoms 0.060 ± 0.005 eV, dimers 0.09 ± 0.01 eV, trimers 0.13 ± 0.01 eV (after Massobrio and Blandin [135]).

not favored for trimers – they prefered a triangular arrangement in adjacent fcc sites. Both static as well as molecular dynamics estimates were made. In the former, the diffusion barrier for single silver adatoms was 0.05 eV. For long-range motion intercell diffusion has to occur, and this required an activation energy of 0.10 eV for dimers and 0.14 eV for trimers. Molecular dynamics simulations were done over a range of temperatures, and as is clear from Fig. 8.85 show increasing barriers as the cluster size increases. The values obtained from the slope of the Arrhenius plot were 0.060 ± 0.005 eV for single atoms, 0.09 ± 0.01 eV for dimers and 0.13 ± 0.01 eV for

8.13 Clusters on platinum surfaces

trimers. The barrier achieved in this way for single atoms is higher than in the static evaluation, but if only the data up to 300 K are used is reduced to 0.052 ± 0.005 eV, and agreement with the static value was obtained. Dimers were unstable at temperatures $T > 300$ K, and trimers above 400 K.

Diffusion of Co dimers on Pt(111) was recently examined by Goyhenex [136] using RGL interactions. During molecular dynamics simulations for 1 nsec at 620 K he observed 18 rotations and 24 inter-cell motions. Translations in the same cell from fcc-fcc to hcp-hcp sites were also observed, but rarely. The potential energy changes in intra-cell diffusion are shown in Fig. 8.86a, and yielded a diffusion barrier of 0.26 eV for movement from hcp-hcp to fcc-hcp sites; movement continues to the stable fcc-fcc configuration over a barrier of 0.11 eV. For inter-cell motion the potential energy curve in Fig. 8.86b gave a barrier of 0.18 eV. The transition state is very broad and Goyhenex claimed that movement backward and forward from the transition state was possible, which limited the rate of success. Such a broad transition state is associated with a change in the dimer distance from 2.47 to 2.38 Å, due to the big mismatch in the sizes of Co and Pt. Molecular dynamics simulations yielded the Arrhenius plot of the diffusivity in

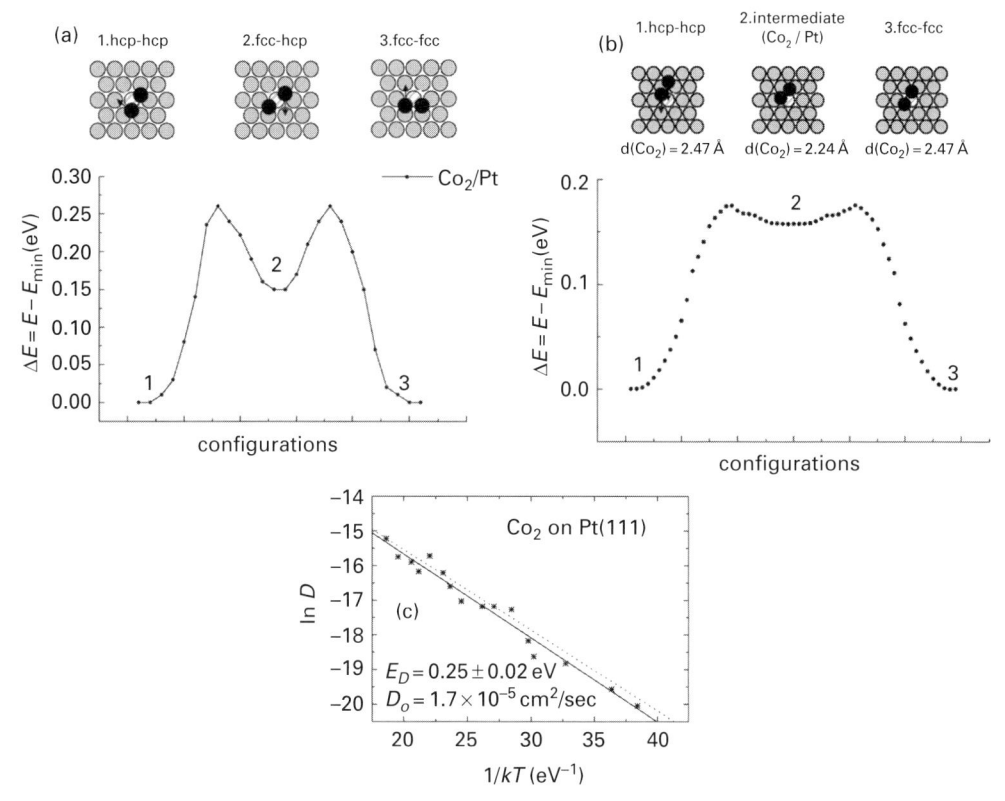

Fig. 8.86 Cobalt dimer diffusion on Pt(111) obtained in RGL calculations. (a) Potential energy change for Co_2 moving in cell defined by six sites around substrate atom. (b) Potential energy curve for the inter-cell motion of Co_2. (c) Arrhenius plot for diffusion of Co_2 on Pt(111). Solid curve gives fit for dimer, dotted curve for monomer (after Goyhenex [136]).

Fig. 8.86c, from which an activation barrier of 0.25 ± 0.02 eV and a somewhat low prefactor of 1.7×10^{-5} cm^2/sec were derived. The energy for dimer diffusion was slightly lower than for diffusion of the monomer. What is very surprising in this study is that intra-cell motion required more energy than inter-cell diffusion.

The self-diffusion of clusters on the Pt(111) surface was widely investigated theoretically; unfortunately, the same cannot be said about hetero-diffusion.

8.13.5 Theoretical investigations: Pt(110)

The leapfrog mechanism has been examined more closely for dimers and larger one-dimensional clusters on Pt(110)-(1 × 2) surfaces by Montalenti and Ferrando [137], who resorted to RGL potentials. In molecular dynamics simulations of dimers, leapfrog transitions were observed most frequently at 1000 K. Much the same was found for trimers and tetramers. For dimers as well as trimers the energy for an atom to step up the {111} wall was 0.84 eV, while stepping down required 0.21 eV for the dimer, and 0.64 eV for the trimer. The crucial difference in leapfrog diffusion of dimers and trimers is caused by the existence of a metastable minimum for LF movement of the trimer, shown in Fig. 7.14; such a metastable minimum is not observed for movement of the LF dimer. The values obtained are close to the experimental estimate of 0.9 eV for up motion and 0.7 eV for down motion [123].

Estimates of platinum dimer diffusion along the channels of Pt(110)-(1 × 2) were made by Feibelman [138] in 2000 using the VASP [21,22] *ab initio* package. For diffusion by the leapfrog mechanism the barrier proved to be only 0.76 eV, significantly less than the estimate for diffusion of single Pt atoms.

Movement on Pt(110)-(2 × 1) surfaces was examined with EAM potentials by Kürpick [42]. Four different diffusion mechanisms were considered and are indicated in Fig. 8.18. She took into account concerted motion of clusters, movement by the leapfrog mechanism as well as by exchange. It is clear that the leapfrog jump up is the limiting step for all the systems treated here. The potential energy experienced by Pt clusters is illustrated in Fig. 8.87. The concerted motion for trimers required a huge energy of 1.55 eV, tetramers needed 2.02 eV, and pentamers 2.48 eV. Exchange was also unlikely over a barrier of 1.30 eV for trimers and 1.36 eV for tetramers and pentamers. Step up in leapfrog motion

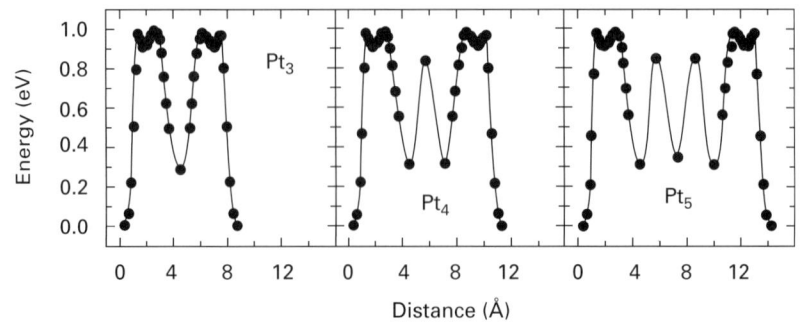

Fig. 8.87 Potential energy of one-dimensional platinum cluster diffusing by leapfrog mechanism on Pt(110)-(2 × 1) plane, calculated by using EAM potentials (after Kürpick [42]).

for chains three to five atoms long required 0.98 eV, a bit higher than estimated by Montalenti and Ferrando [137] and also higher than the experimental value. The values for stepping down, 0.70 eV for trimers and 0.67 eV for tetramers and pentamers, are in excellent agreement with experiment. She also looked at concerted movement on the unreconstructed surface, and for trimers came up with a barrier of 1.48 eV, tetramers 1.93 eV, and pentamers 2.39 eV.

Recently Kim *et al.* [36] used molecular dynamics to look at the energetics of movement of linear clusters two to four atoms long on the unreconstructed Pt(110) surface. For dimers, in-channel hopping is the most probable step over a barrier of 0.842 eV, compared to 1.535 eV for exchange. For trimers, in-channel hopping required 1.195 eV and exchange 1.353 eV; for linear tetramers, however, exchange occurred over a barrier of 1.353 eV, while in-channel hopping was over 1.578 eV. For all of them, cross-channel hopping was a high-energy process. Unfortunately no experiments are available for diffusion on Pt(110) that would allow comparison with the calculated results.

8.14 Clusters on gold surfaces

8.14.1 Gold (100)

Estimates of gold dimer mobility on the reconstructed Au(100) surface were done by Bönig *et al.* [139] drawing on EMT potentials for the interactions. Similar to the real system, in the simulation the top layer was reconstructed, but the second layer had the fcc(100) structure. This configuration caused two kinds of adsorption sites on the surface. They found a diffusion barrier of 0.2 eV for dimer movement, but motion of the dimer did *not* occur by a sequence of one-atom jumps. An important conclusion from this study was that jumps occurred with a bond length practically unchanged. Worth noting is that during adsorption of 2–6 atom clusters, local reconstruction to the fcc(100) structure of the lattice underneath the island was observed.

Bogicevic *et al.* [13] used the VASP simulation package as well as the DACAPO code [14] to look at self-diffusion of Au dimers on the Au(100) unreconstructed plane. The two important processes were shearing, shown in the middle of the second row in Fig. 8.8 and to the left, which in fact can be achieved by a jump to the nearest-neighbor site perpendicular to the dimer axis. The second important process was stretching (middle row, center to right in Fig. 8.8), which can be achieved by a jump in the direction of the dimer axis. For Au_2 as well as for Rh_2 and Al_2, jumping perpendicular to the dimer axis was a lower energy path than stretching with an energy of 0.86 eV compared to 1.11 eV. The authors argue that this is likely to be a general effect for other (100) fcc metals, but did not consider exchange in their investigations. Additionally, in real systems it is known that the Au(100) surface reconstructs.

Molecular dynamics simulations of dimer diffusion have been carried out by Zhuang *et al.* [17] on the Au(100) surface using EAM potentials of Haftel *et al.* [71–73]. The simple exchange mechanism seems to be favorable for Au dimers, with an energy of 0.256 eV compared to 0.782 eV for traditional hopping. The existence of a complicated exchange, as for example a 180 degree rotation-exchange, was also observed on this surface, and was associated with a stress-induced mechanism. Again reconstruction of the surface was not taken into account in this study. Investigation of dimer diffusion was continued by Liu *et al.* [19] with EAM [18] and RGL interactions [52,53]. The static activation energies from EAM potentials indicated that the exchange mechanism, with an energy of 0.2647 eV, was more likely than hopping, which occurred with an energy of 0.8151 eV. The difference in the activation energy calculated with RGL potentials for exchange and hopping was smaller – exchange required 0.3715 eV, hopping 0.5962 eV. However, both potentials show a preference for exchange on this plane. From 2 nsec molecular dynamics simulations with RGL potentials at 650 K, 33 exchange events were observed, but only 3 jumps and 3 complicated exchange mechanisms.

Chang and Wei [20] used the VASP [21,22] package to establish the activation energy for Au dimers on Au(100). Just as in previous studies, they found exchange as the preferred mechanism of movement, with an energy of 0.24 eV, compared to 0.83 eV for hopping. Recently Kim *et al.* [36] used molecular dynamics to find the energetics of atom movement of linear clusters two to four atoms long, elongated along the [110] direction. For Au dimers, exchange required 0.404 eV, while perpendicular hopping demanded slightly more, 0.531 eV, and tangential hopping occurred over a barrier of 0.731 eV. For linear trimers and tetramers only tangential hopping in the [110] direction and exchange in the [100] were considered. Exchange for trimer atoms required 0.512 eV and for tetramer atoms 0.547 eV. Tangential hopping for trimer atoms took place over a barrier of 0.984 eV, and for tetramer atoms over 1.185 eV.

On the Au(100) surface, movement of dimers has gotten most attention. The studies seem to indicate that exchange will be a leading mechanism of movement, and the scatter in activation energies obtained from different potentials is not too bad. After excluding data from RGL potentials, the value is always close to 0.25 eV for exchange and 0.80 eV for hopping, but experiments to allow a comparison are not available and in the real system the surface is known to reconstruct. This will complicate the movement of clusters and only Bönig *et al.* [139] looked at the movement of dimers on such a surface.

8.14.2 Gold (111)

Diffusion on the (111) plane of gold has not been investigated in great detail – there are only three papers which briefly touch this subject. The first one by Chang *et al.* [27] used EAM potentials to calculate diffusion barriers for clusters of up to five atoms. For gold dimers they came up with a barrier of 0.10 eV, for trimers it was 0.25 eV, for tetramers the value was lower at 0.24 eV, but the barrier increased again for pentamers to 0.42 eV.

The close-packed Au heptamer was investigated by Longo *et al.* [40] using EAM potentials parameterized according to Voter and Chen [41]. For a cluster gliding over its own crystal they came up with an energy of 0.64 eV. However in molecular dynamic observations they failed to observe gliding for Au_7, which was attributed to the vertical mobility of the substrate at elevated temperatures.

Recently Kim *et al.* [36] studied the energetics of dimer and trimer atoms moving. For dimers they found that the lowest-energy process was a jump in the direction 30 degrees to

8.14 Clusters on gold surfaces

the dimer axis at 0.100 eV; jumps at 90 degrees required twice as much energy, 0.229 eV, and exchange 1.484 eV. For triangular trimers jumping needed 0.218 eV, exchange 1.109 eV. No experiments have been done to provide an indication of what diffusion barriers for gold clusters should amount to. Additionally, all investigations were done on the unreconstructed surface, although it is known that the Au(111) surface reconstructs.

8.14.3 Gold (110)

The leapfrog mechanism has been examined more closely for dimers and larger one-dimensional clusters on Au(110) and Au(110)-(1 × 2) as well as platinum surfaces by Montalenti and Ferrando [137], who resorted to RGL potentials. In molecular dynamics simulations of dimers on Au(1 × 1) only concerted jumps of the dimer were observed, but for Au(110)-(1 × 2), leapfrog transitions were found most frequently, both at 450 K and 550 K. Much the same was observed for trimers, and tetramers. Concerted jumps of Au dimers on Au(110) required an energy of 0.47 eV, and 0.52 eV on Au(110)-(1 × 2). The energy of dimer atoms stepping up was estimated as 0.45 eV, while stepping down required only 0.14 eV, less than the concerted jump of a dimer. Trimer motion up requires the same energy as for dimers; stepping down, however, was more demanding, needing 0.37 eV, and much less than for the concerted jump of trimers, which was estimated at 0.7 eV. They also made estimates of the lifetimes of adatoms on top of the cluster and for gold concluded that the lifetime would be too short to be observed with dimers, but that it should be feasible to detect it by STM for trimers and higher. There still remains a question about how the atom from atop the cluster descends, whether this occurs by atom exchange or by an ordinary jump. It turned out that exchange events, both going up as well as going down, had a significantly larger activation energy than straight jumps; the exchange descent required 0.55 eV, and going up by exchange needed 0.64 eV. The jumps as dominant mechanisms in both descent and ascent were also confirmed by high-temperature simulations. The conclusion was that the leapfrog mechanism should be common in all sorts of fcc(110)-(1 × 2) surfaces.

Diffusion of various dimers on Au(110)-(1 × 2) was further examined by Montalenti and Ferrando [140] using molecular dynamics with RGL interactions. Concerted jumps of the dimer, shown in Fig. 8.88a, as well as leapfrog diffusion, illustrated in Fig. 8.88b, were considered, but dimer dissociation was also allowed. Examined were the dimers Au_2, Cu_2, and AuCu; the barriers to the different atomic events they obtained are given in Table 8.14. For gold dimers, the energetics for leapfrog diffusion are displayed in

Table 8.14 Activation barriers for leapfrog E_{LF}, concerted jump E_{CJ} and dissociation events E_d on as obtain by quenched molecular dynamics Au(110)-(1×2) [140].

Dimer	E_{LF}(eV)	E_{CJ}(eV)	E_d(eV)
Au_2	0.45	0.52	0.51
Cu_2	0.75	0.66	0.50
AuCu	0.58 (Au-LF)	0.60	0.52 (Au-Di)
	0.67 (Cu-LF)		0.54 (Cu-Di)

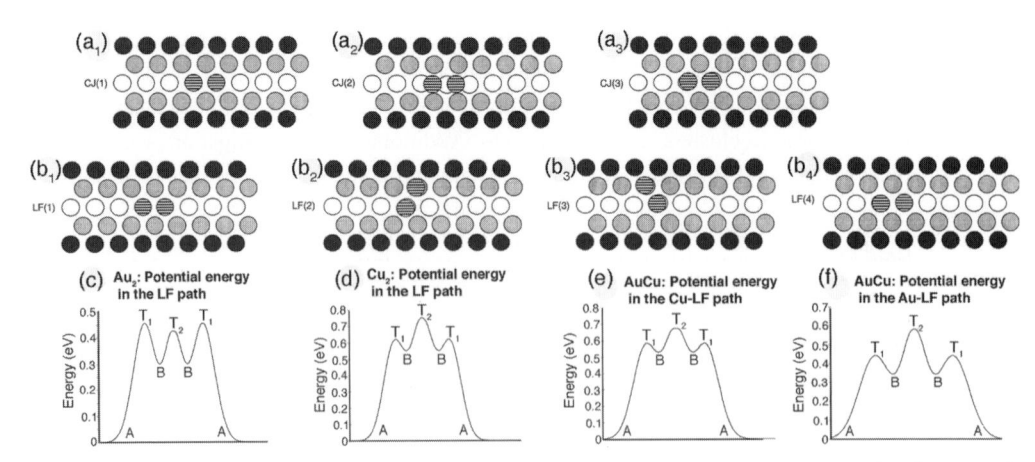

Fig. 8.88 Dimer diffusion on Au(110)-(1 × 2) examined with RGL potentials. (a) Schematic of concerted jump of dimer on Au(110)-(1 × 2). (b) Model for leapfrog mechanism in dimer diffusion on Au(110)-(1 × 2). (c)–(f) Schematic of energy changes in dimer diffusing over Au(110)-(1 × 2). (c) Au_2, (d) Cu_2, (e) AuCu with Cu leapfrog path, (f) AuCu with Au leapfrog path. T_1 denotes saddle point for diffusion (after Montalenti and Ferrando [140]).

Fig. 8.88c. What is shown here is not a universal curve. For Au_2, the first saddle point constitutes the maximum, and therefore corresponds to the activation energy for diffusion. Leapfrog motion plays a major role. For Cu_2 dimers the energy is configured differently, shown in Fig 8.88d, in that the peak lies not at the first barrier but rather at the second, and this is also the situation for AuCu dimers, as is clear in Fig. 8.88e–f. For all three the barrier to dissociation (in Table 8.14) is higher than for single atom diffusion, so that if dissociation takes place the adatoms will move rapidly and recombine with each other. It should be noted that for Cu_2 the barrier to leapfrog motion is higher than for concerted jumps, which therefore dominate. Au dimers are also more stable than Cu dimers. However, for all three dimers, leapfrog diffusion does occur, and this seems to be a general trend on (110)-(1 × 2) surfaces.

Diffusion of nickel atoms and clusters has been simulated at 400 K on the Au(110)-(1 × 2) plane by Fan and Gong [141] using Johnson potentials [18]. Single adatom motion on the reconstructed (110) has already been described. Dimer motion is somewhat similar, as illustrated in Fig. 8.89a. Motion can occur by concerted jumps along the trough with the dimer bonds parallel to the trough, or else with one atom in the trough, the other along the sidewall. Diffusion may also take place with both atoms on the sidewall, or by jumping from one channel to a neighboring one. The last is not a frequent occurrence, however. Occasionally leapfrog diffusion may take place, with one atom jumping to the sidewall and then to the other side of the first atom in the trough, but this is not as frequent as concerted jumping.

The movement of Ni dimers on Au(110)-(1 × 2) is definitely different than movement of Cu or Au dimers. Nickel trimers are in a triangular configuration with two atoms at the bottom of the trough and the third near the local minimum on the sidewall. Some of the diffusion steps are illustrated in Fig. 8.89b. Diffusion may involve concerted jumps

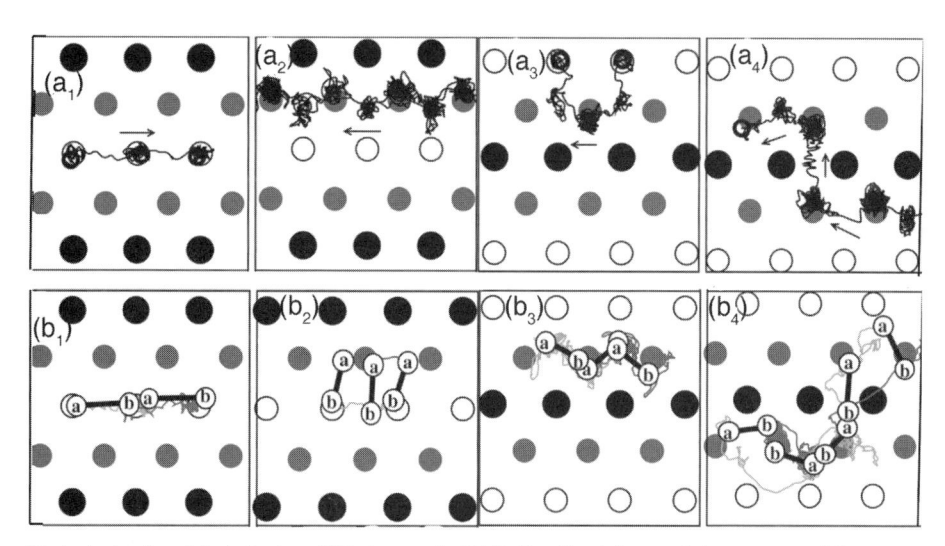

Fig. 8.89 Trajectories for nickel clusters diffusing on Au(110)-(1 × 2) relying on Johnson potentials. (a$_1$) Concerted jumps in trough. (a$_2$) One atom moves in trough, the other one on the side. (a$_3$) Dimer diffusion on side wall. (a$_4$) Jump from one trough to the next. (b) Diffusion of Ni$_3$ on Au(110)-(1 × 2). (b$_1$) Translation in trough. (b$_2$) Rotation. (b$_3$) Translation plus rotation. (b$_4$) Jump over trough by rotation and translation (after Fan and Gong [141]).

along the trough, similar to gliding (Fig. 8.89b$_1$), rotation (Fig. 8.89b$_2$), translation plus rotation (Fig. 8.89b$_3$), and inter-channel motion by rotation and translation (Fig. 8.89b$_4$). Compared to gliding, rotation had quite a low energy because only two atoms move instead of three. Higher clusters were three-dimensional, and moved rapidly by rolling over the surface, but such structures are not considered here. The Ni$_N$ cluster had a sphere-like shape due to stronger Ni-Ni compared with Ni-Au interactions. The work showed that the diffusion mechanism depended on the size of the cluster, but no attention was paid to the energetics of movement. These novel mechanisms of movement should still be confirmed experimentally. However, they showed the importance of in-cluster interactions on diffusion.

Kim *et al.* [36] looked at jumps of linear Au clusters two to four atoms long on an unreconstructed Au(110). For dimers, in-channel jumps were most likely, with an energy of 0.456 eV, exchange required 0.906 eV, and cross-channel jumps 1.187 eV. For trimers, in-channel jumps needed 0.646 eV, exchange 0.756 eV, and cross-channel jumps 1.763 eV. For tetramers, in-channel jumps occurred over a barrier of 0.847 eV, exchange over 0.989 eV, and cross-channel jumps over 2.902 eV.

8.15 Comparisons

Much material has been surveyed here. There is a question that still remains: are there any significant features to indicate trends in the diffusion behavior? Here it is important to notice that considerable work on cluster diffusion has been done on the (110) plane of tungsten. For nickel, palladium, platinum, and iridium, clusters of a few atoms all exist as linear chains oriented along <111>. The activation energies for diffusion rises starting from the lowest value for palladium and going to nickel, and then platinum (see Table 8.10). Presumably the barrier for diffusion of iridium clusters is higher still. The fact that palladium is lowest may be surprising, as this is not in keeping with the trends of the enthalpy of vaporization. However, we should have expected this, as the diffusion of palladium and nickel single adatoms on W(211) behaves in the same way. Rhenium and tantalum clusters are different, and what is interesting is the formation of Re dimers. On the W(110) surface, Re dimers are unstable, but at the same time both in-channel and cross-channel forms are stable on the W(211) surface. The question immediately arises if stability of Re_2 is surface mediated, and is favorable for W(211) and not favorable on the W(110) structure, or if the W(110) structure provides unfavorable bond distances for Re_2 to exist. It will be important to probe this question on different bcc surfaces. What little has been done on rhenium shows trimers both linear and triangular, with bigger clusters two-dimensional. The energetics of diffusion have not been established quantitatively, but mobility increases significantly for tetramers and pentamers, which is probably correlated with the intriguing open structure of these clusters. Palladium clusters have been examined on Ta(110), another bcc metal. Just as on W(110), these clusters exist in the form of chains, but with nine palladium atoms, two-dimensional structures form again, so diffusion should proceed similarly to chains on W(110).

Clusters have been studied in experiments on the close-packed surfaces Pt(111), Ir(111), as well as Re(0001). Iridium and platinum are fcc neighbors in the periodic table, but cluster behavior is not too similar. With iridium, there is a marked drop in the diffusion barrier of iridium tetramers compared to trimers, but that is not seen with platinum clusters on Pt(111). For monomers, the self-diffusion barrier of iridium is higher than for platinum, as expected from the enthalpy of vaporization. However, this relation is not maintained uniformly in going to clusters with a larger number of atoms. Oddly enough, the situation improves in comparing the (111) plane of fcc iridium with hcp Re(0001). As shown in the plot in Fig. 8.90, the barriers for self-diffusion of rhenium clusters up to Re_7 are uniformly higher, but both tetramers have a minimum value. Furthermore, the behavior of rhenium compared to iridium is precisely what is expected from the enthalpies of vaporization of the two materials. What seems to be a common feature of clusters on Ir(111) and Pt(111) is that heptamers are very stable, and have the possibility of gliding over the surface without shape change. However, according to molecular dynamics studies, gliding of heptamers is not observed on all fcc(111) surfaces [40]; this situation should be checked in experiments. For movement and shearing on fcc(111) surfaces, two concerted mechanisms were predicted by simulations – reptation and dislocation motion; however, these should still be probed directly in experiments.

We now turn to clusters on the (100) planes of fcc elements. On platinum (100), dimer diffusion occurs by exchange, trimers seem to move by exchange plus hopping but the clusters are linear. On Rh(100), clusters tend to be two-dimensional, and move with an apparently oscillating activation energy, by atom hopping over the surface. Diffusion of platinum clusters on the (100) plane of rhodium occurs by hopping, but the clusters are

8.15 Comparisons

Table 8.15 Barriers to dimer self-diffusion (eV).

Material	Experiment		Calculation		
Pt(111)	0.37 ± 0.02 [121]	0.35 DFT [130]	0.17 EAM [27]	0.16 EMT [122]	
Ir(111)	0.447 ± 0.013 [120]	0.467 RGL [114]	0.24 EAM [27]		
Cu(111) local	0.018 ± 0.003 [44]	0.009 EAM [58]	0.016 EAM [55]		
Pt(100)	0.41 [33]	0.52 EAM [33]	0.18 VASP [20]	0.372 EAM [17]	
		0.644 EAM [19]	0.292 SEAM [19]	0.799 MD [36]	
Ir(100)	0.88 ± 0.05 [108]	1.078 RGL [114]	0.65 EAM [115]	1.30 VASP [20]	

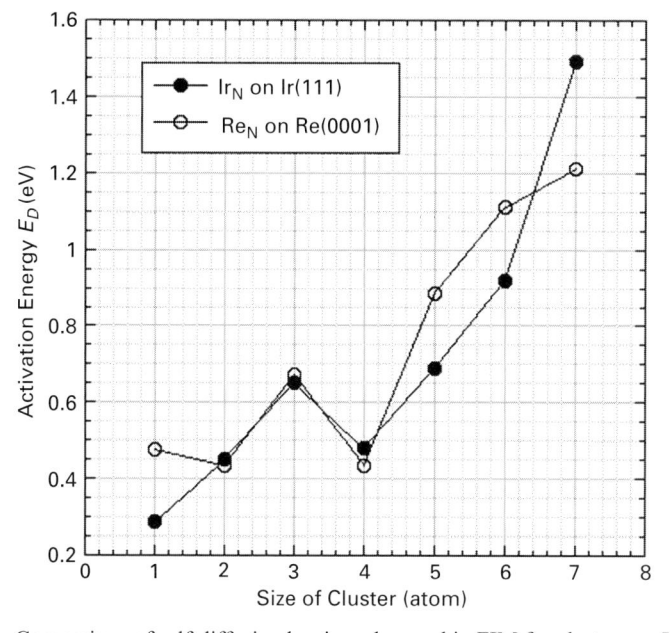

Fig. 8.90 Comparison of self-diffusion barriers observed in FIM for clusters on Ir(111) and Re(0001).

again in the form of linear chains, indicating the importance of long-range interactions. Predictions or comparisons are not easy here. At least three competing mechanisms are involved in the movement on fcc(100) planes – exchange, individual atom hopping and shearing. However, on fcc(100), concerted motion of the whole cluster was *not* observed.

Diffusion of iridium clusters on Ir(100) appears to be different, in that clusters usually appear as either linear chains or two-dimensional islands, the structure depending upon the temperature; low temperatures for linear chains, but at higher temperatures rearrangement to island form occurs. In any event, only little is known about surface diffusion.

In special cases, when clusters are arranged in linear form, diffusion steps can be crudely estimated, but in general, with present knowledge, prediction about the surface diffusion of clusters is difficult.

With the various computational efforts describing cluster motion in place, the question arises – how useful are they. One thing fairly clear is that the calculated barrier heights are not uniformly reliable. That emerges on looking at Table 8.15, where values determined

Table 8.16 Comparison of self-diffusion barriers from experiment and calculation [27] on (111) planes for Pt [121] and Ir [29].

	E_D (eV) for Pt		E_D (eV) for Ir	
Number of Atoms	Expt	EAM	Expt	EAM
1	0.260	0.07	0.290	0.11
2	0.37	0.17	0.45	0.24
3	0.52	0.40	0.65	0.56
4	0.57	0.39	0.48	0.54
5	0.78	0.61	0.69	0.83
6	0.89		~0.92	
7	1.17		1.49	

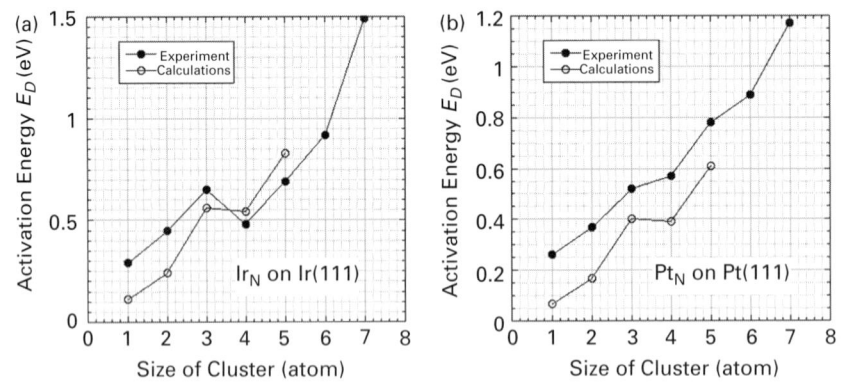

Fig. 8.91 Comparison of diffusion energies for clusters calculated using EAM potentials [27] with experimental values for iridium clusters on Ir(111) [26,120] and platinum clusters on Pt(111) [121].

in direct observations of dimers using the FIM are compared with calculations of different kinds for the same materials. On fcc(111) planes some of the calculational results are in excellent agreement with the experiments, but some are not. For dimers on fcc(100) planes, the calculated diffusion barriers differ quite significantly from the experiments.

It must be emphasized, however, that although the calculations offer a somewhat uncertain guide to absolute barrier heights, they are nevertheless very useful in comparisons of how size affects cluster behavior. This emerges from Table 8.16, as well as from Fig. 8.91, where calculations are compared with experiments for both Pt(111) and Ir(111). The trends in the former are fairly well duplicated by the theoretical efforts.

Most important have been the calculations dealing with atomic jumps in surface diffusion, which are very difficult to derive in experiments. These have made it clear that diffusion of clusters in some instances does not occur by independent single jumps, but proceed by the correlated motion of more than one atom at a time, sometimes by separate motion of cluster parts, that is by dimer and trimer shearing. Calculations have also served to unravel how leapfrog movement of clusters takes place, and have clarified

8.15 Comparisons

conditions under which atom exchange facilitates cluster motion. It is in this area of understanding the mechanisms of diffusive processes that calculations have and will probably make the greatest contributions. It also is worth noting the terrible lack of theoretical efforts to describe diffusion on bcc surfaces, especially for tungsten, where experimental effort has been concentrated. A survey of experimentally determined energetics and prefactors for cluster diffusion is provided in Table 8.17.

Table 8.17 Experimentally measured movement of the center of mass of clusters

System	Mechanism	E_D (eV)	D_o(cm^2/sec)	v_0 (sec^{-1})	Method	Ref.
Cu/Cu(111)	Jump	0.040 ± 0.001		$1 \times 10^{12 \pm 0.5}$	STM	[142]
Cu$_2$/Cu(111)	Jump local	0.018 ± 0.003		$8 \times 10^{11 \pm 0.5}$	STM	[44]
Rh/Rh(100)	Jump	0.84	†10^{-3}		FIM	[46]
Rh$_2$/Rh(100)	Jump	0.97	†10^{-3}		FIM	[46]
Rh$_3$/Rh(100)	Jump	1.02	†10^{-3}		FIM	[46]
Rh$_4$/Rh(100)	Jump	1.05	†10^{-3}		FIM	[46]
Rh$_5$/Rh(100)	Jump	1.01	†10^{-3}		FIM	[46]
Rh$_6$/Rh(100)	Jump	1.14	†10^{-3}		FIM	[46]
Rh$_7$/Rh(100)	Jump	1.04	†10^{-3}		FIM	[46]
Rh$_8$/Rh(100)	Jump	1.14	†10^{-3}		FIM	[46]
Rh$_9$/Rh(100)	Jump	1.22	†10^{-3}		FIM	[46]
Rh$_{10}$/Rh(100)	Jump	1.07	†10^{-3}		FIM	[46]
Rh$_{12}$/Rh(100)	Jump	1.23	†10^{-3}		FIM	[46]
Pt/Rh(100)	Jump	0.92 ± 0.13	$2 \times 10^{-3 \pm 1.9}$		FIM	[65]
Pt$_2$/Rh(100)	Jump	0.91	†1×10^{-3}		FIM	[65]
Pt$_3$/Rh(100)	Jump	1.02	†1×10^{-3}		FIM	[65]
Pt$_4$/Rh(100)	Jump	1.03	†1×10^{-3}		FIM	[65]
Pt$_5$/Rh(100)	Jump	1.03	†1×10^{-3}		FIM	[65]
Pt$_6$/Rh(100)	Jump	1.16	†1×10^{-3}		FIM	[65]
Cu/Ag(111)	Jump	0.065		$1 \times 10^{12 \pm 0.5}$	STM	[57]
Cu$_2$/Ag(111)	Jump	0.073		†10^{12}	STM	[57]
Re/W(211)	Jump	0.86 ± 0.030	$2.2(\times 2.8^{\pm 1}) \times 10^{-3}$		FIM	[82]
Re$_2$/W(211) cross-channel	Jump	0.78 ± 0.013	$4.5(\times 1.7^{\pm 1}) \times 10^{-4}$		FIM	[82]
Re$_2$/W(211) in-channel	Jump	1.01 ± 0.05	$1.8(\times 5.1^{\pm 1}) \times 10^{-3}$		FIM	[82]
Re$_3$/W(211) cross-channel	Jump	0.79 ± 0.02	$5.2(\times 2.3^{\pm 1}) \times 10^{-4}$		FIM	[84]
W/W(110)	Jump	0.90 ± 0.07	$6.2(\times 13^{\pm 1}) \times 10^{-3}$		FIM	[87]
W$_2$/W(110)	Jump	0.92 ± 0.14	$1.4(\times 160^{\pm 1}) \times 10^{-3}$		FIM	[87]
Pt/W(110)	Jump	0.67 ± 0.06	$30.6(\times 20.1^{\pm 1}) \times 10^{-8}$		FIM	[90]
Pt$_2$/W(110)	Jump	0.67 ± 0.06	$9.21(\times 20.1^{\pm 1}) \times 10^{-8}$		FIM	[90]
Pt$_3$/W(110)	Jump	0.79 ± 0.16	$15.2(\times 403^{\pm 1}) \times 10^{-8}$		FIM	[90]
Pt$_4$/W(110)	Jump	0.87 ± 0.16	$45.6(\times 403^{\pm 1}) \times 10^{-8}$		FIM	[90]
Re/W(110)	Jump	0.91 eV	†10^{-3}		FIM	[143]
Re$_3$/W(110)	Jump	~1.22	†10^{-3}		FIM	[93]
Re$_5$/W(110)	Jump	~1.02	†10^{-3}		FIM	[93]

Table 8.17 (cont.)

System	Mechanism	E_D (eV)	D_o(cm^2/sec)	v_0 (sec^{-1})	Method	Ref.
Pd-Re/W(110)	Jump local	0.48 ± 0.02			FIM	[94]
Pd/W(110)	Jump	0.51	†10^{-3}		FIM	[99]
Pd$_2$/W(110)	Jump	0.65	†10^{-3}		FIM	[99]
Pd$_3$/W(110)	Jump	0.68	†10^{-3}		FIM	[99]
Pd$_4$/W(110)	Jump	0.75	†10^{-3}		FIM	[99]
Pd$_5$/W(110)	Jump	0.80	†10^{-3}		FIM	[99]
Pd$_6$/W(110)	Jump	0.84	†10^{-3}		FIM	[99]
Pd$_7$/W(110)	Jump	0.88	†10^{-3}		FIM	[99]
Pd$_8$/W(110)	Jump	0.88	†10^{-3}		FIM	[99]
Pd$_9$/W(110)	Jump	0.95	†10^{-3}		FIM	[99]
W/W(111)	Jump	1.9	†10^{-3}		FIM	[100]
W$_2$/W(111)	Jump	~1.62	†10^{-3}		FIM	[100]
W$_3$/W(111)	Jump	~1.74	†10^{-3}		FIM	[100]
W$_4$/W(111)	Jump	~1.79	†10^{-3}		FIM	[100]
W$_5$/W(111)	Jump	~1.7	†10^{-3}		FIM	[100]
W$_6$/W(111)	Jump	~1.74	†10^{-3}		FIM	[100]
W$_7$/W(111)	Jump	~1.81	†10^{-3}		FIM	[100]
W$_8$/W(111)	Jump	~1.83	†10^{-3}		FIM	[100]
W$_9$/W(111)	Jump	~1.9	†10^{-3}		FIM	[100]
W$_{10}$/W(111)	Jump	~2.01	†10^{-3}		FIM	[100]
Re/Re(0001)	Jump	0.48 ± 0.02	6.13(×2.6$^{±1}$) × 10^{-6}		FIM	[103]
Re$_2$/Re(0001)	Jump	~0.43	†10^{-3}		FIM	[103]
Re$_3$/Re(0001)	Jump	~0.65	†10^{-3}		FIM	[103]
Re$_4$/Re(0001)	Jump	~0.43	†10^{-3}		FIM	[103]
Re$_5$/Re(0001)	Jump	~0.89	†10^{-3}		FIM	[103]
Re$_6$/Re(0001)	Jump	~1.12	†10^{-3}		FIM	[103]
Re$_7$/Re(0001)	Jump	~1.21	†10^{-3}		FIM	[103]
Re$_8$/Re(0001)	Jump	~1.25	†10^{-3}		FIM	[103]
Ir/Ir(100)	Ex	0.84 ± 0.05	6.26(×11$^{±1}$) × 10^{-2}		FIM	[106]
Ir$_2$/Ir(100)	Ex	0.88 ± 0.05	1.42 × 10$^{-3 ± 0.8}$		FIM	[108]
Ir$_4$/Ir(100)	Jump	1.09			FIM	[109]
Rh/Ir(100)	Jump	0.80 ± 0.08	2.09 × 10$^{-3 ± 1.5}$		FIM	[108]
Rh$_2$/Ir(100)	Jump	0.83 ± 0.06	3.53 × 10$^{-4 ± 1.1}$		FIM	[108]
Rh$_4$/Ir(100)	Jump	0.93			FIM	[109]
Re-Ir/Ir(100)α_{Re}	Jump local	0.60 ± 0.07	4 × 10$^{-4 ± 1.5}$	2.0 × 10$^{12.0 ± 1.5}$	FIM	[107]
Re-Ir/Ir(100)β_{Re}	Jump	0.73 ± 0.07	2 × 10$^{-2 ± 1.5}$	2.6 × 10$^{13.0 ± 1.5}$	FIM	[107]
Ir/Ir(111)	Jump	0.29 ± 0.003	3.8(×1.4$^{±1}$) × 10^{-4}		FIM	[29]
Ir$_2$/Ir(111)	Jump	0.45 ± 0.01	2.6(×2.2$^{±1}$) × 10^{-5}		FIM	[29]
Ir$_3$/Ir(111)	Jump	0.65 ± 0.02	4.2(×2.4$^{±1}$) × 10^{-4}		FIM	[29]
Ir$_4$/Ir(111)	Jump	0.48 ± 0.01	1.5(×2.3$^{±1}$) × 10^{-5}		FIM	[29]
Ir$_5$/Ir(111)	Jump	0.69 ± 0.02	6.0(×1.9$^{±1}$) × 10^{-6}		FIM	[29]
Ir$_6$/Ir(111)	Jump	~0.92	~1 × 10^{-5}		FIM	[29]
Ir$_7$/Ir(111)	Jump	1.49 ± 0.03	1.4(×2.4$^{±1}$)		FIM	[29]
Ir$_{18}$/Ir(111)	Jump	1.63 ± 0.07	7.8(×4.5$^{±1}$) × 10^{-4}		FIM	[111]
Ir$_{19}$/Ir(111)	Jump	2.54 ± 0.06	13(×2.6$^{±1}$)		FIM	[111]
Ir/Ir(110)	Jump	0.80 ± 0.04	4 × 10$^{-3 ± 0.8}$		FIM	[144]
Ir/Ir(110)	Ex	0.71 ± 0.02	6 × 10$^{-2 ± 1.8}$		FIM	[144]
Ir$_2$/Ir(110)	Jump	1.05 ± 0.14	3.7 × 10$^{-5.0 ± 1.3}$		FIM	[113]

Table 8.17 (cont.)

System	Mechanism	E_D (eV)	D_o(cm^2/sec)	v_0 (sec^{-1})	Method	Ref.
Ir$_2$/Ir(110)	Ex	1.18 ± 0.12	$2.6 \times 10^{-4.0 \pm 0.7}$		FIM	[113]
Pt/Pt(100)	Ex	0.47	$1.3(\times 10^{\pm 1}) \times 10^{-3}$		FIM	[145]
Pt$_2$/Pt(100)	Ex	0.41			FIM	[33]
Pt$_3$/Pt(100)	Ex+Jump	0.49			FIM	[33]
Pt/Pt(111)	Jump	0.260 ± 0.003	$2.0(\times 1.4^{\pm 1}) \times 10^{-3}$		FIM	[121]
Pt$_2$/Pt(111)	Jump	0.37 ± 0.02	$1.9(\times 4.5^{\pm 1}) \times 10^{-4}$		FIM	[121]
Pt$_3$/Pt(111)	Jump	0.52 ± 0.01	$1.1(\times 2.1^{\pm 1}) \times 10^{-3}$		FIM	[121]
Pt$_4$/Pt(111)	Jump	0.57 ± 0.04	$6.6(\times 6.1^{\pm 1}) \times 10^{-5}$		FIM	[121]
Pt$_5$/Pt(111)	Jump	0.78 ± 0.02	$1.8(\times 2.3^{\pm 1}) \times 10^{-2}$		FIM	[121]
Pt$_6$/Pt(111)	Jump	0.89 ± 0.04	$4.9(\times 5.1^{\pm 1}) \times 10^{-3}$		FIM	[121]
Pt$_7$/Pt(111)	Jump	1.17 ± 0.04	$5.1(\times 3.8^{\pm 1}) \times 10^{-1}$		FIM	[121]

† assumed value

References

[1] D. W. Bassett, M. J. Parsley, Field ion microscope observations of cluster formation in metal deposits on tungsten surfaces, *Nature* **221** (1969) 1046.

[2] D. W. Bassett, Controlled clustering: a technique for use in studying thin film nucleation in the field ion microscope, *J. Phys. E: Sci. Instr.* **3** (1970) 417–418.

[3] D. W. Bassett, Field ion microscope studies of iridium adatom clusters on (110) tungsten surfaces, *Surf. Sci.* **21** (1970) 181–185.

[4] D. W. Bassett, The use of field ion microscopy in studies of the vapour deposition of metals, *Surf. Sci.* **23** (1970) 240–258.

[5] T. T. Tsong, Interaction of individual metal atoms on tungsten surfaces, *J. Chem. Phys.* **55** (1971) 4658–4659.

[6] T. T. Tsong, Direct observation of interactions between individual atoms on tungsten surfaces, *Phys. Rev. B* **6** (1972) 417–426.

[7] D. W. Bassett, D. R. Tice, The stability of W$_2$ and WRe adatom clusters on (110) tungsten surfaces, *Surf. Sci.* **40** (1973) 499–511.

[8] J. C. Tully, G. H. Gilmer, M. Shugard, Molecular dynamics of surface diffusion. I. The motion of adatoms and clusters, *J. Chem. Phys.* **71** (1979) 1630–1642.

[9] D. Ghaleb, Diffusion of adatom dimers on (111) surface of face centered crystals: A molecular dynamics study, *Surf. Sci.* **137** (1984) L103–108.

[10] P. J. Feibelman, Rebonding effects in separation and surface-diffusion barrier energies of an adatom pair, *Phys. Rev. Lett.* **58** (1987) 2766–2769.

[11] C.-L. Liu, J. B. Adams, Structure and diffusion of clusters on Ni surfaces, *Surf. Sci.* **268** (1992) 73–86.

[12] S. Valkealahti, M. Manninen, Diffusion on aluminum-cluster surfaces and the cluster growth, *Phys. Rev. B* **57** (1998) 15533–15540.

[13] A. Bogicevic, S. Ovesson, B. I. Lundqvist, D. R. Jennison, Atom-by-atom and concerted hopping of atom pairs on an open metal surface, *Phys. Rev. B* **61** (2000) R2456–2459.

[14] B. Hammer, Bond activation at monatomic steps: NO dissociation at corrugated Ru(0001), *Phys. Rev. Lett.* **83** (1999) 3681–3684.

[15] O. S. Trushin, P. Salo, M. Alatalo, T. Ala-Nissila, Atomic mechanisms of cluster diffusion on metal fcc(100) surfaces, *Surf. Sci.* **482–485** (2001) 365–369.
[16] F. Ercolessi, J. B. Adams, Interatomic potentials from first-principles calculations: The force-matching method, *Europhys. Lett.* **26** (1994) 583–588.
[17] J. Zhuang, Q. Liu, M. Zhuang, L. Liu, L. Zhao, Y. Li, Exchange rotation mechanism for dimer diffusion on metal fcc(001) surfaces, *Phys. Rev. B* **68** (2003) 113401 1–4.
[18] R. A. Johnson, Analytical nearest-neighbor model for fcc metals, *Phys. Rev. B* **37** (1988) 3924–3931.
[19] Q. Liu, Z. Sun, X. Ning, Y. Li, L. Liu, J. Zhuang, Systematical study of dimer diffusion on metal fcc(001) surfaces, *Surf. Sci.* **554** (2004) 25–32.
[20] C. M. Chang, C. M. Wei, Self-diffusion of adatoms and dimers on fcc(100) surfaces, *Chin. J. Phys.* **43** (2005) 169–175.
[21] G. Kresse, J. Hafner, *Ab initio* molecular dynamics for liquid metals, *Phys. Rev. B* **47** (1993) 558–561.
[22] G. Kresse, J. Hafner, *Ab initio* molecular dynamics simulation of the liquid-metal-amorphous-semiconductor transition in germanium, *Phys. Rev. B* **49** (1994) 14251–14269.
[23] D. M. Ceperley, B. J. Alder, Ground state of the electron gas by a stochastic method, *Phys. Rev. Lett.* **45** (1980) 566–569.
[24] J. P. Perdew, A. Zunger, Self-interaction correction to density-functional approximations for many-electron systems, *Phys. Rev. B* **23** (1981) 5048–5079.
[25] A. Bogicevic, P. Hyldgaard, G. Wahnström, B. I. Lundqvist, Al dimer dynamics on Al(111), *Phys. Rev. Lett.* **81** (1998) 172–175.
[26] S. C. Wang, G. Ehrlich, Structure, stability, and surface diffusion of clusters: Ir$_x$ on Ir(111), *Surf. Sci.* **239** (1990) 301–332.
[27] C. M. Chang, C. M. Wei, S. P. Chen, Self-diffusion of small clusters on fcc metal (111) surfaces, *Phys. Rev. Lett.* **85** (2000) 1044–1047.
[28] C. M. Chang, C. M. Wei, S. P. Chen, Structural and dynamical behavior of Al trimer on Al(111) surface, *Surf. Sci.* **465** (2000) 65–75.
[29] S. C. Wang, Diffusion of Ir clusters on Ir(111), Personal Communication, 1999.
[30] D. Chen, W. Hu, J. Yang, L. Sun, The dynamic diffusion behaviors of 2D small Fe clusters on Fe(110) surface, *J. Phys.: Condens. Matter* **19** (2007) 446009 1–8.
[31] W. Hu, X. Shu, B. Zhang, Point-defect properties in body-centered cubic transition metals with analytic EAM interatomic potentials, *Comput. Mater. Sci.* **23** (2002) 175–189.
[32] W. Hu, H. Deng, X. Yuan, M. Fukumoto, Point defect properties in hcp rare metals with analytic modified embedded atom potential, *Eur. Phys. J. B* **34** (2003) 429–440.
[33] G. L. Kellogg, A. F. Voter, Surface diffusion modes for Pt dimers and trimers on Pt(001), *Phys. Rev. Lett.* **67** (1991) 622–625.
[34] G. L. Kellogg, Surface diffusion of Pt adatoms on Ni surfaces, *Surf. Sci.* **266** (1992) 18–23.
[35] Z.-P. Shi, Z. Zhang, A. K. Swan, J. F. Wendelken, Dimer shearing as a novel mechanism for cluster diffusion and dissociation on metal (100) surfaces, *Phys. Rev. Lett.* **76** (1996) 4927–4930.
[36] S. Y. Kim, I.-H. Lee, S. Jun, Transition-pathway models of atomic diffusion on fcc metal surfaces. I. Flat surfaces, *Phys. Rev. B* **76** (2007) 245407 1–15.
[37] T. Y. Fu, T. T. Tsong, Atomic processes in self-diffusion of Ni surfaces, *Surf. Sci.* **454–456** (2000) 571–574.
[38] J. C. Hamilton, M. S. Daw, S. M. Foiles, Dislocation mechanism for island diffusion on fcc(111) surfaces, *Phys. Rev. Lett.* **74** (1995) 2760–2763.

[39] S. C. Wang, G. Ehrlich, Diffusion of large surface clusters: Direct observations on Ir(111), *Phys. Rev. Lett.* **79** (1997) 4234–4237.

[40] R. C. Longo, C. Rey, L. J. Gallego, Molecular dynamics study of the melting behaviour of seven-atom clusters of fcc transition and noble metals on the (111) surface of the same metal using the embedded atom model, *Surf. Sci.* **459** (2000) L441–445.

[41] A. F. Voter, S. P. Chen, Accurate interatomic potentials for Ni, Al and Ni$_3$Al, *Mater. Res. Soc. Symp. Proc.* **82** (1987) 175–180.

[42] U. Kürpick, Self-diffusion of one-dimensional clusters on fcc(110)(2 × 1) surfaces of Pt, Ag, Cu, and Ni, *Phys. Rev. B* **63** (2001) 045409 1–5.

[43] D. C. Schlösser, K. Morgenstern, L. K. Verheij, G. Rosenfeld, F. Besenbacher, G. Comsa, Kinetics of island diffusion on Cu(111) and Ag(111) studied with variable-temperature STM, *Surf. Sci.* **465** (2000) 19–39.

[44] J. Repp, G. Meyer, K.-H. Rieder, P. Hyldgaard, Site determination and thermally assisted tunneling in homogeneous nucleation, *Phys. Rev. Lett.* **91** (2003) 206102 1–4.

[45] C.-L. Liu, Energetics of diffusion processes during nucleation and growth for the Cu/Cu(100) system, *Surf. Sci.* **316** (1994) 294–302.

[46] G. L. Kellogg, Oscillatory behavior in the size dependence of cluster mobility on metal surfaces: Rh on Rh(100), *Phys. Rev. Lett.* **73** (1994) 1833–1836.

[47] G. Boisvert, L. J. Lewis, Self-diffusion of adatoms, dimers, and vacancies on Cu(100), *Phys. Rev. B* **56** (1997) 7643–7655.

[48] O. Biham, I. Furman, M. Karimi, G. Vidali, R. Kennett, H. Zeng, Models for diffusion and island growth in metal monolayers, *Surf. Sci.* **400** (1998) 29–43.

[49] O. S. Trushin, P. Salo, T. Ala-Nissila, Energetics and many-particle mechanisms of two-dimensional cluster diffusion on Cu(100) surfaces, *Phys. Rev. B* **62** (2000) 1611–1614.

[50] P. Salo, J. Hirvonen, I. T. Koponen, O. S. Trushin, J. Heinonen, T. Ala-Nissila, Role of concerted atomic movements on the diffusion of small islands on fcc(100) metal surfaces, *Phys. Rev. B* **64** (2001) 161405 1–4.

[51] J. R. Sanchez, J. W. Evans, Diffusion of small clusters on metal (100) surfaces: Exact master-equation analysis for lattice-gas models, *Phys. Rev. B* **59** (1999) 3224–3233.

[52] V. Rosato, M. Guillope, B. Legrand, Thermodynamical and structural properties of f.c.c. transition metals using a simple tight-binding model, *Philos. Mag. A* **59** (1989) 321–336.

[53] M. Guillope, B. Legrand, (110) Surface stability in noble metals, *Surf. Sci.* **215** (1989) 577–595.

[54] M. Basham, F. Montalenti, P. A. Mulheran, Multiscale modeling of island nucleation and growth during Cu(100) homoepitaxy, *Phys. Rev. B* **73** (2006) 045422 1–10.

[55] M.-C. Marinica, C. Barreteau, M.-C. Desjonqueres, D. Spanjaard, Influence of short-range adatom-adatom interactions on the surface diffusion of Cu on Cu(111), *Phys. Rev. B* **70** (2004) 075415 1–14.

[56] Y. Mishin, M. J. Mehl, D. A. Papaconstantopoulos, A. F. Voter, J. D. Kress, Structural stability and lattice defects in copper: *Ab initio*, tight binding, and embedded-atom calculations, *Phys. Rev. B* **63** (2001) 224106 1–16.

[57] K. Morgenstern, K.-F. Braun, K.-H. Rieder, Direct imaging of Cu dimer formation, motion, and interaction with Cu atoms on Ag(111), *Phys. Rev. Lett.* **93** (2004) 056102 1–4.

[58] A. Karim, A. N. Al-Rawi, A. Kara, T. S. Rahman, O. Trushin, T. Ala-Nissila, Diffusion of small two-dimensional Cu islands on Cu(111) studied with a kinetic Monte Carlo method, *Phys. Rev. B* **73** (2006) 165411 1–11.

[59] S. M. Foiles, M. I. Baskes, M. S. Daw, Embedded-atom-method functions for the fcc metals Cu, Ag, Ni, Pd, Pt, and their alloys, *Phys. Rev. B* **33** (1986) 7983–7991.

[60] J. Yang, W. Hu, J. Tang, M. Xu, Long-time scale dynamics study of diffusion dynamics of small Cu clusters on Cu(111) surface, *J. Phys. Chem. C* **112** (2008) 2074–2078.

[61] V. Papathanakos, G. A. Evangelakis, Structural and diffusive properties of small 2D Au clusters on the Cu(111) surface, *Surf. Sci.* **499** (2002) 229–243.

[62] V. Chirita, E. P. Münger, J. E. Greene, J.-E. Sundgren, Cluster diffusion and surface morphological transitions on Pt(111) via reptation and concerted motion, *Thin Solid Films* **370** (2000) 179–185.

[63] T. J. Raeker, A. E. DePristo, Molecular dynamic and Monte Carlo simulation of Fe island growth on Cu(111), *Surf. Sci.* **317** (1994) 283–294.

[64] G. Ayrault, G. Ehrlich, Surface self-diffusion on an fcc crystal: An atomic view, *J. Chem. Phys.* **60** (1974) 281–294.

[65] G. L. Kellogg, Diffusion behavior of Pt adatoms and clusters on the Rh(100) surface, *Appl. Surf. Sci.* **67** (1993) 134–141.

[66] A. F. Voter, Classically exact overlayer dynamics: Diffusion of rhodium clusters on Rh(100), *Phys. Rev. B* **34** (1986) 6819–6829.

[67] P. Fernandez, C. Massobrio, P. Blandin, J. Buttet, Embedded atom method computations of structural and dynamic properties of Cu and Ag clusters adsorbed on Pd(110) and Pd(100): evolution of the most stable geometries versus cluster size, *Surf. Sci.* **307–309** (1994) 608–613.

[68] C. Massobrio, P. Fernandez, Cluster adsorption on metallic surfaces: Structure and diffusion in the Cu/Pd(110) and Pd/Pd(110) systems, *J. Chem. Phys.* **102** (1995) 605–610.

[69] E. Z. da Silva, A. Antonelli, Diffusion of Pd clusters on Pd(111) surfaces: A molecular dynamics study, *Surf. Sci.* **452** (2000) 239–246.

[70] A. F. Voter, Simulation of the layer-growth dynamics in silver films: Dynamics of adatom and vacancy clusters on Ag(100), *Proc. SPIE* **821** (1987) 214–226.

[71] M. I. Haftel, Surface reconstruction of platinum and gold and the embedded-atom method, *Phys. Rev. B* **48** (1993) 2611–2622.

[72] M. I. Haftel, M. Rosen, T. Franklin, M. Hettermann, Molecular dynamics observations of the interdiffusion and Stranski-Krastanov growth in the early film deposition of Au on Ag(100), *Phys. Rev. Lett.* **72** (1994) 1858–1861.

[73] M. I. Haftel, M. Rosen, Molecular-dynamics description of early film deposition of Au on Ag(110), *Phys. Rev. B* **51** (1995) 4426–4434.

[74] R. C. Nelson, T. L. Einstein, S. V. Khare, P. J. Reus, Energetics of steps, kinks, and defects on Ag{100} and Ag{111} using the embedded atom method, and some consequences, *Surf. Sci.* **295** (1993) 462–484.

[75] P. R. Schwoebel, G. L. Kellogg, Palladium diffusion and cluster nucleation on Ta(110), *Phys. Rev. B* **38** (1988) 5326–5331.

[76] D. W. Bassett, Field ion microscopic studies of submonolayer films of nickel, palladium and platinum on (110) tungsten surfaces, *Thin Solid Films* **48** (1978) 237–246.

[77] W. R. Graham, G. Ehrlich, Surface self-diffusion of atoms and atom pairs, *Phys. Rev. Lett.* **31** (1973) 1407–1408.

[78] W. R. Graham, G. Ehrlich, Surface diffusion of atom clusters, *J. Phys. F* **4** (1974) L212–214.

[79] W. R. Graham, G. Ehrlich, Surface self-diffusion of single atoms, *Thin Solid Films* **25** (1975) 85–96.

[80] D. A. Reed, G. Ehrlich, Chemical specificity in the surface diffusion of clusters: Ir on W(211), *Philos. Mag.* **32** (1975) 1095–1099.

[81] T. Sakata, S. Nakamura, Surface diffusion of molybdenum atoms on tungsten surfaces, *Surf. Sci.* **51** (1975) 313–317.

[82] K. Stolt, W. R. Graham, G. Ehrlich, Surface diffusion of individual atoms and dimers: Re on W(211), *J. Chem. Phys.* **65** (1976) 3206–3222.

[83] K. Stolt, J. D. Wrigley, G. Ehrlich, Thermodynamics of surface clusters – direct observations of Re_2 on W(211), *J. Chem. Phys.* **69** (1978) 1151–1161.

[84] G. Ehrlich, K. Stolt, in: *Growth and Properties of Metal Clusters*, J. Bourdon (ed.), Surface diffusion of metal clusters on metals, (Elsevier, Amsterdam, 1980), p. 1–14.

[85] D. A. Reed, G. Ehrlich, In-channel clusters: rhenium on W(211), *Surf. Sci.* **151** (1985) 143–165.

[86] P. Cowan, T. T. Tsong, Diffusion behavior of tungsten clusters on the tungsten (110) planes, *Phys. Lett.* **53**A (1975) 383–383.

[87] G. L. Kellogg, T. T. Tsong, P. Cowan, Direct Observation of surface diffusion and atomic interactions on metal surfaces, *Surf. Sci.* **70** (1978) 485–519.

[88] D. W. Bassett, C. K. Chung, D. Tice, Field ion microscope studies of atomic displacement processes on metal surfaces, *La Vide* **176** (1975) 39–43.

[89] H.-W. Fink, G. Ehrlich, Direct observation of three-body interactions in adsorbed layers: Re on W(110), *Phys. Rev. Lett.* **52** (1984) 1532–1534.

[90] D. W. Bassett, Migration of platinum adatom clusters on tungsten (110) surfaces, *J. Phys. C* **9** (1976) 2491–2503.

[91] T. T. Tsong, R. Casanova, Elementary displacement steps in the migration of tungsten diatomic clusters on the tungsten {110} plane, *Phys. Rev. B* **21** (1980) 4564–4570.

[92] T. T. Tsong, R. Casanova, Migration behavior of single tungsten atoms and tungsten diatomic clusters on the tungsten(110) plane, *Phys. Rev. B* **22** (1980) 4632–4649.

[93] H.-W. Fink, G. Ehrlich, Rhenium on W(110): Structure and mobility of higher clusters, *Surf. Sci.* **150** (1985) 419–429.

[94] F. Watanabe, G. Ehrlich, Direct observations of pair interactions on a metal: Heteropairs on W(110), *J. Chem. Phys.* **95** (1991) 6075–6087.

[95] S.-M. Oh, S. J. Koh, K. Kyuno, G. Ehrlich, Non-nearest-neighbor jumps in 2D diffusion: Pd on W(110), *Phys. Rev. Lett.* **88** (2002) 236102 1–4.

[96] T.-Y. Fu, L.-C. Cheng, Y.-J. Hwang, T. T. Tseng, Diffusion of Pd adatoms on W surfaces and their interactions with steps, *Surf. Sci.* **507/510** (2002) 103–107.

[97] R. S. Chambers, Determination of cluster spacings in the FIM: iridium dimers on W(110), *Surf. Sci.* **246** (1991) 25–30.

[98] S. J. Koh, G. Ehrlich, Pair- and many-atom interactions in the cohesion of surface clusters: Pd_x and Ir_x on W(110), *Phys. Rev. B* **60** (1999) 5981–5990.

[99] T.-Y. Fu, Y.-J. Hwang, T. T. Tsong, Structure and diffusion of Pd clusters on the W(110) surface, *Appl. Surf. Sci.* **219** (2003) 143–148.

[100] T.-Y. Fu, W. J. Weng, T. T. Tsong, Dynamic study of W atoms and clusters on W(111) surfaces, *Appl. Surf. Sci.* **254** (2008) 7831–7834.

[101] W. Xu, J. B. Adams, W dimer diffusion on W(110) and (211) surfaces, *Surf. Sci.* **339** (1995) 247–257.

[102] D. Chen, W. Hu, H. Deng, L. Sun, F. Gao, Diffusion of tungsten clusters on tungsten (110) surface, *Eur. Phys. J. B* **68** (2009) 479–485.

[103] J. T. Goldstein, G. Ehrlich, Atom and cluster diffusion on Re(0001), *Surf. Sci.* **443** (1999) 105–115.

[104] P. R. Schwoebel, G. L. Kellogg, Structure of Iridium Cluster Nuclei on Ir(100), *Phys. Rev. Lett.* **61** (1988) 578–580.

[105] C. Chen, T. T. Tsong, Behavior of Ir atoms and clusters on Ir surfaces, *Phys. Rev. B* **41** (1990) 12403–12412.
[106] C. Chen, T. T. Tsong, Displacement distribution and atomic jump direction in diffusion of Ir atoms on the Ir(001) surface, *Phys. Rev. Lett.* **64** (1990) 3147–3150.
[107] C. L. Chen, T. T. Tsong, Observation of two diffusion modes of a Re-Ir dimer-vacancy complex on the Ir(001) surface and their diffusion mechanisms, *Phys. Rev. Lett.* **72** (1994) 498–501.
[108] T.-Y. Fu, T. T. Tsong, Structure and diffusion of small Ir and Rh clusters on Ir(001) surfaces, *Surf. Sci.* **421** (1999) 157–166.
[109] T.-Y. Fu, T. T. Tsong, Structure and diffusion mechanism of Ir and Rh tetramers on Ir(001) surfaces, *Surf. Sci.* **482–485** (2001) 1249–1254.
[110] S. C. Wang, G. Ehrlich, Cluster motion on metals: Ir on Ir(111), *J. Chem. Phys.* **91** (1989) 6535–6536.
[111] S. C. Wang, U. Kürpick, G. Ehrlich, Surface diffusion of compact and other clusters: Ir_x on Ir(111), *Phys. Rev. Lett.* **81** (1998) 4923–4926.
[112] U. Kürpick, B. Fricke, G. Ehrlich, Diffusion mechanisms of compact surface clusters: Ir_7 on Ir(111), *Surf. Sci.* **470** (2000) L45–51.
[113] T. T. Tsong, C. L. Chen, Diffusion of Ir-dimers on Ir(110) surfaces by atomic-exchange and atomic-hopping mechanisms, *Surf. Sci.* **246** (1991) 13–24.
[114] K.-D. Shiang, T. T. Tsong, Molecular-dynamics study of self-diffusion: Iridium dimers on iridium surfaces, *Phys. Rev. B* **49** (1994) 7670–7678.
[115] C. M. Chang, C. M. Wei, S. P. Chen, Modeling of Ir adatoms on Ir surfaces, *Phys. Rev. B* **54** (1996) 17083–17096.
[116] J. C. Hamilton, M. R. Sorensen, A. F. Voter, Compact surface-cluster diffusion by concerted rotation and translation, *Phys. Rev. B* **61** (2000) R5125–5128.
[117] P. R. Schwoebel, S. M. Foiles, C. L. Bisson, G. L. Kellogg, Structure of platinum clusters on Pt(100): Experimental observations and embedded-atom-method calculations, *Phys. Rev. B* **40** (1989) 10639–10642.
[118] G. L. Kellogg, The effect of an externally applied electric field on the diffusion of Pt adatoms, dimers, and trimers on Pt(001), *Appl. Surf. Sci.* **76/77** (1994) 115–121.
[119] D. W. Bassett, M. J. Parsley, The effect of an electric field on the surface diffusion of rhenium adsorbed on tungsten, *Brit. J. Appl. Phys. (J. Phys. D)* **2** (1969) 13–16.
[120] K. Kyuno, A. Gölzhäuser, G. Ehrlich, Growth and the diffusion of platinum atoms and dimers on Pt(111), *Surf. Sci.* **397** (1998) 191–196.
[121] K. Kyuno, G. Ehrlich, Diffusion and dissociation of platinum clusters on Pt(111), *Surf. Sci.* **437** (1999) 29–37.
[122] S. Liu, Z. Zhang, J. Nørskov, H. Metiu, The mobility of Pt atoms and small Pt clusters on Pt(111) and its implications for the early stages of epitaxial growth, *Surf. Sci.* **321** (1994) 161–171.
[123] T. R. Linderoth, S. Horch, L. Petersen, S. Helveg, E. Laegsgaard, I. Stensgaard, F. Besenbacher, Novel mechanism for diffusion of one-dimensional clusters; Pt/Pt(110)-(1 × 2), *Phys. Rev. Lett.* **82** (1999) 1494–1497.
[124] E. P. Münger, V. Chirita, J. E. Greene, J.-E. Sundgren, Adatom-induced diffusion of two-dimensional close-packed Pt_7 clusters on Pt(111), *Surf. Sci.* **355** (1996) L325–330.
[125] D. J. Oh, R. A. Johnson, Simple embedded atom method model for fcc and hcp metals, *J. Mater. Res.* **3** (1988) 471–478.
[126] S. C. Wang, G. Ehrlich, Atom incorporation at surface clusters: An atomic view, *Phys. Rev. Lett.* **67** (1991) 2509–2512.

[127] E. P. Münger, V. Chirita, J.-E. Sundgren, J. E. Greene, Destabilization and diffusion of two-dimensional close-packed Pt clusters on Pt(111) during film growth from the vapor phase, *Thin Solid Films* **318** (1998) 57–60.

[128] V. Chirita, E. P. Münger, J.-E. Sundgren, J. E. Greene, Enhanced cluster mobilities on Pt(111) during film growth from the vapor phase, *Appl. Phys. Lett.* **72** (1998) 127–129.

[129] V. Chirita, E. P. Münger, J. E. Greene, J.-E. Sundgren, Reptation: A mechanism for cluster migration on (111) face-centered-cubic metal surfaces, *Surf. Sci.* **436** (1999) L641–647.

[130] G. Boisvert, L. J. Lewis, Diffusion of Pt dimers on Pt(111), *Phys. Rev. B* **59** (1999) 9846–9849.

[131] K. Albe, M. Müller, Cluster diffusion and island formation on fcc(111) metal surfaces studied by atomic scale computer simulations, *Internat'l Series of Numerical Math.* **149** (2005) 19–28.

[132] K. Albe, K. Nordlund, R. S. Averback, Modeling the metal-semiconductor interaction: Analytical bond-order potential for platinum-carbon, *Phys. Rev. B* **65** (2002) 195124 1–11.

[133] J. Yang, W. Hu, G. Yi, J. Tang, Atomistic simulation of Pt trimer on Pt(111) surface, *Appl. Surf. Sci.* **253** (2007) 8825–8829.

[134] P. Blandin, C. Massobrio, Diffusion properties and collisional dynamics of Ag adatoms and dimers on Pt(111), *Surf. Sci.* **279** (1992) L219–224.

[135] C. Massobrio, P. Blandin, Structure and dynamics of Ag clusters on Pt(111), *Phys. Rev. B* **47** (1993) 13687–13694.

[136] C. Goyhenex, Adatom and dimer migration in heteroepitaxy: Co/Pt(111), *Surf. Sci.* **600** (2006) 15–22.

[137] F. Montalenti, R. Ferrando, Leapfrog diffusion mechanism for one-dimensional chains on missing-row reconstructed surfaces, *Phys. Rev. Lett.* **82** (1999) 1498–1501.

[138] P. J. Feibelman, Ordering of self-diffusion barrier energies on Pt(110)-(1×2), *Phys. Rev. B* **61** (2000) R2452–2455.

[139] L. Bönig, S. Liu, H. Metiu, An effective medium theory study of Au islands on the Au(100) surface: reconstruction, adatom diffusion, and island formation, *Surf. Sci.* **365** (1996) 87–95.

[140] F. Montalenti, R. Ferrando, Dimer diffusion on (110)(1×2) metal surfaces, *Surf. Sci.* **432** (1999) 27–36.

[141] W. Fan, X. G. Gong, Simulation of Ni cluster diffusion on Au(110)-(1×2) surface, *Appl. Surf. Sci.* **219** (2003) 117–122.

[142] N. Knorr, H. Brune, M. Epple, A. Hirstein, M. A. Schneider, K. Kern, Long-range adsorbate interactions mediated by a two-dimensional electron gas, *Phys. Rev. B* **65** (2002) 115420 1–5.

[143] T. T. Tsong, G. Kellogg, Direct observation of the directional walk of single adatoms and the adatom polarizability, *Phys. Rev. B* **12** (1975) 1343–1353.

[144] C. L. Chen, T. T. Tsong, Self-diffusion on the reconstructed and nonreconstructed Ir(110) surfaces, *Phys. Rev. Lett.* **66** (1991) 1610–1613.

[145] G. L. Kellogg, Temperature dependence of surface self-diffusion on Pt(001), *Surf. Sci.* **246** (1991) 31–36.

9 Diffusion of large clusters

Starting in the 1970s, considerable work was done on dimers and trimers and their surface diffusion, but there were no experimental studies of larger clusters, containing twenty or more atoms, since they were assumed to be immobile at the surface. This changed in 1984, with the work of Fink [1–3] using the FIM, in which he assembled a cluster of twenty or more palladium atoms on the (110) plane of tungsten. At 390 K, this large cluster moved over the surface as a unit, as shown in Fig. 9.1, demonstrating its diffusivity. Large clusters turn out to be mobile at relatively low temperatures and their movement needed to be investigated, since it influences the stability of nanostructures and thin film growth kinetics. With the invention of the scanning tunneling microscope, large clusters were rediscovered a few years later, and work began to unravel how diffusion occurred, many of the studies focusing on the dependence of diffusivity on cluster size. This effort will be surveyed, arranged according to the type of the surface. Study of large clusters began with the examination of movement on a bcc surface, on W(110), but this work was not continued later; instead fcc surfaces were investigated in detail.

9.1 Large clusters on fcc(100) surfaces

Theoretical investigations of large clusters on fcc(100) surfaces started in 1980 with the work of Binder and Kalos [4], which initiated a number of discussions of how the cluster diffusivity D was affected by the size and the specific mechanism of diffusion. Binder and Kalos had already derived simple estimates for clusters of radius R_r in a square lattice gas and showed tentatively that the dependence of the diffusivity on cluster size could be explained in terms of a crossover between two mechanisms – peripheral movement and evaporation–condensation, mechanisms yielding different power laws. For small clusters they expected the diffusivity to depend on the number of atoms N in the cluster to the power $1 + 1/d$; for larger clusters the exponent would be $1 - 1/d$, where d is the dimensionality; the location of the crossover depended on the temperature.

The theoretical investigations were continued with Rh clusters on the Rh(100) surface evaluated with Lennard-Jones potentials by Voter [5]. He also observed a crossover in diffusivity between small and large clusters, as is shown in Fig. 9.2a. For clusters larger than 10 atoms the exponent in the size dependence of the diffusivity was 1.76 ± 0.05 and the main mechanism was shown to be edge running. This conclusion was based on

9.1 Large clusters on fcc(100) surfaces

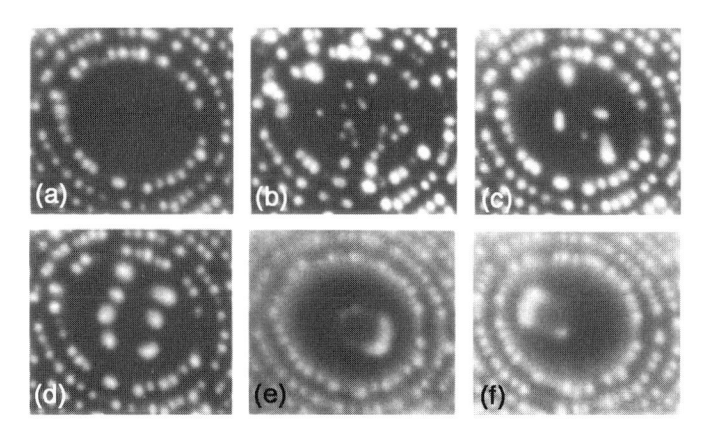

Fig. 9.1 Formation and diffusion of Pd cluster on W(110) plane [3]. (a) Clean surface. (b) Roughly twenty Pd atoms have been deposited. (c) Pd chains have formed after heating to 235 K. (d) After additional deposition and heating, 2D clusters have formed. (e) At 340 K, a large Pd cluster has been created. (f) Cluster has moved over the surface at 390 K.

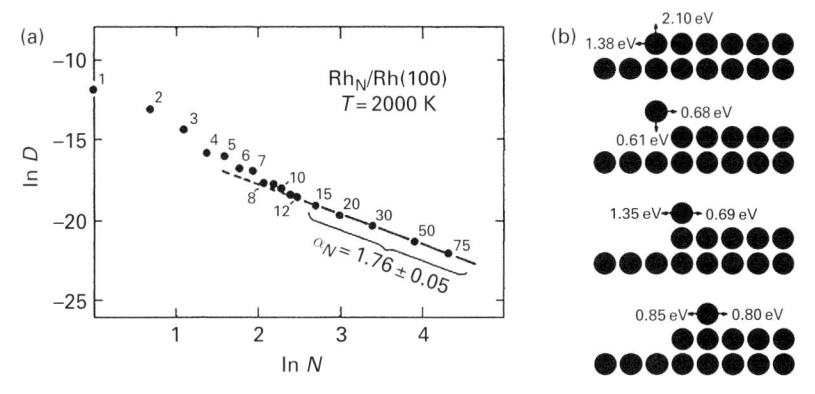

Fig. 9.2 Diffusion of rhodium clusters on Rh(100) obtained from Monte Carlo simulations using Lennard-Jones potentials. (a) Size dependence of Rh cluster diffusivity. (b) Energetics for movement of atom onto a fresh cluster edge (after Voter [5]).

agreement between the energy required for movement of 20 atom clusters (2.08 ± 0.04 eV) and the energy for an atom climbing up to the edge from a kink site (2.10 eV), which is the limiting step for adatom edge running. This as well as other energies calculated for adatom edge running are shown in Fig. 9.2b.

A year later Voter [6] investigated large clusters on Ag(100). The crossover between the diffusivity of small and large clusters was not so clear for this system, as indicated in Fig. 9.3a. However, similar to the previous study, the activation energy for movement of a 100 atom cluster, 0.82 ± 0.09 eV, was in very good agreement with the energy required for the step that limits peripheral movement – an adatom jumping out from a kink site, 0.83 eV. The energetics of other steps are shown in Fig. 9.3b.

Experimental studies on fcc(100) surfaces started with an extensive piece of work by Wen et al. [7] dealing with the diffusion of large silver clusters on the Ag(100) surface.

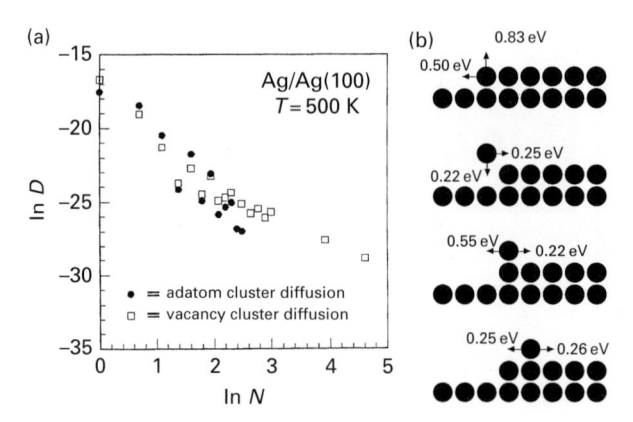

Fig. 9.3 Silver cluster diffusion on Ag(100) surface. (a) Dependence of cluster diffusivity on size. (b) Energetics of atom movement to fresh cluster edge (after Voter [6]).

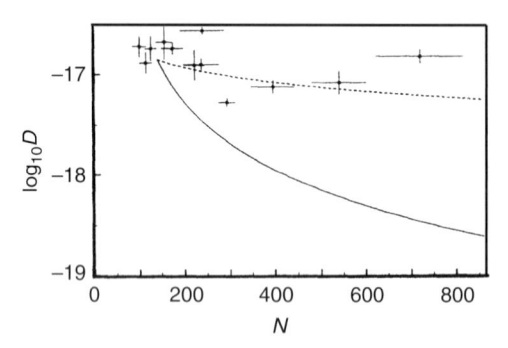

Fig. 9.4 STM observations of room temperature diffusivity of Ag cluster on Ag(100) as a function of the number of cluster atoms N (after Wen *et al.* [7]).

The clusters were made up of from 100 to 800 atoms, and were observed at room temperature with the STM. A plot of the diffusivity against the number of cluster atoms is shown in Fig. 9.4. Wen *et al.* considered two different mechanisms for the motion of the clusters: atom movement around the periphery of the cluster, and evaporation–condensation, in which the cluster is in contact with a dilute two-dimensional gas layer. For peripheral diffusion, Monte Carlo simulations had shown the diffusivity to depend upon the number of cluster atoms N, with

$$D = D^* N^{-\alpha_N}; \tag{9.1}$$

here α_N lies between 1.5 and 2. These results are shown by the solid line in Fig. 9.4, which deviates significantly from the experiments. For the evaporation–condensation mechanism of cluster diffusion they estimated that α_N would be 0.5. This mild dependence upon the number of cluster atoms is given by the dotted line in the figure, and seems to better agree with the experimental data. Direct evidence for evaporation of atoms from the cluster edges was found from observations of the disappearance of small clusters. In short, the mechanism for cluster diffusion was attributed to the evaporation and condensation of cluster atoms.

9.1 Large clusters on fcc(100) surfaces

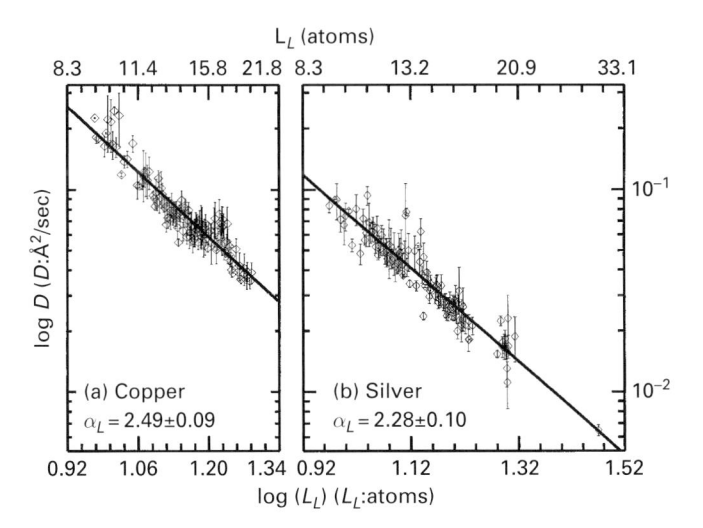

Fig. 9.5 Diffusivity of islands measured with STM in its dependence upon island length L_L (after Pai et al. [8]).

These findings almost immediately stimulated controversy. Pai et al. [8] came out with an STM examination of the diffusivity and coarsening of silver on Ag(100) as well as copper on Cu(100). In the relation between diffusivity and the island length L_L they found

$$D = D^* L_L^{-\alpha_L} \qquad (9.2)$$

with $\alpha_L = 2.28 \pm 0.10$ and $D^* = (18.5 \pm 4.5)$ Å^2/sec for silver and 2.49 ± 0.09 and $D^* = (59.5 \pm 13.1)$ Å^2/sec for copper, as shown in Fig. 9.5, for islands of more than a thousand atoms. These coefficients are reasonably close to what is expected for correlated evaporation–condensation or terrace limited diffusion. However, separate observations revealed essentially no decay of the islands. For copper, hardly any loss was noted during a period of six hours. With silver, the decay rate led to an evaporation–condensation diffusivity of $\approx 2.2 \times 10^{-5}$ Å^2/sec. These values are three orders smaller than the measured diffusivity of $D \approx (3-6) \times 10^{-2}$ Å^2/sec. The conclusion was that movement of atoms around the cluster periphery was responsible for the diffusivity.

The authors claimed that the usual assumption of a structureless periphery was responsible for such disagreement, and to deduce the macroscopic mechanism details of the structure had to be taken into account. They also performed an analysis with a constant kink density, and random attachment and detachment of atoms from kinks. They found that α_L changed from 1 to 3 with increasing radius of the clusters. This study was based on much better statistics than that of Wen et al. [7] and challenged their findings. Wen et al. [9], in their later study on the coarsening mechanism on Ag(100), observed no Ostwald ripening of islands up to a coverage of 0.65 ML, which might indicate the same problem. However they rationalized the lack of Ostwald ripening by a huge probability of recondensation close to the cluster.

Coarsening of large silver clusters on a silver(100) surface was also measured with a scanning tunneling microscope (STM) at 295 K by Stoldt et al. [10] and the Ostwald

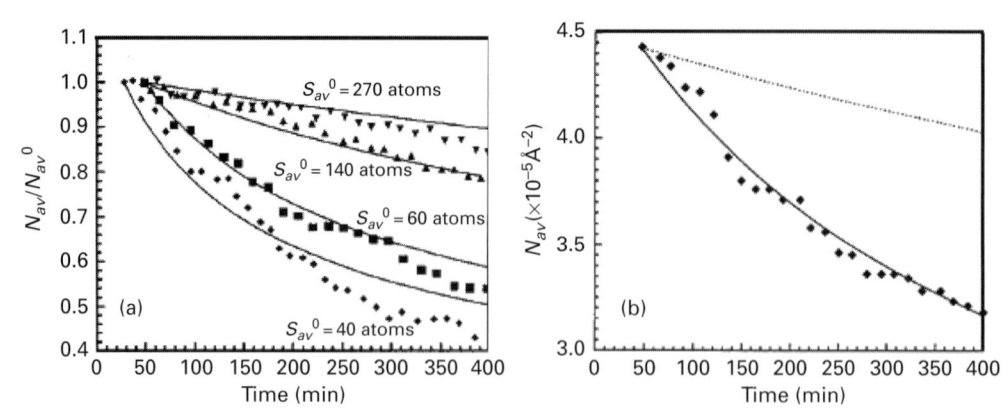

Fig. 9.6 Coarsening kinetics for Ag islands on Ag(100) determined with STM. (a) The unstrained surface, fitted with $D_o = 50$ Å^2/sec and an exponent of 1.5. (b) The strained surface, fitted with $D_o = 250$ Å^2/sec and exponent of 1.5 in Eq. (10.1). Dashed curve for $D_o = 50$ Å^2/sec (after Stoldt et al. [10]).

ripening process was comparable with experimental noise in this study. They found the coarsening rate decreased strongly with increasing size of the islands (coverage), suggesting a strong decrease in the diffusivity with increasing island size. The results of their experiments are shown in Fig. 9.6a, where the data were analyzed based on mean field theory and were fitted with a value of 50 Å^2/sec for D^* and an exponent $\alpha_N = 1.5$ in Eq. (9.1), the value appropriate for perimeter diffusion of atoms around the clusters. They also observed enhanced diffusion rates for islands on a strained Ag(100) surface; their data are shown in Fig. 9.6b, where their findings are fitted with 250 Å^2/sec for D^* and keeping an exponent $\alpha_N = 1.5$ in Eq. (9.1). The origin of the strain was undetermined.

Monte Carlo simulations were carried out by Pal and Fichthorn [11] for the diffusion of large clusters on an fcc(100) surface at 300 K. In Monte Carlo simulations they used a lattice model with periodic boundary conditions and nearest-neighbor interactions on a square lattice. An activation energy of 0.52 eV was assumed for single atom movement, together with a lateral bond strength of 0.8 eV, and a frequency factor of 10^{12} sec^{-1}. Individual atomic steps in cluster diffusion were monitored. Four elementary processes are shown in Fig. 9.7a: atomic diffusion by uncorrelated jumps around the cluster periphery PD, the hopping of a kink atom KD, the attachment KA of an isolated atom to a kink position, and finally the process designated as CPD, a variant of peripheral diffusion, in which an atom jumps from a site where it makes two bonds, to an adjacent position, where it again makes two bonds. The plot for the diffusivity as a function of size is given in Fig. 9.7b, and divides into two sections, the first with a slope of 1.47 for clusters made of fewer than 100 atoms, and the second with a slope of 0.49 for large clusters of more than 100 atoms. For small clusters, diffusion around the edges of the clusters proved to be the dominant step, but for large clusters movement of kink atoms turned out to be the most frequent event, shown in Fig. 9.7c. Evaporation–condensation (EC) suggested from the scaling factor was not found responsible for the observed value of $\alpha_N \approx 0.5$. It was instead the increase in KA/KD events, together with some rise in the CPD process, that was important for the behavior of the cluster. Some oscillations in the

9.1 Large clusters on fcc(100) surfaces

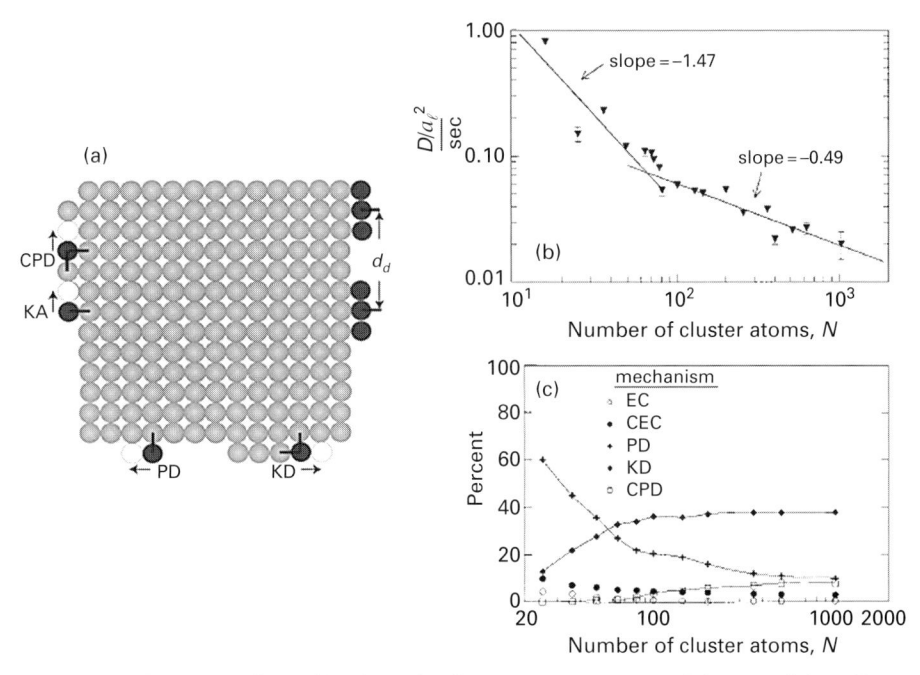

Fig. 9.7 Diffusion on fcc(100) surfaces. (a) Schematic of atom movements around cluster periphery. Empty circles indicate positions that can be filled by adatom. KD – jump of kink atom. PD – peripheral diffusion. KA – kink attachment of isolated adatom. CPD – two bonds broken, two formed. (b) Dependence of diffusivity of clusters on N, the number of cluster atoms. (c) Probability of occurrence for different mechanisms. EC – evaporation–condensation, CEC – evaporation and condensation within two hops (after Pal and Fichthorn [11]).

diffusivity of large clusters were also observed. From a separate analysis of the mechanisms involved in movement, shown in Fig. 9.7c, the condensation–evaporation process was found very rarely, and if an atom detached from the cluster, then reattachment usually was observed within two hops, a process named CEC, evaporation and condensation spatially correlated, in Fig. 9.7c.

Extensive Monte Carlo simulations of diffusion and evaporation for large clusters on fcc(100) surfaces were also done by Lo and Skodje [12], based on a bond breaking model. They allowed for hops only to nearest-neighbor sites, and neglected any concerted gliding, adatom/substrate exchange processes, or dimer shearing. Clusters with as many as 100 000 atoms were examined, at a temperature around 500 K, where both large clusters and monomers were stable. The behavior of the diffusivity as a function of the number N of atoms in the cluster is plotted in Fig. 9.8a. Examination of the diffusion revealed that the primary event was correlated evaporation–condensation, in which an evaporating atom reattaches close to the same site, in agreement with the value found for the α_N exponent. From an Arrhenius plot of the diffusivity in Fig. 9.8b, an activation energy was derived, in agreement with the energy for the removal of an atom from a kink site, shown in the inset. The same value was obtained from the evaporation rate of islands as large as 10 000 atoms. Based on this observation, the authors concluded that removing

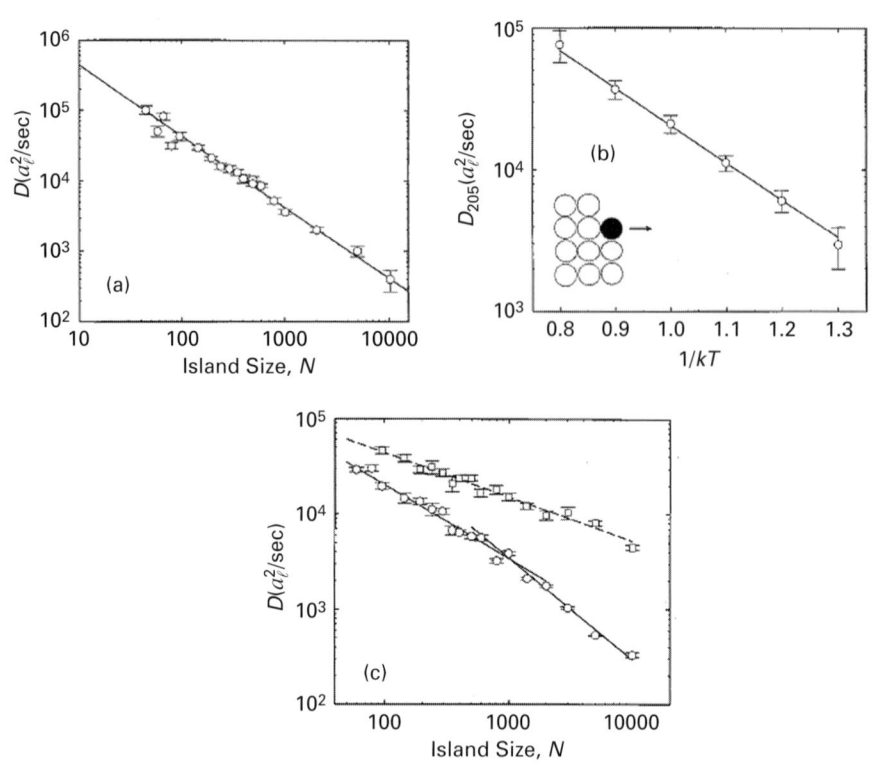

Fig. 9.8 Surface diffusivities on fcc(100) derived in MC simulations. (a) Dependence of cluster diffusivity upon the number of cluster atoms N. Solid line has slope of 1.01 ± 0.05, a_ℓ is the lattice spacing. (b) Arrhenius plot for cluster of 205 ± 10 atoms with an energy of $6.05 \pm 0.08\ E_S$. E_S is the smallest barrier in the system. (c) Dependence of vacancy island diffusivity upon the number of missing cluster atoms; upper plot without substrate, lower with substrate (after Lo and Skodje [12]).

atoms from kink sites was the limiting step in the kinetics of island movement. Here we have to emphasize the different dependence on cluster size from that found by Pal and Fichthorn [11], obtained at lower temperatures and for a smaller range of cluster sizes. The atomic mechanism in the diffusion was also different: for Pal and Fichthorn it was kink movement, while for Lo and Skodje correlated evaporation–condensation. For vacancy islands the dominant mechanism was not derived directly, but from the exponential dependence of the diffusivity on island size in Fig. 9.8c for vacancy islands of 100 to 800 spacings, correlated or uncorrelated evaporation–condensation was suggested, while for bigger islands the limiting step was believed to be terrace diffusion.

Monte Carlo simulations for diffusion of copper islands on Cu(100) were carried out by Heinonen et al. [13] drawing on the EMT energetics of Merikoski and Ala-Nissila [14]. For the diffusion as a function of the number of cluster atoms N they obtained the plot in Fig. 9.9a at a temperature of 1000 K. Islands with more than ten atoms gave a straight line, which yielded the exponent plotted in the inset. This varied from the value for edge diffusion around small clusters to that for terrace diffusion limiting when the cluster became large. That is, the mechanism of diffusion changed with size, but the

9.1 Large clusters on fcc(100) surfaces

Table 9.1 Activation energies for diffusion of silver clusters calculated for Ag(100) plane [15].

Process	Activation energy (eV)	Process	Activation energy (eV)
KD	0.55	M	0.45
TC	0.60	VM	0.55
EA	0.75	TCPD	0.50
PD	0.25	TPD	0.35

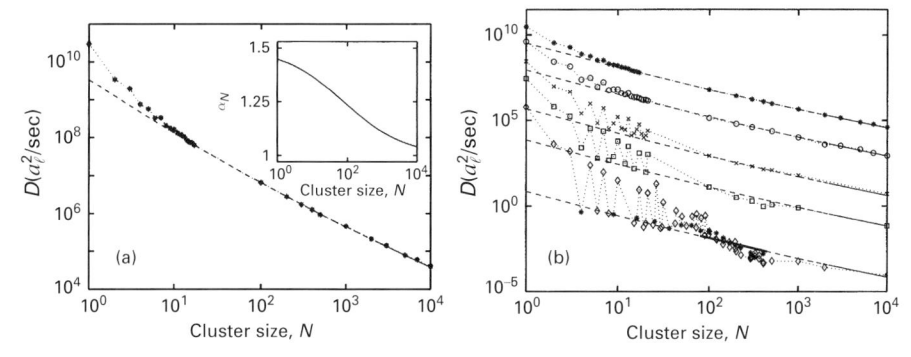

Fig. 9.9 Cluster diffusivity on Cu(100) derived from EMT energetics. (a) As a function of size N at $T = 1000$ K. (b) Diffusivity as function of size at temperatures 1000 K, 700 K, 500 K, 400 K, 300 K, starting at the top (after Heinonen et al. [13]).

coefficient did not strongly depend upon the temperature. For this system the activation energy for diffusion did not change much on going from edge diffusion to terrace diffusion being limiting. They also observed a crossover from periphery dominated transport to vacancy dominated diffusion inside the clusters. At low temperatures, the diffusivity oscillated with increasing size, as shown in Fig. 9.9b. Movement of vacancy islands was found to be very similar to that of adatom islands.

The same year Mills et al. [15] employed Monte Carlo simulations to look at islands on an fcc(100) plane, with parameters to mimic Ag on Ag(100); energetics were obtained from DFT calculation, or from EMT scaled to DFT values. Barriers not available from DFT or EMT were estimated from bond counting, with the strength of one bond at ~0.3 eV. The energetics in the simulations are listed in Table 9.1. It is quite surprising that the energy required for adatom motion along a straight step, 0.25 eV, is much smaller than the energy for isolated adatom movement over the (100) surface, 0.45 eV. This suggests that mass transport across the terrace was less favorable than transport along the edges, so that the evaporation–condensation mechanism would be unlikely to happen. Fig. 9.10a shows the processes considered in this study. The power law dependence of diffusivity D on island size N was described by

$$D \sim \exp(-E_p/kT) N^{-\alpha_N} \qquad (9.3)$$

where E_p denotes the activation energy for movement along steps. They found that the power law was valid only for large clusters, which satisfy the relation $N_l < N < 2000$,

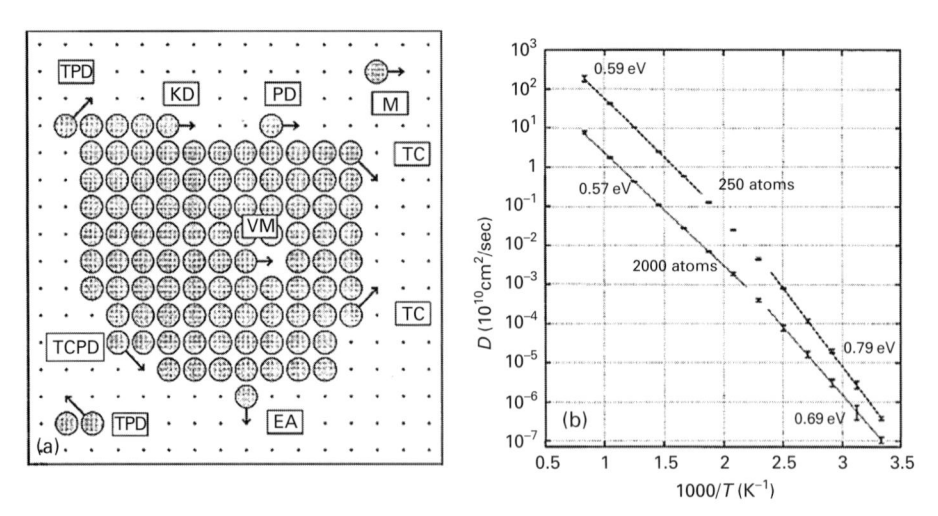

Fig. 9.10 Diffusion of clusters on fcc(100) surface from MC simulations. (a) Illustration of possible atomic steps in diffusion. (b) Arrhenius plots for clusters of 250 and of 2000 atoms. Prefactors increase at low temperatures (after Mills *et al.* [15]).

where N_l depends on temperature; it is 200 at temperatures above 450 K, and 300 below 450 K. They did simulations at 1000 K with an island of 125 atoms and also 2000 atoms with 20 to 30 atoms in the vapor phase under two boundary conditions: first with evaporation–condensation forbidden, second with peripheral motion forbidden. During simulations with evaporation–condensation not allowed, the diffusion coefficient reached 90% of its value when all events were active, showing that evaporation–condensation influences the diffusivity very weakly. At the same time, when peripheral motion was forbidden, the diffusivity reached only 10% of the overall value. This study clearly indicated that diffusion was driven mostly by adatoms moving along the edges.

Mills *et al.* [15] also looked at the temperature dependence of the diffusivity for islands of 125 and 2000 atoms. Their findings are presented in Fig. 9.10b. We clearly see the situation changing from low to high temperatures. Of course the measured diffusivity is an average result over all processes participating. However, two distinct regions in the diffusivity allowed them to speculate that the mechanism will be caused mostly by two main processes, of which one will dominate at low temperatures and a second one at high temperatures. The task would be to decide which processes are involved. At high temperatures, the energy obtained from the Arrhenius plot was 0.57 eV, and a prefactor of 1.7×10^{-7} cm^2/sec for the 2000 atom island, and 0.59 eV, with a prefactor of 5.5×10^{-6} cm^2/sec for the 250 atom island. Both values are comparable to the energy of corner or kink breaking, also called core breaking, which might be the limiting step in the motion. The situation is less clear at low temperatures, where the energy from the Arrhenius plot was much too high for edge hopping or core breaking. The values are 0.69 eV, and the prefactor amounted to 3.6×10^{-6} cm^2/sec for the 2000 atom island and 0.79 eV with a prefactor of 6.3×10^{-4} cm^2/sec for the 250 atom island. A possible rationalization could be nucleation of new rows with an effective energy of 0.81 eV, but it is not clear what really is responsible for such a high energy.

9.1 Large clusters on fcc(100) surfaces

That cluster diffusion occurs by atom movement around the periphery was further confirmed in a subsequent study by Pai et al. [16] of worm-like vacancy clusters. These were created during deposition of an incomplete layer (~0.6 ML) of copper on Cu(100) and silver on Ag(100), after allowing the clusters to collapse to create percolation vacancy islands. Since worm-like vacancy islands are quite irregular, their reshaping allows us to get insight into the diffusion mechanism. The frequent pinch-off observed was found to be distinctive for atom diffusion around the cluster edges, and speaks against any of the other mechanisms. Comparison of the experimentally evaluated shape of the vacancy islands with simulations confirmed that reshaping is a geometry driven process.

The same year Resende and Costa [17] examined the deposition of Cu_{13} clusters on the Cu(100) surface using molecular dynamics and Monte Carlo simulations, relying on interactions based on a second-moment approximation to tight-binding. Following the movement of clusters over the surface they found that clusters maintained their integrity even at 800 K during diffusion and moved by single hops of the peripheral atoms.

Specific attention was paid by Wang et al. [18] to so-called "diagonal atom motion" in the diffusion of clusters, again using Monte Carlo simulation with EMT energetics. Diagonal motion is just the movement of an atom around a cluster corner, as illustrated in Fig. 9.11a[1] together with other hopping processes taken into account in these studies. The diagonal movements are marked in Fig. 9.11a as TPD and TCPD. They occur over an additional barrier E_T that has to be overcome for such movements, compared to E_D for free atom hopping over the substrate. The effect of such transitions on the size dependence of the diffusivity, $D \propto N^{-\alpha_N}$, is shown in Fig. 9.11b at 300 K for clusters of 15–300 atoms. The deviation observed for small clusters can be caused by fluctuations in the cluster size or changes in the shape; that is why Wang et al. took into account only data for $N > 40$. The exponent α_N changes as the relative magnitude of the diagonal barrier E_T changes. As indicated in the inset, $\alpha_N = 1.54 \pm 0.034$ if $E_T = 0$, and it falls to 1.03 ± 0.027 eV when $E_T = 1.55\, E_D$. How the different diffusion mechanisms are affected by changes in the magnitude of the barrier to diagonal motion is indicated in Fig. 9.11c, where T describes diagonal motion, PD perimeter and kink movement, EC evaporation–condensation, and CEC the correlated evaporation–condensation mechanism. The biggest contribution when diagonal motion requires too much energy is made by peripheral diffusion in combination with detachment from kinks, as suggested in the schematic in Fig. 9.11d. From this study it is clear that as long as the energy of diagonal motion is low, $E_T < 0.4\, E_D$, motion around the corners is realized by diagonal movements T and $\alpha_N \rightarrow 1.5$; however, when the energy for diagonal movement increases, $E_T > 0.4\, E_D$, then corner crossing is done by the correlated evaporation–condensation mechanism CEC and $\alpha_N \rightarrow 1.0$.

Jahma et al. [19] used Monte Carlo simulations for copper island diffusion on both Cu(100) and Cu(111) surfaces, based also on energetics from effective medium theory. Results for the diffusivity on Cu(100) as a function of size, given by the number of atoms N, and the temperature are shown in Fig. 9.12b. Here, as in previous studies, there were

---

[1] See also Fig. 9.7a.

674 Diffusion of large clusters

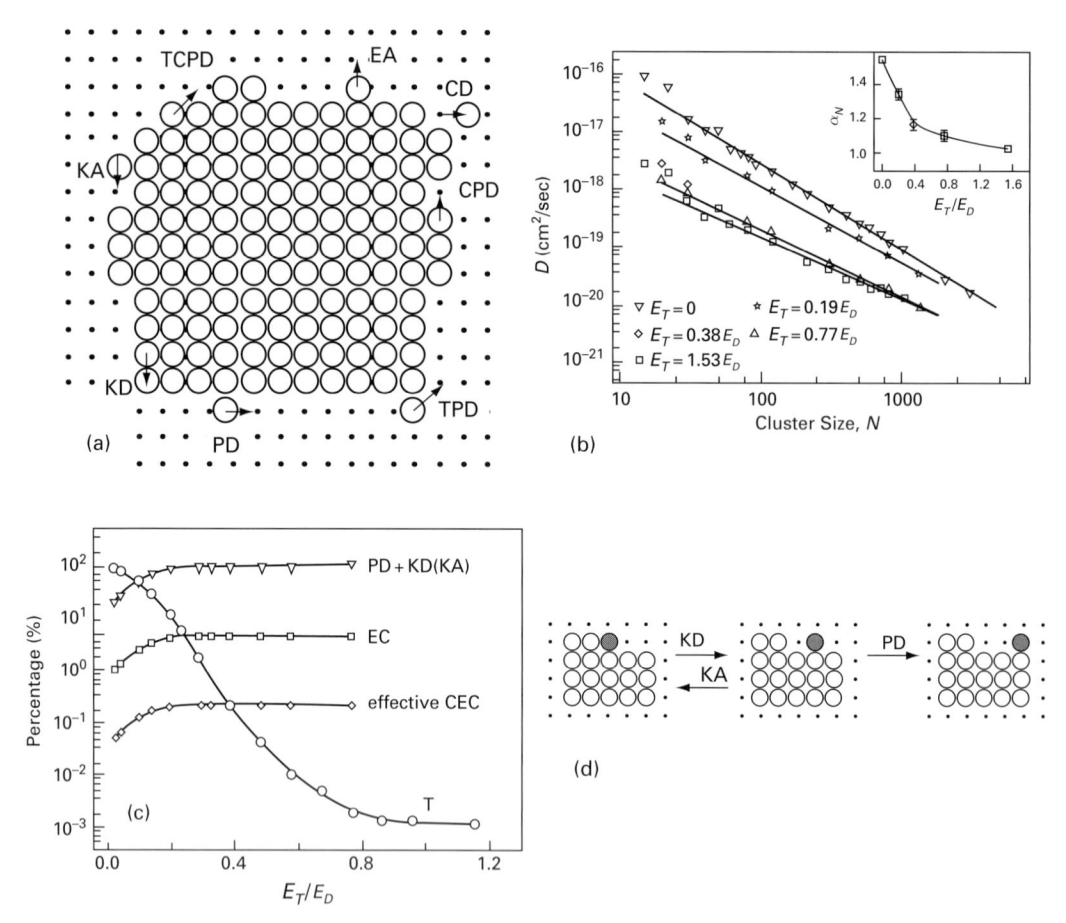

Fig. 9.11 Cluster diffusion on fcc(100) surface derived from MC simulations. (a) Elementary diffusion events on cluster periphery. (b) Dependence of cluster diffusivity at 300 K on cluster size N. Inset shows dependence of scaling coefficient α_N on E_D. (c) Occurrence of different mechanisms as a function of the change in the diagonal motion energy, T – diagonal movement such as TPD and TCPD. (d) Model for rate-limiting step in cluster diffusion (after Wang *et al.* [18]).

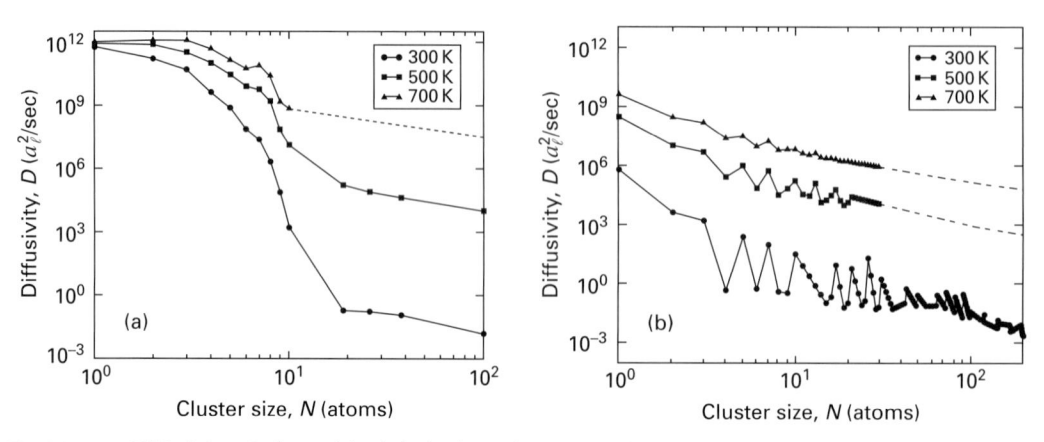

Fig. 9.12 Diffusivity of adatom islands in its dependence upon size N and upon temperature based on EMT. (a) On Cu(111). (b) On Cu(100) (after Jahma *et al.* [19]).

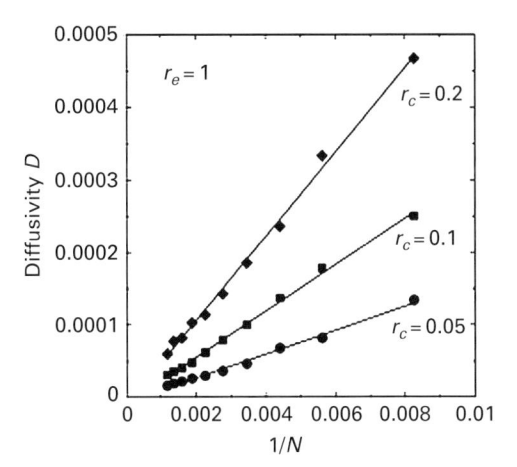

Fig. 9.13 Diffusivities of Ag clusters on Ag(100) surfaces obtained in MC simulations in their dependence upon N^{-1}, where N gives the number of cluster atoms. r_c is the rate of evaporation–condensation, r_e that of peripheral diffusion (after Sánchez [20]).

significant oscillations with size, especially at lower temperatures. At the higher temperature of 700 K, these oscillations became much less significant and the diffusivity was reasonably represented by the curve $D \propto N^{-\alpha_N}$, with $\alpha_N = 1.5$, the value if the limiting step was atom diffusion around the island edges. The activation energy for atom diffusion was 0.4 eV on Cu(100).

Monte Carlo simulations of cluster diffusivities were done for (100) surfaces by Sánchez [20], who examined mobilities as a function of r_c/r_e, the ratio of the rate of evaporation–condensation to the rate of diffusion around the periphery. One interesting result is shown in Fig. 9.13, where the diffusivity is plotted against N^{-1}, N being the number of atoms making up the cluster. For different values of the evaporation–condensation rate, the cluster diffusivity is a linear function of N^{-1}, and has a greater value as r_c/r_e increases. Sánchez concluded that although the predominant process was peripheral diffusion, evaporation–condensation had a pronounced effect on the diffusivity. However, experimental measurements have been obtained mostly around room temperature, at which evaporation could not occur, leaving peripheral diffusion as the dominant mechanism [8,10,16,21].

Studies of large clusters on fcc(100) surfaces predict movement by edge running. The evaporation–condensation mechanism is unlikely, but possible in some specific situations, for example where corner movement is blocked by a huge barrier [18]. Most of the MC simulations relied on the same EMT energetics, which might also lead to misleading results. It is very difficult to judge how all this applies to real materials, since measurements have been mainly done on the Ag(100) surface, and even here deriving the mechanism has not been an easy task. Different materials might require different starting parameters. What is surprising is that there are no investigations concerning the influence of dimer or trimer shearing on the movement of large clusters, mechanisms which are clearly present in the movement of smaller clusters on fcc(100). For fcc(100), there is also

9.2 Large clusters on fcc(111) surfaces

Diffusion of pits and large clusters on Au(111) surfaces in an electrochemical environment was discussed by Trevor and Chidsey [22] in 1991. They used scanning tunneling microscopy and found that "pits, islands and step edges" diffused at room temperature. They proposed a model in which atoms diffuse around the pit edge, moving the center of mass. For this scheme the diffusivity was proportional to R_r^{-3}, where R_r gives the radius of the pit. Indeed, small pits, 20–40 Å in diameter, moved very rapidly, whereas large pits 100 Å in diameter were almost stationary. Based on these observations, they estimated an upper bound for the free energy to create an adatom from the step, together with an energy for adatom diffusion of 0.5 eV.

Vacancy islands on Cu(111) covered with a partial monolayer of cobalt were examined a few years later by de la Figuera *et al.* [23] again using a scanning tunneling microscope under UHV conditions. Vacancy islands were created in a controllable way relying on tip–surface interactions. Islands ranged in size from just a few vacancies to voids with a diameter of 75 Å. In diffusion, the islands maintained their hexagonal shape. The mean-square displacement of individual vacancy islands proved to be linear in time as shown in Fig. 9.14a, in agreement with the notion of Brownian displacements. From measurements on differently sized islands they were able to construct a plot of the mean-square displacement against the island area, shown in Fig. 9.14b. As random motion, the mean-square displacement was approximately inversely proportional to the area of the islands. On a clean Cu(111) surface, however, Cu vacancy islands were essentially not mobile at room temperature, unless strain was present.

Morgenstern *et al.* [24] using the STM looked at vacancy islands on Ag(111). Diffusion of these clusters is shown in Fig. 9.15. They derived an expression for the dependence of the diffusivity for atoms running around the perimeter and for

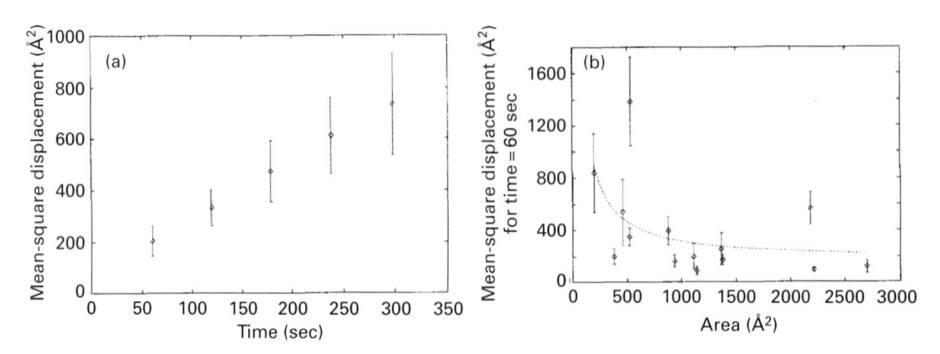

Fig. 9.14 Diffusion of vacancy islands on Cu(111), covered with a partial layer of Co, observed with STM. (a) Mean-square displacement as a function of time. (b) Mean-square displacement at 60 sec in its dependence on area of vacancy islands (after de la Figuera *et al.* [23]).

9.2 Large clusters on fcc(111) surfaces

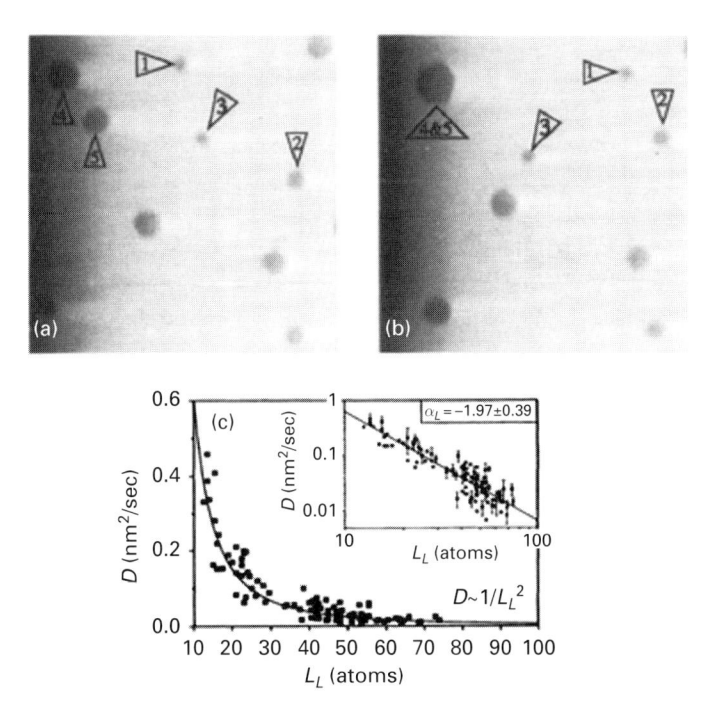

Fig. 9.15 Diffusion of vacancy islands on Ag(111). (a)–(b) Motion as observed with STM. (c) Diffusivity of vacancy clusters as a function of cluster diameter (after Morgenstern et al. [24]).

evaporation–condensation. For cluster motion by atom diffusion around the edges they arrived at a diffusivity proportional to the inverse of the cluster diameter raised to the power three. For evaporation–condensation, this dependence changed to L_L^{-2}, where L_L is the diameter. Measurements were done under UHV conditions, and the mean-square displacement was again found to vary linearly with time, as expected for a random process. Mobility of vacancy islands was already high at room temperature, with bigger islands moving slower. The dependence of the diffusivity D on the size of the vacancy cluster is shown in Fig. 9.15c. As indicated in the inset, the diffusivity is given by $D \propto L_L^{\alpha_L}$, where $\alpha_L = -1.97 \pm 0.39$, in agreement with atom diffusion over the terrace rather than along the edges. What is interesting is that Morgenstern et al. never observed island coalescence, even if the islands came as close as five atomic distances.

These interesting findings have been examined again briefly, in view of a final summary by Rosenfeld et al. [25] about the diffusivity of vacancy clusters on Ag(111). In the above studies, they had found a diffusivity given by $D \propto L_L^{-1.97 \pm 0.39}$, which is expected for evaporation–condensation. However, they subsequently measured the rate of evaporation of atoms from cluster edges and found this was on the order of 1 sec^{-1}; at the same time, they estimated from the prefactor of the diffusivity a rate three orders of magnitude higher, amounting to 750 sec^{-1}. Furthermore, in Monte Carlo simulations of the movement of vacancy clusters on Ag(111), with sizes and energetics comparable to the experiments, they were able to agree with the experimentally found scaling coefficient of two. However, they also discovered that their results were not significantly affected when evaporation was forbidden, so that only diffusion around the perimeter could occur. They concluded that diffusion around the edges of the vacancy cluster was really the mechanism for motion, even though the diffusivity did not vary as $D \propto 1/L_L^3$, as expected for peripheral atom motion, and that further studies were in order.

In 1995 Hamilton et al. [26] introduced a new mechanism for cluster diffusion on fcc(111) planes, already described in Section 7.2. Part of the cluster, in their scheme, moved from fcc to hcp positions, with the two parts separated by a misfit dislocation. This could move across the cluster, resulting in diffusion. This mechanism was shown to be favorable for clusters of 20 to 100 atoms. For movement of clusters smaller than 20 atoms, gliding was predicted as the leading mechanism, while for clusters bigger than 100, movement by kink motion was expected. They showed that the dislocation mechanism was more probable than adatom edge running. No attempt to calculate diffusivities with this new mechanism was made. A year later, Hamilton [27] did further investigations testing his mechanism for the hetero-motion of silver clusters on Ru(0001) surfaces. Ruthenium is an hcp metal, but the (0001) surface looks very similar to fcc(111). Using EAM, he calculated activation energies at 0 K for diffusion of 19, 37, 61, 91, and 127 atom clusters by dislocation movement; the energies were respectively 0.681 eV, 0.729 eV, 0.457 eV, 0.293 eV, and finally 0.786 eV. From a molecular dynamics study he found that diffusion of a 61 atom cluster around room temperature occurred at higher rates than 37 and 91 atom clusters. This happened due to a change of the activation energy for movement with temperature, arising from the temperature dependence of the misfit between the Ag cluster and the Ru(0001) surface, as shown in Fig. 7.28. The minimum activation energy for 61 atom clusters was as low as 0.18 eV at ~309 K.

The diffusion of xenon clusters, a non-metal, on Pt(111) was examined by Monte Carlo simulation in the hands of Sholl and Skodje [28], who showed that evaporation–condensation was the preferred mechanism by which vacancy as well as adatom clusters diffused.

Bitar et al. [29] used Monte Carlo simulations to look at islands 10 to 100 atoms big, and resembling gold on an fcc(111) surface. The diffusivity decreased with the size of the island according to a power law, with an exponent equal to -2.1 ± 0.1 for clusters with $N > 20$; for clusters smaller than 20 atoms the exponent was -1.8 ± 0.1. The authors associated the exponent with the fast movement of atoms along the edges. They also indicated that changes in the exponent between 1.5 and 2 can be observed due to the choice of interactions, including taking into account next-nearest-neighbor effects. From the temperature dependence of the diffusivity the activation energy for movement, E_D^D, was derived, and turned out to be size independent and close to the experimentally derived value for removing an atom from a kink of the Au(110) surface [30]. Values Bitar et al. obtained from Arrhenius plots of the diffusivity, as well as from the cluster velocity, E_D^v, are presented in Table 9.2.

The diffusion of large clusters on surfaces with an fcc(111) structure without misfit was reexamined by Bogicevic et al. [31]; they did Monte Carlo simulations allowing only diffusion of atoms around the cluster edges. In these simulations two different energetics were considered, illustrated in Fig. 9.16a; all rates had a pre-exponential factor of 10^{13} sec^{-1}. The results for the diffusivity at different temperatures as a function of the

9.2 Large clusters on fcc(111) surfaces

Table 9.2 Energetics of clusters from MC simulations on fcc(111) surfaces [29].

Size of island N (atoms)	$E_D^D(eV)$	$E_D^v(eV)$
10	0.59 ± 0.07	0.60 ± 0.09
20	0.59 ± 0.06	0.61 ± 0.09
30	0.61 ± 0.06	0.62 ± 0.08
40	0.59 ± 0.04	0.64 ± 0.07
100	0.60 ± 0.03	0.63 ± 0.05

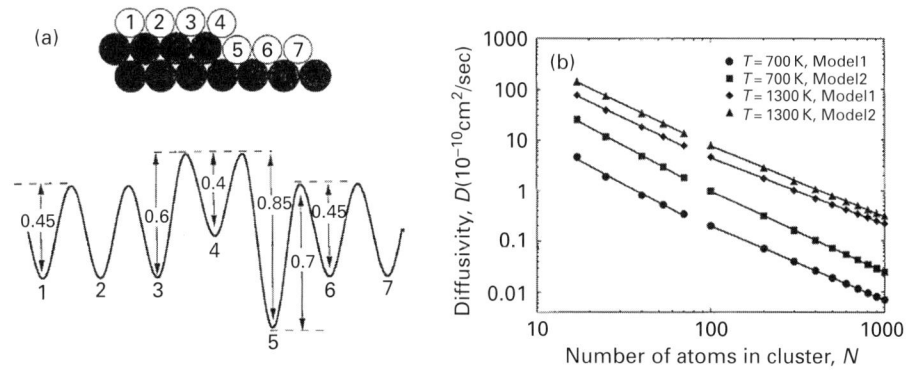

Fig. 9.16 Edge diffusion of adatoms around cluster on fcc(111) surface obtained in MC simulations. (a) Top: model of edge structure. Bottom: Activation energy (in eV) for different atomic movements. Second model arises from different activation energies. (b) Dependence of island diffusivity at 1300 K on the number of atoms in the cluster (after Bogicevic *at al* [31]).

number of cluster atoms are shown in Fig. 9.16b, and it is clear that when the number of cluster atoms exceeds 100 there is an exponential relation with size. However, the exponent is not a constant, as suggested previously. Instead, it varies with the temperature and with the energetics of atom diffusion around the edges, which indicates a material dependence. For clusters much larger, with 1000 to 2000 atoms, the exponent was almost constant and differed by less than 1%.

The same year, Bogicevic *et al.* [32] using density-functional calculations and kinetic Monte Carlo simulations, mapped out the energetics for movement of large clusters on an Al(111) surface. The processes considered are presented in Fig. 9.17a and the energies derived are given in Table 9.3. What is interesting to see is that movement along *B*-type steps proceeded by exchange rather than by hopping; exchange was also the preferred mechanism for kink incorporation. On an *A*-type step hopping is the leading mechanism, but exchange was in operation over this range of temperatures as well. Bogicevic *et al.* also observed a huge anisotropy for corner diffusion; the corner energy barrier to *B*-type steps was 400% higher than to *A*-type steps. Based on the energies derived from DFT they constructed a scale showing the increasing number of processes activated with temperature, as given in Fig. 9.17b. From 0 to 120 K, only terrace and corner diffusion is active, so an atom randomly deposited next to a corner will relax to a position next to the step. Islands are not compact, as movement along edges was suppressed, but a gradual

Diffusion of large clusters

Table 9.3 Energetics of self-diffusion processes from LDA/GGA in Fig. 9.17 as a function of aluminum cluster size N [32].

Process	N	E_h^A	E_e^A	E_h^B	E_e^B
$E_{2 \to 2}$	109–190	0.31/0.31	0.43/0.36	0.45/0.45	0.35/0.26
$K_{3 \to 2}$	135–192	0.47/0.45	0.46/0.38	0.67/0.65	0.49/0.42
$K_{2 \to 3}$	135–192	0.27/0.28	0.26/0.22	0.41/0.42	0.24/0.19
$C_{1 \to 1}$	129	0.18/0.18		0.22/0.22	
$C_{1 \to 2}$	129	0.04/0.05		0.17/0.19	
$C_{2 \to 1}$	129	0.33/0.33		0.32/0.30	
$C_{3 \to 1}$	131	0.62/0.59		0.63/0.60	
$C_{1 \to 3}$	131	0.03/0.04		0.14/0.14	

h-hop, e-exchange

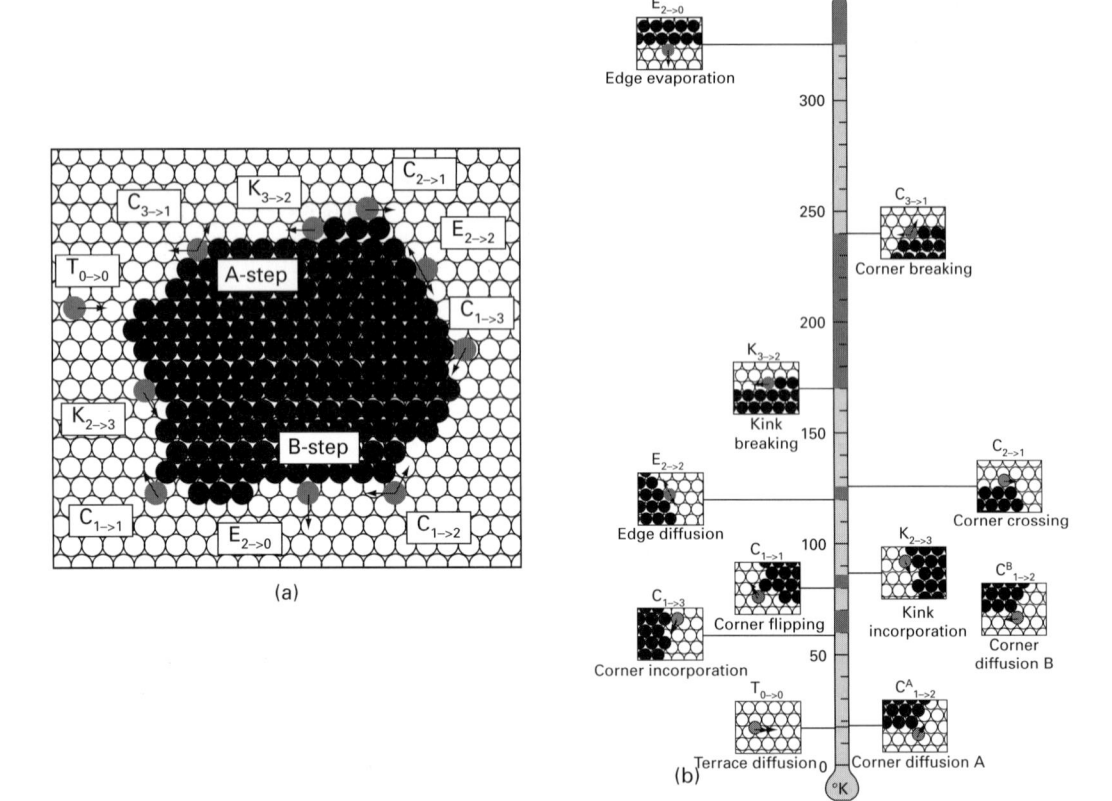

Fig. 9.17 Self-diffusion of clusters on Al(111) derived from density-functional calculations. (a) Schematic of participating diffusion processes. (b) Temperature scale for onset of diffusion events (after Bogicevic *et al.* [32]).

9.2 Large clusters on fcc(111) surfaces

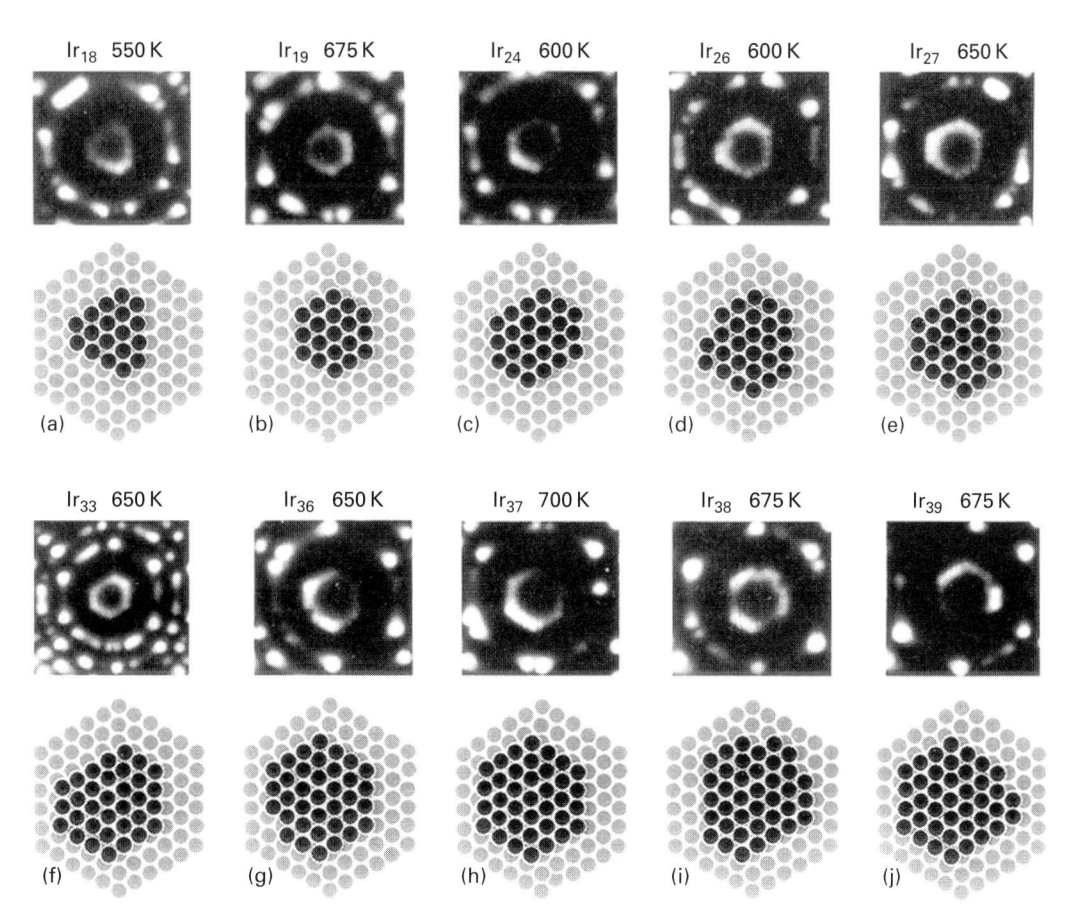

Fig. 9.18 Iridium clusters observed on Ir(111) plane in FIM together with annealing temperature [33].

transition towards less of a fractal shape proceeds. At 120–170 K, a transition begins to compact clusters, arising from edge movement and corner crossing. At 170–325 K the shape changes to islands with an equilibrium form. At higher temperatures, edge evaporation and exchange of atoms between neighboring islands starts playing a role and Bogicevic *et al.* [32] claimed good agreement with unpublished experimental observations by Fischer *et al.* in Kern's laboratory in Lausanne.

Questions concerning cluster diffusion were tackled at roughly the same time by field ion microscopy, which can directly reveal the individual atomic events. Studied by Wang and Ehrlich [21,33] were clusters of iridium on Ir(111), shown in Fig. 9.18 after equilibration, with sides of unequal as well as equal length. In none of these did evaporation intervene in diffusion. With Ir_{18}, Ir_{26} as well as Ir_{36}, diffusion occurred at T ≥ 550 K by atoms moving along the cluster edges, which are seen in Fig. 9.19 to be of unequal length. The diffusive movement proceeded without any additional adatoms on the terrace and with the number of cluster atoms conserved, ruling out the possibility of evaporation–condensation. Changes in the shape of these clusters were directly observed in the FIM; the individual atom jumps, however, could not be detected. It was always two

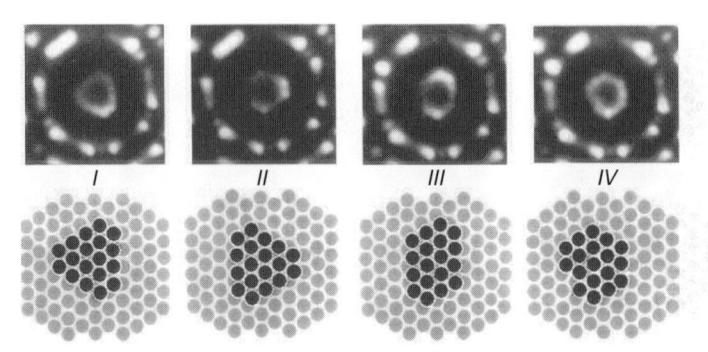

Fig. 9.19 Ir$_{18}$ clusters of non-equivalent form on Ir(111) after equilibration at ~ 550 K observed in FIM [21].

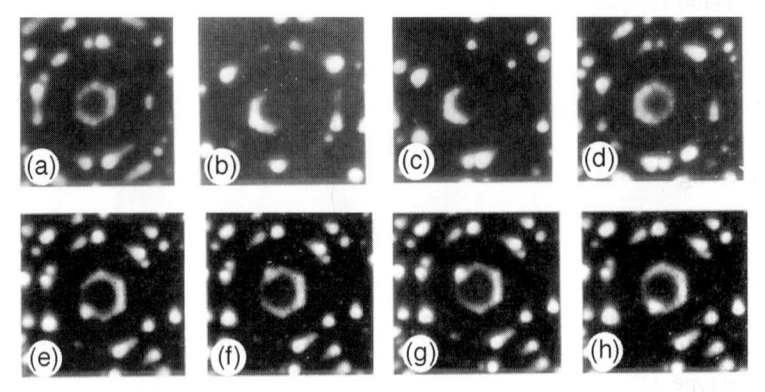

Fig. 9.20 Direct FIM images of Ir clusters diffusing on Ir(111) [33]. (a)–(d) Compact Ir$_{19}$ diffusing at ~675 K without shape change. (e)–(h) Ir$_{38}$ cluster diffusing at 400 K with position of attached Ir atom constantly altering.

atoms with Ir$_{18}$ and three with Ir$_{36}$ clusters that displaced. At a temperature of ~700 K, evaporation of atoms from non-compact clusters was observed; the free atom never came back to the cluster, rather it diffused over the terrace.

Close-packed hexagonal Ir$_{19}$ clusters also moved over the surface, shown in Fig. 9.20a–d. This diffusion differed from Ir$_{18}$, however, in that motion of the hexagonal cluster occurred at temperatures more than 100 K higher, without any apparent movement of atoms around the cluster periphery, and without shape changes. What also became clear from observations of the cluster displacement was the participation of long jumps, that is transitions larger than a displacement between nearest neighbors. The movement of Ir$_{37}$, another compact cluster, also proceeded without any shape changes; it must be noted that this cluster was more stable than non-compact ones. An Ir$_{38}$ cluster created by adding an atom to Ir$_{37}$ is shown in Fig. 9.20e–h. The movement of the additional atom around the stationary, hexagonal core consisting of 37 atoms proceeded on heating to 400 K. The atom was always observed next to type *A* steps, indicating stronger binding there compared with *B*-type steps. The Ir$_{38}$ cluster started changing its shape when the temperature was increased to 700 K, as in Fig. 9.21. Analyzing the frequency at which different cluster shapes appeared, it was possible to draw the

9.2 Large clusters on fcc(111) surfaces

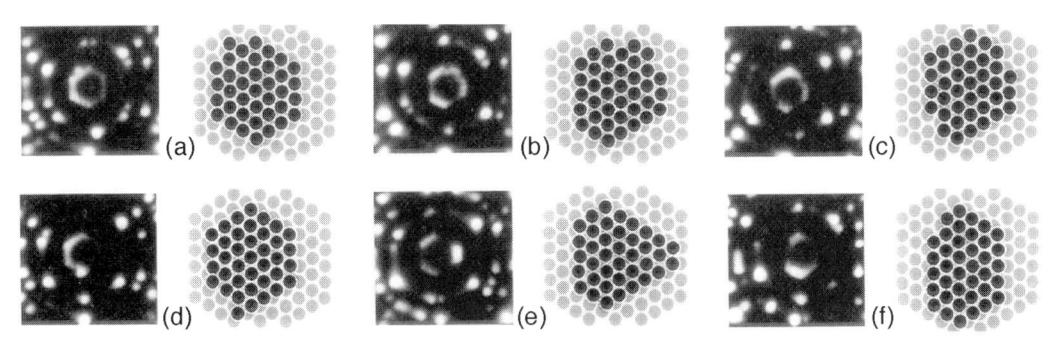

Fig. 9.21 Views of Ir$_{38}$ cluster on Ir(111) in FIM after annealing at ~700 K [33].

conclusion that for other than compact clusters, shapes with longer *B*-type edges are preferred over those with *A*-type edges. That makes binding at *A*-type edges stronger than at steps of type *B*, which agrees with observations of the presence of an atom attached only to step *A* of Ir$_{38}$ at temperature 400 K, shown in Fig. 9.21e–f. From the frequency of observation for different cluster shapes it is possible to derive the free energy change for converting one shape into another, according to

$$\Delta F = -kT \ln[(N_I/\Omega_I)/(N_{II}/\Omega_{II})], \tag{9.4}$$

where N_I and N_{II} denote the frequency of occurrence of forms *I* and *II*, while Ω_I and Ω_{II} denote the degeneracy, the number of equivalent configurations of the two cluster forms.

A rough estimate of the free energy per atom at a specific type of edge showed that at a step of type *A* the average free energy was 0.029 ± 0.008 eV higher than for an adatom at a straight step of type *B*. From the frequency of observation of kinked clusters it was deduced that this structure had the lowest energy for Ir$_{26}$ and Ir$_{39}$. For Ir$_{18}$ and Ir$_{36}$, the difference between the free energy of kinked and more stable forms was very small. For hexagonal clusters, the free energy for other shapes exceeded 0.025 eV.

From a study on Ir(111) surfaces it was possible to divide large clusters into two groups: a hexagonal one, for which the number of atoms N_c can be written

$$N_c = 1 + 3s(s-1), \quad s = 2, 3 \ldots \quad , \tag{9.5}$$

and all the others. Clusters with the number of atoms given above are more stable and remain in a hexagonal shape, while others change their shape and are less strongly bound. For example, Ir$_{19}$ remains hexagonal until it disappears at 700 K, while Ir$_{18}$ frequently changes shape and on heating to 600 K dissociates. Clusters such as Ir$_{39}$ after prolonged heating at 700 K dissociate, leaving a stable Ir$_{37}$ on the surface.

More quantitative studies were subsequently done by Wang *et al.* [34], who measured the mean-square displacement at different temperatures. Shown in Fig. 9.22a,b are plots for diffusion of Ir$_{18}$ as well as Ir$_{19}$ on the Ir(111) plane. What is notable here is that for Ir$_{18}$ clusters with edges of unequal length, the prefactor for the diffusivity is much the same as for a single iridium adatom; the activation energy for diffusion, 1.63 ± 0.07 eV for Ir$_{18}$, is of course much higher than for individual Ir adatoms.

Diffusion was also examined for two hexagonal clusters of different size, Ir$_{19}$ and Ir$_7$. It is clear that both have unusually high prefactors, $13(\times 2.6^{\pm 1})$ cm^2/sec for Ir$_{19}$ and

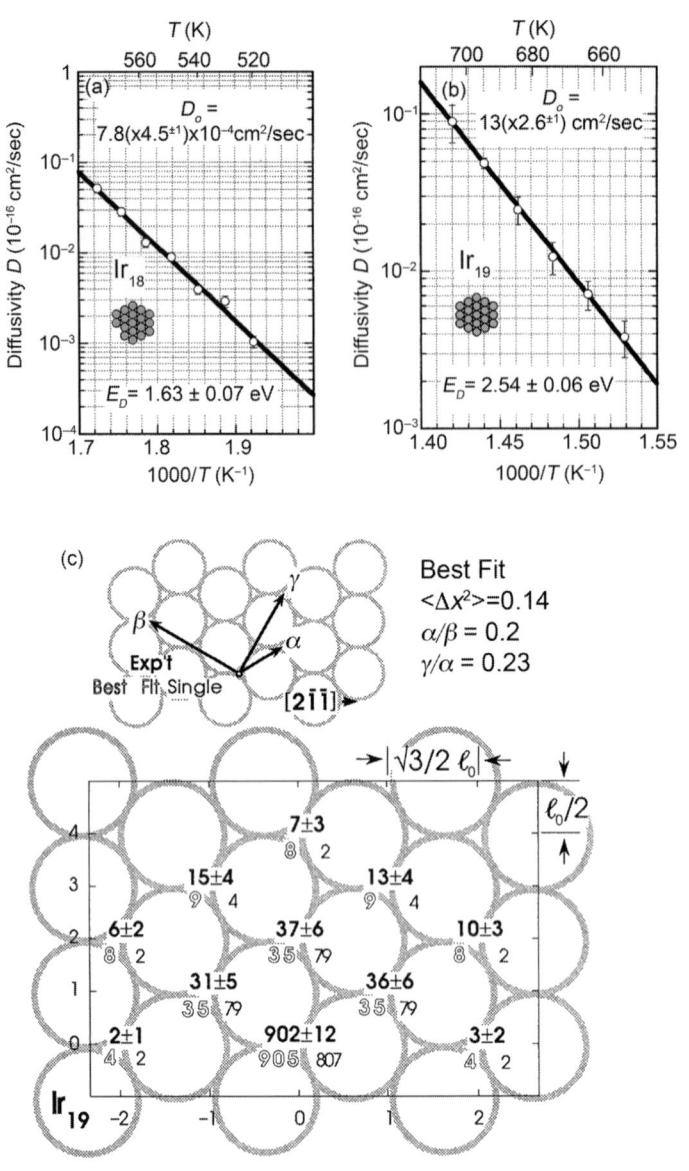

Fig. 9.22 Temperature dependence of the diffusivity for Ir clusters on Ir(111) plane observed in FIM [34]. (a) Arrhenius plot for diffusivity of Ir$_{18}$ cluster. (b) Arrhenius plot for diffusivity of Ir$_{19}$. (c) Distribution of center-of-mass displacements for Ir$_{19}$ cluster during 10 sec at 690 K. Bold letters give observations, best fit in outline numerals below, to the left. Fit with nearest-neighbor jumps only is at the bottom right.

$1.4(\times 2.4^{\pm 1})$ cm^2/sec for Ir$_7$; the diffusion barrier amounted to 2.54 eV for the larger cluster and 1.49 eV for Ir$_7$. The distribution of cluster displacements for Ir$_{19}$ was also measured and is shown in Fig. 9.22c, in which are indicated the three types of jumps identified in diffusion. What is unusual is that the long jumps, double β's and triple γ's occurred at a considerable rate and contributed ~60 % to the diffusivity at 690 K. For Ir$_7$,

9.2 Large clusters on fcc(111) surfaces

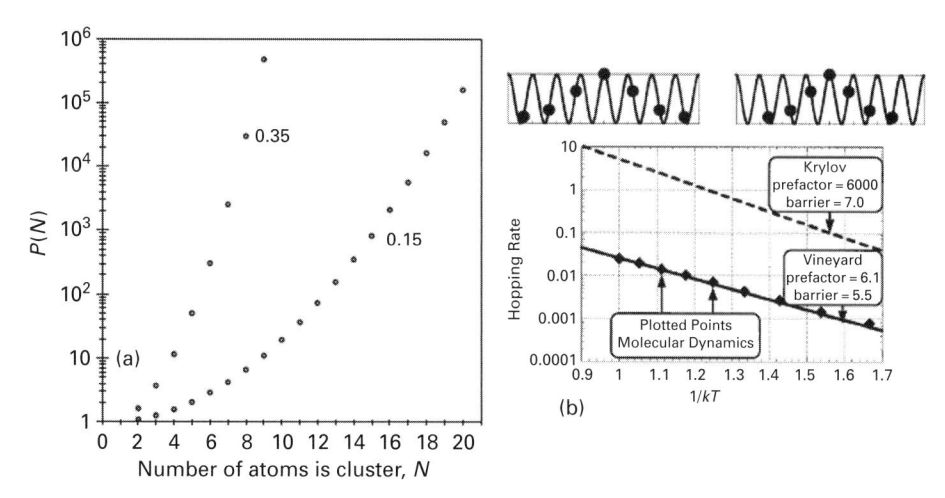

Fig. 9.23 Kinetics of cluster diffusion. (a) Predicted dependence of diffusivity prefactor on the number N of atoms in a cluster for frequency ratios $v^{(c)}/v_0 = 0.15$ and 0.35 (after Krylov [35]).
(b) One-dimensional diffusion of Ir chains. Top: stretched and compressed transition states in one-dimensional diffusion. Bottom: Arrhenius plots for hopping rate of Ir seven atom chain over a periodic potential (after Hamilton and Voter [36]).

long jumps are not of comparable importance. Clearly diffusion of these hexagonal clusters occurs in an unusual manner, and it must be emphasized that the big jump lengths are not responsible for the large values of the prefactors. The mechanism of movement was not revealed. However, it is possible to disallow the evaporation–condensation mechanism, since an Ir_{18} cluster is unstable at the temperature at which compact clusters move, so detachment of one atom will result in dissociation of the entire cluster.

In an attempt to describe the surface diffusion of hexagonal, close-packed clusters, which was found by Wang *et al.* [34] to occur with a very large prefactor and may also have involved long jumps, Krylov [35] came up with a one-dimensional theory for cluster gliding. He was seeking a rationalization for the huge prefactor caused by internal vibrations and arrived at a diffusivity D given by

$$D = D_o P(N) \exp(-E_D/kT), \qquad (9.6)$$

where the additional term $P(N)$ in the prefactor arises from a dynamical misfit. This term is close to unity when the number of atoms N in the cluster is small, but rises exponentially as N increases. A plot of this dependence on the number of atoms is given in Fig. 9.23a, for two values of the vibrational frequency ratio $v^{(c)}/v_0$ upon which $P(N)$ depends, and it is clear that clusters with tens of atoms may indeed have a prefactor much larger than normal. Krylov also advanced a separate kinetic argument for a rise in the number of long transitions in diffusion and associated this effect with size dependent cluster–substrate coupling, caused by interactions of short phonon wave length with intra-cluster vibrations.

This theory has been strongly criticized by Hamilton and Voter [36], who worked out the rate of seven atoms chain hopping by molecular dynamics for one-dimensional motion as well as by resorting to Vineyard's [37] expression for the rate of jumping. Both give a prefactor on the order of unity, as shown in Fig. 9.23b. Krylov [38], however,

Table 9.4 Prefactors for cluster diffusion calculated with GGA potentials on Ir(111) in units of sec^{-1} [40].

Mechanism	1 atom	7 atoms	19 atoms	Reference
Hopping	$2 \times 10^{12}/3 \times 10^{12}$			[41]/ [42]
Bridge glide		$7 \times 10^{13}/1 \times 10^{14}$	$2 \times 10^{16}/7 \times 10^{16}$	[41]/ [42]
Cartwheel shuffle		$7 \times 10^{14}/5 \times 10^{14}$	$8 \times 10^{16}/1 \times 10^{17}$	[41]/ [42]
Experiment	7×10^{11}	3×10^{15}	2×10^{16}	[34]

rejected these arguments, claiming that molecular dynamics simulations fail due to the use of only one spring constant, which might not represent the physics correctly. In the Vineyard approach, the minimum energy barrier was used instead of an averaged effective one for all transitions.

This leaves unsettled an explanation for the magnitude of the cluster prefactors. That question has been taken up in theoretical studies; we first mention work by Kürpick et al. [39], who examined Ir$_7$ clusters on Ir(111) more closely with Sutton–Chen potentials. Three sequences from simulations at 1350 K are shown in Fig. 7.24. The clusters move by slightly different paths from one energy minimum to a neighboring position. For the concerted displacement from fcc to hcp sites, a potential energy increase of 1.23 eV was found with Sutton–Chen potentials and 1.51 eV using RGL; several non-equivalent mechanisms for cluster diffusion were identified with activation energies comparable to the experimental value of 1.49 ± 0.03 eV. Prefactors for the different mechanisms were evaluated and turned out to be in the usual range of 10^{-4} cm^2/sec. Her conclusion was that there exist a number of non-equivalent paths for diffusion, with activation energies comparable to the measured value. Prefactors of the magnitude usual in single adatom diffusion were found in simulations; there was no indication of any unusually large value for any single path. However, the number of mechanisms was large, and they allowed the cluster to move without any significant change in shape. Their superposition created the large overall prefactor found in the experiments.

The intriguing mode of diffusion for compact hexagonal Ir$_7$ and Ir$_{19}$, with very high prefactors, high activation energies, and long jumps, was addressed again by Hamilton et al. [40], using generalized gradient approximations and the nudged elastic band method. They found that a pseudomorphic monolayer had a fairly low energy metastable minimum when in the on-top location on a crystal. Semi-empirical methods, such as EAM, EMT or Sutton–Chen, did not detect this minimum. With this result as a motivation, they considered four different mechanisms of cluster diffusion shown in Fig. 7.29: glide over bridge sites, glide over on-top positions, as well as two combinations of cluster rotation and translation: cartwheel shuffle and cartwheel glide. The results of their calculations for both Ir$_7$ and Ir$_{19}$ are given in Table 9.4, but it should be noted that for Ir$_{19}$, calculations were limited. For Ir$_7$, bridge gliding is in very good agreement with the measured barrier of 1.49 ± 0.03 eV [34]. The situation with Ir$_{19}$ clusters is different, as now bridge glide and cartwheel shuffle have much higher calculated barriers, 3.2 eV and 3.6 eV respectively, compared to the 2.54 ± 0.06 eV found in experiments, but this could have stemmed from calculational approximations. Hamilton et al. believed that

9.2 Large clusters on fcc(111) surfaces

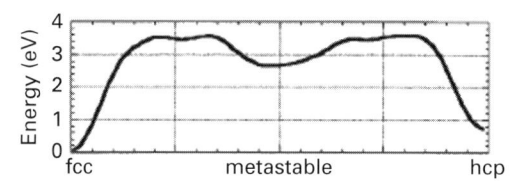

Fig. 9.24 Energy calculated using generalized gradient approximation for translation and rotation in cartwheel shuffle of Ir_{19} cluster on Ir(111) plane (after Hamilton et al [40]).

cartwheel shuffle (possibly combined with bridge glide) should be considered in attempting to account for the long jumps. The energy change during a cartwheel shuffle is shown in Fig. 9.24, and it is clear that there is a very broad metastable position, but that the energy changes rapidly at both fcc and hcp sites. In the Vineyard [37] formulation of rates the prefactor is given by the product of vibrational frequencies in the fcc state over the product in the transition state. Judging from the potential diagram, the latter vibrations should have much lower frequencies, which should give a higher prefactor. To estimate this, calculations were done with Sutton–Chen [41] and Chen [42] potentials, giving the results in Table 9.4. Higher prefactors than usual were in fact obtained.

Monte Carlo simulations of the diffusion of Ir_{18} clusters on Ir(111) have also been made by Zhuang and Wang [43] relying on the energetics of Bogicevic et al. [31]. Presumably diffusion occurred via the peripheral movement of individual atoms around the edges, as is suggested in the diagram in Fig. 9.25a. In field ion microscopic observations, however, two atoms were always seen to transfer from one position at the edge to another [34], probably because single atom events took place too rapidly, making it impossible to observe the intermediate shapes shown in Fig. 9.25a as states a–e. Zhuang and Wang therefore decided to see if the experimentally observed results could be simulated by jumps of an effective dimer, as is suggested in Fig. 9.25b, where the barrier heights were obtained by fitting to the experiments. As shown by the temperature dependence of the diffusivity, plotted in Fig. 9.25c, agreement was very good, as the experimentally derived activation energy was 1.63 ± 0.07 eV with a prefactor $7.8 (\times 4.5^{\pm 1}) \times 10^{-4}$ cm^2/sec. This study shows that dimer shearing movement is most likely responsible for movement of Ir_{18} cluster.

In 2000, Chirita et al. [44] extended their considerations of cluster diffusion on Pt(111) by sub-unit shearing to higher clusters. In their work they relied on the embedded atom potentials of Johnson and Oh [45,46], and worked on platinum clusters of 6 to 25 atoms. An example of their proposed mechanism is shown in Fig. 9.26 for Pt_{11}. Starting with all atoms in fcc sites, 5 atoms glide to hcp sites, which creates a stacking fault. The motion is completed by relaxation of the remainder into hcp positions. For hexamers, they estimate a barrier to diffusion of $\simeq 1.55$ eV, and contend that double (reptation) shearing should "replace concerted cluster gliding as the energetically favored migration mechanism with increasing cluster size." For Pt_6, Pt_7, and Pt_{19} activation energies of 1.22 eV, 1.50 eV, and 4.37 eV were estimated for cluster diffusion, in which all atoms make concerted jumps. If the number of cluster atoms is less than 7, concerted glide and double shearing are competitive, but for larger clusters the latter is favored according to their study. Shearing

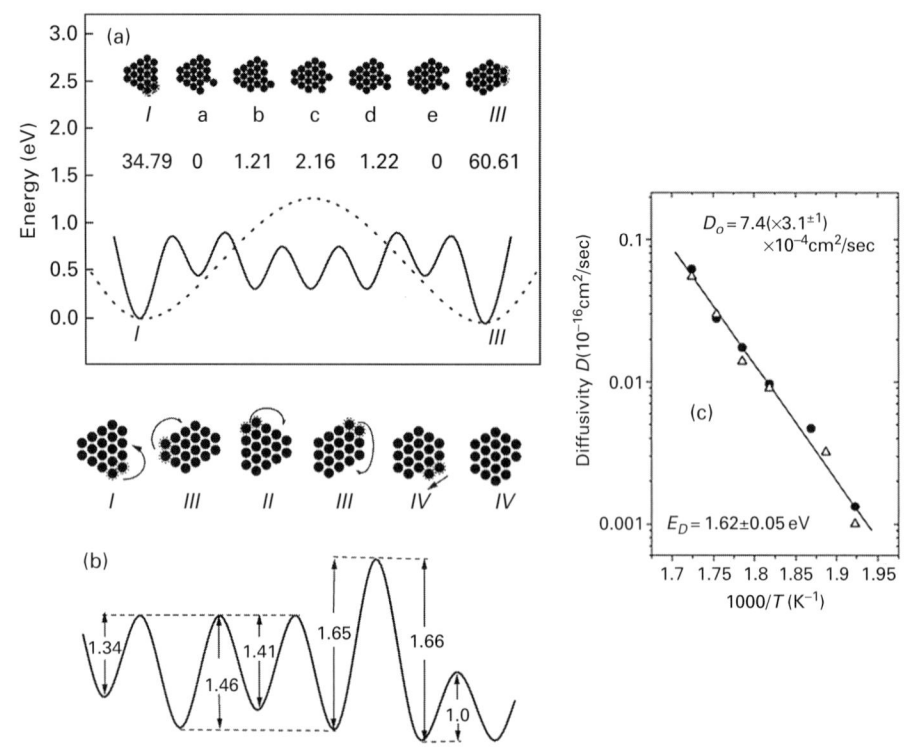

Fig. 9.25 Migration at periphery of Ir_{18} on Ir(111) surface derived from MC studies. (a) Solid curve shows energy for single atom peripheral diffusion. Dashed curve gives effective energy for dimer process. Numerals below a configuration indicate percentage of time spent there. (b) Peripheral diffusion of dimer around Ir_{18}. (c) Arrhenius plot for the diffusivity of Ir_{18} derived from mean-square displacements. Solid circles: from Monte Carlo simulations. Open triangles: experiments of Wang et al. [34] (after Zhuang and Wang [43]).

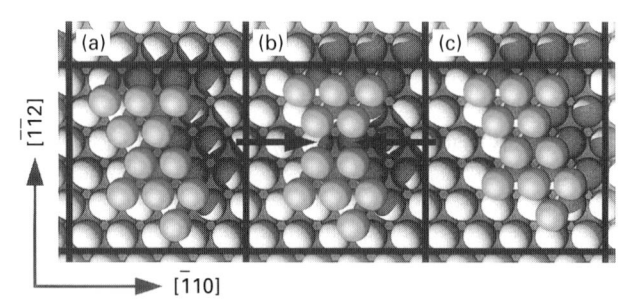

Fig. 9.26 Diffusion of Pt_{11} cluster on Pt(111) by reptation, based on EAM. (a) Starting position. (b) Upper part of cluster glides, forming stacking fault. (c) Form after diffusion (after Chirita et al. [44]).

is also proposed to be important for close-packed clusters of 19 atoms. The value obtained for the heptamer is essentially identical with experiments [47], but that derived for Pt_{19} appear too high.

Diffusion of islands on Cu(111) and Ag(111) was extensively studied with scanning tunneling microscopy by Schlösser et al. [48], who examined the relation between the

9.2 Large clusters on fcc(111) surfaces

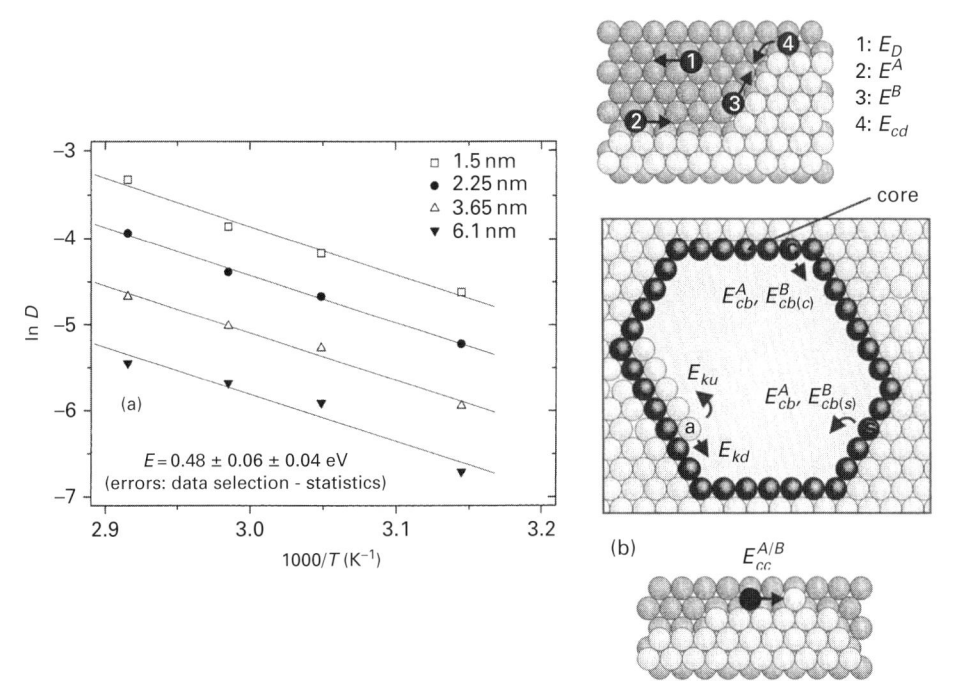

Fig. 9.27 Diffusivity of vacancy islands on Cu(111) derived from STM measurements. (a) Arrhenius plots. (b) Atomic steps in vacancy island motion (after Schlösser et al. [48]).

diffusivity and the island radius R_r. Measurements were made at temperatures from 263 K to 343 K on islands with radii in the range from 1 to 10 nm. For vacancy islands on Cu(111) the exponent α_L in the expression $D \propto R_r^{-\alpha_L}$ for the size dependence was given as 1.37 ± 0.1, and the activation energy for island diffusion derived from four temperatures, shown in Fig. 9.27a, amounted to 0.49 ± 0.01 eV. For similar vacancy islands on Ag(111) the exponent α_L was 1.33 ± 0.11 and the diffusion barrier came to 0.51 ± 0.04 eV. With adatom islands on Ag(111), α_L was 1.63 ± 0.38, and the diffusion barrier amounted to 0.53 ± 0.05 eV. The activation energies for all three systems were much the same. What is most significant, however, is that the exponent α_L in the size relationship for all three differed significantly from the value of three expected for peripheral diffusion of atoms around the island edges as the limiting step. Calculations of the activation energies were made with EMT potentials for the various atomic processes in island diffusion. The processes taken into account are illustrated in Fig. 9.27b, the energetics derived from calculations are listed in Table 9.5. However, Schlösser et al. ended their work with the conclusion that "a simple interpretation of island diffusion in terms of a single rate-limiting process is not possible." The authors concluded that island diffusion arose from a combination of peripheral motion of single atoms plus the breakup of island cores.

Monte Carlo simulations of copper island diffusion have been done both on Cu(100) and Cu(111) surfaces by Jahma et al. [19], based on energetics from effective medium theory. For islands on Cu(111) the diffusivity versus cluster size curves are smooth with no oscillations. There is also relatively little temperature dependence of the diffusivity for small islands, and

Table 9.5 Energetics of cluster self-diffusion on (111) surfaces in eV, calculated using EMT potentials [48].

	Cu(111)	Ag(111)
γ_S	0.211	0.159 [49]
ΔE_{ks}	0.223	0.169
ΔE_{cs}	0.434	0.330
ΔE_{kt}	0.721	0.558
E_D	0.057	0.068
E^A	0.225	0.221
E^B	0.325	0.300
E_{cd}	0.370	0.318
$E_{kd}^{A/B}$	$E^{A/B} + \Delta E_{ks}$	
E_{ku}	$E_{cd} + \Delta E_{ks}$	
E_{cb}^A	$E^B + \Delta E_{cs}$	
$E_{cb(c)}^B$	$E^A + \Delta E_{cs}$	
$E_{cb(s)}^B$	$E_{cb} + \Delta E_{cs}$	
$E_{cc}^{A/B}$	$E^{A/B} + 2\Delta E_{ks}$	

γ_S – formation energy per step atom; ΔE_{ks}, ΔE_{cs} and ΔE_{kt} – binding energy of kink atom, of core atom relative to adatom at the step, and of the kink atom relative to the terrace. The diffusion energies in the table are relative to the atomic motions shown in Fig. 9.27b.

the activation energy for single adatom diffusion was 0.026 eV. For islands made of 19 to 100 atoms the exponent α_N in the size relation was about 1.57, in agreement with the view that large island movement is dominated by diffusion along the perimeter.

Recently Ghosh et al. [50] have carried through a quite thorough Monte Carlo simulation of cluster diffusion on Cu(111) with the NEB method, using EAM interaction potentials and relying on a pattern recognition scheme to identify 49 different diffusion processes. The energetics of the main processes are shown in Table 9.6 for a cluster of 10 atoms. General clusters of 10, 18, and 26 atoms were found to carry out overall random motion. Compact hexagonal clusters, with 19 or 37 atoms, diffused slowly, with occasional large transitions, as shown in Fig. 9.28b. The authors also found another two classes of clusters: with 6s+3 atoms, s being an integer and with a hexagon+1. These were found to execute stick-slip motion, with short transitions between long, essentially stationary periods. At higher temperatures, clusters of all sizes were moving randomly, and vacancies started to form. Arrhenius plots for the diffusivity of general clusters are shown in Fig. 9.28d, and yield activation energies reasonably close together but increasing with size. The most frequent event in these clusters is diffusion along A-type edges, with B-type edges next. Effective barriers for clusters, except for those of 21 and 26 atoms, turned out to be around 0.65 eV. For a cluster made up of 21 atoms the effective barrier for movement of the center of mass proved to be higher, 0.84 eV. For clusters composed of 26 atoms the effective activation energy was derived as 0.76 eV. Hexagonal compact entities diffused so slowly that no barrier heights were evaluated, which is regrettable, as comparison with the

9.2 Large clusters on fcc(111) surfaces

Table 9.6 Energetics based on EAM for main processes in diffusion of 10 atom cluster on Cu(111) [50].

Process	Energy barrier (eV)	Process	Energy barrier (eV)
Step edge A	0.252	Reverse of step rounding B	0.402
Step edge B	0.295	Corner rounding AA at stage 1	0.313
Kink detach along step A	0.519	Corner rounding AA at stage 2	0.143
Kink detach along step B	0.556	Corner rounding AA at stage 3	0.010
Kink incorporation step A	0.220	Corner rounding BB at stage 1	0.374
Kink incorporation step B	0.265	Corner rounding BB at stage 2	0.038
Kink detach out of step A	0.658	Corner rounding BB at stage 3	0.052
Kink detach out of step B	0.590	Corner rounding AB at stage 1	0.317
Kink fall into step A	0.074	Corner rounding AB at stage 2	0.084
Kink fall into step B	0.007	Corner rounding BA at stage 1	0.396
Kink rounding at step A	0.656	Corner rounding BA at stage 2	0.015
Kink rounding at step B	0.678	Rounding chain A	0.063
Reverse of step rounding A	0.374	Rounding chain B	0.019

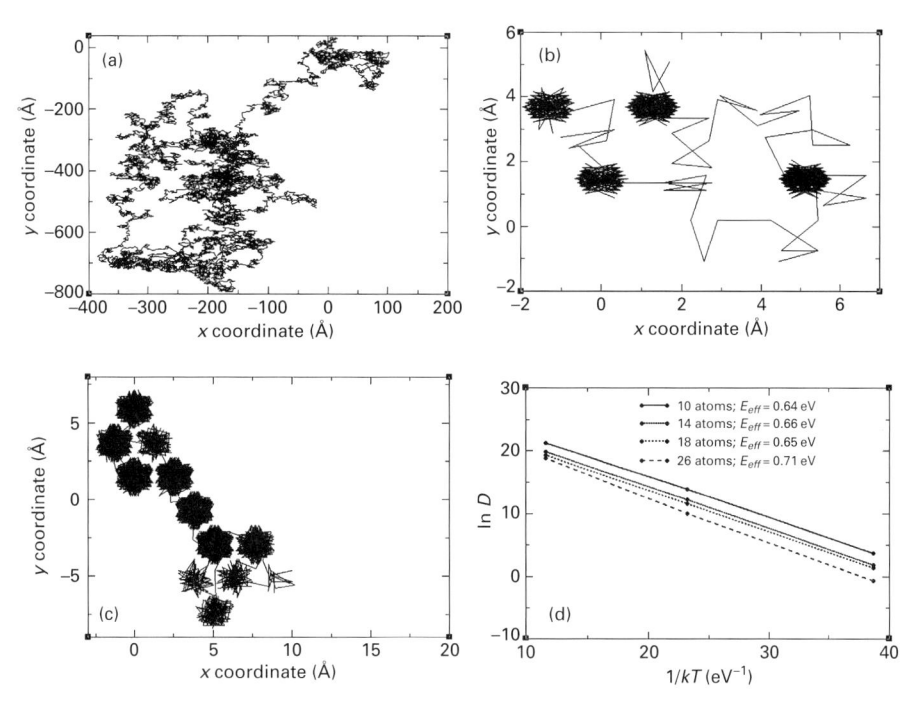

Fig. 9.28 Trace of center-of-mass diffusion obtained from MC simulations using EAM of clusters on Cu (111). (a) Cu$_{10}$ (b) Cu$_{19}$ (c) Cu$_{21}$ (d) Arrhenius plots for diffusion derived from Monte Carlo simulations (after Ghosh et al. [50]).

experiments of Wang et al. [34] would have been interesting. However, there clearly is a similarity in behavior of hexagonal clusters on Ir(111) and Cu(111).

From the information presented for islands of more than 10 atoms on an fcc(111) surface, we can conclude that movement of most clusters proceeds by adatom edge

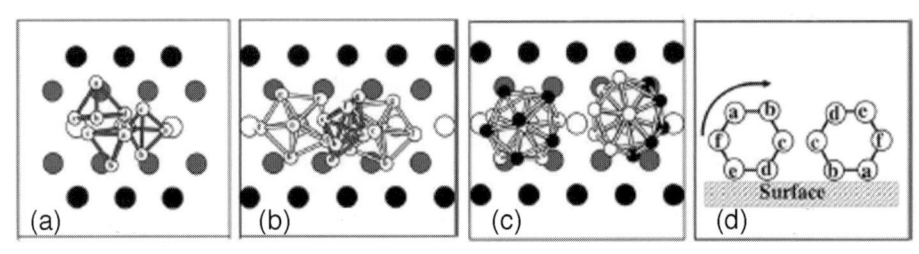

Fig. 9.29 Diffusion of Ni clusters on Au(110)-(1 × 2) simulated with Johnson potentials [46,52]. Illustration of rolling in trough for (a) Ni$_4$ (b) Ni$_7$ (c) Ni$_{13}$ (after Fan and Gong [51]).

running. For some systems, for example like Al(111) [32], movement along edges by exchange should be taken into account. The evaporation–condensation mechanism has been excluded from the picture at least at room temperature. Investigations at higher temperatures are still in order. Movement of hexagonal islands is rather different. The mechanism of movement as well as the reasons for the high prefactors for diffusivity and for the presence of long jumps remain to be assigned with certainty. The small seven atom hexagonal clusters may move by gliding; the mechanism of movement for bigger hexagonal islands is still uncertain. Does it proceed by creation of a dislocation or by shuffle gliding? Individual atom-by-atom movement is certainly unable to explain the observed displacements, and concerted motion should play a role.

Unfortunately experimental data about movement of large clusters on fcc(111) surfaces is limited to three materials: iridium, copper, and silver.

9.3 Large clusters on fcc(110) surfaces

Information about the movement of large clusters on the channeled fcc(110) surface is scarce. For metallic clusters the only study found was for Ni$_N$ on the Au(110)-(1 × 2) reconstructed surface, done by Fan and Gong [51] with EAM and the interactions of Johnson [46,52]. Since Ni-Ni interactions are stronger than Au-Ni, the Ni$_{13}$ cluster prefers the three-dimensional structure of an icosahedron. Such a cluster can roll in the deep channels of the Au surface, shown in Fig. 9.29. A sphere-like shape is also achieved for Ni$_4$ and Ni$_7$ clusters. The diffusion constants are orders of magnitude larger for clusters with spherical structure, compared to clusters which cannot achieve this form. In rolling, only two atoms need overcome an energy barrier. Although rolling seems to be the leading mechanism, gliding and rotation was also observed.

It is of interest to note diffusion studies on rather different materials, decacyclene, and hexa-tert-butyl decacyclene [53]. These flat ring structures have been studied on Cu(110). They are mentioned here because just as in Ir$_{19}$ [34], long jumps as large as 3.9 ± 0.2 and 6.8 ± 0.3 copper nearest-neighbor spacings have been found for the two molecules, with prefactors of $10^{-1.0 \pm 1.0}$ and $10^{0.9 \pm 1.0}$ cm^2/sec. These are again unusually large, and it appears that long jumps may be quite common in the diffusion of such large entities.

Experimental mechanistic studies of large clusters on channeled surfaces have been overlooked, and we have no real clue about the mechanisms of movement or the diffusivities.

9.4 Comments and comparisons

What is clear from this brief survey is that much progress has been made in understanding the diffusion of large clusters over crystal surfaces. Evaporation–condensation does not appear to be a significant process in surface diffusion under the usual conditions. Instead it seems that peripheral diffusion of atoms around the cluster edges is the dominant mechanism. More could be done to define the mechanism of motion for hexagonal clusters with edges of equal length. It is clear, however, that such clusters have unusually large diffusion prefactors, and surprisingly, carry out jumps large by comparison with nearest-neighbor surface spacings. It is evident, moreover, that more quantitative experimental studies will be really desirable.

References

[1] G. Ehrlich, in: *Proceedings of the 9th International Conference on Solid Surfaces*, J. L. de Segovia (ed.), Layer growth – an atomic picture, (A.S.E.V., Madrid, 1983), p. 3–16.
[2] G. Ehrlich, in: *Chemistry and Physics of Solid Surfaces V*, R. Vanselow, R. Howe (eds.), An atomic view of crystal growth, (Springer-Verlag, Berlin, 1984), p. 282–296.
[3] H.-W. Fink, in: *Diffusion at Interfaces – Microscopic Concepts*, M. Grunze, H. J. Kreuzer, J. J. Weimer (eds.), Direct observation of atomic motion on surfaces, (Springer-Verlag, Berlin, 1988), p. 75–91.
[4] K. Binder, M. H. Kalos, "Critical clusters" in a supersaturated vapor: Theory and Monte Carlo simulation, *J. Stat. Phys.* **22** (1980) 363–396.
[5] A. F. Voter, Classically exact overlayer dynamics: Diffusion of rhodium clusters on Rh(100), *Phys. Rev. B* **34** (1986) 6819–6829.
[6] A. F. Voter, Simulation of the layer-growth dynamics in silver films: Dynamics of adatom and vacancy clusters on Ag(100), *Proc. SPIE* **821** (1987) 214–226.
[7] J.-M. Wen, S.-L. Chang, J. W. Burnett, J. W. Evans, P. A. Thiel, Diffusion of large two-dimensional Ag clusters on Ag(100), *Phys. Rev. Lett.* **73** (1994) 2591–2594.
[8] W. W. Pai, A. K. Swan, Z. Zhang, J. F. Wendelken, Island diffusion and coarsening on metal (100) surfaces, *Phys. Rev. Lett.* **79** (1997) 3210–3213.
[9] J. M. Wen, J. W. Evans, M. C. Bartelt, J. W. Burnett, P. A. Thiel, Coarsening mechanisms in a metal film: From cluster diffusion to vacancy ripening, *Phys. Rev. Lett.* **76** (1996) 652–655.
[10] C. R. Stoldt, C. J. Jenks, P. A. Thiel, A. M. Cadilhe, J. W. Evans, Smoluchowski ripening of Ag islands on Ag(100), *J. Chem. Phys.* **111** (1999) 5157–5166.
[11] S. Pal, K. A. Fichthorn, Size dependence of the diffusion coefficient for large adsorbed clusters, *Phys. Rev. B* **60** (1999) 7804–7807.
[12] A. Lo, R. T. Skodje, Diffusion and evaporation kinetics of large islands and vacancies on surfaces, *J. Chem. Phys.* **111** (1999) 2726–2734.
[13] J. Heinonen, I. Koponen, J. Merikoski, T. Ala-Nissila, Island diffusion on metal fcc (100) surfaces, *Phys. Rev. Lett.* **82** (1999) 2733–2736.
[14] J. Merikoski, T. Ala-Nissila, Diffusion processes and growth on stepped metal surfaces, *Phys. Rev. B* **52** (1995) R8715–8718.

[15] G. Mills, T. R. Mattsson, L. Mollnitz, H. Metiu, Simulations of mobility and evaporation rate of adsorbate islands on solid surfaces, *J. Chem. Phys.* **111** (1999) 8639–8650.
[16] W. W. Pai, J. F. Wendelken, C. R. Stoldt, P. A. Thiel, J. W. Evans, D.-J. Liu, Evolution of two-dimensional wormlike nanoclusters on metal surfaces, *Phys. Rev. Lett.* **86** (2001) 3088–3091.
[17] F. J. Resende, B. V. Costa, Molecular dynamics study of copper cluster deposition on a Cu(010) surface, *Surf. Sci.* **481** (2001) 54–66.
[18] X. Wang, F. Xie, Q. Shi, T. Zhao, Effect of atomic diagonal motion on cluster diffusion coefficient and its scaling behavior, *Surf. Sci.* **561** (2004) 25–32.
[19] M. O. Jahma, M. Rusanen, A. Karim, I. T. Koponen, T. Ala-Nissila, T. S. Rahman, Diffusion and submonolayer island growth during hyperthermal deposition on Cu(100) and Cu(111), *Surf. Sci.* **598** (2005) 246–252.
[20] J. R. Sánchez, Metal surface adsorbed clusters: Structure and dynamics, *J. Molec. Catalysis A: Chemical* **237** (2005) 206–209.
[21] S. C. Wang, G. Ehrlich, Diffusion of large surface clusters: Direct observations on Ir(111), *Phys. Rev. Lett.* **79** (1997) 4234–4237.
[22] D. J. Trevor, C. E. D. Chidsey, Room temperature surface diffusion mechanisms observed by scanning tunneling microscopy, *J. Vac. Sci. Technol. B* **9** (1991) 964–968.
[23] J. de la Figuera, J. E. Prieto, C. Ocal, R. Miranda, Creation and motion of vacancy islands on solid surfaces: A direct view, *Solid State Comm.* **89** (1994) 815–818.
[24] K. Morgenstern, G. Rosenfeld, B. Poelsema, G. Comsa, Brownian motion of vacancy islands on Ag(111), *Phys. Rev. Lett.* **74** (1995) 2058–2061.
[25] G. Rosenfeld, K. Morgenstern, G. Comsa, in: *NATO-ASI Surface Diffusion: Atomistic and Collective Processes*, M. C. Tringides (ed.), Diffusion and stability of large clusters on crystal surfaces, (Plenum Press, New York, 1997), p. 361–375.
[26] J. C. Hamilton, M. S. Daw, S. M. Foiles, Dislocation mechanism for island diffusion on fcc(111) surfaces, *Phys. Rev. Lett.* **74** (1995) 2760–2763.
[27] J. C. Hamilton, Magic size effects for heteroepitaxial island diffusion, *Phys. Rev. Lett.* **77** (1996) 885–888.
[28] D. S. Sholl, R. T. Skodje, Diffusion of clusters of atoms and vacancies on surfaces and the dynamics of diffusion-driven coarsening, *Phys. Rev. Lett.* **75** (1995) 3158–3161.
[29] L. Bitar, P. A. Serena, P. García-Mochales, N. García, V. T. Binh, Mechanism for diffusion of nanostructures and mesoscopic objects on surfaces, *Surf. Sci.* **339** (1995) 221–232.
[30] L. Kuipers, M. S. Hoogeman, J. W. M. Frenken, Step dynamics on Au(110) studied with high temperature, high speed scanning tunneling microscope, *Phys. Rev. Lett.* **71** (1993) 3517.
[31] A. Bogicevic, S. Liu, J. Jacobsen, B. Lundqvist, H. Metiu, Island migration caused by the motion of the atoms at the border: Size and temperature dependence of the diffusion coefficient, *Phys. Rev. B* **57** (1998) R9459–9462.
[32] A. Bogicevic, J. Strömquist, B. I. Lundqvist, Low-symmetry diffusion barriers in homoepitaxial growth of Al(111), *Phys. Rev. Lett.* **81** (1998) 637–640.
[33] S. C. Wang, G. Ehrlich, Equilibrium shapes and energetics of iridium clusters on Ir(111), *Surf. Sci.* **391** (1997) 89–100.
[34] S. C. Wang, U. Kürpick, G. Ehrlich, Surface diffusion of compact and other clusters: Ir$_x$ on Ir(111), *Phys. Rev. Lett.* **81** (1998) 4923–4926.
[35] S. Y. Krylov, Surface gliding of large low-dimensional clusters, *Phys. Rev. Lett.* **83** (1999) 4602–4605.

[36] J. C. Hamilton, A. F. Voter, Failure of 1D models for Ir island diffusion on Ir(111), *Phys. Rev. Lett.* **85** (2000) 1580.

[37] G. H. Vineyard, Frequency factors and isotope effects in solid state rate processes, *J. Phys. Chem. Solids* **3** (1957) 121–127.

[38] S. Y. Krylov, Krylov replies, *Phys. Rev. B* **85** (2000) 1581.

[39] U. Kürpick, B. Fricke, G. Ehrlich, Diffusion mechanisms of compact surface clusters: Ir_7 on Ir (111), *Surf. Sci.* **470** (2000) L45–51.

[40] J. C. Hamilton, M. R. Sorensen, A. F. Voter, Compact surface-cluster diffusion by concerted rotation and translation, *Phys. Rev. B* **61** (2000) R5125–5128.

[41] A. P. Sutton, J. Chen, Long-range Finnis-Sinclair potentials, *Philos. Mag. Lett.* **61** (1990) 139–46.

[42] S. P. Chen, Studies of iridium surfaces and grain boundaries, *Philos. Mag. A* **66** (1992) 1–10.

[43] G. Zhuang, W. Wang, The atomic moving process of cluster Ir_{18} diffusion on Ir(111), *Internatl. J. Mod. Phys. B* **14** (2000) 427–434.

[44] V. Chirita, E. P. Münger, J. E. Greene, J.-E. Sundgren, Cluster diffusion and surface morphological transitions on Pt(111) via reptation and concerted motion, *Thin Solid Films* **370** (2000) 179–185.

[45] D. J. Oh, R. A. Johnson, Simple embedded atom method model for fcc and hcp metals, *J. Mater. Res.* **3** (1988) 471–478.

[46] R. A. Johnson, Alloy models with the embedded atom method, *Phys. Rev. B* **39** (1989) 12554–12559.

[47] K. Kyuno, G. Ehrlich, Diffusion and dissociation of platinum clusters on Pt(111), *Surf. Sci.* **437** (1999) 29–37.

[48] D. C. Schlösser, K. Morgenstern, L. K. Verheij, G. Rosenfeld, F. Besenbacher, G. Comsa, Kinetics of island diffusion on Cu(111) and Ag(111) studied with variable-temperature STM, *Surf. Sci.* **465** (2000) 19–39.

[49] P. Stoltze, Simulation of surface defects, *J. Phys.: Condens. Matter* **6** (1994) 9495–9517.

[50] C. Ghosh, A. Kara, T. S. Rahman, Usage of pattern recognition scheme in kinetic Monte Carlo simulations: Application to cluster diffusion on Cu(111), *Surf. Sci.* **601** (2007) 3159–3168.

[51] W. Fan, X. G. Gong, Simulation of Ni cluster diffusion on Au(110)-(1×2) surface, *Appl. Surf. Sci.* **219** (2003) 117–122.

[52] R. A. Johnson, Analytical nearest-neighbor model for fcc metals, *Phys. Rev. B* **37** (1988) 3924–3931.

[53] M. Schunack, T. R. Linderoth, F. Rosei, E. Laegsgaard, I. Stensgaard, F. Besenbacher, Long jumps in the surface diffusion of large molecules, *Phys. Rev. Lett.* **88** (2002) 156102 1–4.

10 Atomic pair interactions

As is clear from the previous chapters, a considerable amount of information has become available about the diffusivity of metal atoms on metal surfaces. This knowledge will have an impact on understanding topics such as crystal growth and dissolution, annealing, sintering, as well as chemical surface reactions. We will not examine any of these important topics, but will instead focus on attempts to measure atomic surface interactions, as an example of effects which have become much clearer through gains in the knowledge of surface diffusion.

Interactions can be divided into two categories, direct, for example van der Waals, dipolar, as well as electronic interactions, and indirect, which are mediated by the substrate, such as coupling by electronic states, elastic effects, or vibrational coupling. In Chapters 8 and 9 we described the movement of clusters created mostly by direct interactions of adatoms. In this chapter we will concentrate on the second type of interactions, mediated by the lattice – indirect interactions. The theory of interactions between adsorbed atoms has been nicely reviewed by Einstein [1], so here we just want to point out that with two atoms in the gas phase, the wave function for a higher state will be confined to the vicinity of the core. Placed on a metal surface, however, quantum interference of wave functions enters and adatom waves combine with metal wave functions from the lattice. The contributions from two atoms may overlap, as shown in Fig. 10.1 [2]. When in phase, attraction occurs between the two atoms, when out of phase, there will be repulsion. In view of the oscillatory nature shown in the figure it is clear that there will also be oscillations in the interactions as the interatomic separation is varied.

The first piece of information of interest in surface interactions is the energy of two adsorbed atoms as a function of their separation. What is the range of these effects? How strong are they? A variety of techniques have been used in the past in an effort to gain an insight into these questions. Such efforts have been briefly reviewed some time ago [3]. With modern methods, difficulties in the measurements are not hard to overcome. What needs to be done is to place two atoms on a surface, and allow them to equilibrate with each other. This is easy to do in the STM or FIM, as two atoms can be directly observed.

How the atoms position themselves will be dependent upon interactions between the adatoms on the surface. Crucial is knowledge about the diffusion characteristics of the atoms, as conditions have to be adjusted to ensure that equilibrium is established. The analysis of the distribution of atom separations to give interactions is straightforward, and a simple derivation [4] proceeds as follows.

Atomic pair interactions

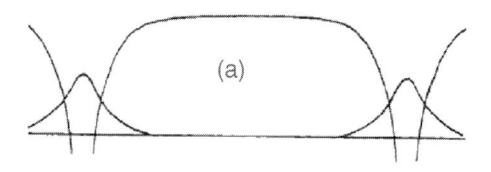

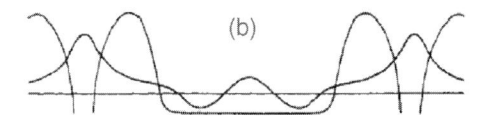

Fig. 10.1 Wave functions for two atoms (a) in the gas phase and (b) adsorbed on a metal surface as a function of separation. For the latter there are interactions between the two adatoms (after Grimley [2]).

Suppose there are two equilibrated atoms located at sites i and j on the surface, and the atoms are in a quantum state ℓ. The probability $P_{ij,\ell}$ of finding these two atoms is given by

$$P_{ij,\ell} = \exp[-E_{ij,\ell}/kT] / \sum_{ij,\ell} \exp[-E_{ij,\ell}/kT] = \exp[-E_{ij,\ell}/kT]/Z; \tag{10.1}$$

here $E_{ij,\ell}$ is just the energy of the two atoms in state ℓ, and Z is the canonical partition function

$$Z = \sum_{ij,\ell} \exp[-E_{ij,\ell}/kT]. \tag{10.2}$$

Only the spatial distribution of the atoms will be measured. Summing over all states ℓ we therefore obtain

$$P_{ij} = \sum_{\ell} \exp[-E_{ij,\ell}/kT]/Z. \tag{10.3}$$

Now the Helmholtz free energy F_{ij} can be written as

$$-F_{ij}/kT = \ln \sum_{\ell} \exp[-E_{ij,\ell}/kT], \tag{10.4}$$

so that the probability P_{ij} of finding a pair at sites i and j is just

$$P_{ij} = \exp[-F_{ij}/kT]/Z. \tag{10.5}$$

Our real interest, of course, is not the probability P_{ij}, but rather the probability $P(\boldsymbol{R})$ of finding atoms at a vector separation $\boldsymbol{R}$ on the surface, for which the total number of pairs possible at this separation is $N_o(\boldsymbol{R})$. This we obtain by summing Eq. (10.5) over all pairs of sites separated by $\boldsymbol{R}$

$$P(\boldsymbol{R}) = CN_o(\boldsymbol{R})\exp[-F(\boldsymbol{R})/kT]; \tag{10.6}$$

Here C is a constant introduced to give normalization, and includes the partition function Z. The probability $P(\boldsymbol{R})$ is given in terms of $N(\boldsymbol{R})$, the number of observations for which a pair of atoms has been found separated by $\boldsymbol{R}$, that is

$$P(\boldsymbol{R}) = N(\boldsymbol{R})/M, \tag{10.7}$$

where M is the total number of observations, so that

$$N(\boldsymbol{R}) = CMN_o(\boldsymbol{R})\exp[-F(\boldsymbol{R})/kT]. \tag{10.8}$$

We can therefore write the free energy of interaction $F(\boldsymbol{R})$ as

$$F(\boldsymbol{R})/kT = C_\infty - \ln[N(\boldsymbol{R})/N_o(\boldsymbol{R})], \tag{10.9}$$

where we have written C_∞ for $\exp(CM)$. This new normalization can be obtained by summing over all separations larger than a critical value R_c for which interactions have vanished, so that

$$C_\infty = \ln\left[\sum_{R>R_c} N(\boldsymbol{R}) / \sum_{R>R_c} N_o(\boldsymbol{R})\right]. \tag{10.10}$$

Gaining an understanding of how the free energy of interaction between two atoms varies with their separation $\boldsymbol{R}$ is now just a matter of measuring $N(\boldsymbol{R})$, the number of times a pair is observed separated by $\boldsymbol{R}$ in an equilibrated system. Such measurements can be made by either field ion microscopy or with the scanning tunneling microscope; the latter technique has only been utilized recently, but most successfully.

10.1 Early measurements

In 1973, measurements of the spatial distribution and of the interactions between rhenium atoms on the (110) plane of tungsten were published by Tsong [5]. It was noticed that two Re atoms migrated over the surface, maintaining a distance of 6–7.5Å, implying that the potential should have two minima at 2.74 and 6.8 Å. The results of more detailed analyses shown in Fig. 10.2 were interesting, but not really correct. Observations were made with five rhenium atoms on a plane, rather than just two, and the statistics were poor. Sites on the surface were not established; instead, observations were analyzed just according to the distance between the atoms, which is not reliable as the magnification of the FIM changes across the plane. Despite these defects, what is worth noting here is a strong attractive interaction, at ~2.5 Å, and an interaction potential that seems to vary between maxima and minima as the distance increases. The latter is, of course, what is expected for Friedel oscillations [6]. It must be noted, however, that no such oscillations were found for tungsten atoms, that conditions for establishing equilibrium were not stated, and the anisotropy of the surface was not considered.

A year later, a more careful study appeared of two tungsten atoms diffusing on the (211) plane of tungsten, separated from each other by an empty channel, as in Fig. 10.3a [7]. For this system, the probability of finding two atoms separated by the distance X along the <111> channels is

10.1 Early measurements

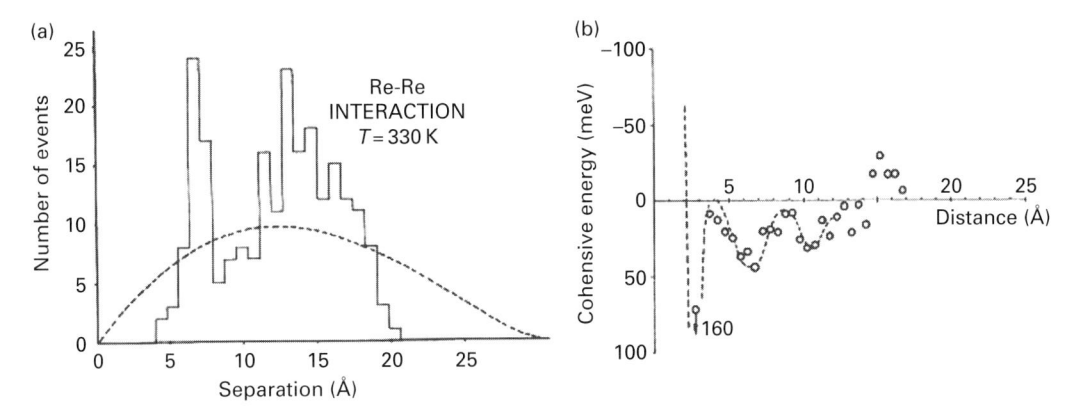

Fig. 10.2 Results obtained from FIM observations of five Re atoms on W(110) plane. (a) Distribution of pair separations at 300 K. (b) Interaction energies between adatoms, derived from (a), as a function of their separation (after Tsong [5]).

$$p(X) = C(2 - \delta_{x0})(L - X) \exp[-F(X)/kT], \tag{10.11}$$

where L is the number of sites in a channel and δ_{x0} is the Kronecker delta.

Tungsten atoms on W(211) always diffuse in the same <111> channel, so that such measurements are easy. The distribution, shown in Fig. 10.3c, does not differ significantly from that of two non-interacting particles, although there may be a slight diminution when the X-separation is zero. The distance between the two rows in these experiments was 8.95 Å. Measurements were also done on the distribution of separations between two tungsten atoms in adjacent rows of W(321), separated by 7.08 Å. Again no interactions were found, but it should be noted that only 300 observations were made, which limits the accuracy. However, two atoms in adjacent channels on W(211), at a distance of 4.48 Å, interact strongly with each other. This study showed that for a pair of W atoms on a bcc channelled surface interactions across the channels become significant once the spacing is shorter than 7Å. Long-range interactions for tungsten atoms in adjacent channels of W(321) were suggested by Nishigaki and Nakamura [8] but this was based on observations of only 10 pairs, and therefore cannot be considered definitive.

More qualitative but also more pleasing experiments were reported by Ayrault [9], who measured the mean-square separation between two rhodium atoms diffusing in adjacent channels of Rh(110), L spacings long. If the atoms are independent of each other, then if atom 1 is held at a given site while 2 moves randomly in the second row,

$$\left\langle (x_2 - x_1)^2 \right\rangle = \frac{1}{L} \int_0^L (x_2 - x_1)^2 dx_2 = \frac{1}{3L}\left[(L - x_1)^3 + x_1^3\right]. \tag{10.12}$$

However, for atom 1, all positions in its channel are really equally likely, so that the mean-square separation is obtained by averaging over all accessible values of x_1, to get

$$\left\langle (x_2 - x_1)^2 \right\rangle = L^2/6. \tag{10.13}$$

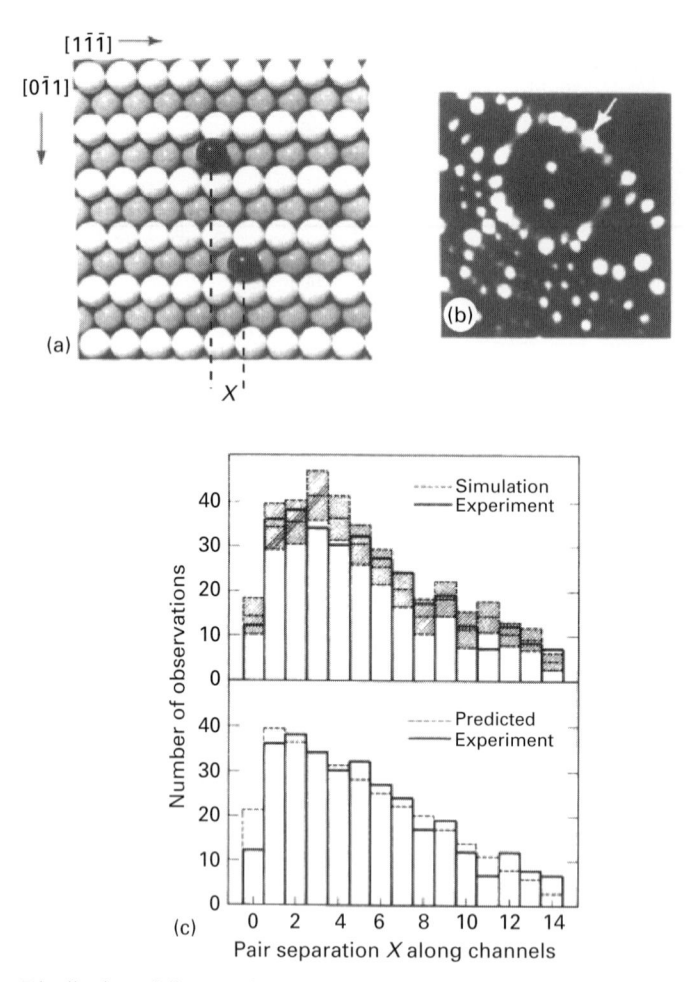

Fig. 10.3 Distribution of distances between two tungsten adatoms at 295 K in rows of W(211) plane separated by an empty channel, as in (a), giving a minimum separation of 8.95 Å [7]. (b) FIM image of two W atoms in rows separated by an empty one on W(211). Arrow gives direction of motion. (c) Distribution of distances between W adatoms. No significant deviations from random distribution are evident.

When there are interactions between the two atoms, the mean-square separation was written as

$$\left\langle (x_2 - x_1)^2 \right\rangle = L^2/6A_i, \tag{10.14}$$

where attractive forces give $A_i > 1$ and repulsive ones $A_i < 1$. Results obtained on the (110) plane as well as on (331) are listed in Table 10.1. It is clear that there are strong attractive interactions between atoms in adjacent channels, and for Rh adatoms they seem to persist even when the atoms are separated by an adjacent channel. This suggests that pair interactions of Rh across channels of the fcc(110) surface are of longer range than for W pairs. Unfortunately, no further energy estimates were made.

10.1 Early measurements

Table 10.1 Correlated Rh atom motion on channeled rhodium planes [9].

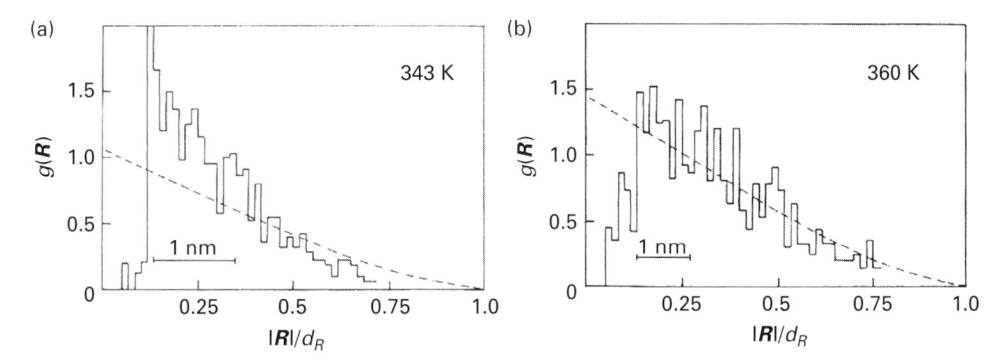

Plane	Channel spacing (Å)	Closest Approach (Å)	Observations	A_i	T (K)
(110)	3.80	7.61	18	6.0	176–198
(110)	3.80	3.80	47	23	190–215
(331)	5.86	11.7	30	1.7	216

Fig. 10.4 Pair distribution $g(R) = P(R)/2\pi|R|$ for Re adatoms on W(110) plane, observed in FIM. Distributions equilibrated at two temperatures (a) 343 K and (b) 360 K. Dashed lines give random distribution on structureless plane. Differences suggest that 343 K was too low a temperature for adequate equilibration; d_R denotes plane diameter (after Bassett and Tice [10]).

Previous estimates of the pair distribution of rhenium on W(110) made by Tsong [5] were criticized in 1975 by Bassett and Tice [10], whose own determinations are shown in Fig. 10.4. Noticeable here is the absence of a strong attraction at small separations, which had been reported by Tsong. Furthermore, they found a significant difference in the distribution on equilibrating at a higher temperature compared with a 20 K lower one, which they suggested showed poor equilibration at the lower temperature of 343 K. They concluded that "the evidence that $U(r)$ for rhenium adatoms on W(110) has the oscillatory character associated with indirect coupling" "is certainly not conclusive". However, the observations were indicative of repulsion at separations less than 5 Å and weak attractions between 7 Å and 10 Å.

Some years later, Fink et al. [11] did quantitative studies with a palladium atom together with a tungsten atom, and later also with a rhenium atom, on W(110). Palladium diffuses at temperatures below 250 K, at which the tungsten and also the rhenium adatom remain stationary, giving a reliable reference point on the surface and confining the atoms to the center of the plane. As seen in the distribution in Fig. 10.5a, consisting of 110 annealing steps at 240 K, palladium most frequently stays at a separation of ~3 Å, the next most frequent position was observed at 11 Å, and possibly a third maximum at 18 Å. For the smaller separation, the atoms were located along the <111> directions, and no atoms were observed between the first and second population peaks. Results obtained for Re-Pd in 51 diffusion intervals at 250 K are shown in Fig. 10.5c. Occupation of nearest-neighbor sites is prevented by repulsive interactions, but there are

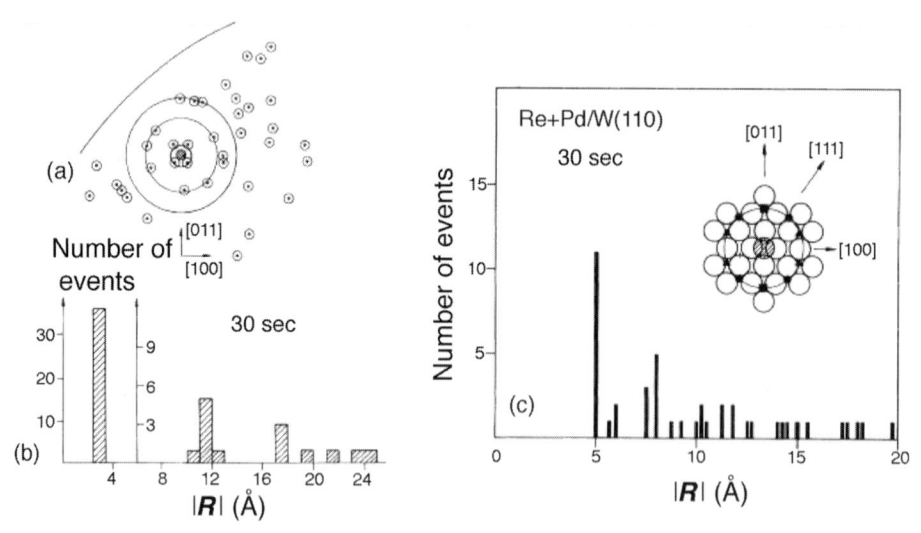

Fig. 10.5 Observations of W-Pd and Re-Pd pairs on W(110) plane in FIM after equilibration. (a) Location of Pd adatoms after diffusion around W adatom at 240 K. (b) Distribution of Pd around W in its dependence upon separation $|R|$ at 240 K. (c) Distribution of separations between stationary Re adatom and Pd, after diffusion at 250 K. Attractive interaction at 5 Å is evident (after Fink et al. [11]).

attractive interactions at 5 Å along <111>. Another separation of 8 Å was not as strongly preferred. From this study it emerged that the <111> direction is preferred for bonding of a Re-Pd pair, in close-bonded dimers as well as for larger interatomic distances. This paper really provided the first experimental indication of long-range interactions.

Attempts to arrive at better evaluations of long-range interactions on the (110) plane of tungsten were again taken up by Tsong and Casanova [12,13]. They examined effects for Re-Re, W-Ir, W-Re, and Ir-Ir with two atoms on the plane, but still maintained a radial analysis rather than endeavoring to identify the binding sites. However, the analysis was improved in that the (110) plane was now assumed to be elliptical [13]. The pair energy for interactions between two iridium atoms is plotted in Fig. 10.6a. There is an attractive minimum at ~ 4 Å. At larger distances, on the order of 7 Å, there is a clear repulsion between the atoms; at bigger separations there is an attraction, followed by what look like weak oscillations.

The interactions for W-Ir are shown in Fig. 10.6b. This diatomic cluster is less stable than Ir-Ir. The pair distribution shows an attractive region around 3 Å, followed by repulsion around 5 Å, and another attraction around 12 Å. Pair energies for Re-Re and Re-W are plotted in Fig. 10.6c and d. During equilibration at 400 K, closely bonded, unstable Re-Re dimers were occasionally created. From their dissociation it was concluded that around 3 Å the cohesive energy was ~260 meV. However, such dimers were not observed in 1045 heating periods of 60 sec at 330 K. The closest interatomic distance found was 6 Å. Based on these observations, the authors concluded that atoms repelled each other up to ~5 Å, and bonded weakly around 7 Å. The situation changed for W-Re dimers, for which observations of closely bonded dimers were frequent. Out of 1111

10.1 Early measurements

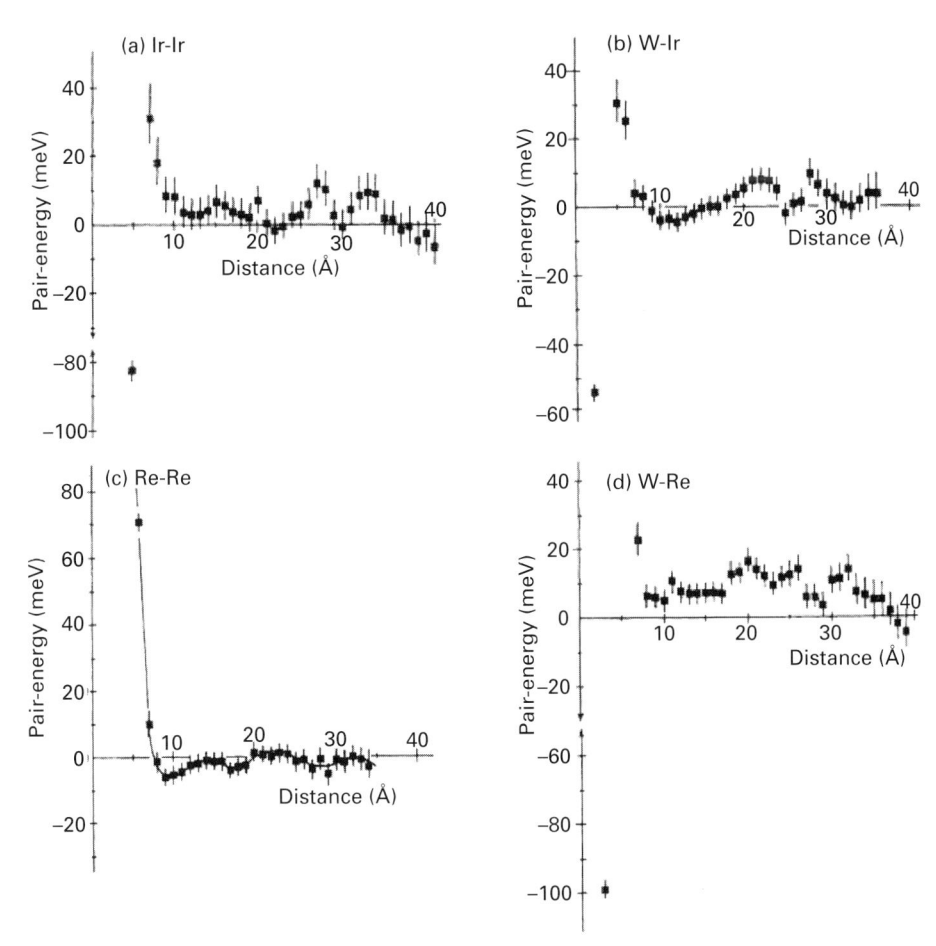

Fig. 10.6 Interaction between two adatoms on W(110) plane, deduced from distance distribution observed with FIM. (a) Ir-Ir pair after diffusion at 330 K. Repulsion is apparent at 7 Å. (b) Interaction energies between W and Ir adatom at different separations on W(110), after diffusion at 300 K. Strong repulsion is seen at ~5 Å. (c) Re-Re energies derived from distance distribution. (d) W-Re interactions (after Tsong and Casanova [13]).

observations, the close dimer was observed as many as 332 times. The repulsive region was between 3 and 6 Å, with an energy around 50 meV. All the systems studied, except for Re-Re, had a strong attractive potential at a nearest-neighbor spacing, indicating interactions of -99.0 ± 0.7 meV for W-Re, -82.0 ± 2.5 for Ir-Ir, and -53.2 ± 3.6 for W-Ir. For W-Re, there is repulsion between 3 and 6 Å, with possible oscillations at higher spacings. Fairly clear oscillations are visible for W-Ir, but were not that well defined for Ir-Ir. The problem with these determinations is that, although the number of observations had been increased to ~1000, no attempt was made to relate interactions to the location of sites on the surface.

In 1983 Stoop [14] investigated pairwise interactions for describing the phase diagram of a silver submonolayer on the W(110) surface. From his work it is clear that pairwise

Table 10.2 Interaction energies (eV) on W(110). ε_1, ε_2, ε_3 – first, second, and third nearest-neigbor pairwise interactions, ξ_1, ξ_2 – trio interactions, ς – quarto interactions.

	ε_1	ε_2	ε_3	ξ_1	ξ_2	ς	Reference
Ag	−0.074	0.037	0.037	−0.037	−0.037		[14]
Cu	−0.222	−0.147	−0.072	0.008	0.052		[16]
Au	−0.293	−0.255	−0.146	0.023	0.158	−0.148	[16]
Ag	−0.177	−0.167	−0.103	0.007	0.094	−0.085	[3]

interactions alone are not enough to describe the experimental phase diagram. He obtained the best fit using attractive first nearest-neighbors, repulsive second and third nearest-neighbor interactions, together with two types of attractive three-body effects. Pairwise interactions extended out to fifth nearest-neighbor distances. This effort was continued by Roelofs and Bellon [15] for Cu and Au based on the cluster calculations of Gollish [16]; their findings for pair, trio and quarto interactions are listen in Table 10.2. Ehrlich and Watanabe [3] followed Roelofs' and Bellon's procedure and derived values for Ag on W(110), also shown in Table 10.2. All pair interactions were attractive and trios repulsive. Even if the data do not agree with the previous findings of Stoop [14], it reproduced the general trends of work functions from experiments by Kolaczkiewicz and Bauer [17]. It was clear that more detailed investigations were needed.

10.2 More recent studies

A start on more careful studies of long-range atomic interactions was made by Fink [18], who mapped out the sites on the W(110) surface at which atoms were held. This was accomplished by allowing a rhenium atom to diffuse over the surface at 380 K, and determining its position after each of 345 diffusion intervals. The grid of binding sites so established is shown in Fig. 2.17, and provides a sound basis for finding distances between two adatoms on the plane; temperature was adjusted so that the root-mean-square displacement of an atom was comparable to the diameter of the plane.

The distribution of separations found between two rhenium atoms on W(110) equilibrated at 400 K is given in Fig. 10.7a, and is compared with the distribution for two non-interacting particles obtained by simulation. The statistics are not great, with only 381 observations, but what clearly emerges again are repulsive interactions at separations less than 10 Å. No attempt to evaluate effects at longer distances was made. Quite surprising here is the lack of dimer formation with two Re atoms, even at low temperatures where atom mobility is reduced. Based on this data we expect the same behavior for the rhenium trimer Re$_3$. Surprisingly enough, this was not the case – stable trimers were formed, as shown in Fig. 10.7b. Clearly three-body interactions are important, with a binding energy of trimers estimated at 0.25 eV.

The real breakthrough in determining long-range interactions between adatoms on a surface was made by Watanabe [19], who adopted the technique of first mapping out the

10.2 More recent studies

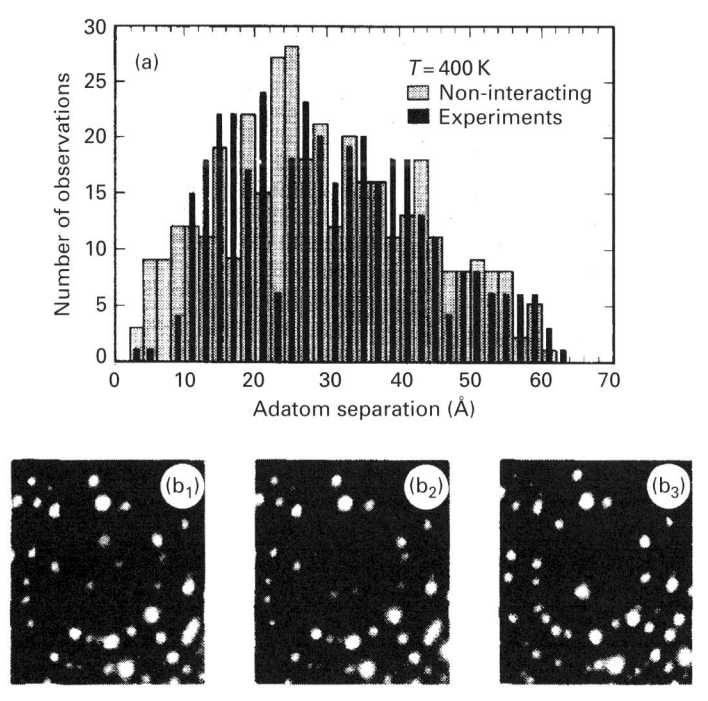

Fig. 10.7 Re atoms on W(110) plane observed in FIM [18]. (a) Distribution of distances between two Re adatoms on W(110) plane after diffusion at 400 K. (b) FIM observations of Re$_3$ trimer formation on W(110). (b$_1$) Three Re atoms have been deposited on (110) plane. (b$_2$) After heating at 420 K for 100 sec. (b$_3$) After one more heating interval at 420 K, trimer has formed.

binding sites on the surface, and then measured the full two-dimensional distribution of separations with a reasonable number of observations, larger than 1000, to reduce scatter. In the first studies, a strongly bound atom was deposited at the center of the (110) plane of tungsten; a second mobile atom was then added [20]. This procedure lowers the total number of observations needed, as atoms are less likely to explore the plane edges. Additionally, if there are no interactions other than between nearest-neighbors the probability of occupation of any site other than next to the central one should be equal. Studied were the pairs Re-Pd as well as W-Pd. The distribution of palladium around a rhenium atom located at the center of the (110) plane is shown in Fig. 10.8a, obtained in 1638 observations after equilibration at 205 K. Here measurements over the entire surface have been replotted in the first quadrant. At this temperature, the root-mean-square displacement of palladium is comparable to the radius of the plane. Striking here are the long-range interactions and the lack of atoms at adjacent sites in the [100] and [1$\bar{1}$0] directions; dimers are orientated along the [1$\bar{1}$1] direction.

The free energy of interaction derived with Eqs. (10.9) and (10.10) from the distribution of atom separations, given in Fig. 10.8b, displays some remarkable features: along the <111> direction, interactions are attractive out to a separation of 10 Å, interactions along <100> and <110> tend to be repulsive. Repulsive interactions are estimated at ~ 45 meV,

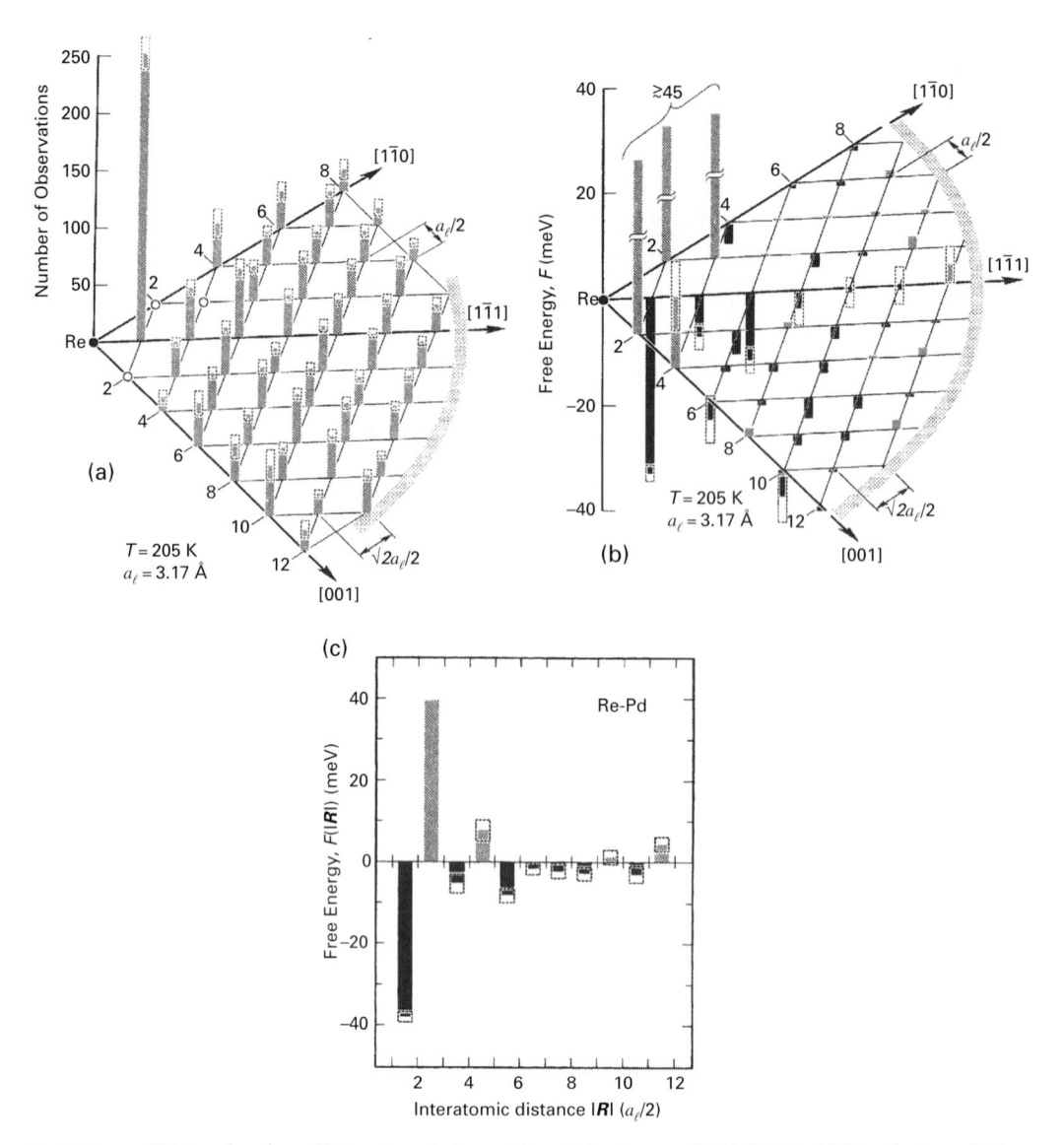

Fig. 10.8 FIM exploration of interactions between Pd and Re atom on W(110) [19]. (a) Distribution of Pd around stationary Re atom after diffusion at 205 K. (b) Interaction between Re and Pd adatom at different sites on W(110) plane. (c) Free energy of interaction as a function of the separation $|R|$ [20].

while attraction at the nearest-neighbor site along <111> amounted to −35 meV. The magnitude of the interactions diminished with distance non-monotonically, and at the third nearest-neighbor site along <111>, attraction amounted to 12 meV. If we plot free energy as a function of the radial separation of adatoms, oscillations are observed in Fig. 10.8c, but these are caused mostly by the surface anisotropy.

Much the same features were found in the interactions of a tungsten atom with palladium on W(110). The distribution of separations is shown in Fig. 10.9a, and the

10.2 More recent studies

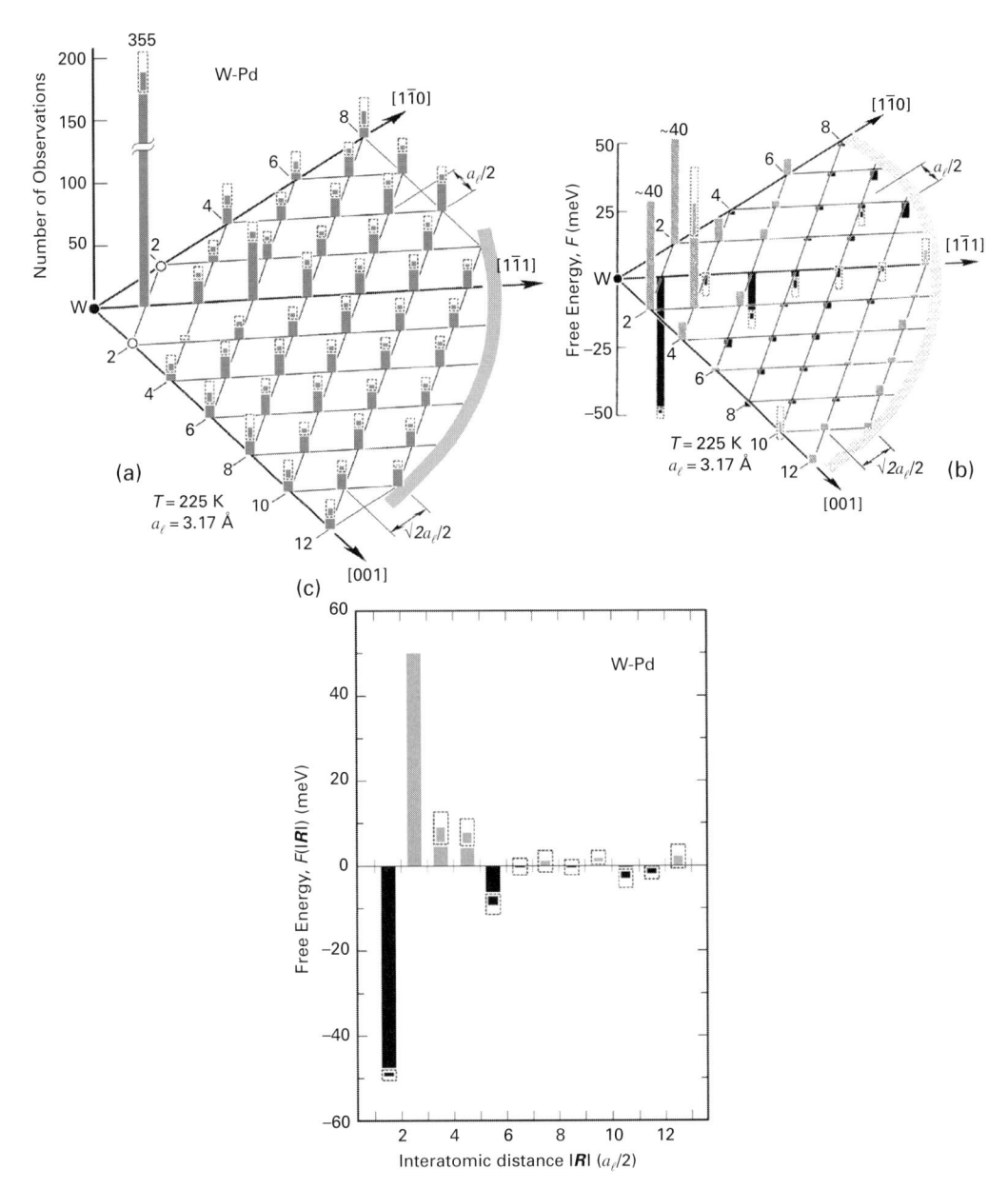

Fig. 10.9 Interactions of Pd atom with stationary W atom on W(110) examined in FIM [20]. (a) Count of Pd adatoms around W adatom on W(110) plane after diffusion at 225 K. (b) Interaction energies between W adatom and Pd atom on W(110). (c) Free energy of interaction for Pd-W analyzed as a function of the separation $|R|$.

free energies of interaction in Fig. 10.9b. Overall features are quite similar to what has been shown before – attraction along <111>, repulsion along <100> and <110>. Interactions oscillate along the <111> direction; at the nearest-neighbor site they are around − 0.050 eV, in the second nearest-neighbor site in this direction they are close to background (−0.0028 eV), then in the third place attraction increases to −0.0175 eV but falls to −0.0064 eV at the next site. In Fig. 10.9c there is shown a superposition of all directions to give the dependence on the radial separation, which shows oscillations caused by the anisotropy of the surface. Nearest-neighbor interactions now are significantly stronger with tungsten (~50 meV) than previously with rhenium.

Some rationalization of what is seen here has been offered by Stoneham [21] and also Kappus [22], who have stressed that interaction of an atom with the surface can induce changes in the position of surface atoms nearby. Kappus has pointed out that if a single adatom causes repulsion between surface atoms along <100> and attraction along <110>, then interactions between two such adatoms will be anisotropic – attractive along <111>, but repulsive along <100> and <110>. However, in his continuous treatment, there are no indications of the oscillations actually found.

Observations were subsequently also made on pairs of like atoms, iridium and rhenium, but with a larger data base of more than 2000 points, shown in Fig. 10.10a [23]. The analysis differs a bit now since both atoms are mobile and the distribution of non-interacting atoms, shown in Fig. 10.10b, is not uniform. The free energy of interaction for Ir-Ir on W(110) is obtained from the ratio of the two distributions in Eq. (10.9) and is plotted in Fig. 10.10c. What is clear here is that it looks quite similar to what was obtained earlier for hetero-pairs, but is stronger. Along <111>, there is first a strong nearest-neighbor attraction, amounting to −0.086 ± 0.002 eV, followed by a small repulsion of 0.0047 eV at a distance two spacings apart; attraction starts again in the third space at −0.0272 eV, grows to −0.0348 eV in the fourth, then diminishes to −0.0156 eV and reaches background (−0.0072 eV) in the sixth space. The interactions are much longer ranged than for hetero-pairs, and extend out to 14 Å. Interactions along <100> and <110> are initially repulsive at close spacings, but then turn to weakly attractive around three spacings. The radial changes in free energy as a function of separation between pairs is presented in Fig. 10.10d. Significantly different from past work [24–26] is the realization, based on studies by Chambers [23,27], that two next to each other iridium atoms resolved in the FIM actually occupy adjacent nearest-neighbor sites, and are not separated by two spacings.

The same procedures as for the Ir pair were followed for rhenium. However, it was shown previously that Re dimers are not created on the W(110) surface [13,18,28], so the pair distribution is spread out over longer distances and a larger number of observations is in order. The Re-Re distribution as well as the free energy of interaction are shown on W(110) in Fig. 10.11, determined in 3145 observations; repulsion is noticeable everywhere at small separations up to 6 Å. There is quite a low atom population at the (1,1) position, and a lack at the close separations along <100> and <110> directions. The interaction starts to be attractive for bigger separations, and the largest number of observations is found at the (4,4) position, at a distance of 11 Å. Along <111>, repulsion amounted to 0.0215 eV at the nearest-neighbor position, is stronger, 0.0554 eV, at the next spacing, and then changes to weak attraction, −0.0164 eV. Attraction is −0.0314 eV at the fourth, and then gradually decreases out to 16 Å and beyond. Along the orthogonal axes, that is along <100> and <110>, strong repulsion at the first spacings is replaced by quite weak attractions further out. What is clear in Fig. 10.11c is that for 6 Å< $|R|$ ≤15 Å interactions

10.2 More recent studies

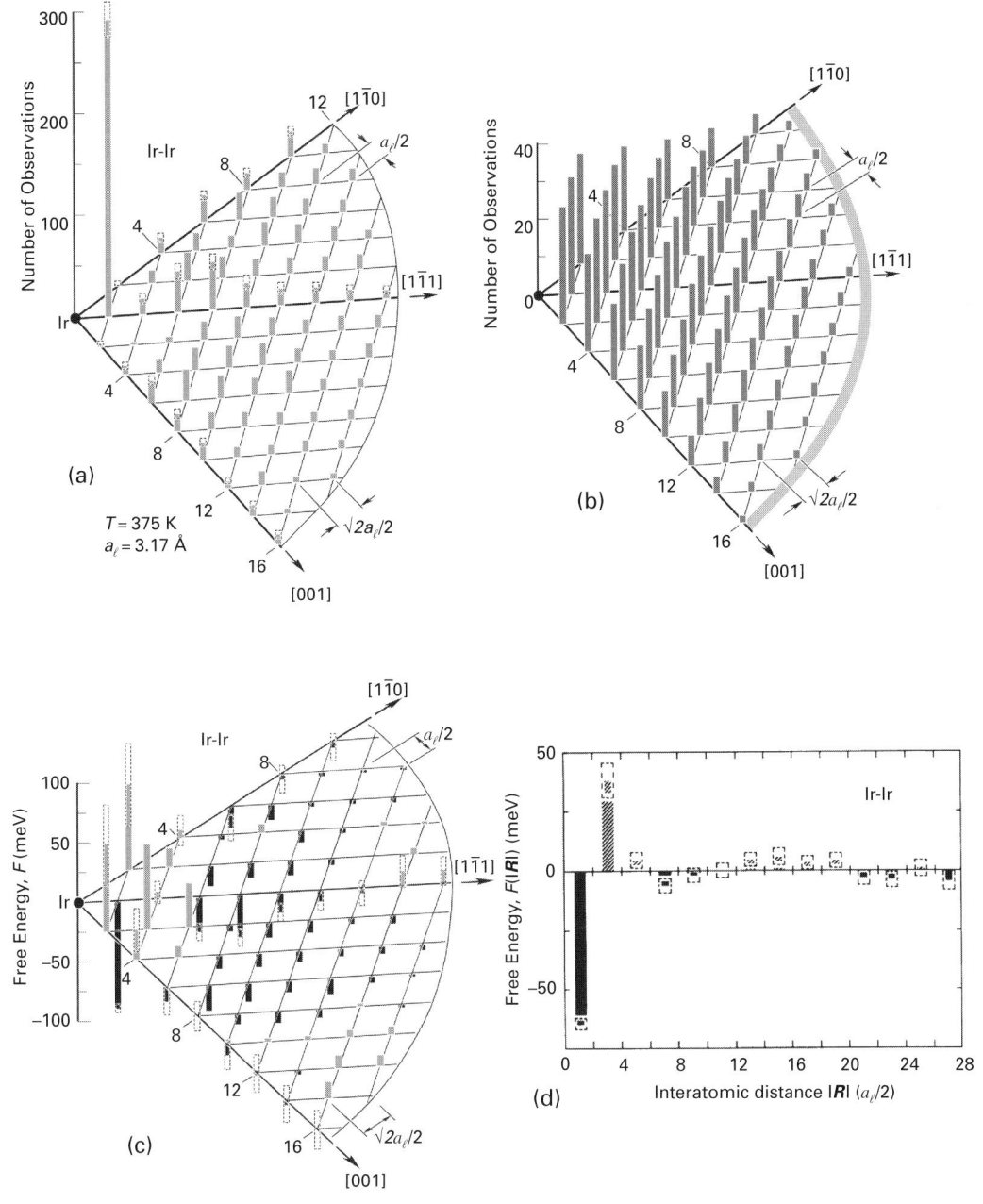

Fig. 10.10 Pair interactions for iridium adatoms on W(110) surface, derived from distribution of interatomic distances measured in FIM [23]. (a) Distribution of distances between two Ir adatoms. (b) Distribution of distances between two identical non-interacting atoms. (c) Free energy of interaction between two Ir atoms after equilibration at 375 K. (d) Free energy of interaction for Ir_2 as a function of the separation $|R|$.

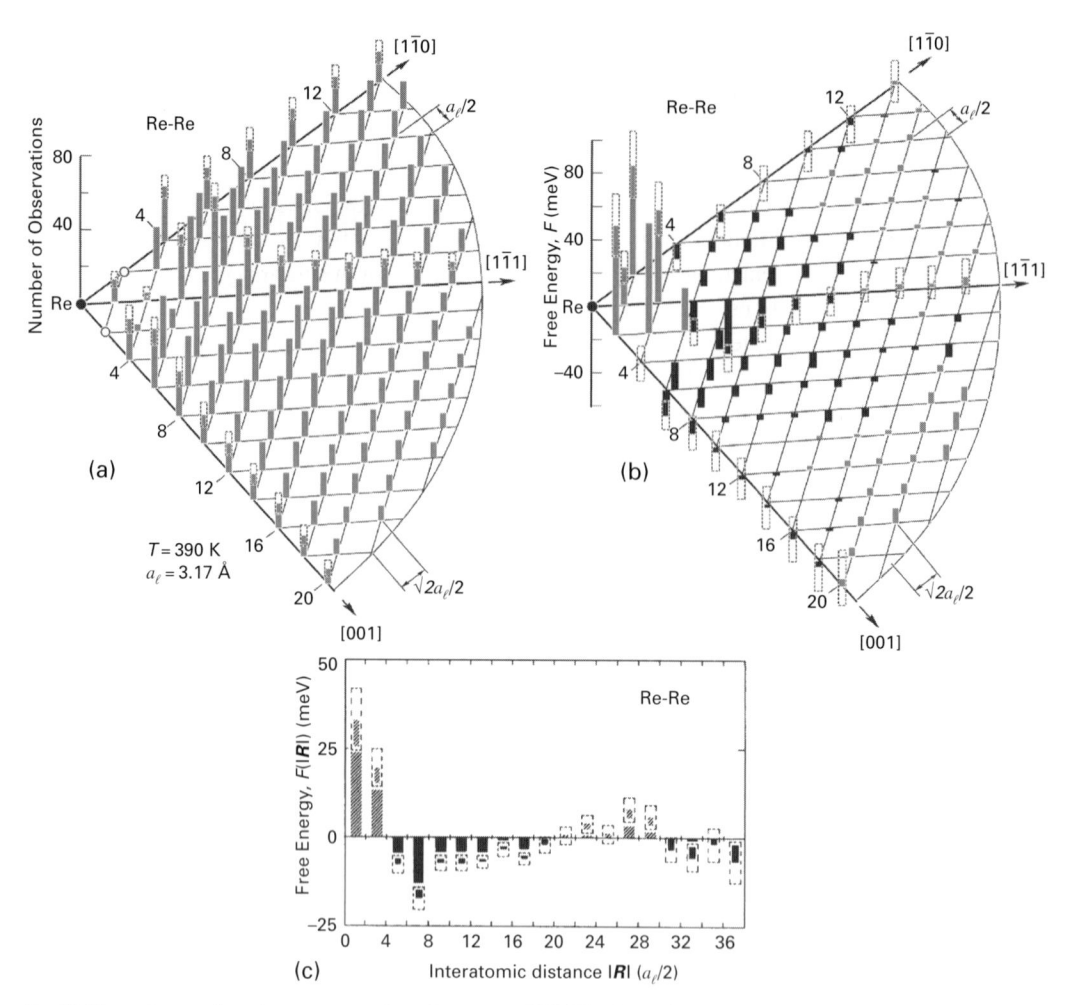

Fig. 10.11 Interactions between two Re adatoms on W(110) plane observed in FIM [23]. (a) Distribution of distances between two Re adatoms. (b) Free energy of interaction for Re$_2$ dimers. Attraction is concentrated along [1$\bar{1}$1]. (c) Free energy of interaction for Re$_2$ as a function of the separation $|\mathbf{R}|$.

are generally attractive but strongest along the close-packed <111> direction. Effects somewhat similar to these findings have been arrived at in calculations by Dreyssé et al. [29]. However, in the experiments the transition from repulsive to attractive interactions occurs over a longer distance and the magnitude of the interactions is overestimated in the calculations. The type of interaction strongly depends on the plane examined: Re-Re pair interactions on the W(211) plane are attractive at close distances [30] with the atoms in adjacent channels or in the same one. This observation directly shows that interactions are surface mediated.

Also studied were Ir and Re trimers, and for both it turned out that pair interactions could not account for the trimers observed at the surface. Based on pair interactions for Ir$_3$, linear clusters should have a free energy of ~ -0.170 eV, while the bent and triangular

10.2 More recent studies

form should be at ~ −0.120 eV. From measurements of the lifetime for dissociation, trio interactions of −0.130 ± 0.070 eV were deduced, indicating that many-body effects amounted to roughly a third of the trimer cohesion. Bigger many-body effects were found for Re trimers. Such trimers were observed stable at the surface, but taking into account only pair interactions, such trimers should have a repulsive free energy of 0.098 eV. However, from separate measurements it was found that the free energy of linear trimers amounted to − 0.240 eV [18] and around 0.030 eV lower for the bent form. Many-body interactions here amouted to − 0.340 eV for the linear and − 0.380 eV for the bent configurations.

A few years later theoretical work on interactions between two tungsten adatoms on W(110) was done by Xu and Adams [31], using a fourth-moment approximation to tight-binding theory. The interaction energies for atoms placed at different sites are shown in Fig. 10.12. There are no detailed experimental studies of tungsten atoms with which to compare these predictions, but the trends in the calculations mirrored experiments on other adatoms. One worrying fact, however, is that two kinds of adsorption sites were seen at the surface, 0.11 Å apart. Two kinds of sites on this plane have never been observed in experiments. Additionally, it was claimed that the most stable adsorption place was three fold, while in experiments atoms sit only in four fold sites. Strong attractive interactions at near sites along <111> are followed by small repulsions, and the same general patterns was also seen along <110> and <100>. However, the magnitude of the energies is too large. Nearest-neighbor energies have typically been measured less than − 0.10 eV, not −2.6 eV found in this study. The strong angular anisotropy of interactions found in experiments on W(110) were, however, duplicated in these theoretical estimates.

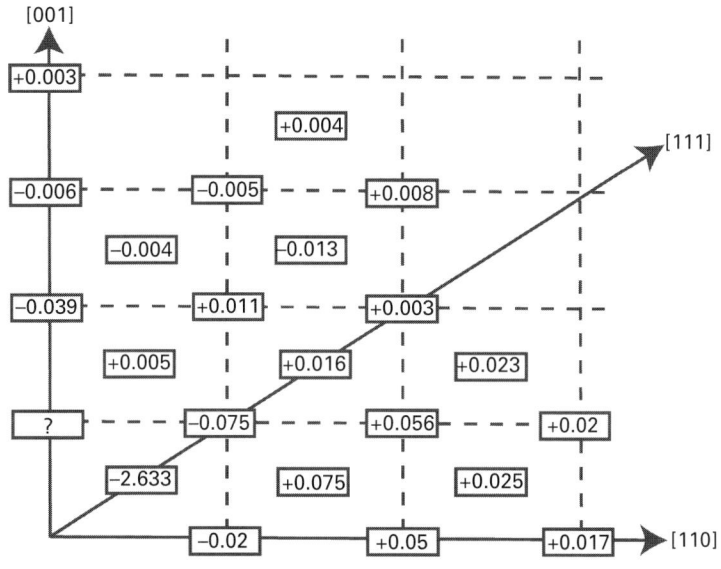

Fig. 10.12 Pair interactions between two W adatoms on W(110) plane, calculated for different locations with a fourth-moment approximation to tight-binding (after Xu and Adams [31]).

Atomic pair interactions

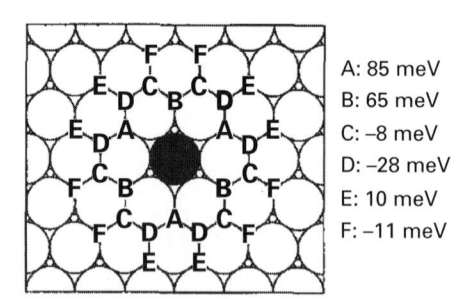

A: 85 meV
B: 65 meV
C: −8 meV
D: −28 meV
E: 10 meV
F: −11 meV

Fig. 10.13 Effect of substitutional In atom (in black) on binding of nearby Cu adatoms, as calculated with EAM. Energies are relative to binding of Cu adatom on a clean Cu(111) surface (after Breeman et al. [32]).

Interesting but different calculations of the binding energy of atoms on Cu(111) were made by Breeman et al. [32] using an embedded atom method devised by Ackland et al. [33]. Their work was concerned with an In atom sitting in the first layer in place of a copper atom; it was found to have a strong effect on the energy of a Cu adatom nearby. As shown in the diagram in Fig. 10.13, the binding energy for a Cu adatom compared to that of a Cu adatom on a clean (111) plane was 0.085 eV higher. At other sites, however, the binding energy was significantly lowered, clearly revealing long-range interactions strongly dependent upon the orientation, and anticipating more elaborate calculations later on.

Indirect estimates of pair interactions based on LEED and AES measurements were done by Kolaczkiewicz and Bauer [34] for Rh, Pt, and Ir on W(110). The values of the interactions were much higher than derived from direct measurements and have no oscillatory dependence on the distance. For Pt, nearest-neighbor interactions amounted to −0.800 eV, while for Rh to −0.600 eV. The second nearest-neighbor interactions were repulsive, amounting to 0.500 eV for Pt and 0.200 eV for Rh. Anisotropy of interactions was not taken into account in those studies, and direct measurements certainly provide more detailed insights.

It is interesting to compare the work on Ir-Ir with more recent FIM studies of palladium interacting with another palladium atom on W(110), done with 1855 observations at $T = 242$ K by Koh [35,36]. The free energies of interaction are displayed in Fig. 10.14 and show the general trends already seen above. Along the close packed <111> we first have strong attraction, −0.0886 eV, followed by a repulsion of 0.0194 eV, and then for the next three spacings by significant attraction. Along <100> and <110> there is now general repulsion; this repulsion extends out significantly farther than for iridium. Interactions extend over a distance of 10 Å. However, as is clear from a comparison of the separation dependence of the free energy in Fig. 10.14c for iridium and palladium, behavior is very similar for the two. These similarities account for the linear chains seen with palladium clusters formed on W(110). What holds for all of the systems studied so far are the strong effects of orientation on interaction, with attraction along the <111> orientation, and repulsion along the orthogonal axes.

Pair interactions do not account for the structure observed for palladium clusters, since for bigger clusters the linear form has a significantly lower energy than do two-dimensional

10.2 More recent studies

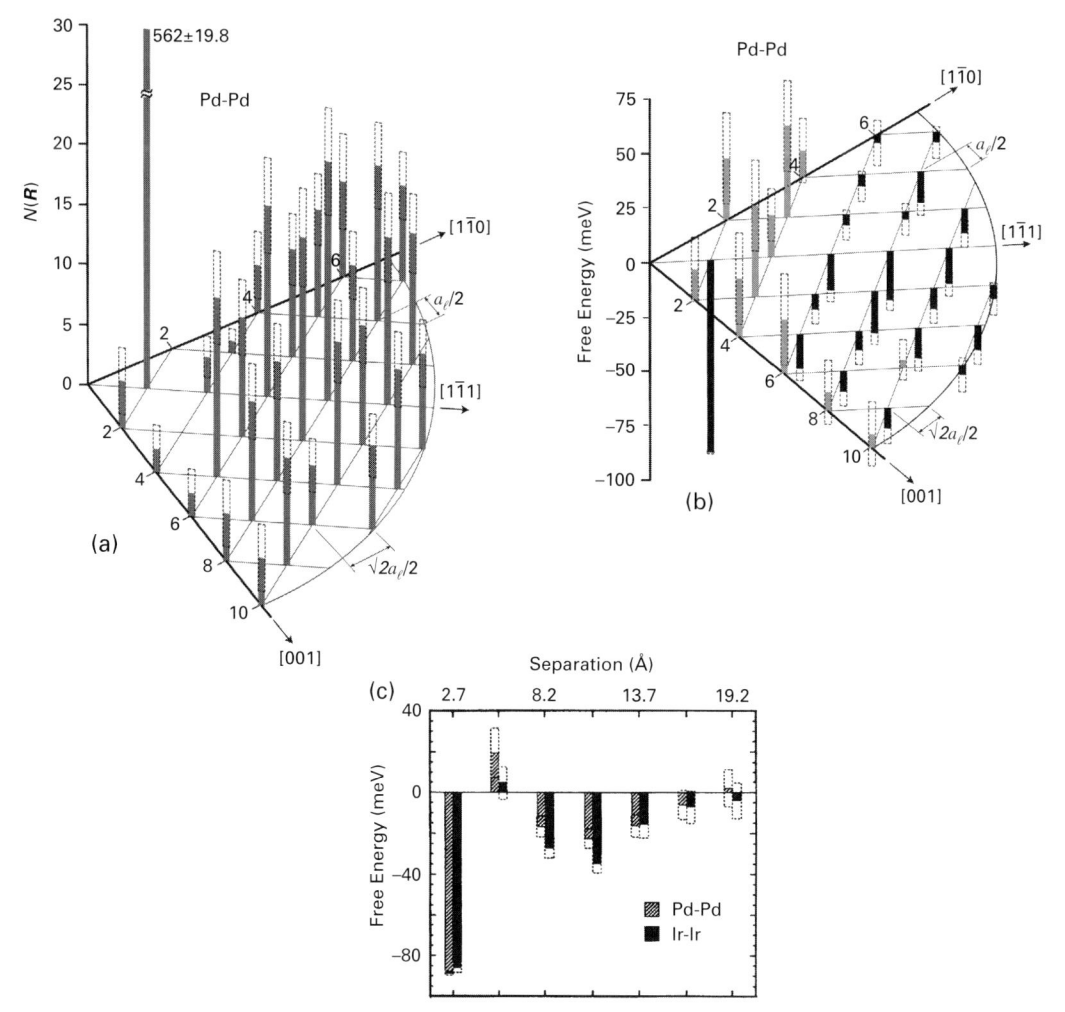

Fig. 10.14 Pair interactions for palladium on W(110) plane observed in FIM after equilibration at 242 K [35]. (a) Number of Pd pairs at different separations. (b) Free energy of interaction for Pd pairs at different locations. (c) Free energies of interaction for Pd-Pd and Ir-Ir as a function of atom separation.

islands. But 2D islands, not chains, were actually observed by FIM for clusters of more than eight atoms. The influence of three-body effects was investigated in detail. Trimers were observed in three configurations, linear, bent, and triangular, shown in Fig. 10.15a, so the free energy can be given as

$$F_{lin} = 2F(1,1) + F(2,2) + F_{3lin} \tag{10.15}$$

$$F_{tri} = 2F(1,1) + F(2,0) + F_{3tri} \tag{10.16}$$

$$F_{bent} = 2F(1,1) + F(0,2) + F_{3bent}, \tag{10.17}$$

714 **Atomic pair interactions**

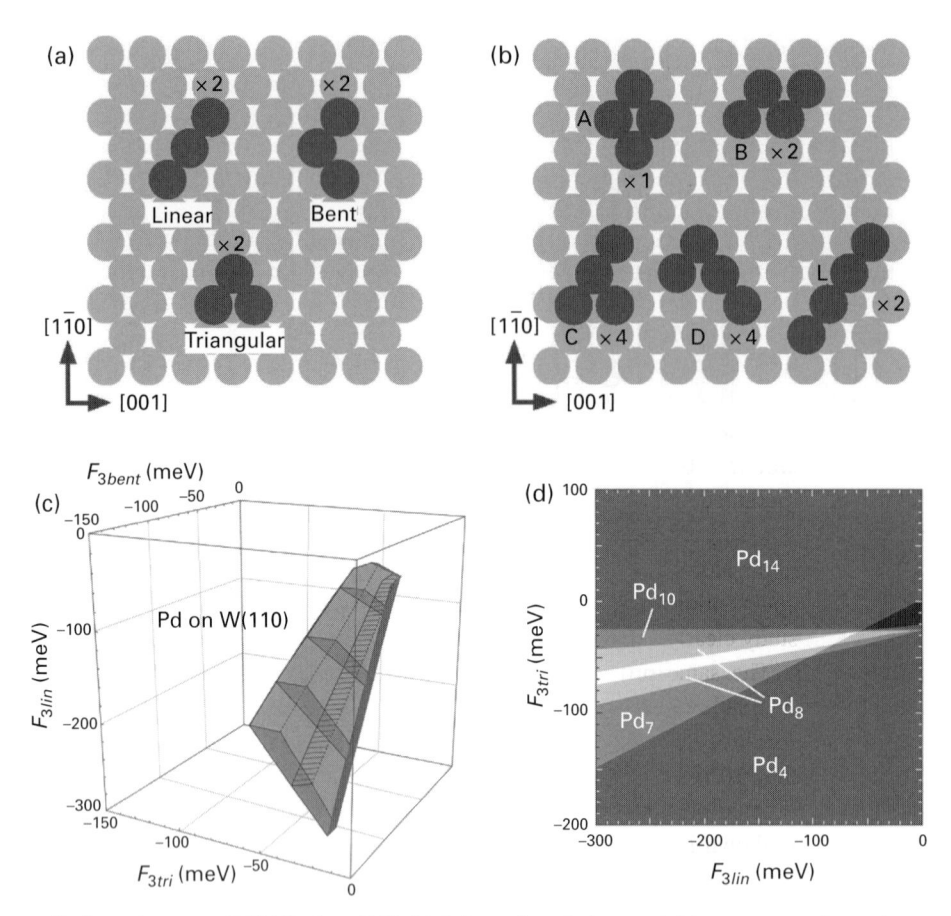

Fig. 10.15 Palladium clusters on W(110) plane [35]. (a) Schematics of trimer and (b) tetramer configurations. (c) Allowed values of F_{3tri}, F_{3bent}, and F_{3lin} for Pd at 242 K. (d) Allowed values of F_{3lin} and F_{3tri}, based on the assumption $F_{3tri} = F_{3bent}$. Only needle-like white zone is allowed.

where F_{lin}, F_{tri}, and F_{bent} denote the free energy of linear, triangular, and bent configurations respectively, $F(x,y)$ the pair free energy for adatoms separated by the distance (x,y); F_{3lin}, F_{3tri}, F_{3bent} give the three-body energy for linear, triangular and bent configurations.

From experiment it is known that the trimer is always observed in a linear configuration. The probability P_{lin} of observing the trimer in the linear configuration is given by

$$P_{lin} > P_{tri} + P_{bent}, \tag{10.18}$$

where P_{tri} and P_{bent} denote the probabilities of observing triangular and bent configurations. The probability of observing a particular form depends exponentially on the free energy,

$$P_{bent} = C\Omega_{bent}\exp(-F_{bent}/kT), \tag{10.19}$$

10.2 More recent studies

where Ω_{bent} is the number of energetically equivalent configurations with the bent shape, and C is given by normalization. Finally putting equations (10.15–19) together we arrive at

$$\Omega_{lin}\exp[-\{2F(1,1)+F(2,2)+F_{3lin}\}/kT] \\ > \Omega_{tri}\exp[-\{2F(1,1)+F(2,0)+F_{3tri}\}/kT] \quad (10.20) \\ +\Omega_{bent}\exp[-\{2F(1,1)+F(0,2)+F_{bent}\}/kT].$$

Information can therefore be deduced about the free energy of the different trimer forms. The same procedure was applied to tetramers in the forms A, B, C, D, and L in Fig. 10.15b, and also to higher clusters containing up to 14 atoms, giving us a set of restrictive equations. Combining all of them together restricted the allowed values of trio-free energies, as shown in Fig. 10.15c for Pd on W(110). F_{3lin} turned out to be attractive around −0.060 eV or higher, a value comparable to the strongest pair interactions measured for this system (−0.0886 eV). Assuming that the trio-energy of triangular F_{3tri} and bent F_{3bent} configurations is the same allowed values for F_{3tri} and F_{3lin} to fall into a narrow region shown in Fig. 10.15d. From this plot it is clear that the smallest trio contribution amounted to −0.06 eV for F_{3lin} and was around −0.025 eV for F_{3tri}. This study clearly showed the big influence of trio-interactions on cluster formation.

The same procedure was applied to Ir clusters, where forming a linear chain is favorable up to 15 atoms. Working out all conditions, Koh again found a small strip for three-body interactions with the total free energy relatively constant, as shown in Fig. 10.16a–b. A comparison of the free energy of clusters obtained by taking into account only pair-interactions, and both pair- and trio-interactions, for Pd as well as Ir is shown in Fig. 10.16c-d. We can see that the fractional influence of trio-interactions increases with the size of the clusters. Trio interactions are attractive and of the same order as the strongest pair-interactions. Allowing even higher-order interactions probably would lower the strength estimated for trio-interactions.

Many such effects have been measured on the bcc(110) plane. What happens on fcc(111) surfaces? The first STM investigation of pair interactions was done by Wahlström et al. [37] on Cu(111). The identification of atoms at the surface was tricky, as atoms emerged from the bulk during annealing at $T > 800K$. Comparison with other sources allowed the authors to suspect that they were dealing with sulphur atoms. Figure 10.17 shows the distribution of distances observed between atoms. For the pair interactions in Fig. 10.17b, at distances 21 ± 3 Å, 36 ± 4 Å and 66 ± 7 Å, three clear peaks were observed in 41 distances between 58 different atoms. The interaction energy was described by

$$\Delta E_{\text{int}} = C_1 \frac{\cos(2k_F|\mathbf{R}|+\varphi)}{|\mathbf{R}|^3} + C_2 \frac{\cos(2k_F|\mathbf{R}|+\varphi)}{|\mathbf{R}|^2}, \quad (10.21)$$

where k_F is the Fermi wave number and $|\mathbf{R}|$ the pair separation. For this system, it meant a period of 15 Å at short distances and 30 Å at longer distances, where the second term dominated. The data are based on quite limited statistics and the segregation of sulphur at higher temperatures suggests the presence of sulphur in the sublayer, which would influence the measured pair interactions. That is why the quantitative data have to be

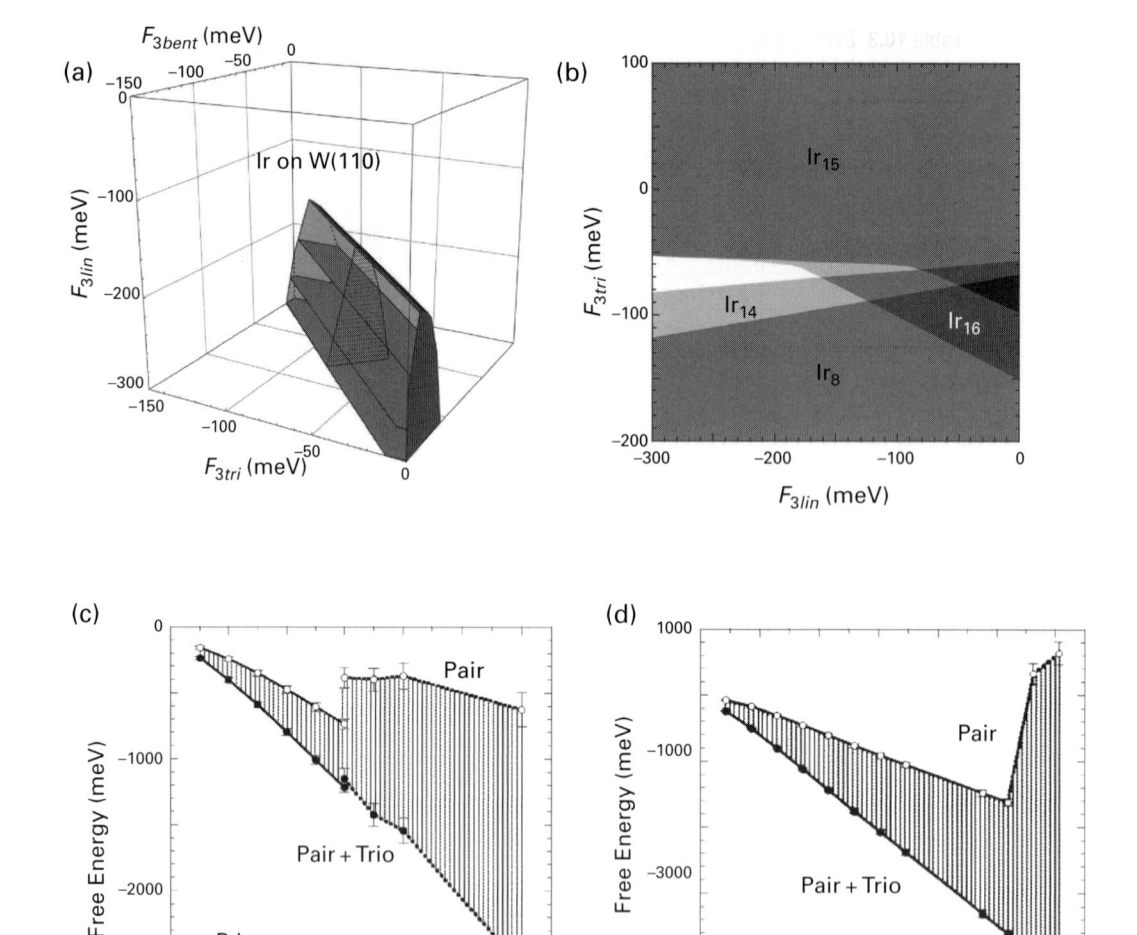

Fig. 10.16 Free energies of Ir clusters on W(110) derived from observations in FIM [35]. (a) Allowed values of F_{3tri}, F_{3bent}, and F_{3lin} for Ir on W(110). (b) Allowed values of F_{3lin} and F_{3tri} for Ir shown in white. Estimates based on the assumption that $F_{3tri} = F_{3bent}$. (c)–(d) Contribution of pair- and trio-interactions to cohesion of clusters on W(110). (c) For Pd, evaluated for $(F_{3lin}, F_{3tri}) = (-0.080$ eV, -0.034 eV). (d) For Ir clusters, evaluated with $(F_{3lin}, F_{3tri}) = (-0.180$ eV, -0.068 eV).

treated carefully. However this study is worth noticing as it is the first STM measurement of pair-interactions.

In 1999, Levanov et al. [38] carried out RGL calculations of interactions between two atoms self-adsorbed on the (100) surface of fcc metals. The surface was modeled with 8 atomic layers consisting of 128 atoms. They evaluated the binding energy E_1 for the atoms at the nearest-neighbor separation, and also at the next-nearest neighbor distance, E_2, as well as E_3 for atoms at third-neighbor positions. The results are given in Table 10.3,

10.2 More recent studies

Table 10.3 Binding energies (eV) between two adatoms self-adsorbed at first-, second-, and third-neighbor positions, obtained in RGL calculations on (100) plane [38].

Atom	E_1	E_2	E_3
Cu	−0.16	−0.02	0.01
Ni	−0.15	−0.07	−0.02
Ag	−0.12	−0.01	0.001
Pd	−0.14	0.005	0.002
Pt	−0.17	0.02	0.003
Au	−0.09	0.01	0.004

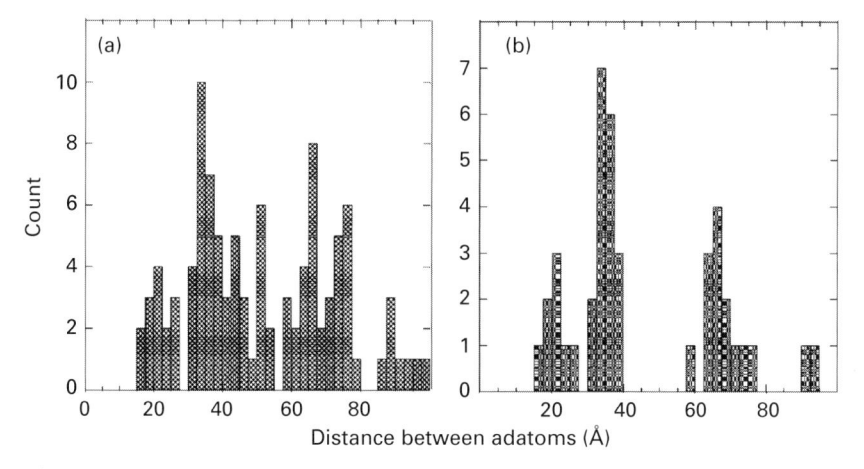

Fig. 10.17 Distribution of interatomic distances for adatoms on Cu(111) seen in STM experiments. (a) All measurements. (b) Distribution for atom pairs only (after Wahlström *et al.* [37]).

and indicate a fairly rapid decay except on Ni(100), where second range interactions (−0.07 eV) are much stronger than for the rest of the materials investigated. The authors associated the behavior of Ni with stronger *d-d* interactions due to the unfilled *d*-shell. It is difficult, however, to judge this study since so far these are the only data for pair interactions on a fcc(100) surface.

Estimates of interactions between silver adatoms on Ag(111) have been made in density-functional calculations by Fichthorn and Scheffler [39]. Work was done on a Ag(111) plane strained so the lattice constant was 4.61% smaller than for bulk silver, as well as on an ordinary Ag(111) surface. Interactions were modeled with a 4 × 4 × 4 slab, and at a fixed distance were assumed independent of the type of adsorption site (fcc or hcp) occupied by the atom. The fcc site was actually favored by 3 meV. Pair- and trio-interactions were considered, with pairs up to 13th neighbor and trios up to 5th. Elastic interactions were found to be small, electronic effects playing the major role. The pair-interactions are plotted in Figs. 2.28 and 10.18, and clearly indicate strong attraction at nearest neighbors but also repulsive interactions ~3 spacings out. The repulsions on the strained surface were bigger than on an ordinary Ag(111), but the diffusion barriers were

718   **Atomic pair interactions**

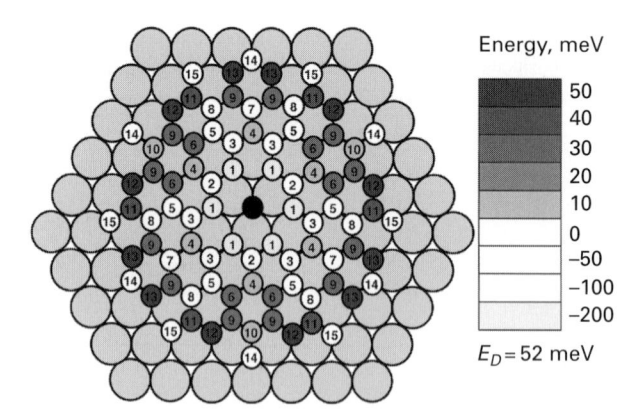

Fig. 10.18   Dependence of pair interaction energy on the separation from the central adatom for Ag on strained Ag(111) surface. Results obtained from density-functional theory calculations (after Fichthorn and Scheffler [39]). See color plate section for color version of this figure.

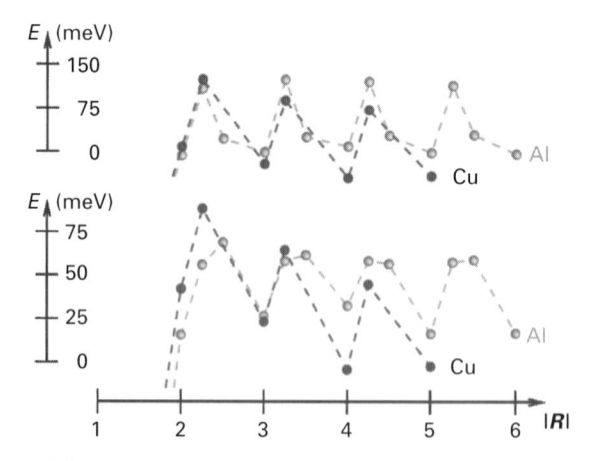

Fig. 10.19   Pair interactions on Al(111) and Cu(111) derived from density-functional calculations. Interactions for Al pairs on Al(111) and Cu on Cu(111). Distance $|R|$ is given in terms of lattice site separation. Top graph gives results for frozen configuration, bottom for entirely relaxed one (after Bogicevic et al. [40]).

smaller. These interactions were found very important in that they led to island densities an order of magnitude larger than predicted in nucleation theory.

Density-functional calculations were also done by Bogicevic et al. [40] using the VASP [41–43] program to examine pair-interactions of aluminum on Al(111) with a supercell $14 \times 4 \times 6$, and copper on Cu(111) with a $12 \times 4 \times 4$ cell. Cu(111) has a surface band of electrons, whereas Al(111) has no surface band states occupied. Results are given in Fig. 10.19 for interactions along the <110> direction at the saddle points for diffusion and at equilibrium binding sites, in both the frozen configuration (at the top of the graph) and the relaxed (at the bottom). Variations in interatomic potential are present even when elastic response is frozen; this indicates

10.2 More recent studies

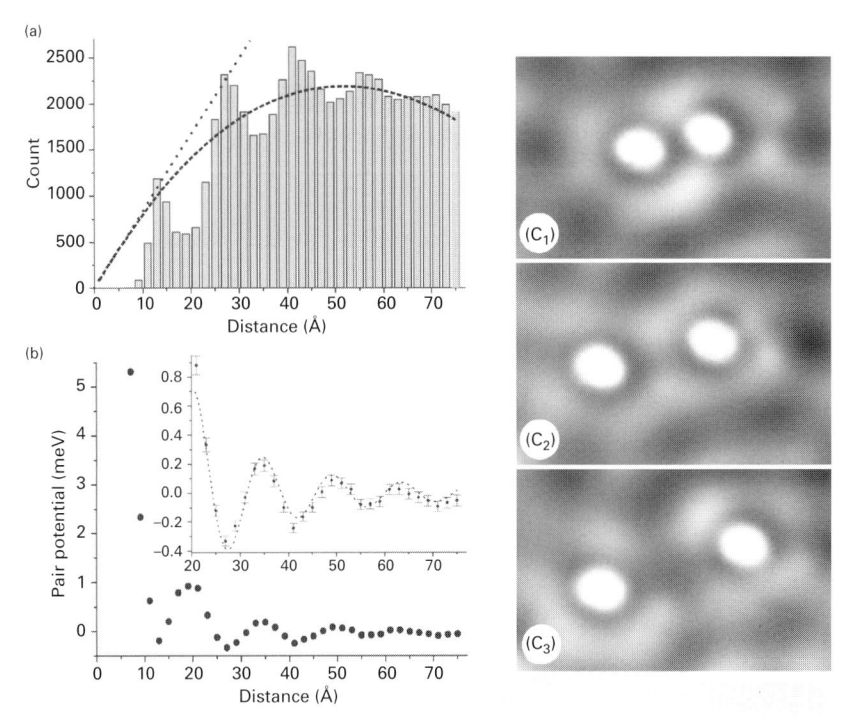

Fig. 10.20 Interactions between Cu adatoms on Cu(111) plane. (a) Distribution of pair separations deduced from STM measurements. (b) Energies for a pair of Cu atoms at different separations. Inset gives results for larger separations. Dotted line is fit of Eq. (10.22). (c) STM images of Cu atoms on Cu(111) at different separations. (c$_1$) 12.9 Å (c$_2$) 20.7 Å (c$_3$) 26.7 Å. Local density of states decreases if there is a potential energy minimum between adatoms (after Repp *et al.* [45]).

that interactions are primarily electronic. Indirect elastic effects should predominate at larger separations, and were found to strongly affect diffusion and aggregation. They also show the island density with and without long-range interactions for Cu on Cu (111); interactions increased the island density by a factor of five. It should be noted, however, that Polop *et al.* [44] later claimed the effects of long-range interactions were overestimated in density-functional theory calculations.

That has been examined by Repp *et al.* [45] in careful low-temperature experiments, using the scanning tunneling microscope, operated from 9–21 K with copper atoms on Cu(111). They measured the diffusion characteristics of copper atoms and were therefore in a position to define proper conditions for equilibration. More than 65 000 distances were tabulated to give the results shown in Fig. 10.20a. A one-dimensional plot of the free energy of interaction is now sufficient. After a peak below a distance of 10 Å, the energy falls and then oscillates steadily out to 70 Å, with a maximum swing of ~1 meV between peaks and valleys. An estimate was also made of a total barrier of 55 ± 5 meV to the formation of more closely spaced pairs. The change in the density of states is illustrated in Fig. 10.20c, showing dimers separated by 12.9 Å, 20.7 Å, and 26.7 Å; the density shows protrusions or depressions due to maximum or minimum free energies between the atoms. There is, however, also the possibility of inelastic Friedel oscillations contributing [46] and the analysis did not touch problems of possible anisotropy in the interactions.

Hyldgaard and Persson [47] at the same time arrived at a theoretical description of adsorbate–adsorbate interactions mediated by a surface-state band, which gave an energy

$$E(\boldsymbol{R}) = -\varepsilon_F \left(\frac{2\sin(\delta_F)}{\pi}\right)^2 \frac{\sin(2q_F|\boldsymbol{R}| + 2\delta_F)}{(q_F|\boldsymbol{R}|)^2}; \qquad (10.22)$$

here ε_F is the Fermi energy measured from the bottom of the band, δ_F the Fermi-level phase shift which characterizes the adsorbate-induced standing surface wave function as observed in the STM, and q_F gives the in-surface Fermi wave vector. Just as in the theory, the energy oscillates with a period equal to $\lambda_F/2$, half the Fermi wave length, the envelope of the magnitude decays as $1/|\boldsymbol{R}|^2$, and the phase shift does not change with separation $|\boldsymbol{R}|$. The adsorbate–adsorbate energy mediated by the surface-state band, calculated from Eq. (10.22) for Cu(111) assuming a Fermi level phase shift $\delta_F = -\pi/2$, is presented in Fig. 10.21, We can clearly see oscillations. However, there were differences between theory and observations, which were mentioned as possibly due to the neglect of electrostatic contributions to the interactions. It was found that scattering of surface state electrons from the adsorbates into the bulk states reduced interactions.

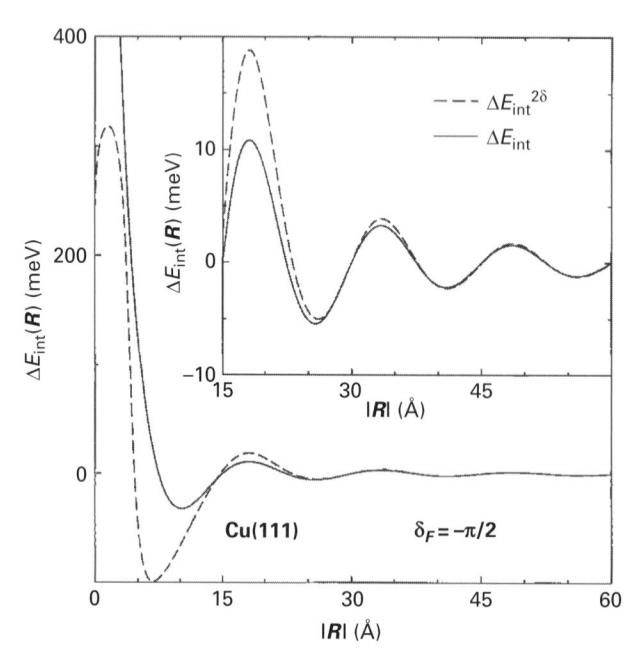

Fig. 10.21 Adatom–adatom interactions mediated by surface-state band. Solid curve gives results from Eq. (10.22), broken curve shows numerical approximation. Long-ranged asymptotic variation is shown in insert (after Hyldgaard and Persson [47]).

10.2 More recent studies

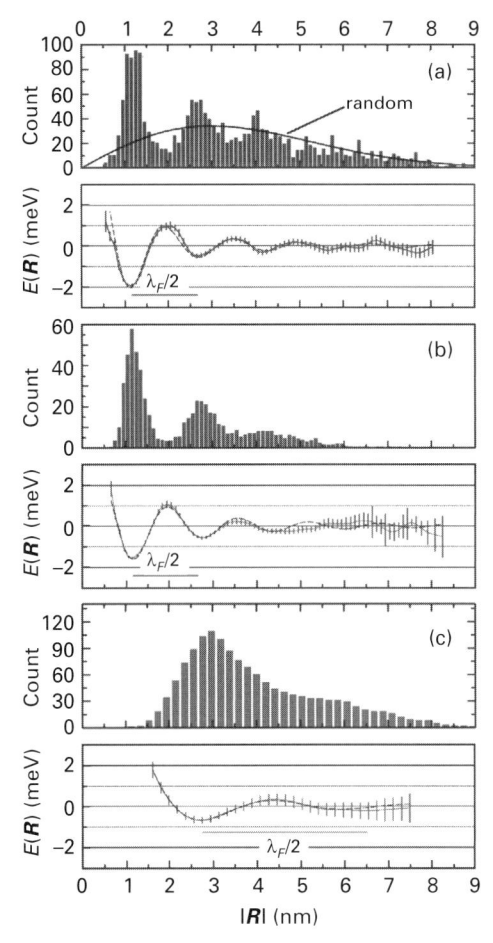

Fig. 10.22 Distribution of distances between adatoms as well as pair interactions on Cu(111) and Ag(111) deduced from observations with STM. (a) Cu adatoms on Cu(111), coverage 1.4×10^{-3} ML, $T = 15.6$ K. (b) Co on Cu(111), coverage 2×10^{-3} ML, $T = 10.2$ K. (c) Co on Ag(111), coverage 4×10^{-4} ML, $T = 18.5$ K. Dashed line gives fit to Eq. (10.22) (after Knorr et al. [48]).

A more extensive and more detailed study was carried out shortly thereafter by Knorr et al. [48], who again used scanning tunneling microscopy to look at the interactions between copper atoms on Cu(111) at 15.6 K, Co on Cu(111) at 10.2 K, and Co on Ag(111) at 18.5 K. Their results for the distribution of separations as well as the interaction energies are shown in Fig. 10.22. The jump rate of copper atoms was measured first as a function of temperature, and a temperature was then picked low enough so that the mean rate of jumps was below the rate of image recording. Furthermore, even though only 1.4×10^{-3} monolayers of copper were deposited, the distance between two atoms was counted only if no third atom or impurity was nearby. As a consequence of these precautions, there was excellent agreement with formula (10.22), indicated by dashed lines in Fig. 10.22, from long distances down to the first minimum, with stronger oscillations than found by Repp et al. [45]. For the maximum in the repulsive interactions between two copper atoms before they formed a copper dimer Knorr et al. estimated a value of ≥ 13 meV.

The experiments with cobalt on Cu(111) were done to establish how the chemical identity of the adatom affected interactions. The results were found to agree with those for copper on Cu(111) within the limit of error. The authors claimed that interactions were insensitive to the chemical identity of the adsorbate, which disagrees with the observation in earlier measurements for Re, Ir, or Pd pairs on W(110) by FIM. Rhenium pair interactions definitely differed from interactions between Pd and Ir pairs. The differences might be connected with the lack of an angular dependence on fcc(111) surfaces. Experiments with Co on Ag(111) were carried out to show that interactions arose from electronic origins, as Ag(111) has a surface band structure different from Cu(111) and $\lambda_F/2$ is 38 Å. It is clear that interactions on Ag(111) are much less pronounced and spread out over much larger spacings, and the "scaling of the interaction period with λ_F of the surface state clearly establishes that the long-range interactions are mediated by the nearly free two-dimensional electron gas of the surface state and excludes mediation via elastic lattice deformation". Evidently the understanding of long-range two atom interactions is in reasonable shape, in large measure based on a knowledge of the diffusion characteristics, and the ability to see individual adatoms. Equation (10.22) turned out to describe quite nicely the observations made by STM, but this equation does not take into account the anisotropy of pair-interactions seen by Watanabe [20].

Hyldgaard and Einstein [49–51] worked out an expression for trio-interactions with experimentally accessible parameters in the case when the leading role came from electrons which scatter at all three locations and traverse the perimeter $d_T = d_{12} + d_{23} + d_{31}$. Assuming that adsorbates couple to the same local environment and have identical phase shifts δ_F

$$\Delta E_{trio}(d_{12}d_{23}d_{31}, \delta_F) \simeq -\varepsilon_F \sin^3(\delta_F) \left(\frac{16\sqrt{2}}{\pi^{5/2}}\right) \gamma_{123} \frac{\sin(q_F d_T + 3\delta_F - 3\pi/4)}{(q_F d_T)^{5/2}}; \quad (10.23)$$

here $\gamma_{123} = d_T^{3/2} \sqrt{d_{12}d_{23}d_{31}}$ is a shape dependent dimensionless ratio. It turns out that trio-interactions have an oscillatory character with slightly weaker amplitude than pair- and bulk-interactions, and decay a bit faster, as $\sim d_T^{-5/2}$. The oscillatory dependence of trio-interactions is illustrated in Fig. 10.23.

Further calculations of the interactions between adatoms at long ranges have been carried out by Stepanyuk et al. [52] using density-functional theory and multiple scattering according to the Korringa–Kohn–Rostoker Green's function method [53–55]. They also used the "force theorem" and self-consistent spin-polarized potentials for single atom calculations. In the fully relaxed geometry, the distance of a Co atom from the surface is reduced by 14% compared to the ideal Cu layer distance. The substrate mediated interactions at large distances were found to be unmodified by relaxation and the calculations yielded the wave length λ_F. The local density of states at the Fermi energy is plotted in Fig. 10.24a for a single cobalt atom on a Cu(111) surface, showing strong oscillations with a period around 15 Å. In Fig. 10.24c is given an STM image of two cobalt atoms ~ 60 Å apart, and in 10.24b the oscillating interactions, determined in STM experiments and from theory, are compared. Theory agreed nicely with experiment, but for distances < 8 Å there are discrepancies with the predictions of Hyldgaard and Persson

10.2 More recent studies

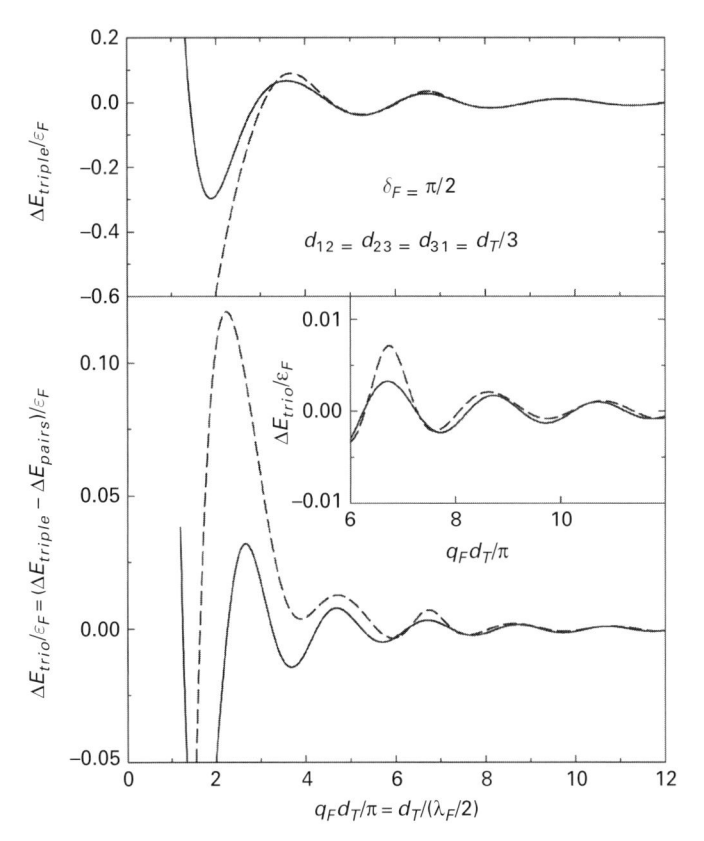

Fig. 10.23 Three-atom interactions mediated by surface-state band. Top: full interaction of three atoms. Bottom: Trio-interaction energy. Solid curves give asymptotic results, dashed curves show numerical calculations. Insert details long-range variations (after Hyldgaard and Einstein [49]).

[47], probably due to their disregarding bulk electronic states in the scattering of surface electrons. The small, long-range interactions shown here have an effect on growth. Cobalt adatoms were found to diffuse alongside a chain of copper atoms, and eventually incorporated into the end of the chain, not into the side, much as has been found in the experiments of Koh [56], described in detail in Section 6.6.

STM observations of cerium adatom layers on an Ag(111) surface were made by Silly et al. [57], who were able to obtain a self-assembled ordered hexagonal array at 3.9 K, shown in Fig. 10.25a. An STM image of a single cerium adatom revealed the oscillating electron density. From measurements of the location of cerium atoms mobile at 4.8 K, as in Fig. 10.25b, and the disappearance of the superlattice at 10 K, they were able to derive the distribution in Fig. 10.25c, which suggests a diffusion barrier of 0.011 eV, assuming a frequency prefactor of 10^{12} sec^{-1}. The interaction potential responsible for positioning the atoms in the array is given in Fig. 10.25d and shows a first minimum at 32 Å with a depth of 0.8 meV, in agreement with the first-neighbor atom position, determined by the scattering properties of the adatoms. The repulsive barrier is in the range 0.3 meV at an adatom distance of 55 Å, so that 4.8 K is high enough to overcome such repulsions. To understand the creation of the superstructure, two- as well as three-body interactions have

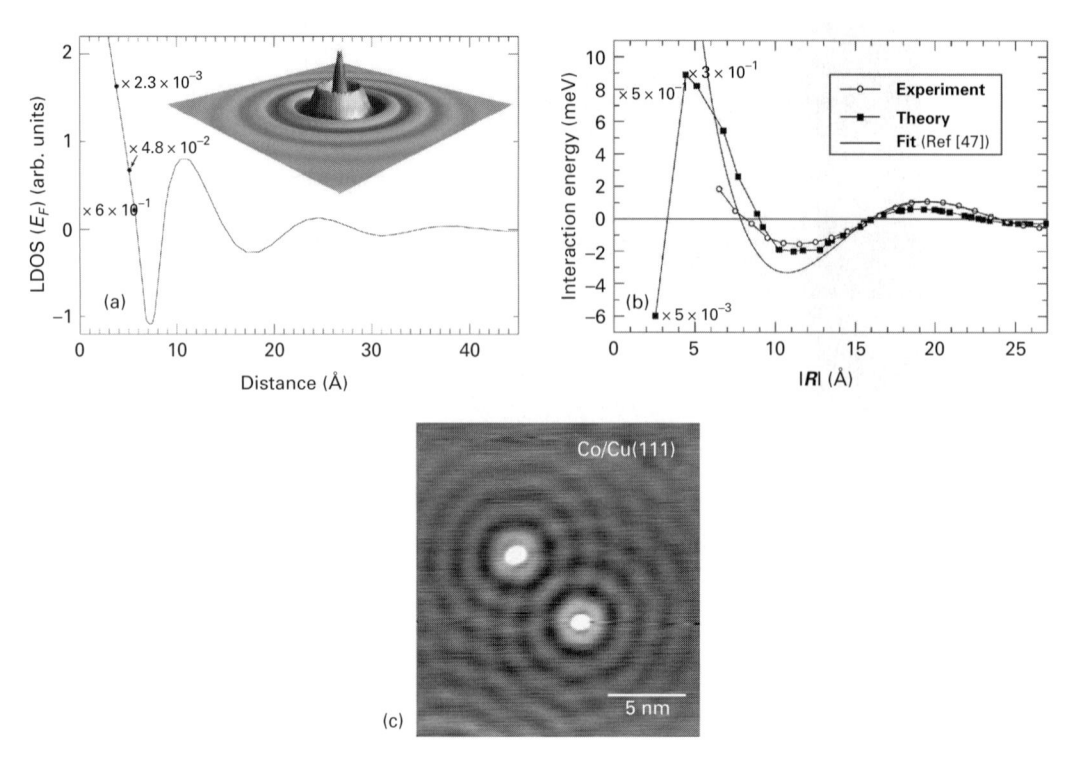

Fig. 10.24 Co adatom interactions on Cu(111). (a) Local density of states for Co adatom on Cu(111), calculated with density-functional and K-K-R method. (b) Co pair interactions as a function of interatomic separation, derived from distribution of distances. Solid curve gives fit to Eq. (10.22). (c) STM image of two Co adatoms, indicating a sharing of density of states oscillations (after Stepanyuk et al. [52]).

to be taken into account. Results are in agreement with the finding of Knorr et al. [48] for Co on Ag(111). Understanding these long-range magnetic interactions is important, and can open up new possibilities for applications. Two years later, Negulyaev et al. [58] performed MC simulations and found a superlattice stabilized by a pair potential well with a depth of 0.6 meV for a coverage between 1% and 1.7% at temperatures < 5 K. At 8 K they observed collapse of the superlattice by formation of dimers. For higher coverages, a local hexagonal order was observed.

Interactions between two palladium adatoms on a W(211) surface have recently been determined in field ion microscopic observations by Fu et al. [59]. Measurements were made of the number of atom pairs at different separations, at temperatures above 200 K, at which palladium diffusion is rapid enough to ensure equilibrium. When the two adatoms were in the same channel, a dimer tended to form with a free energy of -0.0331 ± 0.0007 eV; in the next-nearest site the free energy was 0.1102 eV higher. Rather more impressive results were obtained when the two palladium adatoms were in adjacent channels. Here an attractive minimum was followed by repulsion out to 10 Å, then a small minimum was observed, followed by a rise in repulsion out to ~23 Å, as indicated in Fig. 10.26. The interactions for atoms separated by one or two empty channels were small.

10.2 More recent studies

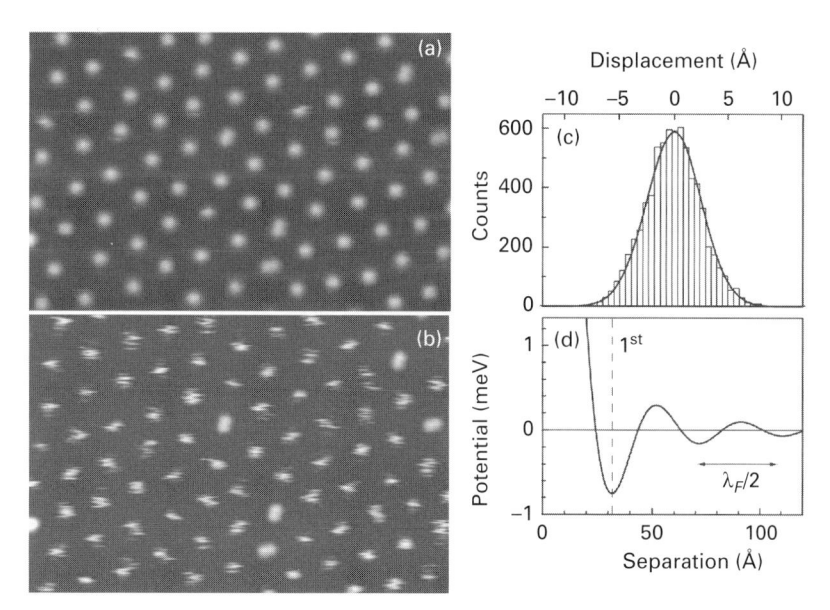

Fig. 10.25 STM images of Ce adatoms on Ag(111) plane self-assembled in a hexagonal superlattice at (a) 3.9 K and (b) 4.8 K. Superlattice is frozen at 3.9 K; at 4.8 K adatoms are mobile around binding sites. (c)–(d) Ce adatoms on Ag(111) surface. (c) Distribution of Ce displacements at 4.8 K around superlattice sites. (d) Pair interaction between Ce adtoms on Ag(111). First-neighbor position at 32 Å is shown by dashed line (after Silly *et al.* [57]).

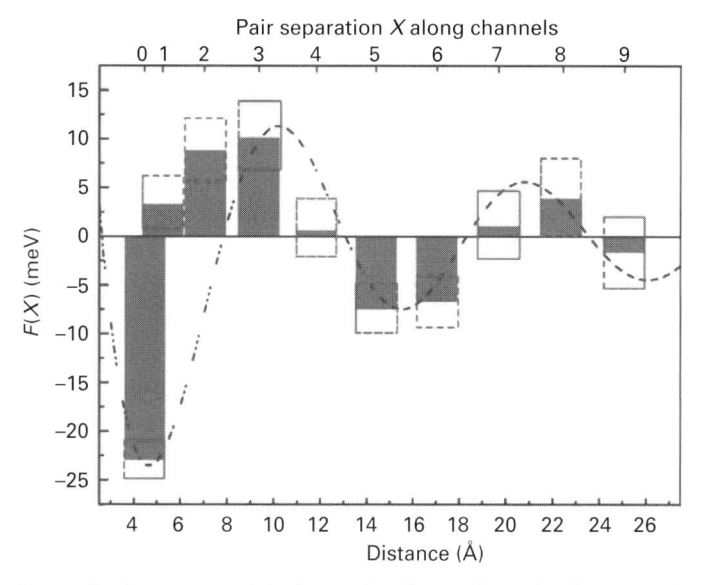

Fig. 10.26 Interaction between two Pd adatoms in adjacent channels of W(211) plane, deduced from FIM observations. Dashed line gives data fit to $|R|^{-n}$ with $n = 1$ (after Fu *et al.* [59]).

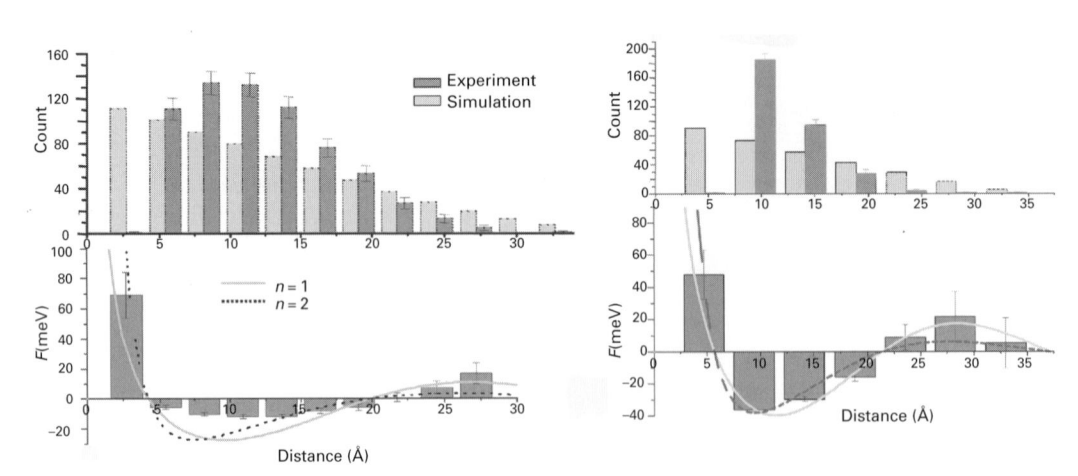

Fig. 10.27 Distribution of pair separations and of interactions for W adatoms on Ir(111) plane, derived from observations in FIM. Left: Along $[2\bar{1}\bar{1}]$ axis. Right: Along $[\bar{1}01]$ axis. Fits with $|R|^{-2}$ and $|R|^{-1}$ dependence shown by dotted and solid curves. Simulations show the distribution for non-interacting W adatoms (after Fu et al. [60]).

A more extensive study by Fu et al. [60] appeared later with results again for two atoms in the same channel and also in neighboring ones. Unfortunately, there was no indication at all of the temperature of the measurements. The free energy of atoms in nearest-neighbor sites in the same channel was now given as -0.070 eV, and in the next nearest-neighbor site the energy was 0.120 eV higher. With the two atoms in adjacent channels, the oscillations in the free energy are quite clear and can be fitted by the relation $(2k_F|R| + 2\delta_F)/(2k_F|R|)^n$, with $k_F = 0.25$ Å^{-1} and $\delta_F = 1.4$. What is interesting here is that the best fit is obtained with n equal to one instead of two as expected from Hyldgaard and Persson [47]. The same kind of behavior was found for interactions between tungsten atoms on the Ir(111) surface. Shown in Fig. 10.27 are results along the $[2\bar{1}\bar{1}]$ axis, and also along $[\bar{1}01]$; both seem to conform to a plot with n equal to one again. In the direction $[2\bar{1}\bar{1}]$, the biggest attraction was observed at 10.88 Å; the biggest repulsion was 0.07 eV. There was no attraction at the nearest-neighbor distance. In the direction $[\bar{1}01]$, the biggest attraction was observed at the next-nearest-neighbor site, at a distance of 9.42 Å; this amounted to 0.090 eV. Again there was no attraction at the closest possible distance. Lack of attraction at nearest-neighbor distances was observed previously for Re pairs on a W(110) surface [18,23].

Density-functional calculations of the interaction energies between silver atoms on a strained Ag(111) surface were carried out by Luo and Fichthorn [61]; the lattice constant was compressed by 4.61% to explore the possibility of adjusting layer formation. Electronic pair interactions obtained are plotted in Fig. 10.28a. They reveal strong attractive effects at the nearest- and next-nearest-neighbor positions. This region is followed by a ring of sites at which interactions are repulsive; this peaks at atomic separations of ~5 Å, and is then followed by a ring of attractive sites, past which, at a separation of ~15 to 16 Å, there is again a repulsive region. Finally, beyond this distance, attraction again sets in once more. Elastic interactions were also calculated, and are listed

10.2 More recent studies

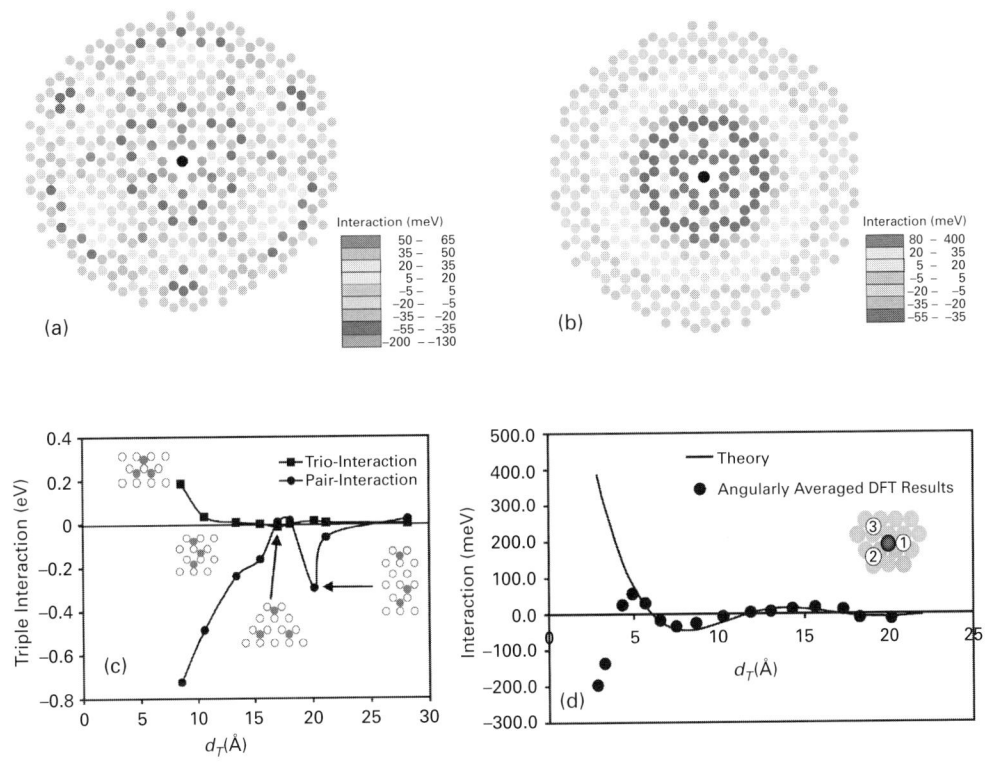

Fig. 10.28 Ag-Ag interactions on Ag(111) plane. (a) Interactions between a Ag adatom and a central Ag on a strained Ag(111) plane as a function of the interatomic separation. Results obtained by density-functional calculations. Lattice constant of silver was 4.16% compressed. (b) Energies from Eq. (10.22) found in four-layer slab in density-functional calculations ($k_F = 0.28$ Å^{-1}, $\varepsilon_F = 0.662 eV$, $\delta_F = 0.45\pi$) for central Ag adatom interacting with a second Ag in different positions. (c) Interactions of three adatoms on Ag(111) in their dependence on trio-perimeter d_T. Total pair-energy indicated by circles. The squares show trio effects. (d) Comparison of interactions from Eq. (10.22) with angularly averaged DFT interactions from (b) (after Luo and Fichthorn [61]). See color plate section for color version of this figure.

together with electronic contributions in Table 10.4. Elastic effects are small, and for the 14th neighbor are below the resolution of the calculations. What is surprising, however, is that the magnitude of the interactions found here on the strained surface are much larger than seen in experiments on the unstrained surface. The period of the interactions amounted to ~11 Å, much smaller than the ~38 Å seen in experiments on an unstrained surface [48], giving an indication of the influence of strain on the interaction character.

Luo and Fichthorn also looked at trio-interactions; Fig. 10.28c shows these interactions as a function of the trio perimeter d_T. Trio-interactions are repulsive at close distances, attractive in the region 15–18 Å, but beyond 13Å are negligible. The calculations show that in this system substrate mediated interactions are primarily pairwise. The energy derived from Eq. (10.22) is given in Fig. 10.28b. The main differences are in the region of first- and second-neighbor distances and are caused by the formation of chemical bonds, not accounted for in Eq. (10.22). DFT calculations also show the anisotropy energy,

Table 10.4 Pair interactions obtained in density-functional calculations for Ag atoms on Ag(111) plane [61].

Neighbor	Distance $|R|$ (Å)	Electronic (eV)	Elastic (eV)
1	2.84	−0.1963	−0.0052
4	4.91	0.0564	0.0007
5	5.67	0.0253	0.002
9	7.50	−0.0249	0.0006
11	8.50	−0.0077	0.0022
14	9.82	−0.0096	−0.0015

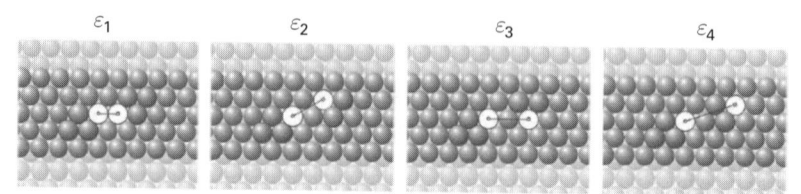

Fig. 10.29 Configurations of adatom pairs on Cu(111) plane, with energy designations at the top (after Stasevich *et al.* [62]).

which is not observed using Eq. (10.22), but was found previously in experiments [23]. After angular averaging the DFT results, agreement with the theoretical equation at longer distances is excellent, as in Fig. 10.28d.

Stasevich *et al.* [62] have recently carried out density-functional calculations using the VASP program [41–43] for the interactions between copper atoms on Cu(111) and Cu(100). The configurations for two copper atoms and their designations are shown in Fig. 10.29. The supercells used in the calculations were (14 × 3 × 2). For this cell fourth-neighbor interactions were ignored and the cell was referred to as (3 × 2); when a (14 × 4 × 2) cell was used, fourth-neighbor interactions were included and it was referred to as (4 × 2). The results on the two planes are given in Table 10.5; only data for the relaxed surface are shown, as the behavior of constrained surfaces is very similar. The results for (3 × 2) and (4 × 2) surfaces do not differ, demonstrating the negligible influence of fourth-neighbor interactions. Second-neighbor interactions are significantly stronger on the (100) plane, but thereafter these effects drop off sharply. On Cu(111), second nearest-neighbor interactions are negligible, but on Cu(100) they are in the range of 1/7 the nearest-neighbor interactions. Interactions at third- and fourth nearest-neighbor distances were found to be negligible on both surfaces. The results are in good accord with the estimates for Cu(100) obtained by Levanov *et al.* [38] using RGL potentials, but do not agree with the experimental data of Knorr *et al.* [48] where longer-range interactions on Cu(111) were detected. The trio-interactions of Stasevich *et al.* for a triangular configuration of atoms on Cu(111) amounted to ~1/3 of nearest-neighbor pair-interactions, (0.117–0.083) ± 0.023 eV. Linear and bent configuration had an even smaller interaction energy, −0.022 ± 0.011 eV and −0.011 ± 0.011 eV. For Cu(100) the trio interaction energy for the linear configuration was about −0.014 ± 0.011 eV, while for the bent one 0.051 ± 0.011 eV.

10.2 More recent studies

Table 10.5 Interaction energies obtained by *ab-initio* calculations between copper atoms in the configurations in Fig. 10.29 [62].

	Cu(100)		Cu(111)	
Interactions energy (eV)	(3 × 2)	(4 × 2)	(3 × 2)	(4 × 2)
ε_1	-0.332 ± 0.016	-0.335 ± 0.012	-0.314 ± 0.019	-0.323 ± 0.011
ε_2	-0.047 ± 0.009	-0.043 ± 0.006	0.004 ± 0.012	0.001 ± 0.012
ε_3	-0.003 ± 0.009	0.013 ± 0.008	0.005 ± 0.006	0.003 ± 0.003
ε_4		0.002 ± 0.004		-0.001 ± 0.003

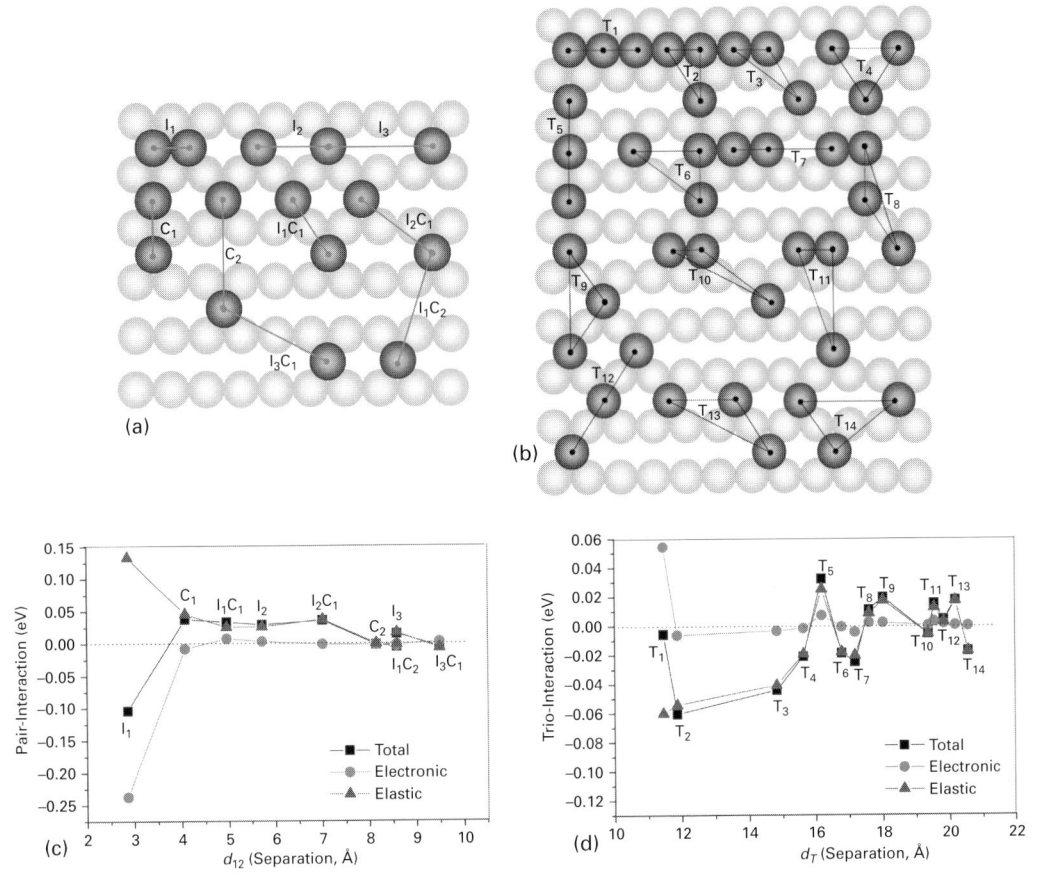

Fig. 10.30 Interactions between Al adatoms on the Al(110) plane derived from density-functional calculations. (a) Designation of pair-interactions. (b) Trio-interactions. (c) Dependence of pair-interactions on separation. (d) Effect of trio perimeter on trio-interactions (after Tiwary and Fichthorn [63]).

Recently, Tiwary and Fichthorn [63] looked at pair- and trio-interactions for aluminum on an Al(110) surface with DFT calculations implemented in the Vienna *ab-initio* simulation package VASP, investigating interactions for the configurations shown in Fig. 10.30a–b. Their data are presented in Table 10.6, and illustrated in Fig. 10.30c for

Table 10.6 Contributions to pair- and trio-energies in Fig. 10.30 calculated using DFT for Al atoms on Al(110) [63].

Pair/Trio	d_{12}/d_T	Total	Electronic	Elastic
I_1	2.86	−0.104	−0.237	0.133
C_1	4.05	0.038	−0.008	0.046
I_1C_1	4.96	0.033	0.008	0.026
I_2	5.72	0.029	0.003	0.026
I_2C_1	7.01	0.036	−0.001	0.037
C_2	8.09	−0.002	−0.002	0.000
I_1C_2	8.58	−0.005	−0.004	−0.001
I_3	8.59	0.015	−0.002	0.017
I_3C_1	9.49	−0.004	0.003	−0.007
T_1	11.45	−0.006	0.054	−0.060
T_2	11.87	−0.060	−0.006	−0.054
T_3	14.83	−0.044	−0.003	−0.041
T_4	15.64	−0.021	−0.002	−0.019
T_5	16.19	0.032	0.007	0.025
T_6	16.78	−0.019	−0.001	−0.018
T_7	17.17	−0.025	−0.004	−0.020
T_8	17.59	0.011	0.002	0.008
T_9	18.00	0.019	0.002	0.017
T_{10}	19.36	−0.005	0.000	−0.006
T_{11}	19.54	0.015	0.003	0.012
T_{12}	19.83	0.004	0.002	0.003
T_{13}	20.17	0.017	0.000	0.017
T_{14}	20.55	−0.017	0.000	−0.017

pair and 10.30d for trio-interactions. For in-channel pair-interactions they observed a strong attraction, −0.104 eV, at a nearest-neighbor separation, followed by weak repulsion of 0.029 eV at the second nearest-neighbor distance. Nearest-neighbor interactions have a mostly electronic origin, while at longer ranges interactions are due to elastic distortions of the lattice and start to play a leading role. Pair-interactions extend out to 8 Å; most trio-interactions, except for collinear trios, are also elastic in origin. Trio-interactions have an oscillatory dependence on distance and were found to be important in understanding homoepitaxy on the Al(110) surface.

10.3 Summary

With knowledge of surface diffusivities in hand, it has been possible to establish conditions for achieving an atom distribution at equilibrium, so the ability to obtain experimental insights into interactions between adatoms has become mature. It is quite clear now from direct experimental observations and also theoretical calculations that interactions between adatoms are not very strong but are of long range. More studies to define interactions can be expected to raise understanding in this area, and investigations of many-body interactions are certainly needed. With more information about the interplay between adatoms available it should also be possible to come to a better understanding of how surface diffusion proceeds with a finite concentration of atoms present on the surface. One thing that has become quite evident is that interactions are *not* short ranged, are likely to be anisotropic and influenced by strain. In addition, it is clear that pair-interactions are not enough to explain the creation of nanostructures on the surface. Moreover, from what has been said here it should be clear how closely surface diffusion and interactions are related. What will be most important in the future are clever experiments to quantitatively establish the magnitude of many-body effects between adatoms.

References

[1] T. L. Einstein, in: *Handbook of Surface Science*, W. N. Unertl (ed.), Interactions between adsorbate particles, (Elsevier, Amsterdam, New York, 1996), p. 577–650.

[2] T. B. Grimley, The indirect interaction between atoms or molecules adsorbed on metals, *Proc. Phys. Soc.* **90** (1967) 751–764.

[3] G. Ehrlich, F. Watanabe, Atomic interactions on crystals: A review of quantitative experiments, *Langmuir* **7** (1991) 2555–2563.

[4] S. J. Koh, Adatom diffusion and interactions: Pd on W(110), Ph. D. Materials Science, University of Illinois at Urbana-Champaign, 1998.

[5] T. T. Tsong, Field-ion microscope observations of indirect interaction between adatoms on metal surfaces, *Phys. Rev. Lett.* **31** (1973) 1207–1210.

[6] J. Friedel, Metallic alloys, *Nuovo Cimento Supplement* **7** (1958) 287–311.

[7] W. R. Graham, G. Ehrlich, Direct measurement of the pair distribution function for adatoms on a surface, *Phys. Rev. Lett.* **32** (1974) 1309–1311.

[8] S. Nishigaki, S. Nakamura, FIM observation of interactions between W atoms on W Surfaces, *Jpn. J. Appl. Phys.* **14** (1975) 769–777.

[9] G. Ayrault, G. Ehrlich, Surface self-diffusion on an fcc crystal: An atomic view, *J. Chem. Phys.* **60** (1974) 281–294.

[10] D. W. Bassett, D. R. Tice, in: *The Physical Basis for Heterogeneous Catalysis*, E. Drauglis, R. I. Jaffee (eds.), Field ion microscopic studies of interactions between atoms adsorbed on tungsten(110) surfaces, (Plenum Press, New York, 1975), p. 231–245.

[11] H.-W. Fink, K. Faulian, E. Bauer, Evidence for nonmonotonic long-range interactions between adsorbed atoms, *Phys. Rev. Lett.* **44** (1980) 1008–1011.

[12] R. Casanova, T. T. Tsong, Pair interactions of metal atoms on a metal surface, *Phys. Rev. B* **22** (1980) 5590–5598.

[13] T. T. Tsong, R. Casanova, Direct measurement of pair energies in adatom-adatom interactions on a metal surface, *Phys. Rev. B* **24** (1981) 3063–3072.

[14] L. C. A. Stoop, A Monte Carlo calculation of lateral interactions in silver monolayers on W (110), *Thin Solid Films* **103** (1983) 375–398.

[15] L. D. Roelofs, R. J. Bellon, Multi–adatom interaction effects in a lattice gas model for Cu and Au adsorption on W(110), *Surf. Sci.* **223** (1989) 585–598.

[16] H. Gollisch, Adsorption of Cu, Ag and Au on W(110): A theoretical study based on a non-additive effective binding potential, *Surf. Sci.* **175** (1986) 249–262.

[17] J. Kolaczkiewicz, E. Bauer, The law of corresponding states for chemisorbed layers with attractive lateral interactions, *Surf. Sci.* **151** (1985) 333–350.

[18] H.-W. Fink, G. Ehrlich, Pair and trio interactions between adatoms: Re on W(110), *J. Chem. Phys.* **81** (1984) 4657–4665.

[19] F. Watanabe, G. Ehrlich, Direct mapping of adatom–adatom interactions, *Phys. Rev. Lett.* **62** (1989) 1146–1149.

[20] F. Watanabe, G. Ehrlich, Direct observations of pair interactions on a metal: Heteropairs on W(110), *J. Chem. Phys.* **95** (1991) 6075–6087.

[21] A. M. Stoneham, Elastic interactions between surface adatoms and between surface clusters, *Solid State Comm.* **24** (1977) 425–428.

[22] W. Kappus, Substrate strain induced interaction of adatoms on W(110), *Z. Phys. B* **38** (1980) 263–266.

[23] F. Watanabe, G. Ehrlich, Direct observation of interactions between identical adatoms: Ir-Ir and Re-Re on W(110), *J. Chem. Phys.* **96** (1992) 3191–3199.

[24] D. W. Bassett, M. J. Parsley, Field ion microscope observations of cluster formation in metal deposits on tungsten surfaces, *Nature* **221** (1969) 1046.

[25] D. R. Tice, D. W. Basset, Nucleation of platinum and iridium monolayer growth on (110) tungsten surfaces, *Thin Solid Films* **20** (1974) S37–40.

[26] H.-W. Fink, G. Ehrlich, Direct observations of overlayer structures on W(110), *Surf. Sci.* **110** (1981) L611–614.

[27] R. S. Chambers, Determination of cluster spacings in the FIM: iridium dimers on W(110), *Surf. Sci.* **246** (1991) 25–30.

[28] D. W. Bassett, C. K. Chung, D. Tice, Field ion microscope studies of atomic displacement processes on metal surfaces, *La Vide* **176** (1975) 39–43.

[29] H. Dreyssé, D. Tomanek, K. H. Bennemann, Multi-adatom interactions on metal surfaces, *Surf. Sci.* **173** (1986) 538–554.

[30] D. A. Reed, G. Ehrlich, In-channel clusters: Rhenium on W(211), *Surf. Sci.* **151** (1985) 143–165.

[31] W. Xu, J. B. Adams, W adatom-adatom interactions on the W(110) surface, *Surf. Sci.* **339** (1995) 241–246.

[32] M. Breeman, G. T. Barkema, M. H. Langelaar, D. O. Boerma, Computer simulation of metal-on-metal epitaxy, *Thin Solid Films* **272** (1996) 195–207.

[33] G. J. Ackland, G. Tichy, V. Vitek, M. W. Finnis, Simple N-body potentials for the noble metals and nickel, *Philos. Mag. A* **56** (1987) 735–756.

[34] J. Kolaczkiewicz, E. Bauer, Desorption energy and lateral interaction energy in the adsorption systems of Rh, Pd, Ir and Pt on W(110), *Surf. Sci.* **374** (1997) 95–103.

[35] S. J. Koh, G. Ehrlich, Pair- and many-atom interactions in the cohesion of surface clusters: Pd_x and Ir_x on W(110), *Phys. Rev. B* **60** (1999) 5981–5990.

[36] S. J. Koh, G. Ehrlich, Atomic interactions and the stability of surface clusters, *Mater. Res. Soc. Symp. Proc.* **528** (1998) 37–43.

[37] E. Wahlström, I. Ekvall, H. Olin, L. Walldén, Long-range interaction between adatoms at the Cu(111) surface imaged by scanning tunnelling microscopy, *Appl. Phys. A* **66** (1998) S1107–1110.

[38] N. A. Levanov, A. A. Katsnel'son, A. É. Moroz, V. S. Stepanyuk, W. Hergert, K. Kokko, Structure and stability of clusters on metal surfaces, *Phys. Solid State* **41** (1999) 1216–1221.

[39] K. Fichthorn, M. Scheffler, Island nucleation in thin-film epitaxy: A first-principles investigation, *Phys. Rev. Lett.* **84** (2000) 5371–5374.

References

[40] A. Bogicevic, S. Ovesson, P. Hyldgaard, B. I. Lundqvist, H. Brune, D. R. Jennison, Nature, strength, and consequences of indirect adsorbate interactions on metals, *Phys. Rev. Lett.* **85** (2000) 1910–1913.

[41] G. Kresse, J. Hafner, *Ab initio* molecular dynamics for liquid metals, *Phys. Rev. B* **47** (1993) 558–561.

[42] G. Kresse, J. Hafner, *Ab initio* molecular dynamics simulation of the liquid-metal-amorphous-semiconductor transition in germanium, *Phys. Rev. B* **49** (1994) 14251–14269.

[43] G. Kresse, J. Furthmüller, Efficient iterative schemes for *ab initio* total-energy calculations using a plane-wave set, *Phys. Rev. B* **54** (1996) 11169–11186.

[44] C. Polop, H. Hansen, C. Busse, T. Michely, Relevance of nonlocal adatom-adatom interactions in homoepitaxial growth, *Phys. Rev. B* **67** (2003) 193405 1–4.

[45] J. Repp, F. Moresco, G. Meyer, K.-H. Rieder, P. Hyldgard, M. Perrson, Substrate mediated long-range oscillatory interaction between adatoms, *Phys. Rev. Lett.* **85** (2000) 2981–2984.

[46] J. Fransson, A. V. Balatsky, Surface imaging of inelastic Friedel oscillations, *Phys. Rev B* **75** (2007) 195337 1–5.

[47] P. Hyldgaard, M. Persson, Long-ranged adsorbate-adsorbate interactions mediated by a surface-state band, *J. Phys.: Condens. Matter* **12** (2000) L13–19.

[48] N. Knorr, H. Brune, M. Epple, A. Hirstein, M. A. Schneider, K. Kern, Long-range adsorbate interactions mediated by a two-dimensional electron gas, *Phys. Rev. B* **65** (2002) 115420 1–5.

[49] P. Hyldgaard, T. L. Einstein, Surface-state-mediated three-adsorbate interaction, *Europhys. Lett.* **59** (2002) 265–271.

[50] P. Hyldgaard, T. L. Einstein, Surface-state mediated three-adsorbate interaction: Electronic nature and nanoscale consequences, *Surf. Sci.* **532–535** (2003) 600–605.

[51] P. Hyldgaard, T. L. Einstein, Surface-state mediated three-adsorbate interaction: Exact and numerical results and simple asymptotic expression, *Appl. Surf. Sci.* **212–213** (2003) 856–860.

[52] V. S. Stepanyuk, A. N. Baranov, D. V. Tsivlin, W. Hergert, P. Bruno, N. Knorr, M. A. Schneider, K. Kern, Quantum interference and long-range adsorbate-adsorbate interactions, *Phys. Rev. B* **68** (2003) 205410 1–5.

[53] V. S. Stepanyuk, W. Hergert, R. Zeller, P. H. Dederichs, Magnetism of $3d$, $4d$, and $5d$ transition-metal impurities on Pd(001) and Pt(001) surfaces, *Phys. Rev. B* **53** (1996) 2121–2125.

[54] V. S. Stepanyuk, W. Hergert, R. Zeller, P. H. Dederichs, Imperfect magnetic nanostructures on a Ag(001) surface, *Phys. Rev. B* **59** (1999) 1681–1684.

[55] K. Wildberger, V. S. Stepanyuk, P. Lang, R. Zeller, P. H. Dederichs, Magnetic nanostructures: $4d$ clusters on Ag(001), *Phys. Rev. Lett.* **75** (1995) 509–512.

[56] S. J. Koh, G. Ehrlich, Self-assembly of one-dimensional surface structures: Long-range interactions in the growth of Ir and Pd on W(110), *Phys. Rev. Lett.* **87** (2001) 106103 1–4.

[57] F. Silly, M. Pivetta, M. Ternes, F. Patthey, J. P. Pelz, W.-D. Schneider, Creation of an atomic superlattice by immersing metallic adatoms in a two-dimensional electron sea, *Phys. Rev. Lett.* **92** (2004) 016101 1–4.

[58] N. N. Negulyaev, V. S. Stepanyuk, L. Niebergall, W. Hergert, H. Fangohr, P. Bruno, Self-organization of Ce adatoms on Ag(111): A kinetic Monte Carlo study, *Phys. Rev. B* **74** (2006) 035421 1–5.

[59] T.-Y. Fu, Y.-J. Hwang, T. T. Tsong, Oscillatory adatom-adatom interactions of Pd adatoms on the W(211) surface, *Surf. Sci.* **566–568** (2004) 462–466.

[60] T.-Y. Fu, T. Y. Wu, T. T. Tsong, Oscillatory adatom–adatom interactions of 1D and 2D surface diffusion systems, *Chin. J. Phys.* **43** (2005) 124–131.

[61] W. Luo, K. A. Fichthorn, First-principles study of substrate-mediated interactions on a compressed Ag(111) surface, *Phys. Rev. B* **72** (2005) 115433 1–8.
[62] T. J. Stasevich, T. L. Einstein, S. Stolbov, Extended lattice gas interactions of Cu on Cu(111) and Cu(001): *Ab initio* evaluation and implications, *Phys. Rev. B* **73** (2006) 115426 1–7.
[63] Y. Tiwary, K. A. Fichthorn, Interactions between Al atoms on Al(110) from first-principles calculations, *Phys. Rev. B* **75** (2007) 235451 1–8.

Appendix

In both field ion microscopy and scanning tunneling microscopy, the preparation of sharp tips is crucial for carrying out observations with atomic resolution. The demands on the tip in scanning tunneling microscopy are not too rigorous; it has to have a sharp region of some kind in order to probe a surface with high resolution. For this task, tips are available commercially. In field ion microscopy, the tip is the surface to be studied, and therefore has to have a well-formed overall shape. As already pointed out in Chapter 2, there are fairly standard methods for preparing and polishing wire samples to the proper shape. We will give a few examples of how this can be accomplished. With the exception of the description for rhodium, these were all worked out by Liu [1], and with some minor changes taken directly from his thesis, but it should be noted that other recommendations are available [2–4].

Preparation of samples for field ion microscopy

"Three conditions have to be met for a good field emitter specimen: (a) The end of the specimen, that is the tip, has to be sharp enough so that strong electric fields can be attained at reasonable voltages; (b) The tapered section, or shank, has to be smooth and devoid of large defects or undercuts, so that the emitter can withstand field stresses; and (c) The whole shank and tip should be smooth and symmetric to avoid serious distortion of the image.

We describe in some detail the techniques for emitter preparation, and compare different techniques following the above guidelines, for five metals: tungsten, molybdenum, rhenium, iridium, and platinum." In addition more recent findings for rhodium are described as well.

A.1 Tungsten

"The easiest and most successful method for preparing tungsten emitters is the drop-off technique. In this method a small section of 0.005 in diameter wire is dipped in 2N NaOH solution, using a ring-shaped platinum or nickel wire as counter-electrode, as shown in Fig. A.1. Approximately 12 V dc is applied across the cell by an automatic shut-off circuit, shown in Fig. 2.11. The tungsten wire is attacked most rapidly at the air-liquid interface, forming a neck there, as shown in Fig. A.2. As polishing proceeds, the portion of the wire below the liquid level is thinned and finally pinched-off at the thinnest place. When this happens the polishing current suddenly drops, as the surface interacting with the solution is greatly reduced; at this moment the potential across the cell is switched off

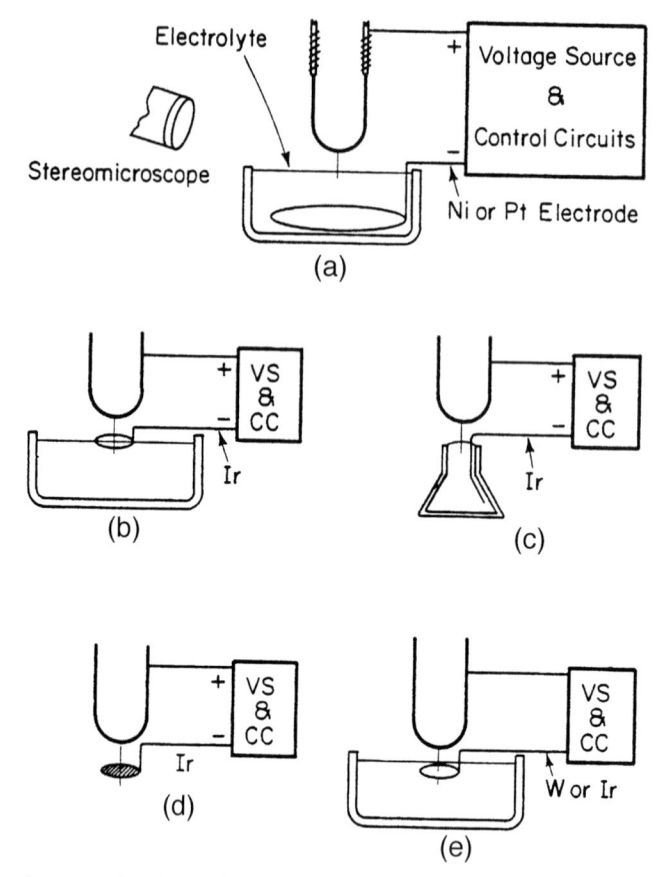

Fig. A.1 Apparatus for electrochemical shaping of samples for field ion microscopy [1]. (a) For tungsten and molybdenum samples; (b) $(11\bar{2}0)$-oriented rhenium wire; (c) arrangement for rhenium emitters; (d) (0001)-oriented rhenium; (e) iridium and platinum specimens.

immediately by the shut-off circuit to stop the back-etching. The wire is then lifted out of the solution and rinsed with distilled water, followed by ethyl alcohol, to remove the residual polishing solution still on the surface. The emitter tips obtained following this procedure ordinarily have a diameter of a few hundred Ångstroms, and the general result is very satisfactory.

A.2 Molybdenum

Molybdenum is located immediately above tungsten in the periodic table; chemically it is quite similar to tungsten. The polishing solution for tungsten (2N NaOH) also polishes molybdenum, but at a much higher rate. When the polishing conditions for tungsten are applied to molybdenum, the large ion currents cause turbulence in the solution which breaks the wire before it is properly dropped off. Attempts to reduce the current by either reducing the potential across the cell or the concentration of the polishing solution for polycrystalline wire lead to uneven etching of the wire.

There is an established method for preparing molybdenum emitters: immerse 1/8 in of 0.003 in diameter molybdenum wire in 3.57N KOH, using nickel as a counter-electrode.

Appendix

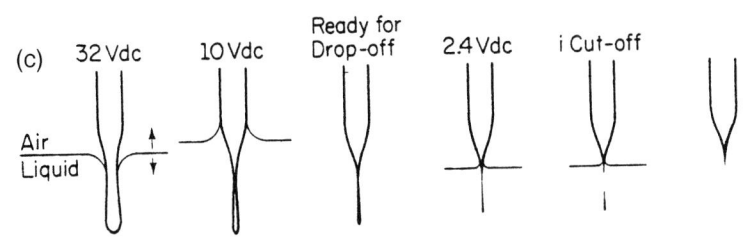

Fig. A.2 Schematics of tip etching for (a) tungsten; (b) zone refined molybdenum; (c) $(11\bar{2}0)$-oriented rhenium [1].

Then apply 4.0 V ac across the cell and polish until the lower section of the immersed wire acquires a long, uniformly thin shape. The potential is then reduced to 1.0 V ac and is applied as intermittent, short pulses while the shape of the wire is monitored with a stereomicroscope. At a magnification of 70X, the end of the wire seems to vanish as the wire is attacked from both the side and bottom. The rate of disappearance of the wire is directly related to the thickness of the wire. By stopping at the highest rate at which the wire disappears from view, a good emitter is usually guaranteed. There is only one disadvantage to this method: for wires of 0.005 in diameter, the section initially immersed has to be 1/4 in long in order to achieve satisfactory results. When high purity zone-refined single crystal wire is used, which is available only in 0.005 in. diameter, this is objectionable from an economic point of view.

For zone-refined wires the following procedure was developed. Immerse 1/32 in of wire in either 2N NaOH or 3.57 KOH solution, using nickel as counter-electrode. Initially the wire is thinned by applying 12 to 15 V dc across the cell. After the wire becomes uniformly thin, as shown in Fig. A.2b, the voltage is lowered to 10 V and the polishing is continued until the lower section of the wire is barely visible in the stereomicroscope at maximum power (70X). The potential is then switched to 1–1.5 V ac and the procedure is finished as described in the previous etching method (Fig. A.2b). It is typical of this method of etching that the emitter does not appear sharp under the optical microscope at 400X. However, about 50% of the emitters prepared this way form a good field ion image at 5–8 kV. This method applies to both 0.003 and 0.005 in diameter wires and is indifferent to previous heat treatments of the wire.

A.3 Rhenium

Rhenium can easily be polished electrochemically using dc potentials and concentrated HNO_3[5]. However, nitric acid tends to attack the air-liquid interface rapidly and good specimens are difficult to obtain consistently. Rhenium does not react readily with acids other than nitric. Most electrochemical polishing solutions cited in the literature [6,7] are fairly slow and time consuming. Besides this, rhenium responds to electrochemical polishing quite differently depending upon its heat treatment, increasing further the difficulty in developing an efficient and versatile etching technique.

To resolve these difficulties, we have developed two methods for etching rhenium. For unheated wires, which are usually $(11\bar{2}0)$ oriented, the polishing solution consists of freshly mixed 20% H_2O, 40% concentrated HNO_3, and 40% HF (48% wt. conc.). In the cell, an iridium counter-electrode in the shape of a small loop about 1/4 in in diameter is immersed in the above electrolyte, as shown in Fig. A.1b. A Petri dish is used as the cell container and is filled with electrolyte up to the level of the iridium loop. When a 32 V dc potential is applied across the cell, polishing is rapid and the attack fastest at the interface. Frequent moving of the wire up and down results in a long, thin neck as shown in Fig. A.2c. The potential is then dropped to ~10 V dc and the polishing becomes uniform along the whole immersed section. Excessive bubbling at the anode (rhenium) is normal at this stage and should not be a matter of concern. After the wire is thinned and appears barely visible under the stereomicroscope, the old solution is abandoned and the final drop-off starts. In this stage a different electrolyte and cell are used: the latter is a narrow necked (5/16 in diameter) flat bottom glass flask containing ~2 cc of the electrolyte; the electrolyte consists of 4 parts concentrated H_3PO_4, 1 part glycerin, 1 part ethyl alcohol, and 1 part 10% aqueous HF. A single iridium wire serves as a counter-electrode, as shown in Fig. A.1c. A dc potential of 2.2 to 2.4 V is applied through the circuit shown in Fig. 2.11. The total current should be restricted to below 100 µA. The position of the air-liquid interface is essential for the success of the drop-off method. The interface should be located just above the thinnest portion of the wire, as shown in Fig. A.2c. Ordinarily a macroscopic undercut is left at the position of the meniscus (Fig. A.2c), since the drop-off solution also attacks the interface preferentially. This does not affect the performance of the tip: we have never lost an emitter due to this.

For (0001) oriented zone-refined wires the final drop-off method is the same as described above. The initial thinning solution is different, however; it is made up from freshly mixed 40% glycolic acid, 30% conc. HNO_3, and 30% HF (10% aqueous solution). The counter electrode is still the 1/4 in iridium loop, but the cell now consists only of a thin layer of electrolyte suspended on the loop, and the rhenium wire. The rhenium wire pierces the electrolyte layer, a small section protruding below it, as shown in Fig. A.1d. Only the section in the solution is etched. The potential across the cell varies from 3 to 4 V dc, depending upon the thickness of the electrolyte layer. Maximum polishing

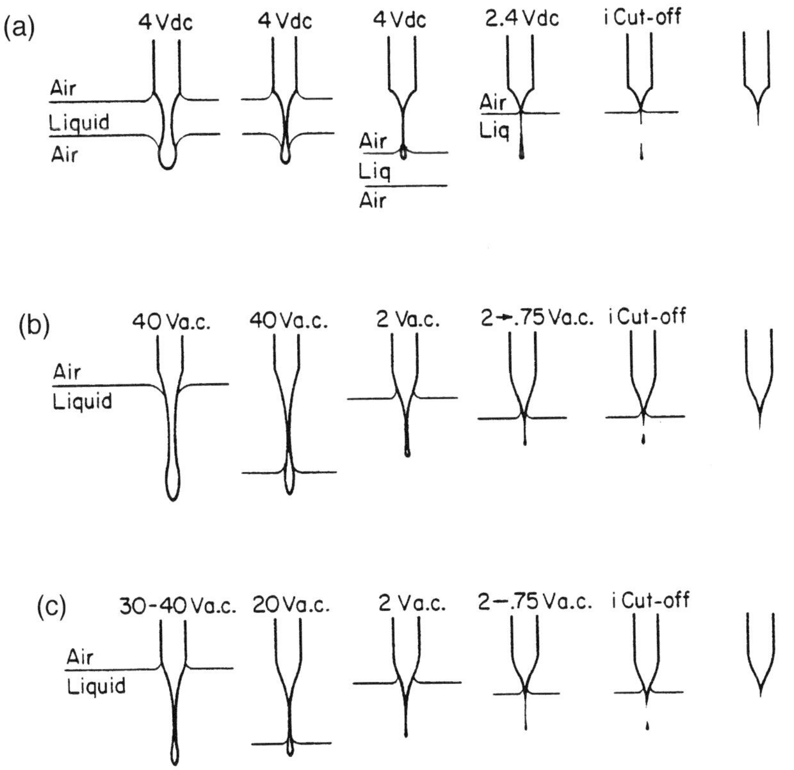

Appendix   739

Fig. A.3   Schematic drawing of tip etching for (a) (0001)-oriented rhenium; (b) iridium; (c) platinum [1].

effect is reached when the ion current is 25 to 30 mA. Polishing is then rapid and smooth; a highly polished surface and a long, narrow section suitable for final drop-off are produced in a couple of minutes. Changes of the wire during polishing are shown schematically in Fig. A.3a.

If incorrect proportions of electrolyte constituents are used, maximum polishing conditions cannot be achieved. This ordinarily results in preferential etching along certain crystallographic directions and hence the creation of deep furrows on the surface. Such undesirable effects usually can be removed by adding some conc. HNO$_3$ to the electrolyte. The field emitters prepared in this way are very sharp. A perfectly smooth surface and clear field ion image can usually be obtained below 5 kV, using He as imaging gas.

A.4 **Iridium**

Iridium is famous for its inertness toward virtually all chemicals, including aqua regia. Reasonably rapid etching of iridium can be accomplished in molten NaCl or KCl salts [5], but control is difficult and the set-up is inconvenient. Chromic acid and ammonium carbonate have been used sequentially to make iridium field emitters [8–10]. This technique is quite slow and the chance of success is rather poor. Quite a simple method was reported by Graham et al. [11]; this uses dilute HF to achieve a thin, narrow taper and

4N HCl to finally sharpen the emitter. The results are generally satisfactory, except that the thinning solution has no effect on heat treated iridium wires.

To etch zone-refined wires we have developed a technique which yields excellent specimens with a high chance of success. This technique employs three different etchants. For convenience, the following solution can be made in advance and stored for indefinitely long periods of time: deposit iridium complex in 4% HF by electrochemically dissolving an iridium wire at 10 V dc and 2–3 A for about 1 hour, until the solution acquires a purplish-bluish color. The initial polishing solution is made of 1 part of the above solution, 1 part distilled water, and 2 parts of concentrated HNO_3. The cell consists of a 1/4 in diameter tungsten loop as counter-electrode, immersed slightly below the electrolyte, as shown in Fig. A.1e. A section 1/16 in in length of 0.005 in diameter iridium wire is inserted into the electrolyte, and a potential ~ 40 V ac is applied across the cell. There should be appreciable noise due to sparking. Raise the voltage if this is not heard. The polishing is generally smooth and relatively fast compared to other methods. After 7 or 8 minutes a slender neck is formed below the interface, as shown in Fig. A.3b. The wire is then raised and the excess wire below the neck etched off. (If this stage is allowed to proceed, etching tends to localize near the meniscus.)

The second step is to continue thinning in saturated $(NH_3)_2CO_3$ solution. This is done simply by replacing the polishing solution and applying 2 V ac for two or three minutes, until the wire is barely visible under 70X magnification. The last stage is to drop off the lower section. This is done with an electrolyte which has been developed for preparing samples for transmission electron microscopy [12]: 1 part conc. H_3PO_4, 1 part conc. HNO_3, and 1 part conc. H_2SO_4. Start out at 2 V ac and gradually decrease to 0.75 V ac as the wire gets thinner. The etching rate of this solution is fairly slow; approximately one hour is needed for the lower section to drop off. A bluish iridium complex tends to form near the anode in this final stage, blocking the view of the iridium wire. This complex does not affect the performance of the emitter, and can usually be removed by blowing air lightly on the solution surface.

The above etching procedures are lengthy.[1] However, the rewards are considerable: if the procedures are followed correctly a good specimen is guaranteed. Also, the sharpness of the tip can be controlled by controlling the size of the portion dropped off. Usually a good helium field ion image can be obtained at voltages ranging from 3 to 25 kV.

20% KCN can also be used for the final drop-off method. This saves considerable time since KCN attacks iridium at a much faster rate than the solution we used. However, KCN tends to leave a rough instead of a finely polished surface. Undercutting of the shank and uneven etching are more frequent and a good specimen is not always achieved. The chance of success is about 30% with KCN.

A.5 Platinum

Platinum is similar to iridium chemically except that it is less resistant to acids. Procedures for etching iridium can be applied to platinum directly. However, we have

---

[1] The time required for the final stage has recently been reduced to ~10 min by applying a continuous ac potential of 0.75 V [13].

discovered a much better and faster method to replace the lengthy procedures used with iridium. The construction of the electrolytic cell is similar to iridium, except that now an iridium counter-electrode is used. The electrolyte is replaced by a solution composed of equal volumes of water and saturated NH$_4$Cl. The surface is highly polished and the etching is smooth and rapid when a potential of 30–40 V ac is applied across the cell. A thin slender neck is, as shown in Fig. A.3c, formed within one or two minutes. The potential is then dropped to 20 V ac and excess wire at the end etched off. The wire is further etched in the same solution at 2 V ac until the thin section becomes barely visible at 70X. Next the wire is taken out and immersed in distilled water to rinse off the ammonium chloride. The final drop-off procedure is the same as with iridium, but since platinum is less resistant than iridium, the time required is only 15 to 30 minutes. Field emitters prepared by this method ordinarily produce excellent helium ion images between 4 to 10 kV.

The mechanical strength of platinum is much less than of tungsten or iridium. In order for the platinum specimen to withstand the stress of the applied field, which corresponds to a negative pressure of 10^5 atmospheres [5], it has to be very smooth and free of mechanical defects. The above procedures yield consistently successful specimens. 20% KCN can also be used for preparing platinum tips. However, the surface finish is far inferior to our method. Due to the roughness of the surface, less than 30% of the field emitters prepared by KCN survive the rigors of field evaporation even when exercising great care."

We should add that iridium etching with three different solutions can end up as quite a long process. For a quick procedure, producing good samples, a dip of the tip into a melt of 3 parts NaNO$_3$ and 7 parts KOH yields good results very quickly.

A.6 Rhodium

Subsequent to Liu's thesis, additional work has been done with rhodium, a material very similar to iridium and platinum. We have found differences in shape for tips etched from single crystal <110> oriented wire and a polycrystalline sample. Both etch rapidly in a solution of 20% KCN and 12.5% NaOH, at a polishing potential of 20 V ac, which is used for creating a neck. For final preparation of the tip, the voltage is lowered to 2 V ac until drop-off occurs. This procedure is good enough to produce really sharp tips which will be imaged in the field ion microscope in a voltage range around ~5 kV. The shape of the tip obtained from polycrystalline and single crystal wire differs, however; the polycrystalline tip is longer. The tips obtained using this method are quite reproducible but it is not clear, at this moment, how mechanically strong they are. We have tried etching rhodium with a solution of molten NaNO$_3$ and NaCl in the ratio of 4 to 1, and this solution is excellent for sharpening the tip in the final stages, but does not work well for creating the initial necking.

References

[1] R. Liu, Chemisorption on perfect surfaces and structural defects, PhD Materials Science, University of Illinois at Urbana-Champaign, 1977.

[2] A. J. Melmed, The art and science and other aspects of making sharp tips, *J. Vac. Sci. Technol. B* **9** (1991) 601–608.

[3] A. J. Nam, A. Teren, T. A. Lusby, A. J. Melmed, Benign making of sharp tips for STM and FIM: Pt, Ir, Au, Pd, and Rh, *J. Vac. Sci. Technol. B* **13** (1995) 1556–1559.

[4] F. Wu, P. Bellon, M. L. Lau, E. J. Lavernia, T. A. Lusby, A. J. Melmed, A new approach to preparing tips for atom probe field ion microscopy from powder material, *Mater. Sci. Eng. A* **327** (2002) 20–23.

[5] E. W. Müller, T. T. Tsong, *Field Ion Microscopy Principles and Applications* (American Elsevier, New York, 1969).

[6] A. J. Melmed, Helium field-ion microscopy of hexagonal close-packed metals, *Surf. Sci.* **8** (1967) 191–205.

[7] W. Kollmar, D. Stark, Zur Feldemission der Metalle mit Hexagonaler Gitterstruktur, *Z. Phys.* **178** (1964) 39–43.

[8] K. M. Bowkett, D. A. Smith, *Field Ion Microscopy* (North-Holland, Amsterdam, 1970) p. 57–61, Appendix 3.

[9] M. A. Fortes, B. Ralph, The occurrence of glissile Shockley loops in field-ion specimens of iridium, *Philos. Mag.* **18** (1968) 787–805.

[10] M. A. Fortes, B. Ralph, The growth of oxide on field-ion specimens of iridium, *Proc. R. Soc. London* **A307** (1968) 431–448.

[11] W. R. Graham, F. Hutchinson, J. J. Nadakavukaren, D. A. Reed, S. W. Schwenterly, Epitaxial deposition of platinum on iridium at low temperatures, *J. Appl. Phys.* **40** (1969) 3931–3936.

[12] C. Cizek, F. Parizek, A. Orlova, J. Tousek, The homogeneization quenching of platinum, *Czech. J. Phys.* **B20** (1970) 56–62.

[13] J. Wrigley, Private Communication.

Index

ab initio, 446, 501
   calculation, 79, 158, 264, 287, 289, 296, 403, 578
   method, 578, 582
   package, 646
   simulation, 476
   technique, 58, 578
   VASP package, 364
   VASP simulation, 326
abstractive
   chemisorption, 169
   dissociation, 170
activated state, 18, 19, 20, 21, 56
activation energy 21
   basic jump, 138
   bridge jump, 268
   cluster diffusion, 542, 559, 577, 581, 586, 589, 593, 609, 617, 618, 619, 635, 637, 644, 652, 665, 669, 671, 678, 687, 689
   compact triangular to linear, 565
   conversion, 138
   corner rounding, 205
   create misfit, 540
   cross-channel jump, 192, 206, 229
   descent, 440, 460, 477, 649
   descent into kink, 441, 448, 473
   detaching corner atom, 428
   difference, 140
   diffusion, 5, 31, 43, 49, 98, 100, 101, 102, 110, 111, 113, 119, 122, 127, 133, 138, 139, 183, 185, 187, 189, 191, 196, 201, 202, 205, 206, 209, 210, 211, 212, 213, 215, 216, 218, 219, 221, 223, 226, 227, 228, 230, 231, 233, 234, 235, 238, 264, 265, 266, 268, 269, 271, 272, 274, 275, 276, 277, 278, 279, 280, 281, 283, 284, 287, 289, 290, 291, 292, 295, 297, 298, 300, 302, 303, 304, 305, 308, 310, 311, 313, 314, 315, 317, 318, 319, 320, 323, 324, 325, 326, 327, 328, 329, 330, 331, 332, 335, 336, 338, 339, 340, 341, 342, 343, 344, 345, 346, 347, 348, 350, 351, 352, 353, 354, 355, 357, 358, 359, 360, 361, 362, 365, 369, 370, 371, 402, 403, 425, 431, 444, 446, 449, 450, 452, 453, 457, 458, 462, 467, 472, 477, 498, 563, 575, 608, 634, 642, 650, 668, 675, 690

activation energy (cont.)
   diffusion with dislocation, 542
   dimer diffusion, 561, 563, 571, 573, 578, 580, 581, 582, 595, 598, 599, 602, 608, 616, 632, 633, 634, 635, 639, 641, 643, 648
   dimer exchange, 563
   double jump, 142, 232, 236
   double-exchange, 326
   effective, 319, 606, 690
   emergence, 199
   escape, 431, 456, 458, 467, 468
   exchange, 89, 90, 185, 192, 197, 206, 218, 226, 228, 236, 262, 264, 279, 286, 287, 294, 309, 323, 325, 327, 331, 334, 335, 363, 364, 367, 368, 371, 447, 455
   exchange descent, 448, 465, 476, 477
   for 18 atom cluster, 683
   for 19 atom cluster, 684
   for 61 atom cluster, 678
   for 7 atom cluster, 683
   gliding, 540, 596, 642
   heptamer diffusion, 630, 643
   hetero-dimer diffusion, 613
   hexamer diffusion, 640
   hopping, 197, 262, 263, 265, 278, 279, 280, 287, 288, 289, 290, 300, 305, 313, 317, 320, 324, 325, 354, 357, 358, 363, 364, 366, 367, 368, 369, 427
   hopping descent, 443, 447, 449, 454, 455, 464, 476
   in-channel ascent, 478
   in-channel dimer diffusion, 617, 632
   in-channel jump, 184, 185, 189, 192, 193, 194, 202, 204, 205, 210, 228, 232, 234, 456
   in-channel motion, 88, 91, 186, 187, 190, 197, 199, 201, 204, 205, 206, 209, 215, 216, 218, 225, 227, 230, 233, 235, 238
   incorporation, 432, 475, 580
   local motion, 564
   long jumps, 138, 139
   long-range diffusion, 566
   motion along step, 671
   motion close to island, 498
   motion cross-channel, 88, 94, 184, 187, 189, 194, 202, 224, 227

Index

activation energy (cont.)
  motion fcc-hcp, 320
  motion hcp-fcc, 350
  motion in-trough, 237
  motion on sidewalls, 237
  motion out of fcc, 350, 352
  motion out of hcp, 352
  motion perpendicular to step, 342
  motion two-dimansional, 93
  motion with misfit, 678
  non-diagonal exchange, 272
  pentamers diffusion, 595
  rebound jump, 145
  short-range movement, 620
  single jump, 127, 141, 232, 236, 278, 335
  spreading, 223
  tetramer diffusion, 623
  three-atom exchange, 263
  trimer diffusion, 602, 607, 622, 640, 642
  zigzag motion, 564
adatom catalyzed double exchange, 109
adatom density, 121
adatom island, 210, 442, 451, 671, 689
additional step-edge barrier, 431, 435, 439, 441, 442, 443, 448, 449, 450, 451, 452, 453, 455, 456, 458, 465, 467, 468, 476, 477, 478, 481, 485, 487
adsorbate-adsorbate mediated energy, 720
ad-trimer state, 108
AES, 323, 334, 368, 712
AFM. *See* Atomic Force Microscopy
AFW
  approximation, 289, 309
  parameters, 184, 190, 194, 195, 210, 235, 261, 278, 324, 357, 363, 366
Ag(001). *See* Ag(100)
Ag(100), 104, 112, 150, 153, 155, 162, 169, 170, 322, 323, 324, 325, 326, 327, 328, 454, 455, 543, 547, 599, 600, 665, 667, 668, 671, 673, 675
  unstrained, 112
Ag(110), 49, 98, 106, 119, 121, 207, 209, 210, 211, 455, 456, 601
  (2 × 1), 601
  (2 × 2), 210
Ag(111), 33, 48, 49, 51, 108, 110, 111, 162, 316, 317, 320, 321, 322, 446, 450, 451, 452, 453, 454, 455, 485, 575, 584, 598, 599, 601, 676, 677, 688, 689, 717, 721, 722, 723, 724, 726
  strained, 717
  unstrained, 51, 717
Al(100), 52, 103, 105, 261, 264, 265, 438, 543, 561, 562, 563
Al(110), 93, 184, 185, 438, 496, 729
Al(111), 52, 75, 164, 166, 168, 169, 265, 266, 267, 268, 269, 436, 438, 536, 563, 564, 566, 567, 679, 692, 718
argon ion scattering, 43
Arrhenius
  curve, 135

Arrhenius (cont.)
  plot, 100, 101, 105, 107, 110, 119, 127, 130, 133, 135, 138, 139, 142, 186, 191, 193, 195, 196, 199, 202, 205, 209, 219, 221, 227, 228, 230, 232, 233, 235, 236, 268, 275, 289, 290, 294, 297, 298, 299, 300, 302, 305, 308, 310, 311, 313, 318, 325, 329, 330, 332, 346, 350, 353, 358, 359, 360, 361, 362, 370, 371, 458, 465, 467, 480, 552, 559, 567, 575, 584, 589, 603, 604, 607, 608, 609, 621, 625, 643, 644, 645, 669, 672, 678, 690
Arrhenius dependence. *See* Arrhenius curve
ARTwork computer program, 317
ascending
  lattice step, 456, 498, 506
  step edge, 505
ascent mechanism
  atom exchange, 504
  in-channel exchange, 649
asymmetric
  bridge site, 168
  distribution, 15
  motion, 13, 14, 15, 211
atom edge running, 596, 598, 664, 672, 675, 678, 691
atom pairs, 164, 167, 168, 169, 261, 724
Atom Probe. *See* Atom Probe microscopy
Atom Probe microscopy, 27, 28, 29, 90, 96, 227
Atomic Force Microscopy, 34
atomic resolution, 34, 352, 469, 735
attempt frequency, 4, 87, 281, 290, 320, 321
  angular, 321
attractive ring, 502
Au(100), 104, 112, 153, 367, 368, 425, 647, 648
  (26 × 5), 425
  hex, 365, 366
  reconstructed, 647
Au(110), 107, 235, 236, 237, 478, 649, 651, 678
  (1 × 2), 127, 158, 162, 232, 235, 236, 237, 649, 650, 692
Au(111), 368, 369, 370, 478, 590, 649, 676
  (23 × √3), 114
  (22 × √3), 370
  reconstructed, 370, 372
  strained (23 × √3), 370
  unreconstructed, 114, 370, 372
autocorrelation function, 30, 291

bakeout, 24, 35
ballistic
  energy, 109
  transition, 121
barrier
  diffusion, 458, 539, 542, 561, 567, 569, 571
  double jump, 288, 340
  height, 20, 112, 113, 194, 197, 200, 205, 233, 235, 317, 324, 339, 364, 370, 372, 653, 654, 687
  peak, 5
  reflective, 41, 456, 458, 480, 482
  single jump, 200, 236

Index

barrier (cont.)
  step-edge, 49, 430, 431, 432, 433, 435, 439, 440, 441, 442, 443, 448, 450, 451, 452, 453, 454, 455, 456, 457, 458, 465, 467, 469, 470, 477, 478, 479, 480, 481, 482, 483, 484, 485, 495
  step-facet, 494
Be(0001), 426
Bi(111), 372
binding energy, 41, 49, 57, 78, 111, 113, 114, 115, 261, 351, 369, 423, 502, 617, 704, 712, 716
binomial theorem, 7, 8
Boltzmann term, 20
Boltzmann–Matano method, 201
bond breaking model, 669
Born–Mayer core repulsion, 57
bridge gliding, 633
bridge site energy, 304
Brownian particle, 118
bulk electronic states, 723

$c(2 \times 2)$ net, 103
cannon-ball mechanism, 168
cannon-ball-like path, 168
cannon-like trajectory, 166
canonical partition function, 697
capture number, 48, 53, 269
cartwheel
  glide, 542, 686
  shuffle, 542, 686
CASTEP numerical code, 274
CEM. *See* corrective effective medium
center of mass, 520, 521, 524, 525, 526, 529, 536, 547, 556, 561, 569, 573, 581, 582, 587, 590, 591, 594, 596, 602, 604, 606, 607, 625, 634, 635, 643, 690
channel plate, 25, 35, 36
chemical potential, 18
chemical surface reactions, 696
chemisorption, 164
classical partition function, 19
cluster
  atom-by-atom jump, 583, 599, 600
  atom-by-atom motion, 583
  compact, 536, 568, 599, 610, 630, 681, 682, 683, 685
  hexagonal, 536, 538, 542, 573, 640, 682, 683, 685, 690, 691, 692, 693
  large, 49, 517, 547, 550, 552, 553, 554, 664, 665, 668, 669, 671, 675, 676, 678, 679, 683, 692, 693
  linear form, 590, 611, 653, 712
  non-compact, 682
  reshaping, 640, 673
  rolling, 692
  rotation, 686, 692
  single jump, 654
  stable, 49, 576, 676
  translation, 686
cluster diffusivity prefactor, 553, 559, 568, 590, 617, 630, 672, 677, 683, 685, 686, 687, 692, 693

cluster frequency prefactor, 575, 578, 581, 589
cluster size, 84, 86, 441, 445, 473, 483, 499, 536, 547, 553, 569, 571, 578, 586, 596, 619, 637, 640, 644, 664, 670, 673, 687, 689
  critical, 49, 188, 304, 309, 365, 370, 452
cluster–substrate coupling, 685
coalescence, 517
coarsening, 667
coefficient
  diffusion. *See* diffusivity
  friction, 119
  scaling, 677
  self-diffusion, 198
  sticking, 443
cohesion, 27, 97
  approximation, 272, 280, 288, 298, 310, 319, 326, 328, 331, 335, 336, 339, 340, 341, 348, 351, 354, 358, 364, 367, 370
  energy, 303
cohesive energy, 57, 263, 272, 702
composition, 24, 27, 28, 29
compressibilities, 56
concentration, 2, 4, 31, 48, 49, 53, 54, 158, 169, 198, 201, 209, 223, 224, 291, 314, 327, 332, 333, 334, 336, 338, 342, 346, 347, 453
  fluctuations, 30
  gradient, 1, 4, 183
  profile, 4, 53
concerted
  gliding, 587, 669, 687
  hopping, 601
  long-range motion, 566
  motion, 478, 517, 536, 538, 562, 574, 584, 585, 586, 587, 588, 589, 592, 595, 599, 601, 633, 640, 646, 653, 692
  sliding, 564, 583, 587
  translation, 542, 583, 633, 640, 642
  two-atom jumps, 599
configuration
  bent, 711, 714, 728
  linear, 728
  linear trimer, 565, 566, 569, 581, 584, 598, 600, 602, 623, 636, 639, 648
  non-compact triangular trimer, 565
  straight, 603
  triangular, 714
  triangular trimer, 565, 569, 576, 598, 601, 608, 623, 633, 635, 639, 640, 649
core breakup, 536, 550, 672
core-core pairwise repulsion, 57
corner
  breaking, 535
  rounding, 205, 207, 550
corrected effective medium, 195, 206, 210, 286, 309, 325, 357
  method, 57
  theory, 118, 155, 276, 278, 297, 305, 313, 441, 444, 450, 453, 473, 475, 478
correlated motion, 107, 341, 404, 567, 578, 593, 654
coupling electronic states, 696

covalent bonding, 104, 262
cross-channel
  ascent, 456, 478
  barrier, 207
  climb, 496
  cluster, 528
  descent, 456
  diffusion, 185, 187
  dimer, 525, 528, 533, 603, 607, 608, 630
  direction, 439, 449
  event, 90, 91, 210
  form, 652
  hopping, 647
  mechanism, 230
  motion, 89, 90, 91, 94, 186, 189, 194, 195, 204, 206, 210, 215, 216, 223, 230, 233, 234, 235, 236, 238, 455, 532, 632
  movement, 90, 93, 94, 96, 98, 184, 187, 190, 194, 202, 206, 207, 211, 216, 230, 233, 237, 532, 630
  process, 184
  transition, 90, 94, 97, 98, 185, 209, 215, 496, 528, 532, 630
  trimer, 527, 528, 607
crossing
  inner-corner, 496
  outer-corner, 496
crowdion, 109, 110
crystal growth, 81, 156, 163, 431, 435, 442, 451, 467, 506, 599, 623, 696
Cu(100), 105, 107, 109, 119, 150, 155, 162, 272, 280, 283, 284, 285, 286, 287, 288, 290, 291, 293, 294, 295, 402, 429, 443, 447, 448, 455, 495, 496, 517, 543, 550, 575, 577, 578, 579, 582, 667, 670, 673, 675, 689, 728
Cu(110), 167, 194, 195, 197, 198, 200, 448, 449, 593, 692
  (2×1), 592
Cu(111), 47, 48, 52, 78, 79, 80, 86, 109, 115, 121, 145, 155, 295, 296, 297, 298, 299, 300, 301, 402, 442, 443, 444, 445, 446, 487, 501, 502, 504, 575, 582, 584, 585, 586, 589, 590, 591, 673, 676, 688, 689, 690, 691, 712, 715, 718, 719, 720, 721, 722, 728
Cu(311), 194, 195
Cu(331), 194, 195
CY-EAM, 358
CY-XEAM1, 358
CY-XEAM2, 358

DACAPO code, 79, 296
dangling bonds, 169
de Broglie wavelength, 19
decacyclene, 145
  hexa-tert-butyl, 145
decay rate, 451
defect, 46, 99, 155
  concentration, 338
degeneracy, 683
density of islands, 49
density of states, 32, 57, 197, 504, 720, 722
density-functional

density-functional (cont.)
  calculation, 104, 112, 129, 210, 232, 263, 266, 268, 280, 309, 319, 334, 344, 358, 369, 426, 428, 449, 456, 473, 496, 504, 563, 580, 595, 600, 634, 639, 640, 679, 717, 718, 726, 728
  estimates, 111, 234, 276, 280, 298, 361, 478
  method, 288, 301, 307, 325, 357
  theory, 56, 79, 112, 166, 185, 262, 264, 265, 268, 274, 289, 294, 298, 313, 319, 354, 357, 360, 364, 367, 368, 370, 436, 454, 504, 563, 564, 575, 679, 719, 722, 728, 729
descent
  multilayer step, 438, 487
  prefactor, 431, 432, 465
descent barrier
  atom exchange, 443, 444, 445, 446, 447, 448, 449, 453, 454, 455, 456, 463, 464, 475, 476, 477, 478
  kink exchange, 441, 463
  vacancy exchange, 465
descent mechanism
  atom exchange, 194, 434, 435, 436, 438, 439, 440, 441, 442, 443, 444, 445, 446, 447, 448, 449, 450, 453, 454, 455, 458, 460, 465, 469, 472, 473, 475, 476, 477, 478, 487, 496, 504, 505
  corner exchange, 447, 470
  hopping over the edge, 194
  in-channel exchange, 649
  kink exchange, 441, 446, 447, 448, 449, 456, 470, 473
  vacancy exchange, 465
desorption energy, 223
DFT. See density-functional theory
diagonal motion, 584, 673
diffusion
  ad-dimer, 108, 364
  cross-channel, 29, 89, 93, 98, 187, 190, 210, 216, 227, 230, 440
  impact cascade, 162
  in-channel, 91, 94, 98, 193, 195, 206, 207, 210, 218
  inter-channel, 89
  leapfrog, 529
  long-range, 563, 564, 566, 569, 577, 582, 587, 594, 620, 623, 626, 644
  non-local, 565
  one-dimensional, 2, 44, 122, 123, 127, 225, 610
  piecewise, 546
  sequential atom jumps, 517
  subsurface, 327
  terrace limited, 517, 536, 667, 671
  thermal, 146, 499, 575
  transient, 146, 149, 151, 153, 155, 157, 158, 161, 162, 163, 164, 171
  two-dimensional, 38, 44, 45, 91, 94, 117, 131, 184, 225
diffusion barrier
  atom exchange, 115, 191, 192, 194, 195, 197, 199, 200, 204, 205, 206, 210, 211, 234, 235, 236, 237, 238, 261, 262, 263, 264, 265, 278, 279, 280, 282, 285, 286, 287, 288, 289, 294, 295, 300, 307, 309, 323, 324, 325, 326, 327, 328, 331, 334, 335, 344, 354, 358, 364, 366, 368, 370, 449

Index

diffusion barrier (cont.)
  cross-channel, 630
  dimer exchange, 563, 571, 578, 581, 595, 598, 600, 601, 632, 634, 637, 639, 648, 649, 651
  hopping, 191, 195, 200, 206, 262, 263, 265, 272, 273, 276, 279, 280, 282, 285, 286, 287, 288
  horizontal jump, 330
  in-channel, 191, 195, 200, 207, 210, 230
  tetramer exchange, 571
  trimer exchange, 573, 593
  vertical jump, 330
diffusion mechanism
  180° dimer exchange, 542, 543, 637, 648
  270° dimer exchange, 543, 637
  360° dimer exchange, 543
  along step exchange, 679, 692
  atom exchange, 29, 51, 89, 90, 94, 96, 98, 99, 100, 101, 102, 103, 104, 105, 107, 108, 109, 112, 115, 117, 118, 162, 184, 185, 186, 189, 190, 191, 192, 193, 194, 195, 197, 198, 199, 200, 202, 205, 206, 207, 209, 211, 216, 218, 226, 227, 228, 229, 232, 233, 234, 235, 261, 262, 263, 264, 265, 268, 272, 276, 277, 278, 279, 280, 281, 283, 285, 286, 287, 288, 289, 290, 293, 297, 298, 299, 300, 301, 305, 307, 308, 309, 310, 311, 312, 314, 320, 321, 323, 324, 325, 326, 327, 328, 331, 334, 335, 341, 344, 353, 354, 358, 360, 361, 362, 363, 364, 365, 366, 367, 368, 370, 371, 372, 402, 403, 423, 425, 427, 428, 433, 456, 460, 478, 528, 531, 570, 571, 622, 634, 635
  ballistic exchange, 162
  between islands exchange, 49, 681
  cluster exchange, 567, 570, 571, 576, 582, 593, 596, 600, 601, 653, 655, 669
  concerted dimer exchange, 545, 630
  concerted exchange, 272, 543
  concerted hopping, 195
  cooperative dimer exchange, 545
  cooperative hopping, 545
  correlated evaporation-condensation, 517, 536, 553, 669, 670, 673
  cross-channel hopping, 190, 191, 195, 197, 229
  detachment-attachment, 517, 549
  diagonal exchange, 272
  dimer exchange, 531, 532, 533, 534, 535, 542, 543, 544, 545, 561, 563, 564, 571, 573, 575, 578, 580, 581, 589, 593, 596, 598, 600, 622, 630, 632, 633, 634, 637, 639, 643, 647, 648, 652
  dimer shearing, 517, 539, 550, 563, 566, 569, 570, 571, 577, 578, 579, 581, 588, 599, 600, 617, 623, 654, 669, 675, 687
  double exchange, 107, 108, 288, 326
  evaporation-condensation, 517, 536, 552, 664, 666, 667, 668, 671, 672, 673, 675, 676, 677, 678, 681, 685, 692, 693
  gliding, 517, 586
  hetero-dimer exchange, 620, 621
  hopping, 103, 104, 109, 112, 115, 117, 121, 191, 192, 193, 194, 195, 197, 200, 202, 205, 211, 229, 235, 261, 262, 264, 265, 272, 275, 277, 278, 279, 280, 281, 282, 285, 286

diffusion mechanism (cont.)
  hopping rotation, 545
  in-channel cluster exchange, 574, 593, 601, 646
  in-channel hopping, 185, 186, 194, 197, 200, 202, 205, 210, 228, 232
  internal-row shearing, 579
  island exchange, 536, 548
  jump-exchange, 106
  jump-exchange-jump, 107
  kink exchange, 679
  leapfrog, 517, 574, 601, 637, 646, 649, 650, 654
  long-range exchange, 265, 279, 325, 358, 360
  multi-particle exchange, 263
  multiple exchange, 105, 107, 109, 110, 493
  non-diagonal exchange, 272
  perimeter, 552, 668
  quadruple exchange, 107, 288
  rotation dimer exchange, 543, 545, 563, 580, 600, 637, 639, 647
  shearing, 576, 577, 652, 653
  sliding. *See* diffusion mechanism gliding
  sub-unit shearing, 687
  tetramer exchange, 575, 582, 593, 598, 600, 601, 639, 651
  three-atom exchange, 263
  trimer exchange, 571, 575, 581, 589, 593, 596, 598, 600, 601, 635, 639, 643, 646, 647, 648, 649, 651, 652
  trimer shearing, 517, 539, 550, 563, 570, 571, 577, 578, 600, 654, 675
  triple exchange, 107, 288
  two-atom dimer exchange, 639
  vacancy-mediated exchange, 301
diffusivity, 1, 2, 4, 5, 7, 21, 22, 30, 41, 42, 46, 48, 49, 51, 53, 54, 86, 590, 596
  cluster, 672
  prefactor, 5, 43, 49, 58, 86, 88, 90, 91, 94, 98, 100, 101, 107, 122, 135, 139, 183, 184, 185, 186, 189, 190, 191, 192, 193, 194, 195, 196, 197, 199, 201, 202, 203, 204, 205, 206, 209, 210, 211, 212, 213, 215, 216, 218, 219, 220, 221, 223, 224, 225, 226, 227, 228, 229, 230, 231, 232, 233, 234, 235, 236, 237, 238, 261, 262, 263, 264, 265, 266, 268, 269, 272, 274, 275, 276, 277, 278, 279, 280, 281, 283, 285, 287, 288, 289, 296, 297, 298, 299, 300, 301, 302, 303, 304, 305, 306, 307, 308, 309, 310, 312, 313, 314, 315, 317, 318, 320, 324, 325, 326, 328, 329, 330, 331, 332, 333, 335, 336, 338, 339, 340, 342, 344, 345, 346, 347, 348, 350, 352, 353, 354, 357, 358, 360, 361, 362, 363, 364, 365, 366, 369, 370, 371, 402, 403, 404, 425, 428, 432
dimer
  fcc-fcc, 79, 575, 587, 641, 645
  fcc-hcp, 79, 575, 587
  gliding, 578
  hcp-hcp, 575, 645
  mixed fcc-hcp, 564
  rotation, 517, 531, 534, 610, 626, 633, 645
  single jump, 602, 607, 608
  translation, 531, 535, 564, 585, 610, 621, 626, 645

dimer configuration
  staggered, 603, 604, 606, 617
  straight, 604, 606, 617
dimer diffusivity
  prefactor, 602, 603, 604, 606, 607, 608, 616, 617, 620, 621, 627, 630, 646
dimer rotation, 535, 562, 564, 599
directional movement, 37
discommensuration line, 370, 372
dislocation, 28, 33, 506, 507, 509, 541, 542
  core, 507, 508
  mechanism, 304, 678
  misfit, 678
  movement, 517, 540, 573, 652, 678
  nucleation, 539, 541, 692
displacement, 6, 9, 42, 43, 44, 45, 100, 103, 115, 122, 123, 133, 144, 156, 269, 304, 308, 321, 335, 372, 508, 524, 527, 531, 565, 606, 607, 611, 622, 627, 692
  average, 13, 14, 165
  ballistic, 162
  Brownian, 676
  center of mass, 520, 526, 531
  cluster, 637, 682
  concerted, 517, 570, 630, 686
  fluctuation, 7, 9, 21, 41, 42, 43, 121, 522, 523, 524, 525, 526, 527, 528, 530, 531, 604, 607, 608
  horizontal, 133
  individual, 4
  length, 6
  mean, 521, 522
  mean-square, 6, 15, 16, 41, 94, 100, 122, 223, 300, 305, 358, 360, 361, 371, 428, 507, 530, 531, 549, 567, 594, 607, 617, 625, 676, 677, 683
  nearest-neighbor sites, 44, 682
  overall, 5, 46
  peripheral, 517, 535
  random, 6
  rate, 637
  root-mean-square, 704, 705
  sequential, 568
  square, 6
  subsequent, 622
  third moment, 13
  variance, 522
  vector, 5
dissipation, 118, 119
dissipation energy, 159, 162
dissociation, 162, 164, 165, 166, 167, 168, 169, 170, 171, 437, 533, 606, 607, 608, 610, 615, 617, 618, 619, 627, 650, 702
  barrier, 650
  cluster, 579, 685
  dimer, 102, 649
  dumbbell, 534
  energy, 170, 569, 571, 606, 641
  island, 51
  lifetime, 711
  rate, 48

dissociation (cont.)
  thermal, 51
dissociative adsorption, 169
dissolution, 696
distribution
  binomial, 10, 11, 15, 16
  cluster displacements, 684
  displacements, 7, 14, 43, 44, 45, 46, 94, 121, 122, 123, 125, 127, 133, 139, 141, 144, 232, 525, 531, 611, 622, 630
  electron density, 56
  Gaussian, 11, 15, 16
  non-interacting atoms, 708
  one-dimensional, 45
  pair, 171, 701, 702, 708
  random, 75, 148, 162, 163, 164, 166
  separations, 699, 704, 706, 721
  terrace length, 323
double jump
  frequency prefactor, 142
  prefactor, 107
double rotation, 535
double shear glide, 640
double-hop, 107
double-shearing, 640
downward funneling, 51, 151, 155, 162
d-shell, 57, 402, 717
dumbbell, 89, 94, 97, 533, 534
dynamical misfit, 685
DYNAMO code, 121

EAM. *See* embedded atom method
edge
  breakup, 548
  diffusion, 596, 670, 671
effective barrier, 587
  cluster movement, 690
  descent, 478
  diffusion, 291
  dimer movement, 586
  metastable walk, 236
  tetramer movement, 588
  trimer movement, 587
effective hopping integral, 57
effective medium
  approximation, 165, 639
  theory, 57, 111, 195, 206, 235, 276, 279, 287, 309, 313, 315, 318, 325, 357, 359, 440, 443, 444, 447, 448, 449, 453, 454, 455, 475, 478, 492, 550, 561, 564, 579, 637, 670, 671, 673, 675, 686, 689
Ehrlich–Schwoebel barrier. *See* additional step-edge barrier
Einstein relation, 7, 21, 121
elastic
  constant, 56, 57
  effects, 696, 727
  indirect effects, 719
  lattice deformation, 722
  theory, 509

Index

electron-charge distribution, 505
electro-polishing, 34
electrostatic attraction, 436
embedded atom method, 56, 58, 105, 109, 159, 190, 192, 197, 207, 228, 261, 265, 288, 297, 304, 309, 312, 317, 318, 325, 354, 358, 364, 440, 445, 446, 447, 448, 449, 454, 472, 473, 478, 517, 538, 541, 566, 578, 582, 600, 617, 639, 678, 686, 692
  approximation, 57, 578
  function, 57
  simulation, 328
emission current, 30, 34
  density, 29
empty zone, 41, 84, 85, 86, 157, 158, 467, 498, 499, 502
EMT. *See* effective medium theory
enthalpy of sublimation, 403
enthalpy of vaporization, 403, 652
entropy, 5, 21, 75, 607
  activation, 87, 122
equilibrium fluctuations, 553
evaporation-condensation, 672

FBD
  approximation, 275, 279
  parameters, 275, 288
Fe(100), 269, 271, 272, 439, 440
Fe(110), 271, 272, 273, 439, 440, 567
Fe(111), 272
FEM. *See* field electron microscopy
  resolution, 30
Fermi
  energy, 720
  level phase shift, 720
  wave number, 715
  wave vector, 720
  wavelength, 720
Fick's law
  first, 1
  second, 2
field desorption, 102
field electron microscopy, 215, 217, 335, 338, 343
field emission current, 28, 342
  fluctuation, 215, 216, 223, 225, 332, 333, 345
field emission microscopy, 53
field emission of electrons, 29
field evaporation, 27, 28, 36, 37, 39, 40, 41, 68, 98, 147, 187, 281, 353, 424, 425, 457, 507, 607, 630, 741
field ion micrograph, 147, 603
field ion microscope, 24, 29, 32, 34, 35, 36, 37, 41, 72, 86, 96, 123, 146, 148, 149, 152, 184, 232, 277, 281, 307, 336, 338, 356, 361, 432, 507, 623, 634, 741
field ion microscopy, ix, 27, 28, 33, 34, 43, 46, 48, 53, 64, 90, 98, 102, 123, 186, 187, 188, 192, 194, 202, 211, 212, 215, 217, 218, 219, 220, 221, 223, 225, 226, 227, 228, 230, 274, 277, 304, 305, 307, 328, 329, 335, 336, 338, 339, 340, 342, 345, 347, 348, 352, 354, 355, 402, 430, 434, 435, 456, 457, 458, 460, 467, 469, 504, 525, 527,

field ion microscopy (cont.)
  537, 556, 593, 601, 602, 608, 610, 613, 614, 618, 634, 654, 664, 681, 698, 708, 712, 713, 722, 735
FIM. *See* field ion microscopy
  image, 39
  magnification, 26, 39, 41, 698
  resolution, 26, 64
Fireball, 58
Fokker–Planck formalism, 331
fourth moment, 13
Fowler–Nordheim relation, 29
free energy, 5, 81, 432, 433, 603, 604, 623, 676, 683, 706, 708, 710, 711, 712, 713, 714, 715, 724, 726
  average, 683
  change, 5, 21, 683
  clusters, 715
  Helmholtz, 697
  interaction, 698, 705, 708, 719
  pair, 714
Frenkel–Kontorova model, 541
frequency
  prefactor, 20, 21, 110, 111, 183, 196, 197, 207, 235, 266, 268, 269, 271, 278, 279, 281, 282, 284, 286, 288, 291, 293, 296, 297, 305, 308, 309, 311, 312, 314, 317, 318, 319, 320, 321, 322, 323, 324, 325, 327, 350, 352, 356, 359, 364, 365, 366, 368, 369, 370, 456, 458
frequency factor
  long jumps, 138, 139
  single jumps, 127
frequency of occurrence, 683
frictional force, 127
Friedel oscillation, 698, 720
full-potential linear muffin-tin-orbital method, 318, 325, 351, 354, 366, 369

generalized gradient approximations. *See* GGA approximation
getter, 35, 36
GGA
  approximation, 263, 264, 274, 276, 288, 289, 298, 301, 325, 357, 367, 454, 496, 564
ghost particle simulation, 116
gliding, 572, 573, 596, 598, 601, 630, 642, 648, 651, 652, 676, 678, 685, 686, 692
  cluster row, 540
  concerted row, 598
glue model, 57

half-chain process, 493
harmonic approximation, 320, 325
head-on-collision, 156
heat of condensation, 150
heat of sublimation, 56
helium atom scattering, 53, 54, 55, 98, 158
  high resolution, 290
  quasielastic, 194, 360
heptamer diffusivity prefactor, 637, 643

Index

heptamers gliding, 637
hetero-dimer, 530, 558, 613, 620
  rotation, 612
hexagonal
  array, 723
  core, 682
hollow
  four fold, 72
  sites, 446, 561
  three fold, 72
hopping
  between fcc sites, 276
  in-channel, 226
  short bridge, 272
hopping rate
  fcc-to-hcp, 296
  hcp-to-fcc, 79, 296
horizontal
  dimer, 549
  intermediate, 530, 531
  states, 549
  transition, 131, 132, 340, 530
"hot" atoms, 166
HRLEED. See LEED high resolution

in-channel
  ascent, 478
  barrier, 224, 226, 234
  climb, 496
  cluster, 575
  diffusion, 210, 440
  dimer, 525, 532, 593, 607, 608, 630
  direction, 441
  hopping, 197, 456, 575, 593, 596, 598, 632, 647
  motion, 89, 187, 190, 191, 194, 195, 197, 202, 205, 206, 211, 218, 223, 227, 230, 233, 234, 235, 236, 237, 455, 456
  movement, 91, 187, 193, 197, 211, 216, 226, 234, 236, 575, 630
  tetramer, 528, 593
  transition, 185, 229
  trimer, 528, 529, 575, 607
incorporation, 433, 434, 435, 436, 439, 441, 443, 444, 445, 450, 455, 456, 458, 460, 461, 462, 463, 464, 472, 473, 475, 479, 480, 500, 502, 504, 506, 509
  ascending step, 504
  barrier, 461, 462, 476, 487, 505
  descending step, 431
  energetics, 436
  lifetime, 481, 483
interactions
  adatom–adatom, 51, 267
  adatom–step, 505
  adsorbate–adsorbate mediated, 720
  AFW, 233
  attractive, 504, 698, 700, 701, 710, 711

interactions (cont.)
  bulk, 722
  corrected effective medium, 325
  cross-channel, 608
  dipolar, 696
  direct, 696
  effective medium theory, 234, 317, 447
  elastic, 436, 717, 726
  electronic, 111, 696, 717, 726
  embedded atom method, 156, 189, 193, 194, 195, 205, 235, 238, 288, 290, 298, 299, 317, 320, 325, 358, 360, 428, 441, 500, 543, 570, 571, 575, 585, 596, 600, 634, 639, 648
  energy, 721
  FBD, 190
  indirect, 696
  Lennard–Jones, 117, 192, 204, 262, 357, 363, 547, 595
  long-range, 33, 48, 49, 52, 188, 205, 266, 300, 321, 502, 504, 593, 611, 653, 699, 702, 704, 705, 712, 719, 722, 723
  long-range magnetic, 724
  MD/MC-CEM, 293, 325, 327, 364
  Morse, 363
  pairwise, 189, 190, 703, 704, 727
  quarto, 704
  repulsive, 51, 425, 446, 504, 701, 704, 705, 717, 721
  RGL, 109, 193, 195, 197, 200, 206, 210, 211, 228, 229, 234, 280, 282, 295, 300, 301, 320, 322, 326, 358, 360, 361, 368, 372, 449, 455, 462, 476, 544, 580, 590, 600, 630, 645, 648, 649
  step-adatom, 51
  step–step, 446
  substrate–mediated, 696
  tip-sample, 47
  trio, 704, 722, 727, 728, 729, 730
  van der Waals, 696
inter-cell
  concerted movement, 566
  concerted sliding, 564
  jump, 626
  movement, 563, 582, 585, 599, 626, 631, 632, 641, 644, 645, 646
  rotation, 631
inter-facets diffusion
  atom exchange, 492, 493, 494
  chain exchange, 493
interior barrier, 440, 480, 481, 482, 483, 484, 485
interlayer
  atom exchange, 478
  diffusion, 441, 444, 448, 451
  jump, 478
  mass transport, 443, 448
  motion, 447, 454, 478, 496
  movement, 49, 442, 443, 445, 449, 450, 476, 478, 487
  process, 445
  transport, 430, 442, 443, 476

Index

intermediate state, 262, 549
internal dimer rotation, 517, 579
internal three-atom shearing, 562
intra-cell
  concerted movement, 566
  concerted rotation, 564
  concerted sliding, 564
  movement, 563, 564, 582, 631, 641, 645, 646
  rotation, 564, 631
  translation, 631
ionization potential, 27
Ir(100), 41, 100, 102, 103, 104, 353, 354, 458, 460, 619, 621, 622, 632, 633, 634, 653
Ir(110), 29, 90, 97, 195, 227, 228, 229, 634
Ir(111), 25, 38, 41, 72, 74, 86, 145, 152, 157, 272, 348, 350, 351, 352, 402, 434, 435, 436, 458, 460, 461, 462, 465, 467, 476, 478, 480, 487, 498, 499, 502, 542, 565, 567, 573, 577, 586, 589, 619, 627, 628, 630, 631, 633, 634, 640, 641, 652, 654, 681, 683, 686, 687, 691, 726
Ir(311), 227, 228
Ir(331), 228
island
  average separation, 49
  coalescence, 121, 677
  decay rate, 667
  density, 49, 51, 52, 53, 121, 188, 205, 266, 267, 271, 278, 295, 302, 312, 316, 317, 320, 322, 355, 359, 365, 370, 452, 719
  dissociation, 51
  fluctuation, 49, 51, 429, 439
  mean separation, 323
  recombination, 51
  rotation, 207
  separation, 271, 284

jump
  basic, 137, 142
  collective, 572
  coordinated, 299
  correlated, 121, 199, 272, 314, 319
  cross-channel, 90, 94, 185, 192, 193, 195, 197, 204, 205, 206, 210, 211, 227, 229, 230, 234, 236, 237, 450, 598, 651
  double, 12, 58, 107, 118, 119, 122, 123, 124, 127, 129, 131, 141, 142, 145, 187, 196, 210, 221, 223, 232, 236, 330, 501, 684
  fcc-hcp-fcc, 299
  horizontal, 140, 141, 330
  in-channel, 185, 191, 193, 211, 233, 234, 236, 598, 601, 633, 651
  inter-channel, 230
  leapfrog, 593, 646
  long, 117, 118, 119, 121, 122, 123, 124, 125, 126, 127, 128, 129, 130, 131, 133, 135, 138, 139, 145, 210, 212, 215, 218, 236, 269, 319, 336, 339, 404, 630, 682, 684, 685, 686, 687, 692

jump (cont.)
  multiple, 200
  nearest-neighbor. *See* single jump
  random, 64, 89, 147
  rebound, 58, 121, 145, 215, 299
  recrossing. *See* rebound jump
  single, 14, 16, 21, 22, 99, 107, 118, 121, 122, 123, 125, 126, 127, 129, 133, 135, 137, 138, 139, 141, 142, 145, 187, 199, 200, 209, 210, 215, 219, 269, 277, 330, 336, 339, 361, 626
  triangular trimer, 589
  triple, 126, 141, 220, 684
  uncorrelated, 668
  vertical, 131, 133, 140, 141, 330, 336
jump length, 1, 4, 5, 6, 87, 274, 685
  characteristic, 121
  mean, 116
  mean-square, 6, 21, 122, 183
  nearest-neighbor, 7, 12, 87, 122
jump rate, 4, 9, 13, 17, 20, 21, 22, 41, 42, 45, 123, 135, 139, 144, 196, 232, 311, 356, 504, 522, 524, 525, 526, 530, 550, 575, 584, 604, 607, 608, 721
transients, 44

kinetic Monte Carlo simulation, 207, 210, 267, 289, 300, 370, 581, 585, 679
kink
  breaking, 672
  escape, 550
  motion, 540, 678
KMC simulations. *See* kinetic Monte Carlo simulation
Kolmogorov relation, 520
Korringa–Kohn–Rostoker Green's function method, 722
Kronecker delta, 604, 699

large cluster
  power law, 664, 671, 678
lateral translation, 150
lattice strain, 111
layer-by-layer growth, 150, 151, 152, 443, 448
LDA approximation. *See* local-density-functional approximation
leapfrog
  jump-up, 574
  transition, 574, 646
LEED, 43, 64, 151, 309, 311, 334, 712
  high resolution, 284, 323
  spot profile, 295, 355, 372, 429
LEIS, 328
Lennard–Jones
  crystal, 89, 116, 117, 157
  fcc(111) plane, 156
  minimum, 560
  system, 149

lifetime, 17, 223, 479, 480, 481, 482, 483, 485, 649
    activated state, 17
    metastable configuration, 637
L-J potentials. *See* Lennard–Jones potentail
local-density-functional
    approximation, 262, 264, 288, 289, 325, 357, 367, 454, 563
    calculation, 359, 561
    method, 355
    theory, 185, 261
long jump prefactor, 135, 138, 139
long-distance transition, 299
long-range
    effect, 503
    mechanism, 493
    transition, 288
    translation, 575
low-energy ion scattering, 211

map
    (1×1), 102, 308
    adsorption places, 39, 41, 72, 103
    binding sites, 72, 78, 623
    sites occupied, 38, 427, 634
    sites visited, 278
MC simulation. *See* Monte Carlo simulation
MD simulation. *See* molecular dynamics simulation
MD/MC-CEM, 57, 195, 206, 210, 279, 281, 282, 286, 293, 308, 309, 311, 319, 324, 366, 368, 590
mean free path, 26
mean-field theory, 48, 49, 668
mean-square separation, 699, 700
melting
    point, 89, 106, 116, 126, 141, 238, 262
    simulations, 596
    temperature, 121, 573, 582, 596, 642
metastable
    location, 72
    walk, 47, 128, 129, 227, 236, 428, 529
Meyer–Neidel rule, 318
microchannel plate. *See* channel plate
mixed hcp-fcc dimer, 587
Mo(100), 303, 322
Mo(110), 273, 302, 303
Mo(111), 303
modified Bessel function, 14, 15, 30, 122
molecular dynamics, 57, 106, 117, 195, 263, 265, 276, 277, 279, 288, 305, 306, 309, 312, 315, 317, 320, 325, 358, 403, 426, 453, 501, 572, 596, 600, 643, 647, 648, 649, 673, 685
    action-driven, 109, 571, 575, 581, 589
    calculation, 118, 364
    classical, 358
    estimates, 89, 200, 207, 237, 275, 364, 644
    investigation, 331
    Langevin, 293
    results, 350
    semi-empirical, 357

molecular dynamics (cont.)
    simulation, 89, 98, 105, 107, 115, 119, 121, 127, 142, 145, 149, 159, 162, 165, 185, 195, 197, 199, 200, 206, 211, 232, 234, 236, 237, 262, 272, 276, 279, 280, 282, 286, 287, 288, 289, 291, 293, 295, 297, 299, 300, 301, 305, 306, 310, 311, 313, 314, 318, 319, 320, 321, 322, 325, 326, 328, 331, 358, 360, 361, 365, 366, 368, 369, 370, 372, 465, 473, 542, 545, 559, 560, 563, 566, 578, 580, 584, 585, 586, 588, 589, 590, 593, 599, 600, 630, 631, 637, 639, 640, 642, 643, 644, 645, 646, 647, 648, 649, 686
    study, 97, 108, 193, 199, 350, 354, 369, 600, 678
    VASP package, 357
molecular statics, 197, 200, 289, 298, 438, 440, 449, 494, 501, 635
    constrained, 108
moment generating function, 12, 15, 525
Monte Carlo
    bond-counting model, 446
    calculation, 42, 550
    estimates, 44, 475
    modeling, 197
    simulation, 41, 52, 57, 123, 147, 158, 159, 162, 235, 264, 268, 271, 278, 298, 300, 305, 309, 317, 320, 323, 355, 363, 365, 403, 429, 439, 443, 446, 451, 480, 501, 502, 553, 578, 579, 604, 666, 668, 669, 670, 671, 673, 675, 677, 678, 687, 689, 690, 724
    TST, 218
Morse
    calculations, 194
    function, 56, 217
movement
    atom-by-atom, 534, 535, 587, 588, 589, 599, 626, 692
    concerted, 566, 585, 590, 596, 647
    fcc-hcp, 572, 582, 588, 590, 596, 686
    hcp-fcc, 587
    hcp-hcp, 265, 299
    hcp-hcp to fcc-hcp sites, 645
    local, 564, 575, 617, 620, 633, 641
    peripheral, 547, 573, 617, 627, 664, 665, 666, 668, 672, 673, 675, 678, 687, 689, 693
    step edge, 589
    vacancy, 552
multilayer
    barrier, 496
    step, 495
    transitions, 498
multilayer descent, 446, 495
    atom exchange, 494, 496

nearest-neighbor transition, 15, 98, 119, 122, 127, 135, 138
NEB method. *See* nudged elastic band method
negative step-edge barrier, 455, 487
net (1×1), 103
neutron scattering, 54
Ni(100), 49, 102, 103, 104, 105, 153, 278, 279, 280, 281, 282, 283, 428, 440, 441, 442, 543, 561, 569, 570, 571, 717
    paramagnetic, 280

Index

Ni(110), 91, 155, 168, 186, 187, 188, 189, 190, 191, 192, 193, 425, 428, 440, 441, 442, 575, 601
Ni(111), 72, 274, 275, 276, 277, 279, 428, 440, 441, 442, 501, 540, 571, 572, 573
Ni(311), 91, 193, 194
Ni(331), 91
non-equivalent
 paths, 686
 processes, 630, 686
non-uniform step-edge barrier, 451
nuclear resonance scattering, 54, 334
nucleation theory, 51, 53, 184, 266, 267, 269, 281, 284, 296, 304, 312, 315, 316, 317, 320, 359, 372, 452, 718
 mean-field, 429
nudged elastic band method, 109, 264, 272, 300, 326, 358, 493, 690

one-dimensional
 arrangement, 632
 chain, 502
 cluster, 619, 646
 cluster diffusion, 601
 crystal, 146
 island, 204, 541
 lattice, 146
 metastable chain, 502
 model, 146
 motion, 21, 520, 530, 685
 movement, 89, 186, 193, 195, 228, 231, 235
 plane, 39, 141, 480
 process, 22
 surface, 183, 592
 system, 123
 theory, 685
 trimer diffusion, 526
Ostwald ripening, 667

PACS. *See* perturbed γ–γ angular correlation spectroscopy
partition function, 18, 19, 20, 21, 698
Pb(110), 238, 478
Pb(111), 478
Pd(100), 104, 151, 153, 158, 309, 310, 311, 312, 598
Pd(110), 204, 205, 206, 207, 596, 598
Pd(111), 171, 312, 313, 314, 450, 596, 598
Pd(311), 205
Pd(331), 205
pentamer
 rotation, 626
 translation, 626, 633
perimeter energy, 553
perturbed γ–γ angular correlation
 spectroscopy, 53, 54, 281, 290
phase shifts, 722
photoemission electron microscopy, 314

piecewise rotation, 546
plane-wave pseudo-potential, 325
potential
 Ackland. *See* potential ATVF
 AFW, 189, 193, 194, 195, 205, 233, 235, 278, 285, 297, 366, 369
 ATVF, 190, 279, 344, 712
 bond-order, 642
 Ceperley–Alder, 563
 corrected effective medium, 152, 286
 EAM5, 191
 effective medium theory, 191, 210, 236, 263, 265, 280, 285, 296, 297, 317, 321, 357, 366, 575, 647, 689
 embedded atom method, 75, 79, 107, 115, 121, 159, 184, 190, 192, 197, 203, 205, 209, 210, 211, 228, 233, 234, 272, 275, 276, 279, 280, 286, 287, 288, 289, 297, 298, 299, 300, 310, 311, 317, 320, 324, 325, 326, 327, 331, 350, 357, 358, 363, 364, 366, 367, 369, 372, 403, 436, 444, 446, 447, 448, 453, 477, 496, 501, 539, 540, 542, 563, 566, 567, 569, 571, 572, 573, 578, 581, 582, 589, 592, 596, 599, 600, 601, 632, 637, 642, 643, 646, 647, 648, 690
 Ercolessi–Adams glue, 263
 FBD, 325, 454
 Finnis-Sinclair, 444, 447
 glue, 114, 370
 hybrid tight-binding-like, 162
 Lennard–Jones, 56, 89, 112, 117, 118, 185, 189, 193, 197, 203, 204, 206, 210, 218, 229, 233, 234, 236, 264, 265, 268, 275, 276, 278, 280, 289, 299, 300, 305, 306, 307, 310, 313, 320, 326, 331, 351, 352, 354, 357, 358, 363, 364, 368, 370, 559, 560, 664
 many-body, 185, 189, 204, 205, 217, 218, 226, 229, 234, 235, 276, 287, 297, 305, 306, 319, 327
 MD/MC–CEM, 191, 287, 309
 Morse, 118, 119, 189, 190, 193, 194, 203, 204, 205, 209, 217, 218, 225, 226, 228, 230, 231, 233, 234, 235, 272, 275, 278, 279, 285, 287, 297, 300, 303, 305, 306, 309, 312, 317, 324, 328, 330, 331, 344, 345, 350, 354, 355, 357, 363, 366, 369, 546
 Oh and Johnson. *See* potential OJ
 OJ, 190
 one-dimensional, 17, 20, 118
 pair, 56, 268, 306
 RGL, 57, 106, 107, 127, 192, 197, 199, 204, 206, 207, 210, 211, 229, 232, 234, 236, 237, 277, 288, 289, 294, 297, 298, 300, 307, 310, 312, 313, 314, 315, 318, 319, 328, 351, 352, 354, 358, 364, 365, 367, 370, 448, 455, 463, 493, 501, 529, 582, 596, 630, 631, 646, 648, 649, 728
 semi-empirical, 56, 98, 357, 403, 448
 surface-embedded atom, 326, 543, 600, 639
 Sutton-Chen, 275, 279, 306, 364, 537, 686, 687
 VC, 57, 185, 189, 190, 193, 205, 233, 234, 235, 261, 285, 288, 309, 312, 325, 363, 366, 369, 454
precursor, 167, 170

Index

probability, 7, 9, 10, 74, 81, 96, 122, 467, 498, 520, 521, 522, 523, 524, 527, 531, 533, 604, 697, 698, 714
  converting, 137
  finding a displacement x, 15
  incorporating, 498
  long jump, 118, 119
  occupation, 705
  recondensation, 667
  reconfiguration, 610
  rotation, 610
Pt(100), 100, 101, 104, 107, 109, 115, 153, 215, 361, 362, 363, 364, 365, 427, 429, 475, 542, 543, 634, 637, 639
  hex, 362, 363
Pt(110), 230, 231, 232, 233, 234, 235, 475, 647
  (1 × 2), 127, 129, 232, 234, 428, 528, 529, 637, 646
Pt(111), 49, 78, 84, 86, 110, 111, 112, 148, 155, 156, 162, 163, 166, 167, 169, 317, 319, 355, 356, 357, 358, 359, 360, 361, 402, 426, 429, 450, 453, 457, 467, 468, 469, 472, 473, 475, 476, 477, 480, 485, 486, 487, 500, 639, 640, 641, 642, 643, 645, 646, 652, 654, 687
Pt(311), 232, 233, 234
Pt(331), 232, 233, 234
pump
  mechanical, 35
  mercury diffusion, 35

quantum confinement of electrons, 502
quantum state, 697
quasi-harmonic approximation, 325
quenched molecular dynamics
  simulation, 210

random walk, 16, 121, 525, 547
rate
  basic jump, 137
  coarsening, 668
  core breakup, 548, 549
  corner rounding, 548
  cross-channel motion, 94
  deposition, 469
  dimer nucleation, 483
  double jump, 135, 142
  escape, 431, 456, 467, 468, 479, 480
  evaporation, 669, 677
  evaporation–condensation, 675
  horizontal jump, 129, 135, 140
  impingement, 49
  in-channel motion, 94
  incorporation, 48, 431, 460
  kink escape, 548
  long jump, 137
  nearest-neighbor transition, 122
  rotation, 626
  shearing, 550

rate (cont.)
  single jump, 122, 135, 137, 138, 139, 141, 142
  vertical jump, 135, 140
ratio
  charge-to-mass, 90
  double-to-single, 123, 125, 126, 127, 223
Re(0001), 80, 263, 347, 348, 404, 457, 618, 619, 652
reaction coordinate, 17
rebound
  energy, 156
  transition, 144
rebound jump frequency prefactor, 145
reflection high energy electron diffraction. *See* RHEED
region
  fcc-like, 114
  hcp-like, 114
relaxation
  adsorbate-induced, 436
  step-induced, 436
reptation, 517, 539, 640, 652, 687
repulsive
  distance, 51
  ring, 501
resonant enhanced multi-photon ionization, 169
RGL, 197, 207, 228, 235, 263, 326, 350, 354, 600, 630, 686, 716
  approximation, 228, 236
Rh(100), 117, 153, 162, 305, 306, 307, 308, 423, 426, 539, 547, 577, 581, 593, 595, 596, 652
Rh(110), 87, 94, 98, 168, 203, 204, 596, 699
Rh(111), 304, 305, 306, 307, 449, 596
Rh(311), 87, 202, 426
Rh(331), 87
RHEED, 271, 323, 439, 451
rolling-over motion, 478
rotation atom-by-atom, 582
row shearing at the edges, 562
Ru(0001), 112, 304, 542, 678
Rutherford backscattering, 290
  spectrometry, 198

saddle point, 20, 89, 113, 272, 546, 579, 622, 635, 650, 718
saturation island density, 110, 163, 184, 207, 268, 317
scanning electron microscopy, 315, 322
scanning tunneling microscopy, 24, 31, 32, 33, 34, 39, 46, 47, 48, 49, 53, 64, 75, 110, 112, 158, 162, 163, 166, 167, 168, 183, 188, 202, 205, 207, 209, 232, 235, 265, 271, 296, 301, 304, 312, 317, 321, 322, 323, 328, 350, 355, 356, 357, 363, 368, 372, 425, 428, 439, 440, 442, 450, 451, 455, 469, 478, 504, 575, 590, 598, 649, 667, 676, 688, 715, 720, 721, 722, 723, 735
  atom tracking, 47
  fast, 296, 637
  image, 127, 163, 169, 528
  low-temperature, 167
  variable temparature, 317

Index

scattering
  elastic, 47
  inelastic, 47
  quasielastic, 207
Schlömilch relation, 14
screw dislocation, 506, 507
second-range transition, 138, 141
shearing, 538, 562, 687
  motion, 626, 627, 628
shuffle gliding, 692
Si(100)–(2 × 1), 169
single jump
  frequency prefactor, 141
  prefactor, 107
single transition, 119, 142, 210, 233
sintering, 696
site
  adsorption, 39, 54, 69, 71, 72, 73, 74, 75, 78, 80, 81, 129, 151, 261, 265, 277, 296, 300, 307, 314, 331, 347, 348, 350, 370, 477, 561, 601, 631, 643, 647, 711, 717
  binding, 54, 68, 69, 72, 77, 81, 84, 105, 113, 114, 117, 152, 156, 262, 265, 275, 304, 344, 347, 357, 358, 598, 610, 636, 702, 704, 718
  bridge, 68, 111, 573, 630, 686
  deep, 225, 226
  equilibrium, 17, 18, 46, 86, 98, 111, 537
  fault, 68, 148, 261, 344
  fcc, 22, 72, 73, 74, 75, 77, 78, 79, 80, 111, 115, 152, 166, 261, 265, 276, 277, 296, 299, 300, 301, 304, 306, 307, 313, 314, 317, 318, 319, 320, 321, 347, 348, 350, 351, 352, 353, 355, 357, 360, 370, 371, 428, 444, 472, 476, 477, 539, 564, 565, 566, 573, 575, 582, 585, 586, 599, 601, 623, 626, 636, 639, 640, 641, 642, 643, 644, 687, 717
  four fold, 69, 80, 477, 632, 711
  hcp, 22, 72, 73, 74, 75, 78, 80, 112, 115, 152, 261, 265, 277, 298, 299, 300, 301, 306, 307, 314, 317, 318, 319, 320, 321, 347, 348, 350, 352, 353, 355, 357, 360, 370, 371, 428, 444, 472, 473, 477, 539, 564, 565, 566, 573, 586, 587, 589, 598, 601, 623, 633, 636, 639, 640, 641, 643, 687
  lattice, 68, 98, 148, 344, 345, 610, 611, 619
  metastable, 477
  normal, 5, 56, 98
  shallow, 69, 226
size distribution, 49, 164, 311
sky-hook mechanism, 426, 428
smart Monte Carlo simulation, 287, 306
SPA-LEED. See LEED spot profile
sphere-like shape, 651, 692
sputtering, 28, 33, 35, 96, 187, 210
stable location, 72
static potential barrier, 56
step
  A-type, 435, 437, 442, 446, 449, 453, 455, 456, 458, 460, 463, 464, 465, 467, 475, 476, 477, 478, 487, 501, 679

step (cont.)
  B-type, 434, 435, 437, 440, 441, 442, 445, 446, 449, 453, 458, 463, 465, 467, 473, 475, 476, 477, 478, 487, 500, 536, 679, 682
step flow, 442, 451, 467
step vacancy, 465
step-edge trap, 457, 479, 480, 481
STM. See scanning tunneling microscopy
stochastic trajectory calculations, 148
strain, 86, 105, 111, 112, 113, 114, 115, 269, 272, 273, 281, 286, 295, 299, 316, 317, 323, 360, 450, 451, 453, 455, 509, 668, 727
  biaxial, 115
  compressive, 110, 111, 113, 295, 509, 564
  local, 114, 115
  surface, 111, 113
  tensile, 111, 113, 115, 509
strain-induced
  event, 543
  exchange, 326
stress, 33, 112, 299, 501, 590, 741
  compressive, 453
  tensile, 104, 112, 453
stress-induced mechanism, 648
stretching dimer, 546, 562, 595, 647
sub-lattice c(2 × 2), 103
sublimation energy, 57, 263
superposition technique, 68
surface defect, 201, 209, 217, 269, 314, 315, 332
surface ionization microscope, 223
surface-embedded atom method, 108, 600, 639
surface-state band, 720
Sutton-Chen model, 57

Ta(100), 328
Ta(110), 328, 404, 601, 652
Ta(111), 328
terrace width, 442
tetramer
  rotation, 623, 626, 633
  translation, 623, 626, 633
thermal
  bath, 127
  desorption, 149
  energy, 135
  equilibrium, 146
thermal energy He atom scattering, 283, 295, 316
thermal energy helium atom diffraction, 162
thermally
  assisted tunneling, 79
  associated tunneling, 79
  equilibrated surfaces, 110
  equilibrated system, 162
third moment, 13
three-body effect, 713
  attractive, 704

tight-binding theory, 57
  fourth-moment approximation, 218, 226, 331, 615, 711
  RGL model, 325
  second-moment approximation, 57, 195, 263, 268, 276, 300, 358, 370, 371, 372, 443, 448, 571, 673
TiO$_2$(110)., 168
transient
  cooling, 604
  heating, 604
  hopping, 147
  mobility, 155, 156, 159, 162, 163, 164, 166, 168, 169, 171
  period, 44
  state, 539
  temperature, 42, 43, 44, 131, 134, 139, 221
  time, 42
transition state, 17, 18, 58, 89, 104, 111, 112, 477, 544, 573, 574, 645, 687
  barrier, 288
  calculations, 276, 289
  estimates, 286
  theory, 4, 17, 122, 235, 236, 263, 266, 275, 279, 288, 293, 305, 309, 311, 317, 327, 331, 364, 366, 369, 595
translation, 19
translation with rotation, 651
translational motion, 573, 582, 596, 642
trimer
  bent form, 710, 713
  linear form, 611, 626, 633, 643, 713
  rotation, 626, 633, 640, 651
  translation, 565, 566, 626, 639, 640, 643, 651
  triangular form, 566, 569, 573, 611, 626, 633, 639, 643, 710, 713
trimer diffusivity prefactor, 643
trio-energy, 715
two-dimensional
  arrangement, 619
  cluster, 483, 556, 613, 652
  distribution of separations, 705
  electron gas, 722
  form, 593
  island, 323, 468, 542, 569, 593, 602, 634, 653, 712
  motion, 21, 88
  movement, 93, 224, 530, 556
  nucleation, 442, 451, 467
  pentamer, 557
  plane, 39
  potential, 546
  process, 22
  shape, 613
  spreading, 290
  structure, 619, 652
  surface, 100, 123, 183, 261, 596, 609

two-dimensional (cont.)
  system, 99
  trimer, 622, 627

uncorrelated motion, 567
upstanding molecule, 168
upward exchange, 449

vacancy, 102, 210, 211, 324, 465, 469, 620, 671, 678
  cluster, 553, 677, 678
  island, 296, 451, 469, 478, 575, 598, 670, 671, 673, 676, 677, 689
  worm-like cluster, 673
van der Waals forces, 148
VASP, 58, 200, 234, 272, 289, 301, 310, 496, 561, 563, 571, 580, 595, 598, 600, 634, 639, 646, 647, 648, 718, 728
  code, 264
VC
  approximation, 184, 210, 233, 275, 279, 280, 286, 288
  parameters, 288, 325
  parametrization, 77
vertical
  dimer, 549
  intermediate, 530, 531
  mobility, 601, 648
  states, 549
  transition, 129, 131, 132, 133, 340
vibrational coupling, 696
vibrational frequency, 87, 685
  effective, 215
Vienna *ab initio* package. *See* VASP
Vineyard, 20, 686
  expression, 685, 687
Vycor, 27, 35, 36

W(100), 272, 341, 342, 344
W(110), 31, 41, 45, 72, 83, 129, 130, 133, 138, 140, 141, 142, 145, 263, 315, 316, 319, 330, 331, 332, 333, 334, 335, 336, 337, 339, 340, 341, 347, 402, 403, 404, 423, 431, 432, 456, 502, 504, 507, 508, 557, 558, 608, 609, 611, 612, 613, 615, 617, 652, 664, 701, 703, 704, 706, 708, 711, 712, 715, 722, 726
W(111), 25, 68, 73, 148, 344, 345, 347, 456, 614
W(211), 38, 41, 45, 68, 81, 82, 87, 125, 126, 129, 141, 142, 145, 211, 212, 215, 216, 217, 218, 219, 221, 223, 224, 225, 226, 263, 402, 403, 433, 456, 525, 527, 558, 602, 603, 604, 606, 607, 610, 652, 699, 700, 710, 724
W(321), 68, 80, 87, 225, 226, 227, 429, 431, 603, 699
waiting time, 613
  prefactor, 613
wave function, 696, 720
wedding cake structure, 429, 451
WIEN97, 294
work function, 28, 30, 53, 201, 704
  change, 24, 200, 224, 278, 302, 332, 335, 338
Wulff polyhedron, 319, 326, 367

Index

XPD, 328
XPS, 328

"zero-time"
   distribution, 44
   experiment, 43, 44, 45, 133, 141

"zero-time" (cont.)
   measurement, 44, 45, 139
   results, 141

zigzag
   motion, 564, 599
   step, 453
   transition, 564